AF598605

Encyclopedia of Applied Physics

SPONSORS

AMERICAN INSTITUTE OF PHYSICS
DEUTSCHE PHYSIKALISCHE GESELLSCHAFT
JAPAN SOCIETY OF APPLIED PHYSICS
PHYSICAL SOCIETY OF JAPAN

DEUTSCHE
PHYSIKALISCHE
GESELLSCHAFT

JAPAN SOCIETY
OF APPLIED PHYSICS

PHYSICAL SOCIETY
OF JAPAN

ENCYCLOPEDIA OF APPLIED PHYSICS

VOLUME 17
Scientific Computing by Numerical Methods to Separation Processes

Edited by
GEORGE L. TRIGG

Associate Editors
EDUARDO S. VERA
WALTER GREULICH

Managing Editor
EDMUND H. IMMERGUT

Assistant Managing Editor
CHRISTOPHER THOMAS MORAN

George L. Trigg
275 Beaver Dam Road
Brookhaven, New York 11719

Walter Greulich
Ziegeleiweg 30
D-69488 Birkenau
Federal Republic of Germany

Eduardo S. Vera
Science and Technology Information Center (ICT)
University of Chile
Beaucheff 850
Santiago, Chile

Edmund H. Immergut
Christopher Thomas Moran
2 Sidney Place
Brooklyn, New York 11201

Library of Congress Cataloging-in-Publication Data

Encyclopedia of applied physics.
(Revised for Vol. 17)

"Sponsors, American Institute of Physics . . . [et al]."
Includes bibliographical references and index.
1. Physics—Encyclopedias. 2. Engineering—Encyclopedias. I. Trigg, George L. II. Vera, Eduardo S. III. Greulich, Walter. IV. American Institute of Physics.
QC5.E543 1991 530'.05 91-8738
ISBN 1-56081-076-9 (VCH Publishers)

British Library Cataloguing in Publication Data

Encyclopedia of applied physics.
Vol. 17
I. Trigg, George L. (George Lockwood) II. Vera, Eduardo S. III. Greulich, Walter
621
ISBN 1-56081-058-0 (set)
ISBN 1-56081-076-9 (vol. 17)

Printed in the United States of America.

ISBN 3-527-28139-8 (Volume 17) VCH Verlagsgesellschaft mbH
ISBN 3-527-26841-3 (set) VCH Verlagsgesellschaft mbH

Printing History:
10 9 8 7 6 5 4 3 2 1

Published jointly by:

VCH Publishers, Inc.
333 7th Avenue
New York, NY 10001

VCH Verlagsgesellschaft mbH
P.O. Box 10 11 61
69451 Weinheim
Federal Republic of Germany

VCH Publishers (UK) Ltd.
8 Wellington Court
Cambridge CB1 1HZ
United Kingdom

ADVISORY BOARD

EDITORIAL CONSULTANTS

MAIN ENTRIES

The subject matter in the *Encyclopedia of Applied Physics* is presented in approximately 500 individual articles, arranged alphabetically. The topics can be classified into 20 sections, similar to the AIP Physics and Astronomy Classification Scheme (PACS):

01	General Aspects: Mathematical, Computational, and Information Techniques
02	Measurement Science, General Devices and/or Methods
03	Nuclear and Elementary Particle Physics
04	Atomic and Molecular Physics
05	Electricity and Magnetism
06	Optics (classical and quantum)
07	Acoustics
08	Thermodynamics and Properties of Gases
09	Fluids and Plasma Physics
10	Condensed Matter A: Structure and Mechanical Properties
11	Condensed Matter B: Thermal, Acoustic, and Quantum Properties
12	Condensed Matter C: Electronic Properties
13	Condensed Matter D: Magnetic Properties
14	Condensed Matter E: Dielectrical and Optical Properties
15	Condensed Matter F: Surfaces and Interfaces
16	Materials Science
17	Physical Chemistry
18	Energy Research and Environmental Physics
19	Biophysics and Medical Physics
20	Geophysics, Meteorology, Space Physics, and Aeronautics

Each article has been assigned a code number consisting of two digits which denotes the section, and a letter which gives the type of article. There are six types: A = Devices, Equipment; B = Materials; C = Methods, Processes; D = Phenomena, Effects; E = Scientific or Technological Fields; F = Institutions, Companies, Societies and other organizations.

CONTRIBUTORS

Joseph S. Accetta, 2812 Bent Tree Drive, Dexter, MI 48130
Sensors, Infrared

Stephen F. Ackley, CREL-RS, 72 Lyme Road, Hanover, NH 03755
Sea Ice

Duncan C. Agnew, Institute of Geophysics and Planetary Physics, University of California, La Jolla, CA 92093-0225
Seismographs

Franco Bassani, Scuola Normale Superiore, Pisa 56126, Italy
Semiconductors, Elemental—Electronic Properties

Jonathan Berger, Institute of Geophysics and Planetary Physics, University of California, La Jolla, CA 92093-0225
Seismographs

P. Capper, GEC-Marconi Infra-Red Ltd., Southampton, 509 7QG, England
Semiconductors, Compound—Material Properties

Christina C. Christara, Computer Science Department, University of Toronto, Toronto, Ontario M5S 1A4, Canada
Scientific Computing by Numerical Methods

M. Dobrowolska, Department of Physics, University of Notre Dame, Notre Dame, IN 46556
Semiconductors, Diluted Magnetic

S. R. Elliott, Department of Chemistry, University of Cambridge, Lensfield Road, Cambridge, CB2 1EW, England
Semiconductors, Amorphous—Material Properties

H. Fritzsche, The James Franck Institute, The University of Chicago, 5640 S. Ellis Avenue, Chicago, IL 60637
Semiconductors, Amorphous—Electronic Properties

J. K. Furdyna, Department of Physics, University of Notre Dame, Notre Dame, IN 46556
Semiconductors, Diluted Magnetic

Roland Grisar, Fraunhofer Institut für Physikalische Meßtechnik, Heidenhofstr. 8, 79110 Freiburg im Breisgau, Germany
Sensors, Optical

W. S. Winston Ho, Corporate Research, Exxon Research and Engineering Company, Annandale, NJ 08801-0998
Separation Processes

Howard R. Huff, SEMATECH, 2706 Montopolis Drive, Austin, TX 78741
Semiconductors, Elemental—Material Properties

R. I. G. Hughes, Department of Philosophy, Logic and Scientific Method, The London School of Economics and Political Science, Houghton Street, London WC2A 2AE, England
Semantic View of Theories

Kenneth R. Jackson, Computer Science Department, University of Toronto, Toronto, Ontario M5S 1A4, Canada
Scientific Computing by Numerical Methods

Carl Kisslinger, CIRES/Campus Box 216, University of Colorado, Boulder, CO 80309-0216
Seismology

Christine E. Koltermann, 3555 Berry Way, Santa Clara, CA 95051-1901
Sedimentary Basin Evolution

Giuseppe La Rocca, Scuola Normale Superiore, Pisa 56126, Italy
Semiconductors, Elemental—Electronic Properties

Norman N. Li, Research and Technology, Allied Signal Inc., 50 E. Algonquin Road, Box 5016, Des Plaines, IL 60017-5016
Separation Processes

H. Luo, Department of Physics, University of Notre Dame, Notre Dame, IN 46556
Semiconductors, Diluted Magnetic

Lawrence C. Lynnworth, Panametrics, Inc., 221 Crescent Street, Waltham, MA 02154-3497
Sensors, Acoustic

Robert Melville Metzger, Department of Chemistry, University of Alabama, Tuscaloosa, AL 35487-0336
Semiconductors to Superconductors: Organic Lower-Dimensional Systems

Martin Pope, Chemistry Department, New York University, 24 Waverly Pl. Rm. 453, New York, NY 10003
Semiconductors, Organic—Electronic Properties

Christian Rückauer, Richard Wagner Str. 13, 76706 Dettenheim, Germany
Sensors, Optical

Hideo Sunami, Central Research Laboratory, Hitachi, Ltd., 1-280 Higashi-koi-gakubo, Kokubunji, Tokyo 185, Japan
Semiconductor-Device Integration

V. Swaminathan, Bell Laboratories, Lucent Technologies, 1600 Osgood Street, MS/21-2E39, North Andover, MA 08145
Semiconductors, Compound—Material Properties

Charles E. Swenberg, Physics Department, George Mason University, Fairfax, VA 22030-4444
Semiconductors, Organic—Electronic Properties

Maurus Tacke, Fraunhofer Institut für Physikalische Meßtechnik, Heidenhofstr. 8, 79110 Freiburg im Breisgau, Germany
Sensors, Optical

Sandip Tiwari, IBM/Watson Research Center, Yorktown Heights, NY 10598
Semiconductors, Compound—Electronic Properties

Toru Toyabe, Department of Information & Computer Sciences, The Faculty of Engineering, Toyo University, 2100 Kujirai, Kawagoe, Saitama 350, Japan
Semiconductor-Device Modeling

Elmar Wagner, Fraunhofer Institut für Physikalische Meßtechnik, Heidenhofstr. 8, 79110 Freiburg im Breisgau, Germany
Sensors, Optical

James Zhou, Research and Technology, Allied Signal Inc., 50 E. Algonquin Road, Box 5016, Des Plaines, IL 60017-5016
Separation Processes

SCATTERING OF LIGHT

See RAMAN SCATTERING

SCIENTIFIC COMPUTING BY NUMERICAL METHODS

Christina C. Christara and Kenneth R. Jackson, *Computer Science Department, University of Toronto, Toronto, Ontario, Canada*

3-527-28139-8/96/$5.00 + .50

INTRODUCTION

Numerical methods are an indispensable tool in solving many problems that arise in science and engineering. In this article, we briefly survey a few of the most common mathematical problems and review some numerical methods to solve them.

As can be seen from the table of contents, the topics covered in this survey are those that appear in most introductory numerical methods books. However, our discussion of each topic is more brief than is normally the case in such texts and we frequently provide references to more advanced topics that are not usually considered in introductory books.

Definitions of the more specialized mathematical terms used in this survey can be found in the Glossary at the end of the article. We also list some mathematical symbols and abbreviations used throughout the survey in the two sections following the Glossary.

1. FLOATING-POINT ARITHMETIC

In this section, we consider the representation of floating-point numbers, floating-point arithmetic, rounding errors, and the effects of inexact arithmetic in some simple examples. For a more detailed discussion of these topics, see Wilkinson (1965) and Goldberg (1991) or an introductory numerical methods text.

1.1 The IEEE Standard

The approval of the IEEE (Institute of Electrical and Electronics Engineers) Standard for Binary Floating-Point Arithmetic (IEEE, 1985) was a significant advance for scientific computation. Not only has this led to cleaner floating-point arithmetic than was commonly available previously, thus greatly facilitating the development of reliable, robust numerical software, but, because many computer manufacturers have since adopted the standard, it has significantly increased the portability of programs.

The IEEE standard specifies both single- and double-precision floating-point numbers, each of the form

$$(-1)^s \times b_0.b_1b_2 \cdots b_{p-1} \times 2^E, \tag{1}$$

where $s = 0$ or 1, $(-1)^s$ is the *sign* of the number, $b_i = 0$ or 1 for $i = 0, \ldots, p-1$; $b_0.b_1b_2 \cdots b_{p-1}$ is the *significand* (sometimes called the *mantissa*) of the number, and the *exponent* E is an integer satisfying $E_{\min} \le E \le E_{\max}$. In single-precision, $p = 24$, $E_{\min} = -126$ and $E_{\max} = +127$, whereas, in double-precision, $p = 53$, $E_{\min} = -1022$, and $E_{\max} = +1023$. We emphasize that a number written in the form (1) is binary. So, for example, $1.100 \cdots 0 \times 2^0$ written in the format (1) is equal to the decimal number 1.5.

A *normalized* number is either 0 or a floating-point number of the form (1) with $b_0 = 1$ (and so it is not necessary to store the leading bit). In single-precision, this provides the equivalent of 7 to 8 significant decimal digits with positive and negative numbers having magnitudes roughly in the range $[1.2 \times 10^{-38}, 3.4 \times 10^{+38}]$. In double-precision, this is increased to about 16 significant decimal digits and a range of roughly $[2.2 \times 10^{-308}, 1.8 \times 10^{+308}]$.

An *underflow* occurs when an operation produces a nonzero result in the range $(-2^{E_{\min}}, +2^{E_{\min}})$. In the IEEE standard, the default is to raise an underflow exception flag and to continue the computation with the result correctly rounded to the nearest *denormalized* number or zero. A denormalized floating-point number has the form (1) with $E = E_{\min}$ and $b_0 = 0$. Because denormalized numbers use some of the leading digits from the significand to represent the magnitude of the number, there are fewer digits available to represent its significant digits. Using denormalized numbers in this way for underflows is sometimes referred to as *gradual underflow*. Many older non-IEEE machines do not have denormalized numbers and, when an underflow occurs, they either abort the computation or replace the result by zero.

An *overflow* occurs when an operation produces a nonzero result outside the range of floating-point numbers of the form (1). In the IEEE standard, the default is to raise an overflow exception flag and to continue the computation with the result replaced by either $+\infty$ or $-\infty$, depending on the sign of the overflow value. Many older non-IEEE machines do not have $+\infty$ or $-\infty$ and, when an overflow occurs, they usually abort the computation.

The IEEE standard also includes at least two NaNs (Not-a-Number) in both precisions, representing indeterminate values that may arise from invalid or inexact operations such as $(+\infty) + (-\infty)$, $0 \times \infty$, $0/0$, ∞/∞, or $\sqrt{x}$ for $x < 0$. When a NaN arises in this way, the default is to raise an exception flag and to continue the computation. This novel feature is not available on most older non-IEEE machines.

It follows immediately from the format (1) that floating-point numbers are discrete and finite, whereas real numbers are dense and infinite. As a result, an arithmetic operation performed on two floating-point numbers may return a result that cannot be represented exactly in the form (1) in the same precision as the operands.

A key feature of the IEEE standard is that it requires that the basic arithmetic operations $+$, $-$, $\times$, $/$, and $\sqrt{\ }$ return *properly rounded* results. That is, we may think of the operation as first being done exactly and then properly rounded to the precision of the result. An operation with $\pm\infty$ is interpreted as the limiting case of the operation with an arbitrary large value in place of the ∞, when such an interpretation makes sense; otherwise the result is a NaN. An operation involving one or more NaNs returns a NaN.

The default rounding mode is *round-to-nearest;* that is, the exact result of the arithmetic operation is rounded to the nearest floating-point number, where, in the case of a tie, the floating-point number with the least significant bit equal to 0 is selected. The standard also provides for directed roundings (round-towards-$+\infty$, round-towards-$-\infty$, and round-towards-0), but these are not easily accessed from most programming languages.

Another important feature of the IEEE standard is that comparisons are exact and never overflow or underflow. The comparisons $<$, $>$, and $=$ work as expected with finite floating-point numbers of the form (1) and $-\infty < x < +\infty$ for any finite floating-point number x. NaNs are unordered and the comparison of a NaN with any other value—including itself—returns false.

The IEEE standard also provides for *extended* single- and double-precision floating-point numbers, but these are not easily accessed from most programming languages, and so we do not discuss them here.

As noted above, the IEEE standard has been widely adopted in the computer industry, but there are several important classes of machines that do not conform to it, including Crays, DEC Vaxes, and IBM mainframes. Although their floating-point numbers are similar to those described above, there are important differences. Space limitations, though, do not permit us to explore these systems here.

1.2 Rounding Errors

Since IEEE standard floating-point arithmetic returns the correctly rounded result for the basic operations $+$, $-$, $\times$, $/$, and $\sqrt{\ }$, one might expect that rounding errors would never pose a problem, particularly in double-precision computations. Although rounding errors can be ignored in many cases, the examples in the next subsection show that they may be significant even in simple calculations. Before considering these examples, though, we need to define an important machine constant and describe its significance in floating-point computation.

Machine epsilon, often abbreviated *mach-eps,* is the distance from 1 to the next larger floating-point number. For numbers of the form (1), mach-eps $= 2^{1-p}$. So for IEEE single- and double-precision numbers, mach-eps is $2^{-23} \approx 1.19\,209 \times 10^{-7}$ and $2^{-52} \approx 2.22\,045 \times 10^{-16}$, respectively.

A common alternative definition of machine epsilon is that it is the smallest positive floating-point number ϵ such that $1 + \epsilon > 1$ in floating-point arithmetic. We prefer the definition given in the previous paragraph, and use it throughout this section, because it is independent of the rounding mode and so is characteristic of the floating-point number system itself, while the alternative definition given in this paragraph depends on the rounding mode as well. We also assume throughout this section that round-to-nearest is in effect. The discussion below, though, can be modified easily for the alternative definition of mach-eps and other rounding modes.

It follows immediately from the floating-point format (1) that the absolute distance between floating-point numbers is not uniform. Rather, from (1) and the definition of mach-eps above, we see that the spacing between floating-point numbers in the intervals $[2^k, 2^{k+1})$ and $(-2^{k+1}, -2^k]$ is mach-eps $\times\, 2^k$ for $E_{\min} \le k \le E_{\max}$. Thus, the absolute distance between neighboring nonzero normalized floating-point numbers with the same exponent is uniform, but the floating-point numbers near $2^{E_{\min}}$ are much closer together in an absolute sense than those near $2^{E_{\max}}$. However, the *relative* spacing between all nonzero normalized floating-point numbers does not vary significantly. It is easy to see that, if x_1 and x_2 are any two neighboring nonzero normalized floating-point numbers, then

$$\frac{\text{mach-eps}}{2} \le \left|\frac{x_1 - x_2}{x_1}\right| \le \text{mach-eps}. \qquad (2)$$

As a result, it is more natural to consider relative, rather than absolute, errors in arithmetic operations on floating-point numbers, as is explained in more detail below.

For $x \in \mathbb{R}$, let $fl(x)$ be the floating-point number nearest to x, where, in the case of a tie, the floating-point number with the least significant bit equal to 0 is selected. The importance of mach-eps stems largely from the observation that, if $fl(x)$ does not overflow or underflow, then

$$fl(x) = x(1 + \delta) \quad \text{for some } |\delta| \le u, \qquad (3)$$

where u, the *relative roundoff error bound,* satisfies u = mach-eps/2 for round-to-nearest and u = mach-eps for the other IEEE rounding modes. Rewriting (3) as

$$\delta = [fl(x) - x]/x,$$

we see that δ is the relative error incurred in approximating x by $fl(x)$, and so (3) is closely related to (2).

If op is one of $+$, $-$, $\times$, or / and x and y are two floating-point numbers, let $fl(x \text{ op } y)$ stand for the result of performing the arithmetic operation x op y in floating-point arithmetic. If no arithmetic exception arises in the floating-point operation, then

$$fl(x \text{ op } y) = (x \text{ op } y)(1 + \delta) \quad \text{for some } |\delta| \le u, \qquad (4)$$

where the (x op y) on the right side of (4) is the exact result of the arithmetic operation. Similarly, if x is a nonnegative normalized floating-point number, then

$$fl(\sqrt{x}) = \sqrt{x}(1 + \delta) \quad \text{for some } |\delta| \le u. \qquad (5)$$

Again, the u in either (4) or (5) is the relative roundoff error bound and the δ is the relative error incurred in approximating x op y by $fl(x \text{ op } y)$ or $\sqrt{x}$ by $fl(\sqrt{x})$, respectively.

Although the relations (4)–(5) are not quite as tight as the requirement that the basic operations $+$, $-$, $\times$, /, and $\sqrt{\ }$ return the correctly rounded result, they are very useful in deriving error bounds and explaining the effects of rounding errors in computations.

1.3 The Effects of Inexact Arithmetic: Some Illustrative Examples

As noted at the start of the last subsection, since IEEE standard floating-point arithmetic returns the correctly rounded result for the basic operations $+$, $-$, $\times$, /, and $\sqrt{\ }$, one might expect that rounding errors would never pose a problem, particularly in double-precision computations. This, though, is not the case. In this subsection we consider a simple example that illustrates some of the *pitfalls* of numerical computation.

Suppose that we compute the expression

$$1 + 10^{10} - 10^{10} \qquad (6)$$

in single precision from left to right. We first compute $fl(1 + 10^{10}) = 10^{10}$, the correctly rounded single-precision result. Then we use this value to compute $fl(10^{10} - 10^{10}) = 0$, without committing an additional rounding error. Thus, $fl((1 + 10^{10}) - 10^{10}) = 0$, whereas the true result is 1.

The key point to note here is that $fl(1 + 10^{10}) = 10^{10} = (1 + 10^{10})(1 + \delta)$, where $|\delta| = 1/(1 + 10^{10}) < 10^{-10} < u = 2^{-24}$. So the rounding error that we commit in computing $1 + 10^{10}$ is small relative to $1 + 10^{10}$, the true result of the first addition, but the absolute error of 1 associated with this addition is not small compared with the true final answer, 1, thus illustrating the **Rule for Sums:**

> Although a rounding error is always small relative to the result that gives rise to it, it might be large relative to the true final answer if intermediate terms in a sum are large relative to the true final answer.

The Rule for Sums is important to remember when computing more complex expressions such as the truncated Taylor series $T_k(x) = 1 + x + x^2/2 + \cdots + x^k/k! \approx e^x$, where k is chosen large enough so that the *truncation error*

$$e^x - T_k(x) = \sum_{i=k+1}^{\infty} x^i/i!$$

is insignificant relative to e^x. It is easy to prove that, if $x \geq 0$ and k is large enough, then $T_k(x)$ is a good approximation to e^x. However, if $x < 0$ and of moderate magnitude, then the rounding error associated with some of the intermediate terms $x^i/i!$ in $T_k(x)$ might be much larger in magnitude than either the true value of $T_k(x)$ or e^x. As a result, $fl(T_k(x))$, the computed value of $T_k(x)$, might be completely erroneous, no matter how large we choose k. For example, $e^{-15} \approx 3.05\,902 \times 10^{-7}$, while we computed $fl(T_k(x)) \approx 2.12\,335 \times 10^{-2}$ in IEEE single-precision arithmetic on a Sun Sparcstation.

A similar problem is less likely to occur with multiplications, provided no overflow or underflow occurs, since from (4)

$$fl(x_1 \cdot x_2 \cdots x_n) = x_1 \cdot x_2 \cdots x_n(1 + \delta_1) \cdots (1 + \delta_{n-1}), \tag{7}$$

where $|\delta_i| \leq u$ for $i = 1, \ldots, n - 1$. Moreover, if $nu \leq 0.1$, then $(1 + \delta_1) \cdots (1 + \delta_{n-1}) = 1 + 1.1n\delta$ for some $\delta \in [-u,u]$. Therefore, unless n is very large, $fl(x_1 \cdot x_2 \cdots x_n)$ is guaranteed to be a good approximation to $x_1 \cdot x_2 \cdots x_n$. However, it is not hard to find examples for which $nu \gg 1$ and $fl(x_1 \cdot x_2 \cdots x_n)$ is a poor approximation to $x_1 \cdot x_2 \cdots x_n$.

The example (6) also illustrates another important phenomenon commonly called *catastrophic cancellation:* all the digits in the second sum, $fl(10^{10} - 10^{10})$, cancel, signaling a catastrophic loss of precision. Catastrophic cancellation refers also to the case that many, but not all, of the digits cancel. This is often a sign that a disastrous loss of accuracy has occurred, but, as in this example when we compute $fl(1 + 10^{10}) = 10^{10}$ and lose the 1, it is often the case that the accuracy is lost before the catastrophic cancellation occurs.

For an example of catastrophic cancellation in a more realistic computation, consider calculating the roots of the quadratic $ax^2 + bx + c$ by the standard formula

$$r_{\pm} = \frac{-b \pm \sqrt{b^2 - 4ac}}{2a}. \tag{8}$$

We used this formula to compute the roots of the quadratic $x^2 - 10^4x + 1$ in IEEE single-precision arithmetic on a Sun Sparcstation. The computed roots were 10^4 and 0, the larger of which is accurate, having a relative error of about 10^{-8}, but the smaller one is completely wrong, the true root being about 10^{-4}. A similar result usually occurs whenever $|ac|/b^2 \ll 1$. The root of larger magnitude is usually computed precisely, but the smaller one is frequently very inaccurate as a result of catastrophic cancellation, since $b^2 - 4ac \approx b^2$ and so $|b| - \sqrt{b^2 - 4ac} \approx 0$. Although the second of these relations signals catastrophic cancellation, the loss of precision occurs in the first.

There is an easy remedy for the loss of precision due to catastrophic cancellation in this case. Use (8) to compute r_1, the root of larger magnitude, and then use the alternative formula $r_2 = c/ar_1$ to compute the smaller one. The relative error in r_2 is at most $(1 + u)^2$ times larger than the relative error in r_1, provided that no overflows or underflows occur in computing c/ar_1.

Another point to note is that, if we compute (6) from right to left, instead of left to right, then

$$fl(1 + (10^{10} - 10^{10})) = fl(1 + 0) = 1.$$

It is not particularly significant that this computation gives the correct answer, but what is important is that it illustrates that floating-point addition is not associative, although it is commutative, since $fl(a + b)$ and $fl(b + a)$ are both required to be the correctly rounded value for $a + b = b + a$. Similar results hold for multiplication and division.

Many other fundamental mathematical relations that we take for granted do not hold for floating-point computations. For example, the result that $\sin(x)$ is strictly increasing for $x \in (0,\pi/2)$ cannot hold in any floating-point system in which x and $\sin(x)$ are in the same precision, since there are more floating-point numbers in the domain $(0,\pi/2)$ than there are in $(0,1)$, the range of $\sin(x)$.

Finally, we end this section by noting that overflows and underflows often cause problems in computations. After an overflow, $\pm\infty$ or NaN frequently propagates through the computation. Although this can sometimes yield a useful result, it is more often a signal of an error in the program or its input. On the other hand, continuing the computation with denormalized numbers or zero in place of an underflow can often yield a useful numerical result. However, there are cases when this can be disastrous. For example, if x^2 underflows, but y^2 is not too close to the underflow limit, then $fl(\sqrt{x^2 + y^2})$, the computed value of $\sqrt{x^2 + y^2}$, is still accurate. However, if both x^2 and y^2 underflow to 0, then $fl(\sqrt{x^2 + y^2}) = 0$, although $\sqrt{x^2 + y^2} \geq \max(|x|,|y|)$ may be far from the underflow limit.

It is often possible to ensure that overflows do not occur and that underflows are harmless. For the example considered above, note that $\sqrt{x^2 + y^2} = s\sqrt{(x/s)^2 + (y/s)^2}$ for any scaling factor $s > 0$. If we choose $s = 2^k$ for an integer $k \approx \log_2[\max(|x|,|y|)]$, then neither $(x/s)^2$ nor $(y/s)^2$ can overflow, and any underflow that occurs is harmless, since one of $(x/s)^2$ or $(y/s)^2$ is close to 1. Moreover, in IEEE floating-point arithmetic, multiplying and dividing by $s = 2^k$ does not introduce any additional rounding error into the computation.

A similar problem with overflows and underflows occurs in formula (8) and in many other numerical computations. Overflows can be avoided and underflows can be rendered harmless in computing $\sqrt{b^2 - 4ac}$ in (8) by scaling in much the same way as described above for $\sqrt{x^2 + y^2}$.

2. THE DIRECT SOLUTION OF LINEAR ALGEBRAIC SYSTEMS

In this section, we consider the *direct* solution of linear algebraic systems of the form $Ax = b$, where $A \in \mathbb{R}^{n\times n}$ (or $\mathbb{C}^{n\times n}$) is a nonsingular matrix and x and $b \in \mathbb{R}^n$ (or $\mathbb{C}^n$). A numerical method for solving $Ax = b$ is direct if it computes the exact solution of the system when implemented in exact arithmetic. Iterative methods for nonsingular linear systems and methods for overdetermined and underdetermined systems are considered in Secs. 4 and 5, respectively.

The standard direct methods for solving $Ax = b$ are based on, or closely related to, *Gaussian elimination* (GE), the familiar variable elimination technique that reduces the original system $Ax = b$ to an *upper-triangular system,* $Ux = \tilde{b}$, which has the same solution x. We present a simple form of GE in Sec. 2.1 and show how $Ux = \tilde{b}$ can be solved easily by *back substitution* in Sec. 2.2. We then explain how the simple form of GE presented in Sec. 2.1 for $Ax = b$ relates to the *LU* factorization of the coefficient matrix A in Sec. 2.3 and to *forward elimination* in Sec. 2.4. Enhancements to this simple form of GE to make it an efficient, robust, reliable numerical method for the solution of linear systems are outlined in Sec. 2.5. The closely related Cholesky factorization for symmetric positive-definite matrices is presented in Sec. 2.6. We consider how to adapt these methods to banded and sparse linear systems in Sec. 2.7. We end with a discussion of the effects of rounding errors on the direct methods in Sec. 2.8 and of *iterative improvement,* a technique to ameliorate these effects, in Sec. 2.9. See Sec. 13 for a discussion of sources of high-quality numerical software for solving systems of linear algebraic equations.

GE can also be applied to singular systems of linear equations or overdetermined or underdetermined linear systems of m equations in n unknowns. However, it is not as robust as the methods discussed in Sec. 5 for these problems, and so we do not present these generalizations of GE here. In addition, we note that there are several mathematically equivalent, but computationally distinct, implementations of GE and the factorizations discussed here. The reader interested in a more comprehensive treatment of these topics should consult an advanced text, such as Golub and Van Loan (1989).

Finally we note that it is generally inadvisable to solve a system $Ax = b$ by first computing A^{-1} and then calculating $x = A^{-1}b$. The techniques discussed in this section are usually both more reliable and more cost effective than methods using A^{-1}. We also note that, although Cramer's rule is a useful theoretical tool, it is an extremely ineffective computational scheme.

2.1 Gaussian Elimination

First, to unify the notation used below, let $A_0 = A$ and $b_0 = b$.

GE for $Ax = b$ proceeds in $n - 1$ stages. For $k = 1, \ldots, n - 1$, we begin stage k of GE with the reduced system $A_{k-1}x = b_{k-1}$, where columns $1, \ldots, k - 1$ of A_{k-1} contain 0's below the main diagonal. That is, for $A_{k-1} = [a_{ij}^{(k-1)}]$, $a_{ij}^{(k-1)} = 0$ for $j = 1, \ldots, k - 1$ and $i = j + 1, \ldots, n$. This corresponds to the variables $x_1, \ldots, x_{i-1}$ having been eliminated from equation i of $A_{k-1}x = b_{k-1}$ for $i = 2, \ldots, k - 1$ and the variables $x_1, \ldots, x_{k-1}$ having been eliminated from the remaining equations $k, \ldots, n$. Moreover, x is the unique solution of both $A_{k-1}x = b_{k-1}$ and $Ax = b$. Note that all the assumptions above hold vacuously for $k = 1$, since, in this case, the "reduced" system $A_0x = b_0$ is just the original system $Ax = b$ from which no variables have yet been eliminated.

During stage k of GE, we further the reduction process by eliminating the variable x_k from row i of $A_{k-1}x = b_{k-1}$ for $i = k + 1, \ldots, n$ by multiplying row k of this system by $m_{ik} = a_{ik}^{(k-1)}/a_{kk}^{(k-1)}$ and subtracting it from row i.

Note that the multipliers m_{ik} are not properly defined by $m_{ik} = a_{ik}^{(k-1)}/a_{kk}^{(k-1)}$ if the *pivot element* $a_{kk}^{(k-1)} = 0$. In exact arithmetic, this can happen only if the $k \times k$ leading principal minor of A is singular. We consider how to deal with zero (or nearly zero) pivots in Sec. 2.5. For now, though, we say that this simple form of GE "breaks down" at stage k and we terminate the process.

After the last stage of GE, the original system $Ax = b$ has been reduced to $Ux = \tilde{b}$, where $\tilde{b} = b_{n-1}$ and $U = A_{n-1}$. Note that $U = [u_{ij}]$ is an upper-triangular matrix (i.e., $u_{ij} = 0$ for $1 \leq j < i \leq n$) with $u_{kk} = a_{kk}^{(k-1)}$ for $k = 1, \ldots, n$. Therefore, if A and all its leading principal minors are nonsingular, then $u_{kk} = a_{kk}^{(k-1)} \neq 0$ for $k = 1, \ldots, n$, so U is nonsingular too. Moreover, in this case, x is the unique solution of both $Ax = b$ and $Ux = \tilde{b}$. The latter system can be solved easily by back substitution, as described in Sec. 2.2.

The complete GE process can be written in pseudocode as shown in Table 1. Note that at stage k of GE, we know the elements $a_{ik}^{(k)} = a_{ik}^{(k-1)} - m_{ik} \cdot a_{kk}^{(k-1)} = 0$ for $i = k + 1, \ldots, n$, and so we do not need to perform this calculation explicitly. Consequently, instead of j running for $k, \ldots, n$ in Table 1, as might be expected, j runs from $k + 1, \ldots, n$ instead.

To reduce the storage needed for GE, the original matrix A and vector b are often overwritten by the intermediate matrices A_k and vectors b_k, so that at the end of the GE process, the upper-triangular part of A contains U and b contains $\tilde{b}$. The only change required to the algorithm in Table 1 to implement this reduced storage scheme is to remove the superscripts from the coefficients a_{ij} and b_i. As explained below, it is also important to store the multipliers m_{ik}. Fortunately, the $n - k$ multipliers $\{m_{ik} : i = k + 1, \ldots, n\}$ created at stage k of GE can be stored in the $n - k$ positions in column k of A below the main diagonal that are elimi-

Table 1. Gaussian elimination (GE) for the system $Ax = b$.

```
for k = 1, . . . , n − 1 do
    for i = k + 1, . . . , n do
        m_ik = a_ik^(k−1)/a_kk^(k−1)
        for j = k + 1, . . . , n do
            a_ij^(k) = a_ij^(k−1) − m_ik·a_kj^(k−1)
        end
        b_i^(k) = b_i^(k−1) − m_ik·b_k^(k−1)
    end
end
```

nated in stage k. Thus, in many implementations of GE, the upper-triangular part of A is overwritten by U and the strictly lower part of A is overwritten by the multipliers m_{ik} for $1 \le k < i \le n$.

A straightforward count of the operations in Table 1 shows that GE requires $n(n-1)/2 \approx n^2/2$ divisions to compute the multipliers m_{ik}, $n(2n-1)(n-1)/6 \approx n^3/3$ multiplications and subtractions to compute the coefficients of U, and $n(n-1)/2 \approx n^2/2$ multiplications and subtractions to compute the coefficients of $\tilde{b}$. Since multiplications and subtractions (or additions) occur in pairs so frequently in matrix calculations, we refer to this pair of operations as a *flop,* which is short for *floating-point operation.* Thus, the computational work required to reduce a system $Ax = b$ of n equations in n unknowns to $Ux = \tilde{b}$ is about $n^3/3$ flops. We show in Sec. 2.2 that $Ux = \tilde{b}$ can be solved by *back substitution* using n divisions and about $n^2/2$ flops.

2.2 Back Substitution

Let $U = [u_{ij}]$ be an $n \times n$ nonsingular upper-triangular matrix. That is, $u_{ij} = 0$ for $1 \le j < i \le n$ and $u_{ii} \ne 0$ for $1 \le i \le n$. Then the linear algebraic system $Ux = \tilde{b}$ can be solved easily by back substitution, as shown in Table 2.

It is easy to see from Table 2 that back substitution requires n divisions and $n(n-1)/2 \approx n^2/2$ multiplications and subtractions. So the computational work is about $n^2/2$ flops.

2.3 The *LU* Factorization

Applying Gaussian elimination (GE) as described in Sec. 2.1 to solve the linear algebraic system $Ax = b$ of n equations in n unknowns is closely related to computing the

Table 2. Back substitution to solve $Ux = \tilde{b}$.

for $i = n, \dots, 1$ do

$$x_i = \left(\tilde{b}_i - \sum_{j=i+1}^{n} u_{ij}x_j\right) \Big/ u_{ii}$$

end

LU factorization of the matrix A, where $L_1 = [l_{ij}^{(1)}]$ is a unit lower-triangular matrix (i.e., $l_{ii}^{(1)} = 1$ for $i = 1, \dots, n$ and $l_{ij}^{(1)} = 0$ for $1 \le i < j \le n$) and $U_1 = [u_{ij}^{(1)}]$ is an upper-triangular matrix (i.e., $u_{ij}^{(1)} = 0$ for $1 \le j < i \le n$) satisfying

$$A = L_1 U_1. \tag{9}$$

The factorization (9) exists and is unique if and only if all the leading principal minors of A are nonsingular. In this case, it can be shown that the matrix U_1 in (9) is the same as the upper-triangular matrix U produced by GE, and the elements in the strictly lower-triangular part of $L_1 = [l_{ij}^{(1)}]$ satisfy $l_{ij}^{(1)} = m_{ij}$ for $1 \le j < i \le n$, where the m_{ij} are the multipliers used in GE. Moreover, the U_1 in (9) is nonsingular if A is; L_1 is always nonsingular if the factorization exists.

From the discussion in Sec. 2.1, it follows that computing the *LU* factorization of an $n \times n$ matrix A in this way requires $n(n-1)/2 \approx n^2/2$ divisions and $n(2n-1)(n-1)/6 \approx n^3/3$ multiplications and subtractions. Thus, the computational work required to calculate it is about $n^3/3$ flops.

If we need to solve $m > 1$ systems $Ax_i = b_i$, $i = 1, \dots, m$, (or $AX = B$, where X and $B \in \mathbb{R}^{n\times m}$ or $\mathbb{C}^{n\times m}$), we may obtain significant computational savings by computing the *LU* factorization of A once only and using the factors L_1 and U_1 to solve each system $Ax_i = b_i$ for $i = 1, \dots, m$ by first solving $L_1\tilde{b}_i = b_i$ for $\tilde{b}_i$ by *forward elimination,* as described in Sec. 2.4, and then solving $U_1 x_i = \tilde{b}_i$ for x_i by back substitution, as outlined in Sec. 2.2. This procedure is essentially the same as performing GE to reduce A to U once only, saving the multipliers $\{m_{ik}\}$ used in the process, and then, for each system $Ax_i = b_i$, using the multipliers to perform the same transformation on b_i to produce $\tilde{b}_i$ and solving $Ux_i = \tilde{b}_i$ for x_i by back substitution. With either of these procedures, the computational work required to solve all m systems $Ax_i = b_i$ is about $n^3/3 + mn^2$ flops, whereas, if we apply GE as outlined in Sec. 2.1 to each system $Ax_i = b_i$, recomputing U each time, the computational work required to solve all m systems $Ax_i = b_i$ is about $m(n^3/3 + n^2)$ flops, which is much greater if m and/or n is large.

Finally, note that we intentionally used the same symbol $\tilde{b}_i$ for the solution of $L\tilde{b}_i =$

b_i and the transformed right-side vector produced by GE, since these vectors are identical.

2.4 Forward Elimination

Let $L = [l_{ij}]$ be an $n \times n$ lower-triangular matrix (i.e., $l_{ij} = 0$ for $1 \le i < j \le n$). If L is nonsingular too, then $l_{ii} \neq 0$ for $1 \le i \le n$ and so the linear algebraic system $L\tilde{b} = b$ can be solved easily by forward elimination, as shown in Table 3.

It is easy to see from Table 3 that forward elimination requires n divisions and $n(n-1)/2 \approx n^2/2$ multiplications and subtractions. So the computational work is about $n^2/2$ flops.

If $L = [l_{ij}]$ is unit lower-triangular (i.e., $l_{ii} = 1$ for $i = 1, \ldots, n$ as well as L being lower-triangular), as is the case for the L produced by the LU factorization described in Sec. 2.3, then the division by l_{ii} in Table 3 is not required, reducing the operation count slightly to about $n^2/2$ flops. However, we have presented the forward-elimination procedure in Table 3 with general $l_{ii} \neq 0$, since other schemes, such as the Cholesky factorization described in Sec. 2.6, produce a lower-triangular matrix that is typically not unit lower-triangular.

The name "forward elimination" for this procedure comes from the observation that it is mathematically equivalent to the forward-elimination procedure used in GE to eliminate the variables $x_1, \ldots, x_{i-1}$ from equation i of the original system $Ax = b$ to produce the reduced system $Ux = \tilde{b}$.

2.5 Scaling and Pivoting

As noted in Sec. 2.1, the simple form of Gaussian elimination (GE) presented there may "break down" at stage k if the pivot $a_{kk}^{(k-1)} = 0$. Moreover, even if $a_{kk}^{(k-1)} \neq 0$, but $|a_{kk}^{(k-1)}| \ll |a_{ik}^{(k-1)}|$ for some $i \in \{k+1, \ldots, n\}$, then $|m_{ik}| = |a_{ik}^{(k-1)}/a_{kk}^{(k-1)}| \gg 1$. So multiplying row k of A_{k-1} by m_{ik} and subtracting it from row i may produce large elements in the resulting row i of A_k, which in turn may produce still larger elements during later stages of the GE process. Since, as noted in Sec. 2.8, the bound on the rounding errors in the GE process is proportional to the largest element that occurs in A_k for $k = 0, \ldots, n-1$, creating large elements during the GE reduction process may introduce excessive rounding error into the computation, resulting in an unstable numerical process and destroying the accuracy of the LU factorization and the computed solution x of the linear system $Ax = b$. We present in this section *scaling* and *pivoting* strategies that enhance GE to make it an efficient, robust, reliable numerical method for the solution of linear systems.

Table 3. Forward elimination too solve $L\tilde{b} = b$.

for $i = 1, \ldots, n$ do

$$\tilde{b}_i = \left(b_i - \sum_{j=1}^{i-1} l_{ij}\tilde{b}_j\right)\Big/ l_{ii}$$

end

Scaling, often called *balancing* or *equilibration,* is the process by which the equations and unknowns of the system $Ax = b$ are scaled in an attempt to reduce the rounding errors incurred in solving the problem and improve its conditioning, as described in Sec. 2.8. The effects can be quite dramatic.

Typically, scaling is done by choosing two $n \times n$ diagonal matrices $D_1 = [d_{ij}^{(1)}]$ and $D_2 = [d_{ij}^{(2)}]$ (i.e., $d_{ij}^{(1)} = d_{ij}^{(2)} = 0$ for $i \neq j$) and forming the new system $\hat{A}\hat{x} = \hat{b}$, where $\hat{A} = D_1 A D_2^{-1}$, $\hat{x} = D_2 x$, and $\hat{b} = D_1 b$. Thus, D_1 scales the rows and D_2 scales the unknowns of $Ax = b$, or, equivalently, D_1 scales the rows and D_2^{-1} scales the columns of A. Of course, the solution of $Ax = b$ can be recovered easily from the solution of $\hat{A}\hat{x} = \hat{b}$, since $x = D_2^{-1}\hat{x}$. Moreover, if the diagonal entries $d_{11}^{(1)}, \ldots, d_{nn}^{(1)}$ of D_1 and $d_{11}^{(2)}, \ldots, d_{nn}^{(2)}$ of D_2 are chosen to be powers of the base of the floating-point number system (i.e., powers of 2 for IEEE floating-point arithmetic), then scaling introduces no rounding errors into the computation.

One common technique is to scale the rows only by taking $D_2 = I$ and choosing D_1 so that largest element in each row of the scaled matrix $\hat{A} = D_1 A$ is about the same size. A slightly more complicated procedure is to scale the rows and columns of A so that the largest element in each row and column of $\hat{A} = D_1 A D_2^{-1}$ is about the same size.

These strategies, although usually helpful, are not foolproof: it is easy to find examples for which row scaling or row and column scaling as described above makes the numerical solution worse. The best strategy is to scale on a problem-by-problem basis depending on what the source problem says about the significance of each coefficient a_{ij} in $A = [a_{ij}]$. See an advanced text such as Golub and Van Loan (1989) for a more detailed discussion of scaling.

For the remainder of this section, we assume that scaling, if done at all, has already been performed.

The most commonly used pivoting strategy is *partial pivoting*. The only modification required to stage k of GE described in Sec. 2.1 to implement GE with partial pivoting is to first search column k of A_{k-1} for the largest element $a_{ik}^{(k-1)}$ on or below the main diagonal. That is, find $i \in \{k, \ldots, n\}$ such that $|a_{ik}^{(k-1)}| \geq |a_{\mu k}^{(k-1)}|$ for $\mu = k, \ldots, n$. Then interchange equations i and k in the reduced system $A_{k-1}x = b_{k-1}$ and proceed with stage k of GE as described in Sec. 2.1.

After the equation interchange described above, the pivot element $a_{kk}^{(k-1)}$ satisfies $|a_{kk}^{(k-1)}| \geq |a_{ik}^{(k-1)}|$ for $i = k, \ldots, n$. So, if $a_{kk}^{(k-1)} \neq 0$, then the multiplier $m_{ik} = a_{ik}^{(k-1)}/a_{kk}^{(k-1)}$ must satisfy $|m_{ik}| \leq 1$ for $i = k + 1, \ldots, n$. Thus no large multipliers can occur in GE with partial pivoting.

On the other hand, if the pivot element $a_{kk}^{(k-1)} = 0$, then $a_{ik}^{(k-1)} = 0$ for $i = k, \ldots, n$, whence A_{k-1} is singular and so A must be too. Thus, GE with partial pivoting never "breaks down" (in exact arithmetic) if A is nonsingular.

Partial pivoting adds a little overhead only to the GE process. At stage k, we must perform $n - k$ comparisons to determine the row i with $|a_{ik}^{(k-1)}| \geq |a_{jk}^{(k-1)}|$ for $j = k, \ldots, n$. Thus, GE with partial pivoting requires a total of $n(n - 1)/2 \approx n^2/2$ comparisons. In addition, we must interchange rows i and k if $i > k$, or use some form of indirect addressing if the interchange is not performed explicitly. On the other hand, exactly the same number of arithmetic operations must be executed whether or not pivoting is performed. Therefore, if n is large, the added cost of pivoting is small compared with performing approximately $n^3/3$ flops to reduce A to upper-triangular form.

Complete pivoting is similar to partial pivoting except that the search for the pivot at stage k of GE is not restricted to column k of A_{k-1}. Instead, in GE with complete pivoting, we search the $(n - k) \times (n - k)$ lower right block of A_{k-1} for the largest element. That is, find i and $j \in \{k, \ldots, n\}$ such that $|a_{ij}^{(k-1)}| \geq |a_{\mu\nu}^{(k-1)}|$ for $\mu = k, \ldots, n$ and $\nu = k, \ldots, n$. Then interchange equations i and k and variables j and k in the reduced system $A_{k-1}x_{k-1} = b_{k-1}$ and proceed with stage k of GE as described in Sec. 2.1. Note that the vector x_{k-1} in the reduced system above is a reordered version of the vector of unknowns x in the original system $Ax = b$, incorporating the variable interchanges that have occurred in stages $1, \ldots, k - 1$ of GE with complete pivoting.

After the equation and variable interchanges described above, the pivot element $a_{kk}^{(k-1)}$ satisfies $|a_{kk}^{(k-1)}| \geq |a_{ik}^{(k-1)}|$ for $i = k, \ldots, n$. So, if $a_{kk}^{(k-1)} \neq 0$, the multiplier $m_{ik} = a_{ik}^{(k-1)}/a_{kk}^{(k-1)}$ must satisfy $|m_{ik}| \leq 1$ for $i = k + 1, \ldots, n$, as is the case with partial pivoting. However, with complete pivoting, the multipliers tend to be even smaller than they are with partial pivoting, since the pivots tend to be larger, and so the numerical solution might suffer less loss of accuracy due to rounding errors, as discussed further in Sec. 2.8.

After the row and column interchanges in stage k, the pivot element $a_{kk}^{(k-1)} = 0$ only if $a_{ij}^{(k-1)} = 0$ for $i = k, \ldots, n$ and $j = k, \ldots, n$ in which case A_{k-1} is singular and so A must be too. Thus, like GE with partial pivoting, GE with complete pivoting never "breaks down" (in exact arithmetic) if A is nonsingular.

Moreover, if A is singular and $a_{ij}^{(k-1)} = 0$ for $i = k, \ldots, n$ and $j = k, \ldots, n$, then the GE process can be terminated at this stage and the factorization computed so far used to advantage in determining a solution (or approximate solution) to the singular system $Ax = b$. However, this is not as robust a technique as the methods discussed in Sec. 4 for overdetermined problems, and so we do not discuss this further here. The reader interested in this application of GE should consult an advanced text such as Golub and Van Loan (1989).

Complete pivoting, unlike partial pivoting, adds significantly to the cost of the GE process. At stage k, we must perform $(n - k + 1)^2 - 1$ comparisons to determine the row i and column j with $|a_{ij}^{(k-1)}| \geq |a_{\mu\nu}^{(k-1)}|$ for $\mu = k, \ldots, n$ and $\nu = k, \ldots, n$. Thus, GE with partial pivoting requires a total of $n(n - 1)(2n + 5)/6 \approx n^3/3$ comparisons. On the other hand, exactly the same number of arithmetic operations must be executed whether or not pivoting is performed. So the cost of determining the pivots is comparable to the cost of performing approximately $n^3/3$ flops required to reduce A to upper-triangular form. In addition, we must interchange rows i and k if $i > k$ and columns j and k if $j > k$ or use some form of indirect addressing if the interchange is not performed explicitly. Thus even though GE with complete pivoting has better roundoff-error properties than GE with partial pivoting, GE with partial pivoting is used more often in practice.

As is the case for the simple version of GE presented in Sec. 2.1, GE with partial or complete pivoting is closely related to computing the *LU* factorization of the matrix A. However, in this case, we must account for the row or row and column interchanges by extending (9) to

$$P_2A = L_2U_2 \tag{10}$$

for partial pivoting and

$$P_3AQ_3^{\mathrm{T}} = L_3U_3 \tag{11}$$

for complete pivoting, where P_2 and P_3 are permutation matrices that record the row interchanges performed in GE with partial and complete pivoting, respectively; Q_3^{T} is a permutation matrix that records the column interchanges performed in GE with complete pivoting; L_2 and L_3 are unit lower-triangular matrices with the $n - k$ multipliers from stage k of GE with partial and complete pivoting, respectively, in column k below the main diagonal, but permuted according to the row interchanges that occur in stages $k + 1, \ldots, n - 1$ of GE with partial and complete pivoting, respectively; and U_2 and U_3 are the upper-triangular matrices produced by GE with partial and complete pivoting, respectively.

A permutation matrix P has exactly one 1 in each row and column and all other elements equal to 0. It is easy to check that $PP^{\mathrm{T}} = I$ so that P is nonsingular and $P^{\mathrm{T}} = P^{-1}$. That is, P is an orthogonal matrix. Also note that we do not need a full $n \times n$ array to store P: The information required to form or multiply by $P = [P_{ij}]$ can be stored in an n-vector $p = [p_i]$, where $p_i = j$ if and only if $P_{ij} = 1$ and $P_{ik} = 0$ for $k \neq j$.

The factorizations (9), (10), and (11) are all called *LU factorizations;* (10) is also called a *PLU factorization.* Unlike (9), the *LU* factorizations (10) and (11) always exist. P_2, P_3, Q_3, L_2, and L_3 are always nonsingular; U_2 and U_3 are nonsingular if and only if A is nonsingular. Moreover, the factorizations (10) and (11) are unique if there is a well-defined choice for the pivot if more than one element of maximal size occurs in the search for the pivot and if there is a well-defined choice for the multipliers in L if the pivot is zero.

The *LU* factorization (10) can be used to solve the linear system $Ax = b$ by first computing $\hat{b}_2 = P_2b$, then solving $L_2\tilde{b}_2 = \hat{b}_2$ by forward elimination, and finally solving $U_2x = \tilde{b}_2$ by back substitution. The steps are similar if we use the *LU* factorization (11) instead of (10), except that the back substitution $U_3\hat{x}_3 = \tilde{b}_3$ yields the permuted vector of unknowns $\hat{x}_3$. The original vector of unknowns x can be recovered by $x = Q_3^{\mathrm{T}}\hat{x}_3$. We have used $\tilde{b}_2$ and $\tilde{b}_3$ for the intermediate results above to emphasize that this is the same as the vector $\tilde{b}$ that is obtained if we perform GE with partial and complete pivoting, respectively, on the original system $Ax = b$.

As noted earlier for (9), if we need to solve $m > 1$ systems $Ax_i = b_i$, $i = 1, \ldots, m$ (or $AX = B$, where $X, B \in \mathbb{R}^{n\times m}$ or $\mathbb{C}^{n\times m}$), we may obtain significant computational savings by computing the *LU* factorization of A once only. The same observation applies to the *LU* factorizations (10) and (11).

We end by noting that not having to pivot to ensure numerical stability can be a great advantage in some cases—for example, when factoring a banded or sparse matrix, as described in Sec. 2.7. Moreover, there are classes of matrices for which pivoting is not required to ensure numerical stability. Three such classes are complex Hermitian positive-definite matrices, real symmetric positive-definite matrices, and diagonally dominant matrices.

2.6 The Cholesky Factorization

In this subsection, we present the *Cholesky factorization* of a real symmetric positive-definite $n \times n$ matrix A. It is straightforward to modify the scheme for complex Hermitian positive-definite matrices.

Recall that $A \in \mathbb{R}^{n \times n}$ is symmetric if $A = A^T$, where A^T is the transpose of A, and it is positive definite if $x^T A x > 0$ for all $x \in \mathbb{R}^n$, $x \neq 0$. The Cholesky factorization exploits these properties of A to compute a lower-triangular matrix L satisfying

$$A = LL^T. \tag{12}$$

The similar *LDL factorization* computes a unit lower-triangular matrix $\tilde{L}$ and a diagonal matrix D satisfying

$$A = \tilde{L} D \tilde{L}^T. \tag{13}$$

We present the *dot product* form of the Cholesky factorization in Table 4. It is derived by equating the terms of $A = [a_{ij}]$ to those of LL^T in the order (1,1), (2,1), . . . , $(n,1)$, (2,2), (3,2), . . . , $(n,2)$, . . . , (n,n) and using the lower-triangular structure of $L = [l_{ij}]$ (i.e., $l_{ij} = 0$ for $1 \leq i < j \leq n$). Other forms of the Cholesky factorization are discussed in advanced texts such as Golub and Van Loan (1989).

It can be shown that, if A is symmetric positive-definite, then

$$a_{jj} - \sum_{k=1}^{j-1} l_{jk}^2 > 0$$

(in exact arithmetic) each time this expression is computed in the Cholesky factorization. Therefore, we may take the associated square root to be positive, whence $l_{jj} > 0$ for $j = 1, \ldots, n$. With this convention, the Cholesky factorization is unique.

Moreover, it follows from

$$l_{jj} = \left(a_{jj} - \sum_{k=1}^{j-1} l_{jk}^2\right)^{1/2}$$

Table 4. The Cholesky factorization of a symmetric positive-definite matrix A.

for $j = 1, \ldots, n$ do

$$l_{jj} = \sqrt{a_{jj} - \sum_{k=1}^{j-1} l_{jk}^2}$$

for $i = j + 1, \ldots, n$ do

$$l_{ij} = \left(a_{ij} - \sum_{k=1}^{j-1} l_{ik} l_{jk}\right) \Big/ l_{jj}$$

end

end

that

$$a_{jj} = \sum_{k=1}^{j} l_{jk}^2.$$

So the elements in row j of the Cholesky factor L are bounded by $\sqrt{a_{jj}}$ even if we do not pivot. Consequently, as noted in Sec. 2.5, it is customary to compute the Cholesky factorization without pivoting.

Note that the method in Table 4 accesses the lower-triangular part of A only, and so only those elements need to be stored. Moreover, if we replace l_{jj} and l_{ij} by a_{jj} and a_{ij}, respectively, in Table 4, then the modified algorithm overwrites the lower-triangular part of A with the Cholesky factor L.

A straightforward count of the operations in Table 4 shows that the Cholesky factorization requires n square roots, $n(n - 1)/2 \approx n^2/2$ divisions, and $n(n - 1)(n + 1)/6 \approx n^3/6$ multiplications and subtractions. This is approximately half the arithmetic operations required to compute the LU factorization of A. Of course, the storage required for the Cholesky factorization is also about half that required for the LU factorization.

The Cholesky, LDL, and LU factorizations are closely related. Let $D_1 = [d_{ij}^{(1)}]$ be the diagonal matrix with the same diagonal elements as $L = [l_{ij}]$ (i.e., $d_{jj}^{(1)} = l_{jj}$ for $j = 1, \ldots, n$ and $d_{ij}^{(1)} = 0$ for $i \neq j$). D_1 is nonsingular, since, as noted above, $l_{jj} > 0$ for $j = 1, \ldots, n$. Moreover, $\tilde{L} = LD_1^{-1}$ is unit lower-triangular. If we also let $D = D_1 D_1 = D_1 D_1^T$, then

$$A = LL^T = (\tilde{L}D_1)(\tilde{L}D_1)^T = \tilde{L}D_1 D_1^T \tilde{L}^T = \tilde{L}D\tilde{L}^T,$$

where $\tilde{L}D\tilde{L}^T$ is the LDL factorization of A. Furthermore, if we let $U = D\tilde{L}^T$, then $\tilde{L}U$ is the LU factorization of A.

The LDL factorization can be computed directly by equating the terms of A to those of $\tilde{L}D\tilde{L}^T$, just as we did above for the Cholesky factorization. This leads to a scheme similar to that shown in Table 4, but without any square roots, although it has $n(n - 1)/2 \approx n^2/2$ more multiplications. Thus the cost of computing the factorization remains about $n^3/6$ flops and the storage requirement remains about $n^2/2$.

An advantage of the LDL factorization is that it can be applied to a symmetric indefinite matrix. The Cholesky factorization is not applicable in this case, since LL^T is always symmetric positive-semidefinite. However, since LDL^T is always symmetric, pivoting must be restricted to ensure that the reordered matrix PAQ^T is symmetric. The simplest way to maintain symmetry is to use *symmetric pivoting* in which $Q = P$. This, though, restricts the choice of the pivot at stage k of the LDL factorization to a_{jj} for $j = k, \ldots, n$. As a result, in some cases, the LDL factorization may incur much more rounding error than GE with partial or complete pivoting.

2.7 Banded and Sparse Matrices

Significant savings in both computational work and storage can often be obtained in solving $Ax = b$ by taking advantage of zeros in the coefficient matrix A. We outline in this subsection how these savings may be realized for banded and more general sparse matrices.

To begin, note that an $n \times n$ matrix A is said to be *sparse* if the number of nonzero elements in A is much less than n^2, the total number of elements in A. *Banded matrices* are an important subclass of sparse matrices in which the nonzero elements of the matrix are restricted to a band around the main diagonal of the matrix. The *lower bandwidth* of a matrix $A = [a_{ij}]$ is the smallest integer p such that $a_{ij} = 0$ for $i - j > p$, the *upper bandwidth* of A is the smallest integer q such that $a_{ij} = 0$ for $j - i > q$, and the *bandwidth* of A is $1 + p + q$. Clearly, if p and $q \ll n$, then A is sparse, since A has at most $(1 + p + q)n \ll n^2$ nonzero elements.

Banded and more general sparse matrices arise in many important applications. For example, quadratic spline interpolation, as described in Sec. 8.6.2, requires the solution of a linear system $Tc = g$, where T is a tridiagonal matrix (i.e., banded with $p = q = 1$) and c is the vector of coefficients for the quadratic spline interpolant. Banded matrices also arise in the solution of boundary-value problems for ordinary differential equations. See Sec. 11.3.1 for an example of a system $Tu = g$, where T is a symmetric positive-definite tridiagonal matrix. More general sparse matrices arise in the numerical solution of partial differential equations. See Sec. 11.3.2 for an example of the matrix associated with the standard five-point difference scheme for Poisson's equation.

If A is large and sparse, it is common to store only the nonzero elements of A, since this greatly reduces the storage requirements. This is easy to do if A is banded, since we can map the elements from the band of A to a $(1 + p + q) \times n$ array representing A in a *packed format*, several of which are commonly used in practice. If $A = [a_{ij}]$ is a general sparse matrix, then we require a more general *sparse-matrix data structure* that stores each nonzero element a_{ij} of A along with some information used to recover the indices i and j.

We consider Gaussian elimination (GE) for a banded matrix first. To begin, note that the reduction process described in Sec. 2.1 maintains the band structure of A. In particular, $a_{ik}^{(k-1)} = 0$ for $i - k > p$, whence the multipliers $m_{ik} = a_{ik}^{(k-1)}/a_{kk}^{(k-1)}$ in Table 1 need to be calculated for $i = k + 1, \ldots \min(k + p, n)$ only and the i loop can be changed accordingly. Similarly $a_{kj}^{(k-1)} = 0$ for $j - k > q$, and so the reduction $a_{ij}^{(k)} = a_{ij}^{(k-1)} - m_{ij}a_{kj}^{(k-1)}$ in Table 1 needs to be calculated for $j = k + 1, \ldots, \min(k + q, n)$ only and the j loop can be charged accordingly. It therefore follows from a straightforward operation count that GE modified as described above for banded matrices requires $np - p(p + 1)/2 \approx np$ divisions to compute the multipliers m_{ik}, either $npq - p(3q^2 + 3q + p^2 - 1)/6 \approx npq$ (if $p \leq q$) or $npq - q(3p^2 + 3p + q^2 - 1)/6 \approx npq$ (if $p \geq q$) multiplications and subtractions to compute the coefficients of U, and $np - p(p + 1)/2 \approx np$ multiplications and subtractions to compute the coefficients of $\tilde{b}$. Furthermore, note that, if we use this modified GE procedure to compute the LU factorization of A, then the lower-triangular matrix L has lower bandwidth p and the upper-triangular matrix U has upper bandwidth q. As noted in Sec. 2.1, it is common to overwrite A with L and U. This can be done even if A is stored in packed format, thereby achieving significant reduction in storage requirements.

The back-substitution method shown in Table 2 and the forward-elimination method shown in Table 3 can be modified similarly so that each requires n divisions, the back-substitution method requires $nq - q(q + 1)/2 \approx nq$ multiplications and subtractions, while the forward-elimination method requires $np - p(p + 1)/2 \approx np$ multiplications and subtractions. In addition, recall that the n divisions are not needed in forward elimination if L is unit lower-triangular, as is the case for the modified LU factorization described here.

A similar modification of the Cholesky method shown in Table 4 results in a procedure that requires n square roots, $np - p(p + 1)/2 \approx np$ divisions, and $(n - p)p(p + 1)/2 + (p - 1)p(p + 1)/6 \approx np^2/2$ multiplications and subtractions. In deriving these operation counts, we used $p = q$, since the matrix A must be symmetric for the Cholesky factorization to be applicable. Moreover, the Cholesky factor L has lower bandwidth p. Thus, as for the general case, the Cholesky factorization of a band matrix requires about half as many arithmetic operations and about half as much storage as the LU factorization, since packed storage can also be used for the Cholesky factor L.

If partial pivoting is used in the LU factorization of A, then the upper bandwidth of U may increase to $p + q$. The associated matrix L is a permuted version of a lower-triangular matrix with lower bandwidth p. Both factors L and U can be stored in packed format in a $(1 + 2p + q) \times n$ array. The operation count for the LU factorization, forward elimination, and back solution is the same as though A were a banded matrix with upper bandwidth $p + q$ and lower bandwidth p. Thus, if $p > q$, both computational work and storage can be saved by factoring A^{T} instead of A and using the LU factors of A^{T} to solve $Ax = b$. If complete pivoting is used in the LU factorization of A, then L and U may fill in so much that there is little advantage to using a band solver. Consequently, it is advantageous not to pivot when factoring a band matrix, provided this does not lead to an unacceptable growth in rounding errors. As noted in Sec. 2.5, it is not necessary to pivot for numerical stability if A is complex Hermitian positive definite, real symmetric positive definite, or column-diagonally dominant. If pivoting is required, it is advantageous to use partial, rather than complete, pivoting, again provided this does not lead to an unacceptable growth in rounding errors.

Extending GE to take advantage of the zeros in a general sparse matrix is considerably more complicated than for banded matrices. The difficulty is that, when row k of A_{k-1} is multiplied by m_{ik} and added to row i of A_{k-1} to eliminate $a_{ik}^{(k-1)}$ in stage k of GE, as described in Sec. 2.1, some zero elements in row i of A_{k-1} may become nonzero in the resulting row i of A_k. These elements are said to *fill in* and are collectively referred to as *fill-in* or *fill.*

However, pivoting can often greatly reduce the amount of fill-in. To see how this comes about, the interested reader may wish to work through an example with the *arrowhead* matrix

$$A = \begin{pmatrix} 5 & 1 & 1 & 1 & 1 \\ 1 & 1 & 0 & 0 & 0 \\ 1 & 0 & 1 & 0 & 0 \\ 1 & 0 & 0 & 1 & 0 \\ 1 & 0 & 0 & 0 & 1 \end{pmatrix}.$$

Since A is a real symmetric positive-definite matrix, there is no need to pivot for numerical stability. It is easy to see that the Cholesky and LU factors of A completely fill in. Interchanging the first and last rows and the first and last columns of A corresponds to forming the permuted matrix $B = PAP^{\mathrm{T}}$, where $P = I - [1,0,0,0,-1]^{\mathrm{T}}[1,0,0,0,-1]$ is a permutation matrix. Since B is also a real symmetric positive-definite matrix, there is no need to pivot for numerical stability when factoring B. However, in this case, the Cholesky and LU factors of B suffer no fill-in at all. This small example can be generalized easily to arbitrarily large arrowhead matrices having the similar properties that the LU factors of A completely fill in while those of the permuted matrix $B = PAP^{\mathrm{T}}$ suffer no fill-in at all.

If we use an appropriate sparse-matrix data structure to store only the nonzeros elements of a sparse matrix A and its LU factors, then reducing the fill-in reduces both the storage needed for the LU factors and the computational work required to calculate them. It also reduces the computational work required for forward elimination and back substitution, since the methods shown in Tables 2 and 3 can be modified easily so that they use the nonzero elements in L and U only, thereby avoiding multiplications by zero and the associated subtractions.

Therefore, the goal in a sparse LU or Cholesky factorization is to reorder the rows and columns of A to reduce the amount of fill-in.

If we need to pivot to ensure numerical stability, then this might conflict with pivoting to reduce fill-in. Unfortunately, even without this complication, finding the optimal reordering to minimize the fill-in is computationally too expensive to be feasible in general. There are, though, many good *heuristic* methods to reorder the rows and columns of A that greatly reduce the fill-in and computational work in many important cases. The reader seeking a more complete description of sparse-matrix factorizations should consult a text on this topic, such as Duff *et al.* (1986) or George and Liu (1981).

We end this subsection with an example illustrating the importance of exploiting the zeros in a sparse matrix A. Consider the linear system derived in Sec. 11.3.2 by discretizing Poisson's equation on an $m \times m$ grid. A is an $n \times n$ symmetric positive-definite matrix with $n = m^2$ and $5m^2 - 4m \approx 5m^2 = 5n$ nonzero elements, out of a total of n^2 elements in A. So, if m is large, A is very sparse. Moreover, if we use the *natural ordering* for the equations and variables in the system, then A, as shown in Sec. 11.3.2, is a banded matrix with lower and upper bandwidth $m = \sqrt{n}$.

The computational work and storage required to solve this linear system are shown in Table 5. We consider three cases:

1. dense, the zeros in A are not exploited at all;
2. banded, we use a band solver that exploits the band structure of A;
3. sparse, we use a general sparse solver together with the *nested dissection ordering* (see George and Liu, 1981) for the equations and variables in the system.

The columns labeled factor and solve, respectively, give the approximate number of flops needed to compute the Cholesky factorization of the matrix A and to solve the linear system given the factors. The columns labeled store A and store L, respectively, give the approximate number of storage locations needed to store A and its Cholesky factor L.

Table 5. Computational work and storage required to solve the linear system $Ax = b$ derived from discretizing Poisson's equation on an $m \times m$ grid.

	Factor	Solve	Store A	Store L
Dense	$m^6/6$	m^4	$m^4/2$	$m^4/2$
Banded	$m^4/2$	$2m^3$	m^3	m^3
Sparse	$O(m^3)$	$O(m^2 \log m)$	$5m^2$	$O(m^2 \log m)$

2.8 Rounding Errors, Condition Numbers, and Error Bounds

In this subsection, we consider the effects of rounding errors in solving $Ax = b$. In doing so, we use vector and matrix norms extensively. Therefore, we recommend that, if you are not acquainted with norms, you familiarize yourself with this topic before reading this subsection. Most introductory numerical methods texts or advanced books on numerical linear algebra contain a section on vector and matrix norms.

The analysis of the effects of rounding errors in solving $Ax = b$ usually proceeds in two stages. First we establish that the computed solution $\tilde{x}$ is the exact solution of a perturbed system

$$(A + E)\tilde{x} = b + r \tag{14}$$

with bounds on the size of E and r. Then we use (14) together with bounds on the size of A and A^{-1} to bound the error $x - \tilde{x}$.

The first step is called a *backward error analysis,* since it casts the error in the solution back onto the problem and allows us to relate the effects of rounding errors in the computed solution $\tilde{x}$ to other errors in the problem, such as measurement errors in determining the coefficients of A and b. Another advantage of proceeding in this two-stage fashion is that, if $\tilde{x}$ is not sufficiently accurate, it allows us to determine whether this is because the numerical method is faulty or whether the problem itself is unstable.

Throughout this section we assume that A is an $n \times n$ nonsingular matrix and $nu < 0.1$, where u is the *relative roundoff error bound* for the machine arithmetic used in the computation (see Sec. 1.2).

If we use Gaussian elimination (GE) with either partial or complete pivoting together with forward elimination and back substitution to solve $Ax = b$, then it can be shown that the computed solution $\tilde{x}$ satisfies (14) with $r = 0$ and

$$\|E\|_\infty \leq 8n^3\gamma\|A\|_\infty u + O(u^2), \tag{15}$$

where

$$\gamma = \max_{i,j,k} \frac{|a_{ij}^{(k)}|}{\|A\|_\infty}$$

is the *growth factor* and $A_k = [a_{ij}^{(k)}]$ for $k = 0, \ldots, n - 1$ are the intermediate reduced matrices produced during the GE process (see Secs. 2.1 and 2.5). It can be shown that $\gamma \leq 2^{n-1}$ for GE with partial pivoting and $\gamma \leq [n(2 \times 3^{1/2} \times 4^{1/3} \times \cdots n^{1/(n-1)})]^{1/2} \ll 2^{n-1}$ for GE with complete pivoting. Moreover, the former upper bound can be achieved. However, for GE with both partial and complete pivoting, the actual error incurred is usually much smaller than the bound (15) suggests: it is typically the case that $\|E\|_\infty \propto \|A\|_\infty u$. Thus, GE with partial pivoting usually produces a computed solution with a small backward error, but, unlike GE with complete pivoting, there is no guarantee that this will be the case.

If A is column-diagonally dominant, then applying GE without pivoting to solve $Ax = b$ is effectively the same as applying GE with partial pivoting to solve this system. Therefore, all the remarks above for GE with partial pivoting apply in this special case.

If we use the Cholesky factorization without pivoting together with forward elimination and back substitution to solve $Ax = b$, where A is a symmetric positive-definite matrix, then it can be shown that the computed solution $\tilde{x}$ satisfies (14) with $r = 0$ and

$$\|E\|_2 \leq c_n \|A\|_2 u, \tag{16}$$

where c_n is a constant of moderate size that depends on n only. Thus the Cholesky factorization without pivoting produces a solution with a small backward error in all cases.

Similar bounds on the backward error for other factorizations and matrices with special properties, such as symmetric or band matrices, can be found in advanced texts, such as Golub and Van Loan (1989).

It is also worth noting that we can easily compute an *a posteriori* backward error estimate of the form (14) with $E = 0$ by calculating the *residual* $r = A\tilde{x} - b$ after computing $\tilde{x}$. Typically $\|r\|_\infty \propto \|b\|_\infty u$ if GE with partial or complete pivoting or the Cholesky factorization is used to compute $\tilde{x}$.

Now we use (14) together with bounds on the size of A and A^{-1} to bound the error $x - \tilde{x}$. To do so, we first introduce the *condition number* $\kappa(A) = \|A\| \, \|A^{-1}\|$ associated with the problem $Ax = b$. Although $\kappa(A)$ clearly depends on the matrix norm used, it is roughly of the same magnitude for all the commonly used norms, and it is the magnitude only of $\kappa(A)$ that is important here. Moreover, for the result below to hold, we require only that the matrix norm associated with $\kappa(A)$ is submultiplicative (i.e., $\|AB\| \leq \|A\| \, \|B\|$ for all A and $B \in \mathbb{C}^{n \times n}$) and that it is consistent with the vector norm used (i.e., $\|Av\| \leq \|A\| \, \|v\|$ for all $A \in \mathbb{C}^{n \times n}$ and $v \in \mathbb{C}^n$). It follows immediately from these two properties that $\kappa(A) \geq 1$ for all $A \in \mathbb{C}^{n \times n}$. More importantly, it can be shown that, if $\|E\|/\|A\| \leq \delta$, $\|r\|/\|b\| \leq \delta$, and $\delta\kappa(A) = r < 1$, then $A + E$ is nonsingular and

$$\frac{\|x - \tilde{x}\|}{\|x\|} \leq \frac{2\delta}{1 - r} \kappa(A). \tag{17}$$

Moreover, for any given A, there are some b, E, and r for which $\|x - \tilde{x}\|/\|x\|$ is as large as the right side of (17) suggests it might be, although this is not the case for all b, E, and r. Thus we see that, if $\kappa(A)$ is not too large, then small relative errors $\|E\|/\|A\|$ and $\|r\|/\|b\|$ ensure a small relative error $\|x - \tilde{x}\|/\|x\|$, and so the problem is *well-conditioned*. On the other hand, if $\kappa(A)$ is large, then $\|x - \tilde{x}\|/\|x\|$ might be large even though $\|E\|/\|A\|$ and $\|r\|/\|b\|$ are small, and so the problem is *ill-conditioned*. Thus, as the name suggests, the condition number $\kappa(A)$ gives a good measure of the conditioning—or stability—of the problem $Ax = b$.

Combining the discussion above with the earlier observation that typically $\|r\|_\infty \propto \|b\|_\infty u$ if GE with partial or complete pivoting or the Cholesky factorization is used to compute $\tilde{x}$, we get the general rule of thumb that, if $u \approx 10^{-d}$ and $\kappa_\infty = \|A\|_\infty \|A^{-1}\|_\infty \approx 10^q$, then $\tilde{x}$ contains about $d - q$ correct digits.

Many routines for solving linear systems provide an estimate of $\kappa(A)$, although most do not compute $\|A\| \, \|A^{-1}\|$ directly, since they do not compute A^{-1}.

There are many other useful inequalities of the form (17). The interested reader should consult an advanced text, such as Golub and Van Loan (1989).

2.9 Iterative Improvement

The basis of *iterative improvement* is the observation that, if x_1 is an approximate solution to $Ax = b$, we can form the residual $r_1 = b - Ax_1$, which satisfies $r_1 = A(x - x_1)$, and then solve $Ad_1 = r_1$ for the difference $d_1 = x - x_1$ and finally compute the improved solution $x_2 = x_1 + d_1$. In exact arithmetic, $x_2 = x$, but, in floating-point arithmetic, $x_2 \neq x$ normally. So we can repeat the process using x_2 in place of x_1 to form another improved solution x_3, and so on. Moreover, if we have factored A to compute x_1, then, as noted in Secs. 2.3 and 2.5, there is relatively little extra computational work required to compute a few iterations of iterative improvement.

The catch here is that, as noted in Sec. 2.8, typically $\|r\|_\infty \propto \|b\|_\infty u$ if GE with partial or complete pivoting or the Cholesky factorization is used to compute $\tilde{x}$. So, if we compute $r_1 = b - Ax_1$ in the same precision, then r_1 will contain few if any correct digits. Consequently, using it in iterative improvement usually does not lead to a reduction of the error in x_2, although it may lead to a smaller E in (14) in some cases. However, if we compute $r_k = b - Ax_k$ in double precision for $k = 1, 2, \ldots$, then iterative improvement may be quite effective. Roughly speaking, if the relative roundoff error bound $u \approx 10^{-d}$ and the condition number $\kappa(A) = \|A\|\,\|A^{-1}\| \approx 10^q$, then after k iterations of iterative improvement the computed solution x_k typically has about $\min(d, k(d - q))$ correct digits. Thus, if $\kappa(A)$ is large, but not too large, and, as a result, the initial computed solution x_1 is inaccurate, but not completely wrong, then iterative improvement can be used to obtain almost full single-precision accuracy in the solution.

The discussion above can be made more rigorous by noting that iterative improvement is a basic iterative method of the form (19)–(20) for solving $Ax = b$ and applying the analysis in Sec. 3.1.

3. THE ITERATIVE SOLUTION OF LINEAR ALGEBRAIC SYSTEMS

In this section, we consider *iterative* methods for the numerical solution of linear algebraic systems of the form $Ax = b$, where $A \in \mathbb{R}^{n\times n}$ (or $\mathbb{C}^{n\times n}$) is a nonsingular matrix, x, and $b \in \mathbb{R}^n$ (or $\mathbb{C}^n$). Such schemes compute a sequence of approximations $x_1, x_2, \ldots$ to x in the hope that $x_k \to x$ as $k \to \infty$. Direct methods for solving $Ax = b$ are considered in Sec. 2.

Iterative methods are most frequently used when A is large and sparse, but not banded with a small bandwidth. Such matrices arise frequently in the numerical solution of partial differential equations (PDEs). See, for example, the matrix shown in Sec. 11.3.2 that is associated with the standard five-point difference scheme for Poisson's equation. In many such cases, iterative methods are more efficient than direct methods for solving $Ax = b$: They usually use far less storage and often require significantly less computational work as well.

We discuss *basic iterative methods* in Sec. 3.1 and the *conjugate-gradient acceleration* of these schemes in Sec. 3.2. A more complete description and analysis of these methods and other iterative schemes are provided in Axelsson (1994), Golub and Van Loan (1989), Hageman and Young (1981), Young (1971), and Varga (1962). See Sec. 13 for a discussion of sources of high-quality numerical software for solving systems of linear algebraic equations.

We discuss multigrid methods in Sec. 11.8, because these iterative schemes are so closely tied to the PDE that gives rise to the linear system $Ax = b$ to which they are applied.

3.1 Basic Iterative Methods

Many iterative methods for solving $Ax = b$ are based on splitting the matrix A into two parts, M and N, such that $A = M - N$ with M nonsingular. M is frequently called the *splitting matrix*. Starting from an initial guess x_0 for x, we compute $x_1, x_2, \ldots$ recursively from

$$Mx_{k+1} = Nx_k + b. \tag{18}$$

We call such a scheme a *basic iterative method*, but it is often also referred to as a *linear stationary method of the first degree*.

Since $N = M - A$, (18) can be rewritten as

$$Mx_{k+1} = (M - A)x_k + b = Mx_k + (b - Ax_k),$$

which is equivalent to

$$Md_k = r_k, \tag{19}$$

$$x_{k+1} = x_k + d_k, \tag{20}$$

where $r_k = b - Ax_k$ is the residual at iteration k. Although (18) and (19)–(20) are mathematically equivalent, it might be computationally more effective to implement a method in one form than the other.

Clearly, for either (18) or (19)–(20) to be effective,

1. it must be much easier to solve systems with M than with A, and
2. the iterates $x_1, x_2, \ldots$ generated by (18) or (19)–(20) must converge quickly to x, the solution of $Ax = b$.

To address point **2**, first note that, since $Ax = b$ and $A = M - N$, $Mx = Nx + b$. So, if the sequence $x_1, x_2, \ldots$ converges, it must converge to x. To determine whether the sequence $x_1, x_2, \ldots$ converges and, if so, how fast, subtract (18) from $Mx = Nx + b$ and note that the error $e_k = x - x_k$ satisfies the recurrence $Me_{k+1} = Ne_k$, or equivalently $e_{k+1} = Ge_k$, where

$$G = M^{-1}N = I - M^{-1}A$$

is the associated *iteration matrix*. So

$$e^k = G^k e_0. \tag{21}$$

Using (21), we can show that, starting from any initial guess x_0, the sequence $x_1, x_2, \ldots$ generated by (18) converges to x if and only if $\rho(G) < 1$, where

$$\rho(G) = \max\{|\lambda| : \lambda \text{ an eigenvalue of } G\}$$

is the *spectral radius* of G. Moreover, $\rho(G)$ is the "asymptotically average" amount by which the error e_k decreases at each iteration. Consequently, $(\log \epsilon)/\log \rho(G)$ is a rough estimate of the number of iterations of (18) required to reduce the initial error e_0 by a factor ϵ. Thus, it is common to define

$$R(G) = -\log \rho(G) \tag{22}$$

to be the *rate of convergence* (sometimes called the *asymptotic rate of convergence* or the *asymptotic average rate of convergence*) of the iteration (18).

One useful general result is that, if $A = M - N$ is Hermitian positive-definite and if the Hermitian matrix $M^{\mathrm{H}} + N$ is positive-definite too, then $\rho(G) < 1$ and the associated iteration (18) converges.

Possibly the simplest iterative scheme is the *RF method* (a variant of Richardson's method) for which $M = I$ and $N = I - A$, whence (18) reduces to

$$x_{k+1} = x_k + r_k,$$

where $r_k = b - Ax_k$ is the residual at iteration k. From the general discussion above, it follows that this scheme converges if and only if $\rho(I - A) < 1$. Because of this severe constraint on convergence, this scheme is not often effective in its own right, but it can be used productively as the basis for polynomial acceleration, as discussed in Sec. 3.2.

We describe the Jacobi, Gauss–Seidel, SOR, and SSOR methods next, and then consider their convergence. In describing them, we use the notation $A = D - L - U$, where D is assumed to be nonsingular and consists of the diagonal elements of A for the point variant of each method or the diagonal submatrices of A for the block variant. L and U are the negatives of the strictly lower- and upper-triangular parts of A, respectively, either point or block, as the case may be. Typically, the block variant of each method converges faster than the point version, but requires more computational work per iteration. Thus, it is usually not clear without additional analysis which variant will be more effective.

The *Jacobi iteration* takes $M_{\mathrm{J}} = D$ and $N_{\mathrm{J}} = L + U$, resulting in the recurrence

$$Dx_{k+1} = (L + U)x_k + b. \tag{23}$$

The associated iteration matrix is $G_{\mathrm{J}} = D^{-1}(L + U)$. The *Gauss–Seidel iteration* takes $M_{\mathrm{GS}} = D - L$ and $N_{\mathrm{GS}} = U$, resulting in the recurrence

$$(D - L)x_{k+1} = Ux_k + b. \tag{24}$$

The associated iteration matrix is $G_{\mathrm{GS}} = (D - L)^{-1}U$. Although (24) may at first appear a little more complicated than (23), it is in fact easier to implement in practice on a sequential machine, since one can overwrite x_k when computing x_{k+1} in (24), whereas this is generally not possible for (23). However, for the Gauss–Seidel iteration, the jth component of x_{k+1} might depend on the ith component of x_{k+1} for $i < j$, because of the factor L on the left side of (24). This often inhibits vectorization and parallelization of the Gauss–Seidel iteration. Note that the Jacobi iteration has no such dependence, and so might be more effective on a vector or parallel machine.

Relaxation methods for $Ax = b$ can be written in the form

$$x_{k+1} = x_k + \omega(\hat{x}_{k+1} - x_k), \tag{25}$$

where $\omega \neq 0$ is the relaxation parameter and $\hat{x}_{k+1}$ is computed from x_k by some other iterative method. The best known of these schemes is *successive overrelaxation* (SOR) for which

$$D\hat{x}_{k+1} = Lx_{k+1} + Ux_k + b. \tag{26}$$

Equations (25) and (26) can be combined to give

$$\left(\frac{1}{\omega}D - L\right)x_{k+1} = \left(\frac{1-\omega}{\omega}D + U\right)x_k + b, \tag{27}$$

which is an iteration of the form (18) with $M_{\text{SOR}}(\omega) = (1/\omega)D - L$ and $N_{\text{SOR}}(\omega) = [(1 - \omega)/\omega]D + U$. It follows immediately from (24) and (27) that the SOR iteration reduces to the Gauss–Seidel method if $\omega = 1$. Moreover, because of the similarity between (24) and (27), the SOR iteration shares with the Gauss–Seidel method the implementation advantages and disadvantages noted above.

Overrelaxation corresponds to choosing $\omega > 1$ in (25) or (27), while underrelaxation corresponds to choosing $\omega \in (0,1)$. Historically, $\omega > 1$ was used in SOR for the solution of elliptic PDEs—hence the name successive *over*relaxation—but underrelaxation is more effective for some problems. See for example Young (1971) for a more complete discussion.

The *symmetric SOR* (SSOR) method takes one half step of SOR with the equations solved in the standard order followed by one half step of SOR with the equations solved in the reverse order:

$$\left(\frac{1}{\omega}D - L\right)x_{k+1/2} = \left(\frac{1-\omega}{\omega}D + U\right)x_k + b, \tag{28}$$

$$\left(\frac{1}{\omega}D - U\right)x_{k+1} = \left(\frac{1-\omega}{\omega}D + L\right)x_{k+1/2} + b. \tag{29}$$

These two half steps can be combined into one step of the form (18) with

$$M_{\text{SSOR}}(\omega) = \frac{\omega}{2-\omega}\left(\frac{1}{\omega}D - L\right)D^{-1}\left(\frac{1}{\omega}D - U\right)$$

and

$$N_{\text{SSOR}}(\omega) = \frac{(\omega-1)^2}{\omega(2-\omega)}D + \frac{1-\omega}{2-\omega}(L + U) + \frac{\omega}{2-\omega}LD^{-1}U.$$

Note that, if $A = D - L - U$ is a real, symmetric, positive-definite matrix and $\omega \in (0,2)$, then $M_{\text{SSOR}}(\omega)$ is a real, symmetric, positive-definite matrix too, since, in this case, $\omega/(2 - \omega) > 0$, both D and D^{-1} are symmetric positive-definite, and $(1/\omega)D - L = [(1/\omega)D - U]^{\text{T}}$ is nonsingular. This property of $M_{\text{SSOR}}(\omega)$ plays an important role in the effective acceleration of the SSOR iteration, as discussed in Sec. 3.2.

We now consider the convergence of the Jacobi, Gauss–Seidel, SOR, and SSOR methods. It is easy to show that the Jacobi iteration (23) converges if A is either row- or column-diagonally dominant. It can also be shown that the Gauss–Seidel iteration (24) converges if A is Hermitian positive-definite. Furthermore, if A is *consistently ordered*, a property enjoyed by a large class of matrices, including many that arise from the discretization of PDEs [see, for example, Young (1971) for details], it can be shown that the Gauss–Seidel iteration converges twice as fast as the Jacobi iteration if either one converges.

The iteration matrix associated with the SOR iteration (27) is

$$G_{\text{SOR}}(\omega) = [M_{\text{SOR}}(\omega)]^{-1}N_{\text{SOR}}(\omega) = \left(\frac{1}{\omega}D - L\right)^{-1}\left(\frac{1-\omega}{\omega}D + U\right).$$

For any nonsingular A with nonsingular D, it

can be shown that $\rho(G_{SOR}(\omega)) \geq |\omega - 1|$ with equality possible if and only if all eigenvalues of $G_{SOR}(\omega)$ have magnitude $|\omega - 1|$. So, if $\omega \in \mathbb{R}$, as is normally the case, a necessary condition for the convergence of SOR is $\omega \in (0,2)$. It can also be shown that, if A is Hermitian, D is positive-definite, and $\omega \in \mathbb{R}$, then the SOR iteration converges if and only if A is positive-definite and $\omega \in (0,2)$. Furthermore, if A is consistently ordered and all the eigenvalues of $G_J = D^{-1}(L + U)$ are real and lie in $(-1,1)$, then the optimal choice of the SOR parameter ω is

$$\omega_0 = \frac{2}{1 + \sqrt{1 - [\rho(G_J)]^2}} \in (1,2)$$

and

$$\begin{aligned}\rho(G_{SOR}(\omega_0)) = \omega_0 - 1 &= \left(\frac{\rho(G_J)}{1 + \sqrt{1 - [\rho(G_J)]^2}}\right)^2 \\ &= \min_{\omega} \rho(G_{SOR}(\omega)) < \rho(G_1) \\ &= \rho(G_{GS}) = [\rho(G_J)]^2 < \rho(G_J).\end{aligned} \tag{30}$$

In many cases, though, it is not convenient to calculate the optimal ω. Hageman and Young (1981) discuss heuristics for choosing a "good" ω.

If A is a real, symmetric, positive-definite matrix, then the SSOR iteration (28)–(29) converges for any $\omega \in (0,2)$. Moreover, determining the precise value of the optimal ω is not nearly as critical for SSOR as it is for SOR, since, unlike SOR, the rate of convergence of SSOR is relatively insensitive to the choice of ω. However, for SSOR to be effective,

$$\rho(D^{-1}LD^{-1}U) \leq \tfrac{1}{4}$$

should be satisfied—or nearly so. If this is the case, then a good value for ω is

$$\omega_1 = \frac{2}{1 + \sqrt{2[1 - \rho(G_J)]}}$$

and

$$\rho(G_{SSOR}(\omega_1)) \leq \frac{1 - \{[1 - \rho(G_J)]/2\}^{1/2}}{1 + \{[1 - \rho(G_J)]/2\}^{1/2}}, \tag{31}$$

where $G_{SSOR}(\omega) = [M_{SSOR}(\omega)]^{-1}N_{SSOR}(\omega)$ is the SSOR iteration matrix.

For a large class of problems, including many that arise from the discretization of elliptic PDEs, $\rho(G_J) = 1 - \epsilon$ for some ϵ satisfying $0 < \epsilon \ll 1$. For such problems, if (30) is valid and (31) holds with $=$ in place of $\leq$, then

$$\rho(G_J) = 1 - \epsilon,$$

$$\rho(G_{GS}) = (\rho(G_J))^2 \approx 1 - 2\epsilon,$$

$$\rho(G_{SSOR}(\omega_1)) \approx 1 - 2\sqrt{\epsilon/2},$$

$$\rho(G_{SOR}(\omega_0)) \approx 1 - 2\sqrt{2\epsilon},$$

whence the rates of convergence for these schemes are

$$R(G_J) \approx \epsilon,$$

$$R(G_{GS}) = 2R(G_J) \approx 2\epsilon,$$

$$R(G_{SSOR}(\omega_1)) \approx 2\sqrt{\epsilon/2},$$

$$R(G_{SOR}(\omega_0)) \approx 2\sqrt{2\epsilon} \approx 2R(G_{SSOR}(\omega_1)),$$

showing that the Gauss–Seidel iteration converges twice as fast as the Jacobi iteration, the SOR iteration converges about twice as fast as the SSOR iteration, and the SOR and SSOR iterations converge much faster than either the Gauss–Seidel or Jacobi iteration. However, the SSOR iteration often has the advantage, not normally shared by the SOR method, that it can be accelerated effectively, as discussed in Sec. 3.2.

Another class of basic iterative methods is based on *incomplete factorizations*. For brevity, we describe only the subclass of *incomplete Cholesky factorizations* (ICFs) here; the other schemes are similar. See an advanced text, such as Axelsson (1994), for details.

For an ICF to be effective, A should be symmetric positive-definite (or nearly so), large, and sparse. If A is banded, then the band containing the nonzeros should also be sparse, as is the case for the discretization of Poisson's equation shown in Sec. 11.3.2. The general idea behind the ICFs is to compute a lower-triangular matrix L_{ICF} such that $M_{ICF} = L_{ICF}L_{ICF}^T$ is in some sense close to A and L_{ICF} is much sparser than L, the true Cholesky factor of A. Then employ the iteration (19)–(20) to compute a sequence of approximations $x_1, x_2, \ldots$ to x, the solution of $Ax =$

b. Note that (19) can be solved efficiently by forward elimination and back substitution as described in Secs. 2.4 and 2.2, respectively, since the factorization $M_{\mathrm{ICF}} = L_{\mathrm{ICF}} L_{\mathrm{ICF}}^{\mathrm{T}}$ is known. This scheme, or an accelerated variant of it, is often very effective if it converges rapidly and L_{ICF} is much sparser than L.

A simple, but often effective, way of computing $L_{\mathrm{ICF}} = [l_{ij}]$ is to apply the Cholesky factorization described in Table 4, but to set $l_{ij} = 0$ whenever $a_{ij} = 0$, where $A = [a_{ij}]$. Thus, L_{ICF} has the same sparsity pattern as the lower-triangular part of A, whereas the true Cholesky factor L of A might suffer significant fill-in, as described in Sec. 2.7. Unfortunately, this simple ICF is not always stable.

As noted in Sec. 2.9, iterative improvement is a basic iterative method of this form, although, for iterative improvement, the error $N = M - A$ is due entirely to rounding errors, whereas, for the incomplete factorizations considered here, the error $N = M - A$ is typically also due to dropping elements from the factors of M to reduce fill-in.

For a more complete discussion of ICFs and other incomplete factorizations, their convergence properties, and their potential for acceleration, see an advanced text such as Axelsson (1994).

We end this section with a brief discussion of *alternating-direction implicit* (ADI) methods. A typical example of a scheme of this class is the Peaceman–Rachford method

$$(H + \alpha_n I)x_{n+1/2} = b - (V - \alpha_n I)x_n, \qquad (32)$$

$$(V + \alpha_n' I)x_{n+1} = b - (H - \alpha_n' I)x_{n+1/2}, \qquad (33)$$

where $A = H + V$, $\alpha_n > 0$, $\alpha_n' > 0$, and A, H, and V are real, symmetric, positive-definite matrices. For many problems, it is possible to choose H, V, and $\{\alpha_n, \alpha_n'\}$, so that the iteration (32)–(33) converges rapidly, and it is much cheaper to solve (32) and (33) than it is to solve $Ax = b$. For example, for the standard five-point discretization of a separable two-dimensional elliptic PDE, H and V can be chosen to be essentially tridiagonal and the rate of convergence of (32)–(33) is proportional to $1/\log h^{-1}$, where h is the mesh size used in the discretization. In contrast, the rate of convergence for SOR with the optimal ω is proportional to h. Note that $h \ll 1/\log h^{-1}$ for $0 < h \ll 1$, supporting the empirical evidence that ADI schemes converge much more rapidly than SOR for many problems. However, ADI schemes are not applicable to as wide a class of problems as SOR is.

For a more complete discussion of ADI schemes, see an advanced text such as Varga (1962) or Young (1971).

3.2 The Conjugate-Gradient Method

The *conjugate-gradient* (CG) *method* for the solution of the linear system $Ax = b$ is a member of a broader class of methods often called polynomial acceleration techniques or Krylov-subspace methods. (The basis of these names is explained below.) Although many schemes in this broader class are very useful in practice, we discuss only CG here, but note that several of these schemes, including Chebyshev acceleration and GMRES, apply to more general problems than CG. The interested reader should consult an advanced text such as Axelsson (1994), Golub and Van Loan (1989), Hageman and Young (1981) for a more complete discussion of polynomial acceleration techniques and Krylov-subspace methods. The close relationship between CG and the Lanzcos method is discussed in Golub and Van Loan (1989).

The *preconditioned conjugate-gradient* (PCG) *method* can be viewed either as an acceleration technique for the basic iterative method (18) or as CG applied to the preconditioned system $M^{-1}Ax = M^{-1}b$, where the splitting matrix M of (18) is typically called a preconditioning matrix in this context. We adopt the second point of view in this subsection.

An instructive way of deriving PCG is to exploit its relationship to the minimization technique of the same name described in Sec. 7.5. To this end, assume that the basic iterative method (18) is *symmetrizable*. That is, there exists a real, nonsingular matrix W such that $S = WM^{-1}AW^{-1}$ is a real, symmetric, positive-definite (SPD) matrix. Consider the quadratic functional $F(y) = \frac{1}{2}y^{\mathrm{T}}Sy - y^{\mathrm{T}}\hat{b}$, where $\hat{b} = WM^{-1}b$ and $y \in \mathbb{R}^n$. It is easy to show that the unique minimum of $F(y)$ is the solution $\hat{x}$ of $S\hat{x} = \hat{b}$. It follows immediately from the relations for S and $\hat{b}$ given above that $x = W^{-1}\hat{x}$ is the solution of both the preconditioned system $M^{-1}Ax = M^{-1}b$ and the original system $Ax = b$. If we take $M = W = I$, then $S = W(M^{-1}A)W^{-1} = A$ and, if we assume that A is a real SPD matrix, PCG reduces to the standard (unpreconditioned) CG method.

It is easy to show that if both A and M are real SPD matrices, then $M^{-1}A$ is symmetrizable. Moreover, many important practical problems, such as the numerical solution of self-adjoint elliptic PDEs, give rise to matrices A that are real SPD. Furthermore, if A is a real SPD matrix, then the splitting matrix M associated with the RF, Jacobi, and SSOR [with $\omega \in (0,2)$] iterations is a real SPD matrix too, as is the M given by an incomplete Cholesky factorization, provided it exists. Hence, these basic iterative methods are symmetrizable in this case and so PCG can be used to accelerate their convergence. In contrast, the SOR iteration with the optimal ω is generally not symmetrizable. So PCG is not even applicable and more general Krylov-subspace methods normally do not accelerate its convergence.

We assume throughout the rest of this subsection that A and M are real SPD matrices, because this case arises most frequently in practice and also because it simplifies the discussion below. Using this assumption and applying several mathematical identities, we get the computationally effective variant of PCG shown in Table 6. Note that W does not appear explicitly in this algorithm. Also note that, if we choose $M = I$, then $\tilde{r}_k = r_k$ and the PCG method reduces to the unpreconditioned CG method for $Ax = b$.

Many other mathematically equivalent forms of PCG exist. Moreover, as noted above, a more general form of PCG can be used if $M^{-1}A$ is symmetrizable, without either A or M being a real SPD matrix. For a discussion of the points, see an advanced text such as Hageman and Young (1981).

Table 6. The preconditioned conjugate-gradient (PCG) method for solving $Ax = b$.

choose an initial guess x_0
compute $r_0 = b - Ax_0$
solve $M\tilde{r}_0 = r_0$
set $p_0 = \tilde{r}_0$
for $k = 0, 1, \ldots$ until convergence do
  $\alpha_k = r_k^T\tilde{r}_k / p_k^T A p_k$
  $x_{k+1} = x_k + \alpha_k p_k$
  $r_{k+1} = r_k - \alpha_k A p_k$
  solve $M\tilde{r}_{k+1} = r_{k+1}$
  $\beta_k = r_{k+1}^T\tilde{r}_{k+1} / r_k^T\tilde{r}_k$
  $p_{k+1} = \tilde{r}_{k+1} + \beta_k p_k$
end

Before considering the convergence of PCG, we introduce some notation. The *energy norm* of a vector $y \in \mathbb{R}^n$ with respect to a real SPD matrix B is $\|y\|_{B^{1/2}} = (y^TBy)^{1/2}$. The *Krylov subspace* of degree $k - 1$ generated by a vector v and a matrix W is $\mathcal{K}_k(v,W) = \text{span}\{v, Wv, \ldots, W^{k-1}v\}$.

It is easy to show that the r_k that occurs in Table 6 is the residual $r_k = b - Ax_k$ associated with x_k for the system $Ax = b$ and that $\tilde{r}_k = M^{-1}r_k = M^{-1}b - M^{-1}Ax_k$ is the residual associated with x_k for the preconditioned system $M^{-1}Ax = M^{-1}b$. Let $e_k = x - x_k$ be the error associated with x_k. It can be shown that the x_k generated by PCG is a member of the shifted Krylov subspace $x_0 + \mathcal{K}_k(\tilde{r}_0, M^{-1}A) \equiv \{x_0 + v : v \in \mathcal{K}_k(\tilde{r}_0, M^{-1}A)\}$. Hence, PCG is in the broader class of Krylov-subspace methods characterized by this property. Moreover, it can be shown that the x_k generated by PCG is the unique member of $x_0 + \mathcal{K}_k(\tilde{r}_0, M^{-1}A)$ that minimizes the energy norm of the error $\|e_k\|_{A^{1/2}} = (e_k^TAe_k)^{1/2}$ over all vectors of the form $e_k' = x - x_k'$, where x_k' is any other member of $x_0 + \mathcal{K}_k(\tilde{r}_0, M^{-1}A)$. Equivalently, $e_k = P_k^*(M^{-1}A)e_0$, where $P_k^*(z)$ is the polynomial that minimizes $\|P_k(M^{-1}A)e_0\|_{A^{1/2}}$ over all polynomials $P_k(z)$ of degree k that satisfy $P_k(0) = 1$. This result is the basis of the characterization that PCG is the optimal polynomial acceleration scheme for the basic iterative method (18).

In passing, note that the iterate x_k generated by the basic iteration (18) is in the shifted Krylov subspace $x_0 + \mathcal{K}_k(\tilde{r}_0, M^{-1}A)$ also and that, by (21), the associated error satisfies $e_k = (I - M^{-1}A)^k e_0$. So (18) is a Krylov-subspace method too and its error satisfies the polynomial relation described above with $P_k(z) = (1 - z)^k$. Thus, the associated PCG method is guaranteed to accelerate the convergence of the basic iteration (18)—at least when the errors are measured in the energy norm.

The characterization of its error discussed above can be very useful in understanding the performance of PCG. For example, it can be used to prove the *finite termination property* of PCG. That is, if $M^{-1}A$ has m distinct eigenvalues, then $x_m = x$, the exact solution of $Ax = b$. Since $M^{-1}A$ has n eigenvalues, $m \leq n$ always and $m \ll n$ sometimes. We caution the reader, though, that the argument used to prove this property of PCG assumes exact arithmetic. In floating-point arithmetic, we rarely get $e_m = 0$, although we frequently get $e_{\tilde{m}}$ small for some $\tilde{m}$ possibly a little larger than m.

The proof of the finite termination property can be extended easily to explain the rapid convergence of PCG when the eigenvalues of $M^{-1}A$ fall into a few small clusters. So a preconditioner M is good if the eigenvalues of $M^{-1}A$ are much more closely clustered than those of the unpreconditioned matrix A.

Because of the finite termination property, both CG and PCG can be considered direct methods. However, both are frequently used as iterative schemes, with the iteration terminated long before $e_m = 0$. Therefore, it is important to understand how the error decreases with k, the iteration count. To this end, first note that the characterization of the error ensures that $\|e_{k+1}\|_{A^{1/2}} \leq \|e_k\|_{A^{1/2}}$ with equality only if $e_k = 0$. That is, PCG is a *descent* method in the sense that some norm of the error decreases on every iteration. Not all iterative methods enjoy this useful property.

The characterization of the error can also be used to show that

$$\|e_k\|_{A^{1/2}} \leq 2\left(\frac{\sqrt{\lambda_n/\lambda_1} - 1}{\sqrt{\lambda_n/\lambda_1} + 1}\right)^k \|e_0\|_{A^{1/2}}, \tag{34}$$

where λ_n and λ_1 are the largest and smallest eigenvalues, respectively, of $M^{-1}A$. (Note that $\lambda_n, \lambda_1 \in \mathbb{R}$ and $\lambda_n \geq \lambda_1 > 0$ since $M^{-1}A$ is symmetrizable.) It follows easily from the definition of the energy norm and (34) that

$$\|e_k\|_2 \leq 2\sqrt{\kappa_2(A)}\left(\frac{\sqrt{\lambda_n/\lambda_1} - 1}{\sqrt{\lambda_n/\lambda_1} + 1}\right)^k \|e_0\|_2, \tag{35}$$

where $\kappa_2(A) = \|A\|_2\|A^{-1}\|_2$ is the condition number of A in the 2-norm (see Sec. 2.8). Although (35) is generally not as tight as (34), it may be more relevant to the practitioner. It follows from either (34) or (35) that, in this context, a preconditioner M is good if λ_n/λ_1 is (much) closer to 1 than is the ratio of the largest to smallest eigenvalues of A.

Unlike many other iterative methods, such as SOR, PCG does not require an estimate of any parameters, although some stopping procedures for PCG require an estimate of the extreme eigenvalues of A or $M^{-1}A$. See an advanced text such as Axelsson (1994), Golub and Van Loan (1989), Hageman and Young (1981) for details.

4. OVERDETERMINED AND UNDERDETERMINED LINEAR SYSTEMS

A linear system $Ax = b$, with $A \in \mathbb{R}^{m\times n}$ and $b \in \mathbb{R}^m$ given and $x \in \mathbb{R}^n$ unknown, is called *overdetermined* if $m > n$. Such a system typically has no solution. However, there is always an x that minimizes $\|b - Ax\|_2$. Such an x is called a *least-squares* solution to the overdetermined linear system $Ax = b$. Moreover, if $\text{rank}(A) = n$, then the least-squares solution is unique.

A linear system $Ax = b$, as above, is called *underdetermined* if $m < n$. Such a system typically has infinitely many solutions. If $Ax = b$ has a solution, then there is an x of minimum Euclidean norm, $\|x\|_2$, that satisfies $Ax = b$. Such an x is called a *least-squares* solution to the underdetermined linear system $Ax = b$. It can be shown that, if $\text{rank}(A) = m$, then the least-squares solution is unique.

In the following subsections, we describe methods to compute the least-squares solution of overdetermined and underdetermined linear systems, assuming that the matrix A has full rank, i.e., $\text{rank}(A) = \min(m,n)$. For the more general case of $\text{rank}(A) < \min(m,n)$, see an advanced text such as Golub and Van Loan (1989).

4.1 The Normal Equations for Overdetermined Linear Systems

Let $Ax = b$ be an overdetermined linear system with m equations in n unknowns and with $\text{rank}(A) = n$. The x that minimizes $\|b - Ax\|_2$ satisfies

$$\frac{\partial[(b - Ax)^T(b - Ax)]}{\partial x_j} = 0 \quad \text{for } j = 1, \ldots, n.$$

These relations can be rewritten in matrix form as $A^T(b - Ax) = 0$ or equivalently

$$A^TAx = A^Tb, \tag{36}$$

which is called the system of *normal equations* for the overdetermined linear system $Ax = b$. It can be shown that the matrix A^TA is symmetric and positive-definite if rank(A) $= n$. So the system (36) has a unique solution, which is the least-squares solution to $Ax = b$.

Note that the matrix A^TA is $n \times n$. Computing the matrix A^TA requires about $mn^2/2$ flops (floating-point operations), while computing A^Tb requires about mn flops, and solving (36) requires about $n^3/6$ flops, if we assume that the Cholesky factorization (see Sec. 2.6) is used.

The condition number of the matrix A^TA is often large. More specifically, it can be shown that if A is square and nonsingular, then the condition number of A^TA is the square of the condition number of A (see Sec. 2.8). As a result, the approach described above of forming and solving the normal equations (36) often leads to a serious loss of accuracy. Therefore, in Secs. 4.4 and 4.7, we discuss more stable alternatives for solving least-squares problems.

4.2 The Normal Equations for Underdetermined Linear Systems

Let $Ax = b$ be an underdetermined linear system with m equations in n unknowns and with rank(A) $= m$. It can be shown that the least-squares solution x to $Ax = b$ can be written in the form A^Ty for some $y \in \mathbb{R}^m$ that satisfies

$$AA^Ty = b. \tag{37}$$

This is a linear system of size $m \times m$, called the system of *normal equations* for the underdetermined linear system $Ax = b$. It can be shown that the matrix AA^T is symmetric and positive-definite if rank(A) $= m$. So the system (37) has a unique solution. The unique least-squares solution to $Ax = b$ is $x = A^Ty$.

Computing the matrix AA^T requires about $nm^2/2$ flops, while computing A^Ty requires about mn flops, and solving (37) requires about $m^3/6$ flops, if the Cholesky factorization (see Sec. 2.6) is used.

As is the case for overdetermined systems, the method of forming and solving the normal equations (37) to compute the least-squares solution to $Ax = b$ is numerically unstable in some cases. In Secs. 4.5 and 4.8, we discuss more stable alternatives.

4.3 Householder Transformations and the *QR* Factorization

An *orthogonal transformation* is a linear change of variables that preserves the length of vectors in the Euclidean norm. Examples are a rotation about an axis or a reflection across a plane. The following is an example of an orthogonal transformation $y = (y_1, y_2)$ to $x = (x_1, x_2)$:

$$x_1 = 0.6y_1 + 0.8y_2,$$

$$x_2 = 0.8y_1 - 0.6y_2.$$

It is easy to see that $\|x\|_2 = \|y\|_2$.

An *orthogonal matrix* is an $m \times n$ matrix Q with the property $Q^TQ = I$. Note that, if (and only if) Q is square, i.e., $m = n$, the relation $Q^TQ = I$ is equivalent to $QQ^T = I$ and so $Q^T = Q^{-1}$. An orthogonal transformation of y to x can be written as $x = Qy$, where Q is an orthogonal matrix. In the above example, the corresponding orthogonal matrix is

$$Q = \begin{pmatrix} 0.6 & 0.8 \\ 0.8 & -0.6 \end{pmatrix}.$$

If Q is an orthogonal matrix, then

$$\begin{aligned} \|x\|_2^2 &= \|Qy\|_2^2 = (Qy)^T(Qy) = y^T(Q^TQ)y \\ &= y^Ty = \|y\|_2^2, \end{aligned}$$

whence $\|x\|_2 = \|y\|_2$. This property is exploited in the methods described below to solve least-squares problems. Moreover, the numerical stability of these schemes is due at least in part to the related observation that orthogonal transformations do not magnify rounding error.

A *Householder transformation* or *Householder reflection* is an orthogonal matrix of the form $H = I - 2ww^T$, where $\|w\|_2 = 1$. Note that, when H is 2×2, the effect of H on a vector x is equivalent to reflecting the vector x across the plane perpendicular to w and passing through the origin of x. A Householder reflection can be used to transform a nonzero vector into one containing mainly zeros.

An $m \times n$ matrix $R = [r_{ij}]$ is *right trian-*

gular if $r_{ij} = 0$ for $i > j$. Note that if (and only if) R is square, i.e., $m = n$, the terms right triangular and upper triangular are equivalent.

Let A be an $m \times n$ matrix with $m \geq n$. The QR factorization of A expresses A as the product of an $m \times m$ orthogonal matrix Q and an $m \times n$ right-triangular matrix R. It can be computed by a sequence $H_1, H_2, \ldots, H_n$ of Householder transformations to reduce A to right-triangular form R. More specifically, this variant of the QR factorization proceeds in n steps (or $n - 1$ if $n = m$). Starting with $A_0 = A$, at step k for $k = 1, \ldots, n$, H_k is applied to the partially processed matrix A_{k-1} to zero the components $k + 1$ to m of column k of A_{k-1}. $Q = H_1H_2 \ldots H_n$ and $R = A_n$, the last matrix to be computed. For the details of the QR factorization algorithm using Householder transformations or other elementary orthogonal transformations, see Hager (1988) or Golub and Van Loan (1989).

4.4 Using the *QR* Factorization to Solve Overdetermined Linear Systems

Assume that A is an $m \times n$ matrix, with $m \geq n$ and rank$(A) = n$. Let $Ax = b$ be the linear system to be solved (in the least-squares sense, if $m > n$). Let $A = QR$ be the QR factorization of A. Note that

$$\begin{aligned}\|Ax - b\|_2 &= \|QRx - b\|_2 = \|Q(Rx - Q^{\mathrm{T}}b)\|_2 \\ &= \|Rx - Q^{\mathrm{T}}b\|_2.\end{aligned}$$

Therefore, we can use the QR factorization of A to reduce the problem of solving $Ax = b$ to that of solving $Rx = Q^{\mathrm{T}}b$, a much simpler task. We solve the latter by first computing $y = Q^{\mathrm{T}}b$. Let $\hat{y}$ be the vector consisting of the first n components of y and $\hat{R}$ the upper-triangular matrix consisting of the first n rows of R. Now solve $\hat{R}x = \hat{y}$ by back substitution (see Sec. 2.2). Then x is the least-squares solution to both $Rx = Q^{\mathrm{T}}b$ and $Ax = b$.

It can be shown that the QR factorization algorithm applied to an $n \times n$ linear system requires about $2n^3/3$ flops, which is about twice as many as the LU factorization needs. However, QR is a more stable method than LU and it requires no pivoting. The QR factorization algorithm applied to an $m \times n$ linear system requires about twice as many flops as forming and solving the normal equations. However, QR is a more stable method than solving the normal equations.

4.5 Using the *QR* Factorization to Solve Underdetermined Linear Systems

Assume that A is an $m \times n$ matrix, with $m < n$ and rank$(A) = m$. Let $Ax = b$ be the linear system to be solved (in the least-squares sense). Obtain the QR factorization of A^{T} by the QR factorization algorithm: $A^{\mathrm{T}} = QR$, where Q is an $n \times n$ orthogonal matrix and R is an $n \times m$ right-triangular matrix. Let $\hat{R}$ be the upper-triangular matrix consisting of the first m rows of R. Solve $\hat{R}^{\mathrm{T}}\hat{y} = b$ by back substitution (see Sec. 2.2) and let $y = (\hat{y}^{\mathrm{T}}, 0, \ldots, 0)^{\mathrm{T}} \in \mathbb{R}^n$. Note that y is the vector of minimal Euclidean norm that satisfies $R^{\mathrm{T}}y = b$. Finally compute $x = Qy$ and note that x is the vector of minimal Euclidean norm that satisfies $R^{\mathrm{T}}Q^{\mathrm{T}}x = b$ or equivalently $Ax = b$. That is, x is the least-squares solution to $Ax = b$.

4.6 The Gram–Schmidt Orthogonalization Algorithm

The *Gram–Schmidt orthogonalization* algorithm is an alternative to QR. The modified version of the algorithm presented below is approximately twice as fast as QR and more stable than solving the normal equations.

Assume that $A \in \mathbb{R}^{m \times n}$ with $m \geq n$ and that A has n linearly independent columns $\{a_j\}_{j=1}^n$. The Gram–Schmidt algorithm applied to A generates an $m \times n$ orthogonal matrix Q and an $n \times n$ upper-triangular matrix R satisfying $A = QR$. In Table 7, we present a stable version of the algorithm, often called the *modified Gram–Schmidt algorithm*.

Table 7. The modified Gram–Schmidt algorithm.

```
for j = 1, . . ., n do                    (q_j and a_j are
    q_j = a_j                              columns j of Q
    for i = 1, . . ., j − 1 do             and A,
        r_ij = q_i^T q_j                   respectively)
        q_j = q_j − r_ij q_i
    end
    r_jj = ||q_j||_2
    q_j = q_j/r_jj
end
```

4.7 Using Gram–Schmidt to Solve Overdetermined Linear Systems

Assume that A is an $m \times n$ matrix, with $m \geq n$ and $\text{rank}(A) = n$. Let $Ax = b$ be the linear system to be solved (in the least-squares sense, if nonsquare). The method for solving $Ax = b$ using the Gram–Schmidt algorithm is similar to that described in Sec. 4.4 for the QR factorization.

First compute the QR factorization of A by the Gram–Schmidt algorithm. Next compute $y = Q^{\mathrm{T}}b$ and solve $Rx = y$ by back substitution (see Sec. 2.2). Then x is the least-squares solution to $Ax = b$.

It can be shown that the Gram–Schmidt algorithm applied to an $n \times n$ linear system requires about $n^3/3$ flops, which is about the same number of arithmetic operations as the LU factorization algorithm needs and about half of what the QR factorization algorithm requires. Moreover, the modified Gram–Schmidt algorithm, as presented in Table 7, is relatively stable. The Gram–Schmidt algorithm applied to an $m \times n$ linear system requires about the same number of flops as that needed to form and to solve the normal equations, but the Gram–Schmidt algorithm is more stable.

4.8 Using Gram–Schmidt to Solve Underdetermined Linear Systems

Assume that A is an $m \times n$ matrix, with $m < n$ and $\text{rank}(A) = m$. Let $Ax = b$ be the linear system to be solved (in the least-squares sense). The method for solving $Ax = b$ using the Gram–Schmidt algorithm is similar to that described in Sec. 4.5 for the QR factorization.

First compute the QR factorization of A^{T} by the Gram–Schmidt algorithm. Next solve $R^{\mathrm{T}}y = b$ by back substitution (see Sec. 2.2) and set $x = Qy$. Then x is the least-squares solution to $Ax = b$.

5. EIGENVALUES AND EIGENVECTORS OF MATRICES

Given an $n \times n$ matrix A, $\lambda \in \mathbb{C}$, and $x \in \mathbb{C}^n$ satisfying $Ax = \lambda x$, λ is called an *eigenvalue* of A and x is called an *eigenvector* of A. The relation $Ax = \lambda x$ can be rewritten as $(A - \lambda I)x = 0$, emphasizing that λ is an eigenvalue of A if and only if $A - \lambda I$ is singular and that an eigenvector x is a nontrivial solution to $(A - \lambda I)x = 0$. It also follows that the eigenvalues of A are the roots of $\det(A - \lambda I) = 0$, the *characteristic equation* of A. The polynomial $p(\lambda) = \det(A - \lambda I)$ of degree n is called the *characteristic polynomial* of A and plays an important role in the theory of eigenvalues.

An $n \times n$ matrix A has precisely n eigenvalues, not necessarily distinct. It also has at least one eigenvector for each distinct eigenvalue. Note also that, if x is an eigenvector of A, then so is any (scalar) multiple of x and the corresponding eigenvalue is the same. We often choose an eigenvector of norm one in some vector norm, often the Euclidean norm, as the representative.

Two matrices A and B are similar if $A = W^{-1}BW$ for some nonsingular matrix W. The matrix W is often referred to as a *similarity transformation*. It is easy to see that, if $Ax = \lambda x$, then $B(Wx) = \lambda(Wx)$. Thus, similar matrices have the same eigenvalues and their eigenvectors are related by the similarity transformation W. Similarity transformations are often used in numerical methods for eigenvalues and eigenvectors to transform a matrix A into another one B that has the same eigenvalues and related eigenvectors, but whose eigenvalues and eigenvectors are in some sense easier to compute than those of A.

Eigenvalues play a major role in the study of convergence of iterative methods (see Sec. 3). Eigenvalues and eigenvectors are also of great importance in understanding the stability and other fundamental properties of many physical systems.

The matrix A is often large, sparse, and symmetric. These properties can be exploited to great advantage in numerical schemes for calculating the eigenvalues and eigenvectors of A.

A common approach to calculate the eigenvalues (and possibly the eigenvectors) of a matrix A consists of two stages. First, the matrix A is transformed to a similar but simpler matrix B, usually tridiagonal, if A is symmetric (or Hermitian), or Hessenberg, if A is nonsymmetric (or non-Hermitian). Then, the eigenvalues (and possibly the eigenvectors) of B are calculated. An exception to this approach is the power method (see Sec. 5.1).

A standard procedure for computing the eigenvectors of a matrix A is to calculate the eigenvalues first then use them to compute the eigenvectors by inverse iteration (see Sec. 5.4). Again, an exception to this approach is the power method (see Sec. 5.1).

Before describing numerical methods for the eigenvalue problem, we comment briefly on the sensitivity of the eigenvalues and eigenvectors to perturbations in the matrix A, since a backward error analysis can often show that the computed eigenvalues and eigenvectors are the exact eigenvalues and eigenvectors of a slightly perturbed matrix $\hat{A} = A + E$, where E is usually small relative to A. In general, if A is symmetric, its eigenvalues are well-conditioned with respect to small perturbations E. That is, the eigenvalues of $\hat{A}$ and A are very close. This, though, is not always true of the eigenvectors of A, particularly if the associated eigenvalue is a multiple eigenvalue or close to another eigenvalue of A. If A is nonsymmetric, then both its eigenvalues and eigenvectors may be poorly conditioned with respect to small perturbations E. Therefore, the user of a computer package for calculating the eigenvalues of a matrix should be cautious about the accuracy of the numerical results. For a further discussion of the conditioning of the eigenvalue problem, see Wilkinson (1965).

5.1 The Power Method

The power method is used to calculate the eigenvalue of largest magnitude of a matrix A and the associated eigenvector. Since matrix–vector products are the dominant computational work required by the power method, this scheme can exploit the sparsity of the matrix to great advantage.

Let λ be the eigenvalue of A of largest magnitude and let x be an associated eigenvector. Also let z_0 be an initial guess for some multiple of x. The power method, shown in Table 8, is an iterative scheme that generates a sequence of approximations z_1, $z_2, \ldots$ to some multiple of x and another sequence of approximations $\mu_1, \mu_2, \ldots$ to λ. In the scheme shown in Table 8, z_k is normalized so that the sequence $z_1, z_2, \ldots$ converges to an eigenvector x of A satisfying $\|x\|_\infty = 1$. Normalizations of this sort are frequently used in eigenvector calculations.

Table 8. The power method.

Pick z_0
for $k = 1, 2, \ldots$ do
    $w_k = Az_{k-1}$
    Choose $m \in \{1, \ldots, n\}$ such that
        $|(w_k)_m| \geq |(w_k)_i|$ for $i = 1, \ldots, n$
    $z_k = w_k/(w_k)_m$
    $\mu_k = (w_k)_m/(z_{k-1})_m$
    test stopping criterion
end

The power method is guaranteed to converge if A has a single eigenvalue λ of largest magnitude. The rate of convergence depends on $|\lambda_2|/|\lambda|$, where λ_2 is the eigenvalue of A of next largest magnitude. With some modifications the power method can be used when A has more than one eigenvalue of largest magnitude.

After the absolutely largest eigenvalue of A has been calculated, the power method can be applied to an appropriately deflated matrix to calculate the next largest eigenvalue and the associated eigenvector of A, and so on. However, this approach is inefficient if all or many eigenvalues are needed. The next sections describe more general-purpose methods for eigenvalue and eigenvector computations. For an introduction to the power method, see Atkinson (1989). For further reading, see Golub and Van Loan (1989) or Wilkinson (1965).

5.2 The *QR* Method

The *QR* method is widely used to calculate all the eigenvalues of a matrix A. It employs the *QR* factorization algorithm presented briefly in Sec. 4.4. Here, we recall that, given an $n \times n$ matrix A, there is a factorization $A = QR$, where R is an $n \times n$ upper-triangular matrix and Q an $n \times n$ orthogonal matrix.

The *QR* method, shown in Table 9, is an iterative scheme to compute the eigenvalues of A. It proceeds by generating a sequence

Table 9. The *QR* method.

Set $A_0 = A$
for $k = 1, 2, \ldots$ do
    Compute the *QR* factorization of $A_{k-1} = Q_kR_k$
    Set $A_k = R_kQ_k$
    test stopping criterion
end

$A_1, A_2, \ldots$ of matrices, all of which are similar to each other and to the starting matrix $A_0 = A$. The sequence converges either to a triangular matrix with the eigenvalues of A on its diagonal or to an almost triangular matrix from which the eigenvalues can be calculated easily.

For a real, nonsingular matrix A with no two (or more) eigenvalues of the same magnitude, the QR method is guaranteed to converge. The iterates A_k converge to an upper-triangular matrix with the eigenvalues of A on its diagonal. If A is symmetric, the iterates converge to a diagonal matrix. The rate of convergence depends on $\max\{|\lambda_{i+1}/\lambda_i| : i = 1, \ldots, n - 1\}$, where $|\lambda_1| > |\lambda_2| > \cdots > |\lambda_n|$.

If two or more eigenvalues of A have the same magnitude, the sequence $A_1, A_2, \ldots$ may not converge to an upper-triangular matrix. If A is symmetric, the sequence converges to a block-diagonal matrix, with blocks of order 1 or 2, from which the eigenvalues of A can be calculated easily. If A is nonsymmetric, the problem is more complicated. See Wilkinson (1965) or Parlett (1968) for details.

In many cases, the QR method is combined with a technique known as *shifting* to accelerate convergence. A discussion of this procedure can be found in Atkinson (1989), Golub and Van Loan (1989), Parlett (1980), and Wilkinson (1965).

The QR method for eigenvalues requires the computation of the QR factorization of a matrix and a matrix–matrix multiplication at each iteration. For a large matrix A, this is an expensive computation, making the method inefficient. To improve its efficiency, the matrix A is normally preprocessed, reducing it to a simpler form. If A is symmetric, it is reduced to a similar tridiagonal matrix, as described in Sec. 5.3. If A is nonsymmetric, it is reduced to a similar upper Hessenberg matrix by a comparable algorithm. It can be shown that, if $A_0 = A$ is in Hessenberg or tridiagonal form, then all the iterates A_k generated by the QR method will also be in Hessenberg or tridiagonal form, respectively. With appropriate implementation techniques, the QR factorization and the matrix–matrix products applied to matrices with these special structures require fewer operations than would be needed for arbitrary dense matrices.

With the modifications discussed above, the QR method is an efficient, general-purpose scheme for calculating the eigenvalues of a dense matrix. The eigenvectors can also be calculated if all the similarity transformations employed in the QR process are stored. Alternatively, the eigenvectors can be calculated by inverse iteration, as described in Sec. 5.4.

5.3 Transforming a Symmetric Matrix to Tridiagonal Form

As noted above, the eigenvalues of a symmetric (or Hermitian) matrix A are often calculated by first transforming A into a similar tridiagonal matrix T. Householder transformations (see Sec. 4.3) are usually employed to perform this task.

The algorithm to obtain the matrix T, given A, resembles the QR factorization algorithm described briefly in Sec. 4.3. It proceeds in $n - 2$ steps and generates a sequence $H_1, H_2, \ldots, H_{n-2}$ of Householder transformations to reduce A to tridiagonal form T. Starting with $A^{(0)} = A$, at step k for $k = 1, \ldots, n - 2$, we form $A_k = H_k A_{k-1} H_k$, where H_k is chosen to zero the components $k + 2$ to n of both row k and column k of A_{k-1}. T is A_{n-2}, the last matrix computed by this process. Note that $T = H_{n-2} \cdots H_1 A H_1 \cdots H_{n-2}$ is similar to A because each H_k is symmetric and orthogonal.

It can be shown that the reduction to tridiagonal form by Householder transformations is a stable computation in the sense that the computed T is the exact T for a slightly perturbed matrix $\hat{A} = A + E$, where E is usually small relative to A. As a result, it can be shown that the eigenvalues of A and T differ very little. For a brief introduction to tridiagonal reduction, see Atkinson (1989). For further reading, see Golub and Van Loan (1989) or Wilkinson (1965). For other methods to reduce A to tridiagonal form, such as planar rotation orthogonal matrices, see Golub and Van Loan (1989).

Similar schemes can be used to reduce a nonsymmetric (or non-Hermitian) matrix to Hessenberg form.

5.4 Inverse Iteration

Inverse iteration is the standard method to calculate the eigenvectors of a matrix A, once its eigenvalues have been calculated.

This scheme can be viewed as the power method (see Sec. 5.1) applied to the matrix $(A - \tilde{\lambda}I)^{-1}$ instead of A, where $\tilde{\lambda}$ is an (approximate) eigenvalue of A.

To see how inverse iteration works, let $\tilde{\lambda}$ be an approximation to a simple eigenvalue λ of A and let x be the associated eigenvector. Also let z_0 be an initial guess to some multiple of x. Inverse iteration, shown in Table 10, is an iterative method that generates a sequence of approximations $z_1, z_2, \ldots$ to some multiple of x. In the scheme shown in Table 10, z_k is normalized so that the sequence $z_1, z_2, \ldots$ converges to an eigenvector x of A satisfying $\|x\|_\infty = 1$. As noted already in Sec. 5.1, normalizations of this sort are frequently used in eigenvector calculations.

Note that, if $\tilde{\lambda} = \lambda$, then $A - \tilde{\lambda}I$ is singular. Moreover, if $\tilde{\lambda} \approx \lambda$, then $A - \tilde{\lambda}I$ is "nearly" singular. As a result, we can expect the system $(A - \tilde{\lambda}I)w_k = z_{k-1}$ to be very poorly conditioned (see Sec. 2.8). This, though, is not a problem in this context, since any large perturbation in the solution of the ill-conditioned system is in the direction of the desired eigenvector x.

The sequence of approximate eigenvectors $z_1, z_2, \ldots$ is typically calculated by first performing an LU decomposition of $A - \tilde{\lambda}I$, often with pivoting, before the start of the iteration and then using the same LU factorization to perform a forward elimination followed by a back substitution to solve $(A - \tilde{\lambda}I)w_k = z_{k-1}$ for $k = 1, 2, \ldots$ (see Sec. 2). We emphasize that only one LU factorization of $A - \tilde{\lambda}I$ is needed to solve all $(A - \tilde{\lambda}I)w_k = z_{k-1}$, $k = 1, 2, \ldots$.

For a brief discussion of the inverse iteration, including its stability and rate of convergence, see Atkinson (1989). For further reading, see Wilkinson (1965).

5.5 Other Methods

Another way to calculate the eigenvalues of A is to compute the roots of the characteristic polynomial $p(\lambda) = \det(A - \lambda I)$. The techniques discussed in Sec. 6.9 can be used for this purpose. However, this approach is often less stable than the techniques described above and it is usually not more efficient. Therefore, it is not normally recommended.

Table 10. Inverse iteration.

```
Pick z_0
for k = 1,2,... do
    Solve (A - λ̃I)w_k = z_{k-1}
    z_k = w_k/||w_k||_∞
    test stopping criterion
end
```

An obvious method for calculating an eigenvector x, once the corresponding eigenvalue λ is known, is to solve the system $(A - \lambda I)x = 0$. Since this system is singular, one approach is to delete one equation from $(A - \lambda I)x = 0$ and replace it by another linear constraint, such as $x_j = 1$ for some component j of x. However, it can be shown [see Wilkinson (1965)] that this method is not always stable and can lead to very poor numerical results in some cases. Thus, it is not recommended either.

Several other methods for calculating the eigenvalues and eigenvectors of a matrix have been omitted from our discussion because of space limitations. We should, though, at least mention one of them, the Jacobi method, a simple, rapidly convergent iterative scheme, applicable to symmetric matrices, including sparse ones.

The eigenvalue problem for large sparse matrices is a very active area of research. Although the QR method described in Sec. 5.2 is effective for small- to medium-sized dense matrices, it is computationally very expensive for large matrices, in part because it does not exploit sparsity effectively. The Lanczos and Arnoldi methods are much better suited for large sparse problems. For a discussion of these methods, see Cullum and Willoughby (1985), Parlett (1980), Saad (1992), and Scott (1981).

6. NONLINEAR ALGEBRAIC EQUATIONS AND SYSTEMS

Consider a nonlinear algebraic equation or system $f(x) = 0$, where $f : \mathbb{R}^n \to \mathbb{R}^n$, $n \geq 1$. A *root* of $f(x)$ is a value $\alpha \in \mathbb{R}^n$ satisfying $f(\alpha) = 0$. A nonlinear function may have one, many, or no roots.

Most numerical methods for computing approximations to roots of a nonlinear equation or system are *iterative* in nature. That is, the scheme starts with some initial guess x_0 and computes new successive approximations x_k, $k = 1, 2, \ldots$, by some formula until a *stopping criterion* such as

$\|f(x_k)\| \leq \epsilon,$

$\|f(x_k)\| \leq \epsilon\|f(x_0)\|,$

$\|x_k - x_{k-1}\| \leq \epsilon,$

$\|x_k - x_{k-1}\| \leq \epsilon\|x_k\|,$ or

$k >$ maxit

is satisfied, where ϵ is the error tolerance and maxit is the maximum number of iterations allowed.

6.1 Fixed-Point Iteration

A *fixed point* of a function $g(x)$ is a value α satisfying $\alpha = g(\alpha)$. A *fixed-point iteration* is a scheme of the form $x_k = g(x_{k-1})$ that uses the most recent approximation x_{k-1} to the fixed point α to compute a new approximation x_k to α. In this context, the function g is also called the *iteration function.*

One reason for studying fixed-point iterations is that given a function $f(x)$, it is easy to find another function $g(x)$ such that α is a root of $f(x)$ if and only if it is a fixed point of $g(x)$. For example, take $g(x) = x - f(x)$. Many root-finding methods can be viewed as fixed-point iterations.

Given an iteration function g, a fixed-point scheme starts with an initial guess x_0 and proceeds with the iteration as follows:

for $k = 1, 2, \ldots$ do
  $x_k = g(x_{k-1})$
  test stopping criterion
end

6.2 Newton's Method for Nonlinear Equations

We consider scalar equations (i.e., $n = 1$) first, and extend the results to systems of equations (i.e., $n > 1$) in Sec. 6.7.

Newton's method is a fixed-point iteration based on the iteration function $g(x) = x - f(x)/f'(x)$, where $f'(x)$ is the first derivative of f. More specifically, the new approximation x_k to the root α of f is computed by the formula

$$x_k = x_{k-1} - \frac{f(x_{k-1})}{f'(x_{k-1})},$$

which uses the previous approximation x_{k-1} to α. Geometrically, Newton's method approximates the nonlinear function $f(x)$ by its tangent (a straight line) at the current approximation x_{k-1} and takes x_k to be the intersection of the tangent with the x axis. That is, x_k is the root of the local linear approximation to $f(x)$. Newton's method is applicable if and only if f is differentiable and f' is nonzero at the point of approximation.

6.3 The Secant Method

The secant method is applicable to scalar equations only and is not a fixed-point iteration. The new approximation x_k to the root α of f is computed using two previous approximations x_{k-1} and x_{k-2} by the formula

$$x_k = x_{k-1} - f(x_{k-1})\frac{x_{k-1} - x_{k-2}}{f(x_{k-1}) - f(x_{k-2})}.$$

The secant method can be viewed as a variant of Newton's method in which $f'(x_{k-1})$ is approximated by $[f(x_{k-1}) - f(x_{k-2})]/[x_{k-1} - x_{k-2}]$. Geometrically, the secant method approximates the nonlinear function $f(x)$ by the chord subtending the graph of f at the two most recently computed points of approximation x_{k-1} and x_{k-2} and takes x_k to be the intersection of the chord with the x axis. That is, x_k is the root of the local linear approximation to $f(x)$. The secant method is applicable if and only if $f(x)$ is continuous and takes different values at x_{k-1} and x_{k-2}. To start, the secant method requires initial guesses for x_0 and x_1. These are usually chosen close to each other and must satisfy $f(x_0) \neq f(x_1)$.

6.4 The Bisection and *Regula Falsi* Methods

The bisection method is not a fixed-point iteration. It is applicable to a scalar equation $f(x) = 0$ if and only if $f(x)$ is continuous and there are two points L and R for which $f(L)f(R) \leq 0$. These conditions guarantee the existence of at least one root of $f(x)$ in the interval $[L,R]$. Without loss of generality, let $L < R$. To start, the bisection method approximates the root by the midpoint $M = (L + R)/2$ of $[L,R]$ and halves the interval at each iteration as follows.

```
forever do
    M = (L + R)/2
    if f(L)f(M) ≤ 0   then R = M
        else L = M
    test stopping criterion
end
```

Note that this iteration maintains the property $f(L)f(R) \leq 0$, as L and R are changed. So, when the algorithm terminates, a root of f is guaranteed to be in $[L,R]$. $M = (L + R)/2$ is often taken as the approximation to the root.

Several root-finding methods are similar to bisection. For example, *regula falsi* chooses

$$M = \frac{f(R)L - f(L)R}{f(R) - f(L)} \tag{38}$$

but is otherwise the same as bisection. Note that the M computed from (38) is the intersection of the chord subtending the graph of $f(x)$ at L and R with the x axis and so is guaranteed to lie in $[L,R]$, since the property $f(L)f(R) \leq 0$ is maintained throughout the iteration even though L and R are changed.

6.5 Convergence

Iterative methods for nonlinear equations can be guaranteed to converge under certain conditions, although they may diverge in some cases.

The bisection method converges whenever it is applicable, but if $f(x)$ has more than one root in the interval of application, there is no guarantee which of the roots the method will converge to.

The convergence of a fixed-point iteration depends critically on the properties of the iteration function $g(x)$. If g is smooth in an interval containing a fixed point α and $|g'(\alpha)| < 1$, then there is an $m \in [0,1)$ and a neighborhood I around the fixed point α in which $|g'(x)| \leq m < 1$. In this case, the fixed-point iteration $x_k = g(x_{k-1})$ converges to α if $x_0 \in I$. To give an intuitive understanding why this is so, we assume that $x_0, \ldots, x_{k-1} \in I$ and note that

$$x_k - \alpha = g(x_{k-1}) - g(\alpha) = g'(\xi_k)(x_{k-1} - \alpha),$$

where we have used $x_k = g(x_{k-1})$, $\alpha = g(\alpha)$ and, by the mean-value theorem, ξ_k is some point in I between x_{k-1} and α. Thus, $|g'(\xi_k)| \leq m < 1$ and so $|x_k - \alpha| \leq m|x_{k-1} - \alpha| \leq m^k|x_0 - \alpha|$, whence $x_k \in I$ too and $x_k \to \alpha$ as $k \to \infty$.

A more formal statement of this theorem and other similar results giving sufficient conditions for the convergence of fixed-point iterations can be found in many introductory numerical methods textbooks. See for example Conte and de Boor (1980); Dahlquist and Björck (1974); Johnson and Riess (1982); Stoer and Bulirsch (1980).

Newton's method converges if the conditions for convergence of a fixed-point iteration are met. [For Newton's method, the iteration function is $g(x) = x - f(x)/f'(x)$.] However, it can be shown that Newton's method converges quadratically (see Sec. 6.6) to the root α of f if the initial guess x_0 is chosen sufficiently close to α, f is smooth, and $f'(x) \neq 0$ close to α. It can also be shown that Newton's method converges from any starting guess in some cases. A more formal statement of these and other similar results can be found in many introductory numerical methods textbooks. See for example Conte and de Boor (1980); Dahlquist and Björck (1974); Johnson and Riess (1982); Stoer and Bulirsch (1980). For a deeper discussion of this topic, see Dennis and Schnabel (1983).

6.6 Rate of Convergence

The rate of convergence of a sequence $x_1, x_2, \ldots$ to α is the largest number $p \geq 1$ satisfying

$$\|x_{k+1} - \alpha\| \leq C\|x_k - \alpha\|^p \quad \text{as } k \to \infty$$

for some constant $C > 0$. If $p = 1$, we also require that $C < 1$. The larger the value of p the faster the convergence, at least asymptotically. Between two converging sequences with the same rate p, the faster is the one with the smaller C.

A fixed-point iteration with iteration function g converges at a rate p with $C = |g^{(p)}(\alpha)|/p!$ if $g \in \mathscr{C}^p$, $g^{(i)}(\alpha) = 0$, for $i = 0, 1, 2, \ldots, p - 1$, and $g^{(p)}(\alpha) \neq 0$, where $g^{(i)}(x)$ is the ith derivative of $g(x)$. Thus, Newton's method usually converges quadratically, i.e., $p = 2$ with $C = |g''(\alpha)|/2$, where $g(x) = x - f(x)/f'(x)$. If $f'(\alpha) = 0$, Newton's method typically converges linearly. If $f'(\alpha) \neq 0$ and $f''(\alpha) = 0$, Newton's method converges at least cubically, i.e., $p \geq 3$. The secant method converges at a superlinear rate of $p = (1 + \sqrt{5})/2 \approx 1.618$, i.e., faster than linear but slower than quadratic. Bisection converges linearly, i.e., $p = 1$ and $C = 1/2$.

6.7 Newton's Method for Systems of Nonlinear Equations

Newton's method for a scalar nonlinear equation (see Sec. 6.2) can be extended to a system of nonlinear equations with $f : \mathbb{R}^n \to \mathbb{R}^n$, for $n > 1$. In this case, the new iterate x_k is computed by

$$x_k = x_{k-1} - [J(x_{k-1})]^{-1} f(x_{k-1}),$$

where $J(x)$ is the *Jacobian* of f, an $n \times n$ matrix with its (i,j) entry equal to $\partial f_i(x)/\partial x_j$. To implement this scheme effectively, the matrix–vector product $z = [J(x_{k-1})]^{-1} f(x_{k-1})$ should *not* be calculated by first computing the inverse of the Jacobian and then performing the matrix–vector multiplication, but rather by first solving the linear system $[J(x_{k-1})]z_k = f(x_{k-1})$ by Gaussian elimination (see Sec. 2.1) or some other effective method for solving linear equations (see Secs. 2 and 3) and then setting $x_k = x_{k-1} - z_k$.

6.8 Modifications and Alternatives to Newton's Method

Solving the linear system $[J(x_{k-1})]z_k = f(x_{k-1})$ requires first evaluating all the partial derivatives of all components of f at the point x_{k-1} and then solving the linear system by performing Gaussian elimination (see Sec. 2.1) or some other effective method for solving linear equations (see Secs. 2 and 3). In the unmodified Newton's method, this procedure is repeated at every iteration, requiring $O(n^3)$ flops (floating-point operations) if Gaussian elimination is used. There exist variants of Newton's method that reduce the computational work per iteration significantly. Even though these schemes typically converge more slowly, they often dramatically reduce the cost of solving a nonlinear system, particularly if $n \gg 1$.

The *chord Newton* method, often called the *simplified Newton method,* holds $J(x_{k-1})$ fixed for several steps, thus avoiding many Jacobian evaluations and *LU* factorizations. However, it still requires one f evaluation and both a forward elimination and a back substitution (see Secs. 2.4 and 2.2, respectively) at each iteration.

Some other variants approximate the Jacobian by matrices that are easier to compute and simpler to solve. For example, the Jacobian may be approximated by its diagonal, giving rise to a *Jacobi-like Newton's* iteration, or by its lower-triangular part, giving rise to a *Gauss–Seidel-like Newton's* scheme. (See Sec. 3 for a discussion of Jacobi and Gauss–Seidel iterations for linear equations.)

Quasi-Newton schemes, which avoid the computation of partial derivatives, are alternatives to Newton's method. Some quasi-Newton schemes approximate partial derivatives by finite differences. For example,

$$[J(x)]_{ij} = \frac{\partial f_i}{\partial x_j}(x) \approx \frac{f_i(x + \delta e_j) - f_i(x)}{\delta},$$

where δ is a small nonzero number and $e_j \in \mathbb{R}^n$ is the jth unit vector, with a 1 in component j and 0's in all other entries.

Possibly the best-known quasi-Newton scheme is *Broyden's* method. It does not require the computation of any partial derivatives nor the solution of any linear systems. Rather, it uses one evaluation only of f and a matrix–vector multiply, requiring $O(n^2)$ flops, per iteration. Starting with an initial guess for the inverse of the Jacobian, $J(x_0)$, it updates its approximation to the inverse Jacobian at every iteration. Broyden's method can be viewed as an extension of the secant method to $n > 1$ dimensions. For a brief description of the algorithm, see Hager (1988).

6.9 Polynomial Equations

Polynomial equations are a special case of nonlinear equations. A polynomial of degree k has exactly k roots, counting multiplicity. One may wish to compute all roots of a polynomial or only a few select ones. The methods for nonlinear equations described above can be used in either case, although more effective schemes exist for this special class of problems. The efficient evaluation of the polynomial and its derivative is discussed below in Sec. 6.10.

Deflation is often used to compute roots of polynomials. It starts by locating one root r_1 of $p(x)$ and then proceeds recursively to compute the roots of $\hat{p}(x)$, where $p(x) = (x - r_1)\hat{p}(x)$. Note that, by the fundamental theorem of algebra, given a root r_1 of $p(x)$, $\hat{p}(x)$ is uniquely defined and is a polynomial of degree $k - 1$. However, deflation may be unstable unless implemented carefully. See an introductory numerical methods textbook for details.

Localization techniques can be used to identify regions of the complex plane that contain zeros. Such techniques include the Lehmer–Schur method, Laguerre's method, and methods based on Sturm sequences [see Householder (1970)]. Localization can be very helpful, for example, when searching for a particular root of a polynomial or when implementing deflation, since in the latter case, the roots should be computed in increasing order of magnitude to ensure numerical stability.

The roots of a polynomial $p(x)$ can also be found by first forming the *companion matrix* A of the polynomial $p(x)$—the eigenvalues $\{\lambda_i : i = 1, \ldots, n\}$ of A are the roots of $p(x)$—and then finding all, or a select few, of the eigenvalues of A. See Sec. 5 for a discussion of the computation of eigenvalues and an introductory numerical methods book, such as Hager (1988), for a further discussion of this root-finding technique.

6.10 Horner's Rule

Horner's rule, also called *nested multiplication* or *synthetic division,* is an efficient method to evaluate a polynomial and its derivative. Let $p(x) = a_0 + a_1x + a_2x^2 + \cdots + a_nx^n$ be the polynomial of interest, and α the point of evaluation. The following algorithm computes $p(\alpha)$ in z_0 and $p'(\alpha)$ in y_1 efficiently.

$z_n = a_n$
$y_n = a_n$
for $j = n - 1, \ldots, 1$ do
  $z_j = \alpha z_{j+1} + a_j$
  $y_j = \alpha y_{j+1} + z_j$
end
$z_0 = z_1\alpha + a_0$

7. UNCONSTRAINED OPTIMIZATION

The *optimization problem* is to find a value $x^* \in \mathbb{R}^n$ that either *minimizes* or *maximizes* a function $f : \mathbb{R}^n \to \mathbb{R}$. We consider only the minimization problem here, since maximizing $f(x)$ is equivalent to minimizing $-f(x)$.

Sometimes the minimizer must satisfy constraints such as $g_i(x) = 0$, $i = 1, \ldots, m_1$, or $h_j(x) \geq 0$, $j = 1, \ldots, m_2$, where g_i and h_j: $\mathbb{R}^n \to \mathbb{R}$. Thus, the general minimization problem can be written as

$$\min_{x \in \mathbb{R}^n} f(x)$$

subject to

$$g_i(x) = 0, \quad i = 1, \ldots, m_1$$

$$h_j(x) \geq 0, \quad j = 1, \ldots, m_2.$$

If any of the functions f, g_i, or h_j are nonlinear, then the minimization problem is *nonlinear;* otherwise, it is called a *linear programming* problem. If there are no constraints, the minimization problem is called *unconstrained;* otherwise, it is called *constrained.*

In this section, we present numerical methods for nonlinear unconstrained minimization problems (NLUMPs) only. For a more detailed discussion of these schemes, see an advanced text such as Dennis and Schnabel (1983).

7.1 Some Definitions and Properties

Let $x = (x_1, x_2, \ldots, x_n)^T \in \mathbb{R}^n$ and $f : \mathbb{R}^n \to \mathbb{R}$. The *gradient* ∇f of the function f is the vector of the n first partial derivatives of f:

$$\nabla f(x) = \left(\frac{\partial f}{\partial x_1}, \frac{\partial f}{\partial x_2}, \ldots, \frac{\partial f}{\partial x_n}\right)^T.$$

Note $\nabla f(x) : \mathbb{R}^n \to \mathbb{R}^n$.

A point x^* is a *critical point* of f if $\nabla f(x^*) = 0$. A point x^* is a *global minimum* of f if $f(x^*) \leq f(x)$ for all $x \in \mathbb{R}^n$. A point x^* is a *local minimum* of f if $f(x^*) \leq f(x)$ for all x in a neighborhood of x^*. If x^* is a local minimum of f, then it is also a critical point of f, assuming $\nabla f(x^*)$ exists, but the converse is not necessarily true.

The *Hessian* $\nabla^2 f(x) = H(x)$ of f is an $n \times n$ matrix with entries

$$H_{ij}(x) = \frac{\partial^2 f}{\partial x_i \partial x_j}(x).$$

If f is smooth, then

$$\frac{\partial^2 f}{\partial x_i \partial x_j} = \frac{\partial^2 f}{\partial x_j \partial x_i},$$

and so the Hessian is symmetric. A critical point x^* is a local minimum of f if $H(x^*)$ is symmetric positive-definite.

Numerical methods for solving NLUMPs are all iterative in character. Some try to compute roots of the gradient, i.e., critical points of f, which are also local minima. These methods are related to schemes used to solve nonlinear equations described in Sec. 6.

Minimization methods can be classified into three main categories:

- *direct-search methods,* which make use of function values only,
- *gradient methods,* which make use of function and derivative values, and
- *Hessian methods,* which make use of function, derivative, and second-derivative values.

Methods from each class are discussed below.

7.2 The Fibonacci and Golden-Section Search Methods

The Fibonacci and golden-section search methods belong to the class of direct-search methods. They are similar to the bisection method for nonlinear equations described in Sec. 6.4, although important differences exist. They are used in one-dimensional minimization only.

The Fibonacci and golden-section search methods are applicable if $f(x)$ is continuous and unimodal in the interval of interest $[a,b]$, where by *unimodal* we mean that $f(x)$ has exactly one local minimum $x^* \in [a,b]$ and $f(x)$ strictly decreases for $x \in [a,x^*]$ and strictly increases for $x \in [x^*,b]$.

The main idea behind these methods is the following. Let x_1 and x_2 be two points satisfying $a < x_1 < x_2 < b$. If $f(x_1) \geq f(x_2)$, then x^* must lie in $[x_1,b]$. On the other hand, if $f(x_1) \leq f(x_2)$ then x^* must lie in $[a,x_2]$. Thus, by evaluating f at a sequence of test points $x_1, x_2, \ldots, x_k$, we can successively reduce the length of the interval in which we know the local minimum lies, called the *interval of uncertainty*.

The Fibonacci and golden-section search methods differ only in the way the sequence of test points is chosen. Before describing the two methods, we give two more definitions.

The *coordinate* ν of a point x *relative to an interval* $[a,b]$ is $\nu = (x - a)/(b - a)$.

The *Fibonacci numbers* are defined by the initial condition $F_0 = F_1 = 1$ and the recurrence $F_k = F_{k-1} + F_{k-2}$ for $k \geq 2$.

The Fibonacci search method applied to a function f on an interval $[a,b]$ starts with two points x_k and x_{k-1} satisfying $(x_k - a)/(b - a) = F_{k-1}/F_k$ and $(x_{k-1} - a)/(b - a) = F_{k-2}/F_k = 1 - F_{k-1}/F_k$, whence $a < x_{k-1} < x_k < b$. It then computes the sequence $x_{k-2}, x_{k-3}, \ldots, x_1$ as follows. After evaluating $f(x_k)$ and $f(x_{k-1})$, the interval of uncertainty is either $[a,x_k]$ or $[x_{k-1},b]$.

- If the interval of uncertainty is $[a,x_k]$, then x_{k-1} belongs to it and $(x_{k-1} - a)/(x_k - a) = F_{k-2}/F_{k-1}$. In this case, x_{k-2} is chosen to satisfy $(x_{k-2} - a)/(x_k - a) = F_{k-3}/F_{k-1}$. The method proceeds to the next iteration with the two points x_{k-1} and x_{k-2} in the interval $[a,x_k]$, with relative coordinates F_{k-2}/F_{k-1} and F_{k-3}/F_{k-1}, respectively. Note that $f(x_{k-1})$ is already computed, so that only $f(x_{k-2})$ needs to be evaluated in the next iteration.
- If the interval of uncertainty is $[x_{k-1},b]$, then x_k belongs to it and $(x_k - x_{k-1})/(b - x_{k-1}) = F_{k-3}/F_{k-1}$. In this case, x_{k-2} is chosen to satisfy $(x_{k-2} - x_{k-1})/(b - x_{k-1}) = F_{k-2}/F_{k-1}$. The method proceeds to the next iteration with the two points x_k and x_{k-2} in the interval $[x_{k-1},b]$, with relative coordinates F_{k-3}/F_{k-1} and F_{k-2}/F_{k-1}, respectively. Note that $f(x_k)$ is already computed, so that only $f(x_{k-2})$ needs to be evaluated in the next iteration.

Thus, at the start of the second iteration, the situation is similar to that at the start of the first iteration, except that the length of the interval of uncertainty has been reduced. Therefore, the process described above can be repeated.

Before the last iteration, the interval of uncertainty is $[c,d]$ and x_1 is chosen to be $(c + d)/2 + \epsilon$, for some small positive number ϵ.

As noted above, the Fibonacci search method requires one function evaluation per iteration. The main disadvantage of the method is that the number of iterations k must be chosen at the start of the method. However, it can be proved that, given k, the length of the final interval of uncertainty is the shortest possible. Thus, in a sense, it is an optimal method.

The golden-section search method also requires one function evaluation per iteration, but the number of iterations k does not need to be chosen at the start of the method. It produces a sequence of test points $x_1, x_2, \ldots$ and stops when a predetermined accuracy is reached.

Let $r = (\sqrt{5} - 1)/2 \approx 0.618$ be the positive root of the quadratic $x^2 + x - 1$. Note that $1/r = (\sqrt{5} + 1)/2 \approx 1.618$ is the famous *golden ratio*. For the Fibonacci search method, it can be proved that, for large k, the coordinates of the two initial points x_k and x_{k-1} relative to $[a,b]$ are approximately r and $1 - r$, respectively. Thus, if, at each iteration, the two points are chosen with these coordinates relative to the interval of uncertainty, the resulting method, called the golden-section search method, is an approximation to the Fibonacci search. Moreover, if a point has coordinate $1 - r$ relative to $[a,b]$, then it has coordinate r relative to $[a, a + (b - a)r]$. Similarly, if a point has coordinate r relative to $[a,b]$, then it has coordinate $1 - r$ relative to $[a + (b - a)(1 - r), b]$. This property is exploited in the golden-section search method, enabling it to use one function evaluation only per iteration.

Both methods described above are guaranteed to converge whenever they are applicable and both have a linear rate of convergence (see Sec. 6.6). On the average, the length of the interval of uncertainty is multiplied by r at each iteration. For a further discussion of these methods, see an introductory numerical methods text such as Kahaner *et al.* (1989).

7.3 The Steepest-Descent Method

The steepest-descent (SD) method belongs to the class of gradient schemes. It is applicable whenever the partial derivatives of f exist and f has at least one local minimum. Whenever it is applicable, it is guaranteed to converge to some local minimum if the partial derivatives of f are continuous. However, it may converge slowly for multidimensional problems. Moreover, if f possesses more than one local minimum, there is no guarantee to which minimum SD will converge.

At iteration k, the SD method performs a search for the minimum of f along the line $x_k - \alpha\nabla f(x_k)$, where α is a scalar variable and $-\nabla f(x_k)$ is the direction of the steepest descent of f at x_k. Note that, since α is scalar, minimizing $f(x_k - \alpha\nabla f(x_k))$ with respect to (w.r.t.) α is a one-dimensional minimization problem. If α^* is the minimizer, x_{k+1} is taken to be $x_k - \alpha^*\nabla f(x_k)$. A brief outline of SD is given in Table 11. See Buchanan and Turner (1992), Johnson and Riess (1982), or Ortega (1988) for further details.

7.4 Conjugate-Direction Methods

The definition of conjugate directions is given w.r.t. a symmetric positive-definite (SPD) matrix A: the vectors (directions) u and v are A-conjugate if $u^TAv = 0$. Thus, conjugate directions are orthogonal or perpendicular directions w.r.t. an inner product $(u,v) = u^TAv$ and, as such, are often associated with some shortest-distance property.

The conjugate-direction (CD) methods form a large class of minimization schemes. Their common characteristic is that the search direction at every iteration is conjugate to previous search directions. Proceeding along conjugate search directions guarantees, in some sense, finding the shortest path to the minimum.

The CD methods are guaranteed to converge in at most n iterations for the SPD quadratic function $f(x) = c + b^Tx - \frac{1}{2}x^TAx$, where A is an $n \times n$ SPD matrix.

There are several techniques to construct conjugate directions, each one giving rise to

Table 11. The steepest-descent method.

Pick an initial guess x_0 and a tolerance ϵ
for $k = 1, \ldots,$ maxit do
  $s_{k-1} = -\nabla f(x_{k-1})$
  if $\|s_{k-1}\| \le \epsilon$ exit loop
  find $\alpha^* \in \mathbb{R}$ that minimizes $f(x_{k-1} + \alpha s_{k-1})$
  $x_k = x_{k-1} + \alpha^* s_{k-1}$
end

a different CD method. The best-known is *Powell's* method [see Buchanan and Turner (1992) or Ortega (1988)].

7.5 The Conjugate-Gradient Method

As the name implies, at each iteration, the conjugate-gradient (CG) method takes information from the gradient of f to construct conjugate directions. Its search direction is a linear combination of the direction of steepest descent and the search direction of the previous iteration. A brief outline of CG is given in Table 12.

The CG method is guaranteed to converge in at most n iterations for the SPD quadratic function $f(x) = c + b^T x - \frac{1}{2}x^T Ax$, where A is a $n \times n$ SPD matrix. See Sec. 3.2, Golub and Van Loan (1989), or Ortega (1988) for a more detailed discussion of this minimization technique applied to solve linear algebraic systems.

There exist several variants of the CG method, most of which are based on slightly different ways of computing the step size β.

7.6 Newton's Method

Newton's method for minimizing f is just Newton's method for nonlinear systems applied to solve $\nabla f(x) = 0$. The new iterate x_k is computed by $x_k = x_{k-1} - [H(x_{x-1})]^{-1} \times \nabla f(f_{k-1})$, where $H(x_{k-1})$ is the Hessian of f at x_{k-1}. See Sec. 6.7 for further details.

If f is convex, then the Hessian is SPD and the search direction generated by Newton's method at each iteration is a descent (downhill) direction. Thus, for any initial guess x_0, Newton's method is guaranteed to converge quadratically, provided f is sufficiently smooth.

For a general function f, there is no guarantee that Newton's method will converge for an arbitrary initial guess x_0. However, if started close enough to the minimum of a sufficiently smooth function f, Newton's method normally converges quadratically, as noted in Sec. 6.6. There exist several variants of Newton's method that improve upon the reliability of the standard scheme.

Table 12. The conjugate-gradient (CG) method.

Pick an initial guess x_0 and a tolerance ϵ
Initialize $s_0 = 0$ and $\beta = 1$
for $k = 1, \ldots,$ maxit do
  if $\|\nabla f(x_{k-1})\| \le \epsilon$ exit loop
  $s_k = -\nabla f(x_{k-1}) + \beta s_{k-1}$
  find $\alpha^* \in \mathbb{R}$ that minimizes $f(x_{k-1} + \alpha s_k)$
  $x_k = x_{k-1} + \alpha^* s_k$
  $\beta = \|\nabla f(x_k)\|^2 / \|\nabla f(x_{k-1})\|^2$
end

7.7 Quasi-Newton Methods

At every iteration, Newton's method requires the evaluation of the Hessian and the solution of a linear system $[H(x_{k-1})]s_k = -\nabla f(x_{k-1})$ for the search direction s_k. Quasi-Newton methods update an approximation to the inverse Hessian at every iteration, thus reducing the task of solving a linear system to a simple matrix–vector multiply. The best-known of these schemes are the *Davidon–Fletcher–Powell* (DFP) and the *Broyden–Fletcher–Goldfarb–Shanno* (BFGS) methods. See Buchanan and Turner (1992) or Dennis and Schnabel (1983) for further details.

8. APPROXIMATION

It is often desirable to find a function $f(x)$ in some class that approximates the data $\{(x_i, y_i) : i = 1, \ldots, n\}$. That is, $f(x_i) \approx y_i$, $i = 1, \ldots, n$. If f matches the data exactly, that is, f satisfies the *interpolation relations* or *interpolation conditions* $f(x_i) = y_i$, $i = 1, \ldots, n$, then f is called the *interpolating function* or the *interpolant* of the given data.

Similarly, it is often desirable to find a simple function $f(x)$ in some class that approximates a more complex function $y(x)$. We say that f *interpolates* y, or f is the *interpolant* of y, at the points x_i, $i = 1, \ldots, n$, if $f(x_i) = y(x_i)$, $i = 1, \ldots, n$. The problem of computing the interpolant f of y at the points x_i, $i = 1, \ldots, n$, reduces to the problem of computing the interpolant f of the data $\{(x_i, y_i) : i = 1, \ldots, n\}$, where $y_i = y(x_i)$.

An interpolant does not always exist and, when it does, it is not necessarily unique. However, a unique interpolant does exist in many important cases, as discussed below.

A standard approach for constructing an interpolant is to choose a set of *basis functions* $\{b_1(x), b_2(x), \ldots, b_n(x)\}$ and form a *model*

$$f(x) = \sum_{j=1}^{n} a_j b_j(x),$$

where the numbers a_j are unknown coefficients. For f to be an interpolant, it must satisfy $f(x_i) = y_i$, $i = 1, \ldots, n$, which is equivalent to

$$\sum_{j=1}^{n} a_j b_j(x_i) = y_i, \quad i = 1, \ldots, n.$$

These n conditions form a system $B\mathbf{a} = \mathbf{y}$ of n linear equations in n unknowns, where $\mathbf{a} = (a_1, a_2, \ldots, a_n)^{\mathrm{T}}$, $\mathbf{y} = (y_1, y_2, \ldots, y_n)^{\mathrm{T}}$, and $B_{ij} = b_j(x_i)$, $i, j = 1, \ldots, n$. If B is nonsingular, then the interpolant of the data $\{(x_i, y_i) : i = 1, \ldots, n\}$ w.r.t. the basis functions $b_1(x)$, $b_2(x), \ldots, b_n(x)$ exists and is unique. On the other hand, if B is singular, then either the interpolant may fail to exist or there may be infinitely many interpolants.

8.1 Polynomial Approximation

Polynomial approximation is the foundation for many numerical procedures. The basic idea is that, if we want to apply some procedure to a function, such as integration (see Sec. 9), we approximate the function by a polynomial and apply the procedure to the approximating polynomial. Polynomials are often chosen as approximating functions because they are easy to evaluate (see Sec. 6.10), to integrate, and to differentiate. Moreover, polynomials approximate well more complicated functions, provided the latter are sufficiently smooth. The following mathematical result ensures that arbitrarily accurate polynomial approximations exist for a broad class of functions.

Weierstrass Theorem: If $g(x) \in \mathscr{C}[a,b]$, then, for every $\epsilon > 0$, there exists a polynomial $p_n(x)$ of degree $n = n(\epsilon)$ such that $\max\{|g(x) - p_n(x)| : x \in [a,b]\} \le \epsilon$.

8.2 Polynomial Interpolation

Techniques to construct a polynomial interpolant for a set of data $\{(x_i,y_i) : i = 1, \ldots, n\}$ are discussed below. Here we state only the following key result.

Theorem: If the points $\{x_i : i = 1, \ldots, n\}$ are distinct, then there exists a unique polynomial of degree at most $n - 1$ that interpolates the data $\{(x_i,y_i) : i = 1, \ldots, n\}$. (There are no restrictions on the y_i's.)

8.2.1 Monomial Basis One way to construct a polynomial that interpolates the data $\{(x_i, y_i) : i = 1, \ldots, n\}$ is to choose as basis functions the monomials $b_j(x) = x^{j-1}$, $j = 1, \ldots, n$, giving rise to the model $p_{n-1}(x) = a_1 + a_2 x + a_3 x^2 + \cdots + a_n x^{n-1}$. As noted above, the interpolation conditions take the form $B\mathbf{a} = \mathbf{y}$. In this case, B is the *Vandermonde matrix* for which $B_{ij} = x_i^{j-1}$, $i, j = 1, \ldots, n$, where we use the convention that $x^0 = 1$ for all x.

It can be shown that the Vandermonde matrix B is nonsingular if and only if the points $\{x_i : i = 1, \ldots, n\}$ are distinct. If B is nonsingular, then, of course, we can solve the system $B\mathbf{a} = \mathbf{y}$ to obtain the coefficients a_j, $j = 1, \ldots, n$, for the unique interpolant of the data.

It can also be shown that, although the Vandermonde matrix is nonsingular for distinct points, it can be ill-conditioned, particularly for large n. As a result, the methods described below are often computationally much more effective than the scheme described here.

8.2.2 Lagrange Basis An alternative to the monomial basis functions discussed above is the Lagrange basis polynomials

$$b_j(x) = l_j(x) = \prod_{\substack{i=1 \\ i \neq j}}^{n} \frac{x - x_i}{x_j - x_i} \quad j = 1, \ldots, n,$$

which are of degree $n - 1$ and satisfy

$$b_j(x_i) = \begin{cases} 1 & \text{if } i = j, \\ 0 & \text{if } i \neq j. \end{cases}$$

Hence $B = I$ and the system $B\mathbf{a} = \mathbf{y}$ of interpolation conditions has the obvious unique solution $a_j = y_j$, $j = 1, \ldots, n$.

Note that changing the basis from monomials to Lagrange polynomials does not change the resulting interpolating polynomial $p_{n-1}(x)$, since the interpolant is unique. It only affects the representation of $p_{n-1}(x)$.

8.2.3 Newton Basis and Divided Differences Another useful basis is the set of Newton polynomials

$$b_j(x) = \prod_{i=1}^{j-1} (x - x_i) \quad j = 1, \ldots, n.$$

The coefficients a_j of the interpolating polynomial written with the Newton basis are relatively easy to compute by a recursive algorithm using *divided differences.* Before describing this form of the interpolating polynomial, though, we must introduce divided differences.

Given a function f with $f(x_i) = y_i$, $i = 1, \ldots, n$, define the divided difference with one point by

$$f[x_i] = y_i \quad i = 1, \ldots, n.$$

If $x_{i+1} \neq x_i$, define the divided difference with two points by

$$f[x_i,x_{i+1}] = \frac{y_{i+1} - y_i}{x_{i+1} - x_i} \quad i = 1, \ldots, n - 1.$$

If $x_i \neq x_{i+k}$, define the divided difference with $k + 1$ points by

$$f[x_i,x_{i+1}, \ldots ,x_{i+k}] = \frac{f[x_{i+1},x_{i+2}, \ldots ,x_{i+k}] - f[x_i,x_{i+1}, \ldots ,x_{i+k-1}]}{x_{i+k} - x_i} \quad i = 1, \ldots, n - k.$$

We can extend this definition of divided differences to sets $\{x_i : i = 1, \ldots, n\}$ with repeated values by noting that

$$\lim_{x_{i+1} \to x_i} f[x_{i+1},x_i] = \lim_{x_{i+1} \to x_i} \frac{y_{i+1} - y_i}{x_{i+1} - x_i} = f'(x_i).$$

So, if $x_i = x_{i+1}$, we define $f[x_i,x_{i+1}] = f'(x_i)$. Similarly, it can be shown that

$$\lim_{\substack{x_{i+1} \to x_i \\ \vdots \\ x_{i+k} \to x_i}} f[x_i,x_{i+1}, \ldots ,x_{i+k}] = \frac{f^{(k)}(x_i)}{k!}.$$

So, if $x_i = x_{i+1} = \cdots = x_{i+k}$, we define $f[x_i,x_{i+1}, \ldots ,x_{i+k}] = f^{(k)}(x_i)/k!$.

Using divided differences, the coefficients a_j can be computed as $a_j = f[x_1, \ldots ,x_j]$. An advantage of the divided-difference form is that it extends easily to the interpolation of derivatives as well as function values, as discussed below. Another advantage is that if the coefficients a_j, $j = 1, \ldots, n$, are computed from n data points, it is easy to add more data points and construct a higher-degree interpolating polynomial without redoing the whole computation, since the new data can be added easily to the existing divided-difference table and the interpolating polynomial extended.

8.3 Polynomial Interpolation with Derivative Data

In this section, we consider *Hermite* or *osculatory interpolation,* which requires that a function and its first derivative be interpolated at the points $\{x_i : i = 1, \ldots, n\}$. The key result is stated below.

Theorem: Given the data $\{(x_i,y_i,y_i') : i = 1, \ldots, n\}$, where the points $\{x_i : i = 1, \ldots, n\}$ are distinct, there exists a unique polynomial interpolant $p_{2n-1}(x)$ of degree at most $2n - 1$ that satisfies $p_{2n-1}(x_i) = y_i$, $i = 1, \ldots, n$, and $p'_{2n-1}(x_i) = y_i'$, $i = 1, \ldots, n$.

The techniques used to construct such a polynomial are similar to those described in Sec. 8.2. The following choices of basis functions are often used.

1. Monomials: $b_j(x) = x^{j-1}$, $j = 1, \ldots, 2n$;
2. Generalized Lagrange basis polynomials: $b_j(x) = [1 - 2(x - x_j)l_j'(x_j)][l_j(x)]^2$, $j = 1, \ldots, n$, and $b_{n+j}(x) = (x - x_j)[l_j(x)]^2$, $j = 1, \ldots, n$;
3. Newton basis polynomials:

$$b_j(x) = \prod_{i=1}^{j-1} (x - x_i)^2, \quad j = 1, \ldots, n,$$

and

$$b_{n+j}(x) = \left(\prod_{i=1}^{j-1} (x - x_i)^2\right)(x - x_j), \quad j = 1, \ldots, n.$$

More general forms of polynomial interpolants are discussed in some numerical methods books. See for example Davis (1975).

8.4 The Error in Polynomial Interpolation

Two key results for the error in polynomial interpolation are given in the following two theorems. For their proofs, see an introductory numerical methods text such as Dahlquist and Björck (1974), Johnson and Riess (1982), or Stoer and Bulirsch (1980). Before stating the results, though, we must define $\mathrm{spr}[x,x_1,x_2,\ldots,x_n]$ to be the smallest interval containing $x, x_1, x_2, \ldots, x_n$.

Theorem: Let $g(x) \in \mathscr{C}^n$, and let $p_{n-1}(x)$ be the polynomial of degree at most $n - 1$ that interpolates $g(x)$ at the n distinct points $x_1, x_2, \ldots, x_n$. Then, for any x,

$$g(x) - p_{n-1}(x) = \frac{g^{(n)}(\xi)}{n!} \prod_{i=1}^{n} (x - x_i),$$

where $g^{(n)}(x)$ is the nth derivative of $g(x)$ and ξ is some point in $\mathrm{spr}[x,x_1,x_2,\ldots,x_n]$.

Theorem: Let $g(x) \in \mathscr{C}^{2n}$, and let $p_{2n-1}(x)$ be the polynomial of degree at most $2n - 1$ that interpolates $g(x)$ and $g'(x)$ at the n distinct points $x_1, x_2, \ldots, x_n$. Then, for any x,

$$g(x) - p_{2n-1}(x) = \frac{g^{(2n)}(\xi)}{(2n)!} \prod_{i=1}^{n} (x - x_i)^2,$$

where $g^{(2n)}(x)$ is the $2n$th derivative of $g(x)$ and ξ is some point in $\mathrm{spr}[x,x_1,x_2,\ldots,x_n]$.

Note that there is a close relationship between the error in polynomial interpolation and the error in a Taylor series. As a result, polynomial interpolation is normally effective if and only if g can be approximated well by a Taylor series.

More specifically, the polynomial interpolation error can be large if the derivative appearing in the error formula is big or if $\mathrm{spr}[x,x_1,x_2,\ldots,x_n]$ is big, particularly if the point x of evaluation is close to an end point of $\mathrm{spr}[x_1,x_2,\ldots,x_n]$ or outside this interval.

8.5 Piecewise Polynomials and Splines

Given a set of *knots* or *grid points* $\{x_i : i = 1, \ldots, n\}$ satisfying $a = x_0 < x_1 < \cdots < x_n = b$, $s(x)$ is a *piecewise polynomial* (PP) of degree N w.r.t. the knots $\{x_i, i = 0, \ldots, n\}$ if $s(x)$ is a polynomial of degree N on each interval (x_{i-1},x_i), $i = 1, \ldots, n$. A polynomial of degree N is always a PP of degree N, but the converse is not necessarily true.

A *spline* $s(x)$ of degree N is a PP of degree N. The term spline usually implies the continuity of $s(x), s'(x), \ldots, s^{(N-1)}(x)$ at the knots $\{x_0,x_1,\ldots,x_n\}$. In this case, $s \in \mathscr{C}^{N-1}$, the space of continuous functions with $N - 1$ continuous derivatives. Sometimes, though, the terms PP and spline are used interchangeably.

Let $s(x)$ be a PP of degree N, and assume that $s(x), s'(x), \ldots, s^{(K)}(x)$ are continuous at the knots $\{x_0,x_1,\ldots,x_n\}$. Since $s(x)$ is a polynomial of degree N on each of the n subintervals (x_i,x_{i-1}), $i = 1, \ldots, n$, it is defined by $D = n(N + 1)$ coefficients. To determine these coefficients, we take into account the continuity conditions that s and its K derivatives must satisfy at the $n - 1$ interior knots $\{x_1,\ldots,x_{n-1}\}$. There are $K + 1$ such conditions at each interior knot, giving rise to $C = (n - 1)(K + 1)$ conditions. Thus, we need $M = D - C = n(N - K) + K + 1$ properly chosen additional conditions to determine the coefficients of s.

In the following we give examples of PPs and splines and their associated basis functions.

8.5.1 Constant Splines The constant PP model function

$$\phi(x) = \begin{cases} 1 & \text{for } 0 \le x \le 1, \\ 0 & \text{elsewhere,} \end{cases}$$

is a constant spline w.r.t. the knots $\{0,1\}$. Note that it is not continuous at the knots, so that $\phi \in \mathscr{C}^{-1}$. The constant PP functions

$$\phi_i(x) = \phi((x - a)/h - i + 1), \quad i = 1, \ldots, n,$$

are constant splines w.r.t. the evenly spaced knots $\{x_i = a + ih : i = 0, \ldots, n\}$, where $h = (b - a)/n$, and form a set of basis functions for the space of constant splines w.r.t. these knots.

8.5.2 Linear Splines The linear PP model function

$$\phi(x) = \begin{cases} x & \text{for } 0 \le x \le 1, \\ 2 - x & \text{for } 1 \le x \le 2, \\ 0 & \text{elsewhere,} \end{cases}$$

is a linear spline w.r.t. the knots $\{0,1,2\}$. Note that ϕ is continuous, but ϕ' does not exist at the knots, so that $\phi \in \mathscr{C}^0$. The linear PP functions

$$\phi_i(x) = \phi((x - a)/h - i + 1), \qquad i = 0, \ldots, n, \tag{39}$$

are linear splines w.r.t. the evenly spaced knots $\{x_i = a + ih : i = 0, \ldots, n\}$, where $h = (b - a)/n$, and form a set of basis functions for the space of linear splines w.r.t. these knots.

8.5.3 Quadratic Splines The quadratic PP model function

$$\phi(x) = \begin{cases} x^2/2 & \text{for } 0 \le x \le 1, \\ [x^2 - 3(x-1)^2]/2 & \text{for } 1 \le x \le 2, \\ [x^2 - 3(x-1)^2 + 3(x-2)^2]/2 & \text{for } 2 \le x \le 3, \\ 0 & \text{elsewhere,} \end{cases}$$

is a quadratic spline w.r.t. the knots $\{0,1,2,3,\}$. Note that ϕ and ϕ' are continuous, but ϕ'' does not exist at the knots, so that $\phi \in \mathscr{C}^1$. The quadratic PP functions

$$\phi_i(x) = \phi[(x - a)/h - i + 2], \quad i = 0, \ldots, n + 1 \tag{40}$$

are quadratic splines w.r.t. the evenly spaced knots $\{x_i = a + ih : i = 0, \ldots, n\}$, where $h = (b - a)/n$, and form a set of basis functions for the space of quadratic splines w.r.t. these knots.

8.5.4 Quadratic Piecewise Polynomials The quadratic PP model functions

$$\phi(x) = \begin{cases} x(2x-1) & \text{for } 0 \le x \le 1, \\ (x-2)(2x-3) & \text{for } 1 \le x \le 2, \\ 0 & \text{elsewhere,} \end{cases}$$

$$\psi(x) = \begin{cases} -4x(x-1) & \text{for } 0 \le x \le 1, \\ 0 & \text{elsewhere,} \end{cases}$$

are continuous, but neither ϕ' nor ψ' exists at their knots $\{0,1,2\}$ and $\{0,1\}$, respectively. So ϕ and $\psi \in \mathscr{C}^0$. The functions

$$\phi_i(x) = \begin{Bmatrix} \phi((x-a)/h - i/2 + 1) & \text{for } i \text{ even} \\ \psi((x-a)/h - (i+1)/2 + 1) & \text{for } i \text{ odd} \end{Bmatrix} \quad i = 0, \ldots, 2n$$

are quadratic PPs w.r.t. the evenly spaced knots $\{x_i = a + ih : i = 0, \ldots, n\}$, where $h = (b - a)/n$. Note that ϕ_i is continuous, but ϕ_i' does not exist at the knots, and so $\phi_i \in \mathscr{C}^0$. These functions form a basis for the space of quadratic PPs in $\mathscr{C}^0$ w.r.t. these knots.

8.5.5 Cubic Splines The cubic PP model function

$$\phi(x) = \begin{cases} x^3/6 & \text{for } 0 \le x \le 1, \\ [x^3 - 4(x-1)^3]/6 & \text{for } 1 \le x \le 2, \\ [x^3 - 4(x-1)^3 + 6(x-2)^3]/6 & \text{for } 2 \le x \le 3, \\ [x^3 - 4(x-1)^3 + 6(x-2)^3 - 4(x-3)^3]/6 & \text{for } 3 \le x \le 4, \\ 0 & \text{elsewhere,} \end{cases}$$

is a cubic spline w.r.t. the knots $\{0,1,2,3,4\}$. Note that ϕ, ϕ', and ϕ'' are continuous, but ϕ''' does not exist at the knots, and so $\phi \in \mathscr{C}^2$. The cubic PP functions

$$\phi_i(x) = \phi((x - a)/h - i + 2), \quad i = -1, \ldots, n + 1 \tag{41}$$

are cubic splines w.r.t. the evenly spaced knots $\{x_i = a + ih : i = 0, \ldots, n\}$, where $h = (b - a)/n$, and form a set of basis functions for the space of cubic splines w.r.t. these knots.

8.5.6 Cubic Hermite Piecewise Polynomials The cubic PP model functions

$$\phi(x) = \begin{cases} x^2(3-2x) & \text{for } 0 \le x \le 1, \\ (2-x)^2(2x-1) & \text{for } 1 \le x \le 2, \\ 0 & \text{elsewhere,} \end{cases}$$

$$\psi(x) = \begin{cases} x^2(x-1) & \text{for } 0 \le x \le 1, \\ (2-x)^2(x-1) & \text{for } 1 \le x \le 2, \\ 0 & \text{elsewhere,} \end{cases}$$

are in $\mathscr{C}^1$ since ϕ, ϕ', ψ, and ψ' are all continuous at the knots $\{0,1,2\}$, but neither ϕ'' nor ψ'' exists at the knots. The functions

$$\phi_i(x) = \begin{Bmatrix} \phi((x-a)/h - i/2 + 1) & \text{for } i \text{ even} \\ \psi((x-a)/h - (i-1)/2 + 1) & \text{for } i \text{ odd} \end{Bmatrix}, \quad i = 0, \ldots, 2n + 1 \tag{42}$$

are cubic PPs w.r.t. the evenly spaced knots $\{x_i = a + ih : i = 0, \ldots, n\}$, where $h = (b - a)/n$. Note that ϕ_i and ϕ_i' are continuous, but ϕ_i'' does not exist at the knots, and so $\phi_i \in \mathscr{C}^1$. These functions form a basis for the space of cubic PPs in $\mathscr{C}^1$ w.r.t. these knots.

8.6 Piecewise Polynomial Interpolation

Piecewise polynomials, including splines, are often used for interpolation, especially for large data sets. The main advantage in using PPs, instead of polynomials, to interpolate a function $g(x)$ at n points, where $n \gg 1$, is that the error of a PP interpolant does not depend on the nth derivative of $g(x)$, but rather on a low-order derivative of $g(x)$. Usually, if $g(x) \in \mathscr{C}^{N+1}[a,b]$ and $s_N(x)$ is a PP of degree N in $\mathscr{C}^K$ that interpolates $g(x)$ at $n(N - K) + K + 1$ properly chosen points, then the interpolation error at any point $x \in [a,b]$ satisfies

$$|g(x) - s_N(x)| \leq Ch^{N+1} \max_{a \leq \xi \leq b} |g^{(N+1)}(\xi)|$$

for some constant C independent of $h = \max\{x_i - x_{i-1} : i = 1, \ldots, n\}$. Another advantage of PP interpolation is that it leads either to simple relations that define the PP interpolant or gives rise to banded systems that can be solved easily for the coefficients of the PP interpolant by the techniques described in Sec. 2.7. In the following subsections, we give examples of PP interpolation. We assume throughout that $h = (b - a)/n$ and $x_i = a + ih$ for $i = 0, \ldots, n$, but these restrictions can be removed easily.

An introduction to PP interpolation is presented in many introductory numerical analysis textbooks, such as Johnson and Riess (1982). For a more detailed discussion, see an advanced text such as Prenter (1975) or de Boor (1978).

8.6.1 Linear Spline Interpolation Let $\phi_i(x)$ be defined by (39). The function

$$s_1(x) = \sum_{i=0}^{n} c_i\phi_i(x)$$

with $c_i = g(x_i)$ is the unique linear spline that interpolates $g(x)$ at the knots $\{x_i : i = 0, \ldots, n\}$.

8.6.2 Quadratic Spline Interpolation Let $\phi_i(x)$ be defined by (40). The function

$$s_2(x) = \sum_{i=0}^{n+1} c_i\phi_i(x)$$

is the unique quadratic spline that interpolates $g(x)$ at the $n + 2$ points x_0, $(x_{i-1} + x_i)/2$, $i = 1, \ldots, n$, and x_n, if the coefficients c_i, $i = 0, \ldots, n + 1$, are chosen so that they satisfy the $n + 2$ conditions $s_2(x_0) = g(x_0)$, $s_2((x_{i-1} + x_i)/2) = g((x_{i-1} + x_i)/2)$, $i = 1, \ldots, n$, and $s_2(x_n) = g(x_n)$. These conditions give rise to a tridiagonal linear system of $n + 2$ equations in the $n + 2$ unknowns c_i, $i = 0, \ldots, n + 1$:

$$\begin{pmatrix} 0.5 & 0.5 & & & & \\ 0.125 & 0.75 & 0.125 & & & \\ & 0.125 & 0.75 & 0.125 & & \\ & & \ddots & \ddots & \ddots & \\ & & & 0.125 & 0.75 & 0.125 \\ & & & & 0.5 & 0.5 \end{pmatrix} \times \begin{pmatrix} c_0 \\ c_1 \\ c_2 \\ \vdots \\ c_n \\ c_{n+1} \end{pmatrix} = \begin{pmatrix} g(x_0) \\ g((x_0 + x_1)/2) \\ g((x_1 + x_2)/2) \\ \vdots \\ g((x_{n-1} + x_n)/2) \\ g(x_n) \end{pmatrix}.$$

Since this system is column diagonally dominant, it is nonsingular and has a unique solution. Moreover, if the first and last equations are multiplied by 0.25, the resulting system is symmetric positive-definite. So the coefficients $\{c_i\}$ of the associated quadratic spline interpolant can be computed easily by the techniques described in Sec. 2.7.

8.6.3 Cubic Spline Interpolation Let $\phi_i(x)$ be defined by (41). The function

$$s_{32}(x) = \sum_{i=-1}^{n+1} c_i\phi_i(x)$$

is a cubic spline that interpolates $g(x)$ at the $n + 1$ points x_i, $i = 0, \ldots, n$, if the coefficients c_i, $i = -1, \ldots, n + 1$, are chosen so that they satisfy the $n + 1$ conditions $s_{32}(x_i) = g(x_i)$, $i = 0, \ldots, n$. These conditions give rise to a linear system of $n + 1$ equations in the $n + 3$ unknowns c_i, $i = -1, \ldots, n + 1$. To determine the unknown coefficients uniquely, two more appropriately chosen conditions (equations) are needed. The standard choices are

1. $s''_{32}(x_0) = 0$ and $s''_{32}(x_n) = 0$, giving rise to the *natural* or *minimum-curvature* cubic spline interpolant;
2. $s'_{32}(x_0) = g'(x_0)$ and $s'_{32}(x_n) = g'(x_n)$, giving rise to a more accurate cubic spline interpolant, but one that requires knowledge of $g'(x)$ (or a good approximation to it) at both end points;
3. the *not-a-knot* conditions, which force $s_{32}(x)$ to have a continuous third derivative at the knots x_1 and x_{n-1}.

In all three cases, the resulting linear system of $n + 3$ equations in $n + 3$ unknowns is almost tridiagonal. More specifically, it is tridiagonal with the exception of the two rows corresponding to the extra conditions. So the coefficients $\{c_i\}$ of the associated cubic spline interpolant can be computed easily by the techniques described in Sec. 2.7.

8.6.4 Cubic Hermite Piecewise Polynomial Interpolation Let $\phi_i(x)$ be defined by (42). The function

$$s_{31}(x) = \sum_{i=0}^{2n+1} c_i\phi_i(x)$$

with $c_{2i} = g(x_i)$, $i = 0, \ldots, n$, and $c_{2i+1} = g'(x_i)$, $i = 0, \ldots, n$, is the unique Hermite PP that interpolates $g(x)$ and its derivative at the $n + 1$ points x_i, $i = 0, \ldots, n$.

8.7 Least-Squares Approximation

It is possible to construct a function $f(x)$ that approximates the data $\{(x_i, y_i) : i = 0, \ldots, m\}$ in the sense that $f(x_i) - y_i$ is "small" for $i = 0, \ldots, m$, but $f(x_i) \neq y_i$ in general. One way is to construct an f in some class of functions that minimizes

$$\sum_{i=0}^{m} w_i[f(x_i) - y_i]^2$$

for some positive *weights* w_i, $i = 0, \ldots, m$. Such an f is called a *discrete least-squares* approximation to the data.

Similarly, it is possible to construct a function $f(x)$ that approximates a continuous function $y(x)$ on an interval $[a,b]$ in the sense that $|f(x) - y(x)|$ is "small" for all $x \in [a,b]$. One way is to construct an f in some class of functions that minimizes

$$\int_a^b w(x)[f(x) - y(x)]^2dx$$

for some positive and continuous *weight function* $w(x)$ on (a,b). Such an f is called a *continuous least-squares* approximation to y.

Often, f is chosen to be a polynomial of degree $n < m$. (For the discrete least-squares problem, if f is a polynomial of degree $n = m$, then f interpolates the data.)

Let f and $g \in \mathscr{C}[a,b]$. We denote the *discrete inner product* of f and g at distinct points x_i, $i = 0, \ldots, m$, w.r.t. the weights w_i, $i = 0, \ldots, m$, by

$$(f,g) = \sum_{i=0}^{m} w_i f(x_i)g(x_i)$$

and the *continuous inner product* of f and g on $[a,b]$ w.r.t. weight function $w(x)$ by

$$(f,g) = \int_a^b w(x)f(x)g(x)dx.$$

Given an inner product $(\cdot,\cdot)$, as above, we denote the *norm* of f by $\|f\| = (f,f)^{1/2}$. Thus we have the *discrete norm*

$$\|f\| = \left(\sum_{i=0}^{m} w_i f(x_i)^2\right)^{1/2}$$

and the *continuous norm*

$$\|f\| = \left(\int_a^b w(x)f(x)^2dx\right)^{1/2}.$$

Therefore, the problem of constructing a least-squares approximation $f(x)$ to a given set of data $\{(x_i,y_i)\}$ or to a given function $y(x)$ is to construct an $f(x)$ that minimizes the norm, discrete or continuous, respectively, of the error $\|f - y\|$.

We note that the "discrete inner product" is not strictly speaking an inner product in all cases, since it may fail to satisfy the property that $(f,f) = 0$ implies $f(x) = 0$ for all x. However, if we restrict the class of functions to which f belongs appropriately, then $(\cdot,\cdot)$ is a true inner product. For example, if we restrict f to the class of polynomials of degree at most n and if $n < m$, then $(f,f) = 0$ implies $f(x) = 0$ for all x. Similar remarks apply to the discrete norm.

Before giving the main theorem on how to construct the least-squares polynomial approximation to a given set of data or to a given function, we introduce orthogonal polynomials and the Gram–Schmidt process to construct them.

8.7.1 Orthogonal Polynomials A set of $n + 1$ polynomials $\{q_i(x) : i = 0, \ldots, n\}$ is *orthogonal* w.r.t. the inner product $(\cdot,\cdot)$ if $(q_i,q_j) = 0$ for $i \neq j$. A set of $n + 1$ orthogonal polynomials $\{q_i(x) : i = 0, \ldots, n\}$ is *orthonormal* w.r.t. the inner product $(\cdot,\cdot)$ if in addition $(q_i,q_i) = 1$ for $i = 0, \ldots, n$.

8.7.2 The Gram–Schmidt Orthogonalization Algorithm The Gram–Schmidt algorithm applied to a set of $n + 1$ linearly independent polynomials $\{p_j : j = 0, \ldots, n\}$ generates a set of $n + 1$ orthonormal polynomials $\{q_i(x) : i = 0, \ldots, n\}$ and a set of $n + 1$ orthogonal polynomials $\{s_i(x): i = 0, \ldots, n\}$. Often the set of $n + 1$ linearly independent polynomials $\{p_j : j = 0, \ldots, n\}$ is chosen to be the set of monomials $\{x^j : j = 0, \ldots, n\}$.

The Gram–Schmidt algorithm for polynomials is similar to the Gram–Schmidt algorithm for matrices described in Sec. 4.6. The reader may refer to an introductory numerical methods text such as Johnson and Riess (1982) for more details. Here we note only that the role of the inner product of vectors in the algorithm in Sec. 4.6 is replaced by the inner product, discrete or continuous, of functions as defined in Sec. 8.7.1.

8.7.3 Constructing the Least-Squares Polynomial Approximation The following result is proved in many introductory numerical methods books. See for example Johnson and Riess (1982).

Theorem: Assume that we are given either a continuous function $y(x)$ on $[a,b]$ or a data set $\{(x_i,y_i) : i = 0, \ldots, m\}$. Let $\{q_j : j = 0, \ldots, n\}$ be a set of orthonormal polynomials w.r.t. an inner product $(\cdot,\cdot)$ appropriate for the given data and assume that $\{q_j : j = 0, \ldots, n\}$ spans $\{x^i : i = 0, \ldots, n\}$, where $n < m$ for the discrete problem. Then

$$p^*(x) = \sum_{j=0}^{n} (y,q_j)q_j(x)$$

is the least-squares polynomial approximate of degree at most n. It is optimal in the sense that, if $p(x)$ is any other polynomial of degree at most n, then $\|y - p^*\| < \|y - p\|$, where $\|\cdot\|$ is the norm associated with the inner product $(\cdot,\cdot)$.

As noted in Sec. 8.7.2, a set of orthonormal polynomials $\{q_j : j = 0, \ldots, n\}$ that spans $\{x^i : i = 0, \ldots, n\}$ can be constructed by the Gram–Schmidt algorithm applied to the monomial basis polynomials $\{x^i : i = 0, \ldots, n\}$.

9. NUMERICAL INTEGRATION—QUADRATURE

In this section, we consider formulas for approximating integrals of the form

$$I(f) = \int_a^b f(x)dx.$$

Such formulas are often called quadrature rules. We assume that a and b are finite and that f is smooth in most cases, but we briefly discuss infinite integrals and singularities in Sec. 9.5.

In many practical problems, $f(x)$ is given either as a set of values $f(x_1), \ldots, f(x_n)$ or $f(x)$ is hard or impossible to integrate exactly. In these cases, the integral may be approximated by numerical techniques, which often take the form

$$Q(f) = \sum_{i=1}^{n} w_i f(x_i).$$

Such a formula is called a *quadrature rule,* the $\{w_i\}$ are called *weights,* and the $\{x_i\}$ are called *abscissae* or *nodes.*

Most quadrature rules are derived by first approximating f by a simpler function, frequently a polynomial, and then integrating the simpler function. Thus, the area under the curve f, which is the exact value of $I(f)$, is approximated by the area under the curve of the simpler function.

For a more detailed discussion of the topics in this section, see an introductory numerical methods text such as Conte and de

Boor (1980), Dahlquist and Björck (1974), Johnson and Riess (1982), or Stoer and Bulirsch (1980).

9.1 Simple Quadrature Rules

Several simple quadrature rules follow:

- The *rectangle rule* approximates $I(f)$ by the area under the constant $y = f(a)$ or $y = f(b)$.
- The *midpoint rule* approximates $I(f)$ by the area under the constant $y = f((a + b)/2)$.
- The *trapezoidal rule* approximates $I(f)$ by the area under the straight line joining the points $(a,f(a))$ and $(b,f(b))$.
- *Simpson's rule* approximates $I(f)$ by the area under the quadratic that interpolates $(a,f(a))$, $(m,f(m))$ and $(b,f(b))$, where $m = (a + b)/2$.
- The *corrected trapezoidal rule* approximates $I(f)$ by the area under the cubic Hermite that interpolates $(a,f(a),f'(a))$ and $(b,f(b),f'(b))$.
- Newton–Cotes rules are discussed in Sec. 9.1.1 below.
- Gaussian rules are discussed in Sec. 9.1.2 below.

The formula $Q(f)$ and the associated error $I(f) - Q(f)$ for each quadrature rule listed above are given in Table 13, where n is the number of function and derivative evaluations, d is the polynomial degree of the quadrature rule (see Sec. 9.1.1 below), η is an unknown point in $[a,b]$ (generally different for each rule), $m = (a + b)/2$ is the midpoint of the interval $[a,b]$, and C and K are some constants. For the derivation of these quadrature rules and their associated error formulas, see an introductory numerical methods text such as Conte and de Boor (1980); Dahlquist and Björck (1974); Johnson and Riess (1982); Stoer and Bulirsch (1980).

9.1.1 Some Definitions A quadrature rule that is based on integrating a polynomial interpolant is called an *interpolatory rule*. All the simple quadrature rules listed in Table 13 are interpolatory. Writing the polynomial interpolant in Lagrange form and integrating it, we see immediately that the weights w_i do not depend on the function f, but only on the abscissae $\{x_i : i = 1, \ldots, n\}$.

Quadrature rules are, in general, not ex-

Table 13. Simple quadrature rules.

Quadrature rule	n	d	Interpolant	$Q(f)$	$I(f) - Q(f)$
Rectangle	1	0	Constant	$(b - a)f(a)$	$\frac{(b-a)^2}{2} f'(\eta)$
Midpoint	1	1	Constant	$(b - a)f(m)$	$\frac{(b-a)^3}{24} f''(\eta)$
Trapezoidal	2	1	Linear	$\frac{b-a}{2}[f(a) + f(b)]$	$-\frac{(b-a)^3}{12} f''(\eta)$
Simpson's	3	3	Quadratic	$\frac{b-a}{6}[f(a) + 4f(m) + f(b)]$	$-\frac{(b-a)^5}{2880} f^{(4)}(\eta)$
Corrected trapezoidal	4	3	Cubic	$\frac{b-a}{2}[f(a) + f(b)] + \frac{(b-a)^2}{12}[f'(a) - f'(b)]$	$\frac{(b-a)^5}{720} f^{(4)}(\eta)$
Newton–Cotes	n	$\geq n - 1$	Deg. $n - 1$	See Sec. 9.1.1	$\frac{(b-a)^{d+2}}{K} f^{(d+1)}(\eta)$
Gaussian	n	$2n - 1$	Deg. $n - 1$	See Sec. 9.1.2	$\frac{(b-a)^{d+2}}{C} f^{(d+1)}(\eta)$

act. An error formula for an interpolatory rule can often be derived by integrating the associated polynomial interpolant error. Error formulas for some simple quadrature rules are listed in Table 13.

A quadrature rule that is exact for all polynomials of degree d or less, but is not exact for all polynomials of degree $d + 1$, is said to have *polynomial degree d*. An interpolatory rule based on n function and derivative values has polynomial degree at least $n - 1$.

Quadrature rules that include the end points of the interval of integration $[a,b]$ as abscissae are called *closed rules,* while those that do not include the end points are called *open rules.* An advantage of open rules is that they can be applied to integrals with singularities at the end points, whereas closed rules usually can not.

Interpolatory rules based on equidistant abscissae are called *Newton–Cotes* rules. This class includes the rectangle, midpoint, trapezoidal, corrected trapezoidal, and Simpson's rules. Both open and closed Newton–Cotes quadrature rules exist.

9.1.2 Gaussian Quadrature Rules A quadrature rule

$$Q(f) = \sum_{i=1}^{n} w_i f(x_i)$$

is fully determined by n, the abscissae $\{x_i : i = 1, \ldots, n\}$, and the weights $\{w_i : i = 1, \ldots, n\}$. Gauss showed that, given n and the end points a and b of the integral,

1. there exists a unique set of abscissae $\{x_i\}$ and weights $\{w_i\}$ that give a quadrature rule—called the Gaussian quadrature rule—that is exact for all polynomials of degree $2n - 1$ or less;
2. no quadrature rule with n abscissae and n weights is exact for all polynomials of degree $2n$;
3. the weights $\{w_i\}$ of the Gaussian quadrature rule are all positive;
4. the Gaussian quadrature rule is open;
5. the abscissae $\{x_i\}$ of the Gaussian quadrature rule are the roots of the shifted Legendre polynomial $q_n(x)$ of degree n, which is the unique monic polynomial of degree n that is orthogonal to all polynomials of degree $n - 1$ or less w.r.t. the continuous inner product

$$(f,g) = \int_a^b f(x)g(x)dx$$

(see Sec. 8.7.1);
6. the Gaussian quadrature rule is interpolatory, i.e., it can be derived by integrating the polynomial of degree $n - 1$ that interpolates f at the abscissae $\{x_i\}$.

Thus, Gauss derived the class of open interpolatory quadrature rules of maximum polynomial degree $d = 2n - 1$. As noted above, these formulas are called *Gaussian quadrature rules* or *Gauss–Legendre quadrature rules.*

9.1.3 Translating the Interval of Integration The weights and abscissae of simple quadrature rules are usually given w.r.t. a specific interval of integration, such as $[0,1]$ or $[-1,1]$. However, the weights and abscissae can be transformed easily to obtain a related quadrature rule appropriate for some other interval.

One simple way to do this is based on the linear change of variables $\hat{x} = \beta(x - a)/(b - a) + \alpha(b - x)/(b - a)$, which leads to the relation

$$\int_\alpha^\beta f(\hat{x})d\hat{x} = \int_a^b f\left(\beta\frac{x - a}{b - a} + \alpha\frac{b - x}{b - a}\right) \times \frac{\beta - \alpha}{b - a}dx. \quad (43)$$

So, if we are given a quadrature rule on $[a,b]$ with weights $\{w_i : i = 1, \ldots, n\}$ and abscissae $\{x_i : i = 1, \ldots, n\}$, but we want to compute

$$\int_\alpha^\beta f(\hat{x})d\hat{x},$$

we can apply the quadrature rule to the integral on the right side of (43). An equivalent way of viewing this is that we have developed a related quadrature rule for the interval of integration $[\alpha,\beta]$ with weights and abscissae

$$\hat{w}_i = \frac{\beta - \alpha}{b - a}w_i \quad \text{and}$$

$$\hat{x}_i = \beta\frac{x_i - a}{b - a} + \alpha\frac{b - x_i}{b - a}$$

for $i = 1, \ldots, n$. Note that, because we have used a linear change of variables, the original rule for $[a,b]$ and the related one for $[\alpha,\beta]$ have the same polynomial degree.

9.1.4 Comparison of Gaussian and Newton–Cotes Quadrature Rules We list some similarities and differences between Gaussian and Newton–Cotes quadrature rules:

1. The weights of a Gaussian rule are all positive, which contributes to the stability of the formula. High-order Newton–Cotes rules typically have both positive and negative weights, which is less desirable, since it leads to poorer stability properties.
2. Gaussian rules are open, whereas there are both open Newton–Cotes rules and closed Newton–Cotes rules.
3. Gaussian rules attain the maximum possible polynomial degree $2n - 1$ for a formula with n weights and abscissae, whereas the polynomial degree d of a Newton–Cotes rule satisfies $n - 1 \leq d \leq 2n - 1$ and the upper bound can be obtained for $n = 1$ only.
4. The abscissae and weights of Gaussian rules are often irrational numbers and hard to remember, but they are not difficult to compute. The abscissae of a Newton–Cotes rule are easy to remember and the weights are simple to compute.
5. The set of abscissae for an n-point Gaussian rule and for an m-point Gaussian rule are almost disjoint for all $n \neq m$. Thus we can not reuse function evaluations performed for one Gaussian rule in another Gaussian rule. Appropriately chosen pairs of Newton–Cotes rules can share function evaluations effectively.
6. Both Gaussian and Newton–Cotes rules are interpolatory.

9.2 Composite (Compound) Quadrature Rules

To increase the accuracy of a numerical approximation to an integral, we could use a rule with more weights and abscissae. This often works with Gaussian rules, if f is sufficiently smooth, but it is not advisable with Newton–Cotes rules, for example, because of stability problems associated with high-order formulas in this class.

Another effective way to achieve high accuracy is to use *composite* quadrature rules, often also called *compound* quadrature rules. In these schemes, the interval of integration $[a,b]$ is subdivided into *panels* (or subintervals) and on each panel the same simple quadrature rule is applied. If the associated simple quadrature rule is interpolatory, then this approach leads to the integration of a piecewise polynomial (PP) interpolant. Thus, using a composite quadrature rule, instead of a simple one with high polynomial degree, leads to many of the same benefits that are obtained in using PP interpolants compared with high-degree polynomial interpolants (see Sec. 8.5).

Table 14 summarizes the composite quadrature rules and the associated error formulas. In the table, n is the number of function and derivative evaluations, d is the polynomial degree of the quadrature rule, η is an unknown point in $[a,b]$ (in general different for each rule), $h = (b - a)/s$ is the step size of each panel, s is the number of panels, and PP stands for piecewise polynomial. Composite quadrature rules based on Gaussian or Newton–Cotes rules can also be used, although they are not listed in Table 14.

9.3 Adaptive Quadrature

We see from Table 14 of composite quadrature rules that the smaller the step size h of a panel, the smaller the expected error. It is often the case that an approximation to the integral

$$I(f) = \int_a^b f(x)dx$$

is needed to within a specified accuracy ϵ. In this case, *adaptive quadrature* is often used. Such schemes refine the grid (or collection of panels) until an estimate of the total error in the integration is within the desired precision ϵ. Adaptive quadrature is particularly useful when the behavior of the function f varies significantly in the interval of integration $[a,b]$, since the scheme can use a fine grid where f is hard to integrate and a coarse grid where it is easy, leading to an efficient and accurate quadrature procedure.

Table 15 gives a simple general recursive procedure for adaptive quadrature. We as-

Table 14. Composite quadrature rules.

Quadrature Rule	n	d	PP interpolant	Formula	Error
Rectangle	s	0	Constant	$h \sum_{i=0}^{s-1} f(a + ih)$	$\frac{h}{2}(b - a)f'(\eta)$
Midpoint	s	1	Constant	$h \sum_{i=1}^{s} f[a + (i - 1/2)h]$	$\frac{h^2}{24}(b - a)f''(\eta)$
Trapezoidal	$s + 1$	1	Linear	$\frac{h}{2}[f(a) + f(b) + 2 \sum_{i=1}^{s-1} f(a + ih)]$	$-\frac{h^2}{12}(b - a)f''(\eta)$
Simpson's	$2s + 1$	3	Quadratic	$\frac{h}{6}\left[f(a) + f(b) + 2 \sum_{i=1}^{s-1} f(a + ih) + 4 \sum_{i=1}^{s} f(a + (i - 1/2)h)\right]$	$-\frac{h^4}{2880}(b - a)f^{(4)}(\eta)$
Corrected trapezoidal	$s + 3$	3	Cubic Hermite	$\frac{h}{2}\left[f(a) + f(b) + 2 \sum_{i=1}^{s-1} f(a + ih)\right] + \frac{h^2}{12}[f'(a) - f'(b)]$	$\frac{h^4}{720}(b - a)f^{(4)}(\eta)$

Table 15. Adaptive quadrature procedure.

```
subroutine AQ(a,b,ε)
   (Q,E) = LQM(a,b)
   if (E ≤ ε) then
      return (Q,E)
   else
      m = (a + b)/2
      return AQ(a,m,ε/2) + AQ(m,b,ε/2)
   end
end
```

sume that we can make use of a routine LQM (Local Quadrature Module) that implements a quadrature rule in some interval $[a,b]$ and returns Q, an approximation to the integral, and E, an estimate of the error. In the next section, we discuss how an error estimate may be obtained.

We note that the adaptive quadrature procedure shown in Table 15 does not illustrate how to reuse function evaluations where possible. An effective adaptive quadrature routine should do this, since function evaluations are often the most computationally expensive part of the procedure.

9.4 Romberg Integration and Error Estimation

As discussed in the last subsection, adaptive quadrature requires an error estimate. As an illustration, we consider how one may be obtained for the composite trapezoidal rule.

Let $T_s(f)$ denote the composite-trapezoidal-rule approximation to

$$I(f) = \int_a^b f(x)dx$$

using s panels and let $E_s = I(f) - T_s(f)$ be the associated error. On the basis of the error formula in Table 13, we expect E_{2s} to be about four times smaller than E_s, assuming that $f''(x)$ does not vary too much. That is,

$$I(f) - T_s(f) = E_s,$$

$$I(f) - T_{2s}(f) = E_{2s} \approx \tfrac{1}{4} E_s.$$

Subtracting these two equations, we get

$$T_{2s}(f) - T_s(f) = E_s - E_{2s} \approx \tfrac{3}{4} E_s.$$

So $E_s \approx 4[T_{2s}(f) - T_s(f)]/3$ and $E_{2s} \approx [T_{2s}(f) - T_s(f)]/3$. Thus, by applying the composite trapezoidal rule first with s and then with $2s$ panels, we obtain estimates of the error in both $T_s(f)$ and $T_{2s}(f)$.

If we add the estimate of the error $E_s = I(f) - T_s(f)$ to $T_s(f)$ we often obtain a better approximation to $I(f)$ than either $T_s(f)$ or $T_{2s}(f)$. That is, $\hat{T}_s(f) = T_s(f) + 4[T_{2s}(f) - T_s(f)]/3 = [4T_{2s}(f) - T_s(f)]/3$ is often a better approximation to $I(f)$ than either $T_s(f)$ or $T_{2s}(f)$, particularly if the function f is smooth and the grid is fine. Thus, by applying the compound trapezoidal rule with s and $2s$ panels and taking an appropriate linear combination of the two approximations, we construct a better approximation to $I(f)$. To be more specific, it can be shown that this eliminates the lowest-order term in the error E_s or E_{2s}. This process can be repeated to eliminate the next higher-order term in the error and so on. In addition, it can be generalized easily to other quadrature rules.

This is the basic idea behind *Romberg integration*. By applying a quadrature rule repeatedly with more panels each time, we can eliminate the leading terms of the error expansion, and thereby obtain better and better approximations to $I(f)$.

9.5 Infinite Integrals and Singularities

If one or both end points of the interval of integration $[a,b]$ are infinite, the integral is called infinite. We restrict the discussion of infinite integrals to the case

$$I(f) = \int_a^\infty f(x)dx,$$

sometimes called a semi-infinite integral, since only one end point is infinite. Other cases can be handled similarly.

Under the assumption that

$$I(f) = \int_a^\infty f(x)dx$$

exists, one way to approximate the integral is to *truncate* $I(f)$, and compute instead

$$\hat{I}(f) = \int_a^b f(x)dx,$$

for some sufficiently large b, by a standard quadrature rule $Q(f)$. The error in this approach is $I(f) - Q(f) = [I(f) - \hat{I}(f)] + [\hat{I}(f) - Q(f)]$. It is often possible to choose b so that

$$I(f) - \hat{I}(f) = \int_b^\infty f(x)dx$$

is small and to choose Q so that $\hat{I}(f) - Q(f)$ is also small.

$I(f)$ can also be approximated by first performing a change of variables to *transform* the infinite integral to a standard one. More specifically, let $x = g(t)$. Then

$$I(f) = \int_a^\infty f(x)dx = \int_{g^{-1}(a)}^{g^{-1}(\infty)} f(g(t)) \cdot g'(t)dt,$$

where $g^{-1}(x)$ is the inverse of $g(x)$. If we can choose g so that $g^{-1}(a)$ and $g^{-1}(\infty)$ are both finite, then $I(f)$ is transformed to a finite integral that can be evaluated by a standard quadrature rule. However, this procedure may introduce singularities (discussed below). If so, it might not lead to a computationally easier problem to solve.

Infinite integrals can also be approximated by special quadrature formulas that are directly applicable to infinite intervals of integration. For further details concerning this approach, see an introductory numerical methods text such as Conte and de Boor (1980), Dahlquist and Björck (1974), Johnson and Riess (1982), or Stoer and Bulirsch (1980).

A singular integral

$$I(f) = \int_a^b f(x)dx$$

is one in which f is singular (i.e., becomes infinite) at some point in $[a,b]$. Singular and infinite integrals are closely related: A change of variables often transforms one into the other.

In the computing of singular integrals by a quadrature rule, the value of f might be required at or close to a point of singularity, and so the quadrature rule may be either inapplicable or inaccurate. It often happens that the singularity in f occurs at the end point a or b, in which case, an open formula, such as a Gaussian rule, may be effective.

The performance of a quadrature rule applied to a singular integral might be improved by a change of variables. A transformation $x = g(t)$ that often helps to remove or lessen the effect of a singularity is $g(t) = b - (b - a)u^2(2u + 3)$ for $u = (t - b)/(b - a)$.

9.6 Monte-Carlo Methods

Monte-Carlo methods (*q.v.*) are of a statistical nature. For the sake of simplicity, we briefly present them for one-dimensional integrals only, but they can be extended easily to multidimensional integrals and are most useful in this context.

To begin, choose n random points $\{U_i : i = 1, \ldots, n\} \subset [0,1]$ and scale each U_i to $[a,b]$ by $u_i = a + U_i(b - a)$. Then

$$Q_n(f) = \frac{(b-a)}{n} \sum_{i=1}^{n} f(u_i)$$

is a Monte-Carlo approximation to

$$I(f) = \int_a^b f(x)dx.$$

If we consider $Q_n(f)$ to be a random variable, then it can be shown that its mean is $I(f)$ and its standard deviation is $|b - a| \times \rho(f)/\sqrt{n}$, where $\rho(f)$ is a constant that depends on f, but not n. Assuming that $Q_n(f)$ is close to being normally distributed, we are led to statistical statements about the error, such as that $|Q_n(f) - I(f)| \leq 2|b - a|\rho(f)/\sqrt{n}$ nineteen times out of twenty.

Note that the error bound above decreases like $1/\sqrt{n}$, much more slowly than the bounds for the standard compound quadrature rules given in Table 14. This suggests that Monte-Carlo methods are not very effective for one-dimensional integrals of smooth functions. However, an error formula similar to that given above continues to hold for multidimensional integrals, while extensions of standard methods become increasingly less efficient as the dimension of the integral to be approximated increases. As a result, Monte-Carlo methods are among the best schemes available for approximating high-dimensional integrals.

10. ORDINARY DIFFERENTIAL EQUATIONS

In this section, we consider numerical methods for the solution of ordinary differential equations (ODEs). We begin by introducing some simple schemes for the initial-value problem (IVP)

$$y'(x) = f(x,y(x)) \quad x \in [a,b],$$
$$y(a) = y_0, \tag{44}$$

where $y : \mathbb{R} \to \mathbb{R}^m$ and $f : \mathbb{R} \times \mathbb{R}^m \to \mathbb{R}^m$. We assume throughout that the IVP (44) has a unique solution for $x \in [a,b]$. We also discuss more sophisticated adaptive methods and explain the terms *stiff* and *nonstiff* problems and how to choose methods appropriate for these two classes of problems. We then briefly consider the boundary-value problem (BVP)

$$y'(x) = f(x,y(x)) \quad x \in [a,b],$$
$$g(y(a), y(b)) = 0, \tag{45}$$

which we assume has a locally unique solution.

For both IVPs and BVPs, we consider systems of first-order ODEs only, since most commonly available codes are for first-order systems and any higher-order ODE can be reduced to a system of first-order equations. However, using a method designed for higher-order equations directly may lead to a more efficient solution of the problem.

Because of space constraints, we do not discuss many important related problems such as differential-algebraic equations or delay differential equations. For a more comprehensive discussion of these and other related topics, see an advanced text such as Ascher *et al.* (1988), Butcher (1987), Hairer *et al.* (1987), Hairer and Wanner (1991), Lambert (1991), or Shampine (1994).

10.1 Initial-Value Problems (IVPs)

10.1.1 Two Simple Formulas Most standard numerical methods for the IVP (44) start with the initial value y_0 at $x_0 = a$ and then compute approximations $y_n \approx y(x_n)$ for $n = 1, \ldots, N$ on a discrete grid $a = x_0 < x_1 < \ldots < x_N = b$. The distance between adjacent grid points, $h_n = x_{n+1} - x_n$, $n =$

$0, \ldots, N-1$, is referred to as the *step size* at step $n+1$. Schemes are often presented with a constant step size $h = h_n$ for all n, but this is generally not required. Moreover, as discussed below, variable–step-size methods are often much more efficient.

Possibly the simplest numerical scheme for (44) is *Euler's method,* sometimes called the *forward Euler method:*

$$y_{n+1} = y_n + h_n f(x_n, y_n). \tag{46}$$

This formula is motivated from the observation that the true solution of a scalar IVP of the form (44) satisfies

$$\begin{aligned} y(x_{n+1}) &= y(x_n) + h_n y'(x_n) + \frac{h_n^2}{2} y''(\eta_n) \\ &= y(x_n) + h_n f(x_n, y(x_n)) + \frac{h_n^2}{2} y''(\eta_n) \end{aligned} \tag{47}$$

for some point $\eta_n \in [x_n, x_{n+1}]$, which follows from standard Taylor-series theory. Thus, the true solution of the IVP (44) satisfies an equation that is very similar to (46). This argument can be extended easily to systems.

The approximations $y_n \approx y(x_n)$ are computed in the order $n = 1, \ldots, N$ using the formula (46). To be more specific, on the first step of Euler's method from x_0 to $x_1 = x_0 + h_0$, we substitute the initial value (x_0, y_0) into the right side of (46) to compute $y_1 \approx y(x_1)$. Thus, at the end of the first step, we have (x_1, y_1). On the second step of Euler's method from x_1 to $x_2 = x_1 + h_1$, we substitute (x_1, y_1) into the right side of (46) to compute $y_2 \approx y(x_2)$. The procedure continues in a similar way for $n = 2, \ldots, N$. Note that this evaluation process applies equally well to systems of equations (i.e., $m > 1$). In this case, h_n is a scalar, but y_{n+1}, y_n and $f(x_n, y_n)$ are all m-vectors.

Euler's method is an *explicit formula* in the sense that the evaluation process described above does not require the solution of any linear or nonlinear algebraic equations. The backward Euler formula

$$y_{n+1} = y_n + h_n f(x_{n+1}, y_{n+1}) \tag{48}$$

is a typical example of an *implicit formula.* It can be motivated from a Taylor-series expansion of the true solution $y(x)$ of the IVP (44) about x_{n+1} similar to the expansion (47) above for $y(x)$ about x_n. Moreover, we again compute the approximations $y_n \approx y(x_n)$ in the order $n = 1, \ldots, N$. However, note that, on step $n+1$ from x_n to $x_{n+1} = x_n + h_n$, we start with (x_n, y_n) and must solve Eq. (48) for y_{n+1}.

We will return to the question of how to solve for y_{n+1} shortly, but first we explain briefly in the next subsection why we may wish to use an implicit scheme (such as the backward Euler formula) rather than an explicit one (such as the forward Euler formula) even though the former clearly requires more work per step than the latter.

10.1.2 Stiff IVPs Roughly speaking, a stiff IVP is one in which some terms in the solution decay rapidly with respect to the length of the integration, while others vary slowly on this time scale. To illustrate this concept, consider the linear constant-coefficient problem $y' = Ay$, $y(0) = y_0$, for $x \in [0,1]$, where

$$A = \tfrac{1}{2}\begin{pmatrix} -10^6-1 & 10^6-1 \\ 10^6-1 & -10^6-1 \end{pmatrix}, \quad y_0 = \begin{pmatrix} 2 \\ 0 \end{pmatrix}.$$

If we let $z = Py$, where

$$P = \tfrac{1}{2}\begin{pmatrix} 1 & -1 \\ 1 & 1 \end{pmatrix}, \quad P^{-1} = \begin{pmatrix} 1 & 1 \\ -1 & 1 \end{pmatrix},$$

$$D = PAP^{-1} = \begin{pmatrix} -10^6 & 0 \\ 0 & -1 \end{pmatrix},$$

then $z' = Py' = PAy = PAP^{-1}z = Dz$ and $z(0) = Py(0) = (1,1)^T$. Therefore, $z(x) = (e^{-10^6x}, e^{-x})^T$ and so $y(x) = P^{-1}z(x) = (e^{-x} + e^{-10^6x}, e^{-x} - e^{-10^6x})^T$. The term e^{-10^6x} that occurs in both $z(x)$ and $y(x)$ gives rise to an initial transient that decays rapidly with respect to the interval of integration [0,1], while the term e^{-x} is associated with a slowly varying smooth term on this scale. The fast and slow terms occur in different components of $z(x)$, but they are mixed in $y(x)$, which is often the case in practice. After the e^{-10^6x} term dies out, both components of $y(x)$ vary smoothly like e^{-x}.

If we apply the forward Euler formula to $y' = Ay$, we get $y_{n+1} = y_n + h_nAy_n = (I + h_nA)y_n$. If we multiply the last equation through by P and perform the change of variables $z_n = Py_n$, we get $z_{n+1} = (I + h_nD)z_n$. The two components of z_n satisfy $z_{n+1}^{(1)} = (1 - 10^6h_n)z_n^{(1)}$ and $z_{n+1}^{(2)} = (1 - h_n)z_n^{(2)}$. If we let $h_n > 2 \times 10^{-6}$ at any point during the integration, then $(1 - 10^6 h_n) < -1$ and $z_n^{(1)}$ will grow in magnitude and oscillate in sign as n increases. Since $y_n = P^{-1} z_n$, this will cause both components of y_n to oscillate about e^{-x_n} with growing amplitude as n increases. Since this instability is undesirable, we must restrict $h_n < 2 \times 10^{-6}$ throughout the integration, even though, after the initial transient, this is likely a much smaller step size than would be required to integrate the slowly varying e^{-x} component of the solution accurately.

On the other hand, if we apply the backward Euler formula to $y' = Ay$, we get $y_{n+1} = y_n + h_n A y_{n+1}$ and so $y_{n+1} = (I - hA)^{-1} y_n$. If we multiply this equation through by P and perform the change of variables $z_n = P y_n$, we get $z_{n+1} = (I - h_n D)^{-1} z_n$. Therefore, $z_{n+1}^{(1)} = z_n^{(1)}/(1 + 10^6 h_n)$ and $z_{n+1}^{(2)} = z_n^{(2)}/(1 + h_n)$. Hence, no matter how large $h_n > 0$ is, $z_n^{(1)}$ decays as n increases. Consequently, after the initial transient, we can choose $h_n > 0$ to integrate $z_n^{(2)}$ accurately without fear of $z_n^{(1)}$ becoming unstable. Since $y_n = P^{-1} z_n$, the same conclusion applies to y_n.

The example above can be generalized to larger systems of equations $y' = Ay$. If A is diagonalizable, then the performance of the method on $y' = Ay$ can be deduced from its performance on the scalar test problems $y' = \lambda y$, where the λ's range over the eigenvalues of A. If the real part of each λ is negative, then $y(x) \to 0$ as $x \to \infty$. We would like the numerical solution to have the same behavior without having to restrict h_n outside the transient region. Methods with this property are called *A-stable*. Generalizing the example above, it is easy to see that the backward Euler formula is A-stable, while the forward Euler formula is not.

The importance of the example above and the scalar test problem $y' = \lambda y$ in particular is that the performance of methods on these simple problems is indicative of their behavior on more general nonlinear stiff problems. A nonrigorous, but intuitive, justification of this follows from the local linearization of $y' = f(x,y)$ at (x_n, y_n):

$$y'(x) \approx f(x_n,y_n) + f_x(x_n,y_n)(x - x_n) + f_y(x_n,y_n)(y - y_n),$$

where $f_y(x,y) = \partial f(x,y)/\partial y \in \mathbb{R}^{m \times m}$ is the Jacobian of f. This problem is usually stiff if

1. some eigenvalue of $f_y(x_n,y_n)$ has a large negative real part with respect to the interval of integration,
2. no eigenvalue of $f_y(x_n,y_n)$ has a large positive real part with respect to the interval of integration, and
3. no eigenvalue of $f_y(x_n,y_n)$ has a large imaginary part unless it also has a relatively large negative real part.

An IVP that is not stiff is called *nonstiff.*

Stiff IVPs arise in many applications, such as chemical kinetics and electrical circuits. As noted earlier, they are characterized by components that vary on vastly different time scales: Some terms in the solution decay rapidly to steady state while others vary slowly.

The observation above that the explicit forward Euler formula is not appropriate for a stiff problem, while the implicit backward Euler formula is, can be generalized. All commonly used formulas that are suitable for stiff problems are implicit in some sense.

See the survey article of Shampine and Gear (1979) or an advanced text such as Butcher (1987), Hairer and Wanner (1991), Lambert (1991), or Shampine (1994) for a more detailed discussion of stiffness.

10.1.3 Solving Implicit Equations We return now to methods for solving for y_{n+1} in an implicit formula such as (48). One common approach is the predictor–corrector technique, which is just a fixed-point iteration as described in Sec. 6.1. For this scheme (and most others), we need in initial approximation $y_{n+1}^{(0)}$ to y_{n+1}. This could be computed, for example, from the forward Euler formula or some other explicit scheme, or simply by taking $y_{n+1}^{(0)} = y_n$. In the terminology of predictor–corrector techniques, the formula for computing $y_{n+1}^{(0)}$ is referred to as the *predictor* formula. For a predictor–corrector method based on the backward Euler formula (48), the corrector formula would be

$$y_{n+1}^{(l+1)} = y_n + h_n f(x_{n+1}, y_{n+1}^{(l)}), \qquad l = 0, 1, \ldots \quad (49)$$

We substitute $y_{n+1}^{(0)}$ into the right side of (49) to compute $y_{n+1}^{(1)}$, which we in turn substitute into the right side of (49) to compute $y_{n+1}^{(2)}$, and so on. It is easy to show that $y_{n+1}^{(l)} \to y_{n+1}$ as $l \to \infty$ if f satisfies the Lipschitz condition

$$\|f(x_{n+1},y) - f(x_{n+1},z)\| \leq L\|y - z\|$$

for some constant L and all y, z in a convex domain containing y_{n+1} and $y_{n+1}^{(l)}$ for $l = 0, 1, \ldots$ and $h_n L < 1$. In most codes, one or two corrections only are needed, since the initial guess $y_{n+1}^{(0)}$ is normally a good approximation to y_{n+1}. Consequently, this scheme is not much more expensive to implement than the explicit forward Euler formula. However, a predictor–corrector implementation of an A-stable method (such as the backward Euler formula) is not A-stable. On the contrary, because of the requirement $h_n L < 1$, it will suffer a step size restriction on a stiff problem similar to that of an explicit formula.

Alternatively, we could rewrite the backward Euler formula (48) as

$$F(y) = y - y_n - h_n f(x_{n+1},y) = 0, \tag{50}$$

where we have replaced the unknown y_{n+1} by y, and then apply one of the other techniques described in Sec. 6 for finding roots of equations to compute the solution $y = y_{n+1}$ of (50). The most commonly used root-finding technique in this context is Newton's method—or a variant of it. As noted in Sec. 6.7, for systems of equations, Newton's method takes the form

$$\begin{aligned}[I - h_n f_y(x_{n+1},y_{n+1}^{(l)})]\Delta_l \\ = y_n + h_n f(x_{n+1},y_{n+1}^{(l)}) - y_{n+1}^{(l)},\end{aligned} \tag{51}$$

$$y_{n+1}^{(l+1)} = y_{n+1}^{(l)} + \Delta_l, \tag{52}$$

where $f_y(x,y) = \partial f(x,y)/\partial y \in \mathbb{R}^{m\times m}$ is the Jacobian of f. Note that we must solve a linear system of m equations in m unknowns to compute the Newton update vector Δ_l in (51). Typically, Gaussian elimination with partial pivoting (see Sec. 2.5) is used to solve such linear systems. A band or sparse solver (see Sec. 2.7) may dramatically decrease the cost of solving (51) if $I - h_n f_y(x_{n+1},y_{n+1}^{(l)})$ is large and sparse. Similarly, iterative methods, such as the preconditioned conjugate-gradient method (see Sec. 3.2), may significantly reduce the cost of solving some large sparse problems. See Sec. 13 for a discussion of sources of high-quality numerical software, including stiff-ODE solvers that incorporate sparse and iterative linear equation solvers.

The computational work required to solve (51) can be decreased significantly by using a chord Newton method, often called a simplified Newton method, that holds the Newton iteration matrix $I - h_n f_y(x_{n+1},y_{n+1}^{(l)})$ constant over several iterations and possibly several steps of the integration, thus avoiding the necessity to factor the Newton iteration matrix on each iteration (see Sec. 6.8). However, even with this savings, the cost per step of the Newton iteration may be much larger than a predictor–corrector method. However, it has the advantage for formulas appropriate for stiff problems that it avoids the step size restriction associated with the predictor–corrector technique or explicit formulas. Thus, even though the Newton iteration might make the scheme much more expensive per step, the step sizes that can be used might be so much larger that the total cost of the integration is significantly less. Finally note that, as for predictor–corrector methods, an initial guess for $y_{n+1}^{(0)}$ is required. It can be computed by the techniques described above.

10.1.4 Higher-Order Formulas A numerical method for ODEs is said to be *of order p* or *p*th *order* or *p*th-*order convergent* if $y_n = y(x_n) + O(h^p)$ for some integer $p \geq 1$, where $O(h^p)$ is any quantity (in this case, the global error) that can be bounded by h^p times a constant that is independent of h, but that may depend on the IVP and the numerical method. Most standard texts on the numerical solution of ODEs show that both the forward and backward Euler methods are first-order convergent.

Higher-order methods are frequently used in practice because they offer the potential of significantly reducing the computational work required to generate an accurate solution to the IVP (44). To get an intuitive feeling for this, suppose that the length of integration $b - a = 1$, that we use a constant step size h throughout the numerical integration, that the global error for a pth-order method satisfies $y_n - y(x_n) = h^p$, and that we require this error to be of size 10^{-10}. Under these assumptions, the optimal step size for the method is $h = 10^{-10/p}$, and the resulting number of steps needed to integrate from a to b is $N = 10^{+10/p}$. To be more specific, for $p = 1, 2, 5, 10$, the number of steps required is $N = 10^{10}, 10^5, 10^2, 10^1$, respectively. Thus, even though a higher-order method may require more work per step than a lower-order scheme, the dramatic reduction in the number of steps required frequently makes a higher-order method much more efficient than a lower-order one—particularly for problems with stringent error tolerances.

Two common second-order formulas are the *trapezoidal rule*

$$y_{n+1} = y_n + \tfrac{1}{2}h_n[f(x_n,y_n) + f(x_{n+1},y_{n+1})] \tag{53}$$

and the *implicit midpoint rule*

$$y_{n+1} = y_n + f(x_n + h_n/2, (y_n, + y_{n+1})/2), \tag{54}$$

each of which is implicit, since one clearly needs to solve for y_{n+1}. Note that neither formula requires much more work per step than the backward Euler formula. Moreover, both formulas are A-stable and effective for solving stiff problems at relaxed error tolerances.

10.1.5 Runge–Kutta Formulas Runge–Kutta (RK) formulas are a general class of methods containing many higher-order schemes. The general form of an s-stage RK formula is

$$k_i = f(x_n + c_i h_n, y_n + h_n \sum_{j=1}^{s} a_{ij}k_j), \qquad i = 1, \ldots, s,$$

$$y_{n+1} = y_n + h_n \sum_{i=1}^{s} b_i k_i. \tag{55}$$

That is, we must first compute the s function values k_i and then form a weighted average of the k_i's to compute y_{n+1} from y_n.

RK formulas are one-step schemes in the sense that all the information required to compute y_{n+1} from y_n is generated on the current step from x_n to x_{n+1}. That is, unlike multistep formulas discussed in the next subsection, a RK formula does not require any information from past steps.

If the stages of the RK formula can be ordered so that $a_{ij} = 0$ for all $j \geq i$, then the formula (55) is explicit in the sense that the k_i's can be computed in the order $i = 1, \ldots, s$ without having to solve any linear or nonlinear equations. In what follows, we assume that the formula has been so ordered if possible. If the RK formula (55) is not explicit, then it is implicit and at least one linear or nonlinear equation must be solved to compute the k_i's.

The coefficients of a RK formula are frequently displayed in a tableau as

$$\begin{array}{c|cccc} c_1 & a_{11} & a_{12} & \cdots & a_{1s} \\ c_2 & a_{21} & a_{22} & \cdots & a_{2s} \\ \vdots & \vdots & \vdots & & \vdots \\ c_s & a_{s1} & a_{s2} & \cdots & a_{ss} \\ \hline & b_1 & b_2 & \ldots & b_s \end{array}$$

If all the elements in the tableau on the diagonal and above it are zero, then the associated formula is explicit.

All the methods considered so far are in fact RK formulas. The RK tableaux for the forward Euler formula, backward Euler formula, implicit midpoint rule, and trapezoidal rule are listed below in that order:

$$\begin{array}{c|c} 0 & 0 \\ \hline & 1 \end{array}, \qquad \begin{array}{c|c} 1 & 1 \\ \hline & 1 \end{array}, \qquad \begin{array}{c|c} 1/2 & 1/2 \\ \hline & 1 \end{array},$$

$$\begin{array}{c|cc} 0 & 0 & 0 \\ 1 & 1/2 & 1/2 \\ \hline & 1/2 & 1/2 \end{array}.$$

Note that the first three are one-stage RK formulas and the final one, the trapezoidal rule, is a two-stage RK formula. The tableau for the classical four-stage fourth-order explicit RK formula is

$$\begin{array}{c|cccc} 0 & 0 & 0 & 0 & 0 \\ 1/2 & 1/2 & 0 & 0 & 0 \\ 1/2 & 0 & 1/2 & 0 & 0 \\ 1 & 0 & 0 & 1 & 0 \\ \hline & 1/6 & 1/3 & 1/3 & 1/6 \end{array}$$

This formula has been widely used since it was published by Kutta in 1901. In the days of hand calculation, the zero coefficients below the diagonal were a distinct benefit, but this is no longer a significant advantage on a modern computer. Moreover, there are now many better formulas, both of order four and of higher order. The interested reader should consult an advanced text such as Butcher (1987), Hairer *et al.* (1987), Hairer and Wanner (1991), Lambert (1991), or Shampine (1994) for further details.

As the sample formulas above suggest, a high-order RK formula requires more stages than a low-order one. The minimum number of stages that an explicit RK formula requires to attain orders 1 to 8 are listed below.

Order	1	2	3	4	5	6	7	8
Stages	1	2	3	4	6	7	9	11

On the other hand, implicit s-stage RK formulas of order $2s$ exist for all $s \geq 1$. Moreover, it can be shown that this is the maximal order possible.

Explicit RK formulas are frequently used to solve nonstiff IVPs. Some implicit RK formulas are A-stable (or nearly so) and are suitable for solving stiff IVPs. Formulas with four or fewer stages are quite effective for problems with relaxed error tolerances, while formulas with five or more stages are suitable for problems with more stringent accuracy requirements. Since RK formulas are one-step schemes, unlike linear multistep formulas (LMFs) discussed in the next subsection, RK formulas are more suitable than LMFs for problems that require rapid changes in step size, such as problems with discontinuities. See Sec. 13 for a discussion of sources of high-quality numerical software, including routines based on RK formulas.

10.1.6 Linear Multistep Formulas Linear multistep formulas (LMFs) can be written in the form

$$y_{n+1} = \sum_{i=1}^{k} \alpha_i y_{n+1-i} + h \sum_{i=0}^{k} \beta_i f(x_{n+1-i}, y_{n+1-i}), \qquad (56)$$

where we assume that at least one of α_k or β_k is nonzero [otherwise we can reduce k in (56)]. Formula (56) is in fact a k-step method, since it uses values over k steps to compute y_{n+1}. Therefore, we assume that, at the start of step $n + 1$, we have $y_{n+1-k}, \ldots, y_n$ and our task is to compute y_{n+1} by (56). In this case, if $\beta_0 = 0$, then the $f(x_{n+1}, y_{n+1})$ term can be dropped from the right side of (56), and so we can evaluate the right side of (56) to compute y_{n+1} without having to solve any linear or nonlinear equations. That is, if $\beta_0 = 0$, the formula is explicit. On the other hand, if $\beta_0 \neq 0$, then y_{n+1} occurs on both sides of (56), and so a linear or nonlinear equation must be solved to compute y_{n+1}. Therefore, the formula is implicit. Depending on the context, a predictor–corrector method or some variant of Newton's method is typically used to solve for y_{n+1}.

Of course, at the start of the integration, $n = 0$ and $y_{1-k}, \ldots, y_{-1}$ are typically not available for $k > 1$. One solution to this problem is to compute $y_1, \ldots, y_{k-1}$ by a one-step method (such as a RK formula) of the same order as the k-step LMF and then start using the k-step LMF at step k. Alternatively, we could use a one-step LMF on step 1, a two-step LMF on step 2, and so on, until we reach step k, after which we can use (56) on that step and all subsequent steps. Most LMF codes employ the latter strategy and adjust the step size so that the accuracy obtained by the formulas with smaller k is comparable to that obtained by formulas with larger k.

If we ignore stability, then it is possible to obtain a k-step LMF of order $2k$ for any $k \geq 1$. Moreover, it is easy to show that this is the maximal order possible. However, these maximal-order formulas are unstable for $k \geq 3$ and so are not useful in practice. It can be shown that, for any $k \geq 1$, the maximal order of a stable k-step LMF is $k + 1$ if k is odd and $k + 2$ if k is even.

We have presented the LMFs in this section for fixed step size only, since variable–step-size formulas are considerably more complicated. However, a variable–step-size variable-order scheme may be far more efficient in practice. Such schemes are discussed in advanced texts on the numerical solution of ODEs, such as Hairer *et al.* (1987), Hairer and Wanner (1991), Lambert (1991), and Shampine (1994), but not usually in introductory numerical methods books.

10.1.7 Adams Formulas Adams formulas are a subclass of LMFs that have the form

$$y_{n+1} = y_n + h \sum_{i=0}^{k} \beta_i f(x_{n+1-i}, y_{n+1-i}). \qquad (57)$$

In the explicit Adams–Bashforth formulas, β_0

= 0 and the remaining k β_i's are chosen to obtain the maximal possible order k. In the implicit Adams–Moulton formulas, $\beta_0 \neq 0$ and the $k + 1$ β_i's are chosen to obtain the maximal possible order $k + 1$. These coefficients are listed in most advanced texts on numerical methods for ODEs and in many introductory numerical methods books. It turns out that the forward Euler formula is the one-step Adams–Bashforth formula, and the trapezoidal rule is the one-step Adams–Moulton formula. Moreover, note that the order of the Adams–Moulton formulas is optimal for k odd and nearly optimal for k even.

The implicit Adams–Moulton formulas have somewhat better numerical characteristics than the explicit Adams–Bashforth formulas. Consequently, Adams formulas are usually implemented in a predictor–corrector fashion, with the k- or $(k + 1)$-step Adams–Bashforth formula used for the predictor and a k-step Adams–Moulton formula used for the corrector. Adams predictor–corrector schemes are the basis for several very effective variable–step-size variable-order codes for nonstiff IVPs. See Sec. 13 for a discussion of sources of high-quality numerical software, including routines based on Adams formulas.

10.1.8 Backward Differentiation Formulas Backward differentiation formulas (BDFs), sometimes called *Gear formulas,* are another subclass of LMFs that have the form

$$y_{n+1} = \sum_{i=1}^{k} \alpha_i y_{n+1-i} + h\beta_0 f(x_{n+1}, y_{n+1}), \tag{58}$$

where $\beta_0 \neq 0$ and so the BDFs are implicit. The $k + 1$ coefficients of a k-step BDF are chosen to obtain the maximal possible order k. However, the BDFs are stable for $1 \leq k \leq 6$ only. They are A-stable for $k = 1$ and 2 and nearly A-stable for $k = 3$, 4, and 5, with the loss of A-stability increasing with k. For $k = 6$ the loss of A-stability increases to such an extent that this formula is frequently excluded from use.

The coefficients for the BDFs are listed in most advanced texts on numerical methods for ODEs and in many introductory numerical methods books. It turns out that the backward Euler formula is the one-step BDF.

Because the BDFs are usually used to solve stiff problems, the implicit equation is normally solved by Newton's method or some variant of this root-finding scheme.

BDFs are the basis for several very effective variable–step-size variable-order codes for stiff IVPs. See Sec. 13 for a discussion of sources of high-quality numerical software, including routines based on BDFs.

10.1.9 Other Methods Taylor-series methods and extrapolation schemes are two other classes of formulas that are sometimes used in practice, but much less frequently than Runge–Kutta or linear multistep formulas. See an advanced text such as Butcher (1987), Hairer *et al.* (1987), Hairer and Wanner (1991), Lambert (1991), or Shampine (1994) for a discussion of these and other classes of methods.

10.1.10 Adaptive Methods Most good programs for the numerical solution of ODEs vary their step size—and possibly their order—in an attempt to solve the problem as efficiently as possible subject to a user-specified error tolerance, *tol.* The error that is controlled is usually the local error on each step, rather than the global error $y_n - y(x_n)$ that the user might at first expect. However, in most good programs the global error is at least roughly proportional to *tol,* so that reducing *tol* usually reduces the global error. A few codes report an estimate of the global error as well. If such an estimate is available, it is often optional, since estimating the global error frequently increases the cost of the integration significantly.

A useful way to interpret *tol* and the associated local error is in the *backward error* sense. (See Sec. 2.8 for a discussion of backward error analysis in the context of solving linear algebraic systems $Ax = b$.) When called to solve the IVP (44), many good programs generate a numerical solution that is the exact solution of the slightly perturbed problem

$$z'(x) = f(x, z(x)) + \delta(x), \quad z(a) = y_0, \tag{59}$$

where $\|\delta(x)\| \lessapprox tol$. A few codes compute $\delta(x)$ explicitly and attempt to ensure that it is bounded by *tol,* but most satisfy (59) indirectly (some less reliably than others) by controlling some measure of the local error. For a more complete discussion of global errors, local errors, and their relationship to the perturbed equation (59), see an advanced text such as Butcher (1987), Hairer *et al.* (1987), Hairer and Wanner (1991), Lambert (1991), or Shampine (1994).

We believe that this backward error approach is often the most natural way to view the error in the numerical integration of an IVP. In many practical problems, we know $f(x,y)$ approximately only, possibly because of measurement errors or neglected terms in the model. Therefore, the true solution of the system satisfies an equation of the form (59), where in this case $\delta(x)$ is the error in the model. So, any solution to an IVP of the form (59) may be equally good provided $\|\delta(x)\|$ is less than the error in the model.

Programs for the numerical solution of ODEs often contain many other useful features. For example, some routines for nonstiff IVPs warn the user if the problem is stiff, while others automatically switch between stiff and nonstiff methods depending on the characteristics of the problem. Some programs contain sophisticated strategies to integrate problems with discontinuities in f or its derivatives much more efficiently and reliably than programs that do not attempt to detect discontinuities. Also, some programs return an interpolant for the numerical solution or allow the user to evaluate the numerical solution at very closely spaced points more efficiently than if the integration method itself produced all these output points. This can greatly increase the efficiency of codes when used to produce graphical output or to detect when the numerical solution satisfies some condition [such as $y(x) = c$ for some constant c].

10.2 Boundary-Value Problems (BVPs)

10.2.1 Shooting Methods *Shooting* is conceptionally one of the simplest numerical techniques for solving the boundary-value problem (BVP) (45). In its simplest form, often called *simple shooting,* we guess an initial condition $y(a) = y_0$ for the IVP (44) for the same ODE as the BVP (45), solve the IVP (44), and test whether the boundary condition $g(y_0,y(b;a,y_0)) = 0$ is satisfied, or nearly so, where $y(b;a,y_0)$ is the solution at $x = b$ of the IVP (44) with the initial condition $y(a) = y_0$. In most cases, the first guess for the initial condition $y(a) = y_0$ does not yield a $g(y_0,y(b;a,y_0))$ that is close enough to 0. So we must apply some root-finding technique to adjust the initial condition $y(a) = y_0$ until $g(y_0,y(b;a,y_0)) = 0$ is satisfied, or nearly so, assuming that there is a solution to the BVP.

Each time we adjust the initial condition, we must solve the IVP (44) again with the new initial condition $y(a) = y_0^{(l)}$ to compute $y(b;a,y_0^{(l)})$ and then $g(y_0^{(l)},(b;a,y_0^{(l)}))$. For a scalar ODE ($m = 1$), we could try a simple technique such as bisection (see Sec. 6.4) to solve $g(y_0,y(b;a,y_0)) = 0$, but this converges slowly and so requires many solutions of the IVP (44) with different initial conditions $y_0^{(l)}$. Moreover, bisection is not applicable to systems of ODEs ($m > 1$).

The usual approach is to apply a variant of Newton's method (see Sec. 6.7) to solve $g(y_0,y(b;a,y_0)) = 0$. However, this requires that we compute an approximation to the Newton iteration matrix

$$\frac{dg(y_0^{(l)},y(b;a,y_0^{(l)}))}{dy_0} = \frac{\partial g(y_0^{(l)},y(b;a,y_0^{(l)})}{\partial y_a} + \frac{\partial g(y_0^{(l)},y(b;a,y_0^{(l)}))}{\partial y_b}\frac{\partial y(b;a,y_0^{(l)})}{\partial y_0},$$

where $\partial g(y_0^{(l)},y(b;a,y_0^{(l)}))/\partial y_a$ is the partial derivative of g with respect to its first argument, $\partial g(y_0^{(l)},y(b;a,y_0^{(l)}))/\partial y_b$ is the partial derivative of g with respect to its second argument, and $\partial y(b;a,y_0^{(l)})/\partial y_0$ is the partial derivative of $y(b;a,y_0^{(l)})$ with respect to the initial condition $y(a) = y_0^{(l)}$. It can be shown that $\partial y(b;a,y_0^{(l)})/\partial y_0 = Y_l(b)$ for $Y_l : \mathbb{R} \to \mathbb{R}^{m\times m}$ the solution of the *variational equation*

$$Y_l'(x) = f_y(x,y_l(x))Y_l(x), \quad Y_l(a) = I, \tag{60}$$

where $y_l(x)$ is the solution of the associated IVP (44) with initial condition $y(a) = y_0^{(l)}$ and $f_y(x,y) = \partial f(x,y)/\partial y \in \mathbb{R}^{m\times m}$ is the Jacobian of f. Therefore, on each iteration of Newton's method, we must solve the IVP (44) with initial condition $y(a) = y_0^{(l)}$ for $y_l(x)$ as well as the variational equation (60) associated with $y_l(x)$. Since it may take many iterations before we find a $y_0^{(l)}$ for which $g(y_0^{(l)},y(b;a,y_0^{(l)}))$ is sufficiently close to 0, this is often a computationally expensive process.

Moreover, the associated IVP (44) may be unstable even though the BVP (45) is stable. As a result, simple shooting may break down or perform poorly. One way around this difficulty is to employ *multiple shooting*. In this scheme, we choose $N + 1$ shooting points $\{x_i : i = 0, \ldots, N\}$ satisfying $a = x_0 < x_1 < \cdots < x_{N-1} < x_N = b$, guess at N initial conditions s_i, $i = 0, \ldots, N - 1$, and solve the N IVPs

$$y_i' = f(x,y_i), \quad x \in [x_i,x_{i+1}] \quad i = 0, \ldots, N-1, \\ y_i(x_i) = s_i. \tag{61}$$

These IVPs are completely independent and so could be integrated simultaneously. Hence, this scheme is often called *parallel shooting*.

We need to adjust the initial conditions s_i, $i = 0, \ldots, N - 1$, so that

$$y_i(x_{i+1}) = s_{i+1}, \quad i = 0, \ldots, N-2, \tag{62}$$

$$g(s_0,y_N(b)) = 0, \tag{63}$$

where the first set of conditions (62) ensures $y_i(x_{i+1}) = y_{i+1}(x_{i+1})$ at the $N - 1$ interior shooting points $x_1, \ldots, x_{N-1}$, thus allowing us to patch the functions $y_i(x)$ together into a continuous function $y(x)$ on $[a,b]$, and the second condition (63) enforces the boundary condition for the BVP (45).

A variant of Newton's method (see Sec. 6.7) is usually used to solve (62)–(63). The solution process is similar to, but somewhat more complicated than, that described above for simple shooting. It should be noted that the linear systems associated with Newton's method for (62)–(63) have a very special structure that can be exploited to great computational advantage. See an advanced text such as Ascher *et al.* (1988) for details.

Both simple and multiple shooting simplify significantly when applied to a linear ODE $y' = A(x)y + b(x)$. Newton's method converges in one iteration and the resulting scheme is equivalent to what is frequently called the method of *superposition*. If the boundary conditions are separated, this scheme simplifies still further. See an advanced text such as Ascher *et al.* (1988) for details.

Good shooting programs contain heuristics for choosing the shooting points and adjusting the tolerance for the IVP solver in an attempt to solve the BVP to within a user-specified tolerance. They also contain many other components, similar to those described in Sec. 10.1.10 for IVPs.

10.2.2 One-Step Methods It is common to apply a one-step method, such as a Runge–Kutta (RK) formula, to solve the BVP (45). Since a collocation method applied to an ODE often reduces to a RK formula, this class of methods is broader than it might at first appear.

To simplify the discussion, assume that the one-step method can be written in the form

$$y_{n+1} = y_n + h_n\phi(x_n,y_n,h_n), \tag{64}$$

where $y_n \approx y(x_n)$, $h_n = x_{n+1} - x_n$, and $a = x_0 < x_1 < \cdots < x_N = b$. Note that the RK formula (55) is of this form with

$$\phi(x_n,y_n,h_n) = \sum_{i=1}^{s} b_i k_i, \\ k_i = f\left(x_n + c_i h_n, y_n + h_n \sum_{j=1}^{s} a_{ij} k_j\right). \tag{65}$$

To apply the one-step formula (64) to the BVP (45), we simply combine the equations (64) together with the boundary conditions to get a large system of equations

$$\Phi(y_0, \ldots, y_N) = \\ \begin{Bmatrix} y_{n+1} - y_n - h_n\phi(x_n,y_n,h_n), \; n = 0, \ldots, N-1, \\ g(y_0,y_N) \end{Bmatrix} \\ = 0. \tag{66}$$

It is usual to apply a variant of Newton's method (see Sec. 6.7) to solve (66). As for shooting, the main difficulty here is to compute the $(N + 1)m \times (N + 1)m$ Newton iteration matrix $\partial\Phi(y_0, \ldots, y_N)/\partial(y_0, \ldots, y_N)$ and solve the associated linear system for the update to the approximate solution $y_0^{(l)}, \ldots, y_N^{(l)}$ to (66). See an advanced text such as Ascher *et al.* (1988) for a more complete discussion of this important point.

Good BVP codes contain heuristics for choosing the grid points to solve the BVP to within a user-specified tolerance. They also contain many other components, similar to those described in Sec. 10.1.10 for IVPs.

10.2.3 Other Methods There are several other classes of numerical methods for BVPs for ODEs. Some of these are discussed in Sec. 11 as numerical methods for BVPs for partial differential equations. An important class of methods, not discussed there, consists of *defect correction* schemes, including *deferred correction* as a special case. The basic idea behind these schemes is to apply a simple technique, possibly in the class discussed in the last subsection, and then estimate the *defect* or truncation error in the discretization and solve a related problem again with the same simple technique in an attempt to eliminate the error. See an advanced text such as Ascher *et al.* (1988) for further details.

11. PARTIAL DIFFERENTIAL EQUATIONS (PDEs)

A partial differential equation (PDE) is an equation in which the partial derivative of some order of the unknown function w.r.t. some independent variable occurs. For example,

$$\frac{\partial^2 u}{\partial x^2} + \frac{\partial^2 u}{\partial y^2} = g(x,y) \tag{67}$$

is a PDE, where $u(x,y)$ is an unknown function, $\partial^2 u/\partial x^2$ and $\partial^2 u/\partial y^2$ denote the partial second derivatives of u w.r.t. x and y, respectively (often denoted by u_{xx} and u_{yy}, respectively), and $g(x,y)$ is a given function. The terms that involve u and its derivatives define the (partial differential) *operator* L, where, for example, $L = \partial^2/\partial x^2 + \partial^2/\partial y^2$ in (67), and the rest of the terms (usually the right side of the equation) form the *source term*.

11.1 Classes of Problems and PDEs

PDEs describe many important physical and technological phenomena. These phenomena can be divided into two basic types, which in turn are associated with two basic classes of problems for PDEs.

1. Equilibrium phenomena, elliptic PDEs, boundary-value problems. In steady-state phenomena, the equilibrium configuration u often satisfies

$$Lu = g \quad \text{in } \Omega, \tag{68}$$

$$Bu = \gamma \quad \text{on } \partial\Omega, \tag{69}$$

where Ω is a spatial N-dimensional domain, $\partial\Omega$ is the boundary of Ω, u is the unknown function of N variables, g and γ are known functions of N variables, and L and B are partial differential operators. Such problems are called *boundary-value problems* (BVPs). Often L is an elliptic operator. Equation (69) is frequently referred to as the *boundary condition* (BC). The definition of an elliptic operator in the general case is beyond the scope of this article, but some typical examples are given below.

2. Propagation phenomena, parabolic and hyperbolic PDEs, initial-value problems. In phenomena of a transient nature, the initial state is often given and we wish to predict the subsequent behavior. The function u at some point $t \in (0,T)$ frequently satisfies

$$Lu = g \quad \text{in } \Omega \times (0,T), \tag{70}$$

$$Bu = \gamma \quad \text{on } \partial\Omega \times (0,T), \tag{71}$$

while the initial configuration satisfies

$$Iu = g_0 \quad \text{in } \Omega \cup \partial\Omega, \tag{72}$$

where $(0,T)$ is the time interval of interest, Ω is a spatial N-dimensional domain, $\partial\Omega$ is the boundary of Ω, u is the unknown function of N spatial variables and one-time variable t, g and γ are known functions of N spatial variables and t, g_0 is a known function of N spatial variables, and L, B, and I are partial differential operators. Such problems are called *initial-value problems* (IVPs). L is often either a parabolic or a hyperbolic operator (see below). Equation (72) is often referred to as an *initial condition* (IC).

11.1.1 Some Definitions The *dimension* of a PDE is the number of independent variables in the PDE. The *order* of a PDE is the order of the highest derivative of the unknown function occurring in the PDE. A PDE is called *linear* if there are no nonlinear terms in the equation involving the unknown function or its derivatives; otherwise it is called *nonlinear*.

For example,

$$\sum_{i=1}^{N}\sum_{j=1}^{N} a_{ij}(x)\frac{\partial^2 u}{\partial x_i \partial x_j} + \sum_{j=1}^{N} b_j(x)\frac{\partial u}{\partial x_j} + c(x)u = d(x) \tag{73}$$

is N-dimensional, second order, and linear, where $x = (x_1, \ldots, x_N)$ is an N-dimensional vector of independent variables, $u(x)$ is the unknown function, and $\{a_{ij} : i,j = 1, \ldots, N\}$, $\{b_j : j = 1, \ldots, N\}$, c, and d are given functions of x. If any of the functions a_{ij}, b_j, or c depends on u or its derivatives, or if d is nonlinear in u or its derivatives, then the PDE is nonlinear.

A linear operator L is called *positive-definite* if $(Lu,u) > 0$ for all $u \neq 0$, where $(\cdot,\cdot)$ denotes an inner product (see Sec. 8.7). In addition, L is called *self-adjoint* if $(Lu,v) = (u,Lv)$ for all u and v in the associated function space. For example, the two-dimensional second-order linear PDE

$$au_{xx} + bu_{xy} + cu_{yy} + du_x + eu_y + fu = g \tag{74}$$

is self-adjoint if $d(x,y) = \partial a/\partial x$, $e(x,y) = \partial c/\partial y$ and $b(x,y) = 0$.

Consider the linear differential equation $Lu = g$, with L self-adjoint and positive-definite. Consider also the quadratic functional $F(u)$ defined by $F(u) = (Lu,u) - 2(u,g)$. The *minimum functional theorem* states that the solution of the differential equation $Lu = g$ coincides with the function u that minimizes $F(u)$. In general, when a solution to a differential equation corresponds to an extremum of a related functional, we have a *variational principle*. Numerical methods that construct approximations to the solution of a differential equation by using such a relationship are called *variational methods, Ritz methods, Rayleigh–Ritz methods,* or *energy methods*. The latter term comes from the observation that many variational methods are based on the physical principle of energy minimization.

The two-dimensional second-order linear PDE (74) is *elliptic, parabolic,* or *hyperbolic* if $D = b^2 - 4ac < 0, = 0$, or > 0, respectively. For the definitions of these terms in the general case, the reader is referred to any introductory PDE book, such as Ames (1992), Celia and Gray (1992), or Hall and Porsching (1990). Typical examples of elliptic, parabolic, and hyperbolic PDEs are

- Laplace's equation, $u_{x_1x_1} + u_{x_2x_2} + \cdots + u_{x_{N-1}x_{N-1}} + u_{x_Nx_N} = 0$, which is elliptic;
- the heat equation, $u_{x_1x_1} + u_{x_2x_2} + \cdots + u_{x_{N-1}x_{N-1}} - u_{x_N} = 0$, which is parabolic; and
- the wave equation $u_{x_1x_1} + u_{x_2x_2} + \cdots + u_{x_{N-1}x_{N-1}} - u_{x_Nx_N} = 0$, which is hyperbolic.

Two other classical elliptic PDEs (given in two dimensions) are

- Poisson's equation, $u_{xx} + u_{yy} = f(x,y)$, and
- the Helmholtz equation, $u_{xx} + u_{yy} + \kappa u = f(x,y)$, where κ is a constant.

The *normal derivative* of a surface $u(x,y)$ is the rate of change of u along the direction of the outward normal (i.e., the direction perpendicular to the surface). Let α be the angle that the direction of the outward normal makes with the x axis at a point (x,y) on u. Then the normal derivative $\partial u/\partial n$ of u (often denoted by u_n) at the point (x,y) is $u_n = u_x \times \cos\alpha + u_y \sin\alpha$. The normal derivative can also be written as the inner product of the gradient of u, ∇u, with the unit outward normal vector, n. That is, $u_n = \nabla u \cdot n$. This definition of the normal derivative for two dimensions can be generalized easily to N dimensions.

11.1.2 Boundary Conditions We now list some common types of boundary conditions (BCs) corresponding to the partial differential operator B in (69) or (71).

- *Dirichlet:* $Bu = u$.
- *Neumann:* $Bu = u_n$.
- *General* (for second-order PDEs): $Bu = \alpha(x)u + \beta(x)u_n$.
- *Mixed* (for second-order PDEs): Often, on parts of the boundary we have Dirichlet BCs and on the other parts Neumann ones. (Also, the term "mixed" may sometimes refer to more general types of BCs.)
- *Essential:* For PDEs of order $2m$, essential BCs involve u and its derivatives of order up to $m - 1$.
- *Natural:* For PDEs of order $2m$, natural BCs involve the derivatives of u from order m to $2m - 1$.

For further reading on the classification of PDE problems, operators, and boundary conditions, see Ames (1992), Celia and Gray (1992), or Hall and Porsching (1990).

11.2 Classes of Numerical Methods for PDEs

The two most commonly used methods for approximating the solution of PDEs are next described briefly.

Finite-Difference Methods (FDMs) have the following main steps:

1. Choose a finite-difference (FD) approximation of the derivatives involved in the PDE, BCs, and ICs. The result is a discretized PDE, BCs, and ICs.
2. Choose a set of n data points in the domain and on the boundary, on which the discretized PDE, BCs, and ICs must be satisfied. The result is a set of n equations w.r.t. the approximate values of u at the n data points.
3. Write the n equations of step **2** as a system and solve the system (discrete model). (If the PDE is linear, the system will usually be linear.) The solution is the approximate value of u at each of the n data points.
4. Evaluate the approximation to u at some point(s) of the domain (if needed).

Finite-Element Methods (FEMs) have the following main steps:

1. Choose a finite-element (FE) space, say n-dimensional, in which the approximation u_Δ is constrained to belong, and a set of basis functions that span the space, say $\{\phi_i : i = 1, \ldots, n\}$. Then write

   $$u_\Delta(x) = \sum_{i=1}^{n} \alpha_i \phi_i(x).$$

   The unknown scalars α_i, $i = 1, \ldots, n$, are often called the *degrees of freedom* (DOF), or *coefficients*, of the FE representation of u_Δ.
2. Choose a set of n conditions that the approximation u_Δ must satisfy. The result is a set of n equations w.r.t. the n coefficients of u_Δ.
3. Write the n equations of step **2** as a system and solve the system (discrete model). (If the PDE is linear, the system will usually be linear.) The solution is the vector of coefficients of u_Δ.
4. Evaluate the approximation to u at some point(s) of the domain.

11.2.1 Analysis of Numerical Methods for PDEs Some common techniques used to analyze numerical methods for PDEs are discussed below. The analysis can be used to evaluate a method w.r.t. some chosen criteria or measures. In the following discussion, we use u_Δ to denote the approximation to u computed by the method.

11.2.1.1 Convergence Analysis (for BVPs and IVPs). We study the behavior of the error $u - u_\Delta$ as n increases. Assuming $\|u - u_\Delta\| \to 0$ as $n \to \infty$, we can write $\|u - u_\Delta\| \leq C(1/n)^\alpha$ for some constants C and α. The largest constant α for which this inequality holds is called the *order of convergence* of the method. As a first rough measure, the larger the α the better the method, as $(1/n)^\alpha$ will converge to 0 faster as $n \to \infty$ for larger α. To estimate the order of convergence of a method experimentally, we often devise PDE problems with known solutions and then solve them using the PDE method under investigation, first using n DOF, then $2n$ DOF, etc. We then plot $\|u - u_\Delta\|$ versus n on a log–log scale. The slope of the plotted line is an approximation to the order of convergence of the method.

11.2.1.2 Stability Analysis (for IVPs). We study the behavior of the error $u - u_\Delta$ as a function of t for increasing t. We often say that a method is *stable* if $\|u - u_\Delta\|$ remains bounded as $t \to \infty$. Otherwise, it is called *unstable*. Or, we study how the error at some point in time propagates to the next point in time. In a stable method, the error is not amplified.

11.2.1.3 Time (Computational) Complexity Analysis. We study the time that the method takes to compute the approximate solution to the PDE as a function of the n DOF. The time is usually proportional to the number of floating-point operations, although it also depends on the implementation and the hardware (computer) used. The most time-consuming part of a FDM or FEM is usually the third step (solution of the system), while the second most time-consuming part is usually the second step (generation of the system). For FDMs, the fourth step can also be time consuming, particularly if the value of the approximation at arbitrary points of the domain is required, since this computation requires interpolation, often using piecewise polynomials (PPs) or splines. The data to be interpolated are the approximate values of u at the grid points. Interpolation is not required in step **4** of a FEM, since the approximate solution can be evaluated at any point of the domain by the formula

$$u_\Delta(x) = \sum_{i=1}^{n} \alpha_i \phi_i(x).$$

By studying the particular implementation of a method, we are usually able to derive an approximate formula, such as time $\approx Kn^{\beta}$, for some constant K, or time $= O(n^{\beta})$, relating the computational complexity to n. The smaller the β, the faster the method, and, among methods with the same β, the smaller the K, the faster the method.

11.2.1.4 Memory Complexity Analysis. We study the memory (storage) requirements of a method as a function of the n DOF. These requirements depend on the storage scheme used for the matrix arising in step **2** and the solver used in step **3**.

11.2.1.5 Overall Efficiency Analysis. Often, the most practical way of comparing two methods is to ask

1. if the methods were to run for the same length of time, which one would give the least error, or
2. given a certain error tolerance, which method satisfies that tolerance faster.

To test the overall efficiency of methods, we usually plot the error versus the time required to compute the approximate solution on a log–log scale. The method with the steepest slope is the most efficient.

11.3 Finite-Difference Methods for BVPs

A FD approximation to a derivative of a function u at a point x is a linear combination of values of u at points near x (often including x). Usually, a FD approximation is first derived for some derivative of a function of one variable, and then it is extended to partial derivatives of functions of several variables.

Let x be the point of interest and h, h_E, h_W small step sizes. The following are several examples of FD approximations in one dimension:

$$u_x(x) = \frac{u(x+h) - u(x)}{h} + O(h), \tag{75}$$

$$u_x(x) = \frac{u(x) - u(x-h)}{h} + O(h), \tag{76}$$

$$u_x(x) = \frac{u(x+h) - u(x-h)}{2h} + O(h^2), \tag{77}$$

$$u_x(x) = \frac{h_W^2 u(x+h_E) + (h_E^2 - h_W^2)u(x) - h_E^2 u(x-h_W)}{h_E(h_E+h_W)h_W} + O(h_E h_W), \tag{78}$$

$$u_{xx}(x) = \frac{u(x+h) - 2u(x) + u(x-h)}{h^2} + O(h^2), \tag{79}$$

$$u_{xx}(x) = \frac{2h_W u(x+h_E) - 2(h_E+h_W)u(x) + 2h_E u(x-h_W)}{h_E(h_E+h_W)h_W} + O(h_E - h_W) + O([\max(h_E,h_W)]^2). \tag{80}$$

Let (x,y) be the point of interest and h, h_E, h_W, h_N, h_S small step sizes. The following are several examples of FD approximations in two dimensions.

$$u_{xx}(x,y) = \frac{u(x+h,y) - 2u(x,y) + u(x-h,y)}{h^2} + O(h^2), \tag{81}$$

$$u_{xx}(x,y) + u_{yy}(x,y) = \frac{u(x+h,y) + u(x,y+h) - 4u(x,y) + u(x,y-h) + u(x-h,y)}{h^2} + O(h^2), \tag{82}$$

$$u_{xy}(x) = \frac{u(x+h,y+h) - u(x-h,y+h) - u(x+h,y-h) + u(x-h,y-h)}{4h^2} + O(h^2), \tag{83}$$

$$u_{xy}(x) = \frac{u(x+h_E,y+h_N) - u(x-h_W,y+h_N) - u(x+h_E,y-h_S) + u(x-h_W,y-h_S)}{(h_E+h_W)(h_S+h_N)} + O(\max(h_E,h_W,h_S,h_N)). \tag{84}$$

Note the following.

1. Each FD approximation listed above includes an error term. The actual FD approximation is the right side of the equation excluding the error term.
2. Approximations (75)–(79) and (81)–(83) use *uniform* step sizes, while the rest use *nonuniform* step sizes.
3. Approximations (75), (76), (80), and (84) are of *first order,* while the rest are of *second order.* The order refers to the (lowest) exponent of the step size(s) in the error term.
4. All FD approximation formulas are derived by using appropriate Taylor-series expansions around the point of approximation.
5. Approximations (81)–(84) are derived by using combinations of one-dimensional Taylor series and make use of values of u at points on a *rectangular grid.*
6. It is possible to derive two-dimensional FD approximations that make use of values of u at points on a *triangular, quadrilateral* (but not rectangular), *polygonal,* or *irregular grid* with points positioned arbitrarily. Such approximations can be derived by using two-dimensional Taylor's series.

11.3.1 An Example of a Finite-Difference Method in One Dimension Consider the problem

$$u_{xx} = g(x) \quad \text{in } (0,1), \tag{85}$$

$$u = \gamma(x) \quad \text{at } x = 0 \quad \text{and} \quad x = 1. \tag{86}$$

Using the FD approximation (79), we transform (85) to

$$[u(x+h) - 2u(x) + u(x-h)]/h^2 = g(x) + O(h^2). \tag{87}$$

Let $\{x_i = ih : i = 0, \ldots, n\}$ with $h = 1/n$ be the set of *grid points* and let $U_i \approx u(x_i)$ for $i = 1, \ldots, n$. Without the $O(h^2)$ error term, the discretized PDE (87) at the grid point x_i becomes

$$(U_{i+1} - 2U_i + U_{i-1})/h^2 = g(x_i). \tag{88}$$

From (86), we have for the point x_1

$$(U_2 - 2U_1)/h^2 = g(x_1) - \gamma(x_0)/h^2. \tag{89}$$

By writing equation (89) first, then (88) for $i = 2, \ldots, n-2$, and finally a relation similar to (89) for the point x_{n-1}, and then multiplying each equation by h^2, we get the following linear system:

$$\begin{pmatrix} -2 & 1 & & & & \\ 1 & -2 & 1 & & & \\ & 1 & -2 & 1 & & \\ & & \ddots & \ddots & \ddots & \\ & & & 1 & -2 & 1 \\ & & & & 1 & -2 \end{pmatrix} \begin{pmatrix} U_1 \\ U_2 \\ U_3 \\ \vdots \\ U_{n-2} \\ U_{n-1} \end{pmatrix} = h^2 \begin{pmatrix} g(x_1) \\ g(x_2) \\ g(x_3) \\ \vdots \\ g(x_{n-2}) \\ g(x_{n-1}) \end{pmatrix} - \begin{pmatrix} \gamma(x_0) \\ 0 \\ 0 \\ \vdots \\ 0 \\ \gamma(x_n) \end{pmatrix}. \tag{90}$$

Note that this system is symmetric, diagonally dominant in all rows, and strictly diagonally dominant in the first and last rows. Therefore it is also positive-definite and has a unique solution. By solving it, we obtain $U_i \approx u(x_i)$, $i = 1, \ldots, n-1$. Using interpolation, we can approximate u at any other point of the domain.

It can be proved that $\max\{|u(x_i) - U_i| : i = 1, \ldots, n-1\} = O(h^2)$. That is, the approximation is second order at the grid points.

The computational complexity of the method described is $O(n)$, since the linear system (90) is tridiagonal and of size $n-1$ (see Sec. 2.7).

11.3.2 An Example of a Finite-Difference Method in Two Dimensions Consider the problem

$$u_{xx} + u_{yy} = g(x,y) \quad \text{in } (0,1) \times (0,1), \tag{91}$$

$$u = \gamma(x,y) \quad \text{on } x = 0, \quad x = 1, \quad y = 0, \quad y = 1. \tag{92}$$

Using the FD approximation (82), we transform (91) to

$$\frac{u(x+h,y) + u(x,y+h) - 4u(x,y) + u(x,y-h) + u(x-h,y)}{h^2} = g(x,y) + O(h^2). \tag{93}$$

Let $\{(x_i,y_j) : x_i = ih,\ y_j = jh,\ i, j = 0, \ldots, n\}$ with $h = 1/n$ be the set of grid points and let $U_{ij} \approx u(x_i,y_j)$. Without the $O(h^2)$ error term, the discretized PDE (93) at the grid point (x_i,y_j), $i = 1, \ldots, n-1$, $j = 1, \ldots, n-1$, becomes

$$(U_{i+1,j} + U_{i,j+1} - 4U_{i,j} + U_{i,j-1} + U_{i-1,j})/h^2 = g(x_i,y_j). \tag{94}$$

From (92), we have for the point (x_1,y_1)

$$(U_{1,2} + U_{2,1} - 4U_{1,1})/h^2 = g(x_1,y_1) - [\gamma(x_0,y_1) + \gamma(x_1,y_0)]/h^2. \tag{95}$$

Similar relations hold at the three other corners of the domain. Also, for the points (x_1,y_j), $j = 2, \ldots, n-2$, we have

$$(U_{1,j+1} - 4U_{1,j} + U_{1,j-1} + U_{2,j})/h^2 = g(x_1,y_j) - \gamma(x_0,y_j)/h^2. \tag{96}$$

Similar relations hold for other grid points one grid line away from the boundary.

One way to number the grid points, and also the equations and unknowns, is bottom up then left to right: (1,1), (1,2), (1,3), . . . , $(1, n-2)$, $(1, n-1)$, (2,1), (2,2), (2,3), . . . , $(2, n-2)$, $(2, n-1)$, ... That is, first (95), then (96) for $j = 2, \ldots, n-2$, then relations similar to (95) for the points (x_1,y_{n-1}), (x_2,y_1), then (94) for $i = 2$, $j = 2, \ldots, n-2$, etc. Using this ordering, we get a linear system $AU = \mathbf{g}$, where, after multiplying each equation by h^2, the matrix A has the form

$$\begin{pmatrix}
-4 & 1 & & & & 1 & & & & & & & & & & \\
1 & -4 & 1 & & & & 1 & & & & & & & & & \\
 & \ddots & & \ddots & & & & \ddots & & & & & & & & \\
 & & 1 & -4 & 1 & & & & 1 & & & & & & & \\
 & & & 1 & -4 & & & & & 1 & & & & & & \\
1 & & & & & -4 & 1 & & & & 1 & & & & & \\
 & 1 & & & & 1 & -4 & 1 & & & & 1 & & & & \\
 & & \ddots & & & & \ddots & \ddots & \ddots & & & & \ddots & & & \\
 & & & \ddots & & & & \ddots & \ddots & \ddots & & & & \ddots & & \\
 & & & & 1 & & & & 1 & -4 & 1 & & & & 1 & \\
 & & & & & 1 & & & & 1 & -4 & & & & & 1 \\
 & & & & & & 1 & & & & & -4 & 1 & & & \\
 & & & & & & & 1 & & & & 1 & -4 & 1 & & \\
 & & & & & & & & \ddots & & & & \ddots & \ddots & \ddots & \\
 & & & & & & & & & 1 & & & & 1 & -4 & 1 \\
 & & & & & & & & & & 1 & & & & 1 & -4
\end{pmatrix}.$$

This can be rewritten as

$$\begin{pmatrix}
T & I & & & \\
I & T & I & & \\
 & \ddots & \ddots & \ddots & \\
 & & I & T & I \\
 & & & I & T
\end{pmatrix},$$

where T is a tridiagonal matrix of size $n - 1$ with -4's on the diagonal and 1's on the super- and subdiagonal and I is the identity matrix of size $n - 1$. Thus the matrix is *block-tridiagonal* of size $(n-1)^2$. The vector of unknown is

$$\begin{aligned}
U = (&U_{1,1},U_{1,2},\ldots,U_{1,n-2},U_{1,n-1},\\
&U_{2,1},U_{2,2},\ldots,U_{2,n-2},U_{2,n-1},\ldots\\
&U_{n-2,1},U_{n-2,2},\ldots,U_{n-2,n-2},U_{n-2,n-1},\\
&U_{n-1,1},U_{n-1,2},\ldots,U_{n-1,n-2},U_{n-1,n-1})^{\mathrm{T}}.
\end{aligned}$$

The right-side vector is

$$\mathbf{g} = (h^2 g_{1,1} - \gamma_{0,1} - \gamma_{1,0}, h^2 g_{1,2} - \gamma_{0,2}, \ldots, h^2 g_{1,n-2} - \gamma_{0,n-2}, h^2 g_{1,n-1} - \gamma_{0,n-1} - \gamma_{1,n}, h^2 g_{2,1} - \gamma_{2,0}, h^2 g_{2,2}, \ldots, h^2 g_{2,n-2}, h^2 g_{2,n-1} - \gamma_{2,n}, \ldots, h^2 g_{n-2,1} - \gamma_{n-2,0}, h^2 g_{n-2,2}, \ldots, h^2 g_{n-2,n-2}, h^2 g_{n-2,n-1} - \gamma_{n-2,n}, h^2 g_{n-1,1} - \gamma_{n,1} - \gamma_{n-1,0}, h^2 g_{n-1,2} - \gamma_{n,2}, \ldots, h^2 g_{n-1,n-2} - \gamma_{n,n-2}, h^2 g_{n-1,n-1} - \gamma_{n,n-1} - \gamma_{n-1,n})^{\mathrm{T}},$$

where $g_{ij} = g(x_i,y_j)$ and $\gamma_{ij} = \gamma(x_i,y_j)$. Note that this system is symmetric, diagonally dominant in all rows, and strictly diagonally dominant in all rows corresponding to grid points one grid line away from the boundary. Therefore, it is positive-definite and has a unique solution. By solving the system $AU = \mathbf{g}$, we obtain $U_{ij} \approx u(x_i,y_j)$ for $i = 1, \ldots, n - 1$ and $j = 1, \ldots, n - 1$. Using interpolation, we can approximate the value of u at any other point of the domain.

It can be proved that $\max\{|u(x_i, y_j) - U_{i,j}| : i, j = 1, \ldots, n - 1\} = O(h^2)$. That is, the approximation is second order at the grid points.

The computational complexity of the method described above depends on the method used to solve the linear system $AU = \mathbf{g}$. Note that A has at most five nonzero entries per row, it is banded with lower and upper bandwidth $n - 1$, and its size is $(n - 1)^2$. If a direct band solver is used to solve $AU = \mathbf{g}$, then the computational complexity of the method is $O(n^4)$, but sparse direct solvers are more efficient (see Sec. 2.7). In addition, there exist iterative methods (e.g., multigrid; see Sec. 11.8 and Briggs, 1987) that can solve this system much more efficiently, reducing the computational complexity of the method to almost $O(n^2)$.

Note that the properties of the matrix A, such as symmetry, diagonal dominance, positive-definiteness, and the sparsity pattern (block-tridiagonal with at most five nonzero entries per row), are highly dependent on the simplicity of the differential operator associated with (91) and boundary conditions (92), the choice of uniform and rectangular grid, and the FD approximation (93). For a differential operator with first-order derivative terms and/or Neumann BCs, symmetry is lost. Symmetry may also be lost if a nonuniform grid is chosen, even if it is rectangular. Diagonal dominance depends on the coefficients of the differential operator and on the absence of first-order derivative terms. The block-tridiagonal form will most likely be affected if an irregular grid is chosen. Fast linear solvers, such as multigrid and FFT (fast Fourier transform) solvers, work well on the matrix A, but may not perform as well on more general systems. The development of fast linear solvers for such matrices is an open and active area of research. See, for example, Van Loan (1992) or Hackbusch (1994) and the references therein. For further reading on FDMs, see Strikwerda (1989).

11.4 Finite-Element Methods for BVPs

The first step in a FEM is to choose a FE approximation space and a basis for it. The most commonly used spaces are piecewise polynomials (PPs) or splines (see Sec. 8.5). Let n be the dimension of the approximation space and let $\{\phi_j(x) : j = 1, \ldots, n\}$ be a set of basis functions for the space.

Consider the problem (68)–(69). Let

$$u_\Delta(x) = \sum_{j=1}^{n} \alpha_j \phi_j(x)$$

be the approximation to u. The next step in a FEM is to choose n conditions that the approximation must satisfy. A FEM is characterized by these conditions. The most common FEMs are the *Galerkin method* and the *collocation method*.

11.4.1 The Galerkin Method Given an inner product $(\cdot,\cdot)$, usually defined by

$$(f,g) = \int_\Omega f(x)g(x)dx,$$

we require that u_Δ satisfies

$$(\phi_i, Lu_\Delta - g) = 0, \quad i = 1, \ldots, n, \tag{97}$$

forcing the *residual* $Lu_\Delta - g$ to be orthogonal to the approximation space, and making it, in a sense, as "small" as possible. If L is a linear operator, then the relations (97) are equivalent to

$$\sum_{j=1}^{n} \alpha_j(\phi_i, L\phi_j) = (\phi_i, g), \quad i = 1, \ldots, n, \tag{98}$$

which can be written in the form $A\boldsymbol{\alpha} = \mathbf{g}$, where A is an $n \times n$ matrix with entries $A_{ij} = (\phi_i, L\phi_j)$, $i = 1, \ldots, n$, $j = 1, \ldots, n$, $\boldsymbol{\alpha} = (\alpha_1, \ldots, \alpha_n)^{\mathrm{T}}$ is the vector of coefficients, and $\mathbf{g}$ is a vector with entries $g_i = (\phi_i, g)$, $i = 1, \ldots, n$. We usually use numerical integration to compute the entries of A and $\mathbf{g}$ (see Sec. 9).

As an example of the Galerkin method in one dimension, consider the problem (85)–(86) and the set of grid points $\{x_i = ih : i = 0, \ldots, n\}$ with $h = 1/n$. Let $\{\phi_i : i = 0, \ldots, n\}$ be the set of linear spline basis functions w.r.t. the knots (grid points) $\{x_i\}$, as defined by (39). Then

$$u_\Delta(x) = \sum_{i=0}^{n} \alpha_i \phi_i(x)$$

is the linear spline approximation to u. From the BC (86), we get $u_\Delta(x_0) = \gamma(x_0)$, which implies $\alpha_0 = \gamma(x_0)$. Similarly, $\alpha_n = \gamma(x_n)$. The remaining unknowns $\{\alpha_i : i = 1, \ldots, n - 1\}$ are determined by the $n - 1$ Galerkin conditions

$$\sum_{j=0}^{n} \alpha_j(\phi_i, L\phi_j) = (\phi_i, g), \quad i = 1, \ldots, n-1,$$

which, for the particular L associated with (85), are equivalent to

$$\sum_{j=0}^{n} \alpha_j(\phi_i, \phi_j'') = (\phi_i, g), \quad i = 1, \ldots, n-1.$$

Writing the inner product (ϕ_i, ϕ_j'') as an integral and applying integration by parts, these conditions reduce to

$$\begin{aligned} \sum_{j=1}^{n-1} \alpha_j \int_0^1 \phi_i' \phi_j' dx &= \left[\phi_i \sum_{j=0}^{n} \alpha_j \phi_j' \right]_0^1 \\ &- \int_0^1 \phi_i g dx - \alpha_0 \int_0^1 \phi_i' \phi_0' dx \\ &- \alpha_n \int_0^1 \phi_i' \phi_n' dx, \quad i = 1, \ldots, n-1. \end{aligned} \tag{99}$$

Note that the term

$$\left[\phi_i \sum_{j=0}^{n} \alpha_j \phi_j' \right]_0^1 = 0,$$

since $\phi_i(0) = \phi_i(1) = 0$ for $i = 1, \ldots, n - 1$. Relations (99) form a linear system of size $n - 1$; the associated matrix A has elements

$$A_{i,j} = \int_0^1 \phi_i' \phi_j' dx.$$

Since the basis functions $\{\phi_i\}$ are nonzero on at most two subintervals, A is tridiagonal with elements

$$A_{i,i} = \int_{x_{i-1}}^{x_{i+1}} \phi_i' \phi_i' dx, \quad i = 1, \ldots, n-1,$$

$$A_{i,i-1} = \int_{x_{i-1}}^{x_i} \phi_i' \phi_{i-1}' dx, \quad i = 2, \ldots, n-1,$$

$$A_{i,i+1} = \int_{x_i}^{x_{i+1}} \phi_i' \phi_{i+1}' dx, \quad i = 1, \ldots, n-2.$$

It can be proved that this matrix is also symmetric positive-definite. Thus, the associated system has a unique solution. By solving the system, we obtain the coefficients $\{\alpha_i : i = 0, \ldots, n\}$ of u_Δ, which we can evaluate at any point of the domain (0, 1).

It can be proved that $\max\{|u(x) - u_\Delta(x)| : x \in [0, 1]\} = O(h^2)$. That is, the approximation is second order on the whole domain.

The computational complexity of the method described is $O(n)$, since the linear system that has to be solved is tridiagonal and of size $n - 1$ (see Sec. 2.7).

Relations (99) can also be derived using a variational method (see Sec. 11.1.1). Thus, for problem (85)–(86), there is a variational method equivalent to the Galerkin method. This is true for every differential equation problem with a self-adjoint positive-definite operator. There exist differential equation problems, though, that are not characterized by variational principles. In such cases, the Galerkin method is applicable, while the variational method is not.

As an example in two dimensions, consider problem (91)–(92) with the grid points $\{(x_i, y_j) : x_i = ih, y_j = jh, i, j = 0, \ldots, n\}$ for $h = 1/n$. A common way to define an approximation space for two-dimensional problems is to choose a tensor product of approximation spaces in each dimension. Let $\{\phi_i(x) : i = 0, \ldots, n\}$ be the linear spline basis functions w.r.t. the knots $\{x_i : i = 0, \ldots, n\}$ and let $\{\phi_j(y) : j = 0, \ldots, n\}$ be the linear spline basis functions w.r.t. the knots $\{y_j : j = 0, \ldots, n\}$, as defined in (39). Then

$$u_\Delta(x,y) = \sum_{i=0}^{n} \sum_{j=0}^{n} \alpha_{ij}\phi_i(x)\phi_j(y)$$

is the bilinear spline approximation to u. Continuing as in the one-dimensional case, we derive a system of $(n + 1)^2$ equations in $(n + 1)^2$ unknowns. The associated matrix A is block-tridiagonal, with at most nine nonzero entries per row and bandwidth $n + 2$. It is also symmetric positive-definite. Thus, the associated system has a unique solution. Moreover, it can be proved that the approximation u_Δ is second order.

Note that, if instead of a rectangular subdivision of the domain and bilinear elements, we choose a triangular subdivision and linear elements (w.r.t. x and y), we would get a system similar to that of Sec. 11.3.2.

An important property of the Galerkin method is that, for any self-adjoint positive-definite differential operator, the resulting matrix is symmetric positive-definite, even if the grid is irregular. As stated before, for every differential equation problem with a self-adjoint and positive-definite operator, there is a variational method equivalent to the Galerkin method. This holds for higher-dimension problems too. Thus, large, sparse, symmetric, positive-definite matrices arise from the application of variational methods.

For an introduction to the FEM, including its computer implementation, see Becker *et al.* (1981). An error analysis is carried out in Strang and Fix (1973).

11.4.2 The Collocation Method We first pick n *collocation points* $\{t_i : i = 1, \ldots, n\}$ in Ω and on $\partial\Omega$. We then require that u_Δ satisfies

$$Lu_\Delta(t_i) - g(t_i) = 0, \quad \text{if } t_i \in \Omega, \tag{100}$$

$$Bu_\Delta(t_i) - \gamma(t_i) = 0, \quad \text{if } t_i \in \partial\Omega, \tag{101}$$

forcing the residuals $Lu_\Delta - g$ and $Bu_\Delta - \gamma$ to be zero at the collocation points, and making them, in a sense, as "small" as possible. If L and B are linear, relations (100)–(101) are equivalent to

$$\sum_{j=1}^{n} \alpha_j L\phi_j(t_i) = g(t_i), \quad \text{if } t_i \in \Omega, \tag{102}$$

$$\sum_{j=1}^{n} \alpha_j B\phi_j(t_i) = \gamma(t_i), \quad \text{if } t_i \in \partial\Omega, \tag{103}$$

which can be written in the form $A\boldsymbol{\alpha} = \mathbf{g}$, where A is an $n \times n$ matrix with entries $A_{ij} = L\phi_j(t_i)$, $j = 1, \ldots, n$, for all $t_i \in \Omega$, and $A_{ij} = B\phi_j(t_i)$, $j = 1, \ldots, n$, for all $t_i \in \partial\Omega$; $\boldsymbol{\alpha} = (\alpha_1, \ldots, \alpha_n)^T$ is the vector coefficients; and $\mathbf{g}$ is a vector with entries $g_i = g(t_i)$ for all $t_i \in \Omega$ and $g_i = \gamma(t_i)$ for all $t_i \in \partial\Omega$.

The choice of collocation points is critical to the success of the method. It affects not only the solvability and other properties (such as symmetry, diagonal dominance, bandedness) of the matrix A but also the accuracy of the approximation u_Δ. Depending on the FE approximation space that u_Δ belongs to, some standard choices of collocation points in one dimension are listed below.

1. If the FE approximation space is the space of quadratic splines (quadratic PPs in $\mathscr{C}^1$), the collocation points are chosen to be the midpoints of the subintervals (x_{i-1},x_i), $i = 1, \ldots, n$, and the two boundary points. The same choice of collocation points is effective if the FE approximation space is composed of any other even-degree splines, with the exception that some additional collocation conditions may be required at boundary points or points close to the boundary.
2. If the FE approximation space is the space of cubic splines (cubic PPs in $\mathscr{C}^2$), the collocation points are chosen to be the grid points $\{x_i : i = 0, \ldots, n\}$. At each of the boundary points, x_0 and x_n, both conditions (100) and (101) are imposed. The same choice of collocation points is effective if the FE approximation space is composed of any other odd-degree splines, with the exception that some additional collocation conditions may be required at boundary points or points close to the boundary.
3. If the FE approximation space is the space of cubic PPs in $\mathscr{C}^1$ (cubic Hermite PPs), the collocation points are chosen to be the two Gauss points $x_{i-1} + (3 \pm \sqrt{3}) \times (x_i - x_{i-1})/6$ in each subinterval (x_{i-1},x_i), $i = 1, \ldots, n$, and the two boundary grid points.

As an example in one dimension, consider

problem (85)–(86) and the set of grid points $\{x_i = ih : i = 0, \ldots, n\}$ with $h = 1/n$. Let the collocation points be the midpoints $t_i = (x_{i-1} + x_i)/2$, $i = 1, \ldots, n$, and the end points $t_0 = x_0$ and $t_{n+1} = x_n$. Let $\{\phi_i : i = 0, \ldots, n + 1\}$ be the quadratic spline basis functions w.r.t. the knots (grid points) $\{x_i\}$, as defined in (40). Then

$$u_\Delta(x) = \sum_{i=0}^{n+1} \alpha_i \phi_i(x)$$

is the quadratic spline approximation to u. Relation (100) for the PDE (85) becomes $u''_\Delta(t_i) = g(t_i)$, and so relation (102) becomes

$$\alpha_{i-1}\phi''_{i-1}(t_i) + \alpha_i\phi''_i(t_i) + \alpha_{i+1}\phi''_{i+1}(t_i) = g(t_i),$$

which reduces to

$$(\alpha_{i-1} - 2\alpha_i + \alpha_{i+1})/h^2 = g(t_i), \quad i = 1, \ldots, n. \tag{104}$$

Relation (101) for the BC (86) becomes $u_\Delta(t_0) = \gamma(t_0)$, and so relation (103) becomes

$$\alpha_0\phi_0(t_0) + \alpha_1\phi_1(t_0) = \gamma(t_0),$$

which reduces to

$$(\alpha_0 + \alpha_1)/2 = \gamma(t_0). \tag{105}$$

Similarly, the collocation condition at $t_{n+1} = 1$ reduces to

$$(\alpha_n + \alpha_{n+1})/2 = \gamma(t_{n+1}). \tag{106}$$

Writing (105) first, then (104) for $i = 1, \ldots, n$, and finally (106), we get a tridiagonal system of equations w.r.t. the coefficients $\{\alpha_i : i = 0, \ldots, n + 1\}$. The system is diagonally dominant and it can be proved that it has a unique solution. It can also be scaled so that it is symmetric positive-definite.

It can be proved that $\max\{|u(x) - u_\Delta(x)| : x \in [0, 1]\} = O(h^2)$. That is, the approximation is second order on the whole domain. There exists a variant of this method, though, that is fourth order at the grid points and midpoints and third order on the whole domain (Houstis *et al.*, 1988).

The computational complexity of the method described above is $O(n)$, since the linear system that must be solved is tridiagonal and of size $n + 1$ (see Sec. 2.7).

As an example in two dimensions, consider problem (91)–(92) and the set of grid points $\{(x_i, y_j) : x_i = ih, y_j = jh, i, j = 0, \ldots, n\}$ with $h = 1/n$. A common approximation space for two-dimensional problems is a tensor product of approximation spaces in each dimension. Let $\{\phi_i(x) : i = 0, \ldots, n + 1\}$ be the quadratic spline basis functions w.r.t. the knots (grid points) $\{x_i : i = 0, \ldots, n\}$ and let $\{\phi_j(y) : j = 0, \ldots, n + 1\}$ be the quadratic spline basis functions w.r.t. the knots $\{y_j : j = 0, \ldots, n\}$, as defined in (40). Then

$$u_\Delta(x,y) = \sum_{i=0}^{n+1}\sum_{j=0}^{n+1} \alpha_{ij}\phi_i(x)\phi_j(y)$$

is the biquadratic spline approximation to u. Continuing as in the one-dimensional case, we derive a system of $(n + 2)^2$ equations and unknowns. The associated matrix is block-tridiagonal, with at most nine nonzero entries per row, and has bandwidth $n + 3$. It can be proved that this system has a unique solution and that the approximation u_Δ is second order. With appropriate modifications, though, the order can be improved as in the one-dimensional case (Christara, 1994).

For a general introduction to collocation methods, see Prenter (1975).

11.5 Finite-Difference Methods for IVPs

Consider the problem (70)–(72). Let the temporal grid points be $\{t_j = jh_t : j = 0, \ldots, m\}$ with $h_t = T/m$. Starting with the initial values of u at t_0, given by (72), most FDMs for IVPs compute approximate values of u at each subsequent temporal grid point t_j, in the order $j = 1, \ldots, m$, using previous and/or current approximate values of u at neighboring space points.

If at each temporal grid point t_j a method uses only approximations from previous temporal grid points, it is called *explicit*, as it does not require the solution of a system of equations to proceed from one temporal grid point to the next. If at some temporal grid point t_j a method uses approximations from the current temporal grid point t_j, it is called *implicit*, as it requires the solution of a system of equations to proceed from one temporal grid point to the next. If at the time step from t_{j-1} to t_j a method uses approximations from t_{j-1} and t_j only, it is called *one-step*. Likewise, we can define *two-step* methods, etc. These definitions for PDEs are similar to those given in Sec. 10 for ODEs.

11.5.1 An Example of an Explicit One-Step Method for a Parabolic IVP Consider the problem

$$u_t = u_{xx} \quad \text{in } (0,1) \times (0,T), \tag{107}$$

$$u = \gamma_0(t) \quad \text{on } x = 0, t \in (0,T), \tag{108}$$

$$u = \gamma_1(t) \quad \text{on } x = 1, t \in (0,T), \tag{109}$$

$$u = g(x) \quad \text{on } t = 0, x \in [0,1]. \tag{110}$$

Using the FD approximations (79) for u_{xx} and (75) for u_t, we transform (107) to

$$[u(x, t + h_t) - u(x, t)]/h_t = [u(x + h, t) - 2u(x, t) + u(x - h, t)]/h^2 + O(h_t + h^2). \tag{111}$$

Let $\{x_i = ih : i = 0, \ldots, n\}$ with $h = 1/n$ be the set of spatial grid points and $\{t_j = jh_t : j = 0, \ldots, m\}$ with $h_t = T/m$ be the set of temporal grid points. Also let $U_{i,j} \approx u(x_i,t_j)$ for $i = 0, \ldots, n$ and $j = 0, \ldots, m$. Then the discretized PDE (111) at the point (x_i,t_j), $i = 1, \ldots, n - 1$, $j = 1, \ldots, m$, becomes

$$(U_{i,j+1} - U_{i,j})/h_t = (U_{i+1,j} - 2U_{i,j} + U_{i-1,j})/h^2.$$

Letting $r = h_t/h^2$, we can rewrite this relation as

$$U_{i,j+1} = rU_{i+1,j} + (1 - 2r)U_{i,j} + rU_{i-1,j} \tag{112}$$

for $i = 2, \ldots, n - 2$. For $i = 1$, we have from (108)

$$U_{1,j+1} = rU_{2,j} + (1 - 2r)U_{1,j} + r\gamma_0(t)_j. \tag{113}$$

Similarly, for $i = n - 1$, we have from (109)

$$U_{n-1,j+1} = r\gamma_1(t_j) + (1 - 2r)U_{n-1,j} + rU_{n-2,j}. \tag{114}$$

For $j = 1$, we have from (110)

$$U_{i,1} = rg(x_{i+1}) + (1 - 2r)g(x_i) + rg(x_{i-1}). \tag{115}$$

Thus, we can compute $U_{i,j} \approx u(x_i,t_j)$ from a linear combination of three neighboring spatial approximations at t_{j-1}.

It can be proved that if $r < \frac{1}{2}$, then $\max\{|u(x_i, t_j) - U_{i,j}| : i = 1, \ldots, n, j = 1, \ldots, m\} = O(h^2 + h_t)$; thus the order of convergence is one w.r.t. to h_t and two w.r.t. h. It can also be proved that if $r < \frac{1}{2}$, the method is stable. However, the restriction $r < \frac{1}{2}$ may be impractical for many problems, since it forces h_t to be very small if h is small and so the method must take many steps to integrate the problem.

The computational complexity of the method is $O(nm)$, since for each grid point (x_i,t_j) a constant number of floating-point operations must be performed.

11.5.2 An Example of an Implicit One-Step Method for a Parabolic IVP Consider the problem (107)–(110) once more. Using the FD approximations (79) for u_{xx} and (76) for u_t, we transform (107) to

$$[u(x, t + h_t) - u(x, t)]/h_t = [u(x + h, t + h_t) - 2u(x, t + h_t) + u(x - h, t + h_t)]/h^2 + O(h_t + h^2). \tag{116}$$

Again let $\{x_i = ih : i = 0, \ldots, n\}$ with $h = 1/n$ be the set of spatial grid points and $\{t_j = jh_t : j = 0, \ldots, m\}$ with $h_t = T/m$ be the set of temporal grid points. Also let $U_{i,j} \approx u(x_i,t_j)$ for $i = 0, \ldots, n$ and $j = 0, \ldots, m$. Then, the discretized PDE (116) at the point (x_i,t_j), $i = 1, \ldots, n - 1$, $j = 1, \ldots, m$, becomes

$$(U_{i,j+1} - U_{i,j})/h_t = (U_{i+1,j+1} - 2U_{i,j+1} + U_{i-1,j+1})/h^2.$$

Letting $r = h_t/h^2$ again, we can rewrite this relation as

$$-rU_{i-1,j+1} + (1 + 2r)U_{i,j+1} - rU_{i+1,j+1} = U_{i,j} \tag{117}$$

for $i = 2, \ldots, n - 2$. For $i = 1$, we have from (108)

$$(1 + 2r)U_{1,j+1} - rU_{2,j+1} = U_{1,j} + r\gamma_0(t_{j+1}). \tag{118}$$

Similarly, for $i = n - 1$, we have from (109)

$$-rU_{n-2,j+1} + (1 + 2r)U_{n-1,j+1} = U_{n-1,j} + r\gamma_1(t_{j+1}). \tag{119}$$

For $j = 1$, we have from (110)

$$-rU_{i-1,1} + (1 + 2r)U_{i,1} - rU_{i+1,1} = g(x_i). \tag{120}$$

Thus, at the jth time step, a tridiagonal linear system must be solved to compute $U_{i,j} \approx u(x_i,t_j)$. The diagonal entries of the associated matrix are all equal to $1 + 2r$, while the off-diagonal entries are all equal to $-r$. The system is symmetric positive-definite and strictly diagonally dominant; thus it has a unique solution.

It can be proved that $\max\{|u(x_i, t_j) - U_{i,j}|: i = 1, \ldots, n, j = 1, \ldots, m\} = O(h^2 + h_t)$; thus the order of convergence is one w.r.t. to h_t and two w.r.t. h. It can also be proved that the method is stable without any restrictions on r (except $r > 0$). The computational complexity of the method is $O(nm)$, since at each time step we must solve a tridiagonal linear system of size $n - 1$ (see Sec. 2.7).

Note that for the problem (107)–(110), which is one-dimensional w.r.t. to space, both the explicit and implicit methods have the same computational complexity. This is not true for problems in more space dimensions. For such problems, the solution of a linear system at each time step can be very time consuming, making an implicit method much more expensive per step than an explicit one. However, because there is no restriction on r for some implicit schemes, while there always is for an explicit one, some implicit schemes may be able to take far fewer time steps than an implicit one. As a result, an implicit method may be computationally more efficient than an explicit one.

11.5.3 An Example of an Explicit Two-Step Method for a Hyperbolic IVP Consider the problem

$$u_{tt} = u_{xx} \quad \text{in } (0,1) \times (0,T), \tag{121}$$

$$u = \gamma_0(t) \quad \text{on } x = 0, \quad t \in (0,T), \tag{122}$$

$$u = \gamma_1(t) \quad \text{on } x = 1, \quad t \in (0,T), \tag{123}$$

$$u = \gamma_0(x) \quad \text{on } t = 0, \quad x \in (0,1), \tag{124}$$

$$u_t = \gamma_1(x) \quad \text{on } t = 0, \quad x \in (0,1). \tag{125}$$

Using the FD approximation (79) for u_{xx} and u_{tt}, we transform (121) to

$$[u(x, t + h_t) - 2u(x, t) + u(x, t - h_t)]/h_t^2 = [u(x + h, t) - 2u(x, t) + u(x - h, t)]/h^2 + O(h_t^2 + h^2). \tag{126}$$

Again let $\{x_i = ih : i = 0, \ldots, n\}$ with $h = 1/n$ be the set of spatial grid points and $\{t_j = jh_t : j = 0, \ldots, m\}$ with $h_t = T/m$ be the set of temporal grid points. Also let $U_{i,j} \approx u(x_i,t_j)$ for $i = 0, \ldots, n$ and $j = 0, \ldots, m$. Then, the discretized PDE (126) at the point (x_i,t_j), $i = 1, \ldots, n - 1$, $j = 1, \ldots, m$, becomes

$$(U_{i,j+1} - 2U_{i,j} + U_{i,j-1})/h_t^2 = (U_{i+1,j} - 2U_{i,j} + U_{i-1,j})/h^2.$$

Letting $r = h_t/h$, we can rewrite this relation as

$$U_{i,j+1} = r^2U_{i-1,j} + 2(1 - r^2)U_{i,j} + r^2U_{i+1,j} - U_{i,j-1} \tag{127}$$

for $i = 2, \ldots, n - 1$. For grid points close to the boundary, the approximate values of U are replaced by the values of the functions γ_0 and γ_1 at the appropriate points, as in Secs. 11.5.1 and 11.5.2. Thus, we can compute $U_{i,j+1} \approx u(x_i,t_{j+1})$ from a linear combination of three neighboring spatial approximations at time t_j and one approximation at time t_{j-1}.

Since (127) is a two-step formula, at the initial time point t_0 it cannot be applied as is. At that point, we use the ICs (124)–(125) and the FD approximation (75) to get

$$U_{i,1} = g_0(x_i) + h_t g_1(x_i). \tag{128}$$

It can be proved that, if $r < 1$, then the method is stable. It can also be shown that $\max\{|u(x_i, t_j) - U_{i,j}| : i = 1, \ldots, n, j = 1, \ldots, m\} = O(h^2 + h_t^2)$; thus the order of convergence is two w.r.t. to both h_t and h. Note that the restriction $r < 1$ is not impractical in this case, since it requires only that $h_t < h$. The computational complexity of the method is $O(nm)$, since for each grid point (x_i,t_j) we apply a formula with a constant number of floating-point operations.

11.6 The Method of Lines

The general idea behind the *method of lines* (MOL) is to use an ODE solver along one of the dimensions of the PDE, while using a PDE discretization across the other dimensions. In its most common form for the solution of IVPs for PDEs, an ODE solver is used along the temporal dimension, while a PDE discretization is employed across the spatial dimensions, transforming an IVP for a PDE into a system of IVPs for ODEs.

To see how this is done, consider the problem (107)–(110) again. Let

$$u_\Delta(x,t) = \sum_{j=1}^{n} \alpha_j(t)\phi_j(x)$$

be a FE approximation to the true solution $u(x,t)$. Now apply a FEM condition to u_Δ to discretize the PDE (107) w.r.t. the spatial dimension. For example, collocation at the points x_i, $i = 1, \ldots, n$, yields

$$\sum_{j=1}^{n} \alpha_j'(t)\phi_j(x_i) = \sum_{j=1}^{n} \alpha_j(t)\phi_j''(x_i).$$

Let $\boldsymbol{\alpha}(t) = (\alpha_1(t), \ldots, \alpha_n(t))^{\mathrm{T}}$, Φ be the matrix with entries $\Phi_{ij} = \phi_j(x_i)$, and A be the matrix with entries $A_{ij} = \phi_j''(x_i)$. Then the PDE (107) is approximated by the system of ODEs $\Phi\boldsymbol{\alpha}'(t) = A\boldsymbol{\alpha}(t)$.

To obtain an IC for the ODE, we construct an interpolant g_Δ of g in the same space as that spanned by $\{\phi_j : j = 1, \ldots, n\}$. Let

$$g_\Delta(x) = \sum_{j=1}^{n} \beta_i\phi_i(x)$$

be the FE representation of g_Δ in that space and set $\boldsymbol{\beta} = (\beta_1, \ldots, \beta_n)^{\mathrm{T}}$. Then

$$\Phi\boldsymbol{\alpha}'(t) = A\boldsymbol{\alpha}(t), \qquad (129)$$

$$\boldsymbol{\alpha}(0) = \boldsymbol{\beta} \qquad (130)$$

is a well-defined IVP for ODEs. Thus, the PDE problem (107)–(110) is converted to an IVP for a system of n ODEs. The latter can be solved by the techniques described in Sec. 10.

Note that applying an ODE method to discretize the IVP (129)–(130) results in a discretization for the PDE (107)–(110). That is, the MOL produces a discretization for a PDE. However, it is generally agreed that standard software for ODEs is more highly developed than for PDEs. Thus, using the MOL to decouple the discretization of the spatial and temporal variables allows us to exploit easily sophisticated time-stepping techniques. As a result, the MOL is often the simplest effective method to solve a PDE.

11.7 Boundary-Element Methods

The general idea behind *boundary-element methods* (BEMs) is to transform the PDE to an integral equation in which the integrations take place along the boundary only of the PDE domain, thus eliminating the need for domain discretization and reducing the dimension of the PDE by one. For example, a one-dimensional integral equation is solved instead of an equivalent two-dimensional PDE. The BEM is applicable to BVPs for Laplace's or Poisson's equation, and many other simple PDEs. If applicable, this approach is often very effective, especially when the PDE domain is highly irregular.

11.8 The Multigrid Method

The multigrid method (MM) exploits the connection between a physical problem and its matrix analog to accelerate the convergence of an iterative method (see Sec. 3). For simplicity, we describe the MM for the one-dimensional problem (85)–(86), although the merits of the scheme become apparent for two- and higher-dimensional problems (see Sec. 11.3.2). Also, we illustrate the technique using Jacobi's method as the basic iterative scheme. The MM, though, can be used with many other iterative methods as a preconditioning technique.

Let A be the matrix in (90). Apply Jacobi's method (see Sec. 3.1) with an extra damping factor of 2 to the linear system (90). The associated iteration matrix is $G = I - A/4$. It can be shown that the eigenvalues of G are $\mu_i = \cos^2[i\pi/(2n)]$, $i = 1, \ldots, n - 1$, and that the components of the eigenvector v_i associated with μ_i are $\sin[i\pi(j/n)]$, $j = 1, \ldots, n - 1$. These are also the eigenvectors of A. Since $\{v_i : i = 1, \ldots, n - 1\}$ spans $\mathbb{R}^{n-1}$, we can write the error e_0 associated with the initial guess for the damped Jacobi iteration as

$$e_0 = \sum_{i=1}^{n-1} \alpha_i v_i$$

for some scalars $\{\alpha_i : i = 1, \ldots, n - 1\}$. It then follows easily from the discussion in Sec. 3.1 that the error at iteration k is

$$e_k = \sum_{i=1}^{n-1} \alpha_i \mu_i^k v_i.$$

The terms of the sum corresponding to small values of i are called *low-frequency components,* while those corresponding to large values of i are called *high-frequency components.* Note that $0 < \mu_{n-1} < \mu_{n-2} < \cdots < \mu_2 < \mu_1 < 1$. Moreover, $\mu_1 \approx 1 - (\pi/2n)^2$, while $\mu_{n-1} \approx (\pi/2n)^2$. Consequently, $e_k \to 0$ as $k \to \infty$, but the low-frequency components of the error converge slowly, while the high-frequency components converge rapidly.

To accelerate the convergence of the low-frequency components of the error, consider solving the problem (90) on a coarse grid with $\hat{n} = n/2$ subintervals and $\hat{n} + 1$ grid points, assuming for simplicity that n is even. Although the coarse grid has about half the number of grid points of the fine one, the $\hat{n} - 1$ eigenvectors of the matrix $\hat{A}$ for the coarse grid provide a good representation of the low- to middle-frequency eigenvectors of A. As a result, the solution to the problem (90) on the coarse grid provides a good approximation to the low- to middle-frequency components of the fine-grid solution. This suggests that the coarse-grid solution can be used to provide good approximations to the low- to middle-frequency components of the fine-grid solution, while the damped Jacobi iteration on the fine grid can be used to provide good approximations to the middle- to high-frequency components of the fine-grid solution.

This is the motivation behind the MM. The term "multigrid" refers to the use of several levels of grids (possibly a fine grid, several intermediate-level grids, and a coarse grid), so that each level damps certain components of the error fast.

To view the MM as a preconditioning technique, consider the linear system $Au = g$ corresponding to the discretization of problem (85)–(86) on some fine grid. Apply one (or a few) damped Jacobi iteration(s) to $Au = g$ to obtain an approximate solution vector $\tilde{u}$. Let $r = g - A\tilde{u}$ be the *residual* vector. Project r to a coarse grid to obtain the *coarse-grid residual* vector $\hat{r}$. This can be done by appropriately interpolating the components of r and evaluating the interpolant at the points of the coarse grid (see Sec. 8). Let $\hat{A}$ be the matrix corresponding to the discretization of problem (85)–(86) on the coarse grid. Solve (or approximately solve) $\hat{A}\hat{\tilde{r}} = \hat{r}$. This can be done by applying a few damped Jacobi iterations, or by recursively applying the MM to $\hat{A}\hat{\tilde{r}} = \hat{r}$, or by using a direct solver (see Sec. 2), since $\hat{A}$ is a smaller matrix than A. Now $\hat{\tilde{r}}$ is the *preconditioned coarse-grid residual* vector. Extend $\hat{\tilde{r}}$ to the fine grid to obtain the *preconditioned fine-grid residual* vector $\tilde{r}$. This can be done by appropriately interpolating the components of $\hat{\tilde{r}}$ and evaluating the interpolant at the points of the fine grid. Add $\tilde{r}$ to $\tilde{u}$ to obtain a new approximate solution vector. Repeat the process until convergence. Usually, only a few iterations are needed. Note that this scheme has some similarities to iterative improvement, as described in Sec. 2.9.

The power of the MM lies in the fact that the coarse grid, which acts as a preconditioner, allows the information to pass from a point of the problem domain to another point in a few steps, while the fine grid maintains the accuracy required. Note that the interpolation and the evaluation of the interpolant needed for the projection of a fine-grid vector to a coarse-grid vector and for the extension of a coarse-grid vector to a fine-grid vector often reduce to simple relations, such as averaging neighboring vector components. A practical introduction to the MM, including an error analysis, can be found in Briggs (1987).

12. PARALLEL COMPUTATION

The increasing demand by scientists and engineers to solve larger and larger problems constitutes the primary motivation for parallel computation. Another impetus is the cost effectiveness of computers consisting of many standard CPUs compared with those based on one very fast CPU.

A parallel computer has the ability to execute simultaneously many different processes by having several independent processors concurrently perform operations on different data. A parallel computer in which all processors perform the same operation on different data is called a *single-instruction, multiple-data* (SIMD) machine, while one in which each processor has the ability to perform different operation(s) on data is called a *multiple-instruction, multiple-data* (MIMD) machine. Sometimes a processor may use data computed by another processor, either by exchanging messages with it or by sharing some common memory area. The former is characteristic of *distributed-memory* or *message-passing* machines, while the latter is characteristic of *shared-memory* machines.

A vector computer may be viewed as a restricted form of a parallel machine. On such a computer, one arithmetic operation is performed by several (usually a few) processors, which cooperate in a pipelined manner. Each processor receives an input operand from another processor, performs a part of the operation on it, and passes the result on to the next processor. While a processor performs its part of the operation on an operand, the next processor performs its part of the operation on the previous operand, etc. This pipelining technique is very effective when the same operation is performed on many data—for example, on each component of a long vector.

To utilize parallel computers effectively, we must be able to split the computation into *parallel processes* to be assigned to different processors. To this end, it is desirable to have fully independent computations in each process. For many problems, though, this is impossible, but there are techniques to minimize the dependence of the computation of one process on another. It is also desirable to have almost equal amounts of computation in each process for *load balancing*. This is also hard to accomplish in many cases, but again there exist techniques to achieve reasonable load balancing.

Many computational scientists study the parallelization of existing numerical methods, as well as the development of new methods appropriate for parallel machines. In the next subsection, we discuss, as a simple example, a numerical method that is effective for solving tridiagonal linear systems on parallel computers.

12.1 Cyclic Reduction

Consider solving the tridiagonal system

$$\begin{pmatrix} a_1 & c_1 & & & & \\ b_2 & a_2 & c_2 & & & \\ & b_3 & a_3 & c_3 & & \\ & & \ddots & \ddots & \ddots & \\ & & & b_{n-1} & a_{n-1} & c_{n-1} \\ & & & & b_n & a_n \end{pmatrix} \begin{pmatrix} x_1 \\ x_2 \\ x_3 \\ \vdots \\ x_{n-1} \\ x_n \end{pmatrix} = \begin{pmatrix} d_1 \\ d_2 \\ d_3 \\ \vdots \\ d_{n-1} \\ d_n \end{pmatrix}$$

by the following *LU* factorization algorithm for tridiagonal matrices:

for $k = 1, \ldots, n - 1$ do
  $b_{k+1} = b_{k+1}/a_k$
  $a_{k+1} = a_{k+1} - b_{k+1}c_k$
end

At the end of the computation, the modified b's form the subdiagonal of the unit lower-triangular matrix L and the modified a's and c's form the diagonal and superdiagonal, respectively, of the upper-triangular matrix U (see Secs. 2.3 and 2.7).

Note that the computation proceeds in the order $b_2, a_2, b_3, a_3, \ldots$. Each value computed depends on the previous one. Therefore, it seems that the computation is purely sequential and that there is no easy way to parallelize it. However, there are other ways to solve tridiagonal linear systems, and some of them can be implemented effectively on a parallel machine.

Assume, for simplicity, that $n = 2^q - 1$, where q is a positive integer. Multiply row 1 by b_2/a_1 and subtract it from row 2, eliminating x_1 from row 2. Also multiply row 3 by c_2/a_3 and subtract it from row 2, eliminating x_3 from row 2. The new row 2 involves variables x_2 and x_4 only.

Repeat the process described above for the $(n - 1)/2$ groups of rows (3,4,5), (5,6,7), etc. This eliminates the odd unknowns from the even equations. The even equations form a new tridiagonal linear system of about half the size, $(n - 1)/2 = 2^{q-1} - 1$, called the *reduced* system. This technique is often called *odd–even reduction*.

Now apply odd–even reduction to the reduced system to obtain another reduced system that is again about half as big as the first reduced system. The recursive application of odd–even reduction continues for $q = \log_2(n + 1)$ steps. At each step, the even equations of the previous step form a reduced tridiagonal system of about half the size of the system from the previous step. At the end of step q, one equation in one unknown remains, so that unknown can be computed easily. This recursive technique is often called *cyclic reduction.*

Then the computation continues in the reverse order with a process called *back substitution.* At each step of back substitution, the even variables are known from the solution of the associated reduced system of about half the size. Substituting these values back into the odd equations of the larger system, we can easily compute all the odd variables.

Both the cyclic-reduction algorithm and the back-substitution algorithm require $q = \log_2(n + 1)$ steps each. In cyclic reduction, the number of floating-point operations is divided by 2 at each step, starting with $O(n)$ floating-point operations in the first step. Thus, it requires $O(n \log n)$ arithmetic operations. Similarly, back substitution also requires $O(n \log n)$ arithmetic operations. Therefore, the computational complexity of the full solution is $O(n \log n)$.

Observe that the algorithm described above is highly parallel. The elimination operations applied to a group of three rows to obtain the reduced system at each step are independent of the elimination operations applied to any other group of three rows, and so can be carried out in parallel. Similarly, the substitution operations to compute the odd unknowns in a reduced system given the even ones are independent of each other. Thus, the unknowns of each back-substitution step can also be computed in parallel.

Assume that we have $p = (n - 1)/2$ processors. Initially, processor 1 is assigned rows (1,2,3), processor 2 is assigned rows (3,4,5), processor 3 is assigned rows (5,6,7), and so on. After the first odd–even reduction step, processor 2 will use equation 2 from processor 1, equation 4 from itself, and equation 6 from processor 3. Similarly, processor 4 will use equation 6 from processor 3, equation 8 from itself, and equation 10 from processor 5, and so on. Only the even processors will continue. The procedure is repeated. The final reduced system is solved by one processor. For the back substitution, one processor works first, then two, then four, and so on.

Thus, the algorithm requires $2 \log_2(n + 1)$ steps with a constant amount of computation done on each processor per step, So the parallel computational complexity of the algorithm is $O(\log n)$, which is a factor of $O(n) = O(p)$ improvement over the $O(n \times \log n)$ computational complexity of the serial version of the algorithm, and a little less than $O(n)$ improvement over the $O(n)$ serial computational complexity of the standard *LU* factorization algorithm for tridiagonal systems. So we can say that, asymptotically, the algorithm has perfect *speedup.*

Note that when a processor uses rows computed by another processor, some communication and/or synchronization must take place between processors. This may degrade the parallel performance of the algorithm from the perfect asymptotic performance. The time spent in communication and/or synchronization depends heavily on the way the processors cooperate. More specifically, it depends on the interconnection network between processors and on the implementation of specific hardware instructions.

For an introduction to parallel numerical methods, see Bertsekas and Tsitsiklis (1989), Ortega (1988), or Van de Velde (1994).

13. SOURCES OF NUMERICAL SOFTWARE

Although most of this article has dealt with elementary numerical methods, we strongly recommend that readers do not program these schemes themselves. High-quality software incorporating these—or more sophisticated—numerical methods is readily available. In addition, good library routines often contain many additional strategies and heuristics (not discussed here) to improve their efficiency and reliability. Using such routines, rather than attempting to reprogram them, will likely save readers a significant amount of time as well as produce superior numerical results.

We highly recommend that readers familiarize themselves with the *Guide to Available Mathematical Software* (GAMS) recently developed by the National Institute of Standards and Technology (NIST). GAMS is both an on-line cross-index of available mathematical software as well as a repository of some 9000 high-quality problem-solving modules from more than 80 software packages. It provides centralized access to such items as abstracts, documentation, and source code of the software modules that it catalogs. Most of this software represents FORTRAN subprograms for mathematical problems that commonly occur in computational science and engineering, such as solution of systems of linear algebraic equations, computing matrix eigenvalues, solving nonlinear systems of differential equations, finding minima of nonlinear functions of several variables, evaluating the special functions of applied mathematics, and performing nonlinear regression. Among the packages cataloged in GAMS are

- the IMSL, NAG, PORT, and SLATEC libraries;
- the BLAS, EISPACK, FISHPAK, FNLIB, FFTPACK, LAPACK, LINPACK, and STARPAC packages;
- the DATAPLOT and SAS statistical analysis systems;
- the netlib routines, including the Collected Algorithms of the ACM (see below).

Note that although GAMS catalogs both public-domain and proprietary software, source code of proprietary software is not available through GAMS, although related items such as documentation and example programs often are. Software can be found either by browsing through a decision tree or performing a key-word search. GAMS can be accessed in several ways: telnet gams.nist.gov; gopher gams.nist.gov; or ⟨www browser⟩ http://gams.nist.gov, where ⟨www browser⟩ is a World Wide Web browser such as Mosaic or Netscape. Report any questions or problems to gams@cam.nist.gov. For more details, log in to the system or see Boisvert *et al.* (1985) and Boisvert (1990).

Included in the software cataloged by GAMS are many high-quality public-domain routines available by electronic mail (e-mail) from *netlib*. These routines are now also available through Xnetlib, a more sophisticated X interface to netlib and the NA-Net Whitepages, or through the World Wide Web at the address http://www.netlib.org/index.html. For more information on netlib, see Dongarra and Grosse (1987) or Dongarra *et al.* (1995), send the message "send index" by e-mail to either netlib@ornl.gov or netlib@research.att.com, or access http://www.netlib.org/index.html through the World Wide Web.

The ACM Transactions on Mathematical Software publishes refereed public-domain software. These high-quality routines, covering a broad range of problem areas, are included in the Collected Algorithms of the ACM, available through both GAMS and netlib.

Not mentioned above are the commercial interactive packages MATLAB, Maple, and Mathematica. MATLAB is built upon a foundation of sophisticated matrix software and includes routines for solving many standard mathematical and statistical problems. In addition, "toolboxes" for several application areas, such as control theory, are available. Both Maple and Mathematica are primarily symbolic algebra packages, but contain many high-quality numerical routines as well. For more information on MATLAB, contact The MathWorks Inc., 24 Prime Park Way, Natick, MA 01760; phone: (508) 653-1415; FAX: (508) 653-2997; e-mail: info@mathworks.com. For more information on Maple, contact Waterloo Maple Software, 450 Phillip St., Waterloo, Ontario, Canada, N2L 5J2; phone: (519) 747-2373; FAX: (519) 747-5284; e-mail: info@maplesoft.on.ca. For more information on Mathematica, contact Wolfram Research Inc., 100 Trade Center Dr., Champaign, IL 61820-7237; phone: (217) 398-0700; FAX: (217) 398-0747; e-mail: info@wri.com.

GLOSSARY

Banded Matrix: An $m \times n$ matrix all of whose nonzero elements occur in a band around its main diagonal.

Block Diagonal: See **Diagonal.**

Characteristic Equation of a Matrix: Defined only for an $n \times n$ matrix A, the equation $\det(a - \lambda I)$.

Characteristic Polynomial of a Matrix: Defined only for an $n \times n$ matrix A, the polynomial $p(\lambda) = \det(A - \lambda I)$, of degree n.

Column-Diagonally Dominant: Describing a matrix A if A^T is **row-diagonally dominant.**

Diagonal: Describing an $m \times n$ matrix $D = [d_{ij}]$ for which $d_{ij} = 0$ for $i \neq j$. The matrix D is *block-diagonal* if each d_{ij} in the preceding definition is a submatrix rather than a single number.

Diagonally Dominant: Describing a matrix A if either A or A^{T} is **row-diagonally dominant.**

Eigenvector: Defined only for an $n \times n$ matrix A, a (possibly complex) number λ such that for some nonzero vector x, $Ax = \lambda x$. Any vector satisfying this equation is an *eigenvector* associated with the eigenvalue λ.

Euclidean Norm (of a vector x): The quantity $\|x\|_2 = (x^{H}x)^{1/2}$, where x^{H} is the complex-conjugate transpose of the vector x. Often called the *2-norm.*

Flop: A floating-point operation on a computer; a multiplication and either an addition or a subtraction.

Hermitian: Describing an $n \times n$ matrix A for which $A^{H} = A$, where A^{H} is the complex-conjugate transpose of A.

Hessenberg: Either **upper** or **lower Hessenberg.** A symmetric Hessenberg matrix is tridiagonal.

Leading Principal Minor of a Matrix A: The $k \times k$ submatrix in the top left corner of A.

Lower Hessenberg: Describing a matrix A for which $a_{ij} = 0$ for $i < j - 1$. That is, it is lower-triangular except for a single nonzero superdiagonal.

Lower Triangular: Describing an $n \times n$ matrix $L = [l_{ij}]$ for which $l_{ij} = 0$ for $1 \le i < j \le n$. It is *strictly lower-triangular* if $l_{ij} = 0$ for $1 \le i \le j \le n$, and *block lower-triangular* or *block strictly lower-triangular,* respectively, if each l_{ij} in the preceding definitions is a submatrix rather than a single number.

Orthogonal: Describing an $m \times n$ matrix Q for which $Q^{T}Q = I$.

Permutation Matrix: An $n \times n$ matrix with exactly one element in each row and column equal to 1 and all other elements equal to 0.

Rank of a Matrix: The maximal number of independent rows (or columns) of the matrix.

Right Triangular: Describing an $m \times n$ matrix $R = [r_{ij}]$ such that $r_{ij} = 0$ for $i > j$. If $m = n$, the terms *right triangular* and *upper triangular* are equivalent.

Row-Diagonally Dominant: Describing an $m \times n$ matrix, with $m \le n$, for which

$$\sum_{j=1, j\neq i}^{n} |a_{ij}| < |a_{ii}|$$

for $i = 1, 2, \ldots, m$.

Similar Matrices: Any two matrices A and B such that $B = WAW^{-1}$ for some nonsingular matrix W, which is called the *associated similarity transformation.*

Sparse: Describing a matrix in which the number of nonzero elements is much less than the total number of elements in the matrix.

Spectral Radius of a (Square) Matrix A: The quantity $\rho(A) = \max\{|\lambda| : \lambda$ an eigenvalue of $A\}$.

Strictly Lower Triangular: See **Lower Triangular.**

Strictly Upper Triangular: See **Upper Triangular.**

Symmetric: Describing an $n \times n$ matrix for which $A^{T} = A$, where A^{T} is the transpose of A.

Symmetric Indefinite: Describing a real symmetric matrix A for which $x^{T}Ax > 0$ for some real n-vector x and $y^{T}Ay < 0$ for some real n-vector y.

Symmetric Positive (Negative) Definite: Describing a real symmetric matrix A for which $x^{T}Ax > 0$ ($x^{T}Ax < 0$) for all real n-vectors $x \neq 0$.

Symmetric Positive (Negative) Semidefinite: Describing a real symmetric matrix A for which $x^{T}Ax \ge 0$ ($x^{T}Ax \le 0$) for all real n-vectors $x \neq 0$.

Transpose: For an $m \times n$ matrix $A = [a_{ij}]$, the $n \times m$ matrix $A^{T} = [a^{t}_{ij}]$ where $a^{t}_{ij} = a_{ji}$ for $i = 1, 2, \ldots, n$ and $j = 1, 2, \ldots, m$. The transpose of a column (row) vector is a row (column) vector.

Unit Lower Triangular: Describing a lower-triangular matrix $L = [l_{ij}]$ for which $l_{ii} = 1$ for $i = 1, 2, \ldots, n$.

Upper Hessenberg: Describing a matrix $A = [a_{ij}]$ such that $a_{ij} = 0$ for $i > j + 1$. That is, it is upper-triangular except for a single nonzero subdiagonal.

Upper Triangular: An $n \times n$ matrix $U = [u_{ij}]$ such that $u_{ij} = 0$ for $1 \le j < i \le n$. It is *strictly upper-triangular* if $u_{ij} = 0$ for $1 \le j \le i \le n$, and *block upper-triangular* or *block strictly upper-triangular,* respectively, if each u_{ij} in the preceding definitions is a submatrix rather than a single number.

Tridiagonal: Describing an $m \times n$ matrix $A = [a_{ij}]$ such that $a_{ij} = 0$ for $|i - j| > 1$.

MATHEMATICAL SYMBOLS USED

A^{H}: The complex-conjugate transpose of the matrix A.

A^{T}: The transpose of the matrix A.

$\mathbb{C}$: The set of complex numbers.

$\mathbb{C}^n$: The set of complex vectors with n components.

$\mathbb{C}^{m \times n}$: The set of complex $m \times n$ matrices.

$\mathscr{C}$: The set of continuous functions.

$\mathscr{C}[a,b]$: The set of continuous functions on the interval $[a,b]$.

$\mathscr{C}^p$: The set of continuous functions with p continuous derivatives.

$\mathscr{C}^p[a,b]$: The set of continuous functions with p continuous derivatives on the interval $[a,b]$.

$O(h^p)$: Any quantity that depends on h that can be bounded above by Ch^p for some constant C and all $h \in (0,H]$ for some $H > 0$.

$O(n^p)$: Any quantity that depends on n that can be bounded above by Cn^p for some constant C and all positive integers n.

$\mathbb{R}$: The set of real numbers.

$\mathbb{R}^n$: The set of real vectors with n components.

$\mathbb{R}^{m \times n}$: The set of real $m \times n$ matrices.

x^{T}: The transpose of the vector x.

$\|x\|$ and $\|A\|$: Norms of the vector x and the matrix A, respectively.

$\|x\|_2$ and $\|A\|_2$: Euclidean norms (also called two-norms) of the vector x and the matrix A, respectively.

z^{H}: The complex-conjugate transpose of the vector z.

$\rho(A)$: The spectral radius of a square matrix A.

ABBREVIATIONS USED

ACM: Association for Computing Machinery.

ADI: Alternating direction implicit.

BC: Boundary condition.

BVP: Boundary-value problem.

CD: Conjugate direction.

CG: Conjugate gradient.

DOF: Degrees of freedom.

FD: Finite difference.

FDM: Finite-difference method.

FE: Finite element.

FEM: Finite-element method.

GAMS: Guide to Available Mathematical Software.

GE: Gaussian elimination.

IC: Initial condition.

ICF: Incomplete Cholesky factorization.

IEEE: Institute of Electrical and Electronics Engineers.

IVP: Initial-value problem.

LMF: Linear multistep formula.

MM: Multigrid method.

ODE: Ordinary differential equation.

PCG: Preconditioned conjugate gradient.

PDE: Partial differential equation.

PP: Piecewise polynomial.

RK: Runge–Kutta.

SD: Steepest descent.

SOR: Successive overrelaxation.

SPD: Symmetric positive-definite.

SSOR: Symmetric successive overrelaxation.

w.r.t.: With respect to.

Works Cited

Ames, W. F. (1992), *Numerical Methods for Partial Differential Equations,* New York: Academic.

Ascher, U. M., Mattheij, R. M. M., Russell, R. D. (1988), *Numerical Solution of Boundary Value Problems for Ordinary Differential Equations,* Englewood Cliffs, NJ: Prentice-Hall.

Atkinson, K. E. (1989), *An Introduction to Numerical Analysis,* 2nd ed., New York: Wiley.

Axelsson, O. (1994), *Iterative Solution Methods,* Cambridge, U.K.: Cambridge Univ. Press.

Becker, E. B., Carey, G. F., Oden, J. T. (1981), *Finite Elements,* Vol. I, Englewood Cliffs, NJ: Prentice-Hall.

Bertsekas, D. P., Tsitsiklis, J. N. (1989), *Parallel and Distributed Computation: Numerical Methods,* Englewood Cliffs, NJ: Prentice-Hall.

Boisvert, R. (1990), "The Guide to Available Mathematical Software Advisory System," in: E. Houstis, J. Rice, R. Vichnevetsky (Eds.), *Intelligent Mathematical Software Systems,* Amsterdam: North-Holland, pp. 167–178.

Boisvert, R. F., Howe, S. E., Kahaner, D. K. (1985), "GAMS: A Framework for the Management of Scientific Software," *ACM Trans. Math. Software* **11** (4), 313–355.

Briggs, W. L. (1987), *A Multigrid Tutorial,* Philadelphia: SIAM.

Buchanan, J. L., Turner, P. R. (1992), *Numerical Methods and Analysis,* New York: McGraw-Hill.

Butcher, J. C. (1987), *The Numerical Analysis of Ordinary Differential Equations,* New York: Wiley.

Celia, M. A., Gray, W. G. (1992), *Numerical Methods for Differential Equations,* Englewood Cliffs, NJ: Prentice-Hall.

Christara, C. C. (1994), "Quadratic Spline Collocation Methods for Elliptic Partial Differential Equations," *BIT* **34** (1), 33–61.

Conte, S. D., de Boor, C. (1980), *Elementary*

Numerical Analysis, 3rd ed., New York: McGraw-Hill.

Cullum, J., Willoughby, R. (1985), *Lanczos Algorithms for Large Symmetric Eigenvalue Computations,* Vol. 1, *Theory,* and Vol. 2, *Programs,* Boston: Birkhäuser.

Dahlquist, G., Björck, Å. (1974), *Numerical Methods,* Englewood Cliffs, NJ: Prentice-Hall.

Davis, P. J. (1975), *Interpolation and Approximation,* New York: Dover.

de Boor, C. (1978), *A Practical Guide to Splines,* New York: Springer-Verlag.

Dennis, J. E., Schnabel, R. B. (1983), *Numerical Methods for Unconstrained Optimisation and Nonlinear Equations,* Englewood Cliffs, NJ: Prentice-Hall.

Dongarra, J., Grosse, E. (1987), "Distribution of Mathematical Software via Electronic Mail," *Commun. ACM* **30** (5), 403–407.

Dongarra, J., Rowan, T., Wade, R. (1995), "Software Distribution Using XNETLIB," *ACM Trans. Math. Software* **21** (1), 79–88.

Duff, I., Erisman, A., Reid, J. (1986), *Direct Methods for Sparse Matrices,* Oxford, U.K.: Oxford Univ. Press.

George, A., Liu, J. (1981), *Computer Solution of Large Sparse Positive Definite Systems,* Englewood Cliffs, NJ: Prentice-Hall.

Goldberg, D. (1991), "What Every Computer Scientist Should Know about Floating-Point Arithmetic," *ACM Comput. Surveys* **23,** 5–48.

Golub, G. H., Van Loan, C. F. (1989), *Matrix Computations,* 2nd ed., Baltimore: John Hopkins Univ. Press.

Hackbusch, W. (1994), *Iterative Solution of Large Sparse Systems of Equations,* New York: Springer-Verlag.

Hageman, L. A., Young, D. M. (1981), *Applied Iterative Methods,* New York: Academic.

Hager, W. W. (1988), *Applied Numerical Linear Algebra,* Englewood Cliffs, NJ: Prentice-Hall.

Hairer, E., Nørsett, S. P., Wanner, G. (1987), *Solving Ordinary Differential Equations I: Nonstiff Problems,* New York: Springer-Verlag.

Hairer, E., Wanner, G. (1991), *Solving Ordinary Differential Equations II: Stiff and Differential-Algebraic Problems,* New York: Springer-Verlag.

Hall, C. A., Porsching, T. A. (1990), *Numerical Analysis of Partial Differential Equations,* Englewood Cliffs, NJ: Prentice-Hall.

Householder, A. (1970), *The Numerical Treatment of a Single Nonlinear Equation,* New York: McGraw-Hill.

Houstis, E. N., Christara, C. C., Rice, J. R. (1988), "Quadratic Spline Collocation Methods for Two-Point Boundary Value Problems," *Int. J. Numer. Methods Eng.* **26,** 935–952.

IEEE (1985), *IEEE Standard for Binary Floating-Point Arithmetic,* New York: American National Standards Institute, ANSI/IEEE Std. 754-1985.

Johnson, L. W., Riess, R. D. (1982), *Numerical Analysis,* Reading, MA: Addison-Wesley.

Kahaner, D., Moler, C., Nash, S. (1989), *Numerical Methods and Software,* Englewood Cliffs, NJ: Prentice-Hall.

Lambert, J. D. (1991), *Numerical Methods for Ordinary Differential Equations,* New York: Wiley.

Ortega, J. M. (1988), *Introduction to Parallel and Vector Solution of Linear Systems,* New York: Plenum.

Parlett, B. N. (1968), "Global Convergence of the Basic QR Algorithm on Hessenberg Matrices," *Math. Comput.* **22,** 803–817.

Parlett, B. N. (1980), *The Symmetric Eigenvalue Problem,* Englewood Cliffs, NJ: Prentice-Hall.

Prenter, P. M. (1975), *Splines and Variational Methods,* New York: Wiley.

Saad, Y. (1992), *Numerical Methods for Large Eigenvalue Problems,* New York: Manchester Univ. Press (Wiley).

Scott, D. (1981) "The Lanczos Algorithm," in: I. S. Duff (Ed.), *Sparse Matrices and Their Uses,* London: Academic, pp. 139–160.

Shampine, L. F. (1994), *Numerical Solution of Ordinary Differential Equations,* New York: Chapman & Hall.

Shampine, L. F., Gear, C. W. (1979), "A User's View of Solving Stiff Ordinary Differential Equations," *SIAM Rev.* **21,** 1–17.

Stoer, J., Bulirsch, R. (1980), *Introduction to Numerical Analysis,* New York: Springer-Verlag.

Strang, G., Fix, G. J. (1973), *An Analysis of the Finite Element Method,* Englewood Cliffs, NJ: Prentice-Hall.

Strikwerda, J. C. (1989), *Finite Difference Schemes and Partial Differential Equations,* Pacific Grove, CA: Wadsworth and Brooks/Cole.

Van de Velde, E. F.. (1994), *Concurrent Scientific Computing,* New York: Springer-Verlag.

Van Loan, C. F. (1992), *Computational Frameworks for the Fast Fourier Transform,* Philadelphia: SIAM.

Varga, R. S. (1962), *Matrix Iterative Analysis,* Englewood Cliffs, NJ: Prentice-Hall.

Wilkinson, J. H. (1965), *The Algebraic Eigenvalue Problem,* Oxford, U.K.: Oxford Univ. Press.

Young, D. M. (1971), *Iterative Solution of Large Linear Systems,* New York: Academic.

Further Reading

Ciarlet, P. G. (1989), *Introduction to Numerical Linear Algebra and Optimization,* Cambridge, U.K.: Cambridge Univ. Press.

Forsythe, G. E., Malcolm, M. A., Moler, C. B. (1977), *Computer Methods for Mathematical Computations,* Englewood Cliffs, NJ: Prentice-Hall.

Forsythe, G. E., Moler, C. B. (1967), *Computer*

Solution of Linear Algebraic Systems, Englewood Cliffs, NJ: Prentice-Hall.

Golub, G. H., Ortega, J. M. (1992), *Scientific Computing and Differential Equations,* New York: Academic.

Golub, G. H., Ortega, J. M. (1993), *Scientific Computing: An Introduction with Parallel Computing,* New York: Academic.

Isaacson, E., Keller, H. B. (1966), *Analysis of Numerical Methods,* New York: Wiley.

Press, W. H., Flannery, B. P., Teukolsky, S. A., Vetterling, W. T. (1986), *Numerical Recipes: The Art of Scientific Computing,* Cambridge, U.K.: Cambridge Univ. Press.

Schultz, M. H. (1973), *Spline Analysis,* Englewood Cliffs, NJ: Prentice-Hall.

Stewart, G. W. (1973), *Introduction to Matrix Computations,* New York: Academic.

Wilkinson, J. H. (1963), *Rounding Errors in Algebraic Processes,* Englewood Cliffs, NJ: Prentice-Hall.

SCINTILLATION DETECTORS

See DETECTORS, SCINTILLATION

SEA ICE

STEPHEN F. ACKLEY, *Cold Regions Research and Engineering Laboratory, Hanover, New Hampshire, U.S.A.*

INTRODUCTION

Sea ice is one form of natural ice in the environment, along with glaciers, seasonal snow, frozen ground, and the ice formed on lakes and rivers. It is distinguished by its formation from bodies of water that contain significant concentrations of salts, the oceans and seas. Ice formed in salt water is significantly different from fresh-water ice in both its growth behavior and structure. Sea ice is more akin to a multiphase alloy than it is to the single–solid-phase, purer ice formed in other environments.

The area covered by sea ice, seasonally or perennially, is restricted to higher latitudes of the northern and southern hemispheres (Fig. 1). In lower latitudes, isolated areas are also found in estuaries, embayments, and sounds when exceptionally cold winters occur. The ice covers about 10% (35×10^6 km^2) of the area of the ocean at its maximum seasonal extent, in a relatively thin veneer varying in thickness from a few tens of centimeters to several meters. Deformed ice can occasionally be tens of meters thick, and in actively deforming regions deformation can alter the mean thickness by several meters of ice.

The climatic and oceanic influences of the ice covers are even greater than might be expected, even considering their substantial area, since the ice can strongly change the air–sea interaction in its presence. For example, the ice albedo reflects 80% or more of the incoming solar radiation compared with 10% for non-ice-covered water, and contributes to lower air temperatures. Sea-ice formation influences oceanic processes: Salts are rejected into the underlying water when the ice forms, changing its salinity and density and causing convective overturning. These denser waters then flow outward from the polar regions at great depths and are replaced by other water masses flowing in from temperate latitudes. Sea-ice growth therefore plays a significant role in the large-scale thermohaline circulation of the world ocean (called the thermohaline conveyor belt). Sea ice can influence biological pro-

3-527-28139-8/96/$5.00 + .50

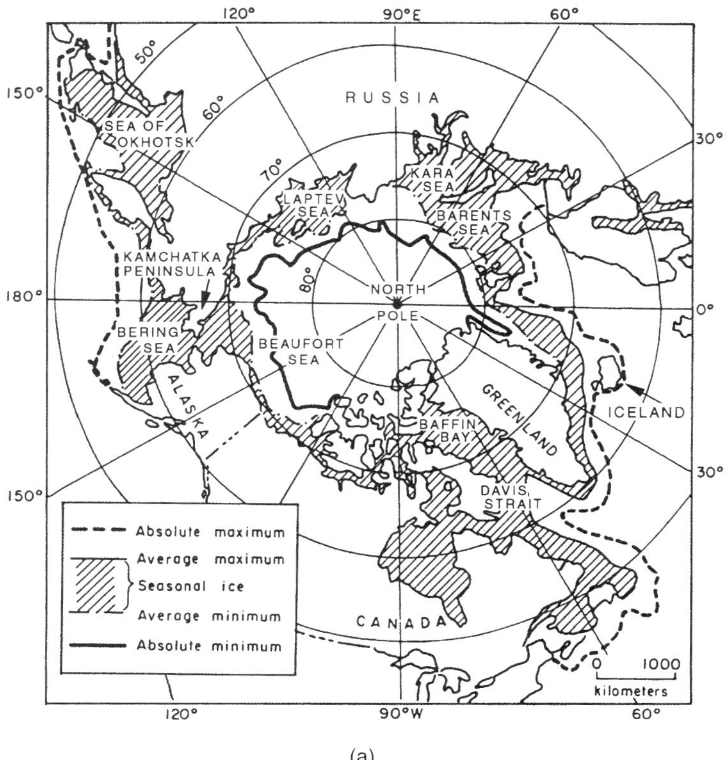

(a)

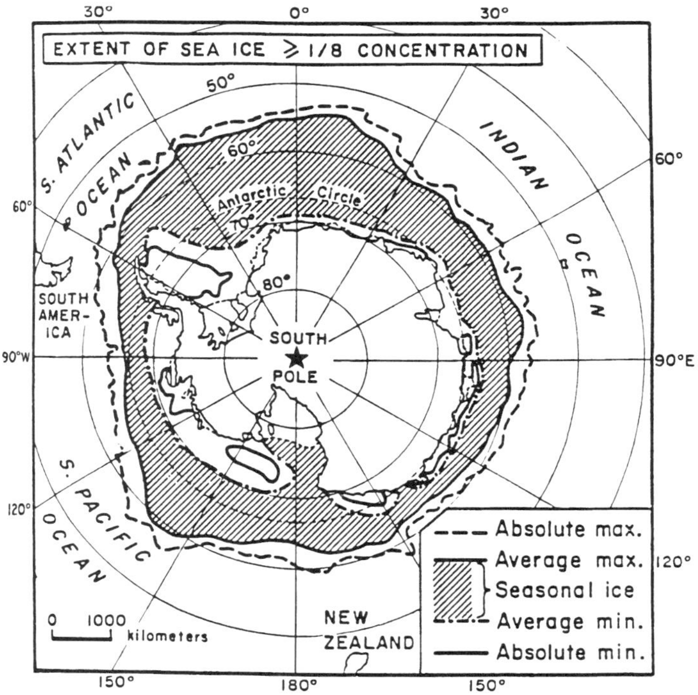

(b)

FIG. 1. Sea-ice extents in the (a) northern and (b) southern hemispheres (after Barry, 1986).

duction in the ocean as well. Ice melt is a mechanism for stabilizing the upper layer of the ocean by lowering the salinity and, therefore, the density of the surface layer. This stabilizing effect allows biological organisms to remain longer in this upper layer, or photic zone, where photosynthesis is greater, enhancing primary production in these ice edge areas.

The properties of sea ice are sensitive to the growth conditions and, after formation of the initial ice cover, its thermal and dynamic history. These conditions are both spatially and temporally variable, leading to significant differences in sea-ice behavior from location to location and season to season. Our theme is to develop the relationship between these processes and the resulting ice properties and ice distribution. We then review how the sea ice interacts with the geophysical and biological environments.

1. SEA-ICE FORMATION AND GROWTH

Sea ice, at temperatures generally seen in the natural environment, is dominated by two phases: salt-free ice, and liquid brine with a high concentration of salts. Depending on the temperature, history of development, and constituents of the sea water, it can also contain varying amounts of gases (primarily air), solid salts, organic inclusions (such as phytoplankton and bacteria), and inorganic inclusions (such as bottom sediments and wind-transported dust).

The two-phase nature of sea ice arises from the relative insolubility of salts in ice, as opposed to water, and a process of brine entrapment during ice growth. The equilibrium solute partition coefficient (k_0) for most solutes in ice (the ratio of the mole fraction of solute in the solid to that in the liquid phase) is about $k_0 < 10^{-4}$. This value suggests that the number of salt impurities that enter substitutionally into the ice lattice is small, so that the ice can be considered a pure phase (Weeks and Ackley, 1986) for most applications. This relative purity and the properties of the ice in the sea-ice matrix arise from its lattice structure.

1.1 Atomic Structure of Ice

A well-known attribute of water, unusual relative to most substances, is that its solid phase is less dense than its liquid phase: i.e., ice floats. Other properties, such as ice's unusually high melting point, and its high strength at high homologous temperatures (temperature relative to its melting point) compared with hydrogen compounds of similar molecular weight (HF, H_2S, CH_4), all result from the hydrogen bonds between water molecules in the ice-crystal lattice. Water is a polar molecule consisting of two relatively mobile protons (hydrogen atoms) located at the two positively charged sites on the molecule and two negative charge clouds associated with the oxygen atom, all distributed around the oxygen atom at roughly equal angles to each other. Four bonds, corresponding to these charge clouds, "point" to the vertices of a tetrahedron with the oxygen atom at the center (Fig. 2), an arrangement identified as tetrahedral coordination (Hobbs, 1974). In the ice lattice, the protons, because of their small size and mobility, can arrange to point to the negative ends of adjacent water molecules, forming hydrogen bonds. Hydrogen bonds are intermediate in bond strength, falling between a high-strength covalent bond (where electrons, rather than protons, are shared between atoms) and a weak ionic bond (where electrons are given up to the other atom rather than shared).

Ice, as a hydrogen-bonded structure, obeys the following rules:

1. Each of the four bonds in ice that connect it to other water molecules has only one hydrogen on it; and
2. of the four hydrogens around an oxygen

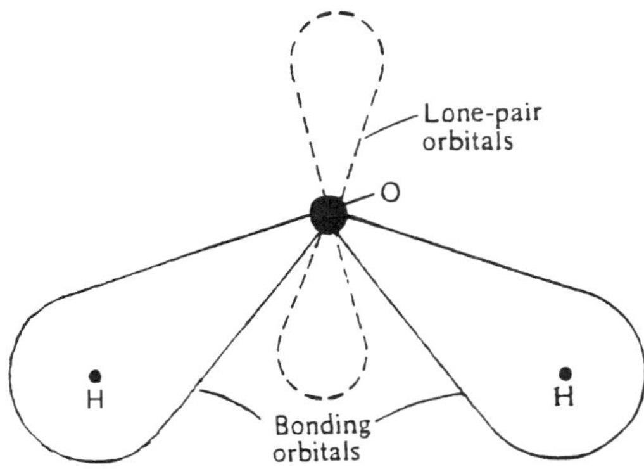

FIG. 2. Tetrahedral bond structure of the water molecule (after Hobbs, 1974).

atom, only two are close to the oxygen atom, so at each lattice site it "looks" like a water molecule (Weeks and Ackley, 1986; Hobbs, 1974).

The hydrogen bond is linear, giving ice its open crystal-lattice structure and its lower density relative to water, in which the hydrogen bonds are broken and the molecules can pack more closely. The tetrahedral coordination gives ice a crystal structure with hexagonal symmetry, affecting many of the large-scale characteristics of both pure and sea ice.

While several crystal structures (polymorphs) of ice exist, the structure of the ice in sea ice, and the other ices that exist on the Earth's surface, is known as ordinary ice (ice I). The ice I crystal structure is such that the oxygen atoms are concentrated close to a series of parallel planes referred to as the basal or (0001) plane in Miller indices. A vector normal to the basal plane is referred to as the crystallographic or *c* axis. Within the basal plane, three equivalent axes (six directions $\langle 11\bar{2}0\rangle$) form the *a* axes, which correspond to three equivalent close-packed rows of atoms (Fig. 3). The *a*-axis directions correspond to the arms of snowflakes growing from the vapor and, for our application, the arms of dendritic ice crystals growing in unstable or kinetic growth from a salt solution.

The relative insolubility of impurities in ice can be understood as a result of hydrogen bonding. Hydrogen bonding applies to the sharing of a proton between pairs of electronegative atoms. These are found to be oxygen, nitrogen, or fluorine (Hobbs, 1974). Because the electronegative impurities found in sea water, e.g., the chloride ion, do not partake in hydrogen bonds, they cannot substitute for oxygen in the ice lattice, and are rejected during the phase transition. Other electropositive impurities (e.g., the sodium ion) are much less mobile than the hydrogen atom and cannot bond between oxygen atoms, and so they too are rejected. The inclusion of gases in the interstices between the ice-lattice sites only occurs under much higher pressures than are found during the freezing of sea water. Air, for example, concentrates and nucleates into bubbles at a

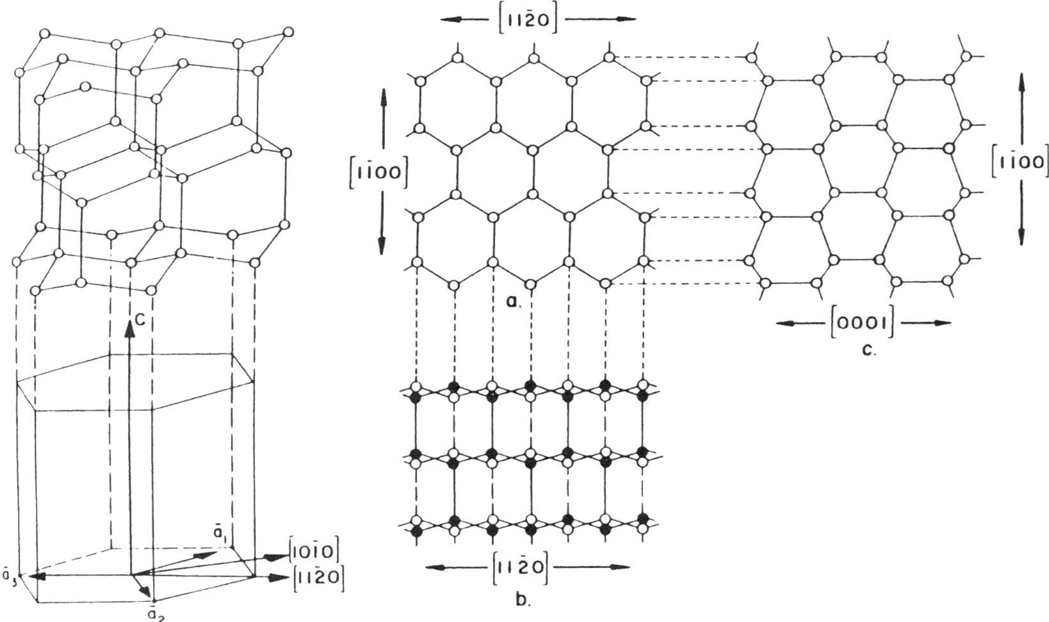

FIG. 3. Structure of ice; crystallographic axes are indicated. Views (a), (b), and (c) are planar sections of the lattice along the directions and planes indicated (after Weeks and Ackley, 1986).

growing ice interface and is not found within the ice lattice.

The impurities in sea water are therefore relatively insoluble in ice at lattice scales, and so the presence of liquid brine in sea ice arises from a macroscopic trapping process controlled by the growth behavior of ice.

1.2 Freezing of Sea Water

Ice growth from the liquid phase can take place under two thermodynamic conditions: stable, equilibrium, or planar growth; and unstable, kinetic, or dendritic growth. Kinetic growth commonly takes place in a supercooled condition of the liquid phase. It occurs in an episodic rapid way, triggered by a nucleation event, and uses the local supercooling as a heat sink for the latent heat released in the transition from the liquid to the solid phase. Dendritic growth is the dominant process in sea-ice growth, while planar growth occurs without supercooling and is seen only in fresh-water bodies. Because of the rapid growth of ice as dendrites and the relatively pure nature of the ice phase, spaces containing the liquid phase are also created, trapping brine within sea ice as it grows.

1.2.1 Constitutional Supercooling In a salt solution the equilibrium freezing point is a function of its composition; for example, sea water at 35 ppt (35 parts per thousand, or 35 g dissolved solids per kg of sea water) of salt has a freezing point of −1.8 °C. But the solution is supercooled if it is cooled to below its freezing point without nucleating ice crystals or changing its composition. At a growing ice interface, most of the salts are rejected. In the solution itself, however, the rate at which heat can be taken away is about ten times that at which salt can diffuse from a region of high to a region of low concentration. While the thermal diffusion lowers the temperature, the salt diffusion, necessary to lower the equilibrium freezing point of the rest of the solution, cannot keep pace with the thermal cooling effect, so that the solution becomes supercooled. Constitutional supercooling is, therefore, inherent in salt solutions.

1.2.2 Kinetic Crystal Growth The crystal anisotropy in ice, because of its structure as a hexagonal material, leads to a geometric selection in its growth from the melt. Crystal-growth theory (Weeks and Ackley, 1986) predicts that atoms arriving from solution at a crystal surface will have a much higher probability of remaining if they arrive at a step in the surface than if they arrive at an atomistically smooth face. Steps are common in ice where line lattice-defect sites (screw dislocations) intersect the crystal surface (Weeks and Ackley, 1986). Dislocations in ice are much more likely to occur in the basal plane (the plane normal to the *c* axis), serving as the step sites for preferential growth. In the early stages of ice growth, disks quickly form with their *c* axes normal to the plane of the disk, rather than as spheres that we might expect if the growth took place isotropically. The lateral growth rate of the disc is about 100 times greater than that which thickens it. These disks can then develop dendritic arms, still in the basal plane, pointing in the close-packed or *a* directions. Supercooling in salt solutions ensures that ice growth takes place kinetically and results in the preferred orientations of the disks and dendrites. This behavior occurs whether the ice forms as free-floating disks or is attached to the bottom of an existing ice sheet. The increase in thickness of a sea-ice sheet is by addition or extension of these dendrites into the supercooled solution below. If these dendrites are attached to an existing ice sheet, the resulting system is one of platelike structures with their basal planes in the growth direction or, equivalently, with their *c* axes perpendicular to the growth direction (Fig. 4).

1.2.3 Brine Entrapment As these ice platelets grow by "stabbing" into the supercooled solution nearby, salt is rejected by the growing pure-ice phase. The resulting solution near the ice crystal is increased in concentration by the rejection process. Since these salty regions are more dense than the surroundings, they descend (as plumes or blobs) and are replaced by upwelled water at the lower ambient concentration, and at the equilibrium freezing point for the solution. This fluid fills the spaces between the dendritic plates and is trapped as an inclusion, either when a new dendrite grows between the existing plates or as the ice plates gradually thicken and neck off the brine inclusion. These brine pockets, if there is a pre-

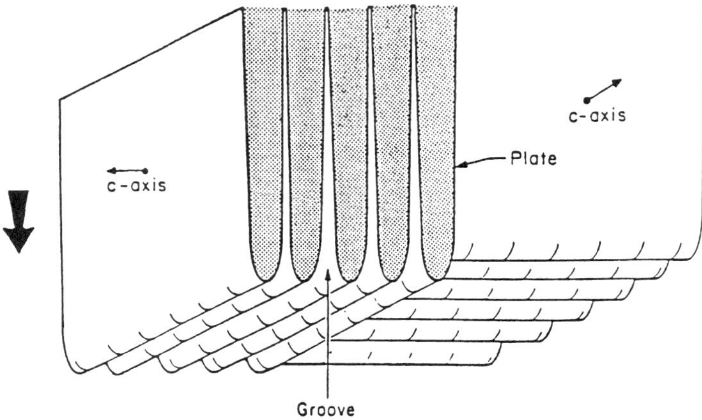

FIG. 4. Dendritic structure of columnar sea-ice crystals (after Gow and Tucker, 1991).

ferred orientation of the ice platelets, can then develop into parallel rows and reflect the local crystal structure in their orientation (Fig. 5). Once trapped, the brine pockets freeze or melt according to the phase relationships of the salt solution in ice. For example, the salt concentration in brine pockets in sea ice at −10 °C is approximately 115 ppt compared with 35 ppt at the freezing point for sea water (−1.8 °C).

1.3 Growth of Sea-Ice Sheets

The majority of sea-ice sheets observed in nature are formed by the kinetic growth of ice in sea water, either initially as free-floating disks, or as dendrites attached to an existing ice sheet, but usually as a combination of the two.

The initial ice formation (Fig. 6) that occurs at or near the surface of the sea water in the form of disks, needles, and platelets is called *frazil*. Continued freezing results in a soupy mixture—grease ice—consisting of sea water and unconsolidated frazil crystals. Wind-induced turbulence or wave action inhibits the immediate development of a solid ice cover. For example, in the presence of a sustained wave field, pancake ice (Fig. 7) forms by the packing and freezing together of the unconsolidated frazil crystals. The constant bumping of the pancakes then pushes frazil crystals up around their perimeters. When the wave action dies off, the pancakes consolidate by freezing together to form a continuous ice sheet. Pancake ice consolidation, up to about 1 m thickness, accounts for about 60% of the sea-ice cover at maximum extent in the Antarctic seas (Weeks and Ackley, 1986). The necessary turbulence at the seasonally advancing ice edge is provided by waves generated in the stormy oceans surrounding Antarctica.

The resulting ice structure of the consolidated sheet retains the crystal sizes and shapes from the initial turbulent formation. When viewed in thin section using polarized light, the texture of the ice sample, showing the individual crystals, exhibits a granular appearance (Fig. 8). Although each crystal probably grew as a disk in its basal plane, the final arrangement is relatively randomly oriented since the disks are pushed into their final configuration in no particular order or orientation. As the individual crystals freeze together, they trap off pockets of liquid between the individual crystal grains of pure ice, at separations corresponding to the crystal diameters, varying from a few tenths of a millimeter to a few millimeters (Weeks and Ackley, 1986). Granular ice texture, as a product of pancake-ice formation, therefore accounts for 60% of the volume of sea ice seen around Antarctica, corresponding to the area of pancake growth.

Under calm conditions, however, the frazil crystals in the grease ice quickly freeze together to form a solid, continuous ice cover with thickness between 1 and 10 cm. Once a continuous ice sheet forms, the underlying ocean is isolated from the cold air. The latent heat of freezing is lost upward through the ice sheet by thermal conduction, so that the growth rate is now determined by

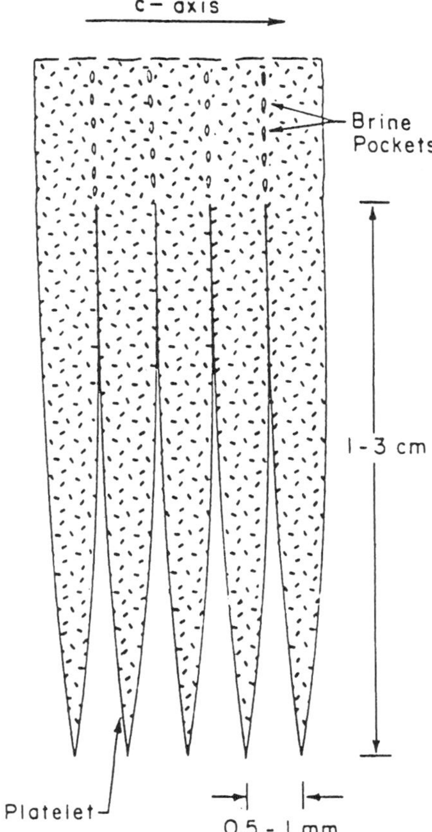

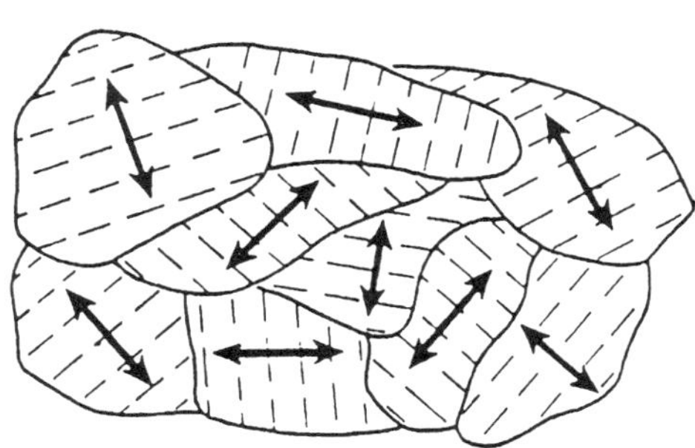

FIG. 5. Necking off of brine pockets from initial brine layers (grooves shown in Fig. 4) as ice growth and cooling take place (after Horner, 1985).

FIG. 6. Initial disks during the freezing of sea water (after Weeks and Ackley, 1986).

FIG. 7. Pancake ice forming in a wave field (after Gow and Tucker, 1991).

FIG. 8. Horizontal thin sections of granular ice. Scale divisions are mm. (After Gow and Tucker, 1991.)

the temperature gradient in the ice and the ice's effective thermal conductivity. In addition, once a continuous sheet has formed, ice crystals lose a degree of growth freedom, i.e. they are attached to the sheet, and so thickening proceeds by freezing at the crys-

FIG. 9. Vertical thin section of columnar ice. Scale divisions are mm.

tals' dendritic ice-growth interfaces at the lower boundary (Fig. 4). Since the direction of growth is defined, the dendrites are oriented to the maximum growth velocity, i.e., so that their basal planes penetrate edge-on into the supercooled water below. Unlike the frazil ice in the upper layers, large sections of the plates (crystal domains) are oriented in the same direction, with layers of brine interspersed between them. In the vertical this yields regions of columnar structure (Fig. 9), each of the columns consisting of an array of ice platelets with a common crystal orientation, alternating with brine layers (Fig. 10). This substructure is sometimes called "intragranular" (within grain) as opposed to the "intergranular" (between grain) location of the brine substructure in frazil or granular ice. The spacing of these plates, or equivalently, the brine-layer separation, varies from a few tenths of a millimeter to about 1 mm, with the wider spacings occurring at the slowest growth rates (Weeks and Ackley, 1986).

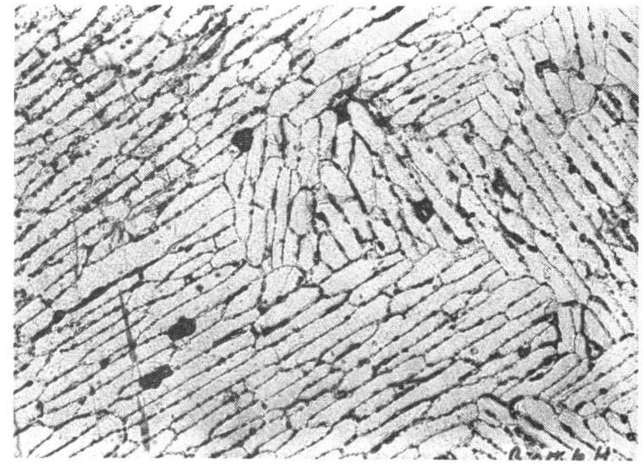

FIG. 10. Brine layers alternating with ice platelets, shown in horizontal thin section. The platelets are of the order of mm sizes, while the brine layers are ~0.1 mm.

In the crystal sense, the size of the individual ice plates is similar for sea ice of frazil or columnar origin, since the distance between the brine inclusions, representing the breakdown of the ice-crystal lattice spacing, is of the order of a few tenths to 1 or 2 mm in both structures. However, the regions over which arrays of these ice plates are ordered vary significantly between these two ice-growth regimes, as comparison of Figs. 8 and 9 indicates. Examination of ice structures by use of these sections therefore is a convenient way to determine whether the growth regime was turbulent (granular structure) or quiescent (columnar structure).

Sea ice in the Arctic Ocean, because of its isolation from surrounding seas by continental land masses, develops under calmer conditions than in the Antarctic seas. Structurally, therefore, the frazil or granular ice is confined to the top few centimeters of the sea ice, and the rest of the ice sheet (about 90%) consists of columnar ice. Because the Arctic Ocean is centered at the pole, while the Antarctic seas surround a continent at the pole, thermal conditions for sea ice formation are generally more severe in the Arctic, resulting in greater mean thickness: 2 m annually, compared with about 1 m in the Antarctic.

1.4 Incidental Processes of Sea-Ice Growth

Together the frazil and columnar structures account for 80%–90% of the initial growth of sea ice in the world's oceans. Other structures, identified as snow ice, platelet ice, and ice stalactites, account for the remaining 10%–20%. In some regions, however, these other forms can account for 30% or more of the ice structure, as, for example, platelet ice does within 50 km of the Antarctic coast.

1.4.1 Snow Ice Snow ice on sea-ice covers develops from the loading of the ice sheet by a snow cover. The snow cover depresses the ice-sheet surface below sea level and, if a crack occurs, sea water can laterally intrude at the snow–ice interface and flood the base of the snow cover. The snow–sea-water slurry or slush can then refreeze as a layer of imposed snow ice on the existing ice surface. Since the refreezing occurs in the slurry, the texture of snow ice is granular. It bears some strong similarities to frazil ice and is difficult to differentiate from frazil ice on structural grounds alone. However, oxygen isotope ratios in snow are quite different than in sea ice because snow forms by vapor diffusion while sea ice forms by freezing of sea water. A measurement of the relative oxygen isotope content in sea ice can therefore provide a well-defined measure of whether a sample is derived from snow or purely from sea water. These measurements in Antarctic sea ice have shown that snow ice accounts for a few percent of the total volume of the ice cover. However, it is estimated that 50% of the surface area of Antarctic sea ice is flooded at some time during its history, and so the surface layer of the sea ice there is in many instances dominated by snow-ice formation. Since remote sensing from space is heavily influenced by the surface condition, determination of ice properties from satellite imagery depends strongly on the degree of surface flooding at the time of the measurement.

1.4.2 Platelet Ice Platelet ice is formed at depth in the ocean and arises from the circulation of ocean waters, primarily under continental ice shelves, and their upwelling at the edges of the shelves. Water circulated at depth can be cooled to temperatures below the surface freezing point, since the freezing point is depressed under pressure. If the ice shelf bottom is sloping, dynamic effects can cause this water to upwell. As the water is brought to a shallower depth adiabatically, it is potentially supercooled, i.e., its actual temperature is lower than the freezing point for water of that composition at its new, shallower depth. If nucleation can occur, then ice-crystal growth can take place since the temperature difference between the higher freezing point and the actual temperature is a heat sink for the latent heat released by crystallization. These ice crystals develop as disks up to several centimeters in diameter and float upward. The disks either consolidate into a sheet at the surface or form cloudlike billow accumulations, up to several meters in thickness, at the base of an existing ice cover. These accumulations occur beneath both the ice shelves and adjacent sea-ice sheets. Their loosely stacked, unconsolidated nature gives them a relatively high proportion of water to ice (3 to 1) and provides a more freely exchanging environment, both for ocean water and biological organisms, than the more consolidated ice overhead. The association with potentially supercooled waters, however, confines platelet-ice development to near-shore regions, particularly in the Antarctic, where waters at depth can flow out from underneath the numerous ice shelves.

1.4.3 Ice Stalactites During the consolidation of an ice cover, brine at a temperature below the freezing point of the surrounding ocean water is rejected from the growing ice sheet. The brine is also more saline than the ocean water, as some of its fresh water has been removed by conversion to ice. As this cold plume descends into the ocean water, both saline and thermal diffusion take place between the plume and the surrounding water. Since the diffusivities for heat and salt are different, water surrounding the plume is cooled faster than it is compositionally changed, leading to the constitutional supercooling observed at the base of an ice sheet. Ice formation therefore takes place in the area adjacent to the plume. Continued rejection of the cold brine leads to the development of a feature similar in appearance to stalactites in caves. Ice stalactites are long features (10–100 cm) with a hollow core through which the brine plume flows, attached to the base of the ice sheet. They have been observed in new ice growth in cracks and under existing ice sheets in both the Arctic and Antarctic (Weeks and Ackley, 1986). While their contribution to the total ice growth is probably small, their appearance is striking (Fig. 11), and they may be dynamically important for fluid exchange from within the ice to the ocean mixed layer.

1.5 Phase Relations in Sea Ice

Once the ice cover has formed, the relationships between the ice, liquid brine, and solid salts are governed by the thermodynamic relationships shown in the phase diagram in Fig. 12 (Weeks and Ackley,

FIG. 11. Photograph of an ice stalactite.

1986). The phase diagram is a useful construct since it allows a computation of, for example, the amount of liquid in ice of a given temperature and composition. Since many of the properties of ice—mechanical, thermal, electromagnetic, and optical—depend on the liquid fraction, its computation is necessary to compare with measurements of these properties or to predict them. Since most naturally occurring sea ice is of lower salinity than the "standard" base for the phase diagram, the amounts of brine and ice are usually computed using equations that fit the slope of the phase diagram by a polynomial, and are then scaled by the ratio of the actual salinity of the ice sample to that of the standard. The measured density of the sample is also used to compute simultaneously the air volume, as the phase diagram gives the masses of ice, brine, and salts in a given mass of sea ice, and so the density is required to convert to volumetric units. The density, temperature, texture, and salinity of the ice sample, with use of the relationships given by the phase diagram, therefore provide a relatively complete set of information on its physical-chemical state.

2. SEA-ICE BIOLOGY

2.1 Sea Ice as Habitat

While our superficial view suggests that ice in general is an inhospitable environment, a closer examination shows otherwise, at least for sea ice. The two-phase nature of sea ice, with sea water, usually enriched in salts and nutrients, entrapped within it, actually provides a microenvironment. Organisms that can fit, tolerate the cold, and adapt to the saline and light changes typical of the ice cover will survive and sometimes thrive there. Most common are marine algae and diatoms (the photosynthetically active organisms), but protists and zooplankton that feed on algae, and the bacteria that decompose dead algal cells and other detritus, are also significant components of the biology in sea ice (Horner, 1985). Especially in the Antarctic, ice habitat has developed in a number of forms falling into three basic categories: those that are found on the top surface, in the interior, or at the bottom surface of the ice (Fig. 13) (Horner *et al.*, 1992). The type of habitat depends on the physical conditions during ice-cover formation and evolu-

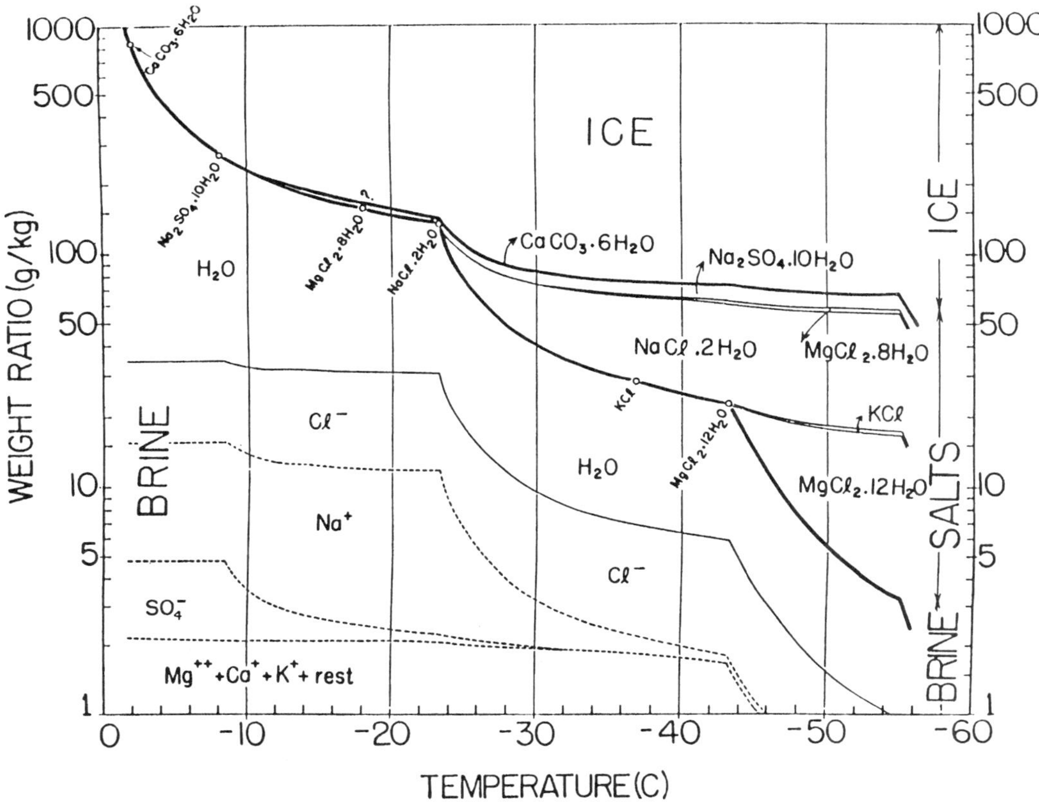

FIG. 12. Phase diagram for sea ice (after Weeks and Ackley, 1986).

tion; so there are significant relationships between the ice properties and the specific biology associated with it.

Advantages of ice habitat, relative to the under-ice marine environment, include closer proximity to the light available for photosynthesis; the presence of solid ice surfaces to attach to near the light, so that energy is not consumed by the organism to stay near the surface; and some protection from predation by larger organisms that cannot fit into the interstices in the ice.

2.1.1 Surface Communities Surface communities develop in water or slush pools created by the flooding of the top surface of the ice cover. Three processes have been identified that contribute to surface flooding. These are melt-pond formation from summer snow melt, snow-load depression of the ice cover so that it floods, and flooding induced by ice-ridge loading on the surface.

While melt ponds are a widespread feature on the Arctic pack, they are less important biologically than sea-water flooding since snow melt, being fresh water, is relatively inhospitable to marine organisms.

Depression of the ice surface, either by snow loading or by the weight of ice blocks produced during ice deformation, can lower the top surface below sea level, which is followed by flooding by sea water through cracks or pores in the ice. The sea water can contain a seed population of organisms as well as the nutrients necessary to sustain photosynthesis. These communities then bloom in the higher-light environment on the top surface of the ice, giving it a distinct color because of the high concentration of organisms (Horner, 1985). Typically, the concentration of marine algae is several hundred times higher per unit volume of flooded snow than it is in the ocean waters directly below the ice. Snow ice and the related bio-

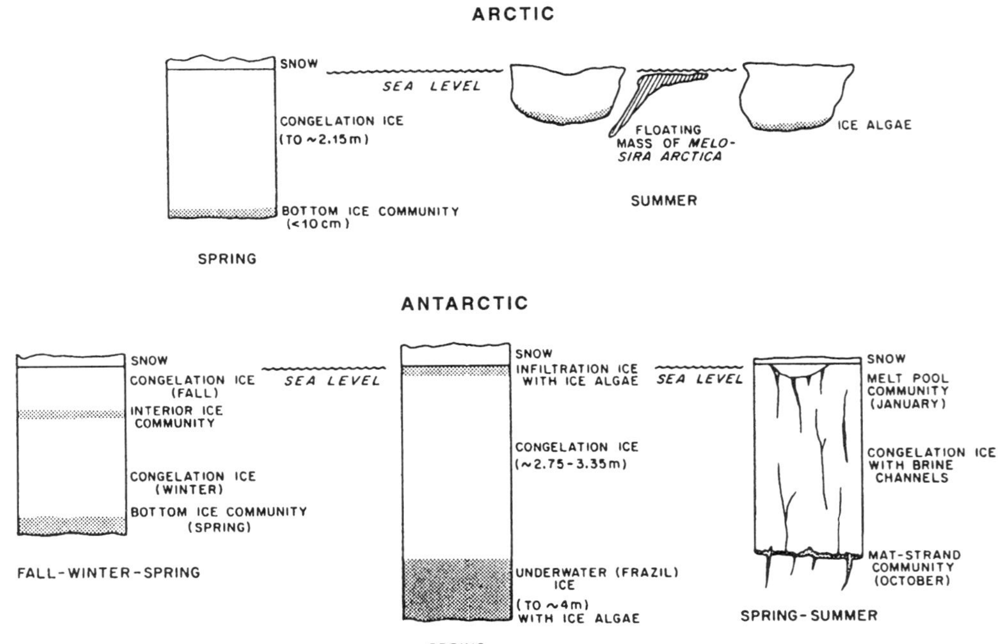

FIG. 13. Habitats identified in sea ice, divided into surface, interior, and bottom (after Horner, 1985).

logical community, resulting from the flooding of the top surface, are common over large areas of the Antarctic sea ice. The thicker ice in the Arctic requires relatively deeper snow to initiate flooding, and so the flooding-produced biological community occurs less frequently there.

2.1.2 Interior Communities Two main interior communities have been identified: frazil-ice communities that develop during ice formation, and the freeboard-layer community that develops in older ice.

The frazil-ice communities were first discovered in newly forming pancake ice in Antarctica, where concentrations of marine organisms were ten to one hundred times those found in the underlying water. It was also discovered, by examination of the nutrient levels in this newer ice, that the organisms were usually there at these higher concentrations before any significant photosynthesis occurred in the ice. Unlike the surface communities, they were physically incorporated at high levels rather than multiplying after incorporation. Mechanisms have been identified that concentrate organisms within the ice, including water pumping and scavenging by floating ice particles. Pumping, where sea water containing organisms is surged through a layer of frazil or grease ice, is thought to concentrate the organisms. As sea water is brought upward into the layer of grease ice, the organisms stick preferentially to the ice surfaces, perhaps because of some biological adhesion. Wave passage or water circulation in closed-loop cells can bring new water containing organisms to the interior of the grease-ice layer where some organisms are deposited, so that over even a few hours, significant enrichment of the ice with organisms can take place. Rafting of the ice (where two adjacent sheets slide over one another) can distribute organisms to the interior of the ice. Since frazil ice is particularly common in the Antarctic, these frazil-ice–associated interior communities are found widely there.

Scavenging of particles by freely floating frazil crystals as they move through the water column, either horizontally or vertically, also can concentrate particles. This process may account, more than wave pumping, for the high concentrations of sediments in ice seen in near-shore regions of the Arctic where wave action is not as common as in the open-ocean regions of the Antarctic.

Freeboard-layer interior communities are generally found between 5 and 10 cm below the ice surface in ice that lasts through the summer period. The condition of this layer, which is usually rich in biology, is a honeycomblike structure of partially melted ice filled with sea water and fresher ice melt. Communication of this layer with the surface sea water is shown by the presence of larger zooplankton, such as the shrimplike krill, that are often found in this layer. The layer may develop through a biophysical feedback process, in which organisms bloom and absorb solar radiation at this near-surface level. Some ice melting, caused by the trapped solar radiation, is followed by intrusion of sea water, containing the larger organisms, through the holes and gaps created by the melting. This layer has been observed in the limited area of Antarctic perennial ice at the end of the summer period. As the ice enters its second year, freezing-induced convection can also lead to nutrient enrichment of the pore water and an autumn bloom of algae inside the ice.

2.1.3 Bottom Communities Bottom communities have been identified in both the Arctic and Antarctic, although the platelet-layer habitat, below the bottom surface of the consolidated ice, appears to be restricted to the Antarctic. While the interior community shows a strong association with frazil ice, the bottom community forms in the platelike dendritic structure seen at the base of columnar ice. Since the overlying ice limits the light penetrating to the base, the bottom community develops in the fall and spring seasons when incident total light levels are so high that the small fraction that is transmitted is sufficient to sustain photosynthesis by algae that attach to the ice dendrites at the base. Even during these periods, the level of snow cover on the ice is a critical factor. It appears that the maximum productivity is attained when a thin snow cover (rather than no snow or a thick snow cover) is present, perhaps optimizing the light condition for these somewhat shade-adapted organisms. Because the bottom surface is in constant communication with the under-ice sea water, nutrient exchange is constantly taking place, unlike at higher levels, within the ice, where fluid flow is more restricted. As a result, the bottom ice communities are able to sustain production at hundreds of times the levels seen elsewhere within the ice. This results in the accumulation of biomass, in the lower 20 cm of the ice, as high as 56 g/m^2 of carbon, a figure similar to hayfield production on land surfaces, and rarely found elsewhere in oceanic waters.

A community also develops in the platelet layer seen in the under-ice coastal region of Antarctica. Even higher levels of biologic activity are seen here than in the overlying ice. Factors contributing to this production are the 3-to-1 sea water-to-ice ratio (which optimizes the space available for colonization), the stability of the layer relatively close to the light, and a high capacity for nutrient exchange over the several-meter thickness of the layer.

In both the Arctic and Antarctic, a thick snow cover impedes the development of these bottom communities, since light limitation restricts their photosynthetic capacity. Bottom communities are relatively sparse, for example, in some drifting pack-ice areas where the snow cover is apparently too thick to allow light penetration sufficient to sustain photosynthesis.

3. DISTRIBUTION OF SEA ICE

At the largest scales, sea ice exhibits features that depend on the full set of geophysical conditions for its formation. The lifetime of the ice cover, for example, can vary from days in the ice's furthest extension from the high latitudes, to a season in its most common form, to several years in the highest latitudes. The ice-thickness distribution and how it evolves best describes these processes and their effects. The categories described here are additionally classified as perennial ice zones and seasonal ice zones, the latter subdivided into fast-ice zones and marginal-ice zones.

3.1 Ice-Thickness Distribution

Plate tectonics changes the visible topographic features of the Earth's crust, cracking and roughening the original, more level, land or ocean-bottom surfaces initially formed by sedimentary or accretionary processes. Similarly, though at different scales, the sea-ice cover undergoes a form of tectonic activity and changes from a relatively more uniform slab, formed by freezing or frazil accumulation, to a rough, uneven surface characterized by ice pileups (pressure ridges) (Fig. 14) and cracks (leads) (Fig. 15). Relative ice motion and stress in the ice arise from the forces exerted on the upper and lower surfaces by wind and water drag, the motion of ocean currents, and the Coriolis force on the ice (Hibler, 1986). If the stress is high enough, the ice breaks, creating either the open-water areas that can then freeze quickly under subfreezing atmospheric conditions, or broken ice that is piled into thick ice ridges as the ice is shoved together. Coastal boundaries can also be an "immovable object" that moving ice breaks and piles against, resulting in thick ridges (Fig. 16). Or coasts may act as source regions for new ice, as fast-freezing open-water regions (polynyas) form when offshore winds, ocean currents, or tides pull the ice cover away instead of pushing it into the coast (Fig. 17).

The ice-thickness changes or, more correctly, the ice-thickness distribution resulting from these processes is the most important property of the ice cover at scales from centimeters (vertically) to hundreds of kilometers (horizontally). One reason for this importance is that the thickness distribution tells us the mass of the ice cover (Rothrock, 1986). The thickness variations also govern, to a large degree, the interaction of the ocean and atmosphere. On ice-covered oceans, the rate of energy transfer (heat loss or gain) to the atmosphere and the ice strength (its resistance to motion and deformation) both depend directly on ice thickness. Ice-growth rate, which governs the salt flux to the ocean, depends itself directly on ice thickness.

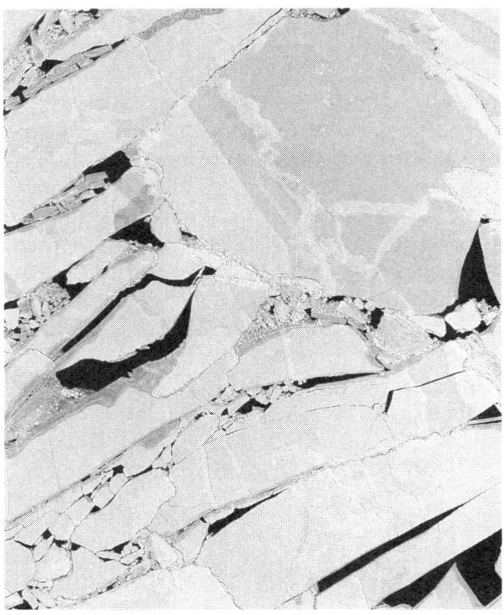

FIG. 15. Photograph of cracks, or leads, in sea ice. Area pictured is 145×180 m^2. (From Steffen, 1986.)

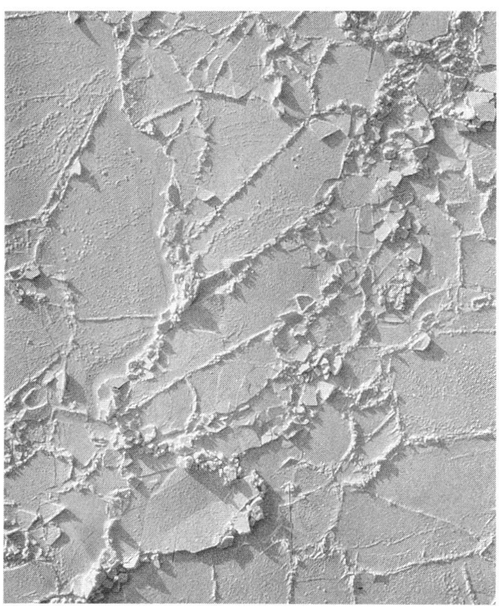

FIG. 14. Photograph of pressure ridges on sea ice. Area pictured is 100×120 m^2. (From Steffen, 1986.)

Regions of modest size can contain ice of many different thicknesses. In a statistical sense we can represent this variation by the ice-thickness distribution $g(h)$ (Rothrock, 1986; Thorndike, 1986). This function can be used to represent the fraction of a region R with thickness between h and $h + dh$ as

$$g(h)dh = \frac{dA(h, h + dh)}{R}, \qquad (1)$$

where $dA(h, h + dh)$ is the area within R

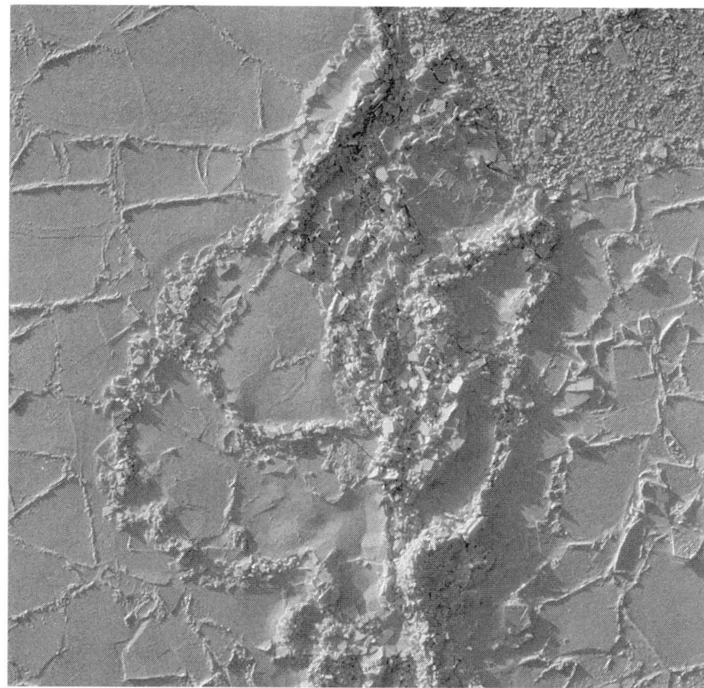

FIG. 16. Photograph of sea ice piling in compression, as in against a coastal boundary. Area pictured is 180×220 m^2. (From Steffen, 1986.)

FIG. 17. Photograph of a coastal polynya. Bottom width of picture is 200 m. (From Steffen, 1986.)

covered by ice in the thickness range $(h, h + dh)$. $g(h)$ itself has the properties of a probability density function with dimensions of length^{-1}.

Note also that when a finite fraction of the region R is ice free, g will have a delta function at $h = 0$, and so g includes information about ice concentration [defined as the areal fraction of ice cover, or $1 - dA(0,dh)/R$] as well as thickness. The mean ice thickness in R is

$$h_m = \int_0^{\infty} hg(h)dh \tag{2}$$

and includes areas of open water.

Taking into account spatial and temporal variations, we can write the ice-thickness distribution more generally as $g(h,\mathbf{x},t)$.

The general thickness distribution is governed by the equation (Thorndike, 1986; Rothrock, 1986) that relates the time and space evolution of the statistical function g to physical forcing terms as

$$\frac{\partial g}{\partial t} = \operatorname{div}(\mathbf{v}g) - \frac{\partial(fg)}{\partial h} + \Psi, \tag{3}$$

where $f(h,\mathbf{x},t) = \partial h/\partial t$ is the thermodynamic growth rate of ice of thickness h, $\mathbf{v}$ is the horizontal velocity vector, $\operatorname{div}(\mathbf{v}g)$ is the flux divergence, and Ψ is a function that accounts for the mechanical redistribution of ice from one thickness to another, modeling the formation of both leads and pressure ridges. Figure 18 shows the behavior of each of these terms (arrows represent delta functions at zero thickness) and the resulting ice thickness distribution. The shapes of the distributions reflect the importance of the various processes. For example, a high thickness tail on the distribution is only possible through the mechanical redistribution that converts thin ice into thick ice (pressure ridging). Purely thermodynamic processes, shown by f, tend to thicken thin ice quickly but only up to an equilibrium thickness, where the growth rate f becomes zero. For thick ice, the growth rate is negative, indicating that melting occurs below very thick ice. This thermodynamic effect is caused by the poor heat conduction through thick ice and the presence of a small, but significant, heat flux from the ocean in most ice-covered regions.

Limited thickness-distribution information

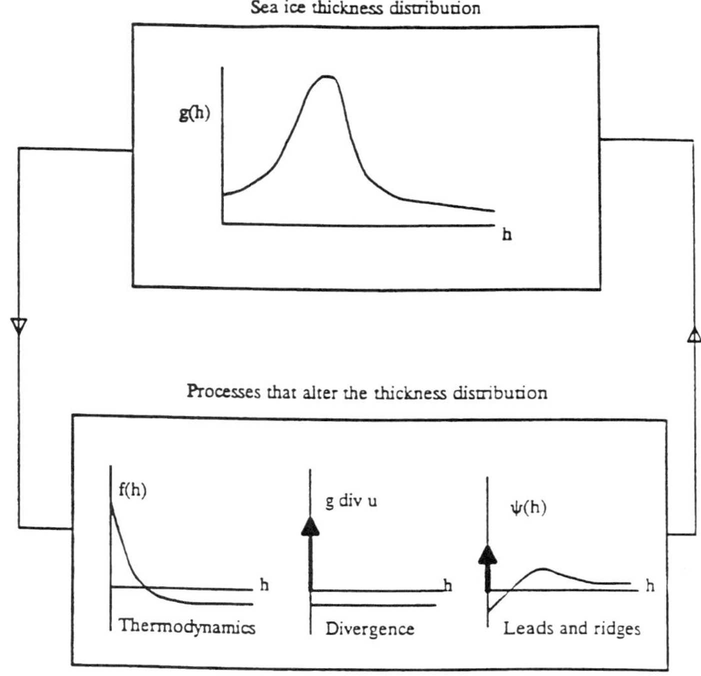

FIG. 18. Ice-thickness distribution and its dependence on growth [$f(h)$], deformation (Ψ), and divergence [$\operatorname{div}(\mathbf{v}g)$] (after Thorndike *et al.*, 1992).

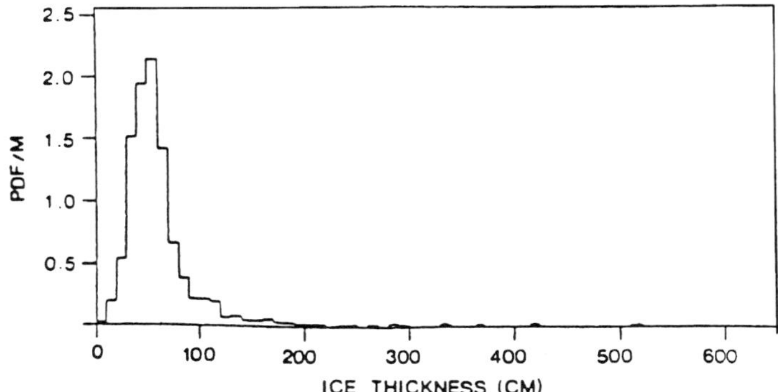

FIG. 19. Antarctic sea-ice–thickness distribution (after Gow and Tucker, 1991).

is available from the polar regions. One from the Antarctic sea ice is shown in Fig. 19, and one for the Arctic is shown in Fig. 20. Even the few distributions available, however, reveal the dominant ice processes and main characteristics of the ice cover on a large scale. The Antarctic ice thickness has a single peak (one mode) and a mean thickness of about 0.6 m. This distribution reflects the generally warm conditions under which the Antarctic pack ice grows, and its formation by primarily thermodynamic processes, shown by the absence of a deformation tail in the distribution. Low divergence, by the absence of a peak at the thicknesses less than the mean, is also indicated. In contrast, a high tail in the Arctic thickness distribution indicates a more robust redistribution of thin ice into thick ice, and significant ridging is in fact present. Similarly, the bimodal (two peaks) distribution indicates that divergence effects have occurred more recently resulting in the thin-ice peak at 0.8 m, a measure of ongoing dynamic activity. The high mean ice thickness (3 m) indicates a generally cold climate, rapid thermodynamic thickening of thin ice, and little ice-thickness loss in the summer.

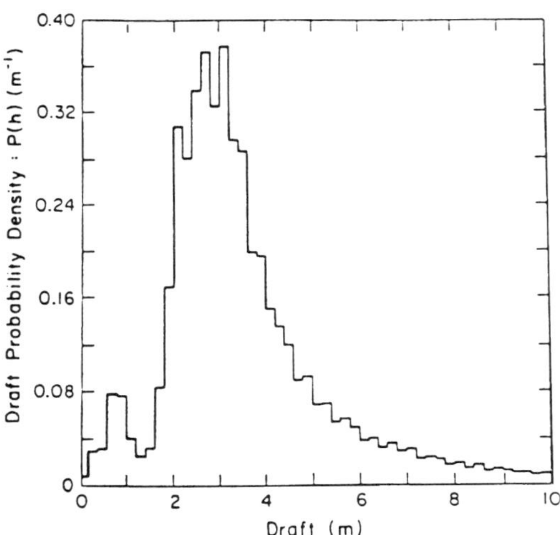

FIG. 20. Arctic sea-ice–thickness distribution. (After Gow and Tucker, 1991.)

3.2 Perennial Ice

Ice covers that last through at least one summer season and into the next growing period are described as perennial ice. Figure 1 shows the distribution of perennial ice in the Arctic and Antarctic, given as the regions inside the seasonal minimum ice edge.

The amounts and distributions of perennial ice in the two regions are highly contrasted; two-thirds of the ice cover in the Arctic is perennial compared with only one-fifth in the Antarctic. The two major influences that lead to this difference are the higher latitudes of the ice-covered ocean in the Arctic relative to the ice-covered ocean in the Antarctic, and the complete open-ocean sea-ice boundary in the Antarctic compared with the Mediterranean character (ocean surrounded by land masses) of the Arctic basin.

The higher latitudes of the Arctic Basin give a longer winter season, and thicker ice develops that only partially ablates away during the summer season. In the Antarctic, the ice is thinner and the open-ocean boundary provides a broad melting zone that extends around the full circumference of the sea-ice zone. Ice is advected into the open ocean from higher latitudes and warmer water destroys it quickly. In contrast, the advected ice from the Arctic can pass only through a relatively narrow passage to be destroyed in the warmer waters of the North Atlantic. This passage at Fram Strait amounts to 10% of the Arctic Basin's circumference. The life of Arctic Basin ice is of the order of 3 to 7 years before it is advected out of the Basin and destroyed, while Antarctic perennial ice is only two years old because the motion through the perennial zone (e.g., the Western Weddell Sea) is relatively quick. Through exposure to several summers with more intense radiation, Arctic perennial ice is relatively low in salinity compared with the Antarctic, where only one summer exposure is common. Flushing by surface snow-melt water also helps desalinate the upper layers of Arctic ice, while the sea-water–flooded snow cover remains relatively intact on the Antarctic ice through the single summer period, ensuring that little salt is lost from the Antarctic ice in its second year.

3.3 Seasonal Sea Ice

3.3.1 Fast-Ice Zones Fast ice is the sea ice that grows seaward from a coast and stays in place (attached, or "fast," to the coast) during winter. In spring and summer it can either melt in place or break away from the coast and then melt (Wadhams, 1986). A small fraction lasts through the summer as perennial fast ice. Fast ice is found in both the Arctic and Antarctic. In some areas, primarily Arctic coasts, the fast ice is stabilized by grounded ice-deformation features (pressure ridges), and can extend out to the 20–30-m depth contour that is the draft limit of such ridges. In constricted channels, such as in the Canadian Archipelago or McMurdo Sound, Antarctica, the fast ice fills the whole width, regardless of water depth, but can be subjected to some movement and deformation if strong winds occur before the ice is stabilized. Fast ice in Antarctica is only 1% of the total area of ice-covered ocean, but is important in the local ecology, providing habitat for some varieties of seals and penguins. It also modifies the oceanic light regime in the coastal region, important in the overall biological productivity, and the ice itself also supports an abundant biological community.

3.3.2 Marginal-Ice Zone The marginal-ice zone is the part of the ice cover that is affected by the open-ocean boundary. Several cold drift currents that contain ice can be considered part of the marginal-ice zone (East Greenland and Labrador) as is ice lying within 150–200 km of the northern Antarctic ice edge. The majority of seasonal ice that drifts is in the marginal-ice zone in its formation period, in its decay period, or both. Generally, ice in the marginal zone has much greater freedom of movement than ice in the central pack and consequently a number of ice–ocean–atmosphere interaction phenomena occur that are peculiar to this zone.

A principal effect on ice processes in this zone is that of ocean waves. As described before, the interaction of the wave field with a growing ice cover accounts for the high proportion of frazil or granular ice seen in the Antarctic. Wave fields also have important influences on the decay of seasonal ice. As large-diameter (kilometer-scale) floes are advected into the zone where waves can penetrate, they are broken by flexure into pieces only a few to 10 m in diameter, greatly increasing the perimeter length that is in contact with warm water and accelerating the ice decay. The lower ice concentration and smaller floes seen here also respond differently to wind and ocean-current forcing than the interior pack, influencing the decay of the ice cover as well.

Some marginal-ice zones, such as the East Greenland Current, coincide with frontal zones in the ocean. Satellite data have shown that complex ocean-eddy structures are revealed by the pattern of the ice floes on the surface. Modeling studies and measurements have shown that the ice is an active participant in ocean dynamics, influencing the kinematics of the eddies and how they form, move, and decay as well as providing a tracer of the eddy motions.

3.4 Geophysical Effects on Ice Distribution

The presence of an ice cover in a region is a product of several forces or factors. Some of these are the surface-energy balance (temperature and radiation) (Maykut, 1986); the ocean thermal structure; and the wind and ocean currents that drive the motion and deformation of the ice cover (Hibler, 1986). Interactions that add levels of complexity to ice conditions include the wave-field effects that cause pancake-ice formation and the sub–ice-shelf ocean dynamics that lead to platelet-ice formation. Wind-driven effects and the deformation of the ice cover can, for example, change the ice-thickness distribution so that the thickest ice occurs in lower latitudes than expected from the location of the coldest conditions (Hibler, 1986). And open water is often found continuously at the far southern boundary of the pack ice nearest the Antarctic continent because of the strong winds that blow ice away from the continent.

3.4.1 Ice Circulation and Deformation in the Arctic and Antarctic Two dominant features of the ice circulation in the Arctic Basin are the Beaufort Sea Gyre, a clockwise circulation pattern in the Beaufort Sea, and the transpolar drift stream, a basin-wide feature that transports ice from the western to the eastern side of the basin (Fig. 21). The ice circulation mimics the wind field that drives it, caused by the mean sea-level pressure field. In the mean, a large high-pressure area exists over the Beaufort Sea and isobars over the remainder of the basin trend west to east. With large daily and seasonal variability, mean velocities of the ice are about 2 km day^{-1} (± 7 km day^{-1}). As a result of this circulation pattern, ice floes can remain in the basin, especially the Beaufort Sector, for many years. The ice dynamics also lead to a buildup of ice along the coasts of Greenland and the Canadian Archipelago. Ice with average thickness in excess of 6 m is found along these coasts in contrast to the 3- to 4-m ice found over the North Pole. Figure 22 shows ice-thickness contours obtained from submarine up-looking sonar data. Model results, where no ice dynamics is allowed, have shown that the thickest ice would be expected over the

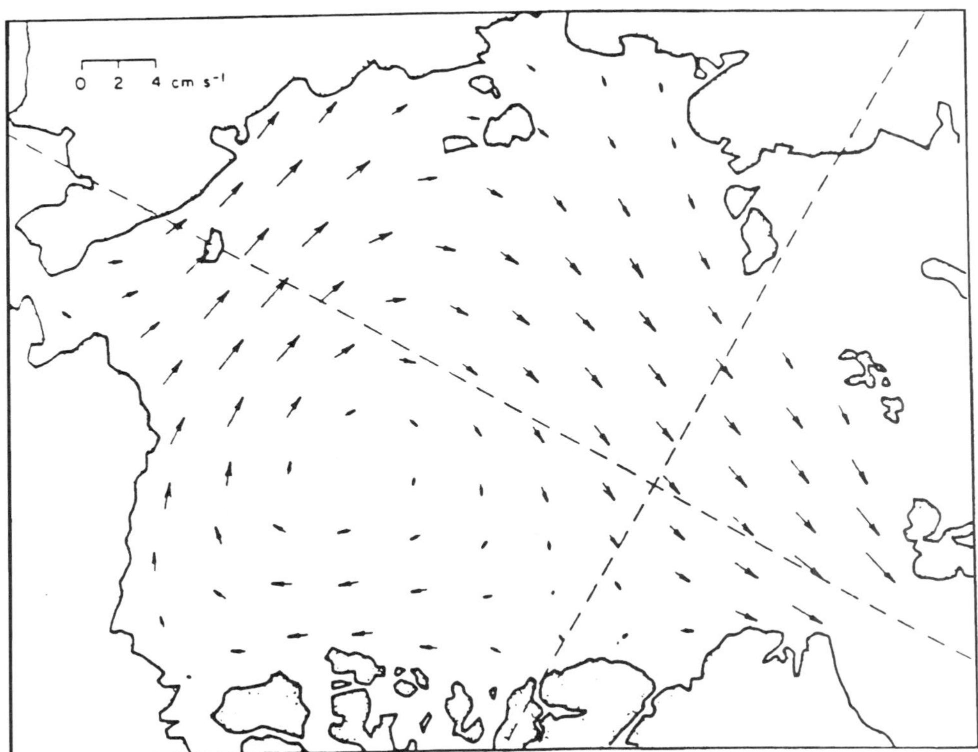

FIG. 21. Mean field of ice motion in the Arctic Basin (after Gow and Tucker, 1991). Greenland is at the bottom right.

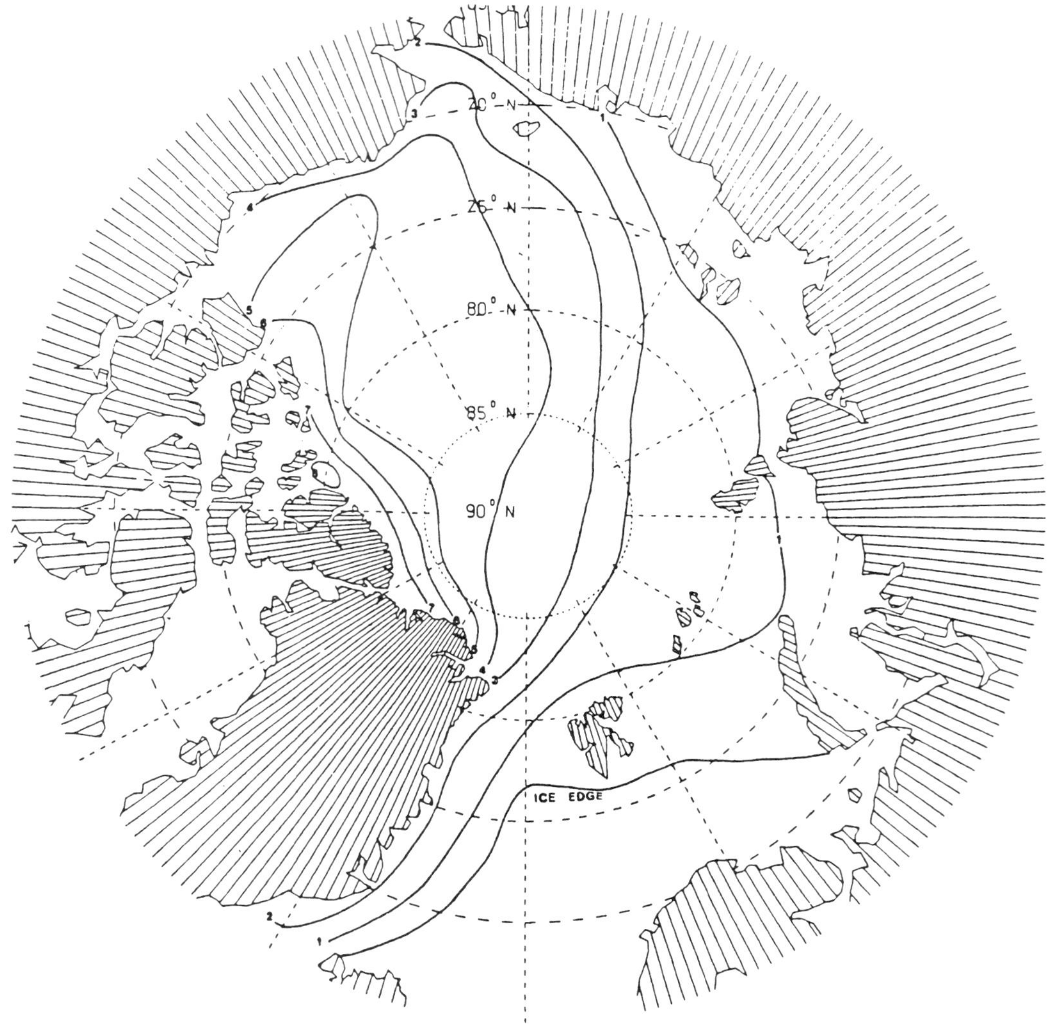

FIG. 22. Contours of Arctic mean ice draft obtained from submarine sonar (after Gow and Tucker, 1991).

North Pole where the coldest conditions prevail. As the data show, however, the thickest ice actually forms elsewhere, as a result of ice dynamics (Hibler, 1986).

In the Antarctic the major drift features are the wind-driven circulation at the north near the boundary with the Antarctic Circumpolar Current, and the East Wind Drift flowing in the opposite direction in the south immediately adjacent to the continent (Fig. 23) (Gow and Tucker, 1991). This coastal flow is directed north by protrusions of the continent in the Weddell and Ross Seas. The ice in the northward flow joins the Circumpolar Current, which transports it into warmer water where it eventually melts.

Because of the northward flow in the Weddell Sea, the ice at the northern end lasts longer than ice elsewhere, thickened by deformation during its journey. It is thicker than ice further south, despite generally colder conditions in the south. In this case, the ice circulation and deformation invert the thickness relationships expected from thermodynamics alone. Mean drift rates (4–6 km day^{-1}) are two to three times those of Arctic sea ice, and ice-drift rates up to 20 km day^{-1} have also been observed, reflecting the

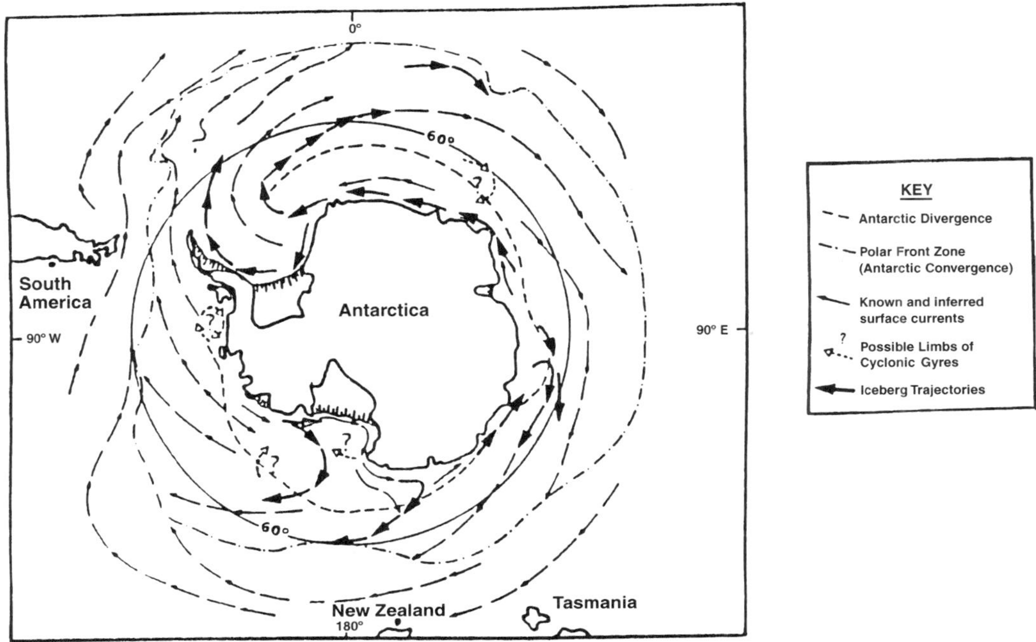

FIG. 23. Surface circulation (of ice and water) around Antarctica (after Gow and Tucker, 1991).

more energetic atmospheric and ice circulations in the Southern Hemisphere.

4. CONCLUSION—SEA ICE, COMPLEX RESPONSE IN A CHANGING CLIMATE

While cold conditions are necessary for ice-cover formation, many other factors in the atmosphere and ocean modify the distribution, thickness, and properties of the ice cover. Some of these include the snow cover, the ocean thermal state, and the wind and ocean currents. Because of the ice cover's capability to influence also the ocean and atmosphere, through feedback processes, the response of the ice cover to climate-change effects would be difficult to predict, even assuming the climate-change effects are themselves predictable.

In the outer areas of the pack ice—in the seas surrounding the Arctic Basin or at the northern limit of the Antarctic pack ice—atmospheric and oceanic thermal effects may determine the presence or absence of the ice covers. Even there, however, certain regions are affected by the advection of ice formed in more polar latitudes. Ice-cover changes may in that case be more sensitive to changes in the wind and ocean circulation fields than they would be to atmospheric temperature changes. It appears, for example, that the paleoclimatic extension of the sea-ice cover into the North Atlantic was in response to a decrease in salinity of the upper ocean there, affecting the deep convection that brings heat to the surface. The expanded sea-ice field then cooled the atmosphere near it and then influenced how weather systems moved across the North Atlantic. Lower temperatures in northern Europe may then have resulted from the sea-ice effects, rather than causing them. Studies on Antarctic sea-ice conditions have also suggested that local temperature changes at the Antarctic coast are related to sea-ice extent, while the ice extent itself is more controlled by the wind fields prevailing at the time than by the temperature.

The sea-ice fields also interact strongly with the biological environment. Changes in the ice cover can drive populations to different regions as the food supply associated with, for example, melting ice edges, or sea-ice biology, responds to external changes.

Deep-ocean areas are generally low in productivity except in special zones, like the ice edges, where melt, wave damping, and other processes occur that enhance biological production. A reduction in sea-ice area or a decrease in the circumference of marginal ice would then potentially reduce the ice edge available for biological production.

The sea-ice system combines the vagaries of atmospheric and oceanic circulation with nuances unique to its state as a more solid rather than strictly fluid medium. Predictions of those circulations are somewhat tractable in their full dimension only by numerical modeling with the full computational power of the world's fastest and largest computers. Further computational development, as well as a better understanding of sea-ice processes, is necessary before the combined atmosphere–ice–ocean–biosphere interaction can be fully understood.

GLOSSARY

Brine: Sea water enriched in salts by ice freezing.

Columnar Ice: Ice grown by one-dimensional cooling by thermal conduction, which produces ice crystals with the appearance of long columns parallel to the growth direction.

Fast Ice: Ice, formed in a water body, that is attached to a shoreline.

Frazil Ice: Ice formed as small disks, plates, and dendritic arms in suspension in a water body, usually under turbulent flow conditions.

Ice Stalactite: An elongated tube of ice formed at and attached to the base of an ice sheet, similar in appearance to a cave stalactite.

Leads: Cracks in a sea-ice cover, of the order of 5 to 100 m in width, and sometimes exceeding many kilometers in length.

Marginal-Ice Zone: The region of the ice-covered ocean that is close to the open ocean where no sea-ice cover is present. Ice floes in the marginal-ice zone are usually broken up by the wave action.

Perennial Ice: Ice that has survived one or more summer seasons.

Platelet Ice: A type of ice crystal formed in and associated with the outflow of cold supercooled water from beneath large continent-based ice shelves.

Polynyas: Large irregularly shaped bodies of open water, either surrounded by sea ice or bounded on one side by coast and on the other by sea ice ("coastal polynyas"). (Originally Russian.)

Pressure Ridges: Ice pileups formed by ice-floe collisions; can be many meters thicker than the original ice sheet.

Seasonal Ice: Ice that forms during the winter and melts by the end of summer.

Snow Ice: Ice formed by the freezing of water-saturated snow.

Works Cited

Barry, R. G. (1986), in: N. Untersteiner (Ed.), *The Geophysics of Sea Ice,* NATO Advanced Study Institutes Series B, Physics, Vol. 146, New York: Plenum, Chap. 15.

Gow, A. J., W. B. Tucker III (1991), *Physical and Dynamic Properties of Sea Ice in the Polar Oceans,* CRREL Monograph 91-1, Hanover, NH: Cold Regions Research and Engineering Laboratory.

Hibler, W. D. III, (1986), in: N. Untersteiner, (Ed.), *The Geophysics of Sea Ice,* NATO Advanced Study Institutes Series B, Physics, Vol. 146, New York: Plenum, Chap. 9.

Hobbs, P. (1974), *Ice Physics,* Oxford: Clarendon Press, p. 13.

Horner, R. (1985), *Sea Ice Biota,* Boca Raton, FL: CRC Press.

Horner, R., S. F. Ackley, G. S. Dieckmann, B. Gulliksen, T. Hoshiai, I. A. Melnikov, W. S. Reeburgh, M. Spindler, C. W. Sullivan (1992), "Ecology of Sea Ice Biota. 1. Habitat and Terminology," *Polar Biol.* **12,** 417–427.

Maykut, G. A. (1986), in: N. Untersteiner (Ed.), *The Geophysics of Sea Ice,* NATO Advanced Study Institutes, Series B, Physics, Vol. 146, New York: Plenum, Chap. 5.

Rothrock, D. A. (1986), in: N. Untersteiner (Ed.), *The Geophysics of Sea Ice,* NATO Advanced Study Institutes Series B, Physics, Vol. 146, New York: Plenum, Chap. 8.

Steffen, K. (1986), *Atlas of the Sea Ice Types, Deformation Processes and Openings in the Ice,* Eidgenössische Technische Hochschule, Zurich, Switzerland, Report No. 20.

Thorndike, A. S. (1986), in: N. Untersteiner (Ed.), *The Geophysics of Sea Ice,* NATO Advanced Study Institutes Series B, Physics, Vol. 146, New York: Plenum, Chap. 7.

Thorndike, A. S., Parkinon, C., Rothrock, D. A. (Eds.) (1992), Report of the Sea Ice Thickness Workshop, Polar Science Center, Univ. of Washington, Seattle, WA.

Wadhams, P. (1986), in: N. Untersteiner (Ed.), *The Geophysics of Sea Ice,* NATO Advanced Study Institutes, Series B, Physics, Vol. 146, New York: Plenum, Chap. 14.

Weeks, W. F., S. F. Ackley (1986), in: N. Untersteiner, (Ed.), *The Geophysics of Sea Ice,* NATO Advanced Study Institutes Series B, Physics, Vol. 146, New York: Plenum, Chap. 1.

Further Reading

Horner, R. (Ed.) (1985), *Sea Ice Biota,* Boca Raton, FL: CRC Press.

Untersteiner, N. (Ed.) (1986), *The Geophysics of Sea Ice,* NATO Advanced Study Institutes, Series B, Physics, Vol. 146, New York: Plenum.

SEDIMENTARY BASIN EVOLUTION

Christine E. Koltermann, *Pegasus Geoscience, Santa Clara, California, U.S.A.*

INTRODUCTION

A sedimentary basin is a large topographic depression in the Earth's crust that accumulates a substantial thickness of sediment plus pore fluids over geologic time. The sinking or vertical downward displacement of the Earth's surface that forms sedimentary basins is called *subsidence.* Subsidence initially is caused by tectonic mechanisms that create vertical and horizontal motions in the Earth's crust and later is augmented by the weight of accumulating sediments, water, and ice, and sediment compaction. Principal causative mechanisms for tectonic subsidence include (Ingersoll, 1988; Busby and Ingersoll, 1995)

1. *extension* (stretching or pulling apart) of the Earth's crust and *rifting* (large-scale faulting and tilting);
2. thermal cooling and contraction of heated crust; or
3. *tectonic loading,* which is depression of the lithosphere adjacent to mountains formed by crustal compression.

As a result of the unique tectonic setting of each basin, basins vary widely in size and shape. However, basins are typically at least tens of kilometers from side to side and often have a circular, semicircular, or elongated trough-shaped geometry in plan view (Einsele, 1992; Miall, 1984).

Important aspects of sedimentary basins in a continental setting are illustrated in Fig. 1. The inset shows basin formation mechanisms, while the larger figure shows a mountainous source area that provides fluid and sediment to an adjacent sedimentary basin. Processes operating in the source area include erosion and the chemical and mechanical *weathering,* or breakdown, of rock into smaller particles. Climate variation in the source area affects the intensity of sediment weathering as well as the magnitude and frequency of storms. Wind, water, and mass movements transport sediment from highland source areas into basins. Additional sediment is produced by biologic activity (the growth of organisms) or chemically precipitated from sea water or lakes. Depositional processes, chemical and biological activity, environmental conditions, and the rise and fall of sea level all affect the spatial patterns of sediment grain size, mineralogical composition, porosity, and permeability that develop within the sedimentary basin fill.

Figure 1 also illustrates the sedimentary rock cycle, a continuous cycle in which the Earth's crust is *uplifted,* or raised in eleva-

3-527-28139-8/96/$5.00 + .50

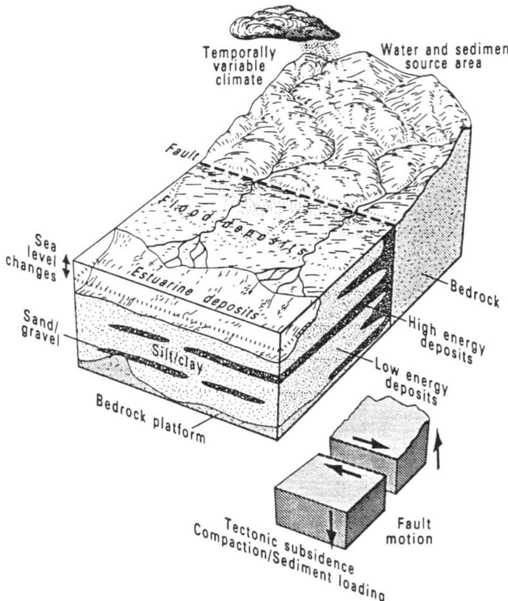

FIG. 1. Example of sedimentary basin-formation and basin-filling processes. Three sets of processes must be considered in the formation and filling of sedimentary basins. Processes in the source area (shown as the mountains) cause weathering and erosion of rock to provide sediment to the receiving basin, into which sediments are transported and accumulate (on the coastal plain and below sea level). Basin-formation processes (shown in the inset) such as tectonics (here shown as motion along faults), sediment loading, and compaction cause subsidence to form a topographic depression. After Koltermann and Gorelick (1992). Reprinted by permission of the American Association for the Advancement of Science.

tion, to form source-area mountains; rocks are weathered and eroded; and sediment is transported and deposited in basins. Sedimentation is not continuous in time. In any given basin, there may be gaps in the sedimentary record of long or short duration, in which either no deposition took place or erosion removed previously deposited sediments. As basin-fill sediments are continually buried by younger deposits, compaction and increasing temperatures as well as precipitation of dissolved minerals from pore fluids cause the sediment grains to cement together and become rock. Physical, chemical, and biological changes to sediments after deposition, collectively called *diagenesis,* include precipitation of cements from circulating subsurface fluids in fractures and pore spaces, and compaction of sediment grains. Diagenetic changes cause sediment particles to *lithify,* that is, to become rock. Folding and faulting of basin-fill sediments can occur during sedimentation as well as after sedimentation ceases. The rock cycle eventually comes full circle, as sedimentary rocks are uplifted into mountains.

Sedimentary rocks contain most of the world's metals, minerals, fossil fuels, and potable underground water supplies. To exploit and manage the Earth's natural geologic resources (groundwater, hydrocarbons, coal, minerals, etc.), we must know the distribution and properties of oil reservoirs, aquifers, and rocks of economic value in the subsurface. Understanding the controls on sedimentary basin formation and filling is critical for successful exploration and production of the Earth's mineral and energy resources as well as prediction of subsurface fluid flow and contaminant migration. Sedimentary rocks also tell us much about the Earth's geologic history, from the progression of life through time to the forces that cause earthquakes.

This article discusses how sedimentary basins evolve; how they are classified, form, and fill with sediment; the typical environments in which sediment is deposited; and common patterns in basin fill. The article is organized around the three components affecting a sedimentary basin and the rock cycle, as illustrated in Fig. 1: the source area, basin-formation mechanisms, and basin-filling processes. First, a brief source-area section discusses the factors affecting fluid and sediment inputs to sedimentary basins. Then, the ways in which sedimentary basins have been classified on the basis of basin-formation tectonics are described. The basin-formation section focuses on mechanisms of basin creation. In the basin-filling section, common depositional environments are first described. Then, the section addresses the transport and accumulation of *clastic* sediment, defined as particles eroded from a previously existing geologic formation, as well as sediment of biological or chemical origins. Chemical and physical processes affecting sediments after deposition, called diagenesis, are then described. A discussion of geologic history analysis—which uses local and regional geologic and climatic data to constrain the structural, sea-level, climatic, and hydrologic histories of a sedimentary basin—is presented. The ways in which sedimentary basin formation and filling and source areas have been modeled mathematically are discussed within each section. Finally, some conclusions are presented. (See also PLATE TECTONICS; STRUCTURAL AND PHYSICAL GEOLOGY; EARTH, INTERIOR STRUCTURE OF THE; VISUALIZATION, SIMULATION, AND MODELING BY COMPUTER.)

1. THE SOURCE AREA

A *source area* is a feature of positive relief that constitutes the drainage area for fluid and sediment transported into a basin. As the source area is uplifted, it is simultaneously eroded, although uplift rates are almost never exactly balanced by erosion. Physical, chemical-mineralogical, and biological weathering of source-area rocks at and just beneath the land surface breaks rocks down into smaller particles, alters minerals into more stable minerals such as clays, forms soils, and dissolves some minerals into water. Source-area *denudation* is defined as the combination of natural processes that result in the progressive lowering of the Earth's surface by weathering, erosion, mass wasting, and transport by wind and water. The rate at which fluid and sediment is transported from a source area into a sedimentary basin is a function of the source-area size, climate, hydrology, and vegetation, and the rock composition. In continental settings, runoff of surface water into streams transports most of the particulate matter and dissolved minerals, but wind and ice can also transport sand and debris, and groundwater transports some dissolved minerals out of source areas and into sedimentary basins. The rocks of the source area are known as the *parent rocks,* or *provenance,* of the basin sediments. The grain sizes and minerals present in the parent rocks are directly reflected in the basin sediments. An extensive discussion of source-area denudation and techniques for estimation of sediment production from source areas is provided by Slingerland *et al.* (1994).

Source-area mountains form when the Earth's crust is deformed as driven by the motion of the giant plates that cover the surface of the Earth, and where volcanic activity creates individual mountains or mountain ranges. Situations in which mountains are built include the coming together or convergence of two continents; the convergence of an oceanic tectonic plate with a continental plate that results in *subduction,* or a diving underneath, of oceanic crust beneath continental crust; the faulting and tilting of large blocks of rock due to crustal extension (stretching); volcanic activity; and the faulting and folding of continental crust in the interior of a continental plate due to compression.

Physical and chemical processes operating in source areas can be explicitly represented to predict the development of stream patterns (called *drainage networks*), the magnitude and frequency of stream flows, patterns of weathering and erosion, and the amount of sediment transported out of the source area. The source area can be mathematically defined through representing the physics of mountain building, called uplift models, as well as through the climatic and hydrologic processes that create drainage networks, cause precipitation in source areas, and transport sediment. These modeling approaches are discussed in Slingerland *et al.* (1994), Koltermann (1993), and Turcotte and Schubert (1982). Models of uplift are mathematical models of tectonic forces that predict areas of high elevation relative to surrounding areas. An example of a model of uplift due to convergence has been presented by Beaumont *et al.* (1991), who explored the role of wind direction and precipitation patterns on erosion of mountain ranges during their uplift. They modeled the physics of two continental plates converging in which rocks were folded and faulted to form a mountain range, which then exerts a tectonic load on the Earth's crust and is subject to fluvial erosion by precipitation that collects on the land surface as streams. They considered the dynamic interaction between tectonic compression, wind (storm) direction, orographic effects, stream drainage-network patterns, erosion, and isostatic compensation (Fig. 2). Modeling the deformation of the continental plates requires assumptions about the relations between compressional forces (the applied stress) and the rate of deformation (the *strain rate*). It was assumed that the lithosphere is a rigid-plastic material that does not deform until a critical threshold stress is

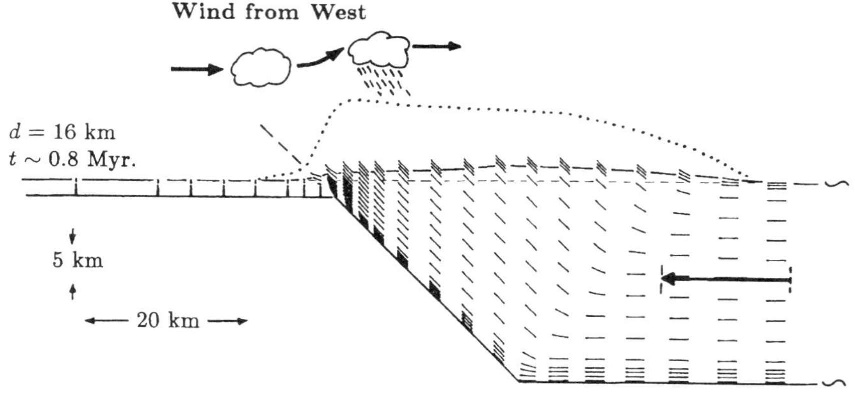

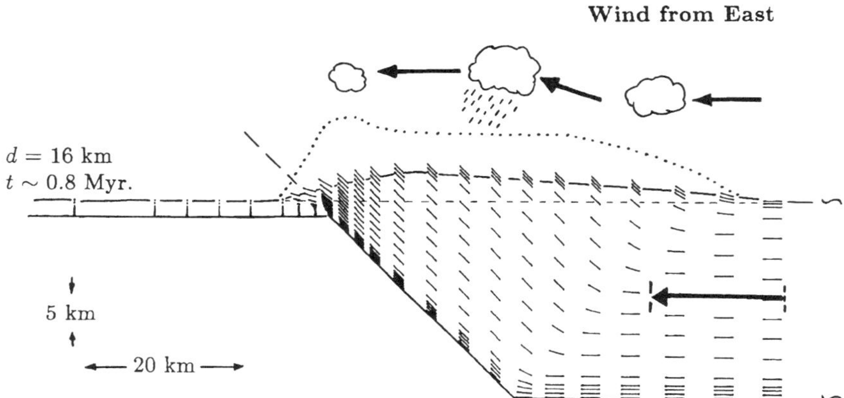

FIG. 2. Model of uplift caused by compressive forces. Results from two simulation experiments with different wind directions (west or east) that affect the spatial distribution of precipitation and therefore the location of fluvial erosion. Both simulations were conducted for 0.8 million years (t = 0.8 Myr) and with the same amount of crustal shortening from tectonic compression (d = 16 kms), shown by the bold horizontal arrows. Small dashed lines are velocity vectors. The dotted lines beneath the clouds show the projected position of the land surface if denudation had not taken place. Model results predict stream-channel network patterns, mountain-range topography, and erosional unloading of the Earth's crust. After Beaumont *et al.* (1991). Reprinted by permission of Chapman and Hall, ITPS Ltd.

achieved. For stress values greater than the critical value, a plastic deformation relation is used, in which the strain rate is proportional to the applied stress. Beaumont *et al.* (1991) represented this situation with the Navier–Stokes equation for conservation of momentum for viscous fluid flow, neglecting the inertial term because over geologic time scales, deformation of rocks is an extremely slow process:

$$\eta_e \nabla^2 \mathbf{v} - \nabla p + \rho_L \mathbf{g} = 0. \qquad (1a)$$

Deformation is assumed to be incompressible:

$$\nabla \cdot \mathbf{v} = 0. \qquad (1b)$$

Here η_e is an effective viscosity, $\mathbf{v}$ is the velocity field (three components), p is pressure, ρ_L is the density of the lithosphere, and $\mathbf{g}$ is gravitational acceleration. Results of this modeling effort show that the mountain topography, orographic effects, and wind direction produce spatially variable precipitation, which preferentially erodes the mountain range and collects in an evolving stream drainage network. Fluvial erosion leads to erosional "unloading" of the mountain range and additional uplift through isostatic rebound, in which the Earth's crust responds to a lightening of the rock "load" on the underlying mantle. The combination of tectonic compression and rebound produces spatial and temporal changes in uplift rates.

A detailed treatment of source-area hydrologic processes is provided by *rainfall-runoff* models, also known as *watershed* or *catchment* models, that represent the physics of hydrologic processes. All aspects of the hydrologic (water) cycle can be considered, including weathering, bedrock and soil erosion, mass movements, hillslope processes, infiltration of water into the subsurface, the uptake of water by plants, overland flow, and flow within stream channels. The amount of precipitation that infiltrates into the soil versus the amount that becomes stream flow is affected by the initial topography, soil types, soil moisture conditions, soil infiltration capacity, soil depth, groundwater flow conditions, groundwater storage properties, and the sequence of storm patterns, durations, and rainfall intensities. An example of a rainfall-runoff model is provided by Willgoose *et al.* (1991), who developed a source-area model that focuses on the long-term evolution of source-area topography and drainage-network development. Other methods of estimation of ancient stream flows from source areas, without explicit modeling of hydrologic processes or uplift, called *paleohydrologic* techniques, are discussed in Koltermann (1993).

2. CLASSIFICATION OF SEDIMENTARY BASINS

Sedimentary basins vary in size, shape, age, morphology, cause of formation, and patterns of basin fill. The location of fluid and sediment sources, rates of subsidence, patterns of sediment deposition, and areas of thickest sediment accumulation also vary. Some basins have been structurally deformed (folded and faulted), while others have been uplifted and eroded to form mountains. Despite the numerous factors that can cause variations between basins, geologists have tended to classify sedimentary basins on the basis of tectonic mechanisms of basin formation, proximity to tectonic plate boundaries, and type of plate boundaries (Dickinson, 1974, 1976; Ingersoll, 1988; Mitchell and Reading, 1986; Busby and Ingersoll, 1995). Table 1 contains an abbreviated basin classification system based on tectonic origin to illustrate some of the numerous basin types, some examples of which are shown in Fig. 3. Basin classification is useful for purposes of comparing common characteristics of basins and causes of basin subsidence, and for understanding the features observed in basins affected by basin-formation mechanisms.

Einsele (1992) points out that while the basin shape and size are controlled by tectonic processes, the morphology of the basin-fill sediment surface is strongly controlled by both tectonic and sedimentation processes. Factors such as locations, rates, and grain-size populations of sediment input and depositional, chemical, and hydrodynamic processes affect the spatial distribution of sediment types and ocean or lake water depths. To address these factors, studies of sedimentary basins account for depositional processes and environmental conditions affecting sedimentation as well as tectonic processes. These latter issues are discussed in the section on basin filling.

3. BASIN FORMATION

Sedimentary basins are formed by a downward sinking of the land surface known as *subsidence*. Three factors contribute to subsidence: tectonics, sediment loading, and compaction. Tectonic subsidence initiates basin formation. Tectonic subsidence is caused primarily by forces that thin the crust and by heating followed by thermal cooling and contraction, and to a lesser extent by compressive forces. As a basin evolves, the weight of accumulating sediments, ice, and water, called *loading*, becomes an additional driving force on subsidence. Tectonics and sediment loading cause subsidence of the basement rock that forms the floor of sedimentary basins as well as subsidence of the sediment surface. *Compaction* operates on the sediments themselves, causing subsidence of the sediment surface but not the

Table 1. Classification of sedimentary basins based on tectonic style [modified from Kingston *et al.* (1983), Mitchell and Reading (1986), and Einsele (1992)].

Basin category	Type of crust	Style of tectonics	Common basin characteristics
Continental or interior sag basins[a]	Continental[b]	Divergence[c]	Basins of regional extent, slow subsidence
Continental or interior fracture basins	Continental	Divergence	Narrow basins, fault-bounded valleys, rapid initial rift-stage subsidence
Basins on passive continental margins, margin sag basins	Transitional[d]	Divergence and shear	Asymmetric basins, outbuilding of sediment, slow subsidence during later stages
Oceanic sag basins	Oceanic[e]	Divergence	Asymmetric basins of regional extent, slow subsidence
Basins related to subduction	Oceanic or transitional/ Oceanic	Convergence[f] or dominantly divergence	Basins can be asymmetric, widely varying depth and subsidence rates
Basins related to collision	Oceanic or continental	Convergence or crustal flexure[g] with local convergence or transform motions[h]	Tendency to increasing subsidence through time, asymmetric basins, rapid sedimentary loading
Strike-slip/wrench basins[i]	Continental and/or oceanic	Transform motion, possible divergence or convergence	Relatively small elongated basins, rapid subsidence

[a]A sag basin is a broad, shallow basin of regional extent, with gently sloping sides.
[b]Continental crust underlies continents and continental shelves.
[c]Divergence occurs when two lithospheric plates move away from one another.
[d]Transitional crust occurs at the boundary between the oceans and the continents.
[e]Oceanic crust underlies oceans. It is thinner, denser, and of lower seismic velocity than continental crust.
[f]Convergence occurs when two lithospheric plates move toward one another.
[g]Flexure is regional bending of the lithosphere due to compression or loading.
[h]Transform motion occurs along transform faults, where plates slide past one another.
[i]A strike-slip or wrench basin occurs along transform margins.

basin floor. Subsidence occurs both above and below sea level, forming nonmarine and marine sedimentary basins, respectively. Subsidence rates can range from 10 to 1000 meters per million years.

3.1 Tectonics

Tectonic (or thermomechanical) subsidence mechanisms are driven by heat production from radioactive elements within the Earth. Horizontal motions of the Earth's crust drive vertical motions caused by changes in crustal thickness, temperature, and loading or unloading of the crust. Vertical motions form sedimentary basins and uplift source areas, and alter the path of sediment transported into and throughout basins. Tectonic mechanisms of basin formation are (Busby and Ingersoll, 1995)

1. *crustal thinning,* primarily caused by crustal extension (stretching or pulling apart) and rifting (large-scale faulting and tilting), but also caused by magmatic withdrawal;
2. *cooling and contraction* of lithosphere that was previously heated by underlying hot asthenosphere;
3. *tectonic loading,* in which subsidence occurs as the lithosphere compensates for the weight of mountains formed by crustal compression;

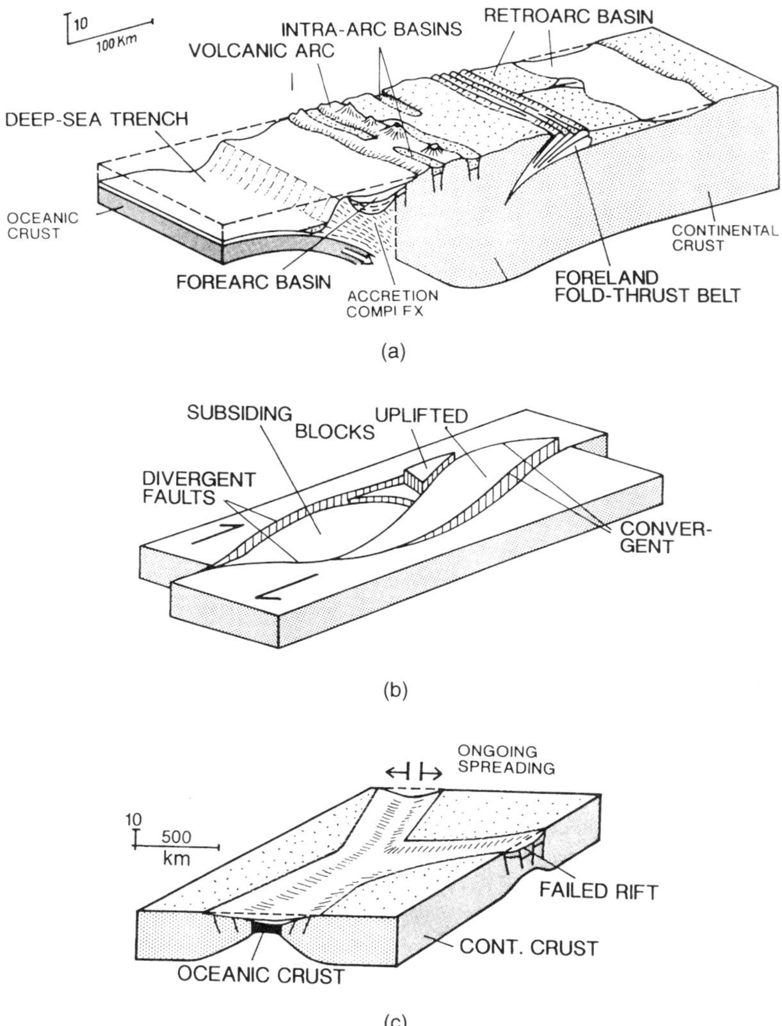

FIG. 3. Tectonic basin classification examples. (a) Continental margin basins formed at a *convergent* tectonic plate margin. This margin is subject to subduction as an oceanic plate dives beneath a continental plate, forming a deep-sea trench offshore, a string of volcanos (called a *volcanic arc*) onshore and parallel to the trench, and numerous associated sedimentary basins. A basin forms in front of the volcanic arc system (*forearc* basin). Sediments beneath this basin (the *accretion complex*) have been scraped off of the subducting oceanic plate and pile up (accrete) against the continental plate. Small basins form within the volcanic arc system from faulting, called *intra-arc* basins. Farther inland, convergence has caused faulting and folding, forming a mountain range known as a *foreland fold-thrust belt.* An elongate basin develops adjacent and parallel to the foreland fold-thrust belt, called a *foreland basin,* or here, a *retroarc* basin because it is behind the volcanic arc system. (b) Strike slip/wrench basins form along *transform* margins, where two tectonic plates slide past one another along a fault or a system of closely spaced faults. Any kinks in the fault(s) or splitting or joining of faults produces compressive or extensional forces that form sedimentary basins, where some blocks are uplifted while others subside. (c) A rift-zone basin, formed at a *divergent* tectonic plate margin. In this example, a new ocean basin floored by hot oceanic crust, shown in black, is being formed at a *spreading center,* along two arms of the rift zone, while a third arm has ceased to spread, and so is known as a *failed rift.* After Einsele (1992). Reprinted by permission of Springer-Verlag New York, Inc.

4. *volcanic loading,* in which subsidence occurs to compensate for the weight of a growing volcano;
5. *subcrustal loading,* in which the lithosphere experiences subsidence due to bending (*flexure*) from thickening due to faulting in the lithosphere caused by compressive forces;
6. *asthenospheric flow,* primarily caused by the descent of a subducting plate; and
7. *crustal densification,* in which the density of the crust increases because of changing pressure and temperature conditions, and the increased density causes sinking.

Each plate tectonic setting has its own combination of controls on basin formation and subsidence (Dickinson, 1974, 1976). Extension and thermal contraction and cooling are characteristic of locations where tectonic plates diverge, such as rift zones. Tectonic and subcrustal loading is characteristic of locations where tectonic plates converge, such as subduction zones. In plate interiors and where plates slide past one another along large-scale faults, called transform faults, subsidence is caused by a combination of extensional and compressional mechanisms.

Crustal extension that produces rifting is the first stage in the formation of an ocean basin [Fig. 3(c)]. Rifts are long, narrow troughs bounded by faults that form because crustal extension ruptures the crust. Rifting occurs on continents or on thinned continental crust beneath oceanic waters. Extension of the lithosphere causes thinning of the crust, an upward vertical movement (*upwelling*) of upper mantle material, and a change in the lithosphere temperature profile (Fig. 4). The result of mantle upwelling is high temperatures brought close to the mantle–crust boundary and the land surface, an increased temperature gradient, some thermal expansion of thinned crustal rocks, and a downward isostatic adjustment of the crust. The thinned crust is riddled with faults and subject to rapid initial subsidence, forming a sedimentary basin. After extension ceases, heat is conducted upward and lost to the atmosphere, and the rocks cool, increase in density, and contract, causing thermal subsidence in addition to the subsidence caused by extension and rifting.

Thermal contraction and cooling causes thermal subsidence, controlled in part by the thermal properties of the *lithosphere,* a layer of strength that includes the Earth's crust and the upper part of the mantle, approximately 100 km in thickness. Thermal subsidence always follows heating of rocks. For example, the eruption of molten rock at plate boundaries, such as the mid-Atlantic ridge, forms a mountain range of high-temperature rocks. As the rocks cool, they contract, increase in density, and consequently subside. Cooling and contraction takes tens of millions of years; therefore, thermal subsidence is a very long-term process, commonly represented in basin-formation models by a decay-type equation.

A more rigorous modeling approach was taken by McKenzie (1978) and Jarvis and McKenzie (1980), who developed time-dependent analytical models for the dependence of heat flow and thermal subsidence on the rate and amount of crustal stretching during extension and rifting. The time history of the thermal anomaly developed beneath the rifted area is represented by a heat-flow equation:

Pre-Extension Configuration

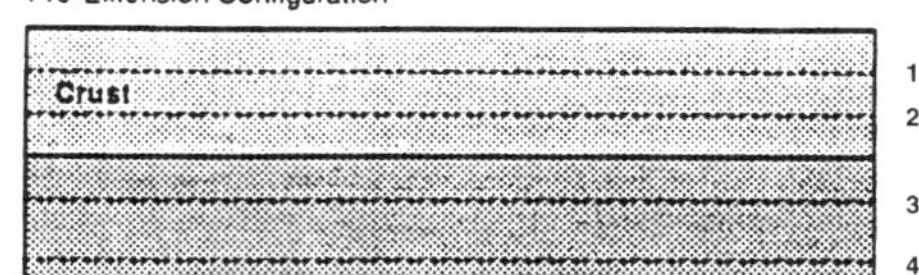

(a)

Post-Extension Configuration

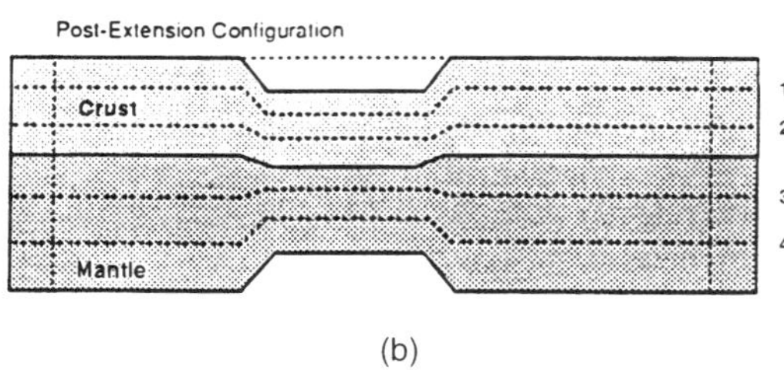

(b)

FIG. 4. Extension, rifting, and subsidence. (a) Pre-extension configuration of crust overlying mantle. Temperature isotherms are labeled 4 (hottest) to 1 (coolest). (b) Post-extension configuration of crust overlying mantle. Extension caused initial subsidence, an upwelling of mantle material, and hotter temperatures brought closer to the land surface. Gradual cooling of the heated rocks causes contraction and additional subsidence. Modified from Slingerland *et al.* (1994). Reprinted by permission of Prentice-Hall, Upper Saddle River, New Jersey.

$$\frac{\partial T}{\partial t} + G(Z - z)\frac{\partial T}{\partial z} = k\frac{\partial^2 T}{\partial z^2}, \tag{2}$$

where T is temperature, t is time, k is the thermal diffusivity, Z is the elevation of the land surface, z is the elevation above the lithosphere–asthenosphere boundary, and G is the vertical velocity gradient.

Crustal *compression* results in an increase in crustal thickness as rocks are folded and faulted on top of their neighbors to form mountains. This increased weight of rocks on the Earth's crust, called *tectonic loading* of the crust, causes subsidence in front of the mountain belt, which, together with subsidence caused by crustal compression, forms a basin adjacent to the mountain range. Subsidence adjacent to loads of rock rather than just beneath such loads is known as *flexural* subsidence. The long, trough-shaped sedimentary basins formed in front of mountain ranges as a result of tectonic loading of rocks in the mountain range are known as *foreland* basins. Jordan (1981) used a model of flexural subsidence to show that tectonic loading coupled with erosion and redistribution of the mountain range can produce foreland basins that are hundreds of kilometers in width. Similarly, subcrustal loading occurs because of a thickening of the lithosphere due to faulting caused by compressive forces. The increased weight of the lithosphere is compensated for by subsidence. Regardless of the cause of the loading, models of loading account for both local and regional subsidence, as discussed in the next section.

3.2 Loading of the Lithosphere

Subsidence due to loading on the Earth's crust has been represented as either local *isostatic compensation* or *flexural bending* of the lithosphere. Mathematical models of isostatic and flexural subsidence are presented by Turcotte and Schubert (1982). Isostatic compensation assumes that the mass of sediments, lithosphere, and water above a certain depth of compensation within the mantle are balanced by an equal pressure from below (Fig. 5). The lithosphere, sediments, water, and ice are treated as one-dimensional vertical loads directly above the corresponding mantle. The isostatic response of the crust can be calculated

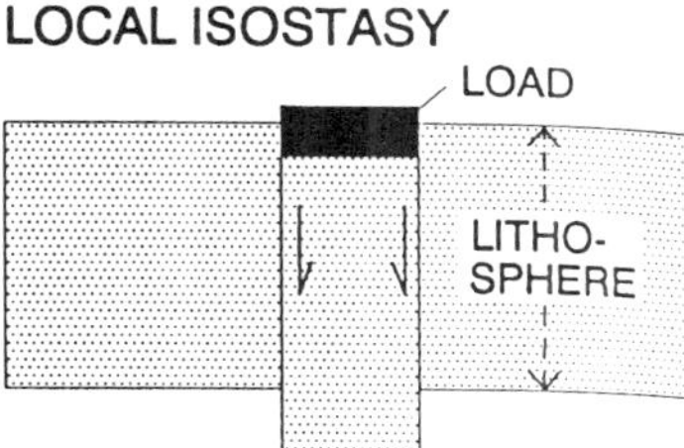

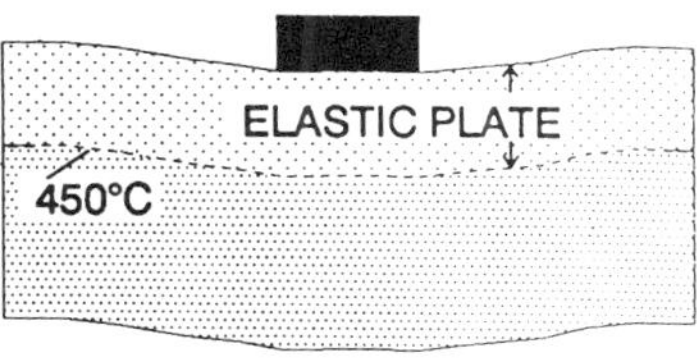

FIG. 5. Isostatic and flexural loading. Local isostatic response of the lithosphere to a load occurs directly beneath the load. Regional flexural response of the lithosphere to a load causes less subsidence beneath the load but greater subsidence over a broad area. After Einsele (1992). Reprinted by permission of Springer-Verlag New York, Inc.

with an Airy model, which assumes that the sediment load is entirely compensated at the location of the load when the load occurs (Turcotte and Schubert, 1982):

$$s = \frac{(\rho_m - \rho_w)}{(\rho_m - \rho_s)}(W_0 - W_f), \tag{3}$$

where ρ_m is the density of the mantle, ρ_s is the density of sediment, ρ_w is the density of water, s is the thickness of the sediment load, and W_0 and W_f are the initial and final water depths.

In addition to the local deformation of the lithosphere beneath loads, the lithosphere adjacent to loads responds to either loading or unloading by subsidence or uplift, respectively. This is because the lithosphere has a strength, and loads from accumulating sediments or growing highlands are compensated regionally by flexure (Fig. 5). The strength of the lithosphere and the manner in which it deforms affect the distance over which loads are compensated and the resulting sedimentary basin geometry. Flexural models mathematically describe the strength of the crust to propagate and support loads of sediment and rock regionally. Flexural models typically represent the lithosphere as an elastic beam that has a specified flexural rigidity, in which deformation is recoverable and the mantle (treated as a denser fluid) supports the lithosphere and modifies its downward flexure (Turcotte and Schubert, 1982):

$$D\frac{d^4w}{dx^4} = q(x) - P\frac{d^2w}{dx^2}, \tag{4a}$$

$$D = Eh^3/12(1 - n), \tag{4b}$$

where D is the flexural rigidity, a measure of the elastic strength of the lithosphere and a function of the elastic constants of the rock, i.e., E (Young's modulus) and n (Poisson's ratio); h is the elastic thickness of the lithosphere; $q(x)$ is the downward force per unit area; P represents horizontal forces; and w is the vertical deflection. The thickness of the elastic lithosphere, h, is estimated either by a strength envelope, which designates the depths at which the lithosphere behaves in a brittle and ductile manner with increasing stress, or by the depth of the 450 °C isotherm, which has been found to produce a good approximation to the base of the elastic lithosphere. The strength and flexural rigidity of the lithosphere are temperature dependent, with both increasing as the lithosphere cools.

3.3 Compaction

The third subsidence mechanism, compaction, occurs as sediments are buried and subject to pressure from the increasing weight of overlying sediments. Compaction is the reduction in bulk volume, thickness, and void space of a volume of sediments. It is a post-depositional, or diagenetic, change to the original sediments. The amount of compaction varies with grain size, shape, and composition as well as with the initial porosity and with increasing pressure as sediments are buried under increasing loads of younger sediments. For example, clays compact more than sand grains. Clays often have extremely high porosities of 70–80% when first deposited, but they can compact to have porosities that are less than half of the original value. Although compaction is typically the least important subsidence mechanism in terms of magnitude, differential subsidence in sands and clays due to differing amounts of compaction strongly affects the land surface topography and consequently affects stream migration patterns. Compaction mechanisms include grain rearrangement to make a tighter particle packing; dissolution of minerals at grain-to-grain contacts because of pressure, called *pressure solution;* grain deformation and fracturing; and the release of water bound to clay minerals. These mechanisms are typically represented empirically in basin-formation models with curves showing how porosity decreases with increasing depth of burial for different sediment types.

Subsurface fluid pressures also exert control on compaction. Most curves of porosity versus depth of burial are developed from field data under hydrostatic pressures; i.e., pore fluids expelled during burial are allowed to escape through the overlying sediments, so that there is an equilibrium between increasing pressure from sediment accumulation, compaction, and pore-water expulsion. However, rapid deposition of low-permeability fine-grained sediments inhibits pore-fluid expulsion. Then, the pore fluid bears some of the weight of the overlying sediment pile, producing an overpressured zone, where pore-water pressures are greater than hydrostatic. In these zones, compaction is reduced from expected values. This situation has been modeled by Bethke *et al.* (1990), who focused on overpressured zones in the Gulf Coast of the United States.

4. BASIN FILLING

Basin filling is strongly controlled by factors governing basin formation, as well as the source area, sea-level changes, energy conditions within the basin, and conditions conducive to the formation of chemical and biological sediments. As a basin fills, the horizontal and vertical land-block movements caused by thermal or mechanical deformation of the Earth's crust, sediment loading, and compaction provide a continuously moving land surface upon which sediments accumulate. The size, shape, and subsidence rates of the basin sediment container control how much of the sediment influx is trapped within a basin and where it accumulates. The basin geometry and rates of subsidence versus sedimentation dictate whether the sediment pile is preserved or whether there are erosion-caused gaps in the sedimentary record. The basin-fill characteristics strongly depend on the rate at which sediment is received and the composition of the sediment. The source area controls the volume and grain-size distribution of sediment and the dissolved mineral load supplied to a basin. In depositional systems above sea level, the dissolved mineral load that infiltrates into the groundwater flow system plays an important role in post-depositional, or diagenetic, changes to sediments, such as the chemical precipitation of minerals in the pore spaces. The location of the shoreline and energy conditions above and below sea level affect the transport and deposition of different sediment grain sizes and biological activity. While the basin-formation processes of uplift and subsidence are critical to the production of clastic sediments and development of the basin container, the basin-filling processes of sediment erosion, transport by water and wind, deposition, water circulation, and chemical and biological activity control most processes occurring within a basin. All of these factors work together to create spatial patterns in different sediment types throughout basin fill.

The spatial patterns and the fossil content of basin-fill sediments are diagnostic of the depositional and environmental conditions under which sediments were deposited, called *depositional environments.* Each depositional environment has a unique set of physical (especially hydrodynamic), chemical, and biological conditions and particular depositional processes, thus producing a characteristic sedimentary deposit (Reineck and Singh, 1980). Within each depositional environment, sedimentary *facies* can be distinguished, defined as the "sum of all the primary characteristics of a sedimentary unit" produced as a result of deposition under given environmental conditions (Reineck and Singh, 1980). These common characteristics include fossil content, sediment chemistry, grain-size distribution, and small-scale sedimentary features. Common characteristics of rocks and sediments are compiled in *facies models,* which visually show, through maps, cross sections, block diagrams, and one-dimensional vertical columns, the typical configuration of sediments deposited within each depositional environment. Facies models are used to interpret and predict the grain size, mineralogical composition, and spatial arrangement of stratigraphic units. Of special interest to earth scientists interested in extraction of hydrocarbons, groundwater, and contaminants from the subsurface are the sand- and gravel-body geometries predicted by facies models, because interconnected sands and gravels produce connected high-permeability pathways for fluid flow.

4.1 Depositional Environments and Facies Models

Depositional environments and facies models have been described in detail by many authors (for example, Reineck and Singh, 1980; Blatt *et al.,* 1980; Reading, 1986; Einsele, 1992). A brief description of common environments based on these references is provided here. Here, depositional environments are divided into the following general categories:

1. nonmarine (fluvial, glacial, desert, and lacustrine);
2. coastal and shallow-sea sediments (beaches, estuaries, deltas, and continental shelf); and
3. marine (continental margin, slope, and deep ocean basins).

It should be noted that sediment can be transported from one depositional environment to another, such as from source-area highlands on continents to marine basins. In addition, sediments of volcanic origin occur in many depositional environments.

4.1.1 Nonmarine Sediments Nonmarine depositional environments are those occurring on land or in surface-water bodies located above sea level. These sediments can be of stream, wind, lake, volcanic, or swamp origin. Sediments are deposited under fluvial, glacial, lacustrine, and desert conditions. *Fluvial* sediments are produced by the action of streams or rivers that erode, transport, and deposit sediment from highlands to lowlands. *Glacial* sediments are deposited by the action of ice or glaciers, while *lacustrine* deposits are deposited in or formed in a lake. Within desert environments, *eolian* sediments are transported and deposited by the wind. Commonly, fluvial, glacial, and eolian sediments are not conducive to the preservation of fossils. A remarkable exception to this statement is the recent discovery of a fossil nesting dinosaur, whose preservation is attributed to burial in a sandstorm (Norell *et al.,* 1995).

Rivers are the most important geomorphic agent in transporting fluid and sediment from highland source areas on the continents to lakes and the ocean. The most notable feature of fluvial systems is the channel pattern, that is, the geometry of channels in map view. The number of channels and their width, depth, slope, sinuosity, and type of sediment load are diagnostic. *Sinuosity* is a measure of the length of a stream channel versus the length of the valley it flows through. Sinuous channels have many bends; straight channels have few or no bends. Sediment load is composed of bedload and suspended load, the proportions of which are affected by sediment availability and stream hydrodynamics. Streams transport the larger, heavier grains, such as gravel and sand, as *bedload,* by rolling, bouncing, and sliding on the stream bottom. In contrast, finer-grained sediments, such as silts and clays, are transported in suspension within the water column by turbulence and called *suspended load.* Fluvial depositional environments include meandering rivers, braided streams, anastomosing rivers, and alluvial fans (Fig. 6). Within each fluvial depositional environment, there are subenvironments, including the channel, its levees, and flood-plain sediments.

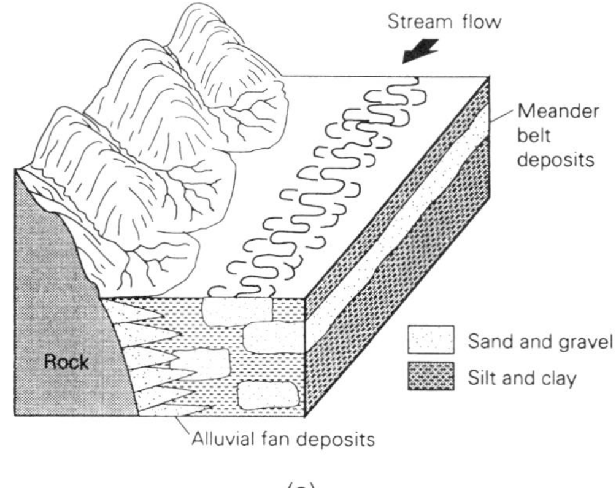

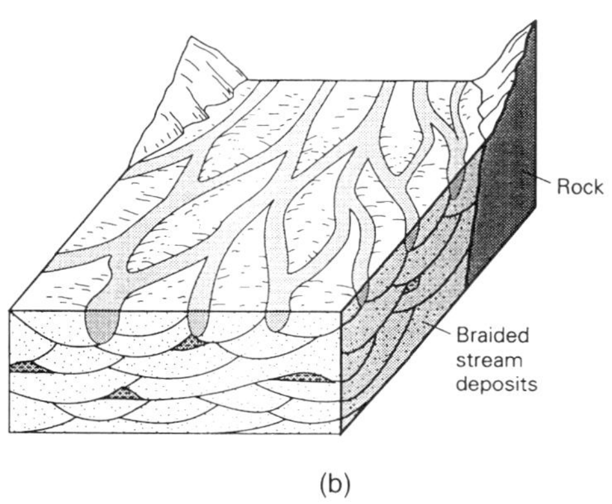

FIG. 6. Fluvial depositional environments. (A) Sands deposited within alluvial fans and by a meandering stream. The wedge-shaped alluvial-fan sands are adjacent to the mountain range, while the sands deposited from the meandering stream lie within and are parallel to the valley axis. (B) Sands from braided-stream deposits with little fine-grained sediment. Sands appear as sheets, ribbons, wedges, or tabular elongate bodies. The legend to the right of (A) also applies to (B); that is, sand and gravel are shown as lightly stippled, while silt and clay are shown as dark-stippled areas. Modified from Allen (1965). Reprinted by permission of Blackwell Science Ltd.

Meandering rivers have one highly sinuous primary channel. These rivers typically form where the land surface slope is low, the stream banks are composed of cohesive materials, the flow discharge is relatively steady, and the suspended-load/bedload ratio is high. When a meandering stream overflows its banks, suspended load in the flood waters is transported out of the stream channel to settle out over the flood plain. Coarse-grained sediments remain behind, transported on the stream channel bottom. The sediments deposited by meandering streams are coarse-grained stream channel deposits embedded in fine-grained flood-plain deposits. In areal view, a *braided* stream looks like a wide, shallow channel containing many interwoven small channels separated by small sand and gravel islands. Channels repeatedly divide to flow around the islands and then rejoin. Channels shift position frequently. Braided streams typically transport coarse-grained bedload sediments, have a steep stream-bottom gradient, and are relatively straight and wide. In sedimentary deposits formed by fluvial processes, the channel type and proportions of bedload versus suspended load affect the geometry of the channel deposits preserved in the sedimentary record (Fig. 7). *Alluvial fans* occur where a stream flows out from a steep, narrow mountain canyon onto a lower-gradient valley or coastal plain, depositing sediment in a characteristic shape that looks like a segment of a cone, where the apex of the cone is at the highest elevation. The fan shape is produced by the decrease in stream power as the gradient abruptly decreases, causing rapid deposition of transported sediments. The flows can be either sediment-laden water flows or debris flows. An alluvial fan that extends into an ocean or a lake is called a *fan delta.*

Glacial sediments are formed by the action of ice and occur both on the continents and in shallow marine waters. The most common sediments in this environment are

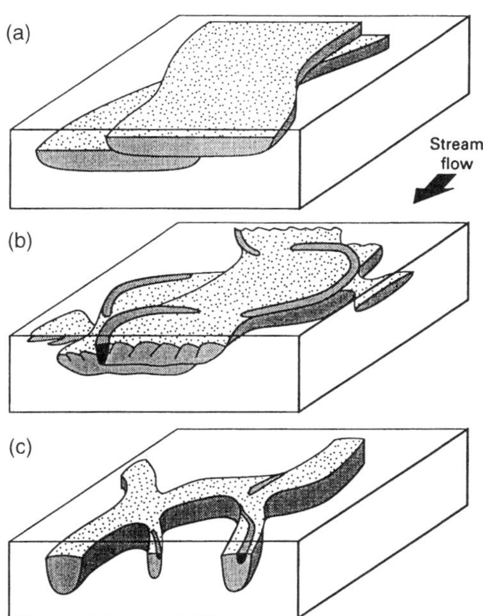

FIG. 7. Fluvial sand-body geometries. A schematic of the three-dimensional geometry of sand bodies typically seen in (A) bedload, (B) mixed-load, and (C) suspended-load fluvial systems. The channel type and proportions of bedload versus suspended load affect the geometry of the channel deposits preserved in the sedimentary record. Modified from Galloway and Hobday (1983). Reprinted by permission of Springer-Verlag New York, Inc.

produced by continental ice sheets, which transport debris from upland source areas, erode sediment at the ice sheet base, and deform the underlying rock. When the ice melts and the glacier retreats, the transported debris is dropped. In contrast to fluvial deposits, the processes of sediment transport within, beneath, and in front of ice sheets do not separate sediments by grain size. Deposits from glaciers are a mixture of grain sizes ranging from boulders to clay-sized sediments, with large boulders or *clasts* appearing to occur at random in a matrix of silt and clay-sized material. Sediments deposited by glacial action are called *till.* The morphology of glacial deposits is indicative of depositional processes. For example, *moraines* are elongated hills that occur at the sides and front of glaciers resulting from debris pushed at the edges of the ice. *Drumlins* and *eskers* form beneath a glacier where there are fissures in the ice. They are teardrop-shaped hills and narrow ridges on the land surface, respectively, left behind after a glacier has melted. The large volume of water produced by a melting glacier forms *meltwater streams* in front of glaciers that transport sediment. *Glacial outwash* deposits occur adjacent to ice sheets and are primarily composed of gravels and coarse sands transported short distances. As meltwater streams flow farther from a glacier, they deposit *glaciofluvial* sediments, with characteristics similar to alluvial fans and braided streams.

Lacustrine sediments, or sediments deposited in lakes, include both clastic sediments transported into lakes and sediments created by biological or chemical activity within lakes. Lakes typically contain fresh water, but some have salinities greater than that of sea water. Incoming streams carrying clastic sediment are affected by the density difference between the lake water and the incoming mixture of water and sediment transported by a stream. Flows denser than lake water occur along the lake bottom, flows of similar density mix within the water column, and less dense flows float at the lake water surface. Coarse-grained materials are primarily deposited near the shoreline, with finer-grained materials deposited successively farther from the shoreline. *Varves* are alternating, horizontal, thin sequences of lacustrine sediment reflecting settling from suspension of fine-grained particles during seasonal changes in temperature or from seasonal precipitation of minerals caused by differences in evaporation. Typically, summer layers are light-colored, and winter layers are dark-colored. The proportion of lake sediments derived from biological activity is strongly controlled by the lake water chemistry and the influx of nutrients. In particular, the salinity content and chemical composition of lakes exert strong control on biological activity and therefore on sediments composed of fossil plant and animal remains. Lacustrine sediments of chemical origin provide deposits of salts that are mined for their economic value, including carbonates, sulfates, chlorides, borates, and nitrates.

Desert environments occur where rainfall is insufficient to produce much vegetation, mean annual evaporation far exceeds precipitation, and sediment transport by wind is significant. Eolian deposits are composed of wind-blown particles, typically fine to medium sand and silt. During transport, silt and some sand particles are carried in suspension within turbulent air, while larger, heavier sand particles may be bounced or rolled along the land surface in a process called *saltation*. During deposition, the action of wind blowing on dry sediment particles produces ripples, dunes, ridges, and large hills characteristic of eolian sediments (Fig. 8). While eolian deposits are typically thought of as occurring only in deserts and on sandy beaches, they also occur within fluvial environments. When rainstorms occur in the desert, they are often violent downpours. Coupling of intense rain with a lack of vegetation produces sediment-laden flash floods and debris flows that produce fan-shaped sedimentary deposits in desert valleys.

4.1.2 Coastal and Shallow-Sea Sediments Coastal and shallow-sea environments include beaches, tidal flats, lagoons, estuaries, deltas, and continental shelves. Sediments deposited in shallow coastal waters and on the continental shelf are affected by ocean energy conditions as well as by processes that control the location, timing, volume, and grain-size distribution of sediment influx. Waves, tidal action, and bottom and surface currents serve to redistribute and rework sediments pouring in from continental sources, removing finer-grained ma-

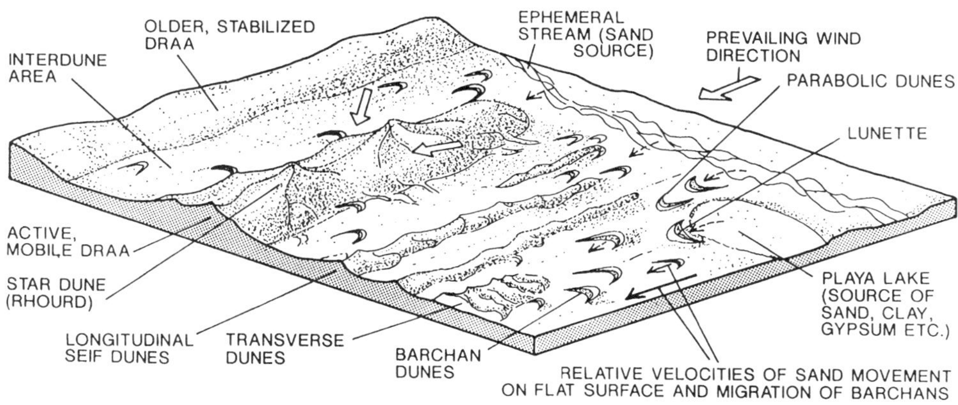

FIG. 8. Eolian depositional environment. The action of wind transporting sediment particles produces many different types of dunes and *draas*, which are very large-scale dunes. In eolian environments, infrequent rains produce infrequent (ephemeral) streams and *playa lakes* that produce evaporitic minerals as they dry up. *Barchan dunes* are isolated, crescent-shaped sand dunes that lie transverse to the prevailing wind direction and are convex in the upwind direction. The wings or horns of barchan dunes point downwind. Barchan dunes can grow to heights of more than 30 m and widths up to 350 m from horn to horn. They form where the sand supply is limited and the wind is constant and of moderate velocity. In contrast, *parabolic dunes* have a long, scoop-shaped (parabolic) form that is convex in the downwind direction, so that its horns point upwind. Parabolic dunes are commonly covered with sparse vegetation and occur along coastal areas where strong onshore winds supply abundant sand. After Einsele (1992). Reprinted by permission of Springer-Verlag New York, Inc.

terials and transporting them into deeper water. Wave energy and the *longshore currents* produced by waves (currents along the coastline) produce commonly observed features, such as cliff erosion, barrier islands, spits, and lagoons. Storms, particularly those with large waves, alter the sediments and sediment geometries produced under normal depositional conditions.

The coast forms the transition between marine ocean and nonmarine continental conditions. Coastal sediments are typically sand but occasionally gravel. Finer-grained silts and clays are transported in suspension into the ocean, whereas the coarser-grained sand and gravel bedload transported by streams is deposited at and near the shoreline. Wave energy and shallow marine currents along the coast redistribute sediment deposited by rivers and exert control on beach morphology. Beach sand can also be reworked by the wind into coastal dunes.

Estuaries are the funnel-shaped seaward end of rivers where fresh water mixes with sea water. Estuary salinity and water level is affected on a daily basis by tidal fluctuations and mixing between fresh water and sea water. The water in an estuary is often stratified, with denser sea water flowing beneath less dense fresh water. Sediments deposited in estuaries are typically clays to sands of both marine and fluvial origin mixed with decomposed organic matter.

Deltas occur where a stream enters a body of standing surface water, depositing sediment that forms a large expanse of land of low relief that builds out into a lake or the ocean. Formation of a delta requires a high sediment load transported in the stream plus subsidence in the location of deltaic sediment deposition. A portion of the delta is above lake or sea level, while the rest is below. Commonly, the areal shape of a delta resembles a segment of a cone, or a triangle; however, the shape is affected by tidal action, coastline shape, the degree of sediment transport along the coastline, and wave action. The stream flowing over the delta branches into numerous smaller streams, called *distributaries.* Marine deltas occur in shallow ocean waters along coastlines where the sediment supply is greater than the amount of sediment waves and currents can erode. When the volume of sediment supplied to a delta is large, the delta builds out (*progrades*) into the standing body of surface water, producing a characteristic sediment geometry in vertical section. The geometry of delta bedding includes relatively flat-lying, usually coarse-grained sediments at and above the lake or ocean (*topset* beds); steeply inclined sandy sediments from near the water surface to near the base of the slope (*foreset* beds); and gently inclined fine-grained sediments at the bottom (*bottomset* beds). Sea-level fluctuations affect the position of the shoreline and, thus, the position of finer-grained marine versus coarser-grained nonmarine deposition and sediments of biological origin. The relative contributions of energy from waves, tides, storms, and marine currents affect the geometry of coarse- versus fine-grained sediments within deltas (Fig. 9).

The common features and depositional environments of a common continental margin are shown in Fig. 10. The *continental shelf* refers to the gently sloping underwater zone from the shoreline down to about 200 m ocean water depth. The magnitude of shelf currents caused by tides, density differences, and waves is highly variable. Currents can distribute high volumes of sediment on the shelf and produce large, migrating ripple structures on the sea floor. Large storms also play a significant role in distributing shelf sediment. Commonly, sediment grain size varies as the water depth increases, from sands at the coast, to clayey silt to silty sand in a transition zone that starts just below the wave base, to mud in the deepest waters of the shelf. Where currents are not strong, there is biological activity. Sediments are *bioturbated* (churned up) by the action of bottom-dwelling organisms, often obliterating sedimentary features formed during sediment deposition.

4.1.3 Marine Sediments There are numerous marine depositional environments, affected by water depth, temperature, salinity, and energy conditions. Beyond the continental shelf lies the *continental slope,* which has a much steeper gradient than does the shelf. The base of the slope lies in deep water and is called the *continental rise.* Sediments on the continental rise are largely from material that has slumped off the continental slope, called *slope apron deposits,* or sediments transported as density currents

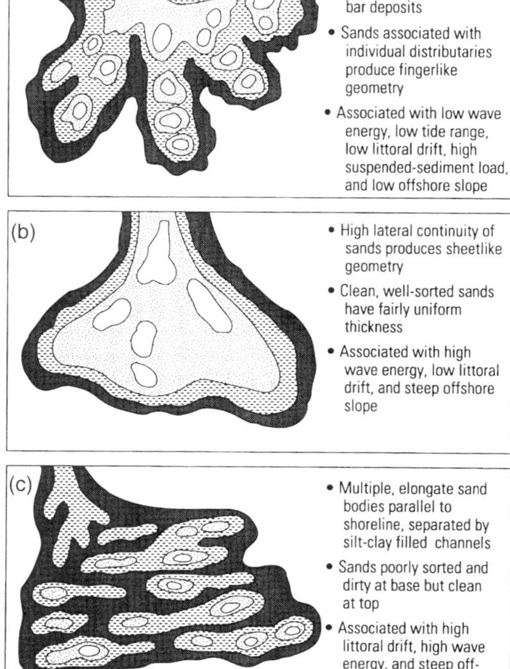

FIG. 9. Deltaic depositional environments. The relative contributions of energy from waves, tides, storms, and marine currents affects the geometry of coarse- versus fine-grained sediments within deltas. Modified from Coleman and Wright (1975). Reprinted by permission of Houston Geological Society.

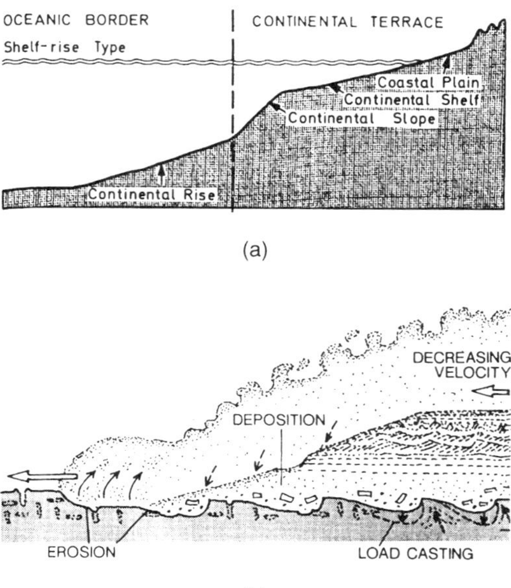

FIG. 10. Marine depositional environments and processes. (A) The common features and depositional environments of a common continental margin include the continental shelf, continental slope, continental rise, and abyssal plains (or ocean basins). After Curray (1969). Reprinted by permission of the American Geological Institute. (B) Turbidity currents are a common sediment transport mechanism in the ocean. As a turbidity current travels down the steeply dipping continental slope and reaches the more gently sloped continental rise and ocean basin, its velocity decreases and deposition starts, here shown as the classic depositional sequence (*Bouma* sequence). Deposition from a turbidity current onto unconsolidated mud produces *load casts,* formed by partial sinking of the deposited materials into the mud. After Einsele (1992). Reprinted by permission of Springer-Verlag New York, Inc.

through submarine canyons to form deep-sea fans. The deepest depositional environments are the *abyssal plains,* or *ocean basins,* which lie beyond the continental rise. Ocean basins are monotonous in their extent and usual lack of topographic features, except for occasional volcanic cones. Their depth is exceeded only by the depth of deep-sea trenches, which are about 2 km deeper than the surrounding ocean floor. Areas of higher elevation within marine environments include sediments deposited on top of seamounts and submarine ridges or platforms. In these locations, sediment sequences are typically thin and composed largely of sediments of biological or chemical origin. Although the continental shelf, slope, and rise type margin has been discussed here, it should be noted that the margins of some continents do not have a well-developed shelf, slope, and rise. Rather, deep-sea trenches, folded and faulted rocks, or submerged steep cliffs are present.

Oceanic sediments are composed of materials transported into the sea by water, wind, or mass movements as well as by sediments of chemical or biological origin produced within the sea. By far, rivers are the greatest contributors of sediment to the oceans. However, continental sediments delivered to the oceans by rivers can be efficiently transported and deposited far from their source areas and at substantial depths within the oceans. While continental margins accumulate sediment largely from deposition by flowing sediment/water mixtures, the portions of ocean basins farthest from the continents accumulate sediments primarily by vertical settling of very fine-grained sediment. Sediment can be redistributed into deeper water by the force of gravity or transported long distances by ocean currents. For example, slope failure on the continental shelf produces mass movements that travel at high velocities down submarine canyons to form deep-sea fans. Transport of coarse sediment long distances into deep ocean waters occurs by *turbidity currents,* which are bottom-flowing density currents containing high concentrations of suspended sediment that travel at high velocities and spread areally when not confined by an underwater canyon. Turbidity currents occur in lakes as well as the ocean. High volumes of sediment are also transported downslope by large-scale gravity slides and slumps. Salinity, temperature, and water pressure affect the density of ocean waters and thus exert control on ocean circulation. Denser water sinks and flows beneath less-dense water to form an ocean bottom current. The result of these density currents is that a global pattern of ocean circulation is caused by the temperature difference between the equator and the poles. The other strong control on ocean circulation is global wind patterns that control wind-driven surface currents.

Sediments of biological origin commonly occur in sedimentary basins with low sedimentation rates, because the growth of organisms is adversely affected by high volumes of clastic sediments. Sediments of biological origin are produced either within the water column (*planktonic*) or on the sea floor (*benthonic*). To thrive, each organism has its own set of required environmental conditions regarding water salinity, temperature, depth, and energy conditions. For example, formation of *carbonate deposits* of biological origin depends on conditions favorable for carbonate-producing marine life, including warm water temperatures, sunlight, and a limited influx of gravel, sand, silt, and clay. Warm, calm, shallow marine waters provide a good environment for the development of carbonates. Carbonates of biological origin are the skeletons or shells of carbonate-producing animals that live in warm, shallow marine waters, typically at depths less than 10 to 20 m and in low latitudes. A reef is an example of a carbonate deposit formed from the structures of frame-building animals such as corals. Sediments formed by slow vertical settling of particles of biological origin include muds containing significant amounts of continental, volcanic, or calcium carbonate materials, oozes, chalk, limestones, and clays. For example, carbonates can result from the skeletons of carbonate-producing animals such as *foraminifera* that settle to the sea floor after the animal dies. Other sediments of biological origin produced in the ocean include the siliceous skeletons of organisms such as diatoms, sponges, and radiolaria. These siliceous skeletons accumulate to form bands and layers of a colorful mineral called *chert.* The proportion of deep marine sediments attributed to the continents or volcanic origins versus the proportion derived from organic skeletal remains of micro-organisms is diagnostic of depositional environments. Sedimentary deposits in which less than 25% of the coarse fraction is composed of continental, volcanic, or larger organisms of continental shelf origin are typically located beyond the continental margin and on the abyssal plain. In contrast, sedimentary deposits in which more than 25% of the coarse fraction is predominantly composed of continental, volcanic, and skeletons of shallow-water organisms usually accumulate near the continental margin.

Shallow seas and other areas of shallow marine waters are the typical sites for sediment of chemical precipitation. Sediments of chemical origin are formed under specific temperature, salinity, *p*H, and oxidation-reduction (Eh) conditions. Carbonate minerals of chemical origin are inorganically precipitated from sea water and include a suite of common rock-forming minerals, such as calcium and magnesium carbonate. Carbonates formed by chemical precipitation require particular environmental conditions causing evaporation and chemical supersaturation of carbonate constituents. The production of carbonate minerals in the ocean is affected by the amount of calcium transported into the oceans from rivers, the ability of animals to extract calcium from sea water, and the availability of other nutrients required by carbonate-producing organisms (Gross, 1980).

Small oceanic basins connected to the world oceans by restricted inlets/outlets can be subject to conditions that differ from those of the open ocean. Examples include the Gulf of California and the Persian Gulf. The salinity of these small oceanic basins can be the same as, greater than, or less than that of the open ocean, depending on the cross-sectional area of the opening to the ocean, the amount of fresh-water influx, and climate conditions, such as conditions amenable to evaporation. Where there is little freshwater influx from precipitation and streams, these basins can attain salinities far greater than average seawater salinities and are therefore good candidates for sediments of chemical origin. In contrast, in areas of large freshwater influx, these basins can have salinities less than that of sea water.

4.1.4 Volcanic Sediments Sediments of volcanic origin can contribute substantially to the total volume of basin fill. The size of volcanic rock fragments ranges from microscopic dust and ash particles to huge blocks. Initial transport and deposition of volcanic rocks is accomplished by fallout and flows. *Fallout* occurs from an ash cloud produced by an eruption. *Flows* are hot mixtures of melted volcanic rock, gas, and solid particles that flow downhill under the influence of gravity and collect, cool, and solidify in topographic depressions.

4.2 Basin-Filling Processes

In addition to an understanding of depositional environments, study of basin filling requires an understanding of how sediment is transported and deposited on a small scale, as well as how large-scale patterns in sedimentary basin fill are produced. For example, knowledge of how sediment is dispersed throughout a basin by wind, water, and mass movements is the key to understanding many features observed in the rock record. Sedimentation patterns, particularly cycles of sedimentation, can be controlled by processes that operate within a sedimentary basin, called internal or *autocyclic* processes, and by processes external to a sedimentary basin, called *allocyclic* processes (Miall, 1984; Einsele, 1992). External factors, such as tectonics, climate change, and sea-level change, affect sedimentation patterns over a large areal extent and therefore produce regional patterns in basin-fill sediments. Tectonic forces exert strong control over the locations of source areas versus receiving basins and rates of uplift and subsidence. Tectonics affects the basin geometry, the elevation difference between the source area and receiving basin, and can influence hydrologic processes through the source-area size, elevation, and topography. Sea level controls the location of marine versus nonmarine depositional environments and associated sediment grain sizes, depositional features such as cross bedding, and fossil content, while the basin energy conditions control how sediment is distributed once it arrives. Climate change affects local lake levels, the magnitude and frequency of stream flows, the amount and composition of sediment transported by streams, weathering and sediment production in the source area, and rates of erosion in the source area. In contrast, internal processes, such as the switching of a stream channel to a lower elevation within a stream valley (*avulsion*) and switching of areas of lobe-shaped deposition on marine deltas (*delta lobes*), contribute to variability in sedimentation patterns over a limited areal extent.

Sediment is transported into a sedimentary basin by water, wind, or the force of gravity. For mixtures of water and sediment, there is a spectrum of transport processes in which the proportion of fluid to sediment grades from relatively clear streams to flows dominated by mud and debris. One must distinguish between *subaqueous* flows, in which sediment is transported beneath the surface of a standing body of water (a lake or an ocean), and *subaerial* flows, in which the flow is in contact with the atmosphere. In subaerial flows, the same fluvial processes that transported sediment into the basin also serve to distribute sediment throughout the basin. In contrast, while fluvial processes may transport sediment into a lake or ocean, subaqueous processes take over the transport and distribution of sediment beneath the water surface. In subaqueous flows, sediment transport is also affected by gravity, density, wind-produced waves and currents, and ocean circulation.

When water is the transporting medium in a mixture of fluid and sediment, the flow is known as a *fluid gravity flow*. These flows obey Newton's law of viscosity. The sediment concentration is small enough (less than 5%) that it does not appreciatively affect the rheological properties of the mixture, and both grain-to-grain interactions and the displacement of fluid by falling grains within the water column are not significant (Blatt *et al.*, 1980). Therefore, the effects of sediment concentration on fluid density, viscosity, particle settling velocities, flow dynamics, and sedimentation mechanics need not be considered in modeling of fluid gravity flows (Tetzlaff and Harbaugh, 1989).

The processes that transport sediment are discussed in detail in Allen (1985) and Blatt *et al.* (1980). In fluid gravity flows, the processes by which sediment is transported are quite different for coarser sediment transported on the stream bed, called bedload, and for finer sediment transported in the water column, called suspended load. Bedload grains are transported by rolling, bouncing, and sliding on the stream bed, whereas suspended-load grains remain in transport by turbulence in the water column. Particles that can be supported in suspension by turbulence are transported from the bedload upward into the suspended load. Deposition occurs when the flow can no longer support particles in suspension or can no longer move particles on the bed. Stream valley deposits of gravels and sands embedded in silts and clays are an example of a sedimentary deposit produced by fluid gravity flows.

When a stream flows into a lake or ocean, the flow cross-sectional area suddenly expands, the stream velocity drops, and the flow turbulence decreases, reducing the flow's capacity to transport both suspended sediment and bedload. Consequently, bedload sediment is rapidly deposited at the mouth of the stream. Clay particles start to flocculate as they are affected by the change in water chemistry from fresh stream water to saline (alkaline) ocean water. *Flocculation* is the process by which small particles suspended in the water column clump together to make bigger particles with higher settling velocities.

Rapid deposition seaward of the stream mouth causes oversteepened slopes on the floor of the lake or ocean. Such slopes are unstable and therefore provide sources of sediment to deeper portions of basins through mechanisms such as slumping, sliding, and flows of dense sediment–water mixtures. The offshore, steeply sloped foreset beds of marine deltas are an example of structures formed by these processes.

A viscous mixture of fluid and sediment that moves downslope under the force of gravity rather than by fluid transport is called a *mass movement.* Mass movements typically occur where there are both a steep topographic gradient and sufficient rock or sediment for slope failure. Rock falls, slides, slumps, creep, and sediment gravity flows form the basic types of mass movements (Blatt *et al.*, 1980). In *rock falls,* a pile of angular rock fragments is deposited adjacent to the source of the fragments. Rock falls occur on steep slopes either on dry land or beneath the sea. *Slides* are characterized by the quick downslope movement of a mass of soil and rock over a surface at the base of the slide, called a *shear surface.* The mass is subject to little internal deformation but may rotate. *Slumps* are quick downslope mass movements that experience rotation and much internal deformation. If the sliding or slumping mass experiences *liquefaction,* in which the solid sediment mass is transformed to a fluid mass because of increased pore pressures and loss of strength, a debris or mudflow develops. *Creep* is the slow downslope movement of soils and unconsolidated sediments.

The mass movements most important in terms of total volume of sediments observed in the geologic record are dense mixtures of fluid and sediment called *sediment gravity flows.* In contrast to fluid gravity flows, in sediment gravity flows the sediment concentration and grain interactions affect the particle settling velocities and the fluid–sediment mixture density and viscosity. Four transport mechanisms have been identified (Blatt *et al.*, 1980). First, turbulence within the fluid–sediment mixture exerts an upward force on sediment grains. Second, grain-to-grain collisions cause grains to support each other. Third, as grains settle under the force of gravity, fluid is displaced, rises, and exerts an upward force on other sediment grains. Fourth, a mixture of fluid and fine-grained sediment has high strength and can therefore support larger grains within the flow. As a result, coarse-grained particles are transported well above the base of the flow, not just at the base of the flow as in fluid gravity flows. Two common forms of sediment gravity flows are *debris* or *mud flows* and *turbidity currents* (Fig. 10). Debris flows are composed of rock fragments, soil, and mud, with large fragments dispersed in and supported by a finer-grained cohesive matrix. Mud flows are primarily fine-grained materials, with few larger fragments. Turbidity currents are high-velocity, short-duration, underwater, turbulent suspension currents that occur as a result of either suspended-sediment–laden floodwaters or, more commonly, a failed sediment mass that entrains ocean or lake water as it proceeds downslope. Sediments deposited by turbidity currents are known as *turbidites*. The grain size and composition of turbidites preserved in the geologic record is diagnostic of source-area parent sediment type, turbidity-current density, sediment transport mechanisms, and proximity to source area. Sediment gravity flows commonly form fan-shaped deposits located seaward of submarine canyons and large rivers, known as submarine fans. Deltas and deep-sea fans are common settings for widespread creep, slumps, debris and mud flows, and turbidity currents.

4.3 Modeling of Basin Filling

Models of basin filling represent spatial and temporal changes in the volume and distribution of sediments accumulating in a basin. These models seek to reproduce sedimentary geometries, locations of depositional environments, and facies relations within a subsiding sedimentary basin. To untangle the relative contribution of each geologic process to the observed geology, all relevant external and internal forcing factors as well as fluid flow and sediment transport processes must be included. There are two categories of basin-fill models: process-based and geometric. Process-based models incorporate the physics of fluid dynamics and sedimentation mechanics to predict sediment erosion, transport, and deposition. In contrast, geometric models are based on algebraic constraints and empirical rules developed from conceptual depositional (facies) models. These models help assess the effects of multiple geologic, hydrologic, and climatic processes on observed basin-fill patterns. Basin-fill models must be linked to source-area and basin-formation models. Source-area models serve to estimate the influx of fluid and sediment into a basin. Basin-formation models predict the basin geometry and subsidence, and assist in predicting the topography.

4.3.1 Process-Based Models Process-based basin-fill models use a combination of governing equations for conservation of mass, momentum, and energy for fluid; conservation of mass for sediment; relations for fluid and sediment rheological properties; empirical relations for sediment transport and entrainment; and relations between rates of erosion or deposition and the size distribution of sediments transported. The equations predict how water, wind, and mass movements transport sediment into and throughout basins and how sediments are distributed during deposition.

A complete account of unsteady flow in three dimensions for an isotropic, Newtonian fluid is provided by the mass-conservation and the Navier–Stokes force-momentum equations (Bird *et al.*, 1960). Conservation of fluid mass mathematically describes the rate of change of density at a fixed point resulting from changes in the mass velocity vector $\rho\mathbf{v}$:

$$\frac{\partial \rho}{\partial t} = -\nabla \cdot \rho\mathbf{v}. \tag{5a}$$

If the fluid is assumed to be incompressible,

$$\nabla \cdot \mathbf{v} = 0, \tag{5b}$$

where t is time, $\mathbf{v}$ is the fluid-flow velocity vector (three components), and ρ is the flow density.

Conservation of fluid momentum mathematically describes how the fluid accelerates or decelerates. The motion of the fluid changes because of the forces (pressure, viscous, and gravitational) acting on it:

$$\frac{\partial(\rho\mathbf{v})}{\partial t} = -(\mathbf{v}\cdot\nabla)\rho\mathbf{v} - \nabla p - [\nabla\cdot\boldsymbol{\tau}] + \mu\nabla^2\mathbf{v} + \rho\mathbf{g}, \tag{6}$$

where $\mathbf{g}$ is gravitational acceleration, p is pressure, $\boldsymbol{\tau}$ is the Navier–Stokes stress tensor, a function of flow viscosity and flow velocity gradients (nine components), and μ is the flow viscosity.

The equation for conservation of energy

must be included when vertical flow-velocity variations, such as flow turbulence, are important. Conservation of fluid energy mathematically describes how the accumulation of internal and kinetic energy changes because of changes in the flow rate, heat conduction, and work done on the fluid by forces (gravitational, pressure, and viscous) acting on it:

$$\frac{\partial}{\partial t}\left(\tfrac{1}{2}\rho\mathbf{v}^2\right) = -(\mathbf{v}\cdot\nabla)\tfrac{1}{2}\rho\mathbf{v}^2 + \rho(\mathbf{v}\cdot\mathbf{g}) - \nabla\cdot p\mathbf{v} - p(-\nabla\cdot\mathbf{v}) - \nabla\cdot[\boldsymbol{\tau}\cdot\mathbf{v}] - (-\boldsymbol{\tau}:\nabla\mathbf{v}), \tag{7}$$

where all variables are as previously defined.

The Navier–Stokes equations as shown are suitable for subaerial flows. In contrast, subaqueous flows require consideration of flow density, buoyancy, and additional frictional forces. First, the density difference between sediment-laden fresh water and sea water caused by differences in sediment concentration, salinity, and temperature must be considered. This density difference is accounted for by multiplying gravity by the relative density of the flow with respect to the surrounding fluid, $(\rho_{ws} - \rho_{sea})/\rho_{ws}$, where ρ_{ws} is the density of the water–sediment mixture, and ρ_{sea} is the density of the surrounding fluid (sea water or lake water). The density difference affects where the flow is located, the location beneath sea level. For flows that float at sea level, friction occurs at the freshwater–seawater interface at the bottom of the flow. In contrast, for submarine flows denser than sea water, friction occurs between the flow and the sea floor as well as at the freshwater–seawater interface at the top of the flow. Second, entrainment of sea water into the sediment-laden flow by turbulent mixing must be considered. Additional important considerations below sea level include the transport of sediment by longshore currents and ocean circulation patterns. Martinez and Harbaugh (1993) provide examples of simulation of longshore transport and sediment transport by wind-driven waves.

Solving the Navier–Stokes equations in three dimensions is computationally intractable for the large spatial and temporal domains of sedimentary basins. However, with a few simplifying assumptions, realistic large-scale problems have been addressed (Tetzlaff and Harbaugh, 1989; Koltermann and Gorelick, 1992). Typically, all flow variables are vertically integrated to eliminate vertical flow-velocity variations and internal shear within the flow. Assuming a homogeneous, incompressible, isothermal fluid allows fluid density and viscosity to be treated as constant for fluid gravity flows. Coriolis forces are neglected unless the effects of ocean currents must be simulated.

Sediment-transport models require an equation of conservation of sediment mass as well as equations relating the fluid mechanics to the ability of the flow to transport sediment, whether the transporting medium is wind or water. Allen (1985) and Blatt *et al.* (1980) provide discussions of sediment transport. Conservation of sediment mass mathematically describes topographic changes due to erosion or deposition and the resulting changes in the sediment concentration in the flow:

$$h_f\left(\frac{\partial c}{\partial t} + [\bar{\mathbf{v}}\cdot\nabla c]\right) = -\frac{\partial z_s}{\partial t}(1 - \phi), \tag{8}$$

where ϕ is porosity, h_f is flow depth, z_S is the sediment surface elevation, $\bar{\mathbf{v}}$ is the vector of depth-averaged flow velocity, and c is the sediment concentration. The thickness of sediment deposited or eroded is reflected in the change in the land surface elevation multiplied by the packing concentration $(1 - \phi)$ of the deposited sediment.

Sediment-transport equations developed from field and laboratory studies have been based either on statistical regression fits to data or on theoretical analysis of the forces required to initiate particle motion and *entrain* particles into the flow. The *entrainment* of particles occurs when grains are dislodged from a position of rest. The critical or threshold shear stress required to initiate particle motion and entrain particles into the flow is related to particle size, density, shape, and cohesion, and to the roughness of the surface the grain rests upon (Komar, 1988; Blatt *et al.*, 1980).

Many equations have been proposed to relate the bedload transport capacity of a stream to flow hydraulic characteristics and the physical properties of both fluid and sediment. Reviews are provided by the American Society of Civil Engineers (ASCE, 1976) and Dawdy and Vanoni (1986). These equations typically estimate bedload transport capacity as a function of the difference between the *boundary shear stress* (the shear stress at the interface between the fluid and the stream bottom) and the *critical shear stress:*

$$q_s = K(\tau_0 - \tau_c)^{3/2}, \quad (9)$$

where q_s is the bedload transport rate, τ_0 is the boundary shear stress [the product of the streambed slope, the stream hydraulic radius (area divided by wetted perimeter), and the fluid specific weight], τ_c is the critical shear stress, and K is an empirically derived function of particle characteristics.

In contrast to bedload, the amount of suspended sediment transported often is supply limited. That is, the capacity of the stream to transport sediment is greater than the supply available in the source area. For a given magnitude of flood, the suspended sediment concentration can vary widely as a result of factors affecting the source area, such as vegetation, precipitation, temperature, parent rock type, and soil moisture conditions. For example, Koltermann and Gorelick (1992) showed that in the coastal mountain ranges of California, streams with similarly sized drainage basins but different climate conditions had suspended sediment concentrations that varied over several orders of magnitude. These kinds of studies show that suspended sediment loads must be measured rather than predicted on the basis of flow hydraulic variables.

In all cases, the particle settling velocities are critical. Settling velocities are greatest in clear water and decrease with increasing sediment concentration because particle collisions and interactions hinder particle settling (Mirza and Richardson, 1979). In particular, the presence of a high concentration of fine-grained particles impedes the settling of coarser particles.

In fluid gravity flows, the spatial patterns of different grain sizes throughout a sedimentary basin is partially controlled by *downstream fining,* the processes that cause the average grain size within a stream channel to diminish with distance from the source area. Two mechanisms control downstream fining. First, particle abrasion breaks bedload particles down into smaller sizes as they are bounced, slid, and rolled along a stream bottom. Second, selective deposition of heavier and larger grains with higher settling velocities results in preferential deposition of these grains closer to the source area. The division of sediment into different grain-size populations during transport or deposition is called *fractionation.* To date, in basin-fill modeling particle abrasion is neglected, while selective deposition is enforced by a hierarchy of grain-size classes in order of decreasing settling velocities.

Examples of fluid–gravity-flow and sediment-transport models used to make predictions about spatial patterns in sedimentation are not numerous but do exist. To predict spatial variations in mean grain size and sorting in one bend of a meandering stream, Bridge (1977, 1992) developed a model of nonuniform stream flow, bedload transport, and changes in bed geometry with erosion and deposition (Fig. 11). Simulation results reproduce the trends in grain size observed in a natural channel. Bitzer and Pflug (1990) simulated steady-state flow and suspended-sediment transport in small ocean basins. An example simulation shows the geometry and distribution of grain sizes in terms of percentages of gravel, sand, and shale predicted from flow through a harbor (Fig. 12). Fluid-flow and sediment-transport models for purposes of geologic predictions can get quite complicated. For example, Ericksen *et al.* (1990) simulated wind- and tide-driven circulation, sedimentation, and erosion in oceanic basins by accounting for continuity of fluid, salt, heat, and sediment as well as sediment transport and density-dependent fluid flow. They applied their model to the late Devonian Catskill Sea to estimate ancient wind and tide conditions.

To date, most process-based basin-fill modeling has been of fluid gravity flows because modeling of sediment gravity flows is more complex and difficult. As a result of the dependence of the flow hydraulic characteristics on the flow rheology, the governing equations for fluid and sediment in sediment gravity flows must be solved simultaneously at each time step, rather than sequentially. In addition, the equation for conservation of energy must be included to represent turbulence. This type of modeling requires consideration of sedimentation from highly concentrated multicomponent suspensions, and prediction of sediment-concentration–dependent fluid viscosity, density, and particle set-

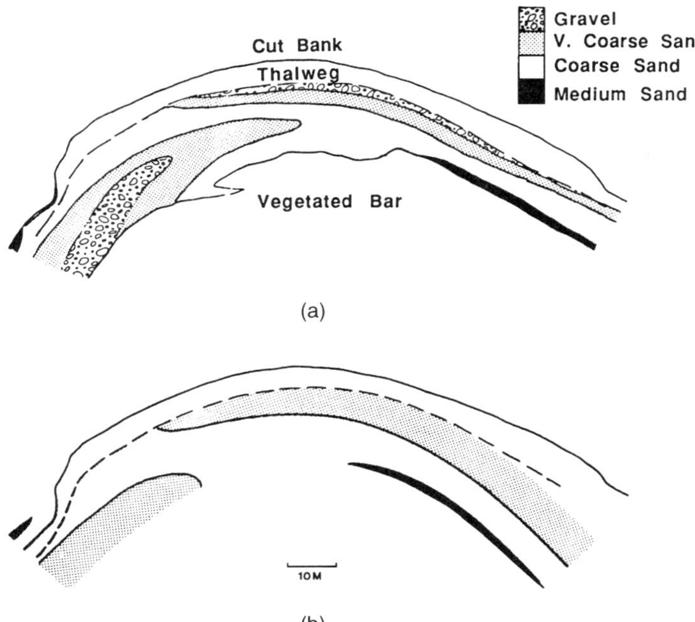

FIG. 11. Process-based image of a meander (stream channel bend). The predicted grain size (lower figure) compares favorably to the observed grain size (upper figure) in a natural channel. After Bridge (1977). Reprinted by permission of John Wiley and Sons, Ltd.

tling velocities (Zeng and Lowe, 1992). In sediment gravity flows such as turbidity currents, the bulk of the coarse sediment is transported within the flow rather than as bedload, sediment size fractionation is affected by particle interactions and fluid displacements by falling grains, deposition is dominated by deposition from suspended load rather than bedload, and sediment size fractionation occurs not during transport but during deposition. Unlike fluid gravity flows in which water entrains sediment particles, the suspended load in a turbidity current results from entrainment of water into a failed sediment mass. Therefore, the sediment concentration in the flow and the volume of the flow are highly variable over short time periods and distances. Modeling of subaqueous sediment gravity flows and density currents has been performed primarily to analyze

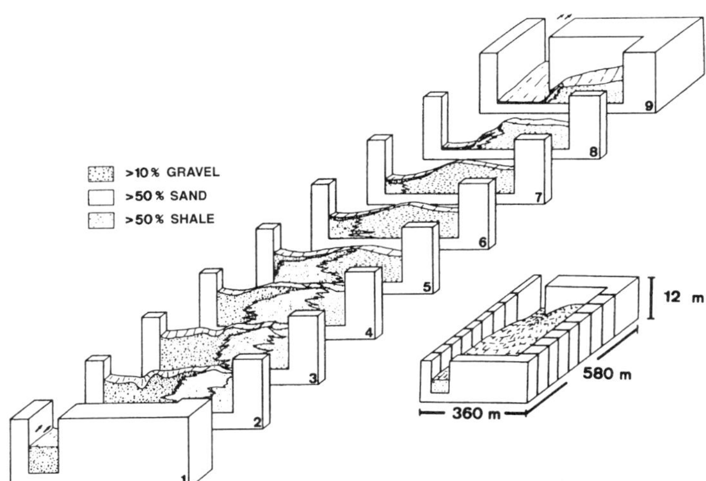

FIG. 12. Process-based image of harbor sedimentation. The geometry and distribution of grain sizes in terms of percentages of gravel, sand, and shale predicted from flow through a harbor is shown. Flow is from lower left to upper right. Downstream fining is evident in the distribution of grain sizes in the simulated harbor, where coarser grain sizes are deposited closer to the harbor inlet, and finer grain sizes are deposited closer to the outlet. The inset shows the size and geometry of the harbor. After Bitzer and Pflug (1990). Reprinted by permission of Prentice Hall, Upper Saddle River, New Jersey.

highly erosive turbidity currents in submarine canyons (Fukushima *et al.*, 1985), rather than to examine sedimentation patterns, although research is progressing in the latter area.

The other important sediment–gravity-flow category is debris flows, the physics of which differs from turbidity currents. Because a debris flow has strength that must be overcome, a debris flow starts moving only after the critical shear stress required to initiate motion is exceeded (Johnson, 1970). The strength of a debris flow is so great that transport of even boulder-size rock fragments occurs. A debris flow remains in motion until the shear stress at the base of the flow drops below the critical shear stress, at which time the entire mass ceases to move. The base of the debris flow deforms as it flows, while the top flows as a rigid mass. Debris-flow sediments show that the upper parts of a debris flow contain coarser materials dispersed in a finer-grained matrix, without the grain-size sorting, stratification, and small sedimentary features produced by flowing fluid or sediment seen in fluvial flows and turbidity currents. Although grain-size variations within debris-flow deposits occur, they are not important to fluid-flow properties such as permeability, which is controlled by the fine-grained matrix. Often, the predicted location and geometry of debris-flow deposits are of greatest concern. Therefore, they have been numerically modeled using a Coulomb continuum approach (Savage and Hutter, 1989), in which a debris flow is represented as one undifferentiated mass that changes shape as it flows from the initiation of movement to its final location. Results of simulations compare favorably to both wet and dry experimental debris flows (Iverson and LaHusen, 1992).

Models of fluid gravity flows have been used in process-based basin-fill models that include external forcing factors, links to a source-area model, and a model of basin formation. The first large-scale process-based model to represent basin fill in three dimensions was that of Tetzlaff and Harbaugh (1989). Their model predicts three-dimensional grain-size variations at a large scale produced in fluvial and deltaic environments. Figure 13 shows a plan and cross-sectional view from a simulation of braided-stream deposits. The general form of the model has lent it to simulation of alluvial fans, deltas, and braided-stream deposits. Realistic sedimentary sequences have been produced for the Golden Oil field, Louisiana, the National Petroleum Reserve, Alaska, and the Ivishak formation, Alaska. They began with a simple model of tectonic subsidence. More recent work has included sediment transport by waves (Martinez and Harbaugh, 1993) and links to simulation of hydrocarbon migration in sedimentary deposits produced by the basin-fill model (Wendebourg and Harbaugh, in press).

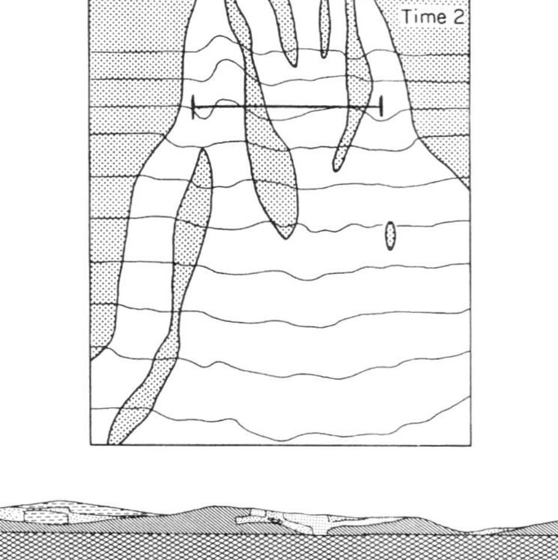

FIG. 13. Process-based image of braided-stream deposits. Areal and cross-sectional views of simulation results from four source streams flowing over a planar surface. The thick horizontal bar located 1/4 of the way down from the top of the areal view shows the location of the cross-sectional view. Lines crossing the figure from left to right are elevation contour lines. Coarser deposits are stippled. In the cross-sectional view, bedrock is shown as cross-hatched, while changes in stippling indicate sediments deposited at successively later times. Modified from Tetzlaff and Harbaugh (1989). Reprinted by permission of Van Nostrand Reinhold publishers.

Koltermann and Gorelick (1992) extended the model of Tetzlaff and Harbaugh (1989) to represent sea-level change, horizontal and vertical fault motion, sediment loading, numerous tectonic subsidence mechanisms, compaction, paleoclimate-driven fluctuations in floods and sediment loads, and realistic sequences of ancient floods. The model was adapted to a supercomputer to simulate sedimentary basin evolution over geologic time scales (hundreds of thousands to millions of years). Simulation of the geologic history of the Alameda Creek alluvial fan in north-central California showed how paleoclimate fluctuations create large-scale patterns in sedimentation by affecting ancient flood magnitudes and the volume and composition of sediment loads transported into a basin. These large-scale patterns produced highly permeable gravel layers separated by thick, low-permeability clay layers (Fig. 14). Results show that basin evolution models can be used to distinguish between large-scale spatial patterns in sedimentary basin fill caused by tectonics, sea-level change, and climate change. The grain-size distributions were used to develop a realistic model of porosity and permeability variations (Koltermann and Gorelick, 1995) and were linked to a groundwater flow simulator (Koltermann, 1993) to illustrate how sedimentary-basin evolution models can be useful predictive tools for subsurface fluid flow.

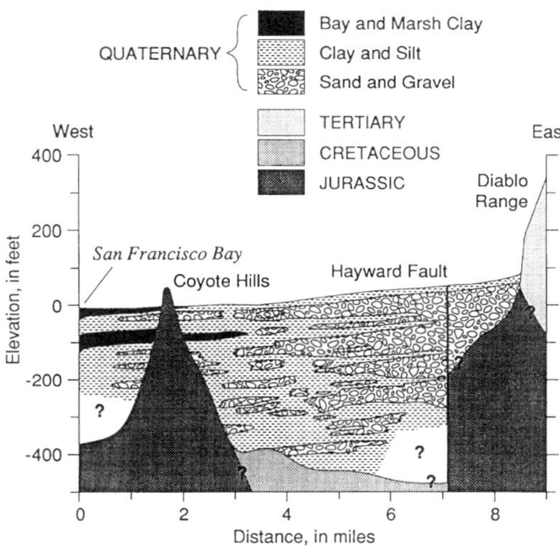

FIG. 14. Alluvial-fan cross section showing cycles of sedimentation. Simulation of the geologic history of the Alameda Creek alluvial fan in north-central California showed how paleoclimate fluctuations create large-scale patterns in coarse, high-permeability gravels and sands versus fine, low-permeability silts and clays by affecting ancient flood magnitudes and the volume and composition of sediment loads transported into a basin. Results show that coupling of a source-area model, a basin-formation model, and a basin-fill model helps geologists distinguish between spatial patterns in sedimentary basin fill due to tectonics, sea-level change, and climate change.

4.3.2 Geometric Models Geometric models predict large-scale grain-size patterns in sedimentary basin fill using rules consistent with observed arrangements of depositional environments and geologic laws, such as the *principle of superposition,* which states that younger sediments are successively deposited on top of older sediments and rock. The rules constrain patterns in sediment grain size, the location of marine versus nonmarine sediments, the location of carbonate minerals, and the slope of the land surface and sea floor. Some geometric methods produce results at fine spatial detail, corresponding to the width of braided channels, whereas others produce coarser results, corresponding to the large-scale sediment geometries resulting from the rise and fall of sea level.

Geometric modeling methods include stratigraphy, analytical, random-walk, and random-avulsion (stream-channel migration) approaches. The most useful and widely used of the geometric modeling methods are stratigraphy models. These models produce cross-sectional images of large-scale lithologic sequences that show depositional environments, grain-size distributions, ages of deposition, and *unconformities,* which are gaps in the rock record resulting from a cessation of deposition or from erosion (Fig. 15). *Stratigraphy* models are commonly applied in shallow marine settings to decipher the effects of subsidence mechanisms, sediment-supply fluctuations, basin geometry, and sea-level changes on the geologic record. These models have received a lot of attention in the petroleum industry and have been used to examine the geometry, extent, and continuity of sand bodies in petroleum reservoirs. Stratigraphy models include an extensive set of rules to achieve a realistic result. These rules

1. constrain the flow direction to be from one side of a cross section (a vertical image) to the other, forcing sediment to move downhill;
2. control topographic gradients;
3. control where sedimentation and erosion occur, often as a function of the location of the shoreline;
4. use downstream fining rules to deposit coarse sediment near the source area and progressively finer sediment away from the source area;

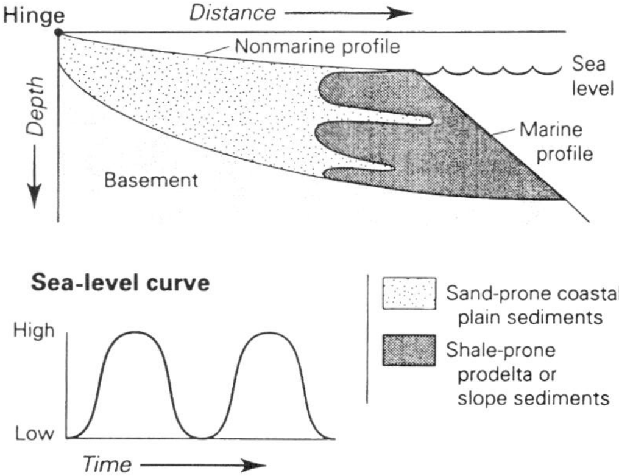

FIG. 15. Stratigraphy model image. Stratigraphy predicted for steady subsidence, two cycles of a sinusoidal fluctuation in sea level, and a constant sediment supply. Results show that sea-level fluctuations produce cyclicity in coarse-grained deposits, with high sea-level stands trapping coarse materials close to the shoreline. Modified from Jervey (1988). Reprinted by permission of the Society for Sedimentary Geology.

5. control the amount of sediment deposited in any given location during an interval of time, usually as a simple function of the distance from the shoreline; and
6. consider the production of carbonate minerals as empirical functions of water depth.

Results produced by these models look like geologic interpretations from seismic sections.

The other geometric methods are of less importance in terms of practical applications. *Analytical* models predict the lateral extent and thickness of high-permeability gravels given a basin size, subsidence rate, and sediment influx. Paola *et al.* (1992) developed an analytical model to examine the effects of four external factors on gravel geometries: sediment influx, subsidence, gravel fraction in the sediment supply, and water flux (Fig. 16). Their work showed that a variety of mechanisms produce similar-looking cycles in sedimentary deposits forming groundwater aquifers and petroleum reservoirs.

Random-walk models predict braided-stream channel network configurations by allowing particles of fluid to travel over a to-

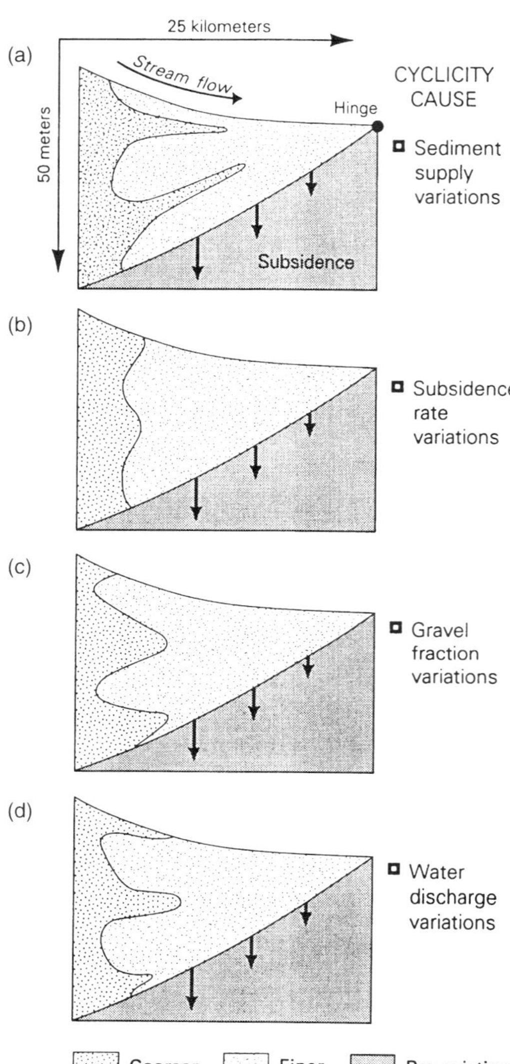

FIG. 16. Analytical images of cyclicity mechanisms. Hypothetical cross sections of a basin showing the effects of rapid variation in (a) sediment supply, (b) tectonic subsidence, (c) gravel fraction in the sediment supply, and (d) water flux. Each of the four cyclicity mechanisms is varied independently in a sinusoidal fashion. Variations follow a sinusoidal curve through time. Results show that multiple mechanisms can be responsible for observed cyclicity in alluvial basins. Modified from Paola *et al.* (1992). Reprinted by permission of Blackwell Science, Ltd.

pography from a source area to a discharge point using probabilistic rules for the location of fluid particles through time (Webb, 1995). Given hydraulic relations between flow discharge, velocity, and channel cross-sectional area, a channel topology can be predicted. The type of sediment deposited throughout the channels can be predicted by applying rules for sediment transport given the flow hydraulics.

Random-avulsion models mimic the abrupt change of location (avulsion) of a meandering stream channel to the lowest topographic elevation within a subsiding valley, using probabilistic rules to govern when avulsion occurs. Images show the large-scale cross-sectional architecture of channel sands encased in clays (Fig. 17). Random-avulsion models have been used to examine the distribution of sand bodies deposited by meandering streams as functions of subsidence rate and the proportions of coarse- versus fine-grained sediment transported by streams (Bridge and Leeder, 1979).

5. DIAGENESIS

Important changes occur after sediments are deposited. As loose, unconsolidated sediments are buried, they are subject to increasing pressure and temperatures, bathed by fluids, subject to chemical reactions, compacted, and finally *lithified,* or turned into rock. As these processes occur, the sediments experience a tighter packing, an increase in density, and a decrease in both porosity and permeability. The post-depositional processes causing chemical/mineralogical, biological, and physical changes to sediments and rocks are collectively referred to as *diagenesis.* Readers are referred to Blatt *et al.* (1980) for a discussion of how diagenesis affects different minerals and to Einsele (1992) for a chapter on diagenesis and fluid flow, thermal histories, and hydrocarbon generation. Diagenetic processes cause compaction and expulsion of pore fluids, precipitation of mineral cements and concretions, growth of minerals such as clays in pore spaces, replacement of primary minerals by secondary minerals, reactions (such as dissolution) of minerals with pore fluids, recrystallization and formation of new minerals, incorporation of water into a mineral or release of water from a mineral, and breakdown of minerals due to the action of bacteria. Chemical processes cause soluble or thermodynamically unstable minerals to dissolve and be reprecipitated or recrystallize as more stable minerals. Diagenesis can overprint original depositional features, such as fossils and small-scale primary sedimentary structures such as cross bedding, as well as larger features such as cycles of sedimentation. Through time, diagenetic processes tend to reduce drastically the original porosity and permeability of sediments. However, fracturing of rock can increase the porosity and drastically increase the permeability. This is referred to as *secondary* porosity and permeability.

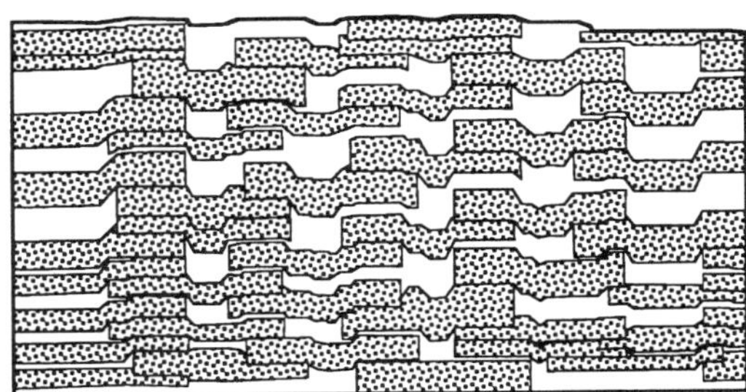

FIG. 17. Random-avulsion image of a meander belt. Cross section from a random-avulsion simulation in which the channel-belt sands (stippled) are encased in fine-grained flood-plain deposits. Compaction produces the distortion of the originally horizontal channel sands. Modified from Bridge and Leeder (1979). Reprinted by permission of Blackwell Science, Ltd.

Near the land surface, an important post-depositional change is the mechanical and chemical weathering of surface sediments and rock and the formation of soil horizons, complete with roots, biological activity, and breakdown of sediment grains into clay minerals. This part of the sedimentary rock cycle produces weathered materials for erosion and sediment transport.

Sediments of chemical and biological origin are particularly susceptible to diagenesis. Carbonate and siliceous minerals experience dramatic changes after deposition. Minerals dissolve, creating additional pore space. New carbonate minerals precipitate, forming cement in the voids of existing rocks. The originally deposited carbonate minerals can be replaced by other, more stable, carbonate minerals, such as the commonly observed calcite (calcium carbonate) and dolomite (calcium magnesium carbonate). Sediments of biological origin rich in silica are highly soluble. Dissolution of one siliceous mineral and reprecipitation as another siliceous mineral is common. Finally, as the temperature and pressure increase with depth of burial, silicon oxide is recrystallized to become the familiar mineral *quartz*.

Geologists tend to distinguish between diagenetic processes that occur when sediments have not been buried very deeply, and processes that occur after deep burial when temperatures have substantially increased. Changes in the temperature of basin sediments cause diagenetic changes because where sediments of organic origin are present, the thermal history of a basin exerts control on the formation of hydrocarbons and coal. This thermal history is recorded by temperature-sensitive organic and inorganic indicators within the sediments. Thermal histories record increasing depth of burial as well as uplift and erosion (Fig. 18). Recall that in basins formed by extension and rifting, high-temperature mantle material is brought closer to the overlying crust, causing increased temperatures in crustal rocks that then cool slowly. These higher temperatures affect the sediments in terms of diagenetic processes. Deep burial also is important to the diagenetic processes that produce ore deposits. Deep subsurface fluids can be of sufficiently high salinity and metal concentrations to precipitate ore minerals (Einsele, 1992). It is important to note that as the temperature increases beyond about 200 °C, diagenesis gives way to low-grade metamorphism (Einsele, 1992).

Modeling of diagenesis includes representation of the chemistry of subsurface waters in contact with sediments and rock, as well as the flow of subsurface fluids and transport of solutes. Freeze and Cherry (1979) and Anderson and Woessner (1992) provide texts on groundwater flow and modeling. Wendebourg and Harbaugh (in press) treat modeling of hydrocarbon migration in sedimentary basins. Readers are referred to Bethke *et al.* (1990) for examples of modeling basin thermal histories, fluid flow, and diagenesis on a large scale in sedimentary basins. Models of diagenesis have not yet been coupled with models of sedimentary basin formation and filling. The only diagenetic process currently commonly represented in sedimentary basin formation and filling models is compaction.

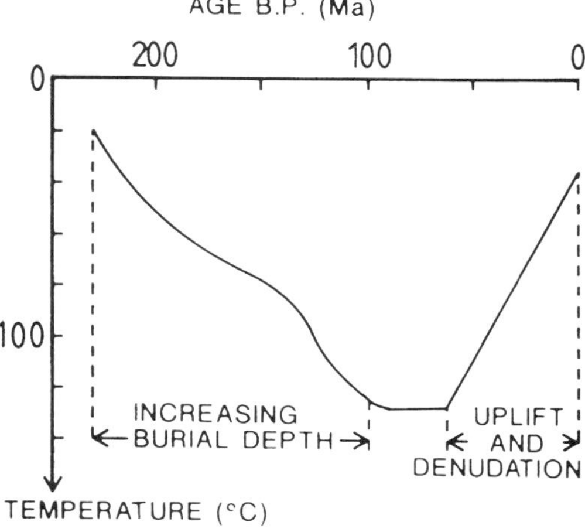

FIG. 18. Thermal history diagram. At this location in the Colorado Plateau, the thermal history, as reflected by temperature-sensitive indicators in the sediments, records an increasing depth of burial until 100 million years (Ma) before present (B.P.), as well as more recent uplift and denudation. After Meyer *et al.* (1989). Reprinted by permission of Springer-Verlag New York, Inc.

6. GEOLOGIC HISTORY ANALYSIS

We have both direct and indirect views of the Earth's subsurface that provide data used in the analysis of a sedimentary basin's history. Direct measurements of rock or sediment properties can be obtained by "field" observation of outcrops, such as in valleys, cliffs, and road cuts, mapping, and the drilling of wells into the sedimentary basin fill. Specific information can include lithologic logs, soil borings, cores, paleontologic information, age data, and well-pressure test data. Indirect measurements include geophysical data, such as seismic profiles and borehole geophysical logs. Seismic data provide images of large-scale geometric relations in sediments, whereas borehole data produce small-scale information on petrophysical properties used to infer lithology. All of the data are combined to produce interpretive cross sections and maps of a basin.

All available data must be compiled and analyzed through techniques of sedimentary basin analysis, which examine the structural, climatic, and sea-level history of a sedimentary basin (Miall, 1984). Through such analysis one can understand the basin tectonic style and geometry; estimate the rates of subsidence, sedimentation, and sea-level change; determine whether subsidence is accelerating or decelerating; estimate the climatic conditions under which the basin formed; and develop facies relations. Several authors provide a list of questions to guide a comprehensive analysis of the geologic history of a basin (Table 2). Basin evolution models require the following information from a geologic history analysis:

1. the paleogeography, including the timing of basin formation, relation to plate boundary and margin type, basin type, the subsidence history, location and motion on faults, reconstruction of horizontal and vertical land-block movements, and the topography through time;
2. the paleobathymetry, including the history of sea-level change, shoreline location through time, and paleocurrents indicating past flow directions;
3. the paleohydrology, including a magnitude–frequency relation for paleofloods, a relation for flood velocity versus discharge, and a relation for the effects of paleoclimate change on paleohydrology; and
4. the paleosediment discharge, including a relation for the amount of bedload and suspended load carried by each flood, a function of climate and hydrology as well as the parent rock type, and the grain-size distribution in the source sediment supply.

The quantity and quality of the geologic history data base is limited by the infrequent or recent nature of our observations (such as sparse well data and stream-flow records that seldom have more than 100 years of record) and difficulties in age dating. The age of rocks and sediments is established on the basis of radiometric dating techniques, dated periodic reversals in the Earth's mag-

Table 2. Questions about the geologic and climatic history of a basin (Modified from Miall, 1984; Pettijohn *et al.*, 1972; Reading, 1986).

1. Basin age:
   How old is the basin under study?
   What does this tell us about the regional history of subsidence and uplift?
2. Tectonism:
   What type of basin is under consideration, in terms of plate-tectonic theory?
   Where within the basin are the rocks or sediments (aquifer or reservoir) of interest?
   What was the tectonic style of the basin during sedimentation?
   Does this information or fossil evidence or radiometric age dates suggest the length of time the basin was active and its rate of sediment accumulation?
   Are there gaps in the rock record, or folds or faults that occurred during deposition that provide evidence of variations in the rate of basin subsidence?
   Are there widespread lithologic units or cyclic sequences that could have been caused by episodic basin subsidence or episodic source area uplift or climatic changes?
3. Climate:
   What were the global and regional climate conditions during basin formation and filling?
   Is there any internal evidence of climate change (abundance or type of flora and fauna, diagnostic minerals, or presence of glacial features)?
   What does this evidence tell us about probable controls on river discharge (perennial or ephemeral flow), bank cohesiveness (vegetation cover), or grain size (chemical vs mechanical weathering)?
   Are there local variations in climate within the sedimentary basin or between the sedimentary basin and source area?
4. Regional depositional trends:
   What are the main facies variations in the basin, and over what scale do the variations occur, both vertically and laterally?
   What do the spatial patterns in regional depositional trends look like?
   What are the patterns of deposition above sea level versus below sea level?
5. Local depositional trends:
   What is the thickness range, lateral variability, and grain-size composition of cycles in the sediments?
   Can a pattern or sequence of stream channel migration or avulsion be detected?
   Can specific morphological units be identified?
   What small-scale sedimentary structures are observed?

netic field, and fossil animal and plant remains. Age dating is particularly difficult in terrestrial sediments (land-based sediments deposited above sea or lake water), where the fossil content may be negligible and where volcanic ash beds may be few and far between. Given some age dates in a basin, the ages of the remaining strata can be estimated relative to the dated strata. Climate data come from cores of sediments from the ocean floor, from modern and ancient lakes, and from ice caps. These data provide records of the isotopic composition of sea water, sedimentation rates, the magnitude and frequency of fluctuations in global ice volume, and records of pollen and plant assemblage fluctuations associated with changes in precipitation and temperature.

7. CONCLUSIONS

Because the spatial and temporal scales of sedimentary basin evolution are far beyond the scales we can examine in the laboratory or within a human lifespan, earth scientists have turned to numerical models to examine indirectly and make predictions about that which we cannot observe directly. In recent years, efforts have been made to quantify the processes through which sedimentary basins evolve, and to integrate descriptive geologic studies with mathematical models of source areas, basin formation and filling mechanisms, and post-depositional diagenetic changes. Such modeling studies have three objectives:

1. to increase our understanding and test our hypotheses about the Earth;
2. to predict the effect of geologic processes for engineering or environmental purposes; and
3. to provide information for resource management decisions, such as three-dimensional quantitative descriptions of porosity and permeability fields used in computer models of subsurface fluid flow to predict well locations for hydrocarbon extraction or groundwater contamination cleanup.

Basin evolution modeling provides a quantitative means to evaluate the effects of basin geometry, factors governing basin formation, source-area fluid and sediment supply histories, and sea-level and climatic histories on the architecture of sedimentary deposits. This approach provides a rigorous framework to help interpret the relative effects of competing forcing factors affecting basin fill and improves our ability to interpret the geologic record and make predictions about basin genesis and subsurface fluid flow. Models allow us to study geologic processes in areas that are relatively inaccessible, such as in submarine canyons, and over extremely long time periods.

Basin evolution studies must address four major components. First, models of the source area predict the time sequence and volume of fluid and sediment transported into the basin. Second, models of basin formation predict the tectonic and loading controls on basin geometries and rates of subsidence. Third, models of basin filling predict the grain-size distribution of the sediments deposited within a basin, the location and type of sediments of chemical or biological origin, and the distribution of depositional environments, particularly related to ocean water depths. Fourth, models of diagenesis predict post-depositional changes. Much research has been devoted to studying and modeling each of these components in isolation rather than as part of an entire system. For example, many geodynamic models have been constructed to examine tectonic subsidence alone and neglected basin filling and the additional subsidence caused by sediment loading. However, quantitative basin evolution modeling is in its infancy because earth scientists have only just begun to merge descriptive geologic studies with available quantitative tools. This approach to model simultaneously all components of basin evolution appears destined to become more common as issues involving complex and interrelated tectonic, sedimentologic, hydrologic, climatic, and diagenetic histories are increasingly addressed.

GLOSSARY

Avulsion: The abrupt switching in location of a stream channel, usually during flooding.

Clastic sediment: Particles eroded from a previously existing geologic formation.

Convergence: The motion of two lithospheric plates toward each other to join together at a common boundary.

Denudation: The combination of natural processes that result in the progressive lowering of the Earth's surface by weathering, erosion, mass wasting, and transport by wind and water.

Depositional environment: The set of physical (especially hydrodynamic), chemical, and biological conditions, and depositional processes, occurring in a given location to produce a characteristic sedimentary deposit.

Diagenesis: The physical, chemical, and biological changes to sediments after deposition, including precipitation of cements from circulating subsurface fluids in fractures and pore spaces, and compaction of sediment grains into a tighter packing arrangement.

Extension: The stretching or pulling apart of the Earth's crust that causes the crust to thin and fracture.

Facies: Geologic units of common characteristics that can be distinguished by the "sum of all the primary characteristics of a sedimentary unit" produced as a result of deposition under given environmental conditions.

Flexural Subsidence: Depression of the Earth's surface adjacent to loads of rock and sediment rather than just beneath such loads. It is a consequence of the fact that the Earth's crust has a strength and can propagate loads.

Geodynamic Models: Models of basin formation.

Geologic History Analysis: A collection of techniques that use local and regional geologic and climatic data to constrain the structural, sea level, climatic, and hydrologic histories of a sedimentary basin.

Isostatic Compensation: The vertical adjustment of the lithosphere to maintain an equilibrium among rocks, sediments, and water as loads are either added to or subtracted from the lithosphere.

Lithification: The processes by which recently deposited, loose sediments are transformed into rock.

Lithosphere: A layer of strength from the surface of the Earth downward. It includes the crust and part of the upper mantle and is approximately 100 km thick.

Paleowater Depths: The depths of ancient seas at an earlier point in time.

Rifting: The large-scale faulting and tilting of the Earth's crust occurring after extension of the crust.

Source Area: A feature of positive relief that constitutes the drainage area for fluid and sediment transported into a basin.

Subduction: The process by which one lithospheric plate descends beneath another.

Subsidence: The sinking or vertical downward displacement of the Earth's surface.

Tectonic Loading: Depression of the lithosphere adjacent to mountains formed by crustal compression.

Uplift: The positive vertical motion of the Earth's surface by tectonic forces or isostatic compensation.

Upwelling: An upward vertical movement of upper mantle material.

Weathering: The group of chemical and physical processes that cause decomposition of geologic materials *in situ* into smaller particles, individual minerals, and dissolved constituents, making materials available for transport by wind and water.

Works Cited

Allen, J. R. L. (1965), "A review of the origin and characteristics of recent alluvial sediments," *Sedimentology* **5,** 89–191.

Allen, J. R. L. (1985), *Principles of Physical Sedimentology,* London: George Allen and Unwin, 272 pp.

American Society of Civil Engineers (ASCE) (1976), in V. A. Vanoni (Ed.), *Sedimentation Engineering,* Manuals and Reports on Engineering Practice No. 54, New York: American Society of Civil Engineers, 745 pp.

Anderson, M. P., Woessner, W. (1992), *Applied Groundwater Modeling. Simulation of Flow and Advective Transport,* San Diego: Academic Press, 381 pp.

Beaumont, C., Fullsack, P., Hamilton, J. (1991), "Erosional Control of Active Compressional Orogens," in: K. R. McClay (Ed.), *Thrust Tectonics,* New York: Chapman and Hall, 447 pp.

Bethke, C. M., Harrison, W. J., Upson, C., Altaner, S. P. (1990), "Supercomputer Analysis of Sedimentary Basins," *Science* **239,** 261–267.

Bird, R. B., Stewart, W. E., Lightfoot, E. N. (1960), *Transport Phenomena,* Chichester: Wiley and Sons, 780 pp.

Bitzer, K., Pflug, R. (1990), "DEPO3D: A Three-Dimensional Model for Simulating Clastic Sedimentation and Isostatic Compensation in Sedimentary Basins," in: T. A. Cross (Ed.), *Quantitative Dynamic Stratigraphy,* Englewood Cliffs, NJ: Prentice Hall, pp. 335–348.

Blatt, H., Middleton, G., Murray, R. (1980), *Origin of Sedimentary Rocks,* Englewood Cliffs, NJ: Prentice Hall, 782 pp.

Bridge, J. S. (1977), "Flow, Bed Topography, Grain Size and Sedimentary Structure in Open Channel Bends: A Three-Dimensional Model," *Earth Surf. Proc.* **2,** 401–416.

Bridge, J. S. (1992), "A Revised Model for Water Flow, Sediment Transport, Bed Topography, and Grain Size Sorting in Natural River Bends," *Water Resources Res.* **28** (4), 999–1013.

Bridge, J. S., Leeder, M. R. (1979), "A Simulation Model of Alluvial Stratigraphy," *Sedimentology* **26,** 617–644.

Busby, C. J., Ingersoll, R. V. (1995), *Tectonics of Sedimentary Basins,* Cambridge, MA: Blackwell Science, 579 pp.

Coleman, J. M., Wright, L. D. (1975), "Modern River Deltas: Variability of Process and Sand Bodies," in: M. L. Broussard (Ed.), *Deltas, Models for Exploration,* Houston: Houston Geological Society, pp. 99–149.

Curray, J. R. (1969), "Shallow Structure of the Continental Margin," in: D. J. Stanley (Ed.), *The New Concepts of Continental Margin Sedimentation,* JC-12, Washington, DC: American Geological Institute, pp. 1–22.

Dawdy, D. R., Vanoni, V. A. (1986), "Modeling Alluvial Channels," *Water Resources Res.* **22** (9), 71S–81S.

Dickinson, W. R. (1974), "Plate Tectonics and Sedimentation," in: W. R. Dickinson (Ed.), *Tectonics and Sedimentation,* SEPM Special Publication 22, Tulsa, OK: Society of Economic and Paleontologic Mineralogists, p. 1–27.

Dickinson, W. R. (1976), *Plate Tectonic Evolution of Sedimentary Basins,* AAPG Continuing Education Course Notes Series No. 1, Tulsa, OK: American Association of Petroleum Geologists, 62 pp.

Einsele, G. (1992), *Sedimentary Basins,* New York: Springer-Verlag, 628 pp.

Ericksen, M. C., Masson, D. S., Slingerland, R., Swetland, D. W. (1990), "Numerical Simulation of Circulation and Sediment Transport in the Late Devonian Catskill Sea," in: T. A. Cross (Ed.), *Quantitative Dynamic Stratigraphy,* Englewood Cliffs, NJ: Prentice Hall, pp. 293–305.

Freeze, R. A., Cherry, J. A. (1979), *Groundwater,* Englewood Cliffs, NJ: Prentice-Hall, 604 pp.

Fukushima, Y., Parker, G., Pantin, H. M. (1985), "Prediction of Ignitive Turbidity Currents in Scripps Submarine Canyon," *Marine Geol.* **67,** 55–81.

Galloway, W. E., Hobday, D. K. (1983), *Terrigenous Clastic Depositional Systems,* New York: Springer-Verlag, 423 pp.

Gross, M. G. (1980), "Variations in the O^{18}/O^{16} and C^{13}/C^{12} ratios of diagenetically altered limestones in the Bermuda Islands," *J. Geol.* **72,** 170–194.

Ingersoll, R. (1988), "Tectonics of Sedimentary Basins," *Geol. Soc. Am. Bull.* **100,** 1704–1719.

Iverson, R. M., LaHusen, R. G. (1992), "Momentum Transport in Debris Flows: Large-Scale Experiments," *EOS, Trans. Am. Geophys. Union* **73** (43), 227.

Jarvis, G. T., McKenzie, D. P. (1980), "Sedimentary Basin Formation with Finite Extension Rates," *Earth Planet. Sci. Lett.* **48,** 42–52.

Jervey, M. T. (1988), "Quantitative geological modeling of siliciclastic rock sequences and their seismic expression," in: C. K. Wilgus, B. S. Hastings, C. G. Kendall, H. M. Posamentier, C. A. Ross, J. C. Van Wagoner (Eds.) *Sea Level Changes—An Integrated Approach,* SEPM Special Publication 42, Tulsa, OK: Society of Economic and Paleontologic Mineralogists, pp. 47–69.

Johnson, A. (1970), *Physical Processes in Geology,* Geology, San Francisco: Freeman, Cooper, and Co.

Jordan, T. E. (1981), "Thrust Loads and Foreland Basin Evolution, Cretaceous, Western United States," *Am. Assoc. Pet. Geol. Bull.* **65,** 2506–2520.

Kingston, D. R., Dishroon, C. P., Williams, P. A. (1983), "Global basin classification system," *Am. Assoc. Pet. Geol. Bull.* **67,** 2175–2193.

Koltermann, C. E., Gorelick, S. M. (1992), "Paleoclimate Signature in Terrestrial Flood Deposits," *Science* **256,** 1775–1782.

Koltermann, C. E., Gorelick, S. M. (1995), "Fractional packing model for prediction of hydraulic conductivity in sediment mixtures," *Water Resources Res.* **31,** 3283–3297.

Koltermann, C. E. (1993), *Geologic Modeling of Spatial Variability in Sedimentary Environments for Groundwater Flow Simulation,* Ph.D. Dissertation, Stanford University, 314 pp.

Komar, P. D. (1988), "Sediment Transport by Floods," in: V. R. Baker, R. C. Kochel, P. C. Patton (Eds.), *Flood Geomorphology,* New York: Wiley, pp. 97–111.

Martinez, P. A., Harbaugh, J. W. (1993), *Simulating Nearshore Processes,* New York: Pergamon, 265 pp.

McKenzie, D. P. (1978), "Some Remarks on the Development of Sedimentary Basins," *Earth Planet. Sci. Lett.* **40,** 25–32.

Meyer, A. J., Landais, P., Brosse, E., Pagel, M., Carisey, J. C., Krewdl, K. (1989), "Thermal History of the Permian Formations from the Breccia Pipes Area (Grand Canyon Region, Arizona)," *Geol. Rundsch.* **78,** 427–438.

Miall, A. D. (1984), *Principles of Sedimentary Basin Analysis,* Heidelberg: Springer-Verlag, 490 pp.

Mirza, S., Richardson, J. F. (1979), "Sedimentation of Suspension of Particles of Two or More Sizes," *Chem. Eng. Sci.* **34,** 447–454.

Mitchell, A. H. G., Reading, H. G. (1986), Sedimentation and Tectonics, in: H. G. Reading (Ed.),

Sedimentary Environments and Facies, 2nd ed., Oxford: Blackwell, pp. 471–519.

Norell, M. A., Clark, J. M., Chiappe, L. M., Dashzeveg, D. (1995), *Science* **378,** 774–776.

Paola, C., Heller, P. L., Angevine, C. L. (1992), "The Large-Scale Dynamics of Grain-Size Variation in Alluvial Basins, 1: Theory," *Basin Res.* **4,** 73–90.

Pettijohn, F. J., Potter, P. E., Siever, R. (1972), *Sand and Sandstone,* New York: Springer-Verlag, 618 pp.

Reading, H. G. (Ed.) (1986), *Sedimentary Environments and Facies,* New York: Elsevier, 557 pp.

Reineck, H. E., Singh, I. B. (1980), *Depositional Sedimentary Environments,* Heidelberg: Springer-Verlag, 549 pp.

Savage, S. B., Hutter, K. (1989), "The motion of a finite mass of granular material down a rough incline," *J. Fluid Mech.* **199,** 177–215.

Slingerland, R., Harbaugh, J. W., Furlong, K. (1994), *Simulating Clastic Sedimentary Basins,* Englewood Cliffs, NJ: Prentice Hall, 220 pp.

Tetzlaff, D. M., Harbaugh, J. W. (1989), *Simulating Clastic Sedimentation,* New York: Van Nostrand Reinhold, 202 pp.

Turcotte, D. L., Schubert, G. (1982), *Geodynamics: Applications of Continuum Physics to Geological Problems,* Wiley: New York, 450 pp.

Webb, E. K. (1995), "Simulation of Braided Channel Topology and Topography," *Water Resources Res.* **31,** 2603–2611.

Wendebourg, J., Harbaugh, J. (in press), *Simulating Oil Migration and Accumulation in Stratigraphic Traps,* Oxford: Pergamon, 350 pp.

Willgoose, G. R., Bras, R. L., Rodriguez-Iturbe, I. (1991), "A Coupled Channel Network Growth and Hillslope Evolution Model," *Water Resources Res.* **27** (7), 1671–1702.

Zeng, J., Lowe, D. R. (1992), "A Numerical Model for Sedimentation from Highly-Concentrated Multi-sized Suspensions," *Math. Geol.* **24,** 393–415.

Further Reading

Harbaugh, J. W., Bonham-Carter, G. (1970), *Computer Simulation in Geology,* New York: Wiley, 575 pp.

Kendall, C. G. St. C., Strobel, J., Moore, P., Cannon, R., Bezdek, J., Biswas, G. (1989), "Simulation of the Sedimentary Fill of Basins," in: E. K. Franseen, W. L. Watney (Eds.), *Sedimentary Modeling: Computer Simulation of Depositional Sequences,* Kansas Geological Survey Subsurface Geology Series 12, Lawrence, KS: Kansas Geological Survey, pp. 1–5.

Kleinspehn, K. L., Paola, C. (Eds.) (1988), *New Perspectives in Basin Analysis,* New York: Springer-Verlag.

Lawrence, D. T., Doyle, M., Aigner, T. (1990), "Stratigraphic Simulation of Sedimentary Basins—Concepts and Calibrations," *Am. Assoc. Pet. Geol. Bull.* **74,** 273–295.

Ritter, D. F. (1986), *Process Geomorphology,* New York: Wm. C. Brown, 579 pp.

Simpson, J. E. (1987), *Gravity Currents: In the Environment and the Laboratory,* New York: Wiley, 244 pp.

Walker, R. G. (1984), "General Introduction: Facies, facies sequences and facies models," in: R. G. Walker (Ed.), *Facies Models,* 2nd ed., Geoscience Canada Reprint Series 1, St. John's, Newfoundland: Geological Association of Canada, pp. 1–9.

Wilgus, C. K., Hastings, B. S., Kendall, C. G., Posamentier, H. W., Ross, C. A., Van Wagoner, J. C. (Eds.) (1988), *Sea-Level Changes: An Integrated Approach,* SEPM Special Publication 42, Tulsa, OK: Society of Economic and Paleontologic Mineralogists.

SEISMOGRAPHS

JONATHAN BERGER AND DUNCAN C. AGNEW, *Institute of Geophysics and Planetary Physics, Scripps Institution of Oceanography, University of California, San Diego, San Diego, California, U.S.A.*

INTRODUCTION

Seismographs are systems for recording the motion of the ground; very similar principles apply to systems for measuring the vibrations of structures. The interest in making such records comes primarily from the study of the waves radiated from a sudden source of stress change, such as an earthquake or explosion. These waves provide information, often the only information, both about the source and about the material properties of the parts of the Earth through which they propagate. Using artificially generated seismic waves to study Earth structure has led to a vast commercial application of seismic methods in the exploration for hydrocarbons and other mineral resources. The development of products for use by that industry has had some influence on what is used in academic studies, but less than might be supposed: Most academic studies are of waves with lower frequency (which travel to greater distances and depths), and from earthquake sources. We focus here on the instruments and procedures used primarily outside the exploration industry.

The seismic system consists of two parts: the seismometer, or seismic sensor, a transducer for converting ground motion into some other recordable quantity (almost always an electrical signal), and a recording system. (Note that in the exploration industry the customary term for seismometer is "geophone.") While the recording system shares some properties with other types of data recording, the nature of seismology imposes special requirements. When earthquake-generated waves are used, the times at which these will occur are unknown. This means that seismographs must be operated continuously; but since the signals of interest occupy only a small fraction of the recording time, the recorder needs a high capacity.

Because each earthquake is a unique and nonreproducible event, geophysicists often use data from instruments that would now be obsolete, extracting new results from old data. It is therefore appropriate to consider the historical development of seismic systems in some detail. The first instruments capable of producing data of useful quality were developed around 1900 (Dewey and Byerly, 1967); studies of earthquakes before

3-527-28139-8/96/$5.00 + .50

this time are characterized as "preinstrumental." These early systems, like those in use today, measured the motion of a suspended mass relative to the ground, but were limited in their ability to magnify this motion, since this was done through a system of levers, either mechanical or optical. Such "direct-recording" systems predominated until the 1950s; one of them, the Wood–Anderson seismometer (Anderson and Wood, 1925), was the basis for the earthquake magnitude scale. An advance on this was the introduction of electromagnetic sensors (discussed below) driving galvanometer recorders. Although developed soon after 1900 by B. Golitzin, these were not used extensively in global seismology until the introduction of the World-Wide Standardized Seismographic Network in 1963. The latest advance, in the mid-1970s, was the introduction of active electronics and feedback into the sensor to shape the response and reduce noise.

The mid-1970s also saw the beginning of routine digital recording in earthquake seismology. Almost all systems before this (except in exploration geophysics) recorded a time record of the signal as a "wiggly" line, purely in analog form. The rapid advance of computer technology since 1970 has meant that digital storage devices have reached the capacity needed for seismology, and virtually all new seismographs are digital in their recording, though the sensors use analog electronics.

1. SEISMIC SENSORS

Except for the limited class of instruments discussed in Sec. 1.8, all seismic sensors are inertial: They measure the motion of a suspended mass relative to the instrument frame, the motion of which can usually be taken to be the same as that of the ground. In all modern instruments the mass is free to travel only in one direction, giving a single degree of freedom, and a single direction of ground motion to which the sensor is (ideally) sensitive. The suspension must also exert some restoring force to keep the mass in a fixed position in the absence of shaking. This restoring force is, again ideally, proportional to the distance of the mass from its stable position, and this makes the equation of motion of the mass that of a standard harmonic oscillator:

$$m\ddot{q} + d\dot{q} + kq = m\ddot{x}, \tag{1}$$

where q is the displacement of the mass relative to the frame and x the motion of the frame in "inertial" space (actually, to a frame corotating with the Earth). The mass is m, the restoring force is kq, and $d\dot{q}$ is a velocity-dependent force included to account for damping. This can be written in a more standard form as

$$\ddot{q} + 2\gamma\omega_0\dot{q} + \omega_0^2 q = \ddot{x}, \tag{2}$$

where ω_0 is the frequency of the undamped mechanical mass/suspension combination and γ is the damping factor.

This equation shows that what a seismic sensor measures is ground acceleration (with some complications to be discussed below). If the frequency of ground motion is much less than ω_0, the mass motion q will increase as ω_0^{-1}. In early instruments, when magnifying the motion was difficult, this made long mechanical periods attractive, which was achieved by minimizing the restoring force. The difficulty this introduces is that the lower the restoring force the more susceptible the instrument is to forces introduced from other sources (such as temperature effects), and the more fragile it becomes: Seismometers for use under rough conditions thus have to have shorter periods than those fixed in observatories. Note that if the frequency of ground motion is much greater than ω_0, the mass displacement is simply proportional to the ground displacement, so that the sensor can be termed a "displacement meter."

One complication that can be of importance is that horizontal accelerations caused by local translation of the ground cannot be distinguished from the apparent accelerations caused by rotations about a horizontal axis (tilts). These accelerations appear because any tilt gives the local gravity vector a component along any originally horizontal instrument axis. At long periods, the acceleration caused by a given translation becomes smaller and the importance of tilt-induced accelerations larger. While it is possible to measure tilts separately using a gyroscope or suspended bar to define a fixed direction, these techniques do not have adequate stability for geophysical use. All devices constructed to measure tilt in fact use the gravity vector to define a direction and are thus affected by accelerations. Whether a device is called a seismometer or a tiltmeter depends on the emphasis placed on one aspect or another: What is measured is the same for both (Agnew, 1986). Similarly, since a vertical seismometer measures changes in vertical acceleration, it is in some sense equivalent to a gravimeter for measuring changes in the acceleration of gravity: Which name is used depends upon the application.

1.1 Sensor Requirements: Dynamic Range, Sensitivity, Bandwidth

What has made seismographs such demanding instruments to construct has been the very wide range, in both frequency and amplitude, in what is to be measured. Figure 1 shows the range of some signals and the background noise against which they must be measured.

Partly because of the peak in the noise spectrum (Fig. 1), known as the microseism peak, and partly for historical reasons, seismic waves (and seismic sensors) have been divided into short period (SP) and long pe-

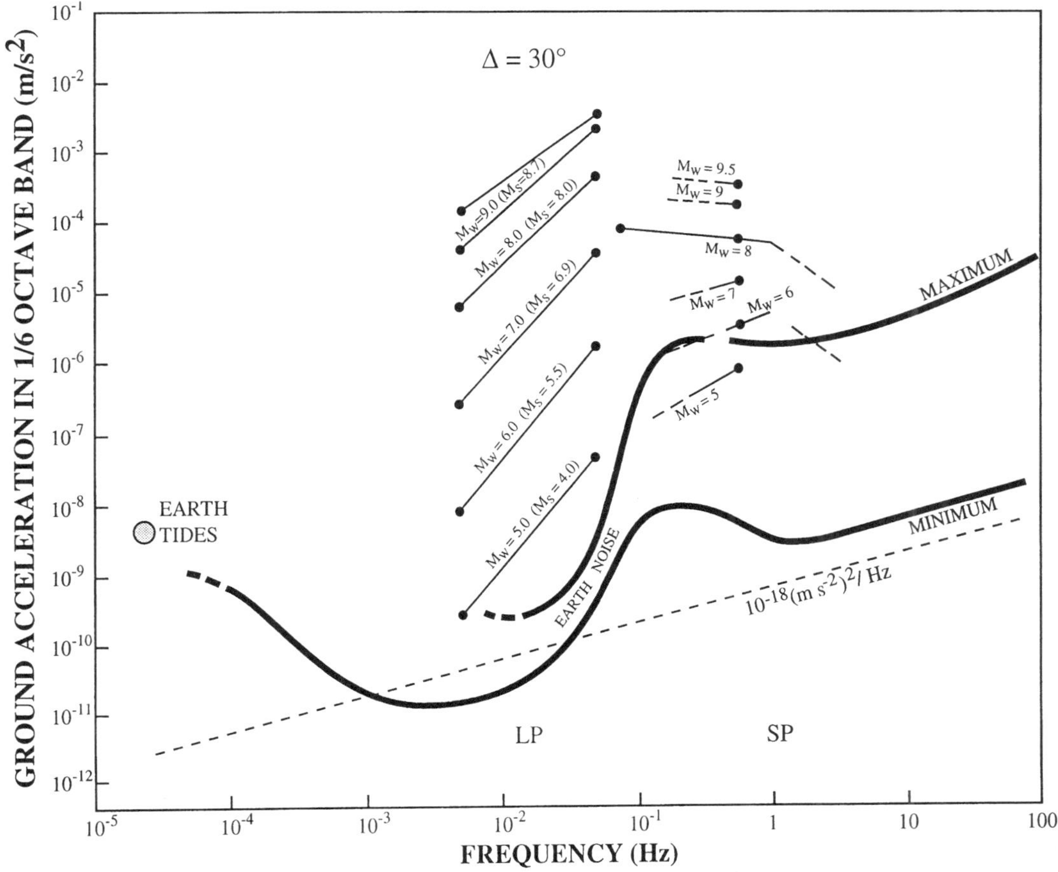

FIG. 1. Ambient seismic noise. The heavy lines show the range of expected ambient noise at a continental site. The light lines show the signals expected from earthquakes of different magnitudes at a distance of 30°. The dashed line indicates a level of constant acceleration with a rms value of 3×10^{-9} m s^{-2} over a 10-Hz bandwidth.

riod (LP), with the dividing period being about 6 s. This is, however, a relative term: In exploration geophysics, a long-period sensor might be sensitive to wave periods longer than 0.5 Hz.

Earthquake signals that propagate as elastic waves extend over a bandwidth from about 3×10^{-4} Hz, the frequency of the gravest normal mode of the whole Earth, to frequencies in excess of 50 Hz. The attenuation in the Earth limits the maximum frequency present to about 100 Hz within tens of kilometers of the source and 1–10 Hz at distances beyond 1000 km.

Ground motions generated by earthquakes can reach from nearly 1 m $s^{-2}/Hz^{1/2}$ in the near field (on the order of an earthquake fault dimension) down to the smallest detectable signals just above the ambient background noise. At teleseismic distances (1000 km or more from the source) or greater, peak accelerations will rarely exceed 5×10^{-3} m $s^{-2}/Hz^{1/2}$, and those will be for surface waves with a frequency of about 5×10^{-2} Hz. Figure 1 illustrates the range of signals expected at about 30° (3300 km) from an earthquake source. The background noise, representing the limit of detectability of earthquake signals, is a composite of the quietest readings observed at different sites around the world (Peterson, 1993). The dynamic range of teleseismic signals to be measured peaks at some 160 dB at about 3×10^{-2} Hz.

1.2 Suspension Designs

As noted above, the mass in an inertial seismometer is attached to the frame through some mechanical system, called the "suspension," which has to allow it to move with only one degree of freedom and provide a restoring force to keep it in position in the absence of acceleration. This is especially difficult for instruments that measure vertical motion (in whole or in part) because the restoring force must then be, not zero, but an amount equal to the local force of gravity upon the mass. The difficulty is then in keeping this nonzero restoring force constant to 10^{-10} (equivalent to the ambient ground noise over a 10-Hz bandwidth). Nearly all inertial seismometers use some combination of elastic materials and geometrical constraints on the mass to provide a suspension. The simplest example of an elastic suspension would be a mass hanging from a coil spring; a simple suspension for horizontal motion would be a mass on the end of a rod, forming a simple pendulum. In the second case the restoring force is provided by gravity. Most actual suspensions are elaborations and combinations of these.

For measurements of horizontal motion the mass may be constrained to move in a straight line, with the restoring force provided by springs: This is the design used in most short-period sensors. For longer-period instruments, an old, but still common, suspension remains the horizontal pendulum (or "garden gate"), in which the mass is constrained to move in a nearly horizontal circle, making the gravitational restoring force small (Rodgers, 1968). Still another way to achieve this is the inverted pendulum, in which the mass rotates above its pivot point; this is, of course, unstable, so that springs are used to provide the necessary restoring forces.

For the shortest-period vertical sensors, the mass hanging on a spring is a satisfactory design, but if the period is (in practice) much more than 1 s, the equilibrium extension of the spring under gravity becomes too long. The usual solution to this is the LaCoste "zero-length spring" suspension (LaCoste, 1934), in which a specially designed spring is used in a particular geometry that can give an arbitrarily small restoring force linear in mass displacement. Other designs are possible; for example the STS-1 seismometer (Wielandt and Streckeisen, 1982) used in many modern installations uses a design based on the elastica (the shape assumed by a thin strip of metal bent elastically to large deflection).

One complication in an actual seismometer design is that whatever constraints are applied to the mass must not apply any additional restoring force to it; thus, for example, if the mass pivots about some point, the pivots must not apply any forces (unless designed to). As the period increases the limits on allowable extraneous forces diminish, and the need for careful design increases. Most seismometers thus require fine mechanical design and construction. A further complication is that many such forces are produced by thermal expansion of the materials of which the sensor is constructed, so that the best operation of a seismometer will take place only in a stable environment.

1.3 Position Sensors

The mechanical design of the sensor provides a mass whose displacement with respect to the instrument frame indicates Earth motion according to Eq. (1). The next step, in any modern instrument, is to convert this motion into an electrical signal that can be recorded. This is the purpose of the position sensor. While a very large number of such sensors have been designed for industrial use, only a few types have been employed in seismometers.

By far the commonest sensor is the moving-coil system, in which a coil moving through the field of a permanent magnet produces a voltage whose magnitude is proportional to the velocity of the mass. This system is in common use for sensors with periods of 1 s and less, and is simple, robust, and linear. In conjunction with modern electronic amplifiers such a sensor can resolve ambient ground motion for periods from 10 s to many hertz (Rodgers, 1994). It does not, however, have adequate sensitivity at longer periods. For longer-period systems it is necessary to use a position sensor whose output is proportional to mass displacement; essentially the only systems used in seismometers are capacitive and inductive sensors [both reviewed in Agnew (1986)]. The capacitive sensor is capable of resolving much smaller motions, but is environmentally less robust than the inductive systems. Both types of sensors require specialized electronics with phase-sensitive detection in order to achieve low noise, and also need careful design to assure a linear response over their full range of motion.

1.4 Feedback

One way of greatly improving sensor performance, now utilized in nearly all "observatory-grade" seismometers (though not in geophones), is to incorporate feedback of some sort. What this means is that the output of the mass-position sensor is not just recorded, but is also used to generate a force that is applied to the mass [in addition to those described by Eq. (1)]. There are a variety of ways of doing this; the most common is to make the force proportional (over some frequency range) to the mass displacement away from a null position: This is called force feedback, or sometimes a force-balance system. This requires low-noise electronics, but has the merit of reducing the motion of the mass significantly, thus reducing the linearity required of the mechanical system and replacing it by that needed in the electronics (and the voltage-to-force converter). The two main methods of applying a feedback force (electromagnetically, through a coil in a magnetic field, and electrostatically, through a capacitor) can both be made very linear.

Figure 2 illustrates schematically the principles of force feedback. In the Fourier domain, the quantities $A_g(s)$ and $E_o(s)$ are the ground acceleration and the output voltage, respectively. The quantities T_p, T_1, and T_2 represent the responses of the seismometer pendulum described, the response of the displacement transducer, and the feedback portion of the system, respectively. The feedback signal, converted to a force, is summed with the force of the ground acceleration by the mass itself with the positive and negative signs indicating that these forces are in opposition.

The overall response of the system is given by

$$E_o(s) = \frac{T_p T_1}{1 + T_p T_1 T_2} A_g(s), \tag{3}$$

where

$$T_p(s) = \frac{1}{s^2 + 2\gamma\omega_0 s + \omega_0^2} \tag{4}$$

from Eq. (2).

By careful design of the responses of the displacement transducer and feedback circuit elements, it is possible to alter the frequency response of the system, that is $E_o(s)/$

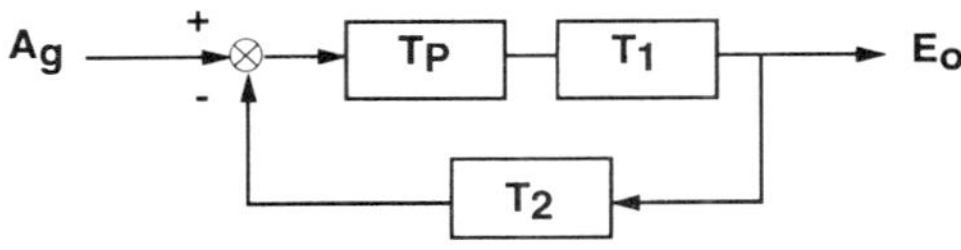

FIG. 2. Block diagram of feedback seismometer.

$A_g(s)$, from that available from the purely mechanical system described by Eq. (2). For example, with a mechanical system of relatively short period, one can obtain a system response that is flat with respect to ground velocity over an extended frequency range (Fels and Berger, 1994).

1.5 Instrumental Noise Sources

When seismologists refer to ground noise, they usually mean motions of the ground caused by something other than the waves of interest. Distinct from this is noise in the seismic sensor itself. Two important sources of noise that are peculiar to such a sensor are the thermal Brownian noise, and apparent tilts.

The thermal Brownian noise is the irreducible minimum noise level of a sensor, and arises from the motion of the mass caused by Brownian motion of the surrounding materials. It can be shown (Melton, 1976) that, for a freely suspended mass whose motion is governed by Eqs. (1) and (2), the noise level is flat in acceleration power spectral density, and is equal to

$$8kT\gamma\omega_0/m, \tag{5}$$

where T is the absolute temperature of the system and k is Boltzmann's constant. As an example, a mass of 0.1 kg with a restoring force that gives a period of 2 s, and critical damping ($\gamma = 1$), will have a noise level at room temperature of 10^{-18} m^2 s^{-4} Hz^{-1}, comparable to the lowest level of ground noise. Since the temperature cannot easily be lowered, the two main ways to reduce the noise are to increase the mass or to decrease the damping force. The latter can usefully be done if the system uses feedback to provide damping, since this does not increase noise; this approach is followed in several modern sensors.

Human activities also are a source of seismic noise, and sensors should be located in an area removed from sources of cultural noise such as towns or major roads. For island settings, the site should be chosen as far from the coast or breaking reef as practical to avoid noise caused by the surf.

As noted above, horizontal seismometers depend on the direction of local gravity for a reference, and thus cannot distinguish between tilts and horizontal motions. At periods greater than about 10 s, the tilts of the ground caused by loading from atmospheric turbulence created by wind produce a noise level on horizontal instruments that exceeds that present in the vertical by a considerable amount. Thus, the seismometer should be well insulated and shielded from the effects of temperature-induced tilts and from the effects of wind- and pressure-generated Earth noise at the surface (Sorrels and Goforth, 1973). Additionally, most sensors are temperature sensitive, and a stable temperature environment is required.

Traditionally, seismometers have been installed in basements, vaults, tunnels, or other places where a good coupling to the local rocks can be achieved, where the sensors can be isolated from common sources of man-made vibrations, and far enough below the surface to reduce the temperature-induced tilts.

In the 1960s much research was devoted to studying the efficacy of installing seismometers in boreholes. It was quickly discovered that while the background noise decreased with depth, so too did the signal strength as a result of destructive interference from surface reflections. Empirically, it was found that not much improvement was achieved beyond a depth of 100 m, and this is now the accepted standard for quiet continental sites.

1.6 Calibration

In order to obtain useful information from the recorded seismogram, it is necessary to have the seismic system calibrated: It must be known, for all frequencies of interest, how large a ground motion in a given direction will produce a given recorded signal. This problem, in most modern systems, is divided into two parts. Relative calibration is the problem of determining the frequency response of the system, which may be done by injecting a known signal into the feedback loop (or an auxiliary forcing system) and performing a cross-spectral analysis between this known input and the recorded output (Berger *et al.*, 1976; Rodgers *et al.*, 1995). This may, with modest care, provide relative responses to better than 1%.

The more difficult problem of absolute calibration is then to relate the output to actual ground motion (perhaps only at one frequency). Providing a known displacement (from putting the seismometer on a shake table) is possible at short periods but becomes difficult at longer ones: Most long-period systems are calibrated by tilting them (so that the change in gravity mimics an apparent acceleration) or by moving masses around them.

Another method of calibrating seismometers is to provide a way to induce a force on the seismometer mass electromagnetically, usually by the application of a voltage to a forcing coil. In practice, it is considerably easier to calibrate the forcing coil than it is to calibrate the overall seismometer and hence this technique is widely used.

Finally there is the apparently simple problem of establishing the actual orientation of the seismometer. Since this often has to be done in boreholes or other underground locations this is not as straightforward as it might appear; and experience has shown that it is often done incorrectly.

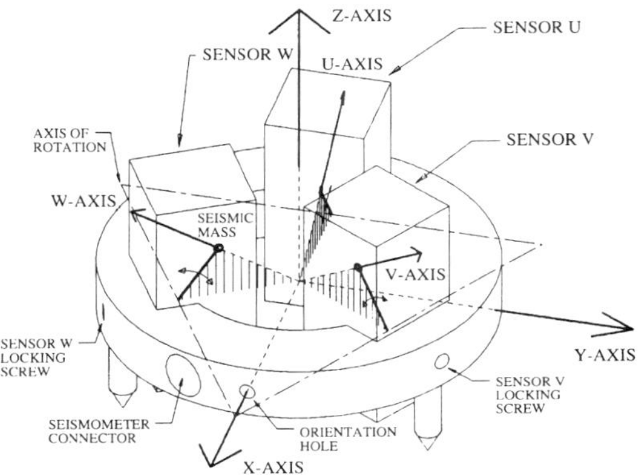

FIG. 3. Streckeisen STS-2 three-component seismometer.

1.7 Examples of Modern Seismometers

Two examples of modern three-component seismometers are shown in Figs. 3 and 4; one is designed for quick and simple installation in a vault or other environment accessible to the installer and the other is designed to be emplaced in an underground borehole. Both instruments employ force feedback and have electronics to shape the response so that the output is proportional to ground velocity over most of their bandwidth, which extends from less than 10^{-2} to about 50 Hz.

The STS-2 manufactured by G. Streckeisen AG of Pfungen, Switzerland, is a second generation of the leaf-spring design of Wielandt and Streckeisen (1982). It is meant to be easily portable and capable of operation over a wide temperature range. The unit employs three identical sensors, which are obliquely oriented, with their sensitive axes 45° to the vertical. Analog circuitry combines the sensor outputs to provide signals proportional to the ground motion in the coordinate system of the local vertical. The sensors and control electronics are mounted in a vacuum-tight cylindrical case of 24-cm diam designed to minimize the effects of distortion by changes in barometric pressure. The top and bottom covers are secured to a massive base plate in such a way as to let the covers compress without stressing the entire package. As with other modern feedback seismometers, it will resolve the minimum ground noise (Fig. 1) over a frequency band from about 2×10^{-3} to 15 Hz.

The CMG-1, manufactured by Guralp Systems, Ltd, of Reading, England, is a borehole seismometer designed to be installed in a casing of at least 15-cm inside dimension. It consists of a vertically oriented and two horizontally oriented sensors housed, together with control electronics, in a package of 12-cm diam by 1.55-m length. Two external cylindrical metal bars are affixed to the casing to fit the inside diameter of the borehole into which the package is installed. After lowering the unit to the desired depth, a spring-loaded arm is released by motor drive, which locks the package firmly against the borehole walls.

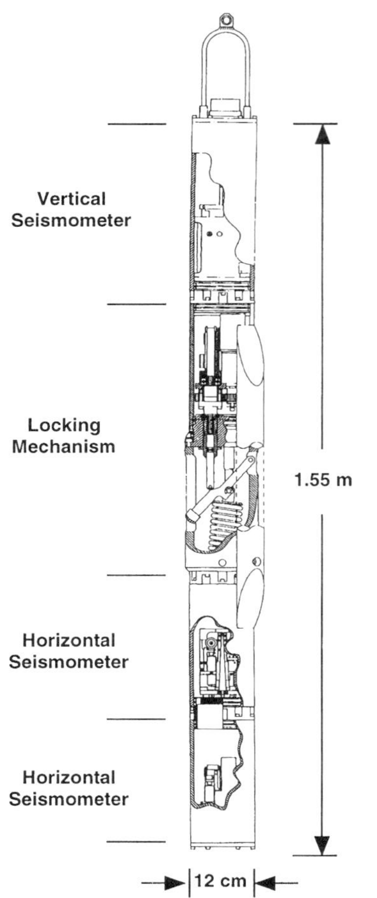

FIG. 4. Guralp CMG-1 three-component borehole seismometer.

The CMG-1 will level itself in boreholes tilted up to 3.5° from vertical. The horizontal components can be rotated relative to the borehole *in situ* to a precision of 0.2°. It has an output that is proportional to ground velocity from about 2.8×10^{-3} to 50 Hz with an internal noise lower than the lowest ambient ground noise (Fig. 1) from about 1.5×10^{-3} to 20 Hz.

1.8 Strain Seismometers

One method of recording seismic waves does not rely on measuring the ground motion, but rather the strains associated with it. One way of doing this is to measure the displacement between two relatively closely spaced points: Since this is (for seismic waves with wavelengths longer than this distance) proportional to the extensional strain, such instruments are also called extensometers. The first strain seismometer of sufficient sensitivity was developed by Benioff in 1935: This used a velocity transducer, one part of which was attached to one point and the other to a 30-m rod extending to the other point; such solid-rod strainmeters have been built in a variety of modes (Agnew, 1986). A greater separation between fiducial points can be achieved by using a laser system to measure the relative displacements. An alternative way to measure strain is to cement a flexible capsule in the ground, and measure the change in volume through the motion of a liquid in and out of it; such systems, called dilatometers, are in some use for looking for very-long-period strain changes. Neither extensometers nor dilatometers have seen much use in recording seismic waves; it does not appear that strain measurements usually provide much information not available from measuring displacements with inertial sensors.

2. RECORDING SYSTEMS

2.1 Recording Systems—Historical

In the earliest seismometers the motion of the mass was magnified mechanically and recorded directly, one of the lowest-friction ways of doing this being to have a stylus scratching paper coated with a layer of soot. Direct optical magnification, recording on photographic paper or film, came into common use somewhat later. Both methods remain in occasional use, the former being an inexpensive and robust method for field operations. The usual way of providing an immediate visual record remains a pen-and-ink recorder, almost always recording on a drum, but the majority of seismic recorders (before the introduction of digital recording) used photographic recording of the light spot reflected off a mirror attached to a galvanometer. The galvanometer provided a low-noise method for detecting the small voltages generated by most seismic sensors (especially those using a moving-coil position sensor), as well as some flexibility in determining the frequency response (Willmore, 1979).

2.2 Modern Data-Acquisition Systems

The general requirements of the seismographic data-acquisition system are dictated by the purpose of the station: Those employed for seismic exploration will differ considerably from those employed for teleseismic monitoring, for example. Many elements are common, however, and the spectra of signals to be recorded and background noise will determine the details. For illustration, we describe the requirements for a teleseismic station. Figure 1 depicts the spectrum of signals to be measured and a model of the ambient ground noise. From this figure, the required bandwidth and dynamic range can be determined.

The main functions of the data-acquisition system are

1. amplification or electrical isolation of each input analog signal,
2. antialias filtering,
3. analog-to-digital conversion,
4. formatting and time stamping of the data,
5. signal processing, mainly filtering and decimation,
6. event detection,
7. digital data storage on some high-capacity medium,
8. data transmission, via a modem or hard wire, to a data-collection center, and
9. digital-to-analog conversion for generating calibration signals or to monitor the digital signals.

Because the sensors are usually placed in a relatively inaccessible location such as an underground vault or borehole, the data-acquisition systems are usually divided into two parts. The amplifiers and analog-to-digital conversion units (ADC) are located close to the sensors to minimize possible electrical interference, while the control and recording units are located in a place with convenient access for operators. A generic seismological data-acquisition system designed primarily for use at earthquake observatories is outlined in Fig. 5.

The analog amplifiers and filtering subsystem provide the interface between the sensors and the ADC. Their purpose is to provide any analog amplification that may be required to match the sensor output levels with the available ADC noise level. Additionally, the unit must low-pass filter the analog signals to insure there will be no aliasing in the analog-to-digital conversion process. In general the ambient ground noise passed through the sensor should produce a voltage slightly greater than the combined noise level of the amplifiers, analog filters, and ADC in order to maximize dynamic range.

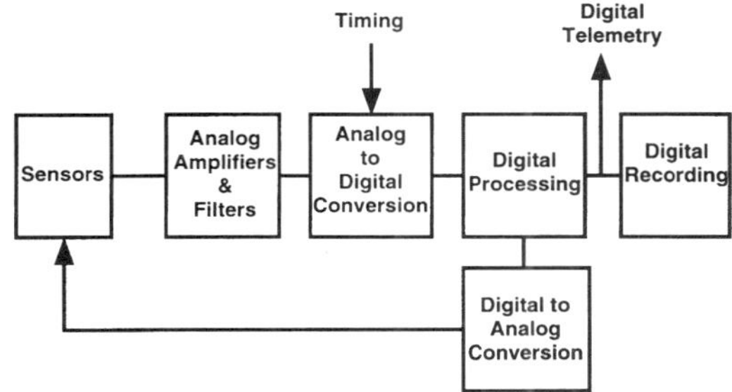

FIG. 5. Generic seismic data-acquisition system.

Quite often, a variety of sensors will be used to measure the entire spectrum of seismic signals and other interesting variables. A typical unit might be designed to accept the following inputs:

1. three axis broadband channels,
2. three axis high-frequency channels,
3. three axis broadband-seismometer mass-position channels.
4. three axis strong-motion channels,
5. one channel for barometric pressure,
6. one channel for temperature sensor, and
7. three axis anemometer channels.

2.2.1 Analog-to-Digital Conversion

The analog-to-digital conversion subsystem is the heart of the data-acquisition system. As the name implies, it performs the conversion of the signal from its continuous analog form into its sampled digital representation. Additionally, it must "attach" to each digital sample the time of sampling.

The bandwidth of the seismic signals of interest (~100 Hz) is not a challenge to modern ADCs. The dynamic range required, however, is available only in the highest-resolution digitizers currently available. As discussed in Sec. 1.1, the dynamic range of teleseismic signals to be measured peaks at about 0.03 Hz at some 160 dB.

The dynamic range can be defined as the ratio of

full scale (FS) to rms noise in $\frac{1}{6}$ decade band ($\Delta f/f = 0.386$).

If the spectral density of the system noise is $N(f)$ per f_n, where f_n is the digitized Nyquist frequency, then the dynamic range per $\frac{1}{6}$ decade band will be *frequency dependent* and given by

$$\text{Dynamic range} = \frac{\text{FS}}{\sqrt{0.386N(f)f/f_n}}. \qquad (6)$$

The process of analog-to-digital conversion introduces an uncertainty into the signal as a result of the finite resolution. Assuming that there is a uniform probability of the actual signal voltage being anywhere in the interval $-\frac{1}{2}\Delta V$ to $+\frac{1}{2}\Delta V$, where ΔV is the resolution or least count (lc) of the digitizer, this is equivalent to adding a noise of variance $\frac{1}{12}(\text{lc})^2$ evenly distributed over the recorded bandwidth. Thus, the perfect digitizer adds a noise of power spectral density

$$N(f) = (\text{lc})^2/12f_n, \qquad (7)$$

where f_n is the digitized bandwidth. (From this formula, it can be seen that the noise of the digitizer can be reduced by increasing the digitizer bandwidth—that is, sampling faster.) For an n-bit ADC, where FS $= 2^n$, the dynamic range for a perfect n-bit encoder in decibels (dB) is then

$$\text{Dynamic range} = 6.02n + 17.4 - 10\log(f/f_n)\ \text{dB}. \qquad (8)$$

In Table 1, the dynamic range of perfect n-bit encoders is given at the Nyquist frequency and at 1/2000 of the Nyquist frequency. For a unit sampling at 200 samples per second these frequencies would be 100 and 5×10^{-2} Hz respectively. Also listed is the sample rate required to produce 160-dB dynamic range at 20 Hz, matching the performance of the seismometer illustrated in Fig. 1. From this table it can be seen that to achieve the desired dynamic range with a single digitizer, 24-bit precision is required.

Table 1. Dynamic range for perfect n-bit encoders.

Number of bits	Dynamic range at f_n (dB)	Dynamic range at $5\times10^{-2}f_n$ (dB)	Sample rate for 160 dB at 20 Hz (s^{-1})
12	89.64	122.64	4.58×10^8
16	113.72	146.72	1.82×10^6
24	161.88	194.88	40

In practice, the noise of the digitizer usually exceeds these theoretical figures as a result of a variety of internal electronic sources. A typical noise spectrum of a digitizer is illustrated in Fig. 6 (Given *et al.*, 1992). In general the shape is common to most such devices: a flat spectrum down to some frequency f_0, and below that a $1/f$ behavior:

$$N(f) = N_0 \quad \text{for } f > f_0 \qquad (9)$$

and

$$N(f) = N_0 f_0/f \quad \text{for} < f_0. \qquad (10)$$

In this case, for $f < f_0$ the dynamic range will be a constant at

$$\text{Dynamic range} = 6.02n + 6.6 - 10\log N_0 - 10\log(f_0/f_n)\ \text{dB}. \qquad (11)$$

2.2.2 Timing Timing of seismographic data is a critical element of station design, as the analysis of the data depends upon a common time base for widely separated sites. Generally speaking, the time *accuracy* should be a small fraction of a sample interval so that time series from across the network can be aligned. For data sampled at 20 Hz, this translates to a desired accuracy specification of about 5 ms.

There are a number of systems that broadcast Universal Coordinated Time (UTC) of sufficient precision for seismographic purposes. A number of national authorities support radio transmissions such as those broadcast on WWVB in the United States and MSF in Northern Europe, but the coverage of these systems is quite limited (for a global network) and reception is often poor. The OMEGA navigational system offers practically global coverage but sometimes reception is adversely affected by local topography. The system provides an accuracy estimated to 5 ms by clock manufacturers, but UTC can only be determined to within ±5 s because of the nature of the broadcast signals. An operator must set the time to within this bound manually. The recently introduced Global Positioning System (GPS), however, offers a very accurate global timing

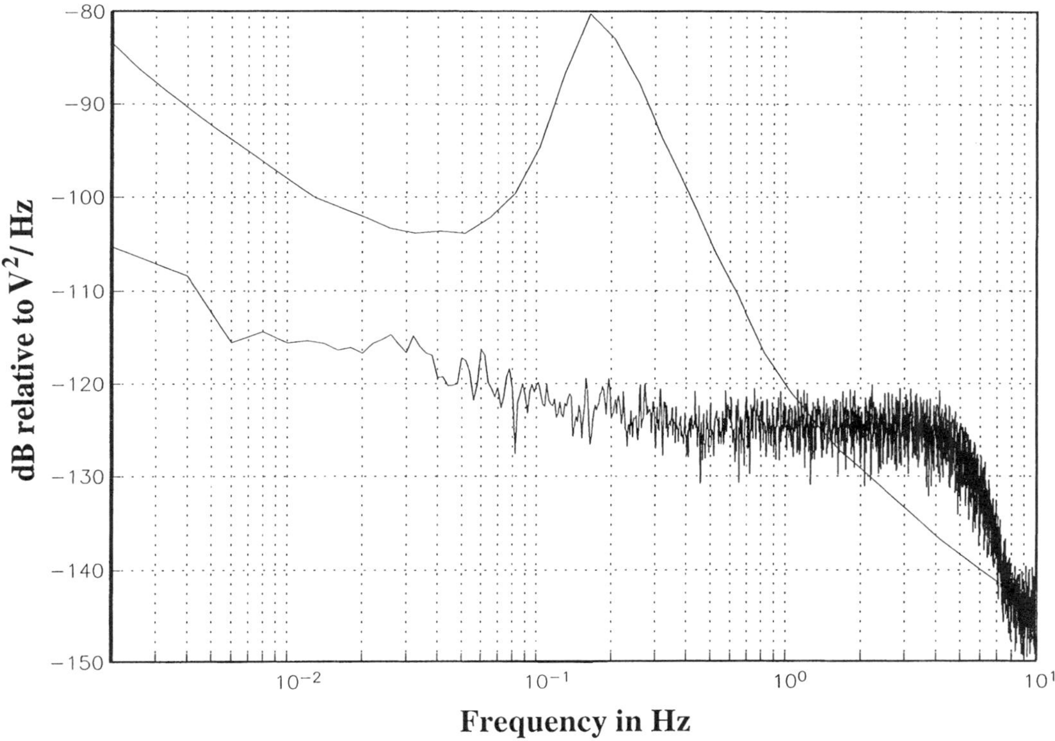

FIG. 6. Performance of a 24-bit Hewlett-Packard analog-to-digital converter. Full scale is 40 V and the sample rate is 20 samples s^{-1}. For frequencies above 0.15 Hz, the noise figure is about 6 dB higher than a perfect 24-bit encoder. For comparison, the smooth curve represents a model spectrum of ambient ground noise that has been passed through a hypothetical seismometer with a conversion factor of 2400 V m^{-1} s^2.

system that is rapidly replacing the other systems completely.

The clocks typically produce two signals: a 1-Hz pulse precisely synchronized to UTC and a time string containing the day, hour, minute, and second, and a quality factor character. The quality factor indicates if the clock is locked to the broadcast standard or, if not, how long it has been out of lock.

2.2.3 Digital Processing As discussed above, it is common to oversample the analog data to achieve better resolution through averaging. The averaging is done as part of the data-acquisition process by an internal processor, which performs digital filtering and decimation. For data-analysis purposes, it is desirable to have a variety of data streams recorded from a single sensor. Typically, sample rates of 100, 40, 20, 1, and 0.1 Hz are recorded.

In addition to the filtering and decimation, digital processing is often required to provide a "detection" filter for selected data streams. This allows, for example, strong-motion sensors to be recorded at a high rate only when a signal is detected. Data are bandpass filtered in a frequency range appropriate for the sensors, and then passed to a trigger algorithm.

A typical detector for strong-motion sensors is based on the comparison between a short-term average (STA) and a long-term average (LTA) of the sensor signal (or channel) passed through a 1.0–30.0-Hz bandpass filter. The LTA is actually computed as an average of STAs and therefore "lags" the STA by a period of time. When the ratio of the STA to the LTA is above a threshold level then the detector triggers for that channel. A trigger is maintained until the STA/LTA ratio drops below some fraction of the trigger ratio, or until the maximum event time is exceeded. Multiple channels may be used as detector inputs; an "event" occurs when a predefined number of input channels trigger (referred to as the "voting threshold"). This helps to decrease the number of false triggers due to a noisy channel. While an event is occurring any streams designated as "triggered" will be saved (otherwise they are kept only temporarily as pretrigger records). A record is usually kept of event start and end times (and approximate amplitudes). Typical event detector parameters might be as follows:

1. STA length = 0.05 s,
2. LTA length = 10 s,
3. trigger ratio = 5,
4. maximum event time = 4 min,
5. detector input channels = three short-period channels, and
6. voting threshold = 2.

2.2.4 Digital Recording It is usually required to record all data streams at the station as a local archive. The data rates from a typical teleseismic station often amount to about 2000 bytes/s of raw data, and so the daily volume of data to be recorded amounts to nearly 175 megabytes. That rate can be reduced by about a factor of 3 by various forms of data compression. A variety of digital recording media such as digital audio tape (DAT) or helical-scan 8-mm tape are readily available to accommodate a week's data on a single volume.

2.2.5 Digital Telemetry It is common to require a method of near real-time recovery of at least some of the data via telecommunication circuits. While data rates recorded at the station may amount to 2000 bytes per second, when required this rate can easily be reduced. Table 2 indicates that the overall data rate can be reduced to about 200 bytes per second for each station.

Using modern error-correcting modems, the capacities of two types of commercial telecommunication circuits are indicated in Table 3. The lower-speed circuits are available virtually worldwide via the commercial "dial-up" telephone system. The higher-speed circuits are becoming more available as services such as the Integrated Services Digital Network spread globally.

Table 2. Typical data rates.

Data streams	Date rates
Continuous seismic	3 channels × 40 samples s^{-1} × 4 bytes $sample^{-1}$ = 480 bytes s^{-1}
Other data	20 channels × 1 samples s^{-1} × 4 bytes $sample^{-1}$ = 80 bytes s^{-1}
Total data	= 560 bytes s^{-1}
Total with compression	Assume factor of ~3 = 200 bytes s^{-1}

3. NETWORKS AND ARRAYS

3.1 Local and Regional Networks

Many areas of the world are subject to frequent damaging earthquakes. In these areas, it is common to deploy a local or regional network for the purpose of locating earthquakes within a small, well-defined region.

An example of a regional network is the Southern California Network (Fig. 7), which has the prime purpose of detecting and locating earthquakes in southern California, and particularly in the urban areas around the Los Angeles Basin. Over 100 stations consisting of short-period, vertical-component seismometers are telemetered to a central site in real time, where the data are processed to detect and locate earthquakes in the region.

3.2 Global Networks

The first truly global seismographic network was the World-Wide Standardized Seismographic Network (WWSSN), deployed in the early 1960s (Oliver and Murphy, 1971) primarily to study the problems associated with detection and identification of underground nuclear testing (though not to monitor such tests). The network data, however,

Table 3. Telecommunication circuit capacities.

Nominal circuit bandwidth (bits s^{-1})	Nominal circuit bandwidth (bytes s^{-1})	One-minute volume (kilobytes)	All data compressed (min day^{-1})
9 600	1200	72	240
64 000	8000	480	36

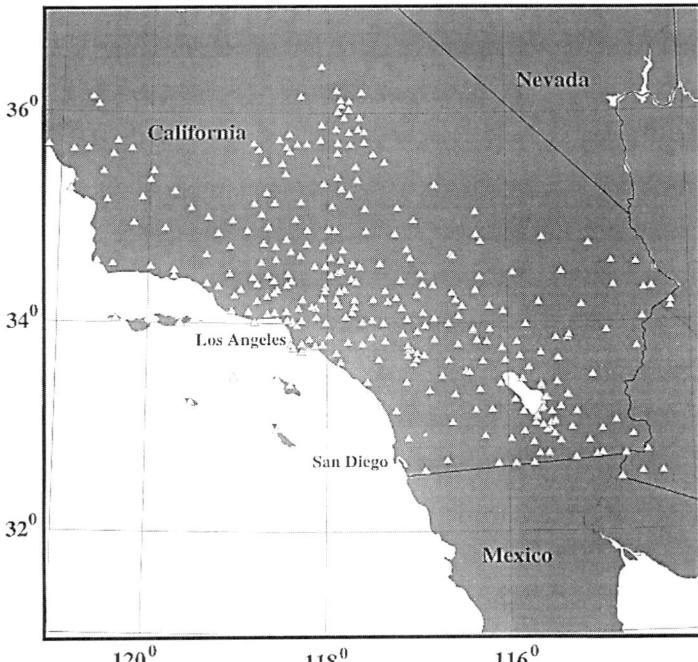

FIG. 7. The Southern California Network.

led to a revolution in solid-earth geophysics. Not only did these data lead to a new understanding of earthquake source mechanisms, but they also led to the seismological evidence that was important to the development of plate tectonics.

Among other things that were learned by an analysis of the WWSSN data were better ways of detecting and identifying clandestine nuclear testing, and this led to a decline in the support of the network and its gradual deterioration. Other less dense networks were deployed, such as GEOSCOPE, the French worldwide digital network, but it was not until the mid-1980s that a concerted plan to replace the WWSSN with modern instrumentation was developed by a consortium of university seismologists called the Incorporated Research Institutions for Seismology (IRIS).

By 1995, some 92 of a planned 147 stations of the Global Seismographic Network were operational (Fig. 8). The stations are not all equipped with identical equipment but they all meet common standards for bandwidth, dynamic range, and linearity. Data from all stations are collected and archived in a common format at a central data center, which distributes data to users upon request.

3.3 Arrays

Seismic arrays consist of two-dimensional groups of sensors deployed over dimensions of a few to tens of kilometers as illustrated in Fig. 9. The basic idea of a seismic array is that the background noise tends to be incoherent over spatial dimensions for which the signals are coherent, and hence processing of the sensor data can enhance the signal-to-noise ratio of such signals (Green *et al.*, 1965). Additionally, processing can "steer" the array in both azimuth and velocity, which can be used to help locate the source of the signals.

If a plane wave in the surface plane impinges upon the array with a velocity c in a direction ϑ, the arrival time at the ith seismometer of the array will be

$$t_i = t_0 + \frac{\cos\vartheta}{c}(x_i - x_0) + \frac{\sin\vartheta}{c}(y_i - y_0), \tag{12}$$

where t_0 is the arrival time at a reference

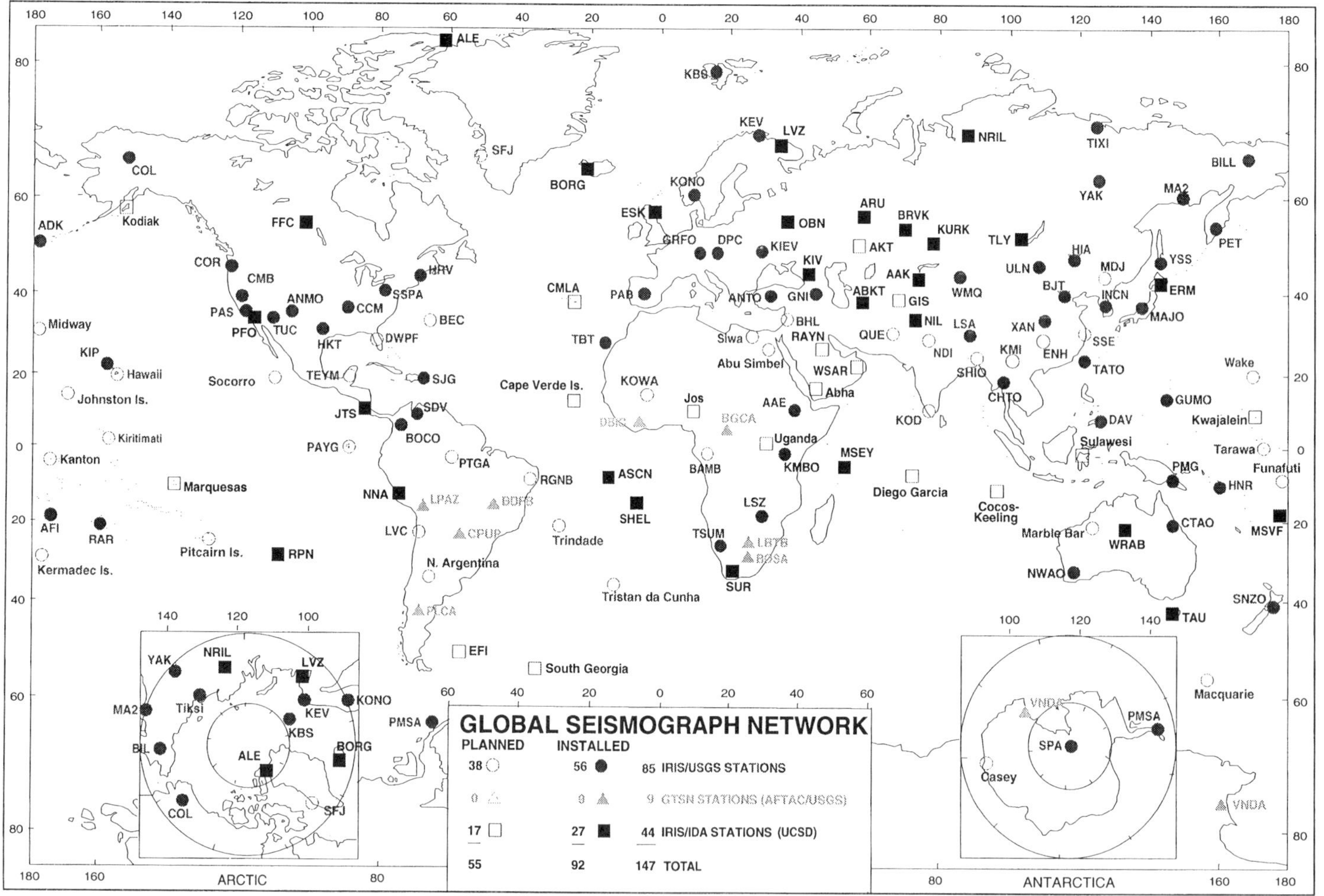

FIG. 8. The Global Seismographic Network. The IRIS/USGS stations are operated on behalf of IRIS by the United States Geological Survey (USGS) while the IRIS/IDA stations are operated by the University of California, San Diego (UCSD). The GTSN refers to stations of the Global Telemetered Seismographic Network operated on behalf of the Air Force Technical Applications Command (AFTAC) by the USGS.

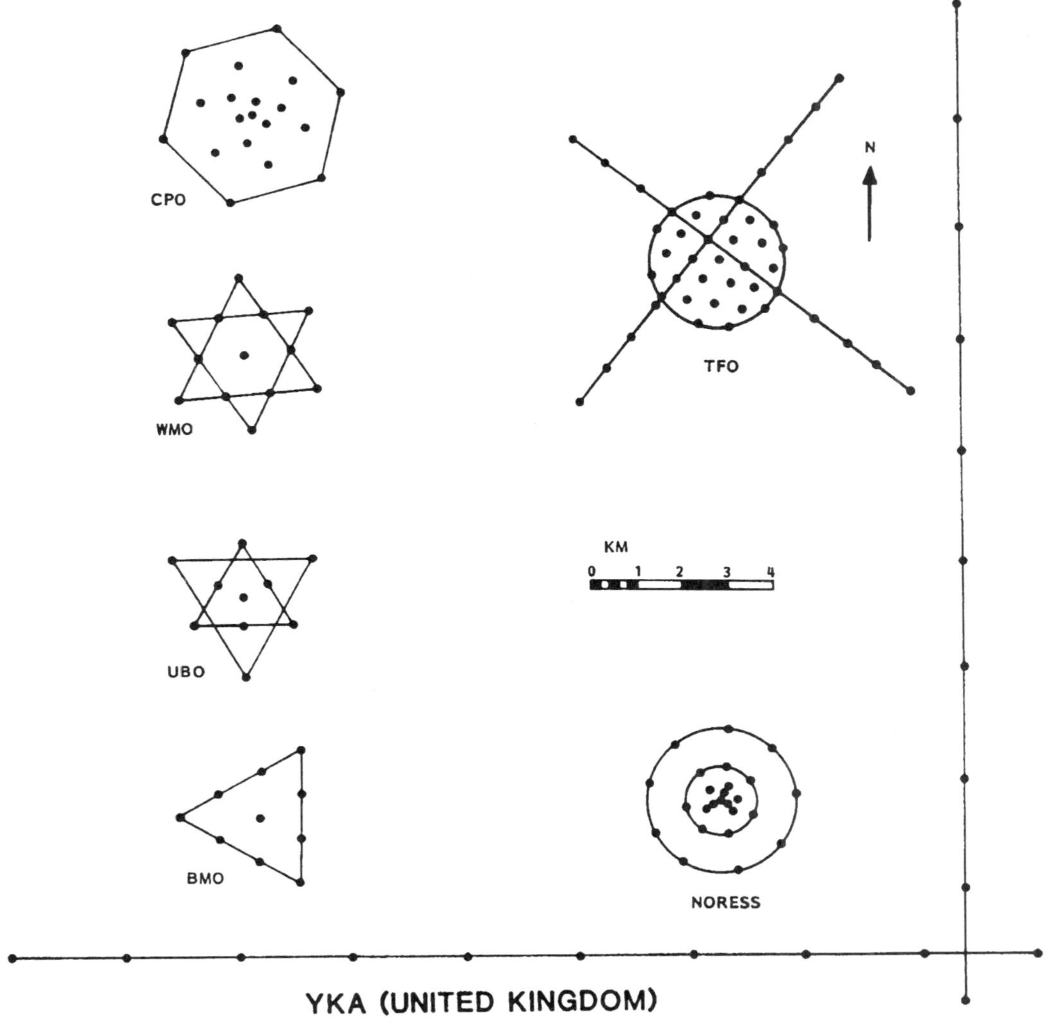

FIG. 9. Some array layouts (after Farrell, 1985). The arrays in the left column all had apertures of about 4 km. TFO was a larger experimental array and NORESS is a modern array of about 3-km aperture and 25 elements. The YKA array is a typical crossed-line array, but in this figure the intersection has been moved.

point (x_0, y_0). The array can be "steered" by delaying each seismometer output by an amount $t_i(\vartheta c)$ and summing the signals.

Plate 1 shows an example of such processing for an array of aperture about 1.5 km shown in the upper left of the figure. The plots of beam amplitude in "slowness" $(\cos\vartheta/c, \sin\vartheta/c)$ space—essentially the reciprocal velocity across the array—are shown for two different signals identified in the beam. Circles indicate constant velocity, with infinite velocity (a vertically traveling wave) at the center. From this analysis it can be seen that the first signal comes from an azimuth of 53° with an apparent velocity of 2.5 km/s—most likely a noise pulse from the nearby city of Ashgabat—while the second comes from an azimuth of 211° with a velocity of 8.6 km/s, appropriate for an earthquake at the location shown.

GLOSSARY

ADC: Analog-to-digital converter.

Geophone: The term used for a seismometer in exploration geophysics. Typically a small unit with a natural frequency of several hertz.

lc: Least count. One digital unit output from an ADC.

LP: Long period—referring to a wave period greater than about 6 s.

LTA: Long-term average.

SP: Short period—referring to a wave period less than about 6 s.

STA: Short-term average.

Teleseismic: Referring to source-to-receiver distances greater than about 2000 km.

UTC: Universal Coordinated Time.

WWSSN: World-Wide Standardized Seismographic Network.

Works Cited

Agnew, D. C. (1986), "Strainmeters and Tiltmeters," *Rev. Geophys.* **24,** 579–624.

Anderson, J. A., Wood, H. O. (1925), "Description and Theory of the Torsion Seismometer," *Bull. Seism. Soc. Am.* **15,** 1–72.

Berger, J., Agnew, D. C., Parker, R. L., Farrell, W. E. (1976), "Seismic System Calibration: 2. Cross-Spectral Calibration Using Random Binary Signals," *Bull. Seism. Soc. Am.* **69,** 271–288.

Dewy, J., Byerly, P. (1967), "The Early History of Seismometry," *Bull. Seism. Soc. Am.* **59,** 183–227.

Farrell, W. E. (1985), "Sensors, Systems and Arrays," in: Ann V. Kerr (Ed.), *Seismic Instrumentation under VELA-Uniform. The VELA Program, A Twenty-Five Year Review of Basic Research,* Washington, DC: Defense Advanced Research Projects Administration.

Fels, J.-F. Berger, J. (1994), "Parametric Analysis and Calibration of the STS-1 Seismometer of the IRIS/IDA Seismographic Network," *Bull. Seism. Soc. Am.* **84,** 1580–1592.

Given, H. K., Berger, J., Fels, J.-F., Horwitt, D., Winther, C. (1992), "The IRIS-3 High Resolution Data Acquisition System," *Technical Report,* Scripps Institution of Oceanography Reference Series No. 92-32.

Green, P. E., Frosch, R. A., Romney, C. F. (1965), "Principals of an Experimental Large Aperture Seismic Array (LASA)," *Proc. IEEE* **53,** 1821–1833.

LaCoste, L. J. B. (1934), "A New Type, Long Period Vertical Seismograph," *Physics* **5,** 178–180.

Melton, B. (1976), "The Sensitivity and Dynamic Range of Inertial Seismographs," *Rev. Geophys.* **14,** 93–116.

Oliver, J., Murphy, L. (1971), "WWSSN: Seismology's Global Network of Observing Stations," *Science* **174,** 254–261.

Peterson, J. (1993), "Observations and Modeling of Seismic Background Noise," *Technical Report,* United States Department of Interior Geological Survey Open-File Report No. 93-322.

Rodgers, P. W. (1968), "The Response of the Horizontal Pendulum Seismometer to Rayleigh and Love Waves, Tilt, and Free Oscillations of the Earth," *Bull. Seism. Soc. Am.* **58,** 1384–1406.

Rodgers, P. W. (1994), "Self-Noise Spectra for 34 Common Electromagnetic Seismometer/Preamplifier Pairs," *Bull. Seism. Soc. Am.* **84,** 222–228.

Rodgers, P. W., Martin, A., Robertson, M., Hsu, M. (1995), "Signal-Coil Calibration of Electromagnetic Seismometers," *Bull. Seism. Soc. Am.* **85,** 845–850.

Sorrels, G. G., Goforth, T. T. (1973), "Low-Frequency Earth Motion Generated by Slowly Propagating Partially Organized Pressure Fields," *Bull. Seism. Soc. Am.* **63,** 1583–1601.

Weilandt, E., Streckeisen, G. (1982), "The Leaf-Spring Seismometer: Design and Performance," *Bull. Seism. Soc. Am.* **72,** 2349–2367.

Willmore, P. L. (1979), *Technical Report, Manual of Seismological Observatory Practice,* World Data Center A Report No. SE-20, Boulder, CO: U.S. Department of Commerce.

Further Reading

Agnew, D. C. (1986), "Strainmeters and Tiltmeters," *Rev. Geophys.* **24,** 579–624.

Aki, K., Richards, P. (1980), *Quantitative Seismology: Theory and Methods,* San Francisco, CA: W. H. Freeman.

Carpenter, E. W. (1965), "An Historical Review of Seismometer Array Development," *Proc. IEEE* **53,** 1816–1821.

Lee, W. H. K., Stewart, S. W. (1981), *Principles and Applications of Microearthquake Networks,* New York: Academic.

Usher, M. J., Burch, R. F., Guralp, C. (1979), "Wide-Band Feedback Seismometers," *Phys. Earth Planet. Inter.* **18,** 38–50.

Willmore, P. L. (1960), "The Detection of Earth Movements," in: S. K. Runcorn (Ed.), *Methods and Techniques of Geophysics,* Vol. 1, New York: Interscience, pp. 230–276.

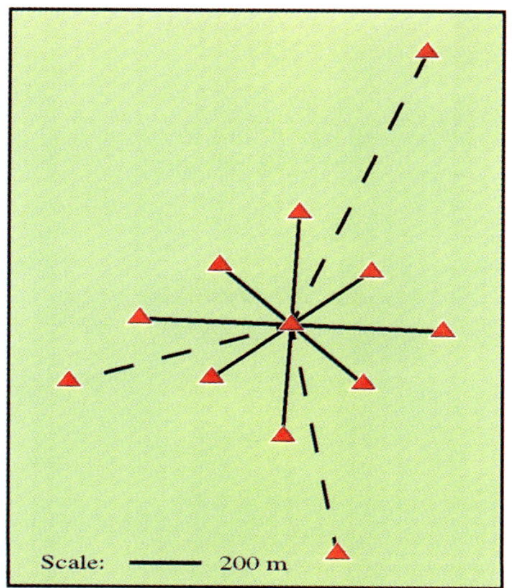

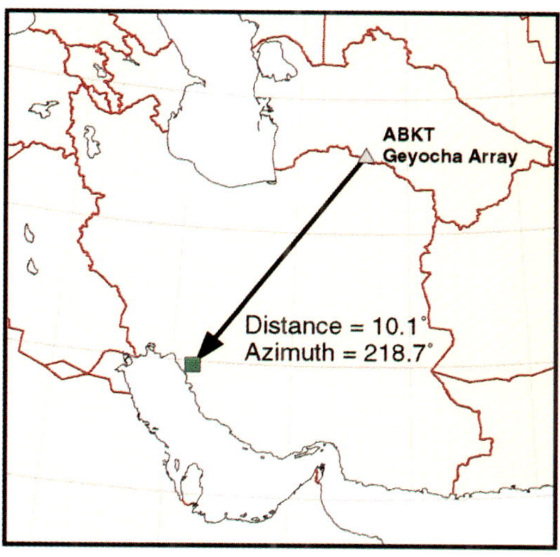

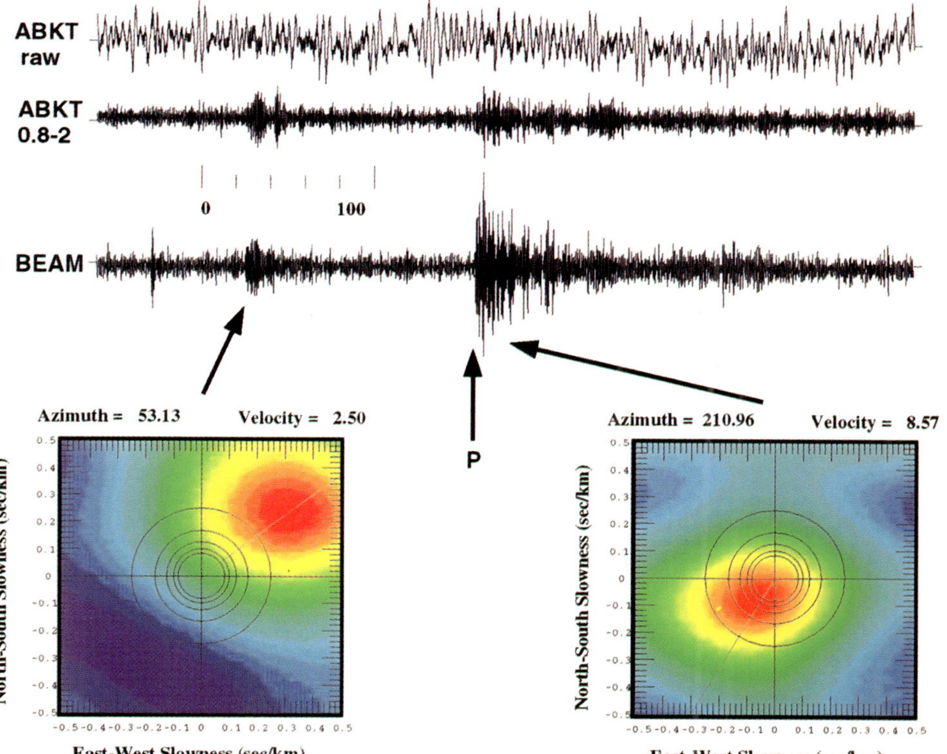

Plate 1. Example of array processing. These results were obtained from an experimental broadband array near the city of Ashgabat, Turkmenistan. The time series labeled ABKT is the signal on a single sensor. ABKT 0.8–2 is the same signal bandpass filtered between 0.8 and 2 Hz. The time series labeled BEAM is simply the undelayed sum of all elements.

SEISMOLOGY

CARL KISSLINGER, *Department of Geological Sciences and Cooperative Institute for Research in Environmental Sciences, University of Colorado, Boulder, Colorado, U.S.A.*

INTRODUCTION

Seismology is the study of ground shaking (from Greek: *seien* = to shake + *ology*). Ground vibrations from a variety of sources, natural (earthquakes, storms) and artificial (explosions, rock bursts in mines, machinery), have been investigated as part of this science. The term is now used primarily for the study of earthquakes and related phenomena. Modern seismology consists of two principal aspects: the study of earthquakes as a geophysical phenomenon and the study of the internal structure of the Earth on the basis of data provided by the observations of waves generated by seismic sources (see EARTH, INTERIOR STRUCTURE OF). For studies of earth structure, the data may come from natural earthquakes or from artificial sources, such as explosions.

Theoretical seismology is based primarily on the mechanics of continuous media, especially solid materials (see MECHANICAL PROPERTIES OF SOLIDS; MECHANICAL AND ELASTIC WAVES IN SOLIDS; ACOUSTICS, LINEAR). The mechanics of failure in solids under stress and the theory of wave propagation in solids and liquids provide the foundation. The materials, in the general case, are lossy and possibly elastically anisotropic.

An earthquake may be defined as a vibration of the Earth's surface caused by a transient disturbance of the equilibrium of rocks near or beneath the surface (Macelwane, 1947, p. 68; Kostrov and Das, 1988). This definition excludes more or less steady-state sources of observable vibrations, such as storms over the ocean. However, the study of the continuous waves generated by such sources, called microseisms, is a traditional topic in seismology. The definition admits sources like explosions as "artificial earthquakes" (Richter, 1958, p. 151).

The point within the Earth at which the disturbance occurs is called the *hypocenter* or *focus* of the earthquake. Because earthquake sources are spatially extended, and in view of the way in which earthquakes are routinely located from instrumental data, the term hypocenter is usually interpreted as the point at which the disturbance initiates. The point on the Earth's surface vertically above

3-527-28139-8/96/$5.00 + .50

the hypocenter is the *epicenter.* The *epicentral distance* to another point on the surface is measured as the arc of the great circle connecting the two points and may be expressed in either linear units (km) or angular units (deg).

1. SEISMIC SOURCE MECHANISMS

As understanding of Earth physics evolved, various energetic transient phenomena were proposed as the cause of earthquakes. Collapse of underground cavities, impacting meteorites, and processes within the chambers of volcanoes (see VOLCANICS) all produce measurable ground motion that fits the definition of "earthquake." The total energy released globally in such processes is negligible compared with that from the dominant cause, the sudden slip of rock masses along geologic faults. Earthquakes caused by fault slip are called *tectonic,* i.e., caused by the forces acting in the interior of the Earth to form geologic structures. More than a century of analysis of field and instrumental observations supports the hypothesis that almost all earthquakes, and all strong ones, are of tectonic origin (Scholz, 1990).

1.1 Geologic Faults

A fault is the interface between two rock masses along which there has been tangential displacement of one side relative to the other. The fault surface is pictured as a plane that intersects the Earth's surface along a line called the *fault trace.* Faults in nature may have complex geometries, which can be represented by a combination of planes with varying orientations. The spatial orientation of a fault surface is given by two angles: the dip, the angle measured downward between the horizontal plane and the fault plane, and the strike, the azimuth of the fault trace measured from the north. The mode of faulting is described by the direction of the slip vector, which lies in the fault plane and describes the motion of the block on one side of the fault relative to the other side. *Dip slip* occurs when the slip vector is parallel to the dip direction, and may be *normal* faulting, in which the block above the fault plane (the hanging wall) moves down relative to the block below (the foot wall), or *reverse,* in which the hanging wall moves up. Reverse faulting on a fault with small dip is called *thrust* faulting. *Strike slip* occurs when the slip vector is horizontal, parallel to the strike of the fault. The fault slip during an earthquake is often a combination of dip and strike slip.

The mode of faulting is directly related to the orientation of the ambient principal tectonic stresses. When the maximum compressive stress (the P axis) is vertical and the minimum compressive stress (the T axis) is horizontal (extensional tectonics), normal faulting dominates. Reverse faulting occurs when the maximum compressive stress is horizontal and the minimum compressive stress is vertical (compressional tectonics). When the maximum and minimum principal stresses are both horizontal, strike-slip faulting results. The axis of the intermediate principal stress lies in the fault plane. It follows that if the orientation of the plane on which slip occurred and the slip direction can be determined from either field or instrumental observations of an earthquake, important information about the regional stress field can be recovered.

A set of values of strike, dip, and slip direction is called a *focal-mechanism solution* for the earthquake. The determination of these basic geometric parameters is the first step in the analysis of the process that caused an event. Other characteristics of the fault can be derived from the recorded ground motion, such as the length and width of the surface that ruptured, the speed with which the rupture propagated across the surface, and the distribution of slip on the surface.

Few earthquakes, usually only strong, shallow ones on land, produce surface evidence of the faulting from which a focal mechanism can be developed from observations in the field. Several procedures have been developed for extracting this information from the records of ground motion produced by the earthquake. They all depend on the known relations between the geometric properties of the source and the pattern of radiation of seismic waves. The fault-slip model of earthquake origin is accepted as valid, but variations of this model that include displacements other than simple shearing can be accommodated. It is necessary, therefore, to examine the properties of the waves generated by an earthquake before considering how they may be used to retrieve its focal mechanism. Because waves represent one form in which the energy released in an earthquake is accounted for, a consideration of the flow of energy in an earthquake is discussed first.

1.2 Energy Flow in an Earthquake: The Elastic-Rebound Model

The original form of the elastic-rebound model of earthquake occurrence was formulated by Reid (1910, pp. 29–31). Working from the detailed data provided by the investigation of the 1906 California earthquake, he proposed the following description of the flow of energy in the process of producing fault rupture:

1. The energy released in an earthquake is present just before the event in the form of elastic strain energy in the rocks around the fault.
2. This energy has accumulated slowly as the rocks deform under the action of tectonic forces. The origin of these forces was not known to Reid (1910, pp. 27–28) nor understood until the development of the theory of plate tectonics (*q.v.*).
3. The strain accumulates until the resulting stresses on the fault exceed the shear strength of the fault surface, at which time sudden slip occurs to produce the earthquake. The rocks surrounding the fault move toward a position of minimum strain (rebound).

The fundamental concept that seismic energy is stored slowly by elastic deformation and released suddenly by fault rupture has held up well with time. The details of the model have been modified by better understanding of the material properties in fault zones and advances in theoretical fracture mechanics (Kostrov and Das, 1988; Scholz, 1990).

If the strain field can be represented by homogeneous simple shear, the energy per unit volume available to produce an earthquake is $e = \frac{1}{2}\mu\gamma^2$, where μ is the shear modulus and γ is the angle of shear. For typical crustal values of the shear modulus and of the shear strains associated with earthquakes, e is of the order of 100 J/m^3.

Several mechanisms are available to absorb the energy released in the event. Energy goes into the creation of a newly fractured surface and possibly the uplift of rock masses, energy appears as heat due to the friction on the sliding surfaces, and energy is radiated away from the source in the form of waves. The energy budget of an earthquake is not well determined, but a large fraction of the released energy goes into waves.

2. SEISMIC WAVES

2.1 Body Waves

The theory of waves in solids demonstrates that two types of waves can exist. The well-known differential equation expressing motion with infinitesimal strains in an unbounded perfectly elastic, isotropic body governs the behavior of these waves:

$$\frac{\partial^2 \mathbf{u}}{\partial t^2} = \mathbf{f} + \frac{(k + \frac{4}{3}\mu)}{\rho}\nabla\nabla\cdot\mathbf{u} - \frac{\mu}{\rho}\nabla \times (\nabla \times \mathbf{u}), \quad (1)$$

where $\mathbf{u}(\mathbf{x},t)$ is the particle displacement, $\mathbf{f}$ is the body force per unit mass, k is the incompressibility modulus, μ is the shear modulus, and ρ is the density of the medium (Aki and Richards, 1980).

The two wave types are the manifestation of the propagation from the source of the two fundamental types of deformation, volumetric strain, given by $\nabla\cdot\mathbf{u}$, and shear strain, given by $\nabla \times \mathbf{u}$. The former, termed compressional waves, gives rise to waves that are longitudinally polarized in an isotropic medium, labeled *P* waves (for primary) in seismology. Shear deformation propagates as shear waves, transversely polarized, labeled *S* waves (for secondary). In a perfectly elastic, isotropic solid, these waves travel with the velocities

$$V_P = \sqrt{(k + \tfrac{4}{3}\mu)/\rho} \qquad V_S = \sqrt{\mu/\rho}. \quad (2)$$

It may be seen from these relations that the *P*-wave velocity is greater than the *S*-wave velocity in any material; hence the designations primary and secondary to describe the order of their arrival at an observing point. The ratio V_P/V_S in common crystalline rocks is about 1.8. It is also seen that shear waves travel only in solid materials, not in fluids, for which the shear modulus is zero.

The situation is more complicated for anisotropic media, in that the polarization is no longer strictly longitudinal or transverse, the wave velocity depends on the direction of propagation, and the velocity of *S* waves depends on the polarization as well as the propagation direction (Musgrave, 1970). Anisotropy does exist in parts of the Earth's crust and upper mantle, and may exist at greater depths. In the following, we assume isotropy.

2.2 Generalized Snell's Law

A complexity encountered in seismology that does not occur in acoustics or electromagnetic wave studies is the conversion of part of the energy in an incident wave of either *P* or *S* type into the other type at boundaries between materials of different properties. The converted waves may be both reflected back from the boundary and transmitted through it. Thus, a wave of either type will, in general depending on the angle of incidence and the contrast in properties of the two materials, give rise to four waves at a boundary, a reflected and transmitted wave of the same type as well as the converted waves. For a *P* wave incident in medium 1, angle of incidence i_{inc}, the directions of propagation of the four derived waves are given by a generalization of Snell's law (see MECHANICAL AND ELASTIC WAVES IN SOLIDS):

$$\frac{\sin i_{inc}}{V_{P1}} = \frac{\sin i_1}{V_{P1}} = \frac{\sin j_1}{V_{S1}} = \frac{\sin i_2}{V_{P2}} = \frac{\sin j_2}{V_{S2}}. \quad (3)$$

In Eq. (3), all angles are measured between the normal to the boundary and the ray. The angle i_1 is the angle of reflection of the *P* wave and is equal to i_{inc}; j_1 is the angle of reflection of the *S* wave converted from the incident *P* wave. The angle i_2 is the angle of refraction of the transmitted *P* wave; j_2 is the angle of refraction of the transmitted *S* wave. For an incident *S* wave, i_{inc} is replaced by j_{inc} and V_{P1} by V_{S1} in the first term. The reflection and transmission coefficients for each of the four derived waves, as functions of the angle of incidence and contrast in material properties, are given in Aki and Richards (1980, pp. 133–151).

One consequence of these relations is that a critical angle of incidence exists whenever the velocity of the incident wave is less than the velocity of the derived ray. Because $V_S < V_P$ in all materials, a critical angle for *S*-to-*P* conversion by reflection always exists, for which $\sin j_{inc} = V_S/V_P$. A frequently occurring case in the Earth is $V_{P1} < V_{P2}$ for waves incident on a boundary from above. Here, for $i_{inc} = \sin^{-1}(V_{P1}/V_{P2})$, the critically refracted energy travels along the boundary with velocity V_{P2} and radiates energy back into the first medium at the critical angle. These *head waves* can be quite prominent and provide a powerful means of mapping subsurface structure.

In an unbounded, homogeneous, perfectly elastic Earth, the signal from an earthquake would consist of a *P* wave followed by an *S* wave, and nothing else. As seen in Fig. 1, an actual seismogram is much more complex. Much of the complexity is due to the reflection, refraction, and conversion from one type to the other of the original *P* and *S* waves at boundaries between layers of material with differing properties. The dominant wave arrivals on seismograms depend on the epicentral distance of the station. Local and regional records show primarily high-frequency phases that have traveled in the layers of the crust. At intermediate distances, 20° to 80°, body waves that have traveled through the mantle along various paths dominate. At greater distances, waves that have traveled through the core become important.

The development of the wave train with increasing distance is illustrated in Fig. 2, in which a few of the many distinct arrivals are identified. The initial *P* arrival at station ZOBO is very small because at this distance the wave has encountered the core of the Earth and the arrival seen has been diffracted at the core–mantle boundary. The arrival at station CTAO marked SKS is a wave that left the source as an *S* wave, entered the core as a *P* wave, then left the core after conversion back to an *S* wave. An epicentral distance of about 80° is the shortest at which a wave that has traveled part of its path in the core arrives. The travel time of the direct *P* wave through the center of the Earth to the antipodes is about 20 minutes.

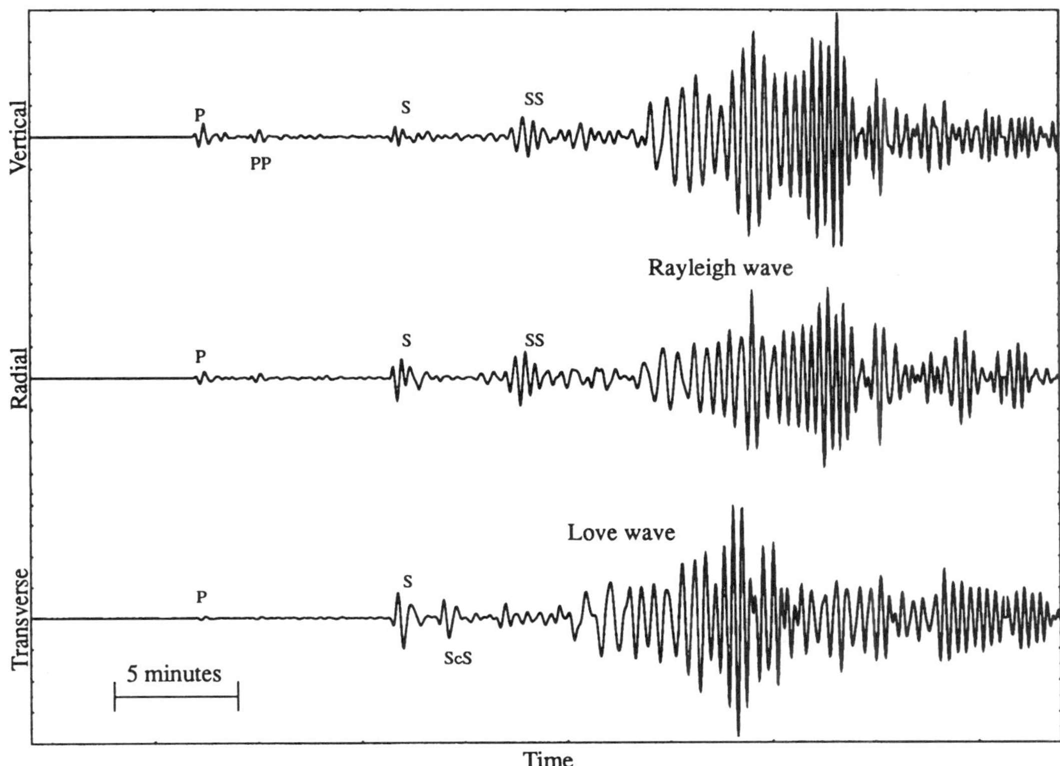

FIG. 1. Long-period seismograms recorded at Glen Almond, Quebec, Canada, from an earthquake in the Fox Islands, Alaska. The earthquake occurred on 5 January 1987, with body-wave magnitude $m_b = 6.0$, surface-wave magnitude $M_S = 6.6$. The epicentral distance to the station was 57.6° or 6400 km. The horizontal components have been rotated to correspond to radial and transverse motion relative to the great circle from the epicenter to the station. The body-wave arrivals labeled are as follows: the direct *P* and *S* wave from the source; PP and SS, *P* and *S* waves reflected one time from the Earth's surface at the midpoint between the epicenter and the station; ScS, *S* wave reflected from the core–mantle boundary. The dispersion in the surface-wave trains is seen. The Rayleigh-wave train is seen on the vertical and radial components, the Love-wave train on the transverse component. Digital wave-form data provided by the National Earthquake Information Center, U.S. Geological Survey.

The boundaries marking the most pronounced contrasts in properties are at the Earth's outer surface, rock against air (PP,SS in Fig. 1), and at the core–mantle boundary, solid silicates against melted iron alloy (ScS in Fig. 1) (see EARTH, INTERIOR STRUCTURE OF THE). Strong signals are received as reflections, including multiple reflections, from these boundaries, as well as from waves that have been refracted through the core. Because no *S* waves that have traveled directly through the core can be detected on seismograms, it is concluded that the core is, at least in part, fluid. Numerous other boundaries marking less sharp changes in properties have been identified from their signatures in the seismic wave train. The most detailed information we have about the internal constitution of the Earth has come from the determination of *P*- and *S*-wave velocities at all depths within the Earth by analysis of the observed travel times of wave energy along the many possible paths.

2.3 Surface Waves

In addition to the body waves that travel through the crust, mantle, and core, wave energy also becomes trapped in waveguides and spreads out in two dimensions. Prominent waveguides are formed by the Earth's outer surface in combination with layers in

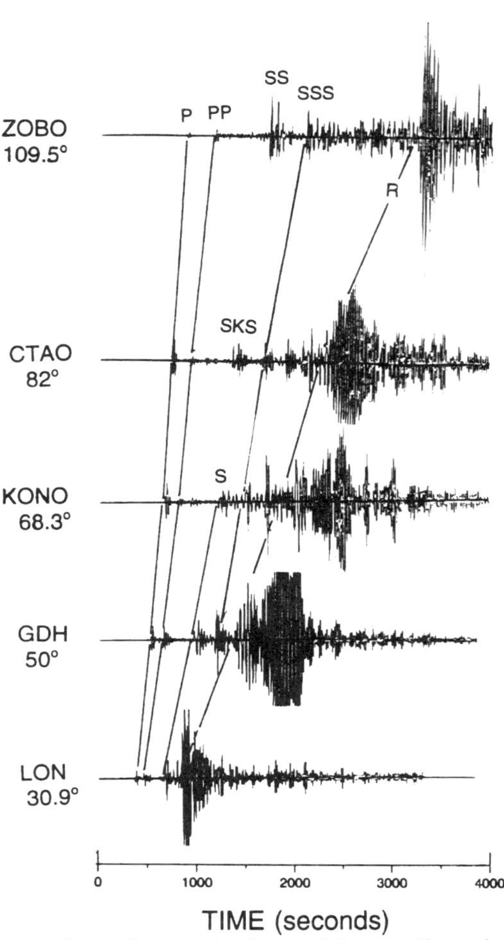

FIG. 2. The long-period vertical-component seismograms from the same earthquake shown in Fig. 1 over a large range of epicentral distances, given in degrees of arc below the station designator. The stations are Longmire, Washington, LON; Godhaven, Greenland, GDH; Kongsberg, Norway, KONO; Charter Towers, Australia, CTAO; and Zongo Valley, Bolivia, ZOBO. The amplitude scale at each station is arbitrary and has been adjusted for convenience in plotting. Only a few of the many phase arrivals are identified. The small-*P* arrival at ZOBO has been diffracted at the core–mantle boundary. CTAO is at about the shortest distance from the epicenter at which a wave that has traveled part of its path in the core, SKS, arrives. The time between phase arrivals at a station is a function of its epicentral distance. Digital wave-form data provided by the National Earthquake Information Center, U.S. Geological Survey.

the crust and upper mantle. The resulting waves, called *surface waves,* produce the largest signals on typical seismograms of earthquakes with hypocentral depths less than 100 km, Figs. 1 and 2. Surface waves are of two general types, one polarized in the plane of incidence (vertical and radial components of displacement), the other polarized horizontally, with displacement perpendicular to the plane of incidence. The former are called *Rayleigh waves,* the latter *Love waves.* Surface waves propagate with little attenuation in the waveguide, and die off exponentially with depth outside the waveguide (Bullen and Bolt, 1985).

Surface waves are *dispersed,* so that the wave velocities depend on the wavelength. Observations of either phase or group velocity as a function of wave period are useful for deriving the Earth structure that controls the propagation along the path from source to seismograph station. Each surface-wave type propagates in an infinite number of modes, manifested as the number of nodal planes of the motion within the waveguide. Love-wave velocities are controlled by the densities and shear-wave velocities and the thickness of the layers making up the waveguide. Rayleigh-wave velocities depend on these variables, as well as on the *P* wave velocities.

2.4 Free Oscillations

In addition to the propagating body and surface waves, earthquakes may also generate *free oscillations* of the Earth, i.e., the normal modes of vibration of the planet as a bounded, layered elastic body (Bullen and Bolt, 1985). Free oscillations were first detected only from extremely powerful earthquakes, but with modern broadband seismographs (see SEISMOGRAPHS) they may be seen from events as small as magnitude 6 (see Magnitude, Sec. 4.2).

Free oscillations are of two fundamental types, spheroidal and toroidal, corresponding to the two fundamental types of surface waves. Spheroidal modes may be recorded on instruments sensitive to the vertical component of motion, including gravimeters, while toroidal modes are found only on the horizontal components. Free oscillations are studied by computing the frequency spectrum of a long sample of the ground motion following an earthquake. They show up as a sequence of spectral lines at well-defined frequencies.

The frequencies of many identified modes have been measured. The theory of normal modes of an elastic body is used to calculate the expected frequencies for a given model of the Earth, specified in terms of a distribution with radius of the density and elastic moduli. The model can then be adjusted to give the best fit to the suite of measured frequencies.

The width of a given spectral line is determined by the anelastic attenuation in the Earth of the particular mode. This provides a powerful tool for exploring anelastic properties within the Earth. The spectral lines are converted into multiplets by the rotation of the Earth and by its ellipticity, and so these effects are taken into account in the calculations.

2.5 Attenuation of Seismic Waves

The amplitudes of seismic waves decrease with distance traveled, primarily because of geometric spreading. For all except plane waves, the amplitude in a perfectly elastic medium must decrease as the wave front expands, in such a way that the total energy flux through the wave front is conserved. Because the energy flux per unit area of the wave front varies as the square of the amplitude, this purely geometrical effect is easily accounted for. The amplitude of a spherically spreading wave decreases as r^{-1}, while a cylindrically spreading wave amplitude decreases as $r^{-1/2}$.

Other causes of seismic-wave attenuation are elastic scattering from inhomogeneities in the medium and anelastic attenuation. Shear-wave energy removed from the advancing wave front by scattering is thought to be the principal source of the short-period coda on seismograms of local and regional earthquakes. Anelastic attenuation is of fundamental importance, because the departures from elasticity of the materials within the Earth provide an important clue to composition and ambient conditions at depth (Aki and Richards, 1980, pp. 167–185).

A number of processes have been proposed as the physical basis for anelastic attenuation in the Earth (Jackson and Anderson, 1970). Attenuation is measured by the quality factor, Q, defined by the fractional loss of energy per cycle as

$$2\pi Q^{-1} = -\Delta E/E. \tag{4}$$

Q is directly related to the viscous and elastic parameters in any of the various models of viscoelastic materials that have been analyzed (e.g., Maxwell solid, Voigt solid, standard linear solid).

The decrease in amplitude with distance traveled, for a wave traveling with velocity V, with geometric spreading removed, may be expressed as $\exp[-\pi fr/VQ]$. The decrease of amplitude with time for a standing wave, such as a free oscillation, is $\exp[-\pi ft/Q]$. Q may be measured either from observations of amplitudes of propagating waves, or from the linewidths in the spectra of free oscillations. A measure of attenuation can also be derived from the rate of decay of the coda of earthquakes recorded at small distances. On the assumption that this coda consists of single backscattered shear waves, a measure called the coda Q is calculated.

The frequency dependence of Q is a fundamental question of seismic-wave propagation (Anderson and Given, 1982). Laboratory and field studies of the attenuation of P and S body waves, surface waves, and free oscillation provide the basic data. Q for both P and S waves is essentially independent of frequency over a wide portion of the seismic spectrum and increases outside of that band, so that the Earth behaves as an absorption-band filter. Anderson and Given (1982) presented evidence that the characteristics of the absorption band vary with depth within the Earth.

Anelastic attenuation introduces frequency dependence in the velocities of seismic waves. Therefore, Earth models derived from the velocities of waves with greatly different frequencies can be reconciled only by taking into account the dispersion introduced by anelasticity.

2.6 Locating Earthquakes

Earthquakes are located by analysis of the time of arrival of the various waves at seismograph stations. Hundreds of such stations have been established at known positions on the Earth. The mean travel times of seismic waves along the many possible paths through the Earth as functions of epicentral distance and hypocentral depth have been established after more than a century of observations. The location process consists of adjusting the latitude, longitude, and depth of the hypocenter and the origin time of the event so that the observed arrival times, especially of the direct *P* and *S* waves, at all of the stations that recorded the event are in the best agreement, in the least-squares sense, with the times calculated from the standard travel-time curves. Adjustments to the standard travel-time curves to take into account local or regional variations in wave velocities from the standard Earth model at the source or at the recording stations may be required for accurate solutions. Solutions are compiled into global earthquake catalogs, such as the Bulletin of the International Seismological Center and the Preliminary Determinations of Epicenters of the U.S. Geological Survey, as well as national and regional catalogs (Bullen and Bolt, 1985, Chap. 10).

3. FOCAL MECHANISMS FROM THE WAVE RADIATION PATTERNS

Slip on a fault is accepted as the cause of almost all earthquakes, yet very few earthquakes produce surface slip that can be observed in the field. The fault origin is accepted because the pattern of radiation of seismic waves, as recorded by seismographs, can, for most earthquakes, be quantitatively explained by this model. The polarity and amplitude of the waves recorded at a station depend on the distance and azimuth of the station from the epicenter, which are known once the earthquake is located, and the dip, strike, and slip direction of the fault, the unknowns for which a solution is sought. The equations describing the radiation patterns of both body and surface waves are given in Aki and Richards (1980, Chaps. 4, 7).

The standard method, still widely used for routine investigations, is based on the distribution of polarities of the first motion of *P* waves at many points on the Earth's surface. Because it is the first-arriving wave, the polarity of the *P* wave (compression or dilatation) from sufficiently strong earthquakes can be read with less ambiguity than for later arrivals.

The analysis begins by approximating the ruptured fault as a point source of the waves. The length of the actual fault and the speed of rupture propagation can later be estimated by extending the analysis. The point source can be modeled as a dislocation or by the equivalent distribution of body forces. For a pure shear dislocation, the body force equivalent is a double couple without moment. The pattern of radiation of the *P*-wave displacement u_p at an observing point at $\mathbf{R}$ from this type of source located at the origin is given by

$$u_p(\mathbf{R},t) = \frac{2(\boldsymbol{\delta}\cdot\mathbf{n})(\boldsymbol{\delta}\cdot\mathbf{V})\mu A\delta}{4\pi\rho V_p^3 r}, \tag{5}$$

where $\boldsymbol{\delta}$ is the unit vector in the direction from the source to the observing point, $\mathbf{n}$ is unit normal to the fault plane, $\mathbf{V}$ is the average vector velocity of slip on the fault plane, A is the area of the fault, and r is the distance to the point of observation (Aki and Richards, 1980, p. 113).

For an observing point in the $(\mathbf{n},\mathbf{V})$ plane, the first factor in the radiation pattern is $\sin\alpha$, where α is the angle between the slip vector and the direction to the station; the second factor is proportional to $\cos\alpha$. The result is a four-lobed radiation pattern of alternating compressions and dilatations, separated by nodal lines at $\alpha = 0$ (along the strike of the fault) and $\alpha = 90°$. When embedded in the Earth with an arbitrary orientation, the pattern divides the planet into four segments, separated by two nodal planes. One of the planes is the fault, the other, the auxiliary plane, normal to the first.

To apply this result, the earthquake must first be located. Then, a small "focal sphere" is imagined to surround the (point) source. The earthquake energy leaves the focal sphere along ray paths that are refracted through the Earth's interior to emerge at points distributed over its surface. From the known distribution of wave velocity with depth, the ray emerging at any point on the Earth's surface can be traced back to the point at which it left the focal sphere. The corresponding direction of the first motion, compression or dilatation, is then assigned to that point. By convention, the distribution of first motions on the lower hemisphere of the focal sphere is mapped by equal-area stereographic projection. The analyst then decides the position of the two nodal curves, constrained to be orthogonal, that separate the quadrants of compressions and dilatations. The two nodal curves are maps of the intersection of the two nodal planes with the focal sphere. The strike, dip, and direction of slip on each of these two planes can then be read from the resulting graph (Lee and Stewart, 1981, pp. 139–148). The entire process is now usually done by computer.

The resulting focal-mechanism solution is fundamentally ambiguous, in that there is no way, in principle, to decide from these data which of the two nodal planes is the fault plane and which the auxiliary plane. Independent information, such as knowledge of the dominant structural trends in the region and the distribution of aftershocks of the event, can help resolve the ambiguity.

Examples of focal-mechanism solutions obtained for one month, August 1992, by the U.S. Geological Survey are plotted at the positions of the earthquake epicenters in Fig. 3. The type of faulting and regional stress patterns associated with various tectonic settings can be inferred from such maps.

Fault-plane solutions do yield useful information about the orientation of the principal stress axes that correspond to the solution. The axis of maximum compression (P axis)

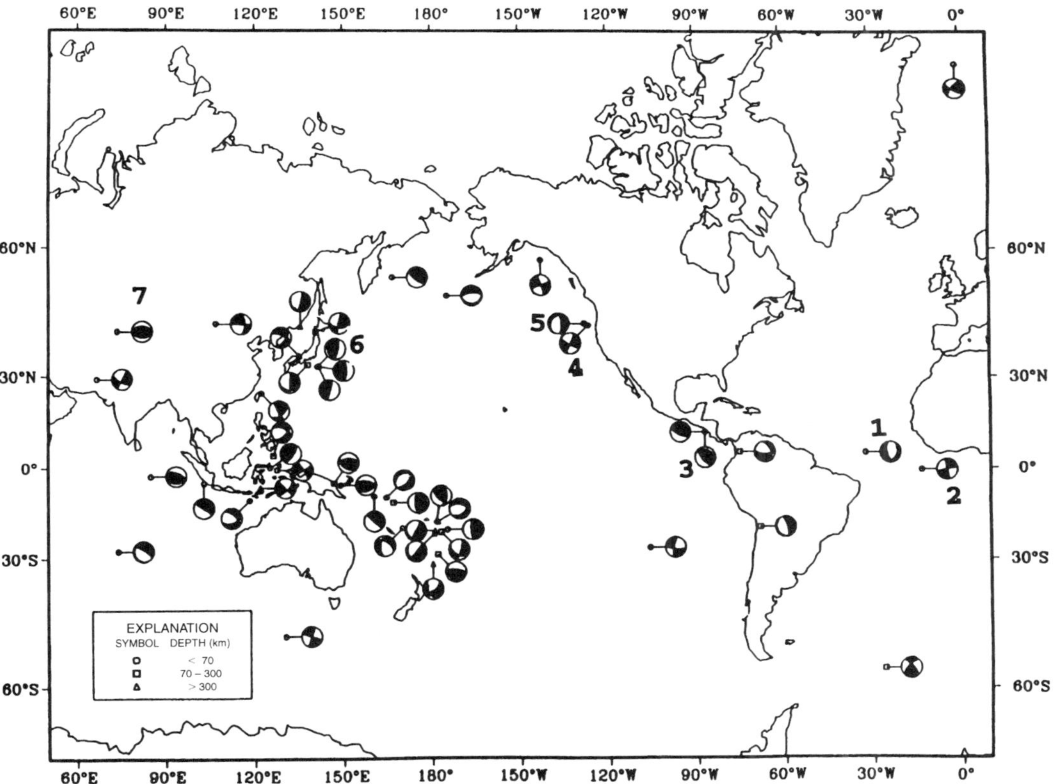

FIG. 3. A world map with focal mechanisms of earthquakes that occurred during August 1992. Each solution is connected to the epicenter of the earthquake by a line segment. The circles are equal-area stereographic maps of the lower hemisphere of the focal sphere. The two nodal planes of each solution, one of which is the fault plane, and the quadrants of compressional first motions (black) and dilatational first motions (white) are shown. A few typical mechanisms are marked with numbers: 1 is a normal fault on the Mid-Atlantic Ridge, 2 is a strike-slip earthquake on a transform fault that offsets segments of the Mid-Atlantic Ridge, 3 is a thrust event at the Middle America subduction zone, 4 and 5 are a strike-slip and normal fault, respectively, associated with the undersea ridge offshore of northern California, 6 is a thrust event associated with subduction of the Pacific Plate under Eurasia at the Japan Trench, and 7 is an intraplate thrust event, reflecting the north–south compression resulting from the collision of the Indian subcontinent into the Asian land mass. From U.S. Geological Survey, Preliminary Determination of Epicenters, Monthly Listing, August 1992.

lies in the center of the quadrant of dilatations, the axis of minimum compression (T axis) lies in the center of the quadrant of compressional first motions, and the axis of intermediate stress lies along the intersection of the two nodal planes. The directions may not be the same as the orientation of the current regional tectonic stress field, as the rupture may have occurred along some previously established plane of weakness in the rocks at an arbitrary angle to that stress field. Techniques have been developed by which the focal-mechanism solutions of many earthquakes in a region can be inverted for the contemporary stress field that best fits the ensemble of data (Gephardt and Forsyth, 1984).

Occasionally the distributions of first motions cannot be separated into four quadrants by great circles on the focal sphere. The implication is that the radiation is not described by Eq. (5), that is, the source was not equivalent to a simple double couple, but also contained mutually orthogonal components of extensional and compressional displacement. Solutions are obtained in this case by extended analysis, on the assumption that the net volumetric change at the source is zero.

4. MEASURES OF EARTHQUAKE SIZE

4.1 Intensity

The size or strength of an earthquake may be measured in several ways. The oldest measure is *intensity,* which is expressed in terms of the visible effects of the earthquake on natural and manmade features (Richter, 1958, Chap. 11). Descriptions of earthquake strength in terms of effects such as human perception, landslides, ground displacement, and damage to buildings were a natural preinstrumental approach. Intensity is still determined and mapped because of its direct applicability to earthquake-resistant design of structures. Several intensity scales have been devised, each of which assigns a numerical value to each level of damage or other observable effects. The most commonly used scale in the United States is the Modified Mercalli scale, which ranges from I (felt by only a few persons under favorable conditions) to XII (total destruction). Slight damage occurs at MM V (Roman numerals are conventionally used for intensities to avoid confusion with earthquake magnitude). The distribution of intensities is often plotted on an *isoseismal map,* on which areas in which the intensity is judged to be constant are separated by contours called *isoseismals.*

Intensity is a useful measure for engineering purposes and because it is possible to assign distributions of intensities to earthquakes from the preinstrumental era from written descriptions of the effects. Intensity is, however, of limited value in the scientific study of earthquakes. Earthquakes in unpopulated areas, including the numerous earthquakes under the oceans, cannot be assigned an intensity on the basis of observed effects. Also, the assignment of an intensity value is often subjective, so that different observers may report somewhat different values. For these reasons and because of the need for an objective universal measure of earthquake strength that could be used for statistical analysis of earthquake catalogs, C. F. Richter developed the magnitude scale in 1935.

4.2 Magnitude and Seismic Moment

The *magnitude* of an earthquake is a measure of its strength based on the recorded amplitude of a specified wave type (Richter, 1958, Chap. 22). The original Richter scale was based on the maximum wave amplitude produced by the earthquake, as recorded on a standard seismograph (fixed magnification, period, and damping) then used in the network of stations operated by the California Institute of Technology. The scale was intended for use only for local and regional earthquakes (epicentral distances less than about 600 km) in southern California, but was so successful that the basic idea has been extended for global use.

The logarithm (base 10) of the amplitude in micrometers of the largest wave on the seismograms was taken as the measure. The logarithm was used because of the vast range of strengths of earthquakes. Because wave amplitudes decrease with distance from the source, it was necessary to adjust the observed values to what they would have been at some standard epicentral distance, chosen as 100 km. Richter developed an empirical attenuation curve of wave amplitude vs distance for southern California, so that this adjustment could be made. There remained only to set the zero value on the scale. Richter decided that an event that produced a maximum recorded amplitude on the standard seismograph of 10^3 μm at 100 km would be a magnitude 3 earthquake.

Magnitudes determined by the original process, using data equivalent to those from the original standard seismograph, are called *local magnitudes,* M_L. The concept was so successful that scales applicable to seismograms recorded with any calibrated seismograph, at any distance from the epicenter, were subsequently developed. The *surface-wave magnitude,* M_S, is calculated from the logarithm of the largest amplitude in the Rayleigh-wave train at periods near 20 s. The recorded amplitude is converted into true ground amplitude by use of the known magnification of the instrument. Standard correction is made for epicentral distance.

Because surface waves are not strongly excited by earthquakes deeper than about 100 km, the surface-wave magnitude scale is of limited applicability. Therefore, Gutenberg and Richter, in 1956, developed a scale based on the seismic body waves, which are generated by all earthquakes. The most widely used scale is the *body-wave magnitude,* m_b, based on the amplitude of *P* waves with periods near 1 s. This scale is based on the logarithm of the ratio of amplitude of the ground motion (in micrometers) to period (in seconds) of the largest amplitude in the first few cycles of the recorded *P* wave. A correction is made for epicentral distance and hypocentral depth.

All of the magnitude scales discussed so far saturate at large magnitudes. That is, the energy release in the earthquake increases without a corresponding increase in the measured magnitude. Saturation is a consequence of the shape of the typical earthquake amplitude–frequency spectrum. The spectrum, shown schematically in Fig. 4, is characterized by a flat segment at low frequencies, a corner frequency, which depends on the physical dimensions of the fault segment that ruptured, and a high-frequency falloff. The corner frequency moves to lower values as the size of the fault and the energy released in the earthquake increase. If the earthquake is so big that the corner frequency is below the frequency of the waves used to calculate magnitude, saturation occurs. A solution to this problem is the use of the flat, low-frequency part of the spectrum, as proposed by Kanamori (1977). The very low frequencies involved were not well recorded by earlier seismographs, but are by modern broadband instruments. The level of the low-frequency part of the spectrum is proportional to the seismic moment (see below), and the magnitude scale based on the moment, called M_W by Kanamori, is now widely used.

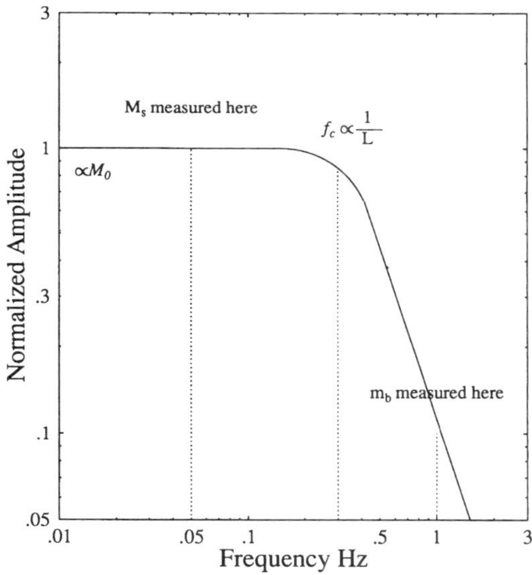

FIG. 4. Schematic earthquake displacement spectrum. The details of the theoretical spectrum are model dependent, but the main features are a flat low-frequency part that is proportional to the seismic moment, a corner frequency f_c that is inversely proportional to the characteristic dimension *L* of the rupture surface, and a high-frequency falloff, usually taken as ω^{-2}, though ω^{-3} is preferred by some seismologists. The body-wave magnitude is measured at about 1 Hz, the surface-wave magnitude at about 0.05 Hz. As the earthquake dimension increases, the corner frequency moves to lower frequencies. If the corner frequency falls below the frequency at which a particular magnitude is measured, the magnitude will not increase in accord with the increase in energy or moment. That magnitude scale is saturated.

An increase of one unit on any of the magnitude scales signifies an increase by a factor of 10 of the amplitude of the wave used. However, studies of the relation of wave amplitude and energy release have led to the empirical relation log*E* (joules) = $1.5M_S + 4.8$, so that an increase in M_S of one unit corresponds to an increase in released seismic energy by a factor of about 30. Thus, it takes almost 1000 $M_S = 4$ earthquakes to release the energy in one $M_S = 6$.

No maximum or minimum possible magnitude follows from the definition of the scale. Nature seems to have an upper limit of about $M_W = 9$, presumably governed by the maximum amount of strain energy rocks are capable of storing and the maximum dimensions of fault rupture surfaces.

The *seismic moment* is a measure of earthquake strength based on the fault-slip model of earthquake generation that is more directly related to the physics of the source than either intensity or magnitude. The seismic moment is defined as $M_0 = \mu AD$, where μ is the shear modulus of the material in which the fault occurs, A is the area of the rupture surface, and D is the average displacement on the surface during the earthquake (Aki and Richards, 1980, p. 49). The seismic moment, in newton meters, is related to the moment magnitude, M_W (Hanks and Kanamori, 1979), by $\log M_0 = 1.5M_W + 9.05$, so that typical values would be a seismic moment of 1.1×10^{18} N m for a moment magnitude 6 earthquake.

5. EARTHQUAKE GEOGRAPHY

The geographic distribution of earthquakes is basic information about the Earth that is to be interpreted in terms of the dynamic processes that shape major geologic features. Figure 5 shows the global distribution of earthquakes for the years 1981–1990. One conclusion from this map is that earthquakes are not randomly or uniformly distributed over the planet, but that most occur in distinct, well-defined seismic zones or belts. This information is essential in assessing earthquake hazards, in that we know where the probability of occurrence of strong earthquakes is high and where they are rare or may never occur. The broad features of the distribution are revealed by observations during a short time, like ten years. However, a short interval will not yield information about some significant characteristics. The strongest earthquake to be expected at any site and the mean time between events of a given magnitude cannot be determined. Also, some areas of infrequent, but significant seismicity may not show activity during the interval.

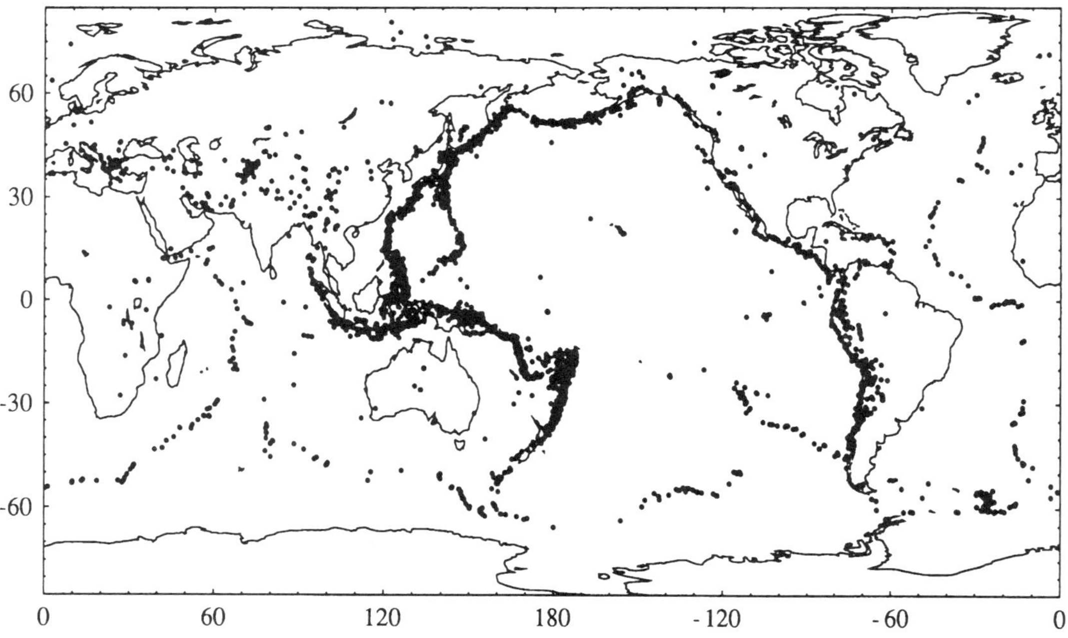

FIG. 5. Global distribution of earthquakes, 1981–1990. All events (3950) with magnitude greater than 5.5 are plotted. The data are taken from the catalog, Preliminary Determination of Epicenters, produced by the U.S. Geological Survey. At this rather high minimum magnitude some areas of lower-level activity, such as the eastern United States, show no activity during the selected decade.

The classical, pre-plate-tectonics designation of the principal concentrations of seismicity was as the Circum-Pacific, Alpide-Himalaya, and Oceanic Active Belts (Gutenberg and Richter, 1949). We now understand that those concentrations exist at the active boundaries between mobile lithospheric plates (Gubbius, 1990) (see PLATE TECTONICS). Some of the most compelling evidence in support of the plate-tectonics model came from the analysis of seismological observations. For example, focal-mechanism solutions of the strong earthquakes show that subduction zones are compressive regimes (thrust faulting) and undersea rifts are tensional regimes (normal faulting), Fig. 3. The sense of strike-slip on transform faults is that expected from the plate-tectonics model. Earthquakes at plate boundaries are called *interplate* earthquakes; those away from plate boundaries are *intraplate*.

The forces driving the plates and plate interactions provide the tectonic forces called for by the elastic-rebound theory. It is presumed that plate motions are the ultimate source of the stresses driving both interplate and intraplate earthquakes, but the relation is not as obvious for the latter. In particular, the reasons for the localization of intraplate events are often obscure. For example, the diffuse distributions of seismicity in eastern North America and in western China are almost certainly the consequence of the westward motion of North America from the Mid-Atlantic Ridge in the first case and the continuing intrusion of the Indian subcontinent into Asia in the second. The reasons for the local details of the stress distribution and the relation of the active sites to local and regional geologic features are, however, not well understood.

Events along the Pacific coast of the coterminous 48 United States are clearly related to the interaction between the Pacific/Juan de Fuca and North American plates. Events farther inland have a variety of focal mechanisms associated with the geologically diverse settings. Activity in the western intermountain zone, in the central Mississippi Valley, and at apparently isolated sites along the east coast do not easily fit into a plate-tectonics explanation. They are interpreted as the result of complex plate interactions over long periods of geologic time. The configuration of the faulting responsible for great earthquakes in the central Mississippi Valley in 1811–1812, and numerous smaller events since, has been clarified by detailed geophysical and geological studies. The problem is more difficult along the Atlantic coast because even strong earthquakes cannot be readily associated with individual faults or fault systems, even when faults in the vicinity of the activity have been mapped.

Earthquakes in Alaska, the largest within the territory of the United States, are directly caused by the relative motion of the Pacific and North American plates, primarily subduction along the Aleutian Trench and under southern Alaska. Earthquakes in Hawaii are concentrated under the "big island," Hawaii, and have a different kind of origin. Some are apparently caused by stresses in the upper part of the crust induced by pressure within the chambers and conduits of the volcanoes. Those on the south flank of Kilauea volcano are thought to be related to the seaward motion of the volcanic pile over the buried oceanic crust, also under the influence of volcano-induced stress and probably gravity (Klein *et al.*, 1987).

The geographic distribution of the zones of high seismicity is an obvious starting point for the assessment of earthquake hazards, one important application of seismology.

6. EARTHQUAKE HAZARD ASSESSMENT AND EARTHQUAKE PREDICTION

The quest for methods to foretell the occurrence of destructive earthquakes goes back to antiquity, but has been the subject of a major international scientific effort only since the mid-1960s. The distinction between hazard assessment and prediction is primarily with regard to the specificity of the projected time of occurrence. An earthquake hazard assessment is a statement of the probability of occurrence in a defined area within a given time interval of either an earthquake with magnitude equal to or exceeding a specified value, or the occurrence of earthquake-generated ground motion exceeding a specified level at a given site. In addition to general concern for public safety, efforts to develop improved hazard-assessment methodologies have been motivated by the need for earthquake-resistant design and construction of critical facilities, such as dams, hospitals, and power plants, the failure of which because of an earthquake might produce catastrophic effects.

An earthquake prediction is a statement of the time, place, and magnitude of a specific future earthquake, with estimates of the uncertainty of each of those parameters. No reliable, generally applicable methods for formulating such predictions have yet been developed, though many approaches have been proposed and tested, with some limited success. Because of differences in the kinds of observations used, it is convenient to classify predictions by the time to the predicted event: long-term (decades to a few years); intermediate-term (a few years to a few weeks); short-term (a few weeks to a few hours). Long-term prediction efforts have become almost the same as hazard assessments, except for the implication that the event is expected to occur near the end of the long interval rather than with equal likelihood at any time within the interval.

6.1 Hazard Assessment

Earthquake hazard assessment starts from knowledge of the geographic distribution of earthquakes (Reiter, 1990). From catalogs of earthquakes located by instruments or from historical records, and geological evidence of prehistorical earthquakes, estimates can be made of the most likely sites for future earthquakes and the largest earthquakes that can be expected in each seismic source region. Uncertainties in this process arise from the possible incompleteness of catalogs for the time interval they cover, the short historical record in many parts of the world, and limited understanding of earthquake genesis, particularly in intraplate settings.

The next step is the estimation of the mean time between occurrences of earthquakes of a given magnitude in a particular seismic source region. The empirical Gutenberg–Richter relation between $N(M)$, the number of earthquakes with magnitude equal to or greater than M, and the magnitude is used for this purpose: $\log_{10}N(M) = a - bM$ (Reiter, 1990, Chap. 10). This relation approximately describes seismicity distributions on scales from global to local, with the value of b near 1. The parameters a and b are calculated as the maximum-likelihood estimates from the available catalog. If the data are normalized with respect to time to 1 year, the reciprocal of $N(M)$ is the average time in years between events at or above magnitude M within the specified area. The longer the time interval covered by the catalog and the wider the range of included magnitudes, the more reliable are the estimates of the parameters.

A problem with the application of this form is the recent discovery that the relation is most likely not a straight line over all magnitudes. Extrapolation from the frequent smaller events to estimates of the recurrence interval of the largest ones may be in serious error. One kind of behavior is the leveling off of the relation at larger magnitudes, with a relatively large number of stronger events with about the same magnitude, and few or none larger (the characteristic earthquake case). In this case extrapolation leads to an underestimate of the frequency of the largest events. Another possibility is a sharp decrease in the number of larger events, so that extrapolation overestimates the frequency of the largest magnitudes. This type of behavior has been explained by the geometry of the faults, with the falloff occurring for events larger than the smallest one that breaks through the entire brittle, seismogenic layer.

The estimation of the strongest earthquake, M_{max}, to be expected in a source region is a difficult problem (Reiter, 1990, Chap. 5). At what magnitude should the Gutenberg–Richter relation be terminated at its upper end? Experience demonstrates that taking the strongest known earthquake of the past as a measure of the strongest to be expected in the future is dangerous, especially in intraplate settings with short earthquake histories of only a few centuries. Mapping of the geologic structures in the region may reveal the dimensions of the largest units, the longest faults in particular, that are likely to participate in a single earthquake. For very long, through-going fault systems of 1000 km or more, it becomes necessary to identify barriers to slip, such as bends in the fault or offsets, that may serve to limit the length of any single break. A variety of statistical methods for estimating the extreme value of future magnitude from the known past events have been tested, but none has been proven to give reliable estimates.

Because of the limitations on hazard assessment imposed by the brevity of recorded human history relative to geologic time, efforts have been made to extend the record through *paleoseismic* investigations. Geologic evidence of prehistoric great earthquakes takes the form of preserved effects of such events. Under favorable circumstances, the amount of slip in the event and the time of its occurrence can be estimated.

A variety of evidence has been invoked for paleoseismic studies, from the offset of ancient shorelines of lakes to the interruption of the growth cycle of old trees on faults by tearing of their roots. Systematic searches for paleoseismic evidence on accessible faults are made by trenching. By careful mapping of the distribution of materials in the walls of a trench dug across or along a fault, the offset of layered beds may be detected, the beds disrupted in a single event identified, and direct evidence of strong ground shaking, such as deposits of sand liquefied by an ancient earthquake, seen. If the disturbed material in the trench walls includes organic matter, such as peat, the age of the disturbance can be determined radiometrically.

Trenching has been especially successful for the southern San Andreas fault in California (Sieh, 1984). The mean time interval between great earthquakes during the past 2000 years is now well determined. However, important for efforts at hazard assessment, clear evidence for clustering of large earthquakes in time has been found (Sieh *et al.*, 1989). Several groups of earthquakes have happened on the southern San Andreas fault with the time between individual events much shorter than the mean interevent time. Then a long time may pass before the next cluster begins. The implication is that the long-term mean time interval may not be a reliable input to probabilistic assessments of the occurrence of the next great earthquake, at least on this fault. The approach taken is to use the best-determined value of the mean time interval and incorporate the statistical uncertainty of this value in the assessment.

Once the probability of occurrence of an earthquake of a given magnitude within a seismic source region has been calculated, the estimation of the damage potential of that event in its surroundings is the final step in hazard assessment. Experience shows that damage depends strongly on local geologic conditions, especially the thickness and properties of poorly consolidated materials near the surface. For general hazard assessments, the peak value of some ground-motion parameter, such as peak acceleration or particle velocity, is estimated for hard-rock conditions as a function of distance from the event of specified magnitude. These estimates, which embody their own uncertainties, are made from empirical curves of ground motion vs distance that have been assembled for various geographic regions.

If recordings of strong ground motion from previous damaging events in the region are available, or if the use of such recordings from other regions can be justified, a response spectrum for the anticipated earthquake can be computed. This measure is the peak response (usually acceleration or velocity) of a simple oscillator with specified viscous damping at any time during the event as a function of the period of the oscillator. The response spectrum contains much more information about the damage potential than a single value of, say, peak acceleration.

The important effects of local ground conditions must be incorporated into the assessment process on a site-specific basis. A well-known example of the effects of near-surface geology in controlling damage is the pattern of destruction in Mexico City as a result of the Michoacán earthquake, 19 September 1985. The heavy damage and loss of life were concentrated hundreds of kilometers from the epicenter, with almost no damage in between. The severe effects were the direct result of the amplification of the ground motion in the unconsolidated lake-bottom sediments on which the damaged part of Mexico City is built. The heavy damage caused in the Marina District of San Francisco by the Loma Prieta, California, earthquake, 18 October (UTC) 1989, is further evidence of local-site effects. The damage at a distance of 100 km from the epicenter was due to a variety of responses, including soil liquefaction and ground-motion amplification, in the artificial fill, mostly sand, on which the structures were built.

6.2 Earthquake Prediction

Research on earthquake prediction has been organized as a serious scientific effort only since about 1960 (Asada, 1982; Scholz,

1990, Chap. 7). Early work was done in the former Soviet Union, Japan, and the People's Republic of China. A formal program began in the United States in 1977 as part of the National Earthquake Hazards Reduction Program. This broadly based international effort has not yet resulted in any reliable techniques for establishing the time of occurrence of a specific earthquake at a given place. Enough cases of observations of anomalies in geophysical parameters prior to earthquakes have been reported to lend support to the concept that at least some earthquakes in some geological situations may be predictable on the basis of physical precursors. On the other hand, developments in nonlinear continuum mechanics leave open the question as to whether it is, in principle, possible to know the conditions within a fault zone and the surrounding region at any time well enough to be able to predict the future state of that zone more than a short time into the future.

The search for prediction methods is conducted within the framework provided by the plate-tectonics model of geologic processes and the elastic-rebound model of earthquake energetics. According to the latter, the goal should be to observe evidence that the rocks in a seismic source zone are in a highly strained state, with the corresponding stresses approaching the shear strength of the fault-zone materials. The possibility of making the necessary observations is limited by the location of the fault, with the best chances for shallow faults on land. Because many of the world's greatest earthquakes happen under the sea in subduction zones (see PLATE TECTONICS), the attempts to predict these events must depend on the interpretation of data collected at a distance from the source region.

Two principal approaches to prediction are the search for specific precursors of earthquakes and the application of pattern-recognition techniques. A precursor is an anomaly in some geophysical observable that precedes an earthquake and can be interpreted as diagnostic of the approaching occurrence. Anomalies in many physical variables that are potentially indicative of a state of high strain or stress have been noted and tested as possible precursors (Asada, 1982; Wyss, 1991). Many of these reported anomalies either have not been reproduced in reevaluation of the supporting data or in additional cases, have been poorly documented, or have not been demonstrated in a convincing way to be truly related to the subsequent earthquake.

The fully developed case for a claimed precursor must include not only the evidence that the phenomenon preceded some earthquakes, but also tests of the false-alarm rate resulting from the use of this phenomenon (anomaly with no earthquake) and the failure-to-predict rate (earthquake with no anomaly). In addition, a quantitative definition of what constitutes an "anomaly" is required. Because some attempts at prediction have been offered on the basis of pseudoscience, it is essential that rigorous standards be applied to the evaluation of claims of success.

The most promising phenomena for further investigation appear to be crustal deformation in the vicinity of faults and changes in the spatial and temporal distributions of background seismicity. Crustal deformation is appealing because it is directly related to the accumulation of strain in rocks. Crustal deformation may be monitored by direct observations of uplift and subsidence relative to some base, changes in the distances between points a few kilometers or tens of kilometers apart along lines that may or may not cross active faults, local measurements of tilt and strain in the rocks, and creep along active faults. Indirect evidence of crustal deformation can be derived from gravity measurements, which are sensitive to elevation changes (but also to possible variations in mass beneath the observing point), and changes in water level in wells in response to strain changes. Pronounced elevation changes have been reported as occurring years to months prior to some strong earthquakes. Tilt and strain anomalies have also been reported as precursors. Observation systems in regions selected for prediction attempts routinely monitor such parameters.

Changes in the distribution of background seismicity are indicators of changes in the stress state or strength distribution on the fault. Patterns of both seismic activation and seismic quiescence have been suggested as intermediate-term precursors of a strong earthquake. If earthquakes above some minimum magnitude for which the catalog is thought to be complete are counted as a function of time over some extended period, an average rate of occurrence for the region covered by the catalog may be computed. Intervals of clustering and decreased activity are interspersed through the time series, but a meaningful long-term mean rate can be established.

Seismic quiescence occurs when the rate of background seismicity becomes significantly lower, in a statistical sense, than the long-term mean rate, and stays lower for a time long compared with the typical fluctuations that characterize the sequence. Quiescence in the future epicentral region has been identified in a number of cases as occurring months to years before strong earthquakes. The subject remains controversial, however, because of imperfections in earthquake catalogs. Some changes in the rate of events derived from catalogs have been demonstrated as artificial, due to changes in the monitoring network or data-analysis procedures. As with all searches for precursors, only the observation of many more cases of carefully documented quiescence in clear association with a subsequent strong earthquake can lead to the acceptance of this phenomenon as reliably diagnostic of an impending event.

Many strong earthquakes are preceded by foreshocks. The final warning in the most famous case of a successful prediction, Haicheng, China, 4 February 1975 (Asada, 1982, pp. 255–259), was based on the observations of foreshocks that began three days before, increased to a peak of about 60 events per hour on the morning of the event, then suddenly stopped some eight hours before the mainshock. However, no characteristics of foreshocks have been identified that can reliably distinguish them as such until after the main shock has occurred. One approach is to assume that every moderate (magnitude 4 to 6) event in a selected area is a foreshock whose occurrence increases the probability of a stronger earthquake during the next few days. This approach is useful for alerting emergency response services and public administrators, as long as it does not lead to an unacceptable false-alarm rate.

Seismicity patterns are among the traits that are considered in pattern-recognition approaches to prediction. As currently implemented, pattern-recognition/prediction algorithms are almost entirely the products of Russian geophysicists, led by Keilis-Borok (Keilis-Borok *et al.*, 1988). The investigator may define as traits any number of phenomena that may conceivably be associated with a coming earthquake. Each trait is then scored in terms of its association with a set of known past earthquakes. Those traits that have high scores are retained as diagnostic. The resulting algorithm can then be tested against another set of known past earthquakes, not part of the original set on which the learning was done. The finally accepted version is then applied to the prediction of future events by yielding Times of Increased Probability (TIPs) in prescribed geographic regions. No attempt is made to understand the physical basis for the success of a particular trait; the approach is purely phenomenological. Although the output of the procedure is the identification of TIPs in those places selected by the algorithm, the amount by which the probability of occurrence is increased over the long-term average probability is not stated quantitatively. The approach is now being evaluated by a number of investigators.

In addition to crustal deformation and seismicity changes, changes in several stress- and strain-dependent properties of rocks have been examined as possible intermediate- or short-term precursors. These include changes in the compressional-wave velocity due to the formation of microcracks in highly stressed materials followed by the filling of the cracks with ground water, changes in electrical conductivity and direction of magnetization, and changes in ground-water chemistry, especially the concentration of radon gas in the water (Scholz, 1990, Chap. 7). Although none of these anomalies has been verified as a reliable precursor, research continues, and it is conceivable that some combination of anomalies, each of which by itself increases the probability of a strong earthquake only slightly, may be found that can be used as a predictive tool.

7. APPLICATIONS OF SEISMOLOGY TO SOCIETAL NEEDS

The reduction of the disastrous effects of earthquakes in populated areas is one direct application of knowledge gained through seismological research. In addition to providing the input to the hazard-assessment procedures described above, seismology also provides structural engineers with data about the properties of strong motion that are needed for the development of building codes and design criteria. The ground-motion properties of primary engineering interest are the dominant frequencies in the motion, the peak values of the motion, typically of ground acceleration, and the duration of the motion.

The mitigation of the disastrous effects of extreme natural events, including earthquakes, is not purely, perhaps even primarily, a technical problem. Effective mitigation measures require actions governed by powerful economic and sociological factors. Strict land-use regulation (avoidance of the hazard) and rigorously enforced building codes can reduce future losses. However, the best technological decisions on such matters are often modified to accommodate other societal needs, including the need to accommodate ever-increasing populations.

The application of seismological techniques to mapping in detail the geological structure of the upper 10 km or so of the Earth's crust is the principal tool for exploration for oil and gas. The principles employed are the same as those used to investigate the planet on a global or regional scale; the difference is that the frequency band of the waves observed and the source–receiver distances are selected to yield the high resolution needed. Controlled sources, such as charges of high explosives, compressed-air guns, and mechanical vibrators, provide the wave energy. Large numbers of receivers are deployed and grouped to provide high signal-to-noise ratios and spatial filtering that enhances the wave arrivals that are useful. Modern techniques of seismic exploration use extensive processing of digitally recorded signals, including both travel-time and amplitude information, to produce the cross sections of subsurface structure that can be interpreted in terms of potential for yielding hydrocarbons (see GEOPHYSICS).

Another application of seismology is monitoring underground nuclear explosions. The problem consists of several parts. An unknown source must be detected by suitably deployed instruments, recognized as a seismic event, and located. It must then be identified as a natural event, an earthquake, or an explosion. This is the discrimination problem. If the decision is that it was an explosion, the yield must be estimated. Support for research to develop reliable monitoring technology has led to major advances in both theoretical and observational seismology. Although the end of the Cold War has eliminated incentives for the known nuclear powers to monitor test programs of each other, seismological monitoring of clandestine explosions has an important role to play in oversight of the nonproliferation of nuclear weapons.

GLOSSARY

Body Wave: A seismic wave that travels through the interior of the Earth, propagating in three dimensions.

Dip of a Fault: The angle between the horizontal plane and a fault plane, usually taken as less than 90°.

Earthquake: Vibration of the Earth's surface due to a transient disturbance of the equilibrium of the rocks in the interior.

Epicenter: Point on the surface vertically above the point of initiation of an earthquake rupture.

Fault: Surface separating two blocks of rock, along which there has been tangential relative motion of the blocks.

Fault Trace: Line of intersection of a fault plane, projected upward if necessary, with the Earth's surface.

Focal Mechanism: A description of the geometric properties of the fault rupture that generated an earthquake. The essential elements are the strike, dip, and direction of slip on the fault surface.

Focal Sphere: An imaginary sphere of small radius surrounding an earthquake source, on which the direction of first motion is mapped.

Hypocenter: The point in the Earth at which the energy in an earthquake is released. It is the point of initial release in the case of an extended fault rupture.

Interplate Earthquake: An earthquake at the interface between two lithospheric plates.

Intraplate Earthquake: An earthquake in the interior of a lithospheric plate.

Intensity: A measure of earthquake strength based on effects on natural features, manmade structures, and human reactions.

Isoseismal: A curve separating zones of equal intensity on a map of earthquake intensities.

Love Wave: A surface wave that is a horizontally polarized shear wave. Love waves exist only in layered systems.

Magnitude: A measure of earthquake strength based on the recorded amplitude of selected waves generated by the earthquake.

Moment, Seismic: A measure of earthquake strength based on the area of the ruptured surface and the mean relative displacement across that surface.

Microseisms: Continuous vibrations of the surface produced by sources such as wind or storms at sea.

Normal Fault: A fault on which the block above the fault plane moves downward in the direction of dip.

Precursor: An anomalous change in some observable, that is diagnostic of an impending earthquake.

***P* Wave:** An elastic wave related to propagating volumetric strain. It is longitudinally polarized in an isotropic medium.

Quiescence: A significant decrease in the rate of occurrence of background seismicity from the long-time average rate.

Rayleigh Wave: A surface wave for which the particle motion is polarized in the plane of incidence. In the absence of layering, Rayleigh waves are nondispersive.

Reverse Fault: A fault on which the block above the fault plane moves upward in the direction of dip.

Strike of a Fault: The angle between the fault trace and geographic north, conventionally defined so that the fault dips down to the right ($<90°$) when one looks in the direction of the strike.

Strike-Slip Fault: A fault on which the relative motion of the blocks is horizontal.

Surface Wave: A seismic wave that travels as a guided wave in layers defined by the Earth's outer surface and prominent boundaries in the interior. Surface waves propagate in two dimensions and exhibit geometric dispersion. They die off exponentially with distance away from the waveguide.

***S* Wave:** An elastic wave related to propagating shear strain. It is transversely polarized in an isotropic medium.

Tectonic Earthquake: An earthquake driven by the forces in the Earth that form geologic structures. Tectonic earthquakes are caused by sudden slip on a fault.

Works Cited

Aki, K., Richards, P. G. (1980), *Quantitative Seismology,* San Francisco: W. H. Freeman.

Anderson, D. L., Given, J. W. (1982), *J. Geophys. Res.* **87,** 3893–3904.

Asada, T. (Ed.) (1982), *Earthquake Prediction Techniques,* Tokyo: University of Tokyo Press.

Bullen, K. E., Bolt, B. A. (1985), *An Introduction to the Theory of Seismology,* 4th ed., Cambridge, U.K.: Cambridge Univ. Press.

Gephardt, J. W., Forsyth, D. W. (1984), *J. Geophys. Res.* **89,** 9305–9320.

Gubbins, D. (1990), *Seismology and Plate Tectonics,* Cambridge, U.K.: Cambridge Univ. Press.

Gutenberg, B., Richter, C. F. (1949), *Seismicity of the Earth,* Princeton: Princeton Univ. Press.

Hanks, T., Kanamori, H. (1979), *J. Geophys. Res.* **84,** 2348–2352.

Jackson, D. D., Anderson, D. L. (1970), *Rev. Geophys. Space Phys.* **8,** 1–63.

Kanamori, H. (1977), *J. Geophys. Res.* **82,** 2981–2987.

Keilis-Borok, V. I., Knopoff, L., Rotwain, I. M., Allen, C. R. (1988), *Nature* **335,** 690–694.

Klein, F. W., Koyanagi, R. Y., Nakata, J. S., Tanigawa, W. R. (1987), in R. W. Decker, T. L. Wright, P. H. Stauffer (Eds.), *Volcanism in Hawaii,* U.S. Geological Survey Professional Paper 1350, Washington, D.C.: U.S. GPO, pp. 1019–1185.

Kostrov, B. V., Das, S. (1988), *Principles of Earthquake Source Mechanics,* Cambridge, U.K.: Cambridge Univ. Press.

Lee, W. H. K., Stewart, S. W. (1981), *Principles and Applications of Microearthquake Networks,* New York: Academic.

Macelwane, J. B. (1947), *When the Earth Quakes,* Milwaukee: Bruce.

Musgrave, M. J. P. (1970), *Crystal Acoustics,* San Francisco: Holden-Day.

Reid, H. F. (1910), *The California Earthquake of April 18, 1906,* Vol. II, *The Mechanics of the Earthquake,* Report of the State Earthquake Investigation Commission, Washington, DC: Carnegie Institution of Washington.

Reiter, L. (1990), *Earthquake Hazard Analysis,* New York: Columbia Univ. Press.

Richter, C. F. (1958), *Elementary Seismology,* San Francisco: W. H. Freeman.

Scholz, C. H. (1990), *The Mechanics of Earthquakes and Faulting,* Cambridge, U.K.: Cambridge Univ. Press.

Sieh, K. (1984), *J. Geophys. Res.* **89,** 7641–7670.

Sieh, K., Stuiver, M., Brillinger, D. (1989), *J. Geophys. Res.* **94,** 603–624.

Wyss, M. (Ed.) (1991), *Evaluation of Proposed Earthquake Precursors,* Washington, DC: American Geophysical Union.

Further Reading

Bolt, B. A. (1993), *Earthquakes,* New York: W. H. Freeman.

Kennett, B. L. N. (1983), *Seismic Wave Propagation in Stratified Media,* Cambridge, U.K.: Cambridge Univ. Press.

Mogi, K. (1985), *Earthquake Prediction,* Tokyo: Academic.

SEMANTIC VIEW OF THEORIES

R. I. G. Hughes, *Department of Philosophy, Logic and Scientific Method, The London School of Economics and Political Science, London, U.K.*

INTRODUCTION

The use of "the semantic method" in "the logical analysis of theories" was proposed by the Dutch philosopher Evert Beth in 1949. The method he advocated was that developed in Tarski's investigation of formal languages in the 1930s. The phrases *the semantic view* and *the semantic conception* now cover a variety of approaches to the study of physical theories. They have in common the rejection of an earlier conception of theories, variously called *the syntactic view, the statement view,* and (to mark its predominance during the first half of this century) *the received view.*

1. THE SYNTACTIC VIEW

Between about 1930 and 1960 the statement view was articulated by the logical empiricists, notably Hans Reichenbach, Rudolf Carnap, and Carl Hempel, who took as a model for their work Hilbert's (1971) formalization of the foundations of geometry. On their view, as adumbrated by Carnap (1939), a physical theory consists of a formal calculus together with a (partial) interpretation of that calculus. The formal calculus is a theory in the logician's sense: it is the deductive closure of a set of axioms. The language in which the axioms are stated and the principles governing deductions are to be those of modern logic. The language contains connectives (e.g., "$\rightarrow$" for "If . . . , then . . ."), quantifiers (e.g., "$\forall x$" for "For any x"), variables ("x", "y", etc.), and predicates ("P", "Q", etc.); the sentence "$\forall x(Px \rightarrow Qx)$" may be read as "For any x, if x is a P, then x is a Q", or, more succinctly, as "All Ps are Qs." This abstract syntactical structure is given content by an interpretation. Only the elements of the language that can be interpreted as referring to observable quantities and their magnitudes (the *observable predicates*) can be "cashed out," and so the formal language is only partially interpreted. Although theoretical predicates receive no direct interpretation, the theory contains correspondence rules that relate them to observational terms.

It was no part of the logical empiricists' program that this articulation of a theory should correspond to the way that it would be presented and used by a practicing physicist. Rather, their aim was to isolate and display the logical structure and the cognitively significant features of a theory. To this end, one of the few fully developed analyses of this kind, Reichenbach's (1957) axiomatization of relativity, "was aimed at exhibiting the roles of experience and of convention in physical theorizing about space, time, and motion" (Hempel, 1977). A detailed account of the development of the received view, and also of the criticisms leveled against it, is provided by Suppe (1977).

The crucial move made by proponents of the semantic view is to reject the thesis that a theory is primarily a linguistic entity. Instead of concentrating, as the logical empiricists did, on the language in which the theory is articulated, they focus attention on the structures that the axioms of a theory define. In the formal version of the semantic approach these structures are models of the theory, in the sense in which that word is

3-527-28139-8/96/$5.00 + .50

used in logic and formal semantics. Hence the phrase, "the semantic view." While an emphasis on the models of a theory is common to all versions of the semantic view, different versions contain different proposals concerning the way these models should be specified. To some extent these differences also correspond to different views about the function that the analysis should serve. In the examples that follow, the demand that the specifications should have a canonical form is progressively relaxed. At the same time, not only do the analyses move closer to scientific practice, but a close match between the two becomes a desideratum.

2. THE SET-THEORETIC APPROACH

The set-theoretic approach was initiated by Suppes (1969) in the 1960s and taken up in the next decade by Sneed (1979) and Stegmüller (1979). Like the logical empiricists, Suppes aimed at clarifying the foundations of physical theories, rather than at an analysis that matched scientific practice. He too emphasized the axiomatic method, but he differed from the empiricists, first, by using this method to define set-theoretical structures and, second, by allowing informal set-theoretical methods of proof where the empiricists had insisted on a rigid adherence to first-order logic. (A logic is *first order* if it allows quantification over individuals but not over predicates. In the sentence, "Every light bulb was defective in some way or other," for example, "every" involves quantification over individuals, but "some," quantification over predicates.) Instead of looking to Hilbert's work in mathematics as a model, Suppes and those who followed him looked to the group of French mathematicians collectively known as Bourbaki.

Sneed (1979) claims that the approach yields a "logical clarification of theories"; his account of classical particle mechanics offers an example of a Bourbaki-style presentation of a scientific theory. Sneed here uses the set-theoretic predicates "is a PM" (is a particle mechanics) and is a "CPM" (is a classical particle mechanics) where other writers would talk of a structure being a system of a particular kind. The predicates are defined as follows.

x is a PM if and only if there exists a P, T, $\mathbf{s}$, m, $\mathbf{f}$, such that

1. $x = \langle P,T,\mathbf{s},m,\mathbf{f}\rangle$;
2. P is a nonempty finite set;
3. T is an interval of real numbers;
4. $\mathbf{s}$ is a function from $P \times T$ into the set of ordered triples of real numbers such that, for all $p \in P$, $t \in T$, $D^2\mathbf{s}(p,t)$ exists;
5. m is a function from P into the real numbers such that, for all $p \in P$, $m(p) > 0$; and
6. $\mathbf{f}$ is a function from $P \times T \times I$ into the set of ordered triples of real numbers such that, for $p \in P$, $t \in T$,

$$\sum_{i \in I} \mathbf{f}(p,t,i) \text{ is absolutely convergent.}$$

A system P of point masses, moving in three-dimensional space during a time interval T under the influence of a collection of forces, constitutes a realization of a PM; clause **4** (in which the operator D is the time derivative) stipulates that the acceleration of each particle is at all times well defined; clause **5**, that the mass m of each particle is positive and never varies; clause **6**, that the total (vector) sum of the forces acting on each particle is well defined.

The class of models is then further restricted by adding Newton's second law, to define a classical particle mechanics.

x is a CPM if and only if

1. x is a PM; and
2. for all $p \in P$ and $t \in T$,

$$m(p)D^2\mathbf{s}(p,t) = \mathbf{f}(p,t,i)$$

Sneed goes on to define the subclass of CPMs that are also governed by Newton's third law (Newtonian classical particle mechanics). In any theory the specification of set-theoretic structures is accompanied by a set of intended applications; an application is a physical system (e.g., our solar system) that may be treated as a particle mechanics. Sneed is unwilling, however, to regard statements of the form "Q is a CPM" as expressing the empirical content of the theory (where Q is an intended application), because of difficulties involving theoretical terms. The problem, in the case of particle mechanics, lies with the mass and force functions. If Q is a putative application of classical particle mechanics, there may be no way to determine the values of these functions for Q that does not already presuppose that Q is a CPM. Sneed acknowledges that he gives only a partial resolution of this problem. In a later work co-authored with Wolfgang Balzer and Ulises Moulines (Balzer *et al.*, 1987), Sneed suggests that the occurrence of the same theoretical term (e.g., mass) in more than one theory offers a different solution.

In this work Balzer *et al.* show how versatile the set-theoretic approach is by applying it, not only to classical particle mechanics, classical collision mechanics, and relativistic collision mechanics, but also to decision theory, Daltonian stoichemistry, simple equilibrium thermodynamics, Lagrangian mechanics, and pure exchange economics. The authors also apply the set-theoretic approach to a general theory of theories, as does Stegmüller (1976, 1979).

3. THE STATE-SPACE APPROACH

The state-space approach stays much closer to orthodox formulations of physical theories than does the set-theoretic approach. The most influential presentation of it was made by van Fraassen (1983), who acknowledged debts to Hermann Weyl and (especially) to Evert Beth. The approach is also implicit in the work of von Neumann (1955) and Mackey (1963) on quantum mechanics and suggests an approach to the interpretive problems raised by that theory. It also offers a way of comparing theories as different as classical and quantum mechanics, and these theories will serve as examples. (See Hughes, 1989.) It has also been successfully applied to evolutionary theory in biology; see Thompson (1989) and Lloyd (1988).

On the state-space approach a theory is characterized by three things:

1. the way in which the state of a system is specified;
2. the way in which the state evolves in time; and
3. the relation the theory postulates between the state of a system and the values of physical quantities associated with the system.

In the Hamilton-Jacobi formulation of classical mechanics, a classical system consists of a set of n point masses. The state of the system at a given time is specified by a point ω in a space Ω of $6n$ dimensions. The coordinates of ω are the three components q_x, q_y, q_z, of position and the three components p_x, p_y, p_z, of momentum, for each of the n particles. The evolution of the state ω through time is determined by the Hamiltonian function $H{:}\Omega \to R$ for the system, whose value for a given state is the system's total energy in that state. The evolution of ω through time is then given by the $3n$ pairs of equations

$$\frac{dq_i}{dt} = \frac{\partial H}{\partial p_i}, \qquad \frac{dp_i}{dt} = -\frac{\partial H}{\partial q_i}.$$

In contrast, the state space of a quantum system is a Hilbert space $\mathcal{H}$, possibly infinite-dimensional. A pure state of the system is represented by a normalized vector $\boldsymbol{\psi}$ in $\mathcal{H}$; more generally, a state is representable by a density operator D (a trace-class operator of trace 1) on $\mathcal{H}$. Here, however, only pure states will be considered. In quantum mechanics, the energy of a system is associated with an operator H on $\mathcal{H}$ (the Hamiltonian operator); the evolution of $\boldsymbol{\psi}$ through time is given by Schrodinger's equation

$$ih\, \partial\boldsymbol{\psi}/\partial t = H\boldsymbol{\psi}.$$

Radical differences between classical and quantum mechanics appear in the relations the two theories postulate between a system's state and the measurable values of physical quantities associated with the system. As in the special case of energy, whereas in classical mechanics each physical quantity A is associated with a real-valued function $f_A{:}\Omega \to R$, in quantum mechanics a physical quantity is associated with a Hermitian operator $A{:}\mathcal{H} \to \mathcal{H}$. In the classical case, the value of f_A for a given state gives the value of the corresponding physical quantity. These values, obviously, will be restricted to the range of f_A. The predictions of the theory take the form "In state ω the measured value of A will be $f_A(\omega)$."

In quantum mechanics, on the other hand, the possible values of a physical quantity are restricted to the spectrum of the corresponding Hermitian operator, A. This spectrum may be discrete (as in the case of a spin operator), in which case the spectrum consists of the eigenvalues of A, or continuous (as in the case of the momentum operator p). In the discrete case, to each eigenvalue a_i of the operator is associated a projection operator P_i from the spectral decomposition of A, so that

$$A = \sum a_i P_i.$$

P_i projects onto the ray spanned by the eigenvectors corresponding to the eigenvalue a_i.

In the continuous case, to the Hermitian operator A we assign a spectral measure $P(x)$, such that

$$A = \int_{-\infty}^{\infty} x dP(x).$$

We then define, for each semi-closed interval $\Delta = \{x: a < x \leq b\}$ of $\mathbb{R}$, the operator

$$P_\Delta^A = P(b) - P(a).$$

In both cases, the theory predicts, not the value of the quantity that measurement will yield, but the probability that a particular outcome will result. In the discrete case the probability p_i that a measurement of A will yield the result a_i is given by

$$p_i = \langle \psi | P_i \psi \rangle.$$

In the continuous case, the probability p that a measurement of A will yield a result in the interval Δ is given by

$$p_i = \langle \psi | P_\Delta^A \psi \rangle.$$

In both the classical and quantum cases, van Fraassen (1983) suggests that the observable magnitudes are to be identified with elements in the "empirical substructures" within the models defined by the theory. For example, in quantum mechanics the set of subspaces of the Hilbert space (each associated with a value, or set of values, of a physical quantity) is the empirical substructure of the Hilbert space.

In this way, the state-space approach may suggest a specific interpretive stance with respect to a theory. For, in the case of quantum mechanics, the set of subspaces has the algebraic structure of a coherent orthomodular lattice (equivalently a transitive partial Boolean algebra: see Hughes, 1989; QUANTUM LOGIC). The "logic" of these structures is nonstandard; within them the Boolean distributive law does not hold, in general. Thus the state-space representation of quantum theory leads naturally to the "quantum logical" approach to its interpretation.

4. THE REPRESENTATIONAL APPROACH

The set-theoretic and state-space versions of the semantic account emphasize foundational theories of physics, like classical mechanics and quantum mechanics. They offer alternative ways of representing the internal structure of these theories. They do not examine the question of how a theory is applied. In contrast, the representational approach accommodates both foundational theories and their applications. For this reason, of the three versions of the semantic view examined, it is the one that most nearly conforms to scientific practice.

On this account, a physical theory involves two things:

1. the specification of a class of models (the *theoretical definition*); and
2. the postulate that a particular part of the world, or of the world as we describe it, can be adequately represented by a model from that class (the *theoretical hypothesis*).

This two-part analysis is due to Giere (1984). It is explicitly endorsed by van Fraassen (1989) in his recent writings and implicitly relied on by (e.g.) Suppe (1989) and Cartwright (1983). These four authors differ, however, in their description of the relation between the model and its subject, the part of the world that it represents. Van Fraassen (1989) takes the subject–model relation to be "some embedding, or approximate embedding." This accords with his suggestion that the empirical content of the model is provided by the empirical substructures of the theory; he describes these as "candidates for the direct representation of observable phenomena." In doing so he limits himself to those comparatively rare phenomena to which a foundational theory can be directly applied. Of the four authors mentioned above, he alone pays no attention to the theoretical models used in specific applications of a theory. While the behavior of such "local models," be they models of atomic structure or of the phenomenon of superconductivity, may be governed by the laws of a foundational theory, the model itself is crafted to fit the individual phenomenon.

As an example of a local model, Giere uses the ideal harmonic oscillator whose behavior conforms to the laws of classical mechanics. No "real oscillating system," no weight suspended from a spring, behaves exactly like the ideal oscillator. Giere (1985) describes the subject–model relation in terms of similarity; the subject is "similar to the proposed model in specified respects, and to a specified degree."

For Suppe (1989) the model is a "kind of replica" of its subject; he goes on to distinguish between replicas produced by *abstraction* and those produced by *idealization.* In abstraction we ignore certain influences that affect actual systems (e.g., friction, air resistance); in these cases it is causally possible (Suppe's term) to realize the conditions under which the behavior of the subject and model agree. In idealization we postulate abstract, unrealizable models, as in particle mechanics where the model consists of a set of point masses.

In terms of the relative importance they assign to foundational theory, Cartwright (1983) stands at the opposite end of the spectrum from van Fraassen. In criticizing van Fraassen's version of the semantic account, she writes: "Schrödinger's equation, even when coupled with principles which tell us what Hamiltonians to use for square-well potentials, two-body Coulomb interactions, and the like, does not constitute a theory of anything. To have a theory of the ruby laser, or of bonding in the benzene molecule, one must have models for these phenomena which tie them to descriptions in the mathematical theory." These models, which Cartwright calls *simulacra* to emphasize their fictional nature, do the work of physics. They are models of *phenomena.* On this view, theorizing is a four-level process: The behavior of phenomena is described in terms of simulacra (models of phenomena); that of the simulacra, in terms of models of a theory (infinite potential wells, linear harmonic oscillators, and the like); and that of these "local" models, in terms of the foundational theory in question. As an example she cites Louisell's treatment of a gas laser. The laser, and its behavior, is the phenomenon to be modeled. It is represented as a collection of three-level atoms in interaction with a quantized electromagnetic field; both these components also interact with a damping reservoir. The behavior of this simulacrum is in turn modeled by representing it in a way that is amenable to treatment by the relevant foundational theory, in this case quantum theory.

Physics offers examples of theoretical representation that conform to the views of each of the four authors cited. Indeed, one virtue of this approach to the analysis of theory is that it draws attention to the many different ways in which theories are applied. It also offers a new approach to vexed philosophical issues such as the status of scientific laws and the nature of scientific explanation (see van Fraassen, 1989; Hughes, 1993). For these reasons it underlies much contemporary work in the philosophy of science.

Works Cited

Balzer, W., Moulines, C. U., Sneed J. D. (1987), *An Architectonic for Science,* Dordrecht: Reidel.

Carnap, R. (1939), *Foundations of Logic and Mathematics,* Chicago: Univ. of Chicago Press.

Cartwright, N. (1983), *How the Laws of Physics Lie,* Oxford: Clarendon Press.

Giere, R. (1984), *Understanding Scientific Reasoning,* 2nd ed., Fort Worth, TX: Holt, Rinehart and Winston.

Giere, R. (1985), "Constructive Realism," in: P. M. Churchland, C. A. Hooker (Eds.), *Images of Science,* Chicago: Univ. of Chicago Press.

Hempel, C. G. (1977), "Formulation and Formalization of Scientific Theories," in: F. Suppe (Ed.), *The Structure of Scientific Theories,* 2nd ed., Urbana, IL: Univ. of Illinois Press.

Hilbert, D. (L. Under, transl.) (1971), *The Foundations of Geometry,* La Salle, IL: Open Court (orig. pub. 1899).

Hughes, R. I. G. (1989), *The Structure and Interpretation of Quantum Mechanics,* Cambridge, MA: Harvard University Press.

Hughes, R. I. G. (1993), "Theoretical Explanation," in: P. A. French, T. E. Uehling, H. K. Wettstein (Eds.), *Philosophy of Science,* Midwest Studies in Philosophy Vol. XVIII, Notre Dame, IN: Univ. of Notre Dame Press.

Lloyd, E. A. (1988), *The Structure and Confirmation of Evolutionary Theory,* Westport, CT: Greenwood Press.

Mackey, G. W. (1963), *The Mathematical Foundations of Quantum Mechanics,* New York: Benjamin.

Reichenbach, H. (M. Reichenbach, J. Freund, transl.) (1957), *Axiomatization of the Theory of Relativity,* Berkeley, CA: Univ. of California Press (orig. pub. 1924).

Sneed, J. D. (1979), *The Logical Structure of Mathematical Physics,* 2nd ed., Dordrecht: Reidel.

Stegmüller, W. (1979), *The Structuralist View of Theories,* New York: Springer Verlag.

Suppe, F. (Ed.) (1977), *The Structure of Scientific Theories,* 2nd ed., Urbana, IL: Univ. of Illinois Press.

Suppe, F. (1989), *The Semantic Conception of Theories and Scientific Realism,* Urbana, IL: Univ. of Illinois Press.

Suppes, P. (1969), *Studies in the Methodology and Foundations of Science,* Dordrecht: Reidel.

Thompson, P. (1989), *The Structure of Biological Theories,* Albany: SUNY Press.

van Fraassen, B. C. (1983), *The Scientific Image,* Oxford: Clarendon Press.

van Fraassen, B. C. (1989), *Laws and Symmetries,* Oxford: Clarendon Press.

von Neumann, J. (R. T. Beyer, transl.) (1955), *Mathematical Foundations of Quantum Mechanics,* Princeton: Princeton University Press (orig. pub. 1932).

Further Reading

The Syntactic View

Carnap, R. (1939), *Foundations of Logic and Mathematics,* Chicago: Univ. of Chicago Press.

Hempel, C. G. (1977), "Formulation and Formalization of Scientific Theories," in: F. Suppe (Ed.), *The Structure of Scientific Theories,* 2nd ed., Urbana, IL: Univ. of Illinois Press.

Hilbert, D. (L. Under, transl.) (1971), *The Foundations of Geometry,* La Salle, IL: Open Court (orig. pub. 1899).

Nagel, E. (1961), *The Structure of Science,* New York: Harcourt, Brace, and World.

Reichenbach, H. (M. Reichenbach, J. Freund, transl.) (1957), *Axiomatization of the Theory of Relativity,* Berkeley, CA: Univ. of California Press (orig. pub. 1924).

Critical Perspectives on the Syntactic View

Grünbaum, A., Salmon, W. C. (Eds.) (1988), *The Limits of Deductivism,* Berkeley: Univ. of California Press.

Suppe, F. (Ed.) (1977), *The Structure of Scientific Theories,* 2nd ed., Urbana, IL: Univ. of Illinois Press. The editor's introduction offers a detailed account and critique of the syntactic view.

The Set-Theoretic Approach

Balzer, W., Moulines, C. U., Sneed, J. D. (1987), *An Architectonic for Science,* Dordrecht: Reidel.

Sneed, J. D. (1979), *The Logical Structure of Mathematical Physics,* 2nd ed., Dordrecht: Reidel.

Stegmüller, W. (1976), *The Structure and Dynamics of Theories,* New York: Springer Verlag.

Stegmüller, W. (1979), *The Structuralist View of Theories,* New York: Springer Verlag. Provides a good introduction to this approach.

Suppes, P. (1969), *Studies in the Methodology and Foundations of Science,* Dordrecht: Reidel.

The State-Space Approach

Hughes, R. I. G. (1989), *The Structure and Interpretation of Quantum Mechanics,* Cambridge, MA: Harvard University Press.

Lloyd, E. A. (1988), *The Structure and Confirmation of Evolutionary Theory,* Westport, CT: Greenwood Press.

Mackey, G. W. (1963), *The Mathematical Foundations of Quantum Mechanics,* New York: Benjamin.

Thompson, P. (1989), *The Structure of Biological Theories,* Albany: SUNY Press.

van Fraassen, B. C. (1991), *Quantum Mechanics: An Empiricist View,* Oxford: Clarendon Press.

von Neumann, J. (R. T. Beyer, transl.) (1955), *Mathematical Foundations of Quantum Mechanics,* Princeton: Princeton University Press (orig. pub. 1932).

All the above apply the state-space approach, explicitly or implicitly, to particular theories. A more general account appears in van Fraassen, B. C. (1983), *The Scientific Image,* Oxford: Clarendon Press.

The Representational Approach

Cartwright, N. (1983), *How the Laws of Physics Lie,* Oxford: Clarendon Press.

French, P. A., Uehling, T. E., Wettstein, H. K. (Eds.) (1993), *Philosophy of Science,* Midwest Studies in Philosophy Vol. XVIII, Notre Dame, IN: Univ. of Notre Dame Press.

Giere, R. (1984), *Understanding Scientific Reasoning,* 2nd ed., Fort Worth, TX: Holt, Rinehart and Winston. Chapter 5 gives a straightforward introduction to this approach.

Hughes, R. I. G. (1993), "Theoretical Explanation," in: French, P. A., Uehling, T. E., Wettstein, H. K. (Eds.) (1993), *Philosophy of Science,* Midwest Studies in Philosophy Vol. XVIII, Notre Dame, IN: Univ. of Notre Dame Press.

Suppe, F. (1989), *The Semantic Conception of Theories and Scientific Realism,* Urbana, IL: Univ. of Illinois Press.

van Fraassen, B. C. (1989), *Laws and Symmetries,* Oxford: Clarendon Press.

SEMICONDUCTOR-DEVICE INTEGRATION

Hideo Sunami, *Memory Business Operation, Semiconductor & Integrated Circuits Division, Hitachi, Ltd., Kodaira, Tokyo, Japan*

INTRODUCTION

Since the transistor was invented in the late 1940s, various kinds of semiconductor devices have been developed. They are rectifiers, amplifiers using small-signal or power transistors, microwave-generating diodes, light-emitting diodes (LEDs), laser diodes (LDs), line and area light sensors, infrared sensors, pressure sensors, and *p*H sensors. Recently, very tiny mechanical devices have been fabricated also using silicon process techniques.

Since an electronic circuit integrating more than two devices was invented and demonstrated in 1958 by J. S. Kilby (Kilby, 1976), the integrated-circuit (IC) industry has grown very rapidly at an average annual rate of greater than 10%. It is predicted that worldwide IC sales will surpass $\$200 \times 10^9$ by the year 2000. The growth rate of the IC industry is greater than that of the whole electronics industry. This is partly because the proportion of IC components in products has been gradually increasing. There are very few electronic products that do not take advantage of ICs today.

Even though there are several other potential semiconductor materials, such as gallium arsenide, gallium phosphide, germanium, and silicon carbide, most integrated circuits commercially available in volume production have been made from silicon. The major reasons are that

1. silicon is a very stable material;
2. there exist appropriate dopant atoms for forming *n*-type and *p*-type regions;
3. the silicon-dioxide film formed by oxidation of the silicon itself is chemically and physically very stable; and
4. a planar technology (Hoerni, 1960) has been invented that is very effective for fabrication of monolithic integrated circuits.

It is also very fortunate for the silicon IC industry that silicon material is plentiful on Earth.

Silicon integrated circuits are categorized according to device use or function. As regards devices, the leading technology has been shifting from bipolar transistor, to *p*-channel MOSFET (metal-oxide-semiconductor field-effect transistor), to *n*-channel MOSFET, and then to CMOS (complementary MOS).

Bipolar devices are still used in ultrahigh-speed logic and memory for supercomputers and main-frames, in linear ICs for consumer products, as microwave transistors for telecommunication, and for high-power devices. BiCMOS (bipolar CMOS) devices are now available combining the advantages of bipolar and CMOS transistors.

3-527-28139-8/96/$5.00 + .50

As for their functions, there are memories, logics, and microprocessors. The memories include DRAM (dynamic random-access memory, SRAM (static RAM), Mask ROM (read-only memory), and nonvolatile ROM such as EPROM (erasable and programmable ROM) and EEPROM (electrically erasable PROM). Flash memory is one kind of EEPROM. The logics include G/A (gate array), signal processors, and standard logics such as TTL (transistor-transistor logic). Most microcomputers incorporate both logics and very high-speed cache memory. As the integration level increases, various kinds of devices are inevitably integrated in response to the new functions required.

The basic configuration and operation of major integrated circuits that are now widely manufactured in semiconductor industry are briefly described in the following.

DRAM (Dynamic Random-Access Memory): A 4-megabit DRAM incorporates 4 194 304 (= 2^{22}) memory cells. The memory cell consists of one MOS transistor and one capacitor. Stored charge in the capacitor is continuously leaking out, resulting in loss of stored "memory"; therefore, a restoring operation is required typically every 100-ms interval. Thus this is called "dynamic" RAM. Since the memory access time is around 100 ns and relatively slower than that of SRAM, DRAM is typically used for the main memory of the computer.

SRAM (Static RAM): Its memory cell principally consists of six transistors configuring a flip-flop. Since the memory cell is several times larger than that of DRAM, one-quarter as much memory volume is available in the same technology generation. As no additional restoring operation is required, this is called "static" RAM. The access time is generally much faster than that of DRAM; thus, SRAM is used for cache memory of the computer and main memory of the supercomputer. At the same time, extremely low-power SRAM is also available making full use of no need for a restoring operation. Beside MOS SRAM, bipolar SRAM is still being produced for ultrahigh-speed operation.

EPROM (Erasable and Programmable Read-Only Memory): Even though there are various kinds of EPROM and EEPROM, the memory cell principally consists of one or two MOS transistors having a storage site of a floating gate or an interface of different materials inside a gate oxide. The writing operation is done by charging the storage site, while erasing is performed by UV-light exposure. Since the erasing time is typically around 10 min with UV exposure, EPROM is in general used as one-time writable ROM.

EEPROM (Electrically Erasable and Programmable Read-Only Memory): The memory cell is principally the same as that of EPROM, while erasing is done by hot-electron injection or tunneling instead. More than 10^4 cycles of erasing is guaranteed. EEPROM is used as user-programmable ROMs in various consumer electronics.

Flash EEPROM: One type of EEPROM. This is so called because erasing proceeds simultaneously across all memory cells like a "flash."

Mask ROM: This is the first read-only memory. It consists of a memory cell with one simple MOS transistor. The "write" operation is no more than a patterning process that makes or does not make an electrical connection to certain transistors at the memory site. This is the highest density memory of all resulting in the least expensive and highest volume usage.

Logic LSI: A generic term for LSI incorporating logic circuits mainly used for logic operation. Usually it contains certain ROM and SRAM required for the logic operation. There are various kinds of logic LSIs in the semiconductor market.

MPU (Micro Processor Unit): One kind of logic LSI incorporating various kinds of logic and memory circuits used for calculation for computing. The memories are principally very fast SRAM, used as a level-1 cache memory, and ROM. The relative area of total memories has become larger and close to almost one-half of total chip area in advanced MPUs.

The market trend in these silicon devices is shown in Fig. 1. The growth rates of memory and microprocessors are bigger than those of others. In 1994, the market share of DRAM was around 60% in the MOS memory category. The rapid market growth from 1992 to 1994 is mainly due to the growth of the personal computer market.

Figure 2 shows device counts of DRAM and microprocessor chips. Fourfold higher packing has been realized every three years since 1970 when the 1-kilobit DRAM was introduced. As of 1995, a leading-edge, 32-bit

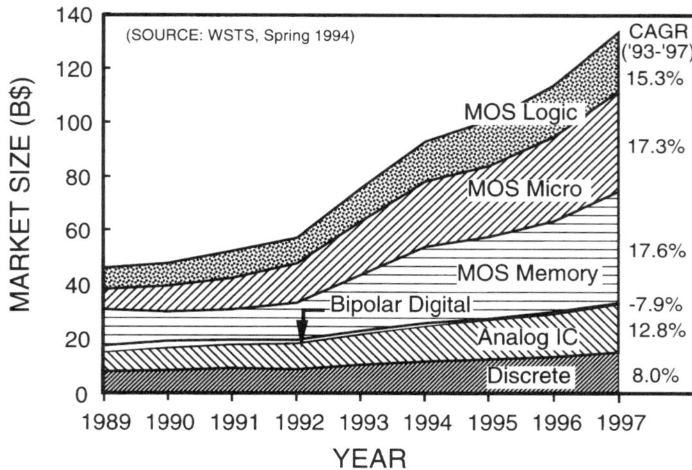

FIG. 1. Market trends in semiconductor products. Trends from 1994 are projected.

microprocessor integrates 3 million transistors on a 300-mm^2 die with 0.8-μm feature size, while a 4-Mbit DRAM has 5 million transistors and 4 million capacitors on an 80-mm^2 die with the same feature size. The packing is limited by memory-cell size for DRAM and by metal interconnect wiring in microprocessors.

A cross section of a typical sub-μm CMOS device identifying critical processing components is shown in Fig. 3. This is representative of a large variety of products, some of which may not use some of the components depending on the function and performance requirements and the manufacturer's experience. Individual elementary processes are not described in this article except for a few selected processes referred to in the next section.

The drive current I_{drive} in the saturation region of an MOS transistor is fundamentally expressed as

$$I_{\text{drive}} = \mu\epsilon_i W_{ch}(V_{gs} - V_t)^2/2T_i L_{ch}, \quad (1)$$

where μ, ϵ_i, W_{ch}, V_{gs}, V_t, T_i, and L_{ch} are carrier mobility, permittivity of gate insulator, channel width, gate-to-source voltage, threshold voltage, gate-insulator thickness, and channel length, respectively. Thus, a thinner gate insulator and a shorter channel length lead to better performance. This is one of

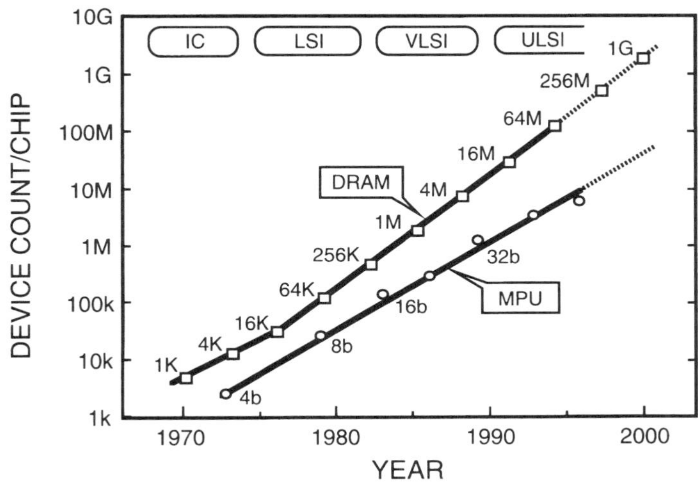

FIG. 2. Device counts of DRAMs and microcomputers. Current DRAM always shows the largest device count among all ICs because of its simple and tiny memory-cell structure.

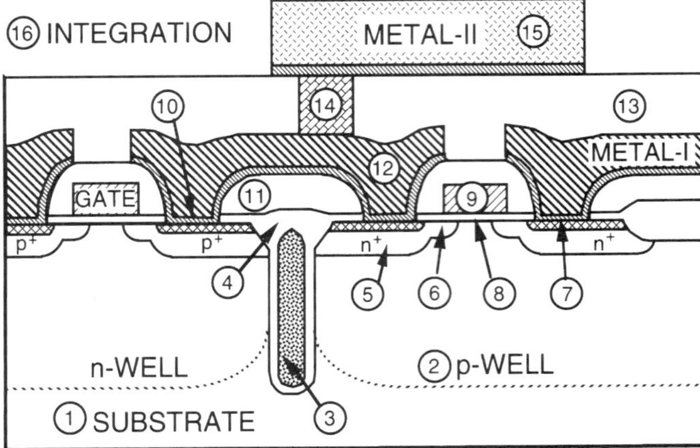

FIG. 3. Elementary components and key issues for submicron CMOS devices. (1) Substrate: epitaxial growth, SOI (Si-on-insulator); (2) well formation: retrograde well, low resistance; (3) deep-trench isolation: etching, side-wall doping, insulator filling; (4) recessed isolation: short bird's beak; (5) junction: low-leakage *p-n* junction, low sheet resistance, silicide; (6) impurity profile: well-defined doping; (7) junction contact: low resistance, barrier material, silicon-cluster precipitation; (8) gate insulator: thinner insulator with good integrity, low defect density, high reliability; (9) gate: low resistance, silicide and/or polycide, refractory metal; (10) self-aligned contact: barrier materials, e.g., titanium nitride, TiN; (11) intermediate insulator I; low-temperature reflow; (12) metallization I; low resistance, heat resistance, free from migration; (13) intermediate insulator II: planarization, less water absorption; (14) via hole: plug, low contact resistance, self-aligning; (15) metallization II: heat resistance, low resistance; (16) fine patterning: delineation accuracy, aligning, and etching.

the biggest motivating forces driving toward smaller devices and resulting in more functions on a die. These characteristics are summarized by the "scaling law" (Dennard *et al.*, 1974).

Since DRAM's greater integration started in 1970, technology innovation has continuously been achieved for these 25 years. The results and projected techniques are shown in Fig. 4 up to the year 2000. Techniques shown in the figure are selected to represent the major components of a typical process flow.

Although a transistor of 0.04-μm gate length has already been demonstrated operating at room temperature, many manufacturing issues remain to be solved for production of 1-gigabit DRAM-class devices and beyond. These are discussed in Sec. 2.3.

1. ELEMENTARY PROCESSES AND DEVICES FOR INTEGRATION

1.1 Elementary Processes

In terms of functions, the elementary processes in wafer fabrication can be categorized into four technology groups as

1. patterning,
2. isolation,
3. device formation, and
4. interconnection.

Each technology group consists of combinations of elementary processing techniques such as crystal growth and wafer preparation, epitaxy, lithography, wet and dry etching, ion implantation and annealing, oxidation, diffusion, dielectric-film deposition, silicidation, and metallization.

In addition to these fabrication techniques, failure analyses and process simulations are essential for realizing higher yield and shortening the design cycle, especially during the development phase. Following fabrication and test, packaging technique (see PACKAGING TECHNOLOGY) is required to realize reliable IC products, which must be resistant to severe environments such as high temperature, high humidity, and mechanical stress. These topics are beyond the scope of this article, which is limited to wafer-fabrication techniques.

1.2 Patterning Technology

As previously shown in Fig. 4, one IC generation of three years duration begins at a feature size that is only 70% of that in the

		1970			1980			1990		FORECAST →		2000
Feature Size (μm)		12	8	5	3	2	1.3	0.8	0.5	0.3	0.2	0.13
DRAM (bit)		1K	4K	16K	64K	256K	1M	4M	16M	64M	256M	1G
Memory Cell	STRUCTURE	3T	1T, PLANAR				STACK or TRENCH					(3D)
	INSULATOR THICK.(nm)*1	120	100	50	35	20	10	8	5	4	3	2
Device	POWER SUPPLY(V)	20	12		5				Int.3.3	3.3 - 1.5		
	TRANSISTOR	pMOS	nMOS				CMOS				(BiCMOS)	
	DRAIN STRUCTURE *2	SD				DD		LDD	(IMPROVED)		(VERTICAL)	
	OXIDE THICK.(nm)	120	100	75	50	35	25	20	15	12	8	5
	CHANNEL LENGTH(μm)	8	5	3	2	1.3	0.8	0.8	0.5	0.3	0.2	0.15
	JUNCTION DEPTH (μm)	1.5	0.8	0.5	0.35	0.3	0.25	0.2	0.15	0.12	0.1	0.08
Process	LITHOGRAPHY *3	CONTACT			1/1PJ	1/10PJ	1/5PJ	(g-LINE)	(i-LINE)	(EXCIMER)		(X/EB)
	ETCHING *4	SOLUTION				PLASMA, RIE			ECR, RIE	HDP		
	ISOLATION	PLANAR	LOCOS				(IMPROVED)		RECESSED			TRENCH
	GATE MATERIAL	Al	POLYSi	DOUBLE POLY		POLYCIDE		(Al-SHUNT)			REFRACTORY	
	METALLIZATION	Al		Al-Si			Al-Si-Cu (TiW, TiN)			W		Cu
	WAFER SIZE ("φ)	2	2.5	3	4	5	6	6-8	8	8-12		(SOI)

*1: SiO_2 equivalent, *2: SD(Single Drain) / DD(Double Drain) / LDD(Lightly-Doped Drain),
*3: PJ(Projection) / X(X-ray) / EB(Electron Beam), *4:RIE(Reactive-Ion Etching) / ECR(Electron Cyclotron Resonance) / HDP(High-Density Plasma)

FIG. 4. Technology road map mainly for MOS ICs. Selected techniques beyond 64-Mbit DRAM are forecast.

previous generation. Therefore patterning is an enabling process for each generation.

Figure 5 shows practical resolutions as a function of wavelength λ for three major lithographies (Okazaki, 1993): optical, x-ray, and EB (electron beam). Since the feature size of integrated circuits that are now under volume production is greater than 0.35 μm, optical reduction projection steppers (step-and-repeat–type projection aligner) with *g*-line (λ = 0.436 μm) and *i*-line (λ = 0.365 μm) light from a mercury-arc source offer enough performance for production.

Two of the most important fundamental performance factors of optical lithography are resolution R and depth of focus DOF. They are expressed as

$$R = k_1\lambda/\mathrm{NA} \tag{2}$$

and

$$\mathrm{DOF} = k_2\lambda/\mathrm{NA}^2, \tag{3}$$

where λ and NA are wavelength of incident light and numerical aperture of lens, respectively, and k_1 and k_2, which are factors defined by photoresist and exposure conditions, are typically 0.6 and 0.5, respectively. It must be noted that increasing NA for the purpose of improving resolution gives rise to an accelerated decrease in DOF. The most advanced stepper using *i*-line light achieves a NA of 0.6 in production today.

Beyond 0.35 μm, a stepper with excimer-laser light source of KrF (λ = 0.248 μm) or ArF (λ = 0.193 μm) should be a potential candidate. Further beyond the 0.2-μm region, F_2 (λ = 0.154 μm), x ray, or EB has a potential performance. Even though an x-ray exposure system with synchrotron-radiation light source has great advantages such as high resolution, high throughput, and large DOF, open issues still remain regarding difficulty in precise mask making and high investment cost. Despite its excellent resolution capability, an x-ray reduction projection system suffers from technical difficulty in implementation of adequate reduction lens and mirror. The electron-beam system has been

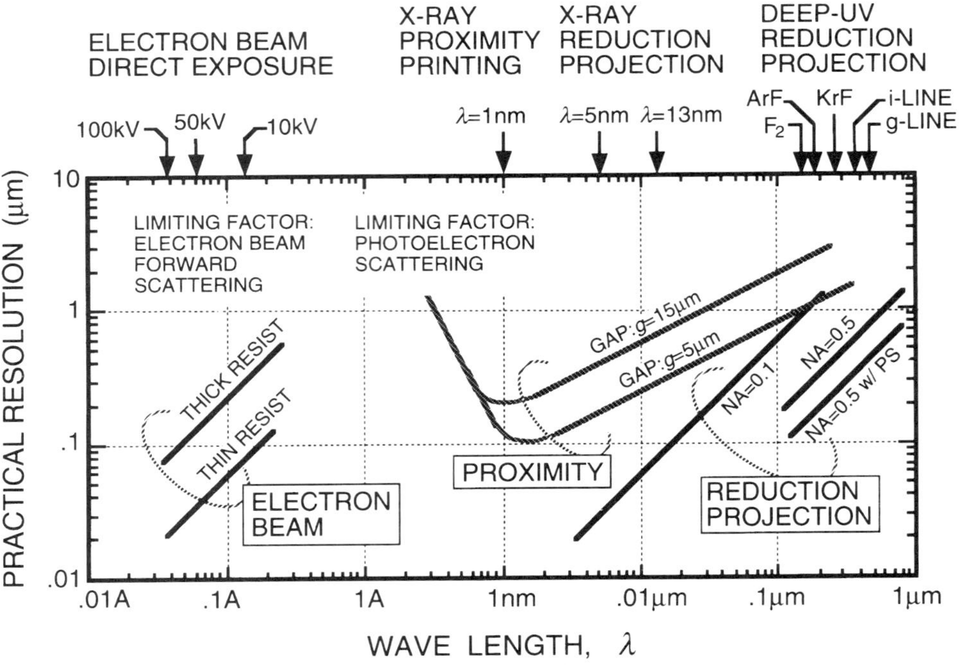

FIG. 5. Practical resolutions of electron-beam, x-ray, and deep-UV lithographies (Okazaki, 1993). Resolution of patterning R is practically proportional to λ/NA for reduction projection system, and to $(\lambda g)^{1/2}$ for proximity printing system, where λ, NA, and g are wavelength of incident light, numerical aperture of lens, and gap length between mask and resist to be exposed. Ultimate attainable resolutions are fundamentally limited by photoelectron scattering for x ray and by forward scattering of electron beam for EB. PS in reduction projection system denotes phase-shift technology, which is discussed later.

widely used to fabricate photomasks for the optical stepper, but its very low throughput is still not capable of volume production.

Besides the technical performance, the cost/performance factor is also very important from the economic viewpoint. Empirically, the practical resolution R of aligners is hyperinversely proportional to its price. For example, ratios of R/price are (2 μm)/(\$0.2 $\times$ 10^6), (0.3 μm)/(\$$10^6$), (0.2 μm)/(\$2 $\times$ 10^6), (0.1 μm)/(\$4 $\times$ 10^6), and 0.1 μm/(10 $\times$ 10^6) for a 1:1 stepper, an i-line stepper, an excimer stepper, an EB machine, and a synchrotron-radiation x-ray system, respectively. Therefore, it is a key issue to extend the resolution limit without changing and/or improving the present aligner. Potential methods are shown in Fig. 6. They are candidates for extending the resolution limit to the next generation. Multilayer resist (MLR), Fig. 6(a), makes use of the better resolution with thin and flat photoresist; contrast-enhanced layer (CEL), Fig. 6(b), accelerated generation in transparent area; focus-latitude enhancement exposure (FLEX), Fig. 6(c), multiple exposures with different focal points for thick and stepped photoresist; and phase-shifting mask (PSM), Fig. 6(d), contrast enhancement by interference between adjacent beams.

Simulated resolution enhancements using a phase-shifting mask are shown in Fig. 7. While keeping a DOF value of ± 0.75 μm, resolution limit size can be reduced to 70% compared with a conventional mask. The coherence factor σ of the incident light is chosen to be 0.5 in this simulation.

1.3 Planarization Technology

As the integration level goes further, multilevel metal wirings have become increasingly important. At present three- or four-level metallization is essential to logic ICs. Since deposition and patterning of metal wires are difficult to carry out on a stepped

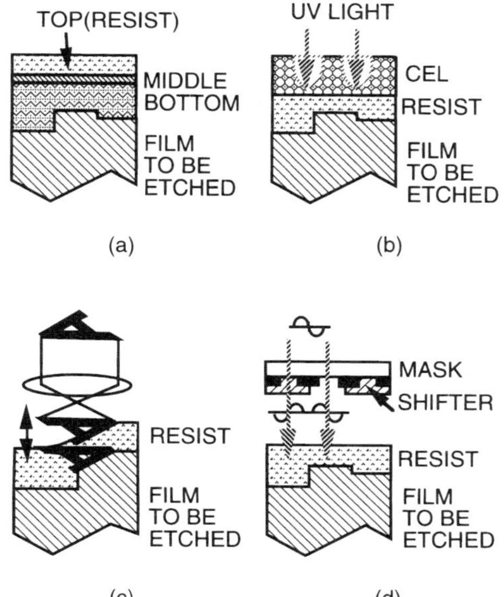

FIG. 6. Proposed methods to enhance resolution: (a) multilayer resist; (b) contrast-enhancement layer; (c) focus-latitude–enhancement exposure; and (d) phase-shifting mask.

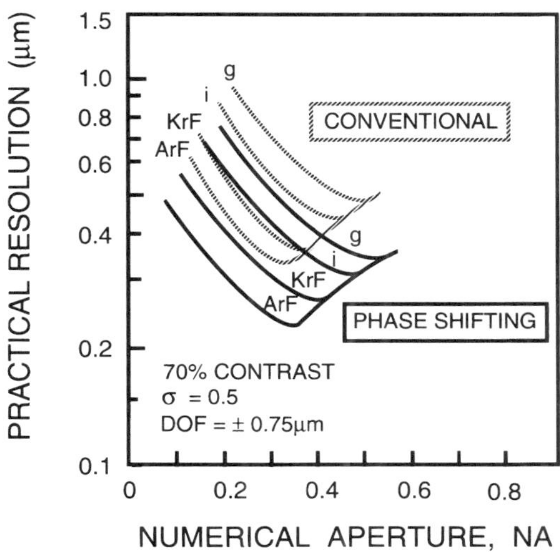

FIG. 7. Simulated resolution enhancement with phase-shifting mask for *g* line of Hg (wavelength λ = 436 nm), *i* line of Hg (λ = 365 nm), KrF excimer (λ = 248 nm), and ArF excimer (λ = 193 nm).

surface, planarization of the insulating film underneath the metal wires is essential. Planarization is also very effective for the finer lithography with its decreased DOF in an advanced stepper having a high-NA lens.

Trends in the number of metallization levels and the aspect ratio of patterns delineated are shown in Fig. 8. As the level increases, the step height on the die surface also increases, resulting in the degradation of focus latitude in lithography and metal thinning due to poor step coverage. To cope with these problems, several techniques are developed to fill narrow gaps and deep vias

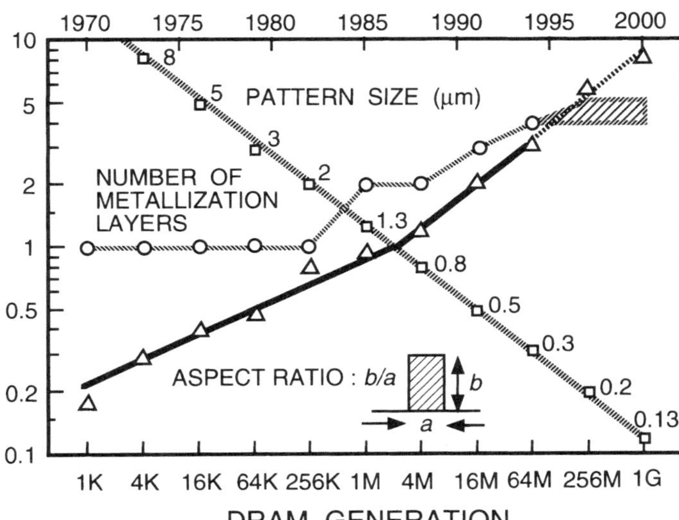

FIG. 8. Increased difficulty in patterning in terms of metallization layers and aspect ratio of patterns to be delineated.

with dielectric films and/or metals. They are the following:

1. ECR (electron cyclotron resonance) sputtering, which results in improved step coverage by taking advantage of high surface mobility of high-energy particles;
2. conformal glass-film deposition with LP-CVD (low-pressure, high-temperature chemical vapor deposition), taking advantage of long mean free path at low pressure;
3. SOG (spin-on-glass), in which an organic or inorganic silicon-compound material in an alcoholic solvent is spin-coated providing excellent gap-filling performance;
4. reflow of BPSG (boro- and phosphosilicate glass, with mole concentrations of boron and phosphorus around 20% and reflow temperature ranging from 800 to 900 °C); and
5. CMP (chemical mechanical polishing).

An example of planarization using CMP, which has increasingly become popular in LSI manufacturing, is shown in Fig. 9. Protruding areas are preferentially polished and simultaneously etched by chemical etchant as shown in the etching-model drawings in the figure. The planarization result is outstanding.

2. PROCESS INTEGRATION TECHNOLOGIES

2.1. Isolation Structure

The isolation structure in LSI has the role of electrical isolation for adjacent transistors. This is "passive performance" as compared with "active performance" of a transistor. The size of a transistor is defined by the performance required on the basis of its drive current, previously described in Eq. (1). While the smaller the better for the isolation region, the isolation region usually becomes bigger than the transistor region.

The most popular isolation technique for more than twenty years has been LOCOS (local oxidation of silicon). As shown in Fig. 10(a), a silicon-nitride (Si_3N_4) film works as a mask against high-temperature oxidation. Oxidation in steam at around 1000 °C for a few tens of minutes produces a several-hundred-nanometer-thick field oxide. Implanted boron forms a p^+ region working as a channel stopper between adjacent transistors.

One major problem associated with LOCOS is "encroachment" of field oxide beneath the nitride mask, causing narrowing of the transistor channel. Since width of the narrowed region, denoted by W_{eff}, is the channel width of the transistor, a pre-enlarged width of W_{mask} should be designed to maintain the minimum channel width. The structures obtained, shown in Fig. 10(c) with "isoplanar" technique, are named "bird's beak" and "bird's head." The isoplanar technique utilizes a recessed silicon surface to keep the entire surface relatively flat as compared with LOCOS.

When the pad oxide beneath the nitride becomes thinner, the bird's beak can be reduced. However, it frequently causes dislocations along field-oxide edges during high-temperature field oxidation due to the huge tensile stresses in the nitride films. The pad oxide works as a stress buffer region between the nitride and the silicon substrate.

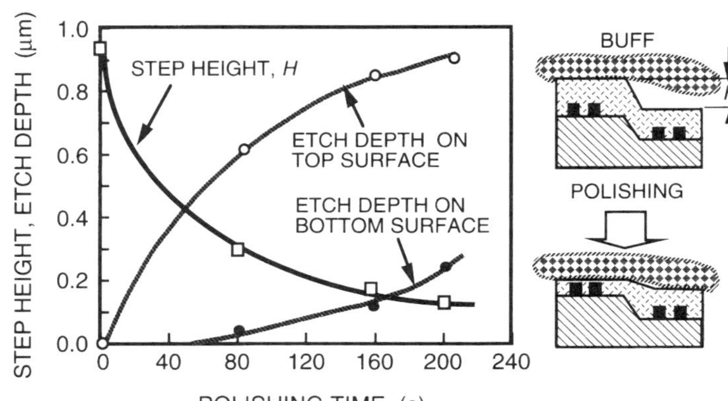

FIG. 9. Step-height reduction with chemical mechanical polishing (CMP).

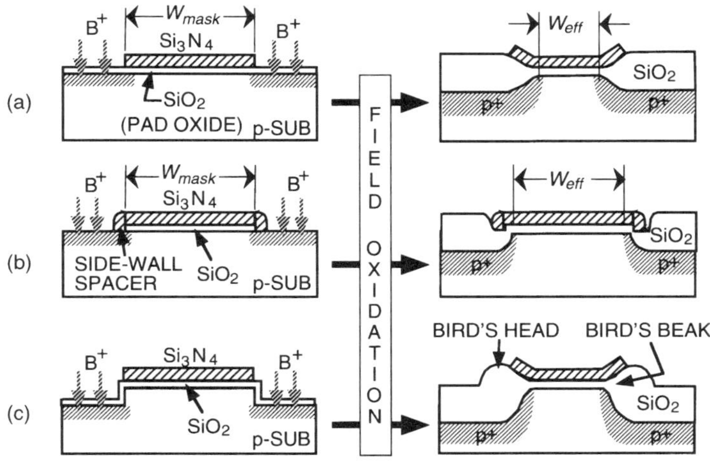

FIG. 10. LOCOS (local oxidation of silicon) structures: (a) conventional LOCOS formation technique; (b) improved LOCOS structure utilizing silicon-nitride side-wall spacer; and (c) conventional isoplanar technique.

An improved LOCOS structure is shown in Fig. 10(b). A side-wall spacer of silicon nitride works as an oxidation stopper resulting in a larger channel width. As the process window of the improved LOCOS is narrower than that of the conventional LOCOS, processing conditions must be carefully chosen to control dislocation formation.

A much improved isolation structure is the deep and narrow trench isolation, previously shown in Fig. 3. There still exist a lot of issues to be solved, such as deep and narrow trench etching, filling materials and techniques, channel-stop implantation into the trench, etc.

2.2 Device Structure

Figure 11 shows a typical process sequence for an advanced submicron n-channel MOS transistor. Key issues are realization of larger effective channel width with an advanced isolation technique and the increase in maximum applicable voltage across source to drain.

Its isolation structure is realized by a technique already shown in Fig. 10(b). The thicker side wall spacer and thicker field oxide cause dislocations beneath the field oxide giving rise to anomalously increased leakage current. To cope with its relatively narrow process window, thicknesses of the pad oxide beneath the side-wall spacer and field oxide have to be controlled.

This device structure is realized with a lightly doped drain (LDD) (Ogura *et al.*, 1980). The relatively lower impurity concentration of the n^+ layer reduces the electric field strength, resulting in increased breakdown voltage across source and drain. A role of the side-wall spacer, shown in Fig. 11(f), is to keep high–impurity-concentration layers of n^{++} away from the front edges of source and drain. Even though this LDD structure improves the breakdown voltage, this brings an adverse effect on the drivability of the transistor as if a series resistance is inserted in the equivalent circuit. It is a trade-off between breakdown voltage and drivability.

2.3 CMOS Process Integration

CMOS ultralarge-scale integration (ULSI) has played an increasingly important role across the entire electronics industry. Major products are memories and microprocessors as previously shown in Fig. 1. Some modern microcomputers incorporate logic and high-speed memory in the same chip. To meet the needs of specific applications, nonvolatile memory can be included to achieve a user-programmable microcomputer.

2.3.1 DRAM Process Integration The equivalent circuit of a DRAM memory cell is very simple, consisting of one MOS transistor and one storage capacitor. The most important performance factor is expressed as C_s/C_b, where C_s is the storage capacitance and C_b is the bit-line capacitance. Ordinarily, 128 or 256 memory cells are connected to one bit line. Historically the value of C_s/C_b

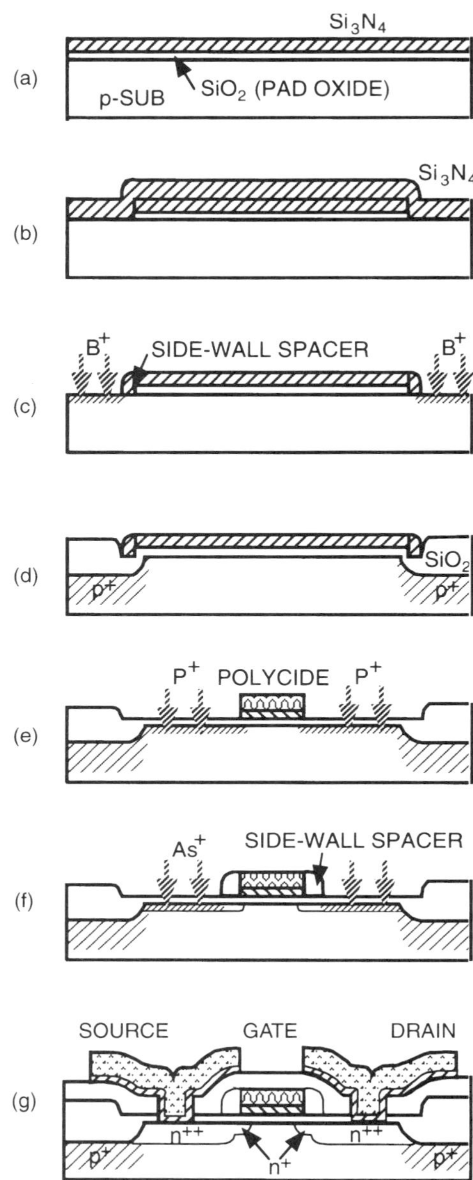

FIG. 11. Process sequence for an advanced *n*-channel MOS transistor: (a) pad oxide and first silicon-nitride film deposition on *p*-type, (100)-oriented silicon substrate; (b) second silicon-nitride film deposition subsequent to delineation of the first for active transistor region; (c) anisotropic dry etching of the second nitride film to leave nitride side-wall spacer and channel-stop implant of boron; (d) field-oxide formation; (e) polycide (silicide/polysilicon) gate delineation and phosphorus implant to form n^+ lightly doped drain; (f) arsenic implant to form n^{++} source and drain with oxide side-wall mask; and (g) glass film is deposited and aluminum/barrier material contacts are formed.

and the signal-voltage swing of the bit line have been chosen to be about 0.1 and 100–200 mV, respectively, over all DRAM generations including considerations for operational margin, soft-error immunity, refresh performance, etc. The storage capacitance C_s is expressed as

$$C_s = \epsilon_i A / T_i, \tag{4}$$

where ϵ_i, A, and T_i are permittivity of the capacitor insulator, area of the capacitor electrodes, and thickness of the capacitor insulator, respectively. Even though the memory cell has been reduced in size, namely, the capacitor area A has been reduced, the storage capacitance cannot be reduced proportionally from operational margin and reliability points of view.

Alpha particles can also be fatal to DRAM operation by causing soft errors. Almost 200 fC of charge, which is generated by an alpha particle having a maximum energy of 5 MeV, is dispersed 25 μm deep into the Si substrate. To keep a sufficiently low soft-error rate, a storage capacitance greater than 20 fF is needed unless an innovative structure is adopted.

Trends in DRAM structural parameters are shown in Fig. 12. From 4 kbit to 1 Mbit, planar capacitors were adequate, whereas three-dimensional (3D) capacitors are needed to keep sufficient storage capacitance beyond 4 Mbit. As clearly shown in the figure, the capacitor area surpasses the cell size after the 4-Mbit generation. A cross-sectional SEM photograph is shown in Fig. 13 for an advanced 16-Mbit stacked-capacitor DRAM cell having double-fin–type CUB (capacitance under bit line) scheme. BPSG reflow under the bit line and SOG over the bit line are utilized for planarization.

The 3D capacitor was introduced in response to the inability to make adequate quality dielectric films that scale with area A according to Eq. (4). Proposed 3D capacitors for a 64-Mbit DRAM are shown in Fig. 14. A simple 3D capacitor with ordinary SiO_2–Si_3N_4–SiO_2 capacitor insulator may not be large enough for the 64-Mbit DRAM and beyond.

Meanwhile, in the 256 Mbit chip and the future, rapid reduction in the thickness is supposed to be required again. One approach is the adoption of a novel insulator

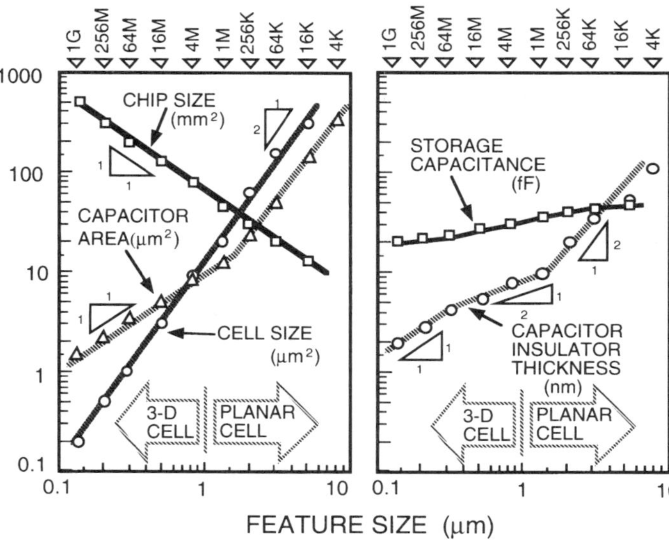

FIG. 12. Device-parameter trends in DRAM. They are divided into two regimes: planar cells before 1 Mbit and three-dimensional cells after 1 Mbit. Reduction of capacitor-insulator thickness is categorized into three regimes in terms of keeping sufficient storage-capacitance value: (1) 4 K to 1 M, capacitance is achieved mainly by the thickness decrease; (2) 1 to 64 M, it is done mainly by structure innovation, such as stacked-capacitor and trench-capacitor cells; (3) 64 M and beyond, both thickness decrease and structure innovation tend to be less effective, and material innovation is likely to be required. Capacitor-insulator thickness stands for SiO_2-equivalent thickness.

having much higher permittivity. Ta_2O_5 and $(Ba,Sr)TiO_3$ films are two of the potential candidates as shown in Fig. 15. The product of dielectric constant and breakdown field strength is nothing but the maximum storable charge per unit area, Q_{max}. It is to be noticed that the relative Q_{max} varies from 1 to 3 over the various kinds of dielectric films plotted inside the shaded region in the figure.

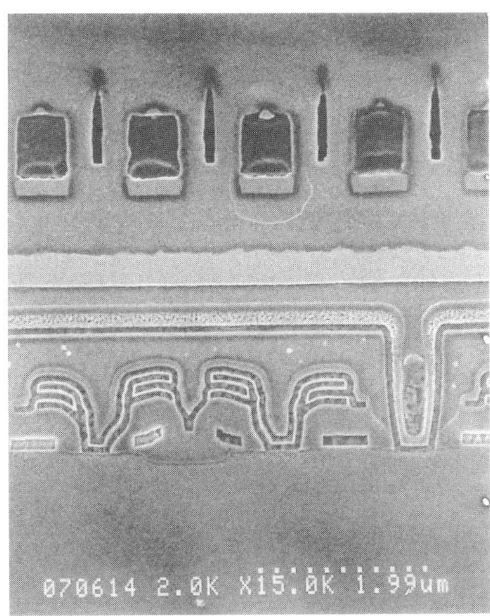

FIG. 13. Cross section of an advanced 16-Mbit DRAM cell that uses double-fin CUB (capacitance under bit line) structure.

A typical R&D-type process sequence for 16-Mbit DRAM is shown in Figs. 16–20. It is no exaggeration to say that almost all manufacturers have been developing process sequences different from each other based on their past experiences. Regarding memory-cell structures, both stacked and trench capacitors are commercially available today. Therefore, there may not be a unique and/or standard solution. Furthermore, a couple of process sequences have been developed for various generations in the same DRAM product, as shown later in Fig. 24.

In this case, eighteen photolithographic steps are defined to complete the sequence. In actual DRAM products, it is natural that more photolithographic steps may be added to produce different kinds of transistors having different threshold voltages and additional metal wiring.

2.3.2 Logic Process Integration Logic ICs have different aspects in their performance requirements compared with memories. In the same 0.8-μm feature-size generation, an average device density is 100 μm^2/device for logics and 10 μm^2/device for DRAM. Wiring density is a key issue to make dense packing in logic LSIs, while memory needs a straightforward increase in number of memory cells. The wiring complexity of a

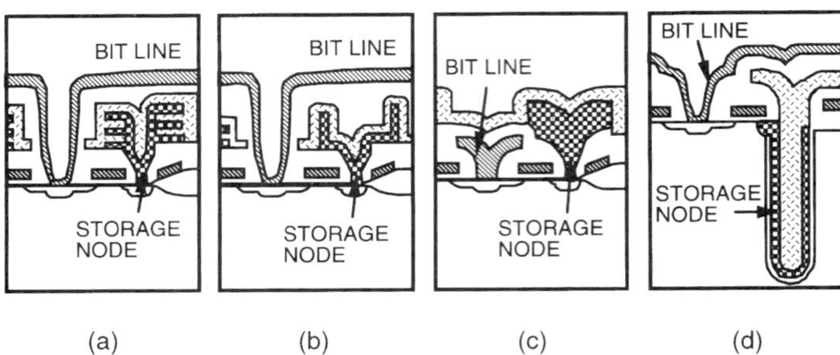

FIG. 14. Proposed 64-Mbit DRAM cells: (a) triple-fin type with a CUB (capacitor under bit line) structure; (b) crown type for CUB; (c) rugged polysilicon with COB (capacitor over bit line); and (d) shielded storage-node–type trench capacitor.

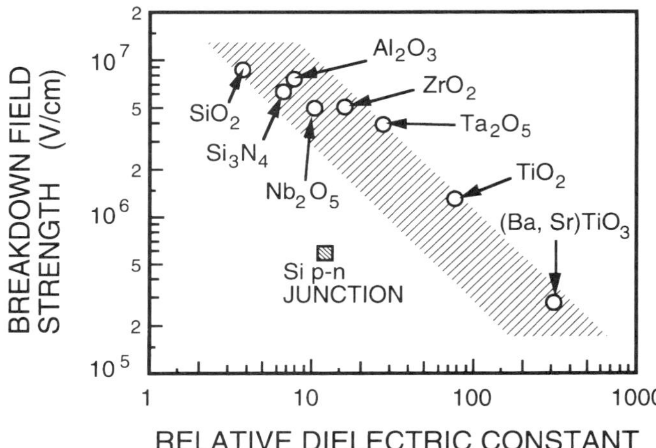

FIG. 15. Potential high-permittivity materials to be used for capacitor insulator in DRAM cells.

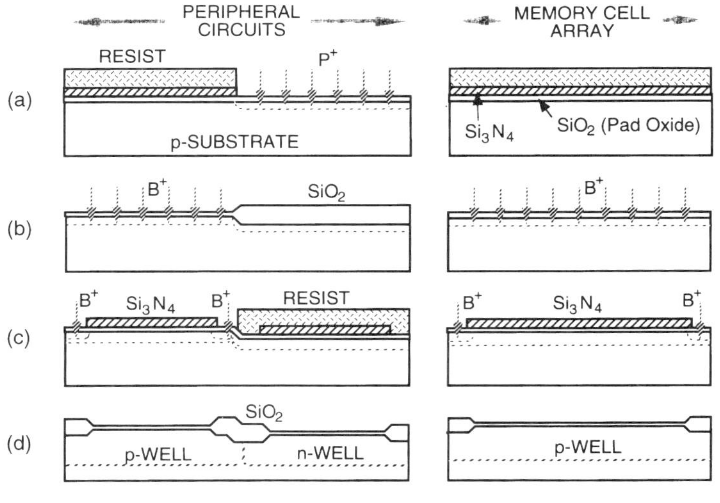

FIG. 16. Twin-well formation step of typical R&D-type process sequence for 16-Mbit DRAM. (a) Selective phosphorus implant into *p*-type, (100)-oriented substrate using photoresist mask (the first photolithography process), for *n*-well formation; (b) field-oxide formation and nonselective boron implant for *p*-well formation; (c) selective nitride deposition with the second photolithography and selective boron implant to form channel stopper using photoresist mask (the third photolithography); and (d) field-oxide formation, *p*- and *n*-wells drive-in, and gate-oxide formation.

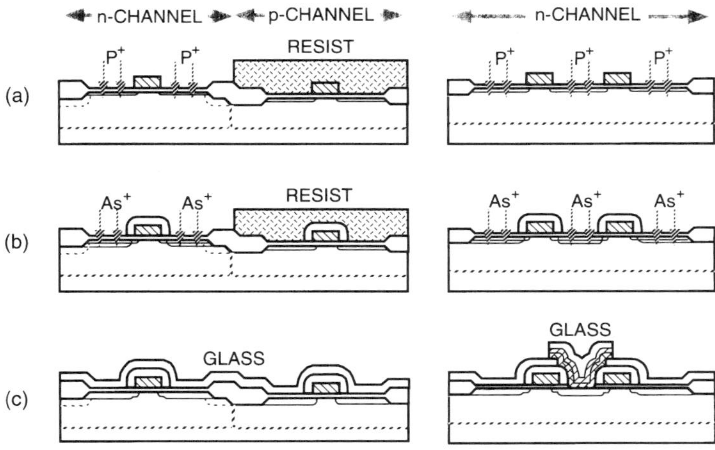

FIG. 17. Transistor-formation step of typical R&D-type process sequence for 16-Mbit DRAM: (a) selective polycide (silicide/phosphorus-doped polysilicon) deposition (the fourth photolithography), selective boron implant into *p*-channel region to form lightly doped drain (fifth photoetching), and selective phosphorus implant into *n*-channel region to form lightly doped drain (the sixth photolithography); (b) selective arsenic implant to form *n*-channel source and drain with oxide sidewall spacer using photoresist mask (the seventh photolithography); and (c) glass-film deposition, the eighth photolithography to form bit-line contact, and selective deposition of polycide bit line (the ninth photolithography).

logic IC theoretically increases with the number of gates as $N^{1.5}$. Therefore, multilayer metallization is very efficient to reduce the die size. For example, the die area of some logic LSIs is reduced to 0.45 by four-level metals and 0.6 by three-level metals as compared with 1 for double-level metals. This die shrinkability can easily overcome the cost increase incurred by more-level metallization.

The key issues for logic IC processes are

1. planarization (see Sec. 1.3) of the underlying insulator and metal films, and
2. low-ϵ dielectric films resulting in higher switching speed and lower power consumption.

Films such as fluorine-incorporated CVD (chemical vapor deposited) SiO_2, organic SOG (spin-on-glass), and polyimide resins are potential candidates for low-ϵ dielectric films.

2.4 Bipolar Process Integration

A typical sequence for the present self-aligned bipolar IC is shown in Fig. 21. One of the key objectives in achieving higher speed is to decrease the base resistance

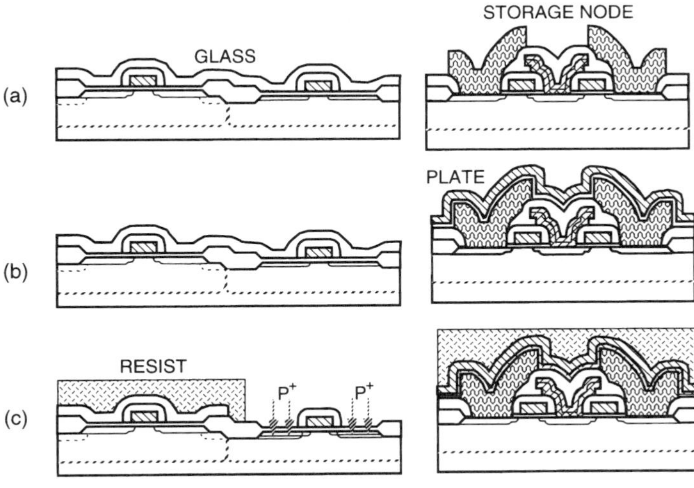

FIG. 18. Memory-cell–formation step of typical R&D-type process sequence for 16-Mbit DRAM: (a) glass-film deposition, storage-node contact formation with tenth photoetching, selective arsenic-doped polysilicon deposition (the eleventh photolithography) to form storage node; (b) very thin capacitor-insulator deposition and selective phosphorus-doped polysilicon plate formation (the twelfth photolithography); and (c) selective boron implant to form *p*-channel source and drain using photoresist mask.

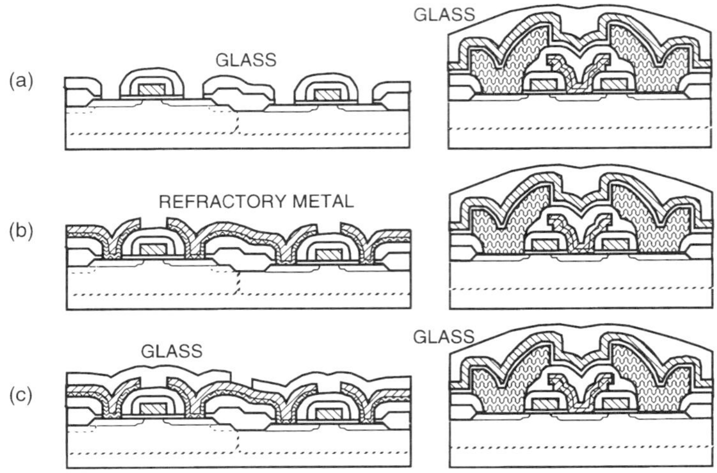

FIG. 19. Metal-wiring formation step of typical R&D-type process sequence for 16-Mbit DRAM: (a) glass-film formation with boro- and phosphosilicate glass (BPSG) reflow and the thirteenth photolithography delineates contact holes to source and drain region; (b) selective first metal wiring of tungsten/titanium-nitride electrode (the fourteenth photolithography); and (c) phosphosilicate glass/spin-on-glass/phosphosilicate glass deposition and subsequent annealing for the densification and first via-hole formation (the fifteenth photolithography).

while keeping the base region thin. Today, bipolar base width can reach below 50 nm and cutoff frequencies approaching 80 GHz have been realized. This high-speed performance is getting rather close to the GaAs microwave transistor. Furthermore, a champion performance of 113-GHz cutoff frequency has recently been achieved with a leading-edge heterobipolar transistor having a SiGe base (Crabbe *et al.*, 1993).

As the demand for downsized computers becomes stronger, very high-speed bipolar ICs have been losing their market share as previously shown as "Bipolar Digital" in Fig. 1. This is partly because they consume huge power to achieve ultrahigh speed and require a very expensive liquid cooling system. Moreover, the high power, sometimes reaching 100 W/chip, does not allow high-density integration, leading to the use of costly multichip modules. Such multichip modules suffer from signal delays due to the relatively long wiring between chips. So, speed and power interact adversely in a high-cost technology, and densely packed CMOS ULSI is expected to outperform the bipolar ICs in the future even if the MOS transistor itself is less speedy than the bipolar transistor.

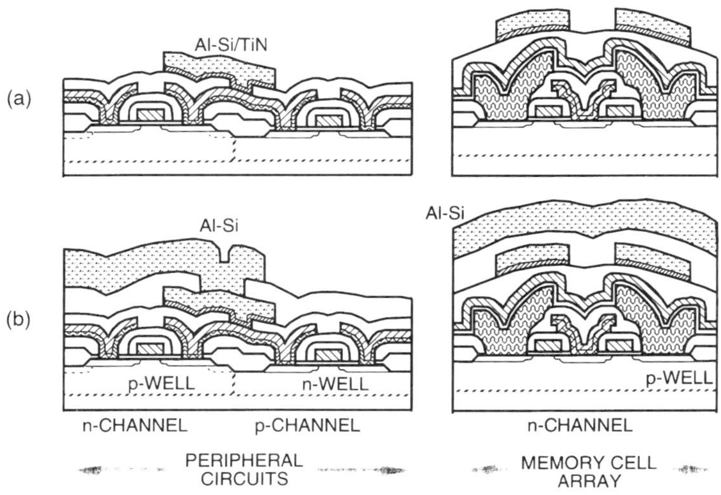

FIG. 20. Multilayer metallization step of typical R&D-type process sequence for 16-Mbit DRAM: (a) Selective deposition of second metal wiring of silicon-contained aluminum/titanium nitride (the sixteenth photolithography); and (b) plasma-oxide/spin-on-glass/plasma-oxide intermediate insulator formation, second via-hole formation with the seventeenth photolithography, and selective aluminum-wiring formation with the eighteenth photolithography.

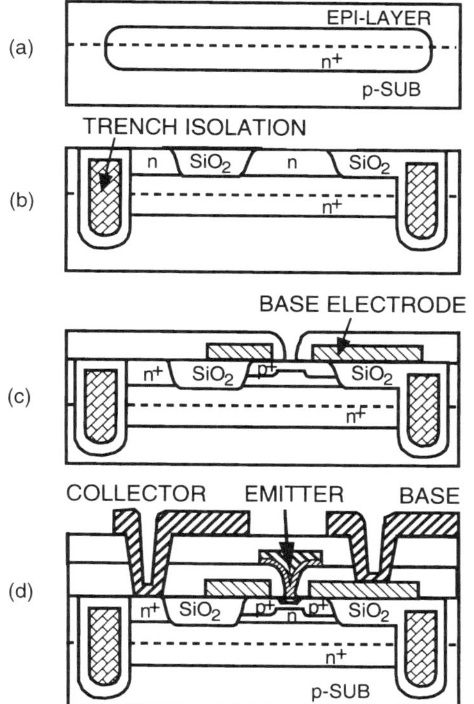

FIG. 21. Process sequence for a self-aligned bipolar IC: (a) forming n^+ buried layer on *p*-type substrate and *n*-type epitaxial growth; (b) etching deep trench, filling with semi-insulating polysilicon, and LOCOS (local oxidation of Si) isolation; (c) boron-doped polysilicon patterning to form the base region and base electrode, glass-film deposition, and anisotropic dry etching to define the emitter window; and (d) arsenic-doped polysilicon deposition to form the emitter electrode, contact-hole etching, and aluminum-electrode formation.

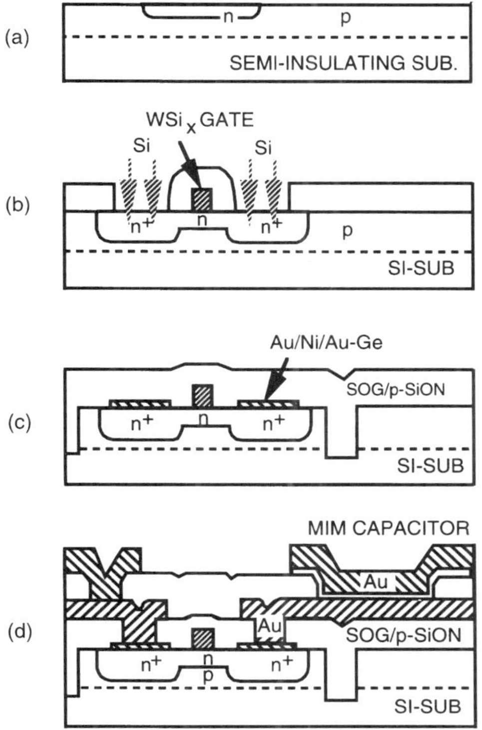

FIG. 22. Process sequence for GaAs MESFET IC: (a) selective *n*-type channel region formation subsequent to Mg-implanted *p*-type layer formation on semi-insulating substrate; (b) transistor region definition, side-wall spacer formation for WSi_x gate, source and drain region formation with Si implantation; (c) Au/Ni/Au–Ge contact formation on source and drain regions, trench etching for isolation, and Au-electrode formation; (d) intermediate insulator deposition, via-hole etching, and simultaneous formation of metal-insulator–metal capacitor and subsequent metallization.

2.5 GaAs Process Integration

Even if a leading bipolar device has already reached the cutoff-frequency region of 100 GHz, the GaAs FET still possesses the advantages of low-noise and ultrahigh-frequency operation. About double the operation speed is realized compared with silicon bipolar LSIs. Figure 22 shows an example of a process sequence for a GaAs MES (metal semiconductor) FET IC.

Unlike silicon, crystalline GaAs neither has a stable insulator nor can it withstand high temperature. Except for a maximum temperature of around 800 °C at impurity activation, ordinary processing temperature is around 400 °C. This is a distinct disadvantage in terms of integrity, reliability, and planarization. For example, a CVD SiO_2 film formed at around 400 °C causes poor step coverage, less resistance to chemical etchant due to its enhanced porosity, and very low dielectric field strength, compared with those of that formed at 800 °C. However, small-scale GaAs ICs have been used in a very limited area where ultrahigh-speed operation is required.

3. FURTHER INTEGRATION TECHNOLOGIES

During the past thirty years, continuous innovation has been a characteristic of silicon IC technology. Despite four times

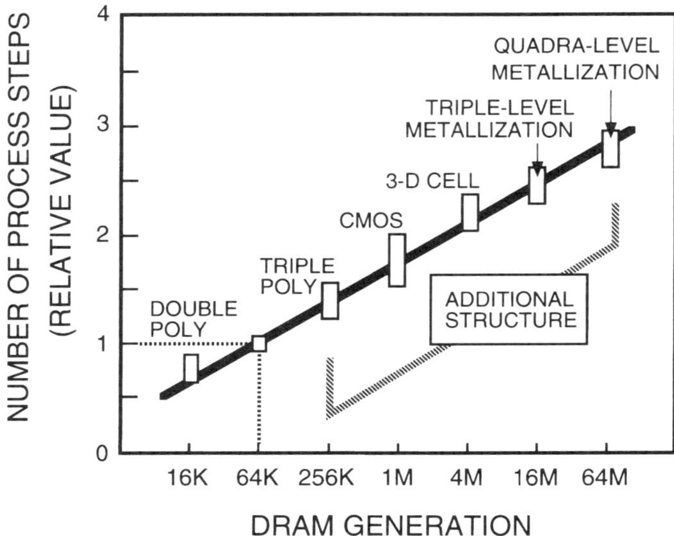

FIG. 23. Increase in number of process steps for DRAMs in response to their structural innovations.

greater integration per generation, prices of DRAMs have remained nearly constant throughout several generations. As a result, bit price has been reduced to 1/1000 from 1 kbit to 1 Mbit. This has certainly led to very rapid growth of the semiconductor industry because of very extensive usage of semiconductor products in various industries.

Increased chip cost due to enlarged die size has been compensated by wafer-size increases. Therefore, if the wafer size ceases to increase, there may not exist any method to compensate for the increase in wafer-fabrication cost caused by process-step increase and the growing investment in new fabrication facilities.

An example of process-step increase is shown in Fig. 23 for the case of DRAM. Difficulty in preventing a decrease in storage capacitance will become fatal in the future. In that era, memories that operate on the basis of different mechanisms will be dominant and may take over the DRAM market assuming that they will be produced less expensively than DRAM.

The candidates might be FRAM (ferroelectric RAM) and/or innovative memories unknown today. Since FRAM utilizes a

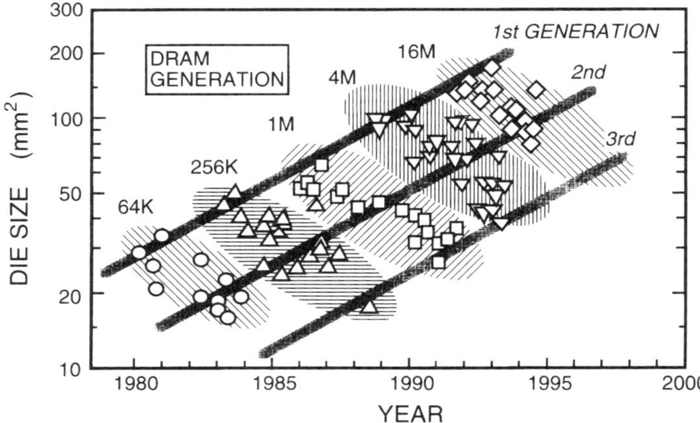

FIG. 24. Die-size reduction for three groupings within a DRAM generation. The third generation delivers almost one-half of the die size as compared with the first.

mechanism of polarization inside a ferroelectric film such as PZT and PLZT, it inherently works as both RAM and ROM. Furthermore, it is switchable by applying a voltage to the plate.

Flash EEPROM has a potential for low-cost memory by virtue of its simple one-transistor memory cell. However, it is not random-access memory but still read-only memory. If its price is low enough, it will be able to take over some part of the mass-storage market. Its operation mechanism is based on charging and discharging of high-energy electrons across thin SiO_2. Macroscopic and microscopic oxide integrities will be key issues for volume production.

4. CONCLUSION

As the semiconductor industry has continuously been growing for these 30 years, device processing has become more and more important. Device count in an IC product has also increased very rapidly resulting in more functions.

Investment in production facilities has increased more rapidly than semiconductor prices. To cope with this dilemma, wafer size has been periodically increased. To get more chips from a wafer, cut-down or die-shrink products have become popular for all IC products. A typical trend is shown in Fig. 24 for DRAMs. Three die-size groupings within each respective DRAM generation are found for products available in the market.

GLOSSARY

Chip Shrink (Model): Reduction of chip size with optically shrunk photomasks to get more chips on a wafer.

Cut Down (Model): Manufacture of a present-level product utilizing redesign and the integration process of the next generation—for example, 4-Mbit DRAM (ordinary with 0.8-μm feature size) manufactured by 0.5 μm (primarily used for 16-Mbit DRAM production). Generally smaller than "chip-shrink" model.

Feature Size: Minimum dimension, which generally denotes the smallest gate length of MOS transistor. It also stands for a level of process generation.

Integration Process: Process sequence to fabricate a semiconductor product, consisting of a specific combination of various elementary processes.

Isolation: Recessed regions on a substrate surface filled with dielectric films such as SiO_2 to isolate active and passive devices in integrated circuits. Typical techniques are LOCOS and trench isolations.

Lithography: Methods to delineate patterns using one of several kinds of aligners such as contact, projection, step-and-repeat, and step-and-scan with ultraviolet light, x rays, or electron beam.

Planarization: Methods to planarize dielectric films underlying metal wires. They include bias sputtering, chemical mechanical polishing (CMP), reflow of CVD oxide, spin-on-glass (SOG), etc.

Plug: Method to fill a via hole with tungsten or other conductive materials to form a contact between two-level conductive layers. Also a metal piece left in the hole.

Polycide: Stacked electrode consisting of a silicide over polysilicon. Polysilicon plays the role of a reliable gate material, and silicide, a low-resistance material.

Process Generation: Representative integration process designated by a feature size such as 8, 5, 3, 2, 1.3, 0.8, or 0.5 μm. DRAM has been a technology driver for 25 years.

Silicidation: Method to form low-resistive layer of silicide consisting of silicon and a metal such as W, Mo, Co, Ti, Ta, etc. A self-aligned silicidation is often called "salicide."

Soft Error: Volatile operation failure caused by an alpha-particle hit. Substantially pronounced in DRAMs; however, the potential also exists in other LSIs.

Three-Dimensional Cell: Vertically formed storage cell for DRAMs such as stacked capacitor and trench capacitor.

Works Cited

Crabbe, E. F., Meyerson, B. S., Stork, J. M. C., Harame, D. L. (1993), "Vertical Profile Optimization of Very High Frequency Epitaxial Si- and SiGE-Base Bipolar Transistor," in: *IEEE International Electron Devices Meeting Technical Digest,* New York: IEEE, pp. 83–86.

Dennard, R. H., Gaensslen, F. H., Yu, H.-N., Rideout, V. L., Bassous, E., LeBlanc, A. R. (1974), "Design of Ion-Implanted MOS FET's with Very Small Physical Dimensions," *IEEE J. Solid-State Circ.* **SC-9,** 256.

Hoerni, J. A. (1960), "Planar Silicon Transistor and Diodes," presented at IRE Electron Devices Meeting, Washington, D.C.

Kilby, J. S. (1976), "The Invention of the Integrated Circuit," *IEEE Trans. Electron Dev.* **ED-23,** 648.

Ogura, S., Tsang, P. J., Walker, W. W., Critchlow, P. L., Shepard, J. F. (1980), "Design and Characteristics of the Lightly Doped Drain-Source (LDD) Insulated Gate Field-Effect Transistors," *IEEE Trans. Electron Dev.* **ED-27,** 1359.

Okazaki, S., (1993), "Lithographic Technologies for Future ULSI," *Appl. Surf. Sci.* **70/71,** 609–612.

Further Reading

IBM J. Res. Develop. (1995), Special Issue, "The IBM CMOS Technology," **39** (1/2).

Kooi, Else (1991), *The Invention of LOCOS,* New York: IEEE.

Sze, S. M. (1988), *VLSI Technology,* New York: McGraw-Hill.

SEMICONDUCTOR-DEVICE MODELING

TORU TOYABE, *Department of Information and Computer Sciences, Faculty of Engineering, Toyo University, Kawagoe, Saitama, Japan*

INTRODUCTION

Since the bipolar transistor was invented by Bardeen, Brattain, and Shockley in 1947, semiconductor devices have developed at an incredible rate. The junction field-effect transistor (JFET) was introduced by Shockley in 1952, and the metal-oxide–semiconductor field-effect transistor (MOSFET) was proposed and fabricated by Kahng and Atalla in 1960. Moreover, the metal-semiconductor field-effect transistor (MESFET), the light-emitting diode, the laser diode, and the charge-coupled device (CCD) were proposed and developed in the succeeding decade.

The advent of the MOSFET was largely based on planar process technology. This has been undoubtedly the most important technology for the development of the integrated circuit in these past 30 years. A fourfold higher packing density has been realized every three years since 1970. As a result the present-day very large-scale integration (VLSI) integrates more than ten million transistors per single chip. Thanks to great advances of fine-pattern technology, 16-Mbit dynamic random-access memory (DRAM) is now in mass production with fabrication technology of minimum feature size of 0.5 μm.

A guiding principle of the MOS LSI development has been the scaling law (Dennard *et al.*, 1974). In early stages of development, the electrical behavior of MOSFETs obeyed well the scaling law that is based on early analytical device modeling, because of their large device size. For miniaturized devices, however, the deviation from the scaling law and classical device modeling has been remarkable. The short-channel effect is a typical example. The device miniaturization has also provoked problems relating to the reliability of LSIs, such as Complementary MOSFET (CMOS) latchup, characteristics degradation due to hot carriers, and alpha-particle–induced soft error (May and Woods, 1979). To cope with these problems, numerical device modeling based on fundamental physical equations for semiconductor devices has become necessary. At present, two-dimensional numerical simulation is commonly used in the development stage of device prototypes.

3-527-28139-8/96/$5.00 + .50

The approach taken in early device modeling was analytic in the sense that closed-form solutions of the basic semiconductor equations are derived for different regions of the interior of the device and combined with boundary conditions between the regions to get a global model. Shockley derived an analytical model for bipolar transistors based on his *p-n*–junction theory in 1949. The bipolar transistors were divided into the emitter, base, and collector regions, and closed-form solutions of the diffusion equations for minority carriers were solved in each region and combined to get an entire bipolar transistor model.

An analytical model for junction field-effect transistors (JFETs) was also given by Shockley in 1952. The operation mechanism of JFETs is two-dimensional, since the direction of the channel current from source to drain is orthogonal to the direction of the controlling gate field. Shockley reduced this two-dimensional problem into two one-dimensional problems. This approach is called the gradual-channel model, since it is assumed that the electric field in the direction of channel is much smaller than the one in the direction of the gate-substrate. A gradual-channel model was also presented for MOSFETs by Ihantolla and Moll (1964). These analytical models were transparent and very useful thanks to the compact forms of their expressions.

The gradual-channel model, however, cannot be applied to recent miniaturized devices, because of the failure of the gradual-channel approximation. Thus, more accurate modeling based on numerical computation is indispensable for understanding current device operation and the design of modern devices. Numerical simulation of semiconductor devices started from the one-dimensional model of bipolar transistors by Gummel in 1964. In his work, it was shown that the coupled system of Poisson and current-continuity equations can be solved by an iterative computation within realistic CPU time. One-dimensional numerical simulation was applied to Gunn diodes and IMPATT diodes to analyze their operational mechanisms in the 1960s.

Two-dimensional numerical simulations that started in the 1970s have successively been applied to MOSFETs, JFETs, MESFETs, and bipolar transistors. Two-dimensional effects on the electrical characteristics, such as short-channel effects in small-geometry MOSFETs and high-level injection phenomena in bipolar transistors, were successfully analyzed. Moreover, three-dimensional numerical simulations became feasible in the 1980s in part thanks to the advent of supercomputers, and have been applied to analysis of narrow-channel effects in miniaturized MOSFETs and alpha-particle–induced soft-error phenomena in DRAM cells.

Silicon VLSI devices have been miniaturized to such a level that devices with the minimum feature size of 0.1 μm will be realized in the near future; however, limits of the conventional device structure and operation principle have become evident. To investigate the relevant physical phenomena, such as hot-carrier, nonlocal impact-ionization, and ballistic effects, and the device limits, various advanced models have been proposed and developed—e.g., the hydrodynamic model, the thermal hydrodynamic model, the Monte Carlo model, and the full-band Monte Carlo model. Moreover, more advanced treatments such as the scattering-matrix method, the cellular-automaton method, and quantum-mechanical analysis are applied to investigate the behavior of electrons in the crucial parts of future devices. In the following, these various device models are described.

1. BASIC EQUATIONS FOR CLASSICAL DEVICE MODELING

The classical analytic device models and the majority of the numerical device models commonly used are based on the drift–diffusion (DD) model. The carrier transport is expressed by the drift current due to the electric field and the diffusion current due to the spatial variation of the carrier density. Thus, the electron current density J_n and the hole current density J_p are expressed as

$$\mathbf{J}_n = qn\mu_n\mathbf{E} + qD_n\nabla n \qquad (1)$$

and

$$\mathbf{J}_p = qp\mu_p\mathbf{E} - qD_p\nabla p, \qquad (2)$$

where q is the electron charge, $\mathbf{E}$ is the electric field, and n, p, μ_n, μ_p, D_n, and D_p are the carrier densities, mobilities, and diffusion constants for electrons and holes, respectively. It is assumed that the Einstein relation holds for the mobility and diffusion constant:

$$D_{n,p} = (k_B T/q)\mu_{n,p}, \tag{3}$$

where T is the lattice temperature.

The basic equations in the DD model are the current-continuity equations for electrons and holes,

$$q\frac{\partial n}{\partial t} = \nabla \cdot \mathbf{J}_n + qG - qR, \tag{4}$$

$$q\frac{\partial p}{\partial t} = -\nabla \cdot \mathbf{J}_p + qG - qR, \tag{5}$$

and the Poisson equation

$$\epsilon\Delta\psi = q(n - p - N_D + N_A), \tag{6}$$

where G and R are the generation rate and the recombination rate of carriers, respectively, ϵ is the dielectric constant of the semiconductor, ψ is the electrostatic potential, and N_D and N_A are the ionized donor and acceptor concentrations, respectively. Physical mechanisms producing G and R are described in Sec. 4.

2. ANALYTIC MODELS FOR SEMICONDUCTOR DEVICES

Analytic models for representative devices such as diodes, bipolar transistors, and MOSFETs, which are clearly the basis for understanding the operational mechanisms, are derived by analytically solving the basic equations given above with appropriate simplifying assumptions.

2.1 *p-n*–Junction Diode

An ideal *p-n*–junction diode has a simple structure as shown in Fig. 1. The diode has rectifying characteristics in which it carries a large current in the forward-biased condition ($V > 0$) and nearly zero current in the reverse-biased condition ($V < 0$). To derive an analytical model for the diode, the basic equations of the drift–diffusion model are

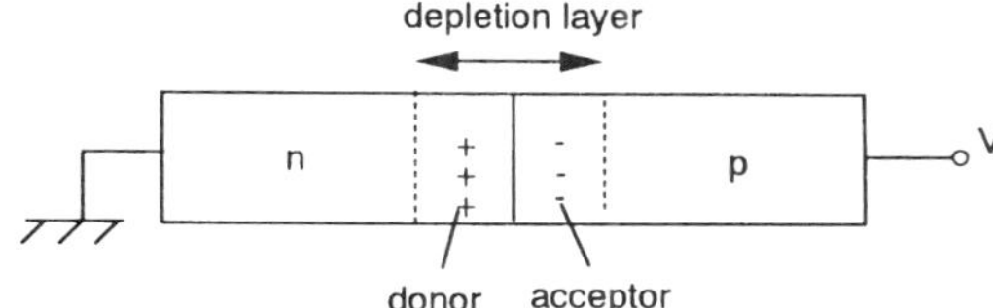

FIG. 1. *p-n*–junction diode.

solved under the following assumptions. In the electrically neutral n and p regions outside of the depletion layer (with no mobile carriers) of the *p-n* junction, the injected minority carriers flow only by diffusion under the forward-biased condition. Then, the current-continuity equations (4) and (5) are simplified into simple diffusion equations by substitution using Eqs. (1) and (2), as

$$D_n\frac{d^2n}{dx^2} = \frac{n - n_{p0}}{\tau_n} \quad \text{in the } p \text{ region}, \tag{7}$$

$$D_p\frac{d^2p}{dx^2} = \frac{p - p_{n0}}{\tau_p} \quad \text{in the } n \text{ region}, \tag{8}$$

where τ_n and τ_p are electron and hole lifetimes characterizing the recombination rate, and n_{p0} and p_{n0} are the equilibrium minority-carrier densities, respectively. The injected minority-carrier densities at the edge of the depletion layer are given by Shockley's law as

$$n = n_{p0}\exp(qV/k_B T), \tag{9}$$

$$p = p_{n0}\exp(qV/k_B T). \tag{10}$$

Solving the diffusion equations (7) and (8) under the boundary conditions (9) and (10), the current density is obtained as

$$J = J_s[\exp(qV/k_B T) - 1], \tag{11}$$

with the saturation current density

$$\mathbf{J}_s = qD_n n_{p0}/L_n + qD_p p_{n0}/L_p, \tag{12}$$

where $L_n = (D_n\tau_n)^{1/2}$ and $L_p = (D_p\tau_p)^{1/2}$ are the diffusion lengths for electrons and holes, respectively. Equation (11) with Eq. (12) is the celebrated Shockley equation, which is the ideal diode law.

2.2 Bipolar Transistor

A one-dimensional *npn* bipolar transistor is shown in Fig. 2. When a positive base–emitter voltage V_{BE} is applied, electrons are injected from the emitter to the base. When a positive collector–emitter voltage V_{CE} is applied, the injected electrons in the base reach the edge of the collector depletion layer and are driven to the collector by the strong electric field in the depletion layer. To derive an analytical model, the following assumptions are made:

1. uniform impurity profile in each region (emitter, base, and collector);
2. low-level injection;
3. constant mobility;
4. no generation–recombination in the depletion layer.

Then, the diffusion equation for the neutral base, emitter, and collector regions under the boundary conditions at the depletion-layer edges given by Shockley's law can be solved, and the emitter, base, and collector current densities can be derived as

$$J_E = q\left(\frac{D_n n_{B0}}{W_B} + \frac{D_p p_{E0}}{L_E}\right)\exp\left(\frac{qV_{BE}}{k_B T}\right), \tag{13}$$

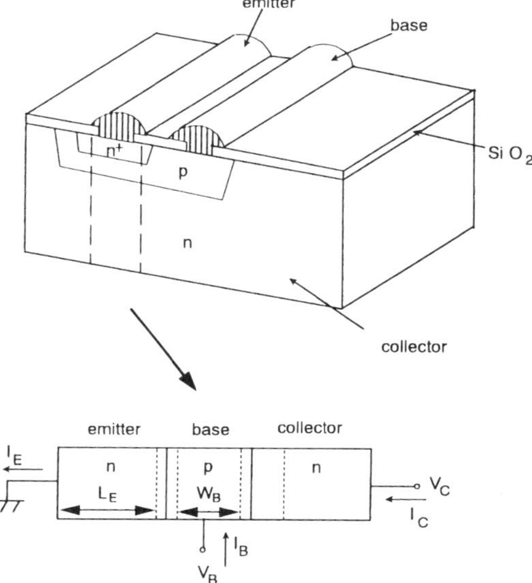

FIG. 2. Schematic structure (upper) and one-dimen-

$$J_B = q\frac{D_p p_{E0}}{L_E}\exp\left(\frac{qV_{BE}}{k_B T}\right), \tag{14}$$

$$J_C = q\frac{D_n n_{B0}}{W_B}\exp\left(\frac{qV_{BE}}{k_B T}\right), \tag{15}$$

where W_B is the neutral base width, L_E is the neutral emitter width, p_{E0} is the equilibrium minority-carrier (hole) density in the emitter (equal to n_i^2/N_E), and n_{B0} is the equilibrium minority-carrier (electron) density in the base (equal to n_i^2/N_B). A formula for the current gain is given as

$$h_{FE} = \frac{I_C}{I_B} = \frac{D_n L_E N_E}{D_p W_B N_B}. \tag{16}$$

It should be noted, however, that actual current gain is smaller than that given in Eq. (16) as a result of band-gap narrowing in the emitter, which is ignored in the above derivation.

2.3 MOSFET

An ideal metal-oxide–semiconductor structure is shown in Fig. 3, where the work-function difference, $q\phi_{ms}$, is equal to zero:

$$q\phi_{ms} = q\phi_m - (q\chi + \tfrac{1}{2}E_g + q\psi_B) = 0, \tag{17}$$

where $q\phi_m$ is the work function of the metal, $q\chi$ is the electron affinity, E_g is the band-gap energy, and $q\psi_B$ is the difference between the Fermi potential and the midgap. The energy-band diagrams for negative and positive voltages applied on the metal are shown in

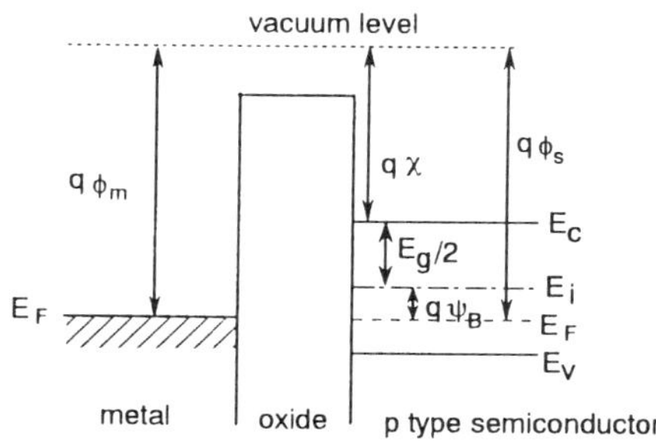

FIG. 3. Energy-band diagram of an ideal MOS diode for zero biasing voltage, where $q\phi_m$ is the metal work function, $q\phi_s$ the semiconductor work function, and χ the semiconductor electron affinity.

Fig. 4. For negative voltage, an accumulation layer of holes appears at the interface; for low positive voltage, a depletion layer extends from the interface; and for high positive voltage, the electron inversion layer appears at the interface. The MOSFET shown in Fig. 5 utilizes these MOS diode characteristics. When the gate voltage is zero or low, a depletion layer appears under the gate and no current flows between the source and the drain. When a high gate voltage is applied, an inversion layer called the channel appears under the gate, and current flows through this channel between the source and the drain for nonzero drain voltage. To derive an analytic model for MOSFET I-V characteristics, the following assumptions are made:

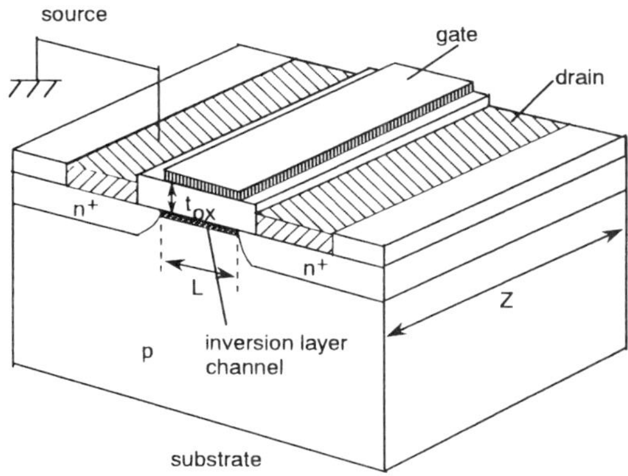

FIG. 5. Structure of a MOSFET.

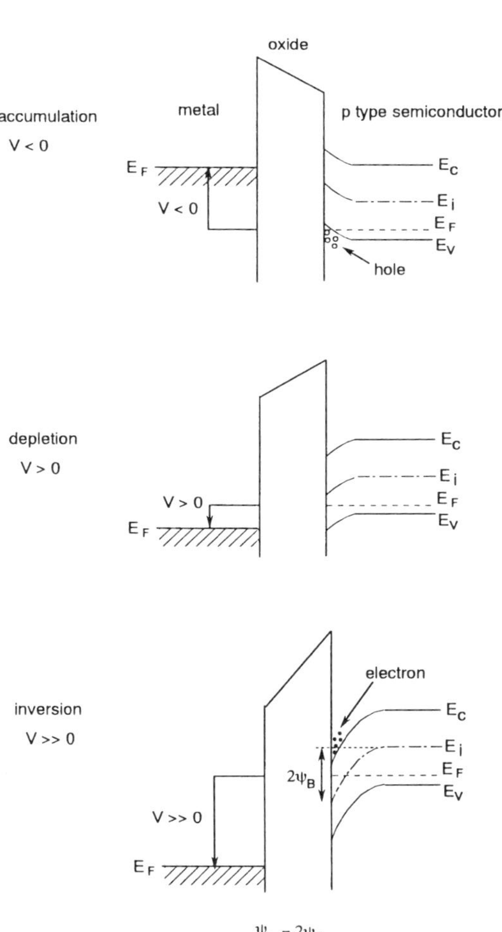

FIG. 4. Energy-band diagram of an ideal MOS diode for nonzero biasing voltage.

1. The MOS diode is ideal.
2. Only drift current is considered (the diffusion current is ignored).
3. Electron mobility in the inversion layer is constant.
4. The impurity density of the substrate is constant.
5. Reverse saturation current is ignored.
6. The electric field component parallel to the channel direction is much smaller than the electric field component normal to the channel direction.

Assumption **6** is called the gradual-channel approximation.

On the basis of these assumptions the following analytic expression for the drain current can be derived:

$$I_D = \frac{W}{L}\mu_n C_{ox}\left\{\left(V_G - 2\psi_B - \frac{V_D}{2}\right)V_D - \frac{2}{3} \times \frac{(2\epsilon_s q N_A)^{1/2}}{C_{ox}}[(V_D + 2\psi_B)^{3/2} - (2\psi_B)^{3/2}]\right\}, \tag{18}$$

where L is the channel length, W is the channel width, C_{ox} is the gate-oxide capacitance ($= \epsilon_{ox}/t_{ox}$), and N_A is the substrate doping density. The drain current [Eq. (18)] of the gradual-channel model is plotted in Fig. 6. When the drain voltage V_D is low, the drain current is given as

$$I_D = (W/L)\mu_n C_{ox}(V_G - V_T)V_D, \tag{19}$$

where V_T is a threshold voltage given by

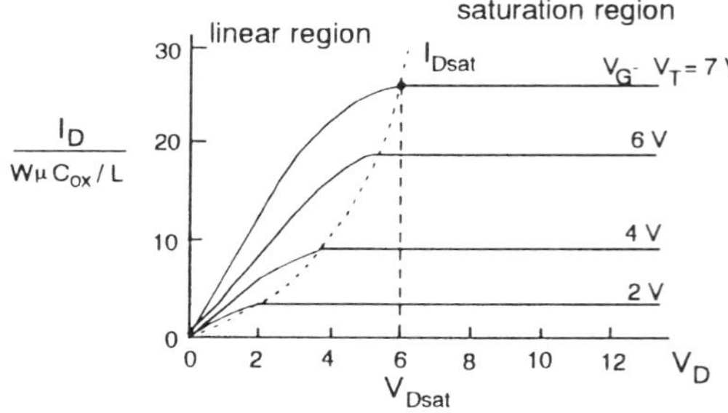

FIG. 6. Drain-current characteristics of a MOSFET derived by the gradual-channel model. The dashed curve indicates the locus of the saturation points (V_{Dsat}, I_{Dsat}).

$$V_T = 2\psi_B + [2\epsilon_s q N_A(2\psi_B)]^{1/2}/C_{ox} + V_{FB}. \tag{20}$$

When the gate voltage is higher or lower than the threshold voltage, the drain current flows or does not flow, respectively. Nonideality of the MOS diode is taken into consideration in Eq. (20) by the flat-band voltage V_{FB}, which is given as

$$V_{FB} = \phi_{ms} - Q_{ox}/C_{ox}, \tag{21}$$

where Q_{ox} is the oxide charge.

3. NUMERICAL MODELS FOR SEMICONDUCTOR DEVICES

The analytical models described in Sec. 2 are based on various assumptions. These assumptions, such as uniform impurity density, constant carrier mobility, or the gradual-channel approximation, inevitably restrict the models' applicability. In contrast, numerical models are general, since they derive device characteristics from solving numerically the basic semiconductor equations, e.g., the drift–diffusion equations in Sec. 1, without the severe assumptions mentioned above. Therefore, the numerical models can be used for quantitative analysis or design of realistic devices. Moreover, the hot-carrier phenomena and the nonlocal impact ionization, which crucially affect current semiconductor devices, and even nonstationary microscopic phenomena such as the ballistic effect, can be treated by the numerical approach based on the appropriate basic equations, e.g., the hydrodynamic equation and the Boltzmann equation.

However, numerical computation techniques to solve the basic semiconductor equations are needed. These techniques should be powerful, since solving partial differential equations or transport equations tends to be computer-expensive. To solve partial differential equations numerically, the finite-difference method (FDM) using orthogonal grid lines, and the control-volume method, using triangular elements in 2D and tetrahedron elements in 3D, are generally used (see SCIENTIFIC COMPUTATION BY NUMERICAL METHODS). A resulting linear system that is expressed by a large sparse matrix can be efficiently solved using recently developed methods such as CGS (conjugate gradient squared) and Bi-CGSTAB (van der Vorst, 1992). At present, device simulation using 10^4 grid points can be executed quite easily in routine design jobs.

At present, the numerical models are divided into several classes: drift–diffusion model, hydrodynamic model, Monte Carlo model, etc. In addition, different approaches to solve the Boltzmann transport equation and various quantum-mechanical treatments have been reported.

The drift–diffusion model is still a standard model from the point of view of the application. For practical use, modeling of physical parameters such as carrier mobility is very important. In the numerical approach, physical-parameter models can be incorporated even if they are rather complicated. Thus, most practical DD model simulators are equipped with sophisticated physical-parameter models.

4. PHYSICAL PARAMETERS

4.1 Carrier Mobility

Carrier mobilities are related to the scattering mechanisms that deflect the carrier motion. Electrons and holes in a semiconductor are scattered by thermal lattice vibrations (phonons), ionized impurities, neutral impurities, and the surface. As a result, carrier mobilities depend on the temperature through phonon scattering, on the impurity density through ionized impurity scattering, and on the electric field through optical phonon scattering.

In the DD model, the carrier mobility is commonly expressed in empirical form, as a function of the temperature, impurity density, and electric field. The temperature dependence of the carrier mobility in Si can be expressed as

$$\mu_L = \mu_0(T/300\ \mathrm{K})^{-\alpha}, \tag{22}$$

where α is about 2.5 and 2.3 for electron and hole, respectively, and μ_0 is the low-field mobility for low impurity density (1450 and 500 $\mathrm{cm^2/V\cdot s}$ for electrons and holes, respectively). The mobility μ_L is regarded as the one for low impurity density and for low electric field.

An empirical model including impurity-density dependence (for Si) is given by Caughey and Thomas (1967) as

$$\mu_{LI} = \mu_{\min} + \frac{\mu_L - \mu_{\min}}{1 + [(N_D + N_A)/N_0]^{\alpha}}, \tag{23}$$

where $\mu_{\min}$, N_0, and α are 80 $\mathrm{cm^2/V\cdot s}$, $1 \times 10^{17}\ \mathrm{cm^{-3}}$, and 0.7, respectively.

For electric fields higher than 5×10^4 V/cm in Si, the carrier drift velocity is nearly constant. This velocity is called the saturation velocity v_{sat}, which is about 1×10^7 cm/s. An empirical expression including this high-field effect is given by Scharfetter and Gumbel (1969) and Thornber (1980) as

$$\mu_{LIE} = \mu_{LI}\left\{1 + \left(\frac{\mu_{LI}E}{v_c}\right)^2\left(\frac{\mu_{LI}E}{v_c} + F\right)^{-1} + \left(\frac{\mu_{LI}E}{v_{\mathrm{sat}}}\right)^2\right\}^{-1/2}, \tag{24}$$

where E is the electric field and v_c and F_0 are parameters.

Carriers in an inversion layer of a MOSFET experience surface scattering besides phonon and impurity scattering. Surface scattering is related to surface roughness, interface charge, and quantum-mechanical effects of a two-dimensional gas. The surface mobility has been measured as an effective mobility determined from the drain conductance for the MOSFET in the low–drain-voltage region (linear region). It has been shown that the surface mobilities follow a universal curve, independent of the substrate impurity concentration, when they are plotted as a function of the effective normal field. Experimental results by Takagi *et al.* (1988) are shown in Fig. 7, which shows a remarkable universality in the wide range from 10^{15} to $10^{18}\ \mathrm{cm^{-3}}$. For MOSFET simulation, implementation of the surface-mobility model is important.

4.2 Generation–Recombination Rate

When a system of electrons and holes deviates from thermal equilibrium, generation–recombination processes take place to restore thermal equilibrium. For indirect–band-gap semiconductors such as Si, the recombination process takes place through indirect transition of electrons and holes between the conduction band and the valence band via trap levels in the band gap.

The single-level recombination model (Shockley *et al.*, 1952) can be derived from four processes: electron capture, electron emission, hole capture, and hole emission. The recombination rate R_{SRH} is given as

$$R_{SRH} = (pn - n_i^2)\left\{\tau_p\left[n + n_i \exp\left(\frac{E_t - E_i}{k_B T}\right)\right] + \tau_n\left[p + n_i \exp\left(-\frac{E_t - E_i}{k_B T}\right)\right]\right\}^{-1}, \tag{25}$$

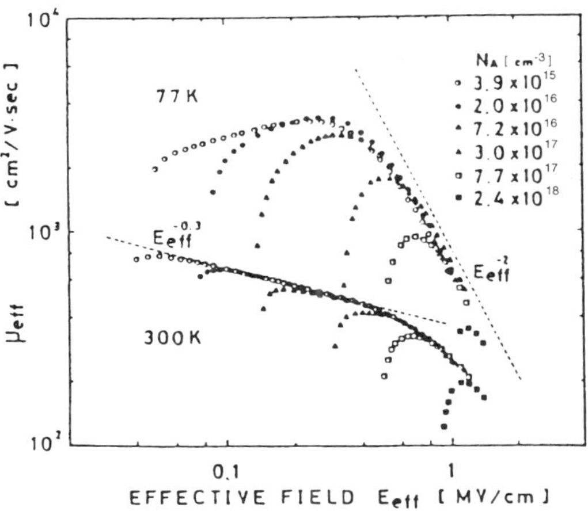

FIG. 7. Effective electron mobility μ_{eff} in n-channel MOSFETs vs normal field E_{eff} with substrate acceptor concentration as a parameter (Takagi *et al.*, 1988).

where τ_n and τ_p are the electron and hole lifetimes, respectively, and E_t is the trap energy level. The carrier lifetimes are determined by the carrier capture cross sections and the trap density. When electrons are injected into p-type Si as minority carriers, the recombination rate R_{SRH} is given approximately as

$$R_{SRH} = (n - n_{p0})/\tau_n, \tag{26}$$

where n_{p0} is the equilibrium minority-electron density. The carrier lifetime is found empirically to depend on the doping density as

$$\tau_{n,p} = \frac{\tau_{n,p0}}{1 + (N_D + N_A)/N_r} \tag{27}$$

or

$$\tau_{n,p} = \tau_{n,p0}[N_r/(N_D + N_A)]^{\alpha}, \tag{28}$$

with α in the range 0.3–0.6.

4.3 Auger-Recombination Rate

Auger recombination involves three-particle transitions. There are three processes:

1. An electron in the conduction band recombines with a hole in the valence band and excites another electron in the conduction band with the generated energy;
2. the recombination energy excites an electron in the light valence band to the heavy valence band; and
3. the recombination energy excites an electron in the spin-orbit–split band to the valence band.

The Auger-recombination rate is given as

$$R_{AUG} = (C_n n + C_p p)(np - n_i^2), \tag{29}$$

where $C_n = 2.8 \times 10^{-31}$ cm^6/s and $C_p = 9.9 \times 10^{-32}$ cm^6/s, at 300 K. Since Auger recombination is dominant for high impurity density, such as the emitter layer in a bipolar transistor, it affects the current gain of the bipolar transistor.

4.4 Impact-Ionization Rate

Impact ionization is a carrier-generation process that takes place in a region of high electric field. In the process, carriers generate electron–hole pairs in their transit, with the energy acquired from the electric field. Since this process is related to the band structure and various scattering mechanisms, accurate analytical treatment is not easy. Empirical models are generally used.

The generation rate of impact ionization, G_{ii}, defined as the number of electron–hole pairs per unit volume and per unit time, can be expressed using the ionization rates α_n and α_p as

$$G_{ii} = \alpha_n |J_n|/q + \alpha_p |J_p|/q. \tag{30}$$

The electron ionization rate α_n is defined as the number of generated electron–hole pairs per unit length of travel and per electron. An empirical model for the ionization rate α_n is given as

$$\alpha_n = A_n \exp(-b_n/E^m), \tag{31}$$

where E is the electric field and A_n, b_n, and m are parameters. The exponent m is equal to 1 or 2. Empirically $m = 1$ is used for Si and Ge, and $m = 2$ is used for GaAs. Different values for A_n and b_n have been reported in the literature. For example, A_n is 6.2×10^5 cm^{-1} and b_n is 1.08×10^5 V/cm for 2.4×10^5 V/cm $< E < 5.3 \times 10^5$ V/cm. For holes, the same expression as Eq. (31) holds.

Solving the Boltzmann transport equation numerically, Baraff (1962) obtained a universal plot for the ionization rate in three-parameter form:

$$\alpha = \frac{1}{\lambda} f\left(\frac{E_r}{E_i}, \frac{E_i}{q\lambda E}\right), \tag{32}$$

where λ is the optical phonon mean free path, E_r is the average energy loss per phonon collision, and E_i is the ionization threshold energy.

For the latest submicron devices, the electric field profile is so steep that nonstationary transport effects are prominent and the impact ionization is not directly determined by the local field. For these cases, the ionization rate is considered as a function of carrier energy or to be dependent on the trajectory of the carrier movement. The impact-ionization rate as a function of the electron energy has recently been calculated on the basis of first principles using realistic band structures (Sano and Yoshii, 1992).

4.5 Band-Gap Narrowing Effect

Highly doped diffusion layers, such as the emitter region in a bipolar transistor, are often used in semiconductor devices. Impurity concentrations in these regions are higher than 10^{20} cm^{-3}. In such cases, the impurity bands are formed below the conduction band and above the valence band and merge with the respective bands. Therefore, the band gap is effectively narrower than for the intrinsic case (very pure with a negligibly small amount of impurities). Band-gap narrowing ΔE_g can be measured and an empirical expression can be given (Slotboom and de Graaff, 1976) as

$$\Delta E_g = qV_1\left\{\ln\left(\frac{N_D + N_A}{N_0}\right) + \left(\left[\ln\left(\frac{N_D + N_A}{N_0}\right)\right]^2 + C\right)^{1/2}\right\} \tag{33}$$

with

$$V_1 = 9 \text{ mV}, \quad N_0 = 10^{17} \text{ cm}^{-3}, \quad C = 0.5. \tag{34}$$

The band-gap narrowing effect brings about an increase in the intrinsic carrier density,

$$n_{ie} = n_i \exp(\Delta E_g/2k_B T), \tag{35}$$

where n_{ie} is called the effective intrinsic carrier density. This effect causes a decrease in the current gain of bipolar transistors from the ideal value, and modeling this effect is indispensable for simulation.

5. AN EXAMPLE USING THE DRIFT–DIFFUSION MODEL

A typical two-dimensional simulation of an n-channel MOSFET is shown. A schematic cross section of a LDD (lightly doped drain) NMOSFET is shown in Fig. 8. The impurity profile was calculated using a process simulator and its bird's-eye-view plot is shown in Fig. 9. The number of grid points is 58 × 61 = 3538. The channel region and the junction region are divided by a fine mesh. The device parameters are as follows:

gate-oxide thickness 105 Å
gate length 0.6 μm

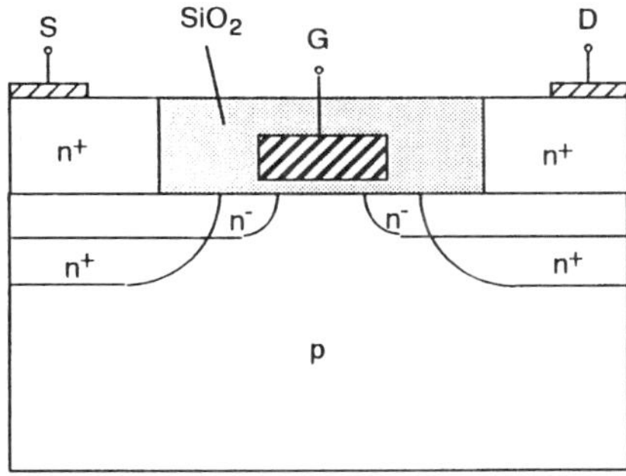

FIG. 8. Cross section of a LDD (lightly doped drain) MOSFET. The source and the drain contacts are replaced by n^+ diffusion layers because of a restriction of the simulator used.

substrate impurity density 2×10^{16} cm^{-3}
channel impurity density 3×10^{17} cm^{-3}
junction depth 0.2 μm.

Calculated drain-current–drain-voltage characteristics are shown with gate voltage as a parameter in Fig. 10. Increase in the drain current due to partial avalanche breakdown is observed for drain voltage larger than 5 V. Potential and electron-density distributions for a typical biasing point of V_G = 3 V and V_D = 3 V are shown in Figs. 11 and 12, respectively.

6. ENERGY-TRANSPORT MODEL

In miniaturized devices, electric fields are high and vary sharply with respect to position. The electrons in these devices acquire high energy from the electric fields, and hot-electron transport plays an important role. In this case the electron energy distribution deviates significantly from the equilibrium distribution, and the drift–diffusion model loses validity. The Boltzmann transport theory presents a way to cope with this problem. Using the energy-transport model, which is also called the hydrodynamic model (Forghieri *et al.*, 1988), is one of the practical approaches to solve the Boltzmann transport equation.

In the energy-transport model, the basic equations for electrons are given as

$$\frac{\partial n}{\partial t} = -\nabla \cdot n\mathbf{v} + G, \tag{36}$$

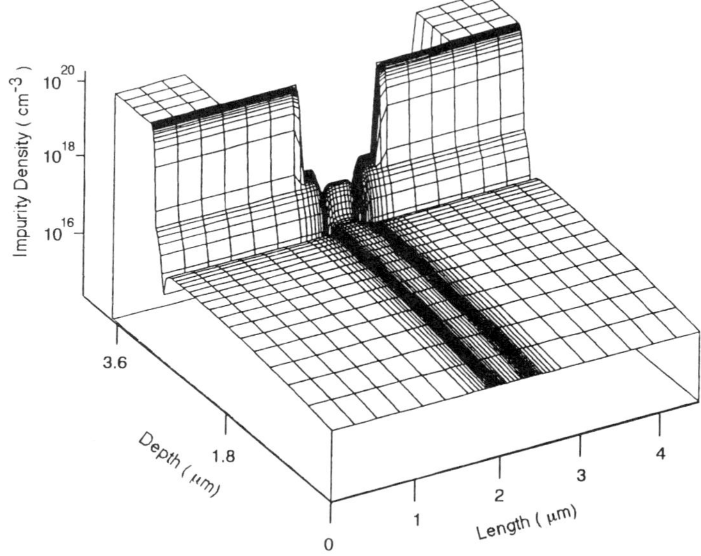

FIG. 9. Impurity-density distribution of the LDD MOSFET.

$$\frac{\partial(m^*n\mathbf{v})}{\partial t} = qn\nabla\psi - \nabla nk_BT_n - \frac{m^*n\mathbf{v}}{\tau_m}, \quad (37)$$

$$\frac{\partial(nw)}{\partial t} = -\nabla\cdot n\mathbf{v}w + qn\mathbf{v}\cdot\nabla\psi - \nabla\cdot n\mathbf{v}k_BT_n + \nabla\cdot\kappa\nabla T_n - \frac{n(w-w_0)}{\tau_w} + Gw, \quad (38)$$

where G is the net generation rate, which is composed of various generation–recombination mechanisms mentioned in the previous section; m^* is the electron effective mass; T_n is the effective electron temperature; τ_m and τ_w are the momentum and energy relaxation times, respectively; w is the average electron energy, which is related to the mean velocity v and the effective electron temperature T_n as

$$w = \tfrac{1}{2}m^*v^2 + \tfrac{3}{2}k_BT_n; \quad (39)$$

and κ is thermal conductivity. Equations (36), (37), and (38) are derived from the moments of the Boltzmann transport equation, and represent mass conservation, momentum conservation, and energy conservation,

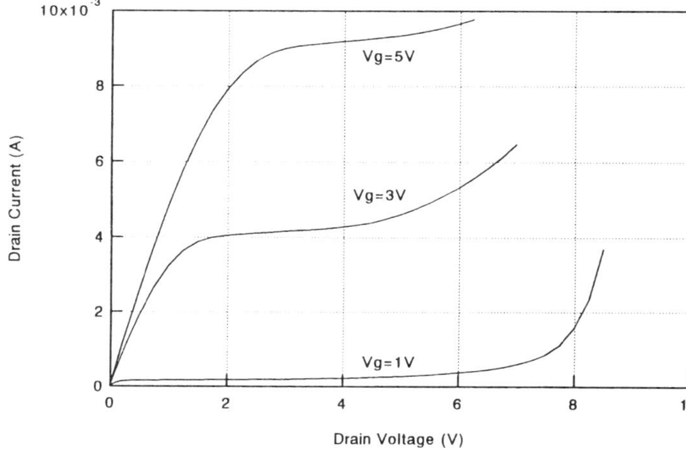

FIG. 10. Drain-current–drain-voltage characteristics of the LDD MOSFET.

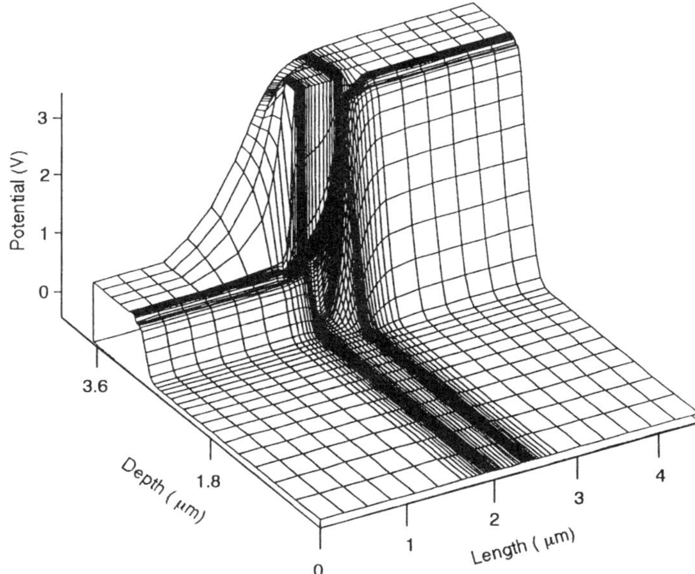

FIG. 11. Potential distribution of the LDD MOSFET for V_D = 3 V and V_G = 3 V.

respectively. The sequence of the moments is truncated at the second moment [Eq. (38)] by assuming that heat flow can be described by $-\kappa \nabla T_n$. It is assumed that the effects of the various scattering mechanisms on momentum and energy are expressed by the relaxation times and can be given as functions of the energy. This approach is also called the relaxation-time approximation. The momentum relaxation time τ_m is related to the mobility by $\mu = (q/m^*)\tau_m$. It is hard to measure the energy relaxation time, and it is often treated as a fitting parameter, though it should be derived from scattering theory.

Hot-carrier phenomena are reflected by high average energy of electrons, resulting in the electron temperature being higher than the lattice temperature, causing a change in

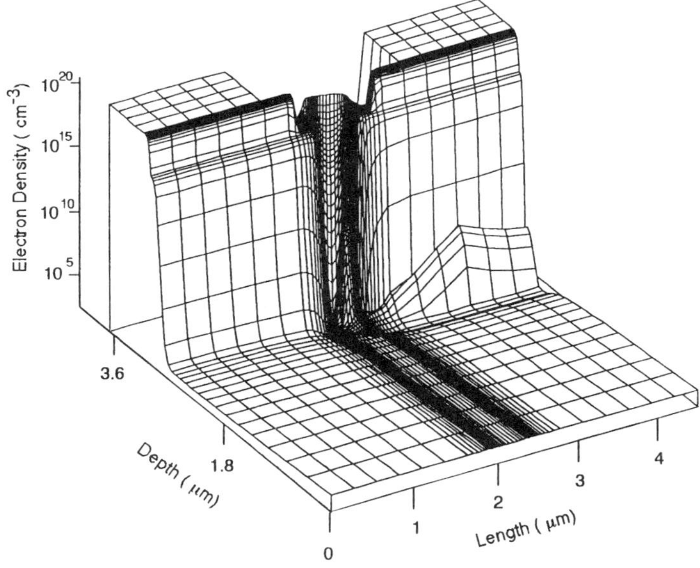

FIG. 12. Electron-density distribution of the LDD MOSFET for V_D = 3 V and V_G = 3 V.

mobility and impact-ionization rate values. The diffusion constant given as $D = (k_B T_n/q)\mu$ indicates a larger diffusion in the "hot" region than in the "cold." This model predicts channel spreading near the drain junction of a miniaturized MOSFET under high drain voltages. Substrate currents in MOSFETs are successfully modeled by the energy-transport model including the energy-dependent impact-ionization rates as shown in Fig. 13 (Katayama and Toyabe, 1989).

7. MONTE CARLO MODEL

7.1 Motivation

As devices shrink into deep submicron and sub–0.1-μm regions, the DD and the hydrodynamic models are no longer valid, since nonequilibrium transport of highly energetic electrons, which affects strongly the electrical characteristics of the devices, must be treated rigorously. For this problem, it is necessary to solve the Boltzmann transport equation (BTE). It is impossible in practice to solve directly the BTE coupled with the Poisson equation self-consistently even by the fastest supercomputer, since the equation is a complicated integro-differential equation with respect to the electron distribution function in 2D or 3D real space and 3D wave-vector space. The Monte Carlo method (*q.v.*), however, is a viable means to solve the BTE.

The Monte Carlo (MC) model can accurately handle various scattering mechanisms and the full band structures of semiconductors. The impact of the full band structures on device physics has been thoroughly investigated by Fischetti and Laux (1988). Since high-field effects predominantly observed in small devices are governed by electrons in the high-energy portion of the bands, not by the ones near the band minima, only the MC model based on the full band structure can give reliable predictions for the electrical characteristics of small devices. Thus, a comparative study has shown that the transconductance of an n-GaAS MOSFET is almost the same as that of an n-Si MOSFET for channel lengths between 0.05 and 0.25 μm, in contrast to the superiority of the low-field characteristics of n-GaAs devices resulting from the high electron mobility at low fields (Fischetti and Laux, 1991).

7.2 Electric Field Calculation

The computation of the Monte Carlo model is composed of the electric field computation, the free-flight computation, and the

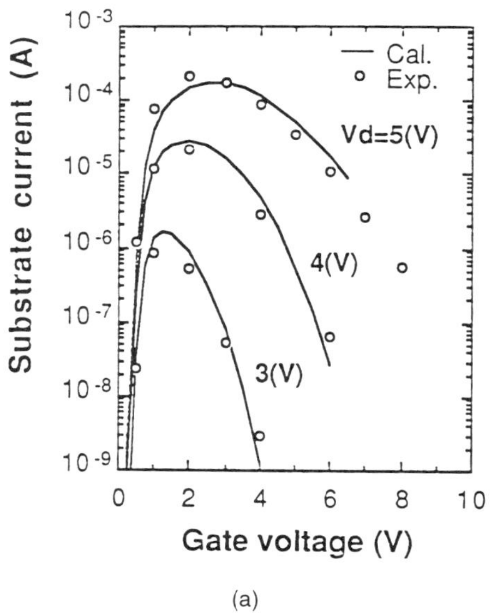

(a)

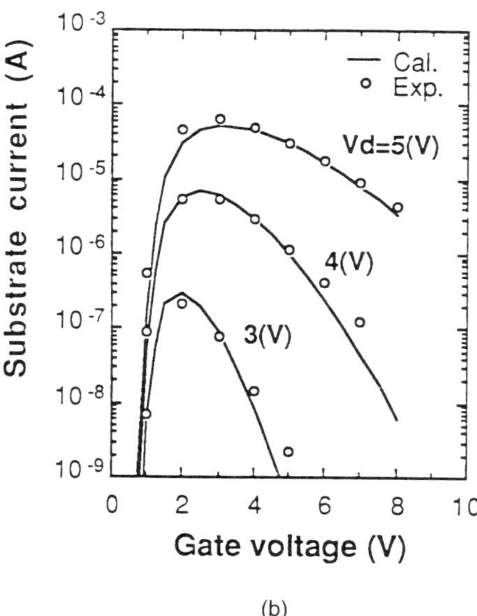

(b)

FIG. 13. Substrate-current characteristics for n-channel MOSFETs, where L_{eff} = 0.99 μm, W = 100 μm. (a) Oxide thickness t_{ox} = 21 nm; (b) t_{ox} = 41 nm. (Katayama and Toyabe, 1989.)

scattering computation as shown in Fig. 14. The electric field can be obtained by solving the Poisson equation, which is usually solved using the finite-difference method (FDM), the finite-element method (FEM), or the fast-Fourier-transform (FFT) method. In these methods, real space is divided by grid points and the Poisson equation is discretized on the grid points.

The particles distributed in space have to be assigned to grid points to form the space-charge–density term of the Poisson equation. A simple way is using the nearest-grid-point method in which each particle is assigned to the nearest grid point. This method produces a discontinuous charge distribution when the number of particles is not large. Another method, called the cloud-in-cell method, assigns particles to neighboring grid points with appropriate weights. This method produces a smoother charge distribution than with the nearest-grid-point method.

7.3 Free-Flight Calculation

Particles travel under the influence of the electric field between collisions. When analytic approximations of the bands are adopted, each free-flight time t_f is determined by a random number r_f that is uniform between 0 and 1 as

$$t_f = -\log(r_f)/\Gamma, \tag{40}$$

where the constant Γ is the total scattering probability including the so-called "self-scattering."

In the MC model adopting the full bands (Fischetti and Laux, 1988), however, the high efficiency of this technique vanishes since the equations of motion,

$$\frac{d\mathbf{r}}{dt} = \frac{1}{\hbar}\nabla_k E_\nu(\mathbf{k}), \tag{41}$$

$$\frac{d\mathbf{k}}{dt} = \frac{q}{\hbar}\nabla_r \psi(\mathbf{r}) = -\frac{q\mathbf{F}(\mathbf{r})}{\hbar}, \tag{42}$$

cannot be integrated analytically, where $E_\nu(\mathbf{k})$ is the νth energy band and $\mathbf{F}$ is the electric field at the electron position $\mathbf{r}$. In this case, a prefixed time step Δt of the order of 0.1 fs is preferable. The band structure of technological interest can be obtained from empirical-pseudopotential calculation. The entire Brillouin zone is divided into irreducible wedges, and each wedge is discretized by mesh points. The energy, group velocity, and effective mass for each mesh point are stored in a lookup table (Fischetti and Laux, 1988).

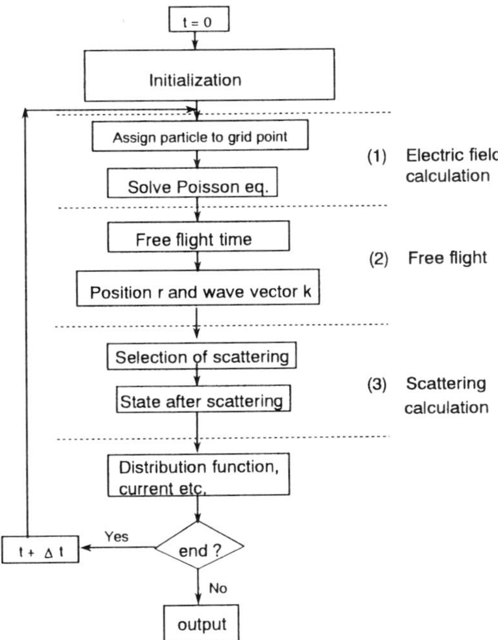

FIG. 14. Flow of Monte Carlo method.

7.4 Scattering Mechanisms

The scattering probability S from a state $|i\rangle$ to a state $|f\rangle$ per unit time is given using the Fermi "golden rule" as

$$S = \frac{2\pi}{\hbar}|\langle i|H'|f\rangle|^2\delta(E_f - E_i), \tag{43}$$

where H' is a perturbation Hamiltonian. In the case of phonon scattering, the scattering probability from a point $\mathbf{k}$ in the νth band to a point $\mathbf{k}'$ in the ν'th band $S(\nu,\mathbf{k},\nu',\mathbf{k}')$ is given as

$$S(\nu,\mathbf{k},\nu',\mathbf{k}') = \frac{2\pi}{\hbar}M(|\mathbf{k}-\mathbf{k}'|)^2|I(\nu,\nu';\mathbf{k},\mathbf{k}')|^2 \times (n_\mathbf{q} + \tfrac{1}{2} \pm \tfrac{1}{2}) \times \delta(E_{\nu',\mathbf{k}'} - E_{\nu,\mathbf{k}} \pm \hbar\omega_\mathbf{q}), \tag{44}$$

where M depends on various parameters of the crystal such as sound velocity, density, deformation potential, etc., I is the overlap integral, $n_{\mathbf{q}}$ is the number of phonons with the wave vector $\mathbf{q}$, and signs + and − correspond to the emission and absorption of a phonon, respectively.

Scattering mechanisms are electron–phonon scattering (acoustic phonon and optical phonon); particle–particle scattering (electron–electron and electron–hole); and other scattering (impurity, alloy, and impact ionization).

In the Monte Carlo method, the successive scattering for each electron is modeled through the selection of a final state after a scattering using random numbers. As can be imagined from the complexity of the band-structure data and the scattering probability expressions, the Monte Carlo model needs a huge amount of CPU time and computer memory. Nevertheless, a large amount of exploring work has been reported by the IBM group, e.g., comparative studies of Si, GaAs, InP, InGaAs, and Ge FET performance (Fischetti and Laux, 1991).

8. OTHER SOLUTION METHODS TO THE BOLTZMANN TRANSPORT EQUATION

8.1 Scattering-Matrix Approach

The scattering-matrix approach (Das and Lundstrom, 1990) begins by dividing a device into a number of small elements (thin slabs) and defining scattering matrices, which relate the fluxes incident upon each element to the emerging fluxes. For a slab shown in Fig. 15, the emerging carrier fluxes are related to those incident on it by

$$\begin{bmatrix} b^+ \\ a^- \end{bmatrix} = \begin{bmatrix} t & r' \\ r & t' \end{bmatrix} \begin{bmatrix} a^+ \\ b^- \end{bmatrix} = S \begin{bmatrix} a^+ \\ b^- \end{bmatrix}, \tag{45}$$

where S is the scattering matrix for the slab. By cascading the individual scattering matrices for the slabs composing a semiconductor device, the composite scattering matrix for the entire device is evaluated.

For actual simulations, the incident and emerging fluxes are resolved into M modes that represent different three-dimensional $\mathbf{k}$ vectors. For this case, the elements of the

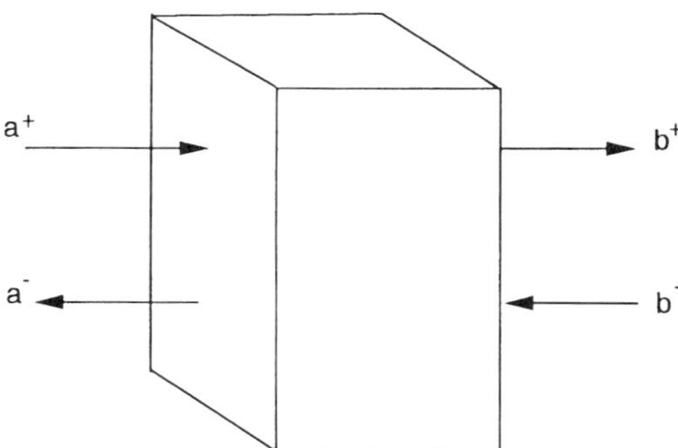

FIG. 15. Characterization of electron transport across a slab of semiconductor in terms of incident and emerging fluxes (Das and Lundstrom, 1991).

scattering matrix, t, r, t', and r', become $M \times M$ submatrices. By assuming local thermodynamic equilibrium at the contacts, the incident fluxes from contacts are evaluated using the Fermi–Dirac distributions. Starting from these fluxes, all the other fluxes can be obtained from the computation using scattering matrices. The carrier current, density, and distribution function in momentum space can be derived from the fluxes. This approach is free from the statistical noise that appears in the MC model. Almost all the computation burden exists in the scattering-matrix computation, which is usually carried out by Monte Carlo simulation in the slabs. Once the scattering matrices are calculated, the subsequent device simulation is carried out very efficiently. The method is capable of treating quantum interference effects as well. Demonstration of this approach was given for high-field transport in bulk Si and nonstationary transport in 1D Si devices with good agreement with Monte Carlo simulations (Das and Lundstrom, 1990).

8.2 Cellular-Automaton Method

In the cellular-automaton method (Kometer *et al.*, 1992), lattices in real ($\mathbf{x}$) and momentum ($\mathbf{k}$) space are defined with the cells. The $\mathbf{x}$ cells and $\mathbf{k}$ cells are populated by fictitious particles—the lattice-gas particles—which obey nondeterministic cellular-automaton (CA) rules. It is shown that the Boltzmann equation for semiclassical transport can be transformed into a Boolean master equation, which represents a cellular automaton with nearest-neighbor interaction in position space. The translation term in **x** and **k** space in the BTE is transformed to an effective collision term resulting in the CA rules. Boolean cell occupancies evolve in time obeying the master equation. The average over an ensemble of CA states (the average over a large number of cells) yields macroscopic quantities such as particle densities. The CA simulations have been verified, with agreement with Monte Carlo results (Kometer *et al.*, 1992). This approach is well suited for carrier systems with pronounced spatial inhomogeneities, large density variations, and two-particle scattering interactions.

9. CONCLUSIONS

Semiconductor-device modeling already has a history of 40 years. From early analytical modeling it has evolved into sophisticated numerical modeling. The classic drift–diffusion model is now a commonly used tool in the design world of the semiconductor industry. For modern devices of small dimensions, advanced approaches of the hydrodynamic model, at least, and the Monte Carlo model are indispensable for understanding the operating mechanism. The full-band Monte Carlo model has been realized to attain rigorous understanding of small devices of 0.1 μm or even smaller. Furthermore, to go beyond the MC model, approaches using the scattering-matrix and cellular-automaton methods have been proposed. The quantum-mechanical treatment will be more important for smaller devices. The continuing increase in the cost of semiconductor development makes numerical modeling of semiconductor devices, with the help of the decrease in the cost of computation, all the more important.

GLOSSARY

Alpha-Particle–Induced Soft Error: Volatile operation failure in memory or logic circuits in LSI caused by an alpha-particle hit.

CG Method: A numerical solution method for linear systems expressed by symmetric positive-definite matrices. CGS and Bi-CGSTAB are extended versions applicable to nonsymmetric matrices.

Drift–Diffusion Model: A traditional model used in device modeling where device operation is described by the potential and the carrier-density distributions governed by the Poisson and current-continuity equations, respectively.

Electron Temperature: The electron temperature is introduced to express a rise of the electron mean energy above the thermal value $3k_BT/2$ in the nonlinear transport caused by a high electric field, assuming the displaced Maxwellian distribution.

Energy-Transport Model: One of the advanced device models, also called hydrodynamic model, where the carrier temperature is introduced to describe the state of the system besides the potential and the carrier-density distributions. The equations for the model are derived by taking the moments of the Boltzmann transport equation.

FET (Field Effect Transistor): A kind of transistor in which the current flowing in the semiconductor is controlled by the electric field produced by the biasing voltage on the gate electrode. Representative types of FETs are junction FET (JFET), metal-oxide–semiconductor FET (MOSFET), and metal-semiconductor FET (MESFET).

Gunn Diode: A microwave solid-state oscillator utilizing the effect of electron transfer in the conduction band from low-energy, high-mobility states to high-energy, low-mobility states in GaAs in a high electric field.

Hot-Carrier Effects: Adverse effects caused by hot electrons or hot holes such as nonlinear conduction, parasitic current of substrate current, and device-performance degradation in MOSFETs.

IMPATT (Impact Avalanche Transit Time) Diode: A microwave solid-state oscillator utilizing the avalanche process in a reverse-biased *p-n* junction for generation of energetic carriers.

Planar Process Technology: Fabrication technology of semiconductor devices with electrodes and active region on the same semiconductor surface.

Relaxation Time: Time constants appearing in the approximation of the collision term of the Boltzmann transport equation. The momentum and energy relaxation times are used in the energy-transport model.

Works Cited

Baraff, G. A. (1962), *Phys. Rev.* **128,** 2507–2517.

Caughey, D. M., Thomas, R. E. (1967), *Proc. IEEE* **52,** 2192–2193.

Das, A., Lundstrom, M. S. (1990), *Solid-State Electron.* **33,** 1299–1307.

Dennard, R. H., Gaensslen, F. H., Yu, H. N., Rideout, V. L., Bassous, E., LeBlanc, A. (1974), *IEEE J. Solid-State Circuits* **SC-9,** 256–268.

Fischetti, M. V., Laux, S. E. (1988), *Phys. Rev. B* **38,** 9721–9745.

Fischetti, M. V., Laux, S. E. (1991), *IEEE Trans. Electron Dev.* **ED-38,** 650–660.

Forghieri, A., Guerrieri, R., Ciampolini, P., Gnudi, A., Rudan, M., Baccarani, G. (1988), *IEEE Trans. Comput. Aided Design* **7,** 231–242.

Gummel, H. K. (1964), *IEEE Trans. Electron Dev.* **ED-11,** 455–465.

Ihantola, H. K. J., Moll, J. L. (1964), *Solid-State Electron.* **7,** 423–430.

Katayama, K., Toyabe, T. (1989), in: *International Electron Devices Meeting Technical Digest,* New York: IEEE Publishing Services, pp. 135–138.

Kometer, K., Zandler, G., Vogl, P. (1992), *Phys. Rev. B* **46,** 1382–1394.

May, T. C., Woods, M. H. (1979), *IEEE Trans. Electron Dev.* **ED-26,** 2–9.

Sano, N., Yoshii, A. (1992), *Phys. Rev. B* **45,** 4171–4180.

Scharfetter, D. L., Gummel, H. K. (1969), *Electron Devices* **ED-16,** 64–77.

Shockley, W., Read, W. T. (1952), *Phys. Rev.* **87,** 835–842.

Slotboom, J. W., de Graaff, H. C. (1976), *Solid-State Electron.* **19,** 857–862.

Takagi, S., Iwase, M., Toriumi, A. (1988), in: *International Electron Devices Meeting Tech. Digest,* New York: IEEE Publishing Services, pp. 398–401.

Thornber, K. K. (1980), *J. Appl. Phys.* **51,** 2127–2136.

van der Vorst, H. A. (1992), *SIAM J. Sci. Statist. Comput.* **13,** 631–644.

Further Reading

Jacoboni, C., Lugli, L. (1989), *The Monte Carlo Method for Semiconductor Device Simulation,* New York: Springer-Verlag.

Selberherr, S. (1984), *Analysis and Simulation of Semiconductor Devices,* New York: Springer-Verlag.

Snowden, C. M. (1986), *Introduction to Semiconductor Device Modelling,* Singapore: World Scientific.

SEMICONDUCTOR DETECTORS

See DETECTORS, SEMICONDUCTOR

SEMICONDUCTOR LASERS

See LASERS, SEMICONDUCTOR

SEMICONDUCTORS TO SUPERCONDUCTORS: ORGANIC LOWER-DIMENSIONAL SYSTEMS

ROBERT MELVILLE METZGER, *Department of Chemistry, University of Alabama, Tuscaloosa, Alabama, U.S.A.*

INTRODUCTION

Physicists tend to think of organic molecules as large, bland, boring, usually colorless assemblages of covalently bonded atoms that are electrically neutral overall, needed for that occasional headache (acetylsalicylic acid), for that pesky mosquito (naphthalene), to dye blue jeans (indigo), to make "plastics," i.e., structural polymers, or to enable the complicated chemical processes of life.

For decades it was unusual to be excited about the solid-state properties of organic molecules and their ions. In part this is because organic solids usually have only covalent chemical bonds within each molecule and van der Waals or London interactions between molecules; partial or complete formal charges on constituent cations and anions, and therefore Coulomb interactions between organic molecules in crystals, are the exception, rather than the rule. However, metallic conduction in organic systems had been dreamed of as early as 1911. Until the mid-1960s the number of serious researchers of the organic solid state remained small. The names of Hideo Akamatsu, Noel S. Bayliss, Melvin Calvin, Daniel D. Eley, Helmuth

3-527-28139-8/96/$5.00 + .50

Kainer, Aleksandr I. Kitaigorodskii, Jan Kommandeur, Yoshio Matsunaga, Harden M. McConnell, Albert Szentgyörgyi, A. N. Terenin, and A. T. Vartanyan come to mind, but the list should include their many students and co-workers, many of whom are still active today.

The discovery of superconductivity in potassium-intercalated graphite by Hannay *et al.* (1965), the development of quasimetallic organic donor–acceptor crystals (charge-transfer, or "two-chain" systems) in the 1960s to early 1970s, with metallic conductivity along the direction of stacking, the discovery of highly conducting doped polyacetylene by Shirakawa *et al.* (1977), the discovery of superconductivity in the polymer polythiazyl, $(SN)_x$, by Greene *et al.* (1975), in organic "one-chain" crystals by Jérome *et al.* (1980), and in alkali-metal salts of C_{60} (buckminsterfullerides) by Hebard *et al.* (1991), and the development of light-emitting diodes based on conducting polymers by Burroughes *et al.* (1990) have all helped to educate the solid-state physics community to the possibilities of organic systems, whose molecular orbitals can be finely "tuned" by chemical synthesis to enhance desired physical properties.

This article reviews the progress made in organic crystalline and polymeric conductors and superconductors, by first introducing relevant issues from solid-state theory and molecular-orbital theory, then summarizing the history of the advances in crystalline conductors, polymers, and superconductors. After that, summaries of molecules, polymers, crystals, and superconductors are presented, and design rules for new systems are discussed.

We start our discussion with concepts from the band theory of solids, discuss what can break the symmetry of one-dimensional systems, introduce electrical conductivity and superconductivity, present the Mulliken charge-transfer theory for solution complexes and its extension to solids, then discuss briefly the simple π-electron theory for long polyenes.

1. ONE-DIMENSIONAL BAND THEORY OF SOLIDS

Consider a one-dimensional (usually almost infinite) set of N atoms, molecules, or point masses, all equally spaced at interparticle distances d along the real-space coordinate x, with Born–von Kármán periodic boundary conditions for the potential energy:

$$V(x) = V(x + Nd). \tag{1}$$

By the theorems of Floquet (in one dimension; 1883) and Bloch (in three dimensions; 1928) the periodicity of the lattice requires that the eigenfunction of the appropriate Hamiltonian must satisfy

$$\psi(x + Nd) = \exp(ikNd)\psi(x), \tag{2}$$

where k is the wave vector. Multiplying this wave vector by Planck's constant h divided by 2π yields the crystal momentum $p = (h/2\pi)k$. Free electrons in vacuum obey the Schrödinger equation

$$\begin{aligned} H\psi(x) &= (p^2/2m)\psi(x) \\ &= (h^2k^2/8\pi^2m)\psi(x) = E\psi(x), \end{aligned} \tag{3}$$

where H is the Hamiltonian, p is the momentum, m is the electron mass, and E is the energy. The free-electron eigenvectors $\psi(x)$ are the standing-wave sine functions $\sin(kx) = \sin(n\pi x/Nd)$ and cosine functions $\cos(n\pi x/Nd)$, or the traveling-wave functions $\exp(\pm in\pi x/Nd)$, where $n = -(N - 1)$ to N.

For the periodic linear lattice one can start from N simple free-electron traveling-wave functions $\exp(\pm ikx)$; the interactions between the N atoms or molecules will break the N-fold degeneracy of the energy E, and create narrowly spaced energy levels that can be considered a semicontinuous energy band of width W; these can fill, lowest energy first, up to the "band edges" $k = \pm\pi/d$. The traveling-wave functions must undergo Bragg scattering at these band edges, and the quadratic dependence of E on k in Eq. (3) must be softened at the band edges, so that an energy gap opens up and $dE/dk = 0$ at $k = \pm\pi/d$. A more realistic dispersion relation $E(k)$ for a periodic linear lattice is given within the tight-binding approximation (Bloch, 1928):

$$E(k) = U - 2t\cos(kd), \tag{4}$$

where U is the on-site energy,

$$U = \langle\psi_i|H|\psi_i\rangle, \tag{5}$$

for the electron on site i, H is the one-electron Hamiltonian, ψ_i is the wave function for the electron localized at site i, and t is the Mulliken transfer integral for an electron moving from site i to the adjacent site $i + 1$:

$$t = \langle \psi_i | H | \psi_{i+1} \rangle. \tag{6}$$

Chemists will recognize Eq. (4) from the simple Hückel molecular-orbital theory for aromatic π-electron systems; t is akin to the Hückel resonance integral β. The bandwidth W is given by

$$W = 4t. \tag{7}$$

This can be verified by defining $W = \max[E(k)] - \min[E(k)]$ in Eq. (4). This band can be filled by electrons (or holes) symmetrically up to the maximum (minimum) Fermi wave vectors k_F ($-k_F$), either with only one electron per site (if the Coulomb electron–electron repulsion discourages more than one electron per site), or with two electrons (spin up and spin down) per site. If the band is filled up to the band edge, then

$$k_F = \pi/d. \tag{8}$$

If the band is only partially filled, then k_F will be some fraction of π/d. Bragg (Umklapp) x-ray scattering can be observed between the minimum ($-k_F$) and the maximum ($+k_F$) of band filling, at the reciprocal wave vector $2k_F$.

For free electrons the Fermi energy ϵ_F is given by

$$\epsilon_F = h^2 k_F^2 / 8\pi^2 m^*, \tag{9}$$

where m^* is the effective electron mass. For bound electrons in an energy band a different expression for ϵ_F can be crafted, e.g., from Eq. (4), as $\epsilon_F = U - 2t\cos(k_F d)$. A "filled band" has two electrons (or holes) per site (with spin up and spin down); a "half-filled band" has only one electron (or hole) per site; a "quarter-filled band" has one electron (or one hole) per two sites.

Very useful for the systems discussed here are the Hamiltonian named after John Hubbard (1963a, 1963b, 1964), but first discussed by Van Vleck (1932):

$$H = -t \sum_{i,\sigma} [a_{i,\sigma}^\dagger a_{(i+1),\sigma} + a_{i,\sigma} a_{(i+1),\sigma}^\dagger] + U \sum_i a_{i,\sigma}^\dagger a_{i,\sigma} a_{i,\sigma}^\dagger a_{i,\sigma} \tag{10}$$

and the extended Hubbard Hamiltonian (Hubbard, 1978):

$$H = -t \sum_{i,\sigma} [a_{i,\sigma}^\dagger a_{(i+1),\sigma} + a_{i,\sigma} a_{(i+1),\sigma}^\dagger] + U \sum_i a_{i,\sigma}^\dagger a_{i,\sigma} a_{i,\sigma}^\dagger a_{i,\sigma} + V \sum_i a_{(i+1)}^\dagger a_{(i+1)} a_i^\dagger a_i, \tag{11}$$

where $a_{i,\sigma}^\dagger$ creates an excitation of spin σ at site i, and $a_{i,\sigma}$ annihilates it, t is the Mulliken transfer integral [Eq. (6)], U is the on-site Coulomb energy [akin to the U of Eq. (5), but here it is a two-particle on-site energy], and V is the nearest-neighbor Coulomb interaction energy. Despite the phenomenological attractiveness of these two Hamiltonians, they cannot be solved for the general case. They have been solved exactly only for one-dimensional systems and for the two-state system, and numerically for a restricted set of conditions.

2. ONE-DIMENSIONAL INSTABILITIES

An organic crystal, or an organic thin film, is a three-dimensional object, but some of its properties are "quasi one-dimensional," i.e., resemble, but cannot coincide with, the predictions of one-dimensional physics, whose characteristics are simple but often peculiar.

An infinite one-dimensional "regular" chain of particles (atoms, electrons, or molecules), separated by equal distances d, is thermodynamically unstable; for instance, entropy considerations imply that it cannot exist as two or more distinct phases, or have a phase transition at finite temperatures (Landau and Lifschitz, 1958).

Peierls (1955) showed that a different, energetic instability of a one-dimensional electron gas, driven by electron–phonon interactions, causes nonuniformities in the electron distribution, and leads to a structural distortion and to a (usually second-order) phase transition, at a temperature T_P (the Peierls temperature). Below T_P either a dimerization into two sets of unequal interparticle distances d' and d'' (such that $d' + d'' = 2d$) or some other structural distortion must occur. The electronic energy of the metallic chain is lowered by the formation of a charge-density wave (CDW) of amplitude $\rho(x)$:

$$\rho(x) = \rho_0[1 + \alpha \cos(2k_F x + \phi)], \tag{12}$$

where x is the coordinate along the chain, ρ_0 is the uniform charge density, and $\alpha\rho_0$ is the charge-modulation amplitude. This phase transition opens up a "Peierls" energy gap 2Δ in the dispersion relation for the energy. Chemists are familiar with a conceptually similar Jahn–Teller distortion in organometallic systems. The Peierls distortion disrupts the periodicity, and can transform a metal (above T_P) to an electrical semiconductor or insulator (below T_P).

Below T_P the static (locked) CDW of the conduction electrons at $2k_F$ (or $4k_F$) couples with the other atomic or molecular electrons in the lattice; the resulting slight lattice distortions give rise to extra x-ray reflections. Above T_P, the CDWs are mobile excitations, with no phase locking between excitations on nearby chains: their x-ray signatures are diffuse reflections, similar to thermal diffuse scattering streaks in reciprocal space. These streaks sharpen as the temperature is lowered and T_P is approached. Below T_P, the static lattice distortion can be observed as new reflections between reciprocal-lattice layer lines. When the band filling is a rational fraction ($\frac{1}{4}$, $\frac{1}{2}$, $\frac{2}{3}$, 1, etc.) then the CDW excitations coincide with certain Bragg reflections of the background lattice, and are more difficult to detect.

When the band filling is not rational, the CDW reflections at wavelength λ are incommensurate with the background lattice of periodicity d, and the extent of charge transfer ρ (defined as the ratio of N_e, the net number of electrons, to N, the total number of sites) can be measured directly by the equation

$$\rho = N_e/N = 2d/j\lambda = 2dk_F/\pi, \tag{13}$$

where $j = 1$ for $2k_F$ and $j = 2$ for $4k_F$ scattering.

A similar magnetic ordering in the spin system can transform an ordered uniform paramagnetic chain ($S = \frac{1}{2}$ per site, $\rho = 1$) above a "spin-Peierls" transition temperature T_{SP} into a chain of spin-paired singlet "dimers" ($S = 0$ per dimer) below T_{SP}. One can also have a spin-density–wave (SDW) instability: many $\rho = \frac{1}{2}$ "one-chain" salts are paramagnetic above a SDW ordering temperature T_{SDW}, and become antiferromagnetic with a trapped SDW state below T_{SDW}. SDW instabilities can also have x-ray signatures.

The theory of CDW instabilities has received much attention: it differentiates between the weak-coupling limit ($U \ll t$), the intermediate-coupling limit, and the strong-coupling limit ($U > t$). There are other possible instabilities. Instabilities in a 1-D system, if driven by a strong on-site electron–electron Coulomb repulsion U, lead to a Mott–Hubbard insulator, particularly for $\rho = 1$ systems: here charge localization ensues, and the crystal becomes an insulator. For ρ values with rational fractions ($\rho = \frac{1}{2}, \frac{2}{3}$), a Wigner crystal can occur: the charges alternate regularly in the crystal, so that one site has $\rho = 0$, the next $\rho = 1$, and so on. Finally, an Anderson metal–insulator transition is driven by even a weak random field due to structural disorder, which can then localize the electronic states.

3. CONDUCTIVITY AND SUPERCONDUCTIVITY

The electrical conductivity σ (for electrons or holes) is defined by

$$\sigma = L/RA, \tag{14}$$

where the sample resistance R is measured between parallel planes L apart for a crystal of cross-sectional area A. For metals σ decreases with increasing temperature, as the mean free path of carriers is increasingly limited by scattering from lattice phonons. For semiconductors, σ increases exponentially with increasing temperature, usually following an Arrhenius-type temperature dependence:

$$\sigma = \sigma_0 \exp(-\Delta E/kT); \tag{15}$$

here ΔE is the experimental activation barrier, T is the temperature, and k is Boltzmann's constant. The conductivity is related to the carrier (electron or hole) mobility μ by

$$\sigma = Ne\mu, \tag{16}$$

where e is the electronic charge and N is the carrier concentration.

For superconductors, some of the electrons close to the Fermi surface form "Cooper pairs" of electrons with opposite spins and opposite crystal momenta, thanks to a "critical" coupling of both electrons to certain lattice phonons. This occurs below a critical temperature T_c, which is affected by both applied pressure and applied magnetic field. For external magnetic fields between zero and a critical field H_c, there is a partial or complete flux exclusion within the superconductor (Meissner–Ochsenfeld effect), so that a bulk superconductor is diamagnetic. In addition to the critical parameters T_c and H_c, there is also the critical current j_c above which the superconductor returns to the normal state.

In the Bardeen–Cooper–Schrieffer (BCS) theory (Bardeen *et al.*, 1957a, 1957b) T_c for the phase transition between the "normal" metal state and the lower-temperature superconducting state of Cooper pairs with infinite electrical conductivity is given by

$$T_c = 1.14\theta_D \exp[-1/U_{ep}D(\epsilon_F)], \tag{17}$$

where U_{ep} is the electron–phonon coupling energy, $D(\epsilon_F)$ is the density of states at the Fermi energy ϵ_F, and θ_D is the Debye temperature of the lattice. For elemental superconductors the product $U_{ep}D(\epsilon_F)$ is typically in the range 0.1 to 0.5, and so T_c is one to five orders of magnitude smaller than θ_D (see SUPERCONDUCTIVITY, LOW-TEMPERATURE). One "chemical" issue is how T_c could be raised, if BCS theory were applicable to new classes of superconductors.

4. MULLIKEN CHARGE-TRANSFER COMPLEXES IN SOLUTION

One of the early concerns was the formation of intense coloration in crystals of weak stoichiometric (1:1) complexes [benzene (**1**) (Fig. 1) with iodine (I_2), or naphthalene (**2**) with trinitrobenzene (TNB, **3**)] and other similar systems. This intense color is formed when the components are mixed in solution, or when they are co-crystallized as stoichiometric solid-state complexes. These solutions and crystals showed, almost unchanged, the full optical absorption spectrum of the neutral components, plus an extra broad, intense absorption band with little or no vibrational structure.

The phenomenon was not understood until Mulliken (1950) proposed his charge-transfer theory. He explained that these 1:1 solution complexes have ground-state wave functions $\Psi_{GS}(D,A)$ and excited-state wave functions $\Psi_{ES}(D,A)$ that are linear combinations of the wave functions ϕ_0 of the neutral donor D and the neutral acceptor A, and of the charge-transfer wave functions ϕ_{CT} of the donor cation D^+ and the acceptor anion A^-, as follows:

$$\Psi_{GS}(D,A) = a\phi_0(DA) + b\phi_{CT}(D^+A^-), \tag{18}$$

$$\Psi_{ES}(D,A) = a\phi_{CT}(D^+A^-) - b\phi_0(DA). \tag{19}$$

Typical values for the coefficients are $a \ll b$, e.g., $a = 0.05$, $b = 0.95$. For the benzene–iodine complex D = one-electron donor = benzene, A = one-electron acceptor = I_2. The charge-transfer transition is between the states $\Psi_{GS}(D,A)$ and $\Psi_{ES}(D,A)$, with charge-transfer band

$$h\nu_{CT} = \langle\Psi_{ES}(D,A)|H|\Psi_{ES}(D,A)\rangle - \langle\Psi_{GS}(D,A)|H|\Psi_{GS}(D,A)\rangle, \tag{20}$$

where H is the appropriate Hamiltonian.

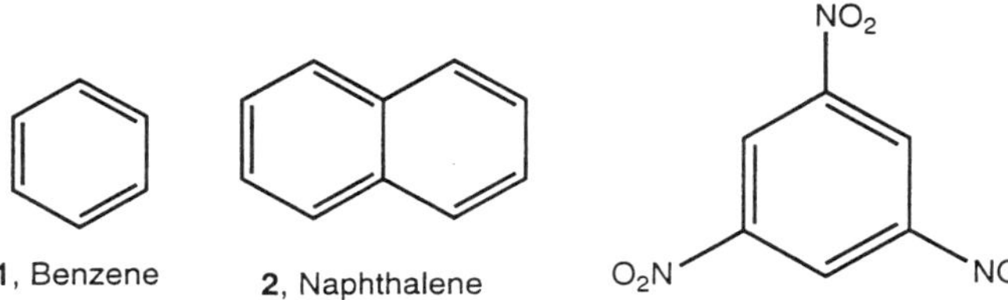

FIG. 1. Molecular structures of the weak one-electron donors benzene (**1**) and naphthalene (**2**), and of the weak one-electron acceptor 1,3,5-trinitrobenzene (**3**).

This transition is strongly allowed, because the D^+A^- state (both molecules at typical van der Waals separations, 3.3–3.8 Å apart) has $S = \frac{1}{2}$ for each ion, but $S = 0$ overall, thanks to the mixing of states in Eqs. (18) and (19). One can quantify the charge transfer ρ ($0 \leq \rho \leq 1$) between adjacent D and A molecules from the amount of mixing of ionic states with the neutral states:

$$\rho = a/(a^2 + b^2)^{1/2}. \tag{21}$$

Nothing in the Mulliken theory prevents a large value of ρ.

The 1:1 complex of benzene (or naphthalene) with trinitrobenzene is called a Mulliken charge-transfer complex, but is also often called a π:π complex, because it is considered to be due to the weak intermolecular interaction of the π electrons of benzene or naphthalene with the π electrons of trinitrobenzene. When discussing molecular orbitals (molecular wave functions), chemists distinguish between σ electrons (which build up electron density along the internuclear axis, and have stronger localization between bonding centers) and π electrons (which have nodes along the internuclear axis, high electron density in the direction perpendicular to the internuclear axis, and greater delocalization between the atoms of the molecule). In planar aromatic molecules, like benzene and naphthalene, there are several π electrons (6 for benzene, 10 for naphthalene) whose wave functions have amplitudes above and below the plane formed by the C and H atoms, and which are considered to be much more mobile than the σ electrons, and almost "separate" from the σ-electron network.

5. CHARGE-TRANSFER CRYSTALS

The arguments used for solution and gas-phase charge-transfer complexes could be extended to a charge-transfer crystal, which Mulliken called an "∞:∞" complex. The first solid charge-transfer crystals were colored, but had low thermodynamic stability, relative to the separate neutral donor and acceptor crystals. However, if the donor D had low gas-phase ionization potential I_D, and if the acceptor had high gas-phase electron affinity A_A—

$$D(g) \rightarrow D^+(g) + e^-, \quad \Delta U = I_D; \tag{22}$$

$$e^- + A(g) \rightarrow A^-(g), \quad \Delta U = -A_A \tag{23}$$

—then the solid complexes tended to show much higher electrical conductivity, and to be paramagnetic. McConnell *et al.* (1965) explained that, if the stacking in the solid state was of planar molecules D atop planar molecules A (later this was called the "mixed-stack" case), then the crystals tended to fall into two distinct classes: the almost ionic case ($\rho \approx 1$), when the crystal Coulomb or Madelung energy E_M was more negative than the difference $I_D - A_A$ was positive, i.e.,

$$E_M + I_D - A_A < 0 \quad \text{(ionic case, } \rho \approx 1\text{)}, \tag{24}$$

or the cost of ionizing the lattice was more than repaid by the ionic binding energy of the crystal; and the almost neutral case ($\rho \approx 0$) when the Madelung energy was not enough to counterbalance the cost of ionizing the lattice:

$$E_M + I_D - A_A > 0 \quad \text{(neutral case, } \rho \approx 0\text{)}. \tag{25}$$

The (almost) ionic complexes would be paramagnetic and have moderate (semi)-conductivity. For "strong" one-electron donors $I_D < 6.5$ eV; for strong one-electron acceptors $A_A > 2.5$ eV. Most strong donors are poor acceptors, and vice versa. A plot of I_D and A_A as a function of number of π electrons converges asymptotically at $I_D = A_A = 4.4$ eV, the value for graphite, which is as good an electron donor as it is an electron acceptor.

In the exact case $\rho = 1$ (one electron per molecule) the electrons (or holes) will be localized, since any transport of charge will put two electrons on the same site, at a huge cost in on-site Coulomb repulsion energy; thus a $\rho = 1$ system is known as a Mott–Hubbard insulator.

The discovery of strong cyanocarbon acceptors TCNE (tetracyanoethylene, **4**) (Fig. 2) by Cairns *et al.* (1957), and TCNQ (7,7,8-8-tetracyanoquinodimethane, **5**) by Acker *et al.* (1960) opened the door to strongly ionic complexes, with interesting magnetic properties. These ion-radical salts ("one-chain" salts) were based either on the the strong

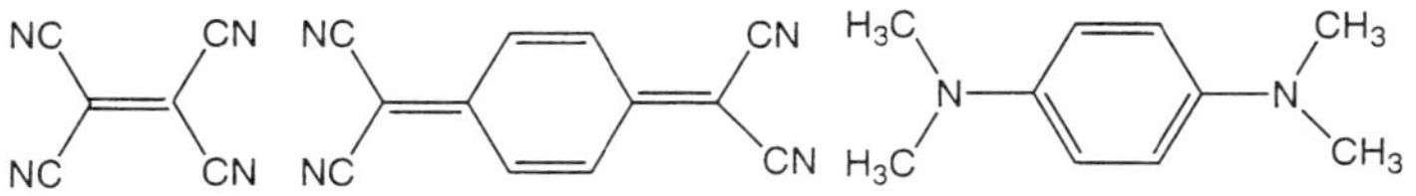

FIG. 2. Molecular structures of the strong one-electron acceptors TCNE (**4**) and TCNQ (**5**), and of the strong one-electron donor TMPD (**6**).

electron acceptor TCNQ, or on the strong electron donor TMPD (*N,N,N′,N′*-tetramethyl-para-phenylenediamine, **6**). The perchlorate salt of TMPD, known as Wurster's blue perchlorate ($TMPD^+ClO_4^-$), led to an understanding of the paramagnetism of linear-chain systems (Thomas *et al.*, 1963), which, at a spin-Peierls ordering temperature T_{SP}, undergo a spin-Peierls transition to a diamagnetic low-temperature ground state.

6. OPTICS OF POLYENES

Kuhn (1949) analyzed the lowest-energy optical absorption band of long linear polyenes, or oligomers of conjugated linear polymers, by free-electron molecular-orbital theory (a slight modification of the particle-in-a-box problem). E_{max}, the energy of maximum absorbance for the lowest optical absorbance band, can be related to n_π, the number of π electrons, to L, the length of the linear "box," and to L_O, the "effective length" of the π-electron chain in the oligomer, defined so that $L = n_\pi L_O$:

$$E_{max} \approx h^2 n_\pi / 8mL^2 = h^2 / 8mL_O^2 n_\pi. \quad (26)$$

Thus a plot of E_{max} versus $1/n_\pi$ is linear for oligomeric substrands of known conducting polymers, and should extrapolate to $E_{max} = 0$ for the perfectly degenerate and conjugated infinite linear polymer, which will be metallic. For instance, $E_{max} = 0$ for graphite and for $(SN)_x$.

7. PROPOSED EXCITONIC SUPERCONDUCTIVITY

Little (1964) made a controversial but very stimulating proposal. If the electron–phonon coupling U_{ep} in Eq. (17) is replaced by the much larger electron–electron coupling matrix element V_{ee}, then T_c should become two orders of magnitude larger, and superconductivity at room temperature becomes plausible. This proposal for "excitonic superconductivity" became a mantra for the hopes for superconductivity at higher T_c in organic systems. No systems to date have been proven to implement the mechanism implied in Little's proposal.

Having discussed the general physics of the subject, we now present brief overviews of inorganic superconductors (the "competition" for the organic case) and of organic quasi one-dimensional metals.

8. INORGANIC SUPERCONDUCTORS

Elemental metallic superconductors have a maximum $T_c = 9.2$ K for Nb, and systems of two metals, such as the A15 crystal Nb_3Ge, reached $T_c = 23.2$ K (Gavaler, 1973). A search for novel superconductors first concentrated on the highly conducting inorganic "Krogmann salts," such as $K_3Pt(CN)_4Br_{0.3}$ $2.3H_2O$, which are stacks of cyanoplatinate ions with Pt ··· Pt nonbonded distances of 2.77 Å (same as in bulk metallic Pt) and "quasi one-dimensional" metallic conduction along the Pt chains (Krogmann, 1969). The CDW states were first found experimentally in the Krogmann salts, but no superconductors were found among them.

As mentioned above, $T_c = 0.55$ K was found in the stage-1 layered graphitic compound KC_8, and superconductivity with $T_c = 0.3$ K was also found for the crystalline polymer polythiazyl, $(SN)_x$.

In the 1980s ceramic perovskite defect oxide superconductors drove T_c very high indeed: 25–40 K in the "214" system $La_{2-x}Sr_xCuO_{4-y}$ (Bednorz and Müller, 1986), 93 K in the "123" system $YBa_2Cu_3O_{7-x}$ (Wu *et al.*, 1987), 110 K in the Bi-1223 system (Maeda *et al.*, 1988), 125 K in the Tl-2134 system (Parkin *et al.*, 1988), and 150 K under pressure in the Hg-1223 system (Chu *et al.*, 1993). Applications are at present limited by relatively low values of the critical current j_c, but intense work is under way to overcome this limitation by better processing and increase of contact between superconducting grains.

The discovery of buckminsterfullerene C_{60} (Kroto *et al.*, 1985) and its practical synthesis (Krätschmer *et al.*, 1990) were followed by the finding that alkali-metal salts of buckminsterfullerenes A_3C_{60} (A = K, Rb, Cs) superconduct from T_c = 18.6 to 40 K (Hebard *et al.*, 1991; Palstra *et al.*, 1995), as does one alkaline-earth salt. These compounds are, however, extremely susceptible to water vapor.

9. TTF-TCNQ AND OTHER QUASI ONE-DIMENSIONAL METALS

After the synthesis of tetrathiafulvalene (TTF, **7a**; see Fig. 3) (Wudl *et al.*, 1970), the first "organic metal," TTF TCNQ, was characterized by Cowan and co-workers (Ferraris *et al.*, 1973). From erroneous measurements in some samples of TTF TCNQ Heeger and co-workers claimed "superconducting fluctuations" at 58 K (Coleman *et al.*, 1973). This stimulated a veritable explosion of interest in organic ionic solids; careful measurement showed that there is no superconductivity in TTF TCNQ (Thomas *et al.*, 1976). TTF TCNQ belongs to a new class of metallic conductors, the quasi one-dimensional metals. These molecular conductors have decreasing electrical conductivity at increasing temperature, and large directional anisotropy in this conductivity: along the molecular and ionic stacks, where there is reasonable intermolecular (interionic) π–π overlap, the conductivity is typically 10 to 1000 times larger than in the transverse directions. These are almost one-dimensional systems, which became real laboratory examples instead of mere theoretical curiosities. Many variants on TTF were studied, and an understanding of what it takes to make organic metals, and what may be needed for organic superconductivity, evolved gradually (Cowan, 1989; Cowan *et al.*, 1990).

Figure 4 shows the room-temperature crystal structure of TTF TCNQ (Kistenmacher *et al.*, 1974), and Fig. 5 shows its conductivity as a function of temperature. At high temperature TTF TCNQ is metallic, with $\sigma(T) \propto T^{-2.3}$; since it has a reasonably high coefficient of thermal expansion, as do most molecular crystals, therefore a more meaningful quantity to consider is the conductivity at constant volume, for which $\sigma_V(T) \propto T^{-1.29}$: This means that one-phonon scattering processes are dominant (Conwell, 1980). It has been determined that a CDW

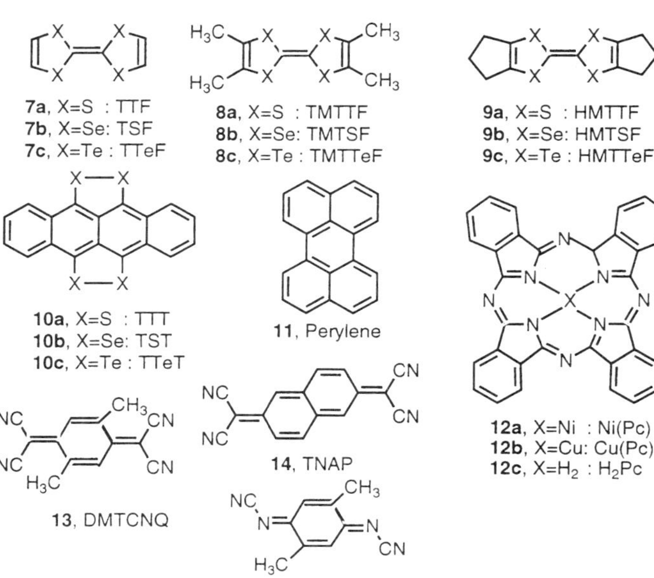

FIG. 3. Molecular structures of TTF (**7a**) and other chalcogen-containing electron donors (**7b–10c**), of perylene (**11**) and phthalocyanines (**12**), and of three electron acceptors (**13–15**) similar to TCNQ.

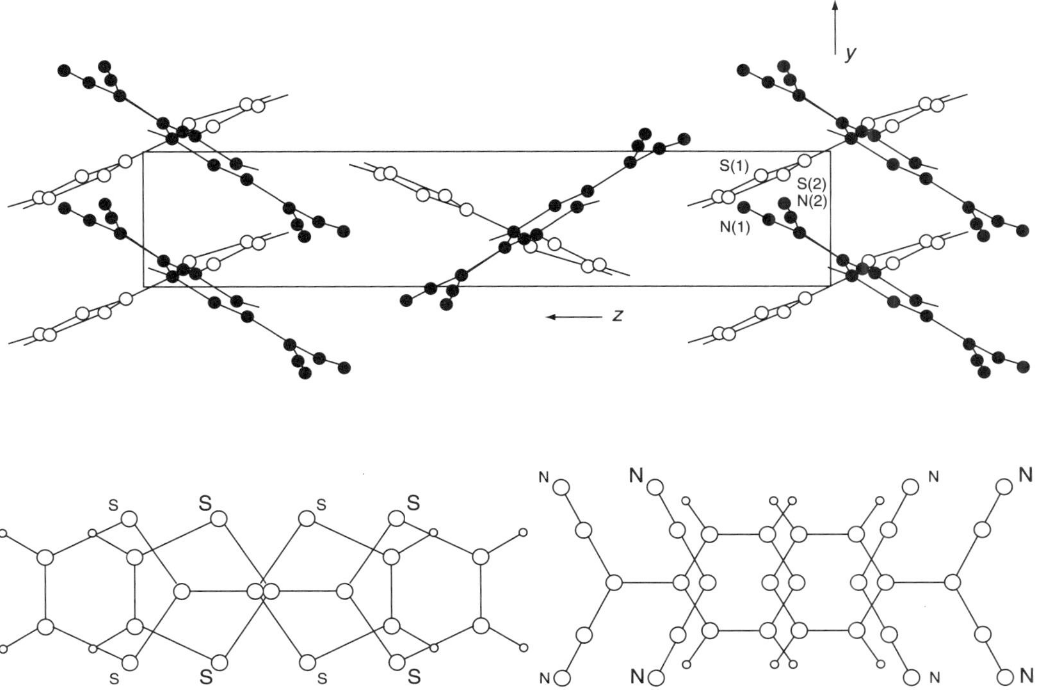

FIG. 4. Room-temperature crystal structure of TTF TCNQ (Kistenmacher *et al.*, 1974). The upper diagram is a projection along [100], with the unit cell axes *a* vertical and *c* horizontal, which shows the stacks of TTF (open circles for the atom positions) and TCNQ (filled circles for the atom positions). The lower diagram shows the intermolecular overlap, projected normal to the least-squares molecular planes, along the TTF stack (left), and the TCNQ stack (right).

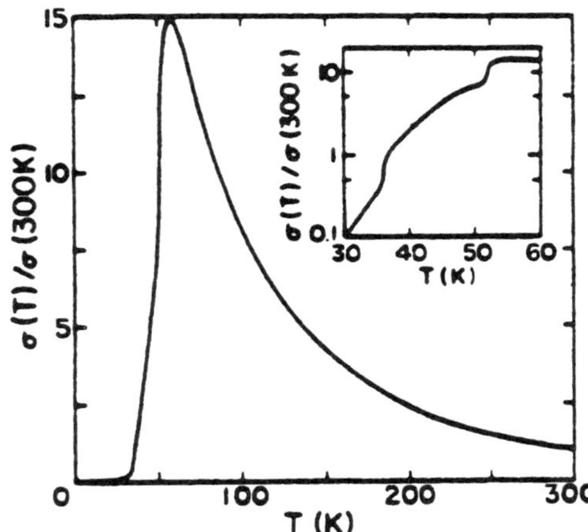

FIG. 5. Temperature dependence of the conductivity σ of TTF TCNQ (Ferraris *et al.*, 1973).

starts on the TCNQ stacks at about 160 K; at 54 K a coupling occurs between CDWs on different TCNQ chains; at 49 K a CDW starts on the TTF stacks, and by 38 K a complete Peierls state is observed. The TTF molecules exhibit a slip of only about 0.034 Å along their long molecular axis, relative to the ordered structure (Coppens *et al.*, 1987); thus the Peierls CDW state involves a very subtle structural distortion.

Table 1 summarizes (Cowan *et al.*, 1990) the electrical conductivity of several charge-transfer crystals, most of which have a definite Peierls transition to a semiconducting state below a temperature T_{max}. In several cases, however, there is instead a very broad maximum in the plot of conductivity versus temperature, and the compound retains its high conductivity to the lowest temperatures measured, without going superconducting: one salt, $Cu(2,5\text{-}DM\text{-}DCNQI)_2$, even reaches at 0.5 K a conductivity of about 5×10^5 S cm^{-1}. Figure 6 shows the temperature de-

Table 1. Selected organic metals (Cowan *et al.*, 1990). None of these compounds superconduct. The temperature of maximum conductivity is in parentheses if it is a broad conductivity maximum. The molecular structures are given in Figs. 2 (molecule **5**) and 3 (molecules **7–15**).

Compound	Molecular structure *D*	*A*	Charge transfer (ρ)	Room-temp. conductivity (σ_{RT}/S cm^{-1})	Maximum conductivity (σ_{max}/S cm^{-1})	Temp. of max. cond. (T_{max}/K)
TTF TCNQ	**7a**	**5**	0.59	500	2×10^4	59
TMTTF TCNQ	**8a**	**5**	0.65	350	5×10^3	60
HMTTF TCNQ	**9c**	**5**	0.72	500	2×10^3	75
TSF TCNQ	**7b**	**5**	0.63	800	1×10^4	40
TMTSF TCNQ	**8b**	**5**	0.57	1200	7×10^3	61
TMTSF DMTCNQ	**8b**	**13**	0.50	500	5×10^3	42
HMTSF TCNQ	**9b**	**5**	0.74	2000	7×10^3	(32)
TTeF TCNQ	**7c**	**5**	0.71	1800	2.5×10^4	<4
HMTTeF TCNQ	**9c**	**5**	n.a.	550	9×10^2	(73)
HMTTeF DMTCNQ	**9c**	**13**	n.a.	460	1×10^3	(83)
HMTSF TNAP	**9b**	**14**	n.a.	2900	2×10^4	50
$(TTT)_2I_3$	**10a**		0.50	1000	3×10^3	(60)
$(TST)_2I$	**10b**		0.50	3900	1×10^4	(35)
$(TST)_2Cl$	**10b**		0.50	2100	2×10^4	(26)
$(Perylene)_2(PF_6)_{1.1}$	**11**		0.55	900	1×10^3	200
Ni(Pc)I	**12a**		0.33	550	5×10^3	(25)
Cu(Pc)I	**12b**		0.33	900	7×10^3	$\ll$(30)
H_2(Pc)I	**12c**		0.33	750	4×10^3	15
Cu(2,5-DM-DCNQI)$_2$	**15**		0.50	800	5×10^5	(3.5)

pendence of the resistivity of four TCNQ salts; HMTSF TCNQ does not undergo a Peierls transition.

10. $(TMTSF)_2X$ OR BECHGAARD SALTS: THE FIRST ORGANIC SUPERCONDUCTORS

By using the donor TMTSF (tetramethyltetraselenafulvalene, **8b**) and electro-crystallization, Bechgaard, Jérome, and co-workers found that $(TMTSF)_2PF_6$ was the first organic superconductor, with $T_c = 0.9$ K at an applied pressure of 10 kbar (Jérome *et al.*, 1980). Here the lateral two-dimensional Se–Se interactions and the applied pressure together defeat the Peierls transition. The first ambient-pressure superconductor, $(TMTSF)_2ClO_4$ ($T_c = 1.4$ K), followed quickly; Figure 7 shows its crystal structure.

Table 2 summarizes all the known organic superconductors, plus the stage-1 intercalated graphite KC_8, the polymeric $(SN)_x$ discussed above, and the fullerene superconductors. Additional relevant molecular structures of donors and acceptors are shown in Fig. 8.

Eight superconducting "Bechgaard" salts $(TMTSF)_2X$ (X = inorganic anions) were found, but almost all have $T_c \approx 1$ K; the highest T_c is 2.1 K for $X = FSO_3^-$. For the Bechgaard salts, an ordering in the tetrahedral anions (ReO_4^-, ClO_4^- under rapid cooling past 24 K) or in the octahedral anions (PF_6^-) can occur; this doubles the unit cell size, opens up a small band gap, and causes antiferromagnetic ordering of the spins in the TMTSF species to a new collective spin-density wave (SDW) state. This ordering makes these salts insulating, not superconducting. In many cases, hydrostatic pressure can prevent the anions from ordering, and a smooth transition from the metallic state to the superconducting state can occur. Figure 9 shows what is understood about the temperature–pressure phase diagram of several TMTSF salts.

11. $(BEDT\text{-}TTF)_2X$ AND OTHER ORGANIC SUPERCONDUCTORS

Cava and coworkers prepared the "all-sulfur" donor BEDT-TTF (bis-ethylenedithiolene-tetrathiafulvalene, or ET for short, **16a**) (Mizuno *et al.*, 1978). Saito and co-workers found that in some $(BEDT\text{-}TTF)_2X$ salts the usual Peierls-driven metal-to-insula-

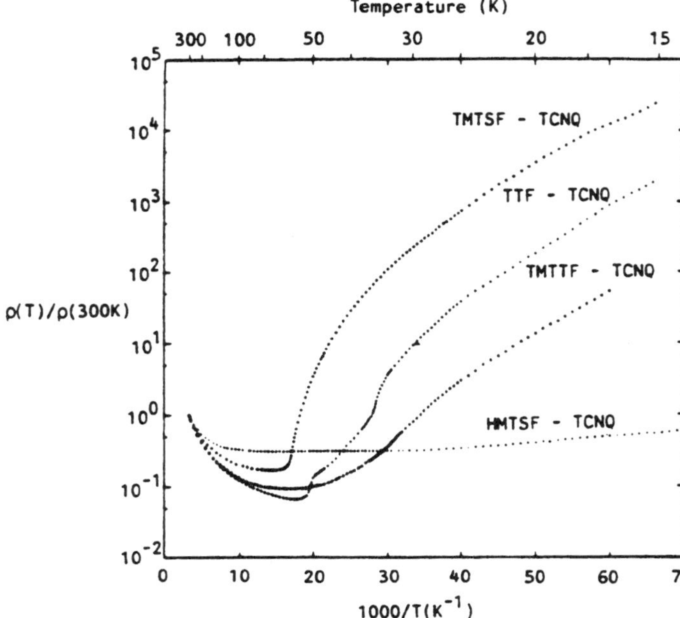

FIG. 6. Temperature dependence of the electrical resistivity ($1/\sigma$) of four TCNQ salts (Bloch *et al.*, 1977).

tor phase transition was suppressed (Saito *et al.*, 1982); then Parkin and co-workers found in $(ET)_2ReO_4$ the first sulfur-based superconductor (Parkin *et al.*, 1983). These salts with formula $(ET)_2X$ (X = inorganic anions) crystallize in a wide variety of crystallographically distinct phases, sometimes with several phases growing from the same solution (Williams *et al.*, 1992). This polymorphism, rare among organic systems, is baffling; for some anions X, certain phases are insulating, while others are highly conducting, and some are superconducting. There seem to be several closely ranked cohesive-energy minima that are available to the growing microcrystal.

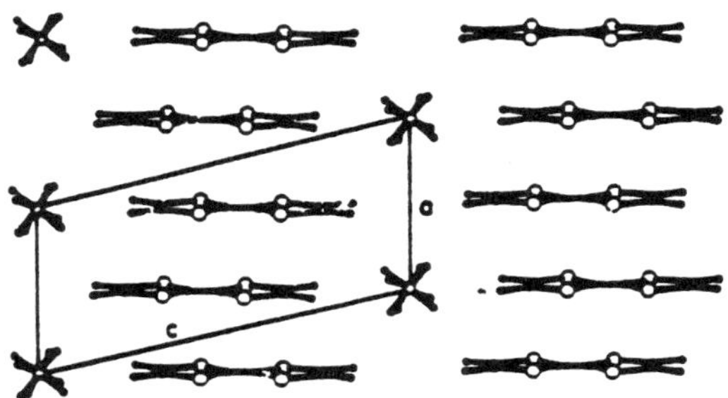

FIG. 7. Crystal structure of $(TMTSF)_2ClO_6$ (Thorup *et al.*, 1981), projected along [010]. The unit cell is shown; the TMTSF stacking is vertical; the stacks are separated by perchlorate anions, which are at the corners of the unit cell.

There are 20 organic superconductors of the ET_2X family; the present record critical temperature is T_c = 12.8 K at 0.3 kbar for κ-(BEDT-TTF)$_2$Cu[N(CN)$_2$]Cl; close ET cousins are one superconducting salt of BEDT-TSF (**16b**), and one superconducting salt of BEDO-TTF (**16c**). Other organic "hole" superconductors are salts of the electron donors MDT-TTF (**17**) (one superconductor), of EDT-TTF (**18**), of MET (**19**), of DMET (**20**) (seven superconducting salts, maximum T_c = 1.9 K), and of TMTTF (**8a**) (one superconductor). Electron superconductors are formed with the electron acceptor anions Ni(dmit)$_2$ (**21a**) (three superconductors, maximum T_c = 5.0 K at 7 kbar) and Pd(dmit)$_2$ (**21b**). Ten alkali-metal salts of C_{60} and one calcium salt of C_{60} are superconductors, but with much higher critical temperatures.

So far, all TMTSF and ET-type superconductors have true mixed valence and $\rho = \frac{1}{2}$ [except $\rho = \frac{2}{3}$ for (BEDT-TTF)$_2$Cl$_3\cdot$2H$_2$O]. The TMTSF salts, and some ET salts, are quasi one-dimensional, but with enough warping of the Fermi surface to make them pseudo 2-D, and to defeat the Peierls transition. The κ-phase (BEDT-TTF) salts are really 2-D systems; this phase, shown in Fig.

Table 2. Molecular structure of donor or acceptor, and superconducting critical temperatures T_c (/kelvin, at 1 bar) or T_c (/kelvin, at applied pressure P/kbar) for organic, intercalated graphite, $(SN)_x$, and fullerene superconductors (updated from Williams *et al.*, 1992). The molecular structures are given in Figs. 3 (molecules **7–8**) and 8 (molecules **16–22**).

Compound	Molecular structure	T_c/K @ P/kbar	Compound	Molecular structure	T_c/K @ P/kbar
KC_8 (stage-1 graphite)		0.55	$(ET)_4Hg_{3-x}Cl_8$	**16a**	1.8 @ 12
$(SN)_x$		0.3	$(ET)_4Hg_{2.89}Br_8$	**16a**	4.3
$(TMTSF)_2PF_6$	**8b**	0.9 @ 10	$(BEDO\text{-}TTF)_3Cu(NCS)_3$	**16c**	1.1
$(TMTSF)_2ClO_4$	**8b**	1.2	λ-$(BEDT\text{-}TSF)_2GaCl_4$	**16b**	8.0
$(TMTSF)_2AsF_6$	**8b**	1.1 @ 12	$(MDT\text{-}TTF)_2AuI_2$	**17**	4.5
$(TMTSF)_2SbF_6$	**8b**	0.4 @ 11	$(DMET)_2I_3$	**20**	0.47
$(TMTSF)_2TaF_6$	**8b**	1.4 @ 12	$(DMET)_2IBr_2$	**20**	0.59
$(TMTSF)_2ReO_4$	**8b**	1.3 @ 9.5	$(DMET)_2AuCl_2$	**20**	0.83
$(TMTSF)_2FSO_3$	**8b**	2.1 @ 6.5	$(DMET)_2AuBr_2$	**20**	1 @ 1.5
$(TMTTF)_2Br$	**8a**	0.8 @ 26	κ-$(DMET)_2AuBr_2$	**20**	1.9
$(ET)_2ReO_4$	**16a**	2.0 @ 4.5	$(DMET)_2AuI_2$	**20**	0.55 @ 5
α_t-$(ET)_2I_3$	**16a**	7–8	$(DMET)_2Au(CN)_2$	**20**	0.8 @ 5
$\alpha\beta$-$(ET)_2I_3$	**16a**	2.5–6.9	$(TTF)[Ni(dmit)_2]_2$	**7a,21a**	5.0 @ 7
β-$(ET)_2I_3$	**16a**	1.4	$[(CH_3)_4N][Ni(dmit)_2]_2$	**21a**	3.0 @ 3.2
β^*-$(ET)_2I_3$	**16a**	8.0 @ 0.5	α-$(TTF)[Pd(dmit)_2]_2$	**21b**	1.7 @ 22
γ-$(ET)_3(I_3)_{2.5}$	**16a**	2.5	α'-$(TTF)[Pd(dmit)_2]_2$	**21b**	6.42 @ 20.7
θ-$(ET)_2(I_3)_{0.98}(AuI_2)_{0.02}$	**16a**	3.6	α-$(EDT\text{-}TTF)[Ni(dmit)_2]$	**18,32a**	1.3
κ-$(ET)_2(I_3)$	**16a**	3.6	Na_2CsC_{60}	**22**	10.5
β-$(ET)_2IBr_2$	**16a**	2.8	K_3C_{60}	**22**	19.6
β-$(ET)_2AuI_2$	**16a**	4.98	K_2RbC_{60}	**22**	22.5
$(ET)_3Cl_2 \cdot 2H_2O$	**16a**	2 @ 16	$K_1Rb_2C_{60}$	**22**	28.0
α-$(ET)_2NH_4Hg(SCN)_4$	**16a**	1.15	$K_{1.5}Rb_{1.5}C_{60}$	**22**	25.1
$\beta(ET)_{1.96}(MET)_{0.04}I_3$	**16a,19**	4.6	Rb_3C_{60}	**22**	29.8
κ-$(ET)_2Cu(NCS)_2$	**16a**	10.4	Rb_2CsC_{60}	**22**	31.3
κ-$(ET)_2Cu(N(CN)_2)Br$	**16a**	11.6	$Rb_1Cs_2C_{60}$	**22**	33
κ-$(ET)_2Cu(N(CN)_2)Cl$	**16a**	12.8 @ 0.3	Cs_3C_{60}	**22**	29.5
κ-$(ET)_2Ag(CN)_2 \cdot H_2O$	**16a**	5.0	Cs_3C_{60}	**22**	40 @ 10
κ_L-$(ET)_2Cu(CF_3)_4 \cdot TCE$	**16A**	4.0	Ca_5C_{60}	**22**	8.4

10, has isolated dimers connected by dispersion interactions to form rough two-dimensional sheets. Figure 11 shows the temperature dependence of the conductivity of κ-$(BEDT\text{-}TTF)_2Cu(NCS)_2$.

Alas, structural correlations of T_c with chemical formula or structure type are limited. The β-$(ET)_2X$ salts with linear anions have a linear dependence of T_c with anion length (but this correlation fails for very long

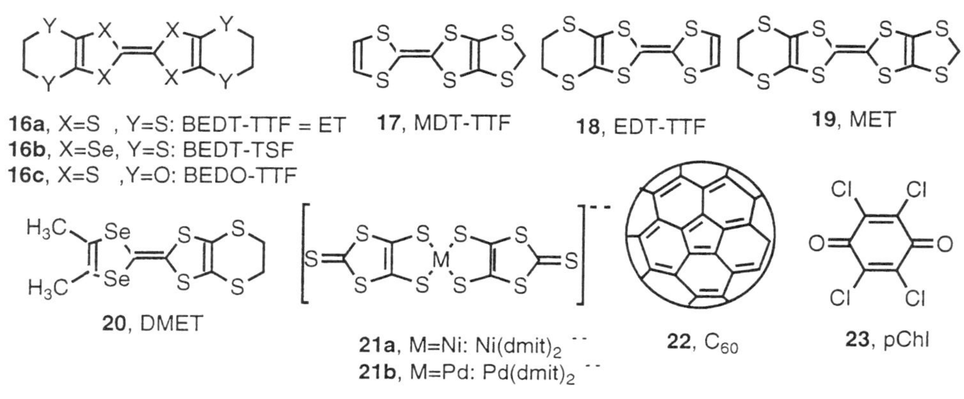

16a, X=S , Y=S: BEDT-TTF = ET
16b, X=Se, Y=S: BEDT-TSF
16c, X=S ,Y=O: BEDO-TTF
17, MDT-TTF
18, EDT-TTF
19, MET
20, DMET
21a, M=Ni: $Ni(dmit)_2^{--}$
21b, M=Pd: $Pd(dmit)_2^{--}$
22, C_{60}
23, pChl

FIG. 8. Molecular structures of more donors (**16–20**) and acceptors (**21–23**).

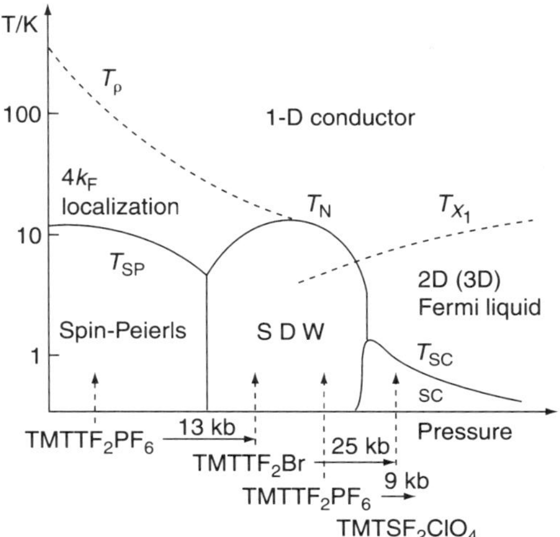

FIG. 9. Temperature-pressure phase diagram of TMTSF salts (Jérome, 1990).

anions, as other phases are formed). The $(TMTSF)_2X$ salts with tetrahedral anions X show a linear dependence of the Peierls metal-to-insulator phase transition temperature with tetrahedral anion radius.

All entries in Table 2 are "hole" superconductors, except for the metal $(dmit)_2$ salts, where the carriers are electrons. The room-temperature conductivities of these compounds are usually about one or two orders of magnitude smaller than those shown in Table 1.

Included in Table 2 are the superconducting (but very air-sensitive) alkali-metal and alkaline-earth fullerides: these are compounds with three-dimensional superconductivity, where the alkali-metal ions are just counterions tucked into tetrahedral and octahedral holes in the cubic fullerene crystal structure (Fig. 12). One can idly debate whether C_{60} is an organic molecule, or just a molecular allotrope of elemental carbon: in fact it is the largest molecule ever "synthesized" by astrophysicists. The critical temperatures of ET salts seem to be stuck at about 13 K, while the fullerene salts have reached 40 K, and the ceramic superconductors are far ahead, at 150 K. For amusement, one can plot the critical temperature T_c as a function of year of discovery (Fig. 13). Such plots do not predict scientific success, but stimulate our hopes.

The coherence length in the organic superconductors is of the order of a lattice constant (as it is for the ceramic-oxide superconductors). The organic superconductors are of type II (there are two critical fields). The "dimensionality" of the (super)conductivity is between 1 and 2 for the organic

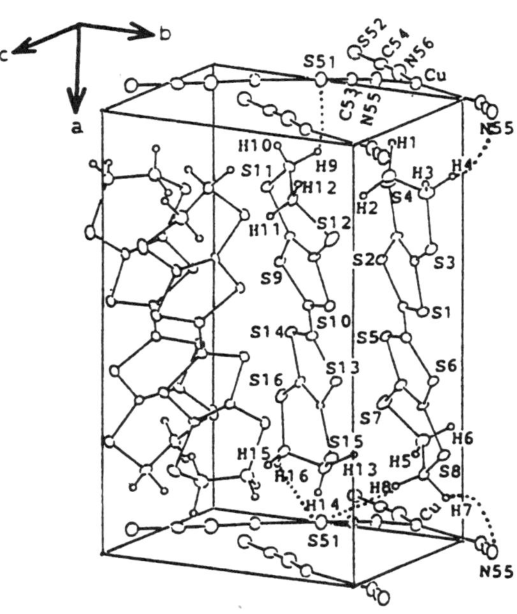

FIG. 10. Crystal structure of κ-$(BEDT\text{-}TTF)_2Cu(NCS)_2$ at 104 K (Oshima *et al.*, 1988).

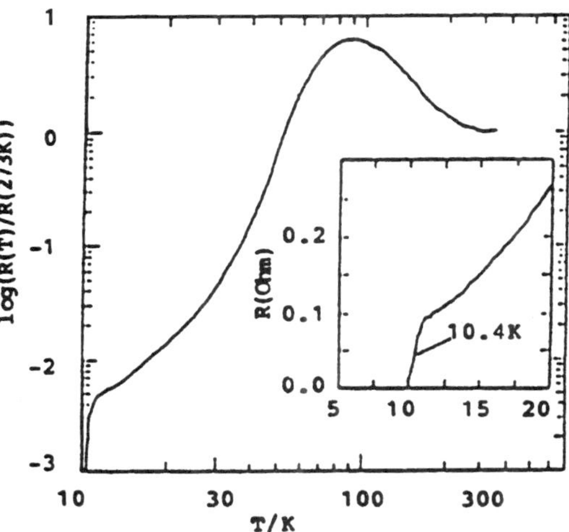

FIG. 11. Temperature dependence of the resistivity ($1/\sigma$) of κ-$(BEDT\text{-}TTF)_2Cu(NCS)_2$ (Saito, 1990).

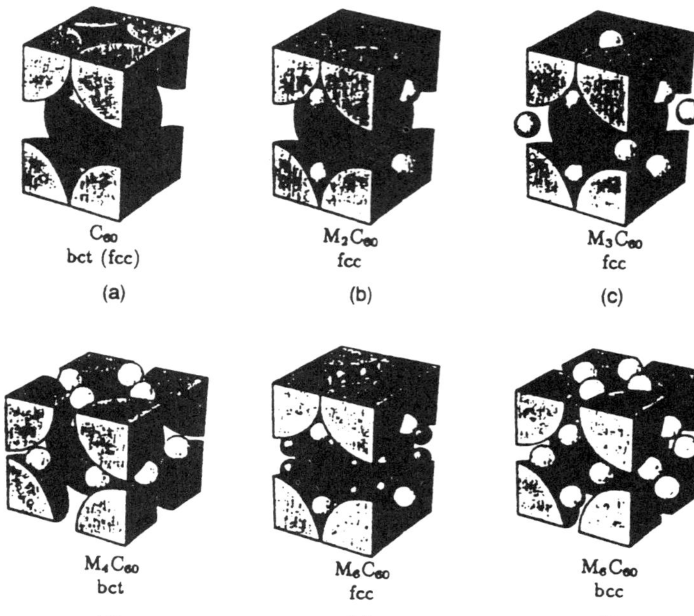

FIG. 12. Crystal structures of alkali fullerides (Dresselhaus *et al.*, 1993).

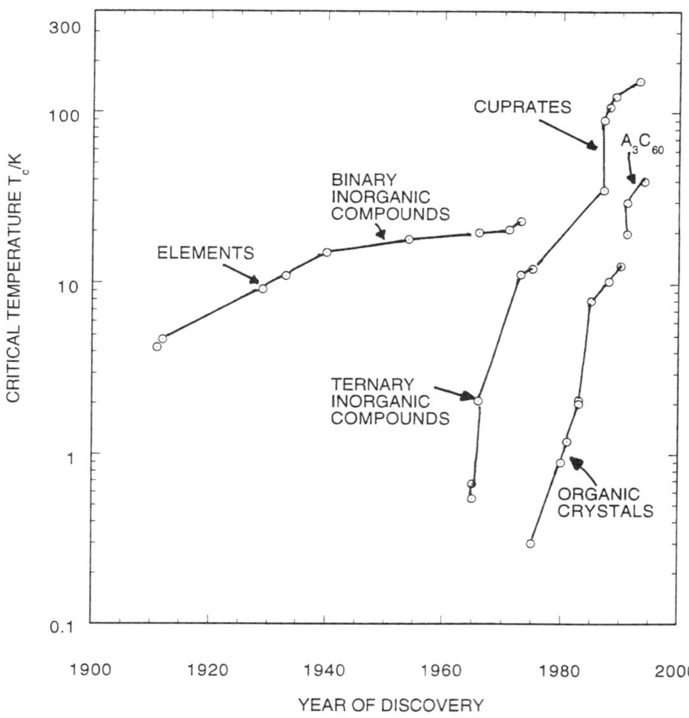

FIG. 13. Logarithm of the critical temperatures T_c of superconductors, plotted against the year of their discovery (adapted from Day, 1990).

superconductors (it is 3 for the fullerides). For organic superconductors, the isotope-effect results are unclear: It is not certain which phonon modes are important for the superconductivity.

For the ceramic-oxide superconductors the mechanism of superconductivity is difficult to explain by standard BCS theory. For the organic superconductors, the superconductivity mechanism is probably of the BCS type.

12. CONDUCTING POLYMERS

After the short-lived interest in polythiazyl, $(SN)_x$ (**24**), and its superconductivity, organic conducting polymers were found when the Shirakawa synthesis (1971) of *cis*- and *trans*-polyacetylene, $(CH)_x$ (**25**), first yielded insulating films of $(CH)_x$, and then Shirakawa and co-workers (1977) doped polyacetylene with electron donors (e.g., sodium) and acceptors (e.g., iodine) to form the first high-conductivity organic polymer. This was followed (see Fig. 14; for a review, see Skotheim, 1986) by the discovery of other *n*-doped or *p*-doped conducting polymers, such as poly-paraphenylene, $(C_6H_4)_x$ (**26**), poly-(paraphenylene vinylene), $(C_8H_6)_x$ (**27**), and the air-stable polymers, such as polypyrrole, $(C_4H_3N)_x$ (**28**), and polythiophene, $(C_4H_2S)_x$ (**29**), and the rediscovery of the properties of emeraldine salt, $(C_{24}H_{22}N_4)_x$, the conducting form of polyaniline (**30**).

The discovery of organic conducting polymers such as polyacetylene raises the question of how long a conducting strand can be, and when the strand can be conducting. When first formed, polyacetylene is in the *cis* form; this can be converted to the all-*trans* form by heating, or by doping with electron donor or acceptor counterions (sodium cation, iodide anion, etc.). Although the band gap in *trans*-$(CH)_x$ is large (4.7 eV), the existence of midgap infrared absorption led to the suggestion that the relevant excitations were neutral solitons (**33**; see Fig. 15) or po-

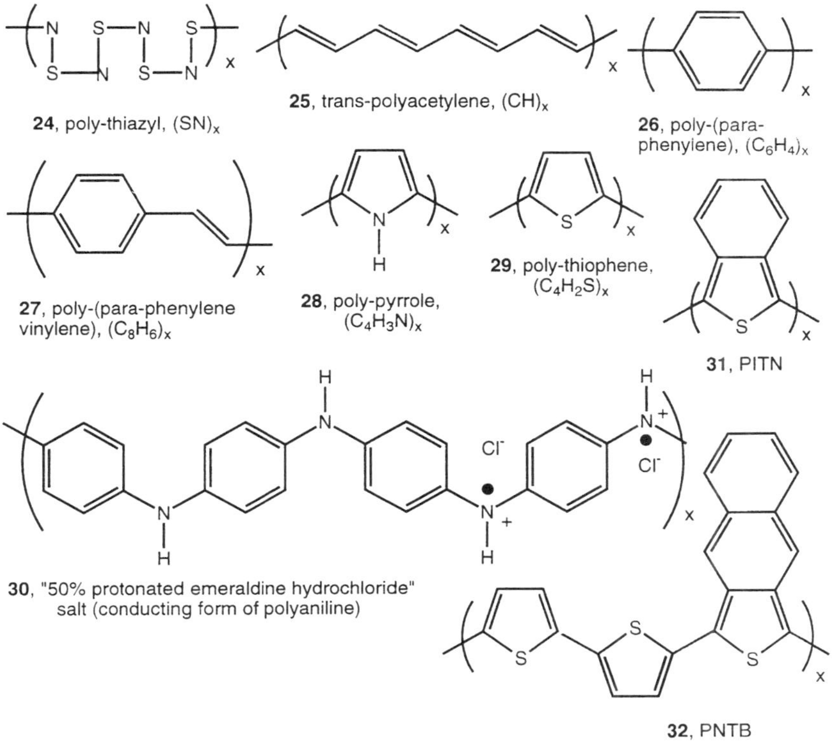

24, poly-thiazyl, $(SN)_x$

25, trans-polyacetylene, $(CH)_x$

26, poly-(para-phenylene), $(C_6H_4)_x$

27, poly-(para-phenylene vinylene), $(C_8H_6)_x$

28, poly-pyrrole, $(C_4H_3N)_x$

29, poly-thiophene, $(C_4H_2S)_x$

31, PITN

30, "50% protonated emeraldine hydrochloride" salt (conducting form of polyaniline)

32, PNTB

FIG. 14. Molecular structure of conducting polymers (**24–32**).

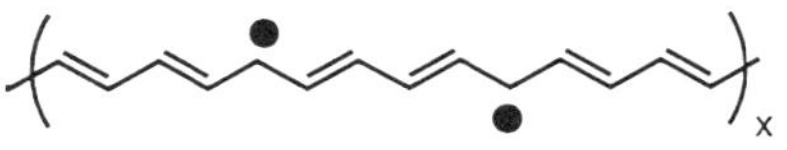

33, Neutral soliton-anti-soliton pair on $(CH)_x$

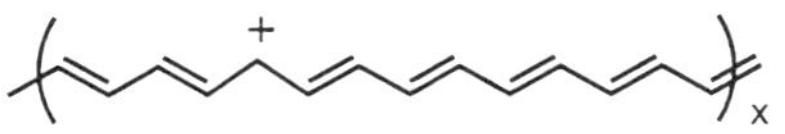

34, Positive polaron on I-doped $(CH)_x$

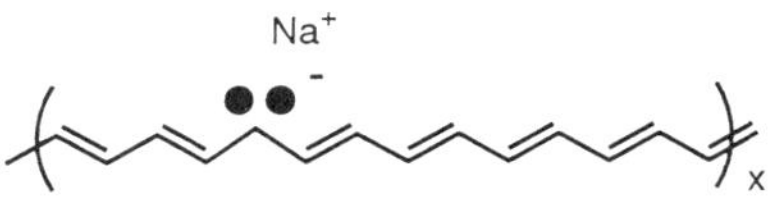

35, Negative polaron on Na-doped $(CH)_x$

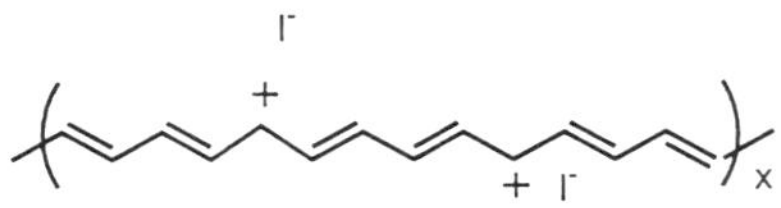

36, Positive bipolaron on heavily I-doped $(CH)_x$

FIG. 15. Depiction of solitons (**33**), polarons (**34–35**), and bipolarons (**36**) in doped and undoped polyacetylene.

larons (**34**, **35**). These solitons are breaks in the conjugation, which occur in pairs (soliton–antisoliton pairs), which can be treated theoretically, and whose experimental existence was proven.

For most (semi)conducting polymers, the "effective conjugation length" L [Eq. (26)] can be estimated from the optical absorption spectrum. For all known conducting polymers, except for graphite and for $(SN)_x$, there is a finite optical gap E_{max}, with contributions from both the σ skeleton and the π electrons. The optical gap is usually close to the band gap: It is assumed that the charge carriers (electrons or holes) have higher mobility in the excited state than in the ground state. The undoped-polymer band gaps range from E_{max} = 4.7 eV for polyacetylene, to 3.8 eV for the emeraldine base form of polyaniline, 3.6 eV for poly-paraphenylene, 3.0 eV for polypyrrole, and 2.58 eV for polythiophene. The single- to double-bond length alternation, which indicates a nonzero E_{max}, is conceptually related to a Peierls gap, and is of the order of 0.08–0.10 Å in polyacetylene.

The band gaps E_{max} are lowered dramatically after doping with electron donors (sodium, potassium, etc.) or electron acceptors (iodine, ferric chloride, arsenic pentafluoride, nitrosonium hexafluorophosphate, etc.); most conducting polymers are dark or black in their conducting state.

For conducting polymers, the monomer site M must accommodate at least one electron (M^-) or hole (M^+), thus forming a mobile polaron (**34**,**35**). A second electron or hole can easily go onto a neighboring site, thus allowing for a new polaron state. If this second electron or hole goes onto the same monomer site, forming M^{++} or M^{--}, then a bipolaron is formed (**36**), which tends to localize the electron transfer and freeze the transport. In conducting polymers, however, the mobility of the polaron is limited by the need to accommodate the movement of the counterion, which is typically a small inorganic ion loosely associated with the polaron site, which must move with the moving charge. This is why conducting polymers have, typically, low mobilities.

Most conducting polymers have room-temperature conductivities σ that vary from 10^{-3} to 10^3 S cm^{-1}; the values depend strongly on the details of the processing. For extremely uniform polyacetylene $\sigma = 5 \times 10^5$ cm^{-1} has been achieved (Naarmann and Theophilou, 1987). For a stereoregular alkylated polythiophene conductivities up to 5.5 $\times$ 10^3 S cm^{-1} have been reported (McCullough and Williams, 1993).

The conductivity is almost always thermally activated—i.e., semiconducting, not metallic. It is thought that the activation energy ΔE [Eq. (15)] is mostly due to interchain hopping barriers. The early conducting polymers were all insoluble, and so molecular-weight determinations and processability were not feasible. With the introduction of solubilizing side chains, the solubility and processability increased dramatically, but at the price of the conductivity, which was presumably lowered by the increased interchain distance. The molecular weights of typical soluble polymers range between 10^3 and 10^4 g mol^{-1}.

13. CONDUCTING LANGMUIR–BLODGETT FILMS

Irving Langmuir and Kathleen Blodgett showed in the 1930s that one can transfer monomolecular layers of amphiphilic molecules (such as cadmium arachidate) from the air–water interface onto solid supports, molecular layer by molecular layer. These LB films are thus very nice, thin, two-dimensional crystalline materials (Roberts, 1990). Large anisotropic conductivities in LB films were demonstrated when Barraud and coworkers showed that a multilayer film of *N*-docosylpyridium TCNQ, deposited on an inert substrate, when doped with iodine vapor, reorganized and exhibited in-plane conductivities of 0.1 S cm^{-1}, and out-of plane conductivities 10 orders of magnitude smaller (Ruaudel-Teixier *et al.*, 1985). Most conducting LB films have activated conductivity, but some are metallic. Recently, Metzger and co-workers showed that K-doped 50-layer LB films of C_{60} superconduct below about 8 K, about 10 K lower than the bulk crystal (Wang *et al.*, 1993).

14. DESIGN RULES FOR CRYSTALLINE ORGANIC METALS AND SUPERCONDUCTORS

Having presented a summary of the recent advances in several areas, we attempt to codify design rules for newer and better conductors and superconductors.

Several semiempirical rules have developed (Cowan, 1989; Cowan *et al.*, 1990) to find new high-conductivity compounds and superconductors. These are summarized and updated as the following five requirements.

14.1 Molecular Requirements

The molecular constituents (electron donors D *and/or electron acceptors* A*) must have stable open-shell (free-radical) states* D^+ *or* A^-, *and stable dication* (D^{++}) *and dianion* (A^{--}) *states.*

For any candidate molecule, one must consider the relative location of the highest occupied molecular orbital (HOMO) of the neutral donor (D) and the lowest unoccupied molecular orbital (LUMO) of the neutral acceptor (A) molecule. The "parent" neutral molecule and the "daughter" anion radical A^- or cation radical D^+ must be both thermodynamically and kinetically stable, and accessible (not too high in energy). Since organic chemists must (or prefer) to work in solution, the species should not oxidize or reduce the solvent. In addition, the second ionization state (D^{++} or A^{--}) must also be stable, since an electron or a hole must also be given a chance to reside on the same, already ionized center. The most convenient experimental test of this stability of radicals is the measurement of the solution cyclic voltammogram. Many relatively stable organic free radicals are known. There are, at present, four methods useful for stabilizing these systems:

1. Force high spin density onto heteroatoms; this works well for TTF and TCNQ.
2. Use the steric hindrance of substituents, so that radical species are stable once made, as in diphenylpicrylhydrazyl, or in nitroxide spin labels.
3. Use push–pull substituent effects.
4. Use derivatives of odd-alternant aromatic hydrocarbons.

Of these four methods, only **1** has been seriously exploited for organic metals; methods **2**, **3**, and **4** do not necessarily stabilize the dication/dianion states.

A representative list of donors and acceptors is given in Figs. 2, 3, and 8. The synthetic variability of these structures is great, but the donors and acceptors must be stable and soluble in reasonable solvents if crystals are to be grown, and so the molecular sizes cannot be increased without limit. Adding solubilizing groups (alkyl chains) to an otherwise insoluble core molecule can help, but often these groups will prevent the efficient packing of molecules in crystals.

14.2 Crystallographic Requirements

Planar molecules D *or* A *with delocalized* π *molecular orbitals are best. Molecular components should be of appropriate or compatible size. One-dimensional metals need segregated stacks of radicals, and not mixed stacks.*

The planarity requirement for the donors and the acceptors is a logical inheritance from the charge-transfer complexes, and caused by the need for reasonable overlap between adjacent centers, to facilitate larger Mulliken transfer integrals t [Eq. (6)], and therefore good electron transport. However, compact molecules or clusters, which have small van der Waals distances between them and large overlaps, may have a moderate amount of intramolecular nonplanarity.

The requirement that components (electron donors D and electron acceptors A) be of compatible size is a crystal-packing requirement. The *horror vacui* that Kitaigorodskii (1973) tried to put on a computational basis in crystal-packing calculations means that, with the best-designed donors and acceptors, if they are of very different sizes, good compact structures with charge transfer are not achieved. This is why there are few interesting salts with TCNE, and so many with TCNQ: TCNE is too small an electron acceptor, compared with most donors, while hundreds of salts are known with TCNQ.

To make good electron acceptors, electron-rich substituent groups like $-SF_5$ or $-SeF_5$ could be attached to molecules, but these bulky groups will usually impede efficient crystal packing. This restricts strong electron acceptors to cyanocarbons; thus there are comparatively few good electron acceptors.

The issue of the "right" organic lattice is important. At one extreme, with very weak donors (high I_D) and very weak acceptors (low A_A), one has slightly colored crystals with so-called "mixed stacks" of alternating D and A molecules spaced at van der Waals distances apart (3.5 Å or higher), forming infinite "*DADA*" stacks, where the π electrons of D overlap with the π electrons of A; these have low ρ values. At the other extreme, with very strong donors (low I_D) and strong acceptors (high A_A), the tendency is to form a so-called mixed-stacking lattice, the organic chemical equivalent of sodium chloride, i.e., a $\rho \approx 1$ lattice with D^+ atop A^-; these crystals with $D^+A^-D^+A^-$ stacking are very deeply colored or shiny black, paramagnetic, and semiconducting.

Torrance pointed out that the segregated stacking that seems essential for high conductivity can be achieved by combining donors of intermediate I_D and acceptors of intermediate to high A_A (Torrance, 1979). Thus, TMPD TCNQ has a mixed stack (I_D = 6.25 eV for TMPD), while TTF TCNQ has a segregated stack (I_D = 6.83 for TTF). Since gas-phase electron affinities A_A are particularly difficult to measure, and gas-phase ionization potentials I_D are not routinely measured, one can seek guidance and solace from solution half-wave reduction potentials $E_{1/2}$, measured typically by cyclic voltammetry. Torrance, and also Saito and Ferraris (1980), have given rules of thumb for which are promising combinations of $E_{1/2}$ values. The interface between neutral and ionic mixed-stack *DA* complexes has been found: The 1:1 crystal complex of TTF (**7a**) with chloranil (**23**) was a *DADA* "neutral" ($\rho \approx$ 0.1) crystal at room temperature, which undergoes a collective transition to an ionic state ($\rho \approx 0.4$) at either low temperature or high pressure (Torrance *et al.*, 1981).

Black mixed-stack crystals and black segregated-stack crystals cannot be told apart by the naked eye, but only by electrical measurements or a solved crystal structure.

An important distinction to be made is between the two-stack systems (like TTF TCNQ, where electrons can migrate along the TCNQ stacks, and holes can migrate along the TTF stacks) and the one-stack systems, or ion-radical salts (or radical-ion salts), such as the $(TMTSF)_2X$ salts, the alkali TCNQ salts M_n(TCNQ), and the $(BEDT\text{-}TTF)_2X$ salts: The holes are localized on the TMTSF or BEDT-TTF sublattice, and the electrons on the TCNQ sublattice, and the rest of the crystal consists of diamagnetic, passive counterions.

However, this business of good stacking should not be exaggerated: The 12.5-K superconductor $\kappa\text{-(BEDT-TTF)}_2\text{Cu[N(CN)}_2\text{]Cl}$ has a layer of donors, which form van der Waals "dimers" of two BEDT-TTF species with charge $\rho = \frac{1}{2}$ each, with good intradimer overlap, but bad interdimer overlap, and yet the system is superconducting!

Finally, the finite crystal size is a problem: Typical organic conductors and superconductors form needle-shaped crystals of reasonable length (0.5 to 2 cm) but very small widths (usually 0.1 to 1 mm, or even as small as 0.025 mm).

14.3 Band Requirements

The HOMO–LUMO gap should be as small as possible, and the bandwidths should be as large as possible. This can be understood in terms of band theory, and of the Hubbard Hamiltonian, Eq. (10). The width of an energy band is $4t$; the effective repulsion of electrons within the same band is U_{eff} (akin to U, but with screening): One wants $U_{eff} < 4t$, and an incompletely filled band. Typical U_{eff} are 1 to 4 eV, while t is typically only

0.1–0.25 eV. U_{eff} can be decreased by increasing the molecular polarizability, by introducing atoms with many electrons (Se, Te, I) into the structure. One could decrease the HOMO–LUMO gap, while increasing the bandwidth, so as to force the LUMO- and HOMO-derived bands to overlap directly: This would yield a semimetallic regime, as for bismuth, but has not yet been realized for organic metals. The "tunability" of molecular orbitals within a certain class of related structures is finite, and the bandwidth of these systems remains relatively small.

14.4 Solid-State Electronic Requirements

Mixed valence, or fractional charge transfer ρ, *should be achieved. Inhomogeneous charge and spin distribution are preferable. A Peierls distortion should be avoided, by hydrostatic pressure, if necessary, or by efficient interchain interactions. Disorder should be avoided in the crystal.*

It was quite clear that intermediate values of the ionicity ρ were possible, indeed necessary for metallic conduction, within "segregated-stack" structures (*D* atop *D*, etc., and *A* atop *A*). One must achieve partial charge transfer (certainly $0 < \rho < 1$, but so far not yet $1 < \rho < 2$), and these species must be mixed valent, i.e., if one of two sites is a cation and the other a neutral molecule, then there must be only one crystallographically distinct species (so that the Franck–Condon rearrangement necessary to move the electron from the cation to the neutral molecule is not large). The theoretical classification into neutral and ionic crystals [Eqs. (24), (25)] by comparing Madelung energies to the cost of ionization works well for mixed-stack crystals, but such calculations did not yield the desired energy minimum for the partial ionicity of mixed-valence crystals (Metzger, 1981), because higher-order polarization and ρ-dependent dispersion energies need to be included in the calculation. TTF TCNQ is thermodynamically stable (Metzger, 1977).

The intermolecular (interionic) distances must be regular. This "mixed valency" requires that there be only one crystallographically unique molecular site, which must share its partial valency with the nearest-neighbor sites along the stack. The many "complex" stoichiometry TCNQ salts, e.g., $Cs_2(TCNQ)_3^{--}$ or triethylammonium$(TCNQ)_2^-$, which exhibit "trimeric" or "tetrameric" units of several crystallographically distinct TCNQ molecules and $TCNQ^-$ anions held at van der Waals separations, cannot conduct well.

14.5 Electron–Phonon Coupling Requirement

The electrons must couple with the crystal phonons so as to enable the right magnitude of coupling for Cooper-pair formation.

This requirement is the most mysterious. It is not known why certain crystals [e.g., HMTSF TCNQ or $Cu(2,5\text{-}DM\text{-}DCNQI)_2$] conduct very well at low temperatures, but do not form Cooper pairs. One can only speculate that certain intramolecular or intermolecular vibrations or rigid-body librational modes must be "right" for superconductivity. Kuzmany (1993) has suggested that strained molecules, e.g., C_{60}, have intramolecular Raman modes that may somehow be very important for superconductivity.

15. DESIGN RULES FOR POLYMERIC ORGANIC CONDUCTORS

Nobody knows how far one can push the conductivity by careful processing, aligning the polymer strands, and removing defects. The tremendous advantage of conducting polymers, in comparison with crystals of organic conductors or superconductors, is that larger samples can be made, at the price of residual catalyst impurities, structural defects, and dependence of physical properties on the processing methods.

It has been suggested (Heeger, 1990) that in a "dirty polymer" such as 10% doped polyacetylene, the charge-carrier concentration can be $N = 4 \times 10^{21}$ cm^{-3}, and the mobility $\mu < 0.2$ cm^2 V^{-1} s^{-1}, whence by Eq. (16) $\sigma < 30$ S cm^{-1}. By increasing the mean free path of charge carriers from about one lattice constant to several hundred lattice constants, the theoretical σ can be increased to hundreds of S/cm. In the "clean limit" of conductivity limited only by phonon scattering, $\sigma = 2 \times 10^6$ S cm^{-1}, i.e., four times higher than that of Cu, was predicted (Heeger, 1991).

Five design rules emerge for metallic conducting polymers.

15.1 Small Band Gaps

Reduce the band gaps by condensing large aromatic rings onto the backbone. The optical band gap E_{max} is reduced if there is a large quinoid contribution (see Fig. 16) to the ground state of the conjugated polymer. This has been accomplished in part in **31** and **32**: A band gap of 0.65 eV has been found for **32** (Lakshmikantham *et al.*, 1993).

15.2 Fiber Alignment

Stress-align the polymer strands. This is a traditional technique in fibrous polymers, and has given excellent results for polyacetylene (Naarmann and Theophilou, 1987). It turns out that for polyacetylene the chain lengths are probably quite long (100 or so C atoms), while for polyaniline and polythiophene 10 to 20 oligomers may be the individual chain length.

15.3 Polymer Solubilization

Increase the polymer solubility by adding long-chain alkyl substituents to the monomer. This brings one gain (solubility and processability of the polymer), but usually with two drawbacks (decreased effective conjugation length because of increased rotational disorder, and decreased interchain overlap). But for the newer soluble polymers one can determine molecular weights, viscosity, and other polymer chemical parameters unavailable, e.g., for polyacetylene.

15.4 Stereoregular Polymerization

Chemically protect all "unwanted" side positions of monomer (McCullough and Lowe, 1992; McCullough and Williams, 1993). This eliminates undesirable branched polymerization at unwanted side positions. For instance, thiophene is naively thought to polymerize preferentially at the α (or 2,2′) positions, but in reality a considerable amount of polymerization at the β (or 3,3′) positions occurs. If one chemically blocks the β positions in thiophene by a bulky substituent, then the polymerization is forced to occur only at the α positions, and to become stereoregular, and the conductivities increase dramatically (McCullough and Lowe, 1992; McCullough and Williams, 1993).

Quinoid state of poly-pyrrole

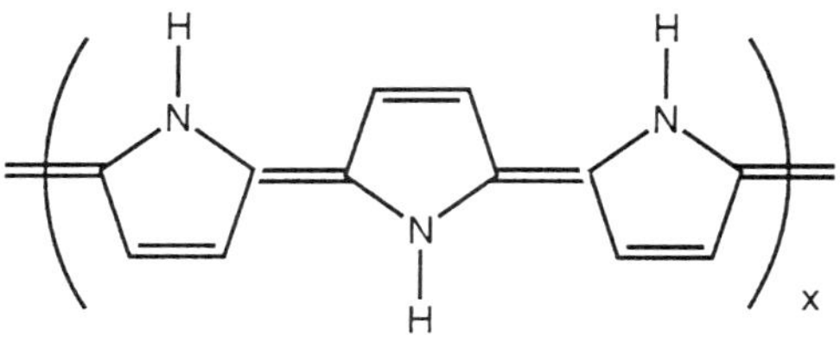

FIG. 16. Structure of the quinoid state for polypyrrole.

15.5 Facilitate Interchain Hopping

Reduce the interchain barrier to hopping by bringing the chains closer together. The conductivity in most polymers is thermally activated, and limited by hopping between conducting strands: The macroscopic behavior is semiconductive, not metallic. If interstrand communication can be improved, maybe metallic conductivity will be achieved. This requires careful molecular modeling, to ensure good packing by exploiting van der Waals forces.

Design rules for superconductivity in inorganic conducting polymers are somewhat premature: So far, $(SN)_x$ remains the only polymeric superconductor. It is likely that the monomer needs to have two stable oxidation or reduction states (see design rule 1 for crystals, above): Then bipolaronic insulating states are neatly avoided.

16. APPLICATIONS

A huge commercial success was registered when International Business Machines Corp. demonstrated in 1964 that a charge-transfer polymeric complex between trinitrofluorenone and polyvinylcarbazole could replace the high-cost selenium disk as the electrostatic imaging surface in electrostatic copiers. The early promise of lightweight batteries based on polyacetylene encountered insurmountable shelf-life problems, because of the intrinsic instability of the polymer. Conducting polymers have been used as electromagnetic shielding, and may have found applications, shrouded in military secrecy, for the wing surfaces in "stealth" bombers. More pedestrian applications, such as thermochromic temperature dating of stored foods in supermarkets and secondary batteries based on polyaniline, have entered the marketplace.

17. CONCLUSION

The fascinating world of organic conductors, superconductors, and polymers has been shown to exhibit not only interesting fundamental lower-dimensional physics, but also some promising commercial applications.

ACKNOWLEDGMENT

This work was supported in part by USAFOSR Grant No. F49620-92-J-0529DEF.

GLOSSARY

Bandwidth: Energy spacing between lowest level in a band and the highest level in the same band.

Bechgaard Salts: The TMTSF (tetramethyltetraselenafulvalene, **8b**) salts $TMTSF_2X$, where X is an inorganic anion; historically the first organic superconductors.

Bipolaron: Pseudoparticle that denotes two close dipoles in a lattice, and that has a diffusive coherent motion in the lattice.

Charge-Density Wave (CDW): Deformation of charge density in a crystal lattice from its equilibrium distribution (which is periodic with the lattice). Can be dynamic or static.

Charge Transfer ρ: Estimate of contribution of ionic state ($\rho = 1$) that replaces, partially or totally, the neutral state ($\rho = 0$). Can be calculated indirectly from $2k_F$ and $4k_F$ CDW x-ray scattering [Eq. (13)], from infrared spectral shifts, or from chemical shifts in nuclear quadrupole resonance spectra, x-ray photoelectron spectra, or (rarely) nuclear magnetic-resonance spectra.

Conduction Band: Energy band that is partially filled, because there are more states available than electrons (or electron pairs) to fill them; the top of the occupied levels is the Fermi level. The electrons at the top of the conduction band are relatively mobile, and can contribute to the electrical conductivity of the crystal.

Cooper Pairing: Pairing of electron states of opposite spin and opposite momentum in a crystal, mediated by coupling of both electrons to a particular lattice phonon.

Counterion: An anion or cation that is not involved in the process of interest (here, conductivity or superconductivity), but is needed as "counterbalance" to achieve electrical neutrality for the stoichiometric unit of a salt.

Effective Conjugation Length: Equivalent length of the π-electron network in a linear polyene, used in a particle-in-a-box model to calculate its lowest-energy optical absorption.

Energy Band: Set of energies for a set of N identical particles (atoms or molecules) that are very closely spaced, because their N-fold degeneracy is lifted by the interparticle interactions, e.g., for molecules in a crystalline solid.

Fermi Energy: Highest occupied energy band in the ground state of a free-electron gas at $T = 0$ kelvin.

Fermi Level: Chemical potential, or partial molar Gibbs free energy, for the electrons in a solid at finite temperatures.

Fermi Wave Vector: Wave vector for electron in the crystal with highest crystal momentum.

Highest Occupied Molecular Orbital (HOMO): Molecular wave function associated with the highest energy level that is occupied by an electron (or electron pair) in a molecule.

Hubbard Hamiltonian: Phenomenological Hamiltonian for N sites in an ordered array (atoms or molecules in a crystal or other lattice), which includes the Mulliken transfer integral t for electron transfer between sites, and the on-site energy U for electron–electron repulsion on the same site.

Lowest Unoccupied Molecular Orbital (LUMO): Molecular wave function associated with the lowest energy level that is not occupied by an electron in a molecule.

Mulliken Charge-Transfer Complex: Weak intermolecular complex, in solution or in the solid state, which mixes the ionic and neutral states of two components, so as to preserve the optical spectrum of these components, yet allow a strong optical transition between the mostly neutral and mostly ionic states.

Mulliken Transfer Integral: Energy involved in moving an electron from one site in a crystal to a nearest-neighbor site.

One-Electron Acceptor: A molecule with relatively high vacuum electron affinity (2 to 3.5 eV), which easily accepts an electron (gets reduced).

One-Electron Donor: A molecule with relatively low vacuum ionization energy (5 to 7 eV), which easily donates an electron (gets oxidized).

Peierls Instability: Structural instability of the one-dimensional uniform lattice toward distortion.

Peierls Transition: Transition between uniform high-temperature lattice and CDW-distorted low-temperature lattice; often separates high-temperature metallic regime from low-temperature semiconducting regime.

π Electron: Electron in a molecular orbital with p (perpendicular) symmetry and a nodal plane in the internuclear region between atoms.

Polaron: Pseudoparticle that denotes a dipole in a lattice, and that has diffusive coherent motion in the lattice.

Quinoid: State of a conjugated system with localized double bonds, as in benzoquinone, rather than delocalized double bonds, as in benzene (opposite: benzenoid).

σ Electron: Electron in a molecular orbital with s (longitudinal) symmetry and finite amplitude along the internuclear axis between atoms.

Soliton: Pseudoparticle representing a localized structural distortion that can diffuse coherently in a lattice.

Spin-Density Wave (SDW): Deformation of spin density in a crystal lattice from its equilibrium distribution (which has the periodicity of the lattice). Can be dynamic or static.

Spin-Peierls Transition: Transition between uniform paramagnet at higher temperature and distorted (lightly dimerized) diamagnet at lower temperature.

Superconductor: Solid with infinite electrical conductivity and diamagnetic flux exclusion.

Tight-Binding Approximation: Assumption that, at a lattice point or its immediate neighborhood, the crystal wave functions can be approximated as the local wave functions of the atom or molecule at that lattice point.

Works Cited

Acker, D. S., Harder, R. J., Hertler, W. R., Mahler, W., Melby, L. R., Benson, R. E., Mochel, W. E. (1960), *J. Am. Chem. Soc.* **82,** 6408–6409.

Bardeen, J., Cooper, L. N., Schrieffer, J. R. (1957a), *Phys. Rev.* **106,** 162–164.

Bardeen, J., Cooper, L. N., Schrieffer, J. R. (1957b), *Phys. Rev.* **108,** 1175–1204.

Bednorz, J. G., Müller, K. A. (1986), *Z. Phys. B* **64,** 189–193.

Bloch, F. (1928), *Z. Phys.* **52,** 555–600.

Bloch, A. N., Carruthers, T., Poehler, T. O., Cowan, D. O. (1977), in: H. J. Keller (Ed.), *Chemistry and Physics of One-Dimensional Metals,* NATO Advanced Study Institutes, Series B: Physics, Vol. 25, New York: Plenum, pp. 47–85.

Burroughes, J. H., Bradley, D. D. C., Brown, A. R., Marks, R. N., Mackay, K., Friend, R. H., Burns, P. L., Holmes, A. B. (1990), *Nature* **347,** 539–541.

Cairns, T. L., Carboni, R. A., Coffman, D. D., Engelhardt, V. A., Heckert, R. E., Little, E. L., McGeer, E. G., McKusick, B. C., Middleton, W. J. (1957), *J. Am. Chem. Soc.* **79,** 2340–2341.

Chu, C. W., Gao, L., Chen, F., Huang, Z. J., Meng, R. L., Xue, Y. Y. (1993), *Nature* **365,** 323–325.

Coleman, L. B., Cohen, M. J., Sandman, D. J., Yamagishi, F. G., Garito, A. F., Heeger, A. J. (1973), *Solid State Commun.* **12,** 1125–1132.

Conwell, E. (1980), *Phys. Rev. B* **22,** 1761–1780.

Coppens, P., Petricek, V., Levendis, D., Larsen, F. K., Paturle, A., Yan, G., LeGrand, A. D. (1987), *Phys. Rev. Lett.* **59,** 1695–1697.

Cowan, D. O. (1989), in: Z. Yoshida, T. Shiba, Y. Ohshiro (Eds.), *New Aspects of Organic Chemistry,* Tokyo: Kodansha.

Cowan, D. O., Fortkort, J. A., Metzger, R. M. (1990), in: R. M. Metzger, P. Day, G. C. Papavassiliou (Eds.), *Lower-Dimensional Systems and Molecular Electronics,* NATO Advanced Study Institutes, Series B: Physics, Vol. 248, New York: Plenum, pp. 1–22.

Day, P. (1990), in: R. M. Metzger, P. Day, G. C. Papavassiliou (Eds.), *Lower-Dimensional Systems and Molecular Electronics,* NATO Advanced Study Institutes, Series B: Physics, Vol. 248, New York: Plenum, pp. 115–128.

Dresselhaus, M. S., Dresselhaus, G., Eklund, P. C. (1993), *J. Mater. Res.* **8,** 2054–2097.

Ferraris, J., Cowan, D. O., Walatka, V. J., Perlstein, J. H. (1973), *J. Am. Chem. Soc.* **95,** 948–949.

Floquet, G. (1883), *Ann. Sci. Ecole Norm. Sup.* **12,** 47.

Gavaler, J. R. (1973), *Appl. Phys. Lett.* **23,** 480–482.

Greene, R. L., Street, G. B., Sutter, L. J. (1975), *Phys. Rev. Lett.* **34,** 577–579.

Hannay, N. B., Geballe, T. H., Matthias, B. T.,

Endres, K., Schmidt, P., Macnair, D. (1965), *Phys. Rev. Lett.* **14,** 225–226.

Hebard, A. F., Rosseinsky, M. J., Haddon, R. C., Murphy, D. W., Glarum, S. H., Palstra, T. T. M., Ramirez, A. P., Kortan, A. R. (1991), *Nature* **350,** 600–601.

Heeger, A. J. (1990), in: R. M. Metzger, P. Day, G. C. Papavassiliou (Eds.), *Lower-Dimensional Systems and Molecular Electronics,* NATO Advanced Study Institutes, Series B: Physics, Vol. 248, New York: Plenum, pp. 293–301.

Hubbard, J. (1963a), *Proc. R. Soc. London A* **276,** 238–257.

Hubbard, J. (1963b), *Proc. R. Soc. London A* **277,** 237–259.

Hubbard, J. (1964), *Proc. R. Soc. London A* **281,** 401–419.

Hubbard, J. (1978), *Phys. Rev. B* **17,** 494–505.

Jérome, D. (1990), in: R. M. Metzger, P. Day, G. C. Papavassiliou (Eds.), *Lower-Dimensional Systems and Molecular Electronics,* NATO Advanced Study Institutes, Series B: Physics, Vol. 248, New York: Plenum, pp. 85–89.

Jérome, D., Mazaud, A., Ribault, M., Bechgaard, K. (1980), *J. Phys. (Paris) Lett.* **41,** L95–L98.

Kistenmacher, T. J., Phillips, T., Cowan, D. O. (1974), *Acta Crystallogr. B* **30,** 763–768.

Kitaigorodsky, A. I. (1973), *Molecular Crystals and Molecules,* New York: Academic.

Krätschmer, W., Lamb, L. D., Fostiropoulos, K., Huffman, D. R. (1990), *Nature* **347,** 354–358.

Krogmann, K. (1969), *Angew. Chem. Int. Ed. Engl.* **8,** 35–42.

Kroto, H. W., Heath, J. R., O'Brien, S. C., Curl, R. F., Smalley, R. E. (1985), *Nature* **318,** 162–163.

Kuhn, H. (1949), *J. Chem. Phys.* **17,** 1198–1212.

Kuzmany, H. (1993), private communication.

Lakshmikantham, M. V., Lorcy, D., Scordilis-Kelley, C., Wu, X.-L., Parakka, J. P., Metzger, R. M., Cava, M. P. (1993), *Adv. Mater.* **5,** 723–726.

Landau, L. D., Lifschitz, L. M. (1958), *Statistical Physics,* Vol. 5 of *Course of Theoretical Physics,* London: Pergamon, p. 482.

Little, W. A. (1964), *Phys. Rev. A* **134,** 1416–1424.

Maeda, H., Tanaka, Y., Fukutomi, M., Asano, T. (1988), *Jpn. J. Appl. Phys. Lett.* **27,** L207–L210.

McConnell, H. M., Hoffman, B. M., Metzger, R. M. (1965), *Proc. Natl. Acad. Sci. U.S.A.* **53,** 46–50.

McCullough, R. D., Lowe, R. D. (1992), *J. Chem. Soc. Chem. Commun.,* 70–72.

McCullough, R. D., Williams, S. P. (1993), *J. Am. Chem. Soc.* **115,** 11608–11609.

Metzger, R. M. (1977), *J. Chem. Phys.* **66,** 2525–2533.

Metzger, R. M. (Ed.) (1981), *Crystal Cohesion and Conformational Energies,* Springer Topics in Current Physics Vol. 26, Berlin: Springer.

Mizuno, M., Garito, A. F., Cava, M. P. (1978), *J. Chem. Soc. Chem. Commun.,* 18–19.

Mulliken, R. S. (1950), *J. Am. Chem. Soc.* **72,** 600–608.

Naarmann, H., Theophilou, N. (1987), *Synth. Met.* **22,** 1–8.

Oshima, K., Mori, T., Inokuchi, H., Urayama, H., Yamochi, H., Saito, G. (1988), *Phys. Rev. B* **38,** 938–941.

Palstra, T. T. M., Zhou, O., Iwasa, Y., Sulewski, T. E., Fleming, R. M., Zagorski, B. R. (1995) in: L. Y. Chiang, P. Bernier, D. S. Bethune, T. W. Ebbesen, R. M. Metzger, J. W. Mintmire (Eds.), *Science and Technology of Fullerene Materials,* Materials Research Society Symposium Proceedings Vol. 359, Pittsburgh, PA: Materials Research Society, pp. 285–288.

Parkin, S. S. P., Engler, E. M., Schumaker, R. R., Lagier, R., Lee, V. Y., Scott, J. C., Greene, R. L. (1983), *Phys. Rev. Lett.* **50,** 270–273.

Parkin, S. S. P., Lee, V. Y., Engler, E. M., Nazzal, A. I., Huang, T. C., Gorman, G., Savoy, R., Beyers, R. (1988), *Phys. Rev. Lett.* **60,** 2539–2542.

Peierls, R. E. (1955), *Quantum Theory of Solids,* Oxford: Clarendon, p. 108.

Roberts, G. (Ed.) (1990), *Langmuir-Blodgett Films,* New York: Plenum.

Ruaudel-Teixier, A., Vandevyver, M., Barraud, A. (1985), *Mol. Cryst. Liq. Cryst.* **120,** 319–322.

Saito, G., Ferraris, J. P. (1980), *Bull. Chem. Soc. Jpn.* **53,** 2141–2145.

Saito, G., Enoki, T., Inokuchi, H. (1982), *Chem. Lett.,* 1345–1348.

Saito, G. (1990), in: R. M. Metzger, P. Day, G. C. Papavassiliou (Eds.), *Lower-Dimensional Systems and Molecular Electronics,* NATO Advanced Study Institutes, Series B: Physics, Vol. 248, New York: Plenum, pp. 67–84.

Shirakawa, H., Ikeda, S. (1971), *Polym. J.* **2,** 231–244.

Shirakawa, H., Louis, E. J., MacDiarmid, A. G., Chiang, C. K., Heeger, A. J. (1977), *J. Chem. Soc. Chem. Commun.,* 578–580.

Skotheim, T. A. (Ed.) (1986), *Handbook of Conducting Polymers,* Vols. 1, 2, New York: Dekker.

Thomas, D. D., Keller, H., McConnell, H. M. (1963), *J. Chem. Phys.* **39,** 2321–2329.

Thomas, G. A., Schafer, D. E., Wudl, F., Horn, P. M., Rimai, D., Cook, J. W., Glocker, D. A., Skove, M. J., Chu, C. W., Groff, R. P., Gillson, J. L., Wheland, R. C., Melby, L. R., Salamon, M. G., Craven, R. A., DePasquali, G., Bloch, A. N., Cowan, D. O., Walatka, V. V., Pyle, R. E., Gemmer, R., Poehler, T. O., Johnson, G. R., Miles, M. G., Wilson, J. D., Ferraris, J. P., Finnegan, T. F., Warmack, R. J., Raaen, V. F., Jérome, D. (1976), *Phys. Rev. B* **13,** 5105–5100.

Thorup, N., Ringdorf, G., Soling, H., Bechgaard, K. (1981), *Acta Crystallogr. B* **37,** 1236–1240.

Torrance, J. B. (1979), *Acc. Chem. Res.* **12,** 79–86.

Torrance, J. B., Girlando, A., Mayerle, J. J., Crowley, J. I., Lee, V. Y., Batail, P., LaPlaca, S. J. (1981), *Phys. Rev. Lett.* **47,** 1747–1750.

Van Vleck, J. H. (1932), *The Theory of Electric and Magnetic Susceptibilities,* Oxford, U.K.: Oxford Univ. Press.

Wang, P., Metzger, R. M., Bandow, S., Maruyama, Y. (1993), *J. Phys. Chem.* **97,** 2926–2927.

Williams, J. M., Ferraro, J. R., Thorn, R. J., Carlson, K. D., Geiser, U., Wang, H. H., Kini, A. M., Whangbo, M.-H. (1992), *Organic Superconductors (Including Fullerenes) Synthesis, Structure, Properties, and Theory,* Englewood Cliffs, NJ: Prentice-Hall.

Wu, M. K., Ashburn, J. R., Torng, C. J., Hor, P. H., Meng, R. L., Gao, L., Huang, Z. J., Wang, Y. Q., Chu, C. W. (1987), *Phys. Rev. Lett.* **58,** 908–910.

Wudl, F., Smith, G. M., Hufnagel, E. J. (1970), *J. Chem. Soc. Chem. Commun.,* 1453–1454.

Further Reading

The interested reader can glean more detail from the several thousand research articles since the 1970s and from the following representative list of references:

Review Articles

Bryce, M. R., Murphy, L. C. (1984), *Nature* **309,** 119–126.

Garito, A. F., Heeger, A. J. (1974), *Acc. Chem. Res.* **7,** 232–240.

Greene, R. L., Street, G. B. (1984), *Science* **226,** 651–656.

Heeger, A. J., Kivelson, S., Schrieffer, J. R., Su, W. P. (1988), *Rev. Mod. Phys.* **60,** 781–850.

Kommandeur, J. (1985), *Nouv. J. Chim.* **9,** 341–345.

Miller, J. S., Epstein, A. J. (1987), *Angew. Chem. Int. Ed. Engl.* **26,** 287–293.

Roth, S., Bleier, H. (1985), *Adv. Phys.* **36,** 385–462.

Schukat, G., Fanghänel, E. (1987), *Sulfur Rep.* **14,** 245–300.

Sólyóm, L. (1979), *Adv. Phys.* **28,** 201–303.

Soos, Z. G. (1983), *Isr. J. Chem.* **23,** 37–48.

Williams, J. M., Schultz, A. J., Geiser, U., Carlson, K. D., Kini, A. M., Wang, H. H., Kwok, W.-K., Whangbo, M.-H., Schirber, J. E. (1991), *Science* **252,** 1501–1508.

Wudl, F. (1984), *Acc. Chem. Res.* **17,** 227–232.

Edited Topical Volumes and Monographs

Billups, W. E., Ciufolini, M. A. (Eds.) (1993), *Buckminsterfullerenes,* New York: VCH.

Briegleb, G. (1960), *Elektronen-Donator-Acceptor Systeme,* Berlin: Springer.

Conwell, E. M. (Ed.) (1988), *Highly Conducting Quasi-One-Dimensional Organic Crystals,* Boston: Academic.

Devreese, J. T., Evrard, R. P., Van Doren, V. E. (Eds.) (1977), *Highly Conducting One-Dimensional Solids,* New York: Plenum.

Ferraro, J. R., Williams, J. M. (1987), *Introduction to Synthetic Electrical Conductors,* San Diego: Academic.

Lieb, E. H., Mattis, D. C. (1964), *Mathematical Physics in One Dimension,* New York: Academic.

Miller, J. S. (Ed.) (1981, 1982, 1983), *Extended Linear-Chain Compounds,* Vols. 1, 2, 3, New York: Plenum.

Monceau, P. (Ed.) (1985), *Electronic Properties of Inorganic Quasi-One-Dimensional Compounds,* Vols. 1, 2, Dordrecht, Holland: Reidel.

Symposium Proceedings:

Ann. N.Y. Acad. Sci. **313** (1978); *Chem. Scripta* **17** (1981); *Mol. Cryst. Liq. Cryst.* **77, 79, 81, 83, 85, 86** (1982); *J. Phys. (Paris) Coll.* **44 C3** (1983); *Mol. Cryst. Liq. Cryst.* **117–120** (1985); *Phys. B* **143** (1986); *Synth. Met.* **17–19** (1987), **27–29** (1989), **41–43** (1991), **57–59** (1993), **69–71** (1995); *Springer Lect. Notes in Phys.* **65** (1977), **95–96** (1979), *Springer Proc. in Phys.* **51** (1990); *Springer Ser. in Solid State Sci.* **88** (1990), **117** (1993); *Mater. Res. Soc. Symp. Proc.* **173** (1990), **247** (1992); *Am. Chem. Soc. Symp. Ser.* **5** (1974), **481** (1992).

Hatfield, W. E. (Ed.) (1979), *Molecular Metals,* New York: Plenum.

Kresin, V., Little, W. A. (Eds.) (1990), *Organic Superconductivity,* New York: Plenum.

Masuda, K., Silver, M. (Eds.) (1974), *Energy and Charge Transfer in Organic Semiconductors,* New York: Plenum.

Ovchinnikov, A. A., Ukrainskii, I. I. (Eds.) (1991), *Low-Dimensional Conductors and Superconductors,* Berlin: Springer.

North American Treaty Organization Advanced Study Institutes (NATO ASI)

Alcácer, L. (Ed.) (1980), *The Physics and Chemistry of Low Dimensional Solids,* NATO ASI Series C: Chemistry, Vol. 56, Dordrecht, Holland: Reidel.

Beeby, J. L. (Ed.) (1991), *Condensed Systems of Low Dimensionality,* NATO ASI Series B: Physics, Vol. 253, New York: Plenum.

Brédas, J.-L. (Ed.) (1991), *Conjugated Polymeric Materials: Opportunities in Electronics, Optoelectronics and Molecular Electronics,* NATO ASI Series E: Materials Science, Vol. 182, Dordrecht, Holland: Kluwer.

Delhaès, P., Drillon, M. (Eds.) (1988), *Organic and Inorganic Low-Dimensional Crystalline Materials,* NATO ASI Series B: Physics, Vol. 168, New York: Plenum.

Jérome, D., Caron, L. G. (Eds.) (1987), *Low-Dimensional Conductors and Superconductors,* NATO ASI Series B: Physics, Vol. 155, New York: Plenum.

Keller, H. J. (Ed.) (1977), *Chemistry and Physics of One-Dimensional Metals,* NATO ASI Series B: Physics, Vol. 25, New York: Plenum.

Metzger, R. M., Day, P., Papavassiliou, G. C. (Eds.) (1990), *Lower-Dimensional Systems and Molecular Electronics,* NATO ASI Series B: Physics, Vol. 248, New York: Plenum.

For lack of space, this review did not discuss certain ancillary areas; the reader is therefore asked to read elsewhere about:

Nonlinear Optical Properties

Prasad, P. N., Williams, D. J. (1990), *Introduction to Nonlinear Optical Effects in Molecules and Polymers,* New York: Wiley.

The Promise of Molecular Electronic Devices

Ashwell, G. J. (Ed.) (1992), *Molecular Electronics,* Taunton, Somerset: Research Studies Press.

Aviram, A. (Ed.) (1992), *Molecular Electronics—Science and Technology,* New York: American Institute of Physics.

Birge, R. R. (Ed.) (1994), *Biomolecular Electronics, American Chemistry Society Advances in Chemistry Series,* Vol. 240, Washington, DC: American Chemical Society.

Carter, F. L. (Ed.) (1982, 1987, 1988), *Molecular Electronic Devices,* Vol. 1, New York: Dekker; Vol. 2, New York: Dekker; Vol. 3, Amsterdam: North-Holland.

Organic and Organometallic Ferromagnets

Gatteschi, D. (1994), *Adv. Mater.* **6,** 635–645.

Miller, J. S., Epstein, A. J. (1994), *Angew. Chem. Int. Ed. Engl.* **33,** 385–415.

SEMICONDUCTORS, AMORPHOUS—ELECTRONIC PROPERTIES

H. FRITZSCHE, *James Franck Institute and Department of Physics, University of Chicago, Chicago, Illinois, U.S.A.*

3-527-28139-8/96/$5.00 + .50

INTRODUCTION

During the past 15 years amorphous semiconductors have spawned a rapidly growing multibillion-dollar industry. Thin-film transistors made of amorphous silicon and amorphous silicon nitride control the pixels in large-area liquid-crystal displays used for flat-panel computer and television screens. Amorphous silicon and its alloys cover the drums of optoelectronic copying machines and are the active ingredients in thin-film photovoltaic modules that have achieved over 10% efficiency in converting sunlight into electricity. Compact-disk optical memories utilize the optical contrast between the amorphous and crystalline phases of chalcogenide materials. In many of these optoelectronic-device applications, use is made of the fact that amorphous semiconductors can be deposited as very thin films on large or curved flexible substrates.

Solid materials that lack the periodic symmetry and regular spatial arrangement of the atoms that are characteristic of crystals are called amorphous or vitreous or more generally noncrystalline. It is useful to distinguish between vitreous and amorphous materials. As vitreous materials (glasses) are cooled from the melt their viscosity increases as the atoms become frozen in place. Below a certain temperature, called the glass-transition temperature, the viscosity is so large that the vitreous material is considered a solid. Its atomic arrangement even though frozen is very similar to that of its liquid state above the glass-transition temperature. Amorphous materials, in contrast, can normally not be prepared by cooling their melt. They can be prepared only in the form of thin films by deposition on substrates that are kept below the crystallization temperature. This article deals with noncrystalline semiconductors, i.e., both vitreous and amorphous materials.

Semiconductors are materials whose electrical conductivity falls between those of metals and of insulators. Moreover, their conductivity can often be changed appreciably by altering the materials composition or adding specific elements that act as electron donors or acceptors. The electron concentration and conductivity of semiconductors can also be changed and controlled by charge injection or by exposure to light.

The study of the electronic properties that specifically distinguish noncrystalline semiconductors from insulators and ionic conductors started with the discovery by Goryunova and Kolomiets (1955) of the semiconductor properties of chalcogenide glasses, the exploration of amorphous Ge and Si by Tauc *et al.* (1966), and the bipolar switching and memory effects found by Ovshinsky (1968) in thin films of chalcogenide semiconductors.

The lack of long-range order has been an obstacle as well as a challenge for scientists trying to understand the electronic and optical properties of noncrystalline semiconductors. Between perfect order and disorder there are innumerable configurational and topological arrangements of atoms for which we do not have even a descriptive terminology. Nonetheless, there are structural similarities between crystalline and noncrystalline semiconductors, particularly in the nearest-neighbor sphere around each atom, because the same chemical forces and the same chemical bonds hold both kinds of solids together. Since the electronic properties of materials are largely governed by the chemical nature of the constituent atoms and the type of bonding between atoms, the same materials that are metals, semiconductors, or insulators as crystals remain so in the noncrystalline state. This leads to a useful classification shown in Table 1, which distinguishes covalent and ionic materials and separates the covalent ones into tetrahedrally bonded semiconductors and chalcogenide glasses. The latter contain as major constituents elements of group VI of the periodic table that are called chalcogens. Unlike crystalline semiconductors, noncrystalline semiconductors may contain many elements in a wide range of compositions since they do not constitute a thermodynamic equilibrium phase. As a consequence, noncrystalline semiconductors tend to be compositionally heterogeneous and contain spatial fluctuations of their band gap or their internal potential, which complicate a description of their physical properties. We restrict the following discussion to a few prototypical materials in each category of Table 1.

Table 1. Classification and examples of noncrystalline semiconductors.

1. Covalently bonded materials
   A. Tetrahedral amorphous films
      Si:H, Ge:H, $SiGe_x$:H, SiC_x:H, SiN_x:H, GaAs, GaSb, ...
   B. Tetrahedral glasses, $A^{II}B^{IV}C_2^V$
      $CdGe_xAs_2$, $CdSi_xP_2$, $ZnSi_xP_2$, $CdSn_xAs_2$, . . .
   C. Chalcogenide glasses
      (i) Elements and compounds
         Se, S, Te, As_2Se_3, As_2S_3, . . .
      (ii) Multicomponent glasses
         Ge–Sb–Se   Si–Ge–As–Te
         Ge–As–Se As_2Se_3–As_2Te_3
         As–Se–Te Tl_2Se–As_2Te_3, . . .
   D. Others
      B, As, P, . . .
2. Quasi-ionic materials
   A. Metal-oxide films
      In_2O_{3-x}, In_2O_{3-x}:Sn, SnO_{2-x}, ZnO_x, $AgSbO_3$, . . .
   B. Oxide glasses
      V_2O_5–P_2O_5 MnO–Al_2O_3–SiO_2
      V_2O_5–GeO_2–BaO TiO_2–B_2O_3–BaO, . . .
3. Dielectric films
   SiO_{2-x}, Al_2O_3, Si_3N_4, BN, . . .

1. ELECTRONIC STATES

1.1 Chemical Bonding

Most amorphous semiconductors are covalently bonded, and the coordinations of their constituent atoms follow simple chemical rules. The solid has the lowest energy when the maximum number of valence electrons participates in covalent bonds, which means that they occupy bonding states and no electrons occupy antibonding states. Each state has two electrons with opposite spins according to Pauli's exclusion principle. Group IV elements form four hybridized sp^3 orbitals and are tetrahedrally coordinated. Group V elements form three covalent bonds with their p orbitals and leave a pair of s electrons nonbonded. Group VI elements, the chalcogens, use only two p electrons for bonding in twofold coordination, leaving two s electrons and two p electrons in nonbonding orbitals. This leads to the widely observed $8 - N$ rule, which relates the coordination number C with the column number N of the element:

$$C = 8 - N \quad \text{for} \quad N \geq 4, \tag{1}$$

$$C = N \quad \text{for} \quad N < 4. \tag{2}$$

The bond energies are further optimized by favoring stronger heteropolar over weaker homopolar bonds. Deviations of the covalent bond lengths and bond angles from the ideal values and stronger interactions of the nonbonding electrons contribute to the excess energy of the amorphous state relative to the energy of the crystalline state.

Materials that easily form glasses have average covalent coordinations C_{av} between 2 and 3. Phillips (1979) suggested that the optimal value is 2.45 for glass forming because the short-range order imposed by bond-stretching and bond-bending forces is just sufficient to exhaust the local degrees of freedom. Materials with C_{av} between 3 and 4 are overconstrained and can only be deposited as amorphous films, except for some compounds like $CdSi_xP_2$, which are called tetrahedral glasses; see Table 1.

1.2 Extended and Localized States

The bonding and nonbonding orbitals that hold all the valence electrons spread into a band of states, the valence band, as the atoms are brought close together to form a covalently bonded solid. The band is several eV wide and the spreading is caused by the interaction of neighboring orbitals, which overlap and form electron wave functions that extend over the whole volume of the solid. Likewise, the antibonding orbitals interact, overlap, and form extended wave functions. Their states spread into a band that is called the conduction band. Because the noncrystalline solid lacks translational symmetry, these extended wave functions are very different from the Bloch functions of periodic structures.

In materials that contain a dominant concentration of group VI elements, i.e., chalcogens, the valence band is formed by the two nonbonding (or lone-pair) p orbitals. These noncrystalline solids are called chalcogenide glasses or lone-pair semiconductors (Kastner, 1972).

The valence and conduction bands of states as well as the energy gap separating them are a consequence of chemical bonding. These bands do not have sharp energetic boundaries as they do in crystalline semiconductors. Instead, the variations in bond angles and bond lengths and the presence of weaker bonds as well as of topological bond structures that exist in noncrystalline solids, such as loops with odd numbers of covalent bonds, produce electronic states that tail from both bands into the gap. These tail states are localized in space with a typical localization radius of a few bond lengths. The energy that separates the localized tail states from the extended band states is called the mobility edge because at very low temperatures an electron should still be mobile in an extended state while it is immobile in a localized state. There is a mobility edge E_V separating band and tail states at the top of the valence band and a mobility edge E_C separating these two kinds of states at the bottom of the conduction band. The conductivity of a semiconductor at a given temperature is governed by the small concentration of electrons that are in the conduction band and the small concentration of missing electrons in the valence band. These missing electrons behave like positive charges and are called holes just as in crystalline semiconductors (see CONDUCTIVITY IN INSULATORS AND SEMICONDUCTORS). Since it costs less energy to produce a missing electron at the top compared with the bottom of the valence band, the energy of the hole is least at the top of the valence band.

The valence and conduction bands are approximately 5–10 eV wide and contain about twice as many electron states as there are atoms in the solid. The factor 2 comes from the two spin directions of the electron. Less than 0.1% of these states are localized tail states. Nonetheless, these tail states are important because they outnumber the concentrations of electrons and holes participating in conduction. One not only needs to know the energy of the states, but the concentration of states that exists in an energy interval dE at an energy E. This quantity is called the density of states $g(E)$ and given the units [$\mathrm{cm}^{-3}\ \mathrm{eV}^{-1}$]. This density-of-states function is sketched against electron energy for a typical noncrystalline semiconductor in Fig. 1. Since this is the energy range of interest, we only show the top of the valence band with its mobility edge E_V and the valence-band tail and the bottom of the conduction band with its tail and mobility edge E_C. Values of the gap $E_C - E_V$ range between 0.5 and 3 eV for semiconductors. Amorphous materials having a larger gap are called insulators. Near the gap center a peak in $g(E)$ is sketched in Fig. 1. This peak is due to localized states originating from defects, which are discussed in the following.

1.3 Electronically Active Defects and Dangling Bonds

A noncrystalline semiconductor whose atoms satisfy their valences by following the $8 - N$ rule of coordination can be considered defect free. Atoms that deviate from the $8 - N$ coordination rule are electronically active defects. If they are overcoordinated, they donate an electron to the host material, thereby changing their number of valence electrons to obey the $8 - N$ rule. For example, a P atom bonded to four neighbors donates an

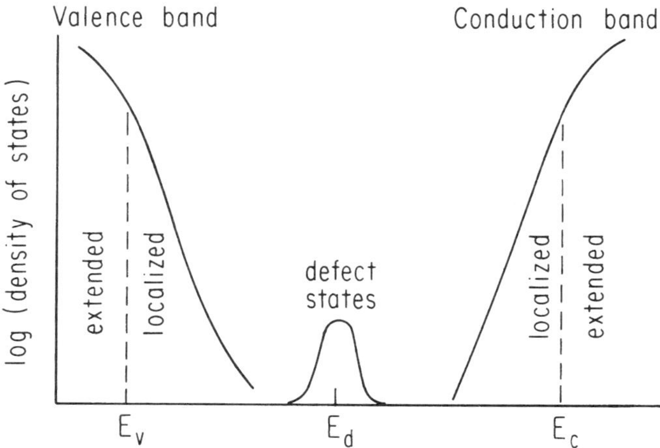

FIG. 1. Density of states distribution in a noncrystalline semiconductor. $E_c - E_v$ is the mobility gap.

electron and becomes positively charged, such that the number of valence electrons comes to agree with $N = 4$. Likewise an undercoordinated atom adjusts its number of valence electrons by accepting an electron and becoming negatively charged.

Coordination defects of the host material are often amphoteric defects that can exist in three charge states, D^+, D^0, and D^-, as they are occupied by zero, one, or two electrons. In chalcogenide glasses, these are undercoordinated (dangling-bond) defects and also overcoordinated chalcogen or pnictogen atoms. In *a*-Si:H, the dominant defect is believed to be a three-fold coordinated Si atom whose fourth sp^3 hybrid orbital remains unbonded. Such a dangling bond is often written as Si_3 where the subscript stands for the coordination number. After accepting an electron and becoming a negatively charged Si_3^-, its tetrahedral bond angles (~109°) might relax toward a p^3 configuration (~90°), whereas the positively charged Si_3^+ might prefer a sp^2 planar bond configuration (~120°). Alternatively, the dominant defect in *a*-Si:H has been pictured as a floating bond, i.e., a five-fold coordinated Si where the wave function of the gap state is distributed on the five neighbors of the central atom (Pantelides, 1986).

1.4 Positive- and Negative-*U* Defects

According to Pauli's exclusion principle, every electronic state can be occupied by up to two electrons. Doubly and singly occupied levels of localized states will not have the same energy because of the repulsive Coulomb energy U_C of the two electrons. Moreover, a change in charge state is usually accompanied by an adjustment of the neighboring bonds, which contributes a relaxation energy W. The two-electron ($2e$) level is then separated from the one-electron ($1e$) level by an effective correlation energy

$$U = U_C - W. \tag{3}$$

The possibility of a negative U occurring for large relaxations, $W > U_C$, was pointed out by Anderson (1975) and such defects were identified by Street and Mott (1975) in chalcogenide glasses.

Fermi statistics are not applicable to correlated electron states. Instead, the occupation probabilities f_0, f_1, f_2 for zero, one, and double occupancy are obtained from the grand partition function as

$$f_0 = z^{-1} = \{1 + 2\exp[-\beta(E - E_F)] + \exp[-\beta(2E - 2E_F + U)]\}^{-1}, \tag{4}$$

$$f_1 = z^{-1}\, 2\exp[-\beta(E - E_F)], \tag{5}$$

$$f_2 = z^{-1}\exp[-\beta(2E - 2E_F + U)]. \tag{6}$$

Here the zero of energy E is placed at the level E_d of the defect and $\beta = (kT)^{-1}$. One finds that for $+U$ defects the $2e$ level exists only when the $1e$ level is occupied. For $-U$ defects, however, the $1e$ level lies above the $2e$ level and remains unoccupied for $kT \ll |U|$. The important difference between $+U$ and $-U$ defects is illustrated in Fig. 2, which shows the shift of the Fermi energy E_F as the defect occupancy n/N of N defects is varied as, for instance, by doping or by inducing a space charge (Adler and Joffa, 1976). The difference in behavior of E_F in Figs. 2(a) and 2(b) can be understood as follows: $E_C - E_F$

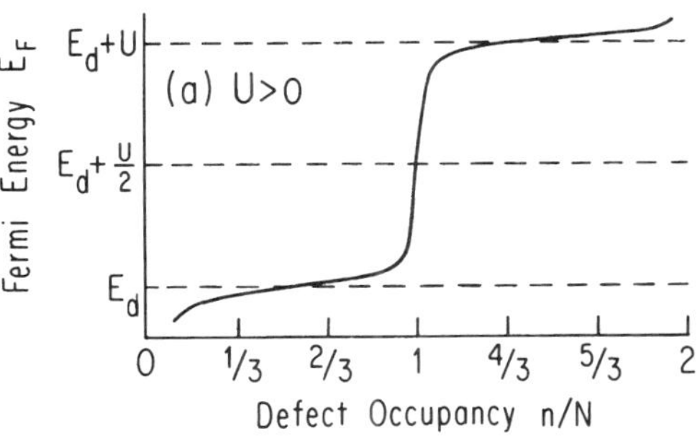

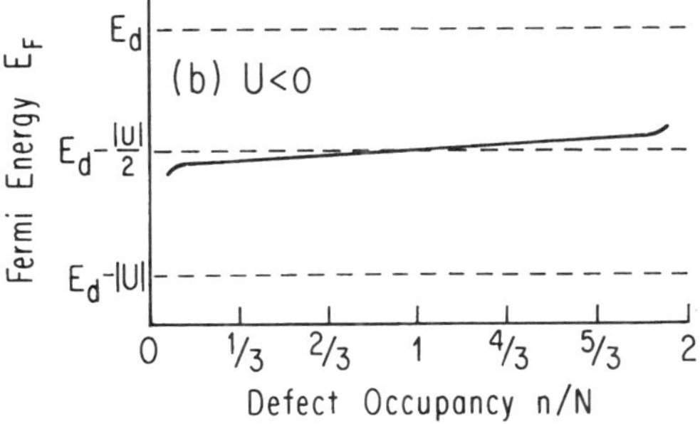

FIG. 2. Fermi energy as a function of electronic occupancy *n*/*N* of defects having (a) positive correlation energy *U* and (b) negative correlation energy, $-|U|$. After Adler and Joffa (1976).

represents the average thermal energy for exciting electrons to the conduction band. This energy remains the same for $-U$ defects; it is the average of the first and second ionization energy, $E_C - E_F = \frac{1}{2}(E_1 + E_2)$, as long as there are some $-U$ defects in the D^- and D^+ states. Hence E_F remains pinned for $0 < n/N < 2$. For $+U$ defects E_F moves quickly from the $1e$ to the higher lying $2e$ level at $n = N$. Figure 1 shows the density of one-electron states. Therefore, the $2e$ level of D^- does not appear in this figure.

1.5 Thermodynamic Equilibrium of Defects

The observation that coordination defects produce electronic levels in the gap or act as donors or acceptors holds for both crystalline and noncrystalline semiconductors. There are important differences, however. The translational symmetry of crystals imposes the same coordination environment on all equivalent atomic sites. The energies of defects of each kind are therefore the same and the doping efficiency of the doping elements is 100 percent. In noncrystalline semiconductors, the coordination is not confined by symmetry, and foreign atoms can achieve their preferred $(8 - N)$-rule coordination. The number of coordination defects is therefore governed above a thermodynamic equilibration temperature T_E by minimizing the free energy, and the doping efficiency can be considerably smaller than unity. In glasses, T_E is the glass-transition temperature, which usually lies between 100 and 300 °C. In hydrogenated amorphous silicon and silicon alloys, the defect reactions are mediated by diffusing hydrogen. The defect reaction can therefore attain thermodynamic equilibrium on a time scale set by hydrogen diffusion and T_E can be taken as the temperature below which the defect equilibration takes longer than about 10 min. This temperature has been found to lie between 150 and 200 °C.

1.5.1 Defect Pool The native dangling-bond defect concentration N_D often lies in the low 10^{15}-cm^{-3} range in undoped a-Si:H because of the presence of about 10 at.% hydrogen, which bonds to most potential dangling bonds. In addition to passivating preexisting dangling bonds, H atoms break strained Si–Si bonds, relieving strain in the material. Why is there still such residue of dangling bonds in the presence of about 5×10^{21} cm^{-3} hydrogen atoms? The residual defect concentration may be the equilibrium value N_D established above T_E from a pool of potential defects, N_{Pool},

$$N_D = N_{\text{Pool}} \exp(-E_D/kT_E). \tag{7}$$

The effective defect-formation energy E_D depends on the position of the Fermi energy E_F because of the energy $E_d - E_F$ gained by transferring an electron from D^0 to E_F or the energy $E_F - (E_d + U)$ gained by transferring an electron from E_F to D^0. The energy in the gap E_d of a dangling bond depends on the bonding environment. It is possible to conceive of a pool of potential defects that, if picked at random, would cover energies over a wide range in the gap. The point is that they are not picked at random but with preference for a low formation energy. Hence, defects with small E_d are favored in n-type material where E_F lies high and defects with large E_d are favored in p-type material where E_F lies low in the gap. According to this defect-pool model, defects having small or large E_d equilibrate at $T \geq T_E$ in response to the energy position of E_F. This happens regardless of whether E_F is determined by doping or by an induced space charge as for instance in a field-effect transistor.

1.5.2 Doping Equilibration We restrict the discussion to amorphous, tetrahedrally coordinated hydrogenated materials because they are essentially the only noncrystalline semiconductors that can be doped. Choosing a-Si:H as an example one finds the following doping and defect reactions (Street, 1985):

$$\text{Si}_4^0 + \text{P}_3^0 \rightleftarrows \text{P}_4^+ + D^-, \tag{8}$$

$$\text{Si}_4^0 + \text{B}_3^0 \rightleftarrows \text{B}_4^- + D^+, \tag{9}$$

$$\text{Si}_4^0 + \text{Si}_4^0 \rightleftarrows D^+ + D^-. \tag{10}$$

The left-hand sides are the normal $(8 - N)$-rule coordinations. Hence, P_3^0 and B_3^0 do not act as donors and acceptors, respectively. The right-hand sides require the formation energies for dangling bonds and for changing the dopant coordinations to P_4 and B_4. These energies depend on the position of E_F

as a result of the energy gain of transferring the electrons to and from E_F as discussed above, and are smallest when E_F lies near the gap center. As a consequence, the doping efficiency is highest in compensated a-Si:H that contains equal concentrations of P and B. The reactions (8)–(10) are mediated by hydrogen diffusion and hence come to rest at $T < T_E$. The P_4^+ donors in n-type material are compensated by dangling-bond defects D^- according to Eq. (8). As a consequence, E_F remains between the donor and D^- gap states and cannot move into the conduction band as it does in heavily doped degenerate crystalline semiconductors. The same defect compensation occurs upon p-type doping according to Eq. (9).

The final charge state of dopants and defects depends on E_F, i.e., the combined effect of all defect reactions and the presence of induced space charges.

1.5.3 Valence-Alternation Defects It sometimes costs less energy to form a pair of coordination defects than a single one. This happens in chalcogenide glasses where a charge transfer from one defect to another leads to a bonding change and a net gain in energy. This means that the reaction

$$2D^0 \rightarrow D^+ + D^- \qquad (11)$$

is exothermic and D is a $-U$ defect (Street and Mott, 1975). Kastner *et al.* (1976) proposed that the D^+-D^- pair of defects derives from alternating the valences of the group VI (chalcogen) and group V (pnictogen) atoms. The $8 - N$ rule is then obeyed by the charged defects in the following manner: The D^+ defects are overcoordinated and the D^- defects are undercoordinated chalcogen or pnictogen atoms. Since the number of covalent bonds is not changed by creating such D^+-D^- valence-alternation pairs (VAP), the formation energy E_{VAP} of a defect pair should be small. $E_{VAP} \sim 0.5$–0.7 eV is approximately the positive Coulomb energy E_C of placing the second electron on the undercoordinated D^- center. The equilibrium concentrations $[D^-]$ and $[D^+]$ of the defects near the glass-transition temperature T_g follow from the law of mass action and are

$$[D^-][D^+] = N_A^2 \exp(-E_{VAP}/kT_g) = N_0^2. \qquad (12)$$

N_A is the concentration of atoms capable of undergoing valence alternations by utilizing the nonbonding orbitals of adjacent atoms. An appreciable density, $N_0 \sim 10^{17}$ cm^{-3}, of intrinsic VAP defects is therefore expected to exist even in well-annealed chalcogenide glasses.

The discussion of the pinning of E_F by $-U$ defects of concentration N (see Fig. 2) suggests that foreign dopant atoms exceeding $2N$ can unpin E_F and produce n-type or p-type chalcogenide glasses. Such attempts fail because Eq. (12) demands that the product $[D^-][D^+]$ reach an equilibrium value N_0^2 at T_g. Large concentrations of charged acceptors $[A^-]$ added to the glass above T_g will force a compensating increase in $[D^+]$ to satisfy the neutrality condition

$$[D^+] = [D^-] + [A^-]. \qquad (13)$$

As a consequence, the total number $[D^+] + [D^-]$ of $-U$ defects increases and always exceeds $[A^-]$ while the product $[D^-][D^+]$ remains fixed obeying Eq. (13) (Fritzsche and Kastner, 1978). This self-compensation mechanism of VAP defects can be circumvented by adding modifying additives to the glass below T_g where the VAP concentration remains frozen (Ovshinsky, 1977).

1.6 Inhomogeneities, Potential Fluctuations, Percolation

Amorphous semiconductors tend to be inhomogeneous. Only glasses having an average coordination close to $C_{av} = 2.4$ can be prepared rather strain-free from the melt (Phillips, 1979). Tetrahedrally coordinated semiconductors, including those that are hydrogenated, form overconstrained networks whose internal strain is often relaxed by the formation of voids and clusters of certain constituents. If the length scale of the inhomogeneities exceeds the tunneling distance of charge carriers, then fluctuations of the band gap or of the internal potential create classical turning points for carriers, which restrict the carrier motion to percolation paths (Zallen, 1983). Compositional inhomogeneities that lead to band-gap fluctuations require analysis on mesoscopic scales. Long-range potential fluctuations arising from random distributions of charged coordination defects can be estimated as follows.

For a concentration N of charged defects,

the statistical deviation ΔZ of their average number in a volume L^3 is

$$\Delta Z = (NL^3)^{1/2}. \tag{14}$$

On a length scale L, these cause potential variations

$$\Delta V = (NL^3)^{1/2}e/4\pi\epsilon\epsilon_0 L. \tag{15}$$

The physics of the semiconductor determines either ΔV or L as illustrated by the following examples. In an n-type sample containing a concentration of N shallow donors and according to Eq. (8) an equal number of compensating $+U$ defects, the length scale in Eq. (15) is the screening length

$$L = (8\pi\epsilon\epsilon_0/ne)^{1/2}, \tag{16}$$

where n is the concentration of trapped and free charges. In a sample compensated with equal concentrations N of shallow donors and acceptors, screening occurs when the fluctuations move E_F into the band tails. Hence $\Delta V \leq E_g/2$ and L follows from Eq. (15). In an undoped semiconductor containing only $+U$ defects, $\Delta V \approx U/2$, and L follows again from Eq. (15). Chalcogenide glasses containing $-U$ defects should be free of potential fluctuations because E_F remains pinned. ΔZ of Eq. (14) is kept zero by charge transfer between the $-U$ defects.

Compositional heterogeneities lead to band-gap and potential fluctuations, which affect the processes that depend on the position of the Fermi energy with respect to the band edges, such as the defect concentration, their charge state, and optical absorption (Bar-Yam *et al.*, 1986; Branz and Silver, 1990; Fritzsche, 1994). Moreover, long-range potential fluctuations affect electronic transport, in particular the activation energy of the thermopower (Overhof and Beyer, 1984) (see Sec. 2.4) and the drift mobility of electrons and holes (Street *et al.*, 1988b; Howard and Street, 1991).

1.7 Determination of Electronic States

The basis for understanding a semiconductor is the knowledge of the concentration and energies of the energy eigenstates. They are difficult to extract from experiments on amorphous semiconductors because there is a lack of sharp demarcation energies such as band edges to serve as reference points. Moreover, defect energies depend on the position of E_F if they arise from a defect pool and are subject to relaxations. In addition, defect and dopant concentrations exhibit a number of metastabilities, discussed in a later section.

1.7.1 Optical Absorption The optical absorption spectra of different noncrystalline semiconductors are very similar except for the magnitude of the optical gap. Because of the lack of long-range order and translational symmetry, there are no momentum-conservation selection rules governing optical transitions. All energy-conserving transitions from occupied to empty states in Fig. 1 are allowed. This explains why materials exhibit stronger absorption when they are amorphous than when they are crystalline. The change in optical properties is used for data storage in phase-change compact-disk optical memories.

A typical absorption spectrum is shown in Fig. 3. The high absorption region I ($\alpha \geq 10^4$

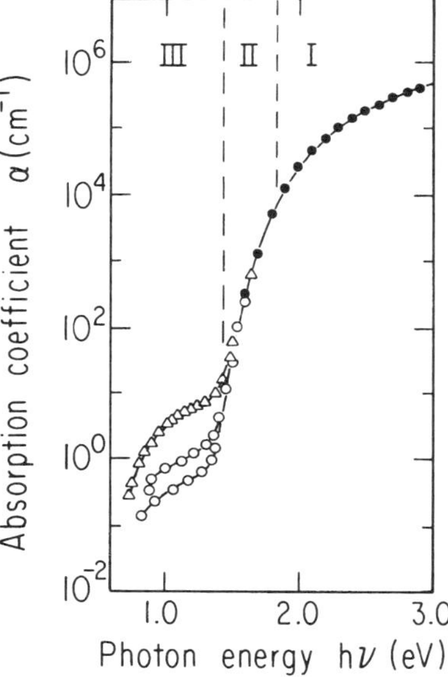

FIG. 3. Absorption spectrum of *a*-Si:H: interband-absorption region I, Urbach-edge region II, and defect-absorption region III. The open symbols are data taken with the photothermal-deflection method. The different curves in region III represent samples with different defect concentrations.

cm^{-1}) reflects the joint density of extended band states with the above assumptions. Assuming parabolic densities of states, as for free particles, Tauc (1974) obtained the relation

$$(\alpha h\nu)^{1/2} = B(h\nu - E_g), \tag{17}$$

which defines the optical Tauc gap E_g. The constant B contains the optical matrix element and the effective electron and hole masses. Its magnitude has been found to lie between 300 and 1000 $(\mathrm{eV\ cm})^{-1/2}$ when the energies are measured in eV and the absorption coefficient in cm^{-1}. Toward lower photon energies $h\nu$ and below $\alpha = 5 \times 10^3$ cm^{-1} the absorption is exponential over one to three decades and is described by

$$\alpha = \alpha_0 \exp(h\nu/E_0). \tag{18}$$

This Urbach-edge region II arises from transitions between the band tails and the opposite bands. E_0 is the exponential slope of the density of states of the valence- or the conduction-band tail, whichever is larger. This is usually the valence-band tail, and the magnitude ranges between $E_0 \sim 0.05$ and 0.08 eV.

Absorption region III toward lower $h\nu$ is due to transitions to or from coordination defects in the gap. Their concentration N can be obtained by integrating over the defect absorption,

$$N = A \int \alpha d(h\nu). \tag{19}$$

The coefficient A is obtained by determining N by other methods or by assuming the same dipole-matrix element as for other optical transitions. The value of A depends on the method used for determining α in region III. For photothermal-deflection spectroscopy $A \approx 8 \times 10^{15}$ cm^{-2} eV^{-1} in *a*-Si:H (Jackson and Amer, 1982), and for the constant-photocurrent method one often uses $A \approx 1.6 \times 10^{16}$ cm^{-2} eV^{-1} because only half of the transitions yield electron photocarriers. The correct value depends on doping and the position of the Fermi energy.

1.7.2 Photoemission Spectroscopies

Direct information about the valence- and conduction-band density of states is obtained from two photoemission spectroscopies (Ley, 1984). The first probes the occupied valence-band states. Photons of known energy excite electrons above the vacuum level so that they can leave the sample to be counted and energy analyzed. The second probes the empty conduction-band states. Electrons of known energy bombard the sample. As they fall into unoccupied states they emit x rays. The conduction band $g(E)$ is obtained by measuring the x-ray intensity at fixed wavelength as the electron-beam energy is varied.

Figure 4 compares the valence- and conduction-band spectra of amorphous hydrogenated silicon with those of crystalline silicon. The similarity of these spectra emphasizes that the chemical bonds have a much stronger effect on the energy states than the presence of long-range order (Jackson *et al.*, 1985).

Optical absorption spectra are essentially the product of the square of the optical dipole matrix element and the joint densities of the valence- and conduction-band states.

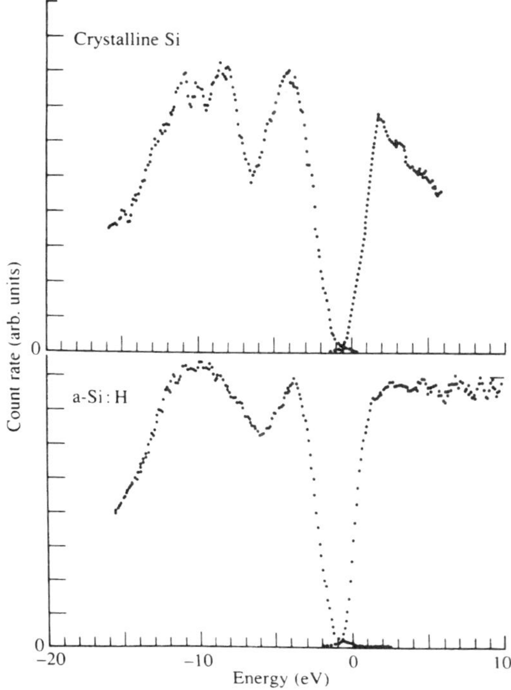

FIG. 4. The valence and conduction bands of crystalline silicon and *a*-Si:H measured by x-ray photoemission spectroscopy. After Jackson *et al.* (1985).

By combining the information from optical absorption and their photoemission spectra of Fig. 4, Jackson *et al.* (1985) were able to determine the dipole matrix element for *a*-Si:H. They found a magnitude of 10 Å times the electronic charge independent of energy below ~3.4 eV. This magnitude is the same for transitions between extended states and between localized and extended states. Transitions between localized states have vanishingly small matrix elements since most of them are spatially uncorrelated. These findings support the interpretation of optical absorption spectra as reflecting joint densities of states, and they indicate that the random-phase approximation is valid for *a*-Si:H.

Further details of $g(E)$ near the valence-band edge are revealed by a photoemission technique that measures the total yield $Y(h\nu)$ of emitted electrons as a function of photon energy (Winer and Ley, 1987):

$$Y(h\nu) \propto M^2(h\nu) \int_{E_{\rm vac}}^{\infty} g_V(E) g_c(E + h\nu) dE. \quad (20)$$

The density of states of the valence band $g_V(E)$ is proportional to $dY/d(h\nu)$ because the probability of photoemission is nearly constant above the work-function energy $E_{\rm vac}$ and the transition-matrix element $M(h\nu)$ is assumed to be a constant. Figure 5 shows the results for *a*-Si:H. One observes an exponential tail region with $E_0 = 0.051$ eV and a band of defect states centered about 0.7 eV above E_V. These measurements cannot identify the mobility edge E_V. Other evidence suggests $g_V(E_V) \approx 10^{21}$ cm^{-3} eV^{-1}.

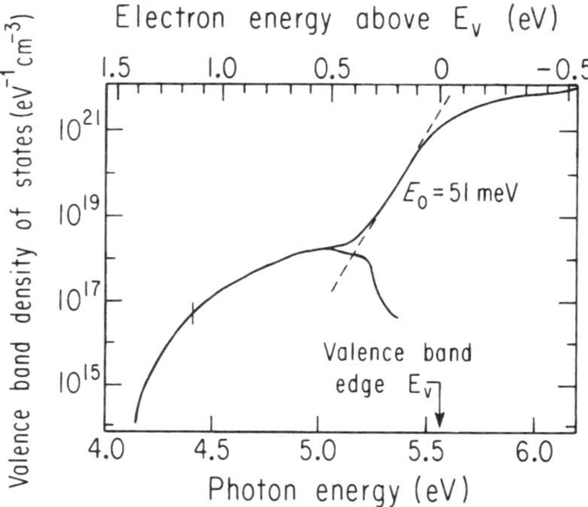

FIG. 5. Photoemission total yield measurements of the valence-band density of states of *a*-Si:H. After Winer and Ley (1987).

1.7.3 Electron Spin Resonance (ESR)

ESR yields important information about the concentration, the structure, and the localization radius of singly occupied localized states. Since the local defects are randomly oriented in an amorphous material, one measures the powder average of the gyromagnetic g tensor and of the nuclear hyperfine interaction. Modern ESR spectrometers can detect as few as 10^{10} spins per gauss of linewidth. By measuring ESR signals as a function of sample thickness, one can separate spin contributions of the bulk of the sample from those of the surface and of interfaces. Moreover, spin orientations play an important role in hopping of charge carriers between localized states and in recombination processes as discussed in Sec. 3.6. As a result, photoconductivity and luminescence become spin dependent and serve as sensitive tools to detect ESR optically (Solomon, 1979; Morigaki, 1984).

The lack of any ESR signal in chalcogenide glasses (Agarwal, 1975) indicates that the localized valence-band-tail states are doubly occupied and the conduction-band tail states are empty and thus have the parentage of their respective bands. Moreover, the general lack of paramagnetic states in these materials suggests that the defects have a negative effective correlation energy (Anderson, 1975) and exist in the D^+ and D^- charge states, which are diamagnetic. By illuminating these glasses at low temperatures, Bishop *et al.* (1977) created the metastable neutral D^0 states and measured the light-induced ESR (LESR). The concentrations (10^{16}–10^{17} cm^{-3}) of the defects and the observed anisotropies of the g tensors and hyperfine interactions agree with those expected for valence-alternation defects (Bishop *et al.*, 1977; Gaczi, 1982).

In evaporated or sputtered *a*-Si, which does not contain hydrogen, the spin signal is always very large, originating from 10^{19}–10^{20} cm^{-3} defects. These have a gyromagnetic ratio $g = 2.0055$ and have been identified by Brodsky and Title (1969) with silicon dangling bonds. In hydrogenated material, *a*-Si:H, the defect concentration is reduced to the low 10^{15}-cm^{-3} range. Even though the defect concentration increases with doping according to Eqs. (8) and (9), the ESR signal vanishes as the Fermi level is shifted with doping because the defects become charged and hence diamagnetic. From these observations one deduces a positive correlation energy $U \approx 0.3$ eV. By using light-induced ESR and creating metastable D^0 centers, one can determine the defect concentration in doped *a*-Si:H.

ESR signals at $g = 2.0080$ and $g = 2.0040$, respectively, are detected for valence-band tail states and conduction-band tail states that are singly occupied either by doping or by exciting electrons and holes into the tail states by light at low temperatures (Stuke, 1977). In strongly doped *a*-Si:H, one also finds an ESR signal of neutral phosphorus or arsenic donors in *a*-Si:H. From the hyperfine splitting, one deduces a localization radius of 10 ± 1 Å for the extra electron bound to the phosphorus and 9 ± 1 Å for the arsenic donor (Stutzmann and Street, 1985). In a similar manner, one has identified the spin signals of the Ge dangling bonds and of the singly occupied tail states in *a*-Ge:H (Stutzmann *et al.*, 1987).

1.7.4 Transient Emission Spectroscopies Several transient emission techniques have been developed to explore the energies and concentrations of deep electronic gap states in semiconductors. The principle of junction transient measurements is illustrated in Fig. 6. A semiconductor is shown with one discrete level of deep states in the space-charge region of a Schottky barrier, which is produced by making a metal contact to an *n*-type semiconductor. In Fig. 6(a) the sample is at equilibrium at a certain reverse bias. In Fig. 6(b) a voltage filling pulse allows gap states in the depletion region to capture electrons from the conduction band. When the bias is returned to its original value, as shown in Fig. 6(c), occupied gap states above E_F remain occupied until their electrons are excited to the conduction band by thermal or optical transition processes. The emitted electrons are then rapidly swept out by the high electric field of the depletion region. The thermal release time is

$$\tau_e = \nu_0^{-1} \exp(\Delta E/kT), \tag{21}$$

which increases exponentially with the energy depth ΔE of the gap state from E_C. The frequency ν_0 is related by detailed balance to the trapping cross sections of the gap states. Its magnitude must be determined independently; $\nu_0 \approx 10^{12}$ to 10^{13} s^{-1} (Lang *et al.*, 1982a; Johnson, 1983), but values near $\nu_0 \approx 10^8$ s^{-1} have also been reported (Okushi *et al.*, 1982). As the extra carriers are released, the original depletion region gets reestablished.

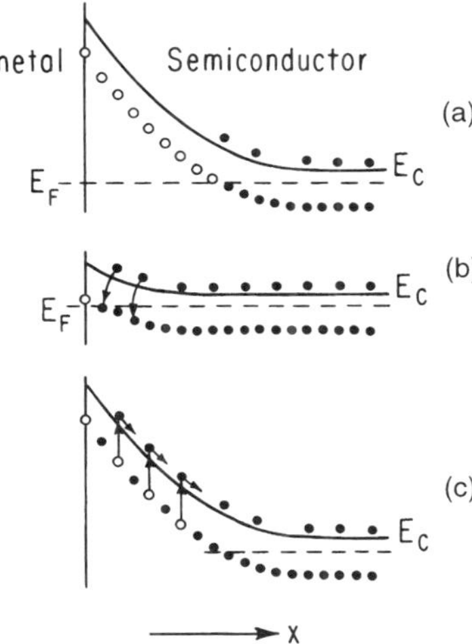

FIG. 6. Sketch of a metal–semiconductor Schottky barrier. (a) Reverse bias at equilibrium, (b) after voltage filling pulse, and (c) return to original reverse bias and thermal emission of nonequilibrium carriers.

The thermal-emission transient can be monitored by measuring the change in the Schottky-barrier capacitance $\Delta C(t)$ as a function of time, which occurs because the depletion region will contract as negative charge is lost and the positive charge density increases. $\Delta C(t)$ can be analyzed to yield $g(E)$, the energy dependence of the density of gap states as shown for an *n*-type *a*-Si:H sample in Fig. 7. According to Eq. (21) extra electrons in deep gap states take a long time to be released. It is therefore often found convenient to study $\Delta C(t_0)$ after a time window (t_0) as a function of temperature. The analysis of such thermal-emission spectra gave the first evidence for the existence of a broad band of defects in *n*-type *a*-Si:H located near the gap center (Cohen *et al.*, 1980; Lang *et al.*, 1982a).

Another variation is the use of subgap light in place of the purely thermal excita-

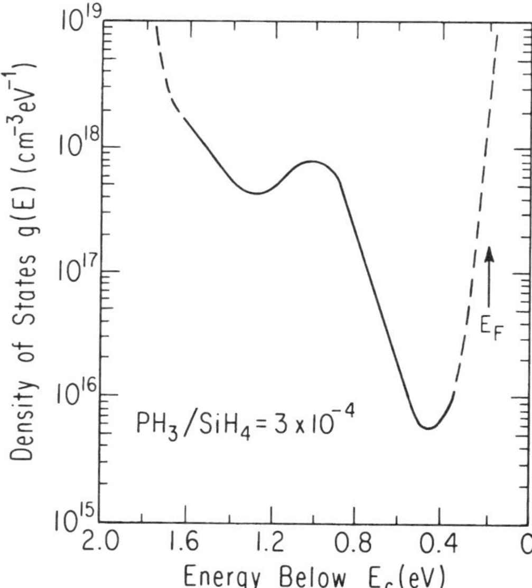

FIG. 7. Density of states in *n*-type *a*-Si:H measured by transient capacitance spectroscopy. The central peak is associated with dangling-bond defects. After Lang *et al.* (1982a).

tion process. This method is called photocapacitance spectroscopy. The photocapacitance signal, defined as the difference in the capacitance transient signals with and without subgap light exposure, reveals the distribution of optical transitions for the nonequilibrium electron population produced by the filling pulse. This spectrum is very similar to that obtained by photothermal-deflection spectroscopy (Johnson and Biegelsen, 1985; Gelatos *et al.*, 1986, 1988). However, because the change in capacitance is sensitive to the sign of the charge left behind in the depletion region, these measurements can be used to distinguish minority- and majority-carrier excitation processes (Gelatos *et al.*, 1988).

Several useful variations of this basic deep-level transient spectroscopy (DLTS) have been developed. For example, by changing the duration of the filling pulse, structural relaxations of the dangling-bond defect following the trapping of an extra electron can be studied (Cohen *et al.*, 1993), though the interpretation of these results remains controversial. By combining ESR measurements with DLTS, the paramagnetic nature of the gap state, either before or after receiving an extra electron, can be identified (Cohen *et al.*, 1982; Johnson and Biegelsen, 1985). In this manner, one can demonstrate that the peak in $g(E)$ located about 0.9 eV below E_C in n-type a-Si:H shown in Fig. 7 originates from D^- dangling-bond states. A deep minimum in $g(E)$ was found between these D^- states and the exponential rise in $g(E)$ of the conduction-band tail.

A final variation on such capacitance methods that has been used to study deep defects in *a*-Si:H with good success has been the drive-level capacitance profiling method (Michelson *et al.*, 1985). This method is an extension of the more familiar *C-V* profiling method that is used to determine the spatial variation of dopants in crystalline semiconductors. In the drive-level method, however, the variation of the junction capacitance is determined not only as a function of the dc bias but also as a function of the measurement frequency (or temperature) as well as the amplitude of the alternating voltage. In this manner it is possible to determine the distribution of deep defects as a function of both thermal energy depth and their spatial position within the sample (Unold and Cohen, 1991).

The transient emission spectroscopies require a manipulation of the band bending near a Schottky contact or a junction. This is not possible in chalcogenide glasses, whose Fermi level is pinned by the $-U$ defect centers (see Secs. 1.4 and 1.5.3).

2. ELECTRONIC TRANSPORT

In dealing with the transport properties of noncrystalline semiconductors one faces some major problems. Many studies are made on thin films whose surfaces and interfaces can have different properties from that of the bulk on account of surface adsorbates, out-diffused hydrogen, or interface states. Furthermore, electronic transport is quite sensitive to material heterogeneities that depend on preparation conditions.

Even when care is taken to alleviate these material problems, it is very difficult to extract microscopic transport parameters from experiments because of the influence of the large number of localized tail states and the fact that the anomaly of the Hall coefficient makes it impossible to learn about the band mobilities of the carriers and their temperature dependence.

Finally one is faced with profound theoretical problems. The disorder potential, which originates from the lack of long-range order in noncrystalline semiconductors, prevents us from solving the Schrödinger equation and from treating electronic transport within the effective-mass approximation. From Anderson's theory of localization (Anderson, 1958) one expects regions of localized states at the edges of the conduction and valence bands as sketched in Fig. 1 and verified experimentally as discussed in previous sections. Even though the distinction between localized and extended states may be conceptually sharp, the electronic transport usually involves both kinds of states.

In this spirit one may associate a differential conductivity $\sigma(E)$ with each energy slice dE at E by using an energy-dependent mobility $\mu(E)$ and write

$$\sigma(E) = e\mu(E)g(E)f(E)[1 - f(E)], \tag{22}$$

$$\sigma = \int \sigma(E)dE. \tag{23}$$

Even this approach (Kubo, 1957; Greenwood, 1958) is based on assumptions that are not easily justified in amorphous semiconductors. First, the use of a single-particle density of states $g(E)$ and the Fermi function $f(E)$ neglects correlation effects of electrons in localized states. Second, Eqs. (22) and (23) assume that conduction in each energy shell can be treated separately, a doubtful procedure when charge motion involves energy exchange with the lattice as in the case of phonon-assisted hopping and polaron conduction. Finally, the above equations assume that the material is homogeneous and that conduction is not spatially restricted to percolation paths.

2.1 Hopping Transport in Localized States

In contrast to extended states, the wave functions of localized states are confined in space, decaying exponentially as $\exp(-r/a)$. The localization radius a presumably depends on the binding energy $E_c - E$ of the state. A reasonable estimate is

$$a(E) = [h^2/m(E_c - E)]^{1/2}. \tag{24}$$

The tunneling or hopping rate of a charge carrier from a state at energy E_i to an unoccupied state at E_F over a separation R is

$$\nu(R) = \nu_0 \exp(-2R/a) \exp[-(E_F - E_i)/kT] \quad \text{for} \quad E_F > E_i, \tag{25a}$$

$$\nu(R) = \nu_0 \exp(-2R/a) \quad \text{for} \quad E_F < E_i. \tag{25b}$$

Both the density of localized tail states and the localization radius a increase as one approaches the mobility edge from the gap. As a consequence, the hopping probability rises rapidly so that the mobility of the charge carriers may lack the sharp discontinuity that gave the mobility edge its name (Cohen *et al.*, 1969).

At low temperatures very few charge carriers are excited into the extended states and conduction proceeds by phonon-assisted hopping first to neighboring localized states and at even lower temperatures to more distant localized states that are close in energy (variable-range hopping) (see Sec. 8 in CONDUCTIVITY IN INSULATORS AND SEMICONDUCTORS). A particular energy level in the band tail of localized states, called the transport energy, plays a crucial role in hopping transport of carriers in equilibrium as well as nonequilibrium conditions for both steady-state and transient phenomena. For a density-of-states distribution that falls off exponentially with a logarithmic slope E_0 as in Fig. 5, the distance of the transport energy E_t from the nearest mobility edge is

$$E_t = 3E_0 \ln\left[\frac{3E_0 a(4\pi N_0/3)^{1/3}}{2kT}\right], \tag{26}$$

where N_0 is the concentration of localized tail states. The transport energy maximizes the hopping rate and is the most probable final energy of hopping carriers regardless of their initial energy (Grünewald and Thomas, 1979; Shapiro and Adler, 1985; Monroe, 1985).

The hopping processes have been studied in a-In_2O_{3-x} at low temperatures (Ovadyahu, 1986, 1990; Bellingham *et al.*, 1991). Evaporated or sputtered films of a-Ge or a-Si also contain a sufficient concentration of defect states near E_F to support hopping conduction at low temperatures; however, lack of homogeneity of these materials makes a quantitative analysis difficult. The photoconductivity in both chalcogenide glasses and *a*-Si:H, at low temperatures (see Sec. 3.4), is due to hopping of photocarriers from higher-energy to lower-energy localized gap states before they recombine.

2.2 dc Conductivity in Extended States

For conduction in extended states Eqs. (22) and (23) yield for electrons

$$\sigma(T) = e\mu_c g(E_C)kT \exp[-(E_C - E_F)/kT] \tag{27}$$

and an equivalent expression for holes. In general both E_F and the mobility edges E_C and E_v depend on temperature. Above 100 K the dependence is linear so that

$$E_C - E_F = E_a - (\delta_c + \delta_F)T. \tag{28}$$

The conductivity follows the Arrhenius law

$$\sigma(T) = \sigma_0 \exp(-E_a/kT) \tag{29}$$

with the prefactor

$$\sigma_0 = e\mu_c g(E_C)kT \exp[(d_c + \delta_F)/k]. \tag{30}$$

The disorder potential severely limits the phase coherence of the extended states. The carrier mobility in the extended states is between 1 and 10 cm^2/(V s). This corresponds to a mean free path of the order of 5–7 Å at 300 K. Simple transport equations that assume weak scattering conditions are obviously not valid for these small scattering lengths. Using the random-phase approximation, Friedman (1971) obtained for the band mobility

$$\mu = \frac{2\pi}{3}\frac{ea^3}{h} C_{av} \frac{J}{kT} a^3 J g(E_V), \tag{31}$$

where $C_{av} = 3$ is an average coordination number, $J \approx 1$ eV is the two-site transfer integral, and $a = 0.25$ nm is the interatomic spacing. Another estimate (Cohen, 1970) treats the charge transport in classical terms as diffusion of carriers jumping between neighboring sites with jump rate $\nu = J/h$:

$$\mu = \frac{1}{6}\frac{ea^2}{kT}\nu. \tag{32}$$

Both expressions (31) and (32) yield $\mu \sim 5$ cm^2/(V s) at 300 K.

2.2.1 Chalcogenide Glasses The conductivity of chalcogenide glasses follows the Arrhenius law (28) with an activation energy E_a that is close to one-half the optical gap: $E_a = E_g/2$. This quite general relationship between E_a and E_g in chalcogenide glasses suggests that E_F is pinned near midgap and that the optical gap E_g is approximately equal to the mobility gap $E_C - E_V$. Figure 8 shows the logarithm of the room-temperature conductivity against E_a for a large number of chalcogenide glasses. The straight lines have the slope $(kT)^{-1}$ with $T = 300$ K, and the intercept with the $E_a = 0$ axis is σ_0. For most of these materials one finds $10^2 < \sigma_0 < 10^4$ V^{-1} cm^{-1}. The positive sign of the thermopower indicates that conduction is by holes in the valence band. From the mid-range value $\sigma_0 = 10^3$ V^{-1} cm^{-1} and Eq. (30) we can estimate the hole mobility μ_v in the following manner. In these glasses $\delta_F = 0$ because E_F is pinned by the coordination defects, which have a negative correlation energy (see Sects. 1.4 and 1.5.3). The temperature dependence of the hole mobility edge E_V is approximately half of that of the gap, which is for these glasses is

$$\frac{dE_g}{dT} \approx 8 \times 10^{-4} \text{ eV/K}. \tag{33}$$

Substituting these values into Eq. (30) and using a reasonable estimate $g(E_V) = 10^{21}$ cm^{-3} eV^{-1} we obtain at 300 K a hole mobility $\mu_v = 2$ cm^2/(V s). This agrees well with the estimates from Eqs. (31) and (32). This shows that between, say, 200 K and the glass-transition temperature T_g the conduction takes place in the extended states of the valence band. Only some Bi-containing chalcogenide glasses show *n*-type conduction.

2.2.2 Hydrogenated Amorphous Silicon The importance of *a*-Si:H arises from the fact that the Fermi energy can be shifted through the gap by *n*-type or *p*-type doping

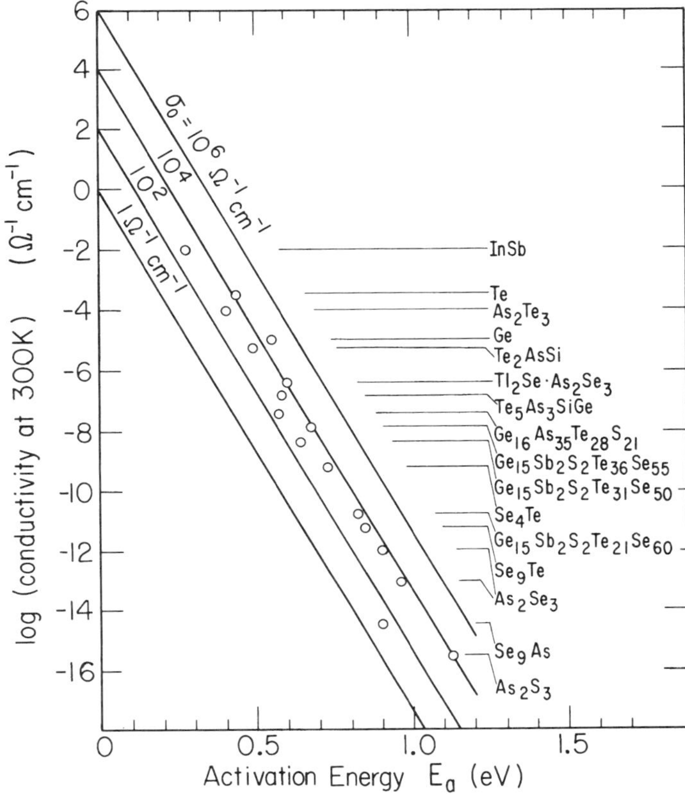

FIG. 8. Relation between conductivity at 300 K and activation energy E_a defined by Eq. (29) for several chalcogenide glasses.

(Spear and LeComber, 1975). Figure 9 shows the dramatic change of the room-temperature conductivity as small fractions of dopant gases B_2H_6 (for *p*-type) and PH_3 (for *n*-type) are added to the SiH_4 gas used for plasma-assisted chemical vapor deposition of these materials.

Over a large temperature range $150 < T < 450$ K the conductivity follows the Arrhenius law (29). In contrast to chalcogenide

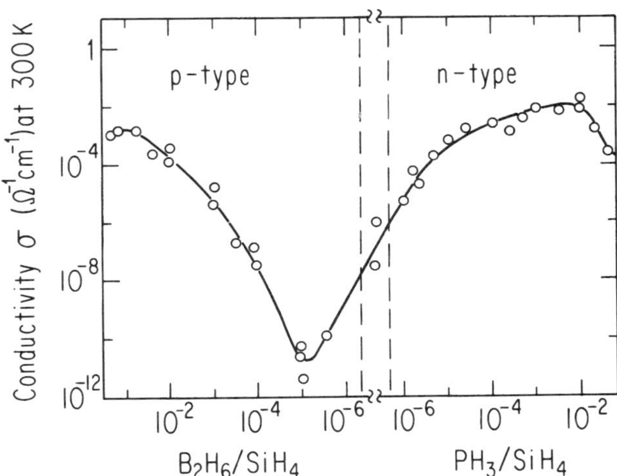

FIG. 9. Conductivity at 300 K of *a*-Si:H as a function of dopant concentration in the gas phase. After Beyer and Overhof (1984).

glasses, however, σ_0 is not constant but changes over five orders of magnitude as E_a is changed by *n*-type or *p*-type doping. This is shown in Fig. 10. This relation can approximately be described by

$$\ln\sigma_0 = \ln\sigma_{00} + E_a/kT_m, \tag{34}$$

where $\sigma_{00} \approx 0.1\ \mathrm{V^{-1}\ cm^{-1}}$ and $T_m \approx 600$ K. T_m is called the Meyer–Neldel temperature because relation (34) was first observed in a number of polycrystalline materials by Meyer and Neldel (1937). Overhof and Thomas (1989) tried to explain this relation in *a*-Si:H in the following manner. In strongly doped material, E_F lies in the steeply rising $g(E)$ of the band tails. E_F moves in general toward lower $g(E)$ with increasing T in order to conserve charge neutrality. This so-called statistical shift of E_F contributes therefore a term δ_F to Eq. (28), which can be of either sign depending on the shape of $g(E)$ near E_F. As a consequence, σ_0 given by Eq. (30) is expected to depend on E_a. The solid curve in Fig. 10 was calculated by them for a reasonable model for $g(E)$. At high doping levels (small E_a) the small values of σ_0 may be due to hopping conduction in the localized tail states.

2.2.3 Amorphous In_2O_{3-x} In most amorphous semiconductors, E_F cannot be shifted into the extended states to produce a degenerate semiconductor. In chalcogenide glasses, E_F is pinned by $-U$ defect centers and in *a*-Si:H doping is limited by the self-compensating increase in dangling-bond defect concentration. In amorphous (and microcrystalline) In_2O_{3-x}, however, the electron concentration is governed by the oxygen deficiency x and E_F can be shifted through the mobility edge E_c to produce a metallic state. A series of $\sigma(T)$ curves obtained by stepwise oxidizing a conducting film of *a*-In_2O_{3-x} is shown in Fig. 11. The lower conductivity curves exhibit the Meyer–Neldel relation with $\sigma_{00} \approx 0.1\ \mathrm{V^{-1}\ cm^{-1}}$ and $T_m \approx 600$ K similar to *a*-Si:H. The films are metallic for $\sigma \geq 150\ \Omega^{-1}\ \mathrm{cm^{-1}}$ (Bellingham *et al.*, 1991; Pashmakov *et al.*, 1993).

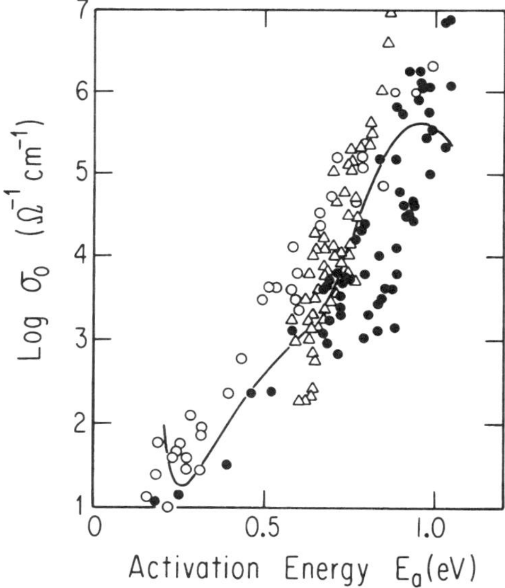

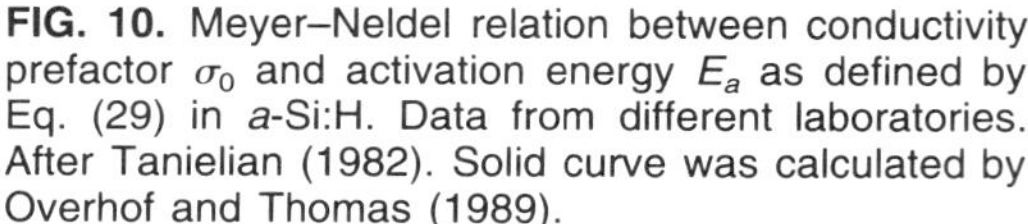
FIG. 10. Meyer–Neldel relation between conductivity prefactor σ_0 and activation energy E_a as defined by Eq. (29) in *a*-Si:H. Data from different laboratories. After Tanielian (1982). Solid curve was calculated by Overhof and Thomas (1989).

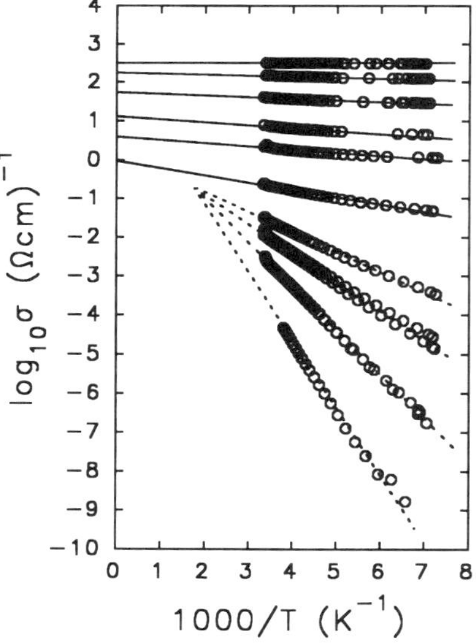

FIG. 11. Temperature dependence of the conductivity of a 25-nm-thick In_2O_{3-x} sample, first made metallic by photoreduction and then reoxidized in steps. Lower σ curves extrapolate to a common point at T_m = 590 K. After Pashmakov *et al.* (1993).

2.3 Hall Effect

When a magnetic field B_y is applied normal to a current density j_x, an electric Hall field E_z appears perpendicular to both the current and the magnetic field driven by the Lorentz force. The Hall coefficient R relates these quantities:

$$E_z = Rj_xB_y. \tag{35}$$

Since the conductivity is proportional to the product of the carrier concentration and mobility, studying both $\sigma(T)$ and $R(T)$ yields the temperature dependence of the carrier concentration and the mobility. In amorphous semiconductors (with one exception) the Hall effect is anomalous and does not give the desired information (Fritzsche, 1974; LeComber *et al.*, 1977; Kakalios, 1989). It has the opposite sign from what is expected from doping and thermopower measurements and the magnitude is about 50 times smaller than expected from conductivity measurements. The origin of this anomaly is still a mystery. The Hall-effect anomaly does not seem to be caused by the small value of the carrier mobilities in the extended band states because crystalline semiconductors with similar or smaller mobilities behave normally.

In contrast to the other noncrystalline semiconductors, the metal-oxide films listed under 2.A. in Table 1 show a normal sign and magnitude of the Hall effect. The carriers are electrons and their mobility μ = 5–20 $cm^2/(V\ s)$ is similar to that of electrons in *a*-Si:H or of holes in chalcogenide glasses. This shows that the Hall-effect sign anomaly of the latter is not due to the small value of the mobility. The carrier concentration determined from R in *a*-In_2O_{3-x} agrees with the optical free-carrier absorption (Bellingham *et al.*, 1990). Moreover, the Hall mobilities of *a*-In_2O_{3-x} and of $AgSbO_3$ are nearly the same in the amorphous and in the crystalline states (Bellingham et al., 1990; Yasukawa *et al.*, 1995).

2.4 Thermopower

A temperature gradient across a conductor causes the charge carriers at one end to be more energetic than at the other. This causes a drift or diffusion of the energetic carriers from the hotter to the cooler end until an opposing electric field, the thermopower, prevents further charge flow and establishes a steady state. The sign of the potential at the cold end is therefore the sign of the dominant charge carriers.

The thermopower or Seebeck coefficient is $S = \Delta V/\Delta T$ where ΔV is the potential difference measured at two points of the sample having a temperature difference ΔT. The interpretation of the thermopower is the same for amorphous and crystalline materials. The Onsager relation associates the thermopower S with the Peltier coefficient:

$$\Pi = ST. \tag{36}$$

This coefficient has the simple meaning that it is the energy transported by the charge carriers weighted by their contribution to the electrical conductivity. The energy transported is measured relative to the Fermi energy E_F. One therefore can write quite generally (Fritzsche, 1971)

$$\Pi = ST = -\frac{1}{e}\frac{\int (E - E_F)\sigma(E)dE}{\int \sigma(E)dE}. \tag{37}$$

All contributions to Eq. (37) from energy states above E_F are from electrons; contributions from states below E_F have the opposite sign and hence are from holes. S has the sign of the dominant charge carrier. The validity of Eq. (37) is limited by the same assumptions mentioned in connection with Eqs. (22) and (23).

For electron conduction above the electron mobility edge E_c, Eq. (37) reduces to

$$S = -k/e\left[\frac{E_C - E_F}{kT} + A\right]. \tag{38}$$

The first term in this expression describes the thermopower when all electrons have the same energy E_c. However, since the electrons have an average energy of about $\frac{3}{2}kT$ above E_c, there is the correction term $A \approx \frac{3}{2}$.

For hole conduction near the mobility edge E_V, the sign of S is positive as E_C in Eq. (38) is replaced by E_V. According to Eqs. (27) and (38), the temperature dependences of $S(T)$ and of $\ln\sigma(T)$ are both governed by the energy separation of E_F from the nearest mobility edge. Hence, the quantity $Q = \ln\sigma + (e/k)S$ should be a constant. Experimentally one finds, however,

$$Q = Q_0 - E_Q/kT \quad (39)$$

with $0.05 < E_Q < 0.2$ eV for *a*-Si:H, *a*-Ge:H, and the chalcogenide glasses. This additional E_Q term depends on preparation parameters and heat treatment and may be due to long-range random potential fluctuations (Beyer and Overhof, 1979; Overhof and Beyer, 1984).

2.5 Conductance Fluctuations in Hydrogenated Amorphous Silicon

Studies of conductance fluctuations have been employed to investigate aspects of defect kinetics and electronic properties in amorphous semiconductors that are not revealed in bulk transport measurements. Observations of random telegraph switching noise (RTSN) in semiconducting devices with small effective volumes and constraining geometries at cryogenic temperatures, which limit the number of free charge carriers in the material, have been attributed to carrier capture and emission processes of individual defects. The recent observation of RTSN, that is, sharp jumps between discrete resistance levels, in coplanar and transverse current measurements in macroscopic volumes of *a*-Si:H at and above room temperature was therefore quite surprising (Rogers *et al.*, 1986; Arce *et al.*, 1989; Choi *et al.*, 1990; Parman *et al.*, 1991). Figure 12 shows a time trace of the resistance of an *n*-type (10^{-3} PH_3/SiH_4) doped *a*-Si:H film with coplanar electrodes (width 0.2 mm, separation 1 mm) at 400 K (Parman *et al.*, 1991). Switching events have been observed for time scales ranging from a few milliseconds to several tens of seconds, with fractional resistance jumps varying from $\Delta R/R \sim 10^{-2}$ to 10^{-4}. Comparison of two-probe and four-probe measurements confirms that the RTSN is not a contact effect.

If the RTSN in *a*-Si:H arose from charge trapping in defects, then for the macroscopic film in Fig. 12 the magnitude of the resistance jumps implies that 10^5 electrons are simultaneously trapped and released. Alternatively, it has been suggested that the RTSN reflects the presence of inhomogeneous current filaments, which arise from the long- or medium-range compositional disorder or potential fluctuations in the amorphous semiconductor. The switching noise indicates that the resistance of these microchannels is time dependent, possibly as a result of hydrogen or electronic motion. If the fractional resistance jump is associated with a change in the volume of a microchannel available to carry current, then the cross-sectional diameter of a current filament is estimated to be on the order of ~1–10 μm, which is consistent with microchannel size estimates based on RTSN studies of transverse current measurements (Arce *et al.*, 1989; Choi *et al.*, 1990; Teuschler *et al.*, 1993). However, it is not obvious how hydrogen motion could influence the conductance of a channel this large, nor how a smaller filament could influence the electronic conductivity of a larger region. If the conductance fluctuations in amorphous silicon do result from inhomogeneous current filaments, then, unlike other

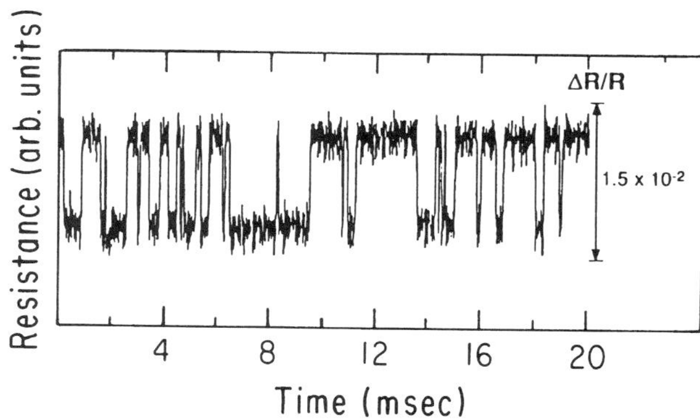

FIG. 12. Resistance fluctuations in *a*-Si:H at 400 K. After Parman *et al.* (1991).

transport measurements, noise studies (in essence an electronic measurement) can reveal information concerning disorder over length scales that are much longer than the inelastic scattering length.

3. PHOTOEXCITATION AND RECOMBINATION

Illumination with band-gap light, $h\nu \geq E_g$, creates a nonequilibrium condition in which excess electrons and holes populate extended and localized states. Radiative as well as nonradiative recombination of excess electron–hole pairs tends to restore equilibrium. The excess carriers give rise to a photoconductivity that can exceed the dark conductivity by a factor of many million, which is used for light detectors, for detector arrays, and in electrophotographic copiers. The excess carriers convert light into electricity in photovoltaic cells and solar panels. The process of light emission during radiative recombination is called photoluminescence.

The thermalization and recombination processes involve a sequence of steps, illustrated in Fig. 13, which are distinguished by different time scales. Thermalization in extended states proceeds with emission of phonons at a cooling rate of about 2 eV/ps, much faster than in crystals, because the crystal-momentum selection rules do not apply to noncrystalline materials. Thermalization into localized states occurs at a rate $\nu_0 \sim 10^{13}\ s^{-1}$, a phonon frequency. Transitions between localized states separated by R occur by phonon-assisted hopping at a rate described by Eq. (25).

3.1 Radiative and Nonradiative Recombination

Figure 13 illustrates a number of recombination processes. They occur from the extended states or from the localized tail states. Since the recombining electron–hole pairs have energies between $E_g/2$ and E_g, an appreciable energy has to be dissipated in the process. This occurs either radiatively by emission of a photon or nonradiatively by an Auger process or the emission of many phonons. Radiative recombination (photoluminescence) is favored at low temperatures and for larger recombination energies. However, the process cannot be faster than $\tau_0 \sim 10^{-8}$ s, the dipole radiation lifetime. Since the carriers get trapped by localized tail states in about $\nu_0 \sim 10^{-13}$ s, one does not observe recombination between extended states. Radiative recombination between localized electrons and holes separated a distance R occurs by tunneling at a rate

$$\nu_r = \tau_0^{-1} \exp(-2R/a). \tag{40}$$

As this rate is relatively slow, nonradiative recombination processes dominate in materials with large electron–phonon interactions. In hydrogenated amorphous silicon, for instance, the Si dangling-bond defects are effective nonradiative recombination centers. In chalcogenide glasses it is possible that the recombination energy is dissipated by a brief

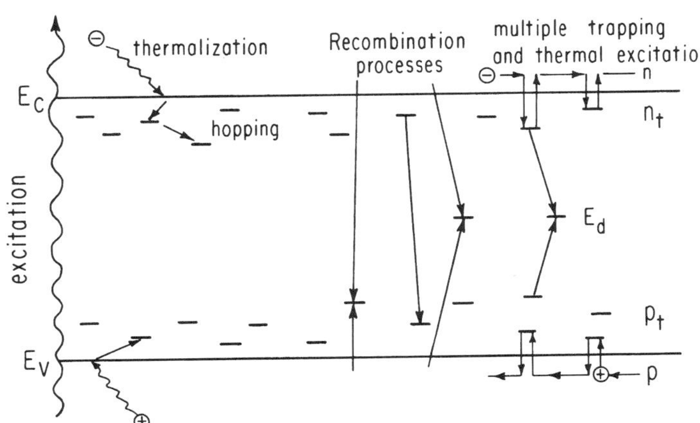

FIG. 13. Schematic representation of excitation, thermalization, and recombination of charge carriers.

configuration change, which is modeled by Street (1977) and illustrated in Fig. 14. This model suggests that an electron or a hole localized in tail states changes temporarily the valency of nearby atoms. This causes a distortion toward the configuration of an intimate valence-alternation pair (see Sec. 1.5.3). In this example the pair consists of overcoordinated and undercoordinated Se atoms. Path I in Fig. 14 corresponds to recombination of the transient exciton, path II to the photoinduced creation of a metastable defect pair, a process to be discussed in Sec. 4.1. Auger processes are probably responsible for the fast nonradiative recombination (a few picoseconds) of very high photocarrier concentrations (10^{19}–10^{21} cm^{-3}) excited by ultrashort laser pulses.

3.2 Geminate Recombination

Recombination is called geminate when the very same electron–hole pair recombines as was created by photoexcitation. It occurs when the pair does not drift far apart, i.e., at low temperatures and low light intensities and in materials having small thermal-carrier concentrations. The geminate recombination lifetime is independent of excitation intensity because the fate of each pair is independent of the others. This distinguishes it from nongeminate recombination.

Onsager (1938) calculated the probability for an electron to escape the attractive Coulomb force of its hole. This model is less applicable to amorphous semiconductors because the electron need not overcome the Coulomb potential but can escape by hopping to localized states in the Coulomb well.

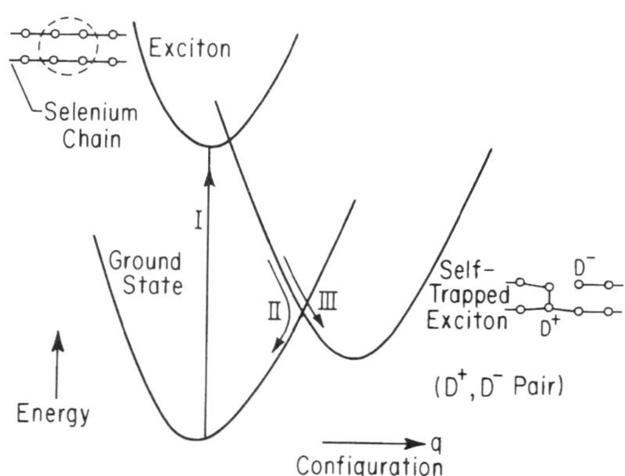

FIG. 14. Energy-configuration diagram showing (I) photoexcitation, (II) recombination via a transient exciton, and (III) the formation of a self-trapped exciton. After Street (1977).

3.3 Photoluminescence

3.3.1 Steady-State PL The light emitted by the radiative recombination processes is called photoluminescence. Figure 15 shows for three chalcogenide glasses the photoluminescence spectrum (PL) as well as the spectrum of photoexcitation light (PLE) that produces the photoluminescence. It is understandable that the PLE spectrum starts with the spectrum of optical absorption α and falls off at higher photon energies where the excitation is limited to a surface layer. More remarkable is the rather large Stokes shift of the PL to about one-half the energy gap. This Stokes-shift energy is understood to be dissipated by a lattice relaxation that is closely related to the relaxation that produces a negative effective correlation energy

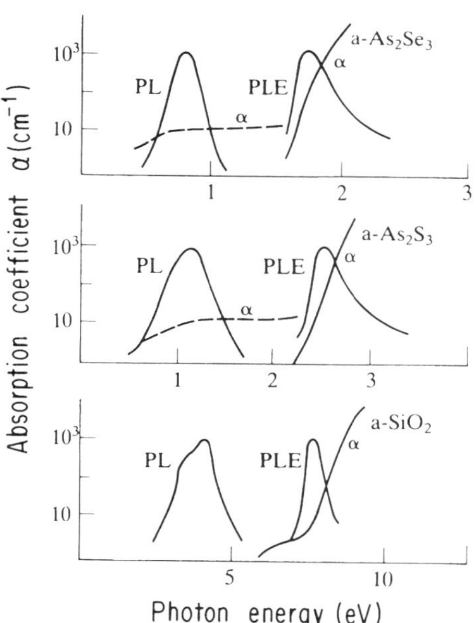

FIG. 15. Photoluminescence (PL) and photoluminescence excitation (PLE) spectra of three chalcogenide glasses. The solid curve α represents light-induced metastable absorption. After Gee and Kastner (1979).

($-U$) of the coordination defects in these glasses (see Secs. 1.4 and 1.5.3).

Prolonged light exposure of chalcogenide glasses decreases the PL efficiency at low temperatures. This PL fatigue is accompanied by the growth of a subgap absorption shown by the dashed lines in Fig. 15. These light-induced effects and the appearance of an ESR signal disappear upon warming the samples to room temperature. These effects arise at low temperatures when an increasing number of normally charged $-U$ defects become neutral and paramagnetic by trapping photoexcited carriers.

In *a*-Si:H, excitation with band-gap light produces a PL peak at low temperatures near $h\nu$ = 1.3 eV with a full width of 0.3 eV. Although the PL energy is again lower than the excitation energy, the difference is believed to be the energy loss of the carriers as they hop down into tail states before recombining radiatively. The defects are predominantly nonradiative recombination centers and produce only a weak PL at 0.9 eV.

3.3.2 Time-Resolved PL After a short (less than 10^{-9} s) excitation pulse, the decay of PL proceeds with diminishing intensity to about 10^{-2} s. The PL decay of *a*-Si:H is shown as an example in Fig. 16. The lower panel shows the distribution of radiative lifetimes calculated from the upper-panel data. One notices a broad (two orders of magnitude wide) distribution of lifetimes centered near $\tau = 10^{-4}$ s or 10^{-5} s depending on the temperature. The broad distribution of lifetimes suggests that the electron–hole pairs recombine radiatively by tunneling over separations R according to Eq. (39). With an increase in temperature the PL intensity decreases because of thermally assisted tunneling of carriers to nonradiative recombination centers. This affects in particular long-lived pairs causing in Fig. 16 the shift of the lifetime distribution to shorter times (Tsang and Street, 1979).

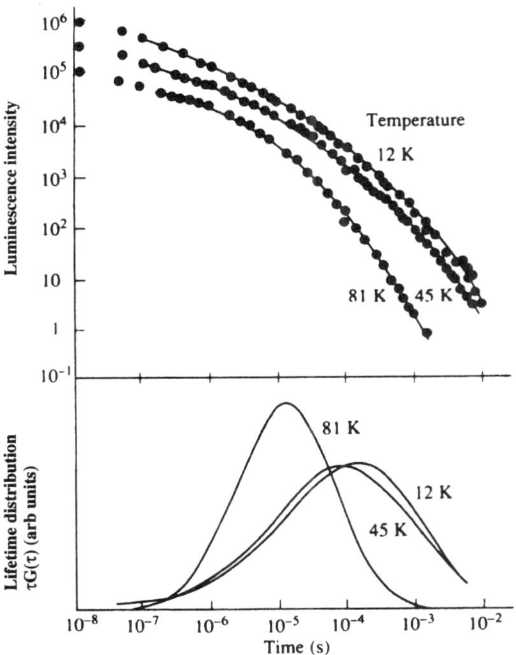

FIG. 16. Decay of photoluminescence of *a*-Si:H following a short excitation pulse. The lower curves are the distribution of luminescence lifetimes calculated from the data. After Tsang and Street (1979).

3.4 Photoconductivity

3.4.1 Primary and Secondary Photoconductivity Ordinary photoconductivity is measured by using Ohmic electrical contacts to the semiconductor, which pass both electrons and holes. The material is supposed to be free of space charge and the photocurrent is proportional to the applied electric field. This is the secondary photoconductivity and the topic of the following section. If the contacts are blocking, one deals with the primary photocurrent. Examples are charge collection in *p-i-n* solar cells or structures with Schottky-barrier metal contacts. The primary photocurrent is mentioned in Sec. 3.5 on time-of-flight drift mobility. The recombination processes are best studied using the secondary photoconductivity because space charges, which distort the applied field, are absent.

3.4.2 Steady-State Photoconductivity In steady state the photoconductivity does not change with time because the constant excitation rate is exactly balanced by the recombination rate. A steady-state nonequilibrium distribution function can be defined (Simmons and Taylor, 1971), which when substituted into Eqs. (22) and (23) yields the photoconductivity σ_p. For electrons in extended states having a mobility μ_C one can write (with equivalent expressions for holes)

$$\sigma_p = en\mu_C, \tag{41}$$

$$n = \eta_n G \tau_n. \tag{42}$$

Here G is the electron–hole pair-generation rate (in units of $cm^{-3}\ s^{-1}$) and τ_n is the electron recombination lifetime. The quantum efficiency η_n is the fraction of electrons that not only get excited to conduction-band states above E_c in Fig. 1 or 13 but that also recombine nongeminately. The quantum efficiency is smaller than unity when carriers get excited to localized states or when pairs recombine geminately and hence are unable to transport a net charge.

The electron recombination lifetime τ_n usually gets shorter with increasing generation rate G because more holes become available for recombination. Consequently Eqs. (41) and (42) yield a sublinear dependence on light intensity,

$$\sigma_p \propto G^\gamma \quad \text{with} \quad \gamma \leq 1. \tag{43}$$

As is illustrated in Fig. 17 the temperature dependence of the photoconductivity has essentially the same shape for many amorphous semiconductors, even though the magnitudes differ. The small values found in chalcogenide glasses may be attributed to a small quantum efficiency due to geminate recombination. The rapid rise of σ_p with increasing T is attributed to the rise in free-carrier concentration n relative to the trapped-carrier concentration n_t. Major factors that decrease the magnitude of σ_p for a given generation rate G are the defect concentration, the concentration of localized tail states, geminate recombination, potential fluctuations, and compositional heterogeneities.

The unusual decrease and minimum in $\sigma_p(T)$ observed in Fig. 17 for a-Si:H near room temperature is called thermal quenching of photoconductivity (Rose, 1963). In this material it can be attributed to the opening of a new recombination channel as, with increasing T, the trapped holes become more mobile and enhance recombination through defect states. In this region one observes supralinearity, $\gamma \geq 1$ in Eq. (43).

The temperature-independent photoconductivity at very low T cannot be described by Eqs. (41) and (42). It is due to the hopping of photocarriers down to lower-energy localized tail states. This energy-loss hopping attains an overall drift component in the applied field, which results in photoconductivity (Shklovskii *et al.*, 1989).

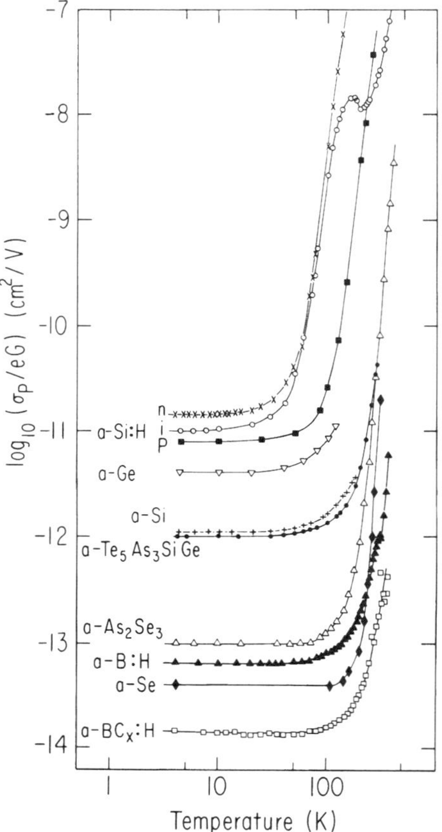

FIG. 17. Temperature dependence of the photoconductivity normalized by the electron charge and the excitation rate for different noncrystalline semiconductors. After Johanson *et al.* (1989).

3.4.3 Photoconductivity Decay Upon terminating illumination the photocurrent decays from its steady-state value as the nonequilibrium distribution of the carrier population returns via thermal excitation and recombination to the equilibrium distribution $f(E)$. The initial logarithmic decay defines the decay time τ_D as

$$\left.\frac{d \ln\sigma_p}{dt}\right|_{t=0} = \frac{1}{\tau_D}. \tag{44}$$

A measurement of τ_D and of $\mu_C\tau_n$ from the

steady-state photoconductivity allows one to calculate the electron-drift mobility μ_d using the relation (Hoheisel *et al.*, 1988)

$$\mu_C \tau_n = \mu_d \tau_D. \tag{45}$$

By measuring the decay of the photocurrent after a short laser pulse (1–10 ns), instead of the steady-state photocurrent, one gains information about the product of the band mobility and the deep trapping or recombination lifetime of the dominant photocarriers (Antoniadis and Schiff, 1992).

3.4.4 Modulated Photoconductivity

The modulated photocurrent method (Oheda, 1981) is used to determine the density of localized states in a certain energy region of the mobility gap. The modulated photoconductivity $\sigma_m(\omega)$ is measured in response to a sinusoidal component of the electron–hole pair-generation rate $G_m(\omega)$. The density of states $g(\Delta E)$ at the energy

$$\Delta E(\omega) = kT \ln(\nu_0/\omega) \tag{46}$$

below the nearest mobility edge is obtained from the magnitude of $\sigma_m(\omega)$ and the phase shift ϕ of the photocurrent relative to the light modulation. According to Brüggemann *et al.* (1990) the relation is

$$g(\Delta E) = \frac{2}{\pi} g(E_C) \frac{\omega}{\nu_0} \left[\frac{e\mu_C G_m(\omega)}{\omega \sigma_m(\omega)} \sin\phi - 1 \right], \tag{47}$$

where $g(E_C)$ and μ_C are the density of states and the mobility at the electron mobility edge E_C. Here it is assumed, as in Eq. (27), that electrons contribute more to the photocurrent than holes. Equation (46) is derived from Eq. (21) by relating the modulation frequency ω with the inverse thermal release time τ_e^{-1}. Their relation is the basis of this experiment because photocarriers at energy $\Delta E(\omega)$ are thermally released and retrapped synchronously with the modulation frequency albeit with a phase lag ϕ.

The gap energy ΔE of Eq. (46) can be scanned by varying either the frequency ω or the temperature T. A larger energy range is explored by changing both. An example of the density of gap states of an *a*-Si:H sample determined by the modulated photoconductivity method is shown in Fig. 18. The energy separation ΔE from the electron mobility edge was determined for each T and ω from Eq. (46). The temperature scale for the 8-Hz modulation measurements is shown on top of the figure. The nearly identical density-of-states distributions obtained from the three modulation frequencies indicate the correctness of the basic assumptions. The peak at $\Delta E = 0.65$ eV is due to dangling-bond defect states.

3.5 Drift Mobility

As sketched on the right-hand side of Fig. 13, each carrier is frequently trapped into localized tail states and thermally emitted back

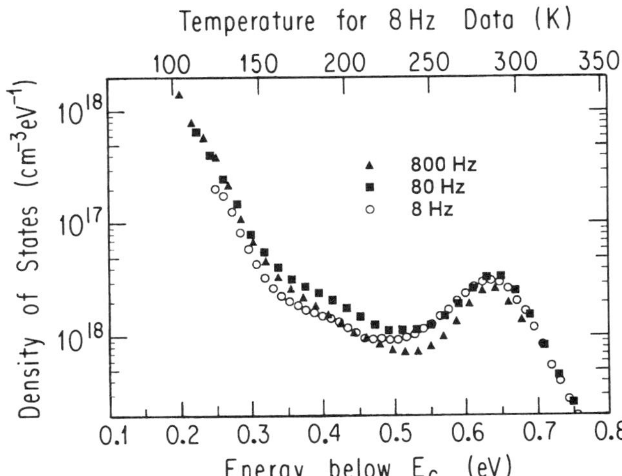

FIG. 18. Density of gap states of *a*-Si:H determined with the modulated photoconductivity method by scanning the temperature for three modulation frequencies. The energy distance ΔE from the electron mobility edge was obtained from Eq. (46) by accounting for a small temperature variation of ν_0. After Zhong and Cohen (1993).

into the extended states. The long-time average mobility of an electron, for example, is not its extended-state mobility μ_C but a smaller drift mobility μ_d which depends on the ratio of the time the electron is free τ_f to the time it is trapped τ_t as

$$\mu_d = \mu_C \tau_f/(\tau_f + \tau_t). \tag{48}$$

One way of measuring μ_d is with the time-of-flight apparatus sketched in Fig. 19(a). An excitation light pulse creates electron–hole pairs near the left transparent contact to the sample. Its photon energy is sufficiently high to have a short absorption depth. Depending on the sign of the applied field F, either electrons or holes are drawn through the sample. In the example of Fig. 19(a) the photoexcited holes drift with an average velocity

$$v_d = \mu_d F \tag{49}$$

through the length L of the sample toward the negative electrode. The drift current I_d is recorded as a function of time. The current I_d should persist until the carriers arrive at the negative electrode after a transit time

$$\tau_T = L/v_d, \tag{50}$$

which with Eq. (49) determines μ_d. In most cases, however, the observed $I_d(t)$ is so featureless that τ_T cannot be identified. Scher and Montroll (1975) discovered that this is to be expected when the motion of the carriers is subjected to a broad distribution of trap release times as is the case here.

They further found that τ_T is defined by the position of a break in the curve when $I_d(t)$ is plotted on a logarithmic plot as shown in Fig. 19(b). This transport is called dispersive and the parameter α obtained from the slopes of the $\log I_d$ vs $\log t$ curves is called the dispersion parameter.

A characteristic feature of dispersive transport is that μ_d depends on the measurement time. The reason is the following. The trapping time τ_t in Eq. (48) increases exponentially with increasing binding energy of the tail-state traps,

$$\tau_t = \nu_0^{-1} \exp[(E_C - E)/kT]. \tag{51}$$

As a consequence, when the measuring time is increased, more deep traps can participate in trapping and releasing drifting electrons. Hence, τ_t increases thereby reducing μ_d. In the time-of-flight experiment, the measurement time is τ_T, which depends on the field F according to Eqs. (49) and (50). This in brief explains the observed field dependence of μ_d shown by the data in Figs. 20(a) and 20(b). Plotted here is the temperature dependence of μ_d of electrons and holes in a-Si:H. These curves and the dispersion parameter α can be fitted with assumed density of states distributions $g(E)$ of localized tail states. Good fits are obtained with $g(E)$ increasing exponentially toward E_C and E_V, respectively, as sketched in Fig. 1. The logarithmic slopes of these tails are $E_0 \sim 0.05$ eV for holes and 0.025 eV for electrons (Tiedje and Rose, 1980; Orenstein and Kastner, 1981).

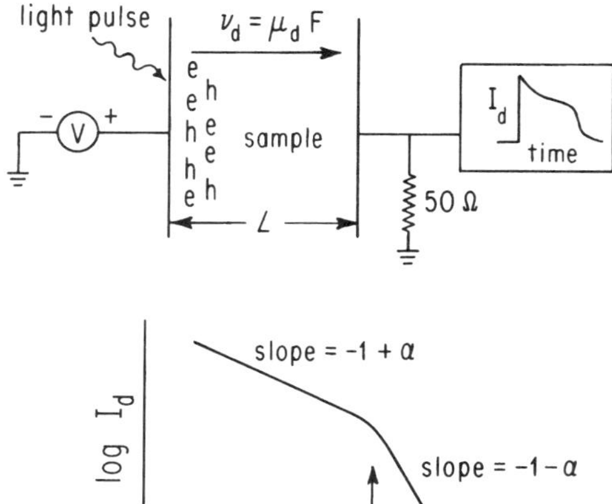

FIG. 19. (a) Sketch of time-of-flight apparatus and (b) time dependence of dispersive drift current.

3.5.1 Deep Trapping In the absence of deep traps or defect states one should eventually collect the total charge Q_0 generated by the light pulse in a time-of-flight experiment. On the other hand, when carriers are lost to deep traps at a rate $1/\tau$, a smaller Q is collected. According to Hecht (1932)

$$\frac{Q}{Q_0} = \frac{\mu_d \tau F}{L}\left[1 - \exp\left(-\frac{L}{\mu_d \tau F}\right)\right]. \tag{52}$$

A measurement of charge collection Q/Q_0 as a function of applied field F in Fig. 19(a)

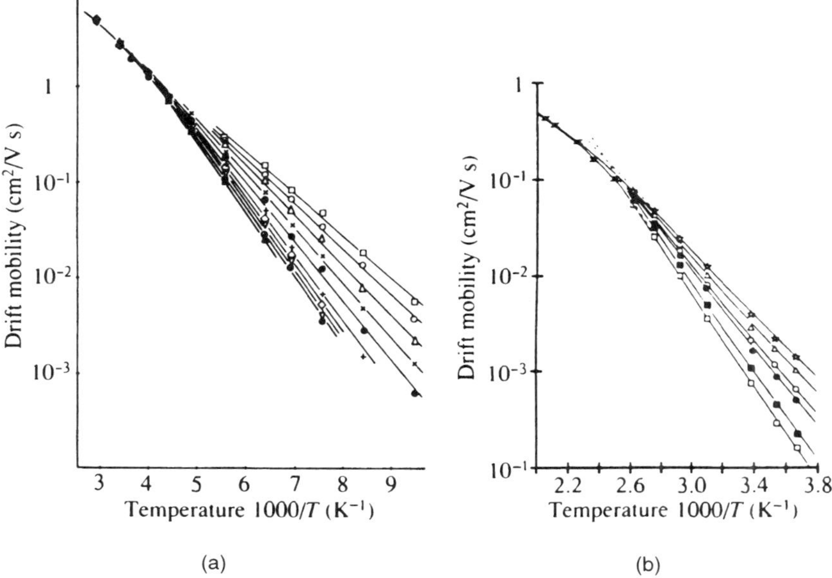

FIG. 20. Temperature dependence of the (a) electron and (b) hole drift mobilities of *a*-Si:H at different electric fields increasing from the bottom from 5 × 10^2 to 5 × 10^4 V/cm. The field dependence is caused by dispersion. After Street (1991b).

yields the product $\mu_d \tau$ of drift mobility and deep trapping time. Figure 21 shows $\mu_d \tau$ at 300 K for electrons and holes in both *n*-type and *p*-type as well as compensated and intrinsic *a*-Si:H (Street *et al.*, 1983). The results confirm the following picture. The dangling-bond defect concentration increases with *n*-type or *p*-type doping according to Eqs. (8) and (9). In *n*-type samples these defects are negatively charged D^- and hence can trap holes but not electrons. The results of Fig. 21 on *n*-type samples confirm this: $\mu_d \tau$ of holes decreases with *n*-type doping but $\mu_d \tau$ of electrons remains constant at 10^{-6} cm^2 V^{-1}. The opposite happens in *p*-type samples where the defects become D^+ and can trap electrons but not holes. In intrinsic *a*-Si:H, in contrast, many defects are neutral so that $\mu_d \tau$ of both electrons and holes are reduced with increasing defect concentration. The results also indicate that the drift mobility μ_d and thus the tail-state slope E_0 are not affected by doping.

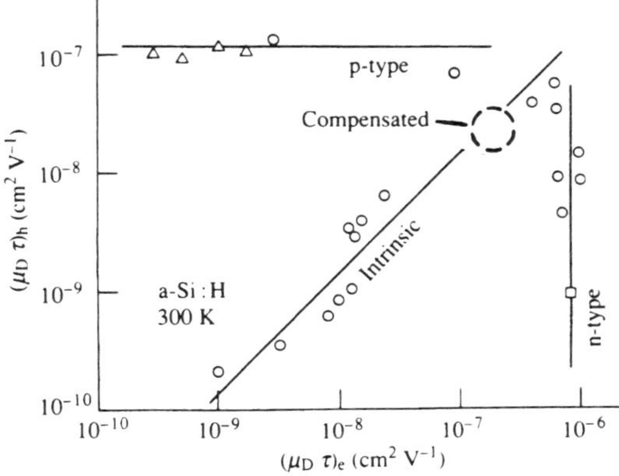

FIG. 21. Product of band mobility and deep trapping lifetime of electrons and holes for undoped, doped, and compensated *a*-Si:H. After Street *et al.* (1983).

3.6 Spin-Dependent Recombination

Recombination of an electron–hole pair is allowed when the spins of the electron and the hole are antiparallel and forbidden when they are parallel because the final state of the recombination process must have zero spin. Geminate recombinations proceed unhindered unless spin-flip processes occur during thermalization. Nongeminate pairs, on the other hand, have random spin alignments. The recombination rates can, therefore, be changed by perturbing the thermal spin alignments by a magnetic field or by changing the population of Zeeman-split levels resonantly with microwave fields (ESR-induced transitions). Since different g values are associated with tail-state electrons, tail-state holes, and neutral defect states, the role of each of these entities in the recombination event can be studied in detail.

A magnetic field produces a spin polarization P_e and P_h of the electrons and holes, respectively, which decreases their recombination rate W:

$$W = W_0(1 - P_eP_h). \tag{53}$$

At resonance, the microwave field tends to equalize the populations of the spin states and to reduce the spin polarizations. This enhances the photoluminescence and reduces the photoconductivity. At 300 K a spin polarization of $P_eP_h \sim 10^{-6}$ is expected for $g \sim 2.0$ spins and a field of 3000 G. Experimentally one finds, however, changes in photoluminescence, photoconductivity, and photoinduced absorption that are several orders of magnitude larger than this value. This enhancement is caused by the fact that a net surplus of spin-aligned electron–hole pairs accumulates because their recombination is forbidden. No macroscopic magnetization occurs, however, because the orientations of the spin-aligned pairs remain random. A spin flip of either partner of the spin-aligned pair at ESR resonance induces recombination. This large spin correlation between recombining electron–hole pairs enhances the sensitivity of optically detected magnetic resonance far beyond the sensitivity of ordinary ESR (Solomon, 1979; Morigaki, 1984).

4. METASTABILITY

Essentially all noncrystalline semiconductors exhibit changes in their electronic and optical properties when they are subjected to electron bombardment, strong light, or other excitations. These changes are metastable and reversible in that the original properties can be restored by annealing the material at a characteristic equilibration temperature T_E that usually lies between 150 and 300 °C.

Metastabilities are well known in glasses whose structure, density, and general properties can be changed by quenching from high temperatures where they are soft or liquid to room temperature where they are solid. The original properties can be restored by annealing near the glass-transition temperature. The metastabilities of noncrystalline semiconductors are quite different, however. They are produced by electronic excitations below the equilibration temperature and others occur in materials such as a-Si:H and a-In_2O_3, which are amorphous films and do not exhibit a softening or glass transition. It appears that the local dissipation of the recombination energy of an electron–hole pair may give rise to a metastable change in bonding configuration.

4.1 Chalcogenide Glasses

4.1.1 Photostructural Changes Exposure to light or other radiation that excites electron–hole pairs produces metastable structural changes in nearly all chalcogenide glasses (DeNeufville *et al.*, 1974; Ka. Tanaka, 1976, 1981; Owen *et al.*, 1985; Ke. Tanaka, 1990; Pfeiffer *et al.*, 1991). As a consequence of light-induced changes in atomic configurations, a variety of physical and chemical properties are altered such as density, hardness, elastic constants, electronic transport and optical properties, and chemical solubility, which allows the use of these materials for photolithography.

A striking light-induced effect is called photodarkening, a shift ΔE of the optical gap E_g and the Urbach edge (see Sec. 1.7.1) toward lower energies. The magnitude of ΔE produced at different illumination temperatures T_i relative to T_g is shown for several glasses in Fig. 22. Glasses containing S show the largest and those containing Te the smallest shift.

Photostructural changes are largest at low temperatures and are not present for illumination near T_g. They are caused by electron–hole pair-producing radiation and are attributed to the accumulated effects of local bonding changes caused by nonradiative electron–hole pair recombinations. More than 10^{26} cm^{-3} recombinations are needed to bring the photostructural changes to saturation. The metastable changes can be eliminated by incorporating certain elements such as Cu into the glass (Liu and Taylor,

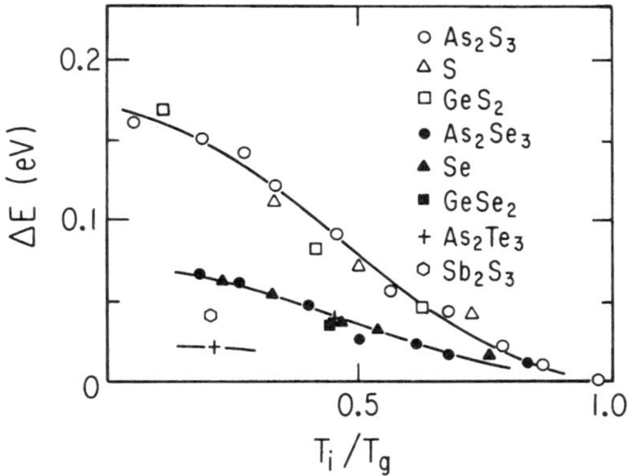

FIG. 22. Photoinduced shift of the absorption edge as a function of the illumination temperature T_i normalized by the glass-transition temperature for several chalcogenide glasses. After Ke. Tanaka (1983).

1987), presumably because these elements provide recombination channels that do not yield structural changes.

Some models relate photostructural changes with specific sites or defects (Ke. Tanaka, 1990), while other models attribute them to bonding changes of the network and a resulting decrease in medium-range order (Elliott, 1986; Pfeiffer *et al.*, 1991; Fritzsche, 1993a). The photodarkening shift ΔE of Fig. 22 is attributed to a broadening of the valence band due to increased lone-pair orbital interactions in the metastable configuration. This accounts for the dependence of the photodarkening ΔE on the chalcogen element (see Fig. 22). The broadening of the valence band giving rise to photodarkening has been observed by x-ray photoelectron spectroscopy (Kolobov *et al.*, 1991).

4.1.2 Metastable Defects The concentration of $-U$ defects in some chalcogenide glasses can be increased more than tenfold by exposure to band-gap light (Biegelsen and Street, 1980). This results in an increase in the subgap absorption and in the light-induced ESR signal, and a decrease in photoluminescence. The equilibrium concentration of $-U$ defects is reestablished by annealing at T_g. The metastable defects are valence-alternation pairs, which may be thought of being produced via path III of the nonradiative recombination scheme of Fig. 14 (Street, 1977).

4.1.3 Optical Anisotropies Exposure of chalcogenide glasses to linearly or circularly polarized band-gap light produces an anisotropy of the dielectric tensor that results in linear or circular dichroism as well as in linear or circular birefringence, respectively (Zhdanov *et al.*, 1979; Lyubin and Tikhomirov, 1989, 1991). These photoinduced anisotropies can be reversibly reproduced after annealing, and the axes defining the optical anisotropy can be turned by turning the polarization direction of the exposing light. The induced dichroism has normally the sign $\alpha_{\parallel} < \alpha_{\perp}$ and the induced birefringence yields $n_{\parallel} < n_{\perp}$, where $\parallel$ and $\perp$ refer to the relative polarizations of the probe light and inducing light. These effects have been attributed to geminate nonradiative recombination events that produce local changes in bond configurations (Fritzsche, 1993b, 1995).

4.2 *a*-Si:H and Related Materials

4.2.1 Light-Induced Defects In 1977 Staebler and Wronski discovered that prolonged exposure to band-gap light decreases both the dark conductivity and the photoconductivity in doped and undoped *a*-Si:H as shown in Fig. 23. The degraded conductivity state is stable at room temperature but the original values are restored by annealing between 150 and 200 °C. Electron-spin-resonance (ESR) experiments demonstrated that this effect is due to the creation of metastable dangling-bond defects (Dersch *et al.*, 1980), which move the Fermi level toward midgap and act as recombination centers reducing the photoconductivity and photoluminescence. Irradiation with ions (Street *et al.*, 1979), electrons (Voget-Grote *et al.*, 1980), or x rays (Pontuschka *et al.*, 1982) also produces these metastable defects. It is believed that they are created during the recombination of irradiation-produced electron–hole pairs (Guha *et al.*, 1983), because the recombination events produced by a forward-bias current in *a*-Si:H pin devices create the same metastable defects (Street, 1991a). The concentration of excess defects ΔN_D saturates near 10^{17} cm^{-3} in undoped *a*-Si:H. The defect creation is relatively inefficient as more than 10^{26} cm^{-3} recombination events are needed to reach that value. ΔN_D can reach values greater than 10^{18} cm^{-3} in

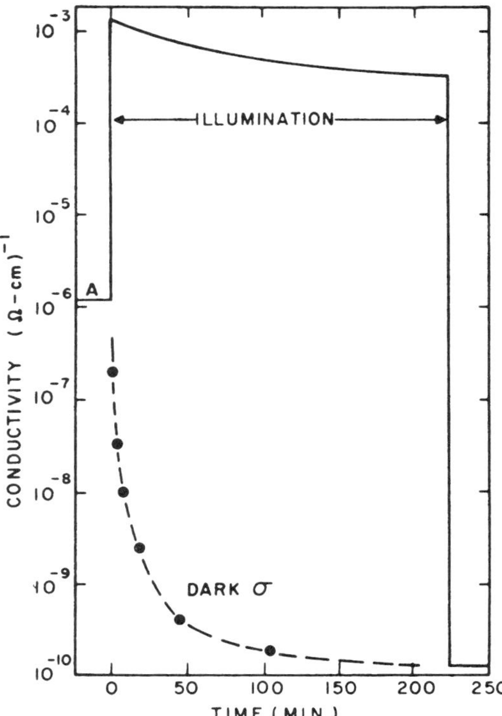

FIG. 23. Decrease of the photoconductivity (solid curve) and of the dark conductivity (dashed curve) of *a*-Si:H with time of exposure to 200-mW/cm² filtered tungsten light. After Staebler and Wronski (1980).

doped *a*-Si:H and with short-pulse laser irradiation.

The defect-creation kinetics has been studied by Stutzmann *et al.* (1985). They found that the defect density N_D increases as

$$\Delta N_D(t) = \text{const} \times G^{0.6}t^{0.33} \tag{54}$$

when ΔN_D is larger than twice the concentration of N_D of native defects. For small ΔN_D, a slower growth is observed. This nonlinear dependence on time and electron–hole pair generation-rate G in Eq. (54) was explained by Stutzmann and co-workers (1985) with the following model. Defects are created by nonradiative recombination of electron–hole pairs at spatially correlated tail states that correspond to the bonding and antibonding states of weak Si–Si bonds. The recombination energy of about 1.4 eV, released locally while the bond breaks, allows a bonded hydrogen to insert itself such that the weak bond cannot reform. These are rare events as most recombinations are assumed to proceed via dangling-bond defects. This creation mechanism yields not single defects but pairs of defects. Such pairing has not been confirmed by ESR, but a defect pair may get separated by further bond switching.

At elevated temperatures, annealing of light-induced defects follows a stretched-exponential time dependence (Jackson and Kakalios, 1988),

$$\Delta N_D = N_D(0) \exp[-(t/\tau)^{\beta}]. \tag{55}$$

The parameters of the decay, the dispersion parameter β, and the temperature dependence of the decay time τ are found to be consistent with the hydrogen-glass model in which hydrogen diffusion governs relaxation toward equilibrium (Street *et al.*, 1987; Kakalios, 1991). According to this model, however, light-induced defect creation and annealing should cease with the diffusion of hydrogen. This is not the case. The defects are created with comparable efficiencies at temperatures between 4.2 and 400 K, and annealing of defects created at low temperatures commences near 150 K where hydrogen cannot diffuse (Stradins and Fritzsche, 1994).

There are many open questions regarding the details of the creation mechanism, the role of impurities, and whether native and light-induced defects have the same energy in the gap and whether defect formation occurs preferentially at internal surfaces of voids. The elimination of light-induced degradation is of great importance for the long-term stability of solar cells.

4.2.2 Metastabilities Produced by Rapid Quenching Dangling-bond defects may be created not only by recombination of photoexcited electron–hole pairs but also by recombination of thermally excited carriers at high temperatures (Smith and Wagner, 1985). This process might explain the presence of $N_D \sim 10^{15}$ cm^{-3} native defects in high-quality *a*-Si:H. More defects should be observed when the material is quenched from higher temperatures. Indeed, rapid quenching from above 250 °C gave the first experimental evidence for a thermodynamic equilibration of the defect and dopant reactions Eqs. (8)–(10) (Ast and Brodsky, 1979; Street *et al.*, 1986; Kakalios *et al.*, 1987).

As shown in Fig. 24, the conductivity of doped *a*-Si:H below an equilibration temperature T_E depends on the thermal history, the quench temperature T_Q, and the rate of cooling. This is typical behavior for a glass, which falls out of equilibrium below a glass-transition temperature. The conductivity of rapidly quenched samples is higher than that of slowly cooled samples because reactions (8) and (9) are endothermic with an activation energy for changing the coordinations of the dopant and Si atoms. Hence, a higher dopant concentration is quenched from higher T_Q. The nonequilibrium excess carrier concentration relaxes below T_E. The relaxation process follows the stretched exponential of Eq. (55) with relaxation parameters that suggest that hydrogen diffusion controls the relaxation process (Kakalios *et al.*, 1987; Street *et al.*, 1988a). Further evidence of the role of hydrogen diffusion in the relaxation process is the observation that samples that have a large H-diffusion rate have a low value of T_E and slower stretched-exponential relaxation. The H-diffusion rate increases as E_F moves away from the gap center and is larger for *p*-type than for *n*-type *a*-Si:H, while the opposite trend is observed for T_E. Hydrogen-glass equilibration above a temperature T_E and metastabilities below T_E not only occur in *a*-Si:H but are generally observed in hydrogenated tetrahedral materials and alloys with N, C, and O.

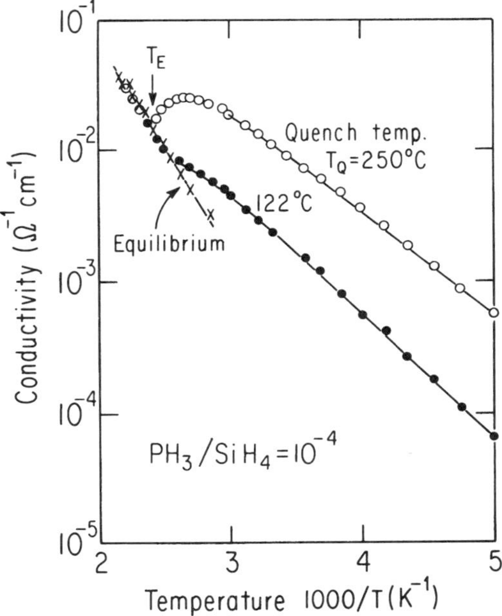

FIG. 24. Temperature dependence of the conductivity of *n*-type *a*-Si:H after annealing (equilibrium) and quenching from different temperatures. The data were taken during subsequent warming. After Street *et al.* (1988b).

4.2.3 Metastabilities Produced by Space Charges The presence of a space-charge layer in junction devices or field-effect transistors produces a change in the defect concentration relative to the equilibrium concentration in the space-charge–free material. Similar effects arise from space charges produced by injected currents (Kruhler *et al.*, 1984), and a metastable, field-induced increase in the doping efficiency was reported by Lang *et al.* (1982b). These effects provide convenient means for studying below T_E the temperature dependence of the creation and annealing kinetics since the space charges can be manipulated with external voltages and measured capacitively (Hepburn *et al.*, 1986; Powell *et al.*, 1987, Jackson and Kakalios, 1989). One finds that the creation and anneal processes are described by stretched exponentials, Eq. (55), with relaxation parameters appropriate for H diffusion in either *n*-type or *p*-type material (Jackson and Kakalios, 1989).

A possible cause for these effects can be that at a given temperature, the equilibrium concentrations of defects and of dopants depend on the formation energies that govern the reactions (8)–(10). These formation energies depend in turn on the Fermi energy E_F since the energy involved in charging the centers is measured relative to E_F. Any change in space charge produces a local change in E_F and causes a relaxation to a new equilibrium state.

4.3 In_2O_{3-x}

Amorphous layers of In_2O_{3-x} exhibit metastabilities that may be related to light-induced changes in the oxygen deficiency *x*, which determines the electron concentration. The family of conductivity curves shown in Fig. 11 was obtained on one sample. The different conductivity states can be reached by photoreduction of the oxide with ultraviolet band-gap light or by reoxidation (Pashmakov *et al.*, 1993). These effects need to be considered when indium oxide or indium tin oxide layers (ITO) are used as transparent conducting electrodes.

5. INTERFACES AND MULTILAYERS

The bulk properties of an amorphous semiconductor can be extracted from measurements on thin films only when the effects of contacts, surfaces, and interfaces are understood. The effects of the latter extend to a depth of the diffusion length or the width of the space-charge region. They can dominate the conduction properties of thin semiconductor films whose defect density is small. Surface states can dominate the electron-spin-resonance signal, measurements of subgap absorption by photothermal-deflection spectroscopy, or the photoemission spectra. For amorphous semiconductors there is no known method to obtain a clean surface equivalent to cleaving crystals in ultrahigh vacuum. Moreover, the atomic layers near the surface and the substrate often differ from the bulk in composition and structure.

These concerns do not apply to measurements on chalcogenide glasses. The failure of attempts to make Schottky-barrier contacts or to bend their bands by a gate voltage in a field-effect geometry confirms the effectiveness of $-U$ defects in pinning the Fermi level (see Sec. 1.4). The following sections refer therefore to high-quality *a*-Si:H.

5.1 Oxide Interface and Gas Adsorption

By measuring the thickness dependence of the spin density and of the subgap defect absorption, Jackson *et al.* (1983) found a surface defect density of about 10^{12} cm^{-2}, which is two orders of magnitude less than the density on freshly cleaved crystalline Si surfaces. This is attributed to a saturation of Si bonds with hydrogen. This surface passivation greatly reduces the oxidation rate of *a*-Si:H relative to crystalline Si (Ponpon and Bourdon, 1982). At 300 K, it takes about 10 hours to grow a 2-Å-thick oxide layer on *a*-Si:H in air. A native oxide layer is negatively charged. A small positive space-charge layer exists adjacent to the surface in *a*-Si:H, which can dominate the sheet conductance of *p*-type samples (Aker and Fritzsche, 1983), while the corresponding electron depletion in *n*-type samples does not significantly alter the film's conductance.

The passivation of surface states with hydrogen makes *a*-Si:H particularly sensitive to effects of gas exposure. Chemisorption of molecules involves a charge transfer to the adsorbing substance. In the absence of surface states, these charges produce a space-charge layer in the amorphous semiconductor, which can dominate its conductance. Adsorbed layers of water and ammonia act as electron donors and increase the conductance of undoped or slightly *n*-type *a*-Si:H by several orders of magnitude. In *p*-type *a*-Si:H these adsorbates compensate for the effect of a negatively charged oxide layer and decrease the conductance manyfold (Tanielian, 1982).

The enhancement of the conductance by many orders of magnitude by controlling the space-charge layer with a gate voltage across a thin insulator is the desired function of *a*-Si:H thin-film transistors, which are used as control elements in flat-panel liquid-crystal displays. Here great care is taken to avoid interface states that would reduce the charge induced in the semiconductor.

5.2 Multilayers

Multilayer structures consist of a large number of ultrathin layers of different semiconductors. A certain sequence of materials can be repeated with any desired periodicity. Because of the layer periodicity they are known as superlattice structures in the field of crystalline semiconductors (see HETEROSTRUCTURES AND SUPERLATTICES, SEMICONDUCTOR). A much larger variety of multilayers can be prepared with amorphous compared with crystalline materials because no restriction for matching lattice constants of adjacent layers limits the choice of amorphous materials. For example, the Si–N bond in *a*-SiN_x:H is 1.7 Å long and the Si–Si bond in *a*-Si:H is 2.3 Å. Nevertheless, these materials can be multilayered with interface state densities less than 10^{11} cm^{-2}. Similarly, *a*-Si:H/*a*-Ge:H multilayers have fewer than 10^{10} cm^{-2} interface states. As a consequence of the periodic change in composition there is a corresponding change in the band gap in these multilayers. In contrast, doping multilayers are made by alternating *n*-type and *p*-type doping in the same semiconductor, *npnp* multilayers, or by inserting an undoped layer in between, producing what are called *nipi* structures. Charge transfer between the differently doped layers causes the internal potential to fluctuate with the periodicity of the doping while the band gap remains constant.

5.2.1 Compositional Multilayers The sharpness of x-ray diffraction patterns indicates that compositional multilayers can be prepared with remarkable flatness and regularity for hundreds of layers (Abeles and Tiedje, 1983). The abruptness of the interfaces depends on the growth sequence; the Si-on-SiO_x interface, for instance, is atomically abrupt while the SiO_x-on-Si interface has a 5-Å-wide region graded in oxygen concentration, according to photoemission measurements (Yang *et al.*, 1989). Optical absorption measurements show an increase in the optical gap of *a*-Si:H/*a*-SiN_x multilayers when the silicon-layer thickness is less than 30 Å, suggesting that the electron coherence length at room temperature is of this order if interpreted as a quantum confinement effect. Although a blue-shift absorption edge can also arise from unintentional alloying of N in the *a*-Si:H layers, Hattori *et al.* (1987) reported the observation of steps in the differential absorption spectra of *a*-Si:H/*a*-SiC_x:H multilayers with silicon quantum well thicknesses between 20 and 50 Å. A coherence length this long is surprising in view of the short mean free path of 5–7 Å suggested by transport measurements (see Sec. 2.2).

5.2.2 Doping-Modulated Multilayers When *npnp* or *nipi* *a*-Si:H multilayers are illuminated, the photoexcited electron–hole pairs find themselves in regions of strong electric fields due to the series of back-to-back *pn* junctions. If the drift length of the carriers is greater than the layer thickness, these charge carriers are spatially separated to different layers. When illumination is stopped, recombination of the excess charge carriers is inhibited by the built-in electric fields, and the excess electrons in the *n* layers and holes in the *p* layers increase the parallel conductance of the multilayer film. This field-induced enhanced recombination lifetime was predicted by Döhler (1972) and confirmed at low temperatures by Hundhausen *et al.* (1984). Room-temperature exposure creates a persistent photoconductivity that takes days to decay (Kakalios and Fritzsche, 1984). This effect appears to be due to the different rates of metastable defect creation in the *n*-type and *p*-type layers, which raises the Fermi level of the multilayer (Hamed, 1991).

6. CONCLUSIONS

Noncrystalline semiconductors are playing an increasing practical role in applications such as large-area displays, compact optical computer memories, electrophotographic processing, photovoltaics, and optical readout devices. This is partly due to their inexpensive thin-film processing, which extends and complements conventional chip processing by allowing multilayer stacking.

Amorphous and vitreous semiconductors exhibit fascinating new properties that differ fundamentally from those of crystals. Whereas the dopant and defect concentrations remain essentially fixed in crystals after they are grown, these concentrations are subject to thermodynamic equilibration laws above the equilibration or glass-transition temperatures in noncrystalline semiconductors. Moreover, while the electronic energies of defect and dopant states are fixed and have definite values in crystals, these energies depend on the position of the Fermi level in noncrystalline semiconductors because the electrochemical potential affects the creation energy of the defects and dopants. Furthermore, because of the greater structural flexibility of noncrystalline materials and the shorter localization radius of charge carriers in their defect and dopant states, there is generally a larger lattice relaxation associated with a change in charge state in these materials than in crystals. The strength of this relaxation can be so strong in chalcogenide glasses that their defects have a negative effective correlation energy. The thermodynamic equilibrium of defects established in noncrystalline semiconductors can easily be disturbed by external electron–hole pair excitations. The ensuing recombination events produce metastable structural changes that produce a variety of new phenomena rarely observed in crystals. Among these are light-induced metastable defects, light-induced changes in the optical gap, and optical anisotropies induced by linearly or circularly polarized light. These phenomena are uncommon in the more constrained structures of crystals.

We have moved far from viewing glasses and amorphous materials as highly imperfect crystals. The study of noncrystalline semiconductors has greatly enriched and will continue to challenge our understanding of materials.

ACKNOWLEDGMENT

Preparation of this article was supported by the National Science Foundation Grant NSF DMR 9408670.

GLOSSARY

Acceptor: Impurity that can accept an electron from the host semiconductor.

Amphoteric Defect: A defect that can either release or accept an electron from the host semiconductor.

Chalcogen: Element of group VI of the periodic table: O, S, Se, Te.

Conduction Band: The lowest unfilled band of electron states.

Correlation Energy: The Coulomb repulsion energy of two electrons of opposite spin in one state. The effective correlation is the Coulomb repulsion minus the relaxation energy of the surrounding in response to the addition of the second electron.

Dangling Bond: Unsatisfied bond or free radical. In amorphous silicon, for example, the paramagnetic defect state associated with a threefold-coordinated silicon atom.

Defect Pool: The reservoir of potential defect sites.

Density of States: Number of states available for an electron per unit volume and per unit energy range.

Donor: Impurity that can give up an electron.

Doping: Addition of impurities that act as either donors or acceptors and hence can increase the conductivity of a semiconductor sample.

Drift Mobility μ_d: The average velocity of a charge carrier in a unit electric field when the carrier spends a fraction of its time immobilized in a trap.

Extended States: Electron states whose wave functions extend over the volume of the sample.

Geminate Recombination: Recombination of the very same electron–hole pair that was excited by a photon.

Hole: A missing electron in a normally occupied state.

Hopping Conduction: Motion of electron between localized states as a result of the finite overlap of the wave functions of these states. Phonons are emitted or absorbed when the states differ in energy.

Hydrogen-Glass Model: Description of a hydrogenated semiconductor in which hydrogen diffusion governs the kinetics of defect and dopant equilibration.

Localized States: Electron states whose wave function describes an electron localized in a small region of the semiconductor.

Lone-Pair Semiconductor: A semiconductor whose valence band originates from filled nonbonding orbitals.

Mobility Edge: The energy that separates localized from extended states near the band edges of a noncrystalline semiconductor.

Mobility μ: The net velocity developed by an electron or hole as a result of one unit of electric field.

Persistent Photoconductivity: A metastable light-induced increase in the dark conductivity.

Phonon: Quantum of vibrational energy of one of the normal modes into which the random thermal motion of the atoms in a solid may be decomposed.

Photoconductivity: The increase in electrical conductivity induced by absorption of light.

Photodarkening: Light-induced metastable change in the structure of chalcogenide materials that causes a red shift of the optical absorption edge.

Photoluminescence: Secondary light emitted from a sample that is excited by primary light.

Pnictogen: Element of group V of the periodic table: N, P, As, Sb.

Staebler–Wronski Effect: Light-induced creation of metastable defects in amorphous silicon and related alloys.

Statistical Shift of Fermi Level: The temperature dependence of the Fermi energy that is demanded by charge conservation in the presence of a density of states changing with energy.

Stokes Shift: Energy difference between the exciting primary light and the emitted secondary light.

Telegraph Noise: Sudden random changes of the conductance of a sample between two or more discrete values.

Thermopower: The potential difference per degree centigrade between the warm end and the cold end of a sample placed in a temperature gradient.

Transport Energy: The energy of predominant hopping conduction in an energy-dependent density of localized states.

Urbach Edge: Region of the optical absorption spectrum in which the absorption coefficient increases exponentially with photon energy.

Valence-Alternation Defects: A defect pair of which one partner is positive and an overcoordinated atom and the other is negative and an undercoordinated atom. They may be close (intimate pair) or distant.

Valence Band: The uppermost filled energy band of electron states.

Works Cited

Abeles, B., Tiedje, T. (1983), *Phys. Rev. Lett.* **51,** 2003–2006.

Adler, D., Joffa, E. J. (1976), *Phys. Rev. Lett.* **36,** 1197–1200.

Agarwal, S. C. (1975), *Phys. Rev. B* **7,** 685.

Aker, B., Fritzsche, H. (1983), *J. Appl. Phys.* **54,** 6628–6633.

Anderson, P. W. (1958), *Phys. Rev.* **109,** 1492–1505.

Anderson, P. W. (1975), *Phys. Rev. Lett.* **34,** 953–955.

Antoniadis, H., Schiff, E. A. (1992), *Phys. Rev. B* **46,** 9482–9492.

Arce, R., Ley, L., Hundhausen, M. (1989), *J. Non-Cryst. Solids* **114,** 696–698.

Ast, D. G., Brodsky, M. H. (1979), in: B. L. H. Wilson (Ed.), *Proceedings of the 14th International Conference on the Physics of Semiconductors,* IOP Conference Series No. 43, Bristol: Institute of Physics, p. 1159.

Bar-Yam, Y., Adler, D., Joannopoulos, J. D. (1986), *Phys. Rev. Lett.* **57,** 467–470.

Bellingham, J. R., Graham, M., Adkins, C. J., Phillips, W. A. (1991), *J. Non-Cryst Solids* **137&138,** 519–532.

Bellingham, J. R., Phillips, W. A., Adkins, C. J. (1990), *J. Phys.: Condensed Matter* **2,** 6207–6221.

Beyer, W., Overhof, H. (1979), *Solid State Commun.* **31,** 1–4.

Beyer, W., Overhof, H. (1984), in: J. I. Pankove (Ed.), *Semiconductors and Semimetals,* Vol. 21C, New York: Academic, p. 257.

Biegelsen, D. K., Street, R. A. (1980), *Phys. Rev. Lett.* **44,** 803–806.

Bishop, S. G., Strom, U., Taylor, P. C. (1977), *Phys. Rev. B* **15,** 2278–2294.

Branz, H. M., Silver, M. (1990), *Phys. Rev. B* **42,** 7420–7428.

Brodsky, M. H., Title, R. S. (1969), *Phys. Rev. Lett.* **23,** 581–585.

Brüggemann, R., Main, C., Berkin, J., Reynolds, S. (1990), *Philos. Mag. B* **62,** 29–45.

Choi, W. K., Owen, A. E., LeComber, P. G., Rose, M. J. (1990), *J. Appl Phys.* **68,** 120–123.

Cohen, J. D., Harbison, J. B., Wecht, K. W. (1982), *Phys. Rev. Lett.* **48,** 109–112.

Cohen, J. D., Lang, D. V., Harbison, J. P. (1980), *Phys. Rev. Lett.* **45,** 197–200.

Cohen, J. D., Leen, T. M., Rasmussen, R. J. (1993), *Phys. Rev. Lett.* **69,** 3358–3361.

Cohen, M. H. (1970), *J. Non-Cryst. Solids* **4,** 391–409.

Cohen, M. H., Fritzsche, H., Ovshinsky, S. R. (1969), *Phys. Rev. Lett* **22,** 1065–1068.

DeNeufville, J. P., Moss, S. C., Ovshinsky, S. R. (1974), *J. Non-Cryst. Solids* **13,** 191–223.

Dersch, H., Stuke, J., Beichler, J. (1981), *Appl. Phys. Lett.* **38,** 456–458.

Döhler, G. H. (1972), *Phys. Status Solidi B* **52,** 79–92.

Elliott, S. R. (1986), *J. Non-Cryst. Solids* **81,** 71–98.

Friedman, L. (1971), *J. Non-Cryst. Solids* **6,** 329–341.

Fritzsche, H. (1971), *Solid State Commun.* **9,** 1813–1815.

Fritzsche, H. (1974), in: J. Tauc (Ed.), *Amorphous and Liquid Semiconductors,* London: Plenum, p. 221.

Fritzsche, H. (1993a), *Philos. Mag. B* **68,** 561–572.

Fritzsche, H. (1993b), *J. Non-Cryst. Solids* **164–166,** 1169–1172.

Fritzsche, H. (1994), *Appl. Phys. Lett.* **65,** 2824–2826.

Fritzsche, H. (1995), *Phys. Rev. B* **52,** 15854–15861.

Fritzsche, H., Kastner, M. (1978), *Philos. Mag. B* **37,** 285–292.

Gaczi, P. (1982), *Philos. Mag. B* **45,** 241–259.

Gee, C. M., Kastner, M. (1979), *Phys. Rev. Lett.* **42,** 1765–1769.

Gelatos, A. V., Cohen, J. D., Harbison, J. P. (1986), *Appl. Phys. Lett.* **49,** 722–724.

Gelatos, A. V., Mahavadi, K. K., Cohen, J. D., Harbison, J. P. (1988), *Appl. Phys. Lett.* **53,** 403–405.

Goryunova, N. A., Kolomiets, B. T. (1955), *Zh. Tekh. Fiz.* **25,** 984–994 and 2069–2078.

Greenwood, D. A. (1958), *Proc. Phys. Soc. London* **71,** 585–596.

Grünewald, M., Thomas, P. (1979), *Phys. Status Solidi B* **94,** 125–133.

Guha, S., Yang, T., Czubatyj, W., Hudgens, S. J., Hack, M. (1983), *Appl. Phys. Lett.* **42,** 588–589.

Hamed, A. (1991), *Phys. Rev. B* **44,** 5585–5602.

Hattori, K., Mori, T., Okamoto, H., Hamakawa, Y. (1987), *Appl. Phys. Lett.* **51,** 1259–1261.

Hecht, K. (1932), *Z. Phys.* **77,** 235–245.

Hepburn, A. R., Marshall, J. M., Main, C., Powell, M. J., van Berkel, C. (1986), *Phys. Rev. Lett.* **56,** 2215–2218.

Hoheisel, M., Fuhs, W. (1988), *Philos. Mag. B* **57,** 411–419.

Howard, J. A., Street, R. A. (1991), *Phys. Rev. B* **44,** 7935–7940.

Hundhausen, M., Ley, L., Carius, R. (1984), *Phys. Rev. Lett.* **53,** 1598–1601.

Jackson, W. B., Amer, N. M. (1982), *Phys. Rev. B* **25,** 5559–5562.

Jackson, W. B., Kakalios, J. (1988), *Phys. Rev. B* **37,** 1020–1023.

Jackson, W. B., Kakalios, J. (1989), in: H. Fritzsche (Ed.), *Amorphous Silicon and Related Materials,* Singapore: World Scientific, p. 247.

Jackson, W. B., Biegelsen, D. K., Nemanich, R. J., Knights, J. C. (1983), *Appl. Phys. Lett.* **42,** 105–107.

Jackson, W. B., Kelso, S. M., Tsai, C. C., Allen, J. W., Oh, S.-J. (1985), *Phys. Rev. B* **31,** 5187–5198.

Johanson, R. E., Fritzsche, H., Vomvas, A. (1989), *J. Non-Cryst. Solids* **114,** 274–276.

Johnson, N. M. (1983), *Appl. Phys. Lett.* **42,** 981–983.

Johnson, N. M., Biegelsen, D. K. (1985), *Phys. Rev. B* **31,** 4066–4069.

Kakalios, J. (1989), *J. Non-Cryst. Solids* **114,** 372–374.

Kakalios, J. (1991), in: J. I. Pankove (Ed.), *Semiconductors and Semimetals,* Vol. 34, New York: Academic, p. 381.

Kakalios, J., Fritzsche, H. (1984), *Phys. Rev. Lett.* **53,** 1602–1605.

Kakalios, J., Street, R. A., Jackson, W. B. (1987), *Phys. Rev. Lett.* **59,** 1037–1040.

Kastner, M. (1972), *Phys. Rev. Lett.* **28,** 355–357.

Kastner, M., Adler, D., Fritzsche, H. (1976), *Phys. Rev. Lett.* **37,** 1504–1507.

Kolobov, A. V., Kostikov, A. A., Yu, P., Lantratova, S. S., Lyubin, V. M. (1991), *Fiz. Tverd. Tela* **33,** 781–785 [*Sov. Phys. Solid State* **33,** 444–447].

Kruhler, W., Pfleiderer, H., Plattner, R., Stetter, W. (1984), in: P. C. Taylor, S. G. Bishop (Eds.), *Optical Effects in Amorphous Semiconductors,* AIP Conference Proceedings No. 120, New York: American Institute of Physics, p. 311.

Kubo, R. (1957), *J. Phys. Soc. Jpn.* **12,** 570–586.

Lang, D. V., Cohen, J. D., Harbison, T. P. (1982a), *Phys. Rev. B* **25,** 5285–5320.

Lang, D. V., Cohen, J. D., Harbison, J. P. (1982b), *Phys. Rev. Lett.* **48,** 421–424.

LeComber, P. G., Jones, D. I., Spear, W. E. (1977), *Philos. Mag.* **35,** 1173–1187.

Ley, L. (1984), in: J. D. Joannopoulous, G. Lucovsky (Eds.), *The Physics of Hydrogenated Amorphous Silicon II,* Berlin: Springer-Verlag, Chap. 3.

Liu, J. Z., Taylor, P. C. (1987), *Phys. Rev. Lett.* **59,** 1938–1941.

Lyubin, V. M., Tikhomirov, V. K. (1989), *J. Non-Cryst. Solids* **114,** 133–135.

Lyubin, V. M., Tikhomirov, V. K. (1991), *J. Non-Cryst. Solids* **137&138,** 993–996.

Meyer, W., Neldel, H. (1937), *Z. Tech. Phys.* **12,** 588–593.

Michelson, C. E., Gelatos, A. V., Cohen, J. D. (1985), *Appl. Phys. Lett.* **47,** 412–414.

Monroe, D. (1985), *Phys. Rev. Lett.* **54,** 146.

Morigaki, K. (1984), in: J. I. Pankove (Ed.), *Semiconductors and Semimetals,* Vol. 21C, Orlando, FL: Academic, p. 155.

Oheda, H. (1981), *J. Appl. Phys.* **52,** 6693–6700.

Okushi, H., Tokumaru, Y., Yamasaki, S., Oheda, H., Tanaka, K. (1982), *Phys. Rev. B* **25,** 4313–4316.

Onsager, L. (1938), *Phys. Rev.* **54,** 554–557.

Orenstein, J., Kastner, M. (1981), *Phys. Rev. Lett.* **46,** 1421–1424.

Ovadyahu, Z. (1986), *J. Phys. C* **19,** 5187–5213.

Ovadyahu, Z. (1990), in: H. Fritzsche, M. Pollak (Eds.), *Hopping and Related Phenomena,* Singapore: World Scientific, p. 193.

Overhof, H., Beyer, W. (1984), *Philos. Mag. B* **49,** L9–L14.

Overhof, H., Thomas, P. (1989), *Electronic Transport in Hydrogenated Amorphous Semiconductors,* Springer Tracts in Modern Physics Vol. 114, Berlin, Heidelberg, New York: Springer-Verlag.

Ovshinsky, S. R. (1968), *Phys. Rev. Lett.* **21,** 1450–1453.

Ovshinsky, S. R. (1977), in: W. E. Spear (Ed.), *Proceedings of the 7th International Conference on Amorphous Semiconductors,* Edinburgh: Centre for Industrial Consultancy and Liaison, Univ. of Edinburgh, p. 519.

Owen, A. E., Firth, A. P., Ewen, P. J. S. (1985), *Philos. Mag. B* **52,** 347–362.

Pantelides, S. T. (1986), *Phys. Rev. Lett.* **57,** 2979–2982.

Parman, C. E., Israeloff, N. E., Kakalios, J. (1991), *Phys. Rev. B* **44,** 8391–8394.

Pashmakov, B., Claflin, B., Fritzsche, H. (1993), *J. Non-Cryst. Solids* **164–166,** 441–444.

Pfeiffer, G., Paesler, M. A., Agarwal, S. C. (1991), *J. Non-Cryst. Solids* **130,** 111–143.

Phillips, J. C. (1979), *J. Non-Cryst. Solids* **34,** 153–181.

Ponpon, J. P., Bourdon, B. (1982), *Solid-State Electron.* **25,** 875–876.

Pontuschka, W. M., Carlos, W. E., Taylor, P. C., Griffith, R. W. (1982), *Phys. Rev. B* **25,** 4362–4376.

Powell, M. J., van Berkel, C., French, I. D., Nicholls, D. H. (1987), *Appl. Phys. Lett.* **51,** 1242–1244.

Rogers, C. T., Buhrman, R. A., Korger, H., Smith, L. N. (1986), *Appl. Phys. Lett.* **49,** 1107–1109.

Rose, A. (1963), *Concepts in Photoconductivity and Allied Problems,* New York: Interscience.

Scher, H., Montroll, E. W. (1975), *Phys. Rev. B* **12,** 2455–2477.

Shapiro, F. R., Adler, D. (1985), *J. Non-Cryst. Solids* **77&78,** 139–142.

Shklovskii, B. I., Fritzsche, H., Baranovskii, S. D. (1989), *Phys. Rev. Lett.* **62,** 2989–2992.

Simmons, J. G., Taylor, G. W. (1971), *Phys. Rev. B* **4,** 502–511.

Smith, Z., Wagner, S. (1985), *Phys. Rev. B* **32,** 5510–5513.

Solomon, I. (1979), in: M. H. Brodsky (Ed.), *Amorphous Semiconductors,* Topics in Applied Physics Vol. 36, Berlin: Springer, Chap. 7.

Spear, W. E., LeComber, P. G. (1975), *Solid State Commun.* **17,** 1193–1196.

Staebler, D. L., Wronski, C. R. (1977), *Appl. Phys. Lett.* **31,** 292–294.

Staebler, D. L., Wronski, C. R. (1980), *J. Appl. Phys.* **51,** 3262–3268.

Stradins, P., Fritzsche, H. (1994), *Philos. Mag. B* **69,** 121–139.

Street, R. A. (1977), *Solid State Commun.* **24,** 363–365.

Street, R. A. (1985), *J. Non-Cryst. Solids* **77&78,** 1–16.

Street, R. A. (1991a), *Appl. Phys.* **59,** 1084.

Street, R. A. (1991b), *Hydrogenated Amorphous Silicon,* Cambridge, U.K.: Cambridge Univ. Press.

Street, R. A., Biegelsen, D. K., Stuke, J. (1979), *Philos. Mag B* **40,** 451–464.

Street, R. A., Mott, N. F. (1975), *Phys. Rev. Lett.* **35,** 1293–1296.

Street, R. A., Zesch, J., Thompson, M. J. (1983), *Appl. Phys. Lett.* **43,** 672–674.

Street, R. A., Kakalios, J., Hayes, T. M. (1986), *Phys. Rev. B* **34,** 3030–3033.

Street, R. A., Kakalios, J., Tsai, C. C., Hayes, T. M. (1987), *Phys. Rev. B* **35,** 1316–1333.

Street, R. A., Hack, M., Jackson, W. B. (1988a), *Phys. Rev. B* **37,** 4209–4224.

Street, R. A., Kakalios, J., Hack, M. (1988b), *Phys. Rev. B* **38,** 5603–5609.

Stuke, J. (1977), in: W. E. Spear (Ed.), *Proceedings of the 7th International Conference on Amorphous and Liquid Semiconductors,* Edinburgh: Centre for Industrial Consultancy and Liaison, Univ. of Edinburgh, p. 406.

Stutzmann, M., Street, R. A. (1985), *Phys. Rev. Lett.* **54,** 1836–1839.

Stutzmann, M., Jackson, W. B., Tsai, C. C. (1985), *Phys. Rev. B* **32,** 23–47.

Stutzmann, M., Biegelsen, D. K., Street, R. A. (1987), *Phys. Rev. B* **35,** 5666–5701.

Tanaka, Ka. (1976), in: G. Lucovsky, G. L. Galeener (Eds.), *Structure and Excitations in Amorphous Solids,* AIP Conference Proceedings No. 31, New York: American Institute of Physics, p. 148.

Tanaka, Ka. (1981), in: F. Yonezawa (Ed.), *Fundamental Physics of Amorphous Semiconductors,* New York: Springer, p. 104.

Tanaka, Ke. (1983), *J. Non-Cryst. Solids* **59&60,** 925–928.

Tanaka, Ke. (1990), *Rev. Solid State Sci.* **2,** (1990), ibid. **3,** 641–659.

Tanielian, M. (1982), *Philos. Mag. B* **45,** 435–462.

Tauc, J. (1974), *Amorphous and Liquid Semiconductors,* London and New York: Plenum.

Tauc, J., Grigorivici, R., Vancu, A. (1966), *Phys. Status Solidi* **15,** 627–637.

Teuschler, T., Hundhausen, M., Ley, L., Arce, R. (1993), *Phys. Rev. B* **47,** 12687–12695.

Tiedje, T. (1984), in: J. I. Pankove (Ed.), *Semiconductors and Semimetals,* Vol. 21C, Orlando, FL: Academic, Chap. 6.

Tiedje, T., Rose, A. (1980), *Solid State Commun.* **37,** 49–52.

Tsang, C., Street, R. A. (1979), *Phys. Rev. B* **19,** 3027–3040.

Unold, T., Cohen, J. D. (1991), *Appl. Phys. Lett.* **58,** 723–725.

Voget-Grote, U., Kümmerle, W., Fischer, R., Stuke, J. (1980), *Philos. Mag. B* **41,** 127–140.

Winer, K. (1989), *Phys. Rev. Lett.* **63,** 1487–1490.

Winer, K., Ley, L. (1987), *Phys. Rev. B* **36,** 6072–6078.

Yang, L., Abeles, B., Eberhardt, W., Slasiewski, H. E., Sondericker, D. (1989), *Phys. Rev.* **39,** 3801–3816.

Yasukawa, M., Hosono, H., Ueda, N., Kawazoe, H. (1995), *Jpn. J. Appl. Phys.* **34,** L281–L284.

Zallen, R. (1983), *The Physics of Amorphous Solids,* New York: Wiley.

Zhdanov, V. G., Kolomiets, B. T., Lyubin, V. M., Malinovskii, V. K. (1979), *Phys. Status Solidi A* **52,** 621–626.

Zhong, F., Cohen, J. D. (1993), *Phys. Rev. Lett.* **71,** 597–600.

Further Reading

Adler, D., Fritzsche, H. (1984), *Tetrahedrally-Bonded Amorphous Semiconductors,* New York: Plenum.

Adler, D., Fritzsche, H., Ovshinsky, S. R. (1984), *Physics of Disordered Materials,* New York: Plenum.

Brodsky, M. H. (Ed.) (1979), *Amorphous Semiconductors,* Topics in Applied Physics Vol. 36, Berlin: Springer.

Elliott, S. R. (1990), *Physics of Amorphous Materials,* 2nd ed., London: Longmans.

Fritzsche, H. (Ed.) (1989), *Amorphous Silicon and Related Materials,* Advances in Disordered Materials Vol. 1, Singapore: World Scientific.

Fritzsche, H. (1990), *Transport, Correlation and Structural Defects,* Advances in Disordered Materials Vol. 3, Singapore: World Scientific.

Kastner, M. A., Thomas, G. A., Ovshinsky, S. (Eds.) (1987), *Disordered Semiconductors,* New York: Plenum.

Mott, N. F., Davis, E. A. (1979), *Electronic Processes in Non-Crystalline Materials,* Oxford: Clarendon.

Pankove, J. I. (Ed.) (1984), *Hydrogenated Amorphous Silicon,* Semiconductors and Semimetals Vol. 21, Orlando, FL: Academic.

Street, R. A. (1991), *Hydrogenated Amorphous Silicon,* Cambridge, U.K.: Cambridge Univ. Press.

Zallen, R. (1983), *The Physics of Amorphous Semiconductors,* New York: Wiley.

SEMICONDUCTORS, AMORPHOUS—MATERIAL PROPERTIES

S. R. ELLIOTT, *Department of Chemistry, University of Cambridge, Cambridge, U.K.*

INTRODUCTION

Amorphous semiconductors have no long-range translational order (periodicity), by definition, and this structural disorder confers a number of unique and characteristic properties in their physical behavior. Thus, both electronic and vibrational states can become spatially localized under certain circumstances, and the transport and other behavior associated with such localized states in amorphous solids can be very different from that exhibited by the corresponding crystalline material counterparts. The influence of structural disorder on electronic transport and related properties is covered by the article SEMICONDUCTORS, AMORPHOUS: ELECTRONIC PROPERTIES. Here, attention is concentrated on optical and vibrational (thermal) properties of amorphous semiconductors, together with a discussion on methods of preparation and classification schemes for such materials.

1. PREPARATION

Amorphous materials differ from their crystalline counterparts by possessing a greater degree of configurational entropy and free energy. In general, such excess entropy and free energy has to be incorporated either sufficiently rapidly in the course of formation of the solid that crystallization is precluded or *post facto* by means of damage mechanisms. Because of the predominantly covalent nature of the bonding in amorphous semiconductors, certain amorphization methods that work for, say, metals are not applicable—e.g., solid-state diffusional amorphization. Nevertheless, there are many general techniques for producing amorphous semiconductors, and these can be divided into three categories, depending on whether amorphization commences in the vapor, liquid, or solid phase.

1.1 Vapor–Solid Amorphization

Vapor deposition is a commonly used technique for preparing amorphous semiconductors in thin-film form. The effective cooling rate when a vapor is condensed on a cold substrate is extremely high, and usually high enough to preclude crystallization of the growing film unless the substrate temperature is sufficiently elevated to give rise to

3-527-28139-8/96/$5.00 + .50

appreciable surface diffusion. A number of different techniques exist for ensuring that the elements needed to produce the amorphous film exist in the vapor phase prior to condensation on a substrate. In addition, the deposition process may simply be physical (collision with, and adhesion on, the substrate by the vapor species), or it may involve a (heterogeneous) chemical reaction at the growing vapor–solid interface.

1.1.1 Evaporation This is perhaps the simplest technique, conceptually. A charge of the material to be deposited is heated *in vacuo* to form a vapor, and a thin film of the material condenses on a substrate held in the vapor stream. For relatively low–melting-point materials, such as chalcogenides, the material may simply be placed in a resistively heated boat. For higher–melting-point materials, e.g., Si or Ge, electron-beam heating may be required in order to vaporize enough material.

Evaporation of multicomponent alloy systems may be troublesome because of preferential vaporization of high–vapor-pressure species (atoms or molecules), leading to deposited films with a composition different from that of the starting material. "Flash" evaporation, whereby grains of the charge are heated extremely rapidly (e.g., by dropping on a preheated element), can obviate such disproportionation problems to a certain extent.

1.1.2 Sputtering An alternative method of producing material in the vapor phase is by sputtering a solid target with charged ions from a plasma established by electrical discharge in a low-pressure gas. Thin-film deposition occurs on a substrate held on the far side of the plasma from the target. When the gas used is inert, e.g., Ar, "physical" sputtering occurs: ablation of the target takes place with subsequent deposition of the ablated material on a substrate, perhaps containing some physically entrained sputtering gas. This process is better at preserving the stoichiometry of multicomponent compounds than is simple evaporation. Alternatively, a reactive gas may be added to the sputtering gas, resulting in its becoming chemically incorporated in the growing film: an example is the use of hydrogen in producing hydrogenated amorphous silicon, *a*-Si:H.

1.1.3 Chemical Vapor Deposition Chemical vapor deposition (CVD) is a general term for the deposition of thin solid films following the chemical reaction between vapor-phase precursor species, usually by means of a heterogeneous reaction at the growing film–vapor interface. In the simplest case, the heterogeneous reaction is promoted by heating the substrate upon which the film is to be deposited. However, the very high temperatures required to cause the reaction to take place often lead to crystal formation; hence, straightforward CVD is not particularly useful for the preparation of films of amorphous semiconductors. Nevertheless, a variant termed the "hot-wire" technique has proved useful in producing device-quality (low defect-density) *a*-Si:H films that have rather low H contents (0.1–1 at.%) (Mahan *et al.*, 1991). In this, a metallic filament (e.g., tungsten) is heated to temperatures in the range 1000–2000 °C, upon which the silane (SiH_4) feedstock gas decomposes catalytically (Matsumura, 1989). The resulting free radicals (SiH_n,H) then impinge on a substrate positioned nearby, and held at a much lower temperature (~200 °C), thereby forming a film of *a*-Si:H.

CVD can also be promoted using energy sources other than thermal to initiate reaction. Perhaps the most common technique is the use of an electrical plasma in so-called plasma-enhanced (PE) CVD or glow-discharge (GD) decomposition. The plasma causes dissociation of the feedstock gas into ions and free radicals that then impinge on a moderately heated substrate (~200 °C) to form a thin film. The processes occurring in the plasma and at the surface are complex and interrelated (Roca i Cabarrocas, 1993). The primary electron-initiated decomposition of, say, silane, in the formation of *a*-Si:H, creates free radicals (e.g., SiH_3). There is a subsequent competition between surface-related processes (deposition, surface mobility, and reaction) and secondary gas-phase reactions leading to clusters, polymers, or even powder depending on the rf power, gas pressure, flow rate, and reactor geometry used. The composition of the feedstock gas is also critical: high dilution of SiH_4 with H_2, for example, leads to the formation of microcrystalline, not amorphous, silicon. The PECVD process is important because not only can it be used to produce device-quality films of *a*-

Si:H (with about 10–15 at.% hydrogen), but such films can be controllably doped *n*-type or *p*-type by incorporating into the gas-flow phosphine (PH_3) or diborane (B_2H_6), respectively (Spear and Le Comber, 1975). A disadvantage of conventional PECVD is that bombardment of the growing film surface by ions from the plasma can be deleterious to film quality. To avoid this effect, the plasma can be operated remote from the growth chamber, in which case only one of the gaseous species (e.g., hydrogen) is excited by the plasma, the resulting ions being extracted and reacting with other gases (e.g., silane) in the growth chamber (Lucovsky and Tsu, 1987).

Finally, an alternative energy source that can be used to initiate CVD is light. Photo-CVD uses photons (typically in the vacuum-ultraviolet) directly to photolyze precursor molecules in the gas prior to deposition. For example, photo-CVD of disilane (Si_2H_6) has been shown to produce high-quality *a*-Si:H films (Yoshida *et al.*, 1990).

1.2 Liquid–Solid Amorphization

The formation of glasses by means of rapid quenching of a melt is perhaps the simplest and most widespread method of producing amorphous materials. Although commonly used for the production of conventional oxide (e.g., silicate) glasses, the melt-quenching technique can also be used to produce semiconducting chalcogenide (S-, Se-, Te-based) glasses, as well as (with difficulty) the tetrahedrally coordinated "chalcopyrite" [II–IV–$(V)_2$] glasses, e.g., $CdGeAs_2$. It is necessary that the cooling rate be sufficiently rapid that crystallization is precluded (i.e., faster than the homo/heterogeneous crystalline nucleation rate). For many chalcogenide glasses, quenching of an ampoule containing the melt (giving a cooling rate of a few hundred degrees per second) is sufficient; however, for those compositions more prone to crystallization (e.g., tellurides or silicon chalcogenides), considerably faster quenching rates are required, necessitating the use of techniques such as twin-roller quenching.

Generally, a prerequisite for vitrification by melt quenching is that the atomic structures of the amorphous solid and of the melt be very similar so that vitrification is favored. This is the case for conventional glass formers, such as oxides and chalcogenides, but not for Si or Ge, where, although the semiconducting solid forms (both amorphous and crystalline) have tetrahedral coordination, the metallic liquid state is characterized by octahedral coordination. Although this change in coordination prevents the amorphous state being formed from the liquid by conventional (relatively slow) quenching methods, nevertheless it has been demonstrated that illumination of crystalline Si by picosecond laser pulses causes melting of the surface layer, and subsequent ultrafast ($\sim 10^{14}$ K s^{-1}) quenching results in an amorphous Si product (Tsu *et al.*, 1979; Liu *et al.*, 1979; Cullis *et al.*, 1982; Smirl *et al.*, 1986). It is not clear at present to what extent the very thin layers of *a*-Si (often associated with *c*-Si) produced by "laser glazing" resemble material produced by vapor-deposition techniques (e.g., evaporation, sputtering, GD decomposition); it is likely that the *a*-Si resulting from laser-quenching experiments is considerably more disordered and defective.

1.3 Solid–Solid Amorphization

There are two methods, with application to semiconductors, by which crystalline materials can be rendered amorphous while remaining in the solid state, namely, pressure quenching and radiation damage. The former topic will be considered first since it bears some passing resemblance to liquid–solid amorphization discussed in the previous section.

The formation of noncrystalline solids by means of pressure quenching is based on the behavior whereby a (high-coordination, often metallic) high-pressure crystalline phase becomes mechanically unstable at a particular (high) pressure and temperature, and the transition to the crystalline modification stable under such circumstances (usually semiconducting with a lower coordination number) is kinetically hindered on release of the pressure and/or reduction in temperature, resulting in a disordered, often amorphous, phase being produced.

This technique is particularly appropriate for the group-IV materials Si and Ge and their III–V analogs, e.g., GaSb, where the melt is octahedrally coordinated, as is a high-pressure phase, although the crystalline (and amorphous) phase stable at ambient pressure and temperature is tetrahedrally coordinated (see Fig. 1). Pressure quenching of the GaSb II octahedral crystalline phase from a pressure of 90 kbar at room temperature produces bulk amorphous GaSb with crystalline inclusions, the proportion of which depends on the melt temperature from which the initial isobaric temperature quench to form the high-pressure crystalline phase (II) is commenced (Demishev *et al.*, 1994). Similar bulk (~mm^3 volume) samples of *a*-Ge and *a*-Si, again with 20–50 vol % crystallite inclusions, can also be made by pressure quenching (at ~100 K) from a high-pressure crystalline phase (II); for *a*-Ge, the

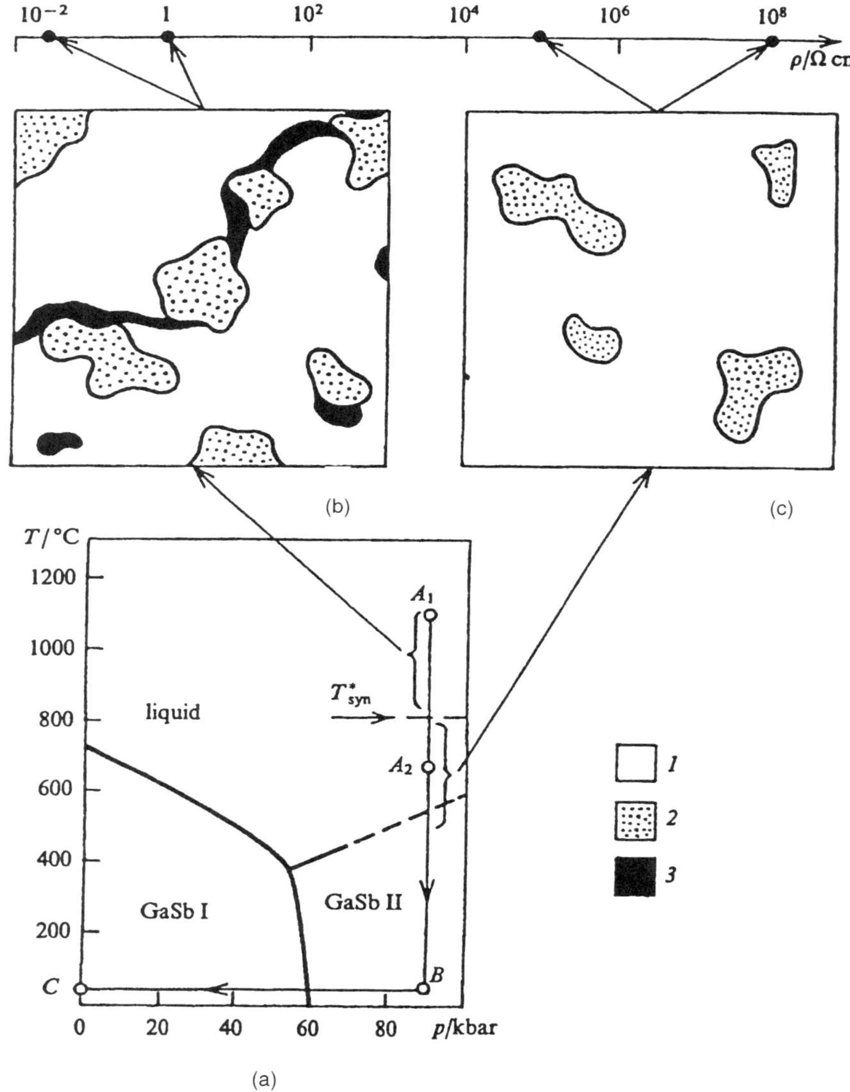

FIG. 1. Illustration of the formation of bulk GaSb by pressure quenching, A → B → C (Demishev *et al.*, 1994), and typical morphologies of the samples obtained by thermal quenching from two temperatures in the melt, A_1 and A_2 (material 1 is bulk *a*-GaSb, 2 is *c*-GaSb, and 3 is a crystalline Ga_xSb_{1-x} phase). Phase I is the low-pressure, tetrahedrally coordinated form, and II, the high-pressure, octahedrally coordinated form, of crystalline GaSb.

transition pressure needed to transform to the amorphous phase is ~1 GPa, and for *a*-Si, ~5 GPa (Brazhkin *et al.,* 1992). The advantage of this technique is that bulk samples of the amorphous phase can be produced that otherwise can only be formed in thin-film form by vapor deposition; the disadvantage is that currently it seems impossible to obtain such pressure-quenched samples completely free of minority crystallite inclusions.

An alternative approach to creating amorphous materials via a solid–solid transformation is by means of radiation damage—e.g., associated with ion implantation. For example, irradiation of crystalline silicon by 2-MeV Si^+ ions causes the material to become amorphous (Hecking *et al.,* 1986). Sufficient energy is deposited into the "spike" of the target material traversed by the impinging ion that the material locally melts. If the surrounding sample temperature is high enough, the atomic mobility is sufficiently large that the melt can recrystallize. However, recrystallization is kinetically hindered at low temperatures, and at sufficiently low temperatures ($T \leq 125$ K for Si), amorphization in the collision-cascade region is complete (Hecking *et al.,* 1986).

Finally, a novel form of radiation-induced amorphization is the athermal photoinduced amorphization found to occur in thin films of crystalline $As_{50}Se_{50}$ (Elliott and Kolobov, 1991; Kolobov and Elliott, 1995). This process is reversible: illumination of a thermally crystallized film of $As_{50}Se_{50}$, having the realgar structure composed of As_4Se_4 molecules, by low-intensity band-gap light causes complete amorphization; subsequent thermal annealing causes the photoamorphized film to crystallize once more, whereupon the whole process can be repeated (see Fig. 2). Although the precise mechanism of photoamorphization is not yet clear, it is believed that the action of the light is to break strained intramolecular As–As bonds with the subsequent reformation of bonds between molecules, thereby forming a (partially) cross-linked amorphous structure. Mechanical strain associated with the substrate–film interface may also play a role (Kolobov and Elliott, 1995).

2. CLASSIFICATION OF AMORPHOUS SEMICONDUCTORS

The definition of a semiconductor is rather imprecise, and the distinction between a semiconductor and an insulator is merely

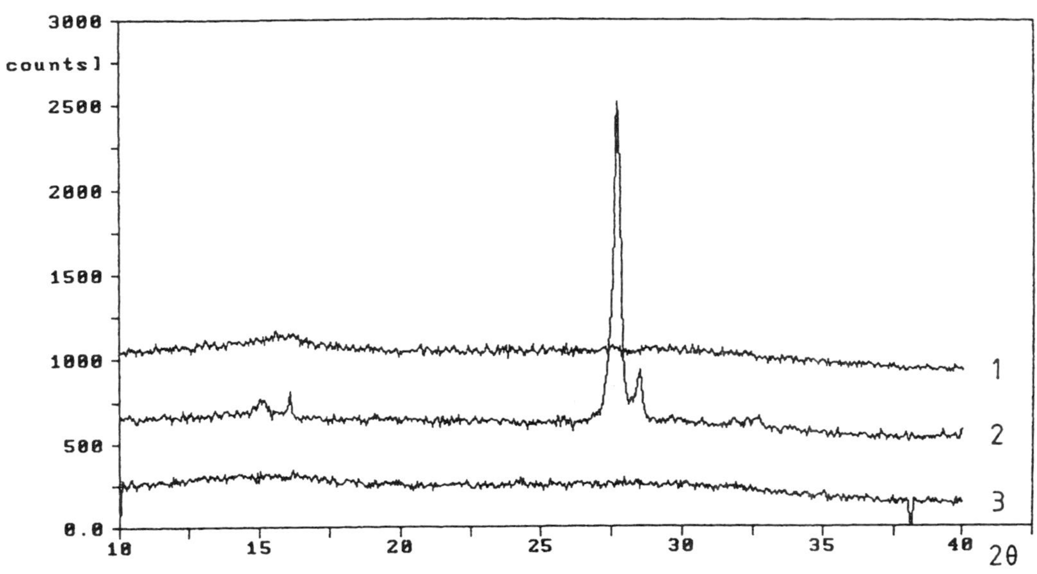

FIG. 2. Reversible photoamorphization of $As_{50}Se_{50}$ (Kolobov and Elliott, 1995). The x-ray diffractograms refer respectively to (curve 1) as-evaporated amorphous film; (curve 2) crystalline film produced by annealing at 185 °C for 24 h; (curve 3) photoamorphized film (after illumination for 4 h.

one of degree, in either magnitude of the band gap between valence and conduction bands or the value of the dc conductivity at, say, room temperature. For the purpose of discussion in this article, a material will be regarded as a semiconductor if its band gap is less than 4 eV and its electrical conduction is intrinsic; i.e., the charge carriers result from thermal excitation across the gap. This arbitrary definition has the advantage that whole classes of material are thereby completely included/excluded. For example, with this definition, all amorphous chalcogenide materials (alloys of S, Se, Te with Si, Ge, As, B, etc.) are deemed to be semiconductors, whereas all oxide glasses are regarded as insulators. [The exceptions to this otherwise appealingly simple picture are those oxide glasses containing mixed-valence (transition-metal) ions, e.g., Fe^{2+}/Fe^{3+}, V^{4+}/V^{5+}, etc.; in such cases, "small-polaron" transport of electrons/holes, involving lattice distortion, can occur between ions with differing charge states (Mott and Davis, 1979), giving rise to a value of intrinsic electronic conductivity comparable to that of a material having a much smaller band gap.]

It is also useful, for the purpose of discussing their properties, to classify amorphous semiconductors further, on the basis of either their atomic or electronic structure, although to some extent these two classifications are linked.

2.1 Structural Classification

It has been conventional in the past, perhaps because of prejudice emanating from the *crystalline* semiconductor field, to divide amorphous semiconductors into two categories, i.e., those that are tetrahedrally coordinated and those that are not. In addition to Si and Ge, other tetrahedrally coordinated amorphous materials include the III–V compounds (e.g., GaAs, GaSb) and the so-called chalcopyrite glasses based on the II–IV–$(V)_2$ composition, e.g., $(CdAs_2)_{1-x}Si_x$ (Houserova and Hruby, 1972). While there is some sense in such a division based on electronic-structure (sp^3-bonding) arguments (see Sec. 2.2), there is also merit in considering a division based on structural/mechanical considerations but, instead, where the average atomic coordination number, r, is greater than, or less than, a critical value, $r_c = 2.67$, rather than equal to, or less than, a value of 4.

Such a structural/mechanical classification is based on the Phillips–Thorpe idea of "rigidity percolation" wherein, for a covalent network constrained by bond-stretching and bond-bending interactions, a mechanical critical point occurs when the number of constraints per atom, n_c, is equal to the dimensionality n_d of the embedding space (Phillips, 1981a):

$$n_c = n_d. \tag{1}$$

The constraints are counted as follows (Thorpe, 1995). A single constraint is associated with each bond, and correspondingly, $r/2$ constraints with each r-coordinated atom. In addition, there are (a maximum of) $2r - 3$ angular constraints per atom. If the total (maximum) number of constraints per atom

$$n_c = r/2 + (2r - 3) \tag{2}$$

is equated to $n_d = 3$ [Eq. (1)], then the critical average coordination so obtained is

$$r_c = 2.4. \tag{3}$$

For average coordinations higher than this value, the structure is said to be overconstrained; for lower coordinations the structure is underconstrained or floppy. Should a fraction m/N of twofold-coordinated (anion) sites have the angular constraint lifted, then the rigidity-percolation threshold occurs at the average coordination number (Zhang and Boolchand, 1994)

$$r_c = 2.4 + 0.4m/N. \tag{4}$$

For AX_2-type glasses with 4:2 coordination (A = Si, Ge; X = O, S, Se), if *all* anions have the angular constraint broken, then $m/N = 2/3$ and Eq. (4) yields for the threshold average coordination number

$$r_c = 2.67 \tag{5}$$

in accord with the stoichiometry AX_2 for chemical stability.

The significance of this approach is that a simple mechanical criterion dictates whether glass formation is to be expected or not. Overconstrained materials, with $r > r_c$, are *not* expected to be glass formers because of the high degree of (homogeneous) mechanical strain expected to be incorporated if they were to form, whereas underconstrained materials, with $r < r_c$, are expected to form glasses readily. This picture has some validity, since chalcogenide and oxide materials, with $r < 2.67$, are indeed good glass formers, whereas As ($r = 3$), Si, and Ge ($r = 4$) do not readily form glasses but can nevertheless be produced in amorphous form by vapor quenching (see Sec. 1.1). However, the approach is somewhat simplistic since it completely ignores weaker interactions, such as van der Waals forces, and the fact that tetrahedrally coordinated materials such as $(CdAs_2)_{1-x}Si_x$, $0 < x < 0.25$, do form glasses, with difficulty, by rapid quenching from the melt (Houserova and Hruby, 1972) [and even *a*-Si is formed by ultrafast glazing (Sec. 1.3)] poses problems for the model. Examples of amorphous semiconductors categorized by this classification scheme are given in Table 1.

2.2 Electronic Classification

Another useful type of classification of amorphous semiconductors is in terms of electronic structure. It has been customary in the past to categorize materials into those that have sp^3-hybrid bonding (i.e., are tetrahedrally coordinated) and those characterized by having lone-pair (i.e., nonbonding) electrons. Examples of lone-pair semiconductors are chalcogenides, where the twofold-coordinated group-VI chalcogen atoms, having an atomic electronic configuration of s^2p^4, in fact give rise to two lone-pair bands in the solid, disparate in energy: one, the *s*-lone-pair band, lies deep within the valence band, and the other, the *p*-lone-pair band, lies at the very top of the valence band, with the intervening band comprising *p*-bonding states. The valence-band density-of-states profiles for these two types of amorphous semiconductors, obtained from photoemission experiments, are shown in Figs. 3(a) and 3(b).

Table 1. Structural classification of amorphous semiconductors in terms of mechanical rigidity.

Material	Average coordination number	Overconstrained (O)/ Underconstrained (U)
Si, Ge	4	O
GaAs	4	O
$CdGeAs_2$	4	O
As	3	O
$GeSe_2$, GeS_2	2.67	—
As_2Se_3, As_2S_3	2.4	U
Se	2	U

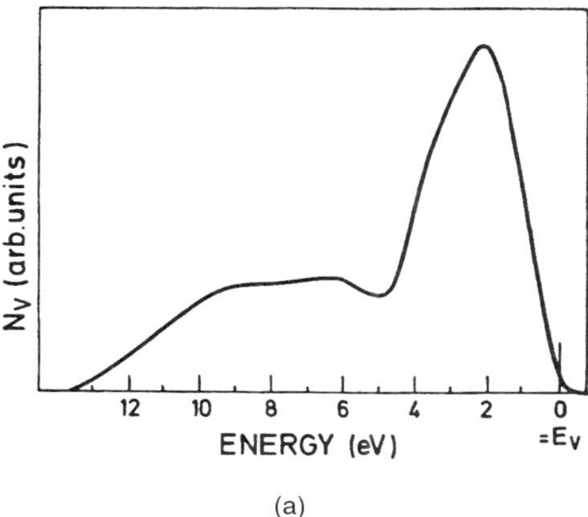

(a)

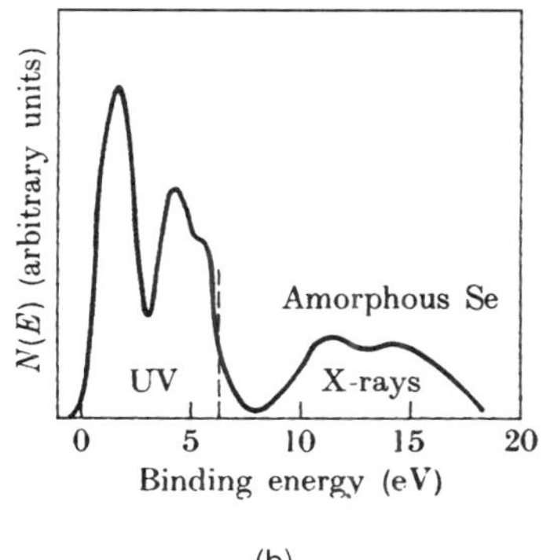

(b)

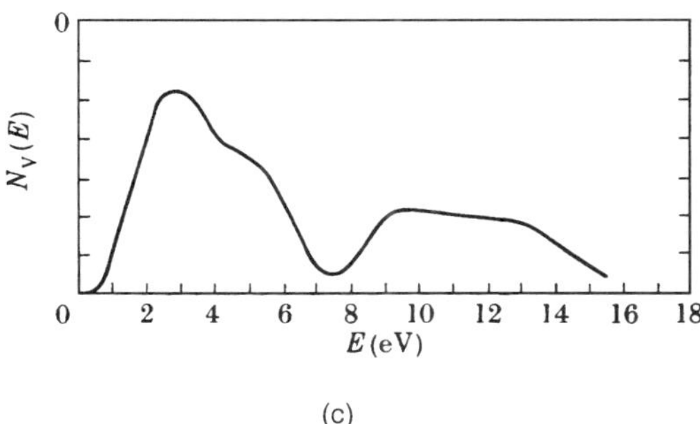

(c)

FIG. 3. Valence-band density of states obtained from photoemission for (a) *a*-Si (Ley, 1984), (b) *a*-Se (Shevchik *et al.*, 1973), (c) *a*-As (Ley *et al.*, 1973).

In fact, the key discriminator is the existence, or otherwise, of p-like lone-pair states at the top of the valence band, since these can be preferentially photoexcited and give rise to a plethora of photoinduced phenomena in the case of amorphous chalcogenides. In the absence of lone-pair states at the top of the valence band, from an electronic point of view it makes little difference, in fact, whether the bonding states forming the valence band derive from sp^3-hybrid orbitals (e.g., Si) or from other states, e.g., p orbitals (i.e., As). Thus, it is perhaps preferable to devise a classification scheme of amorphous semiconductors, based on electronic-structure considerations, in which one category includes materials that have (p-like) lone pairs comprising the top of the valence band, and the other includes materials where the top of the valence band, at least, comprises bonding orbitals. In this way, the awkward position of the group-V semiconductors, As and P, is resolved. With an atomic electronic configuration of s^2p^3, these materials have a deep-lying s-like lone pair but with the upper part of the valence band composed of p-like σ bonding orbitals associated with the threefold covalent coordination characteristic of these materials [see Fig. 3(c)] and so should not properly be regarded as lone-pair semiconductors. Examples of amorphous semiconductors categorized by this classification scheme are given in Table 2.

3. OPTICAL PROPERTIES

3.1 Electronic Structure

The absence of long-range order (translational periodicity) in amorphous semiconductors has a number of ramifications in relation to their electronic structure and cognate properties, e.g., optical behavior. The fact that there is no real-space periodic lattice means that correspondingly there is no reciprocal lattice either for amorphous materials, and so the electron wave vector **k** is no longer a good quantum number and cannot be used as a label for electron states, as in the case of crystalline materials. However, the density of states (DOS) *is* still valid as a description of electron states, even in the presence of disorder. Although the densities of states of amorphous and crystalline forms of the same (semiconducting) material are superficially similar—for example, the bandwidths and band gaps are often comparable (reflecting the marked similarities in the local bonding and short-range order in such cases)—nevertheless there can also be pronounced differences between the densities of states for amorphous and crystalline forms as a result of disorder. In particular, the sharp peaks in the DOS of crystalline semiconductors resulting, say, from Van Hove singularities (where $\nabla_{\mathbf{k}}E = 0$) are absent in the case of amorphous materials (see Fig. 4).

Table 2. Electronic classification of amorphous semiconductors

Material	Lone pair (L)/Bonding (B)
Si, Ge	B
GaAs	B
$CdGeAs_2$	B
As	B
$GeSe_2$, GeS_2	L
As_2Se_3, As_2S_3	L
Se	L

The lack of a reciprocal lattice, and the corresponding inutility of **k** as a label for electron states in amorphous materials, means that the additional selection rule for optical transitions in crystalline semiconductors, based on **k** conservation between initial and final states, breaks down in the case of amorphous semiconductors. In effect, *all* electronic states are accessible in an optical transition for an amorphous semiconductor possessing complete structural disorder. In practice, however, remnants of structural order—e.g., medium-range order—may cause partial reinstatement of the **k**-selection rule for optical transitions, the degree of which is still the subject of some debate.

An additional effect of the presence of structural disorder on electron states and related optical behavior is the broadening caused to the density of states at the band edges, bordering the forbidden gap. Fluctuations in bond lengths and angles result in the formation of "band tails," which appear to have an exponential dependence on energy for most amorphous semiconductors. A model for the ubiquitous existence of exponential band tails is the thermal-equilibration model of Bar-Yam *et al.* (1986), in which the atomic and electronic structures of an amor-

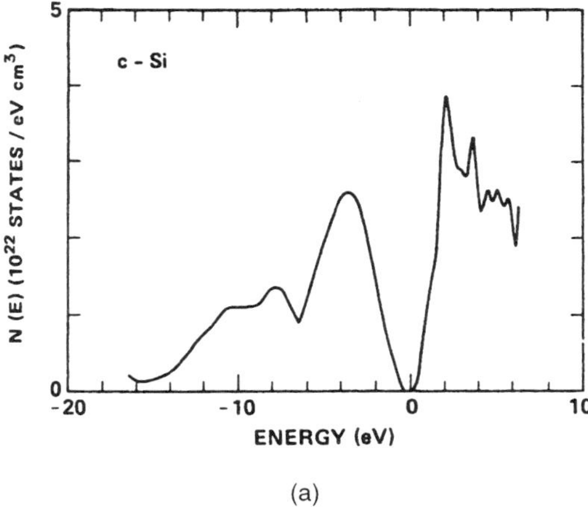

(a)

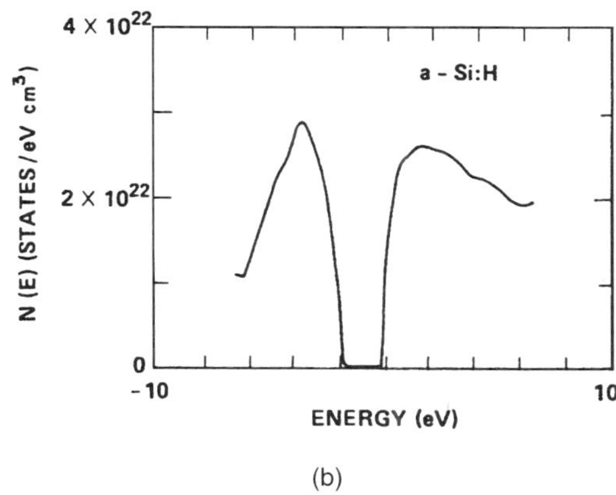

(b)

FIG. 4. Valence- and conduction-band DOS of (a) crystalline Si and (b) amorphous Si:H measured by x-ray photoemission (XPS) and inverse XPS, respectively (Jackson *et al.*, 1985).

phous solid are assumed to be interrelated: energetics determine the structure, which in turn controls the electronic density of states. The starting point is an ideal disordered network, containing no defects, with optimally coordinated atoms and equal bond lengths and angles. Fluctuations in bond lengths and angles, and defects, are then assumed to be generated thermally in such an ideal network, being in thermal equilibrium above, and frozen in below, a critical temperature T^*, which might be the glass-transition temperature of a glass or the film-deposition temperature for an amorphous thin film. The Boltzmann factor involved in the probability of creating above T^*, say, an extra filled electronic state at an energy $E - E_v$ above the valence-band mobility edge at E_v naturally leads to an exponential form for the valence-band–tail density of states (Bar-Yam *et al.*, 1986):

$$N_v(E) = N_0 \exp[-(E - E_v)/kT_v], \tag{6}$$

where

$$T_v = T^*/(1 + \delta_v), \tag{7}$$

and a similar expression for the conduction-band–tail state density T_c, where

$$T_c = T^*/\delta_c, \tag{8}$$

and where δ_v and δ_c are parameters relating the changes in system energy and the disorder-induced raising and lowering of valence- and conduction-band states, respectively. An estimate for the ratio $\alpha = \delta_c/(1 + \delta_v) = T_c/T_v \simeq 0.5$ for Si obtained from tight-binding considerations (Bar-Yam *et al.*, 1986) is in accord with the experimental value of $T_v = 500$ K and $T_c = 300$ K for *a*-Si:H (Tiedje, 1984).

The final consequence of the absence of long-range order on the properties of electron states and associated optical behavior to be discussed concerns the *character* of the states: structural disorder can cause electron states to become spatially *localized* in the Anderson (1958) sense. A critical amount of disorder is necessary to localize electron states, with the consequence that there exist at the edges of the valence and conduction bands demarcation energies, so-called mobility edges, separating extended and localized states, where the magnitude of the electron mobility changes drastically (see, e.g., Elliott, 1990). Although there is no change in the DOS at a mobility edge, the change in character of the states (localized → extended) does have an influence on the optical behavior through the optical transition matrix element, being different for extended–extended (or extended–localized) and localized–localized transitions. With this in mind, it is convenient to divide, somewhat arbitrarily, the discussion about optical properties of amorphous semiconductors into three sections, dealing respectively with interband (valence–conduction band) transitions, transitions involving states in the band tails (the so-called Urbach-edge region), and transitions involving localized defect (and other) states deep in the gap between valence and conduction bands.

3.2 Interband Transitions

The optical properties of amorphous and crystalline semiconductors are almost entirely determined by the imaginary part (ϵ_2) of the complex dielectric constant ϵ^*,

$$\epsilon^* = \epsilon_1 + i\epsilon_2, \tag{9}$$

where ϵ_2 is related to the optical absorption coefficient $\alpha(\omega)$ by

$$\epsilon_2(\omega) = nc\alpha(\omega)/\omega \tag{10}$$

and where n is the refractive index and c is the speed of light. In a conventional optical-absorption experiment, α is obtained from the transmitted light intensity T and the reflectivity R via the Beer–Lambert expression

$$T = (1 - R)^2\exp(-\alpha d), \tag{11}$$

where d is the sample thickness.

For an amorphous semiconductor, where **k** conservation no longer holds because of the presence of structural disorder, ϵ_2 is related to a convolution of occupied valence and unoccupied conduction-band densities of states (the so-called "joint" DOS) and the momentum matrix element $P(\omega)$ (Connell, 1979):

$$\epsilon_2(\omega) = \frac{8\pi^2e^2a^3}{m^2\omega^2}P^2(\omega)\int N_v(E)N_c(E + \hbar\omega)\,dE, \tag{12}$$

where N_v and N_c are the valence-band and conduction-band DOS, respectively, and a is the interatomic spacing. A comparison of experimental curves of $\epsilon_2(\omega)$ for crystalline and amorphous Si is given in Fig. 5. Although superficially rather similar, the curve for a-Si is featureless, whereas that for c-Si exhibits a number of sharp features (Van Hove singularities) associated with the joint DOS. Furthermore, ϵ_2 for c-Si lies below that for a-Si at energies lower than the peak because optical transitions in c-Si at energies ~3 eV are "indirect," i.e., involve absorption/emission of a phonon to conserve wave vector (the band gap of c-Si at 1.1 eV is indirect), and correspondingly have a lower transition probability than the effectively "direct" transitions of a-Si.

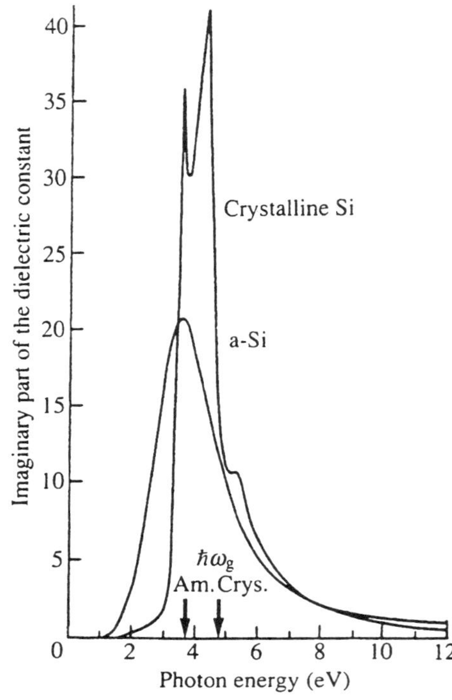

FIG. 5. Comparison of the imaginary part of the dielectric constant, $\epsilon_2(\omega)$, for crystalline and amorphous (a-) Si (Pierce and Spicer, 1972). The energy $\hbar\omega_g$ is the energy of the peak in ϵ_2.

The momentum matrix element $P^2(\omega)$ is related to the dipole-moment matrix element $R^2(\omega)$ via the following relation (Jackson *et al.*, 1985):

$$P^2(\omega) = (m\omega/\hbar)^2R^2(\omega). \tag{13}$$

The energy dependence of $R^2(\hbar\omega)$ can be obtained from the measured spectrum of $\epsilon_2(\omega)$, together with a knowledge of the DOS of valence and conduction bands, from Eq. (12). Figure 6 shows a comparison of $R^2(\hbar\omega)$ for c-Si and a-Si:H (Jackson *et al.*, 1985). The matrix element for c-Si increases rapidly for energies above the direct gap at ~3.4 eV,

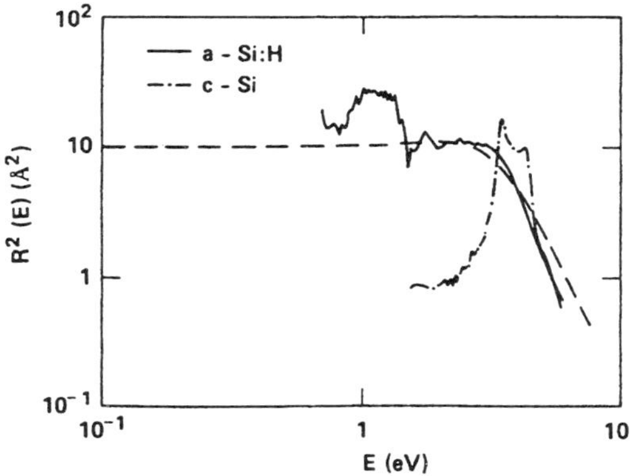

FIG. 6. Energy dependence of the dipole-moment matrix element $R^2(\omega)$ for *a*-Si:H and *c*-Si (Jackson *et al.*, 1985). The dashed line is a fit to the data for *a*-Si:H assuming a damped harmonic-oscillator model.

whereas that for *a*-Si:H is approximately constant below this energy, and decreases rapidly above, and can be fitted quite well with the form of the matrix element expected for a classical damped harmonic oscillator

$$R^2(\hbar\omega) \sim 1/[(\omega^2 - \omega_0^2) + \Gamma^2\omega^2] \qquad (14)$$

with $\hbar\omega_0 \simeq 3.5$ eV and $\Gamma = 4$ eV.

The quantity of primary interest obtainable from interband optical transitions (leading to absorption coefficients in the range $10^3 < \alpha < 10^6$ cm^{-1}) is the optical gap. However, for the case of amorphous semiconductors there is no unique (nor commonly accepted) definition for this quantity. One type of definition is based on an arbitrary level set for the optical absorption coefficient, α, or the density of states, namely, the photon energy corresponding to $\alpha = 10^3$ cm^{-1} (E_{03}) or 10^4 cm^{-1} (E_{04}), or to the transition between the valence- and conduction-band densities of states equal to 10^{20} states/eV cm^3 (E_g^{20}).

Alternatively, an estimate for the optical gap can be obtained by extrapolation, assuming a particular form for the shape of the density of the band edges. For instance, if it is assumed that the conduction-band DOS scales as $N(E) \sim (E - E_c)p$ from the conduction-band edge at E_c, with a similar form for the valence-band DOS with exponent q, then Eq. (12) implies that

$$\omega^2\epsilon_2(\omega) \propto \omega\alpha(\omega) \propto P^2(\omega)(E - E_g)^{p+q+1} \propto \omega^2R^2(\omega)(E - E_g)^{p+q+1}, \qquad (15)$$

where E_g is the (optical) gap defined in terms of the separation of the two extrapolated band edges. With the assumption that the momentum matrix element P^2 is independent of frequency and that the DOS at the band edges have a parabolic energy dependence $N(E) \propto E^{1/2}$, characteristic of free-electron states, then the so-called Tauc plot (Tauc *et al.*, 1966) for the dependence of the optical absorption coefficient on photon energies in the vicinity of the gap is obtained:

$$[\hbar\omega\alpha(\hbar\omega)]^{1/2} = A(\hbar\omega - E_g), \qquad (16)$$

which is widely used to obtain estimates for E_g. However, as seen from Fig. 6 for the case of *a*-Si:H, it is instead the dipole-moment matrix element R^2 that is more nearly constant, and furthermore the energy dependence of the band tails is more nearly linear than parabolic, resulting in an appreciably different estimate for E_g. Table 3 contains a number of estimates for E_g obtained for *a*-Si:H using a variety of assumptions.

3.3 Urbach Edge

The part of the optical absorption tail corresponding to absorption coefficients in the range $1 \leq \alpha \leq 10^3$ cm^{-1} has, for all amorphous semiconductors, a universal exponential energy dependence

$$\alpha(\omega) = \alpha_0 \exp[(E - \hbar\omega)/E_0], \qquad (17)$$

Table 3. Estimates for the energy gap for *a*-Si:H obtained from optical and density-of-states data under a variety of assumptions (Jackson *et al.*, 1985).

Energy gap or extrapolation	DOS energy dependence	Constant matrix element	Gap (eV)
E_g^{μ} (mobility gap)	—	—	1.93
E_g^{20}	—	—	1.61
E_{04}	—	—	1.85
E_{03}	—	—	1.68
$(\epsilon_2)^{1/2}$	$E^{1/2}$	R^2	1.64
$(\epsilon_2)^{1/3}$	E	R^2	1.49
$(E^2\epsilon_2)^{1/2}$	$E^{1/2}$	P^2	1.86
$(E^2\epsilon_2)^{1/3}$	E	P^2	1.55

where the energy E is comparable to the threshold energy obtained from interband optical absorption at higher photon energies (Sec. 3.2) and the slope parameter E_0 typically has values in the range 50–100 meV. This ubiquitous behavior is similar in its energy dependence to that found in alkali and silver halide crystals (Urbach, 1953) and is named after its discoverer; however, the temperature dependence is somewhat different.

The exponential character of the photon-energy dependence of $\alpha(\omega)$ is clearly evident in the data for a-Si:H shown in Fig. 7 and is now believed simply to reflect the exponen-

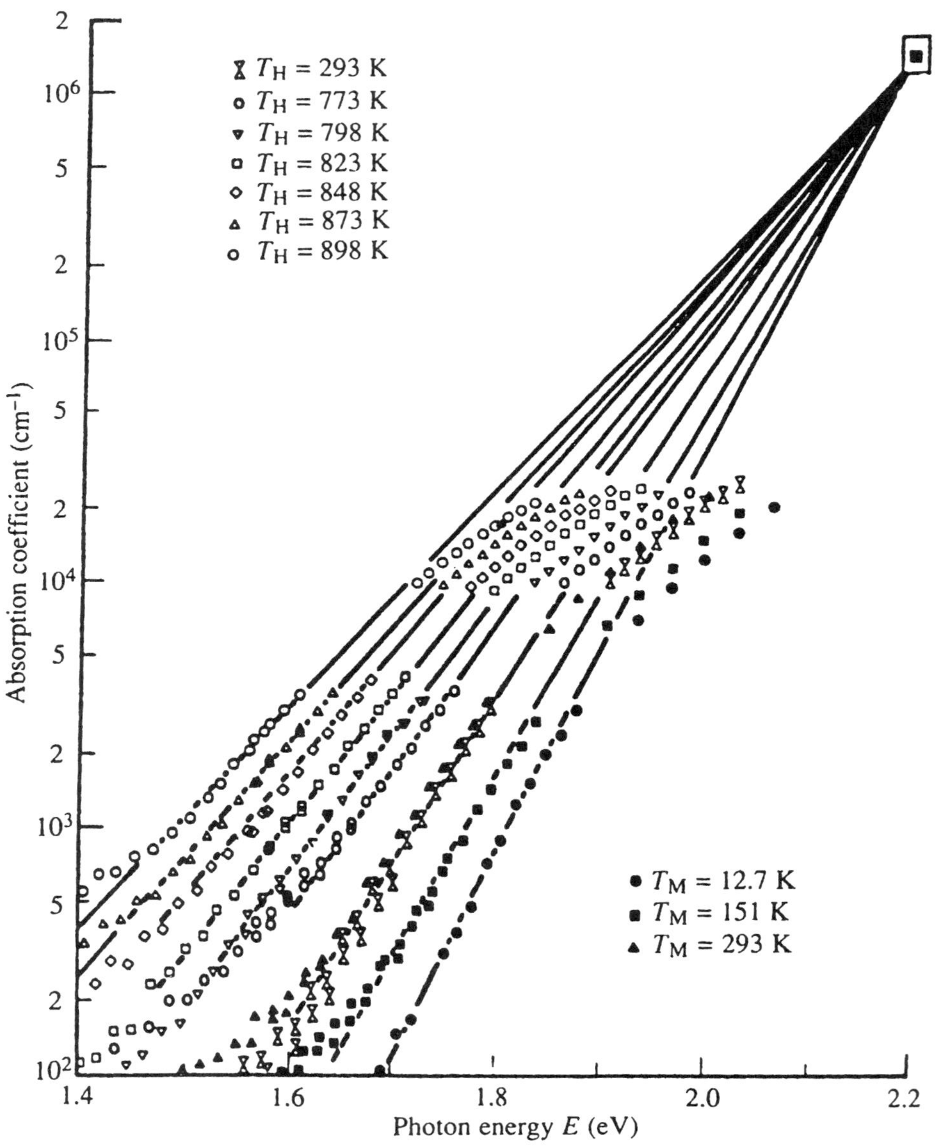

FIG. 7. Optical absorption edge of a-Si:H, showing Urbach-type behavior, measured at various temperatures T_M, and also measured at 300 K after annealing at the temperatures T_H shown (Cody, 1984).

tial energy dependence of the tail-state density of states [Eq. (6)], through the joint density of states [Eq. (12)], caused by disorder (static structural and thermal). However, in the case of *a*-Si:H, the Urbach slope reflects mostly the valence-band–tail state density since this is broader than the conduction-band–tail width.

The temperature dependence of the Urbach absorption edge can be understood theoretically in terms of a model involving both thermal and static disorder (Cody *et al.*, 1981; Street, 1991). The temperature dependence of the optical gap E_g can be regarded as arising predominantly from the electron–phonon interaction via the deformation potential, B (neglecting the thermal-expansion contribution of the network), viz.,

$$E_g(T) = E_g(0) - B[u^2(T) + u_D^2 - u^2(0)], \tag{18}$$

where $u^2(T)$ is the mean square displacement of the atoms due to thermal vibrations, $u^2(0)$ is that due to zero-point motion, and u_D^2 is that due to static disorder. The Urbach theory of optical absorption in crystals proposes that the slope parameter E_0 is proportional to $u^2(T)$. For amorphous semiconductors, there will be an additional contribution due also to static disorder u_D^2, so that

$$E_0(T) = K[u^2(T) + u_D^2], \tag{19}$$

where K is a constant. Combining Eqs. (18) and (19) gives

$$E_g(T) = E_g(0) - Bu^2(0) \times \left[\frac{E_0(T)}{E_0(0)}\left(1 + \frac{u_D^2}{u^2(0)}\right) - 1\right]. \tag{20}$$

Note that Eq. (20) predicts a linear correlation between the magnitude of the optical gap E_g and the Urbach slope parameter E_0 at any given temperature that is borne out experimentally [see Fig. 8(c)].

On the assumption of an Einstein model for the lattice vibrations, characterized by the Einstein temperature θ_E (which is $\frac{3}{4}$ of the Debye temperature θ_D), then

$$u^2(T)/u^2(0) = \coth(\theta_E/2T) \tag{21}$$

when M is the (effective) atomic mass. The

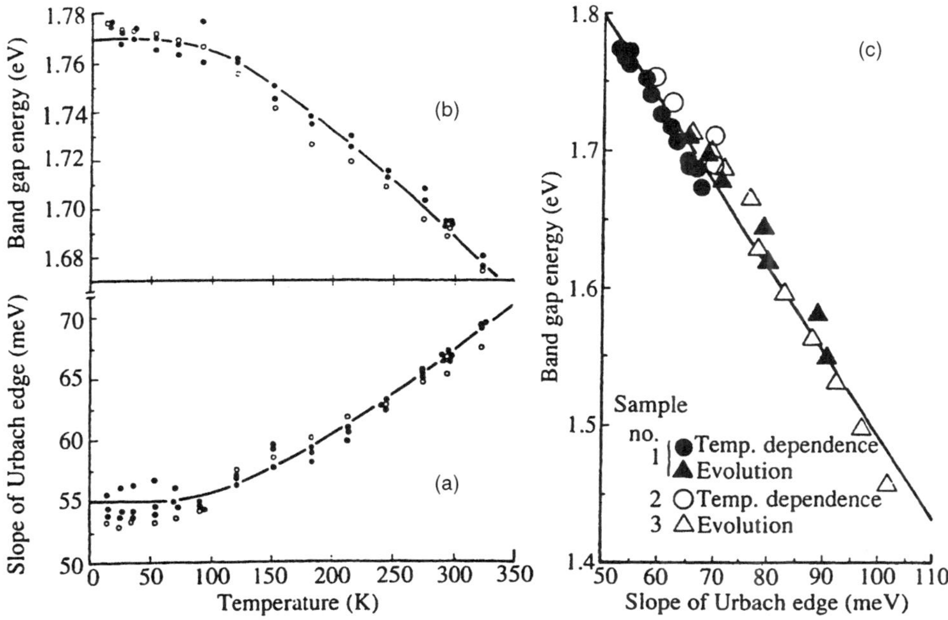

FIG. 8. Behavior of Urbach-edge parameters for *a*-Si:H (Cody *et al.*, 1981). (a) Temperature dependence of the Urbach slope parameter, E_0; (b) temperature dependence of the optical gap, E_g; (c) correlation between E_g and E_0.

explicit form for the temperature dependence of the optical gap, $E_g(T)$, is then obtained by substituting Eq. (21) into Eq. (18). For the case of *a*-Si:H, a fit to the data of $E_g(T)$ [obtained from the Tauc relation, Eq. (16)] is obtained with $\theta_D = 536$ K [see Fig. 8(b)], comparable to the literature value of $\theta_D = 625$ K for *c*-Si. The temperature dependence of the Urbach slope parameter, $E_0(T)$, is obtained in a similar fashion, with the added assumption (to eliminate the constant K) that $E_0 \to kT$ as $T \to \infty$ (Cody *et al.*, 1981), giving

$$E_0(T) = k\,\frac{\theta_E}{2}\left[\coth\left(\frac{\theta_E}{2T}\right) + \frac{u_D^2}{u^2(0)}\right]. \qquad (22)$$

Use of the same value of θ_D (or θ_E) as obtained from the fit to $E_g(T)$ also accounts for the data of $E_0(T)$ remarkably well [see Fig. 8(a)].

3.4 Subgap Transitions and Defect States

Amorphous semiconductors invariably contain structural defects, e.g., dangling bonds, whose electron states lie deep within the forbidden energy gap. These strongly localized states can also be probed optically, albeit with somewhat more difficulty than is the case for transitions involving band-edge or tail states (see Secs. 3.2 and 3.3). Since the absorption coefficient for transitions involving gap states is so low ($\alpha \leq 10\ \mathrm{cm}^{-1}$), reflecting the low level of the defect density of states, conventional absorption measurements to obtain values of α are precluded, particularly for thin-film amorphous semiconductors (e.g., *a*-Si:H). Instead, techniques such as photothermal deflection spectroscopy (PDS) can be used (see Amer and Jackson, 1984, for a review). The PDS detects, as heat, all the optical energy absorbed in a sample, which is proportional to αd when $\alpha d \ll 1$ (d being the sample thickness). Light from a chopped pump beam is absorbed by the sample, and the associated heat is detected by the deflection of a low-intensity probe beam, at grazing incidence to the film surface, caused by the thermally induced modulation of the refractive index of a medium (e.g., a liquid) in thermal contact with the sample.

Results of PDS measurements of defect gap states in *a*-Si:H, made by PECVD and deposited at different rf powers, are shown in Fig. 9(a); the correlation between the integrated gap-state absorption and the dangling-bond defect density as determined from electron-spin resonance is evident in Fig. 9(b).

4. VIBRATIONAL AND THERMAL PROPERTIES

4.1 Vibrational Density of States

In the case of crystalline materials, vibrational excitations can be described in terms of dispersion curves, $\omega(\mathbf{q})$, giving the dependence of the frequency of propagating collective modes (phonons) on the wave vector $\mathbf{q}$. The character of these modes can be distinguished most simply in the long-wavelength limit: "acoustic" branches are characterized by their frequency going to zero as $\mathbf{q}$ tends to 0, whereas "optic" branches have finite frequencies in the $\mathbf{q} = 0$ limit.

For the case of amorphous materials, the picture is complicated by the lack of translational periodicity, as it is for electronic states (Sec. 3.1). Thus, $\mathbf{q}$ is no longer a good quantum number nor a good label of vibrational states in general. Nevertheless, in the low-frequency, long-wavelength limit, amorphous solids appear structurally isotropic and act as elastic solids; as a consequence, vibrational excitations can still be described as acoustic phonons, albeit more heavily damped (with a shorter lifetime) than for crystals. At higher $\mathbf{q}$ values, however, the $\omega(\mathbf{q})$ relation becomes increasingly smeared out (see Fig. 10), and the modes will no longer be plane-wave states (as for crystals) and will be more localized.

However, the vibrational density of states (VDOS) is, like its electronic counterpart (Sec. 3.1), still valid as a description of vibrational states in amorphous materials. The VDOS may be determined experimentally by inelastic neutron scattering (see Elliott, 1990), although IR absorption and Raman scattering spectra can, under certain circumstances, also give an insight into the VDOS; however, the existence of frequency-dependent matrix elements in the latter two cases often precludes a quantitative determination of the VDOS by these means. Figure 11

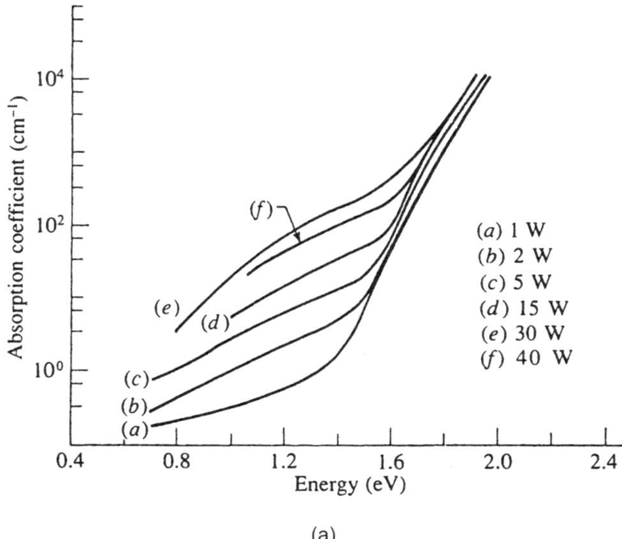

(a)

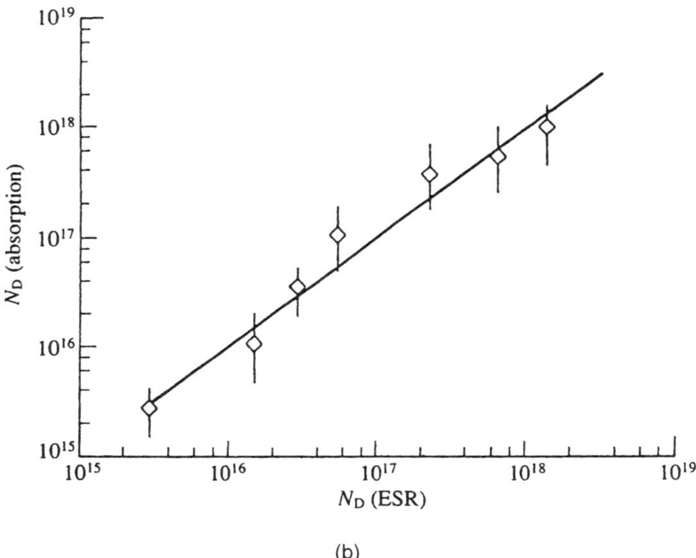

(b)

FIG. 9. (a) PDS measurements of the Urbach edge and gap-state transitions in PECVD *a*-Si:H deposited at the various rf powers shown (Jackson and Amer, 1982). (b) Correlation between the integrated gap-state absorption and the ESR spin density for various samples of *a*-Si:H (Jackson and Amer, 1982).

shows the experimental (inelastic-neutron) VDOS for *a*-Si with, for comparison, that calculated for *c*-Si, as well as Raman spectra for *a*- and *c*-Si. It can be seen that the density-of-states distributions are rather similar for crystalline and amorphous forms, except that, as for the case of the electronic density of states (Sec. 3.1), the sharp cusplike features (Van Hove singularities) characteristic of the crystalline density of states are smeared out in the amorphous case. This overall similarity is not too surprising since the short-range order is the same (tetrahedral) in both cases, with very similar bond lengths and angles, and hence they have correspondingly similar bond-stretching and bond-bending force constants. Note also that, because of the lack of translational periodicity in the amorphous phase, and the corresponding absence of a well-defined reciprocal lattice, the wave vector **q** is no longer a good quantum number and that the

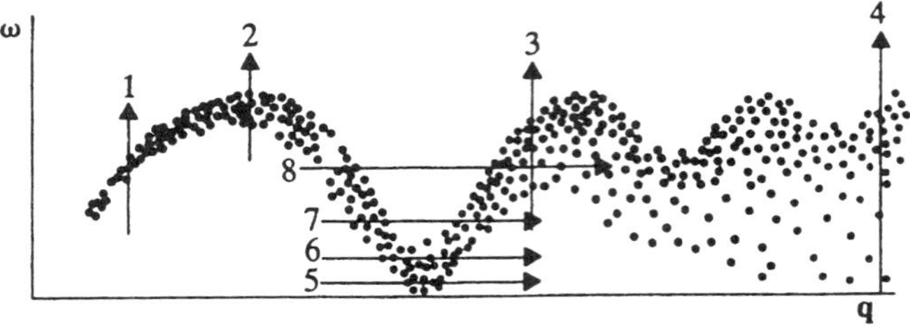

(a)

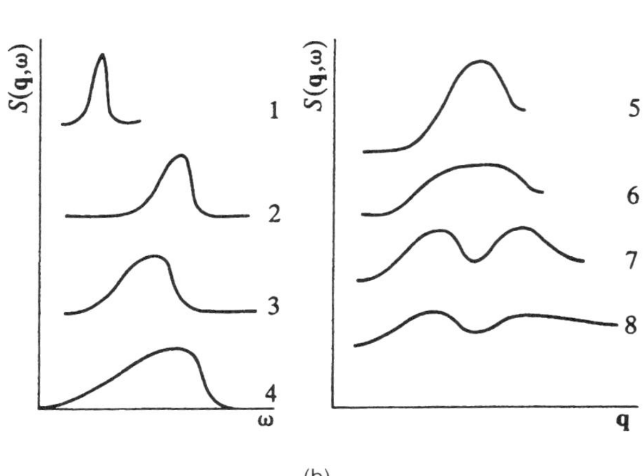

(b)

FIG. 10. (a) Schematic illustration of a phonon dispersion relation $\omega(\mathbf{q})$ for an acoustic branch in an amorphous material. The form of the scattering law, i.e., the dynamic structure factor $S(\mathbf{q},\omega)$, is shown in (b) for various cuts through the curve in (a) (Leadbetter, 1973).

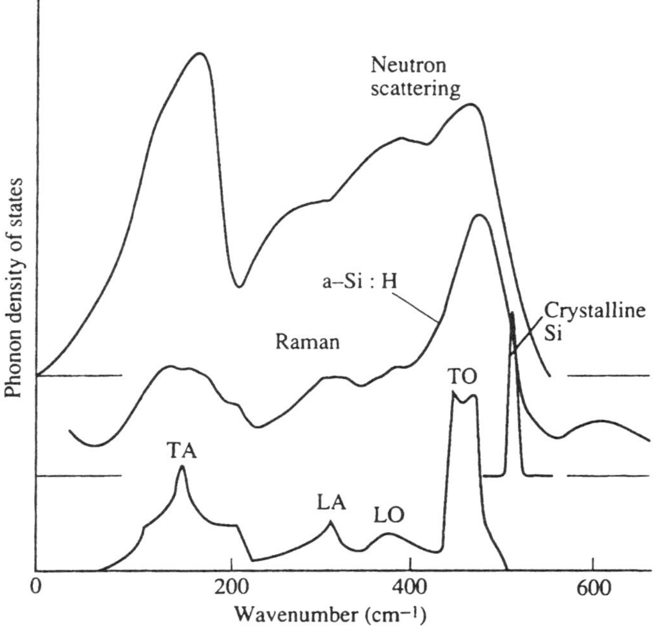

FIG. 11. Vibrational density of states for *a*-Si:(:H) measured by inelastic neutron scattering (Kamitakahara *et al.,* 1984) and Raman scattering (Tsai and Nemanich, 1980), compared with the Raman spectrum for *c*-Si and a calculated VDOS for *c*-Si, where the transverse (T) and longitudinal (L) modes of the acoustic (A) or optic (O) phonons are marked (Street, 1991).

selection rule $\Delta \mathbf{q} = 0$ governing optical transitions between vibrational states in crystals is relaxed in the amorphous state. Thus, although the transition in the case of the Raman spectrum of the crystal is restricted to zone-center phonons (thereby having a frequency corresponding to a transition, between lowest acoustic and highest optic bands, equal to the maximum band width of the VDOS), in the case of amorphous semiconductors, in principle, transitions between *all* vibrational states are allowed because of the breakdown of the **q**-selection rule, and as a result, the Raman spectrum more nearly represents the VDOS (see Fig. 11).

4.2 Propagating and Localized Phonons

Phonons, like electron states, are extended, propagating modes in the case of single crystals (without impurities, defects, etc.). However, the structural disorder characteristic of amorphous materials can serve to localize phonons, in the same way as for electron states, giving rise to a mobility edge at a frequency ω_c, below which phonons are extended and above which they are localized. The existence of such a mobility edge is signaled by the precipitous decrease in the phonon mean free path l (obtained from thermal-conductivity measurements; see Sec. 4.5) as the phonon wavelength decreases, or alternatively as the temperature increases (Zeller and Pohl, 1971)—see Fig. 12. This behavior is observed for all types of nonmetallic amorphous solid, semiconductors (Se) and insulators (GeO_2 and SiO_2) alike. The mean free path of the glasses reaches a minimum saturation value when the inverse phonon wave vector, $k = 2\pi/\lambda$, is comparable to l, the so-called Ioffe–Regel (1960) limit for localization,

$$kl \simeq 1. \tag{23}$$

Additional evidence for the existence of phonon localization comes from the anomalously long phonon lifetime (longer by three orders of magnitude than that for *c*-Si) found in *a*-Si:H for frequencies $\omega \geq 100$ cm^{-1} from time-resolved Raman experiments (Scholten *et al.*, 1993). This result can be understood if, say for the case of localized TO phonons (480 cm^{-1}), decay into two high-frequency extended acoustic phonons

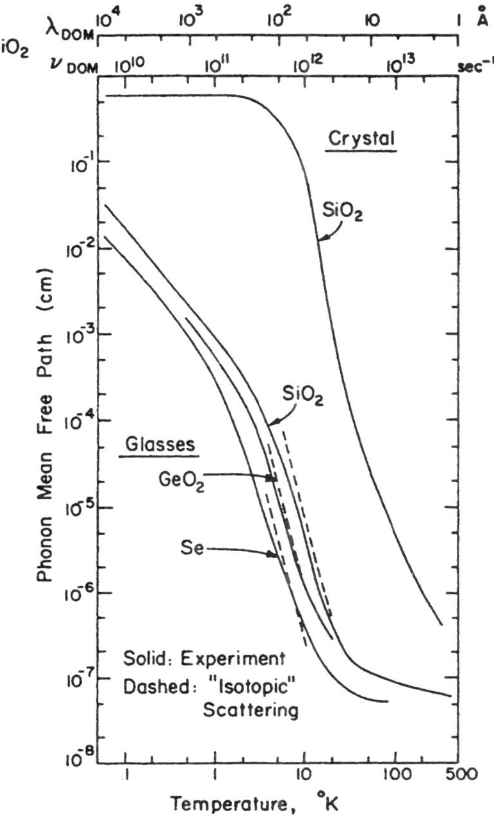

FIG. 12. Variation of phonon mean free path with temperature (or equivalently phonon wavelength) for glassy Se, GeO_2, and SiO_2, compared with that for crystalline SiO_2 (Zeller and Pohl, 1971).

cannot take place as in the crystalline case, but instead a low-probability "hopping" process (Orbach, 1992) occurs, whereby decay can only take place via another high-frequency localized phonon and a low-frequency extended phonon.

The precise cause of the phonon localization is still controversial, although the frequency dependence $l \propto \omega^{-4}$ deduced from Fig. 12 is indicative of Rayleigh scattering controlling the mean free path, and a model invoking elastic Rayleigh scattering from atomic-density fluctuations in amorphous solids on a length scale corresponding to short- and medium-range order has been proposed by Elliott (1992). Scattering of phonons by "soft configurations" (Karpov *et al.*, 1983), groups of atoms having anomalously low-frequency harmonic or anharmonic potentials (see Sec. 4.3), has also been suggested (Buchenau *et al.*, 1992) to be responsible for the behavior of the mean free path for frequencies below the mobility edge.

Although not of direct relevance to *fully dense* amorphous semiconductors, mention should also be made briefly of the localized vibrational excitations known as "fractons" (Orbach, 1986) that are supported by self-similar fractal structures for which the mass scales with distance as

$$M(r) \propto r^{D_H}, \tag{24}$$

where the Hausdorff dimension $D_H \leq d$ (d is the normal Euclidean dimension), over the range of length scales $a < r < \zeta$, where a is a typical interatomic distance and ζ marks the distance beyond which the material stops being self-similar and becomes structurally homogeneous. Thus, at frequencies lower than a crossover frequency ω_{CO}, normal phonon modes exist; at frequencies higher than ω_{CO}, localized fracton modes exist instead, for which the vibrational density of states has the frequency dependence

$$\rho(\omega) \propto \omega^{\tilde{d}-1} \tag{25}$$

where $\tilde{d}(\leq D_H \leq d)$ is the so-called fracton dimension, conjectured to have the universal value of $\frac{4}{3}$ (Alexander and Orbach, 1982), compared with the Debye-type behavior $\rho(\omega) \propto \omega^2$ for plane-wave phonons.

There is no experimental evidence for fractallike structures in fully dense amorphous semiconductors in either bulk-glass or thin-film forms, and so fractons cannot be responsible for the anomalous behavior observed in either the heat capacity (Sec. 4.4) or thermal conductivity (Sec. 4.5) in such materials. Nevertheless, in those materials, such as silica aerogels, that *do* possess a fractallike structure, fracton behavior does appear to be observed (Courtens *et al.*, 1988).

4.3 Soft Configurations and Two-Level (Tunneling) Systems

Analysis of the vibrational response of crystalline materials is almost invariably carried out within the harmonic approximation. However, in the case of amorphous materials, local variations in the atomic structure associated with disorder can lead to varying degrees of anharmonicity in the interatomic potentials, which can have a profound effect on the thermal properties of glasses, such as heat capacity (Sec. 4.4) and thermal conductivity (Sec. 4.5).

A general, albeit phenomenological, model accounting for such anharmonicity is the so-called soft-configuration model (Karpov *et al.*, 1983; Galperin *et al.*, 1989) that considers soft local modes involving an atom, or group of atoms, having the following empirical form for the potential as a function of displacement x in terms of a generalized coordinate:

$$V(x) = E[\eta(x/a)^2 + \xi(x/a)^3 + (x/a)^4]. \tag{26}$$

Here a is a distance of the order of the interatomic spacing, and E sets the energy scale and is given approximately by $E \simeq mv_s^2$, where m is the atomic mass and v_s is the sound velocity. The coefficients η and ξ are assumed to be randomly distributed in amorphous materials and, depending on their values, Eq. (26) describes harmonic or anharmonic single-well, and double-well, potentials.

In this sense, the soft-configuration model is a generalization of the two-level (or tunneling-state) one proposed earlier by Anderson *et al.* (1972) and Phillips (1972) to account for the low-temperature ($T < 1$ K) thermal anomalies of glasses (see Secs. 4.4 and 4.5). The two-level model, too, is a phenomenological model and assumes that certain groups of atoms in amorphous solids can occupy one of two equilibrium positions corresponding to a double-well potential (Fig. 13). At sufficiently low temperatures, atomic tunneling can occur through the potential barrier separating the two potential minima, and the resulting tunnel-split energy levels could be one origin of two-level systems. Although, as will be seen later (Secs. 4.4 and 4.5), the two-level model accounts well for the low-temperature thermal behavior of glasses, nevertheless it is still a phenomenological approach lacking a proper microscopic (atomistic) explanation. Although a motion of coupled (corner-shared) tetrahedra has been proposed to be the mode responsible in the case of v-SiO_2 (Buchenau *et al.*, 1986), this cannot be generally the case for all amorphous materials

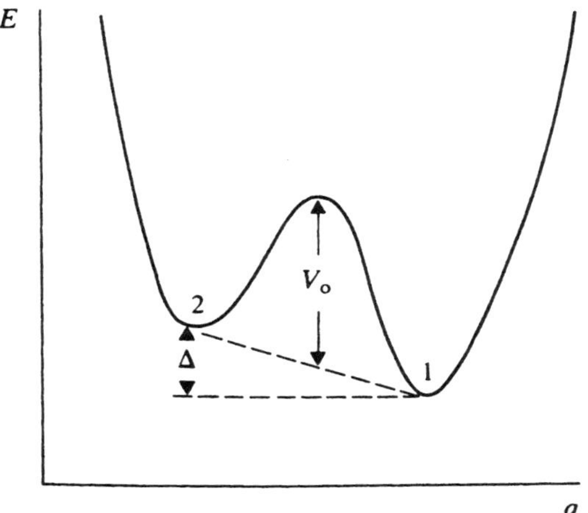

FIG. 13. Schematic illustration of the double-well potential characterizing a two-level system plotted as a function of some configuration coordinate, q. The parameters Δ (asymmetry energy) and V_0 (barrier height) are assumed to be randomly distributed in a glass.

(metals, semiconductors, and insulators) that exhibit the ubiquitous anomalous low-temperature thermal behavior.

4.4 Heat Capacity

The lattice-vibrational contribution to the heat capacity of crystalline solids at low temperatures (less than the Debye temperature θ_D) is well accounted for by the famous Debye law, $C_v \propto T^3$. In the same temperature regime, the behavior of *all* glasses (metallic and nonmetallic, alike) is different (see Phillips, 1981b, for a review): the nonelectronic, atomic-vibrational contribution to the temperature dependence of the heat capacity can be represented as (Zeller and Pohl, 1971)

$$C_v = c_1 T^\alpha + c_3 T^3, \tag{27}$$

where the exponent α has a value close to unity. This behavior (for the case where $\alpha = 1$) can be confirmed most simply by plotting heat-capacity data in the form C_v/T vs T^2: the intercept gives the coefficient c_1 and the gradient gives c_3.

Such a plot for the case of samples of As_2S_3, with varying impurity concentrations, is shown in Fig. 14 (Stephens, 1976). It is apparent that there is a marked extrinsic effect manifested in the heat capacity, although no one-to-one correspondence with the impurity content is evident. Nevertheless, the highest-purity glass does indeed show a temperature dependence for the heat capacity of the form given by Eq. (27), and it is apparent, as is generally the case, that the coefficient c_3 for the cubic term is appreciably larger than the value expected from simple Debye theory. Indeed, at temperatures of the order of 10 K, all glasses exhibit a peak in the temperature dependence of the heat

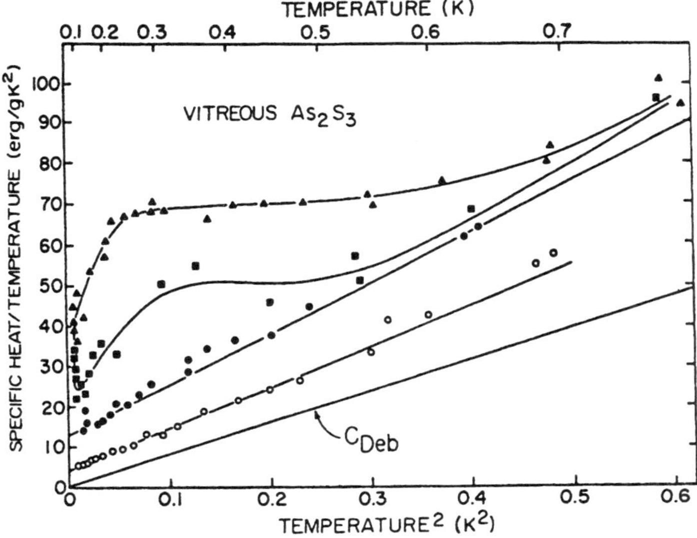

FIG. 14. Specific heat at low temperatures for four samples of a-As_2S_3 with varying impurity contents, plotted as C_v/T vs T^2 (Stephens, 1976). The straight line through the origin is the heat capacity predicted by the Debye model; the presence of the anomalous linear term in the specific heat is marked by the nonzero intercept of the data in such a plot. The purity of the samples (in terms of both metal and water content) increases from the upper to the lower set of data.

capacity normalized to the Debye (T^3) law (see Fig. 15), indicative of the existence of a low-frequency peak in the VDOS (again normalized to the corresponding Debye ω^2 form) that is also widely observed in low-frequency Raman scattering spectra of glasses (the so-called "boson peak").

The anomalous, almost linearly temperature-dependent specific heat observed in glasses at low temperatures (<1 K) was first accounted for by the two-level system model (Anderson *et al.*, 1972; Phillips, 1972). The heat capacity in this case is given by

$$C_v = k_B \int_0^\infty n(E)\left(\frac{E}{2k_BT}\right)^2 \operatorname{sech}^2\left(\frac{E}{2k_BT}\right) dE, \quad (28)$$

where E is the energy separation between the two levels and $n(E)$ is the corresponding density of states. For the case of the tunneling model, the energy is given by

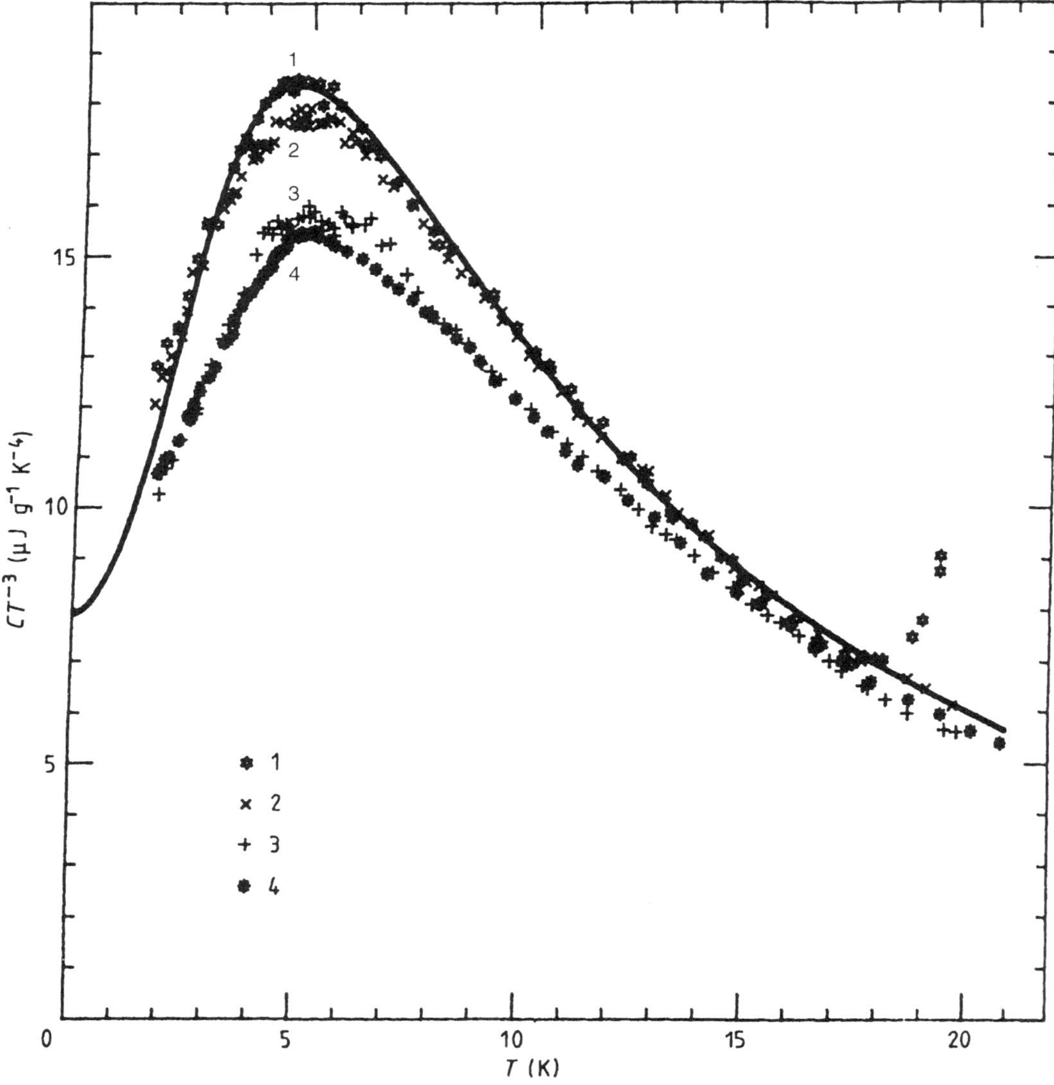

FIG. 15. Heat capacity of four samples of bulk glassy As_2S_3 (as quenched, samples 1 and 2; annealed, samples 3 and 4) normalized with respect to the Debye T^3 form (Ahmad *et al.*, 1986).

$$E^2 = \Delta^2 + \Delta_0^2, \quad (29)$$

where Δ_0^2 is the energy difference between the two lowest tunnel-split states in the symmetric-well case ($\Delta = 0$, see Fig. 13), and

$$\Delta_0 = \hbar\omega_0 e^{-\lambda}, \quad (30)$$

where

$$\lambda = (2mV_0/\hbar^2)^{1/2}d \quad (31)$$

with ω_0 being the oscillation frequency in an individual well. Evaluation of Eq. (28), assuming that the density of states is constant in energy, $n(E) = n_0$, over the energy range of interest (10^{-5}–10^{-4} eV), gives

$$C_v = (\pi^2/6)n_0k_B^2T. \quad (32)$$

If, instead, a weak energy dependence of $n(E)$ is admitted [$n(E) \propto E^m$, with $m = 0.1$–0.3], then a slightly superlinear temperature dependence of the heat capacity results, $C_v \propto T^{1+m}$, in accord with many experimental data (Stephens, 1976).

The more general soft-configuration model (Karpov *et al.*, 1983) treats the low-temperature (<1 K) linear-temperature-dependence regime of the heat capacity in terms of anharmonic excitations (identical to tunneling states), but it can also account for data at higher temperatures (<10 K) in terms of excitations involving soft harmonic potentials (Buchenau *et al.*, 1991).

An interpretation of the low-temperature heat capacity in terms of extra excitations associated with two-level systems or soft configurations is understandable for the case of the amorphous semiconductors *a*-Se or As_2S_3 and the glassy insulators SiO_2 and B_2O_3, since they all contain two-fold coordinated group-VI atoms (O,S,Se) that it is easy to envisage could be involved in the soft modes responsible. However, it is more difficult to understand in this manner that the notionally tetrahedrally coordinated material *a*-Si:H also shows a quasi-linear temperature dependence of the heat capacity, with a value of the coefficient c_1 (Eq. 27) almost equal to that for *v*-SiO_2 (Graebner *et al.*, 1984). It is conceivable that singly coordinated hydrogen, or low-coordinated atoms in the vicinity of the voids responsible for the 14% density deficit (relative to *c*-Si), could be the origin of the tunneling states, but it remains a remarkable coincidence that the densities of two-level systems are the same as between *a*-Si:H and SiO_2. A proper, microscopic, understanding of the heat capacity of amorphous semiconductors (and insulators) is still lacking.

4.5 Thermal Conductivity

Thermal transport in crystalline nonmetallic materials occurs by means of propagating phonons. The thermal conductivity may be written, from kinetic theory, as

$$\kappa = \tfrac{1}{3}\sum_i \int_0^{\omega_D} C_i(\omega)v_i(\omega)l_i(\omega)\,d\omega, \quad (33)$$

where the sum is over the three longitudinal and transverse phonon modes, v and l are the phonon velocity and mean free path, respectively, and $C(\omega)$ is the contribution to the specific heat from phonons of frequency ω. The temperature dependence of κ for a crystal typically exhibits a peak at low temperatures; above, say 10 K, phonons are scattered by intrinsic (phonon–phonon umklapp) processes, and hence l and, consequently, κ decrease with increasing temperature since the number of phonons increases; at lower temperatures, l increases with decreasing temperature and eventually becomes equal to the sample dimensions and hence is constant, and thereafter $\kappa(T)$ follows the temperature dependence of $C_v(T)$ and is therefore proportional to T^3 [see Fig. 16(a)].

In contrast, the behavior of all amorphous nonmetallic materials is completely different in two respects (Zeller and Pohl, 1971; Freeman and Anderson, 1986): no peak in $\kappa(T)$ is evident, but instead there is a plateau in the range 10–100 K for most amorphous materials [see Fig. 16(b)], although this appears to start at 100–200 K for the case of *a*-Si:H (Cahill *et al.*, 1994). The plateau is followed by a further rise at higher temperatures; this is not evident in Fig. 16(b) but is shown in Fig. 17 for the case of the amorphous semiconductor *a*-Se and some insulating oxide glasses. Furthermore, $\kappa(T)$ for amorphous materials also differs from that of crystals at low temperatures: $\kappa(T) \propto T^2$ for amorphous materials in general (although the limiting low-temperature behavior of *a*-Ge shown in

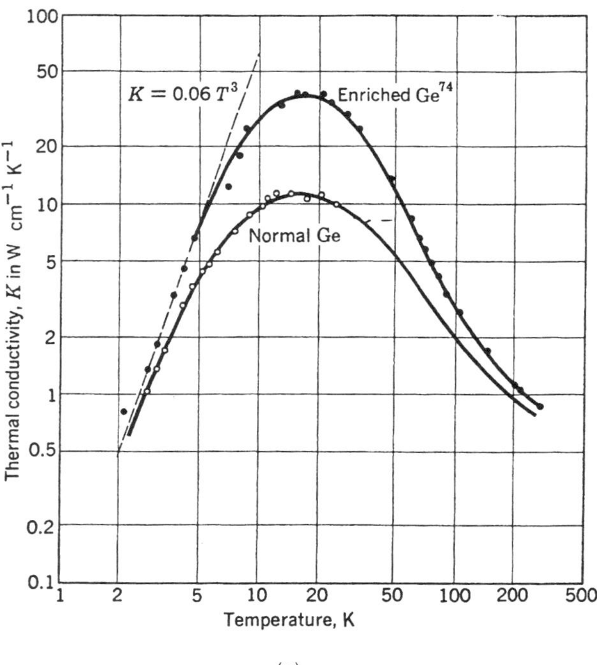

(a)

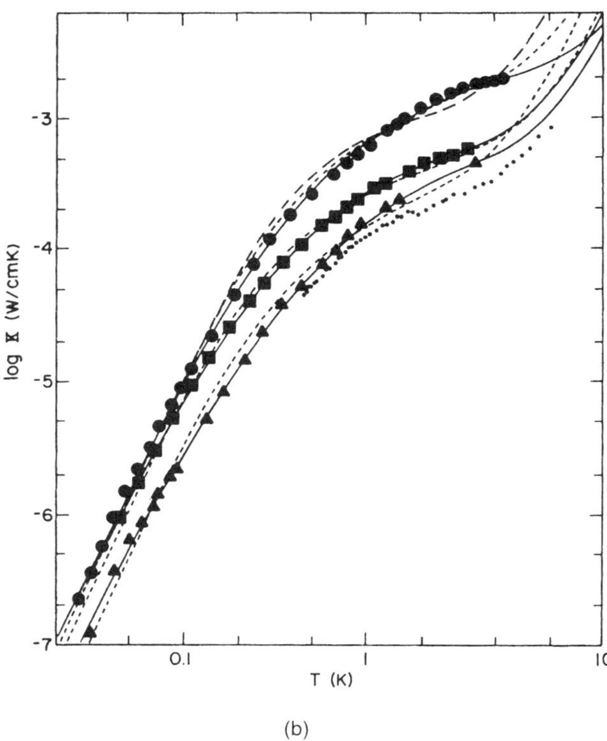

(b)

FIG. 16. Temperature dependence of the thermal conductivity of Ge: (a) *c*-Ge (Kittel, 1976); (b) *a*-Ge (Graebner and Allen, 1983). The three data sets in (b) refer to different samples of *a*-Ge with, from top to bottom, decreasing densities.

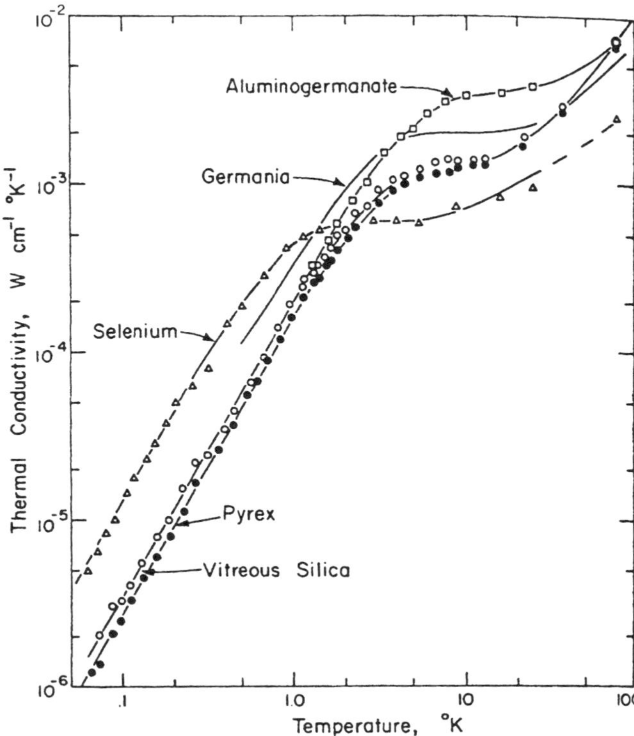

FIG. 17. Temperature dependence of the thermal conductivity of some semiconducting and insulating glasses (Zeller and Pohl, 1971).

Fig. 16(b) appears to be proportional to T^3 as for crystals).

The low-temperature behavior of $\kappa(T)$ for amorphous materials has been explained in terms of the additional phonon-scattering centers controlling the mean free path at low temperatures (<1 K), i.e., two-level systems (see, e.g., Phillips, 1981b; Freeman and Anderson, 1986) or, more generally, soft configurations (Buchenau *et al.*, 1992). Phonons having an energy $\hbar\omega = E$, the tunnel splitting, can be scattered resonantly by a process involving excitation from the ground state of the two-level system, followed by the subsequent decay and emission of an incoherent phonon, for which the mean free path is (Phillips, 1981b)

$$l_{res} = \frac{\rho v^3}{\pi n_0 M^2 \omega} \coth\left(\frac{\hbar\omega}{2k_B T}\right) \tag{34}$$

where M is the coupling constant relating E to the strain. In the dominant-phonon approximation, $\hbar\omega \sim k_B T$, and so Eq. (34) predicts that $l_{res} \propto T^{-1}$, which, taken together with $C_v \propto T^3$ (because the heat is carried by phonons at temperatures below the onset of the plateau in κ), gives from Eq. (33) $\kappa \propto T^2$, as found experimentally (Zeller and Pohl, 1971; Freeman and Anderson, 1986). However, although the phenomenological tunneling-state model can account for the functional form of $\kappa(T)$ for $T \leq 1$ K, it is difficult to understand thereby the reason for the empirical relation

$$l/\lambda = \rho v^2/M^2 n_0 \simeq 150 \tag{35}$$

that is apparently valid for all glasses (Freeman and Anderson, 1986).

At higher temperatures ($T > 1$ K) or higher frequencies ($\omega \geq 10^{11}$ s^{-1}), the phonon mean free path decreases precipitously according to

$$l \propto \omega^{-n} \quad (n = 3\text{–}4). \tag{36}$$

This behavior therefore accounts for the temperature independence of $\kappa(T)$ in the so-called plateau region. Possible physical mechanisms that could give rise to relation (36) are Rayleigh scattering of phonons by density fluctuations, for which $n = 4$ (see Elliott, 1992, for a recent discussion), or alternatively scattering by soft configurations with an assumed density of quasiharmonic states $n(\omega) \propto \omega^4$ (Buchenau *et al.*, 1992).

At the upper frequency/temperature end of the plateau in κ, the mean free path has become reduced to a length of the order of interatomic spacings, $l \sim a$, and also to the dimension of the phonon wavelength (see Fig. 12), suggesting, from Eq. (23), that phonon localization is taking place there. The upturn in κ observed above the plateau region (Fig. 17) is therefore surprising and indicative of another mechanism being responsible for heat transport in this high-temperature/frequency region since the propagating phonons are in the Dulong–Petit limit for which $\kappa(T)$ = constant. Neither the two-level system nor the soft-configuration model can account for this behavior (Buchenau *et al.*, 1992). Orbach (1992, 1993) has proposed that a "phonon-hopping" process occurs in this regime, analogous to the case of variable-range hopping of localized electrons (see, e.g., Mott and Davis, 1979; Elliott, 1990), in which, as a result of *anharmonic* interactions, a localized phonon with frequency ω "hops" to a neighboring site, thereby producing a localized phonon of frequency ω' there, mediated by a (low-frequency) extended phonon of frequency ω_0 ($\omega = \omega' + \omega_0$). A calculation of $\kappa(T)$ appropriate for this situation (taken from a treatment for fractons in fractal structures) gives (Orbach, 1992, 1993)

$$\kappa \propto T \tag{37}$$

for temperatures greater than the plateau region, in qualitative agreement with experimental data (Zeller and Pohl, 1971).

GLOSSARY

Amorphous Solid: A noncrystalline material lacking long-range translational or orientational structural order.

Band Tail: Electron states extending into the forbidden energy (band) gap of an amorphous (*q.v.*) semiconductor, either up from the valence band or down from the conduction band, resulting from the presence of disorder.

Boson Peak: A peak at low frequencies in the vibrational density of states (*q.v.*) of amorphous solids (*q.v.*), most often detected by Raman scattering, whose temperature dependence follows the bose factor.

Chalcogenide: An alloy of a chalcogen (an element from group VI of the periodic table, i.e., S, Se, or Te) with other elements [often Si, Ge, B, As, P, etc. in the case of amorphous materials (*q.v.*)].

Chemical Vapor Deposition (CVD): Method of producing solid materials in thin-film form by means of chemical reactions between precursors in the vapor phase and a solid substrate.

Dangling Bond: A defect comprising an undercoordinated atom, ubiquitous in covalent amorphous materials (*q.v.*).

Density of States (DOS): The number of states (electronic or vibrational) at a particular energy (frequency) per unit energy (frequency) interval.

Fractal: A self-similar structure for which the mass scales with distance as a power law [Eq. (24)] with an exponent less than that (3) characteristic of a fully dense material.

Fracton: A localized (*q.v.*) vibrational excitation supported by a fractal (*q.v.*) structure.

Glass: An amorphous solid (*q.v.*), often prepared by melt quenching, that exhibits a glass transition (discontinuity in the heat capacity).

Glow-Discharge (GD) Decomposition: Synonym for plasma-enhanced chemical vapor deposition (*q.v.*).

Localized State: An electron (or vibrational) state whose wave function is confined to a particular spatial region for all times, as a result of disorder.

Lone Pair: Nonbonding pair of electrons in an orbital.

Medium-Range Order: Structural organization in an amorphous solid (*q.v.*) at distances beyond that characteristic of short-range order (*q.v.*), i.e., 5–20 Å.

Mobility Edge: Demarcation energy separating localized (*q.v.*) from extended (electron or vibrational) states.

Polaron: A charge carrier (electron or hole) associated with lattice distortion.

Short-Range Order: Structural organization in an amorphous solid (*q.v.*) at the shortest length scale (2–5 Å), associated with bond lengths and angles and coordination numbers.

Soft Configuration: A group of atoms in an amorphous solid (*q.v.*) supporting anomalously low-frequency vibrational modes and associated with strongly anharmonic vibrational potentials [Eq. (26)].

Sputtering: Method of thin-film deposition based on ablation of a target by ions from a neighboring gas-phase plasma.

Two-Level System: Double-well vibrational potential for a "soft" configuration (*q.v.*) of a group of atoms, for which atomic tunneling through the potential barrier separating the two wells can take place at low temperatures.

Variable-Range Hopping: Tunneling motion of a localized particle (electron or phonon) between sites with varying separation, such that more distant coupled sites have smaller energy separations, and vice versa.

Urbach Edge: Exponential dependence of optical absorption coefficient on photon energy [Eq. (17)] associated with electronic transitions between band-tail (*q.v.*) electron states distributed exponentially in energy in the case of semiconducting amorphous materials (*q.v.*).

Works Cited

Ahmad, N., Hutt, K. W., Phillips, W. A. (1986), *J. Phys. C* **19,** 3765–3773.

Alexander, S., Orbach, R. (1982), *J. Phys. (Paris) Lett.* **43,** L625–L631.

Amer, N. M., Jackson, W. B. (1984), in: J. Pankove (Ed.), *Semiconductors and Semimetals,* New York: Academic, Vol. 21B, p. 83.

Anderson, P. W., Halperin, B. I., Varma, C. M. (1972), *Philos. Mag.* **25,** 1–9.

Anderson, P. W. (1958), *Phys. Rev.* **109,** 1492–1505.

Bar-Yam, Y., Adler, D., Joannopoulos, J. D. (1986), *Phys. Rev. Lett.* **57,** 467–470.

Brazhkin, V. V., Lyapin, A. G., Popova, S. V., Voloshin, R. N. (1992), *Pis'ma Zh. Eksp. Teor. Fiz.* **56,** 156–159 (*JETP Lett.* **56,** 152–156).

Buchenau, U., Galperin, Yu. M., Gurevich, V. L., Schober, H. R. (1991), *Phys. Rev.* **B43,** 5039–5045.

Buchenau, U., Galperin, Yu. M., Gurevich, V. L., Parshin, D. L., Ramos, M. A., Schober, H. R. (1992), *Phys. Rev.* **B46,** 2798–2808.

Buchenau, U., Prager, M., Nücker, N., Dianoux, A. J., Ahmad, N., Phillips, W. A. (1986), *Phys. Rev.* **B34,** 5665–5673.

Cahill, D. G., Katiyar, M., Abelson, J. R. (1994), *Phys. Rev.* **B50,** 6077–6081.

Cody, G. D. (1984), in: J. Pankove (Ed.), *Semiconductors and Semimetals,* New York: Academic, Vol. 21B, p. 11.

Cody, G. D., Tiedje, T., Abeles, B., Brooks, B., Goldstein, Y. (1981), *Phys. Rev. Lett.* **47,** 1480–1483.

Connell, G. A. N. (1979), in: M. H. Brodsky (Ed.), *Amorphous Semiconductors,* New York: Springer-Verlag, Vol. 36, p. 73.

Courtens, E., Vacher, R., Pelous, J., Woignier, T. (1988), *Europhys. Lett.* **6,** 245–250.

Cullis, A. G., Webber, H. C., Chew, N. G., Poate, J. M., Baeri, P. (1982), *Phys. Rev. Lett.* **49,** 219–222.

Demishev, S. V., Kosichkin, Yu. V., Sluchanko, N. E., Lyapin, A. G. (1994), *Usp. Fiz. Nauk.* **164,** 195–230 (*Sov. Phys. Usp.* **36,** 185–217).

Elliott, S. R., Kolobov, A. V. (1991), *J. Non-Cryst. Sol.* **128,** 216–220.

Elliott, S. R. (1992), *Europhys. Lett.* **19,** 201–206.

Elliott, S. R. (1990), *Physics of Amorphous Materials,* 2nd ed., New York: Longman.

Freeman, J. J., Anderson, A. C. (1986), *Phys. Rev.* **B34,** 5684–5690.

Galperin, Yu, M., Karpov, V. G., Kozub, V. I. (1989), *Adv. Phys.* **38,** 669–737.

Graebner, J. E., Allen, L. C. (1983), *Phys. Rev. Lett.* **51,** 1566–1569.

Graebner, J. E., Golding, B., Allen, L. C., Knights, J. C., Biegelsen, D. K. (1984), *Phys. Rev.* **B29,** 3744–3746.

Hecking, N., Heidemann, K. F., Te Kaat, E. (1986), *Nucl. Instrum. Meth. Phys. Res.* **B15,** 760–764.

Houserova, J., Hruby, A. (1972), *Czech J. Phys.* **B22,** 861–863.

Ioffe, A. F., Regel, A. R. (1960), *Prog. Semicond.* **4,** 237–291.

Jackson, W. B., Amer, N. M. (1982), *Phys. Rev.* **B25,** 5559–5562.

Jackson, W. B., Kelso, S. M., Tsai, C. C., Allen, J. W., Oh, S.-J. (1985), *Phys. Rev.* **B31,** 5187–5198.

Kamitakahara, W. A., Shanks, H. R., McClelland, J. F., Buchenau, U., Gompf, F., Pintchovious, L. (1984), *Phys. Rev. Lett.* **52,** 644–647.

Karpov, V. G., Klinger, M. I., Ignat'ev, F. N. (1983), *Zh. Eksp. Teor. Fiz.* **84,** 760–765 (*Sov. Phys. JETP* **57,** 439–448).

Kittel, C. (1976), *Introduction to Solid State Physics,* 5th ed., New York: Wiley.

Kolobov, A. V., Elliott, S. R. (1995), *Philos. Mag.* **B71,** 1–10.

Leadbetter, A. J. (1973), in: B. T. M. Willis (Ed.), *Chemical Applications of Thermal Neutron Scattering,* Oxford: Oxford Univ. Press, p. 146.

Ley, L. (1984), in: J. D. Joannopoulos and G. Lucovsky (Eds.), *The Physics of Hydrogenated Amorphous Silicon,* Vol. II, New York: Springer Verlag, p. 61.

Ley, L., Pollak, R. A., Kowalczyk, S., McFeely, R., Shirley, D. A. (1973), *Phys. Rev.* **B8,** 641–646.

Liu, P. L., Yen, R., Bloembergen, N., Hodgson, R. T. (1979), *Appl. Phys. Lett.* **34,** 864–866.

Lucovsky, G., Tsu, D. V. (1987), *J. Non-Cryst. Sol.* **97–98,** 265–268.

Mahan, A. H., Carapella, J., Nelson, B. P., Crandall, R. S., Balberg, I. (1991), *J. Appl. Phys.* **69,** 6728–6730.

Matsumura, H. (1989), *J. Appl. Phys.* **65,** 4396–4402.

Mott, N. F., Davis, E. A. (1979), *Electronic Processes in Non-Crystalline Solids,* 2nd ed., Oxford: Clarendon Press.

Orbach, R. (1993), *J. Non-Cryst. Sol.* **164–166,** 917–922.

Orbach, R. (1992), *Philos. Mag.* **B65,** 289–301.

Orbach, R. (1986), *Science* **231,** 814–819.

Phillips, J. C. (1981a), *J. Non-Cryst. Sol.* **43,** 37–77.

Phillips, W. A. (Ed.) (1981b), *Amorphous Solids: Low-Temperature Properties,* New York: Springer-Verlag.

Phillips, W. A. (1972), *J. Low Temp. Phys.* **7,** 351–360.

Pierce, D. T., Spicer, W. E. (1972), *Phys. Rev.* **B5,** 3017–3029.

Roca i Cabarrocas, P. (1993), *J. Non-Cryst. Sol.* **164–166,** 37–42.

Scholten, A. J., Akimov, A. V., Verleg, P. A. W. E., Dijkhuis, J. I., Meltzer, R. S. (1993), *J. Non-Cryst. Sol.* **164–166,** 923–926.

Shevchik, N. J., Cardona, M., Tejeda, J. (1973), *Phys. Rev.* **B8,** 2833–2841.

Smirl, A. L., Boyd, I. W., Boggers, T. F., Moss, S. C., van Driel, H. M. (1986), *J. Appl. Phys.* **60,** 1169–1182.

Spear, W. E., Le Comber, P. G. (1975), *Solid State Comm.* **17,** 1193–1196.

Stephens, R. B. (1976), *Phys. Rev.* **B13,** 852–865.

Street, R. A. (1991), *Hydrogenated Amorphous Silicon,* Cambridge, U.K.: Cambridge Univ. Press.

Tauc, J., Grigorovici, R., Vancu, A. (1966), *Phys. Status Solidi* **15,** 627–637.

Thorpe, M. F. (1995), *J. Non-Cryst. Sol.* **182,** 135–142.

Tiedje, T. (1984), in: J. Pankove (Ed.), *Semiconductors and Semimetals,* New York: Academic, Vol. 21C, p. 207.

Tsai, C. C., Nemanich, R. J. (1980), *J. Non-Cryst. Sol.* **35–36,** 1203–1208.

Tsu, R., Hodgson, R. T., Tan, T. Y., Baglin, J. E. (1979), *Phys. Rev. Lett.* **42,** 1356–1358.

Urbach, F. (1953), *Phys. Rev.* **92,** 1324.

Yoshida, A., Inoue, K., Ohashi, H., Saito, Y. (1990), *Appl. Phys. Lett.* **57,** 484–486.

Zeller, R. C., Pohl, R. O. (1971), *Phys. Rev.* **B4,** 2029–2041.

Zhang, M., Boolchand, P. (1994), *Science* **266,** 1355–1357.

SEMICONDUCTORS, COMPOUND—ELECTRONIC PROPERTIES

Sandip Tiwari, *IBM Research Division, Thomas J. Watson Research Center, Yorktown Heights, New York, U.S.A.*

INTRODUCTION

The term compound semiconductor usually refers to the semiconducting materials formed by compounds of group IIIB elements (Al, Ga, In, e.g.) and group VB elements (N, P, As, Sb, e.g.), group IIB elements (Zn, Cd, Hg, e.g.) and group VIB elements (S, Se, Te, e.g.), and combinations of them. GaAs, $Ga_{1-x}Al_xAs$, $Ga_{1-x}In_xAs$, $Ga_xIn_{1-x}As_yP_{1-y}$, $Al_{1-x}In_xAs$, InP, $Ga_{1-x}In_xP$, GaN, ZnSe, and $Cd_xZn_{1-x}Se$ are some of the more popular examples—binary, ternary, and quaternary—and appealing in device structures because of their electronic or optical properties. The electronic properties that attract the interest are usually the superior transport of carriers—faster than in silicon, the existence of semi-insulating substrates and large breakdown voltages in some of the larger–band-gap semiconductors—each of which is particularly useful in high-frequency microwave applications or even high-speed applications. Many compound semiconductors have direct energy band gaps allowing efficient radiative recombination. Lattice-matched growth allows stacking of different compound semiconductors or grading of their compositions with changes in band gaps, refractive indexes, etc., while maintaining crystalline and interfacial quality. These allow for confinement of carriers and/or photons (wave guiding) to be achieved, resulting in compact lasers as well as very high-mobility structures where donors can be separated from carriers and ionized-impurity scattering minimized. Compound semiconductors have thus been attractive for both engineering and scientific reasons. The group II–VI compound semiconductors have poor mobilities and hence have not been applied to electronic devices in any significant way. They have been employed in optical devices—particularly light-emitting diodes and lasers in the short visible wavelength region. This article will

3-527-28139-8/96/$5.00 + .50

therefore concentrate on the III–V compound semiconductors, and Table 1 is a compilation of properties that we refer to throughout.

1. BAND STRUCTURES

1.1 Mathematical Formulations

Describing the behavior of electrons and holes—the energy-momenta relationship—in a solid such as a compound semiconductor crystal is inherently complex, a many-body problem. The lattice potential is large and thus cannot be treated as a perturbation, and the electrons can best be described by plane waves with lattice periodicity (Bloch functions) whose modulation is very strong (like free atoms) near the ions and very weak (like free electrons) in between. An approximate description of the behavior of electrons in this periodic potential requires a solution to the Schrödinger equation using the crystal Hamiltonian. This crystal Hamiltonian contains energy terms due to kinetic, potential, and magnetic contributions arising from electrons in a lattice structure of the nuclei. The solution can be accomplished in a number of ways using perturbation theory: using tight-binding approximation (i.e., using functions that are linear combinations of either atomic or molecular orbitals), using pseudopotentials, and using various other functional approaches such as Green's-function techniques. Perturbation theory allows the determination of the eigenenergies using the smaller Hamiltonian terms as perturbing functions. Pseudopotentials allow the most convenient means to solve the problem at the present time. The influence of each ion and the environment on electrons in the conduction and the valence bands is incorporated through a potential that approximates realistically, although empirically through experimental matching of band-structure results, the influence of the ion in the important region away from the core while taking into account the crystal symmetry in the form of a Fourier series over the reciprocal lattice vector and the nonlocal effects such as due to different angular momentum components of electrons (for $|s\rangle$ and $|p\rangle$, e.g.) through variations in the pseudopotential. Since the basis functions, involving calculations based on using atomic orbitals near the cores, are $|s\rangle$ and $|p\rangle$ like, the resulting functions are combinations of these at various important points of the Brillouin zone. The band structure is the interesting result of the calculation that describes the relationship between the energy and the wave vector of electrons—the specific values that they can have. There are bands in energy that are allowed and bands that are not allowed. Semiconductors in their pure state are compounds that have these allowed bands either completely filled or completely empty at absolute zero. At a finite temperature, thermal energy allows carriers to be transferred from the fully occupied band to the empty band because the gap between the two, the band gap, is small enough (on the order of electron volts). This makes them semiconducting and allows for carrier transport to take place in both the partially occupied band (electrons in the conduction band) and the partially empty band (holes representing the absence of electrons in the valence band).

1.2 Conduction-Band Structure

Figure 1 shows the conduction- and valence-band structure for GaAs over a large energy range, and for comparison and discussion we include the conduction-band structure of three other popular compound semiconductors: InP (Fig. 2), InAs (Fig. 3), and AlAs (Fig. 4). GaAs is perhaps the most common compound semiconductor in use as a substrate for field-effect transistors and utilized together with AlAs, i.e., in the ternary form $Ga_{1-x}Al_xAs$, in formation of heterojunction discontinuities that have wide applications. Alloy compositions of GaAs with InAs, $Ga_{1-x}In_xAs$, have also been quite common and utilize discontinuities with $Al_{1-x}In_xAs$, InP, and in strained structures with GaAs and its ternary compounds. At the composition of $Ga_{0.47}In_{0.53}As$, it is lattice matched to InP and grown on it. The band structures of these materials also point out a number of interesting features of the variety of compound semiconductors.

Some general observations: GaAs is a direct–band-gap semiconductor with the ordering of minima from low to high energies in the conduction band as Γ ($\langle 000\rangle$), L ($\langle 111\rangle$ and its equivalents), and X ($\langle 100\rangle$ and its equivalents). The lowest constant-energy sur-

Table 1. A compilation of some of the major parameters of compound semiconductors.

Property	AlN	AlP	AlAs	AlSb	GaN	GaP	GaAs	GaSb	InN	InP	InAs	InSb
Lattice[a]	ZB	ZB	ZB	ZB	WZ	ZB	ZB	ZB	WZ	ZB	ZB	ZB
a, Å	a axis 3.112	5.467	5.661	6.136	a axis 3.160	5.451	5.653	6.096	a axis 3.544	5.869	6.058	4.479
	c axis 4.980				c axis 5.125				c axis 5.703			
ρ, g/cm^3		2.40	3.76	4.26		4.138	5.36	5.614		4.81	5.667	5.775
ϵ_0	9.1	9.8	10.06	12.04	10.4_c	11.1	12.91	15.69		12.61	15.15	16.8
ϵ_∞	4.8	7.54	8.16	10.24	5.8_c	9.075	10.1	14.44	9.3	9.61	12.25	15.68
E_g, eV, 300 K	6.2	2.45	2.14	1.63	3.44	2.268	1.423	0.70	2.09	1.3511	0.356	0.18
ΔE_Γ, eV	0	1.237	0.767	0.507		0.496	0	0		0	0	0
ΔE_L, eV		0.824	0.322	0.292		0.415	0.323	0.217		0.832	1.078	1.038
ΔE_X, eV		0	0	0		0	0.447	0.491		1.492	1.607	1.569
E_g, eV, 77 K			2.223	1.666		2.338	1.512		2.21	1.4135	0.414	0.228
$\hbar\omega_{LO}$, meV	112.8	62.1	50.1	42.1	92.4	50.0	35.4		86.1	42.8	29.6	23.7
$\hbar\omega_{TO}$, meV	82.7	54.5	44.8	39.5	69.4	45.5	33.2		59.3	37.7	26.9	22.3
n_i, cm^{-3}				5.0×10^6		3.0×10^6	1.8×10^6	4.3×10^{12}		1.2×10^8	1.3×10^{15}	2.0×10^{16}
m_Γ		0.166	0.149	0.126		0.122	0.067	0.048		0.083	0.031	0.018
m_{Ll}		1.556	1.456	1.601		1.498	1.538	1.400		1.893	1.590	1.700
m_{Lt}		0.162	0.165	0.142		0.358	0.126	0.139		0.425	0.326	0.096
m_{Xl}		1.118	1.664	2.350		1.868	1.987	1.630		0.927	2.499	2.000
m_{Xt}		0.256	0.223	0.255		0.249	0.229	0.212		0.277	0.274	0.268
E_{so}, eV		0	0.29	0.645		0.08	0.340	0.750		0.110	0.410	0.810
m_{hh}		0.630	0.760	0.940		0.790	0.510	0.290		0.850	0.600	0.440
m_{lh}		0.200	0.150	0.290		0.140	0.082	0.047		0.089	0.027	0.016
μ_{el}, cm^2/V·s, 300 K				200		200	8×10^3	5×10^3		5×10^3	3×10^4	
($<10^{15}$ cm^{-3}), 77 K						2.8×10^3	1.5×10^5			6×10^4	3×10^5	4×10^5
μ_{el}, cm^2/V·s, 300 K		50		250	600	160	4.8×10^3			3.2×10^3	2×10^4	8×10^4
($\approx 10^{17}$ cm^{-3}), 77 K		30		800		1.2×10^3	5.2×10^3			4×10^3	3.5×10^4	1.5×10^5
μ_{el}, cm^2/V·s, 300 K			50		180	95	2.2×10^3			1.8×10^3	1.3×10^4	1.0×10^4
($\approx 10^{19}$ cm^{-3}), 77 K			18			450	3×10^3				2×10^4	8×10^4
μ_h, cm^2/V·s, 300 K				450		150	400	1500		180	480	1500
($<10^{15}$ cm^{-3}), 77 K							9×10^3			1×10^3		9×10^3
μ_h, cm^2/V·s, 300 K				375		120	320	880		150		
($\approx 10^{17}$ cm^{-3}), 77 K				1.8×10^3				2.4×10^3		600		3×10^3
μ_h, cm^2/V·s, 300 K				175		30	50	300		28		
($\approx 10^{19}$ cm^{-3}), 77 K								1×10^3		18		700

[a]ZB: zincblende; WZ: wurtzite.

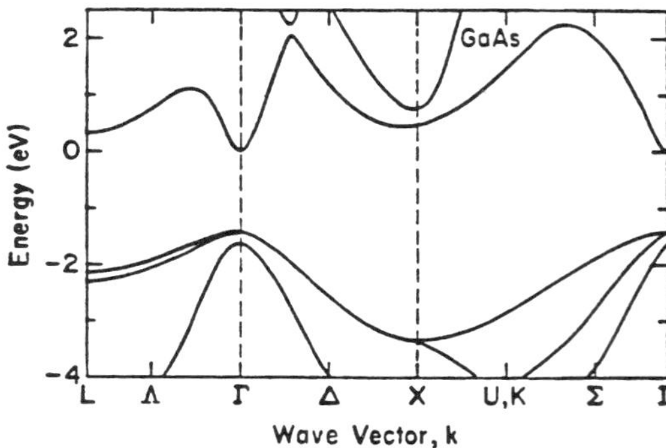

FIG. 1. Conduction- and valence-band structure of GaAs over major points and directions in the first Brillouin zone.

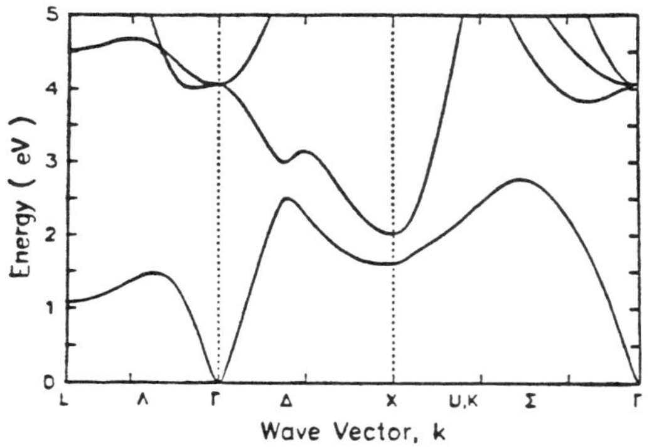

FIG. 3. Conduction-band structure of InAs over major points and directions in the first Brillouin zone.

face, due to the Γ valley, is spherically symmetric. The band gap of intrinsic GaAs, at 300 K, is 1.42 eV; the L valley is 0.29 eV, and the X valley is about 0.48 eV, above the Γ minimum. InP and InAs are also direct–band-gap compound semiconductors with an ordering of conduction-band minima similar to that of GaAs. The band gap of intrinsic InP is 1.35 eV, and that of InAs is significantly smaller, ≈0.36 eV, at 300 K. AlAs and GaP are two of the few compound semiconductors with indirect energy-band gap.

AlAs is an indirect–band-gap semiconductor with an intrinsic band gap of 2.14 eV; the minimum in conduction-band energy occurs nearly 85% of the way to the X point from the Γ point (similar to silicon). This is followed by the L valley. It exhibits a behavior similar to that of silicon, and some of these characteristics may be traced to aluminum, which is a light atom like silicon. Constant low-energy surfaces around this valley, loosely referred to as the X valley, are sixfold degenerate and aligned along orthogonal axes.

In indirect semiconductors, such as AlAs (silicon and germanium are the two common elemental semiconductor examples), the lowest conduction bands have multiple valleys associated with them. A consequence of this is the availability of a large number of states to scatter into. A large change in momentum is required for this process and is available through zone-edge phonons. The result of this is that, at low energies, these semiconductors experience higher scattering than direct–band-gap compound semiconductors.

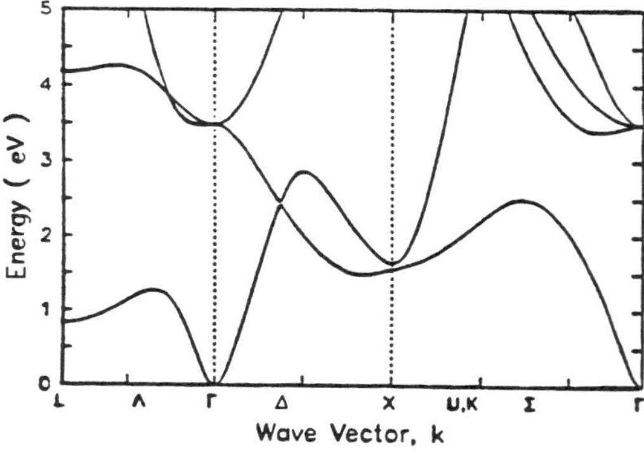

FIG. 2. Conduction-band structure of InP over major points and directions in the first Brillouin zone.

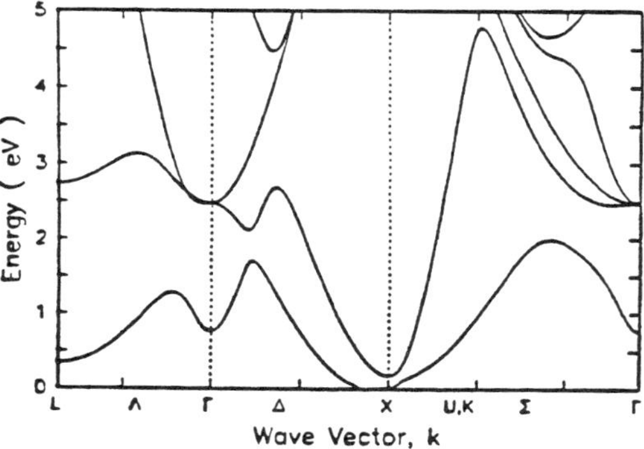

FIG. 4. Conduction-band structure of AlAs over major points and directions in the first Brillouin zone.

1.3 Valence-Band Structure

All compound semiconductors exhibit a universal feature of a degenerate light- and heavy-hole band at zone center, and a split-off band. The split-off band comes about because of the perturbation from the magnetic energy associated with the spin and orbit of the electron. This coupling is generally referred to as spin–orbit coupling or $L \cdot S$ coupling, and the split-off band arising from it is 0.34 eV below the valence-band maximum in GaAs. Silicon, an elemental semiconductor, does not exhibit a large separation of the split-off band. The origin of this is in the spin–orbit energy that gives rise to the formation of the split-off band: Light atoms exhibit small spin–orbit interaction. In silicon, spin–orbit splitting is only 0.044 eV. Spin–orbit energy, however, rises rapidly with atomic mass. Germanium has a spin–orbit splitting of 0.29 eV and does exhibit the split-off band. In InP a split-off band occurs 0.78 eV below. A consequence of multihole bands, split-off bands, large effective masses, etc., is a large hole-scattering rate. This together with the large mass results in poor hole mobilities. Only GaSb, of all the compound semiconductors, exhibits any reasonable hole mobility although still not comparable to the electron mobility.

1.4 Band Gaps

One of the unique attributes of compound semiconductors, exploited in many specialized device structures, is the rich selection of various semiconductors that form heterojunctions of the same lattice constant yet differing in band gap, thus allowing for a broader selection and tuning of desired characteristics. Figure 5 shows the variation of band gap as function of lattice constant for various ternary semiconductors with their binary end points. A broad range of band gaps is possible; for optical devices it is useful in achieving lasing at a variety of useful wavelengths—from infrared to blue at the present time. In electronic devices, its major significance has been in growing single-crystal materials of differing band gap and achieving either discontinuities in band edges at the interfaces or a grading of band gap in the interface region. Discontinuities allow for unique transport and charge control properties that are useful in field-effect transistors. As an example, $Ga_{1-x}Al_xAs$/GaAs and $Al_{1-x}In_xAs/Ga_{1-x}In_xAs$ allow large electron densities in accumulation/inversion layers at the interfaces in the smaller–band-gap material while the doping is located in the larger–band-gap material. This leads to superior mobilities for carriers primarily because of a reduction in ionized-impurity scattering. Grading of band gaps allows for placement of the change in band gap in the conduction band edge or valence band edge. An *n-p-n* $Ga_{1-x}Al_xAs$/GaAs/GaAs bipolar junction transistor with a graded interface at the emitter–base junction allows for different barriers for electrons and holes, a smaller barrier for electrons than for holes leading to superior injection efficiencies.

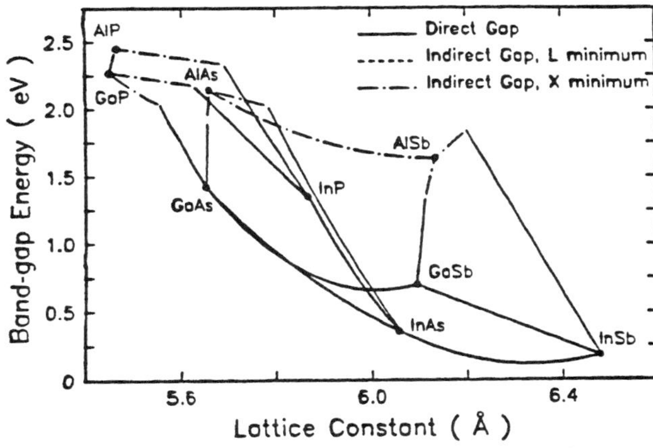

FIG. 5. Band gap versus lattice constant for compound semiconductors. Legend in the figure describes the key to the lowest valley in the various combinations of materials.

2. HETEROJUNCTIONS AND DISCONTINUITIES

2.1 Unstrained Systems

Figure 6 summarizes very approximately the conduction- and valence-band discontinuity of compound semiconductors as a function of the lattice constant, assuming that a combination can be grown in good crystalline quality and that the members are lattice matched, or slightly mismatched to allow pseudomorphic* growth and, hence, allow the fabrication of structures where band-gap changes take place. Figure 6 is an approximate description, based on observed band-gap discontinuities, of the expected conduction- and valence-band discontinuities if one were to make an ideal interface between two semiconductors shown in the figure. Tables 2 and 3 show the discontinuity reference values and bowing and other parameters relevant to the figures of band-gap variation and band-discontinuity variation. The connecting lines in Fig. 6 have been drawn to follow the behavior for the conduction-band edge. The figure is approximate and should be used with caution.

In heterojunctions, the semiconductor band structure may be considered undisturbed right up to the interface. At the interface, in the abrupt heterojunctions, discontinuities occur in the conduction band (ΔE_c) and valence band (ΔE_v), whose magnitude is experimentally determined. Many *ab initio* models and calculations of band discontinuities have been presented, some of which are in fair agreement with experiments. Histori-

*A is a *pseudomorph* of B if it possesses the external form characteristic of B. When a semiconductor is grown on another semiconductor, the grown semiconductor, if it is crystalline, takes the in-plane lattice periodicity of the basis semiconductor. It is said to be pseudomorphic.

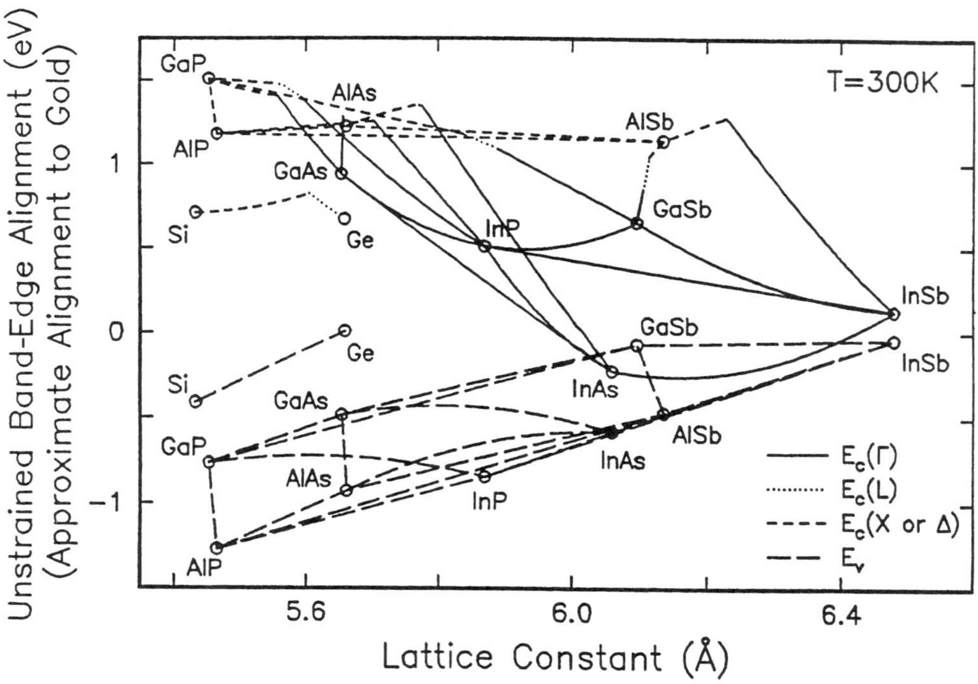

FIG. 6. Discontinuities at heterostructure interfaces for various compound semiconductor systems plotted as a function of lattice constant. This figure is approximate and should be used with great caution. It has been assembled using available and reliable discontinuity data, most of which is for unstrained interfaces. The strained data have also been incorporated approximately, and hence the behavior described by this figure should be considered an approximate description of what one might expect if one were to make an ideal interface of two sets of materials.

Table 2. Discontinuity references at 300 K.

Lattice constant of group (Å)	Material system	ΔE_g (eV)	ΔE_c (eV)	ΔE_v (eV)
5.44	GaP/Si	1.148	0.798	0.35
5.66	GaAs/Ge	0.758	0.268	0.49
	AlAs/Ge	1.48	0.55	0.93
	$Ga_{0.7}Al_{0.3}As/GaAs$	0.395	0.263	0.132
	AlAs/GaAs	0.73	0.29	0.44
	$Ga_{0.51}In_{0.49}P/GaAs$	0.463	0.223	0.24
5.87	$Al_{0.48}In_{0.52}As/InP$	0.106	0.30	−0.194
	$InP/Ga_{0.47}In_{0.53}As$	0.600	0.20	0.40
	$Al_{0.48}In_{0.52}As/Ga_{0.47}In_{0.53}As$	0.706	0.50	0.206
6.10	AlSb/GaSb	0.887	0.487	0.40
	GaSb/InAs	0.370	0.88	−0.51
	AlSb/InAs	1.257	1.367	−0.11

cally, the first of the models was Anderson's electron affinity rule, which assumed continuity of the vacuum level. This predicts $\Delta E_c = \chi_2 - \chi_1$, where χ's are the electron affinities in the two semiconductors. This is in error for most compound semiconductors. For the SiO_2/Si interface, it is relatively close, but does not predict the observed orientation dependence. Atomic-orbital–based calculations predict that when the anion species are common at heterojunctions, such as at $Ga_{1-x}Al_xAs/GaAs$, most of the discontinuity should occur in the conduction band. The argument is that valence-band states are *p*-orbital like and come from the anion that is common. In practice conduction-band discontinuity does not always dominate, showing that significant mixing takes place. For $Ga_{1-x}Al_xAs/GaAs$, the conduction-band discontinuity is $\approx 0.65\Delta E_g$ for aluminum mole fractions of less than 0.4. Semiempirical techniques are therefore usually employed.

Table 3. Bowing parameters for band edges at 300 K.

Material system	c_{Γ_v} (eV)	$c_{\Gamma_c-\Gamma_v}$ (eV)	$c_{L-\Gamma_v}$ (eV)	$c_{X-\Gamma_v}$ (eV)
$Ga_{1-x}Al_xP$	—	—	—	—
$In_{1-x}Al_xP$	—	—	—	—
$In_{1-x}Ga_xP$	0.098	0.77	1.15	0.207
$Ga_{1-x}Al_xAs$	0	0.37	0.055	0.245
$In_{1-x}Al_xAs$	0.479	0.675	—	—
$In_{1-x}Ga_xAs$	0.460	0.431	0.5	0.08
$Ga_{1-x}Al_xSb$	—	0.47	0.7	—
$In_{1-x}Al_xSb$	—	0.43	—	—
$In_{1-x}Ga_xSb$	—	0.413	0.33	0.13
$AlAs_{1-x}P_x$	—	—	—	—
$AlSb_{1-x}As_x$	—	—	—	—
$AlSb_{1-x}P_x$	—	—	—	—
$GaAs_{1-x}P_x$	—	0.19	0.16	0.21
$GaSb_{1-x}As_x$	—	1.15	—	—
$GaSb_{1-x}P_x$	—	—	—	—
$InAs_{1-x}P_x$	—	0.32	—	—
$InSb_{1-x}As_x$	—	0.53	—	—
$InSb_{1-x}P_x$	—	—	—	—
$Ge_{1-x}Si_x$	—	—	0.33	0.17

The discontinuity is, of course, with respect to a specific band minimum. Different minima have different discontinuities that follow different trends, represented, e.g., in the bowing parameters of the alloy. At different specific compositions, different minima may be the lowest. At a heterojunction, which minimum should be considered to be important, and hence which discontinuity should be used, depends on the dominant physical process. All the discontinuities Γ-Γ, *L*-*L*, *X*-*X*, etc., exist at the heterojunction interface. As an example, consider tunneling through a barrier of large–band-gap material such as $Ga_{1-x}Al_xAs$ from GaAs. Electrons in GaAs are largely in the Γ valley; if the $Ga_{1-x}Al_xAs$ barrier is relatively thin, these Γ electrons are most likely to tunnel via Γ-like states in the barrier. However, if the barrier is thick, they may do so through *X*-like states, a process that will require phonon interaction in order to allow for change in momentum. Thus, for thick barriers, the difference in energy between the Γ valley of GaAs and the *X* valley of $Ga_{1-x}Al_xAs$ may be more pertinent. Consider the same for transfer of carriers from donors in $Ga_{1-x}Al_xAs$ to GaAs. The transfer of carriers takes place from whichever is the lowest valley in $Ga_{1-x}Al_xAs$, irrespective of the width of the barrier. This interesting behavior of the discontinuity (Fig. 7) has implications for the charge-transfer process and for perpendicular current flow.

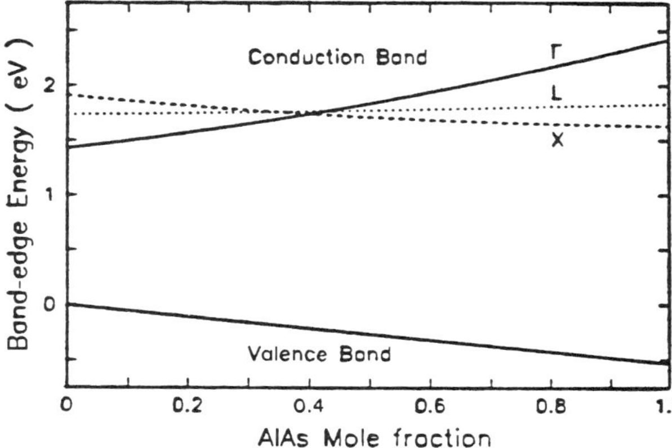

FIG. 7. The conduction- and valence-band discontinuities in the (Ga,Al)As/GaAs system as functions of aluminum mole fraction for the Γ and X valleys.

At aluminum mole fractions of greater than 0.45, the X band is the minimum and forms the lower barrier. So, for $x > 0.45$, the barrier between the X valley in $Ga_{1-x}Al_xAs$ and the Γ valley in GaAs is lower, and in fact for charge transfer, and for tunneling in larger barrier widths, the relevant maximum discontinuity occurs at $x \approx 0.45$. For field-effect transistors based on heterostructures, this results in a larger leakage current between the gate and the channel as mole fractions go toward the AlAs end of the semiconductor composition, counterintuitive to the simplistically expected trend. Suffice it to say, the barrier thickness and barrier height for each of the valleys need to be considered together with elastic and nonelastic tunneling processes to determine the dominant barrier in the current-transport mechanism; for charge-transfer calculations, these are independent of the geometric characteristics of the barrier.

$Ga_{1-x}Al_xAs$ remains closely lattice matched at all mole fractions of aluminum and gallium arsenide (lattice mismatch remains less than 0.1% to both GaAs and AlAs). $Ga_{1-x}In_xP$, GaP_xSb_{1-x}, etc., are lattice matched only at specific mole fractions. For limited thickness of the mismatched layers, however, the resulting strain can be accommodated. Thus, materials with close match only at specific mole fractions can still be grown with a fair degree of reproducibility. Examples of these, of interest, are $Ga_{0.47}In_{0.53}As$ and $Al_{0.48}In_{0.52}As$ to InP, and $Ga_{0.51}In_{0.49}P$ to GaAs. A consequence of this is a tunability of characteristics of the material to achieve desired objectives. A differing barrier height than that of metal-semiconductor junction can be achieved—small heights for microwave detection and large heights for suppressing gate-leakage current in field-effect transistors. The changes in band gap can be used in suppressing injection characteristics of one particular carrier through a suitable use of grading and doping as in heterostructure bipolar transistors. A major usage of the changes in band gap at a spatially abrupt interface has been in obtaining two-dimensional channels of carriers for operation in field-effect transistors.

Practical limitations quite often arise from our ability to grow lattice-mismatched semiconductors. Lattice strain prevents growth of semiconductors with large mismatch because of energy considerations. Thermal energy provides the requisite barrier energy for this relaxation process, and the lattice regains its unstrained lattice constant. A lattice with high strain relaxes by generating defects such as dislocations, etc. A low temperature of growth allows one to make structures that are metastable and have a significant strain due to large lattice mismatch, but they may relax by radiative or nonradiative means during operation of a device. In any case, there is a critical thickness beyond which, for given growth conditions, one cannot grow a suitably high-quality semiconductor. The thicknesses required for heterostructure field-effect transistors are usually sufficiently small that quite significant mismatch can be accommodated. An example of this is the considerable work in the $Ga_{1-x}Al_xAs/Ga_{1-x}In_xAs$ system where low mole fractions of indium arsenide have been used to obtain improvement in amount of charge and transport. The effect of strain in such structures is both in the band structure itself and in scattering due to changes in band structure.

2.2 Strained Systems

During the growth of a structure, the grown epitaxial layer assumes the lattice constant of the crystal and is under strain of $\epsilon = \Delta a/a$, where a is the lattice constant of the unstrained crystal. This strain has an effect on the band structure because it introduces a spatially periodic perturbation differing from that of the unstrained crystal. The strain introduced due to the growth, caused by lattice in the plane perpendicular to the growth orientation, is a biaxial strain.

The symmetry of the band structure has an important relationship with the polarity of the effect on band gap. For direct materials, such as $Ga_{1-x}In_xAs$ grown on GaAs, this biaxially compressive strain leads to a decrease in the band gap. The effect of compressive strain in an indirect-gap crystal—e.g., $Al_xIn_{1-x}P/AlP$—is to increase the band gap from the unstrained situation. So, the polarity of this is dependent on the symmetry of the band-gap minima.

These changes in band gap occur together with changes in the band curvature at the band minima of the conduction band and band maxima of the valence bands, with the changes in the latter being particularly significant. The constant-energy surface for the valence bands is a warped surface. It exhibits differing curvature in different directions. Also, at these energies, we need to consider both the light-hole and the heavy-hole bands. The primary effect of the strain in these is to lift the degeneracy. For the example of $Ga_{1-x}In_xAs$ grown on GaAs, it causes, at the zone center, the heavy-hole band to become the lower-hole energy band, and the light-hole band is pushed significantly farther away. The heavy-hole band also changes considerably in curvature at the zone center, resulting in a lower effective mass for holes at the zone center. This effective-mass change is orientation dependent because the constant-energy surface is warped. Significant complications have now been introduced in the description of the electronic structure by introduction of this strain. Since the masses are different in different orientations and degeneracy has been lifted at the zone center, we would expect interesting consequences for hole transport due to the effect on mass and scattering and for electron–hole interactions, such as in optical processes.

For transport, the primary effect is the lifting of the light-hole and heavy-hole degeneracy, which reduces scattering and hence improves the hole-transport properties. The effect on distortion of the conduction band is smaller than that for hole bands in these examples. The in-plane distortion of the crystal due to the biaxial strain is fourfold symmetric. Band-edge symmetries can therefore be destroyed by this strain. For the examples considered, for electrons, the out-of-plane transport of electrons improves, but the in-plane transport deteriorates.

3. SCATTERING AND TRANSPORT

3.1 Phonon Dispersion and Scattering

At finite temperatures, atoms vibrate around their mean position. The periodic vibration leads to perturbation of the stationary periodic potential and causes changes in the electron wave function in time. The lattice vibrations can be described in a framework that is analogous to the description of electrons. Phonons that represent these vibrations occurring at a frequency ω_q have an energy that is $\hbar\omega_q$, where $\hbar$ is the reduced Planck's constant, and have a momentum that is $\hbar\mathbf{q}$. Phonons describe vibrations just as photons describe electromagnetic waves.

In a three-dimensional crystal, movement of atoms about their mean position may be described by three independent modes of oscillations: one unique longitudinal mode along the wave vector of the oscillation and any two orthogonal directions for transverse oscillations. In the primitive cell, if there are N different types of atoms either of differing mass or ordering in space, $3N$ vibration modes will result. In general, three of these branches, the acoustic branches, will disappear at zone center. The remaining $3N - 3$ branches will be optical branches. For cubic crystals—GaAs, e.g.—$N = 2$, and hence there are three acoustic (low-energy) branches and three optical (high-energy) branches, each

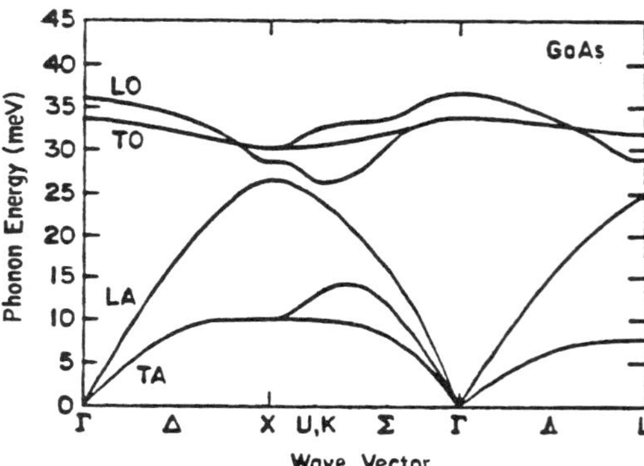

FIG. 8. Phonon dispersion curves for GaAs along major directions in the Brillouin zone.

three comprising one longitudinal and two transverse modes. Scattering resulting from optical and acoustic phonons is commonly referred to as optical-phonon scattering and acoustic-phonon scattering, respectively. Since the energy of the acoustic-phonon mode is smaller, it produces less perturbation than the optical phonons. However, for ⟨100⟩, ⟨110⟩, or ⟨111⟩ directions, the two transverse directions are indistinguishable. So, in the cubic crystal, the two transverse oscillations are degenerate.

Examples of phonon dispersion curves, which describe the energy or frequency dependence as a function of momentum, are shown in Figs. 8 and 9 for GaAs and InAs. The order of magnitude of the optical-phonon energy is between 25 and 50 meV for most semiconductors of interest, on the same order of magnitude as the thermal energy. Heavier ions should be more sluggish; the acoustic branch then should have a lower peak energy. This is reflected in the ascending ordering of peak acoustic energies of InAs, Ge, GaAs, and Si in inverse proportion to the weight of the heaviest element. A similar trend holds for the lowest optical-branch energy for these materials. An additional interesting feature of these figures is that, for compound semiconductors, the longitudinal optical phonon is slightly more energetic than the transverse optical phonon at low momenta because of the presence of ionic charge. At high momenta, in certain directions, they may even cross over because the higher momenta results in a larger reduction in energy of the longitudinal mode due to the effect of the dipole. These figures also show that at specific large momenta, in directions differing from the major crystal axis directions ⟨100⟩, ⟨110⟩, or ⟨111⟩, the degenerate transverse modes split because of the breakdown of symmetry. This occurs for both optical and acoustic modes.

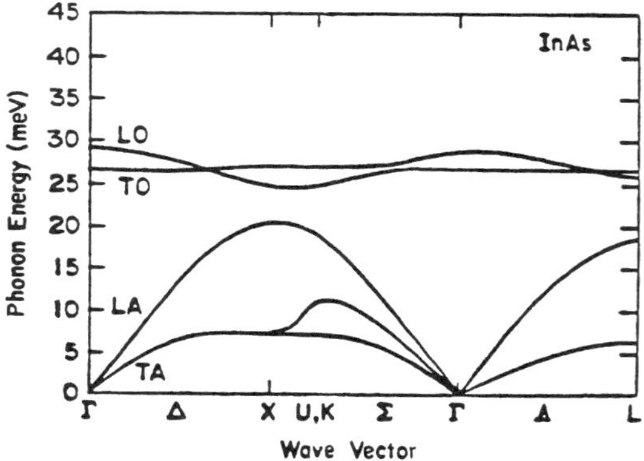

FIG. 9. Phonon dispersion curves for InAs along major directions in the Brillouin zone.

The carrier transport in a medium occurs under the influence of applied forces, which accelerate the carriers in a preferred direction, and carrier interactions with their environment, whose general effect is to render the motion random. Scattering is a general term for the interaction of the carrier with the surroundings.

An electron in the crystal under the influence of an external force, such as due to the electric field, picks up its energy during its acceleration, but it changes energy and/or momentum by various scattering mechanisms that come about because of lattice vibrations, other carriers in its surroundings, and defects of the lattice. The various scattering mechanisms are events of interaction between the carrier and its surroundings that cause it to change its momentum and/or energy. It therefore represents the individual and aggregate effect of the carrier's surroundings that influences the carrier's individual and aggregate transport behavior. If the scattering processes were not to occur, the application of an electric field would increase the velocity linearly with time, an occurrence generally described as ballistic motion. In reality, for long sample lengths, it reaches a limiting value. This limiting value is proportional to the field at low fields (v =

$\mu\mathscr{E}$ where v is the velocity, μ the mobility, and $\mathscr{E}$ the electric field, a constant-mobility region) and saturates to the value v_s at high fields (the saturation-velocity limit). The division of these scattering mechanisms is in three general classes:

1. scatterings that come about from imperfections in the periodicity of the lattice, which we call defect scattering;
2. carrier-related scattering due to interactions between the carriers, a process that becomes important at high carrier concentrations and that can couple with other modes of scattering; and
3. scattering due to interaction of carriers with the perturbing potential produced by lattice vibrations.

Imperfections in the periodic potential of a crystal come about as a result of crystal defects such as vacancies, dislocations, interstitials, etc. These are quite uncommon and the density of them is less than that of unintentional substitutional impurities. Defect scattering also arises from impurity atoms themselves that are used to dope the crystal; this is generally referred to as impurity scattering and originates in the deviation of the local potential around an impurity from the local and periodic potential of the atoms of the host crystal. If the impurity atom is ionized (as is almost always the case), it is called ionized-impurity scattering, and this particular scattering is strong, as shown in Fig. 10 for GaAs. If the impurity atom is not ionized, i.e., it is neutral, it is referred to as neutral-impurity scattering. Neutral-impurity scattering is quite weak because of the small perturbation in potential from a neutral atom on a lattice site. Consider the scattering due to impurity atoms in a crystal as a function of temperature. We assume that impurity density is not so large so as to cause formation of an impurity band and its coalescence with the conduction band. Such a material, therefore, shows freeze-out of carriers; i.e., at low temperatures, the carriers do not have sufficient thermal energy to make the transition to the conduction or valence bands. At temperatures below the freeze-out temperature, the impurities are largely neutral, and hence there is a weak impurity scattering. As the temperature is raised, impurities ionize, and impurity scattering becomes stronger, and hence the carrier mobility decreases. As the temperature is raised further, the thermal energy of the carriers continues to increase, their velocities are larger, and the impurity scattering process becomes weak again. An analogy may be drawn between this and the Rutherford scattering used to analyze the scattering of light atoms by heavier nuclei. A high velocity and energy of the particle causes both a smaller time of interaction and a smaller effect on the velocity of the carrier. Because of the large disparity in masses between the atoms and the moving carriers, this scattering conserves energy and is an example of elastic scattering. At high temperatures, other scattering processes (see Fig. 10) become important, and impurity scattering plays an increasingly smaller role. It is generally important in the 77 K (liquid nitrogen temperature) to 300 K (room temperature) range. The third defect scattering that is im-

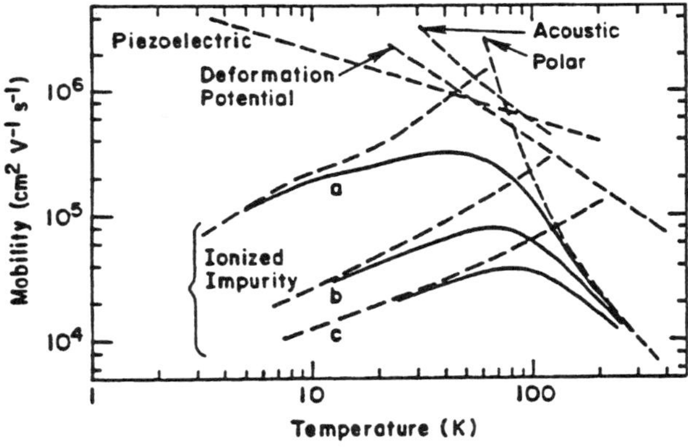

FIG. 10. Electron mobility in *n*-type lightly doped GaAs as a function of temperature together with contributions from the various sources of scattering. After Wolfe *et al.* (1970). The polar scattering referred to in this figure is the component of scattering arising from the perturbation arising from the polar nature of GaAs.

portant occurs in mixed crystals such as $Ga_{1-x}In_xAs$ and $Ga_{1-x}Al_xAs$: alloy scattering. Because of the random variation of the atoms on either the cation or the anion sites of the lattice, a nonperiodic perturbation occurs naturally in such material even when no defect has been added. The perturbations are sufficiently small that alloy scattering generally becomes important only below 300 K.

The last of the defect scattering processes important to devices is related to interactions at the surfaces and interfaces of the lattice (interface scattering). This is a crystal-defect scattering process and is certainly very important at insulator–semiconductor interfaces. It can be important if interfaces are rough and have steps such as in misoriented compound semiconductor growth. Since carriers in two-dimensional systems are confined very close to the interface, any surface-potential perturbation such as through surface atomic steps leads to a decrease in the mobility. The silicon surface mobility is almost half as much as the corresponding bulk mobility.

Steps are not the only way this scattering can cause a net energy or momentum change of the carriers. If there are a number of surface states (surface states result from a number of causes, principal among which are surface reconstruction and nonterminated bonds at the surface), they may cause perturbation effects of their own even for a perfectly smooth surface. For an ideal interface, an incident carrier goes through a reflection without losing its energy. Its momentum parallel to the interface is maintained while its momentum perpendicular to the interface is reversed. An example of this is the $Ga_{1-x}Al_xAs$/GaAs interface, which can be made both atomically smooth as well as with negligible interface state density. A low-energy electron that cannot penetrate the $Ga_{1-x}Al_xAs$/GaAs barrier loses no energy and maintains the momentum parallel to the interface. From the scattering perspective, a nonideal interface is the SiO_2/Si interface. Quite a large fraction of incident carriers lose energy and momentum at the interface because of interface states and atomic roughness. The result of such scattering, called diffuse scattering, is a thermalization of carrier energy distribution and decrease of mobility.

Carrier–carrier scattering conserves the total energy and momentum of the interacting particles. This scattering, also sometimes referred to as plasma scattering, does cause redistribution of the energy and momentum among the interacting particles, and hence its primary effect is in influencing the carrier distribution as a function of energy and momentum. Thus, the scattering influences those parameters that depend on the shape of the distribution of carriers in energy and momentum. The relaxation rate of energy and transfers to other valleys or bands that depend strongly on energy tails do therefore get influenced directly by the carrier–carrier interactions. An additional important way that carrier–carrier scattering influences transport is through its coupling with other scattering mechanisms—the coupling of plasma scattering together with phonons, referred to as plasmon scattering, is an important scattering mechanism in hot-electron devices. In inversion or accumulation layers at high charge densities, this can be quite significant.

Lattice scattering is a broad term used for those scattering mechanisms that arise from the existence of the lattice itself such as due to optical and acoustic phonons. An electron that scatters via optical-phonon scattering can have a large change in momentum since optical phonons can have large momentum. Large momentum changes, most often, occur with changes in the valleys or bands that the carriers occupy. Holes, with their degenerate bands, can easily change bands—e.g., between light-hole and heavy-hole; thus corresponding scattering processes, termed interband scattering processes, can occur even at low energies. These interband scattering processes are unlikely for electrons at low energies. High-energy electrons, such as during an avalanche process, may change bands. All such processes involving changes in bands are called interband scattering processes, while if a process occurs within a band, it is called an intraband scattering process.

Electrons need ≈ 36 meV in GaAs before they can emit an optical phonon, i.e., lose energy in an optical scattering process, since the optical-phonon energy in GaAs is ≈ 36 meV. With lowering of temperature, the consequence of this high energy of optical phonons is a decrease in their numbers, referred to as freeze-out of optical phonons. Corresponding to this smaller likelihood of occurrence of optical phonons, there is a smaller likelihood of absorption of an optical phonon. Hence, at low temperatures there is a reduction in optical-phonon scattering. The emission process can still occur and continues to be the leading cause of loss of energy at high electric fields. Both of these factors are important to the observed velocity–field behavior and operation of high-mobility and hot-electron structures as a function of temperature and bias.

Now consider the nature of the perturbing potential in acoustic mode. A change in atomic spacing, i.e., deformation of the lattice, leads to a disturbance in the potential. The magnitude of the deformation potential is proportional to the strain induced by vibrations, and the corresponding scattering is called deformation-potential scattering. In addition to this spacing effect, if the atoms are slightly ionic as in all of the compound semiconductors, then the displacement of the atoms leads to a potential perturbation due to the charge on the ions. This is called the piezoelectric potential, and the resulting scattering process is called piezoelectric scattering. It is important in compound semiconductors at low temperatures in relatively pure materials (see Fig. 10). Recall that optical scattering is weak at low temperatures and low fields on account of optical-phonon freeze-out, etc., and impurity scattering is weak because of purity and carrier freeze-out, leaving acoustic-phonon scattering as the dominant process. The deformation-potential scattering is weaker than the piezoelectric scattering, and hence piezoelectric scattering dominates.

Optical vibrations also lead to potential perturbation in two ways similar to the case of acoustic vibrations. For one, the perturbation effect of the strain, which is nonpolar in nature, leads to a scattering that is sometimes referred to as nonpolar optical scattering. The second, due to the polarization from the ionic charge, is referred to as polar optical scattering. In GaAs, AlAs, etc., both these scattering processes are important. In general, for compound semiconductors, the polar optical-phonon scattering is the dominant optical scattering mechanism.

3.2 Majority-Carrier Transport

We will now discuss the relationship of these various scattering mechanisms with the steady-state transport behavior of carriers. The operation of a device is interlinked with the transport behavior of carriers. The velocity–field curves for carriers in a semiconductor (Fig. 11) describe the steady-state behavior of the carriers for both low fields, where perhaps an equally useful parameter is the mobility, and high fields. The transport of the carrier is related to the scattering of the carriers. Minority-carrier transport is important in the base of a bipolar transistor. It can be significantly different from a majority-carrier device at comparable background doping and identical temperatures. This comes about because of changes in scattering behavior that result from changes in screening effects of the perturbations. Scattering can also change when the band occupation behavior is changed such as through

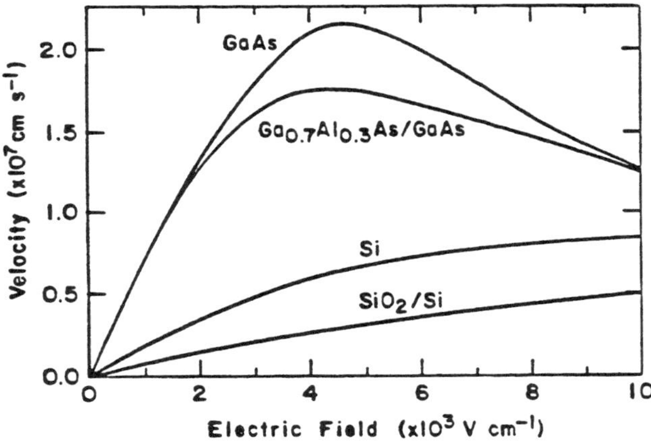

FIG. 11. Steady-state velocity–field characteristics of electrons, in lightly doped bulk Si and at the SiO_2/Si interface, and in lightly doped GaAs and at the $Ga_{0.7}Al_{0.3}As$/GaAs interface at 300 K.

the quantum-mechanical effects of two-dimensional carrier gas systems.

We now consider scattering rates of some common compound semiconductors in order to relate them to the transport characteristics of interest in the operation of devices. Figure 12 shows some of the scattering rates due to phonon processes, at 300 K and low energies, for GaAs, InP, $Ga_{0.47}In_{0.53}As$, and InAs. The minimum in the scattering rates, occurring at very low energies, is due to acoustic scattering. At the low-energy threshold for the optical-phonon modes, scattering rises rapidly because of the large rate of optical-phonon scattering at 300 K. At 77 K and lower temperatures, this rate would be lower. At low energies the scatterings are intravalley events. At higher energies, comparable with or higher than the secondary valley separation, a large density of states is available for intervalley scattering. So, these scattering rates show another threshold where the secondary valley occurs in GaAs, InP, and $Ga_{0.47}In_{0.53}As$. InAs has a secondary valley at higher than 1 eV and does not, therefore, show such a threshold. Note that InP has the highest low-energy scattering; it has both a higher effective mass that leads to higher density of states to scatter into and a stronger coupling coefficient for scattering. Both $Ga_{0.47}In_{0.53}As$ and InAs show favorable low scattering-rate characteristics over the energy range, a behavior partly due to the larger secondary valley separation, which inhibits intervalley scattering over a significant energy range.

The two divisions of lattice scattering that we have considered are intervalley and intravalley scattering. In intravalley scattering, low-momentum phonons are involved. At low fields, this intravalley scattering is important in materials where the conduction-band minimum occurs at the Γ point, e.g., GaAs, InP, and most other compound semiconductors. Intervalley scattering involves large-momentum phonons and is important at low fields in materials with conduction-band minima at nonzero crystal momentum, i.e., non-Γ-minimum semiconductors.

The phonons that take part in intervalley scattering are sometimes referred to as intervalley phonons even though they may be either of the acoustic or of the optical type. When fields are high, electrons pick up enough energy between scattering events that even where the central valley is the minimum energy valley, intervalley scattering becomes important. The negative differential mobility of GaAs, e.g., arises from intervalley scattering from Γ to L and X valleys. Figure 13 shows the velocity–field characteristics for GaAs ($\approx$mid-10^{16} cm^{-3} doping) as a function of temperature. GaAs, InP, etc., actually show a significant region of negative differential mobility arising from the transfer of carriers to secondary valleys at the higher energies. Transfer to the secondary valleys occurs into a high density of states and therefore occurs with a large scattering rate. The resulting pronounced negative differential characteristic has been utilized in achieving negative differential resistance for gener-

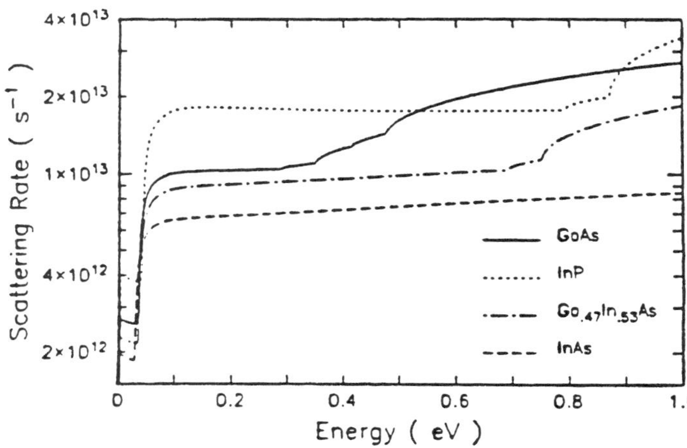

FIG. 12. Steady-state phonon scattering rates as functions of energy, at low energies, in GaAs, InP, $Ga_{1-x}In_xAs$, and InAs at 300 K.

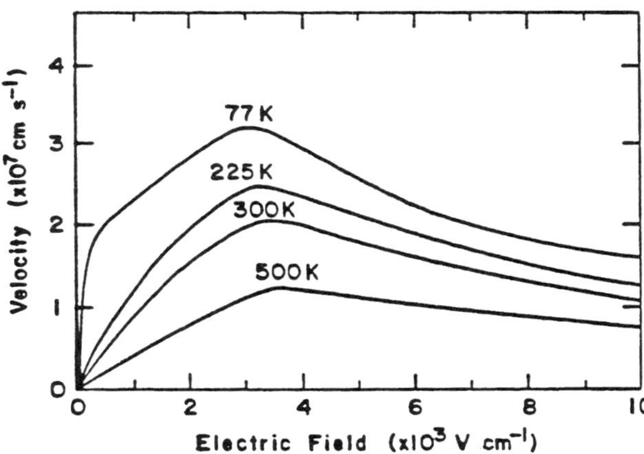

FIG. 13. Velocity–field characteristics for electrons in lightly doped GaAs as a function temperature.

ation of microwave power. The decrease in mobility at low fields is due to intravalley scattering, the decrease in velocity at higher fields is due to intervalley scattering, and the saturation in velocity at very high fields is due to a balance between the energy gain between scattering events and the energy loss due to high intervalley scattering.

In thermal equilibrium, the energy and momentum exchange between electrons and the lattice is balanced out by phonon absorption and emission. With field, the energy and momentum is lost at a higher rate to the lattice, and at moderate fields, the intervalley optical-phonon scattering processes dominate. The momentum loss causes randomization of the electron velocity distribution in momentum space and consequently results in a finite mobility. Note that the momentum-loss rate is higher than the energy-loss rate (also recall that optical phonons involve a loss of ≈36 meV of energy per emission event). These losses are fitted phenomenologically by time constants called relaxation times (τ_p for momentum and τ_w for energy).

We may now describe some general features of the relative importance of different scattering mechanisms in the transport of carriers. Since ionized-impurity scattering is inefficient at higher carrier energies, it becomes important in the 77- to 300-K temperature range. Alloy scattering is important for transport in mixed crystals, in the temperature range below 300 K. Carrier–carrier scattering preserves both the energy and the momentum if it is between carriers of the same type. It therefore causes a substantial difference between the transport of a majority carrier and a minority carrier. Lattice scattering becomes dominant at high temperatures and high carrier energies. Optical phonons, with their large momentum and energy compared to the acoustic phonons, are dominant when carriers have large energies and are dominant in intervalley transfers. Acoustic phonons establish a lower limit in scattering since they can exchange very low energies also. At very low temperatures, therefore, while optical-phonon absorption can be frozen out because of their reduced density, acoustic phonons continue to be important. At low temperatures and low fields, both the acoustic-phonon and the impurity scattering are important. At high fields at very low temperatures, impurity scattering becomes weak because of the higher carrier energy, but optical-phonon emission becomes important. Acoustic-phonon scattering should also be considered. At high temperatures phonon scattering prevails, and both absorption and emission processes become dominant.

Now, let us consider the temperature dependence of the low-field and high-field behavior in the velocity–field curves of Fig. 13. At high fields, as we have already discussed, this velocity is dominated by intervalley scattering processes occurring largely through optical phonons, which are polar in compound semiconductors and nonpolar in silicon and germanium. At low fields, other mostly intervalley scattering processes dominate. The low-field behavior is equally well characterized by the mobility, the drift mobility, which relates the velocity with the low electric field. While drift mobility is the real mobility of interest in most electronic devices, one quite often also uses Hall mobility, which is a far easier and accurate transport parameter to measure.

Drift mobility, which we have discussed thus far, depends on the carrier scattering embodied in the phenomenological carrier relaxation time. To calculate the mobility, we need to determine this relaxation time, including its functional dependence on energy. The average value of this relaxation time then is obtained by averaging over the energy states that the carrier occupies. For electrons, using a parabolic band approximation, we obtain

$$\tau_\mu = \bar{\tau}. \tag{1}$$

The averaging is over energy, which is related to the momentum in the parabolic approximation through a square law. The relaxation time, therefore, is also often written as

$$\tau_\mu = \frac{\langle (E - E_c)\tau \rangle}{\langle E - E_c \rangle} = \frac{\langle v^2 \tau \rangle}{\langle v^2 \rangle}. \tag{2}$$

The time constant for mobility can be calculated given knowledge of what scattering process is dominant. Alternately, if more than one is important, they all have to be evaluated and the resultant determined as a geometric mean. The geometric mean comes about because scattering rates add, and the time constant is inversely proportional to this rate. Thus, the time constant τ_μ is related to the time relating the inverse of scattering rates $\tau_1, \tau_2, \ldots$

$$\frac{1}{\tau_\mu} = \frac{1}{\tau_1} + \frac{1}{\tau_2} + \cdots, \tag{3}$$

and the drift mobility is related to this time constant through*

$$\mu_{\text{drift}} = \frac{q\tau_\mu}{m^*} = \frac{q}{m^*} \frac{\langle v^2 \tau \rangle}{\langle v^2 \rangle}. \tag{4}$$

If one associates $\mu_1, \mu_2, \ldots$ as the mobilities with the various scattering processes corresponding to $\tau_1, \tau_2, \ldots$, then

$$\frac{1}{\mu} = \frac{1}{\mu_1} + \frac{1}{\mu_2} + \cdots. \tag{5}$$

For a given momentum, the velocity of carriers is inversely related to the effective mass. A larger effective mass results in smaller velocity for identical forces on the carrier. Since, in general, the effective mass is a tensor, it has a directional dependence. Let us now consider how the occupation of the bands, the motion of carriers due to nonisotropy of masses, and occupation of more than one degenerate bands are related. In thermal equilibrium, the carriers occupy bands following Pauli's exclusion principle. Constant-energy surfaces are used to show those momentum values that have equal energy and hence an equal likelihood of occupation in thermal equilibrium. The shapes of such surfaces are dependent on the band anisotropy. In cubic crystals, at thermal equilibrium, three types of constant-energy surfaces are usually encountered. They are a spherical surface encountered for Γ valley minima in the conduction band, an ellipsoidal surface encountered for L and X valley minima in the conduction band, and a warped surface encountered for two degenerate valence bands. A parabolic band has an isotropic mass; its constant-energy surface is spherical. L- and X-valley minima, however, have a rotational symmetry around the crystallographic directions and show a differing effective mass in the transverse direction to it than in the longitudinal direction. The constant-energy surface is an ellipsoid. Valence bands are usually doubly degenerate at zone center. For any energy in excess of this zone-center energy, the constant-energy surface is warped.

Thus, even for moderate doping, in cases where the band structure is not a simple parabola, where the band minimum does not occur at the zone center, and where degenerate bands exist, the effective mass for use in conduction calculations has to be derived as a suitable average that reflects the net effect. We first consider the conduction band. For the lowest conduction-band minimum in the Γ band, the bands are quite isotropic, and we may use the usual mass. At higher energy, however, e.g., when a large field is

*This relationship, Mathiessen's rule, has limited, not universal, validity since it is based on the relaxation-time approximation.

applied and intervalley transfer occurs, we may not use this constant isotropic effective mass. Transfer of carriers, in GaAs, occurs first to the L valley and then to the X valley with increasing energy. An L valley has eight equivalent minima in eight different directions. However, since the minimum always occurs at the Brillouin-zone edge, only four equivalent complete valleys need to be considered. For X valleys, the minimum occurs slightly inside the Brillouin zone, and hence all six equivalent minima and six equivalent valleys should be considered. The L and X minima are ellipsoidal surfaces of constant energy; i.e., they can be represented by

$$E - E_c = \frac{\hbar^2}{2m_l^*}(k_l - k_{ol})^2 + \frac{\hbar^2}{2m_t^*}(k_{t1} - k_{t2})^2, \tag{6}$$

where the subscripts l and t denote longitudinal and transverse directions. This is simply a translation of axes from the original Cartesian coordinate in the reciprocal space to a new Cartesian coordinate that is aligned along the longitudinal axis, and because of symmetry, the transverse variations are identical. In terms of the effective mass, if the transformed x axis is now aligned with the longitudinal axis, then the new effective mass tensor is

$$\frac{1}{m^*} = \begin{pmatrix} 1/m_l^* & 0 & 0 \\ 0 & 1/m_t^* & 0 \\ 0 & 0 & 1/m_t^* \end{pmatrix}. \tag{7}$$

If the electric field is applied in the $\langle 100 \rangle$ direction, for X minima, there are four ellipsoid surfaces for which the transverse effective mass is the relevant mass in the determination of the response of the carrier to the field. There are two ellipsoidal surfaces for which the longitudinal mass is relevant. Hence, in the process of approximating into the parabolic band picture, i.e., an expression of $E - E_c = \hbar^2 k^2/2m_c^*$, we must use the mass

$$m_c^* = \tfrac{1}{6}[2m_t^* + 4(m_l^* m_t^*)^{1/2}]. \tag{8}$$

For the L-valley minima with a field in the $\langle 100 \rangle$ direction, all the valleys are equivalent with the effective conduction mass

$$m_c^* = \tfrac{1}{4}(\tfrac{1}{3}m_t^{*2} + \tfrac{2}{3}m_l^* m_t^*)^{1/2}. \tag{9}$$

The valence band is centered in the Brillouin zone, and the light- and heavy-hole bands are degenerate. The equivalent mass can be derived as

$$\frac{1}{m_c^*} = \left(\frac{g_{hh}(E)}{m_{hh}^*} + \frac{g_{lh}(E)}{m_{lh}^*}\right)\frac{1}{g_{hh}(E) + g_{lh}(E)}, \tag{10}$$

where $g_{hh}(E)$ and $g_{lh}(E)$ are the heavy-hole and light-hole densities of states. The ratio of densities of states in the expression above represents the probability of finding the hole in a state with that mass. Since the density of states varies as $\frac{3}{2}$ power of the mass, the conduction mass for holes is

$$\frac{1}{m_c^*} = \frac{m_{hh}^{*1/2} + m_{lh}^{*1/2}}{m_{hh}^{*3/2} + m_{lh}^{*3/2}}. \tag{11}$$

Hall mobility is a mobility parameter that results from Hall-effect measurement. It is a simple and hence very common measurement to gauge the low-field transport characteristic of semiconductors. Our discussion here is to emphasize its relationship with the drift mobility, which is of more immediate interest from a device perspective. The Hall effect is a phenomenon that results from the application of a magnetic field on a sample. Consider an n-type rectangular semiconductor sample with a magnetic field applied perpendicular to the sample, a voltage bias applied to force a current to flow in the plane of the sample, and a voltage being measured perpendicular to this flow of current. A voltage, called the Hall voltage, develops in the perpendicular direction because of the Lorentzian force acting on the charged particle. For the electron, this force is given by

$$\mathbf{F} = -q(\mathbf{v} \times \mathbf{B}). \tag{12}$$

In steady state, carriers accumulate and deplete in the transverse direction causing a transverse electric field, which opposes and cancels the field due to the Lorentzian force, to appear. This allows us to write the electric field in the transverse direction, related to the Hall voltage developed across the transverse direction, in terms of the magnetic field, the current, and a parameter called the Hall coefficient. This field is related as

$$\mathscr{E}_z = R_H J_x B_y, \tag{13}$$

where R_H is the Hall coefficient. The current itself is related to the drift mobility of carriers. The Hall coefficient is related to this drift mobility by

$$R_H = r\left[\frac{n\mu_{dn}^2 - p\mu_{dp}^2}{(n\mu_{dn} + p\mu_{dp})^2}\right] \tag{14}$$

for a sample with both electron and hole conduction. The prefactor r, the Hall factor, usually varies between 1 and 2 for most common scattering processes. This factor is related to the relaxation time for Hall mobility determination, τ_H (analogous to τ_μ, the relaxation time for drift mobility determination), by

$$\tau_H = \langle\tau^2\rangle/\langle\tau\rangle. \tag{15}$$

Note the difference between this and the relaxation time associated with the drift mobility. If only acoustic scattering dominates, and the conduction and valence bands can be assumed to be parabolic, then $r = 3\pi/8$. The motion of carriers in the transverse direction occurs as a result of the applied magnetic field and is determined by the velocity in the longitudinal direction. The proportionality of constant in the longitudinal direction is the drift mobility; however, the proportionality in the transverse direction is a mobility term that depends on the same prefactor as in the Hall factor. Hence, the Hall mobility ratio in samples with unipolar conductivity is

$$\mu_H = r\mu_d. \tag{16}$$

Because the ratio r varies between 1 and 2, the Hall mobility is always larger than the drift mobility. Its significance and ubiquity is largely a result of the ease of its measurement.

The Hall and drift mobility are, of course, related to each other through the characteristics of the dominant scattering processes, i.e., through the ratio r above. Figure 14 shows the Hall mobility as a function of temperature for various dopings of GaAs.

3.3 High-Field Transport of Electrons

The mobility of carriers decreases as a function of particle energy because the scattering rate continues to increase as a function of the energy. Thus, drift and Hall mobilities become negligible at high energies because carriers move short distances between scattering events.

It is the velocity of carriers that is of direct interest to us. At high energies, even though the mobility becomes negligible, the velocity for a high applied field is still high. At low energies, both the velocity and the mobility are useful parameters because they are directly related to each other. However, mobility is used more commonly, both because of the ease of measurement and because of simplifications in many steady-state drift–diffusion calculations in devices.

The saturation of velocity at high fields occurs for the following reason. Carriers drift in the high field between scattering. Any increase in energy resulting from this transport in high field causes an increase in scattering rate also. Increase in electric field,

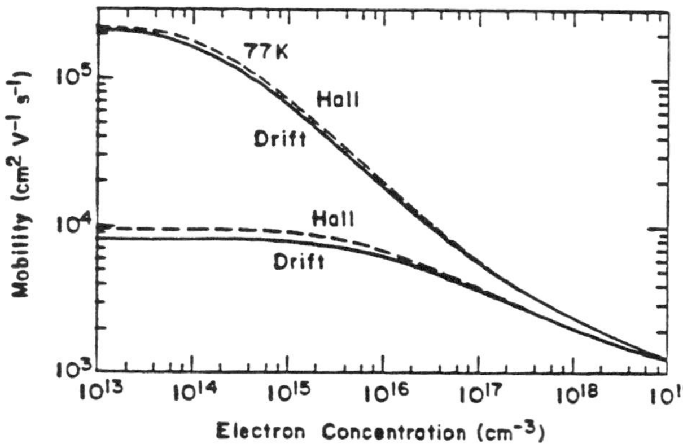

FIG. 14. Drift and Hall mobilities as functions of background electron density at 300 and 77 K for GaAs.

therefore, does not result in significant additional increase in the velocity, and the mobility becomes negligible. An alternative way of looking at this behavior in high fields, where the carrier is in equilibrium with its local conditions, is to note that the mobility is a function of the carrier energy. At low fields, the average carrier energy increases proportionally with the electric field, the scattering rate increases proportionally, and hence the carrier mobility remains constant and independent of the energy and the electric field. At high fields, however, the carrier energy increase causes the mobility to decrease inversely with the electric field and hence velocity saturation.

Optical-phonon scattering is the principal source of scattering in the high-field region. Since the occupation of the optical modes decreases strongly as the temperature decreases, energy relaxation is not as strong, and hence absorption processes are less likely. The drift velocity at high fields also shows a temperature dependence. An example of this is given in Figure 15, where there is nearly a 20% change in the drift velocity at high fields (approximately the saturated velocity) between 300 and 77 K.

4. MINORITY-CARRIER TRANSPORT PROPERTIES

An electron as a majority carrier suffers Coulomb scattering from the donors, other electrons, and any other residual impurities. Usually, the latter two are small; only at high carrier densities, such as at high dopings or in two-dimensional electron gas, do the carrier–carrier interactions also become strong. An electron as a minority carrier suffers Coulomb scattering from the acceptors, the holes, other electrons, and any other residual impurities. The last two, as in the case of electron as a majority carrier, are usually small. Holes are the new twist to the environment of the electron.

The behavior of the electron as a minority carrier depends on the semiconductor and on any residual field in the semiconductor device. It depends on the semiconductor because any interaction between electron and hole—rather the heavy hole since it is far more common—involving exchange of momentum and energy is a function of their relative effective masses. The screening of acceptors by holes (i.e., reduction of Coulomb perturbation by holes in the vicinity of the ionized acceptor) may allow for an increase in the carrier mobility. Holes, if they are considerably heavier than the electron, may cause scattering very similar to that due to ionized acceptors. A slight residual field in the structure causes motion of carriers in opposite directions, which may result in a drag effect on each carrier by the other. The magnitude of the effect on each carrier is a function of their effective mass and the disparity in the effective mass. Scattering is a strong function of temperature; hence, different considerations may apply to different temperatures.

Figure 16 shows a plot, for electrons, of minority- and majority-carrier mobilities in

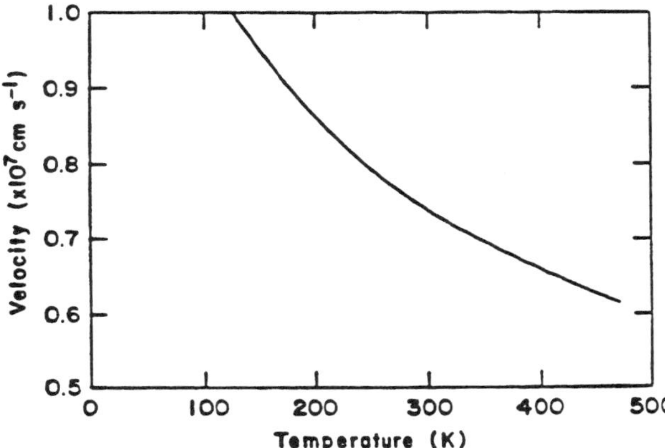

FIG. 15. Velocity of electrons at a field of 50 kV/cm in lightly doped GaAs as a function temperature.

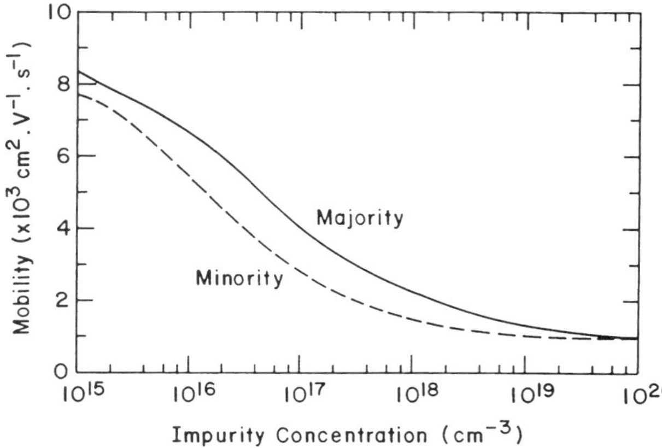

FIG. 16. Majority- and minority-carrier mobilities for electrons for GaAs.

GaAs as functions of background donor and acceptor density. This figure is for mobility at 300 K. GaAs shows a minority-carrier mobility that is nearly half the value of the majority-carrier mobility. Comparatively, silicon shows similar mobility for electrons as a majority and minority carrier. In silicon, the effective masses for electrons and holes are relatively similar. Exchanges involving electrons and holes, and screening of impurities by holes, together contribute to this similarity of mobilities. In GaAs, electrons and holes have widely differing masses. Holes can be considered static for scattering purposes and lead to Coulomb scattering of a similar magnitude as the ionized acceptor. This behavior may change at a different temperature. At 77 K, e.g., for GaAs in the 10^{15}-cm^{-3} doping range, the minority-carrier mobility for electrons is higher than that as a majority carrier.

Minority-carrier devices usually have fields in regions where the electron transits. In the base of the bipolar transistor, e.g., a small field exists even for uniformly doped base structures. This field is the result of the gradient in the hole distribution established to compensate for the electron distribution. It causes a rapid decrease in mobility with the electric field, even though the mobility at zero field is quite comparable to the majority-carrier mobility. The majority-carrier mobility, of course, does not vary substantially. This difference in behavior is usually ascribed to the electron–hole drag effects, which are quite significant for Si because of comparable effective masses. The carrier–carrier scattering conserves momentum. In the absence of electric field, the average drift momentum is zero. In the presence of drift field, electrons and holes move in opposite directions. The interacting electrons and holes that are in the vicinity of each other and are, on an average, moving in opposite directions, exert a Coulomb attractive force on each other. This reduces the effect of the drift field, and externally, appears as the lowering of mobility. The effect is strong and shows up at a low electric field. At higher fields, the drag effect does not remain as important because the Coulomb scattering rate decreases (the electron now has a higher energy).

The consequence of such an effect can be strong in devices such as a bipolar transistor, even if no electric field is built into the structure. In the presence of current, there exists a gradient of electron density in the base. In order to maintain charge neutrality, the hole density also changes as a function of position to compensate for the excess charge due to electrons. The excess hole population is equal to the electron population under quasistatic conditions. The diffusion current due to the gradient in hole density is nearly compensated by the drift current resulting from a small electric field that is established in the base. The magnitude of this electric field can be, approximately, found by setting the hole transport current equal to zero (it is substantially smaller than the electron current for the *n-p-n* bipolar transistor). The current-density equation in the drift–diffusion approximation can be used to model this condition. For an electric field $\mathscr{E}$ and diffusivity $\mathscr{D}_p$ for holes, the absence of hole current implies

$$J_p = q\mu_n\mathscr{E}p - q\mathscr{D}_p\frac{dp}{dz} \approx 0; \tag{17}$$

i.e.,

$$\mathscr{E} \approx \frac{kT}{q}\frac{1}{p}\frac{dp}{dz}. \tag{18}$$

Because the magnitude of dp/dz is negative, the electric field is pointed toward the electron-injecting electrode.

This is a very substantial simplification of a considerably complex many-body problem. It has been argued that under certain conditions, this drag effect should be balanced by the effect of majority-carrier distribution. Hole current being small, the holes are nearly stationary. Thus, they should not transfer much momentum to electrons. Electrons, however, do transfer momentum to holes, resulting in the space charge and electric field to prevent the resulting hole current. This field aids the motion of electrons, just as in the high injection effect, and hence should compensate.

4.1 Diffusion Coefficients

The phenomenon of diffusion is one of the very direct consequences of scattering. Net diffusion takes place from regions of high carrier concentrations to regions of low carrier concentrations. Regions of high concentrations have a higher carrier population, resulting in a higher number of scattering events. Some of these lead to carrier motion toward the low–carrier-concentration region. However, it is not completely compensated for by scattering events that lead to the motion of carriers from the low–concentration-region to the high–concentration-region. The net result is a flow, or diffusion, of carriers determined by the gradient of the carrier concentration and characterized by the diffusion coefficient ($\mathscr{D}_n$ for electrons and $\mathscr{D}_p$ for holes).

Diffusion coefficients are related to mobilities because their underlying basis is in scattering. This relationship, the Einstein relationship, was derived for Brownian motion in the Maxwell–Boltzmann distribution limit of classical gases. In thermal equilibrium (i.e., with constant quasi-Fermi levels E_{Fn} for electrons and E_{Fp} for holes) in an n-type sample,

$$J_n = qn\mu_n\mathscr{E} + q\mathscr{D}_n\frac{dn}{dz} = 0, \tag{19}$$

and its converse. The carrier concentration is given by

$$n = N_c F_{1/2}\left(\frac{E_{Fn} - E_c}{kT}\right), \tag{20}$$

where $F_{1/2}(\ )$ represents the Fermi integral of order $\frac{1}{2}$. This allows us to show that

$$\frac{\mathscr{D}_n}{\mu_n} = \frac{kT}{q}\frac{F_{1/2}((E_{Fn} - E_c)/kT)}{F_{-1/2}((E_{Fn} - E_c)/kT)}, \tag{21}$$

the Einstein relationship for electrons, and following similar arguments,

$$\frac{\mathscr{D}_p}{\mu_p} = \frac{kT}{q}\frac{F_{1/2}((E_v - E_{Fp})/kT)}{F_{-1/2}((E_v - E_{Fp})/kT)}, \tag{22}$$

the Einstein relationship for holes. In the Maxwell–Boltzmann limit, these can be simplified to

$$\frac{\mathscr{D}_n}{\mu_n} = \frac{\mathscr{D}_p}{\mu_p} = \frac{kT}{q}. \tag{23}$$

The diffusion coefficients of electrons in GaAs and other compound semiconductors are larger than in silicon for the same reason as for mobility—largely because of a smaller effective mass. The high diffusion coefficient leads to shorter base transit time in bipolar transistors in GaAs and other compound semiconductors for comparable base widths. It also has implications for noise behavior; a larger diffusion coefficient usually results in smaller magnitude of noise. Since, in general, conduction-band structure can be anisotropic, scattering processes are anisotropic, and since fields exist along specific directions, the diffusion coefficient shows anisotropy. Examples of magnitude of diffusion coefficient with field are given in Fig. 17 for the longitudinal and transverse directions with respect to the field.

The anisotropic nature of polar scattering

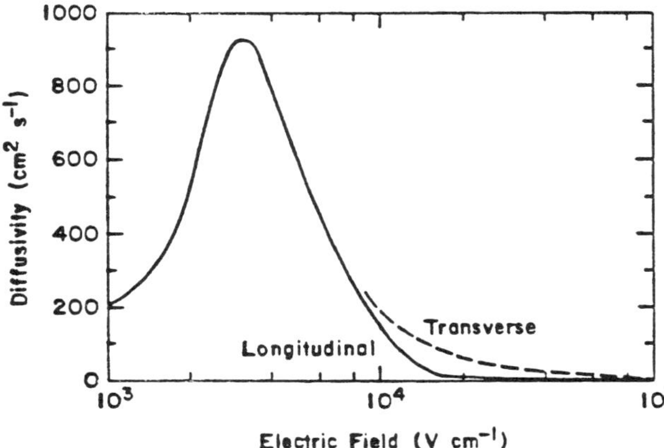

FIG. 17. Diffusion coefficients in the longitudinal and transverse directions for GaAs at 300 K for low doping.

is a particularly strong contributor to this anisotropy. Since polar scattering occurs across the field range so long as it leads to carrier energies above the optical-phonon threshold, the diffusion coefficient also exhibits similar behavior. Anisotropy exists at both low and high fields. At high fields, the *L* and *X* valleys are mostly occupied, the longitudinal effective mass is large, and hence longitudinal diffusion is suppressed. It decreases to very low values compared to the zero-field value, just as in the case of mobility. The pinch-off regions of a field-effect transistor are regions of high fields where the carrier gradient can become large for large–gate-length devices. The low diffusion coefficient reduces the diffusive current in such regions even if carriers move with finite velocity (the saturated velocity) as a result of drift effects.

5. CONFINEMENT EFFECTS

5.1 Transport Consequences of Confinement

In two-dimensional electron gases, scattering due to ionized impurities occurs through a remote process because donors and carriers are separated. The perturbation in potential due to ionized donors (sometimes this is referred to as Coulomb scattering also) is separated from the electron gas, as shown in Fig. 18. Decrease of the ionized-impurity scattering and increase of the carrier screening of the ionized impurities because of large carrier densities in the two-dimensional channel result in a significantly higher mobility. Now, scattering processes such as the piezoelectric scattering become important, at least in part of the temperature range, as shown in Fig. 19. At the lowest temperatures, the residual background doping at the high-mobility interface is still important, although because this is low, the mobility is very high. It is interesting to compare this figure with Fig. 10, which shows the mobility versus temperature for low-doped GaAs. Note that over a fairly broad range of temperature, near and above 77 K, the behavior is very similar. Below that temperature, the improvements due to carrier screening make the mobilities of the two-dimensional electron-gas system significantly larger.

The added complication in the two-di-

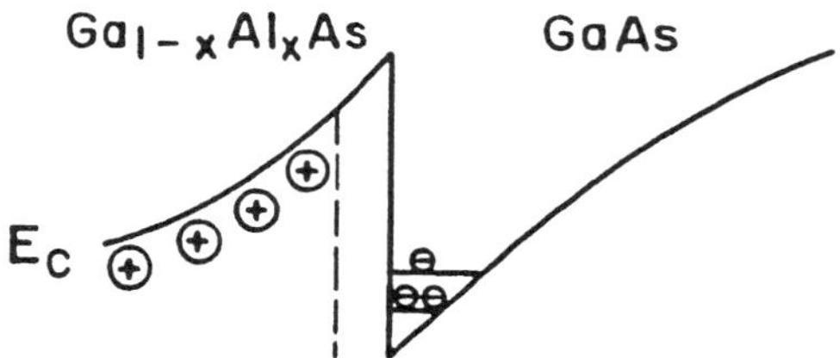

FIG. 18. Separation of two-dimensional electron gas from the ionized donors in a two-dimensional electron gas structure. The larger–band-gap material ($Ga_{1-x}Al_xAs$) from which electron transfer occurs is shown on the left. The right-hand side shows the smaller–band-gap material (GaAs) to which the transfer occurs. An undoped spacer layer, used to decrease remote Coulomb scattering, is also shown in the figure.

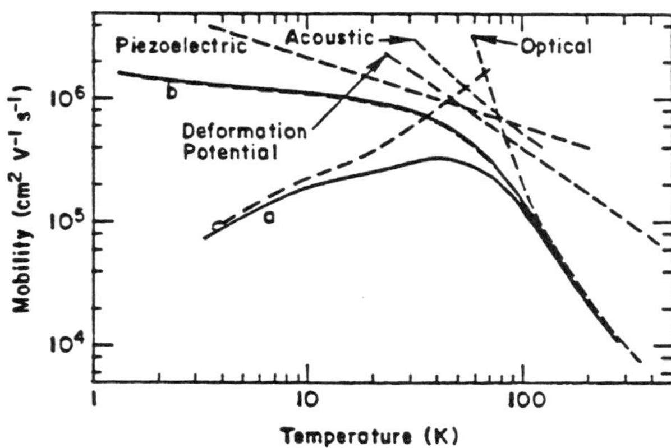

FIG. 19. Theoretical mobility in a two-dimensional electron gas at the $Ga_{0.7}Al_{0.3}As/GaAs$ interface as a function of temperature. The figure also shows components of mobility arising from different scattering mechanisms in bulk pure GaAs.

mensional electron-gas structures is that one should also consider the quantization and its effects on scattering, as well as the effect of carrier–carrier interaction and carrier–dopant interaction. The former leads to a decrease in mobility; the latter tends to screen the perturbation potential associated with the dopant. Thus, there is actually an optimum in carrier concentration in the channel where the two balance and allow the highest mobilities at low temperatures. Note the large increases in low field mobility that can be achieved by a decrease or removal of the ionized-impurity scattering. At 300 K, since ionized-impurity scattering is only one of the two dominant scattering processes, the increase is not very large. At 77 K, mobilities of greater than 70 000 $cm^2\ V^{-1}\ s^{-1}$ are quite common. The behavior of two-dimensional electron gases is quite complicated; it has increased carrier–carrier scattering whose effect, as we have discussed, is to change the distribution function but not to affect the energy and momentum.

The interaction of carrier–carrier scattering processes with other processes are also important in these structures. Coupling between optical phonons and the carrier plasma—the plasmon—becomes stronger as a function of carrier density in GaAs. Significant deviation and large energy effects begin to occur at mid-10^{17} cm^{-3} carrier density. Carrier densities higher than this exist in two-dimensional electron gases in device structures, and therefore mobility is fairly sensitive to carrier scattering. Too low a carrier density leads to reduction of screening of Coulomb scattering from remote and local donors, and too high a density leads to increased carrier–carrier scattering effects.

6. HOT-CARRIER PROPERTIES

There are a number of device-related phenomena that can be intuitively related to the band structures. One example of this is the relationship between scattering rate and density of states. Scattering is proportional to the coupling between the two states connected by a scattering process and the number of states available for scattering into. Thus, at the onset of secondary valley transfer, i.e., when carriers have kinetic energy approximately equal to the intervalley separation, a rapid rise in scattering rate occurs, as seen in Fig. 12 for GaAs, InP, and $Ga_{1-x}In_xAs$. Also note that at low energies, InP has the highest scattering rate. InP has the highest effective mass of the four semiconductors in the figure; it therefore has the highest density of states.

Effective mass, which is inversely proportional to the second derivative of the energy with respect to the wave vector, i.e., related to the band structure, is central to a lot of transport parameters of interest. Mobility is inversely proportional to it; a lower effective mass usually implies faster low-field transport characteristics. It also results in improvement in diffusion coefficient and higher tunneling probability. A smaller effective mass means lower density of states, and hence carriers have a higher velocity for the same energy, and so long as the scattering rate is lower, the nonequilibrium effects are stronger. The highest velocities, the maximum group velocity, that the carriers can attain under the most ideal of circumstances is also limited by the band structure.

Another good example of a strong band-structure relationship exists in the phenomenon of avalanche breakdown due to impact ionization. This is a process in which a carrier picks up enough energy between scattering events to cause an electron to jump from the valence band into the conduction band during a scattering event. The process is both band-structure and scattering-process dependent.

6.1 Avalanche and Other Generation Effects

The breakdown phenomenon of most interest occurs in depletion regions of devices. In bipolar transistors, it is the breakdown in the base–collector junction region. In field-effect transistors, it is the breakdown in the drain-to-substrate and gate-to-drain depletion regions. These occur because there exists a high field, due to the reverse bias at the junction, and this high electric field is capable of accelerating the carrier and imparting to it a large kinetic energy. Indeed, this kinetic energy can become sufficiently large to cause an electron, or even more than one electron under the right conditions, to be removed from the valence band and transferred to the conduction band.

The process involving a single electron–hole pair generation consists of a transfer of energy from the hot carrier to an electron in the valence band, which jumps into the conduction band leaving a hole behind. Also, for small–band-gap semiconductors, such as InAs, even multiple such pairs may be generated.

With complicated band structure, electrons can be from different bands, holes can be from different bands, phonons—particularly optical phonons with their large momentum—may be involved, etc. Indeed, since the band structure is orientation dependent, and impact ionization occurs as a consequence of large energy of the carrier from rapid acceleration in a field, in the event of insufficient randomization, it may even be orientation dependent.

The ionization process is characterized in a semiconductor by the parameter ionization rate. Ionization rate is the relative increase in carrier density per unit length of carrier travel. Thus, the electron ionization rate (α_n) is the rate of increase in number of electrons per unit length of the travel of an electron.

For GaAs, the impact ionization occurs largely from carriers that are not in the lowest bands. Electrons cause ionization from higher bands, and holes generally cause it from the split-off band. At low fields, the ionization rate is also orientation dependent because of lack of randomization. Examples of ionization rates as a function of electric field are given in Figs. 20 and 21.

For most compound semiconductors with greater than 1 eV of band gap, the hole ionization rate is larger than the electron ionization rate. The hole effective mass being larger, for similar velocities in the high-field regions, the hole has significantly higher kinetic energy. At lower fields, electrons are more likely to have a higher energy since they usually have a smaller scattering rate than holes. Thus at low fields, the electron ionization rate tends to be larger. For smaller–band-gap semiconductors, InAs and $Ga_{1-x}In_xAs$, e.g., the hole ionization rate is

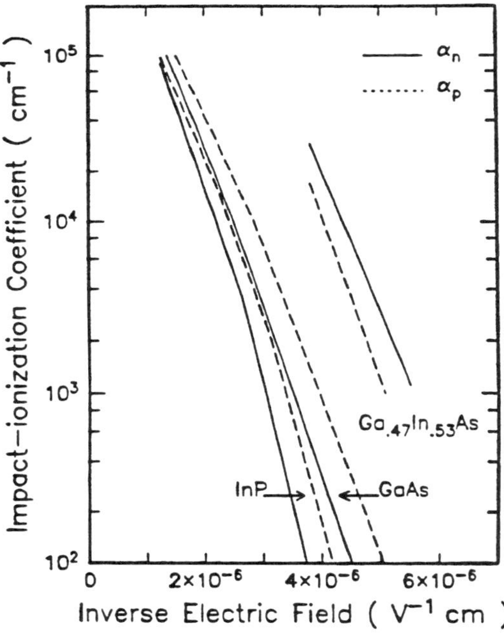

FIG. 20. Impact-ionization rates for electrons and holes for some compound semiconductors.

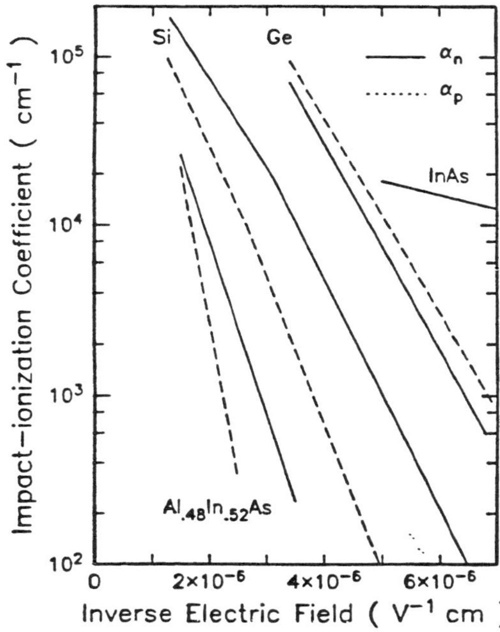

FIG. 21. Ionization rates for electrons and holes for some compound semiconductors, silicon, and germanium.

always low because the threshold energy for ionization is lower.

Impact ionization does not necessarily need an electric field to show a large generation rate for electrons and holes. In compound-semiconductor heterostructures, there exist many smaller–band-gap compounds that can be lattice matched to other semiconductors with a band-gap discontinuity for conduction or valence band that is larger than the band gap. Even in the absence of an electric field, high-energy carriers injected into the smaller–band-gap semiconductor can cause impact ionization. This process is often referred to as Auger generation.

When impact ionization occurs in the presence of an electric field, such as in a *p-n* junction, both electrons and holes get accelerated in opposite directions, and both cause further impact ionization. The process of electron–hole pair creation, acceleration of carriers, and creation of more electron–hole pairs sets up an avalanche, and hence the generation of electron–hole pairs in an electric field is also known as an avalanche process, and the related breakdown of a *p-n* junction, an avalanche breakdown.

The increase in electron or hole current resulting from the ionization caused by electron and hole current flow is

$$\frac{dJ_n}{dz} = \frac{dJ_p}{dz} = -(\alpha_n J_n + \alpha_p J_p), \tag{24}$$

where J_n is the electron current density and J_p is the hole current density. A simple expression for multiplication factor for electrons ($\mathcal{M}_n$), the ratio between electron current at the collecting edge to that at the injecting edge of the high field region, results in the form

$$1 - \frac{1}{\mathcal{M}_n} = \int_{-z_p}^{z_n} \alpha_n \exp\left[-\int_{\zeta}^{z_n} (\alpha_n - \alpha_p)\, d\eta\right] d\zeta, \tag{25}$$

where z_n and z_p are the coordinates of the edges of the depletion region.

The condition for breakdown caused by electron-initiated avalanche is, therefore, the condition where $\mathcal{M}_n$ goes to infinity. This occurs when the integral in this equality goes to unity. Avalanche breakdown, therefore, occurs when

$$\int_{-z_p}^{z_n} \alpha_n \exp\left[-\int_{\zeta}^{z_n} (\alpha_n - \alpha_p)\, d\eta\right] d\zeta = 1. \tag{26}$$

The condition for breakdown with a hole-initiated process can be found following a similar treatment with an analogous mathematical expression of the process.

While generally, at the high fields that dominate avalanche process, the hole ionization coefficient dominates, this is not true for small–band-gap semiconductors, such as $Ga_{0.47}In_{0.53}As$ and InAs. Here hole-initiated processes are less likely, and so even though by virtue of a smaller band gap they are more easily likely to avalanche, because of suppression of hole avalanching, the decrease in breakdown voltage is not as pronounced.

Quite often, the integral on the right of the above equations for multiplication factors is approximated by the relationship $(V/BV)^\nu$, where BV is the breakdown voltage; i.e.,

$$\mathcal{M} = \frac{1}{1 - (V/BV)^\nu}. \tag{27}$$

This relationship, depending on the magnitude of the power ν, shows a soft breakdown or a hard breakdown. Soft breakdown is meant to imply a slow and gradual increase in multiplication factor with increasing bias. It occurs when the ionization rates are slowly varying functions of the electric field. A hard breakdown, with $\nu \geq 4$, leads to a rapid change in multiplication factor and comes about when the ionization rate changes rapidly with electric field. Small–band-gap materials usually have a higher magnitude of the ionization rate and a slow change with electric field in it. They therefore exhibit relatively softer breakdown characteristics.

6.2 Off-Equilibrium Transport

When carriers traverse a region where there is a significant gradient in the electric field, they undergo a velocity overshoot in parts of the region. Velocity overshoot is a term that implies that the local velocity v at a local field $\mathscr{E}$ becomes higher than what it would reach for a slowly varying field $\mathscr{E}$ of the same magnitude. A local off-equilibrium occurs. It arises because the carrier energy and the local electric field are no longer in equilibrium with each other. An example of this local equilibrium at high fields is the existence of velocity saturation. The mobility of carriers is a function of energy. Because of the parallel between the energy and the electric field, the mobility can be written as a function of electric field. When the electric field becomes large, higher carrier energy results. Higher carrier energy causes a larger amount of scattering, which reduces the mobility. Thus, velocity, a product of mobility and electric field with mobility varying inversely with field, becomes a constant. When a large gradient of electric field exists, the carrier does not immediately achieve the energy that would correspond to the local field if the field were slowly varying. It actually has a lower energy and hence suffers less scattering, and yet it is accelerated by the high field. The carrier, therefore, is said to have a velocity overshoot. This large velocity occurs in spite of the fact that it has a lower energy. Another way of saying this is that the relaxation rate for momentum is larger than that for energy. This behavior has become increasingly important as device dimensions have shrunk and is particularly important in compound semiconductors because of their lower scattering rates at low energies. Thus, velocity overshoot occurs in field-effect transistors in the rapidly varying field region near the drain where channel pinch-off is usually said to have occurred. It also occurs in bipolar junction transistors in the base–collector region.

An example of such a velocity overshoot, for the case of GaAs, InP, and $Ga_{1-x}In_xAs$, for a gradient in electric field is shown in Fig. 22. In the case of GaAs, the reduction in kinetic energy occurs when the carriers begin transferring to the L valley. Such a transfer can occur after the carrier has acquired an energy of 0.36 eV (the Γ-L intervalley transfer threshold energy). When such transfer can occur, because the carrier has acquired the requisite energy, the carrier scattering rate becomes larger, and the carriers begin to lose energy. The Γ-L transfer of car-

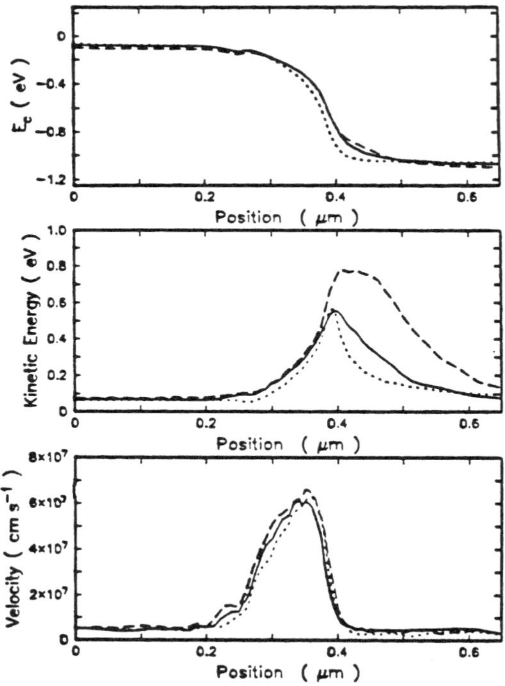

FIG. 22. Conduction-band–edge energy, kinetic energy, and velocity in a short field-effect structure, where electric field (the gradient of the band edge) changes rapidly. The solid lines are for GaAs, the short dashed lines for InP, and the long dashed lines for $Ga_{1-x}In_xAs$.

riers takes place in a time scale of at most 100 fs, and hence the carriers can still transit distances corresponding to this, ≈300 Å, before their energy begins to decrease. InP has a higher scattering rate and higher effective mass at lower energies than either GaAs or $Ga_{1-x}In_xAs$, and hence its velocities are lower toward the source end. However, at higher carrier energies, its velocity overshoot is higher than that of GaAs, because GaAs exhibits a larger scattering rate corresponding to the intervalley transfer. $Ga_{1-x}In_xAs$ has the most favored overshoot characteristics because of its lower scattering rate and large intervalley transfer threshold.

7. RECOMBINATION

The interactions between carriers of opposite types are particularly important in compound semiconductors because most of them are direct-band-gap semiconductors. Additional interactions between electrons and holes through levels within the band gap, due to traps, are also important in compound semiconductors. So, several forms of excitation processes can result either in the recombination of electron and hole pairs or in their generation. Radiative recombination, usually involving band-to-band transitions, with the energy released as photons is the basis of light-emitting diodes and lasers. Radiative recombination can also occur sometimes through discrete levels—e.g., in the Zn–O center combination in GaP. This is a unique example where the Zn–O pair forms an isoelectronic recombination center by replacing a Ga–P pair with an energy level in the band gap. Recombination can also occur without the involvement of radiative processes, such as through traps, through other processes where momentum conservation involves phonons, as well as through Auger processes where excess energy of annihilation or creation of an electron–hole pair is associated with a third carrier. All these processes are important under different conditions in devices.

In order to understand the statistics of these processes and their significance, we use the principle of detailed balance that states that, for a system in thermal equilibrium, the rate of any process and that of its inverse balance each other. The adjective "detailed" emphasizes that this balance occurs in all details of the process. Consider radiative recombination in thermal equilibrium. Detailed balance requires that, in thermal equilibrium, the same fraction of electron–hole recombination is radiative as is the electron–hole pair generation, and with the same spectral characteristics as the radiation that occurs with electron–hole pair recombination.

7.1 Defects and Hall–Shockley–Read Recombination

Traps, because they can exist in more than one charge state, have the ability to act as recombination-generation centers. This ability comes about because these centers can either capture an electron and/or a hole from the conduction band or emit them to the bands. Thus, electrons and holes can recombine or lead to a generation process at these traps. The statistics for this were first analyzed by Hall and by Shockley and Read, and it is known as Hall–Shockley–Read (HSR) recombination.

We will consider a donorlike trap, of density N_T, in an n-type crystal. Being a donorlike trap, it exists either in a positively charged state, and we consider a donor that exists only as a singly ionized donor of density N_T^+, or in an un-ionized state with density N_T^0. The singly ionized state of the trap exists in the band gap at an energy E_T. The net rate of change of excess density can be written as

$$-\frac{dp'}{dt} = -\frac{dn'}{dt} = (np - n_i^2) \times \left\{\left[n + N_c \exp\left(-\frac{E_c - \xi_f}{kT}\right)\right]\frac{1}{c_h N_T} + \left[p + N_v \exp\left(-\frac{E_T - E_v}{kT}\right)\right]\frac{1}{c_e N_T}\right\}^{-1}. \tag{28}$$

This expression relates the net rate of change in carrier concentration with time, i.e., $\mathcal{U} = \mathcal{G} - \mathcal{R} = dp'/dt = dn'/dt$ where $\mathcal{U}$ is the rate of change in carrier concentration and $\mathcal{G}$ and $\mathcal{R}$ are the generation and recombination rates.

Usually, one defines parameters τ_{p0} and τ_{n0} as

$$\begin{aligned} \tau_{p0} &= 1/c_h N_T \\ \tau_{n0} &= 1/c_e N_T, \end{aligned} \tag{29}$$

and hence the minority-carrier lifetime for the n-type semiconductor, for a donorlike trap with a single state, is

$$\begin{aligned} \tau_p &= -\frac{p'}{dp'/dt} \\ &= \frac{p'}{np - n_i^2}\left\{\left[n + N_c \exp\left(-\frac{E_c - \xi_f}{kT}\right)\right]\tau_{p0} \right. \\ &\quad \left. + \left[p + N_v \exp\left(-\frac{E_T - E_v}{kT}\right)\right]\tau_{n0}\right\}, \end{aligned} \tag{30}$$

and majority-carrier lifetime is

$$\begin{aligned} \tau_n &= -\frac{n'}{dn'/dt} \\ &= \frac{n'}{np - n_i^2}\left\{\left[n + N_c \exp\left(-\frac{E_c - \xi_f}{kT}\right)\right]\tau_{p0} \right. \\ &\quad \left. + \left[p + N_v \exp\left(-\frac{E_T - E_v}{kT}\right)\right]\tau_{n0}\right\}. \end{aligned} \tag{31}$$

These equations assume a single-level donorlike trap in the limit of Maxwell–Boltzmann approximation. Traps can be acceptorlike; the equations have the same form as above. A considerably more complex situation occurs when a trap has multiple levels in the band; e.g., transition metals with incomplete d orbitals usually have several levels associated with different states of ionization. Chromium, e.g., has three levels within the band gap of GaAs, and it is among the impurities used to obtain semi-insulating GaAs. Statistics in such situations can be quite complicated, because the presence of one state excludes the other. Figure 23 shows a plot of electron lifetime as a function of acceptor doping for GaAs, showing regions where HSR and Auger recombination, which we discuss next, dominate.

7.2 Auger Recombination

Electrons and holes can recombine, without the HSR-type interaction at a trap or the involvement of photon emission, by releasing the excess energy to another carrier. Such processes involving more than two carriers are usually referred to as Auger processes. Impact ionization, discussed earlier, is actually one example of an Auger process, a generation process that involves the creation of an additional electron–hole pair as a result of the excess energy of a hot carrier.

The number of Auger recombination processes is large. Auger transitions may take place through band-to-band, band–to–shallow-levels, or shallow-levels–to–shallow-levels transitions involving excitons, etc. In all these cases the excess energy released in the transition is picked up by a mobile carrier. Because of the involvement of carriers, shallow levels, and transition across much of the band gap, the process becomes stronger both

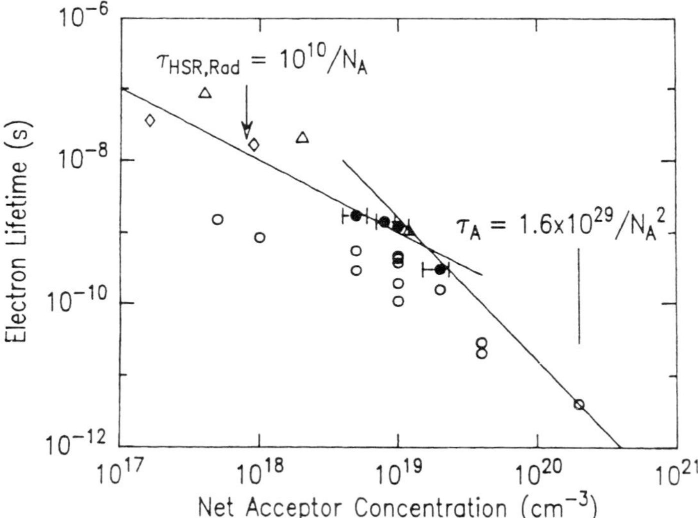

FIG. 23. Lifetime as a function of acceptor concentration in GaAs.

with increases in doping and by reduction in band gap. In small–band-gap materials, it is also strongly temperature dependent. Electrons are also more easily likely to get energetic since they usually have a lighter mass. Light holes are also likely to pick up excess energy; however, most holes exist as heavy holes because of the larger density of states for the heavy-hole band. Band structure is therefore very central to how and which Auger recombination processes are important, as it is for its inverse, the Auger generation process.

The net Auger recombination rate due to electrons and holes is given by

$$\begin{aligned} \mathcal{U}_A &= \mathcal{R}_{Ae} - \mathcal{G}_{Ae} + \mathcal{R}_{Ah} - \mathcal{G}_{Ah} \\ &= (\gamma_n n + \gamma_p p)(np - n_i^2). \end{aligned} \tag{32}$$

Under low-level injection, these expressions reduce to a form where the net recombination rate is proportional to the excess carrier concentration. For *p*-type material, such as the base of a bipolar transistor, the net recombination rate is

$$\mathcal{U}_A = \gamma_p N_A^2 n', \tag{33}$$

where n' is the excess carrier concentration. Using this, in a way similar to that of HSR recombination, an Auger lifetime can be defined whose magnitude is

$$\tau_A = 1/\gamma_n N_A^2. \tag{34}$$

For most semiconductors, where the band-to-band Auger processes dominate, in the limit of heavy doping, the lifetime does appear to follow an inverse dependence with the square root of the doping.

8. TUNNELING

Tunneling is the important example of the wave-particle duality. The de Broglie wavelength *vis-à-vis* the device dimension characterizes an estimate of the importance of the wave aspect. At device dimensions smaller than or comparable to the de Broglie wavelength, it is more appropriate to discuss electrons in crystals as a wave, and the classical picture that assumes a negligible time scale of scattering, drift, and diffusion averaged from a number of scattering events, etc., is increasingly poor. Tunneling, e.g., cannot be explained in the particle description; it requires the wave description. The de Broglie wavelength is related to the effective mass, which because it is generally small in the direct–band-gap compound semiconductors leads to large de Broglie wavelength, and increasing wave effects, tunneling being the most prominent of them.

Tunneling is a manifestation of the wave nature of electrons. If an overlap between the states in the valence band and the conduction band is large, then electrons can tunnel from the valence band to the conduction band in the junction. One may treat the forbidden gap as a potential barrier of the classical barrier-tunneling problem. Zener tunneling is a classical example of this and occurs in a variety of tunnel diodes, and it is a source of breakdown in *p-n* junctions. Zener breakdown is a form of internal emission that takes place by tunneling.

Zener tunneling will be substantial whenever the potential barrier height is small, the potential barrier width is short, or the effective mass is small. Any of these allow for a large overlap of the electron and hole wave functions. For a given band structure, the potential barrier height and width, related as the electric field, are conducive to tunneling at large dopings. Small–band-gap semiconductors have intrinsically a smaller potential barrier, leading to larger Zener tunneling probabilities, and it may take place even at low doping. The process thus has an effect on breakdown voltages both for highly doped semiconductors and for small–band-gap semiconductors. The breakdown voltages can be obtained by determination, again, of the transmission function and the source function, the two characterizing the probability of tunneling and the availability of carriers for tunneling.

Like our discussion of avalanche breakdown, Zener processes may also involve phonon processes. In the classical problem of tunneling through a barrier we ignore it; i.e., we consider only energy-conserving elastic processes. Phonon-assisted tunneling, an inelastic process, however, can be quite strong. During such a process, a longitudinal or transverse optical or acoustic phonon is emitted or absorbed in order to conserve momentum. In fact, for indirect semiconductors, this is the dominant form of tunneling because it can accommodate the large differences in momenta between the bottom of the conduction band and the top of the valence band.

9. HEAVY DOPING EFFECTS

In the presence of a large number of electrons and holes, the strong carrier–carrier interaction leads to a reduction in the band gap. Thus, under heavy doping, the band gap decreases. Figure 24 summarizes ΔE_g at 300 K for GaAs derived from measurements on transistor structures. A parametric fit to these data, optimized for higher dopings, is given by the relation

$$\Delta E_g = 2.00 \times 10^{-11} N_A^{0.50}, \tag{35}$$

where the unit of ΔE_g is eV and that of N_A is cm^{-3}. The magnitude of this shrinkage is quite substantial. At an acceptor doping of $1 \times 10^{19}\ \mathrm{cm}^{-3}$, this shrinkage is ≈ 0.07 eV, nearly 5% of the band gap of GaAs. It is obviously of substantial significance in highly doped base regions of bipolar transistors as well as source and drain regions of field-effect transistors. Figure 24 also includes the effective band-gap shrinkage from a theoretical calculation. We emphasize that none of these is an actual band parameter; they are quantitative fits.

10. PIEZOELECTRIC EFFECTS

An issue unique to compound semiconductors is related to their ionic character and noncentrosymmetry. Stress-induced polarization results in polarization charge, whose effect shows up significantly through electrostatic effects in the operation of the device. The polarization can be viewed to result in additional charge because

$$\nabla \cdot (\epsilon_s \mathscr{E} + \mathscr{P}) = \rho, \tag{36}$$

where $\mathscr{P}$ is the polarization vector and ρ is the charge density.

The polarization results in additional charge and hence device-parameter variation because of stress in the material. This stress can occur in devices because of different expansion coefficients of the materials employed; it may result from the dielectric films employed on semiconductor surfaces or the metalizations of the gate and Ohmic contacts that are in proximity to the device active region.

When a compressive or tensile stress occurs, the angle between bonds for the atoms away from the center increases or decreases,

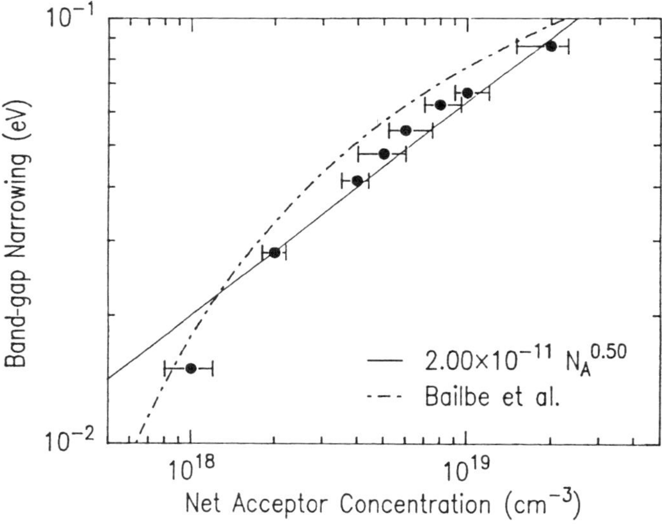

FIG. 24. Band-gap change with doping in GaAs at 300 K.

resulting in a net displacement. Consequently, there is a net dipole moment in the crystal. This polarization, caused by stress, can be determined in general by tensor analysis as

$$\mathcal{P} = \boldsymbol{d}:\boldsymbol{\sigma}, \tag{37}$$

where $\boldsymbol{d}$ is the piezoelectric tensor and $\boldsymbol{\sigma}$ is the stress tensor.

There is an additional requirement for this polarization to lead to a cumulative charge effect and hence a macroscopic variation with position. This is that the crystal be noncentrosymmetric. Noncentrosymmetry causes the polarization of the unit cells to cumulate because the polarizations are additive and have the same direction. Centrosymmetric crystals do not cumulate the polarizations of unit cells since they oppose in alternate unit cells.

Most of the elements of $\boldsymbol{d}$, except six, are zero for compound semiconductors and have identical value. This results in a polarization vector

$$\begin{bmatrix} \mathcal{P}_x \\ \mathcal{P}_y \\ \mathcal{P}_z \end{bmatrix} = d\begin{bmatrix} \sigma_{xy} \\ \frac{1}{2}(\sigma_{xx} - \sigma_{zz}) \\ \sigma_{yz} \end{bmatrix}, \tag{38}$$

where d is the magnitude of the equal nonzero components, the x subscript denotes $\langle 01\bar{1}\rangle$ direction, the y subscript denotes $\langle \bar{1}00\rangle$ direction, and the z subscript denotes $\langle 011\rangle$ direction. The strongest effect arises from the surface dielectric and metal films in close vicinity of the active region of the devices. In a metal-gate–controlled transistor, stress occurs because of both the gate film and the dielectric films used on the surface. The presence of polarization charge can cause electric-parameter (e.g., threshold voltage) change of both signs. In addition, depending on the orientation of the gate or dielectric lines, the polarization results are different. Si_3N_4, one of the dielectric films employed, is usually under compressive stress, while amorphous SiO_2, another dielectric employed, is usually under tensile stress. The resulting polarization from the dielectric film is therefore of the opposite sign, and hence threshold voltage variation is opposite.

In transistors made on (100) surfaces with gates in the $\langle 011\rangle$ or $\langle 01\bar{1}\rangle$ direction, the piezocharge is at either its maximum or the minimum under the gate. Near the tail of the doping density, its effect can be substantial because of reduction in dopant charge. It can, therefore, significantly disturb the low-current behavior of devices—i.e., it affects parameters such as threshold voltage and conductance at low biases with the effect gate-orientation dependent. The problem of changes in threshold voltage in devices that are oriented differently on the same wafer places a significant design constraint in their use.

The threshold shift also becomes more pronounced with a decrease in dimensions, i.e., shortening of the gate length. Shortening of gate length naturally leads to an increase in the stress underneath the gate due to stress from dielectric films. This gate-length effect is in addition to the effect of short-channel phenomenon.

GLOSSARY

Acoustic Phonons: The low-energy mode of lattice vibrations.

Avalanche Process: The cascade process where an energetic carrier loses some of its excess energy by creating an electron–hole pair.

Auger Recombination: The inverse process of avalanche process.

Band Gap: The difference between minimum of conduction-band edge and maximum of valence-band edge.

Discontinuity: The change in conduction- or valence-band edge at an abrupt transition from one composition to another composition of a crystal.

Drift Mobility: The constant of proportionality relating the drift velocity with local electric field.

Hall Mobility: A mobility derived from the Hall voltage that develops transverse to the current flow in the presence of a magnetic field.

Heterojunction: The junction formed between two dissimilar materials.

Majority Carrier: The carrier type that is in majority in thermal equilibrium in a semiconductor.

Minority Carrier: The carrier type that is in minority in thermal equilibrium in a semiconductor.

Optical Phonons: The high-energy modes of lattice vibrations.

Overshoot or Off-Equilibrium Effect: The condition when the local velocity of a carrier is larger than the velocity obtained from the product of mobility and electric field.

Piezoelectric Effect: The effects arising from accumulation of polarization charge in presence of strain in a noncentrosymmetric crystal.

Scattering: The interaction between a carrier and its surroundings.

Work Cited

Wolfe, C. M., Stillman, G. E., Lindley, W. T. (1970), *J. Appl. Phys.* **41,** 3088–3091.

Further Reading

Adachi, S. (1985), "GaAs, AlAs, and $Al_xGa_{1-x}As$: Material Parameters for Use in Research and Device Applications," *J. Appl. Phys.* **58,** R1–R29.

Blakemore, J. S. (1982), "Semiconducting and Other Major Properties of Gallium Arsenide," *J. Appl. Phys.* **53,** R123–R181.

Burstein, E., Lundqvist, S. (Eds.) (1969), *Tunneling Phenomena in Solids,* New York: Plenum.

Conwell, E. M. (1978), "High Field Transport in Semiconductors," in: F. Seitz, D. Turnbull, H. Ehrenreich (Eds.), *Solid State Physics,* Vol. 9, New York: Academic.

Duke, C. B. (1969), "Tunneling in Solids," in: F. Seitz, D. Turnbull, H. Ehrenreich (Eds.), *Solid State Physics,* Vol. 10, New York: Academic.

Harrison, W. A. (1969), *Solid State Theory,* New York: McGraw-Hill.

Hess, K. (1982), "Aspects of High-field Transport in Semiconductor Heterolayers and Semiconductor Devices," in: *Advances in Electronic and Electron Physics,* Vol. 59, New York: Academic.

Jaros, M. (1985), "Electronic Properties of Semiconductor Alloy Systems," *Rep. Prog. Phys.* **48,** 1091–1154.

Nag, B. R. (1980), *Electron Transport in Compound Semiconductors,* Berlin: Springer.

Paul, W. (Ed.) (1982), *Handbook on Semiconductors,* Amsterdam: North-Holland.

Pearsall, T. P. (Ed.) (1990), *Strained Layer Superlattices: Physics,* Semiconductors and Semimetals Vol. 32, San Diego: Academic.

Pearsall, T. P. (1991), *Strained Layer Superlattices: Materials Science and Technology,* Semiconductors and Semimetals Vol. 33, San Diego: Academic.

Ridley, B. K. (1988), *Quantum Processes in Semiconductors,* Oxford: Clarendon.

Sze, S. M. (1981), *Physics of Semiconductor Devices,* New York: Wiley.

Tiwari, S. (1992), *Compound Semiconductor Device Physics,* New York: Wiley.

Wolfe, C. M., Holonyak, N., Jr., Stillman, G. E. (1989), *Physical Properties of Semiconductors,* Englewood Cliffs, NJ: Prentice-Hall.

SEMICONDUCTORS, COMPOUND—MATERIAL PROPERTIES

V. SWAMINATHAN, *Bell Laboratories, Lucent Technologies, North Andover, Massachusetts, U.S.A.*

P. CAPPER, *GEC-Marconi Infra-Red Ltd., Southampton, United Kingdom*

INTRODUCTION

Semiconductors are classified into two broad categories as elemental semiconductors and compound semiconductors. Si and Ge are examples of elemental semiconductors. The compounds are grouped into families depending on the chemical nature of the constituents. III–V compounds represent semiconductor compounds made from elements of group III and group V of the periodic table (e.g., GaAs, InP, GaP, InAs, InSb, GaN). Similarly, II–VI compounds are made from elements of group II and group VI (e.g., CdTe, ZnTe, ZnSe, CdSe, CdS).

While the early research was devoted to the physical properties of the III–V compounds, the emphasis was quickly shifted to technology and applications. Phenomenal progress has been made in the preparation and processing of these compounds, and some of them (e.g., GaAs, InP) are used to make technologically important electronic and photonic devices. In this latter aspect, they have an advantage over Si which, being an indirect–band-gap semiconductor, is not an efficient photonic material. Besides, they (e.g., GaAs, InP) are attractive for high-speed devices since the electron saturation velocity in GaAs is 1.5 times that of Si. The low-field

3-527-28139-8/96/$5.00 + .50

electron mobility in GaAs or InP is also higher than that in Si. The advantages of high speed and low power dissipation of GaAs devices appeal to the communications industry. When compared to Si, III–V compounds have higher radiation hardness and are capable of operating over a wider temperature range from −196 to 300 °C. Because of this, GaAs devices are of special interest for space and military applications.

The major use of GaAs at present is in the area of microwave devices, high-speed digital integrated circuits, and as substrates for epitaxial layer growth to fabricate photonic and electronic devices. For InP, the use is almost exclusively as substrates for growing lattice-matched epitaxial films of alloy semiconductors such as GaInAs and GaInAsP. These ternary and quaternary compounds are the materials of choice for making light sources and detectors for the present-day fiber-optic communication systems. Among other III–V compounds, GaP has been the material of choice for visible-light–emitting diodes. InSb is used for infrared detectors. Both InSb and GaSb are used as substrate material for epitaxial growth of Sb-based compounds. Recently, interest in wide–band-gap (>2 eV) III–V nitrides has increased on account of progress in their growth by epitaxy.

Improvements in both bulk and particularly epitaxial growth techniques in recent years, combined with a deeper understanding of defects and contacting/passivation issues, have led to a resurgence in II–VI compound studies and use. Electrical, optical, and magnetic properties have been utilized to make a wide range of emission and detector devices. Wide-gap II–VI compounds, such as ZnSe, are used in lasers and LEDs, while narrow-gap compounds, such as CdTe and HgCdTe, are used in solar cells, optical devices, room-temperature γ- and x-ray detectors, and infrared detectors and emitters. It is probable that HgCdTe is the third most widely studied semiconductor system, after Si and GaAs, because of the effort expended over the last 30 years or so, with most support coming from military sources. Zinc oxide is used as an acoustic-wave, electro-optic, and acousto-optic material and in some solar-cell structures. IIa–VIa compounds, such as CaS, are indirect-gap semiconductors and are used as electroluminescent materials and as phosphors, when doped with rare-earth elements (e.g., Eu, Ce). Optical modulators and isolators are made from diluted magnetic semiconductors (*q.v.*) (DMS) based on II–VI compounds, such as CdMnTe and HgFeSe. The attraction of these DMS materials lies in the ability to tune—by varying the composition, the energy gap, and the lattice parameter—the purely magnetic properties, including paramagnetic, spin-glass, and antiferromagnetic properties, and those properties, such as giant Faraday rotation of visible and near-infrared light and giant negative magnetoresistance in the semiconductor–semimetal transition region, that are unique to DMS materials.

The following sections summarize the relevant data for the more important binary and ternary III–V and II–VI compounds, which have been distilled from the references. There is insufficient space in this article to provide detailed discussion of the various properties and effects, particularly in view of the ranges given in the literature for many of the parameters. However, a significant bibliography is given at the end, and the reader is referred to books in that list.

1. III–V COMPOUNDS

1.1 Binary Compounds

1.1.1 Bonding and Crystal Structure

Most III–V compound semiconductors crystallize in the cubic zinc-blende (sphalerite) structure. The wide–band-gap nitrides (e.g., AlN, GaN, InN) crystallize in the hexagonal wurtzite structure. In the tetrahedral arrangement of atoms in the sphalerite structure, each group-III atom is bonded to four group-V atoms in the tetrahedron and vice versa. The cubic unit cell of a sphalerite structure is shown in Fig. 1. There are four molecules of the compound *AB* per unit cell.

1.1.1.1 Fractional Ionic Character in Bonding. Although the bonding in the sphalerite structure consists of sp^3 hybridized orbitals, there is some charge transfer between the two types of atoms, giving rise to a partial ionic character to the bonding, unlike in diamond cubic semiconductors. If f_i and f_h denote the fractions, respectively, of ionic or heteropolar character and covalent or homopolar character in the bond, then

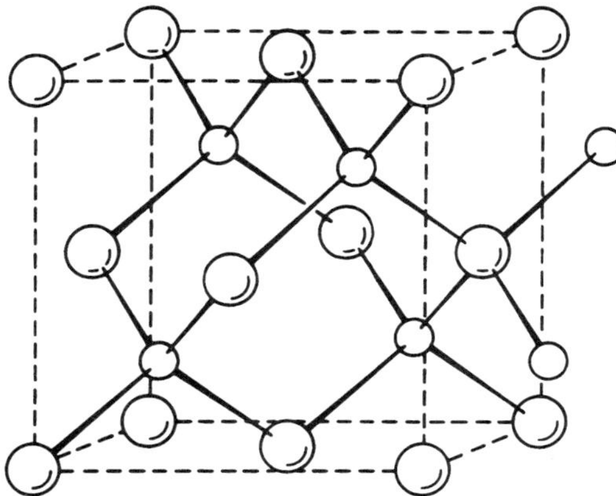

FIG. 1. The cubic unit cell of a sphalerite structure.

Table 1. Fractional ionic character of bonds and cohesive energy of III–V semiconductors (Phillips, 1973).

Crystal	Fractional ionic character, f_i	Cohesive energy[a] (kcal/mole)
BAs	0.002	
BP	0.006	
AlSb	0.250	
BN	0.256	−293.9
GaSb	0.261	−118.3
AlAs	0.274	−157.3
AlP	0.307	−169.5
GaAs	0.310	−135.4
InSb	0.321	−108.5
GaP	0.327	−152.4
InAs	0.357	−125.6
InP	0.421	−132.9
AlN	0.449	
GaN	0.500	

[a]Gibbs free energy of sublimation into neutral atoms at STP.

the sum $f_i + f_h$ equals unity. In elemental semiconductors $f_h = 1$ and $f_i = 0$. A semiempirical theory of fractional ionic or covalent character of the bond has been developed by Phillips (1973), which involves estimating the ionic contribution to the energy gap of the AB semiconductor. Table 1 lists the f_i values for III–V compounds (Phillips, 1973). Also listed in Table 1 are the cohesive energies, denoted by the Gibbs free energies of atomization at standard temperature and pressure (STP), of the respective compounds.

1.1.1.2 Brillouin Zones. The electronic and vibronic states in crystals are best described in reciprocal space or **k** space. The smallest unit cell in reciprocal space is called the first Brillouin zone. While $(a/2)[110]$, $(a/2)[101]$, and $(a/2)[011]$ are the primitive translation vectors of the face-centered cubic (fcc) lattice, $(2\pi/a)[11\bar{1}]$, $(2\pi/a)[1\bar{1}1]$, and $(2\pi/a)[\bar{1}11]$ are the primitive translation vectors in the reciprocal lattice. Incidentally, the latter are the primitive translation vectors of a body-centered cubic (bcc) lattice, so that the bcc lattice is the reciprocal lattice of the fcc lattice. The volume of the primitive cell in the reciprocal fcc lattice is $4(2\pi/a)^3$.

The first Brillouin zone for the cubic semiconductors is the truncated octahedron shown in Fig. 2. The principal symmetry points and lines in **k** space are labeled, using the standard nomenclature. The Cartesian coordinates of these symmetry points are $\Gamma = (0,0,0)$; $X = (2,0,0)$; $L = (1,1,1)$; $\Lambda = (x,x,x)$; $\Delta = (2x,0,0)$; $W = (2,1,0)$; $K = U = (3/2,3/2,0)$; $\Sigma = (3/2x, 3/2x,0)$; where x denotes a number between 0 and 1 and the coordinates are given in units of π/a where a is the lattice constant.

Table 2 lists the lattice constant, density, melting point, Debye temperature, coefficient of thermal expansion, and thermal conductivity of the binary compounds. For the last two parameters, the values at or close to 300 K are given (Madelung *et al.*, 1982, 1984).

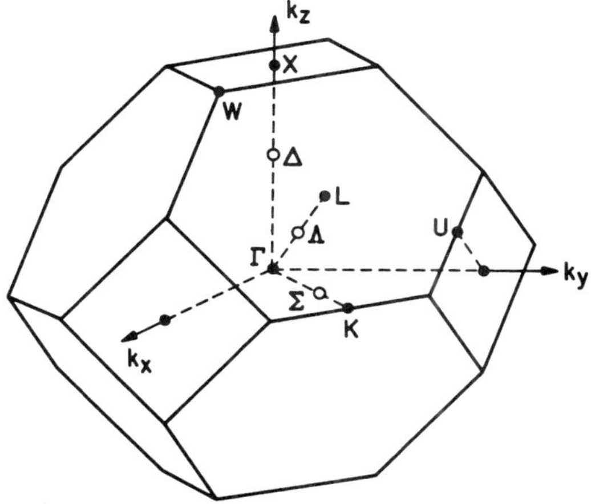

FIG. 2. The first Brillouin zone for the cubic semiconductor showing the symmetry points and axes (from Zallen, 1982).

Table 2. Lattice constant, density, melting point, Debye temperature, coefficient of thermal expansion, and thermal conductivity of III–V compounds.

Compound	Lattice constant (Å)	Density (gm cm^{-3})	Melting point (K)	Debye Temperature[a,b] (K)	Coefficient of thermal expansion[c] (10^{-6}/°C)	Thermal conductivity[a,b,c] (W cm^{-1} K^{-1})
AlN	3.11[e]	3.255	3273		5.27 (⊥)	
	4.98[f]				4.15 (‖)	
AlP	5.467	2.4	2823	588	4.5	0.9
AlAs	5.66	3.7	2013	417	4.9	0.8
AlSb	6.136	4.26	1338	292	4.0	0.57
GaN	3.16[e]	6.095	1973	600	3.17 (⊥)	1.3
	5.125[f]				5.59 (‖)	
GaP	5.4512	4.138	1740	456	4.5	0.77
GaAs	5.6532	5.3161	1513	344	6.86	0.46
GaSb	6.0959	5.6137	985	266	7.75	0.39
InN	3.5446[e]	6.89	1373		3.9 (⊥)	
	5.7034[f]				2.95 (‖)	
InP	5.8687	4.81	1335	321	4.75	0.68
InAs	6.0583	5.667	1215	249	4.52	0.273
InSb	6.4794	5.7747	800	203	5.37	0.166

[a]From Holland (1966).
[b]From Madelung *et al.* (1982).
[c]Values near or at 300 K.
[e]Lattice parameter = a.
[f]Lattice parameter = c.

1.1.2 Energy-Band Structure The $\mathbf{k}\cdot\mathbf{p}$ method is used to calculate the band structure in the vicinity of the most important symmetry points in the Brillouin zone in great detail (Kane, 1982). Semiconductors with tetrahedral coordination have valence-band maxima at the Γ point, and many have conduction-band minima also at the Γ point (e.g., GaAs, InP). The representation of the valence band is Γ_{15}. When electron spin is included, there are altogether six valence-band states. The six states are further split by spin–orbit interaction into a fourfold degenerate $J = 3/2$ and a twofold degenerate $J = 1/2$ states, the $J = 1/2$ band being the lowest band. The $J = 3/2$ states are further characterized by the azimuthal quantum number of J relative to the k axis, $m_j = 3/2$ or $1/2$. They are labeled as the heavy-hole (twofold degenerate, $J = 3/2$, $m_j = 3/2$) band and light-hole (twofold degenerate, $J = 3/2$, $m_j = 1/2$) band. The $J = 1/2$ band is called the spin–orbit split-off valence band. The value of the spin–orbit splitting for III–V semiconductors is given in Table 3 (Phillips, 1973). The conduction band is derived from s-like orbitals and denoted by the representation Γ_1. It is very much like the mirror image of the spin–orbit split-off valence band.

The calculated energy-band structure of GaAs is shown in Fig. 3 (Cohen and Chelikowsky, 1988). The various symmetry points in the Brillouin zone are also indicated. The lowest conduction-band minimum Γ_1 occurs at $k = 0$ with higher minima occurring in the $\langle 100\rangle$ (Δ) and $\langle 111\rangle$ (Λ) directions. The L and X minima are respectively 0.31 and 0.52 eV above the Γ_1 minimum (Cohen and Chelikowsky, 1988). The valence band consists of heavy-hole and light-hole bands degenerate at $k = 0$ and a spin–orbit-split band at 0.34 eV lower energy. The various critical-point energies are also indicated in Fig. 3. These critical points satisfy the condition $\nabla_k E(k) = 0$.

The parameters pertinent to the electronic

Table 3. Spin–orbit splittings in III–V compounds (Phillips, 1973).

Compound	Δ_0 (eV)
GaP	0.127
GaAs	0.34
GaSb	0.80
InP	0.11
InAs	0.38
InSb	0.82

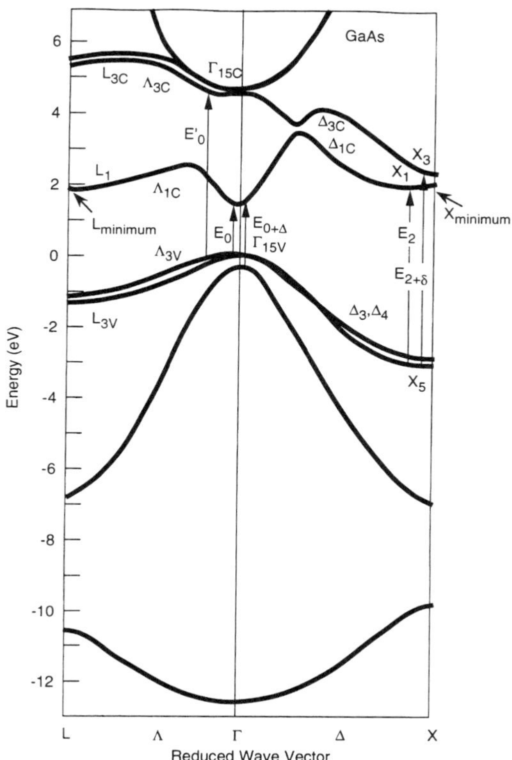

FIG. 3. Calculated energy band structure of GaAs in two principal directions in the Brillouin zone. The various critical point energies are indicated. (From Cohen and Chelikowsky, 1988.)

band structure are summarized in Tables 4 and 5. The direct and indirect energy gaps, their temperature dependences, and the pressure dependence of the minimum energy gap are given in Table 4. For AlP, AlAs, AlSb, and GaP, the conduction-band minimum is located at the Δ axes near the Brillouin-zone boundary. For AlAs the k value of the conduction band minimum is $0.087k_{max}$ from the X point, where $k_{max} = 2\pi/a$, a being the lattice constant. For GaP the corresponding value is $0.043k_{max}$ from the X point.

The temperature dependence of the band gap is generally expressed by (Varshni, 1967)

$$E_g(T) = E_g(0) - \alpha T^2/(T + \beta), \qquad (1)$$

where $E_g(0)$ is the energy gap at 0 K, and α and β are constants, with β having a value close to the Debye temperature. Near room temperature and above, $E_g(T)$ varies linearly with temperature. The values of the linear temperature coefficient of the energy gap are given in Table 4. Table 6 gives the α and β values for obtaining E_g below room temperature (Casey and Panish, 1978).

1.1.2.1 Effective Mass. Table 7 lists the effective masses at conduction- and valence-band edges. Semiconductors that have conduction-band minimum at $\Gamma = 0$ are characterized by one mass only. Those that have minimum near the X point are characterized

Table 4. Energy-band structure of III–V compounds.

Compound	Direct energy gap (eV) $\Gamma_{15} - \Gamma_1$	Indirect energy gap (eV) $\Gamma_{15} - X_1$	Indirect energy gap (eV) $\Gamma_{15} - L_1$	$\frac{dE_d}{dT}$ (10^{-4} eV K^{-1})[a]	$\frac{dE_{ind}}{dT}$ (10^{-4} eV K^{-1})[b]	$\frac{dE}{dP}$ (10^{-6} eV/bar)[c]
AlN	6.2					
AlP	3.62 (77 K)	2.45			−3.6	
AlAs	3.14	2.14[d]		−5.2	−4.0	
AlSb	2.22	1.63[d]			−3.5	−1.5
GaN	3.44			−6.7		4.2
GaP	2.78	2.268[d]		−4.5	−5.2	−1.6
GaAs	1.424	1.804	1.81 (110 K)	−3.9	−2.4[e]	12.0
GaSb	0.7	1.25 (10 K)	0.81	−3.7		14.7
InN	2.09					
InP	1.34	2.04	1.74	−2.9	−3.7[e]	8.8
InAs	0.356			−3.5		10.6
InSb	0.18			−2.8		15.9

[a]Temperature dependence of direct energy gap.
[b]Temperature dependence of indirect energy gap.
[c]Pressure dependence of minimum energy gap.
[d]Minimum situated at the Δ axes near the boundary: $k = (0.903, 0, 0)$ for AlAs; $k = (0.95, 0, 0)$ for GaP.
[e]Temperature dependence of the energy separation between Γ and X minima.

Table 5. Critical-point energies (eV) of III–V compounds. $E_0 = \Gamma_{15v} - \Gamma_{15c}$, $E_1 = \Lambda_{3V} - \Lambda_{1C}$, $E_1 + \Delta = \Lambda_{4V} - \Lambda_{1C}$, $E_2 = X_{5V} - X_{1C}$, $E_2 + \delta = X_{5V} - X_{3C}$.

Compound	E_0	E_1	$E_1 + \Delta$	E_2	$E_2 + \delta$
AlAs	4.34			4.54	4.89
AlSb	3.7 (77 K)	2.78	3.18		
GaN	8.5	6.25		7.65	
GaP	4.8 (80 K)	3.7 (80 K)	3.9 (80 K)		
GaAs	4.49 (4.2 K)	2.9	3.17		
GaSb	3.2 (80 K)				
InP	4.8 (77 K)	3.15	3.3	5.04	5.6
InAs	4.5 (77 K)	2.5	2.75	4.72	5.3
InSb	3.16	1.88	2.38	4.08	4.6 (110 K)

by longitudinal and transverse effective masses, m_l and m_t. Because of lack of inversion symmetry, the degenerate bands at the X point are split into a higher-lying X_3 band and a lower-lying X_1 band. For the indirect–band-gap semiconductors, the density-of-states effective mass is given by $\nu^{2/3}(m_l m_t^2)^{1/3}$, where ν is the number of equivalent conduction-band minima. When the minimum occurs at the X point, $\nu = 3$. The valence band is characterized by three effective masses corresponding to the heavy-hole, light-hole, and spin–orbit bands. The density-of-states effective mass for holes is given by $(m_{hh}^{3/2} + m_{lh}^{3/2})^{2/3}$.

1.1.3 Impurities and Defects Crystalline imperfections play an important role in determining the material quality of III–V compound semiconductors; they include point defects, dislocations, stacking faults, grain boundaries, and interfaces. Point defects can be native defects, such as vacancies and interstitials of the group-III or group-V atoms, as well as foreign atoms. The latter are generally intentionally added to the material to obtain the desired electrical properties. They can also be present as background impurities in the crystals because of contamination from the crystal growth systems. Dislocations and stacking faults are examples of line defects and two-dimensional defects, respectively. They can be present in the as-grown material or can be introduced during device processing steps, such as diffusion and ion implantation. Their presence is, in general, harmful to the devices. Interfaces between semiconductors, as in multilayer structures, or between semiconductors and dielectric films that are deposited on them as part of device processing can be a source or sink for defects. As a result, defect-free interfaces are desired.

Table 6. Temperature dependence of the minimum energy gap. Parameters α and β as defined in Eq. (1) are obtained from Casey and Panish (1978).

Compound	$E_g(0)$ (eV)	α (10^{-4} eV K^{-1})	β (K)
AlP	2.52	3.18	588
AlAs	2.239	6.0	408
AlSb	1.687	4.97	213
GaP	2.338	5.771	372
GaAs	1.519	5.405	204
GaSb	0.810	3.78	94
InP	1.421	3.63	162
InAs	0.420	2.50	75
InSb	0.236	2.99	140
$Ga_{0.47}In_{0.53}As$/InP	0.822	4.5	327[a,b]

[a]From Pearsall *et al.* (1983).
[b]The linear temperature coefficient obtained from absorption measurement is -3.48×10^{-4} eV/K (Zielinski *et al.*, 1986).

In III–V components, the electrical properties are controlled to a large extent by foreign impurities, which are either intentionally added or present as contaminants. Impurities give rise to two kinds of energy levels, which are often denoted as shallow and deep energy levels. Shallow impurities have energy levels that are placed on the order of 0.1 eV or less with respect to the band edge. On the other hand, deep impurities have energy levels toward the middle of the energy gap. The deep impurities are also sometimes referred to as traps.

Shallow impurities are further classified as donors or acceptors depending upon whether they produce electron (n-type) or hole (p-type) conductivity. For impurities that substitute for the cation or anion site in

Table 7. Effective-mass parameters of III–V compounds.

	Conduction band (units of m_0)			Valence band (units of m_0)		
Compound	m_l	m_t	m_e	m_{hh}	m_{lh}	m_{hso}
AlP				0.63	0.2	0.29
AlAs	1.5	0.19	0.79[a]	0.76	0.15	0.24
AlSb	1.64	0.23	0.92[a]	0.94	0.14	0.29
GaN			0.27	0.8		
GaP	7.25	0.313	1.86[b]	0.54	0.16	0.24
GaAs			0.067	0.49	0.08	0.15
GaSb			0.044	0.34	0.044	0.13
InN			0.12	0.5	0.17	
InP			0.075	0.56	0.12	0.12
InAs			0.024	0.37	0.025	0.14
InSb			0.014	0.39	0.016	0.47

[a]For three equivalent minima, m_e is given by $3^{2/3}(m_l m_t^2)^{1/3}$.
[b]Because of the presence of a camel's-back conduction-band structure and extremely high nonparabolicity, there is considerable scatter in the effective-mass values reported in the literature.

the III–V compound, donor or acceptor action is realized depending on the chemical valence difference between the host and the impurity. If the valence difference is greater than zero, donors are created, and if it is less than zero, acceptors are created. If there is no valence difference, then the impurity is called isovalent—eg., N substituting for P in GaP.

For the III–V compounds, group-VI impurities such as S, Se, and Te substitute for group-V atoms and behave as shallow donors. Likewise, group-II impurities such as Be, Mg, Zn, and Cd substitute for group-III atoms and give rise to shallow acceptors. Group-IV impurities like Si, Ge, and Sn, on the other hand, show amphoteric behavior. That is, a group-IV impurity on the anion site behaves as an acceptor, while one on the cation site behaves as a donor.

The energy level associated with a donor (acceptor) represents the binding energy of an electron (hole) to the impurity in the semiconductor host. For shallow impurities, the binding energy can be obtained using the effective-mass approximation (Kohn, 1957) and is given by

$$E_n = \frac{1}{n^2}\left(\frac{m^*}{m_0}\right)\left(\frac{1}{\varepsilon^2}\right)\mathcal{R}, \qquad (2)$$

where $\mathcal{R}$ is 13.6 eV, the ionization energy of the free hydrogen atom, ε is the dielectric constant of the semiconductor, m^* is the effective mass, and m_0 is the free-electron mass. For electrons (donors) m^* is the electron effective mass, and the energies are measured from the bottom of the conduction band. For holes (acceptors) m^* is the hole effective mass and the energies are measured from the top of the valence band. The Bohr radius corresponding to the bound states is given by

$$a = a_{\mathrm{H}}\frac{\varepsilon}{(m^*/m_0)}, \qquad (3)$$

where a_{H} is the Bohr radius of the hydrogen atom.

For direct-gap III–V compounds, such as GaAs and InP with nondegenerate conduction band and isotropic electron effective mass, the effective-mass approximation is a good description for the binding of electrons to donors. The binding energies and Bohr radii of donors in GaAs and InP are respectively E_D = 5 meV, a_D = 102 Å, and E_D = 7 meV, a_D = 84 Å, with ε = 12.9 and m_e = $0.067m_0$ for GaAs and ε = 12.6, m_e = $0.08m_0$ for InP. For acceptors, the situation is somewhat complicated because of the complex valence-band structure (Kohn, 1957), but qualitatively it can be expected that the larger effective mass of the holes compared to that of the electrons gives rise to larger acceptor binding energy and smaller Bohr radius compared to that of donor.

The effective-mass theory predicts the same energy levels for donors and acceptors irrespective of their chemical nature. But in reality, the binding energies vary depending on the impurity, and the variation is significant, especially for the acceptors. The difference between the effective-mass binding energy and the experimental value is referred to as a "chemical shift" (Pantelides, 1975). Phillips (1970) proposed a dielectric model to explain the chemical shifts in terms of the differences in electronegativity between the substitutional impurity and the host atom that it has replaced. When the electronegativity difference is large, the deviation of the binding energy from the effective-mass value is large.

The energy levels of donors and acceptors are usually obtained by observing the transitions of electrons from the impurity state to one of the band continua or to another impurity state or, conversely, from one of the band continua to the impurity state. The ground state and/or the excited states of the impurity may also be involved in such transitions. A large number of electrical, optical, and magnetic resonance techniques have been employed to characterize defects and obtain information of their binding energies and microscopic structure. Many chemical analysis techniques have been used to assess impurity concentrations. Often, special analysis tools with sensitivity in the parts per billion range are required. Secondary-ion mass spectrometry, otherwise known as SIMS, is one of the popular chemical analysis techniques used in the compound semiconductor industry.

Impurities from columns other than II, IV, and VI of the periodic table can also be incorporated in a III–V crystal, and they may have energy level(s) deep within the band gap. The electronic wave functions centered on these impurities are, in general, more localized than is the case for simple donors and acceptors. Further, the effective-mass approximation is not a good description of the binding energy of the electronic particle. 3*d* transition metal atoms such as Ti, Cr, Mn, Fe, Co, Ni, and Cu are examples of deep impurities in III–V compounds. They are used to obtain semi-insulating material with carrier concentrations and resistivities similar to that of the intrinsic semiconductor. Also, these impurities can act as efficient recombination centers and limit the efficiency of light-emitting devices. In some cases, the 3*d* elements have been found to give doubly ionizable donors or acceptors. A junction technique known as deep-level transient spectroscopy, DLTS, is widely used to probe deep impurities and defects and obtain information on ionization energies and emission/capture cross sections for electrons and holes.

There are six possible native defects in an *AB* compound: two types of vacancies (V_A,V_B), interstitials (A_i,B_i), and misplaced atoms (A_B,B_A). Since the *A*/*B* site ratio required by the crystal structure has to be maintained, the native defects occur in sets containing at least two types in a stoichiometric compound (Kroger, 1974). These sets are the Schottky disorder involving V_A and V_B, interstitial disorder involving A_i and B_i, and antistructure disorder involving A_B and B_A. There is also the possibility of combinations of these three disorder types. A disorder involving vacancies and interstitials—for example, $V_A + A_i$ or $V_B + B_i$—is called Frenkel disorder.

Native defects play an important role in determining the electronic properties of III–V compounds. Their concentrations are affected by the conditions of preparation of the crystal (e.g., partial pressure of constituents) that represent the equilibrium between the solid phase and the external liquid or gas phase. Incorporation of substitutional impurities, which include many of the donor and acceptor atoms, depends on the concentration of native defects. Native defects themselves can exist in neutral or charged form, and as charged species they sometimes directly control the conductivity type of the semiconductor. A well-known example is the EL2 deep level in GaAs, which gives rise to the semi-insulating characteristics of undoped material (Martin and Makram-Ebied, 1986). Semi-insulating GaAs required for fabricating integrated circuits such as field-effect transistors can be obtained by doping with Cr. However, Cr redistributes or outdiffuses during annealing steps in processing. Hence, undoped semi-insulating GaAs is very attractive for fabricating integrated circuits, and this is achieved by the deep donor EL2 that has been associated with an As_{Ga} defect. Other examples where native defects directly affect the electronic properties are GaN and GaSb. In undoped form the former is *n* type, and the latter is *p* type, believed to be caused by V_N and V_{Ga}, respectively.

1.1.4 Diffusion Diffusion data in III–V compounds can be separated into two sections, one dealing with the diffusion of the main constituents of the crystal, self-diffusion, and another with the diffusion of impurities. There are two self-diffusion coefficients in a III–V compound: one corresponding to the group-III atom and another to the group-V atom. The self-diffusion coefficients are functions of the deviation of the stoichiometry of the compound, which is controlled by the partial pressure of the constituents during diffusion. Self-diffusion coefficients are generally determined using radioactive tracers. Impurity diffusion occurs under conditions of concentration gradients and is described by a chemical diffusion coefficient. In addition to the component pressures during diffusion, other factors such as ionization state of the impurity, electric fields, mechanical stresses, complex interactions of impurities with other point defects, etc., affect impurity diffusion coefficients. For this reason, data for impurity diffusion coefficients reported in the literature tend to vary over a wide range. Interdiffusion of constituent elements taking place in the presence of a concentration gradient is of importance in superlattice structures (e.g., GaAs/AlAs superlattice).

Diffusion constants are customarily described by the Arrhenius expression, $D = D_0 \exp(-Q/kT)$, where D is the diffusion constant, Q is the activation energy for diffusion, k is the Boltzmann constant, and T is the absolute temperature. In compound semiconductors, D also depends on the component pressure. Table 8 summarizes the self-diffusion data for some of the III–V compounds, and Table 9 summarizes the diffusion coefficients for some common impurities (Madelung *et al.*, 1984). The wide range of values for Q and D_0 for a given constituent of the compound as in self-diffusion coefficients or for a given impurity in a host is caused, to a large extent, by the different diffusion conditions such as component pressure, distribution of defects in the crystal existing prior to the diffusion, etc. For this reason the data in Tables 8 and 9 should be used only to obtain order-of-magnitude estimates of diffusion coefficients.

Table 8. Self-diffusion constants in III–V compounds.

Compound	Element	D_0 ($cm^2\ s^{-1}$)	Q (eV)
GaAs	Ga	1.0×10^{7}	5.6
	Ga	4×10^{-5}	2.6
	As		10.2
	As	5.5×10^{-4}	3.0
GaSb	Ga	3.2×10^{3}	3.15
	Sb	3.4×10^{4}	3.45
	Sb	8.7×10^{2}	1.13
InP	In	1×10^{5}	3.85
	P	7×10^{10}	5.65
InAs	In	6×10^{5}	4.0
	As	3.0×10^{7}	4.45
InSb	In	1.8×10^{-9}	0.28
	In	5.0×10^{-2}	1.82
	In	1.8×10^{13}	4.3
	Sb	1.4×10^{-6}	0.75
	Sb	5.0×10^{-2}	1.94
	Sb	3.1×10^{13}	4.3

1.1.5 Optical Properties Table 10 lists the optical properties—refractive index near the band-gap wavelength, its temperature dependence, the static and high-frequency dielectric constants, and the frequencies of the zone-center LO and TO phonons. The optical absorption of solids can be described either by the complex index of refraction $n + ik$ or by the complex dielectric function $\epsilon + i\epsilon_2$, where n, k, ϵ_1, and ϵ_2 are all real functions of the frequency ω, and

$$\epsilon_1 = n^2 - k^2, \quad \epsilon_2 = 2nk. \tag{4}$$

The functions $\epsilon_1(\omega)$ and $\epsilon_2(\omega)$ are connected by the Kramers–Kronig relations. The absorption coefficient α is related to k by

$$\alpha = 2\omega k/c, \tag{5}$$

where k, the imaginary part of the refractive index, is the extinction coefficient. Values of k and n over a wide range of photon energies for many III–V semiconductors have been compiled by Seraphin and Bennett (1967).

The static and high-frequency dielectric constants are related by

$$\epsilon(0) = \epsilon(\infty) + 4\pi N e_T^2/\omega_{TO}^2 M, \tag{6}$$

where M is the reduced mass, e_T is the effective charge, ω_{TO} is the TO phonon frequency, and N is the number of ion pairs. Note from Table 10 that both $\epsilon(0)$ and $\epsilon(\infty)$ show an inverse dependence on E_g, decreasing with increasing E_g.

Table 9. Chemical diffusion coefficients for common impurities in III–V compounds.

Impurity	Host	D_0 ($cm^2\ s^{-1}$)	Q (eV)	Remarks
Au	GaP	8	2.5 ± 0.3	From A face
		20	2.4 ± 0.2	From B face
	GaAs	1.0×10^{-3}–29	1–2.6	
	InP	1.3×10^{-5}	0.48	
	InAs	5.8×10^{-4}	0.65	
	InSb	7×10^{-4}	0.32	
Be	GaAs	7.3×10^{-6}	1.2	
Cd	GaAs	5.0×10^{-2}	2.43	
	GaSb	1.5×10^{-6}	0.72	
	InP	1.8	1.9	
	InAs	7.4×10^{-4}	1.15	
	InSb	$(1.0\text{–}13) \times 10^{-5}$	1.1–11.2	
Cu	GaAs	3×10^{-2}	0.53	
	InP	3.8×10^{-3}	0.69	
	InAs	3.6×10^{-3}	0.52	
	InSb	9.0×10^{-4}	1.08	
Mn	GaAs	6.5×10^{-1}	2.49	
O	GaAs	2.0×10^{-3}	1.1	
S	GaAs	$(1.6\text{–}12) \times 10^{-5}$	1.63–1.86	
	GaP	3.2×10^{3}	4.7	
	InAs	6.78	2.2	
Se	GaAs	3×10^{3}	4.16	
	InAs	12.6	2.20	
	InSb	1.6	1.87	
Si	GaAs	0.11	2.5	
Zn	GaAs			Complex profiles; extensive studies
	GaSb	4×10^{-2}	1.6	
	InP			Complex profiles; extensive studies
	InAs			Complex profiles; extensive studies

Table 10. Optical properties of III–V compounds.

Compound	Energy gap E_g (eV)	Refractive index n near E_g	$\frac{1}{n}\frac{dn}{dT}$ (K^{-1})	Dielectric constant[a] $\epsilon(0)$	Dielectric constant[a] $\epsilon(\infty)$	Wave number of Raman phonon (cm^{-1}) LO	Wave number of Raman phonon (cm^{-1}) TO
AlN	6.28			9.14[c]	4.84[c]		
AlP	2.45	3.03	3.5×10^{-5}	9.8	7.54		
AlAs	2.14	3.18	4.6×10^{-5}	10.06	8.16	404.1	360.9
AlSb	1.63	3.4	3.5×10^{-5}[b]	12.04	10.24	339.6	318.8
GaN	3.44	2.6	2.6×10^{-5}[c]	12.2	5.2		
GaP	2.27	3.45	2.5×10^{-5}[b]	11.1	9.08	403.0	367.3
GaAs	1.42	3.65	4.5×10^{-5}	12.91	10.9	291.9	268.6
GaSb	0.70	3.82	8.2×10^{-5}	15.69	14.44	233.0	224.0
InN	2.09	3.12			9.3[c]		
InP	1.34	3.41	2.7×10^{-5}	12.61	9.61	345.0	303.7
InAs	0.36	3.52	6.5×10^{-5}[b]	15.15	12.25	238.6	217.3
InSb	0.18	4.00	1.2×10^{-4}	17.7	15.68	190.8	179.8

[a]In units of ϵ_0, the permittivity of free space 8.85×10^{-14} F cm^{-1}.
[b]From Ghosh *et al.* (1986).
[c]From Madelung *et al.* (1982).

In the study of transport and optical properties of polar semiconductors, the coupling between the electron and LO phonons becomes important. The electron–LO-phonon interaction is given by the well-known Fröhlich coupling constant (De Vreese, 1972):

$$\alpha_F = \frac{e^2}{4\pi\epsilon_0}\left(\frac{m^*}{2\hbar^3\omega_{LO}}\right)^{1/2}\left(\frac{1}{\epsilon(\infty)} - \frac{1}{\epsilon(0)}\right), \tag{7}$$

where ω_{LO} is the LO phonon frequency. Equation (7) shows that α_F depends on the ionic polarizability of the crystal, which is related to $\epsilon(\infty)$ and $\epsilon(0)$.

1.1.6 Mechanical Properties The elastic compliances S_{ij} in units of 10^{-12} cm^2 $dyne^{-1}$ are listed in Table 11. For a cubic crystal, there are three independent elastic constants: S_{11}, S_{12}, and S_{44}. The elastic compliance relates strain to stress,

$$\epsilon_i = \sum_j S_{ij}\sigma_j, \tag{8}$$

where the strain ϵ_i and stress σ_j are the components of second-rank tensors and S_{ij}'s are components of a fourth-rank tensor. Equation (8) can be expressed in its covariant form

$$\sigma_i = \sum_j C_{ij}\epsilon_j, \tag{9}$$

where C_{ij}'s are the elastic stiffness constants. The compliance tensor is the reciprocal of the stiffness tensor. For a cubic crystal, C_{ij}'s are related to S_{ij}'s according to (Nye, 1957)

$$\begin{aligned} C_{11} &= (S_{11} + S_{12})/S, \\ C_{12} &= -S_{12}/S, \\ C_{44} &= 1/S_{44}, \end{aligned} \tag{10}$$

where

$$S = (S_{11} - S_{12})(S_{11} + 2S_{12}).$$

From the point of practical utility, the elastic constant of interest is the Young's modulus E. Another elastic property of interest is the Poisson's ratio ν. In cubic crystals both E and ν are anisotropic. The expressions for E and ν for an arbitrary crystallographic direction in a cubic crystal are given by Brantley (1973):

$$\begin{aligned} \frac{1}{E} &= S_{11} - 2(S_{11} - S_{12} - \tfrac{1}{2}S_{44})(l_1^2l_2^2 + l_2^2l_3^2 + l_1^2l_3^2), \\ \nu &= -\frac{S_{12} + (S_{11} - S_{12} - \tfrac{1}{2}S_{44})(l_1^2m_1^2 + l_2^2m_2^2 + l_3^2m_3^2)}{S_{11} - 2(S_{11} - S_{12} - \tfrac{1}{2}S_{44})(l_1^2l_2^2 + l_2^2l_3^2 + l_1^2l_3^2)}, \end{aligned} \tag{11}$$

where l is the longitudinal stress axis, m is an orthogonal direction to l, and the l's and m's are the direction cosines for l and m with respect to the cube axes. For the ⟨100⟩ axes

$$\begin{aligned} 1/E &= S_{11} \\ \nu &= -S_{12}/S_{11}. \end{aligned} \tag{12}$$

From Eq. (11) it follows that the factor $2(S_{11} - S_{12})/S_{44}$ is a measure of deviation from isotropy. For the biaxial plane stress conditions associated with thin films, often encountered in the case of heterostructures, longitudinal stresses and strains parallel to the film substrate are related to each other by $E/(1 - \nu)$. The expressions for this composite elastic constant are easily obtained from Eq. (11).

The bulk modulus for zinc-blende–type crystals is given by

Table 11. Elastic compliances of III–V compounds, in units of 10^{-12} cm^2 $dyne^{-1}$.

Compound	S_{11}	S_{12}	S_{44}
AlP	1.090	−0.350	1.630
AlAs	1.070	−0.320	1.840
AlSb	1.696	−0.562	2.453
GaP	0.973	−0.298	1.419
GaAs	1.176	−0.365	1.684
GaSb	1.582	−0.495	2.314
InP	1.650	−0.594	2.170
InAs	1.945	−0.685	2.525
InSb	2.443	−0.863	3.311

Table 12. Deformation-potential constants of III–V compounds.

Compound	a (eV) Direct gap	a (eV) Indirect gap	b (eV)	d (eV)
AlP				
AlAs				
AlSb	−5.9	2.2	−1.35	−4.3
GaP	−9.6		−1.65	−4.5
GaAs	−9.77		−1.70	−4.55
GaSb	−8.28		−2.0	−4.7
InP	−6.35		−2.0	−5.0
InAs	−6.0		−1.8	−3.6
InSb	−7.7		−2.0	−4.9

$$B = 1/3(S_{11} + 2S_{12}). \tag{13}$$

For ⟨100⟩ axes $B = E/3(1 - 2\nu)$. Knowing the density ρ and C_{ij}, one can obtain the sound velocity v using the relation

$$v = (C_{ij}/\rho)^{1/2}. \tag{14}$$

For a cubic crystal, using $C_{ij} = C_{11}$ in Eq. (14) gives the velocity of a longitudinal wave propagating along a cube axis, and using $C_{ij} = C_{44}$ gives the velocity of a shear wave.

1.1.6.1 Deformation Potential. Application of an external stress has a profound effect on the band structure. Hydrostatic stress causes a shift of the energy states. Uniaxial stress lowers the symmetry of the lattice and splits some degenerate states. In cubic semiconductors, the fourfold degeneracy of the valence band is lifted, resulting in, respectively, $J = 3/2$, $m_j = 3/2$ and $J = 3/2$, $m_j = 1/2$ heavy-hole and light-hole bands. The shift of E_g with hydrostatic stress and the splitting of the valence band δE_v are given by the deformation-potential constants. The values of these constants are listed in Table 12. Variation of direct energy gap with hydrostatic pressure p is described by a constant a given by

$$a = -\tfrac{1}{3}(C_{11} + 2C_{12})\frac{dE_g}{dp}. \tag{15}$$

The splitting of the valence band under [100] stress is described by the deformation-potential constant b and the splitting under [111] stress by constant d, according to the relations

$$\delta E_v = b(S_{11} - S_{12})\sigma$$
$$\delta E_v = (d/2\sqrt{3})S_{44}\sigma, \tag{16}$$

where σ is the applied uniaxial stress.

1.1.7 Heterojunctions Heterostructures have attracted much attention because of scientific and technological interest (Milnes and Feucht, 1982; Kroemer, 1985). The primary reason for the consideration of heterostructures for semiconductor device applications is the abrupt change in the energy-band structure at the heterointerface. This leads to discontinuities, or offsets, in the conduction- and valence-band edges. Such offsets act as potential steps in addition to pure electrostatic potential variation that may be present and as such can be used to control the flow and distribution of electrons and holes in device structures.

Heterojunctions can be classified by their band lineup. The types of band offsets that occur at abrupt semiconductor heterojunctions are illustrated in Fig. 4. In the strad-

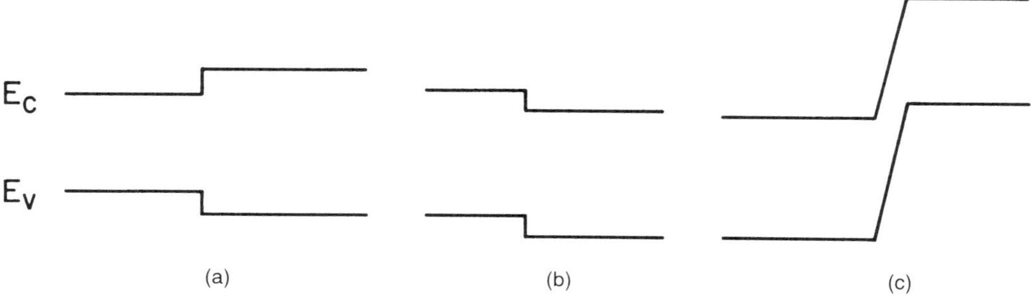

FIG. 4. Schematic band-gap diagrams showing the different heterojunction band lineups: (a) straddling lineup, (b) staggered lineup, and (c) broken-gap lineup.

dling band lineup shown in Fig. 4(a), the band offsets in both the conduction and valence bands act as potential barriers and keep electrons and holes in the smaller–band-gap semiconductors. Heterostructures that have straddling lineups are called type I heterostructures. Notable examples of these are the GaAs/$Al_xGa_{1-x}As$ and the InP/GaInAsP heterojunctions. Some heterojunctions have staggered [Fig. 4(b)] or broken-gap [Fig. 4(c)] band lineups. These are called type II heterostructures. The heterojunction InAs/GaSb is of the broken-gap type in which the top of the valence band of GaSb is higher than the bottom of conduction band of InAs. In this situation, the electrons in the top region of the valence band of GaSb can freely enter the conduction band of InAs when the two semiconductors are brought into contact with each other.

The parameter of utmost importance in heterostructures for device applications is the band-edge offset. The band-edge offsets in the conduction and valence bands are related to the difference in the band gaps of the two components of the heterojunction as

$$\Delta E_V + \Delta E_C = \Delta E_g. \quad (17)$$

Thus, if one of the offsets is calculated, the other is easily obtained since ΔE_g is reasonably well known. The calculation of ΔE_V, for example, requires knowledge of the absolute energies of the valence-band edges of the semiconductors. Further, when the two semiconductors are brought into contact with each other, the charge redistribution in the interface region and the associated electronic interface dipole effects may be significant. Given these two facts, ΔE_V may be written as

$$\Delta E_V = (E_{V0}^A - E_{V0}^B) - \delta E_V, \quad (18)$$

where E_{V0}^A and E_{V0}^B are the energies with respect to the vacuum level of the valence-band maxima of the two semiconductors A and B, respectively, and δE_V is the energy shift in the band edges at the interface caused by the formation of the interface dipole. Apart from charge redistribution at the interface, the interface dipole effect may also be affected by the presence of localized interface defects, particularly in mismatched heterostructures (Tejedor and Flores, 1978). It is also important to include in the calculations of band offset the effect of strain caused by the lattice mismatch of the constituents of the heterostructure.

As shown schematically in Fig. 5, ΔE_C is equal to the difference in electron affinities of A and B (Anderson, 1962),

$$\Delta E_C = \chi_A - \chi_B. \quad (19)$$

ΔE_V is obtained from Eq. (17), neglecting the contribution of δE_V. This is the electron affinity rule (EAR) to calculate the band offset. From Fig. 5 the difference in the ionization energies of A and B is related to the valence-band offset (Harrison, 1977):

$$\Delta E_V = \phi_B - \phi_A = E_{V0}^B - E_{V0}^A. \quad (20)$$

Note that δE_V is neglected in determining ΔE_V. Equation (20) represents the Harrison atomic-orbital (HAO) model to obtain the band offset. Table 13 lists the valence-band offsets for some of the heterostructures in III–V compounds.

1.2 Alloys

1.2.1 Energy-Band Structure of Alloys

III–V alloy semiconductors have assumed an important role in the semiconductor industry because of their utility in many electronic and photonic devices such as heterostructure transistors, light-emitting diodes, and lasers. Let us consider two III–V compounds AC and BC that form a cation alloy $A_xB_{1-x}C$ (e.g., $Ga_xIn_{1-x}As$). Similarly, compounds AC

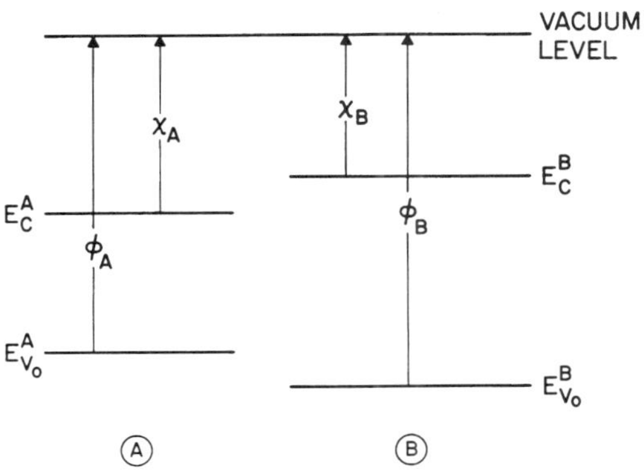

FIG. 5. Schematic band diagram showing the differences in electron affinity and ionization potential of semiconductors A and B.

Table 13. Calculated and experimental valence-band offsets (eV) for GaAs- and InP-based heterostructures[a].

Heterostructure	Experiment	EAR[b]	HAO[c]	Tersoff's model	Effective-dipole model[a]	Total-energy minimization
Ge/InP	0.64	0.90	0.64	0.58	0.67	
Ge/GaAs	0.25–0.65	0.70	0.41	0.52	0.51	
Si/InP	0.57	0.58	0.24	0.40	0.36	
Si/GaAs	0.05	0.38	0.03	0.34	0.22	
GaAs/AlAs	0.19–0.50	0.52	0.04	0.35	0.36	
GaAs/InAs	0.17	0.19	0.32	0.20	0.16	
GaAs/ZnSe	0.96–1.10	1.31	1.05	—	0.95	
InP/CdS	1.63	1.40	1.36	—	0.97	
InSb/InP	—	0.87			0.71	
InSb/GaAs	—	0.67			0.53	
InAs/InP		0.39			0.29	
GaAs/InP	—	0.20			0.13	
$Ga_{0.47}In_{0.53}As/InP$	0.346 ± 0.01[d] 0.39[e]					0.41
$Al_{0.48}In_{0.52}As/InP$	0.36 0.46[f]					0.25
$Ga_{0.47}In_{0.53}As/Al_{0.48}In_{0.52}As$	0.13, 0.20 0.21[f]					0.17

[a]Ruan and Ching (1987).
[b]Electron affinity rule model.
[c]Harrison atomic orbital model.
[d]Lang (1987).
[e]Forrest (1987).
[f]Hybertsen (1990a, 1990b).

and AD form an anion alloy AC_xD_{1-x} (e.g., $InAs_xP_{1-x}$). Mixed compounds of this type are called pseudobinary alloys. If the two compounds do not share a cation or an anion, one gets a quaternary alloy—for example, $Ga_xIn_{1-x}As_yP_{1-y}$.

It has been found experimentally that in many semiconductor alloys the direct energy gap ($\Gamma_{15}^v - \Gamma_1^c$) varies nonlinearly with composition. For an $A_xB_{1-x}C$ compound, the energy gap can be expressed in the form (Van Vechten and Bergstresser, 1970)

$$E_0(x) = a + bx + cx^2, \tag{21}$$

where c is called the bowing parameter and is four times the deviation of $E_0(x)$ from linearity at $x = 0.5$. The other two parameters, a and b, are determined by the values of E_0 observed in the pure binary compounds. The bowing parameter has been calculated using the dielectric method of band-structure calculation (Phillips and Van Vechten, 1969; Van Vechten, 1969). The total bowing parameter has been taken to be the sum of two terms: an intrinsic bowing and an extrinsic bowing. The latter is the result of the aperiodic crystal potential caused by disorder in the alloys and is essentially a short-range (within one unit cell) effect. The calculated values of these two bowing parameters and the experimental value pertaining to the minimum direct gap are given for some III–V compounds in Table 14 (Van Vechten and Bergstresser, 1970).

Figure 6 shows the lattice parameter versus energy gap at room temperature for various III–V compounds and their alloys (Ma-

Table 14. Comparison of theory and experiment for the bowing parameter pertaining to the minimum direct energy gap (Van Vechten and Bergstresser, 1970). c_i and c_e are the intrinsic (virtual crystal) and extrinsic (disorder) bowing parameters. $c_{cal} = c_i + c_e$.

Alloy	c_i (eV)	c_e (eV)	c_{cal} (eV)	c_{exp} (eV)
GaAs-P	0.21	0.09	0.30	0.21
InAs-P	0.15	0.08	0.23	0.20, 0.26
Ga-InSb	0.12	0.24	0.36	0.43
Ga-InAs	0.28	0.29	0.57	0.33, 0.56
Ga-AlAs	0.0	0.03	0.03	~0.2
Ga-InP	0.39	0.31	0.70	0.88

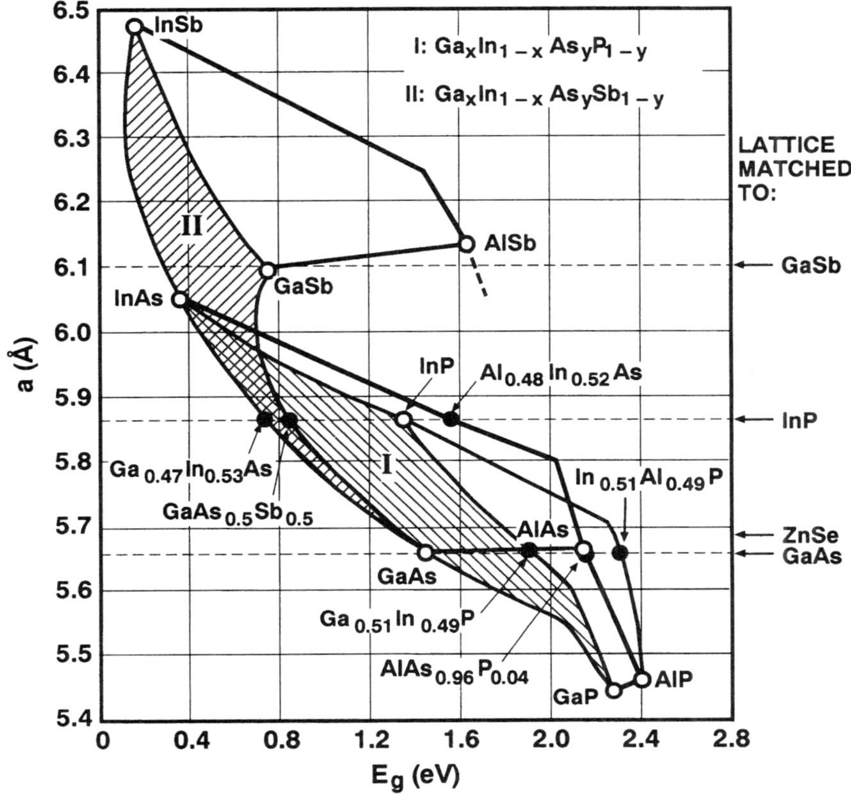

FIG. 6. Lattice parameter versus energy gap at room temperature for various III–V compounds and their alloys (from Madelung and Schulz, 1987). The lines connecting the binary end points represent the effect of the bowing parameter.

delung and Schulz, 1987). The composition of the lattice-matched ternary alloys and the corresponding substrates on which they are grown are indicated in Fig. 6. The availability of good-quality substrates dictates to a large extent which of the lattice-matched heterostructures would be of use for device structures. At present, technologically important III–V heterostructures are limited to those that are grown on GaAs and InP substrates.

The shaded region I in Fig. 6 represents the quaternary alloy $Ga_xIn_{1-x}As_yP_{1-y}$, which can be grown lattice matched on GaAs or InP substrates. When GaAs is used as the substrate, the lattice-matched GaInAsP alloy covers the energy range 1.42–1.91 eV, more or less the same energy range as that covered by the ternary alloy AlGaAs. The GaInAsP alloy grown on InP substrate spans the energy range 0.75–1.35 eV between the band gap of $Ga_{0.47}In_{0.53}As$ (0.75 eV) and that of InP (1.35 eV). The ternary alloy $Al_xIn_{1-x}As$ is lattice matched to InP at $x \approx 0.48$ and has a direct energy gap of 1.45 eV at room temperature. The $Al_xGa_yIn_{1-x-y}As$ alloy covers the spectral range 0.75–1.45 eV and is lattice matched to InP substrate for $x + y \approx 0.47$. This alloy, containing only one group-V atom, covers a wider spectral range than $Ga_xIn_{1-x}As_yP_{1-y}$ and may be advantageous for many heterostructure devices.

1.2.2 Material Parameters of Alloys

In deriving many physical parameters of the alloys, an interpolation scheme is generally adopted using the values of the related binary compounds. For a ternary compound $A_xB_{1-x}C$, the parameter P is derived from

$$P(A_xB_{1-x}C) = xP_{AC} + (1 - x)P_{BC}. \tag{22}$$

Material parameters such as lattice constant vary linearly with composition as per Eq. (22). As noted in the last section, some parameters such as the energy gap deviate from linearity, and in that case, Eq. (21) is more appropriate.

The parameter of a quaternary compound such as $A_xB_{1-x}C_yD_{1-y}$ can be obtained from the respective values of the four binaries, AC, AD, BC, and BD:

$$P(A_xB_{1-x}C_yD_{1-y}) = xyP_{AC} + x(1-y)P_{AD} + (1-x)yP_{BC} + (1-x)(1-y)P_{BD}. \tag{23}$$

If the parameters for the ternary alloy $A_xB_{1-x}C$, $A_xB_{1-x}D$, AC_yD_{1-y}, and BC_yD_{1-y} are available, then Eq. (22) can be modified to give (Glisson *et al.*, 1978)

$$P(A_xB_{1-x}C_yD_{1-y}) = \frac{x(1-x)[(1-y)P_{ABD} + yP_{ABC}] + y(1-y)[xP_{ACD} + (1-x)P_{BCD}]}{x(1-x) + y(1-y)}. \tag{24}$$

The ternary parameters P_{ABD} and so on may include a quadratic dependence given by Eq. (21). This interpolation equation reduces to the average of the four ternary parameters at $x = y = 0.5$. In general, when specific experimental data are unavailable, most physical parameters of alloy semiconductors may be derived using a linear interpolation scheme.

1.2.2.1 Lattice Constant. The lattice constant of ternary III–V alloys generally varies linearly with composition (Vegard's law). This is illustrated in Fig. 7 for the $Ga_xIn_{1-x}As$ alloy (Hocking *et al.*, 1966). Lattice-matched composition for growth on InP substrates occurs at $x \approx 0.468$.

Vegard's law can also be assumed for the quaternary alloys. For the $Ga_xIn_{1-x}As_yP_{1-y}$ alloys, Vegard's law is given by

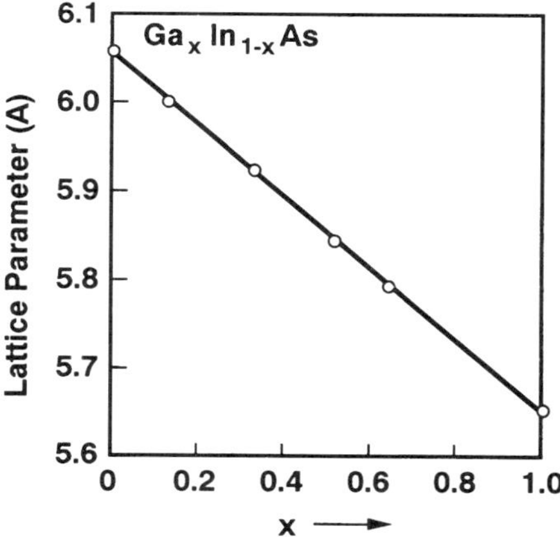

FIG. 7. Room-temperature lattice parameter versus composition for $Ga_xIn_{1-x}As$ (from Hocking *et al.*, 1966).

$$a(x,y) = 0.1896y - 0.4175x + 0.0124xy + 5.8687 = a_{InP}[1 + (\Delta a/a)_0], \tag{25}$$

where $\Delta a/a$ is the relaxed lattice mismatch relative to the substrate. The quaternary alloys grown lattice matched on InP substrates can be considered part of a pseudobinary system between $Ga_{0.47}In_{0.53}As$ and InP. The lattice-matching relation between x and y can be expressed as

$$x \cong 0.47y \qquad (0 < y < 1.0)$$

or, using Eq. (23) more rigidly, as

$$x = \frac{0.1896y}{0.4175 - 0.012y} \qquad (0 < y < 1). \tag{26}$$

The quantity $(\Delta a/a)_0$ in Eq. (25) is related to the measured $(\Delta a/a)_\perp$ for the case of thin epitaxial layers (Hornstra and Bartels, 1978):

$$\left(\frac{\Delta a}{a}\right)_\perp = \frac{1+\nu}{1-\nu}\left(\frac{\Delta a}{a}\right)_0. \tag{27}$$

For {100} crystals according to Eq. (10), Eq. (27) reduces to

$$\left(\frac{\Delta a}{a}\right)_{100} = \frac{S_{11} - S_{12}}{S_{11} + S_{12}}\left(\frac{\Delta a}{a}\right)_0, \tag{28}$$

where S_{ij} is the elastic compliance (see Sec. 1.1.6). Using the S_{ij} values in Table 11, we obtain

$$(\Delta a/a)_{100} = 2.125(\Delta a/a)_0. \tag{29}$$

1.2.2.2 Band Gap. For the alloys, a parameter that is often of interest is the variation of E_g with composition. Besides the variation of the minimum energy gap, the variation of the energy separation between the different conduction-band minima is also of interest. The composition dependence of the energy gap is often represented by Eq. (21). Table 15 gives E_g versus composition for some important ternary alloy semiconductors (Madelung *et al.*, 1982, 1984; Madelung and Schulz, 1987).

1.2.2.3 Quaternary Alloys. For determining the band gap of the quaternary alloy, Moon *et al.* (1974) have used the following equation:

$$E_g(A_xB_{1-x}C_yD_{1-y}) = xE_{ACD} + (1-x)E_{BCD} - \Delta, \quad (30)$$

where Δ is the bowing-parameter term for the quaternary alloy and is given by

$$\Delta = x(1-x)[(1-y)c_{ABD} + yc_{ABC}] + y(1-y)[xc_{ACD} + (1-x)c_{BCD}], \quad (31)$$

where c_{ABD} and so on are the ternary bowing parameters of Eq. (21). For $x = y = 0.5$, Δ is $\frac{1}{8}$ of the sum of the four ternary bowing parameters, which is twice as large as the quaternary bowing parameter given by Eq. (24). A comparison of the two interpolation schemes given by Eqs. (24) and (30) with measured values of the band gap in $Ga_xIn_{1-x}As_yP_{1-y}$ alloys showed that both schemes give comparable errors (Glisson *et al.*, 1978). However, for calculation of the band gap, Eq. (30), which is the quaternary version of Eq. (21), is preferred since it has some theoretical basis. Using the band gaps and lattice constants of the binary compounds and Eqs. (22), (24), and (30), Glisson *et al.* (1978) have derived the lowest-energy band gap and lattice constant of several quaternary alloys of the III–V semiconductors at room temperature. Their calculations are reproduced in Figs. 8 and 9. The composition dependences of energy gaps for the quaternary compounds $Ga_xIn_{1-x}As_yP_{1-y}$ and $Al_xGa_yIn_{1-x-y}As$ are also given in Table 15.

1.2.2.4 Heteroepitaxial Strain and Band Gap. One of the aims in growing device-quality heterostructures is to reduce the lattice mismatch between the epitaxial layer and the substrate as much as possible. Since the layers are usually much thinner than the substrate, the strain caused by lattice mismatch can be accommodated elastically or plastically by the generation of dislocations when the thickness of the layers exceeds a

Table 15. Compositional dependence of the energy gap in III–V ternary and quaternary alloy semiconductors at 300 K.

Alloy	Direct energy gap E_Γ	Indirect energy gap E_X	E_L
$Al_xIn_{1-x}P$	$1.34 + 2.23x$	$2.24 + 0.18x$	
$Al_xGa_{1-x}As$	$1.424 + 1.247x (x < 0.45)$ $1.424 + 1.087x + 0.438x^2$	$1.905 + 0.10x + 0.16x^2$	$1.705 + 0.695x$
$Al_xIn_{1-x}As$	$0.36 + 2.35x + 0.24x^2$	$1.8 + 0.4x$	
$Al_xGa_{1-x}Sb$	$0.73 + 1.10x + 0.47x^2$	$1.05 + 0.56x$	
$Al_xIn_{1-x}Sb$	$0.172 + 1.621x + 0.43x^2$		
$Ga_xIn_{1-x}P$	$1.34 + 0.511x + 0.6043x^2$ $(0.49 < x < 0.55$, VPE layer)		
$Ga_xIn_{1-x}As$	$0.356 + 0.7x + 0.4x^2$		
$Ga_xIn_{1-x}Sb$	$0.172 + 0.165x + 0.413x^2$		
GaP_xAs_{1-x}	$1.424 + 1.172x + 0.186x^2$		
$GaAs_xSb_{1-x}$	$0.73 - 0.5x + 1.2x^2$		
InP_xAs_{1-x}	$0.356 + 0.675x + 0.32x^2$		
$InAs_xSb_{1-x}$	$0.18 - 0.41x + 0.58x^2$		
$Ga_xIn_{1-x}As_yP_{1-y}$	$1.35 + 0.668x - 1.068y + 0.758x^2 + 0.078y^2 - 0.069xy - 0.322x^2y + 0.03xy^2$		
$Al_xGa_yIn_{1-x-y}As$	$0.36 + 2.093x + 0.629y + 0.577x^2 + 0.436y^2 + 1.013xy - 2.0xy(1 - x - y)$		

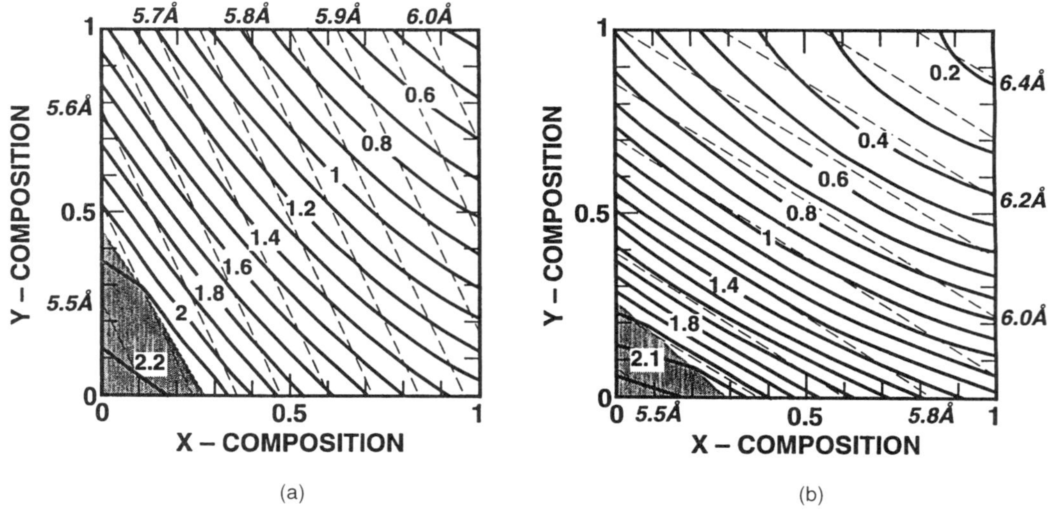

FIG. 8. Energy-band gap (solid curves) and lattice-constant contours (dashed curves) at 300 K for (a) $Ga_{1-x}In_xP_{1-y}As_y$ and (b) $Ga_{1-x}In_xP_{1-y}Sb_y$ alloys. The shaded region shows the compositional range over which the material has an indirect band gap. (From Glisson *et al.*, 1978.)

certain critical value (Asai and Oe, 1983). Both types of strain have a profound influence on device characteristics.

In the elastic regime, for coherent growth (in-plane lattice parameter of the epitaxial layer equals that of the substrate) the lattice mismatch is accommodated by a tetragonal distortion of the layer (Hornstra and Bartels, 1978). For this case, the lattice parameter of the unstrained epitaxial layer is given by Eq. (27). When $(\Delta a/a)_\perp$ is positive (the layer has a larger lattice parameter than the sub-

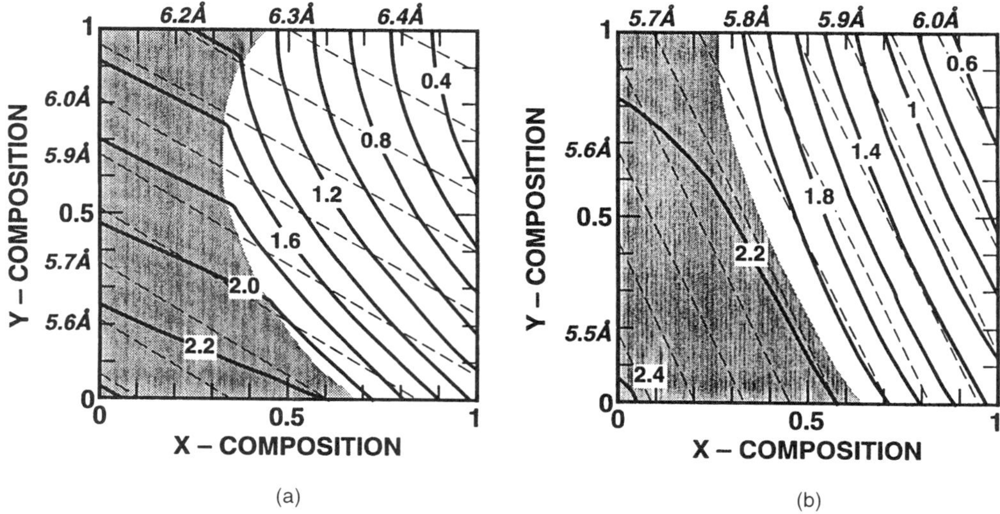

FIG. 9. Energy-band gap (solid curves) and lattice-constant contours (dashed curves) at 300 K for (a) $Al_{1-x}In_xP_{1-y}Sb_y$ and (b) $Al_{1-x}In_xP_{1-y}As_y$ alloys. The shaded region shows the compositional range over which the material has an indirect band gap. (From Glisson *et al.*, 1978.)

strate), there is in-plane compressive strain in the layer and tensile strain normal to the interface. When $(\Delta a/a)_\perp$ is negative, there is in-plane tension and compression normal to the layer. The hydrostatic component of the biaxial strain shifts the band gap while the shear component lifts the degeneracy of the valence band, as shown in Fig. 10. For biaxial compression the valence band closer to the conduction band is the heavy-hole band, whereas it is the light-hole band that is closer for biaxial tension. A similar situation is realized for externally applied uniaxial strain of tension and compression, respectively. The wafer bowing resulting from the biaxial strain is also shown in Fig. 10.

Asai and Oe (1983) derived the transition energies to the $J = 3/2$, $m_j = 3/2$ and $J = 3/2$, $m_j = 1/2$ valence bands using the stress-dependent Hamiltonian (Bir and Pikus, 1974), and they are as follows:

$$E_{(3/2,3/2)} = \left[-2a\,\frac{C_{11} - C_{12}}{C_{11}} + b\,\frac{C_{11} + 2C_{12}}{C_{11}}\right]\varepsilon, \tag{32}$$

$$E_{(3/2,1/2)} = \left[-2a\,\frac{C_{11} - C_{12}}{C_{11}} - b\,\frac{C_{11} + 2C_{12}}{C_{11}}\right]\varepsilon, \tag{33}$$

where ε is the mismatch strain, taken to be positive for compression.

From Eqs. (32) and (33), it follows that for in-plane compressive strain (positive mismatch), the lowest-energy optical transition involves the $m_j = 3/2$ valence band, and for in-plane tensile strain (negative mismatch) it involves the $m_j = 1/2$ valence band. Figure 11 shows the band gap, derived from photoluminescence measurements, of $Ga_xIn_{1-x}As$ grown on InP for lattice mismatch varying from −0.2% to +0.2% (Bassignana *et al.*, 1989). The data clearly illustrate the strain effect on band gap. There is good agreement between the experimental points and the solid lines derived from Eqs. (32) and (33) using the extrapolated values of elastic and deformation-potential constants for $Ga_{0.47}In_{0.53}As$ from the binary values.

1.2.2.5 Effective Mass in Ternary and Quaternary Alloys. According to the **k·p** approximation, the effective mass of the car-

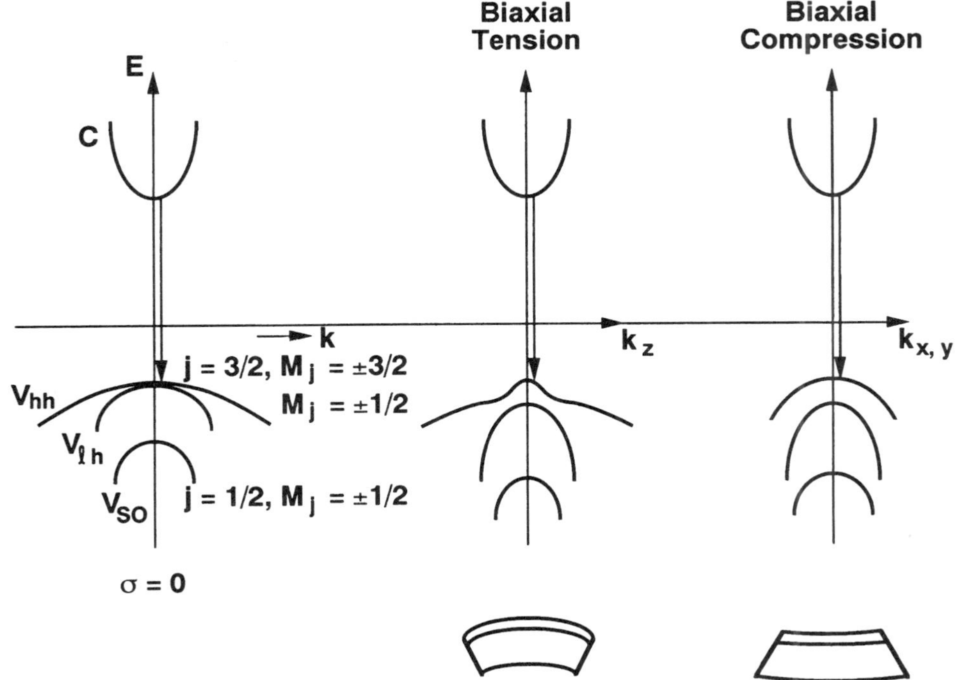

FIG. 10. Valence-band structure of cubic semiconductors under biaxial strain. The fourfold degenerate valence band is split into two twofold degenerate bands. The valence band closest to the conduction band is different depending on the nature of the strain. Wafer bowing resulting from biaxial strain is also shown.

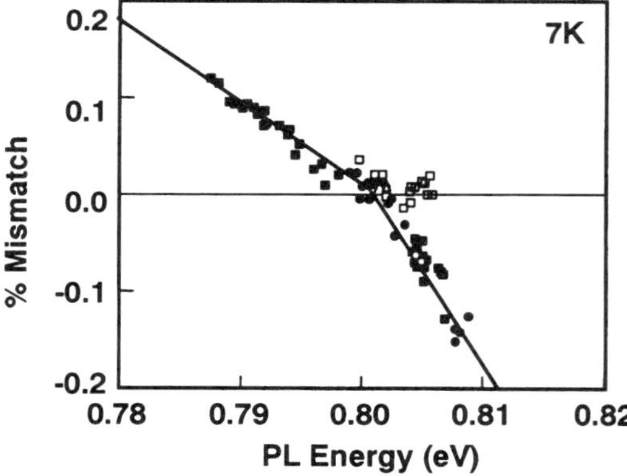

FIG. 11. Photoluminescence peak energy at 7 K of $Ga_xIn_{1-x}As/InP$ as a function of lattice mismatch. The effect of biaxial strain on peak energy is explained on the basis of Eqs. (32) and (33). (From Bassignana *et al.*, 1989.)

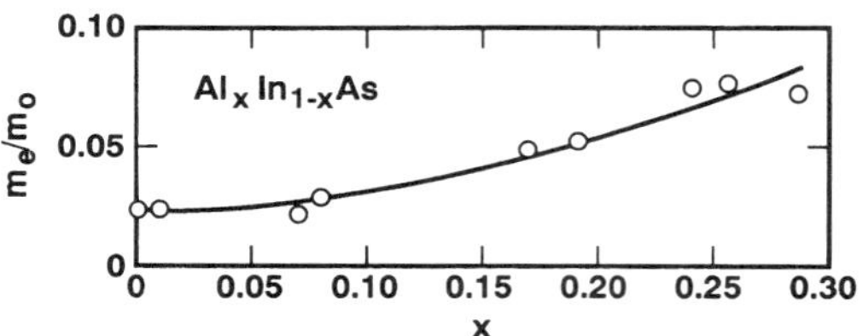

FIG. 13. Electron effective mass versus composition in $Al_xIn_{1-x}As$ (from Matyas and Karoza, 1982.)

rier is related to the square of the matrix element P connecting the conduction band and the light-hole and spin–orbit-split valence bands. It has been found in many ternary alloys that the effective mass calculated under the assumption that P^2 varied linearly between the values for the binary compounds does not agree well with the experimental values. The origin of this discrepancy has been attributed to disorder-induced mixing of the conduction and valence bands at the Γ point (Berolo *et al.*, 1973). The effect of this mixing is to reduce P^2 and thus to increase the effective mass above that which would be obtained by the $\mathbf{k \cdot p}$ calculation.

Figures 12 and 13 show the electron effective mass for $Ga_xIn_{1-x}As$ (Thomas and Woolley, 1971) and $Al_xIn_{1-x}As$ (Matyas and Karoza, 1982), respectively. Figures 14 and 15 show, respectively, the electron effective mass and hole effective masses for lattice-matched $Ga_xIn_{1-x}As_yP_{1-y}$ (Pearsall, 1982). The heavy-hole and split-off band masses can be obtained from the values of the binary compounds using Eq. (23).

1.2.2.6 Phonons in Ternary and Quaternary Alloys. In ternary or quaternary alloys, one can expect phonon frequencies due to the different binary compounds. Such multiple-mode behavior has been observed in the Raman-scattering and infrared-reflectivity measurements on alloys. Figure 16 shows the phonon frequencies in lattice-matched $Ga_xIn_{1-x}As_yP_{1-y}$ (Soni *et al.*, 1986) as functions of y. The low-frequency (220–260 cm^{-1}) phonons are assigned to GaInAs modes. At $y = 1$, they are separated into InAs-type and GaAs-type vibrations. The high-frequency (300–360 cm^{-1}) band ap-

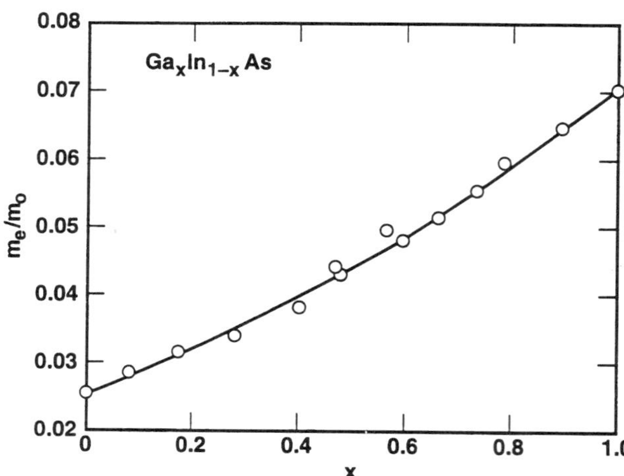

FIG. 12. Electron effective mass versus composition in $Ga_xIn_{1-x}As$ (from Thomas and Woolley, 1971.)

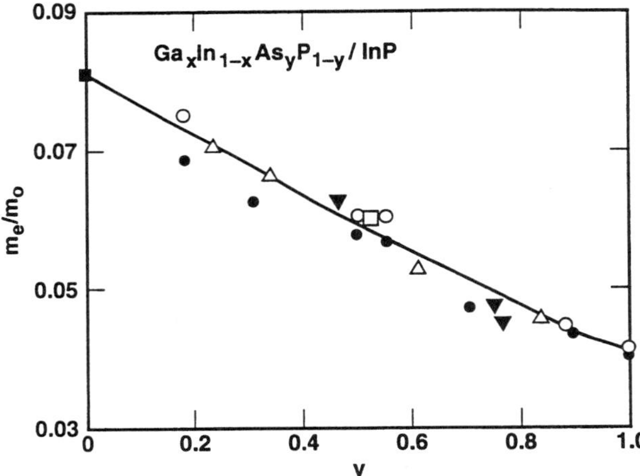

FIG. 14. Electron effective mass versus composition in $Ga_xIn_{1-x}As_yP_{1-y}$/InP (from Pearsall, 1982.)

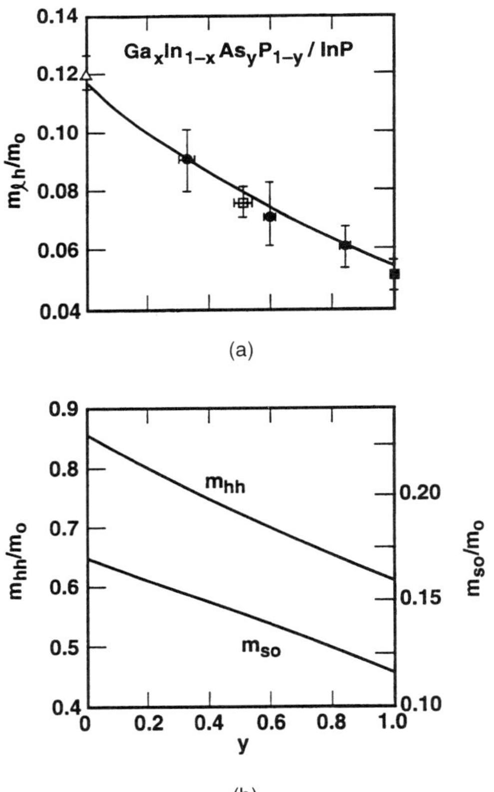

FIG. 15. (a) Light-hole effective mass and (b) heavy-hole effective mass and effective mass of the spin–orbit-split-off valence band versus composition in $Ga_xIn_{1-x}As_yP_{1-y}$ on InP substrates (from Pearsall, 1982.)

pears for all compositions, except for $y = 1$, and is attributed to InP and GaP vibrations.

1.2.2.7 Refractive Index. One of the important factors in the design of heterostructure lasers and optoelectronic devices is the refractive index n. Often, knowledge of n as a function of photon energy near and below the band gap is required. Adachi (1982) proposed a generalized model of dielectric constants of semiconductors based on models of the intraband transitions. According to this model, the real part of the dielectric constant ϵ is expressed as a function of photon energy E as

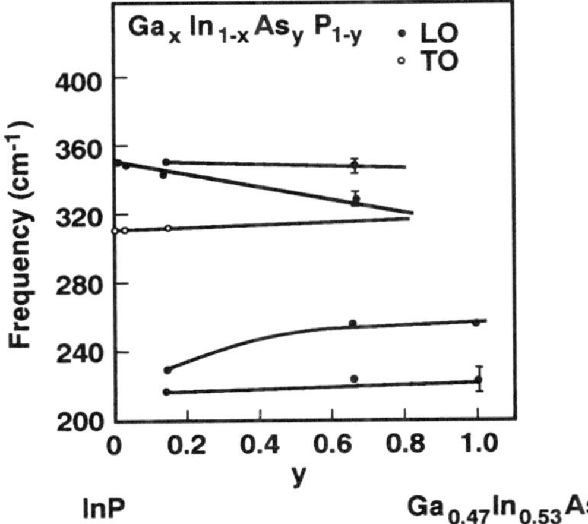

FIG. 16. Measured phonon frequencies of various Raman modes in $Ga_xIn_{1-x}As_yP_{1-y}$ as a function of y. The solid lines emphasize the experimental trends. (From Soni *et al.*, 1986.)

$$\epsilon(E) = A_0\left\{f(x) + \frac{1}{2}\left[\frac{E_g}{E_g + \Delta_0}\right]^{3/2} f(x_0)\right\} + B_0, \quad (34)$$

where A_0 and B_0 are constants obtained by fitting experimental data to Eq. (32), and

$$f(x) = x^{-2}[2 - (1 + x)^{1/2} - (1 - x)^{1/2}],$$
$$x = E/E_g,$$
$$x_0 = E/(E_g + \Delta_0). \quad (35)$$

Figure 17 shows the experimental values of $\epsilon(\approx n^2)$ for $Ga_xIn_{1-x}As_yP_{1-y}$ as a function of wavelength (Chandra *et al.*, 1981) for several compositions. The dispersion curves calculated based on Eqs. (34) and (35) are also shown in Fig. 17. The agreement with experimental curves is quite good over most of the wavelength range.

1.2.2.8 Thermal Resistivity. Thermal resistivity of semiconductors is an important parameter in the design of power-dissipating devices such as semiconductor lasers. It is also important in calculating the figure of merit of thermoelectric devices (e.g., Peltier devices). The thermal resistivity of ternary and quaternary alloys follows a quadratic relation the same as E_g. For the ternary alloy $A_xB_{1-x}C$ it can be written as

$$W(x) = xW_{AC} + (1 - x)W_{BC} + c_{A-B}x(1 - x), \quad (36)$$

where W_{AC} and W_{BC} are the thermal resistivities of the binary compounds. Values of the bowing parameter c_{A-B} for several ternary alloys have been obtained by Adachi (1983) and are given in Table 16. Note that c_{As-P} values of InAsP and GaAsP alloys are considerably smaller than c_{In-Ga} of InGaAs and that $c_{Ga-Al} \approx c_{As-P}$. The mean atomic weight of In-Ga is about two times than that of As-P, which is nearly the same as that of Al-Ga. The anharmonic contribution to thermal resistivity resulting from mass difference between the atoms is supposed to be responsible for the differences and similarities in the c values.

The thermal resistivity of a quaternary alloy $A_xB_{1-x}C_yD_{1-y}$ can be expressed in the same way as Eq. (35) but with two bowing parameters c_{A-B} and c_{C-D}. The thermal resistivity of $Ga_xIn_{1-x}As_yP_{1-y}$ lattice-matched to InP calculated using $c_{In-Ga} = 72$ W^{-1} deg cm and $c_{As-P} \simeq 25$ W^{-1} deg cm is shown in Fig. 18 (Adachi, 1983). The thermal resistivity increases with increasing As content and reaches a maximum value of 24 W^{-1} deg cm at $y \sim 0.75$.

1.2.2.9 Thermal Expansion Coefficient. In double heterostructures consisting of compositionally different layers, differences in the thermal expansion coefficient of the layers can generate elastic stresses during cooling from the growth temperature to room temperature or during thermal cycling of heterostructure devices. In Fig. 19

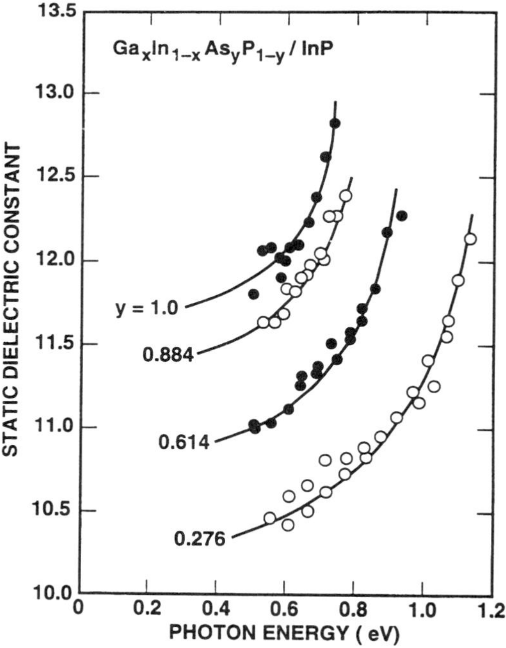

FIG. 17. Dielectric constant versus photon energy for $Ga_xIn_{1-x}As_yP_{1-y}$/InP at various compositions (from Chandra *et al.*, 1981.)

Table 16. The alloy disorder bowing parameter c_{A-B} denoting the deviation from linearity of the thermal resistivity of ternary alloys. The bowing parameter is obtained by fitting experimental data by Eq. (36), From Adachi (1982).

Ternary alloy	c_{A-B} (W^{-1} deg cm)
InGaAs	72
InAsP	30
GaAsP	20
AlGaAs	30
InGaP	72 (est)
InGaSb	72 (est)

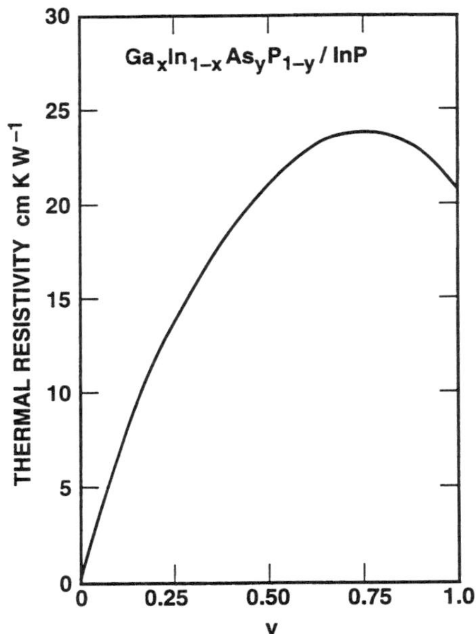

FIG. 18. Thermal resistivity versus composition calculated for $Ga_xIn_{1-x}As_yP_{1-y}$/InP (from Adachi, 1983.)

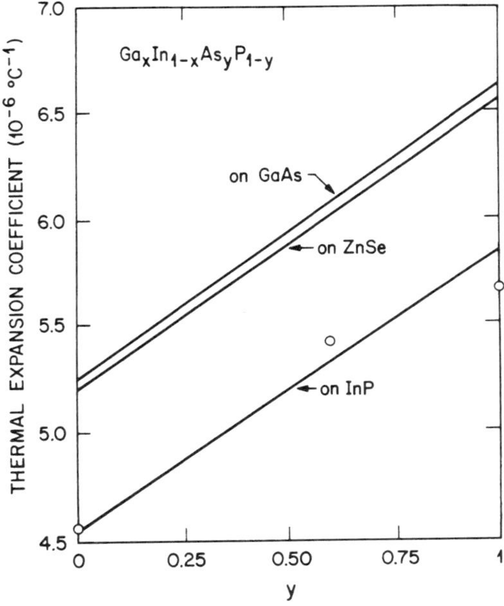

FIG. 19. Thermal expansion coefficient of $Ga_xIn_{1-x}As_yP_{1-y}$ lattice matched to InP, ZnSe, and GaAs as a function of composition. The curves are calculated and the points are experimental. (From Bisaro *et al.*, 1979.)

the thermal expansion coefficient of $Ga_xIn_{1-x}As_yP_{1-y}$ alloy lattice-matched to InP obtained from linear interpolation is shown together with experimental data at selected y values (Bisaro *et al.*, 1979). Pietsch and Marlow (1986) found that the thermal expansion coefficient of the relaxed lattice is about 30% smaller than the values shown in Fig. 19.

2. II–IV COMPOUNDS

2.1 Binary Compounds

2.1.1 Bonding and Crystal Structure II–VI compounds cover the entire range from fully ionic (NaCl structure, in CdTe at high pressure) to fully covalent (cinnabar, α-HgS) bonding. In all the tables of data it will be noted that several gaps occur; this is particularly so for HgS, which has received relatively little study, and ZnO. Table 17 gives the fractional ionic character (Phillips) and the cohesive energy. Table 18 summarizes lattice constants, densities, melting points, thermal expansion coefficients, and thermal conductivities at ~300 K for the binary compounds.

Most mercury chalgocenides crystallize in the cubic zinc-blende (sphalerite) structure with tetrahedral coordination, although the stable phase of HgS, α-HgS, has the hexagonal cinnabar structure. Some of the mercury chalgocenide alloys with other II–VI compounds crystallize in the hexagonal (wurtz-

Table 17. Fractional ionic character and cohesive energy of II–VI binaries.

Compound	Phillips ionicity	Cohesive energy[a] (kcal mole^{-1})
HgS	0.79	—
HgSe	0.68	—
HgTe	0.65	—
CdS	0.69	−120
CdSe	0.7	—
CdTe	0.72	−80
ZnS	0.63	−130
ZnSe	0.68	−110
ZnTe	0.55	−95
ZnO	0.62	−140

[a]Gibbs free energy of sublimation into neutral atoms at STP.

Table 18. Basic properties of the binary II–VI compounds. Values in parentheses are for hexagonal structures.

Compound	Lattice constant (a, c = hexagonal) (Å)	Density (g cm^{-3})	Melting point T_m (K)	Thermal expansion coefficient (10^{-6} °C^{-1})	Thermal conductivity (W cm^{-1} K^{-1})
HgS	5.85	7.7	1723	7.7	—
HgSe	6.086	8.2	1073	8.3	0.016
HgTe	6.462	8.05	943	5.2	0.031
CdS	5.82 (4.13 a, 6.74 c)	(4.82)	1678–1748	(4.2)	(0.2)
CdSe	6.082 (4.3 a, 7.01 c)	5.7 (5.81)	1523	(4.8)	(0.043)
CdTe	6.481 (4.57 a, 7.47 c)	5.67 (5.9)	1365–1371	5.1 (5.9)	0.01
ZnS	5.41 (3.82 a, 6.26 c)	(4.1)	1991–2123	(6.2)	0.026
ZnSe	5.67 (4.00 a, 6.54 c)	5.42 (5.26)	1788–1798	6.8 (6.2–6.5)	0.19
ZnTe	6.13 (4.27 a, 6.99 c)	5.72 (5.7)	1563	(8.2)	0.18
ZnO	(3.25 a, 5.21 c)	(5.28)	—	(7.2)	—

ite) structure when immiscibility is found, but complete miscibility exists if both binaries are of the sphalerite structure. The tetrahedrally coordinated bonds between the metal atoms and the four neighboring chalcogen atoms are formed by sp^3 hybrids, and the bonding is partly covalent and partly ionic. The band structure of the mercury chalcogenides is of the symmetry-induced zero-gap type. The Γ_6 level, which is usually in the conduction band, is below the Γ_8 level, usually in the valence band, giving negative fundamental band gaps; i.e., the compounds are semimetals.

Cadmium-based and zinc-based compounds tend to crystallize in the cubic zincblende structure, at atmospheric pressure, but CdSe, CdS, and ZnS can form in the hexagonal wurtzite structure. At high pressures, CdTe transforms into the NaCl structure and then, at higher pressures still, into the white tin structure. CdTe is the most ionic of the Cd and Zn chalcogenides (see Table 17) with ZnTe the least ionic.

The IIa–VIa binary compounds have the NaCl crystal structure, and these refractory compounds are prepared in a high-temperature floating melt-zone Xe-arc furnace.

2.1.2 Energy-Band Structure The Brillouin zone of cubic crystals was described in Sec. 1.1.1.2. The lattice of wurtzite crystals consists of two interpenetrating hexagonal close-packed lattices, one containing anions and the other cations. The translation vectors are

$$(a/2)(1,-\sqrt{3},0),\ (a/2)(1,\sqrt{3},0),\ \text{and}\ c(0,0,1),$$

where the vectors are given with respect to Cartesian axes. The reciprocal lattice is hexagonal having the translation vectors

$$(2\pi/a)(1,-3^{-1/2},0),(2\pi/a)(1,3^{-1/2},0),(2\pi/c)(0,0,1).$$

The Brillouin zone is shown in Fig. 20. The calculated energy-band structures for the cubic ZnSe and CdTe are given in Figs. 21(a) and 21(b), while the hexagonal ZnS diagram is given in Fig. 21(c) and that of the semimetal HgTe is given in Fig. 21(d). The critical-point energies are shown in Figs. 21(a)–21(d).

The spin–orbit splittings for the II–VI binaries are given in Table 19. Table 19 also shows the energy gaps of the binaries and their temperature and pressure dependences, where known.

Figure 22 shows a plot of the energy gap against lattice parameter for selected II–VI compounds. The lines are simply tie lines,

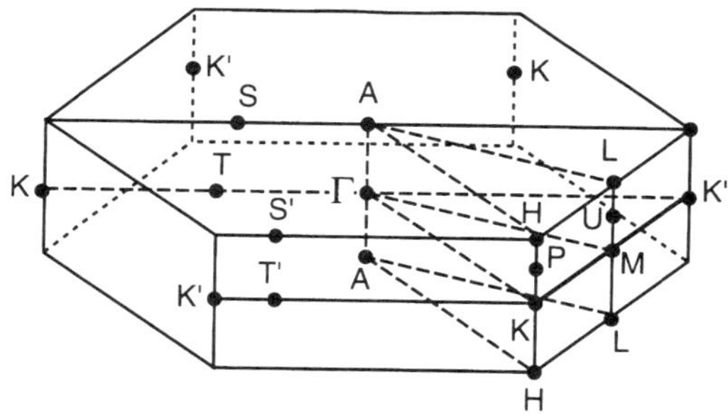

FIG. 20. Brillouin zone for the wurtzite lattice with lines and points of special symmetry (from Aven and Prenner, 1967.)

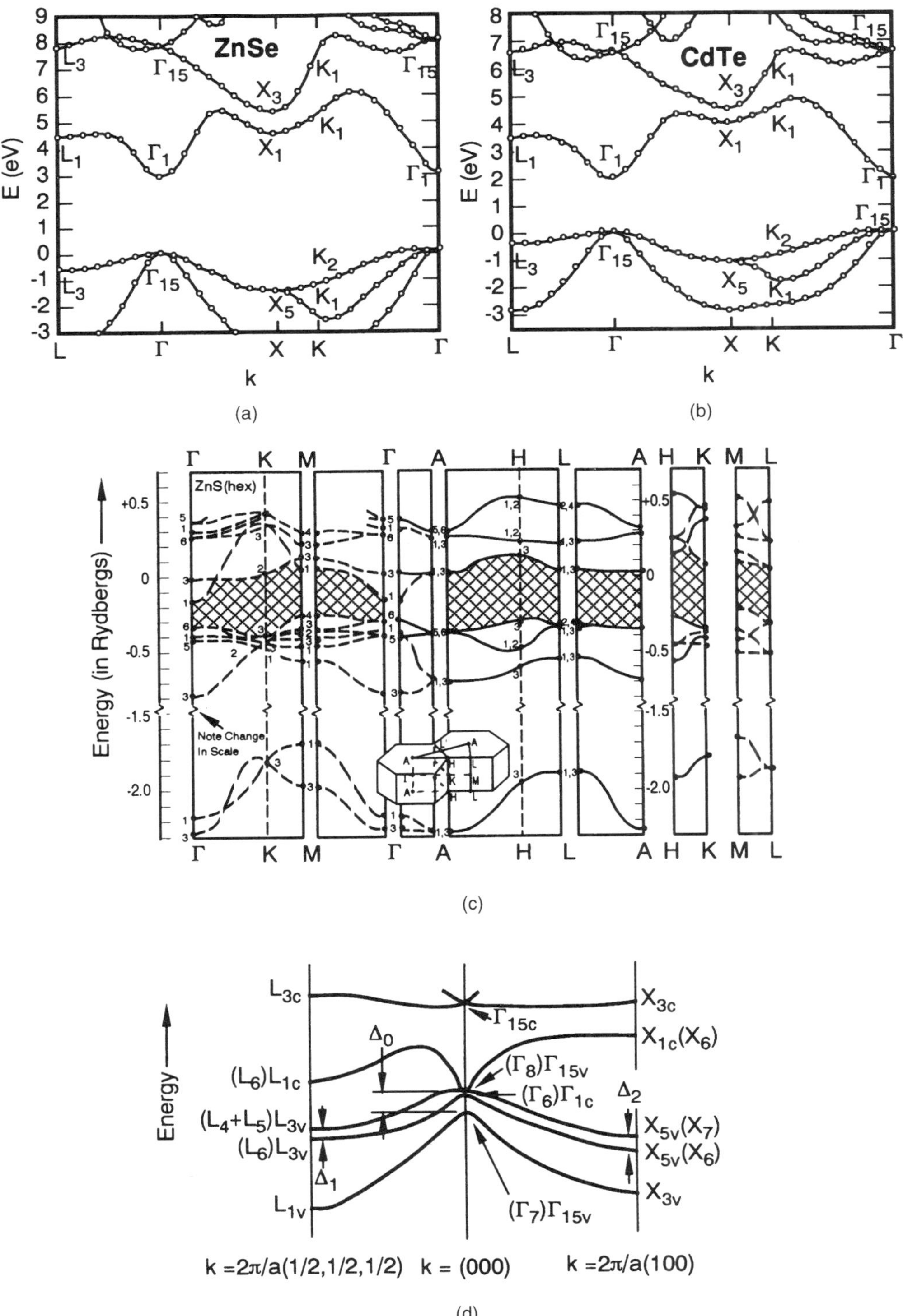

FIG. 21. Energy bands for (a) ZnSe, (b) CdTe, (c) ZnS, and (d) HgTe (from Aven and Prenner 1967.)

Table 19. Energy-band structure parameters for binary II–VI compounds. Values in parentheses are for hexagonal structures.

Compound	Electron affinity	Spin–orbit splitting (e = exp) (eV)	Energy gap (eV)	$\frac{dE_g}{dP}$ (10^{-6} eV kbar^{-1})	$\frac{dE_g}{dT}$ (10^{-4} eV K^{-1})
HgS	—	—	−0.2	—	—
HgSe	—	0.35	−0.22	—	−7
HgTe	—	1.08 (e), 0.78	−0.3	7–14	—
CdS	(4.79)	0.065	2.42, (2.53)	—	−5.2
CdSe	(4.95)	0.42	1.67, (1.9)	—	−9
CdTe	4.28	0.9 (e), 0.86	1.5	8	−3.6
ZnS	3.9	0.27	3.6, (3.78)	—	−5 to −10
ZnSe	4.09	0.43	2.67	—	−7.2
ZnTe	3.53	0.91 (e), 0.89	2.2	—	−5
ZnO	(4.57)	—	(3.2)	—	−10

but, as Vegard's law applies in all the systems, the divisions indicate the relevant proportions of the binaries required to produce a given energy gap. Table 20 gives the effective masses of the binary compounds.

Table 21 gives the energy gaps (at 2 K) of the IIa–VIa binaries, and Fig. 23 shows the theoretical band structure for the sulfide binaries; see Kaneko and Koda (1988). Most of these compounds are predicted to be indirect-gap semiconductors, except for BaO.

2.1.3 Impurities and Defects Point defects are very important in II–VI compounds while line defects, such as dislocations, are probably not significant in affecting transport properties. The most important point defects are vacancies and interstitials of the constituents, either of which can be monovalent or divalent. In CdTe and CdS, interstitials are present in much lower concentrations than residual impurities but could still be responsible for deep trap levels. There is

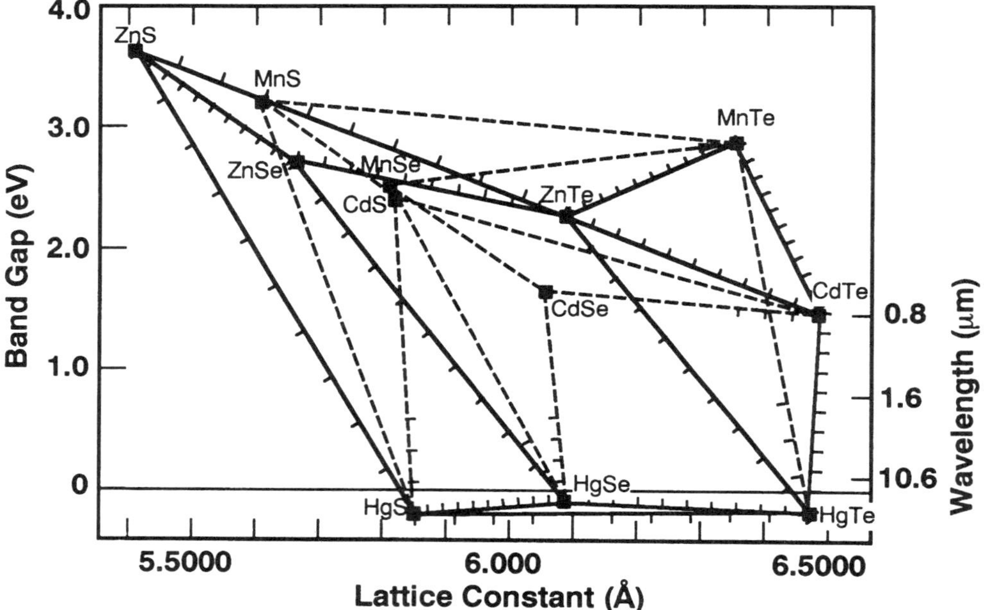

FIG. 22. Energy gap versus lattice parameter for selected II–VI binaries.

Table 20. Effective electron and hole masses of binary II–VI compounds. Values in parentheses are for hexagonal structures.

Compound	m_e/m_0	m_h/m_0
HgSe	0.05	0.02–0.08
HgTe	0.03	0.6
CdS	(0.2)	(0.8)
CdSe	(0.13)	(0.45)
CdTe	(0.14) 0.09	(0.37)
ZnS	0.31	0.5–1.0
ZnSe	(0.17)	(0.6)
ZnTe	(0.16) 0.12	(0.1–0.3)

Table 21. Band-structure parameters (in eV) of binary IIa–VIa compounds.

Compound	Indirect gap X_c–Γ_v (eV)	Direct gap $E_g(X)$ (eV)	Direct gap $E_g(\Gamma)$ (eV)
CaO	—	6.875	—
CaS	4.434	5.343	5.8
CaSe	3.85	4.898	—
SrO	—	5.793	6.08
SrS	4.32	4.831	5.387
SrSe	3.813	4.475	4.57
BaO	—	3.985	8.3
BaS	3.806	3.941	5.229
BaSe	3.421	3.658	4.556

no evidence of S, Se, or Te interstitials, and isolated vacancies of S, Se, and Te appear to be inactive electrically. Vacancies of Zn and Cd have been the most widely studied point defects, either isolated or complexed with impurities. The Zn-vacancy–shallow-donor complex (*A* center) is the major compensating defect in *n*-type ZnSe and ZnS. Similar complexes (Cd-vacancy–shallow donor) occur in CdS and CdSe heat-treated in S and Se atmospheres, respectively. Vacancies of Zn are double acceptors in ZnTe. In general, the

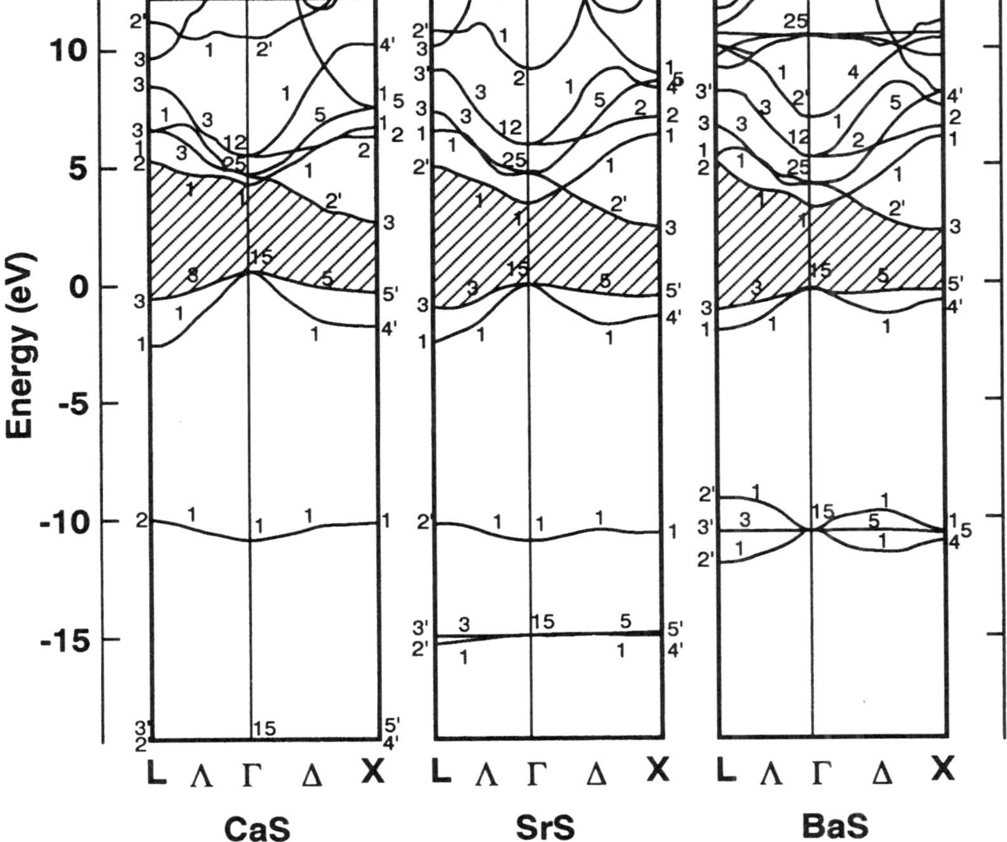

FIG. 23. Energy-band structure for IIa–VIa binaries (from Kaneko and Koda, 1988).

concentration of isolated interstitials and vacancies is quite low, except for cation vacancies in the tellurides.

In early work on wide-gap II–VI compounds, it was found impossible to produce *p*-type ZnSe, CdS, CdSe, or ZnS (even after annealing in the chalcogen vapor), while ZnTe could only be prepared *n* type. This was due to the low solubility of impurities and to compensation problems (probably together with deep trap levels). In general, the higher the ionicity of the compound, the more easily *n*-type material forms and the more difficult it is to obtain *p*-type material. With the advent of the more advanced nonequilibrium epitaxial techniques of metal-organic vapor-phase epitaxy (MOVPE), photo-assisted (PA) MOVPE, molecular beam epitaxy (MBE), PAMBE, atomic-layer epitaxy (ALE), and plasma-assisted epitaxy (PAE), *p*-type doping in ZnSe has been achieved using Li, N, and O, despite the latter being isoelectronic with Se. *n*-type doping has also been accomplished using Al, Cl, Br, and iodine. ZnSe has received more attention than the other wide-gap compounds because of the drive to produce blue lasers.

On account of the lack of good-quality bulk II–VI substrates, most growth is on III–V compound substrates such as GaAs (which is mismatched by ~0.27% with ZnSe and matched to ZnS_xSe_{1-x} with $x = 0.05$), InSb, or InP, or on Si or Ge. The orientation of the substrate can affect the phase of CdS and CdSe epitaxial layers grown. A major step has been the recent production of purer Se from which to grow purer, undoped, high-electrical-resistivity, uncompensated ZnSe.

Low-temperature photoluminescence is the main tool used in the study of impurities and defects, and there is a plethora of complex structures involving two or more defects and extrinsic impurities. Transmission electron microscopy (TEM) studies have revealed dislocation networks at substrate/layer interfaces and the presence of III_2VI_3 interfacial compounds. Stacking faults and microtwins are also often observed [particularly on the (111) orientation]. GaAs epitaxial buffer-layer growth prior to the deposition of the ZnSe can reduce these markedly.

Now that controlled bipolar doping has been achieved in ZnSe, the emphasis is switching to studies of the deep-level centers, which may be limiting the radiative recombination efficiency, and those structural defects that also limit device performance. Clearly reproducibility of both growth and device technologies are being addressed currently. Strained-layer superlattices (SLSs) of ZnSe/ZnSSe and ZnSe/ZnMnSe are thought to show promise for laser structures.

CdTe has received a great deal of attention, and defects can render the material *n* or *p* type, depending on the heat treatment used. Such treatments do, however, affect the amount of second-phase precipitation, the dislocation density, and the strain in the material. Extrinsic doping has met with the same problems of low solubility and compensation as in the wide-gap compounds. Once again, use of nonequilibrium low-temperature growth, such as PAMBE, has led to *n*-type doping using In and iodine and to *p*-type doping with As and Sb. Additions of Zn and/or Se to CdTe hardens the lattice, reduces the dislocation density, and decreases the x-ray rocking-curve widths. Such materials are extensively used as substrates in the epitaxial growth of HgCdTe layers. Defects in epitaxial CdTe grown on GaAs substrates have been delineated with TEM and cathodoluminescence, where misfit dislocations are seen to be numerous. Various buffer layers can be used to decrease these dislocations. Deep-level centers have been studied with deep-level transient spectroscopy (DLTS), electron paramagnetic resonance (EPR), and optically detected magnetic resonance (ODMR), while native metal vacancies have recently come under investigation using positron annihilation techniques.

There is very little information on defects in the narrow-gap Hg compounds, particularly for HgS. HgSe can only be prepared *n* type, as the usual acceptor species from groups I and V are inactive. HgTe can be prepared both *n* and *p* type. HgTe has a very high electron mobility at low temperatures as a result of its low effective mass and high purity. Doping of HgTe with acceptors from group I and donors from groups III and IV has been accomplished. "Doping" of HgSe with Fe produces drastic changes in both the electrical and magnetic properties of the binary. Similar addition of Mn to HgTe produces interesting magnetic properties. These diluted magnetic semiconductor effects were originally studied in CdMnTe but now include Co^{2+}, Fe^{2+}, and Eu^{2+} as "dopants."

The ability to produce zinc-blende structure binaries of MnSe, MnTe, and FeSe by MBE led to the increased interest in these compounds. See SEMICONDUCTORS, DILUTED MAGNETIC.

2.1.4 Diffusion Diffusion data are important in both material and device technologies, particularly in the areas of control of stoichiometry, impurity doping, and compositional interdiffusion. Most data to date have been gathered in bulk samples, as epitaxial material needs much lower temperatures and the material is available in much smaller volumes, making quantitative diffusion studies more difficult. There is a need to monitor or control temperature, nonstoichiometry, and doping/impurity levels during diffusion work. Diffusion occurs in conditions of either chemical equilibrium, with no gradients in chemical potential, or chemical disequilibrium, where diffusion occurs down gradients in chemical potential. Atoms can migrate via nondefect (ring or direct exchange processes) or defect (vacancy, interstitial, or complexes) mechanisms. Diffusion coefficients are related to concentration profiles and are measured using radio-tracer, Rutherford back-scattering (RBS), resistivity or capacitance-voltage (C–V) profiling, and *p*/*n*-junction depth measurements. Diffusion measurements in multiquantum-well (MQW) and SLS structures are becoming more important, but problems are severe because of the small dimensions.

The mechanism of chalcogen self-diffusion is via neutral self-interstitials, under chalcogen-rich conditions, and via chalcogen vacancies at high metal partial pressures. In general, diffusion coefficients $D(X)$ are $\sim 10^{-8}$ cm^2 s^{-1} at 1 atm but fall rapidly with temperature; e.g., for CdTe at $T_m/2$, $D(X) \sim 10^{-12}$ cm^2 s^{-1}. Metal self-diffusion is more complex as a result of nonstoichiometry and doping effects. In ZnSe, ZnTe, and CdTe, $D(M) \sim 10^{-8}$ cm^2 s^{-1} at T_m but falls to $\sim 10^{-16}$ cm^2 s^{-1} at $T_m/2$ and is believed to be due to either neutral associates $[(M_iV_m)$ or $(V_mV_x)]$ or metal interstitials (M_i). Doping has been seen to increase metal self-diffusion in CdTe. In ZnS, CdS, and CdSe diffusion varies with partial pressure in the wurtzite structure but not in the zinc-blende structures. Despite the wurtzite structure of CdS no anisotropy of diffusion is observed. Data in Hg-containing compounds is scarce, but metal diffusion is much faster in HgTe than in CdTe, at a given temperature. Short-circuit path diffusion has been suggested to explain this difference. Chalcogen diffusion rates in HgTe were similar to those in CdTe. Information on chemical self-diffusion is scarce but diffusion coefficients $D(\Delta)$ are in the range 10^{-4}–10^{-6} cm^2 s^{-1} at T_m and $\sim 10^{-12}$ cm^2 s^{-1} at $T_m/3$. In CdTe, $D(\Delta)$ is independent of Cd partial pressure, decreases in doped material, is purity related, but is independent of dislocation density. It is thought to be due to metal interstitials $(Cd_i)^{2-}$ and is important in controlling stoichiometry at the annealing temperatures used and during cooldown from the growth temperatures.

Impurity-diffusion profiles in II–VI compounds are very often complex and do not follow simple erfc or Gaussian distributions. They often have two sections, arbitrarily designated as fast and slow, with the former possibly linked to short-circuit paths—e.g., dislocations or grain boundaries. Elements from groups I, III, IV, V, and VII have all been studied, mainly because of their doping behavior. A vast range of diffusion coefficients D_i exists at a given temperature together with a variety of mechanisms. For Au in CdTe, $D_i \sim 10^{-6}$ cm^2 s^{-1} at T_m, while for Cu in CdTe a value of 10^{-13} cm^2 s^{-1} at $T_m/4$ has been determined. For group-III elements in ZnTe, CdTe, and ZnSe, values of 10^{-6} cm^2 s^{-1} at T_m are typical, decreasing to 10^{-11}–10^{-13} cm^2 s^{-1} at $T_m/2$. Data for other groups are scarce, but it should be noted that diffusion of group-I elements and iodine in CdTe has been reported to be significant even at room temperature. Few data exist in HgTe, or other Hg-containing compounds, but where they do, diffusion coefficients appear to be higher than equivalent ones in CdTe.

Interdiffusion $[\bar{D}(x)]$ in ternary compounds is a function of composition (x), nonstoichiometry, doping, and temperature. Typically $\bar{D}(x)$ values lie in the range 10^{-10}–10^{-13} cm^2 s^{-1} at 600–1100 °C. In common-metal ternaries $\bar{D}(x)$ should be independent of doping, whereas donor and possibly acceptor doping enhances $\bar{D}(x)$ in common-chalcogen systems. The latter could have important consequences for stability in doped MQW and SLS structures, for which there is a dearth of information. Impurity diffusion

Table 22. Optical properties of binary II–VI compounds.

Compound	Refractive index (wavelength in μm)	ε_0	ε_∞	LO phonon (cm^{-1})	TO phonon (cm^{-1})
HgS	n_e = 3.2–2.8 (0.6–11) n_0 = 2.9–2.6 (0.6–11)	—	11.4	—	—
HgSe	—	23	12.2	180	130
HgTe	—	20.8	15.1	148	148
CdS	2.7	9.1 ($E \parallel c$) 8.58 ($E \perp c$)	5.26 ($E \perp c$) 5.31 ($E \parallel c$)	304	240
CdSe	—	8.41 9.38 ($E \parallel c$)	5.98 ($E \perp c$) 6.07 ($E \parallel c$)	210	170
CdTe	2.65	10.36	7.18	170	140
ZnS	2.4–2.3 (0.5–1)	8.32	5.17	352	320
ZnSe	2.43–2.39 (3–12)	8.53	5.7	252	200
ZnTe	2.63 (10.6)	9.1	6.75	200	177
ZnO	—	~8	—	—	—

coefficients in MQW structures are typically $\sim 10^{-18}$ cm^2 s^{-1} at 300 °C.

2.1.5 Optical Properties Refractive indices, static and high-frequency dielectric constants, and the frequencies of zone-center LO and TO phonons of the binaries are given in Table 22. See Sec. 1.1.5 for a full description of the pertinent equations. Both the static and high-frequency dielectric constants decrease with increasing energy gap, as seen in the III–V compounds.

Optical bistability has been observed in CdTe, CdS, ZnSe, and ZnS. Optical nonlinearities for applications in optical-bistability and all-optical circuits have been studied in wide- and narrow-gap II–VI compounds (Miller, 1987, 1992). The polarization P of a material varies in a nonlinear manner with electric field E, as

$$P = \varepsilon_0(\chi^{(1)}E + \chi^{(2)}E^2 + \chi^{(3)}E^3 + \text{etc.}), \quad (37)$$

where ε_0 = permittivity of free space and χ's are susceptibilities. $\chi^{(2)}$ is responsible for second-harmonic generation, sum- and difference-frequency generation, paramagnetic oscillation, optical rectification, and the linear electrolyte effect (Pockels). $\chi^{(3)}$ gives rise to third-harmonic generation, the quadratic electrooptical effect (Kerr), stimulated Brillouin and Rayleigh scattering, stimulated Raman scattering, four-photon mixing, and optical phase conjugation. Measured values of nonlinear optical coefficients range from a nonresonant value of $\sim 5 \times 10^{-12}$ e.s.u. in CdS to >1 e.s.u. in CdHgTe under resonant conditions.

2.1.6 Mechanical Properties Bulk modulus (B), elastic stiffness constants (C_{11},C_{12},C_{44}), and elastic compliances (S_{11},S_{12}) for the binary compounds are given in Table 23. Section 1.1.6 gives the relevant basic equations and the relationships be-

Table 23. Mechanical properties of binary II–VI compounds. Values in parentheses are for hexagonal structures.

Compound	B^a (10^{11} Pa)	C_{11} (10^{11} Pa)	C_{12} (10^{11} Pa)	S_{11} (10^{-11} Pa^{-1})	S_{12} (10^{-12} Pa^{-1})	C_{44} (10^{11} Pa)
HgTe	4.76	0.585	0.408	—	—	0.222
CdS	6.64 (6.61)	0.858	0.5334	—	—	1.49
CdSe	5.57 (5.32)	0.749	0.461	—	—	0.132
CdTe	4.45	0.535	0.367	4.25	−1.7	0.2
ZnS	8.02 (8.21)	1.07	0.67	1.89	−0.72	0.46
ZnSe	6.67	0.826	0.498	2.26	−0.85	0.44
ZnTe	5.09	0.722	0.409	—	—	0.308
ZnO	(13.2)	—	—	—	—	—

[a]For cubic crystals, bulk modulus $B = 1/(3S_{11} + 2S_{12})$.

tween these parameters and Young's modulus E and Poisson's ratio ν ($\sim -S_{12}/S_{11}$). The elastic stiffness constants, together with lattice parameter and hardness values for the binaries, are used to calculate the microhardness of alloys. The formation of alloys is often accompanied by an increase in hardness, e.g., Zn into HgTe or HgCdTe. This is termed solid-solution hardening. Hardness of a material can be related to the type of chemical bonding and can serve as a measure of binding strength.

2.1.7 Heterojunctions Four considerations apply to the choice of a heterostructure system. These are energy gap, lattice mismatch, doping type, and band offsets. Figure 22 gives the relationship between the first two parameters, and Sec 2.1.3 outlined the doping behavior. Strain energy between mismatched binary pairs increases with thickness until a critical value is reached and dislocations are introduced to reduce the strain. Metastable structures, where a layer is sufficiently strained to match the previous layer but not enough to result in dislocations, also exist. Typical values reported for the valence-band offsets for the binary pairs indicated are given in Table 24, along with the electron affinities. It is clear that the differences in electron affinities between binary pairs do not normally agree with the experimental or theoretical (Tersoff) values of valence-band offset, in disagreement with Eq. (18) and

Table 24. Valence-band offsets (VBO) and electron affinities for pairs of binary II–VI compounds.

Binaries	VBO (exp), theory	Electron affinities
HgTe/CdTe	(0.35), 0.37	
HgSe/CdSe	0.57	
HgTe/ZnTe	(0.25), 0.26	
HgSe/ZnTe	0.03	
HgSe/ZnSe	0.6	
CdTe/ZnTe	(0.1), 0.13	4.28/3.53
CdS/CdSe	0.83	4.8/4.95
CdSe/MnSe	0.25	
CdTe/MnTe	0.25	
ZnS/ZnSe	0.82	3.9/4.09
ZnTe/ZnSe	1.0	3.53/4.09
ZnTe/CdSe	0.61	3.53/4.95
ZnSe/CdS	0.54	4.09/4.8
ZnS/CdSe	1.05	3.9/4.95
ZnS/CdS	0.28	3.9/4.8

Fig. 5. This offset is required to be low in solar-cell structures also, and this is thought to be one of the critical parameters in device performance. Figure 4 shows the three types of band offset that occur in heterojunctions. In II–VI compounds type I heterojunctions include ZnTe/ZnSe, while type II heterojunctions include ZnTe/CdSe. Type III semimetal/semiconductor superlattices consist of a semimetal, such as HgTe, and a semiconductor, such as CdTe. The wavelength of the infrared detector made from such structures can be controlled by the layer thickness rather than the composition uniformity, and the larger effective masses reduce the dark currents. The high diffusion coefficients in these systems, however, may lead to interdiffusion, even at the low growth temperature of ~200 °C used in MBE or on high-temperature storage of devices.

These heterojunctions form the basis for numerous SL, SLS, and MWQ structures with useful electrical, optical, and magnetic properties. The last field has benefited from the availability of zinc-blende structure Mn-containing compounds, for use as barrier layers, via nonequilibrium growth techniques.

2.2 Alloys

2.2.1 Energy-Band Structure of Alloys Ternary alloys enable tuning of the band gap and lattice constants by varying the alloy composition, making these compounds very versatile. In pseudobinary alloys and wide-gap II–VI compounds, the fundamental band edge $E_g = \Gamma_6 - \Gamma_8$ varies, in general, nonlinearly with composition and/or temperature. Most measurements are fitted by

$$E_g = [xE_g(AC) + (1 - x)E_g(BC)] - bx(1 - x), \tag{38}$$

where $E_g(AC)$ and $E_g(BC)$ are the band gaps for the binaries and b is a "bowing parameter," but significant temperature dependences are also seen. Empirical relationships (both theoretical and experimental) of $E_g(x,T)$ (with T in K and E_g in eV) are given in Table 25.

The lattice parameter versus energy gap at room temperature for the II–VI binaries is shown in Fig. 22. The availability of good-

Table 25. Variation of energy gap with composition and temperature in ternaries and quaternaries.

Compound	$E_g(x, T)$
$Hg_{1-x}Cd_xTe$	$E_g = -0.313 + 1.787x + 0.444x^2 - 1.237x^3 + 0.932x^4 + (0.667 - 1.714x + 0.76x^2)T/1000$ $(T > 70\ K)$
$Hg_{1-x}Cd_xSe$	$E_g = -0.209(1 - 7.172x - 2.174x^2) + 0.000737\,(1 - 1.277x - 0.151x^2)T + 2.001 \times 10^{-9}\,(1 + 23.45x - 599.4x^2)T^2$
$Hg_{1-x}Zn_xTe$	$E_g = -0.27 + 1.3x + 1.27x^2$ $(T = 77\ K)$ $E_g = -0.17 + 1.1x + 1.27x^2$ $(T = 300\ K)$ $E_g(x, T) = -0.3 + 0.0324x^{1/3} + 2.731x - 0.629x^2 + 0.533x^3 + 0.00053T\,(1 - 0.76x^{1/2} - 1.29x)$
$HgSe_{1-x}Te_x$	$E_g = -0.0687 - 0.484x + 0.408x^2$ $(T = 300\ K)$
$Hg_{1-x}Mn_xSe$	$E_g = -0.27 + 0.044x$ $(x\ 0.37)$ $(T = 10\ K)$
$Zn_{1-x}Hg_xSe$	$E_g = 2.781 - 0.31x + 2.3x^2$
$Cd_{1-x}Zn_xTe$	$E_g = 1.4637 + 0.49613x + 0.2289x^2$ $(T = 300\ K)$ $E_g = 1.586 + 0.5006x + 0.29692x^2$ $(T = 77\ K)$
$CdTe_xSe_{1-x}$	$E_g = 1.78 + 0.937x + 0.755x^2$ $(T = 5\ K)$
$Cd_{1-x}Mn_xTe$	$E_g = 1.59 + 1.4x$ $(T = 4\ K)$
ZnS_xSe_{1-x}	$E_g = [xE_g(ZnSe) + (1 - x)E_g(ZnS)] - 0.456x(1 - x)$
$ZnSe_xTe_{1-x}$	$E_g = [xE_g(ZnSe) + (1 - x)E_g(ZnTe)] - 1.23x(1 - x)$ $E_g = 2.67 - 1.64x + 1.23x^2$ $(T = 300\ K)$
ZnS_xTe_{1-x}	$E_g = [xE_g(ZnS) + (1 - x)E_g(ZnTe)] - 3x(1 - x)$
$Zn_xCd_{1-x}S$	$E_g = 2.43 + 1.28x$ $(T = 300\ K)$
$Zn_xCd_{1-x}Se$ $(x < 0.6)$	$E_g = E_g(CdSe) + [E_g(ZnSe) - E_g(CdSe) - 0.82]x + 0.82x^2$ (300 K)
(W, $x < 0.5$)	$E_g = E_g(CdSe) + [E_g(ZnSe) - E_g(CdSe) - 0.87]x + 0.87x^2$ (80 K)
(ZB, $x > 0.7$)	$E_g = 1.832 + 0.688x + 0.35x^2$ $(T = 77\ K)$
$Zn_{1-y}Cd_ySe_{1-x}Te_x$	$E_g(x, y) = 2.82 - 1.39y - 1.935x + 1.182xy + 0.35y^2 + 1.507x^2 - 0.184xy^2 - 0.752yx^2$ $(T = 5\ K)$

quality, lattice-matched substrates of sufficient size and purity/perfection is still a limitation in the growth of well-developed ternary systems (see Sec. 2.2.2).

2.2.2 Material Properties of Alloys In general, the II–VI binaries form complete solid solutions with each other. Exceptions to this rule are where one of the binaries crystallizes in the hexagonal wurtzite form. In these cases, miscibility gaps exist. Diluted magnetic semiconductor materials, based on Mn, show these effects when Mn(Te,Se,S) combine with the other II–VI binaries. Figure 24 shows an overview of the crystal structure with zinc-blende (Cub) and wurtzite (Hex) regions highlighted. Many ternary alloys follow Vegard's law (see Sec. 1.2.2.1); i.e., the lattice parameter varies linearly with composition between the binaries—e.g., the HgTe/CdTe, HgTe/ZnTe, ZnS/ZnSe, and ZnS/CdS systems. Where lattice-parameter differences are large, the choice of substrate, and its quality and size, is difficult, and recourse is often then taken to growing a II–VI compound on a III–V substrate—e.g., ZnSe on GaAs. Some systems show a bowing in lattice parameter between the end binaries. This general rule applies to many other physical properties of ternary alloys. There is insufficient space in this article to give much detail of ternaries and quaternaries, and so illustrative examples from the $Hg_{1-x}Cd_xTe$ and $Cd_{1-x}Zn_xTe_{1-y}Se_y$ systems will be given.

As mentioned in the Introduction, materials from the CdTe, CdZnTe, and CdTeSe series are used extensively as substrates for the epitaxial growth of $Hg_{1-x}Cd_xTe$, which is then used to fabricate infrared detectors. The properties that make the Cd/Zn/Te/Se materials suitable include high transmission (~65%), lattice matching to the HgCdTe (for x or $y \sim 0.04$ in $Cd_{1-x}Zn_xTe_{1-y}Se_y$), low levels of large-size second-phase precipitation, low dislocation densities (mid-10^4 cm^{-2}), high electrical resistivity (10^9 Ω cm), similar thermal and mechanical properties to HgCdTe, and availability in sufficient single-crystal, twin-free quantities. The addition of Zn and Se to CdTe is known to harden the lattice (as does the addition of Zn to HgTe—see Sec. 2.1.6) as well as improve the lattice matching. Problems remaining in these materials include reproducibility, purity, and uniformity of lattice matching in the Zn-containing compounds due to the strong segre-

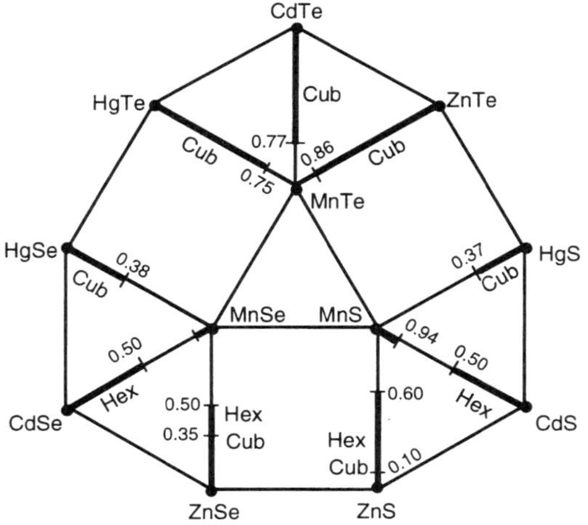

FIG. 24. Overview of $II_{1-x}Mn_xVI$ alloys and their crystal structure. "Hex" and "Cub" indicate wurtzite and zinc blende, respectively. (From Willardson and Beer, 1978.)

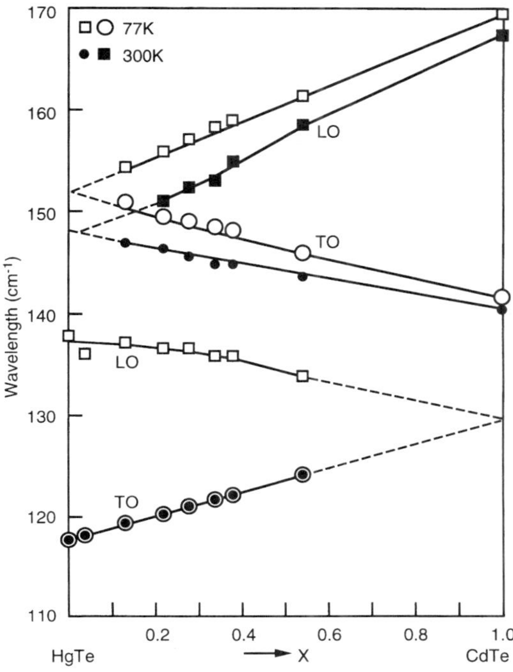

FIG. 25. Longitudinal (LO) and transverse (TO) optical phonon frequencies in HgCdTe at T = 77 and 300 K (from Capper, 1994.)

gation of Zn during growth. Despite these residual difficulties, many suitable high-quality substrates are produced and used, mainly in LPE growth of HgCdTe.

The electron effective mass in CdHgTe varies according to

$$(m_e/m_0) = 1 + 2F + 0.33E_p[2/E_g + 1/(E_g + \Delta)], \quad (39)$$

where F is a constant (~ -0.8), E_p = interband energy ($\sim$19 meV), Δ = spin–orbit splitting ($\sim$1 eV), and E_g = energy gap, which is given in Table 25 as a function of composition x and temperature. For the heavy hole, effective-mass values lie in the range $(0.4–0.7)m_0$. Extrapolation from CdTe to CdZnTe and CdTeSe can be made using

$$(m_e/m_0)(2/E_g) + 1/(E_g + \Delta) = 0.187. \quad (40)$$

In CdHgTe, polar optical phonons arise from CdTe-type and HgTe-type modes and fine structure due to vibration modes of the various combinations of anions and cations in the lattice (from far-IR and Raman-scattering studies). The variations with composition, at 77 and 300 K, of these modes are shown in Fig. 25 (Capper, 1994).

The dielectric constants of CdHgTe vary with composition as

$$\varepsilon_\infty = 15.2 - 15.6x + 8.2x^2 \quad (41)$$

$$\varepsilon_0 = 20.5 - 15.6x + 5.7x^2. \quad (42)$$

ε_0 for CdZnTe varies nonlinearly between the binaries, increasing as Zn content increases up to 0.3, then decreasing. The dispersion of the refractive index of CdHgTe has been fitted by

$$n^2 = a_1 + \frac{a_2\lambda^2}{\lambda^2 - (1/a_3)^2} + \frac{a_4\lambda^2}{\lambda^2 - (a_5)^2}, \quad (43)$$

where a's are constants and λ = wavelength. Experimental data and the fits by this equation are shown in Figs. 26(a) and 26(b) (Capper, 1994). There are no experimental data on n for CdZnTe and CdTeSe, and various extrapolation procedures similar to the above expression have been proposed, but for CdTeSe these are problematic on account of the phase change that occurs as the Se content increases.

Linear thermal expansion coefficients for HgTe and CdTe can be approximated by

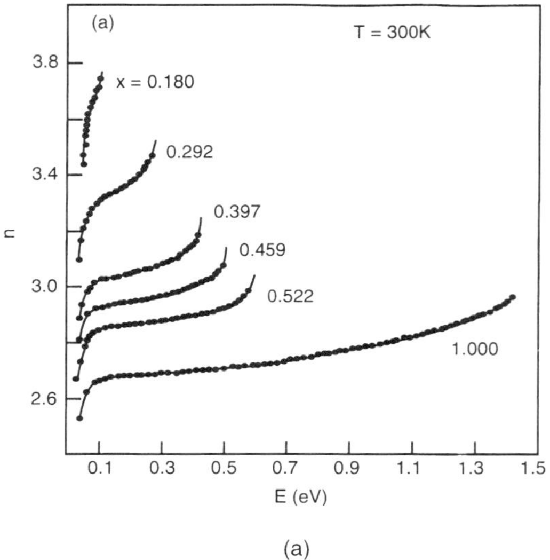

(a)

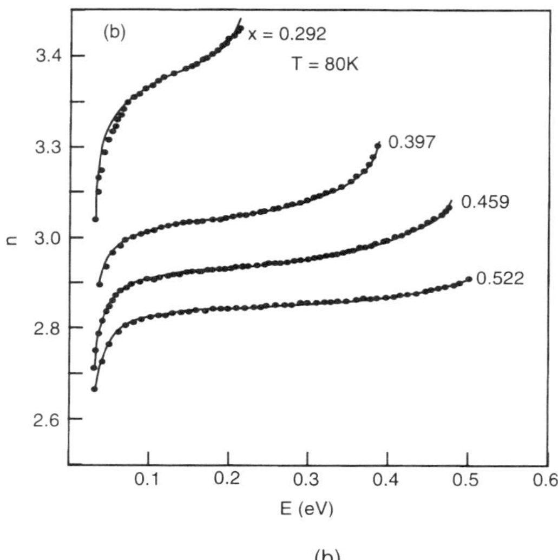

(b)

FIG. 26. Spectral dependences of refractive index of HgCdTe at (a) T = 300 K and (b) T = 80 K (from Capper, 1994.)

$$\alpha = 4 \times 10^{-6} + 4 \times 10^{-9}T \tag{44}$$

and

$$\alpha = 4 \times 10^{-6} + 3.3 \times 10^{-9}\,T, \tag{45}$$

respectively. Values for $x = 0.2$ and 0.7 CdHgTe are 4.26×10^{-6} and 4.5×10^{-6}, respectively, at 290 K. Negative values are found for temperatures below about 60 K. No data exist for CdZnTe and CdTeSe.

The x range of interest for infrared detectors made from $Hg_{1-x}Cd_xTe$ is mainly 0.2–0.3, covering the 3–5 and 8–12 μm bands (Brice and Capper, 1987; Capper, 1994). Most relevant properties vary strongly with both x and temperature (see Capper, 1994, for details). The fundamental properties of high optical absorption coefficient, low intrinsic carrier concentration, high minority-carrier lifetime, and high electron/hole mobility ratio, together with the capability for band-gap engineering (see Table 25), make it almost ideal for a wide range of infrared detector types including photoconductors, Signal Processing In The Element (SPRITEs; see Glossary), photodiodes, and charge-transfer devices. Bipolar doping has been achieved using the lower diffusing elements in most growth techniques following the recognition of the importance of controlling the stoichiometry at the growth temperature. This control is achieved after a low-temperature Hg anneal to minimize the concentration of metal vacancies, which are the principal defects and act as acceptors (possibly doubly ionized). However, the physical properties of the material, which include high mercury vapor pressure, a widely separated phase diagram, high diffusion coefficients, softness, and brittleness, have presented challenging problems to both the material preparation and device fabrication areas. Challengers to its preeminent position include Schottky barriers on silicon, SiGe heterojunctions, SiGe SLs, GaAlAs MQWs, III–V *nipi* structures, high-temperature superconductors, and thermal detectors. It is noteworthy, however, that none of these competitors can really compete in terms of fundamental properties.

3. CONCLUSION

This review has given an overview of some of the main material properties of a wide range of III–V and II–VI binary, ternary, and quaternary compounds. In addition to basic bonding and crystal and energy-band structures, various optical, mechanical, and defect properties have been described. Some of the multitude of device types have been outlined and suitable material combinations detailed. Representatives of the two types of compounds, i.e., GaAs and HgCdTe, are the second and third most studied semiconductors after silicon.

GLOSSARY

Band-Edge Offset: Difference in energy of conduction or valence bands in heterojunctions, which act as potential barriers to carrier movement.

Brillouin Zone: The polyhedron in reciprocal space over which the wave vector **k** is allowed to vary.

Bowing Parameter: Quantity describing the deviation from linearity of an alloy property variation with composition.

Carrier Effective Mass: Related to the matrix element P connecting the conduction band and the light-hole and spin–orbit-split valence bands.

Debye Temperature: A measure of the temperature above which all vibrational modes in a crystal are excited and below which they are frozen out.

Diluted Magnetic Semiconductors (DMS): Semiconductors with a fraction of constituent ions replaced by magnetic ions (e.g., Mn^{+2}, Fe^{2+}, etc.).

Fröhlich Coupling Constant: Constant describing the interaction of electrons with the LO phonon.

Heterojunction: Junction between two binaries with an abrupt change in energy-band structure.

Kramers–Kronig Relation: Relation between the static and high-frequency dielectric constants.

LO: Longitudinal optical phonon.

Multiquantum Well (MQW): Material with many QWs essentially producing a bulk solid with two-dimensional properties.

Phillips Ionicity: Fractional ionic character of binary compound estimated semiempirically from the ionic contribution to the energy gap.

Poisson's Ratio: Elastic constant derived from elastic compliances relating transverse to axial strain.

Quantum Well (QW): A three-layer (minimum) structure producing carrier confinement giving rise to a "two-dimensional" semiconductor.

Signal Processing in the Element (SPRITE): A type of infrared detector where the scene is scanned across the detecting element at the same rate as carriers drift along the element.

Spin–Orbit Splitting: Splitting of orbitally degenerate energy levels in crystals by the spin–orbit interaction.

Strained-Layer Superlattice (SLS): As for SL, but where the binaries have very different lattice parameters.

Superlattice (SL): As the separation of wells in a MQW structure is reduced, the energy levels in one well start to interact with those in neighboring wells, producing an SL.

TO: Transverse optical phonon.

Vegard's Law: Linear variation of lattice parameter with composition in alloy systems.

Young's Modulus: Elastic constant derived from elastic compliances relating stress to strain in the elastic region.

Works Cited

Adachi, S. (1982), *J. Appl. Phys.* **53,** 5863–5869.

Adachi, S. (1983), *J. Appl. Phys.* **54,** 1844–1848.

Anderson, R. L. (1962), *Solid State Electron.* **5,** 341–351.

Asai, H., Oe, K. (1983), *J. Appl. Phys.* **54,** 2052–2056.

Bassignana, I. C., Miner, C. J., Puetz, N. (1989), *J. Appl. Phys.* **65,** 4299–4305.

Berolo, O., Woolley, J. C., Van Vechten, J. A. (1973), *Phys. Rev.* **B8,** 3794–3798.

Bir, G. L., Pikus, G. E. (1974), *Symmetry and Strain Induced Effects in Semiconductors,* New York: Wiley.

Bisaro, R., Mevenda, P., Pearsall, T. P. (1979), *Appl. Phys. Lett.* **34,** 100–102.

Brantley, W. A. (1973), *J. Appl. Phys.* **44,** 534–535.

Brice, J. C., Capper, P. (Eds.) (1987), *Properties of Mercury Cadmium Telluride,* EMIS Datareview Series No. 3, London: INSPEC/IEE.

Capper, P. (Ed.) (1994), *Properties of Narrow-Gap Cadmium-based Compounds,* EMIS Datareview Series No. 10, London: INSPEC/IEE.

Casey, H. C., Jr., Panish, M. B. (1978), *Heterostructure Lasers Part B: Materials and Operating Characteristics,* New York: Academic, p. 9.

Chandra, P., Coldren, L. A., Strege, K. E. (1981), *Electron. Lett.* **17,** 6–7.

Cohen, M. L., Chelikowsky, J. R. (1988), *Electronic Structure and Optical Properties of Semiconductors,* Berlin: Springer-Verlag.

DeVreese, J. T. (Ed.) (1972), *Polarons in Ionic Crystals and Polar Semiconductors,* Amsterdam: Elsevier Science Publishers.

Forrest, S. R. (1987), in: F. Capasso, G. Margaritondo (Eds.), *Heterojunction Band Discontinuities,* Amsterdam: North-Holland, p. 311.

Ghosh, D. K., Samanta, L. K., Bhar, G. C. (1986), *Infrared Phys.* **26,** 111–114.

Glisson, T. H., Hauser, J. R., Littlejohn, M. A., Williams, C. K. (1978), *J. Electron. Mat.* **7,** 1–16.

Harrison, W. A. (1977), *J. Vac. Sci. Technol.* **14,** 1016–1021.

Hocking, E. F., Kudman, I., Seidel, T. E., Schmalz, C. M., Steigmeier, E. F. (1966), *J. Appl. Phys.* **37,** 2879–2887.

Holland, M. G. (1966), in: R. K. Willardson, A. C. Beer (Eds.), *Semiconductors and Semimetals,* Vol. 2, New York: Academic, p. 3.

Hornstra, J., Bartels, W. J. (1978), *J. Cryst. Growth* **44,** 513–517.

Hybertsen, M. (1990a), *J. Vac. Sci. Technol.* **B8,** 773–778.

Hybertsen, M. (1990b), *Phys. Rev. Lett.* **64,** 555–558.

Kane, E. O. (1982), in: W. Paul (Ed.), *Handbook on Semiconductors,* Vol. 1, Amsterdam: North Holland, Chap. 4A.

Kaneko, Y., Koda, T. (1988), *J. Cryst. Growth* **86** 72–78.

Kohn, W. (1957), in: F. Seitz, D. Turnbull (Eds.), *Solid State Physics—Advances in Research and Applications,* Vol. 5, New York: Academic, p. 257.

Kroemer, H. (1985), "Surface and Interface Effects in VLSI," in: N. G. Einspruch, R. S. Bauer (Eds.), *VLSI Electronics Microstructure Science,* Vol. 10, Chap. 4, New York: Academic.

Kroger, F. A. (1974), *The Chemistry of Imperfect Crystals,* Vol. 2, Amsterdam: North-Holland, p. 237.

Lang, D. V. (1987), in: F. Capasso, G. Margaritondo (Eds.), *Heterojunction Band Discontinuities,* Amsterdam: North-Holland, p. 377.

Madelung, O., Schulz, M., Weiss, H. (Eds.) (1982), *Landolt-Börnstein: Numerical Data and Functional Relationships in Science and Technology,* Group III, *Crystal and Solid State Physics,* Vol. 17a, *Physics of Group IV Elements and III–V Compounds,* Berlin: Springer.

Madelung, O., Schulz, M., Weiss, H. (Eds.) (1984), *Landolt-Börnstein: Numerical Data and Functional Relationships in Science and Technology,* Group III, *Crystal and Solid State Physics,* Vol. 17d, *Technology of III–V, II–VI and Non-Tetrahedrally Bonded Compounds,* Berlin: Springer.

Madelung, O., Schulz, M., (Eds.) (1987), *Landolt-Börnstein: Numerical Data and Functional Relationships in Science and Technology,* Group III, *Crystal and Solid State Physics,* Vol. 22a, *Intrinsic Properties of Group IV Elements and III–V, II–VI and I–VII Compounds,* Berlin: Springer.

Martin, G. M., Makram-Ebeid, S. (1986), "The Mid-Gap Donor Level EL2 in GaAs," in: S. Pantelides (Ed.), *Deep Centers in Semiconductors,* Chap. 6, New York: Gordon and Breach.

Matthews, J. W., Blakeslee, A. E. (1974), *J. Cryst. Growth* **27,** 118–125.

Matyas, E. E., Karoza, A. G. (1982), *Phys. Status Solidi B* **111,** K45–K48.

Miller, A. (1987), in: T. C. Brice, P. Capper (Eds.), *Properties of Mercury Cadmium Telluride,* EMIS Datareview Series No. 3, London: INSPEC/IEE, p. 35.

Miller, A. (1992), in: H. E. Ruda (Ed.), *Widegap II–VI Compounds for Opto-Electronic Applications,* London: Chapman & Hall.

Milnes, A. G., Feucht, D. L. (1982), *Heterostructures and Metal Semiconductor Junctions,* New York: Academic.

Moon, R. L., Antypas, G. A., James, L. W. (1974), *J. Electron. Mater.* **3,** 635–644.

Nye, J. F. (1957), *Physical Properties of Crystals,* Oxford: Clarendon Press.

Pantelides, S. T. (1975), in: H. J. Queisser (Ed.), *Festkörperprobleme,* Vol. XV, Berlin: Springer, p. 149.

Pearsall, T. P. (1982), in: T. P. Pearsall (Ed.), *GaInAsP Alloy Semiconductors,* New York: Wiley, p. 295.

Pearsall, T. P., Eaves, L., Portal, J. C. (1983), *J. Appl. Phys.* **54,** 1037–1047.

Phillips, J. C. (1970), *Phys. Rev B* **1,** 1540–1544.

Phillips, J. C. (1973), *Bonds and Bands in Semiconductors,* New York: Academic.

Phillips, J. C., Van Vechten, J. A. (1969), *Phys. Rev. Lett.* **22,** 705–708.

Pietsch, U., Marlow, D. (1986), *Phys. Status Solidi A* **93,** 143–149.

Ruan, Y. C., Ching, W. Y. (1987), *J. Appl. Phys.* **62,** 2885–2897.

Seraphin, B. O., Bennett, H. E. (1967), in: R. K. Willardson, A. C. Beer (Eds.), *Semiconductors and Semimetals,* New York: Academic, p. 499.

Soni, R. K., Abbi, S. C., Jain, K. P., Balkanski, M., Slempkes, S., Benchimol, J. C. (1986), *J. Appl. Phys.* **59,** 2184–2188.

Tejedor, C., Flores, F. (1978), *J. Phys.* **C11,** L19–L23.

Thomas, M. B., Woolley, J. C. (1971), *Can. J. Phys.* **49,** 2052–2060.

Van Vechten, J. A. (1969), *Phys. Rev.* **187,** 1007–1020.

Van Vechten, J. A., Bergstresser, T. K. (1970), *Phys. Rev.* **B1,** 3351–3358.

Varshni, Y. P. (1967), *Physica* **34,** 149–154.

Zallen, R. (1982), in: W. Paul (Ed.), *Handbook on Semiconductors,* Amsterdam: Elsevier Science Publishers, Vol. 1, Chap. 1.

Zielinski, E., Schweizer, H., Streubel, K., Eisek, H., Weimann, G. (1986), *J. Appl. Phys.* **59,** 2196–2204.

Further Reading

Adachi, S. (1985), *J. Appl. Phys.* **58,** R1.

Advanced Materials for Optics and Electronics— European Workshop II–VI (1994), **3.**

Aven, M., Prener, J. S. (Eds.) (1967), *Physics and Chemistry of II–VI Compounds,* Amsterdam: North-Holland.

Bartoli Jr., F. J., Schaake, H. F., Schetzina, J. F. (Eds.) (1990), *Properties of II–VI Semiconductors, Bulk Crystals, Epitaxial Films, Quantum Well Structures and Dilute Magnetic Systems, Mater. Res. Soc. Symp. Proc.* **161.**

Bhargava, R. N., Ruth, R. P., Yno, T., Nurmikko, A. V. (1994), Special issue on II–VI compounds, *J. Cryst. Growth* **138,** 1–1120.

Biefeld, R. M., Gunshor, R. L., Malik, R. J. (Eds.) (1991), *Long-Wavelength Semiconductor Devices, Materials and Processes, Mater. Res. Soc. Symp. Proc.* **216.**

Casselman, T. N. (Ed.) (1986), Proceedings of the 1985 U.S. Workshop on Physics and Chemistry of Mercury Cadmium Telluride, *J. Vac. Sci. Technol.* **A4,** 1955–2238.

Cheung, D. T. (Ed.) (1982), Proceedings of the 1st U.S. Workshop on Physics and Chemistry of Mercury Cadmium Telluride, *J. Vac. Sci. Technol.* **21,** 109–263.

Cheung, D. T. (Ed.) (1983), Proceedings of the 1982 U.S. Workshop on Physics and Chemistry of Mercury Cadmium Telluride, *J. Vac. Sci. Technol.* **A1,** 1586–1764.

Doe, John (Ed.) (1985), Proceedings of the 1984 U.S. Workshop on Physics and Chemistry of Mercury Cadmium Telluride, *J. Vac. Sci. Technol.* **A3,** 55–284.

Dornhaus, R., Nimtz, G. (Eds.) (1985), *Narrow-Gap Semiconductors,* Berlin: Springer-Verlag.

Elliott, C. T., Gordon, N. T. (1993), in: T. S. Moss (Ed.), *Handbook on Semiconductors,* Amsterdam: Elsevier Science Publishers B. V., Vol. 4 (Ed.) C. Hilsum.

Farrow, R. F., Schetzina, J. F., Cheung, J. T. (Eds.) (1987), *Materials for Infrared Detectors and Sources, Mater. Res. Soc. Symp. Proc.* **90.**

Fonash, A. (1981), *Solar Cell Device Physics,* New York: Academic Press.

J. Cryst. Growth **59** (1982); **72** (1985), **86** (1988), **117** (1991).

J. Electron. Mater. **22/8** (1993), p. 801–1112.

Katz, A. (Ed.) (1991), *Indium Phosphide and Related Materials: Processing, Technology, and Devices,* London: Artech House.

Madelung, O., Schulz, M., Weiss, H. (Eds.) (1982), *Landolt-Börnstein: Numerical Data and Functional Relationships in Science and Technology,* Group III, *Crystal and Solid State Physics,* Vol. 17a, *Physics of Group IV Elements and III–V Compounds,* Berlin: Springer.

Løvold, S., Mullin, B. (Eds.) (1992), Proceedings of the NATO Workshop on Narrow-Gap Semiconductors, *Semicond. Sci. Technol.* **6,** C1–C136.

Madelung, O., Schulz, M., Weiss, H. (Eds.) (1984), *Landolt-Börnstein: Numerical Data and Functional Relationships in Science and Technology,* Group III, *Crystal and Solid State Physics,* Vol. 17d, *Technology of III–V, II–VI and Non-Tetrahedrally Bonded Compounds,* Berlin: Springer.

Madelung, O., Schulz, M., (Eds.) (1987), *Landolt-Börnstein: Numerical Data and Functional Relationships in Science and Technology,* Group III, *Crystal and Solid State Physics,* Vol. 22a, *Intrinsic Properties of Group IV Elements and III–V, II–VI and I–VII Compounds,* Berlin: Springer.

Mater. Sci. Eng. **B16** (1993).

McGill, T. C., Sotomayor-Torres, C. M., Gebhardt, W. (Eds.) (1989), *Growth and Optical Properties of Wide-Gap II–VI Low-Dimensional Semiconductors, NATO Advanced Study Institutes, Series B: Physics,* vol. 200, New York: Plenum.

Mullin, B., Stradling, T. (Eds.) (1993), Special issue on narrow-gap semiconductors, *Semicond. Sci. Technol.* **8,** S1–S456.

Neuberger, M. (1972), *Handbook of Electronic Materials,* Vol. 7, *III–V Ternary Semiconducting Compounds—Data Tables,* New York: Plenum.

Ruda, H. E. (Ed.) (1992), *Widegap II–VI Compounds for Opto-Electronic Applications,* London: Chapman & Hall.

Schaake, H. F. (Ed.) (1987), Proceedings of the 1986 U.S. Workshop on Physics and Chemistry of Mercury Cadmium Telluride, *J. Vac. Sci. Technol.* **A5,** 2997–3210.

Schaake, H. F. (Ed.) (1988), Proceedings of the 1987 U.S. Workshop on Physics and Chemistry of Mercury Cadmium Telluride, *J. Vac. Sci. Technol.* **A6,** 2589–2839.

Schaake, H. F. (Ed.) (1990), Proceedings of the 1989 U.S. Workshop on Physics and Chemistry of Mercury Cadmium Telluride, *J. Vac. Sci. Technol.* **A8,** 989–1259.

Seiler, D. G. (Ed.) (1991), Proceedings of the 1990 U.S. Workshop on Physics and Chemistry of Mercury Cadmium Telluride and Novel Infrared Detector Materials, *J. Vac. Sci. Technol.* **B9,** 1611–1901.

Seiler, D. G. (Ed.) (1992), Proceedings of the 1991 U.S. Workshop on Physics and Chemistry of Mercury Cadmium Telluride and Other II–VI Compounds, *J. Vac. Sci. Technol.* **B10,** 1345–1662.

Swaminathan, V., Macrander, A. T. (1991), *Materials Aspects of GaAs and InP Based Structures,* New Jersey: Prentice Hall.

Willardson, R. K., Beer, A. C. (Eds.) (1978), *Semiconductors and Semimetals,* Vols. 13, 25, New York: Academic.

Willardson, R. K., Beer, A. C. (Eds.) (1981), *Semiconductors and Semimetals,* Vols. 16, 18, New York: Academic.

SEMICONDUCTORS, DILUTED MAGNETIC

J. K. FURDYNA, M. DOBROWOLSKA, AND H. LUO, *Department of Physics, University of Notre Dame, Notre Dame, Indiana, U.S.A.*

INTRODUCTION

Diluted magnetic semiconductors (DMSs) are semiconducting alloys whose lattice is made up in part of substitutional magnetic atoms. The most extensively studied and most thoroughly understood materials of this type are the $A^{II}_{1-x}Mn_xB^{VI}$ alloys, in which a fraction of the group-II sublattice is replaced at random by Mn. The worldwide scientific interest and intense research activity that this group of materials enjoys is evidenced by over two thousand papers published on DMSs to date, by an exceptionally large number of invited papers devoted to this subject at international conferences, and by a number of books (see, e.g., Furdyna and Kossut, 1988; Jain, 1991) and survey articles (see, e.g., Furdyna, 1988; Kossut and Dobrowolski, 1993) that have appeared in the last decade.

In recent years, studies involving II-VI semiconductors containing other transition-metal ions (e.g., iron and cobalt) also began to appear, as methods to prepare these new systems became perfected. However, the amount of research carried out on these new compounds, although steadily increasing, is still considerably smaller than that on the "traditional" manganese-based DMSs. Apart from the II-VI materials alloyed with transi-

3-527-28139-8/96/$5.00 + .50

tion-metal chalcogenides, also IV-VI DMS alloys (e.g., $Pb_{1-x}Eu_xTe$, $Pb_{1-x}Mn_xTe$) are beginning to be investigated. Finally, there is a small body of literature already in existence on II-V semiconductor alloys involving transition metals [e.g., $(Cd_{1-x}Mn_x)_3As_2$].

The intense research activity in DMSs has its origin in the fact that the subject represents an interface of two well-established disciplines—semiconductor physics and magnetism—and the "cross products" of these otherwise independent areas of research lead to a host of new electrical, optical, and magnetic phenomena. Specifically, these "cross products" hold promise of modifying the electronic properties of a semiconductor by introducing magnetic interactions. In the $A^{II}_{1-x}Mn_xB^{VI}$ alloys, this change is brought about by the *sp-d* exchange interaction between the spins of band electrons and the localized moments of magnetic ions. The consequences, as we will detail later, are quite dramatic: g factors are effectively enhanced by as much as two orders of magnitude, Faraday rotations become very large, and the magnetoresistance can become negative, reaching gigantic values and leading to an insulator-to-metal transition induced by increasing magnetic field. Further, the DMS alloys are also of fundamental relevance to contemporary problems in disordered magnetism, since they display a low-temperature magnetic phase that strongly resembles a spin glass, as well as various types of antiferromagnetism. These magnetic properties arise from the exchange interaction between the d electrons of the magnetic ions (the *d-d* exchange). Finally, the hope is to harness some of the magnetic-field–related effects in DMS materials for magneto-optical device applications.

Recently the interest in DMS properties has been further increased by the development of methods to prepare DMSs in layer and multilayer forms with a high degree of perfection, including new crystal phases that do not exist in the bulk. Combining the physics of quantum wells and superlattices with the broad range of electronic and magnetic phenomena that exist in DMS materials promises an unusually rich spectrum of entirely new physical phenomena. Alternatively, extending magnetic studies to DMSs at concentrations of magnetic ions not accessible by bulk growth (e.g., $Zn_{1-x}Mn_xTe$ for $x \gtrsim 1$) has led to discoveries of novel forms of long-range antiferromagnetic order. All this is reflected in intense—and a rapidly growing amount of—research activity around the world devoted specifically to layered DMS structures.

Because of its multicomponent nature—semiconductor physics, magnetism, bulk and thin-film phenomena, combined with a large number of possible DMS alloys—a comprehensive presentation of the subject presents a challenge. In addressing this task, our strategy is as follows. We focus primarily on Mn-based DMSs, which by now have been most extensively studied, and whose behavior is representative of the phenomena that we consider most interesting. We also begin with a thorough discussion of bulk DMS systems. Having established the basic DMS properties in this way, it is then a simple matter to extend the discussion also to DMS-based layered structures.

1. OVERVIEW OF BASIC PROPERTIES

1.1 Crystal Structure of Bulk DMS Alloys

In all cases that are of present interest the "magnetic" transition-metal ions are substituted for the cations in a host semiconductor. Since transition metals do not belong to the same column of the periodic table as the atoms that they replace, this fact limits the atomic fraction of magnetic ions that can be incorporated in a stable form of various bulk DMSs.

Table 1 lists the range of molar fractions in which various DMS alloys were obtained, by bulk crystal growth, as well as the crystallographic structure of these alloys. One can see from the table, for example, that $Cd_{1-x}Mn_xTe$ forms a ternary alloy of zinc-blende structure with x up to 0.77, while $Zn_{1-x}Mn_xSe$ exhibits zinc-blende structure for $x < 0.30$ and wurtzite structure for $0.30 < x < 0.57$. Given the fact that the stable crystal structures of MnTe, MnSe, and MnS are neither zinc-blende nor wurtzite, it is remarkable that the $A^{II}_{1-x}Mn_xB^{VI}$ solid solutions shown in Table 1 can reach such high values of x. For example, MnTe itself crystallizes in the NiAs structure, yet in the case of $Cd_{1-x}Mn_xTe$ the zinc-blende structure of the

Table 1. Crystal structures and ranges of composition of bulk DMS alloys (for references see for example Furdyna, 1988, and Kossut and Dobrowolski, 1993).

Compound	Crystal structure	Composition range
$Zn_{1-x}Mn_xS$	Zinc blende	$0 < x < 0.1$
	Wurtzite	$0.1 < x < 0.45$
$Zn_{1-x}Mn_xSe$	Zinc blende	$0 < x < 0.30$
	Wurtzite	$0.30 < x < 0.57$
$Zn_{1-x}Mn_xTe$	Zinc blende	$0 < x < 0.86$
$Cd_{1-x}Mn_xS$	Wurtzite	$0 < x < 0.45$
$Cd_{1-x}Mn_xSe$	Wurtzite	$0 < x < 0.50$
$Cd_{1-x}Mn_xTe$	Zinc blende	$0 < x < 0.77$
$Hg_{1-x}Mn_xS$	Zinc blende	$0 < x < 0.37$
$Hg_{1-x}Mn_xSe$	Zinc blende	$0 < x < 0.38$
$Hg_{1-x}Mn_xTe$	Zinc blende	$0 < x < 0.75$
$(Cd_{1-x}Mn_x)_3As_2$	Tetragonal	$0 < x < 0.12$
$(Zn_{1-x}Mn_x)_3As_2$	Tetragonal	$0 < x < 0.15$
$Pb_{1-x}Mn_xS$	Rocksalt	$0 < x < 0.05$
$Pb_{1-x}Mn_xSe$	Rocksalt	$0 < x < 0.17$
$Pb_{1-x}Mn_xTe$	Rocksalt	$0 < x < 0.40$
$Zn_{1-x}Fe_xS$	Zinc blende	$0 < x < 0.26$
$Zn_{1-x}Fe_xSe$	Zinc blende	$0 < x < 0.21$
$Zn_{1-x}Fe_xTe$	Zinc blende	$0 < x < 0.01$
$Cd_{1-x}Fe_xSe$	Wurtzite	$0 < x < 0.20$
$Cd_{1-x}Fe_xTe$	Zinc blende	$0 < x < 0.03$
$Hg_{1-x}Fe_xSe$	Zinc blende	$0 < x < 0.20$
$Hg_{1-x}Fe_xTe$	Zinc blende	$0 < x < 0.02$
$Zn_{1-x}Co_xS$	Zinc blende	$0 < x < 0.15$
$Zn_{1-x}Co_xSe$	Zinc blende	$0 < x < 0.05$
$Cd_{1-x}Co_xSe$	Wurtzite	$0 < x < 0.08$

"parent" CdTe survives for x as high as 0.77, and even higher for $Zn_{1-x}Mn_xTe$.

The miscibility of manganese with II-VI hosts is particularly large, as compared with other transition-metal ions. The possibility of obtaining crystals in which the amount of the magnetic component can be varied in a continuous fashion within wide limits is one of the reasons why DMSs acquired such great interest in the scientific community.

The study of lattice parameters has been carried out most systematically in the case of DMSs based on II-VI compounds. The lattice parameters were found to vary linearly in proportion with the molar fraction of the magnetic component. (Such linear dependence on concentrations of respective constituents of an alloy is frequently referred to as Vegard's law.) For example, in the case of the zinc-blende structure, the lattice parameter can be expressed in the form

$$a = (1 - x)a_{\text{II-VI}} + xa_{\text{TM-VI}}, \tag{1}$$

where x is the molar fraction of the transition-metal chalcogenide, and $a_{\text{TM-VI}}$ and $a_{\text{II-VI}}$ represent the lattice constants of the end-point binary compounds: a transition-metal chalcogenide in the (hypothetical) zinc-blende phase and the II-VI "parent," respectively.

In view of the diversity of possible lattice structures (zinc-blende and wurtzite) that various II-VI DMSs can acquire, it is useful to introduce the notion of the cation–cation distance d, in terms of which lattice dimensions can be given without specifying the explicit structure of the material. Figure 1 shows the variation of the cation–cation distance in II-VI–based DMSs containing manganese. For extensive information and references on this subject, see Chapter 1 in Furdyna and Kossut (1988).

The usefulness of such a plot is made clear by considering, for example, the linear behavior of d observed for the selenides. Note from Table 1 that $Hg_{1-x}Mn_xSe$ is cubic,

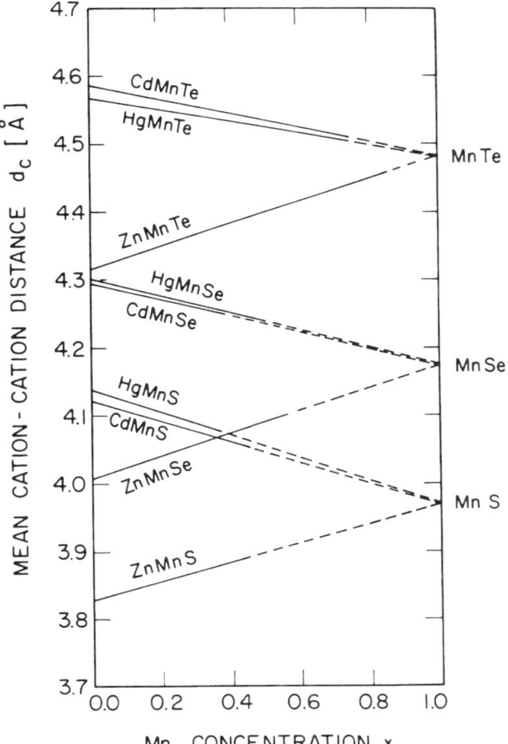

FIG. 1. Mean cation–cation distances d as a function of Mn mole fraction x for $A^{II}_{1-x}Mn_xB^{VI}$ alloys (after Furdyna, 1988).

$Cd_{1-x}Mn_xSe$ is hexagonal, and $Zn_{1-x}Mn_xSe$ changes its crystal structure "in stride" as x increases. However, having established the parameter d for MnSe by extrapolating, e.g., the lattice-constant data for the hexagonal $Cd_{1-x}Mn_xSe$ to $x = 1$, we can predict the lattice constant for the cubic $Hg_{1-x}Mn_xSe$, or for either of the two phases of $Zn_{1-x}Mn_xSe$. Two additional points are worthy of note in Fig. 1. First, it is interesting that the mutual relationship of d for HgSe, CdSe, and ZnSe in this figure closely resembles that for HgTe, CdTe, and ZnTe, despite the differences in structure. We also note that the value of d for $Zn_{1-x}Mn_xSe$ increases smoothly through the cubic-to-hexagonal transition, suggesting that it is the increase in the lattice spacing that forces the onset of hexagonal structure, rather than the other way around.

The actual lattice parameters can be obtained from the parameter d as follows:

$$\text{zinc blende: } a = \sqrt{2}d, \tag{2}$$

$$\text{wurtzite: } a = d, \quad c = (\tfrac{8}{3})^{1/2}d. \tag{3}$$

It should be noted that, apart from its fundamental importance, the precise knowledge of the lattice parameter is of considerable practical interest in that it provides a convenient determination of crystal composition in ternary alloys. It is a fortunate circumstance in this respect that in the case of all $A^{II}_{1-x}Mn_xB^{VI}$ alloys the linear variation of the lattice parameter with x is very large, making this approach to the determination of composition quite reliable.

1.2 Band Structure of DMS Crystals in the Absence of a Magnetic Field

1.2.1 General Band-Structure Features

The band structure of the $A^{II}_{1-x}Mn_xB^{VI}$ DMS alloys can be presented in a unified way by stressing four underlying features:

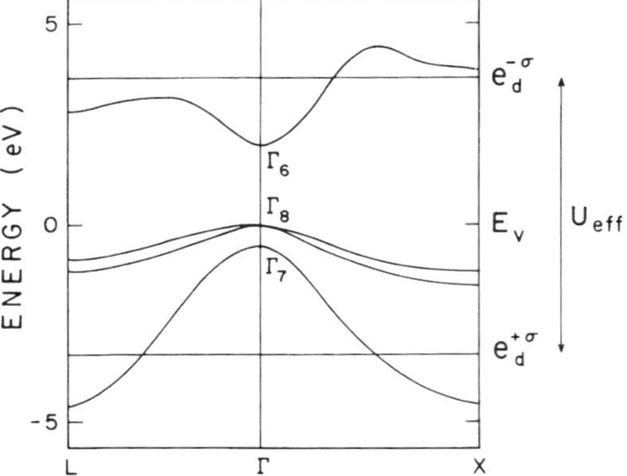

FIG. 2. Schematic band structure for a zinc-blende $A^{II}_{1-x}Mn_xB^{VI}$ alloy. The behavior of the *sp* bands is qualitatively the same as in zinc-blende $A^{II}B^{II}$ semiconductors. The Mn levels are spin split at each site by the energy U_{eff} ($\approx$7 eV). The effect of *p-d* hybridization and of the tetrahedral crystal field on the Mn levels is neglected in this schematic presentation, since the effect is small on the scale of the figure. For wurtzite $A^{II}_{1-x}Mn_xB^{VI}$ alloys the band structure would be qualitatively the same except that the degeneracy seen at Γ_8 in the figure would be removed, resulting in a slight valence-band splitting ($\approx$20 meV) at the Γ point.

1. Qualitatively, the band structure of the $A^{II}_{1-x}Mn_xB^{VI}$ alloys closely resembles that of the $A^{II}B^{VI}$ materials having the same crystal structure. In particular, these alloys are direct-gap semiconductors, with the band extrema occurring at the Γ point. This is illustrated by Fig. 2, where the curves show the characteristic energy (E) vs wave vector k dispersion typical for a zinc-blende $A^{II}B^{VI}$ semiconductor. These bands, common to the $A^{II}B^{VI}$ compounds and to the $A^{II}_{1-x}Mn_xB^{VI}$ alloys alike, originate from the s and p orbitals of the constituent atoms, and we shall refer to them as *sp* bands.
2. Quantitatively, the band structure of the $A^{II}_{1-x}Mn_xB^{VI}$ material transforms in a smooth manner from that of the parent $A^{II}B^{VI}$ compound to the band structure of the tetrahedrally bonded MnB^{VI} "hypothetical" binary compound. Although this evolution is not strictly linear with x, a linear variation (as would be predicted by the virtual-crystal approximation; see, e.g., Hass and Ehrenreich, 1983), is a good first-order guide for a quantitative description of the dependence of the energy gap and other band parameters on x. The band structures of the parent $A^{II}B^{VI}$ binary compounds are well known. Although tetrahedrally bonded MnB^{VI} compounds either do not exist in nature (e.g., MnTe), or are of such poor quality that electronic measurements on these com-

pounds are not reliable, band structures of zinc-blende or wurtzite MnS, MnSe, and MnTe can be established by the various computational schemes commonly used in band-structure calculations. Furthermore, with the advent of epitaxy (see Sec. 1.3), it is now possible to grow some transition-metal chalcogenides (e.g., MnTe, FeSe) in thin-film form using this nonequilibrium growth technique, giving access to band-structure information of these materials. The band structures thus obtained can then serve as the $x = 1$ limit for interpolation between the $A^{II}B^{VI}$ and MnB^{VI}. This procedure, although not always precise, nevertheless serves as a very useful guide for estimating the properties of $A^{II}_{1-x}Mn_xB^{VI}$ alloys for $0 < x < 1$.

3. Superimposed on the above band-structure characteristic of the $A^{II}B^{VI}$ semiconductors is the effect of the $3d^5$ shell of Mn. The energy of the narrow band originating from the ground state of $3d^5$ electrons is approximately 3.5 eV below the valence-band edge (see, for example, Franciosi *et al.*, 1985), i.e., relatively far from the band edges that determine basic properties of a semiconductor. This level is marked as $e_d^{+\sigma}$ in Fig. 2. It is interesting to note that the position of this manganese-derived state is rather insensitive to the II-VI semiconductor "parent." The location of an excited state that corresponds to placing an extra electron on the $3d$ shell is less certain ($e_d^{-\sigma}$ in Fig. 2). Inverse photoemission experiments (i.e., adding a sixth electron to a given Mn site) appear to indicate that it lies some 3.5 eV above the top of the valence band in $Cd_{1-x}Mn_xTe$ (Wall *et al.*, 1989), but more work is needed to establish this conclusively.

4. While the knowledge of the positions of both $e_d^{+\sigma}$ and $e_d^{-\sigma}$ bands, described above, is extremely important for formulating *p-d* hybridization (which underlies both *sp-d* and *d-d* exchange interactions, to be discussed later), it must be emphasized that these levels are determined in the context of a removal of an electron ($3d^5 \rightarrow 3d^4$) from, or of an addition of an electron ($3d^5 \rightarrow 3d^6$) to, a given Mn site. These transitions should not be confused with intra-ion transitions, involving a spin flip within the $3d^5$ level itself. The latter transitions, occurring at energies around 2.1 eV, are of fundamental importance in the optical properties of the $A^{II}_{1-x}Mn_xB^{VI}$ alloys, as will be discussed in Sec. 1.2.3.

In the remainder of this section we focus on the direct-gap and on the intra-Mn transitions, since these features are of primary importance in determining the optical and electrical properties of DMSs in zero magnetic field. For a discussion of the general band structure, including points far from the Γ point and deep-lying bands, we refer the reader to the theoretical and experimental papers dealing explicitly with the subject (see, for example, Larson *et al.*, 1988).

1.2.2 The Energy Gap and its Dependence on Composition All $A^{II}_{1-x}Mn_xB^{VI}$ alloys are direct-gap semiconductors, like their $A^{II}B^{VI}$ parent materials, as is shown in Fig. 3 for three groups of alloys: wurtzite, zinc-blende open gap, and zinc-blende zero gap. As in nonmagnetic semiconductors, the valence band for the zinc-blende case consists of light and heavy holes. For open-gap zinc-blende materials, the heavy-hole and the light-hole bands are degenerate at $k = 0$. The closed-gap zinc-blende case corresponds to a degeneracy of the heavy hole and the conduction band at $k = 0$. For the wurtzite crystals, these degeneracies are lifted by the lower symmetry of this structure.

The effect on the energy gap of replacing the atoms of the group-II element by Mn is illustrated by the case of $Cd_{1-x}Mn_xTe$. As Mn is substituted for Cd in the CdTe lattice, the energy gap increases, to the point that the originally opaque material eventually becomes transparent to visible light. Figure 4 shows the values of the fundamental optical energy gap determined experimentally in CdMnTe at several temperatures. The saturation observed for $x > 0.4$ results from very efficient optical transitions within the Mn^{2+} ions, whereas the gap variation is given by the position of the feature labeled A, observed by Lee and Ramdas (1984) in piezoreflectivity experiments. This technique permits detection of valence-to-conduction-band transitions even in the presence of strong intra-Mn^{++} absorption lines. The latter optical transitions within Mn^{++} ions of course also take place for $x < 0.40$, but for those concentrations they are completely obscured by the fundamental band-edge absorption. As

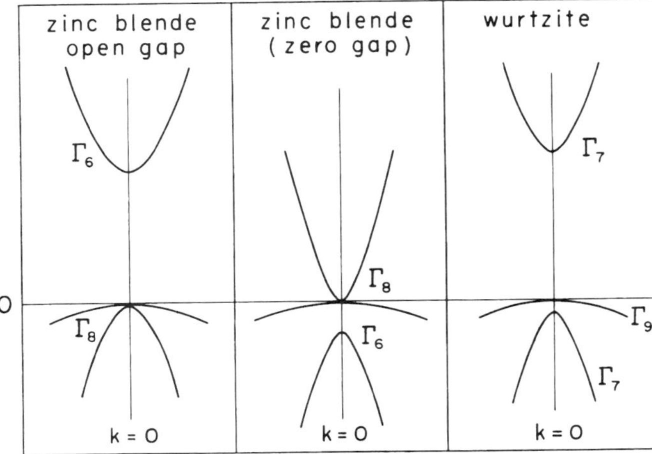

FIG. 3. Band structure near the Γ point for three cases of interest: an open-gap zinc-blende semiconductor (e.g., $Cd_{1-x}Mn_xTe$, or $Hg_{1-x}Mn_xTe$ for $x > 0.07$); a zero-gap zinc-blende semiconductor (e.g., $Hg_{1-x}Mn_xTe$ for $x < 0.05$); and a wurtzite semiconductor (e.g., $Cd_{1-x}Mn_xSe$). In the wurtzite case, the $\Gamma_9 \rightarrow \Gamma_7^{(c)}$ and $\Gamma_7^{(v)} \rightarrow \Gamma_7^{(c)}$ transitions determine the optical absorption edge for the **E** $\perp$ **c** and the **E** $\parallel$ **c** polarizations, respectively, where **E** is the electric field of incident light and **c** is the hexagonal axis. This is the basis of dichroism observed in wurtzite crystals.

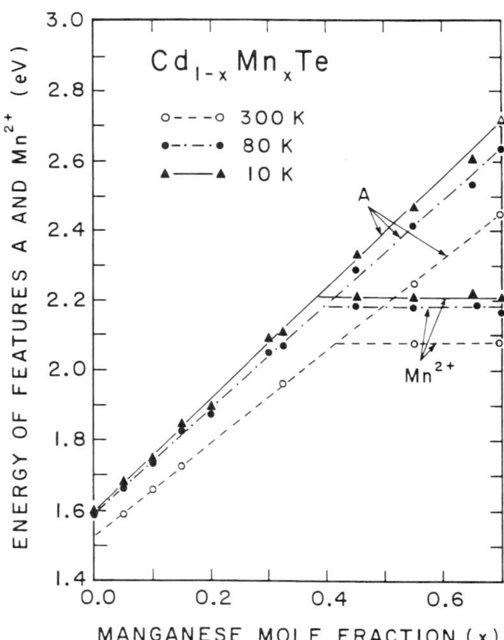

FIG. 4. The energy gap (obtained from the free-exciton transition, marked "A" in the figure) and the ${}^6A_1 \rightarrow {}^4T_1$ intra-Mn transition (marked "Mn^{2+}") as functions of Mn concentration x for $Cd_{1-x}Mn_xTe$ for three temperatures. Note the qualitative similarity in the temperature dependence of both types of transitions (from Lee and Ramdas, 1984).

we can see from the figure, the variation of the fundamental band gap with concentration in diluted magnetic semiconductors can be described, in first approximation, by a linear dependence. When a linear extrapolation to $x = 1$ is made, the value of the energy gap in hypothetical zinc-blende MnTe is obtained. Its value at 4.2 K is 3.2 eV. The same extrapolated value is obtained in the case of $Zn_{1-x}MnTe$ and $Hg_{1-x}Mn_xTe$.

Such "tunability" of the energy gap by composition has an especially profound effect in the case of Hg compounds. Consider $Hg_{1-x}Mn_xTe$ as an example. In this case the role of Mn is similar to that of Cd in $Hg_{1-x}Cd_xTe$. For low values of x this alloy is a zero-gap semiconductor, like the "parent" HgTe. At several atomic percent of Mn ($x \approx 0.07$ at 4.2 K; $x \approx 0.05$ at room temperature) the energy gap of $Hg_{1-x}Mn_xTe$ "opens," and for larger values of x the alloy becomes an ordinary (i.e., positive gap) semiconductor, the gap being extremely sensitive to x, very much like in $Hg_{1-x}Cd_xTe$. A great deal of interest exists in this feature because of potential applications of variable narrow-gap semiconductors as infrared detectors.

The energy gap in DMSs—as in other semiconducting compounds—exhibits considerable temperature dependence, such that the energy gap tends to open wider as the temperature decreases. This behavior is similar to that characteristic of the parent $A^{II}B^{VI}$ compounds, with the result that in the general vicinity of the optical band edge the $A^{II}_{1-x}Mn_xB^{VI}$ alloys tend to be more transparent to shorter and shorter wavelengths as the temperature is decreased. The temperature variation for Cd_xMn_xTe is illustrated in Fig. 4. Detailed studies indicate this variation to be roughly linear between 77 and 300 K, a rather useful fact for making interpolations to intermediate values of T (Furdyna and Kossut, 1988, Chap. 2). In Table 2 we summarize the energy-gap values for the $x = 0$ and $x = 1$ limits of the II-VI DMS alloys for three temperatures. The possibility of tuning the band gap of DMSs, a feature characteristic for many other semiconducting alloy systems, represents an advantageous feature, since by adjusting the composition one can achieve a gap value required by particular application needs. For an extensive discussion of the band structure and references, see Chapter 2 in Furdyna and Kossut (1988).

1.2.3 Intra-ionic Optical Transitions in $A^{II}_{1-x}Mn_xB^{VI}$ Alloys The $A^{II}_{1-x}Mn_xB^{VI}$ systems differ from the $A^{II}B^{VI}$ compounds, and from "orthodox" II-VI ternaries such as $Cd_{1-x}Zn_xTe$, in that the 3d shell of the Mn^{++} ions is only half-filled. This leads to new intra-Mn^{++} electronic transitions, generally in the vicinity of 2.1 eV, which dominate the optical properties of $A^{II}_{1-x}Mn_xB^{VI}$ alloys at high values of x, and are responsible for their characteristic deep-red color in that range of compositions. Furthermore, these transitions within the 3d shell make wide-gap $A^{II}_{1-x}Mn_xB^{VI}$ materials, such as $Zn_{1-x}Mn_xSe$ and $Zn_{1-x}Mn_xS$, of considerable interest in the context of electroluminescent display devices.

Although the Mn^{++} d levels shown in Fig. 2 hybridize with the sp bands, and thus broaden into d bands, the degree of such broadening is in fact quite small. The description of 3d electronic states, therefore, can often be made in terms of atomic orbitals, modified by the presence of a crystal field of an appropriate symmetry. In particular, manganese with its exactly half-filled d shell forms a ground state that is an orbital singlet (orbital moment $L = 0$) denoted by ^6S, which does not split further in the presence of the crystal field nor by the spin-orbit interaction. In the presence of these interactions the ground state of d electrons in Mn ions is traditionally labeled as $^6A_1$. It is, of course, a spin sextet ($S = \frac{5}{2}$) (↑↑↑↑↑). The lowest-lying excited states, which correspond to flipping the spin of one of the five d electrons of manganese (↑↑↑↑↓), are subject to splitting by the crystal-field interaction (see ELECTRON LEVEL SPLITTING), as shown schematically in Fig. 5. The relative position of the crystal-field–split states was studied intensively in the past for various symmetries

Table 2. Energy gaps for $x = 0$ and $x = 1$ limits of the $A^{II}_{1-x}Mn_xB^{VI}$ alloys for liquid helium (LHe), liquid nitrogen (LN_2), and room (R.T.) temperatures (for references see for example Furdyna, 1988, and Kossut and Dobrowolski, 1993).

Compound	T	E_g (eV)	Compound	T	E_g (eV)
ZnS	LHe		HgS	LHe	<0
	LN_2			LN_2	~0
	R.T.	3.8–3.9		R.T.	>0
ZnSe	LHe	2.8	HgSe	LHe	−0.27
	LN_2			LN_2	−0.205
	R.T.			R.T.	+0.10
ZnTe	LHe	2.38	HgTe	LHe	−0.300
	LN_2	2.375		LN_2	−0.252
	R.T.	2.27		R.T.	−0.146
CdS	LHe		MnS	LHe	~3.3
	LN_2			LN_2	~3.0
	R.T.	2.45		R.T.	~3.0
CdSe	LHe	1.83	MnSe	LHe	3.3
	LN_2	1.82		LN_2	3.14
	R.T.	1.75		R.T.	2.9
CdTe	LHe	1.595	MnTe	LHe	3.2
	LN_2	1.586		LN_2	3.05
	R.T.	1.528		R.T.	2.9
			FeSe	LHe	~3.02
				LN_2	
				R.T.	

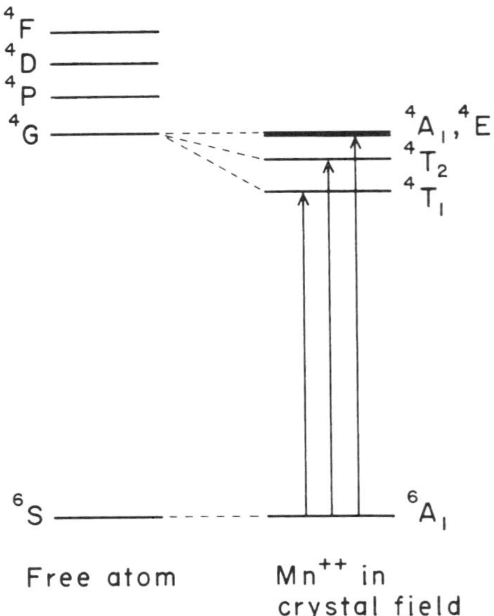

FIG. 5. A schematic diagram of the splitting of the lowest excited state of the $3d^5$ level (^4G) relative to the ground state (^6S) for a Mn^{++} ion in the presence of a tetrahedral crystal field. The arrows indicate possible intra-Mn transitions.

of the host material (see, e.g., Abragam and Bleaney, 1970).

Although optical transitions from the ground ^6S state to lowest excited states are forbidden for a free Mn ion, the spin-orbit interaction and the absence of the inversion symmetry may relax this selection rule when Mn ions are placed into a crystalline host matrix. In fact, optical transitions from the ground state ($^6A_1$) to excited states originating from the ^4G multiplet of the free Mn ion are very probable and correspond to very strong absorption of light. Of these, the $^6A_1 \rightarrow {}^4T_1$ is the lowest-energy (and therefore the most important) transition. It corresponds to approximately 2.1 eV, and constitutes in effect an optical absorption edge in crystals that contain a relatively high concentration of Mn ($x > 0.1$).

The energy of the $^6A_1 \rightarrow {}^4T_1$ transitions in relation to the energy gap is shown for $Cd_{1-x}Mn_xTe$ in Fig. 4, illustrating that either the energy gap or the $^6A_1 \rightarrow {}^4T_1$ transition determines the absorption edge (i.e., the onset of opacity), depending on which of these occurs at the lower energy. It is interesting to note that the temperature dependence of the energy gap and the intra-Mn^{++} transitions has the same qualitative character.

The transitions within the Mn $3d$ shell are the source of an efficient emission of light observed in the electroluminescence in $Zn_{1-x}Mn_xSe$ and $Zn_{1-x}Mn_xS$ (for a review of the energy-transfer mechanisms responsible for the electroluminescent properties of zinc-based DMSs see, e.g., Gumlich, 1981). Because of this intense emission, an application in flat-panel displays has been found for these materials.

Turning to Fe as the constituent of DMSs, let us note that—in contrast to the case of Mn—now even the ground state of the $3d$ electrons is subject to splitting by the crystal field of the $A^{II}B^{VI}$ matrix. This is because Fe, with its six d electrons, possesses a nonvanishing orbital moment $L = 2$. The lowest-lying state is in this case a singlet. Thus the $A^{II}_{1-x}Fe_xB^{VI}$ materials display paramagnetic behavior of the Van Vleck type. The existence of low-lying excited states of Fe completely changes the absorption spectra in the infrared of wide-gap DMSs containing iron, as compared with their Mn counterparts. While, e.g., $Cd_{1-x}Mn_xTe$ is transparent in these ranges of photon energies, compounds containing Fe ions possess very characteristic absorption bands occurring in the near infrared as well as in the far infrared. These absorption bands are due to optical transitions between various Fe levels split by the crystal-field and spin-orbit interactions. From such spectroscopic investigations of absorption by transition-metal impurities in various semiconductor matrices, a great deal of information concerning the crystal-field and spin-orbit interaction parameters can thus be obtained. More details about Fe as the constituent of DMSs can be found, e.g., in the review by Kossut and Dobrowolski (1993).

1.3 Epitaxial DMS Systems

As may be seen from Table 1, there are limits to the composition of DMS alloys in bulk crystal form (i.e., those grown by equilibrium crystal-growth methods). For example, the alloy $Zn_{1-x}Mn_xTe$ is zinc blende and can be obtained in single phase for $x < 0.86$, while $Cd_{1-x}Mn_xSe$ is wurtzite and has a composition range of $x < 0.50$. The higher composition ranges for such materials are not accessible in equilibrium bulk growth, because the equilibrium phases of the end-point materials (for example, rocksalt MnSe and NiAs-structure MnTe) are incompatible with the crystal structure of the parent semiconductor.

The development of epitaxial growth techniques—such as molecular-beam epitaxy (MBE)—allows us to transcend such limitations because of three factors: the nonequilibrium nature of the growth process, the low growth temperature, and the effect of the substrate. These factors allow the growth of crystal phases that would be unstable under typical bulk-growth conditions. The first example of such epitaxial "stabilization" was the growth of zinc-blende $Zn_{1-x}Mn_xSe$ epilayers over the composition range $0 < x \leq 0.66$ (see Kolodziejski *et al.*, 1985). (Bulk-grown $Zn_{1-x}Mn_xSe$ crystallizes in the zinc-blende phase up to $x = 0.3$ and is wurtzite for higher Mn content.) Further, the zinc-blende phase of pure MnSe was also stabilized in the form of thin, strained layers sandwiched between ZnSe layers (see Kolodziejski *et al.*, 1986). Perhaps the most striking example of MBE stabilization of new phases is the growth of single-phase zinc-blende $Cd_{1-x}Mn_xSe$ over the composition range $0 < x \leq 1.00$, even though both the end-point materials CdSe and MnSe have different stable structures for bulk growth (wurtzite and rocksalt, respectively; see Samarth *et al.*, 1989). Finally, both the binary magnetic semiconductors MnTe (Durbin *et al.*, 1989) and FeSe (Jonker *et al.*, 1988) have been grown in the zinc-blende phase by MBE.

2. EXCHANGE INTERACTIONS

As pointed out in the Introduction, DMSs such as $A^{II}_{1-x}Mn_xB^{VI}$ or $A^{II}_{1-x}Co_xB^{VI}$ differ from "normal" semiconductor alloys by the existence of two types of exchange interaction associated with the presence of the magnetic constituent in the lattice. The strong Kondo-type *sp-d* exchange between the spins of band electrons and the localized moments of the magnetic ions has a significant influence on electronic properties, while the weaker Heisenberg inter-ion exchange interaction (the *d-d* exchange) underlies the static and dynamic magnetic properties. Theoretical calculations have shown how the exchange interactions in the $A^{II}_{1-x}Mn_xB^{VI}$ alloys can be related to the band parameters $e_d^{+\sigma}$, U_{eff} (see Fig. 2), and a parameter V_{pd} that describes the degree of hybridization of the p and d orbitals of the alloy system (Ehrenreich *et al.*, 1987; Larson *et al.*, 1988). Similar calculations for exchange interactions in Co- and Fe-based DMSs are not yet available.

2.1 The *sp-d* Exchange Interaction

When a semiconductor contains localized magnetic moments (e.g., $Zn_{1-x}Mn_xTe$), its band structure will be modified by the exchange interaction of these moments (i.e., the $3d^5$ electrons) with band (i.e., *sp*) electrons. It is this *sp-d* exchange coupling that is responsible for most of the electronic peculiarities of the DMSs, e.g., for the greatly enhanced spin splitting of the band states in the presence of a magnetic field (see ELECTRON LEVEL SPLITTING). We start with a description of the physical origin and the form of the *sp-d* exchange interaction, and then relate the modifications of the band structure that result from this mechanism.

The *sp-d* exchange contribution to the band structure can be expressed formally by adding a new exchange term H_{ex} to the original Hamiltonian H_0, so that the total Hamiltonian H_T now reads

$$\begin{aligned} H_T &= H_0 + H_{ex} \\ &= H_0 + \sum_{\mathbf{R}_i} [J^{sp\text{-}d}(\mathbf{r} - \mathbf{R}_i)]\mathbf{S}_i \cdot \sigma, \end{aligned} \tag{4}$$

where $\mathbf{S}_i$ and σ are the spin operators for Mn^{++} and for the band electrons, respectively, $J^{sp\text{-}d}(\mathbf{r} - \mathbf{R}_i)$ is the electron–ion *sp-d* exchange-coupling constant, and $\mathbf{r}$ and $\mathbf{R}_i$ are the coordinates of the band electron and of the Mn^{++} ion, respectively. The summation is only over the lattice sites occupied by the Mn^{++} ions.

Two convenient approximations can be made to simplify H_{ex}. First, since the electronic wave function is very extended, so that the electron "sees" a large number of Mn^{++} ions at any time, we can make use of the molecular-field approximation, replacing $\mathbf{S}_i$ by the thermal average $\langle \mathbf{S} \rangle$ taken over all Mn^{++} ions. For paramagnetic systems, if the applied field $\mathbf{H}$ is in the z direction, we have

$\langle \mathbf{S} \rangle = \langle S_z \rangle$, a quantity intimately related to the magnetization of the system (see Sec. 4).

Second, and again because the electronic wave function spans a large number of lattice sites, we can replace $J^{sp\text{-}d}(\mathbf{r} - \mathbf{R}_i)$ by $xJ^{sp\text{-}d}(\mathbf{r} - \mathbf{R})$, where now $\mathbf{R}$ denotes the coordinate of every site of the fcc (or hcp) cation sublattice, with the summation now carried out over all R. We are thereby expressing the exchange interaction in terms of the virtual-crystal approximation. With these approximations, we have

$$H_{ex} = \sigma_z \langle S_z \rangle x \sum_{\mathbf{R}} J^{sp\text{-}d}(\mathbf{r} - \mathbf{R}), \tag{5}$$

with the summation extending over all cation sites. The major advantage of Eq. (5) is that in that approximation H_{ex} has the periodicity of the lattice. This allows us to use the same wave functions for solving the exchange-interaction problem as those used for diagonalizing H_0.

The *sp-d* exchange modification of the band structure will thus be determined by the values of the exchange constants $J^{sp\text{-}d}$, which are different for the Γ_6 and the Γ_8 bands. Traditionally, the exchange constant characterizing the interaction between the d electrons and the s-like Γ_6 electrons at the center of the Brillouin zone (the Γ point) is designated by the symbol α, and the exchange constant corresponding to the interaction between d electrons and the p-like Γ_8 electrons, also at the Γ point, is designated by β. The constant α originates only from ferromagnetic potential exchange and is hence positive. The parameter β has a larger magnitude than α and is negative. This has contributions mainly from antiferromagnetic kinetic exchange arising from the strong p-d hybridization of valence-band states. Calculations for the $A^{II}_{1-x}Mn_xB^{VI}$ alloys show that the parameter β is related to the band-structure parameters defined in the previous section through the Schrieffer–Wolff expression (Larson *et al.*, 1988)

$$N_0\beta = -32(V_{pd})^2[(e_d^{+\sigma} + U_{\text{eff}} - E_v)^{-1} + (E_v - e_d^{+\sigma})^{-1}], \tag{6}$$

where V_{pd} is the hybridization parameter and N_0 is the number of cations per unit volume.

The experimentally determined values of α and β for various DMSs are listed in Table 3. For comparison, we have also included the latest measurements of these constants in the Fe- and Co-based DMSs. Note that in both latter cases the *sp-d* exchange is significantly larger than in the $A^{II}_{1-x}Mn_xB^{VI}$ alloys.

The consequences of Eq. (5) have been extensively discussed in the literature. It is easy to see that the major consequence of the *s-d* and *p-d* interactions is to modify substantially the band structure of the parent semiconductor in the presence of an external magnetic field. We will illustrate this using the example of wide-gap $A^{II}_{1-x}Mn_xB^{VI}$ alloys. In this case the Zeeman splitting due to H_{ex} is much greater than the orbital (Landau) splitting and the spin splitting predicted by ordinary *sp* band theory, because in wide-gap materials the effective masses m^* are large and the "*sp* band" g factors are of the order of unity. The resulting Zeeman splitting of the conduction- and valence-band edges at the center of the Brillouin zone is schematically depicted in Fig. 6. Note that the splitting is qualitatively identical to "ordinary" spin splitting that takes place in nonmagnetic semiconductors (two spin levels for Γ_6, four for Γ_8), and in this sense the exchange contribution can be naturally absorbed into the concept of an effective g factor whenever the concept of a g factor is meaningful. The situation differs from the case of nonmagnetic semiconductors primarily by the fact that the magnitude of the spin splitting is now extremely large, that its be-

Table 3. Experimental *sp-d* exchange constants $N_0\alpha$ and $N_0\beta$ for the $A^{II}_{1-x}T_xB^{VI}$ alloys (in electronvolts) (N_0 is the number of cations per unit volume).

Alloy	$N_0\alpha$	$N_0\beta$	N_0 $(\alpha - \beta)$
$Zn_{1-x}Mn_xS$	...	...	...
$Zn_{1-x}Mn_xSe$	0.26	−1.11	1.37
$Zn_{1-x}Mn_xTe$	0.18	−1.05	1.23
$Cd_{1-x}Mn_xS$	0.22	−1.80	2.02
$Cd_{1-x}Mn_xSe$	0.26	−1.11	1.37
$Cd_{1-x}Mn_xTe$	0.22	−0.88	1.10
$Hg_{1-x}Mn_xS$	...	...	...
$Hg_{1-x}Mn_xSe$	0.40	−0.70	1.10
$Hg_{1-x}Mn_xTe$	0.40	−0.60	1.00
$Zn_{1-x}Co_xSe$	...	...	2.42
$Cd_{1-x}Co_xSe$	0.32	...	...
$Zn_{1-x}Fe_xSe$	0.22	−1.74	1.96
$Cd_{1-x}Fe_xSe$	0.225	−1.90	2.12

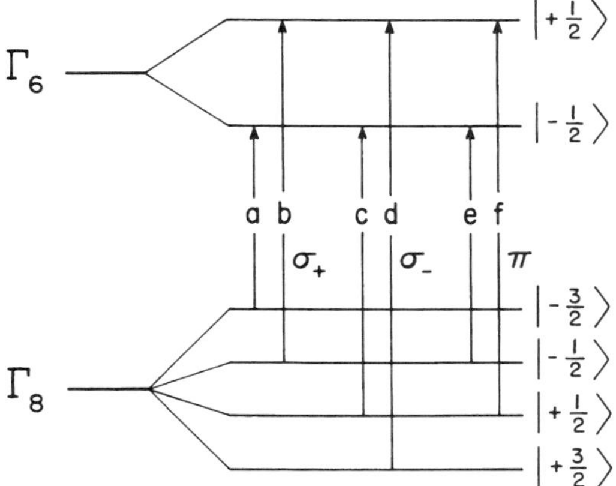

FIG. 6. A schematic picture of the spin splitting of the top of the valence band (Γ_8) and the bottom of the conduction band (Γ_6) for a wide-gap $A^{II}_{1-x}Mn_xB^{VI}$ alloy in a magnetic field. In the fourfold-degenerate valence band, $\pm\frac{3}{2}$ indicates heavy holes, $\pm\frac{1}{2}$ indicates light-hole states. The diagram also shows the electric-dipole–allowed transitions (arrows) for the two circular polarizations rotating transverse to the applied field (usually observed in the Faraday geometry, designated by σ_+ and σ_-), and for the linear polarization parallel to the field (observed in the ordinary Voigt geometry, designated by π). The Faraday-geometry transitions marked *a* and *d* are significantly stronger than those marked *b* and *c*.

havior is determined primarily by the properties of the magnetization (i.e., that it varies with temperature and is not necessarily linear in H), and that—in this approximation—it is almost completely independent of the band structure. However, the selection rules governing transitions between the levels shown in Fig. 6 remain the same as between spin levels of a nonmagnetic semiconductor.

As seen in the figure—and as already anticipated in the preceding discussion of α and β—the splitting of the heavy-hole band (indicated by $\pm\frac{3}{2}$) is considerably larger than for the conduction band, and is opposite in sign (i.e., the energy for spin-up electrons increases, and for spin-up heavy holes it decreases, when a magnetic field is applied). On the other hand, the behavior of light holes (indicated by $\pm\frac{1}{2}$ in the fourfold-degenerate Γ_8 band) is similar to that of electrons, in sign and in magnitude.

The Zeeman splittings in Fig. 6 can be conveniently represented by an effective g factor in the form

$$\Delta E = g_i^{\text{eff}}\mu_B B, \tag{7}$$

where the subscript i refers to the conduction or the heavy-hole band and g_i^{eff} is given by

$$g_e^{\text{eff}} = g_e^* - N_0\alpha x\langle S_z\rangle/\mu_B B, \tag{8}$$

$$g_{hh}^{\text{eff}} = g_{hh}^* - N_0\beta x\langle S_z\rangle/\mu_B B. \tag{9}$$

It can further be shown that for light holes

$$g_{lh}^{\text{eff}} = g_{lh}^* + \tfrac{1}{3}N_0\beta x\langle S_z\rangle/\mu B. \tag{10}$$

Here g_i^* is the g factor obtained from the band structure without the exchange contribution (typically of the order of unity), B is the applied magnetic field, x is the atomic fraction of cations replaced by the magnetic atoms, and $\langle S_z\rangle$ is the thermal average of the z component of Mn^{++} spin. $\langle S_z\rangle$ is related to the dc magnetization M by

$$M = xN_0 g_0 \mu_B\langle S_z\rangle, \tag{11}$$

where μ_B is the Bohr magneton and g_0 is the g factor of the magnetic ion (e.g., $g_{Mn} = 2$). The terms containing α and β are strongly temperature dependent because of the presence of $\langle S_z\rangle$ [see Eq. (5)], and can be as large as 100 or more at low temperatures.

2.2 The *d-d* Exchange Interaction

The interaction between the magnetic constituents of the DMS lattice (e.g., the Mn–Mn interaction in the $A^{II}_{1-x}Mn_xB^{VI}$ family of ions) takes place via the *d* electrons localized on the magnetic ions, and is hence termed the *d-d* exchange. This interaction is expressed via the exchange integral J^{dd}, which determines the magnetic properties of the given DMS system (see Sec. 4).

Experimental measurements of the *d-d* exchange integral have been made using a wide variety of techniques—high-field magnetization, neutron scattering, Raman scattering, and low-field magnetic susceptibility. The results of measurements of the nearest-neighbor exchange J are summarized in Table 4. Once again, note the substantially larger values of the exchange integral in the Co–DMS alloys.

The *d-d* exchange in the $A^{II}_{1-x}Mn_xB^{VI}$ alloys has been calculated perturbatively by Larson *et al.* (1988), who showed that J^{dd} has

Table 4. Nearest-neighbor *d-d* exchange integrals for the $A^{II}_{1-x}T_xB^{VI}$ alloys (in kelvins) (for references, see for example Furdyna, 1988, and Kossut and Dobrowolski, 1993).

Alloy	J^{dd}
$Zn_{1-x}Mn_xSe$	−13.3, −13
$Zn_{1-x}Mn_xTe$	−9.5, −10.0
$Cd_{1-x}Mn_xS$	−10.6
$Cd_{1-x}Mn_xSe$	−7.9, −8.3
$Cd_{1-x}Mn_xTe$	−6.3, −6.9
$Hg_{1-x}Mn_xSe$	−10.9
$Hg_{1-x}Mn_xTe$	−7.15
$Zn_{1-x}Co_xS$	−47 ± 6
$Zn_{1-x}Co_xSe$	−54 ± 8
$Zn_{1-x}Fe_xSe$	−22
$Cd_{1-x}Fe_xSe$	−11.25
$Hg_{1-x}Fe_xSe$	−19
$Hg_{1-x}Fe_xSe$	−18

contributions from three classes of antiferromagnetic interactions: superexchange, the Bloembergen–Rowland interaction (in which the Mn–Mn exchange is mediated by valence-band holes), and RKKY-type interaction between the Mn^{++} *d* electrons and the conduction electrons. In these alloys, the dominant contribution arises from the hybridization-induced superexchange, which involves only the anion-derived upper valence-band states and the Mn *d* states. It may thus be viewed as being mediated by the *sp-d* exchange process (i.e., *d-d* superexchange can be symbolized as a *d-sp-d* interaction), which accounts for the fact that the α's and β's discussed in the preceding section are significantly larger than the J^{dd}'s. Surprisingly, this superexchange seems to be the dominant exchange mechanism even in the narrow-gap Hg compounds where one might have expected the Bloembergen–Rowland interaction to play a significant role (Lewicki *et al.*, 1988). A useful parametrization of the superexchange interaction can be made in terms of the band-structure parameters shown in Fig. 2 (Larson *et al.*, 1988):

$$J^{dd}(R) = -2(V_{pd})^4[(e_d^{+\sigma} + U_{\text{eff}} - E_v)^{-2}(U_{\text{eff}})^{-1} + (e_d^{+\sigma} + U_{\text{eff}} - E_v)^{-3}]f(R). \quad (12)$$

In the wide-gap $A^{II}_{1-x}Mn_xB^{VI}$ alloys, the spatial dependence of the Mn–Mn exchange, $f(R)$, is well described by a Gaussian decay. The *d-d* interaction is primarily a nearest-neighbor interaction: for instance, in $Cd_{1-x}Mn_xTe$, the next-nearest-neighbor exchange constant J_2 is typically about 5 to 10 times weaker than the nearest-neighbor exchange J.

3. EFFECT OF *sp-d* EXCHANGE ON ELECTRONIC PROPERTIES

3.1 Bulk DMS Alloys

The *sp-d* exchange interactions influence physical phenomena that involve electrons in the conduction and valence bands (e.g., magnetotransport, interband and intraband magneto-optics), exciton levels (e.g., Faraday rotation), and impurity levels. This interaction has led to a multitude of new, often spectacular electrical and optical effects. We illustrate this with several representative examples. For a more comprehensive treatment of these effects, as well as for other illustrations, the reader is referred to Furdyna and Kossut (1988) or one of the review articles now available (e.g., Kossut and Dobrowolski, 1993).

3.1.1 Exchange Splitting of Exciton Transitions in DMSs Zeeman splitting of the free-exciton ground state provides one of the most direct and convenient situations for measuring exchange effects in wide-gap DMSs. For simplicity, we shall consider excitons in zinc-blende $A^{II}_{1-x}Mn_xB^{VI}$ alloys. The most important interband optical transitions are those between the band-edge states. They are shown in Fig. 6, together with electric-dipole–allowed transitions for various polarizations of light indicated by vertical arrows. Here σ_- and σ_+ denote transitions elicited by light that is circularly polarized in the plane perpendicular to the applied magnetic field (usually realized in the so-called Faraday geometry), and π corresponds to linear polarization, with the electric field of the photon parallel to the magnetic field (the ordinary Voigt geometry).

The experimentally observed magnetic field dependence of the exciton transition energies is shown in Fig. 7 for $Zn_{1-x}Mn_xTe$. Qualitatively, there is nothing unusual about the nature of the spectrum, in the sense that the same selection rules are involved as in a nonmagnetic semiconductor. What is unique is the very size of the effect—splittings of the order of 100 meV or more, to observe which

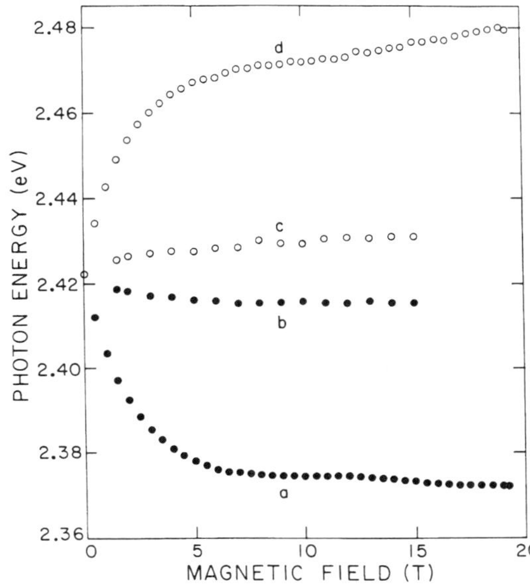

FIG. 7. Magnetic field dependence of the energies of transitions *a*, *b*, *c*, and *d* (see Fig. 6) of the 1*s* exciton in $Zn_{1-x}Mn_xTe$ ($x = 0.05$), observed at 1.4 K in magnetoreflectance in the Faraday configuration. Solid and open circles correspond to transitions observed with the σ_+ and σ_- circular polarizations, respectively (after Aggarwal *et al.*, 1986).

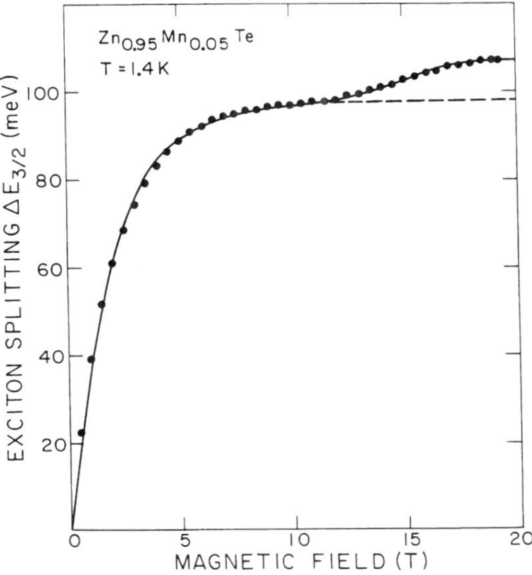

FIG. 8. Magnetic field dependence of the energy difference between the *d* and *a* transitions of the 1s exciton in $Zn_{1-x}Mn_xTe$ ($x = 0.05$), corresponding to the data in Fig. 7. Experimental data are shown by the points. The solid curve is the best theoretical fit to the data, including the contribution of Mn^{++}–Mn^{++} pairs. The dashed curve shows the Brillouin-function fit without the pairs (after Aggarwal *et al.*, 1986).

in a nonmagnetic semiconductor one would require fields of the order of a megagauss. Second, in practical terms the entire Zeeman splitting is caused by the exchange term H_{ex}. Because of the size of the splitting and the ease of measuring sharp exciton lines, this feature provides an exceptionally convenient experimental technique for determining α, β, and the details of magnetization M. This one-to-one correspondence between the splitting of excitonic features and the magnetization enables us to use optical measurements to study magnetic properties. This often is of considerable advantage, since optical measurements are very precise. Moreover, such measurements can be performed on samples whose magnetic properties are otherwise difficult to assess (e.g., because of their small size or mass, as in the case of ultrathin layers). This optical access to M has been exploited, for example, for observing the high-field magnetization steps (see Sec. 4.1.2) using exciton splitting as the measuring technique, as shown in Fig. 8. Measurements of the optical properties were also employed by researchers studying the spin-glass phase in DMSs, in this case exploiting the phenomenon of Faraday rotation (see, e.g., Rigaux *et al.*, 1986).

Zeeman splitting in wurtzite $A^{II}_{1-x}Mn_xB^{VI}$ alloys behaves similarly, with the added minor complication introduced by the fact that the top of the valence band is no longer degenerate at the Γ point. This introduces a small splitting of the exciton spectrum at $H = 0$. Except for this, everything that has been said about zinc-blende $A^{II}_{1-x}Mn_xB^{VI}$ materials applies equally well to the wurtzite members of the DMS family. A similar splitting is also present in DMSs in layer form, again because of the lowering of the cubic symmetry that automatically takes place in such structures as a result of tetragonal distortion via compressive or tensile strain within the layers.

For further details concerning excitons in DMSs, the reader is referred to the excellent and comprehensive treatment of this subject in Chapter 7 of Furdyna and Kossut (1988).

3.1.2 Giant Faraday Rotation in DMS Alloys The exceedingly large Zeeman splitting of the absorption edge and of the exciton level in wide-gap DMSs also results in a proportionately large difference in dispersion (i.e., a large difference in the index of refraction) for light of opposite circular polarizations in the Faraday geometry, particularly in the immediate vicinity of the band edge. This in turn provides the mechanism for Faraday rotation, which attains extremely large values (of the order of 1000°/cm-kG) at liquid-helium temperatures. The immense size of the effect is illustrated in Fig. 9. Because of its size and ease of measurement, Faraday rotation in DMSs—like the spin splitting of the exciton, to which it is closely linked—has frequently been used as a diagnostic tool to map the details of magnetization, both static and dynamic. Its very large size also holds promise of magneto-optical device applications, such as isolators, circulators, and other nonreciprocal devices.

Faraday rotation θ_F can be expressed in a convenient form for the temperature range where Curie–Weiss law holds [see Eq. (18) in Sec. 4.1.2]. This range includes room temperature, which is of particular importance from the point of view of quantitatively assessing the usefulness of this effect in nonreciprocal devices. Faraday rotation can then be expressed as

$$\theta_F = \frac{xPHl}{(T + \theta_0 x)} \frac{N_0(\beta - \alpha)\hbar^2\omega^2}{\{[E_g(0) + x\Delta E_g]^2 - \hbar^2\omega^2\}^{3/2}}, \qquad (13)$$

where θ_0 is the Curie–Weiss temperature defined in section 4.1.2, ΔE_g equals $E_g(1) - E_g(0)$, $E_g(1)$ and $E_g(0)$ are the energy gaps for $x = 0$ and $x = 1$, respectively (see Table 2), and P is a constant. In Eq. (13), θ_F is in degrees, T and θ_0 in kelvins, the energies $N_0(\beta - a)$, $\hbar\omega$, and E_g in electronvolts, H in gauss, and l in centimeters. The constant P has the value

$$P = (35 \pm 2) \text{ K deg/G cm},$$

obtained empirically, and is in principle in-

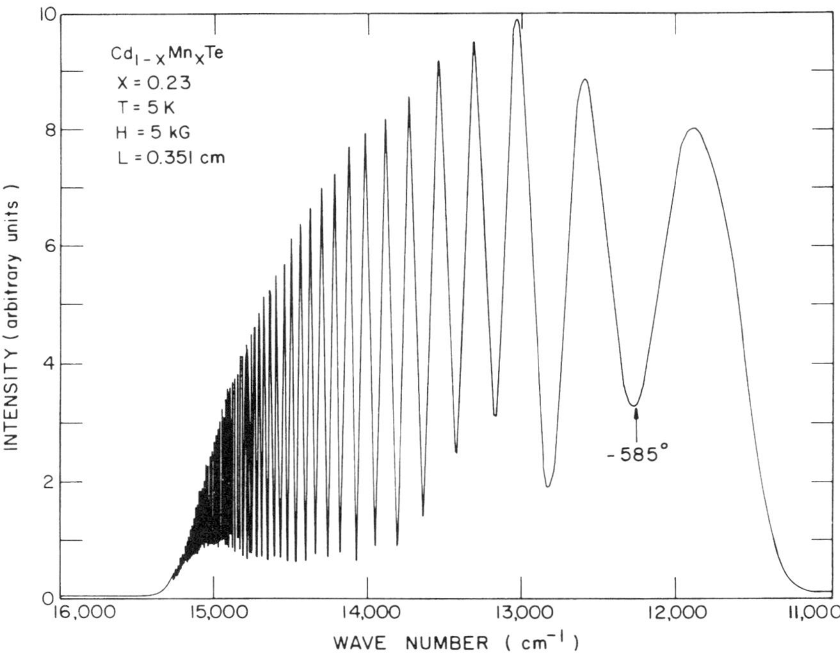

FIG. 9. Transmission as a function of wave number of initially linearly polarized light through a $Cd_{1-x}Mn_xTe$ ($x = 0.23$) slab, sandwiched between two linear polarizers, displaying giant Faraday rotation. At 15 000 cm^{-1} the rotation exceeds −8000 degrees. The experiment is carried out at 5 K and 5 T (after Bartholomew *et al.*, 1986).

dependent of the material. Note, however, that an additional dependence of θ_F on temperature will follow from the temperature dependence of E_g (see Table 2).

3.1.3 Acceptors and Negative Magnetoresistance in *p*-Type $Hg_{1-x}Mn_xTe$ One of the most striking effects of the *sp-d* exchange is the negative magnetoresistance exhibited by *p*-type $Hg_{1-x}Mn_xTe$ (Mycielski and Mycielski, 1980). This phenomenon is due to the mixing of light- and heavy-hole band contributions to the acceptor wave function in the presence of a magnetic field. The admixture arises as follows. In the presence of a magnetic field, the exchange interaction leads to an anisotropy of the constant-energy surfaces associated with the four spin-split components of the valence band near the Γ point. A $\mathbf{k}\cdot\mathbf{p}$ perturbation calculation yields

$$E_{\pm 3/2}(k) = -\tfrac{1}{2}\hbar^2\left[\left(\frac{3}{4m_{lh}} + \frac{1}{4m_{hh}}\right)(k_x^2 + k_y^2) + \frac{1}{m_{hh}}k_z^2\right] \pm 3B, \qquad (14)$$

$$E_{\pm 1/2}(k) = -\tfrac{1}{2}\hbar^2\left[\left(\frac{1}{4m_{lh}} + \frac{3}{4m_{hh}}\right)(k_x^2 + k_y^2) + \frac{1}{m_{lh}}k_z^2\right] \pm B, \qquad (15)$$

where m_{lh} and m_{hh} are the effective masses of heavy and light holes and $B = \frac{1}{6}N_0\beta x\langle S_z\rangle$.

The acceptor ionization energy in $Hg_{1-x}Mn_xTe$ depends on x, and for concentrations of interest here ($x \approx 0.15$) is about 10 meV in the absence of magnetic field (see, e.g., Mycielski, 1983; Furdyna and Kossut, 1988, Chap. 8). For B negative, the valence subband that is closest to the acceptor level is the $E_{-3/2}(k)$ sublevel. This spin sublevel is characterized by a large longitudinal mass (m_{hh}) and a small transverse mass ($\frac{4}{3}m_{lh}$). As the magnetic field is increased, the other subbands move further away from the acceptor level (spin splittings between the subbands can be considerably larger than 10 meV), so that the acceptor wave function becomes increasingly dominated by the properties of $E_{-3/2}(k)$. Thus, with increasing magnetic field, the acceptor Bohr orbit acquires an increasing admixture of the light-hole mass. Because of the decrease in the mass determining the Bohr orbit, the acceptor wave function expands spatially in the direction perpendicular to the applied magnetic field. This leads to a decrease in the ionization energy of the acceptor, hence increasing the free-carrier concentration. The increased spatial extent of the acceptor wave function also results in greater impurity conduction (i.e., hopping). As a result of the disklike shapes of the acceptor wave functions, the effects arising from the swelling of the Bohr orbit are anisotropic. These consequences are manifested in the form of a dramatic negative magnetoresistance of nearly an order of magnitude per tesla at low temperatures, which is anisotropic with respect to the direction of H (Fig. 10) (Wojtowicz and Mycielski, 1983).

3.1.4 Magnetic-Field–Induced Metal–Insulator Transition The negative magnetoresistance described above is also exhibited by *n*-type $Cd_{1-x}Mn_xSe$ (see, e.g., Shapira, 1987). The effect here is not as well understood as in the case of *p*-type $Hg_{1-x}Mn_xTe$, but is obviously very different, since it involves shallow donors and the conduction band. The ionization energy of the donors is in this case typically about 20 meV, and the Zeeman splitting of the donor level is expected to follow closely the behavior of the conduction-band edge. It is likely, however, that the negative magnetoresistance in $Cd_{1-x}Mn_xSe$ arises from the expansion of the impurity wave function in increasing magnetic field due to the destruction of magnetic polarons.

In certain ranges of impurity concentration, the negative magnetoresistance in *p*-$Hg_{1-x}Mn_xTe$ and *n*-$Cd_{1-x}Mn_xSe$ leads to a magnetic-field–induced insulator-to-metal transition (Mott transition) (Wojtowicz *et al.*, 1986). Unlike the phenomenon in ordinary semiconductors, the transition from an insulating phase to a metallic phase occurs in increasing magnetic field. Above a critical field H_c, the conductivity displays the form (Wojtowicz *et al.*, 1986)

$$\sigma(T,H) = \sigma(0,H) + A(H)T^{1/2}. \qquad (16)$$

In other words, the conductivity at absolute zero remains finite, indicating metallic behavior. The values of $\sigma(0,H)$ for *p*-$Hg_{1-x}Mn_xTe$ and *n*-$Cd_{1-x}Mn_xSe$ are shown in Fig. 11. In both these systems, the metal–insulator transition is the direct result of the

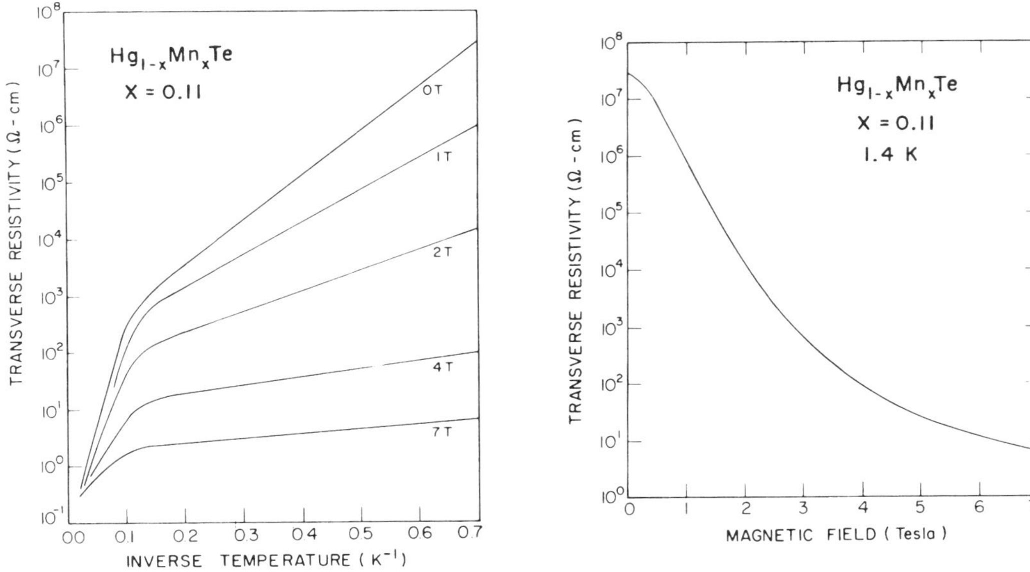

FIG. 10. Large negative magnetoresistance in the impurity-conduction region of p-type $Hg_{1-x}Mn_xTe$ ($x \approx 0.11$, acceptor concentration $\approx 10^{16}$ cm^{-3}) (after Wojtowicz and Mycielski, 1983).

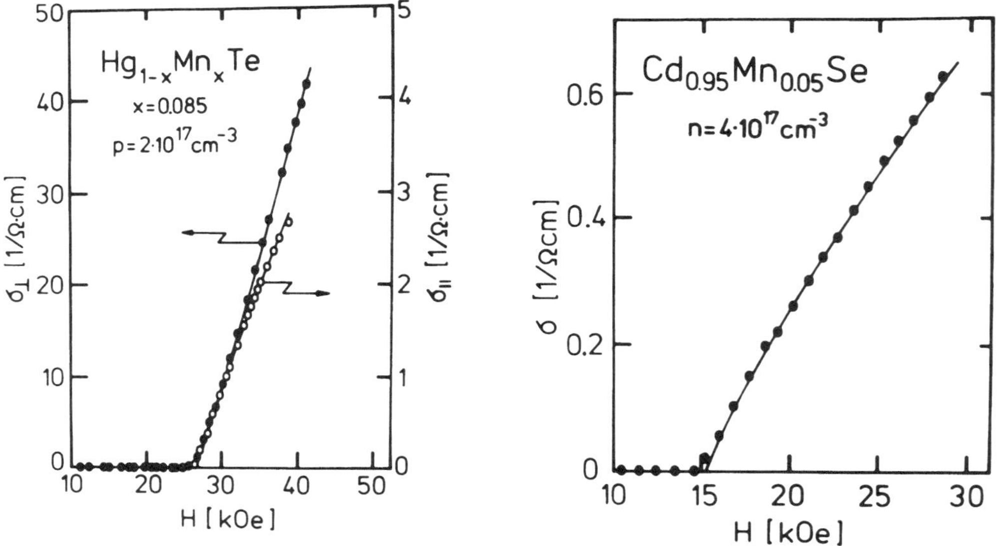

FIG. 11. Magnetic field dependence of the extrapolated zero-temperature conductivity for p-type $Hg_{1-x}Mn_xTe$ and n-type $Cd_{1-x}Mn_xSe$. The $Hg_{1-x}Mn_xTe$ shows that the transverse and longitudinal magnetoconductivities, although different, yield the same value of critical field. (After Wojtowicz *et al.*, 1986.)

giant spin splitting produced by the *sp-d* exchange. Above the critical field H_c, this splitting produces a redistribution of carriers between spin subbands that are effectively decoupled because of a Coulomb gap. The metal–insulator transition is then determined entirely by delocalized carriers within a single spin-polarized subband. This is a unique property of the metal-insulator transition in DMSs, and has led to the definition of a new universality class (Wojtowicz *et al.*, 1986).

3.1.5 Magnetic Polarons All the above effects occur in the presence of external magnetic fields. We finally mention one that is observed when the magnetic field is zero: the bound magnetic polaron (BMP). The BMP phenomenon can be understood as follows: Consider, as an example, an electron bound in a donor level. The hydrogenlike orbit of the electron encloses a large number of Mn^{++} ions (~300 for $x \approx 0.05$), with which the electron interacts through spin–spin exchange. This interaction polarizes the Mn^{++} spins, leading to a finite magnetization on the scale of the donor orbit. In addition, the spontaneous fluctuations of magnetization can also lead to a finite magnetic moment on the same local scale. As a result of both contributions the bound electron perceives a finite magnetization, and energy is therefore required to flip the spin of the donor electron even when there is no external magnetic field. Since the first observation of bound magnetic polarons in *n*-type $Cd_{1-x}Mn_xSe$ (donors) and in $Cd_{1-x}Mn_xTe$ (acceptors), a great deal of attention has been given to this subject in the DMS context.

For additional details and a full theoretical treatment, the interested reader is directed to the review of the bound magnetic polaron problem in DMSs in Chapter 10 of Furdyna and Kossut (1988).

3.2 Quantum Wells and Superlattices

Semiconductor heterostructures owe most of their spectacular properties to our ability to deposit in an atomically coherent fashion a sequence of alternating ultrathin layers of different materials, characterized by different energy gaps (see, e.g., Bastard, 1988; HETEROSTRUCTURES AND SUPERLATTICES, SEMICONDUCTOR). Quantum wells and superlattices represent an extremely important example of such heterostructures. Because of the difference in the energy gaps E_g of the constituent materials, as well as the relative alignment of the band edges (the "band offsets"), electrons and holes become confined along the direction perpendicular to the layer plane. Among numerous semiconductor heterostructures fabricated so far, those incorporating DMSs are in a class by themselves, since they combine many fascinating aspects of semiconductor quantum structures (quantum confinement, efficient radiative recombination, etc.), with the unusual magnetic signatures characteristic of DMS systems discussed in the foregoing sections. For example, the *sp-d* interaction has exciting implications for quantum wells and superlattices consisting of nonmagnetic and DMS layers. Clearly most of the important properties of semiconductor heterostructures—such as localization of electron and hole states (see ELECTRON STATES: LOCALIZED), as well as optical and electrical properties involving those states—arise in one way or another from the existence of band-gap differences (band offsets) between adjacent layers. In DMS/non-DMS heterostructures the large Zeeman splitting of the band edges in the DMS layers thus provides a handle for "tuning" this relative band alignment over a significant scale (typically over several tens of meV) simply by varying an applied magnetic field. By tuning the band alignment, one then automatically has the means of dramatically changing the properties of the heterostructure during the course of the experiment. In this section we review several striking examples of such "Zeeman tuning," unique to DMS-based heterostructures.

3.2.1 Spin Segregation The ability to tune the band alignment by the application of a magnetic field underlies the idea that—under favorable circumstances—a magnetic field can be used to induce a spin-dependent potential in a DMS/non-DMS multilayer system, so as to form a "spin superlattice."

A "spin superlattice" (SSL) is a superlattice in which carriers with opposite spin states are confined in different layers. In order to achieve such spin modulation, we look for a structure in which the energy gaps of the constituent layers are initially (i.e., in the absence of magnetic field) equal, and the band offsets at the interfaces are initially zero, as shown at the top of Fig. 12 (where DMS regions are designated by shading). When a magnetic field is applied, the large Zeeman splitting of the band edges in the DMS layers results in induced band offsets and, consequently, in a spatial separation of the spin-up and spin-down states, as shown in Fig. 12 (bottom). The SSL phenomenon has already been observed experimentally in several DMS/non-DMS multiple-quantum-well systems (see, for example, Dai *et al.*, 1991).

One particularly attractive DMS system for achieving the structure described above—which will be used as an example in this section—is the $ZnSe/Zn_{1-x}Mn_xSe$ superlattice. The energy gap of $Zn_{1-x}Mn_xSe$ exhibits a rather striking bowing with Mn concentration x at low temperatures, first decreasing with x, and then rapidly increasing. It is thus possible to find a value of x at which $Zn_{1-x}Mn_xSe$ has the same energy gap as ZnSe (this occurs for $x \approx 0.04$ at 1.5 K). Furthermore, since the valence-band offsets at $ZnSe/Zn_{1-x}Mn_xSe$ interfaces are close to zero by the common-anion rule (which works reasonably well in wide-gap II-VIs), the value of x that gives equal energy gaps in ZnSe and $Zn_{1-x}Mn_xSe$ automatically leads to zero (or, in practice, to very small) band offsets in the conduction and the valence bands, thus satisfying the properties stipulated in the preceding paragraph.

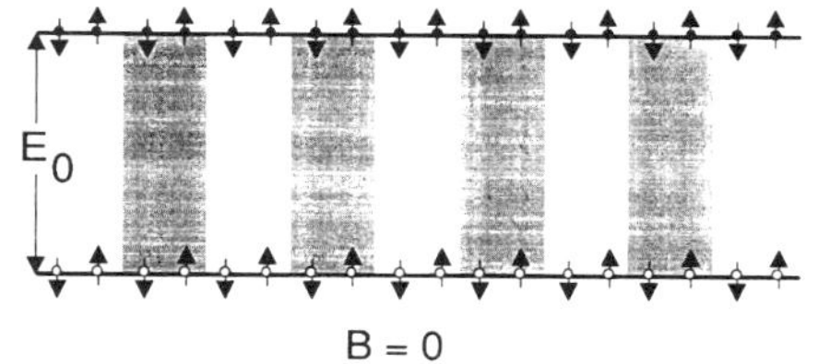

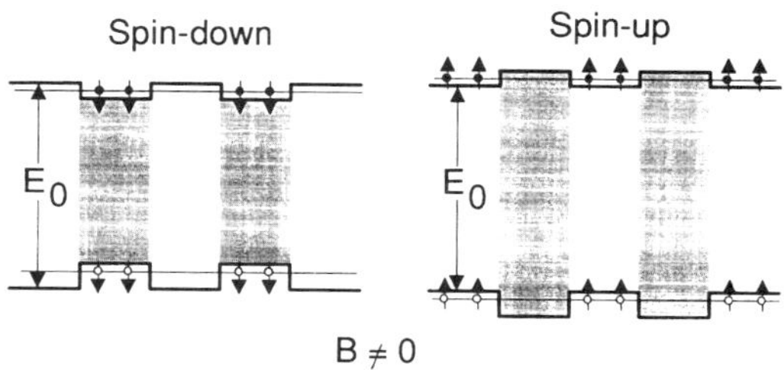

FIG. 12. A schematic diagram of the band structure of a magnetic-field–induced spin superlattice, consisting of DMS (shaded) and non-DMS (unshaded) layers with equal energy gaps at $B = 0$ (upper panel). The lower panel shows the band structure at $B \neq 0$ for spin-down (left) and spin-up (right) states of electrons and heavy holes. The Zeeman splitting in non-DMS layers is negligible. The spin states are indicated by arrows.

Examining Fig. 6 we note that, when the magnetic field is applied, the band edge of $Zn_{1-x}Mn_xSe$ for spin-down electrons will move down, and it will move up in energy for spin-down holes. The $Zn_{1-x}Mn_xSe$ layers thus become wells for holes and electrons of the spin-down orientation. Similarly, for spin-up states the conduction band moves up in energy, and the heavy-hole band moves down, so that for this spin orientation the ZnSe layers become the wells. This magnetic-field–induced band offset has therefore the consequence of spatially separating electron and hole states of different spin orientations. It is particularly convenient to observe the formation of such spin separation in magnetoabsorption experiments with the magnetic field applied perpendicular to the SSL layers (the Faraday geometry), in which only spin-conserving transitions are allowed ($\Delta S = 0$). Using left- and right-handed circular polarizations (designated by σ_L and σ_R), one can then observe separately transitions between spin-up valence- and conduction-band states (corresponding to σ_R), and those between spin-down states (σ_L).

Figure 13 (left) shows the magnetic field dependence of excitonic transitions observed with the σ_L and σ_R polarization in a $ZnSe/Zn_{0.96}Mn_{0.04}Se$ SSL, with DMS and non-DMS layers 112 Å wide. The figure illustrates very nicely the spatial separation of spins occurring in the SSL. Considering first the spin-down (σ_L) transitions, we see the rapid redshift of the σ_L absorption line, since the initial and final states of the transitions are both in the DMS layers at finite magnetic fields, and follow the field dependence of the valence- and conduction-band edges of the DMS material (see Fig. 6). In contrast, the spin-up transitions take place between states localized in the ZnSe layers, which become the wells for this spin orientation when the field is applied. This is responsible for the flatness of the transition energy at higher fields—i.e., once the offset becomes well defined. The small initial rise in energy seen in

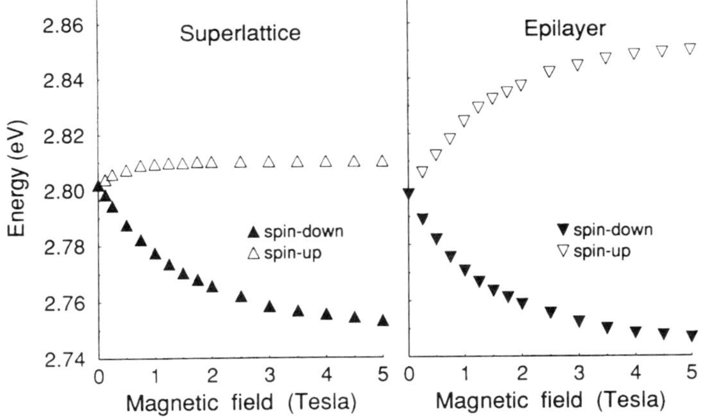

FIG. 13. Experimental energies of the Zeeman-split excitonic ground-state transitions in a $Zn_{0.96}Mn_{0.04}Se$/ZnSe (112 Å/112 Å) SSL for the two spin orientations (left panel). Note the striking asymmetry in the Zeeman shift for the spin-up and spin-down transitions relative to $B = 0$. The right panel shows the energies of the Zeeman splitting of the exciton transition observed on a $Zn_{2.96}Mn_{0.04}Se$ epilayer (after Dai *et al.*, 1991).

Fig. 13 at low fields (i.e., when the offset is just beginning to form) arises from the fact that in that region of extremely shallow wells the state in the ZnSe well is much more sensitive to the height of the (DMS) barrier, which increases with the field. As the barrier continues to increase, the energy of the state confined in the ZnSe well becomes less and less sensitive to the barrier height, and eventually flattens out, reflecting the behavior of the band edges in the nonmagnetic quantum well in which that state is now confined.

For comparison, we show on the right of Fig. 13 the position of free-exciton magnetoabsorption data observed for σ_L and σ_R polarizations in an epilayer grown with the same Mn concentration as the $Zn_{1-x}Mn_xSe$ layers of the SSL. Note that, in contrast with the SSL data, the σ_L and σ_R branches for the epilayer are quite symmetric. The asymmetry seen on the left-hand side is a characteristic signature observed in spin-separating structures.

3.2.2 Wave-Function Mapping One of the most important properties of semiconductor heterostructures is their ability to localize electrons and holes in space. Localization of carriers (as measured by their spatial wave-function distributions) is fairly obvious for low-lying states in deep quantum wells. It is an altogether different matter, however, for states at energies above the barriers, which may be confined in well layers or in barrier layers, or may not be localized at all. Furthermore, below-barrier states with energies close to the barriers (i.e., states in very shallow wells, or high-lying excited states) exhibit significant leakage into the barrier regions. Determination of the degree of leakage in the latter case, and of the region of localization in the case of above-barrier subbands, presents a major challenge.

One can meet this challenge by exploiting the fact that, in heterostructures consisting of DMS and non-DMS layers, the Zeeman splitting of a given state will reflect its weighted probability distribution over these two media. In other words, the Zeeman splitting will be determined effectively by how many localized magnetic moments the electron "sees". In this section we emphasize the application of this idea further, specifically as a tool for mapping the wave function (or probability) distribution in heterostructures.

Figure 14 shows photoluminescence from two single nonmagnetic quantum wells ($Zn_{1-x}Cd_xSe$, $x = 0.12$) of different widths, sandwiched between DMS ($Zn_{1-x}Mn_xSe$, $x = 0.20$) barriers and grown in a monolithic structure. The figure compares the photoluminescence peak positions for the two wells in $B = 0$, $B = 3$ T, and $B = 6$ T for the σ_L (spin-down) polarization. Note that for the narrower well (where the states lie nearer the top) the Zeeman redshift is considerably greater, indicating that the closer the energy of a given state is to the top of the well, the more its wave function penetrates into the barriers. By comparing the observed splitting with that exhibited by an epilayer grown from the barrier material itself, one can thus extract rather precise information on the de-

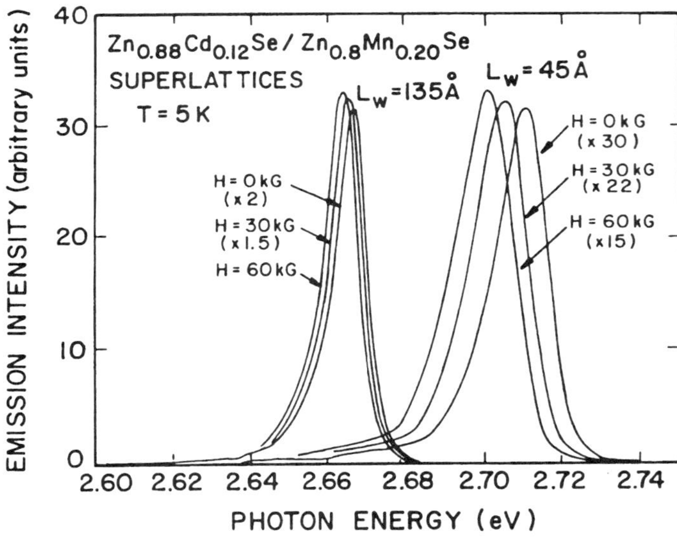

FIG. 14. Photoluminescence signal obtained from two $Zn_{0.88}Cd_{0.12}Se$ quantum wells of different thicknesses (135 and 45 Å) grown in a monolithic structure between thick $Zn_{0.8}Mn_{0.20}Se$ barriers (after Alonso *et al.*, 1991).

gree of penetration of the wave function into the DMS barrier.

This technique has been applied, for example, to pinpoint the localization of above-barrier subbands in superlattices. Using a type-I superlattice (i.e., a structure in which the same layer serves as the potential well for both electrons and holes), Zhang *et al.* (1992) have compared the Zeeman splitting of transitions between below-barrier states (which of course occur in the well layers) and between above-barrier states. On the basis of this comparison, they demonstrated that above-barrier states are localized in the barriers. Type-II superlattices are structures with a potential profile such that a given layer serves as the well for one type of carrier (say, electrons) and as the barrier for the other (say, holes). The ground-state transitions in such a structure are therefore spatially indirect. Using similar techniques it was also shown, however (Zhang *et al.*, 1993), that there exist spatially direct (i.e., type-I-like) excitons in type-II superlattices (formed by one state in the well and one above the barrier). The mechanism for localization of the above-barrier states has been theoretically shown (Luo and Furdyna, 1993) to arise from constructive interference of free-carrier waves at above-barrier energies, but the unambiguous experimental demonstration of how these states are distributed in space could only be accomplished by exploiting the differences in Zeeman splittings in the different layers, as described here.

3.2.3 Spin Lifetimes So far we have focused on the tunability of the band alignments in DMS/non-DMS heterostructures by means of an external magnetic field, and on the effect of such tuning on optical transition energies. In particular, focusing on spin-conserving transitions, we have discussed the band structure for the spin-up states separately from that for spin-down states, as if they represented physically different media. Of course the spin-split energy schemes coexist within the same sample (as in Fig. 12, lower panel), and this provides a unique opportunity for studying the lifetimes of excited spin states, as follows.

Consider an electron in the spin-up state (such as that residing in the non-DMS wells in the lower panel of Fig. 12). This electron can recombine radiatively with a spin-up hole in accord with the spin-conserving selection rule, but it can also make a nonradiative transition to a DMS well by means of a spin flip. The latter process for electrons—and the analogous relaxation of the holes to the energetically favored spin-down states in the valence band—thus compete with the radiative transitions. It is easy to see that, as the band offset increases, the nonradiative transition will become more and more probable, at the expense of the radiative spin-conserving process.

This process manifests inself dramatically in photoluminescence. If the sample is photoexcited so as to populate both spin states in the conduction band (as by linear polarization), the photoluminescence observed at $B = 0$ will contain equal intensities of the two circular polarizations. However, as the field is increased—i.e., as the field-induced energy difference between the spin-up and spin-down states becomes larger—the emitted photoluminescence will contain less and less of the higher-energy photons (the σ_R component). By analyzing the rate of decrease of this component as a function of spin splitting, one can obtain information about the spin lifetimes and, in appropriate geometries, also about the effect of spatial separation on these lifetimes (since in a heterostructure such as the SSL, a spin flip also involves a transfer to an adjacent layer). For more details we refer the reader to Jonker *et al.* (1993).

While photoluminescence contains such information in principle, extracting the values of the lifetimes is in this case rather indirect, involving a number of assumptions. An alternative—and more direct—approach is to use time-resolved methods for this purpose. In either case, the important point to emphasize is that Zeeman tuning provides a unique laboratory for such experiments, by its ability to separate the energies of the two spin states continuously, and over a significant range.

4. MAGNETIC PROPERTIES OF DMS SYSTEMS

We now present an overview of magnetic properties of the DMS alloys as they are presently understood. While the magnetic properties are interesting in their own right, it should be clear from the preceding section that they also bear on the electrical and optical properties of these DMS materials through the mechanism of the *sp-d* exchange, which implicitly involves the magnetization M (i.e., the parameter $\langle S_z \rangle$). This gives additional impetus for the need of obtaining a clear picture of magnetism in DMSs at both microscopic and macroscopic levels. The main role in the determination of magnetic properties of DMSs is played by electrons from the partially filled 3*d* shells of transition-metal atoms. It has to be remembered, however, that semiconductors that constitute host matrices of DMSs are themselves weak diamagnets. Although weak, this diamagnetism has to be included in a quantitative analysis of magnetic data.

4.1 Bulk DMS Alloys

4.1.1 Magnetic Phases Low-field static magnetic susceptibility measurements indicate that the $A^{II}_{1-x}Mn_xB^{VI}$ alloys exist in different magnetic phases, depending on the Mn concentration x and on the temperature T. Figure 15 shows the magnetic phase dia-

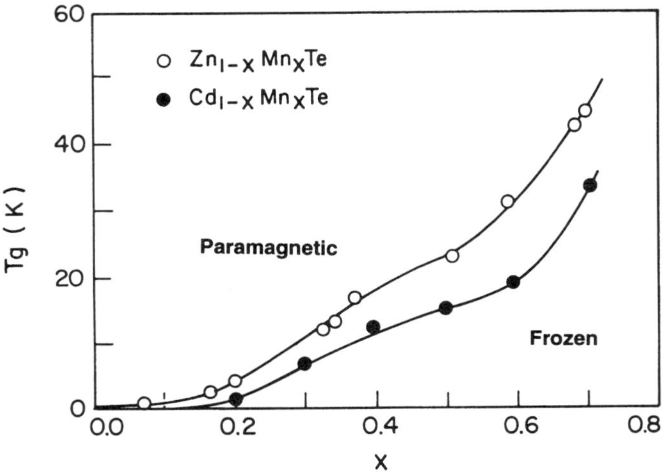

FIG. 15. Magnetic phase diagram for $Zn_{1-x}Mn_xTe$ and $Cd_{1-x}Mn_xTe$ (after Furdyna, 1988). The area above and to the left of each curve denotes the paramagnetic region for the respective materials; below and to the right corresponds to the frozen phase. In mapping out this diagram, the phase boundary is based on the occurrence of a cusp in the temperature dependence of the low-field static magnetic susceptibility. T_g denotes the temperature at which this cusp is observed.

gram for $Zn_{1-x}Mn_xTe$ and $Cd_{1-x}Mn_xTe$, which is qualitatively typical of all DMSs. The phase diagram consists of two regions: a high-temperature paramagnetic phase and a low-temperature frozen phase. The latter phase generally occurs when $x > 0.2$ (as suggested by the major portion of Fig. 15), but recent work has shown that spin freezing can also occur for lower values of x at very low temperatures.

In mapping this diagram, the phase boundary is based upon the occurrence of a cusp in the temperature variation of the low-field static magnetic susceptibility. Neutron-diffraction measurements on bulk DMS crystals (i.e., typically for $x < 0.6$) show that the low-temperature phase does not exhibit any long-range magnetic order in that range of concentrations. Assuming that the anomaly observed in the susceptibility corresponds to a genuine phase transition, we have to conclude that the low-temperature phase is a disordered but "frozen" state, such as a spin glass, with the possibility of short-range antiferromagnetic clusters.

We now proceed to discuss the susceptibility and magnetization behavior in the two phases shown in Fig. 15.

4.1.2 Magnetic Susceptibility of Paramagnetic Phase In the very dilute limit the transition-metal ions can be regarded as isolated entities. In this range of crystal compositions magnetic properties of DMSs are governed by the sequence and the distances between the energy levels of $3d$ electrons in an environment of a specified symmetry. Since the ground state of d electrons in Mn ions is a spin sextet, $S = \frac{5}{2}$ (see Sec. 1.2.3), Mn-containing DMSs are simple paramagnets. However, the lowest-lying state of Fe d electrons in a semiconducting matrix is a singlet. Therefore in Fe-containing DMSs Van Vleck paramagnetism is observed. For example, in $A^{II}_{1-x}Mn_xB^{VI}$ the static magnetic susceptibility displays a rapid variation with temperature, which roughly corresponds to a $1/T$ dependence given by the Curie law

$$\chi = C_0x/T, \tag{17}$$

where $C_0 = N_0(g\mu_B)^2S(S+1)/(3k_B)$. However, the same amount of Fe in exactly the same semiconducting host is characterized by a susceptibility that is practically independent of temperature—a behavior typical for Van Vleck paramagnetism.

When the magnetic transition-metal atoms are incorporated in DMSs in larger proportions, deviations from such simple magnetic behavior become observable. They are connected with the fact that the probability of finding an isolated atom, i.e., with no nearest magnetic neighbors, becomes smaller and smaller, and an increasing number of magnetic ions must be viewed as being members of clusters. The mean size of such clusters grows with the amount of transition-metal atoms present in the crystal. However, at low field, the magnetization M is found to be linear in H, as in the dilute case, again allowing us to define the static susceptibility χ ($M = \chi H$). As shown in Fig. 16, experimental measurements of χ indicate that at high temperatures χ displays a Curie–Weiss behavior,

$$\chi = C(x)/[T - \theta(x)]. \tag{18}$$

The Curie–Weiss temperature $\theta(x)$ obtained

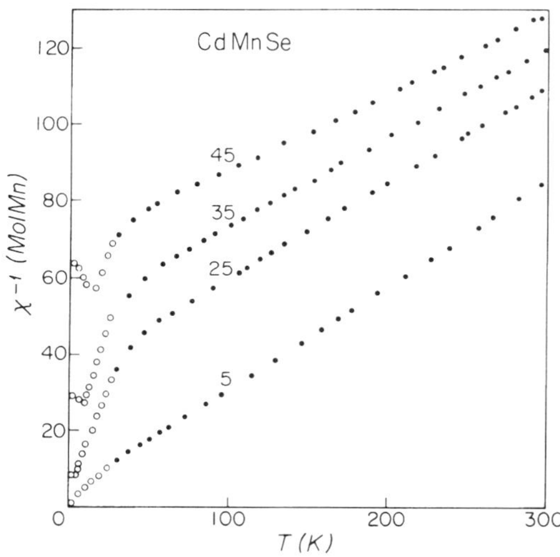

FIG. 16. Inverse magnetic susceptibility as a function of temperature for $Cd_{1-x}Mn_xSe$ of various mole fraction (given in atomic percent in the figure). The low-temperature data (open circles) were taken at 30 gauss for increasing temperature, after zero-field cooling. The high-temperature data (solid circles) were obtained at 8.5 kG. Note the linear Curie–Weiss behavior above 40 K, the characteristic downturn of χ^{-1} at lower temperatures, and the cusp seen in all but the $x = 0.05$ data, signaling the spin-glass transition (after Oseroff, 1982).

from experiment is negative, indicating antiferromagnetic interactions between the spins.

The high-temperature behavior of χ can be analytically derived by representing $A^{II}_{1-x}Mn_xB^{VI}$ alloys as randomly dilute magnetic systems with antiferromagnetic Heisenberg interactions. The calculation of χ in the high-temperature limit yields (Spalek *et al.*, 1986) $C(x) = C_0x$ [where C_0 is the Curie constant already defined in Eq. (17)], and

$$\theta(x) = x\theta_0 = -\tfrac{2}{3}xS(S + 1)ZJ/k_B. \quad (19)$$

Here Z is the number of nearest neighbors ($Z = 12$ for zinc-blende and wurtzite DMSs), and J is the nearest-neighbor d-d exchange integral discussed in Sec. 2.2. Note, then, that J can be estimated from experimental measurements of $\theta(x) = \theta_0x$; and, conversely, that the high-temperature χ can be calculated analytically for any x if J is known for a given alloy family.

For higher values of x at low temperatures, experimental results show a departure from the Curie–Weiss law, in the form of a downturn in the plot of χ^{-1} vs T, as seen in Fig. 16. This downturn is observed in all $A^{II}_{1-x}Mn_xB^{VI}$ alloys. There has been no quantitative analytical explanation of this behavior. However, it can be qualitatively understood by including higher-order expansion terms in the calculation of χ.

4.1.3 Role of Antiferromagnetically Coupled Pairs At very high fields ($B > 10$ T) the magnetization of relatively dilute DMSs ($x < 0.15$) shows a steplike behavior as a function of field (Shapira *et al.*, 1986). This is a direct consequence of the existence of antiferromagnetically coupled nearest-neighbor pairs. In the case of Mn pairs in DMSs interacting via Heisenberg-type exchange, the energy-level scheme is particularly simple. It consists of a ladder of levels, each characterized by the total spin of the pair S_T ($S_T = 0, 1, 2, 3, 4, 5$). The separations between the levels are determined by the magnitude of the exchange constant J. Analytically, the energy-level scheme is given by

$$E = -J[S_T(S_T + 1) - \tfrac{35}{2}] + g\mu_B mH, \quad (20)$$

where J is the nearest-neighbor Mn^{++}–Mn^{++} exchange integral; and for each value of S_T, m takes values $-S_T, -S_T + 1, \ldots, S_T$. For $J < 0$, the ground state is nonmagnetic at zero field, with $S_T = 0$, and it remains the ground state as long as $g\mu_BH/|J|$ is less than 2.

While this condition holds, pairs do not contribute to the total magnetization. However, once the value of $g\mu_BH/|J|$ exceeds 2, the energy for $S_T = 1$, $m = -1$ becomes the ground state, as shown in the upper part of Fig. 17. Suddenly there is a contribution to the magnetization from all the pairs, and this results in a step in the magnetization (lower part of Fig. 17). Other such steps are similarly predicted for higher fields, as also shown in the figure. The magnetic fields at which these steps occur hence provide a direct measure of J. Determination of the exchange constants from the position of the magnetization steps so far has been done only for Mn atoms in various DMSs.

For completeness, it should be remarked that similar information from the energy-level spectrum of a pair of magnetic ions can be obtained from inelastic neutron scattering (Giebultowicz *et al.*, 1987). In fact, in the case of very large exchange constants (such as those in Co^{++} systems), where magnetic fields required to observe magnetization steps would be prohibitively large, inelastic neutron scattering is the primary and most precise source of such information.

4.1.4 Spin-Glass Transition and Short-Range Antiferromagnetic Ordering As we remarked earlier, the low-field magnetic susceptibility of the $A^{II}_{1-x}Mn_xB^{VI}$ DMSs shows a well-defined cusp at a critical temperature T_g (as seen in Fig. 16), signaling a possible phase transition. This serves to define a magnetic phase diagram, showing the boundary between the high-temperature paramagnetic phase and the low-temperature frozen phase, shown in Fig. 15. The low-temperature phase is macroscopically disordered and displays many of the characteristics of a spin-glass state. In particular, the low-temperature specific heat has a linear temperature dependence and—at least in samples with $x < 0.65$—does not show any anomaly at T_g. Furthermore, remanence effects and irreversible phenomena are observed in magnetization studies below T_g. Although the precise nature of this phase remains a subject of

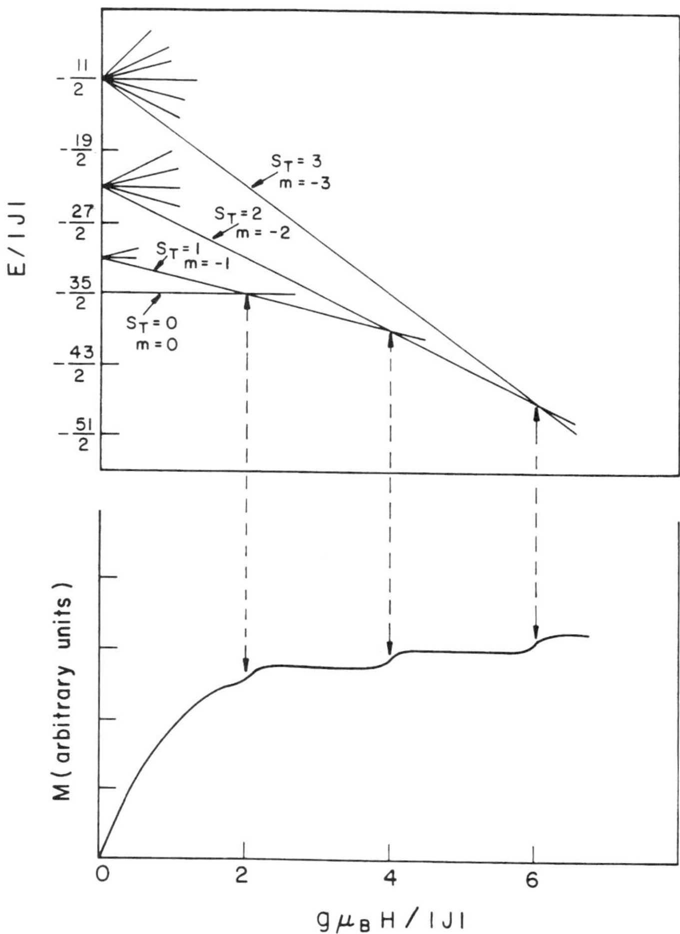

FIG. 17. Energy-level scheme for a pair of Mn^{++} ions, showing how steps occur in the magnetization at high fields. The relative size of the steps in *M* is exaggerated for clarity.

some controversy, we will for the present adhere to the tradition of calling it a spin glass, with the understanding that the rigorous magnetic classification of DMS alloys is yet to come.

The spin-glass behavior in DMSs has frequently been attributed to frustration of the antiferromagnetic interactions between the Mn^{++} ions, arising from the lattice topology of the $A^{II}_{1-x}Mn_xB^{VI}$ alloys. This is based on theoretical arguments showing that, for a diluted magnetic fcc lattice with nearest-neighbor interactions only, a spin-glass state becomes possible when the magnetic ion concentration exceeds a percolation threshold of $x \cong 0.19$. Indeed, an extrapolation in Fig. 15 of all experimental results obtained for temperatures above ~4.2 K tends toward $x \approx 0.2$. This feature has been traditionally taken as evidence that frustration is the underlying mechanism of spin-glass formation in DMSs. However, it is now clear that the spin-glass phase can form for arbitrarily small *x* at sufficiently low temperatures, as has indeed been observed for $Cd_{1-x}Mn_xSe$, $Zn_{1-x}Mn_xTe$, $Zn_{1-x}Mn_xSe$, and $Hg_{1-x}Mn_xTe$. For this reason, and because fcc and hcp lattices do in fact allow for antiferromagnetic order of the third kind (see below), we shall deemphasize the role of frustration as a primary mechanism of the spin-glass formation.

For the highest concentrations of transition-metal atoms attainable in bulk DMSs there are clear indications of magnetically ordered phases appearing at low temperatures. These conclusions were originally inferred from measurements of the susceptibility and the specific heat (where an anomaly

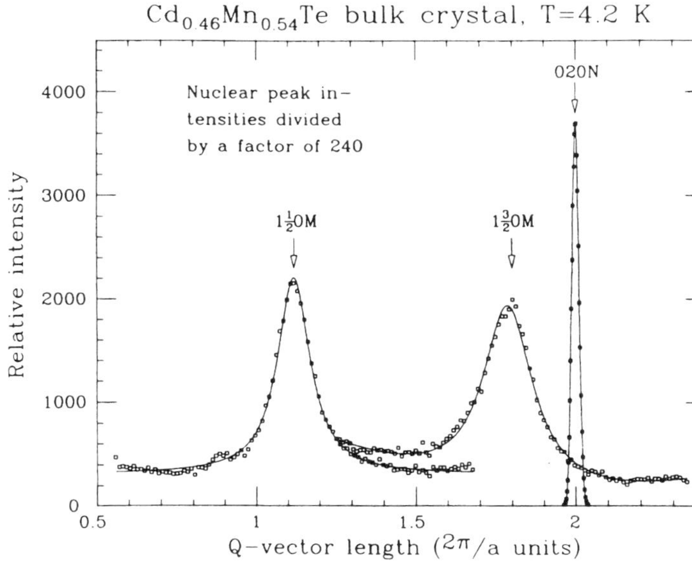

FIG. 18. Neutron-scattering data for a bulk $Cd_{1-x}Mn_xTe$ single crystal. Note that the magnetic peaks are much broader than the nuclear peaks, indicating short-range order.

is found at the transition temperature), and have been conclusively established by neutron-diffraction studies, where magnetic peaks were observed in $Cd_{1-x}Mn_xTe$, $Zn_{1-x}Mn_xTe$, and $Zn_{1-x}Mn_xSe$ (see Chap. 4 in Furdyna and Kossut, 1988). An example of such peaks is presented in Fig. 18 obtained for $Cd_{0.46}Mn_{0.54}Te$. The figure shows a sharp nuclear Bragg peak corresponding to the zinc-blende structure of the alloy, and diffuse magnetic peaks that arise from a magnetically ordered array of Mn spins. The positions of these diffuse peaks are a signature of type-III antiferromagnetic order (AFM-III), where the Mn spins are arranged as shown in Fig. 19.

Neutron-diffraction studies of $Zn_{1-x}Mn_xTe$ and $Zn_{1-x}Mn_xSe$ yield much the same results as for $Cd_{1-x}Mn_xTe$, showing spin-cluster formation, AFM-III short-range order, and satu-

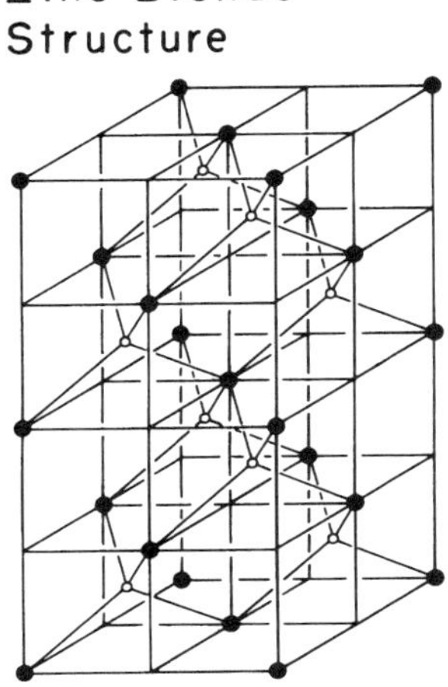

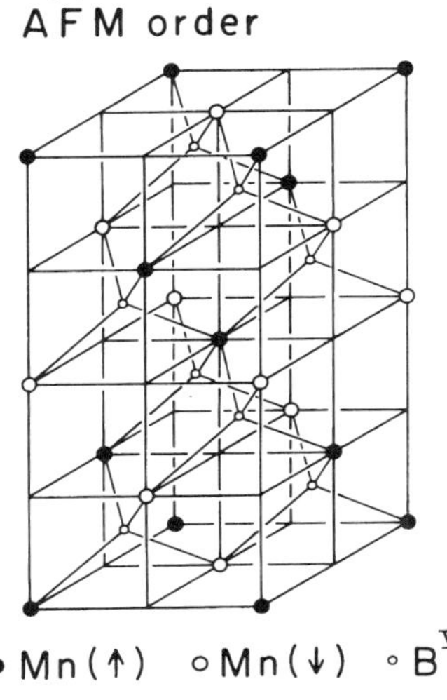

FIG. 19. The tetragonal unit cell for the AFM-III structure of zinc-blende crystals shown alongside two zinc-blende unit cells.

ration of the diffuse magnetic peak widths at low temperatures. The case of $Zn_{1-x}Mn_xSe$ is especially interesting, in that for $x > 0.30$ this material crystallizes in the wurtzite structure. Low-temperature data then reveal the existence of spin clusters correlated in the wurtzite version of the AFM-III antiferromagnetic order (i.e., that displayed by wurtzite β-MnS).

4.2 Magnetic Properties of DMS Layers and Multilayers

The magnetic properties of epitaxial DMS layers and DMS superlattices show remarkable differences compared with the behavior observed in the bulk. Magnetism in these systems has been investigated by direct magnetic probes such as neutron scattering (see, e.g., Giebultowicz *et al.*, 1992), SQUID magnetization measurements (see, e.g., Awschalom *et al.*, 1987), and spin-flip Raman scattering (Arora *et al.*, 1987), as well as indirect methods (Zeeman shift of photoluminescence peaks; Kolodziejski *et al.*, 1986). We begin by addressing the question of magnetic ordering in DMS epilayers and superlattices. Neutron-scattering experiments were performed on epitaxial layers of $Cd_{1-x}Mn_xSe$ ($x \approx 0.7$) and $Zn_{1-x}Mn_xTe$ ($0.70 < x \leq 1.0$), as well as ZnSe/MnSe, ZnTe/MnTe, and ZnTe/MnSe superlattices grown by molecular-beam epitaxy. (Although MnTe and MnSe can hardly be referred to as "diluted," we include them in this discussion because they represent the end points of the DMS families of alloys, and because their study "evolved" in the context of DMS investigations.) The epitaxial layers were 1 to 2 μm thick, and the individual superlattice layers were between typically 10 and 100 Å. All materials were in the zinc-blende phase.

The neutron diffraction observed on $Cd_{0.3}Mn_{0.7}Se$ is shown in Fig. 20 (Giebultowicz *et al.*, 1990). As in lower concentrations investigated on bulk specimens, we again observe clear signatures of AFM-III order [the presence of $(1,\frac{1}{2},0)$ and $(1,\frac{3}{2},0)$ peaks]. We note immediately, however, that—unlike the bulk data—these magnetic peaks are as sharp as the nuclear (0,2,0) peak, indicating long-range AFM-III order. Furthermore, the transition to long-range AFM-III order occurs at a sharply defined Néel temperature, above which there is no sign of any magnetic order. This again is in strong contrast to the bulk case, where the magnetic correlation length increases gradually as the temperature is decreased.

In a cubic crystal (and thus in the epilayers just discussed) the AFM-III structure exists in three "domains" corresponding to the three equivalent orientations of the unit cell shown in Fig. 19, allowed by the cubic symmetry. Studies involving the ZnSe/MnSe and

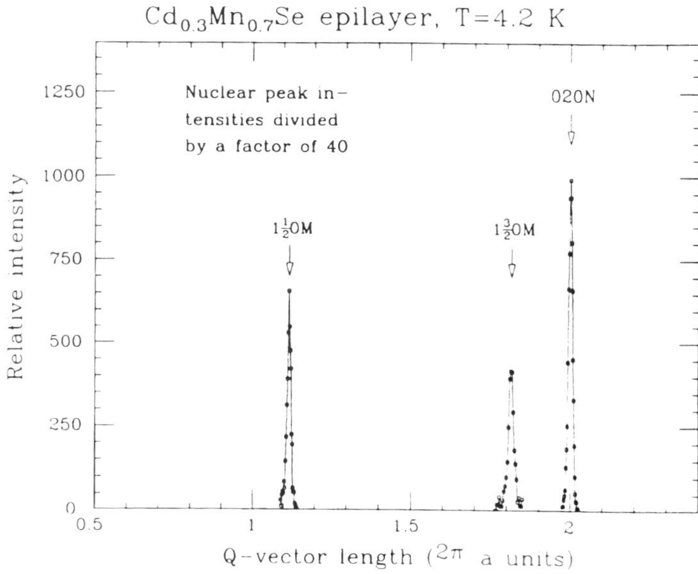

FIG. 20. Neutron-scattering data for a $Cd_{1-x}Mn_xSe$ epilayer (x = 0.70). Note that the magnetic peaks and nuclear peaks have similar widths, indicating that the magnetic order is long range.

ZnTe/MnTe superlattices show similar long-range AFM-III order, but with a major difference: Only one type of AFM-III domain is seen, with the tetragonal axis perpendicular to the plane of the epilayer. This is caused by the fact that the magnetic layer is in both these cases under compression, which lowers the originally cubic symmetry to tetragonal. In both of these superlattice families the magnetic order is long-ranged within the plane of the epilayer, while the correlation range normal to the epilayer corresponds exactly to the MnSe layer thickness. The long-range AFM-III order persists even when the individual MnSe layers are made as small as 20 Å. A totally surprising situation evolves when the magnetic layers are subjected to tension rather than compression, as is the case in ZnTe/MnSe superlattices. In that case AFM-III order disappears and, instead, a long-range helimagnetic phase sets in that is incommensurate with the lattice (Giebultowicz *et al.*, 1992). This system of magnetic semiconductors thus represents a unique system in which the ordering of spins that characterize the magnetic sublattice can be "engineered" by strain.

We mention, finally, that earlier studies of ZnSe/MnSe and $Cd_{1-x}Mn_xTe$/CdTe superlattices—albeit preliminary—have also suggested that dramatic changes should be expected as the DMS layers approach two-dimensionality. Magnetoluminescence and SQUID magnetization measurements in ZnSe/MnSe superlattices have shown that the MnSe layers are antiferromagnetically ordered when layer thicknesses are greater than a few monolayers. However, as the layer thickness approaches the monolayer limit, the MnSe layers remain paramagnetic, at least down to the lowest temperatures investigated so far (2 K). Similarly, SQUID magnetization measurements in CdTe/$Cd_{1-x}Mn_xTe$ superlattices (x in the vicinity of 0.2) have indicated a clear spin-glass transition in relatively thick $Cd_{1-x}Mn_xTe$ layers (80 Å), signaled by a cusp in the susceptibility. However, the cusp disappears when the layers are of the order of 20 Å, suggesting that the layers do not undergo a spin-freezing transition. All these studies strongly indicate that a crossover from three to two dimensions inhibits the formation of both spin-glass and antiferromagnetic order.

5. APPLICATIONS OF DMSs

DMS alloys exhibit large changes of band structure with composition, and thus constitute important candidates for those device applications where band-structure engineering is involved. Thus, for example, $Hg_{1-x}Mn_xTe$ is a viable candidate for infrared detectors, having properties very similar to those of $Hg_{1-x}Cd_xTe$. Since in this context of band-gap engineering the properties of DMSs closely parallel those of their nonmagnetic counterparts, we will not consider this issue beyond pointing out the range of band-gap adjustability implicit in Table 2.

Instead, we focus on the *sp-d* exchange, whereby effects associated with electronic spins—normally ignored in semiconductors—become sufficiently large in DMSs to merit consideration in the device context. Specifically, we will focus on Faraday rotation.

The large size of Faraday rotation and associated dispersion of circularly polarized modes in the alloys offers three types of applications to be considered. Since a rotation of the order of 45° is readily obtained at room temperature in moderate magnetic fields with a high figure of merit (see Dillon *et al.*, 1990), one may contemplate nonreciprocal optical devices (isolators and circulators) as a practical possibility. Faraday rotation may also be readily made use of in optical modulation and switching. Current efforts in this direction have examined the effect of an electric field on Faraday rotation in $Cd_{1-x}Mn_xTe$ films ($x \approx 0.5$) fabricated by ionized-cluster–beam deposition. These experiments have shown a large electric-field–induced change in the Verdet constant for photon energies in the vicinity of the energy gap. Finally, the large size of the Faraday effect might be exploited in highly sensitive optical-fiber–fed dielectric magnetometers, for measuring minute static as well as dynamic (up to gigahertz frequencies; Butler *et al.*, 1986) magnetic fields. Preliminary investigations have indicated that sensors with small dimensions (1 mm diameter) are feasible, and that static magnetic fields of 10 G can easily be resolved.

The large Faraday rotation involves, of course, large phase shifts of two counterrotating circular polarizations. If, instead of working with linear polarization, one employs a single circular polarization traveling along the applied magnetic field, one can then construct a magnetic-field–tunable phase shifter. This is of interest in interferometry as well as in optical cavity tuning (e.g., to optimize laser output).

We should note here that in the area of magneto-optical devices—especially of isolators—the use of ferromagnetic garnets is at a fairly advanced stage. These materials (e.g., Bi-enriched yttrium iron garnets) manifest immense Faraday rotations (and have a built-in magnetic field), but their use is restricted to wavelengths longer than 0.9 μm because of prohibitive absorption at shorter wavelengths. For $\lambda < 0.9$ μm DMS alloys (especially $Cd_{1-x}Mn_xTe$ and $Cd_{1-x}Mn_xSe$) are by far the best candidates for magneto-optical applications, showing larger Faraday rotations and better figures of merit (rotation/loss) than their competitors, such as paramagnetic garnets or rare-earth glasses. It should also be noted that even for $\lambda > 0.9$ μm, garnets suffer from the disadvantages that it is (at least at this stage) not possible to incorporate them monolithically in integrated optoelectronic circuits, the unit cell of a garnet consisting of 124 atoms. DMS alloys, on the other hand, are readily grown on substrates such as GaAs, making DMS films highly suitable for integrated optical-circuit structures. In this context the study of Faraday-effect phenomena in planar geometries, suitable for nonreciprocal devices based on DMS films, still requires considerable attention, since practically all magneto-optical work in DMS materials has been focused on bulklike phenomena.

GLOSSARY

Domain: In a magnetically ordered system, a region that contains a single configuration of the ordered structure.

Donor and Acceptor Levels: Typically refers to a substitutionally introduced atom, which has one more (or less) electron in the outermost shell than the atom that it replaces. The energy position of the extra (or missing) electron is called the donor (or the acceptor) level.

Epilayers: Layers that are grown epitaxially and are sufficiently thick for quantum effects to be negligible.

Exciton: A combination of an electron and a hole (the latter having a positive charge) bound together through Coulomb interaction. Both the form of the interaction and the eigenstates are very similar to those in a hydrogen atom.

Helimagnetic Phase: An ordered magnetic phase in which the spins of nearest neighbors are at an angle different from 180° and, therefore, form a helical pattern along the direction perpendicular to the spins.

Heterostructure: Consists of different materials, grown to form a multilayer coherent monolithic structure.

Magnetoresistance: A change in resistance arising from the application of a magnetic field.

Monolithic Structure: Consists of materials that are grown together without having to break the atomic arrangement of the original crystal structures. They are said to be atomically coherent.

Optical Isolators and Modulators: Devices that can be used to block reflected signals from returning to the source (isolators) and to modulate light intensity (modulators).

Radiative and Nonradiative Recombination: An electron in the conduction band can recombine with a hole in the valence band by emitting a photon, which is called radiative recombination. When the emitted energy is not in the form of photons, the process is referred to as nonradiative recombination.

Spin-Glass Phase: A magnetic phase in which orientations of spins are randomly distributed, but are nevertheless highly correlated. In other words, it behaves as a disordered spin system with individual spins "frozen."

SQUID: Acronym for superconducting quantum interference device, which has extremely high sensitivity in measuring magnetic susceptibility.

Superlattice: A monolithic structure with two or more materials grown as an alternating sequence of layers.

Universality Class: A common behavior (for very different physical systems) that typically depends on only a few pure numbers, such as the dimensionality of the space in which the phenomenon takes place.

Zeeman Splitting: The splitting of the energy levels associated with different spin states in an applied magnetic field.

Works Cited

Abragam, A., Bleaney, B. (1970), *Electron Paramagnetic Resonance of Transition Ions,* Oxford, U.K.: Clarendon.

Aggarwal, R. L., Jasperson, S. N., Becla, P., Furdyna, J. K. (1986), *Phys. Rev. B* **34,** 5894–5896.

Alonso, R. G., Oh, E., Ramdas, A. K., Luo, H., Samarth, N., Furdyna, J. K., Ram-Mohan, L. R. (1991), *Phys. Rev. B* **44,** 8009–8016.

Arora, A. K., Bartholomew, D. V., Peterson, D. L., Ramdas, A. K. (1987), *Phys. Rev. B* **35,** 7966–7972.

Awschalom, D. D., Hong, J. M., Chang, L. L., Grinstein, G. (1987), *Phys. Rev. Lett.* **59,** 1733–1736.

Bartholomew, D. V., Furdyna, J. K., Ramdas, A. K. (1986), *Phys. Rev. B* **34,** 6943–6950.

Bastard, G. (1988), *Wave Mechanics Applied to Semiconductor Heterostructures,* New York: Halsted.

Butler, M. A., Martin, S. J., Baughman, R. J. (1986), *Appl. Phys. Lett.* **49,** 1053–1055.

Dai, N., Luo, H., Zhang, F. C., Samarth, N., Dobrowolska, M., Furdyna, J. K. (1991), *Phys. Rev. Lett.* **67,** 3824–3827.

Dillon, J. F. Jr., Furdyna, J. K., Debska, U., Mycielski, A. (1990), *J. Appl. Phys.* **67,** 4917–4919.

Durbin, S. M., Han, J., Sungki, O., Kobayashi, M., Menke, D. R., Gunshor, R. L., Fu, Q., Pelakanos, N., Nurmikko, A. V., Li, D., Gonsalves, J., Otsuka, N. (1989), *Appl. Phys. Lett.* **55,** 2087–2089.

Ehrenreich, H., Hass, K. C., Larson, B. E., Johnson, N. F. (1987), in: R. L. Aggarwal, J. K. Furdyna, S. Von Molnar (Eds.), *Diluted Magnetic (Semimagnetic) Semiconductors,* Materials Research Society Symposia Proceedings Vol. 89, Pittsburgh: Materials Research Society, pp. 187–196.

Franciosi, A., Chang, S., Reifenberger, R., Debska, U., Reidel, R. (1985), *Phys. Rev. B* **32,** 6682–6687.

Furdyna, J. K. (1988), *J. Appl. Phys.* **64,** R29–R64.

Furdyna, J. K., Kossut, J. (Eds.) (1988), *Diluted Magnetic Semiconductors,* Vol. 25 of R. K. Williamson, A. C. Beer (Eds.), *Semiconductors and Semimetals,* New York: Academic.

Giebultowicz, T. M., Rhyne, J. J., Furdyna, J. K. (1987), *J. Appl. Phys.* **61,** 3537–3539.

Giebultowicz, T. M., Klosowski, P., Samarth, N., Luo, H., Rhyne, J. J., Furdyna, J. K. (1990), *Phys. Rev. B* **42,** 2582–2585.

Giebultowicz, T., Samarth, N., Luo, H., Furdyna, J. K., Klosowski, P., Rhyne, J. J. (1992), *Phys. Rev. B* **46,** 12076–12079.

Gumlich, H. E. (1981), *J. Lumin.* **23,** 73–99.

Hass, K. C., Ehrenreich, H. (1983), *J. Vac. Sci. Technol. A* **1,** 1678–1682.

Jain, M. (Ed.) (1991), *Diluted Magnetic Semiconductors,* Singapore: World Scientific.

Jonker, B. T., Krebs, J. J., Qadri, S. B., Prinz, G. A., Volkening, F. A., Koon, N. C. (1988), *J. Appl. Phys.* **63,** 3303–3305.

Jonker, B. T., Fu, L. P., Yu, W. Y., Chou, W. C., Petrou, A., Warnock, J. (1993), *J. Electron. Mater.* **22,** 489–495.

Kolodziejski, L. A., Gunshor, R. L., Bonsett, T. C., Venkatasubramaniam, R., Datta, S., Byslma, R. B., Becker, W. M., Otsuka, N. (1985), *Appl. Phys. Lett.* **47,** 169–171.

Kolodziejski, L. A., Gunshor, R. L., Otsuka, N., Gu, B. P., Hefetz, Y., Nurmikko, A. V. (1986), *Appl. Phys. Lett.* **48,** 1482–1484.

Kossut, J., Dobrowolski, W. (1993), in: K. H. J. Buschow (Ed.), *Handbook of Magnetic Materials,* Vol. 7, Amsterdam: Elsevier, Chap. 4.

Larson, B. E., Hass, K. C., Ehrenreich, H., Carlsson, A. E. (1988), *Phys. Rev. B* **37,** 4137–4154.

Lee, Y. K., Ramdas, A. K. (1984), *Solid State Commun.* **51,** 861–863.

Lewicki, A., Spalek, J., Furdyna, J. K., Galazka, R. R. (1988), *Phys. Rev. B* **37,** 1860–1863.

Luo, H., Furdyna, J. K. (1993), *Mod. Phys. Lett. B* **7,** 299–305.

Mycielski, A., Mycielski, J. (1980), *J. Phys. Soc. Jpn.* **49,** Suppl. **A,** 807–810.

Mycielski, J. (1983), *Applications of High Magnetic Fields in Semiconductors,* Berlin: Springer.

Oseroff, S. B. (1982), *Phys. Rev. B* **25,** 6584–6594.

Rigaux, C., Mycielski, A., Barilero, G., Menant, M. (1986), *Phys. Rev. B* **34,** 3313–3318.

Samarth, N., Luo, H., Furdyna, J. K., Qadri, S. B., Lee, Y. R., Ramdas, A. K., Otsuka, N. (1989), *Appl. Phys. Lett.* **54,** 2680–2682.

Shapira, Y., Foner, S., Becla, P., Domingues, D. N., Naughton, M. J., Brooks, J. S. (1986), *Phys. Rev. B* **33,** 356–365.

Shapira, Y. (1987), in: R. L. Aggarwal, J. K. Furdyna, S. von Molnar (Eds.) *Diluted Magnetic (Semimagnetic) Semiconductors,* Materials Research Society Symposia Proceedings Vol. 89, Pittsburgh: Materials Research Society, pp. 209–218.

Spalek, J., Lewicki, A., Tarnawski, Z., Furdyna, J. K., Galazka, R. R., Obuszko, Z., (1986), *Phys. Rev. B* **33,** 3407–3418.

Wall, A., Franciosi, A., Gao, Y., Weaver, J. H ., Dow, J. D., Kasowski, R. V. (1989), *J. Vac. Sci. Technol. A* **7,** 656–662.

Wojtowicz, T., Mycielski, A. (1983), *Physica B* **117 & 118,** 476–478.

Wojtowicz, T., Dietl, T., Sawicki, M., Plesiewicz, W., Jaroszynski, J. (1986), *Phys. Rev. Lett.* **56,** 2419–2422.

Zhang, F. C., Dai, N., Luo, H., Samarth, N., Dobrowolska, M., Furdyna, J. K., Ram-Mohan, L. R. (1992), *Phys. Rev. Lett.* **68,** 3220–3223.

Zhang, F. C., Luo, H., Dai, N., Samarth, N.,

Dobrowolska, M., Furdyna, J. K. (1993), *Phys. Rev. B* **47,** 3806–3810.

Further Reading

Aggarwal, R. L., Furdyna, J. K., von Molnar, S. (Eds.) (1987), *Diluted Magnetic (Semimagnetic) Semiconductors,* Materials Research Society Proceedings Vol. 89, Pittsburgh: Materials Research Society.

Brandt, N. B., Moshchalkov, V. V. (1984), *Adv. Phys.* **33,** 193–256.

Goede, O., Heimbrodt, W. (1988), *Phys. Status Solidi B* **146,** 11–62.

Shahzad, K., Samarth, N., Furdyna, J. K., Permogorov, S. (Eds.) (1992), *J. Lumin.* **52,** 183–191.

Taguchi, T., Schulman, J. N. (Eds.) (1993), *Physica B, Condensed Matter* **191.**

SEMICONDUCTORS, ELEMENTAL—ELECTRONIC PROPERTIES

Franco Bassani and Giuseppe La Rocca, *Scuola Normale Superiore, Pisa, Italy*

INTRODUCTION

The existence of solids whose resistivity is intermediate between that of metals and of insulators was known already in the 19th century. This class of materials is very large, since their resistivities at room temperature can vary between 10^{-4} and 10^{10} Ω cm. It was also recognized that the value of the resistivity is not the best criterion to define semiconductors because at absolute zero temperature they are all insulators; moreover, their resistivity is extremely sensitive to the presence of impurities. Semiconductors can rather be defined by the property that their resistivity decreases with increasing temperature in a way that is exponential with the inverse temperature. Another property of semiconductors, opposite to that of metals, is to be transparent to electromagnetic radiation of low frequency, and to absorb totally light above a critical frequency ω_0, which can be in the visible or infrared region.

These were all mysterious properties for a long time, because the classical model of free electrons, which was able to explain the metallic properties, could not be adopted, and the alternative model of electrons bound

3-527-28139-8/96/$5.00 + .50

to their atom in the solid would produce insulators. But even more mysterious was the fact, already known at the beginning of this century, that positive-charge carriers were responsible for electrical currents in some semiconductors, as shown by the sign of the Hall coefficient, and that the number of carriers greatly varied from sample to sample. The clue to an understanding of the semiconductor properties came between the two world wars with the application to solids of the concepts of quantum mechanics and the notion of energy band states. The Pauli exclusion principle led to the introduction of the concept of "hole" and explained the negative Hall constant of some semiconductors. Furthermore, the crucial role of impurities (doping) was understood.

During World War II, much interest was devoted to silicon, particularly to point-contact rectifiers for radar technology. After the war, detailed studies were undertaken of the transport properties of group-IV elements, which crystallize in the diamond structure and are basic semiconductors. Germanium in particular was of interest because its conductivity could be easily controlled by adding impurities. This led to the discovery of the transistor by Bardeen, Brattain, and Shockley in 1947, which can be singled out as the most important milestone in the field of semiconductor physics and technology; since then, solid-state electronics has had a dramatic growth. The development of microdevices also stimulated a great improvement in the growth techniques of monocrystals and in the control of their impurity content (e.g., Czochralski technique, floating-zone refinement; see CRYSTAL GROWTH), so that totally reproducible results could be obtained. A tremendous advance was the production of silicon monocrystals, of good-quality silicon surfaces, and of interfaces of silicon with metals and oxide insulators. In the late 1950s, the control of film formation, lithography, and etching led to the introduction of integrated circuits by Kilby and Noyce. After the 1960s, large-scale integrated circuitry was developed, with the consequences that are now apparent in computer and communications microelectronics, and in optoelectronics.

The theoretical understanding of the phenomena has proceeded parallel to the experimental progress. The group-IV elements are simple crystals, and their electronic structure was understood first. Their band structure has been computed since the 1960s, and it is continuously improved upon. The detailed knowledge of the Ge and Si band structures in fact preceded their verification by optical and photoemission experiments and has given the ideas for new phenomena like the Gunn oscillations.

At present we feel that the field is in full expansion, particularly in the optoelectronic effects: the transformation of light into electrical current (solar cells) and the transformation of electricity into laser radiation (semiconductor lasers). New progress in this field is forthcoming, due to the possibility of growing layers of atoms, or strings and dots of atoms, of one type on a substrate of different material by molecular-beam epitaxy. Also in this field Si and Ge may be the leading materials, because they are so abundant and form crystal alloys in all proportions. Ge_xSi_{1-x} alloys of different thickness and shape can be deposited on a Si substrate to form completely new microstructures, tailored to obtain desired properties.

1. SYMMETRY PROPERTIES OF THE DIAMOND CRYSTAL

The group-IV elemental semiconductors (germanium, silicon, and gray tin) are crystals with the symmetry of diamond; i.e., they belong to the space group $Fd3m$ (O_h^7). The space-group symmetry, along with time-reversal symmetry, determines completely the symmetry properties of a crystal at the macroscopic as well as at the microscopic level. Symmetry profoundly affects all the properties of a material, as will be shown in the following sections. Other elemental semiconductors (such as boron or selenium) have more complex crystal structures; as they are not related to Si and Ge (by far the most important materials in this class), they will not be considered in this article.

The diamond structure can be described as two interpenetrating face-centered cubic (fcc) lattices displaced by $\sqrt{3}a/4$ along the main diagonal of the conventional cubic cell of side a, as shown in Fig. 1. The lattice constants a of diamond (C), silicon (Si), germanium (Ge), and gray tin (α-Sn) are 3.56, 5.43, 5.65, and 6.46 Å, respectively. The transla-

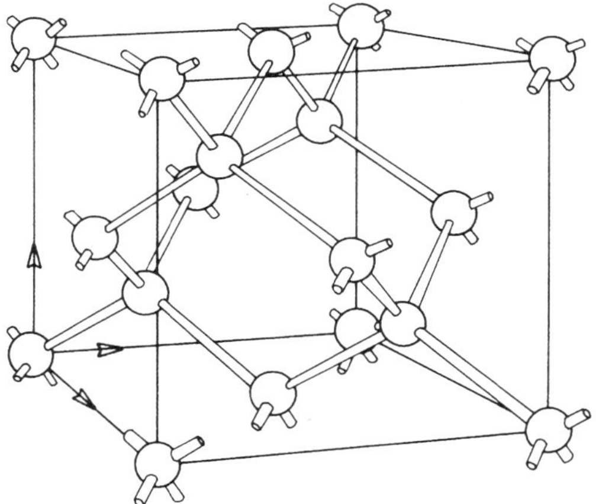

FIG. 1. Conventional (nonelementary) cubic cell of the diamond structure.

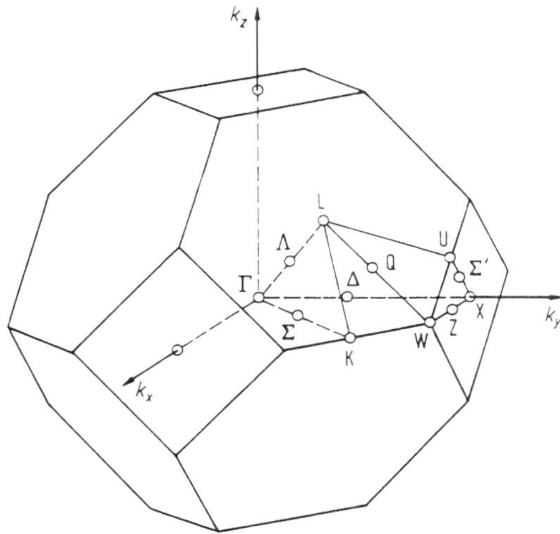

FIG. 2. Brillouin zone of the diamond crystal; the capital letters label lines and points of high symmetry.

tional symmetry is, therefore, described by a Bravais fcc lattice of primitive translations $\boldsymbol{\tau}_1 = (a/2)(0,1,1)$, $\boldsymbol{\tau}_2 = (a/2)(1,0,1)$, and $\boldsymbol{\tau}_3 = (a/2)(1,1,0)$, the crystal being invariant for translations by any vector $\mathbf{R} = n_1\boldsymbol{\tau}_1 + n_2\boldsymbol{\tau}_2 + n_3\boldsymbol{\tau}_3$ with integral values of n_1, n_2, and n_3. The reciprocal lattice is body-centered cubic (bcc) with primitive vectors $\mathbf{h}_1 = (2\pi/a)(\bar{1},1,1)$, $\mathbf{h}_2 = (2\pi/a)(1,\bar{1},1)$, and $\mathbf{h}_3 = (2\pi/a)(1,1,\bar{1})$; the reciprocal-lattice vectors, $\mathbf{G} = m_1\mathbf{h}_1 + m_2\mathbf{h}_2 + m_3\mathbf{h}_3$, with m_1, m_2, and m_3 integers, have the property $e^{i\mathbf{R}\cdot\mathbf{G}} = 1$, for any lattice translation $\mathbf{R}$. The primitive (symmetric) cell of the reciprocal lattice, i.e., the first Brillouin zone (BZ), of the diamond structure is shown in Fig. 2. The physical significance of the reciprocal lattice and of the BZ is that, as a consequence of the periodic translational symmetry (Bloch theorem), all elementary excitations in a crystal (such as phonons and electrons) can be described by wave functions of the form $\Psi(\mathbf{k},\mathbf{r}) = e^{i\mathbf{k}\cdot\mathbf{r}}u(\mathbf{r})$, where the wave vector $\mathbf{k}$ belongs to the BZ and the function u is invariant under lattice translations $[u(\mathbf{r} + \mathbf{R}) = u(\mathbf{r})]$; furthermore, when two or more such elementary excitations interact, their total wave vector is conserved modulo a reciprocal lattice vector; i.e., $\mathbf{k}_i = \mathbf{k}_f + \mathbf{G}$, $\mathbf{k}_i$ and $\mathbf{k}_f$ being the total initial and final wave vectors (processes in which $\mathbf{G} \neq 0$ are called umklapp processes).

Besides translational symmetry, the point-group symmetry characterizes the properties of the diamond structure under rotations and reflections; the point group of diamond is the full cubic group (O_h). At the microscopic level, though, half of the 48 symmetry operations of O_h are accompanied by a fractional translation such that the two interpenetrating fcc lattices are exchanged (the space group of diamond being a nonsymmorphic space group with a basis of two atoms). The point group determines the macroscopic symmetry of a crystal, in particular the form of all the material tensors (such as the polarizability tensor), the relations among elastic constants, etc. Furthermore, at the microscopic level, the point-group symmetry determines the degeneracies of elementary excitations as a function of their wave vector $\mathbf{k}$ in the BZ zone. In particular, elementary excitations at points and lines of high symmetry in the BZ, marked in Fig. 2, are characterized by their transformation properties under point-group symmetry operations that leave their wave vector unchanged (i.e., the little group of their wave vector). Finally, the O_h point-group symmetry allows one to restrict oneself to 1/48 only of the Brillouin zone (irreducible wedge), as the entire zone can be obtained from it by applying rotation and reflection symmetry operations. Standard group-theoretical methods are employed to obtain systematically all the wealth of information that symmetry alone provides (Bir and Pikus, 1974; Bassani and Pastori Parravicini, 1975). Besides space-group symmetry, time-reversal symmetry also affects the properties of a crystal. At a macroscopic level, it constrains the form of material tensors (Onsager's relations); for instance, it ensures that the conductivity tensor is symmetric: $\sigma_{ij} = \sigma_{ji}$. When external perturbations break time-reversal symmetry, such relations are correspondingly modified; for instance, when a magnetic field $\mathbf{B}_0$ is present, $\sigma_{ij}(\mathbf{B}_0) = \sigma_{ji}(-\mathbf{B}_0)$. At a microscopic level, time-reversal symmetry brings about additional degeneracies for the elementary excitation spectra. The most important with regard to electrons is that, when their spins are taken into account, in a crystal with inversion symmetry (as for diamond) at each wave vector **k** the electronic states are (at least) doubly degenerate, because the state $|\mathbf{k},\uparrow\rangle$ with spin up is degenerate with the state $|-\mathbf{k},\uparrow\rangle$ by inversion symmetry, and the latter is degenerate with the state $|\mathbf{k},\downarrow\rangle$ by time-reversal symmetry. The spin degeneracy between the states $|\mathbf{k},\uparrow\rangle$ and $|\mathbf{k},\downarrow\rangle$ is, in general, absent in materials lacking an inversion center (such as in III–V compound semiconductors) or when time-reversal symmetry is lifted, e.g., by an external magnetic field (Kim *et al.*, 1989).

2. ELECTRONIC ENERGY BANDS

Virtually all properties (transport, optical, vibrational) of a semiconductor depend on its electronic structure (see ELECTRON STRUCTURE OF SOLIDS); the latter is described by the energy bands $E_n(\mathbf{k})$, i.e., the energy levels of electronic states having a wave vector **k** and definite transformation properties under the symmetry operations of the little group of **k** (see Sec. 1). The high symmetry of the crystal enormously simplifies the problem of calculating the energy bands and provides rigorous results, such as the degeneracies at special lines and points of the Brillouin zone and the compatibility relations among different lines and points (Bassani and Pastori Parravicini, 1975). Nevertheless, a quantitative computation of the electronic structure of a given semiconductor is a very difficult task, even when basic approximations are made, such as the adiabatic approximation considering the nuclei fixed at their equilibrium sites (see Sec. 3) and the one-electron approximation (neglecting many-body correlations among the electrons). Besides, in the so-called band approximation, each electron moves in the presence of a local potential $V(\mathbf{r})$, including the interactions with the ions as well as with all other electrons in an average (mean-field) sense; the crystal potential $V(\mathbf{r})$ can be either empirically determined or calculated *ab initio*. In any case, $V(\mathbf{r})$ has the full symmetry of the crystal, and the corresponding solutions for the electronic states are Bloch functions $\Psi(\mathbf{k},\mathbf{r})$ with a wave vector **k** in the BZ (see Sec. 1). Ψ indicates a two-component wave function (spinor), one component for each of the two electron-spin orientations. In the band approximation, the fundamental equation for calculating the electronic eigenenergies $E_n(\mathbf{k})$ and eigenfunctions $\Psi_n(\mathbf{k},\mathbf{r})$ is

$$\left[\frac{p^2}{2m} + V(\mathbf{r}) + \frac{\hbar}{4m^2c^2}\boldsymbol{\sigma}\cdot(\nabla V \times \mathbf{p})\right]\Psi_n(\mathbf{k},\mathbf{r}) = E_n(\mathbf{k})\Psi_n(\mathbf{k},\mathbf{r}), \quad (1)$$

where $\mathbf{p}$ is the momentum operator ($\mathbf{p} = -i\hbar\nabla$) and the last term on the left-hand side is the spin–orbit interaction (σ_j are the 2×2 Pauli spin matrices). The spin–orbit coupling is a relativistic correction, which induces energy splittings according to the symmetry properties of spin-$\frac{1}{2}$ particles, i.e., the double-group symmetry (Bassani and Pastori Parravicini, 1975); spin–orbit effects are significant for heavy elements but, in a first approximation, can be neglected in light elements. For each **k**, the equation above has a series of solutions labeled by the band index n, and, for each n, the energy band $E_n(\mathbf{k})$ is a continuous function of **k** (see Figs. 3–5). At zero temperature, the electrons present in a semiconductor fill the levels $E_n(\mathbf{k})$ in increasing order of energy, according to the Pauli exclusion principle, up to the topmost fully occupied band (valence band), which is separated by an energy gap E_g from the lowest empty band (conduction band). For typical semiconductors, the energy gap is of the order of 1 eV: for Ge $E_g = 0.774$ eV, for Si $E_g = 1.17$ eV (at room temperature, respectively 0.66 and 1.1 eV); diamond has $E_g = 5.4$ eV and would rather be considered an insulator instead of a semiconductor; gray tin has a vanishing energy gap (zero-gap semi-

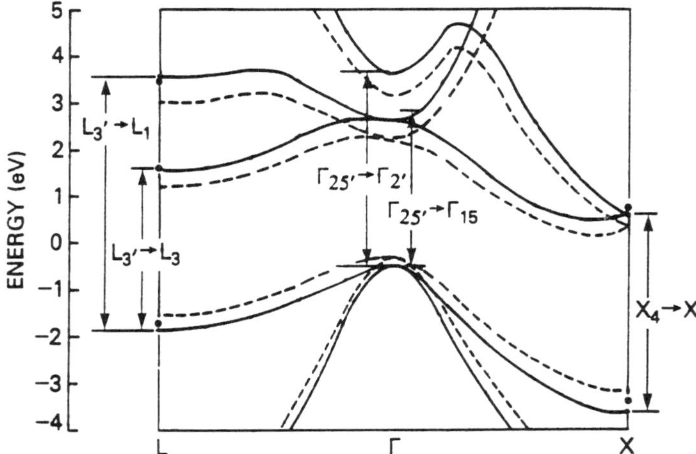

FIG. 3. Band structure of silicon: the dashed lines are from a local-density approximation *ab initio* calculation; the solid lines are from a quasiparticle *ab initio* calculation including many-body correlation effects (Pickett and Wang, 1984); the arrows indicate experimental transition energies (Madelung, 1982).

conductor or ideal semimetal). From the theoretical point of view, an energy gap of the order of 1 eV is the hallmark of semiconductors, from which all their characteristic properties depend.

2.1 Electronic Structure Calculations

The methods employed to calculate the electronic energy bands of semiconductors are numerous and diverse; for instance, while some consider a crystal as a collection of atoms (localized orbital methods), others treat it as a perturbed uniform fluid (plane-wave methods); while some are to a large extent empirical, others instead are *ab initio*. A very popular localized orbital approach is the tight-binding method in which the wave functions are expanded in a basis of linear combination of atomic orbitals (LCAO); usually, only a limited number of atomic orbitals are included, and only the interactions among a limited number of neighbors are considered. This method, at least in its sim-

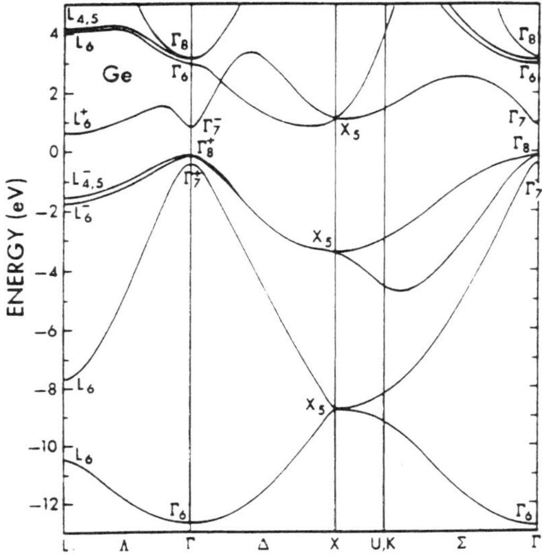

FIG. 4. Band structure of germanium from an empirical pseudopotential calculation (Chelikowsky and Cohen, 1976).

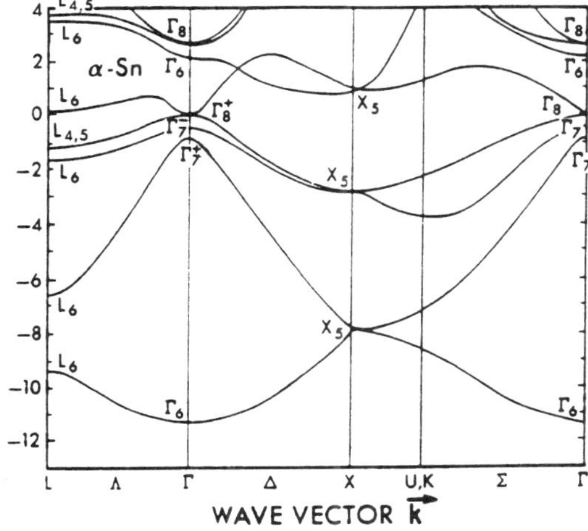

FIG. 5. Band structure of α-Sn from an empirical pseudopotential calculation (Chelikowsky and Cohen, 1976).

pler semiempirical versions, is transparent and economic; it has been very successful in calculating both electronic and structural properties of semiconductors and establishing their chemical trends (Harrison, 1989).

Among the plane-wave methods, the most important is the pseudopotential approach. In a crystal, the valence (or conduction) -band wave functions are rapidly varying near each atomic site, as they are orthogonal to those of the core states; therefore, their expansions in plane waves are very slowly convergent. It is possible, though, to replace them with pseudowave functions that are smooth (nodeless) near the atomic sites while replacing the ionic potential with a much smaller pseudopotential, in such a way that the energies $E_n(\mathbf{k})$ of valence and conduction-band states can be computed using a limited number of plane waves (Bassani and Celli, 1961); furthermore, from the pseudowave functions, the electronic charge density outside the cores can also be obtained. The pseudopotential itself (or only its first few Fourier components) can be empirically chosen to reproduce a set of data (mostly optical transition energies) or can be calculated *ab initio* and self-consistently (Bachelet *et al.*, 1982). The pseudopotential method has been very successful in conjunction with density-functional theory in the local-density approximation (LDA), i.e., using an exchange-correlation potential dependent only on the local electronic charge density. The local-density approximation, though, is not quite satisfactory; in particular, the energy gaps of semiconductors are usually significantly underestimated (even though the shapes of both valence and conduction bands are well described; see Fig. 3). This shortcoming can be mended by going beyond the one-electron approximation and calculating the quasiparticle self-energies including (approximately) many-body effects (Strinati *et al.*, 1980; Hybertsen and Louie, 1985).

2.2 Electronic Bands

Group-IV elemental semiconductors are characterized by tetrahedral bonding with sp^3 hybridization of their four outer-shell electrons. Going from diamond to gray tin, as the ionic cores become larger, the lattice constants increase and the energy gaps decrease. The general appearance of the energy bands (see Figs. 3–5) of elemental semiconductors presents close analogies, especially for valence bands, which are due to the similar chemistry of the bonds and the identical symmetry properties of the crystals. Diamond has the top of the valence band at $\mathbf{k} = 0$ (symmetry Γ'_{25}) and the bottom of the conduction band along the Δ line ($\langle 100 \rangle$ axis). The band structure of silicon, shown in Fig. 3, is very similar, but the conduction band Γ_{15} p-like states are pushed to higher energies relative to the conduction band Γ'_2 s-like states through orthogonalization with respect to the core p levels (absent in C). This effect is further enhanced as one goes from Si to Ge where also orthogonalization to the core d levels pushes the Γ_{15} states (partly d-like) even higher relative to s-like bands. We show in Fig. 4 the band structure of germanium: The effects of spin–orbit coupling (neglected for the light elements above, for which are qualitatively similar) are larger in this case, because the increase in atomic number magnifies all relativistic effects. We notice that the topmost valence band (Γ'_{25} three times degenerate without spin) splits into a Γ_8^+ fourfold degenerate and a Γ_7^+ twofold degenerate band separated by a spin–orbit energy gap $\Delta = 0.3$ eV (for Si, Δ is only 0.04 eV). Besides, the Γ'_2 conduction band (Γ_7^- in double-group notation) is lower than the Γ_{15} band (split into Γ_8^- and Γ_6^-) and, consequently, the lowest conduction state (symmetry L_6^+) is at the L point (boundary of the BZ along the $\langle 111 \rangle$ direction). The band structure of gray tin is shown in Fig. 5: Relativistic effects are even more important, and the Γ_7^- band is lower than the Γ_8^+ band (inverted gap), which turns out to be both the topmost valence band and the lowest conduction band (Groves and Paul, 1963).

All the main features of the band structures presented above are firmly established from both the theoretical and the experimental points of view. These results represent one of the most important achievements in the physics of semiconductors and provide a framework within which almost all material properties (from phonon spectra to optical and transport properties) can be understood.

3. PHONON BANDS

The vibrations of the nuclei around their sites in the perfect crystal affect both the equilibrium properties of a solid and its response to external probes. The phonons are the quantized traveling eigenmodes of vibration. As in the case of electrons, the phonon energies as functions of the wave vector **k** are arranged in bands according to the space-group symmetry of the crystal; for the diamondlike semiconductors, there are three acoustic and three optical phonon branches (Kittel, 1966).

3.1 Lattice-Dynamics Calculations

Since the early days of solid-state physics, the calculation of the lattice dynamics of semiconductors has attracted much attention for both its theoretical and its practical relevance. From the microscopic point of view, the basic problem is the evaluation of the dynamical matrix by diagonalization of which the phonon eigenenergies and eigendisplacements are obtained (Born and Huang, 1954). For this purpose, many phenomenological models have been developed, which successfully yield physical insight in the interpretation of experimental data. With modern computing capabilities, also accurate *ab initio* calculations have become feasible.

The valence-force–field model explains the lattice dynamics on the basis of a set of empirical force constants similar to those employed in the description of molecular vibrations. The potential energy is given in terms of internal coordinates such as bond lengths, bond angles, and their combinations involving a limited number of neighbors. Within such a framework, Tubino *et al.* (1972) have been able to describe well the phonon-dispersion relations of elemental semiconductors using as fitting parameters six force constants corresponding to the reaction of the valence electrons to deformations of the network of covalent bonds (e.g., stretching, bending). Another important phenomenological approach is the adiabatic bond-charge model proposed by Weber (1977) in which, besides the ions, also massless, pointlike bond charges appear in the dynamical equations. Four kinds of forces are considered: long-range Coulomb forces, ion–ion central forces (as for metals), ion–bond-charge central forces (which stabilize the bond charges at the equilibrium positions at the middle of the bonds), and, finally, bond-bending forces (of the valence-force–field type). With only four fitting parameters, the phonon-dispersion curves for Si, Ge, and α-Sn are well described; in particular, the typical flattening of the transverse acoustic branch can be explained. Many other semiempirical models have been developed (very popular are the shell models), but most of them employ a large number of free parameters (ten or more), the interpretation of which is not always transparent; therefore, they provide rather a neat interpolation of the data than physical insight.

Modern *ab initio* methods relating the vibrational properties of solids to their electronic structure are very accurate and computationally demanding. An efficient scheme, based on density-functional linear-response theory, has been developed by Giannozzi *et al.* (1991). It is a self-consistent perturbative approach in which the dynamical matrix is derived from the response of the electronic system to an external static potential representing the ion displacements. The electronic system is described using a plane-wave expansion, *ab initio* pseudopotentials, and the local-density approximation for the exchange-correlation potential (see Sec. 2); then, to compute the response to the perturbation, a Green's-function technique is employed so that the computational burden for the perturbed system is not significantly heavier than that for the unperturbed one. Excellent results have been obtained for the phonon eigenenergies and eigendisplacements of elemental (as well as polar) semiconductors; they are displayed for Si and Ge in Figs. 6 and 7, respectively.

3.2 Phonon Bands and Density of States

From the experimental point of view, the most useful technique to measure the phonon-dispersion curves $\omega_\nu(\mathbf{k})$ is the coherent scattering of thermal neutrons whereby both energy and (pseudo)momentum are exchanged with the lattice with the participation of phonons having any wave vector in the Brillouin zone. One-phonon Raman scattering can only involve phonons with a negligible wave vector (i.e., at the Γ point of the BZ). However, from the folded phonon peaks of a superlattice structure, the bulk phonon dispersion can be reconstructed (Jusserand and Cardona, 1989). The two-phonon Raman scattering spectrum, on the other hand, mimics the phonon density of states. Impu-

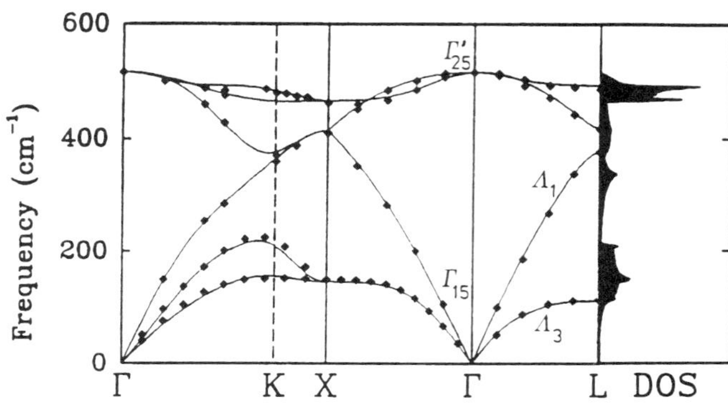

FIG. 6. Phonon-dispersion curves of Si: the data points are from inelastic neutron scattering (Nilsson and Nelin, 1972), and the lines are from an *ab initio* density-functional–perturbation theory (Giannozzi *et al.*, 1991); the rightmost panel shows the calculated phonon density of states.

rity-activated infrared absorption also gives information on the phonon density of states, but it is not always easily interpreted. Figures 6 and 7 show the phonon-dispersion curves of silicon and germanium as measured by neutron scattering, compared to modern *ab initio* calculations (also shown are the corresponding densities of states). Just as in the case for electrons, the phonon-dispersion curves are labeled according to the space-group symmetry of the diamond lattice, and the corresponding critical points clearly show up as Van Hove singularities in the phonon density of states (see Sec. 5).

4. BAND EXTREMA AND EFFECTIVE-MASS APPROXIMATION

Many of the transport and optical properties of semiconductors are dominated by carriers (electron and holes introduced by doping or excited across the energy gap) that occupy the bottom of the conduction band or the top of the valence band. An accurate description of the electronic energy bands near these extrema is given for the unperturbed crystal by the $\mathbf{k}\cdot\mathbf{p}$ method. Then, the response of the carriers to external perturbations, such as electric and magnetic fields, can be dealt with using the effective-mass approximation.

The $\mathbf{k}\cdot\mathbf{p}$ method consists of a perturbative calculation of the electronic states in the proximity of a given $\mathbf{k}_0$ point in the Brillouin zone (usually of high symmetry) corresponding to a maximum in the valence band (hole pocket) or a minimum in the conduction band (electron pocket). If we write a Bloch function at $\mathbf{k}$ in terms of all those at $\mathbf{k}_0$,

$$\psi_j(\mathbf{k},\mathbf{r}) = \sum_n c_n^j(\mathbf{k})\psi_n(\mathbf{k}_0,\mathbf{r})e^{i(\mathbf{k}-\mathbf{k}_0)\cdot\mathbf{r}}$$

the energy eigenvalues $E_j(\mathbf{k})$ at $\mathbf{k}$ as well as the expansion coefficients $c_n^j(\mathbf{k})$ for the respective eigenstates are given by the solution of the secular problem:

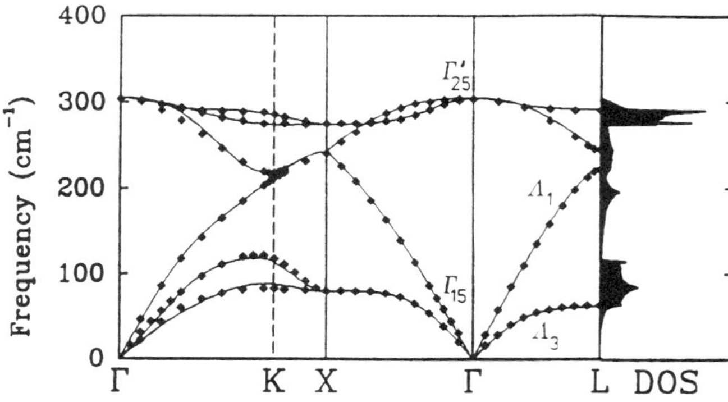

FIG. 7. Phonon-dispersion curves of Ge: the data points are from inelastic neutron scattering (Nilsson and Nelin, 1971), and the lines are from an *ab initio* density-functional–perturbation theory (Giannozzi *et al.*, 1991); the rightmost panel shows the calculated phonon density of states.

$$\sum_{n'}\left\{\left[\frac{\hbar^2}{2m}(\mathbf{k}-\mathbf{k}_0)^2+E_n(\mathbf{k}_0)-E_j(\mathbf{k})\right]\delta_{nn'}+\frac{\hbar}{m}(\mathbf{k}-\mathbf{k}_0)\cdot\mathbf{p}_{nn'}(\mathbf{k}_0)\right\}c^j_{n'}(\mathbf{k})=0, \tag{2}$$

where

$$\mathbf{p}_{nn'}(\mathbf{k}_0)=\int\psi^*_n(\mathbf{k}_0,\mathbf{r})\mathbf{p}\psi_{n'}(\mathbf{k}_0,\mathbf{r})\,d^3\mathbf{r}. \tag{3}$$

For $\mathbf{k}\simeq\mathbf{k}_0$, a limited number of bands with energies $E_n(\mathbf{k}_0)$ close to the energy $E_{n0}(\mathbf{k}_0)$ of the extremum of interest are included in the summations above. For a nondegenerate band, a single-band approximation is usually employed, the interaction with all other bands being treated in perturbation theory up to second order in $\mathbf{k}-\mathbf{k}_0$. In this case, one gets simply

$$E_{n0}(\mathbf{k})\simeq E_{n0}(\mathbf{k}_0)+\frac{\hbar^2}{2m}(\mathbf{k}-\mathbf{k}_0)^2+\frac{\hbar^2}{m^2}\sum_{n'\neq n0}\frac{[(\mathbf{k}-\mathbf{k}_0)\cdot\mathbf{p}_{n0n'}][\mathbf{p}_{n'n0}\cdot(\mathbf{k}-\mathbf{k}_0)]}{E_{n0}(\mathbf{k}_0)-E_{n'}(\mathbf{k}_0)}$$
$$\equiv E_{n0}(\mathbf{k}_0)+\frac{\hbar^2}{2}(\mathbf{k}-\mathbf{k}_0)\cdot\mathbf{m}^{-1}\cdot(\mathbf{k}-\mathbf{k}_0), \tag{4}$$

where the inverse effective-mass tensor $\mathbf{m}^{-1}$ for the extremum of interest has been introduced. Finally, measuring the energy from $E_{n0}(\mathbf{k}_0)$ and the wave vector from $\mathbf{k}_0$, and choosing the $\hat{\mathbf{1}}$, $\hat{\mathbf{2}}$, and $\hat{\mathbf{3}}$ directions along the principal axes of $\mathbf{m}^{-1}$, the standard expression for ellipsoidal pockets is obtained:

$$E(\mathbf{k})=\frac{\hbar^2}{2}\left(\frac{k_1^2}{m_1}+\frac{k_2^2}{m_2}+\frac{k_3^2}{m_3}\right), \tag{5}$$

where the effective-mass parameters can be positive, negative, or infinite. From Eq. (4), it follows that small gaps are in general accompanied by small effective masses.

Then, in the presence of slowly varying external perturbations, the dynamics of electrons and holes can be described as that of free carriers (i.e., without considering the rapidly varying crystal potential) with the (anisotropic) effective mass just discussed. The effective-mass approximation can be generalized to deal with degenerate bands and to consider the spin–orbit interaction (Luttinger and Kohn, 1955; Bir and Pikus, 1974).

4.1 Electron Pockets

The symmetry of the dispersion relation of electron (or hole) pockets as well as the values of the corresponding effective-mass parameters can be measured by cyclotron resonance, i.e., the absorption of radio-frequency radiation in the presence of an external magnetic field $\mathbf{B}$. The resonance frequency is $\omega_c=eB/m^*(\hat{\mathbf{B}})c$, with

$$m^*(\hat{\mathbf{B}})=\left[\frac{m_1m_2m_3}{m_1(\hat{\mathbf{B}}\cdot\hat{\mathbf{1}})^2+m_2(\hat{\mathbf{B}}\cdot\hat{\mathbf{2}})^2+m_3(\hat{\mathbf{B}}\cdot\hat{\mathbf{3}})^2}\right]^{1/2}. \tag{6}$$

From the dependence of the observed absorption on the magnitude and direction of the magnetic field, the symmetry of the pockets of carriers as well as their effective-mass parameters can be determined.

The conduction band of Ge has four electron pockets at the L point of the BZ (see Fig. 4; the half pockets at $\mathbf{k}$ and $-\mathbf{k}$ are to be combined together); they are ellipsoids of revolution elongated along the ⟨111⟩ directions. The values of the longitudinal mass $m_L=m_3\simeq 1.6\ m$ and of the transverse mass $m_T=m_1=m_2\simeq 0.08\ m$ have been determined from cyclotron resonance measurements such as those shown in Fig. 8 (Dresselhaus *et al.*, 1955): with $\mathbf{B}$ along $[\sqrt{3},\sqrt{3},\sqrt{2}]/2$, the four ⟨111⟩ electron pockets split into three groups inequivalently oriented with respect to the magnetic field, each one of these giving rise to a distinct resonance according to Eq. (6). The conduction band of Si has six electron pockets along the ⟨100⟩ directions (see Fig. 3); they are ellipsoids of revolution elongated along the same directions with a longitudinal mass $m_L=m_3\simeq 1.0\ m$ and a transverse mass $m_T=m_1=m_2\simeq 0.2\ m$ (measured from cyclotron resonance; see Fig. 9 where because of the choice of magnetic-field orientation two distinct electronic resonances show up).

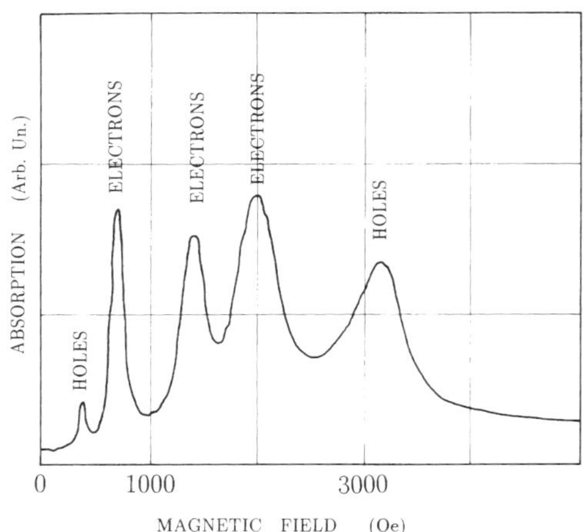

FIG. 8. Cyclotron resonance in Ge (at a frequency of 24 GHz and a temperature of 4 K) with magnetic field in the (110) plane at 60° from the [001] axis (Dresselhaus *et al.*, 1955).

4.2 Hole Pockets

The top of the valence band in both Si and Ge is at the center of the BZ: Γ'_{25} without spin, Γ_8^+, and Γ_7^+ including spin (see above). The Γ_7^+ split-off hole states can be described as a simple band with an effective mass $m_{so} \simeq 0.25m$ in Si and $m_{so} \simeq 0.08m$ in Ge. In a rough approximation, the Γ_8^+ hole states can be described as belonging to two distinct spherical pockets each characterized by an isotropic mass: the heavy-hole pocket (with mass $m_{hh} \simeq 0.5m$ for Si and $m_{hh} \simeq 0.3m$ for Ge) and the light-hole pocket (with mass $m_{lh} \simeq 0.16m$ for Si and $m_{lh} \simeq 0.044m$ for Ge). Also the hole masses have been determined from cyclotron resonance measurements such as those shown in Figs. 8 and 9. This simple picture is not quite correct; even when the hole kinetic energies are larger than $\hbar\omega_c$ and only two fundamental hole cyclotron resonances appear (classical regime), the heavy-hole pocket is significantly anisotropic. Furthermore, when the hole kinetic energies are comparable to $\hbar\omega_c$, a very complex hole cyclotron-resonance spectrum appears (quantum regime), as shown in Fig. 10 (Hensel and Suzuki, 1974). A complete description of the hole states can be given using the multiband effective-mass approach (Luttinger and Kohn, 1955), in which the valence-band degeneracy is fully accounted for.

5. DENSITY OF STATES AND CRITICAL-POINT ANALYSIS

Many material properties are determined to a large extent by the density of states $g(E)$, i.e., the number of phonon (electron, etc.) states per unit volume and unit energy. Since the elementary excitations in a crystal have energy spectra arranged in bands $E_n(\mathbf{k})$, the density of states can be written as

$$g(E) = \sum_n \int_{\mathrm{BZ}} \frac{d^3k}{(2\pi)^3}\, \delta(E_n(\mathbf{k}) - E), \qquad (7)$$

where the integral is over the Brillouin zone and the summation is over all bands (and includes the multiplicity of degenerate bands,

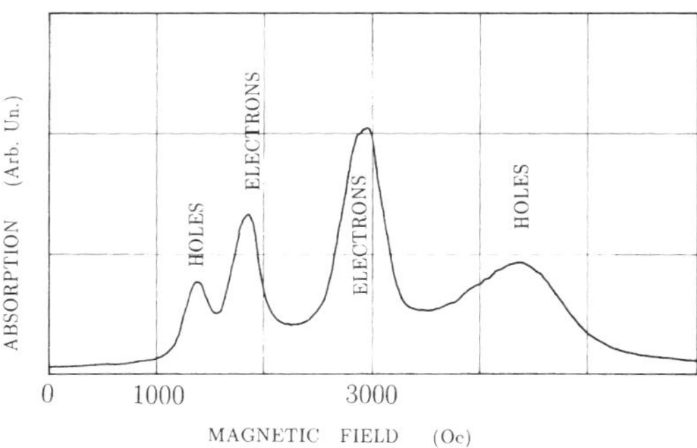

FIG. 9. Cyclotron resonance in Si (at a frequency of 24 GHz and a temperature of 4 K) with magnetic field in the (110) plane at 30° from the [001] axis (Dresselhaus *et al.*, 1955).

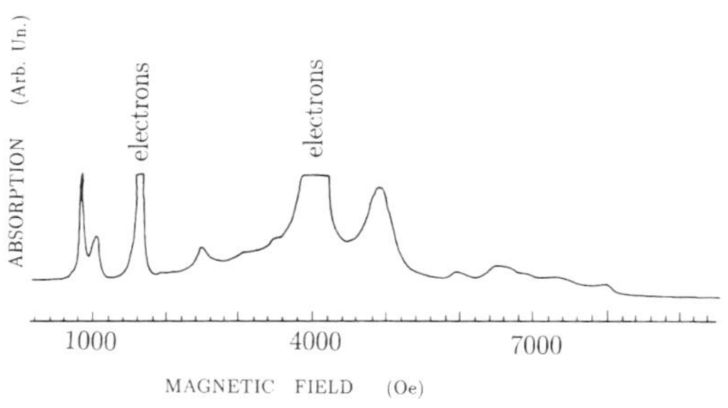

FIG. 10. Cyclotron resonance in Ge (at a frequency of 52.9 GHz and a temperature of 1.2 K) with magnetic field along the [111] axis; the complex spectrum of hole transitions is evident (Hensel and Suzuki, 1974).

at least 2 for electronic states in elemental semiconductors). An equivalent expression is

$$g(E) = \sum_n \frac{1}{(2\pi)^3} \int_{S(E)} \frac{dS}{|\nabla_{\mathbf{k}} E_n(\mathbf{k})|}, \tag{8}$$

where the two-dimensional integral is over a surface in $\mathbf{k}$ space for which $E_n(\mathbf{k}) = E$. The latter expression for $g(E)$ clearly shows that a large contribution to the density of states comes from points where the gradient of $E_n(\mathbf{k})$ with respect to $\mathbf{k}$ vanishes (critical points).

Topological considerations indicate that each band has at least a minimum set of critical points: one maximum, one minimum, and six saddle points (three of type M_1 and three of type M_2). Around a critical point $\mathbf{k}_0$, the energy band $E_n(\mathbf{k})$ has in general a quadratic expansion in $\mathbf{k} - \mathbf{k}_0$, as given by Eq. (5): if all the coefficients (i.e., the effective masses for the electronic case) are positive, the critical point is a minimum (type M_0); if one only is negative, it is a saddle point of type M_1; if two are negative, it is a saddle point of type M_2; if all are negative, it is a maximum (type M_3). The energies of the critical points increase going from those of type M_0 to M_1, M_2, and M_3; the three critical points of type M_1 may correspond to the same energy, and likewise those of type M_2. A critical point having energy E_0 is accompanied by a singularity in the density of states around E_0 having a characteristic shape (shown in Fig. 11) depending on the type of the critical point (van Hove singularities):

$$M_0 \Rightarrow g(E) = \begin{cases} B + O(E - E_0) & \text{if } E < E_0 \\ B + A(E - E_0)^{1/2} + O(E - E_0) & \text{if } E > E_0 \end{cases}$$
$$M_1 \Rightarrow g(E) = \begin{cases} B - A(E_0 - E)^{1/2} + O(E - E_0) & \text{if } E < E_0 \\ B + O(E - E_0) & \text{if } E > E_0 , \end{cases}$$
$$M_2 \Rightarrow g(E) = \begin{cases} B + O(E - E_0) & \text{if } E < E_0 \\ B - A(E - E_0)^{1/2} + O(E - E_0) & \text{if } E > E_0 \end{cases}$$
$$M_3 \Rightarrow g(E) = \begin{cases} B + A(E_0 - E)^{1/2} + O(E - E_0) & \text{if } E < E_0 \\ B + O(E - E_0) & \text{if } E > E_0 \end{cases}$$

where B is a constant and $O(E - E_0)$ is a term of order $E - E_0$ when E tends to E_0. In some cases, the analytical behavior of the density of states at a critical point is different from that just described because for the corresponding $E_n(\mathbf{k})$ the quadratic expansion does not hold; in particular, for the acoustic-phonon branch around the minimum at $E = 0$, the dispersion is linear and $g(E)$ increases as E^2 (Debye spectrum).

The location of critical points in the BZ is mainly (but not completely) determined by symmetry considerations. For the elemental semiconductors with the diamond lattice, the points Γ, X, L, and W in the BZ (see Fig. 2) are always critical points; other critical points usually occur along lines or planes of symmetry, but critical points at a generic wave vector $\mathbf{k}$ cannot be excluded. Photoemission spectra (see Sec. 7) often give direct information on $g(E)$; optical spectroscopies (see Sec. 6) give access to the joint density of states, which has analogous singularities. For phonons, $g(E)$ is obtained from inelastic neutron scattering (see Sec. 3). In an experimental phonon or electron density of states, the van Hove singularities may be difficult to identify; using modulation techniques (see Sec. 10), the derivative spectra can be obtained that show very clearly the stronger singularities in the first (and higher-order) derivatives of $g(E)$. The study of van Hove singularities has been instrumental in establishing the quite detailed picture of phonon and electronic energy bands of elemental semiconductors (see Secs. 2 and 3) on a sound experimental basis (Brust *et al.*, 1962; Cohen and Chelikowsky, 1988).

6. OPTICAL PROPERTIES

The most precise way to investigate the atomic and electronic structures of all substances is to study their response to electromagnetic radiation. This general principle, however, has held for semiconductors only since the 1960s, when the optical quality of

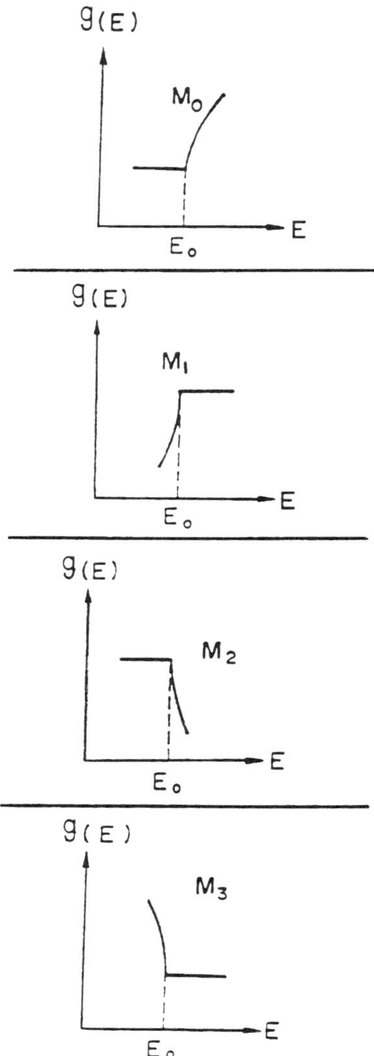

FIG. 11. Schematic representation of the three-dimensional density of states $g(E)$ near critical points of type M_0, M_1, M_2, and M_3; see text for details.

the surfaces made it possible to measure the real and the imaginary parts of the dielectric function as functions of the light frequency. Before, it was not possible to measure light dispersion and absorption because the absorption is so strong when the frequency is higher than the threshold frequency that no radiation goes through the semiconductor, even for very thin layers. This difficulty was finally overcome by measuring the reflectivity over a large frequency range with sufficient accuracy so that application of the Kramers–Kronig (K.K.) relations made it possible to determine the real and the imaginary parts of the reflection coefficient. From this the dispersion and the absorption could be obtained, and the bases for the experimental verification of the electronic structure of semiconductors were established (Philipp and Ehrenreich, 1963; Greenaway and Harbeke, 1968; Bassani and Pastori Parravicini, 1975).

6.1 General Concepts and Methods

We recall that the normal-incidence reflection coefficient, defined as the ratio between the electric fields of reflected and incident waves, is related to the complex index of refraction $\tilde{n}$ by the Fresnel relation

$$r = \frac{E_r}{E_i} = \frac{\tilde{n} - 1}{\tilde{n} + 1} = |r|e^{i\theta} \tag{9}$$

and, taking the logarithm,

$$\log r = \log |r| + i\theta. \tag{10}$$

Since this is an analytic function of the frequency ω in the upper complex plane $\omega + i\eta$, the following K.K. relation holds true:

$$\theta(\omega) = -\frac{2\omega}{\pi} \mathrm{P} \int_0^\infty \frac{\log |r(\omega')|}{\omega'^2 - \omega^2} d\omega', \tag{11}$$

where P denotes the principal value; and from the measured reflectivity $|r(\omega')|^2$, $\theta(\omega)$ can be obtained, and consequently real and imaginary parts of $\tilde{n} = n + ik$ can be derived from Eq. (9). This amounts to the knowledge of dispersion and absorption since the complex dielectric function is related to $\tilde{n}$ by

$$\tilde{\epsilon} = \epsilon_1 + i\epsilon_2 = \tilde{n}^2. \tag{12}$$

The dissipation is related to the imaginary parts, since the absorption coefficient is given by

$$\alpha(\omega) = \frac{2k\omega}{c} = \frac{\epsilon_2 \omega}{nc}, \tag{13}$$

and the real part $\epsilon_1 = n^2 - k^2$ gives the dispersion.

To relate the measured optical constants to the electronic structure, we must consider the coupling dipolar interaction between the electromagnetic radiation and the electrons,

$$H'(\mathbf{r}) = e\mathbf{E}\cdot\mathbf{r}, \qquad (14)$$

where $\mathbf{E} = \mathbf{E}_0 e^{-i\omega t} + \mathbf{E}_0^* e^{i\omega t}$ denotes the electric field of the radiation and $\mathbf{r}$ is the position of the electron that can make a transition between different electronic states. We recall that the transition probability per unit time between initial state $|i\rangle$ and final state $|f\rangle$ is given by the Fermi "golden rule"; in first order

$$P^{(1)}_{i\to f} = \frac{2\pi}{\hbar} |\langle f|H'|i\rangle|^2 \delta(E_f - E_i \pm \hbar\omega), \qquad (15)$$

and in second order

$$P^{(2)}_{i\to f} = \frac{2\pi}{\hbar} \left| \sum_n \frac{\langle f|H'|n\rangle\langle n|H'|i\rangle}{E_n - E_i \pm \hbar\omega} \right|^2 \times \delta(E_f - E_i \pm \hbar\omega \pm \hbar\omega), \qquad (16)$$

where in the energy-conserving delta functions we include the energies of the photons absorbed ($-\hbar\omega$ in first order) or emitted ($+\hbar\omega$ in first order). In the case of semiconductors we must sum over all occupied initial states and empty final states in the unit volume to compute the electromagnetic power dissipated at a given frequency. In first order,

$$W(\omega) = \frac{1}{V} \sum_{i,f} P^{(1)}_{i\to f} \hbar\omega (f_i^0 - f_f^0), \qquad (17)$$

where V is the crystal volume and f_n^0 is the Fermi distribution function for state n:

$$f_n^0 = \frac{1}{\exp[(E_n - \mu)/kT] + 1}. \qquad (18)$$

The absorption coefficient is by definition

$$\alpha(\omega) = \frac{\text{dissipated power per unit volume}}{\text{incoming flux}} = \frac{W(\omega)}{(n^2E^2/2\pi)(c/n)}. \qquad (19)$$

We can observe that it is better to consider the imaginary part of the dielectric function because it does not depend on the real part of the index of refraction. From the above expressions, we obtain in first order for the dissipative function

$$\epsilon_2(\omega) = 4\pi^2 \frac{1}{V} \sum_{i,f} |\langle f|e\mathbf{r}|i\rangle|^2 \delta(E_f - E_i - \hbar\omega), \qquad (20)$$

where the summation is over all occupied initial states and empty final states. The real part $\epsilon_1(\omega)$ is then obtained from the K.K. dispersion relation

$$\epsilon_1(\omega) = 1 + \frac{2}{\pi} \mathrm{P} \int_0^\infty \frac{\omega' \epsilon_2(\omega')}{\omega'^2 - \omega^2} d\omega'. \qquad (21)$$

In second order we can have two-photon absorption or emission, or Raman processes where one photon is absorbed and one emitted.

It is also possible to have second-order processes where only one photon is involved, provided we consider the additional interaction with the vibrations of the lattice, which may be seen as a phonon field, which contributes an additional interaction to (14) of the type

$$H'(\mathbf{r}) = \sum_{\mathbf{q}\nu} (V_\mathbf{q}^{(\nu)} e^{i\mathbf{q}\cdot\mathbf{r}} a_{\mathbf{q}\nu} + V_\mathbf{q}^{(\nu)*} e^{-i\mathbf{q}\cdot\mathbf{r}} a_{\mathbf{q}\nu}^+), \qquad (22)$$

where $a_{\mathbf{q}\nu}$ ($a_{\mathbf{q}\nu}^+$) are destruction (creation) operators, which refer to phonons of mode ν and with wave vector $\mathbf{q}$, such that

$$a_{\mathbf{q}\nu}|n_{\mathbf{q}\nu}\rangle = \sqrt{n_{\mathbf{q}\nu}}|n_{\mathbf{q}\nu} - 1\rangle, \quad a_{\mathbf{q}\nu}^+|n_{\mathbf{q}\nu}\rangle = \sqrt{n_{\mathbf{q}\nu} + 1}|n_{\mathbf{q}\nu} + 1\rangle, \qquad (23)$$

$n_{\mathbf{q}\nu}$ being the phonon number, given by the statistical distribution function of Bose–Einstein:

$$n_{\mathbf{q}\nu} = \frac{1}{\exp[\hbar\omega_\nu(\mathbf{q})/k_BT] - 1}, \qquad (24)$$

where we have denoted with $\hbar\omega_\nu(\mathbf{q})$ the phonon energy. Evidence has been found of all the above processes, which has provided a verification of the electronic and also of the vibrational structure of elemental semiconductors.

6.2 Direct Interband Transitions

Most of the optical absorption in crystals is due to interband transitions, i.e., to promotion of electrons from the valence bands $E_v(\mathbf{k})$ to the conduction bands $E_c(\mathbf{k})$, the values of $\mathbf{k}$ being the same because the photon wave vector is very small (direct transitions). This means that the summations on initial and final states in (20) can be replaced by an integration on the $\mathbf{k}$ vector space, on which the electron energies depend. Because of the δ-function condition in (20), we obtain a singular behavior in the vicinity of critical points of the Brillouin zone, where

$$\nabla_k[E_c(\mathbf{k}) - E_v(\mathbf{k})] = \mathbf{0}. \tag{25}$$

These give different types of singularities in the joint density of states (Van Hove singularities), analogous to those in the density of states discussed in Sec. 5. These structures have been found in all elemental semiconductors, and this has allowed the identification of the critical-point transition peaks and the verification of the band structure.

As a typical example, we show in Fig. 12 the optical excitation spectrum of Ge, measured as explained above and computed using the band-structure calculations described in Sec. 2 and Eq. (20) (Brust *et al.*, 1962). Similar results are obtained for Si and C, but in the case of gray tin the energy gap between valence and conduction bands vanishes because the Γ_7^- state falls below the Γ_8^+ state, which is half full (see Sec. 2). We stress that not only have the critical points been identified but also the full band structure has been put to test by the described analysis of the optical properties. An immediate verification comes from a detailed study of the edge and of the E_1 peaks at 2.1 eV. They turn out to be split by the spin–orbit separation as expected from the band structure described in Sec. 2 ($\Delta \simeq 300$ meV in Ge at the edge and about 2/3 of it at E_1). A further verification comes from the calculation of the dipole matrix elements at all points of the Brillouin zone and for all possible transitions to higher conduction bands. This justifies the fact that almost all of the possible transitions take place to the lowest empty conduction bands, which can also be proved by verifying the f-sum rule in the optical range (Bassani and Altarelli, 1983):

$$\int_0^{\omega_M} \omega' \epsilon_2(\omega')\, d\omega' \simeq \frac{2\pi^2 e^2 n_{\text{eff}}}{m}, \tag{26}$$

where in this case the frequency ω_M is just above the spectrum in the visible region and n_{eff} is about equal to the number of valence electrons that take part in the optical transitions (eight per unit cell in elemental semiconductors).

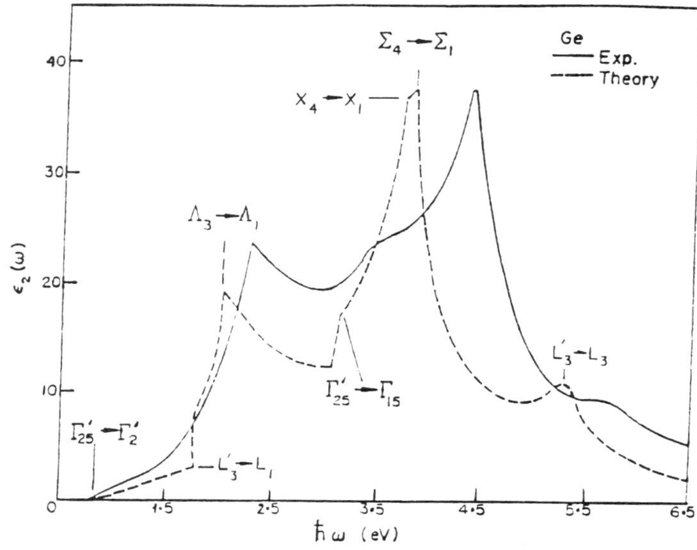

FIG. 12. Imaginary part of the dielectric function of Ge: the solid line is experimental, and the dashed one, theoretical; optical transitions at critical points are explicitly indicated (Brust *et al.*, 1962).

6.3 Indirect Interband Transitions

The band structures of C, Si, and Ge clearly show that, while the top of the valence band is at the center of the Brillouin zone (Γ point), the minimum of the conduction band occurs away from Γ (close to the X point along the Δ line in diamond and Si, at the L point in Ge). This has been verified by cyclotron resonance experiments, but it also shows in the optical properties, because the phonon interaction (22) makes possible an optical transition between different points of the Brillouin zone, the missing (pseudo)momentum being provided by the phonon emitted or adsorbed. We make use of the second-order transition probability (16), taking the perturbation interaction to be the sum of Eqs. (14) and (22), and obtain, assuming a constant matrix element squared (C) involving the states near the extrema that give the strongest contribution,

$$\epsilon_2(\omega) = \frac{4\pi^2 e^2 C}{m^2\omega^2}(n_{\mathbf{q}_0\nu_0} + \tfrac{1}{2} \pm \tfrac{1}{2}) \times \int_{BZ}\int_{BZ} \frac{2d^3k_1 d^3k_2}{(2\pi)^6}\,\delta(E_c(\mathbf{k}_1) - E_v(\mathbf{k}_2) \pm \hbar\omega_{\nu 0}(\mathbf{q}_0) - \hbar\omega), \quad (27)$$

where $+$ $(-)$ refers to creation (destruction) of a phonon of mode ν_0 with wave vector $\mathbf{q}_0$ connecting the valence-band maximum and the conduction-band minimum. The summation over the final electronic states above the edge ($\hbar\omega_0$) gives an energy dependence that is different from that of direct transitions, and precisely

$$\epsilon_2(\omega) = 0 \text{ when } \omega < \omega_0, \quad \epsilon_2(\omega) \sim (\omega - \omega_0)^2 \text{ when } \omega > \omega_0.$$

Furthermore, different edges occur in correspondence to each phonon, and their values depend on the fact that the phonons are absorbed or emitted. At low temperature we can only create phonons, and consequently the edge occurs at higher energy:

$$\hbar\omega_0 = E_g + \hbar\omega_{\nu 0}(\mathbf{q}_0). \quad (28)$$

At higher temperature we can also absorb phonons, with a temperature-dependent probability given by the Bose–Einstein factor, and the edge is shifted to lower energy:

$$\hbar\omega_0 = E_g - \hbar\omega_{\nu 0}(\mathbf{q}_0). \quad (29)$$

Indirect transitions are responsible for the weak absorption tail in C, Si, and Ge, which reflects their conduction-band structures. We show in Fig. 13 the absorption tail of Si; a fit to the profile of the indirect edge gives the frequency of the X phonons that make the transition possible.

Under conditions of electronic excitations, the initial state may be the minimum of the conduction band, and the electron–hole recombination produces luminescence. The recombination luminescence is indirect in ele-

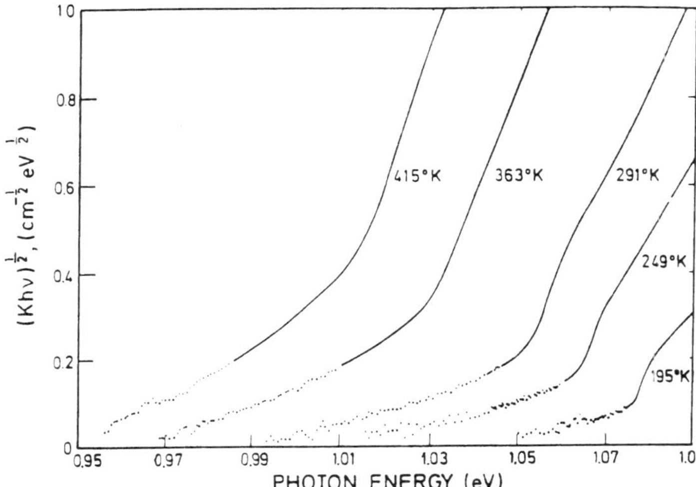

FIG. 13. Indirect absorption edge in Si at different temperatures (MacFarlane *et al.*, 1958).

mental semiconductors and shifted to lower energy at low temperatures because the conservation of energy requires

$$\hbar\omega = E_g - \hbar\omega_{\nu 0}(\mathbf{q}_0). \quad (30)$$

6.4 Inelastic Scattering

Besides the indirect transitions, the phonons are involved in the inelastic scattering of light (Raman scattering for optical phonons and Brillouin scattering for acoustic phonons) whereby one photon is absorbed and a second photon is emitted, the energy difference between the two being due to the phonon created (Stokes process) or absorbed (anti-Stokes process). The Stokes process, being proportional to $n + 1$, is present at all temperatures, while the anti-Stokes process, proportional to n (the Bose–Einstein factor for the relevant phonon), is significant only at a sufficiently high temperature. The Raman spectrum of diamond, from which the energy of the zone-center optical phonon can be precisely measured, is shown in Fig. 14; a typical Brillouin-scattering spectrum of Ge, involving acoustic phonons of appropriate wave vector, is illustrated in Fig. 15. Brillouin- and, especially, Raman-scattering data provide valuable information on the electron-phonon coupling that mediates these processes (resonant conditions with electronic transitions are also frequently employed). Therefore, they are very useful for understanding the electron and phonon band structures (see Sects. 2 and 3) as well as the scattering mechanisms limiting the free-carrier mobilities (see Sec. 11).

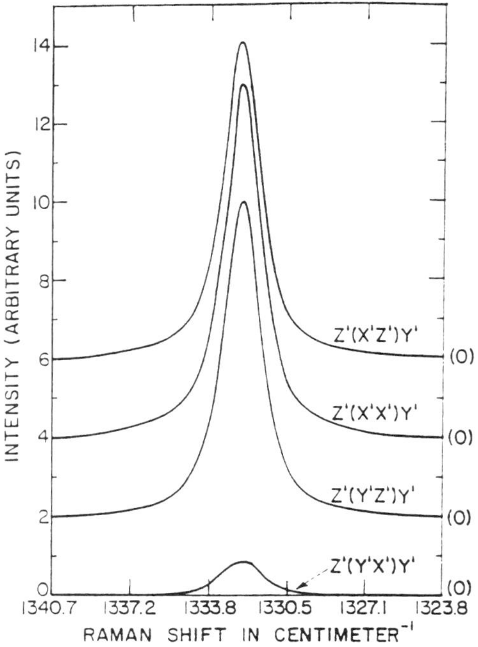

FIG. 14. One-phonon Raman spectrum of diamond; the polarization features correspond to the Γ'_{25} symmetry of the zone-center optical phonon (Solin and Ramdas, 1970).

6.5 Excitonic and Polaritonic Effects

As a result of exchange and correlation effects within the many-electron system, the excitation states are not simply obtained as the sums of the occupied band-state energies of the electrons, as assumed in the preceding sections, but depend on an additional two-body interaction between the electron promoted to the conduction band and the hole that is left in the valence band. This interaction consists of two terms (Bassani and Pastori Parravicini, 1975), an electron–hole Coulomb attraction screened by the dielectric function of the medium $\epsilon(\mathbf{k})$ and a spin-dependent electron–hole exchange of the form

$$2J\delta_S, \quad (31)$$

where $\delta_S = 1$ for singlet states and $\delta_S = 0$

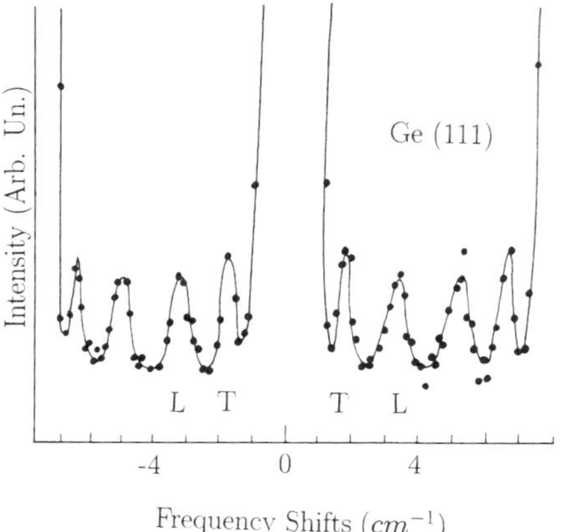

FIG. 15. Brillouin spectrum of Ge showing both transverse acoustic (T) and longitudinal acoustic (L) phonons (Sandercock, 1972).

for triplets. The term J is a two-body contribution, which can be computed from the wave functions of valence and conduction states. It consists of a short-range term J_s that corresponds to a local field correction, and a long-range contribution J_l that is singular in the limit $\mathbf{k} \rightarrow 0$, depending on the fact that the exciton is longitudinal or transverse with respect to the direction of motion. As a consequence we obtain a hydrogen-type series of states below the gap, of the form

$$E_n = E_g - \frac{R^*}{n^2} + 2J_s\delta_S + 2J_l\begin{pmatrix} \frac{8}{3}\pi \\ -\frac{4}{3}\pi \end{pmatrix}\delta_S, \qquad (32)$$

where the coefficient $+8\pi/3$ refers to longitudinal excitons and $-4\pi/3$ to transverse excitons. Since $R^* = (\mu^*/m)(R/\epsilon^2)$ is the effective Rydberg, modified with respect to the atomic value by the reduced effective mass μ^* and by the dielectric screening, and both J_s and J_l are much smaller than R^*, the excitonic series is hydrogen-like with only the lowest states appearing in the excitation spectrum. If we consider the free motion of the center of mass, we obtain a dispersion law for the exciton states of the type

$$E_n(k_{\text{ex}}) = E_n + \frac{\hbar^2 k_{\text{ex}}^2}{2(m_e + m_h)}, \qquad (33)$$

where for indirect gaps $\mathbf{k}_{\text{ex}}$ is measured with respect to $\mathbf{q}_0$, the wave-vector difference between the two band extrema. This produces a frequency dependence above the exciton states that is of the type $[\hbar\omega - E_n \pm \hbar\omega_{\nu 0}(\mathbf{q}_0)]^{1/2}$, rather than the dependence typical of indirect interband transitions. The excitonic effects are almost negligible in elemental semiconductors because of the large values of ϵ but have been duly observed. They become much more important in polar compounds.

Associated to the excitonic states there is also a modification of the electromagnetic waves in the crystal, which produces anomalous dispersion in the frequency region close to the exciton. This is obtained from the usual dispersion law of propagating waves in media

$$\omega^2 = [c^2/\epsilon(\omega,\mathbf{k})]k^2, \qquad (34)$$

where the dielectric function has poles in correspondence to the excitons (33). The solution of (34) for the transverse modes gives an upper polariton and a lower polariton, upper and lower being referred to the transverse exciton energy (Hopfield, 1958). Though polaritonic effects are more important in polar materials, they have been nevertheless observed in elemental semiconductors.

6.6 Core-State Excitations

The sum rule (26) that we mentioned before is strictly valid only when the upper frequency limit is infinite and the number n refers to the total number of electrons. This means that additional absorption occurs as the frequency increases, with edges corresponding to transitions due to electrons from inner shells (1s in C; 1s, 2s, 2p in Si; etc.). The theory of inner-shell excitation spectra is similar to that for valence electrons, but account must be taken of the fact that the energy bands of the "core" states are flat. Consequently, only direct transitions occur, and the structures in the interband transition spectrum depend only on the conduction-band density of states. The exciton states depend only on the electron effective masses and are dispersionless. Measurements of the core-state excitations require frequencies from the vacuum ultraviolet region to the hard x-ray region and can be performed using synchrotron radiation; they fully confirm the conduction-band structures described in Sec. 2 (Bassani and Altarelli, 1983).

7. PHOTOELECTRON SPECTROSCOPY

Besides optical spectroscopy, the most useful experimental technique to measure the electronic energy bands is photoelectron spectroscopy (see PHOTOEMISSION AND PHOTOELECTRON SPECTRA). The dispersion curves of electronic states that are below the Fermi level and, in some cases, above the vacuum level can be studied by a number of variations of photoemission spectroscopy (Cardona and Ley, 1978) in which the energy, wave vector, and polarization (spin) of the incident photon (emitted electron) can be selectively chosen or averaged over. The electronic states with energies between the Fermi level and the vacuum level are not involved in photoemission processes as they can be neither initial states (because they are empty in the ground state) nor final states (because the electrons are unable to escape from the solid once excited). These levels can be investigated by inverse photoemission, which is, in a sense, the time-reversed process of photoemission: An incoming electron falls into an electronic level available in the solid, emitting a photon, which is detected. The main advantage of photoemission and inverse photoemission with respect to optical spectroscopy is the capability of providing absolute energy levels rather than energy differences between levels. Besides lower resolution and (especially for inverse photoemission) lower statistics, the main disadvantage for the study of bulk properties is that photoelectron spectroscopy is surface sensitive because of the short escape or penetration depths of electrons (of the order of 10 Å). The progress in ultrahigh-vacuum and sample-preparation techniques as well as the availability of synchrotron radiation have led to significant developments in photoemission.

7.1 Photoemission

A simple physical picture of the photoemission process is given by the three-step model: First an electron in the solid is promoted to an empty band by absorption of the incoming photon, then it travels toward the surface and, finally, escapes into the vacuum to be detected. The electrons may (secondary electrons) or may not (primary electrons) undergo inelastic scattering processes in the solid before escaping through the surface. The yield I of primary electrons emitted with energy E by photons of energy $\hbar\omega$ is determined by three factors: the probability P of optical excitation, T of reaching the surface, and D of leaving the solid $[I(E,\hbar\omega) = P(E,\hbar\omega)T(E)D(E)]$. T is a smooth function of the electron energy and so is D, beyond the low-energy cutoff due to the potential barrier at the solid-to-vacuum interface. Therefore, the structures in the energy distribution curves measured in angle-integrated photoemission reflect mainly the dependence of P on incoming photon energy and emitted electron energy. The behavior of P is analogous to that of the imaginary part of the dielectric constant ϵ_2 (see Sec. 6) and is determined by two factors: the optical excitation matrix element $M_{fi}(\mathbf{k})$, which depends on the wave functions of initial (valence) and final (conduction) electronic states in the semiconductor, and the energy distribution of the joint density of states, which depends on the dispersion relations of the initial $E_i(\mathbf{k})$ and final $E_f(\mathbf{k})$ levels:

$$P(E,\hbar\omega) \propto \sum_{f,i} \int d^3k M_{fi}(\mathbf{k})\delta(E_f(\mathbf{k}) - E_i(\mathbf{k}) - \hbar\omega)\delta(E - E_f(\mathbf{k})). \quad (35)$$

The approximation of a constant matrix element is often made and, for photon energies higher than about 20 eV, the structures in P correspond to those in the density of states of the occupied states only (the density of final states being practically structureless); in this case, the photoemission spectra mimic the density of states of the valence band.

Photoemission spectra (corrected for the secondary-electron contribution) of diamond, Si, and Ge are shown in Fig. 16 (Cavell *et al.*, 1973). The three main peaks correspond in each case to prominent peaks in the density of states of the valence band: From higher to lower binding energies, they are associated, respectively, to predominantly s-like, mixed

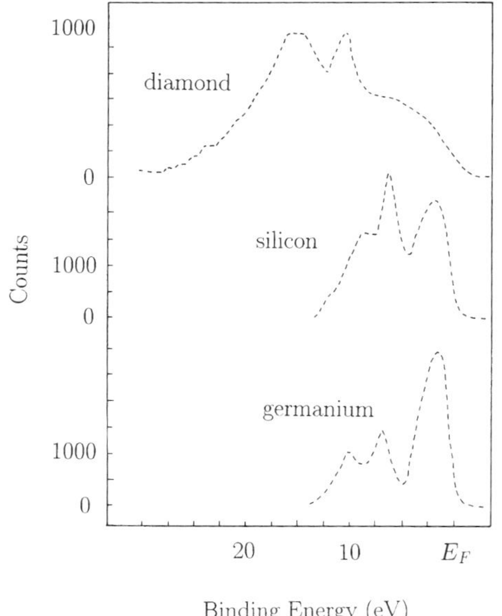

FIG. 16. X-ray photoelectron spectra of diamond, Si, and Ge (Cavell *et al.*, 1973).

s- and *p*-like, and predominantly *p*-like states. Accurate information on the dispersion relations of individual bands can be obtained by angle-resolved photoelectron spectroscopy in which the angle of emission of the electrons is measured. Representative results for Ge are shown in Fig. 17 (Hsieh *et al.*, 1984).

7.2 Inverse Photoemission

Low-lying empty states can be investigated by inverse photoemission; in particular, combining photoemission (whereby an electron is extracted from the valence band) and inverse photoemission (whereby an electron is injected into the conduction band) measurements, the energy gap between single-particle valence and conduction states can be directly determined. In this respect, it is to be noted that the gaps measured by optical spectroscopy (whereby an electron is excited from the valence into the conduction band) are in general smaller than the difference between single-particle states because of the attractive electron–hole interaction. Figure 18 shows the energy dispersion of conduction-band states in Si as derived from angle-resolved inverse photoemission data (Straub *et al.*, 1985) along with the topmost valence band as derived from angle-resolved photoemission (Uhrberg *et al.*, 1984); the energy gaps at the *L* point of the Brillouin zone are compared with the optical E_1 and E_1' gaps (Madelung, 1982): a marked lowering of the E_1 gap (by about 0.5 eV) due to the excitonic effect is evident.

8. DIELECTRIC SCREENING

We have already assumed, when dealing with excitons, that the Coulomb attraction between charges is screened by a wave-vector–dependent dielectric function $\epsilon(\mathbf{k})$. The usual dielectrical constant ϵ_0 corresponds to the screening of a fixed charge at large distance and coincides with the limit of the optical dielectric function (see Sec. 6) $\epsilon(\omega)$ for $\omega \to 0$.

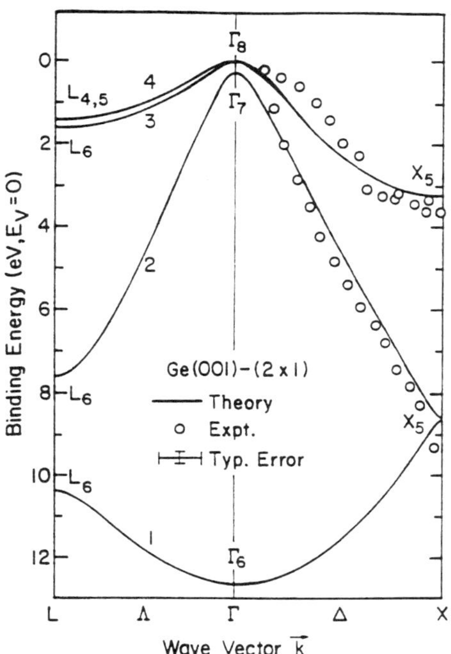

FIG. 17. Valence-band dispersion curves of Ge: the circles are obtained from angle-resolved photoemission (Hsieh *et al.*, 1984), and the solid lines are from a pseudopotential calculation (Chelikowsky and Cohen, 1976).

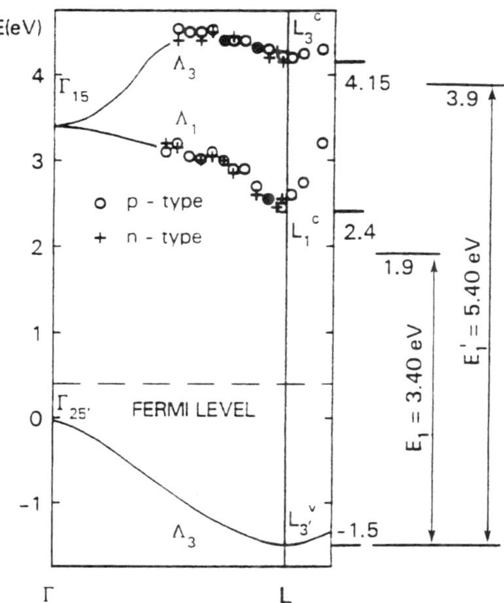

FIG. 18. Energy dispersion of conduction-band states of Si from inverse photoemission (Straub *et al.*, 1985). Also shown are the valence band from photoemission (Uhrberg *et al.*, 1984) and, for comparison, the optical gaps at *L* (Madelung, 1982).

8.1 General Theory

Since in elemental semiconductors we do not have polarization due to phonons, the origin of the dielectric screening is all in the modification of the electronic functions of the crystal produced by the electric charge. This, in principle, can be computed from the detailed knowledge of the band structure, taking into account correlation effects. A general expression (Lindhardt formula), which can be derived from the quantum definition of polarizability (in the random-phase approximation), for the diagonal expression of the dielectric tensor [i.e., neglecting umklapp processes that would produce off-diagonal tensor terms $\epsilon(\mathbf{k}, (\mathbf{k} + \mathbf{G}))$, with $\mathbf{G} \neq 0$] is given by

$$\epsilon(\mathbf{k},\omega) = 1 + \frac{4\pi e^2}{k^2} \sum_{v,c} \int_{\mathrm{BZ}} \frac{2d^3k'}{(2\pi)^3} \frac{|\langle \Psi_{c\mathbf{k}'}|e^{-i\mathbf{k}\cdot\mathbf{r}}|\Psi_{v\mathbf{k}'+\mathbf{k}}\rangle|^2}{E_c(\mathbf{k}') - E_v(\mathbf{k}' + \mathbf{k}) - \hbar\omega - i\hbar\epsilon^+}, \tag{36}$$

ϵ^+ denoting that the positive zero limit is considered (see Bassani and Pastori Parravicini, 1975). It can be shown that for $\mathbf{k} = 0$, the above expression reduces to that described in Sec. 6 for the optical transitions. We can also observe that the static limit ($\omega \to 0$) has a finite value in semiconductors because of the energy gap, and its asymptotic limit for large wave vectors is 1. It also follows from (36) that the values of the dielectric screening increase with decreasing band gaps, which explains the large values of the dielectric constants in semiconductors and their trend from C to Sn ($\epsilon_0 = 12$ in Si, 16 in Ge, 24 in gray tin).

8.2 Simplified Model Calculations

Many authors have performed calculations of the static screening function $\epsilon(\mathbf{k},0)$, using approximate models for the band structure, and also improving on expression (36) by adding off-diagonal terms with $\mathbf{G} \neq 0$ to the polarization propagator. The first of such simplified models is the "free electron with a gap" model due to Penn (1962), who fitted the numerical results with a simple formula that contains all the relevant features:

$$\epsilon(k) \simeq 1 + \left(\frac{\hbar\omega_p}{E_g}\right)^2 \left(1 - \frac{E_g}{4E_F}\right) \times \left[1 + \frac{E_F}{E_g}\left(\frac{k}{k_F}\right)^2 \left(1 - \frac{E_g}{4E_F}\right)^{1/2}\right]^{-2}, \tag{37}$$

E_F being the Fermi energy given by the electron density and $\hbar k_\mathrm{F}$ the corresponding Fermi momentum, and E_g is the gap introduced between the Fermi energy and the empty states.

For the purpose of illustration we show in Fig. 19 the computed values of the static dielectric function for silicon, in a number of approximations. It can be observed that, in the case of excitons or of shallow impurity states (see Sec. 9) in elemental semiconductors, the Bohr radius is so large that the use of the approximation $\mathbf{k} = 0$, appropriate to charges at infinite distance, is justified. This will not be the case in strong insulators, or organic materials, where the interaction is localized inside the unit cell and the limit of large wave vectors is more appropriate.

9. IMPURITY LEVELS

Foreign atoms, in interstitial or substitutional sites, are the most important defect in semiconductor crystals. The ability to control

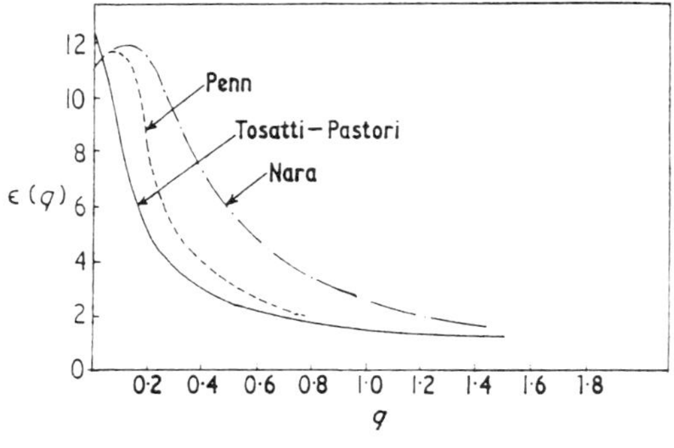

FIG. 19. Dielectric screening function of silicon as a function of the wave vector (in arbitrary units): the dashed-dotted line is from Nara (1965), the dashed one from Penn (1962), and the solid one from Tosatti and Pastori Parravicini (1971).

the content of impurities, and their beneficial as well as pernicious effects, is a key issue in semiconductor technology: The concentration of undesired dopants in Si is typically kept below 10^{12} cm^{-3}. An impurity breaks the translational symmetry of the perfect crystal and may give rise to electronic states, more or less localized around the defect site, with energies lying in the forbidden energy gap of the host material. The main character of these impurity levels is whether they are close to the conduction- or valence-band edges (shallow donor or acceptor states) or rather are close to the middle of the gap (deep states). In the first case, they dominate the transport properties, providing free carriers to the conduction or valence band; in the second case, they mainly affect the recombination of free electron–hole pairs by trapping them at the same site. Furthermore, besides shallow and deep states, an impurity may bring about resonant states that are still partly localized around the defect site but have energy coinciding with that of extended states of the host crystal.

All the properties of the electronic states associated with impurity levels depend on the electronic band structure of the host semiconductor as well as on the impurity potential (for substitutional sites, the difference between the foreign- and replaced-atom potentials). General theoretical schemes have been developed to handle all types of impurity states in a unified way (Bassani *et al.*, 1969); in the following, though, the hydrogenic model based on the effective-mass approximation (see Sec. 4) will be mostly referred to, because it is physically transparent and well suited for the shallow states that are used for doping (see Sec. 11).

9.1 Shallow Levels

In group-IV elemental semiconductors, such as Si and Ge, substitutional group-III impurities, such as Al and B, behave as shallow acceptors, and group-V ones, such as P and As, as shallow donors (see Tables 1 and 2). The extra hole or electron, with respect to the fully occupied valence band, is bound to the defect site by the long-range Coulomb potential due to the charge difference between the impurity and host nuclei. Because of the low effective masses ($m^* \simeq 0.2\, m$) and the high dielectric constants ($\epsilon \gtrsim 10$) typical of semiconductors, the corresponding localized states extend over many unit cells, and these impurities can be described as a solid-state analog of the hydrogen atom with scaled-down Rydberg energy $R^* = m^*e^4/2\hbar^2\epsilon^2$ ($\simeq$20 meV) and Bohr radius $a^* = \hbar^2\epsilon/m^*e^2$ ($\simeq$50 Å).

Table 1. Group-III acceptor levels in silicon and germanium (measured from the valence band in meV).

	B	Al	Ga
Si	45	67	72
Ge	10.4	10.2	10.8

This simple picture can be easily extended to include details of the host band structure, such as the many-valley conduction-band minima and the degeneracy of the valence-band maximum (Ramdas and Rodriguez, 1981). In particular, because the electron pockets in Si and Ge are ellipsoids of revolution with a longitudinal mass different from the transverse mass (see Sec. 4), the three p levels are no longer degenerate but split into a singlet p_0 level and a doublet $p_\pm$ level, as observed in absorption spectroscopy when electric dipole transitions from the 1s ground state of the donor impurity are excited (see Fig. 20). Furthermore, as a result of the interaction between different equivalent valleys, the 1s donor ground states split into one A_1 singlet, one E doublet, and one T_2 triplet in Si (which has six electron pockets) and into one A_1 singlet and one T_2 triplet in Ge (which has four electron pockets).

Aside from complications due to details of the host band structure, the simple hydrogenic model cannot account for significant shifts in binding energies of different impurities (chemical shifts). These are usually important only for the ground state (more localized) and are due to the departure of the impurity potential from a Coulomb potential at short distances from the defect site. This

Table 2. Group-V donor levels in silicon and germanium (measured from the conduction band in meV).

	P	As	Sb
Si	45	54	39
Ge	12	12.7	9.6

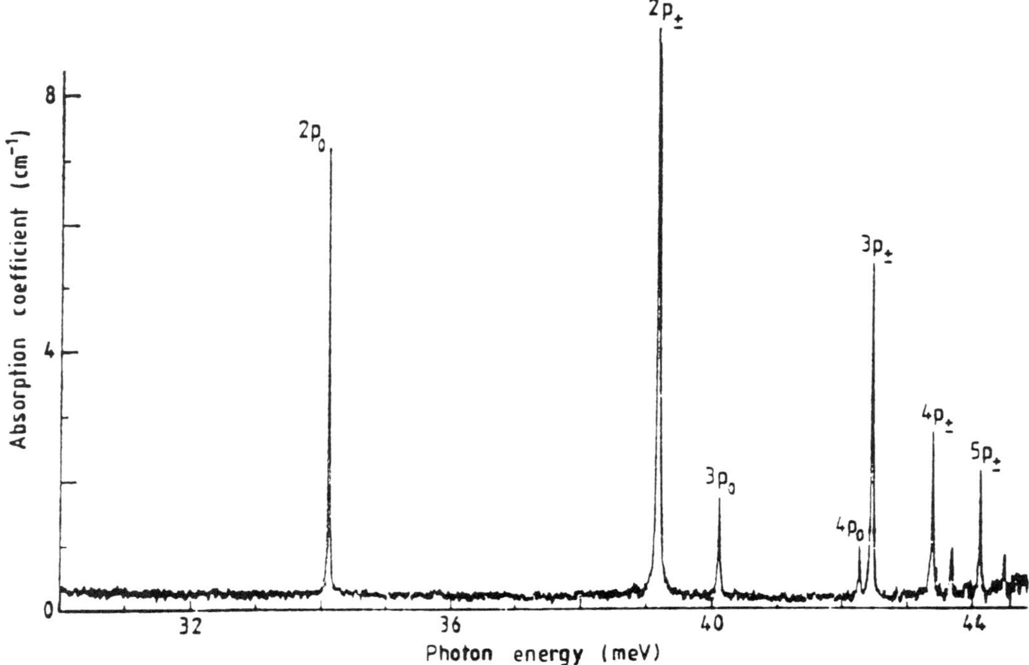

FIG. 20. Excitation spectrum of P donors in Si at liquid-He temperature (Jagannath *et al.*, 1981).

central-cell correction is marked for noniso-coric impurities (i.e., those with a core electron distribution different from that of the host), for which a pseudopotential approach has been used (Pantelides, 1978). Furthermore, at short distances (large wave vectors) the use of a (large) constant dielectric function is questionable as the screening efficiency is decreased (see Sec. 8).

9.2 Resonant Levels

If the host band structure has a secondary extremum (in either the conduction or valence band), the impurity potential may give rise to a localized state associated with it. Such a level is degenerate with extended Bloch states of the host crystals with values of **k** corresponding to the principal extremum. Therefore, the impurity level close to the secondary extremum is broadened by the interaction with the degenerate continuum of the band states belonging to the principal extremum. Such a situation occurs, for instance, for acceptor states at the top of the valence band of Si and Ge with the split-off hole band (Γ_7^+) playing the role of the secondary maximum, as depicted in Fig. 21.

9.3 Deep Levels

When the wave function of an impurity state is localized over a small number of unit cells around the defect site, the effective-mass approximation breaks down, and no simple theoretical description can be given. Such impurities have energy levels deep in the forbidden energy gap. For instance, in silicon, Pd has an acceptor level (at 0.34 eV from the valence band) and O has two donor levels (at 0.16 and 0.51 eV from the conduction band) and two acceptor levels (at 0.41 and 0.74 eV from the valence band); in germanium, Cu has two acceptor levels (at 0.33 and 0.53 eV from the valence band) and Mn has two acceptor levels (at 0.16 and 0.42 eV from the valence band). The corresponding wave functions, when expanded in Bloch states, comprise a large part of the Brillouin zone; they may therefore act as efficient radiative recombination centers for electron–hole pairs in indirect semiconductors, partic-

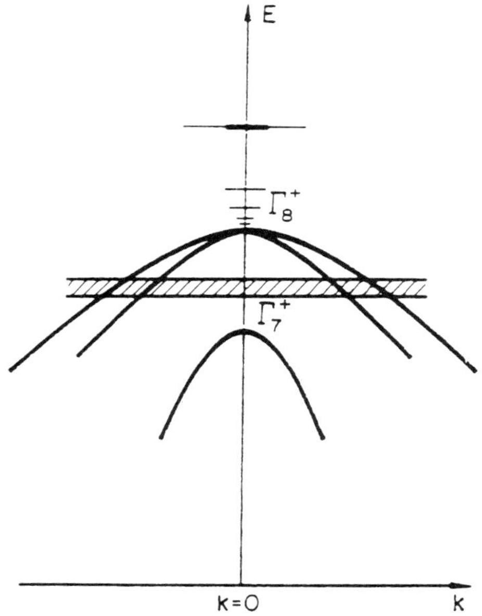

FIG. 21. Acceptor states near the top of the valence band, the shaded level representing a resonant state associated with the secondary maximum of the Γ_7^+ band.

ularly when they have an energy level close to the middle of the band gap. Such is the case, for example, of Au in Si: increasing the Au concentration from 10^{14} to 10^{18} cm^{-3} reduces the lifetime of nonequilibrium free carriers from 1 μs to 0.1 ns. A special case of localized levels is the one due to isoelectronic impurities, such as Ge in Si, for which there is no long-range Coulomb potential, but only an impurity potential localized in the central cell. Such impurities may or may not split a bound state from a band of extended states depending on the ratio between the strength of the impurity potential and the band width: for a larger band width (dependent on the host), a stronger central-cell potential (dependent on the impurity-host difference) is required to introduce a localized state.

10. EXTERNAL PERTURBATIONS

From the experimental point of view, much information on the properties of semiconductors has been obtained by measuring changes induced by externally applied perturbations. The latter can preserve the symmetry of the crystal (as for hydrostatic pressure) or lower it (as for uniaxial stress or electric and magnetic fields), in which case detailed information on the symmetry properties of, say, the critical points of the electronic bands can be obtained. A special case of symmetry-lowering perturbation is the presence of surfaces or interfaces that reduce the translational symmetry of a crystal (Stella, 1993); surface electronic levels in Ge were first observed by Chiarotti *et al.* in 1968.

The possibility of employing a phase-sensitive detection technique in conjunction with a periodic modulation of the external perturbation improves the signal-to-noise ratio by several orders of magnitude on the one hand, and, on the other hand, allows one to enhance the sharp features of, say, the reflectivity with respect to a smoothly varying background. Among the pioneering works in modulation spectroscopy, there are the studies of surface traps in Ge (Chiarotti *et al.*, 1962) and of the Franz–Keldysh effect in Ge and Si (Seraphin and Hess, 1965; Frova and Handler, 1965).

10.1 Hydrostatic Pressure

The application of hydrostatic pressure brings about a lattice-constant change that, through a modification of the crystal (pseudo)potential, may significantly affect the electronic band structure. In particular, the optical gaps between valence and conduction states at various points in the Brillouin zone of semiconductors such as Ge and Si shift in a characteristic way determined by symmetry: the gap between the top of the valence band at Γ (Γ'_{25}) and the conduction band at Γ (Γ'_2) increases at a rate of about 10 meV/kbar, the one between Γ'_{25} and the conduction band along the [111] direction (Λ_1) increases at a rate of about 5 meV/kbar, whereas the one between Γ'_{25} and the conduction band along the [100] direction (Δ_1) decreases at a small rate of about -1.5 meV/kbar. Thus, at elevated pressures (higher than about 50 kbar), the minimum of the conduction band in Ge moves from L_1 to Δ_1; as a consequence, the many-valley structure at the bottom of the conduction band becomes analogous to that of silicon with ellipsoidal electron pockets along the cubic axes.

10.2 Uniaxial Stress

Uniaxial stress lowers the symmetry of cubic semiconductors and, therefore, induces qualitative changes in their properties. Piezospectroscopy has been instrumental in establishing the main features of the electronic bands of Si and Ge. For instance, reflectivity of Ge around 2.2 eV is dominated by interband transitions involving states along the Λ line in the BZ: uniaxial stress along the [001] direction does not split the degeneracy among the eight [111] valleys, whereas uniaxial stress along the [111] direction induces a difference between the two of them along the stress axis and the other six; the latter stress brings about a marked polarization dependence of the reflectivity (Gerhardt, 1965). More striking features are observed in the piezospectroscopy of shallow impurity levels, whereby the bound states closely follow the shifts of the respective band extrema (Ramdas and Rodriguez, 1981).

10.3 Electric Field

The application of an electric field is another powerful technique to probe the electronic structure of semiconductors, particularly in conjunction with modulation spectroscopy (Cardona, 1969). An electric field **E** drastically alters the symmetry of a crystal, breaking the translational invariance of the potential along the direction of the field; as a consequence, the component of the wave vector **k** along the field is no longer a good quantum number, but changes with time according to

$$\mathbf{k}(t) = \mathbf{k}(0) - e\mathbf{E}t/\hbar. \quad (38)$$

Considering the corresponding change in the character of the wave functions, the effects of an electric field on the optical properties can be calculated in the vicinity of a critical point (where they are particularly important). For an absorption edge (M_0 critical point), the main changes in $\epsilon_2(\omega)$ are an exponential tail in the gap and an oscillatory behavior above threshold (Franz–Keldysh effect), as sketched in Fig. 22. The exponential tail is proportional to

$$\epsilon_2(\omega) \propto \exp\left[-\frac{4}{3}\sqrt{\frac{(\omega_0 - \omega)^3 2\mu\hbar}{e^2E^2}}\right], \quad (39)$$

where μ is the reduced mass and ω_0 is the frequency of excitations at the critical point. The changes in optical constants measured by electroreflectance modulation spectros-

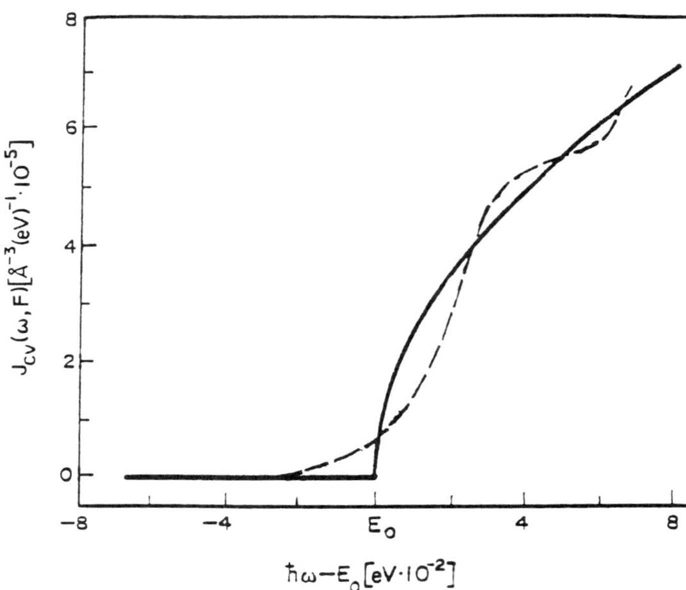

FIG. 22. Electric field effects on the behavior of the joint density of states in the vicinity of an absorption edge (M_0); the solid line represents the zero-field case.

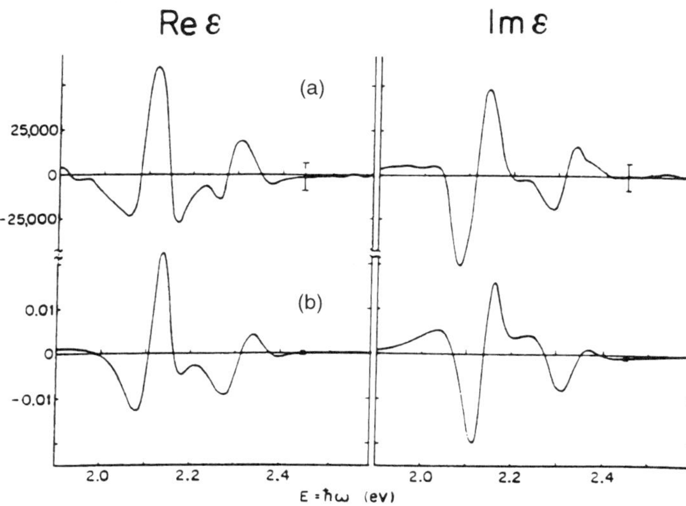

FIG. 23. Optical constants of Ge at the E_1 and $E_1 + \Delta_1$ gaps from low-field electroreflectance data (b) compared to third-order derivatives (a) calculated from ellipsometric data (Aspnes, 1972).

copy have a very simple interpretation in the regime of low applied fields. In this case, the spectra are simply proportional to the third derivative of the respective optical constant with respect to frequency according to

$$\Delta\epsilon(\omega) \propto \frac{e^2E^2}{\mu\hbar}\frac{1}{\omega^2}\frac{d^3}{d\omega^3}[\omega^2\epsilon(\omega)], \tag{40}$$

as shown in Fig. 23 (Aspnes, 1972).

10.4 Magnetic Field

A magnetic field alters the electronic states drastically: The motion in the plane perpendicular to the field is quantized into discrete levels (Landau levels) arranged in a regular ladder (at least for simple bands) with a spacing given by the cyclotron frequency ω_c, whereas the dispersion along the field remains unaffected. Therefore, the three-dimensional density of states collapses into a series of one-dimensional densities of states (each corresponding to a Landau level), as shown (for an absorption edge) in Fig. 24. The Landau quantization is responsible for the cyclotron resonance described in Sec. 4 and for many other phenomena, such as an oscillatory behavior of the magnetization (de Haas–van Alphen effect) or of the conductivity (Shubnikov–de Haas effect) as a function of magnetic field (Kittel, 1966).

Magneto-optical experiments demonstrate very directly the presence of Landau levels. Figure 25 (Aggarwal *et al.,* 1967) shows a

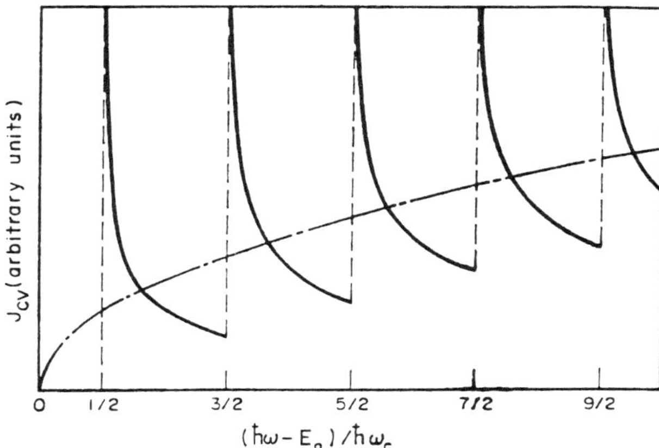

FIG. 24. Joint density of states in the presence of a magnetic field for a critical point M_0 (the dot-dashed line represents the zero-field density of states); the one-dimensional singularities do not take into account broadening effects.

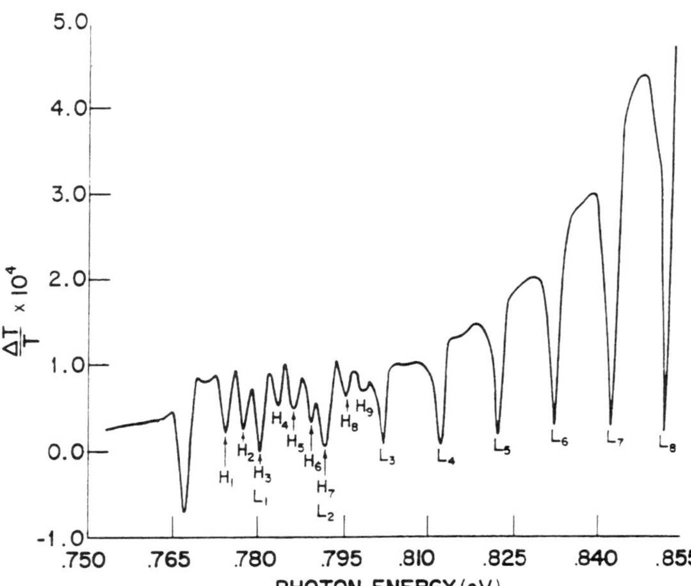

FIG. 25. Magnetopiezotransmission spectrum in Ge (Aggarwal *et al.*, 1967); see text for details.

piezomodulated transmission spectrum of Ge in the presence of a magnetic field along the [110] axis. The two series of sharp features (L_1, L_2, . . .; H_1, H_2, . . .) correspond to transitions between the same level at the top of the valence band and the Landau levels associated with the four electron pockets along the [111] directions: Two of these are perpendicular to the magnetic field and have a higher effective mass and a lower Landau level spacing ($H_{n+1} - H_n$) than the other two pockets (associated to the L series). The Landau level energies increase (approximately linearly) with magnetic field, and the results of magneto-optical experiments are usually presented as a "fan" diagram from which important material parameters (such as effective masses) can be derived (Madelung, 1982).

11. ELECTRICAL CONDUCTIVITY

The most typical property of pure semiconductors is the temperature dependence of their resistivity. From the technological point of view, on the other hand, the possibility of controlling their resistivity through the introduction of a suitable concentration of shallow impurities (doping) is of paramount importance (Sze, 1985). The main factors determining the dc electrical conductivity of homogeneous bulk Si and Ge, pure as well as doped, are discussed in the following.

11.1 Free-Carrier Conductivity

In a cubic semiconductor, the low-field conductivity tensor reduces to a scalar σ given (see INSULATORS AND SEMICONDUCTORS, CONDUCTIVITY IN) by

$$\sigma = e(n\mu_n + p\mu_p), \tag{41}$$

where n and p are the densities of conduction-band electrons and valence-band holes, respectively, and μ_n and μ_p are the corresponding mobilities given by

$$\mu_{n,p} = (e/m_{n,p}^{(\sigma)})\tau_{n,p}, \tag{42}$$

where $m^{(\sigma)}$ is the appropriate effective mass and τ is the appropriate momentum relaxation time. For a degenerate semiconductor, as in the case of a metal, τ corresponds to the momentum relaxation time of carriers at the Fermi energy, whereas for a nondegenerate semiconductor it reduces to

$$\tau = \langle v^2\tau(v)\rangle/\langle v^2\rangle, \tag{43}$$

where the averages are taken over a Maxwell–Boltzmann distribution, v is the carrier thermal velocity, and the scattering is assumed to be energy dependent, but isotropic. For a many-valley band extremum, such as for the electron pockets in Si and Ge, the conductivity effective mass to be used above is an average given by

$$1/m_n^{(\sigma)} = (2/m_T + 1/m_L)/3, \tag{44}$$

and n is the total density of electrons equally distributed among the equivalent valleys. From the values of the electron pocket transverse and longitudinal masses (see Sec. 4), one obtains for Si $m_n^{(\sigma)} \simeq 0.27\ m$ and for Ge $m_n^{(\sigma)} \simeq 0.12\ m$. When more than one electron or hole bands are populated, e.g., light- and heavy-hole bands, their contributions add up in the above expression for σ.

The conductivity of a semiconductor, in contrast to that of a metal, varies strongly (see below) because of the dramatic dependence of the electron and hole densities on the temperature and the doping level; a smoother variation is also brought about by mobility changes.

11.2 Carrier Density

In a nondegenerate semiconductor, the density of electrons in the conduction band, n, and of holes in the valence band, p, can be expressed as

$$n(T,\mu) = N_c(T) \exp[(\mu - E_c)/k_B T], \tag{45}$$

$$p(T,\mu) = N_v(T) \exp[(E_v - \mu)/k_B T], \tag{46}$$

μ being the chemical potential and

$$N_c = \left(\frac{k_B T}{\pi\hbar^2}\right)^{3/2} \frac{(m_c^{(D)})^{3/2}}{\sqrt{2}}, \tag{47}$$

$$N_v = \left(\frac{k_B T}{\pi\hbar^2}\right)^{3/2} \frac{(m_v^{(D)})^{3/2}}{\sqrt{2}}, \tag{48}$$

where $m^{(D)}$ is the density-of-states effective mass given for the conduction band by an average over the equivalent valleys,

$$m_c^{(D)} = g^{2/3}(m_T^2 m_L)^{1/3}, \tag{49}$$

g being the valley degeneracy (6 for Si, 4 for Ge), and for the valence band by an average over heavy and light holes,

$$m_v^{(D)} = (m_{lh}^{3/2} + m_{hh}^{3/2})^{2/3}. \tag{50}$$

The densities of electrons and holes in thermal equilibrium at temperature T satisfy the law of mass action

$$np = N_c N_v e^{-E_g/k_B T}. \tag{51}$$

In a pure (intrinsic) semiconductor, the density of electrons n_i in the conduction band equals the density of holes p_i in the valence band, and they are given by

$$n_i = p_i = \sqrt{N_c N_v} \exp(-E_g/2k_B T);$$

the intrinsic concentration of carriers is dominated by the exponential dependence on $E_g/2k_B T$. From the effective-mass and energy-gap values (see Secs. 2 and 4), one obtains at room temperature (300 K) for Si $N_c \simeq 3 \times 10^{19}\ \text{cm}^{-3}$, $N_v \simeq 1 \times 10^{19}\ \text{cm}^{-3}$, $n_i = p_i \simeq 1.4 \times 10^{10}\ \text{cm}^{-3}$, and for Ge $N_c \simeq 1 \times 10^{19}\ \text{cm}^{-3}$, $N_v \simeq 5 \times 10^{18}\ \text{cm}^{-3}$, $n_i = p_i \simeq 2.3 \times 10^{13}\ \text{cm}^{-3}$. For the nondegenerate regime here considered, the intrinsic Fermi level (chemical potential) is given by

$$\mu_i = \tfrac{1}{2}(E_c + E_v) + \tfrac{3}{4} k_B T \ln(m_v^{(D)}/m_c^{(D)}) \tag{52}$$

and is close to the center of the gap [i.e., $(E_c + E_v)/2$)].

In a (partially compensated) doped semiconductor with a concentration of donors N_d and of acceptors N_a, if $N_d > N_a$ (n type) the electron (majority-carrier) density is approximately $n = N_d - N_a$ and the hole (minority-carrier) density is correspondingly $p = n_i p_i/n$; if $N_d < N_a$ (p-type), similar formulas hold with the role of majority and minority carriers interchanged. In n-type semiconductors the chemical potential μ is shifted a bit closer to E_c with respect to μ_i, as given by

$$\frac{N_d - N_a}{2n_i} = \sinh\left(\frac{\mu - \mu_i}{k_B T}\right); \tag{53}$$

the converse is true for p-type semiconductors. In the above, the temperature has been assumed to be high enough to ionize all the impurities but not so high that the concentration of intrinsic carriers (i.e., those ther-

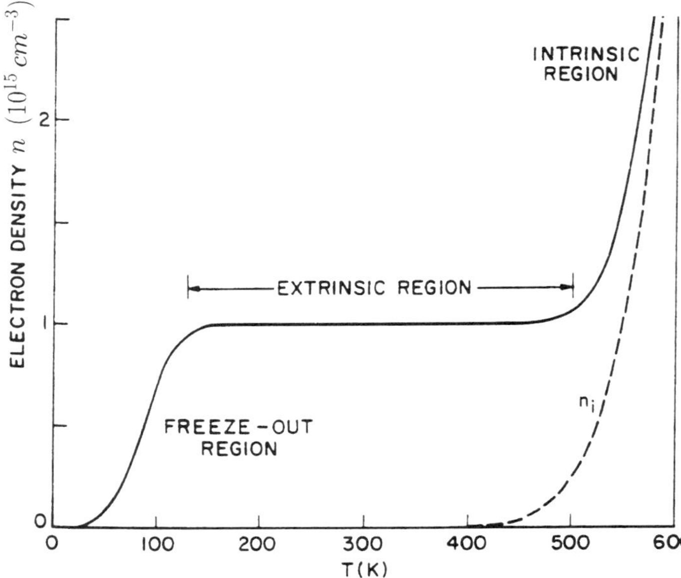

FIG. 26. Typical temperature dependence of majority-carrier concentration in a doped semiconductor. The intrinsic carrier concentration is shown by the dashed line for comparison. The numerical values are for Si with $N_d = 10^{15}$ cm^{-3}.

mally excited across the energy gap) should be comparable to the majority-carrier concentration. In this case, the carrier concentration is independent of temperature and determined by the doping level (extrinsic regime). At lower temperatures, the concentration of ionized impurities decreases because more and more electrons fall into the impurity bound states, and the concentration of free carriers drops (freeze-out regime). The temperature dependence of the majority-carrier concentration in a doped semiconductor over a wide temperature range is depicted in Fig. 26.

In all of the above, it has been assumed that the chemical potential is within the energy gap, so that the thermal distribution of electrons in the conduction band and holes in the valence band is given by the (classical) Boltzmann distribution (nondegenerate regime). When the level of doping is extremely high, the number of electrons may be larger than $(m_c^{(D)}k_BT/\hbar^2)^{3/2}$ or the number of holes larger than $(m_v^{(D)}k_BT/\hbar^2)^{3/2}$, in which case the majority carriers are in the degenerate regime (like the electrons in a metal) and must be described using the Fermi distribution with a chemical potential (possibly) higher than E_c or lower than E_v.

For the case of a p-n junction at equilibrium, the chemical potential is the same on both sides, being closer to the conduction-band edge in the n-type material and to the valence-band edge in the p-type material. This situation implies a bending of the electron energy bands (shown in Fig. 27), which is associated to the electrostatic field appearing in the depletion layer at the junction. The corresponding potential acts as a barrier for the diffusion of the majority carriers; lowering this potential barrier by means of an applied forward bias leads to an exponentially increasing current. The rectifying properties of p-n junctions are discussed at length in the various articles on semiconductor devices.

11.3 Carrier Mobility

The carrier mobility depends on the temperature and the doping level through the average momentum relaxation time τ,

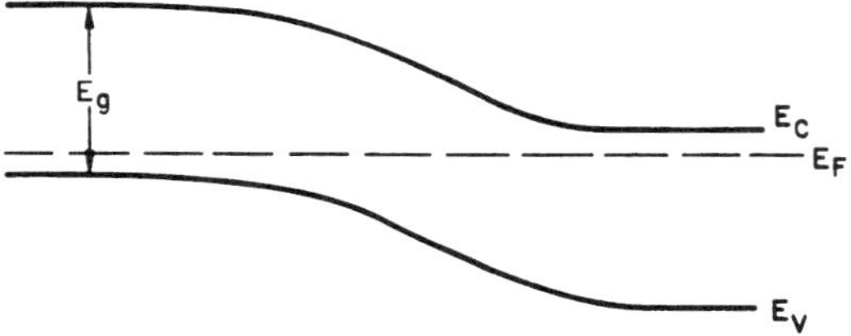

FIG. 27. Energy-band diagram of a p-n junction.

which is determined by the various scattering mechanisms. At room temperature, in pure or moderately doped elemental semiconductors, the mobility is limited mainly by longitudinal acoustic-phonon deformation-potential scattering, whereas at lower temperatures or heavy doping levels, the ionized impurity scattering is dominant (see INSULATORS AND SEMICONDUCTORS, CONDUCTIVITY IN). The acoustic-phonon scattering rate $1/\tau_{ac}$ depends on temperature like $T^{3/2}$. In fact, from perturbation theory (Fermi's "golden rule"), it is proportional to the product of the interaction matrix element squared and the density of final states; the former contains a factor of T coming from the number of phonons present (the energy of an acoustic phonon being much smaller than the thermal energy) and the latter a factor of $T^{1/2}$ as the density of states of the deflected electron scales with the square root of its kinetic energy. Experimentally (Ziman, 1960), the phonon-scattering–limited mobility scales like $T^{-1.6}$ in n-type Ge, $T^{-2.3}$ in p-type Ge, $T^{-2.6}$ in n-type Si, and $T^{-2.3}$ in p-type Si; apart from the first case, there are significant deviations from the $T^{3/2}$ behavior, mostly due to the mixture of acoustic and optical phonon scattering (and, for n-type Si, intervalley scattering). The mobility values in pure materials at room temperature are for Si $\mu_n \simeq$ 1500 cm^2/(V s), $\mu_p \simeq$ 500 cm^2/(V s), and for Ge $\mu_n \simeq$ 3900 cm^2/(V s), $\mu_p \simeq$ 1900 cm^2/(V s).

The impurity Coulomb scattering rate $1/\tau_{im}$, instead, is proportional to $T^{-3/2}N_{im}$, where N_{im} is the density of ionized impurities. In fact, it is given by N_{im} times the product of the Rutherford cross section and the electron velocity; a factor of T^{-2} comes from the energy dependence of the former (i.e., faster electrons are deflected less) and a factor of $T^{1/2}$ from the latter. This result does not change much when one considers the free-carrier screening (which essentially cuts off the small-angle scattering). Experimentally, it is not easy to see the $T^{3/2}$ dependence of the impurity-scattering–limited mobility as the temperature range in which this mechanism is dominant is restricted at low temperatures by the carrier freeze-out and at high temperatures by the onset of phonon scattering. However, when more sources of scattering are competing, the total resistivity is approximately given by the sum of the resistivities obtained from each single-scattering mechanism at a time (Matthiessen's rule). As a result, the mobility of doped semiconductors exhibits a maximum at the temperature corresponding to the crossover from impurity scattering ($\mu_{n,p} \propto T^{3/2}/N_{im}$) to phonon scattering ($\mu_{n,p} \propto T^{-3/2}$). Such behavior is evident in Fig. 28, which shows the electron mobility in Si versus temperature for various doping levels (Sze, 1985).

The temperature dependence of the conductivity of Si at different doping levels is shown in Fig. 29 (Morin and Maita, 1954). The intrinsic regime is evident over most of the temperature range; the flattening of some of the curves at about room temperature corresponds to the extrinsic regime, in which the conductivity is determined by the doping level. The temperature dependence of σ is dominated by that of the carrier concentration in the intrinsic regime, whereas its weaker variation in the extrinsic regime is due to the temperature dependence of the mobility.

11.4 Hall Effect and Magnetoresistance

Measuring the Hall coefficient is a widely used method to characterize the concentration (and type) of free carriers (see INSULATORS AND SEMICONDUCTORS, CONDUCTIVITY IN); the Hall coefficient R_H is given by

$$R_H = \frac{p\mu_p^2 - n\mu_n^2}{ce(p\mu_p + n\mu_n)^2}. \tag{54}$$

In weak fields, the mobilities appearing in the numerator of the above expression are to be multiplied by corrective factors of order unity (i.e., the ratios of Hall mobilities to drift mobilities: μ_H/μ) that depend on the scattering mechanism and the band-structure details. For instance, for an n-type many-valley semiconductor such as Si or Ge:

$$\frac{\mu_H}{\mu} = \frac{\langle v^2\tau^2\rangle\langle v^2\rangle}{\langle v^2\tau\rangle^2}\,\frac{3\gamma(\gamma + 2)}{(2\gamma + 1)^2}, \tag{55}$$

where $\gamma = m_L/m_T$ and the factor $\langle v^2\tau^2\rangle\langle v^2\rangle/\langle v^2\tau\rangle^2$ is 1.18 for acoustic deformation potential scattering and 1.93 for ionized impurity scattering; the mass anisotropy factor is also close to unity: about 0.8 for $\gamma \simeq 20$, which is appropriate for Ge (about 0.9 for Si).

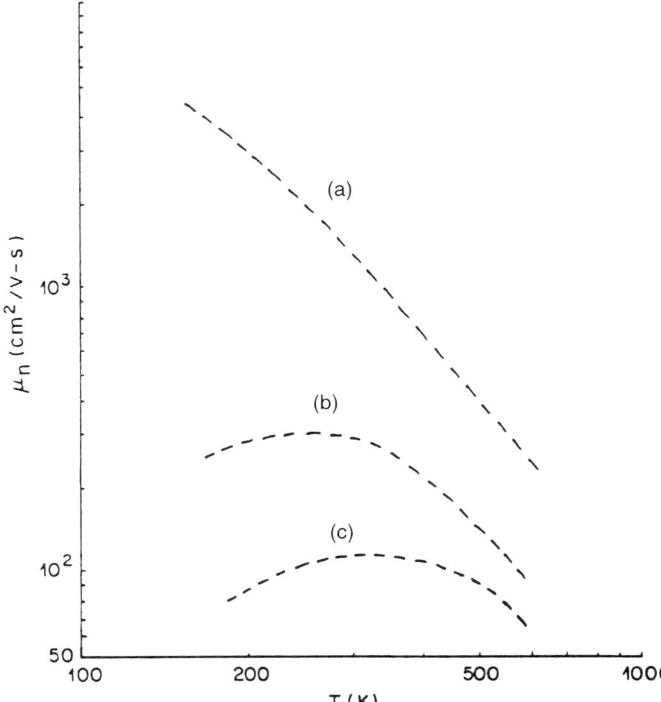

FIG. 28. Electron mobility in Si versus temperature for various doping levels (after Sze, 1985).

The magnetoresistance is much more sensitive to the mass anisotropy and other band-structure features; in particular its strong dependence on the current and magnetic-field directions with respect to the cubic axes (see Fig. 30) allows the identification of the symmetry of the carrier pockets involved (Seeger, 1991). In weak fields, the magnetoresistance $\Delta\rho$ is proportional to the square of the magnetic-field strength and varies with the direction cosines (i,j,k) and (l,m,n), respectively, of the current and the magnetic field according to the inverted Seitz equation

$$\Delta\rho/\rho B^2 = b + c(il + jm + kn)^2 + d(i^2l^2 + j^2m^2 + k^2n^2), \tag{56}$$

where b, c, and d are material-dependent coefficients.

11.5 High-Field Effects

So far we have considered the low-field conductivity corresponding to the Ohmic regime. With increasing field strength, the free-carrier drift velocity approaches their thermal velocity and hot-carrier effects become important (see INSULATORS AND SEMICONDUCTORS, CONDUCTIVITY IN). In particular, in Si and Ge, the high-field mobility (limited mainly by inelastic phonon scattering) eventually decreases linearly with the electric field and the drift velocity saturates, as shown in Fig. 31 (Levinshtein and Simin, 1992). The room-temperature saturation velocities are in Ge, for both electrons and holes, about 5×10^6 cm/s, and in Si, for both electrons and holes, about 10^7 cm/s, whereas the corresponding saturation fields are about 5 kV/cm in n-type Ge, about 15 kV/cm in p-type Ge, about 15 kV/cm in n-type Si, and about 60 kV/cm in p-type Si (but a strongly non-Ohmic behavior is already evident at smaller fields). In the presence of hot electrons, the equivalent valleys need not be equally populated, and, as a consequence, even in a cubic material like n-type Ge, the current need not be parallel to the electric field (Sasaki–Shibuya effect). Moreover, also inequivalent, higher-energy valleys may become populated: For instance, dramatic changes in the Hall coefficient of n-type Ge at electric fields higher than 1 kV/cm are brought about by the transfer of elec-

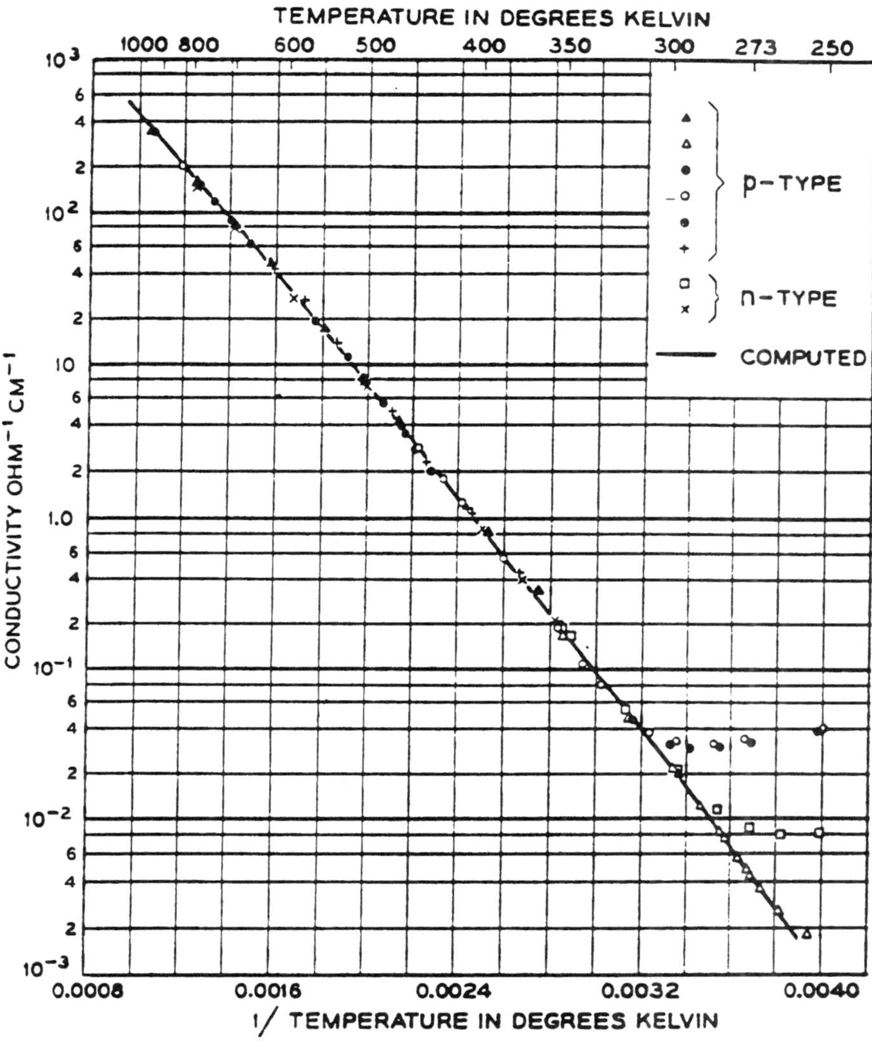

FIG. 29. Temperature dependence of conductivity in Si at various doping levels (Morin and Maita, 1954).

trons from the ⟨111⟩ lowest minima to the ⟨100⟩ higher minima (Heinrich *et al.*, 1970); the Hall coefficient is dominated by the contribution of the ⟨111⟩ high-mobility valleys and starts to increase as their population starts to decrease. In several compound semiconductors such as GaAs, such a carrier transfer to inequivalent valleys is responsible for a decrease of drift velocity at high electric fields, which leads to peculiar current oscillations (Gunn effect). In the presence of hot holes, complex features of the degenerate valence band become important—in particular, the nonparabolicity of the hole dispersion, i.e., the variation with kinetic energy of the effective mass (Costato and Reggiani, 1970). If one increases the electric field strength even beyond the saturation regime, a sudden "runaway" growth of the conductivity is eventually observed (breakdown). This is due to the avalanche multiplication of the number of free carriers: The field is so strong that a carrier can gain between two collisions enough kinetic energy to knock out an electron from the valence to the conduction band (impact ionization). Typical break-

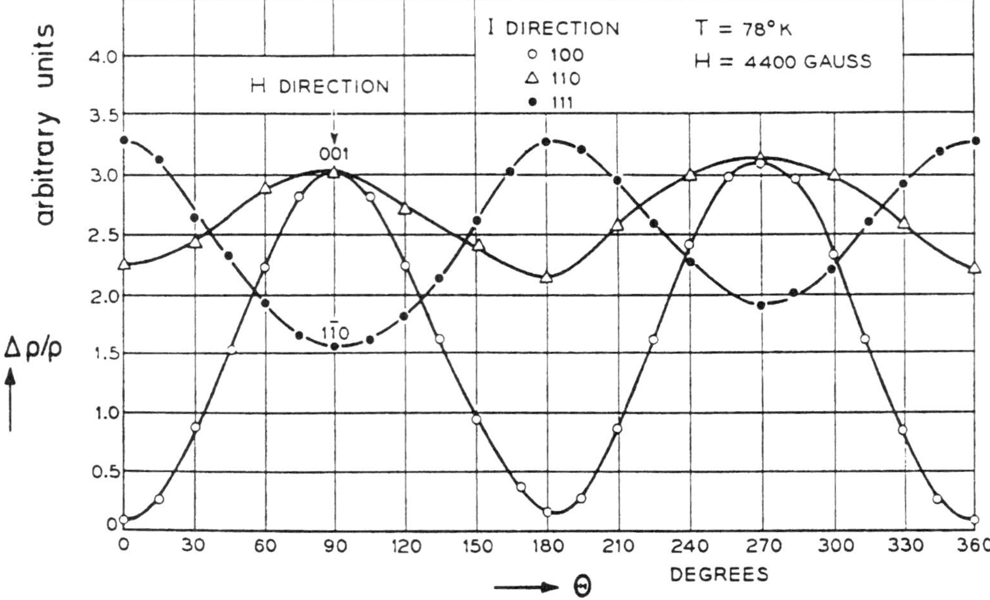

FIG. 30. Magnetoresistance of *n*-type Si as a function of the angle between the current and the magnetic field (in a [010] plane) at liquid-nitrogen temperature (after Pearson and Herring, 1954).

down electric fields are 10^5 V/cm for Ge and 3×10^5 V/cm for Si.

11.6 Impurity-Band Conductivity

In doped semiconductors, when the temperature is lowered so that from the extrinsic regime one enters into the freeze-out regime, the conductivity drops very rapidly because the concentration of free carriers becomes negligible. If the doping level is not too low, though, the resistivity increase flattens out because of the contribution of impurity-

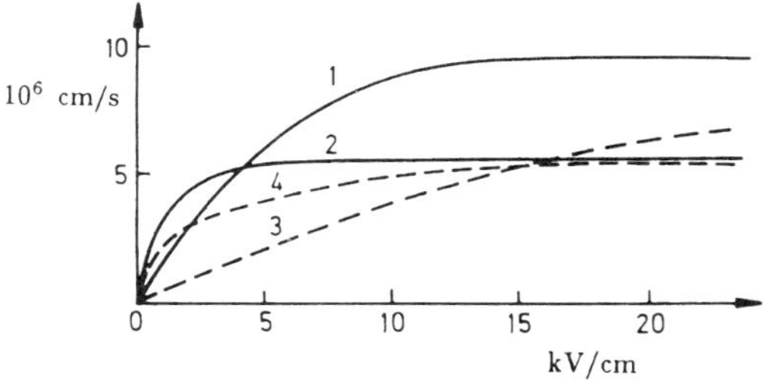

FIG. 31. Drift velocity versus electric field for electrons (solid lines) and holes (dashed lines) in Si (curves 1 and 3) and Ge (curves 2 and 4) at room temperature (after Levinshtein and Simin, 1992).

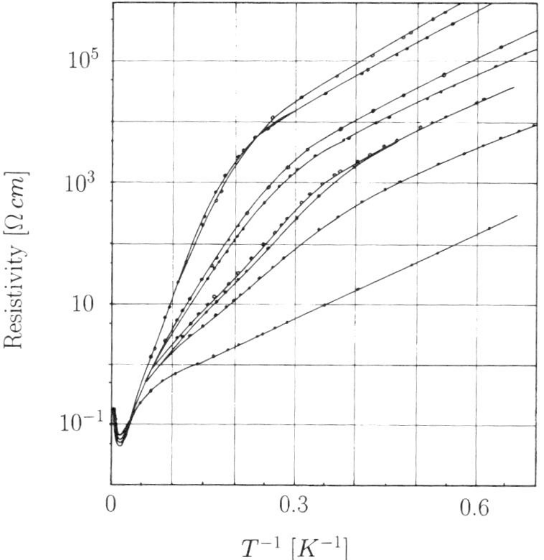

FIG. 32. Impurity-band conductivity in *p*-type Ge; the doping level increases from the top curve ($N_t \simeq 2.1 \times 10^{16}$ cm^{-3}) to the bottom one ($N_t \simeq 7.3 \times 10^{16}$ cm^{-3}) (Fritzsche, 1955).

bound carriers to the conductivity (see Fig. 32). This so-called impurity-band conductivity can be described by two types of fixed-range (nearest neighbor) hopping processes with two different activation energies:

$$\sigma = c_1 e^{-E_1/k_B T} + c_2 e^{-E_2/k_B T}; \tag{57}$$

the one with lower activation energy dominates σ at very low temperatures and is due to the hopping from neutral to ionized impurities (in partially compensated materials), while the other (not always evident in the curves of Fig. 32) involves the double occupancy of impurities. When the doping level is very high ($\simeq 10^{18}$ cm^{-3}), the wave functions pertaining to different impurities overlap each other to the point that their energy levels broaden into an impurity band that merges with the conduction band (impurity metal regime).

Works Cited

Aggarwal, R. L., Zuteck, M. D., Lax, B. (1967), *Phys. Rev. Lett.* **19,** 236–238.

Aspnes, D. E. (1972), *Phys. Rev. Lett.* **28,** 168–171.

Bachelet, G. B., Hamann, D. R., Schlüter, M. (1982), *Phys. Rev. B* **26,** 4199–4228.

Bassani, F., Altarelli, M. (1983), in: N. Koch (Ed.), *Handbook on Synchrotron Radiation,* Vol. 1A, Amsterdam: North Holland.

Bassani, F., Celli, V. (1961), *J. Phys. Chem. Solids* **20,** 64–75.

Bassani, F., Iadonisi, G., Preziosi, B. (1969), *Phys. Rev.* **186,** 735–746.

Bassani, F., Pastori Parravicini, G. (1975), *Electronic States and Optical Transitions in Solids,* Oxford: Pergamon Press.

Bir, G. L., Pikus, G. E. (1974), *Symmetry and Strain-Induced Effects in Semiconductors,* New York: Wiley.

Born, M., Huang, K. (1954), *Dynamical Theory of Crystal Lattices,* Oxford: Oxford University Press.

Brust, D., Phillips, J. C., Bassani, F. (1962), *Phys. Rev. Lett.* **9,** 94–97.

Cardona, M. (1969), "Modulation Spectroscopy," in: F. Seitz, H. Ehrenreich (Eds.), *Solid State Physics,* Suppl. 11, New York: Academic Press.

Cardona, M., Ley, L. (Eds.) (1978), *Photoemission in Solids I and II,* Topics in Applied Physics Vol. 26 and Vol. 27, Berlin: Springer-Verlag.

Cavell, R. G., Kowalczyk, S. P., Ley, L., Pollak, R. A., Mills, B., Shirley, D. A., Perry, W. (1973), *Phys. Rev. B* **7,** 5313–5316.

Chelikowsky, J. R., Cohen, M. L. (1976), *Phys. Rev. B* **14,** 556–582.

Chiarotti, C., Del Signore, G., Frova, A., Samoggia, G. (1962), *Nuovo Cimento* **26,** 403–406.

Chiarotti, C., Del Signore, G., Nannarone, S. (1968), *Phys. Rev. Lett.* **21,** 1170–1172.

Cohen, M. L., Chelikowsky, J. R. (1988), *Electronic Structure and Optical Properties of Semiconductors,* Berlin: Springer-Verlag.

Costato, M., Reggiani, L. (1970), *Lett. Nuovo Cimento Ser. I,* **3,** 239–245.

Dresselhaus, G., Kip, A. F., Kittel, C. (1955), *Phys. Rev.* **98,** 368–384.

Fritzsche, H. (1955), *Phys. Rev.* **99,** 406–419.

Frova, A., Handler, P. (1965), *Phys. Rev. Lett.* **14,** 178–180.

Gerhardt, U. (1965), *Phys. Rev. Lett.* **15,** 401–403.

Giannozzi, P., De Gironcoli, S., Pavone, P., Baroni, S. (1991), *Phys. Rev. B* **43,** 7231–7242.

Greenaway, D. L., Harbeke, H. (1968), *Optical Properties and Band Structures of Semiconductors,* Oxford: Pergamon Press.

Groves, S., Paul, W. (1963), *Phys. Rev. Lett.* **11,** 194–196.

Harrison, W. A. (1989), *Electronic Structure and the Properties of Solids,* New York: Dover.

Heinrich, H., Lischka, K., Kriechbaum, M. (1970), *Phys. Rev. B* **2,** 2009–2016.

Hensel, J. C., Suzuki, K. (1974), *Phys. Rev. B* **9,** 4219–4257.

Hopfield, J. J. (1958), *Phys. Rev.* **112,** 1555–1567.

Hsieh, T. C., Miller, T., Chiang, T. L. (1984), *Phys. Rev. B* **30,** 7005–7008.

Hybertsen, M. S., Louie, S. G. (1985), *Phys. Rev. Lett.* **55,** 1418–1421.

Jagannath, C., Grabowski, Z. W., Ramdas, A. K. (1981), *Phys. Rev. B* **23,** 2082–2098.

Jusserand, B., Cardona, M. (1989), in: M. Cardona, G. Güntherodt (Eds.), *Light Scattering in Solids V,* Berlin: Springer-Verlag.

Kim, N., La Rocca, G. C., Rodriguez, S., Bassani F. (1989), *Rivista Nuovo Cimento* **12** (2), 1–67.

Kittel, C. (1966), *Introduction to Solid State Physics,* New York: Wiley.

Levinshtein, M. E., Simin, G. S. (1992), *Getting to Know Semiconductors,* Singapore: World Scientific.

Luttinger, J. M., Kohn, W. (1955), *Phys. Rev.* **97,** 869–883.

MacFarlane, G. G., McLean, T. P., Quarringron, J. E., Roberts, E. V. (1958), *Phys. Rev.* **111,** 1245–1254.

Madelung, O. (Ed.) (1982), *Landolt-Börnstein: Numerical Data and Functional Relationships in Science and Technology,* Group 3, *Semiconductors,* Vol. 17a, *Physics of Group IV Elements and III-V Compounds,* Berlin: Springer.

Morin, E. J., Maita, J. P. (1954), *Phys. Rev.* **94,** 1525–1529.

Nara, H. (1965), *J. Phys. Soc. Jpn.* **20,** 778–784.

Nilsson, G., Nelin, G. (1971), *Phys. Rev. B* **3,** 364–369.

Nilsson, G., Nelin, G. (1972), *Phys. Rev. B* **6,** 3777–3786.

Pantelides, S. T. (1978), *Rev. Mod. Phys.* **50,** 797–858.

Pearson, G. L., Herring, C. (1954), *Physica* **20,** 975–978.

Penn, D. R. (1962), *Phys. Rev.* **128,** 2093–2097.

Philipp, H. R., Ehrenreich, H. (1963), *Phys. Rev.* **129,** 1550–1560.

Pickett, W. E., Wang, C. S. (1984), *Phys. Rev. B* **30,** 4719–4733.

Ramdas, A. K., Rodriguez, S. (1981), *Rep. Prog. Phys.* **44,** 1297–1387.

Sandercock, J. R. (1972), *Phys. Rev. Lett.* **28,** 237–240.

Seeger, K. (1991), *Semiconductor Physics,* New York: Springer-Verlag.

Seraphin, B. O., Hess, R. B. (1965), *Phys. Rev. Lett.* **14,** 138–140.

Solin, S. A., Ramdas, A. K. (1970), *Phys. Rev. B* **1,** 1687–1698.

Stella, A. (1993), in: A. Stella, L. Miglio (Eds.), *Semiconductor Superlattices and Interfaces,* Proceedings of the International School of Physics "Enrico Fermi", course CXVII, New York: North-Holland.

Straub, D., Ley, L., Himpsel, F. J. (1985), *Phys. Rev. Lett.* **54,** 142–145.

Strinati, G., Mattausch, H. J., Hanke, W. (1980), *Phys. Rev. Lett.* **45,** 290–294.

Sze, S. M. (1985), *Semiconductor Devices,* New York: Wiley.

Tosatti, E., Pastori Parravicini, G. (1971), *J. Phys. Chem. Solids* **32,** 623–626.

Tubino, R., Piseri, L., Zerbi, G. (1972), *J. Chem. Phys.* **56,** 1022–1039.

Uhrberg, R. I. G., Hansson, G. V., Karlsson, U. O., Nicholls, I. M., Persson, P. E. S, Flodström, S. A., Engelhardt, R., Koch, E. E. (1984), *Phys. Rev. Lett.* **52,** 2265–2268.

Weber, W. (1977), *Phys. Rev. B* **15,** 4789–4803.

Ziman, J. M. (1960), *Electrons and Phonons,* Oxford: Oxford University Press.

Further Reading

Besides the works cited, further references are listed in the selected bibliographies of Bassani and Pastori Parravicini (1975) and Cohen and Chelikowsky (1988). Finally, an excellent advanced textbook is Yu, P. Y., Cardona, M. (1995), *Fundamentals of Semiconductor Physics,* New York: Springer-Verlag.

SEMICONDUCTORS, ELEMENTAL—MATERIAL PROPERTIES

HOWARD R. HUFF, *SEMATECH, Austin, Texas, U.S.A.*

INTRODUCTION

Electronics is the world's largest industry with 1995 global yearly revenue of approximately 1.15×10^{12} (U.S.). This is expected to reach approximately 2.0×10^{12} (U.S) by the end of the century. Elemental semiconductor materials constitute a minor component of the electronics-based revenue stream, accounting for about 5.1×10^9 (U.S.) in 1995 and projected to reach about 9.4×10^9 (U.S.) by the year 2000. Although elemental-semiconductor-materials revenue constitutes only about 0.5% of the electronics industry revenue, elemental semiconductor materials are, nevertheless, the cornerstone upon which the industry resides.

The unique characteristic of crystalline semiconducting materials, compared to metals, is their reduction of resistivity with increasing temperature (negative temperature coefficient of resistivity) over selected regions of temperature as schematically illustrated in Fig. 1 (Teal *et al.,* 1985). The electrical resistivity in these regions decreases exponentially with absolute temperature. The decrease of resistivity with increasing absolute temperature in the low-temperature range is due to the thermal excitation of free carriers (i.e., electrons or holes) from the dopant impurities and is referred to as extrinsic conduction. That is, unlike metals where the addition of impurities causes the resistivity to increase, specific dopant impurities in semiconductors can result in drastic decreases in resistivity. The increase in resistivity with further increasing temperature in the low-temperature range is caused by the decrease of the free-carrier mobility. Finally, the decrease in resistivity at higher temperatures is due to the thermal excitation of electron–hole pairs (i.e., intrinsic conduction) dominating the resistivity. The temperature at which intrinsic conduction begins increases

3-527-28139-8/96/$5.00 + .50

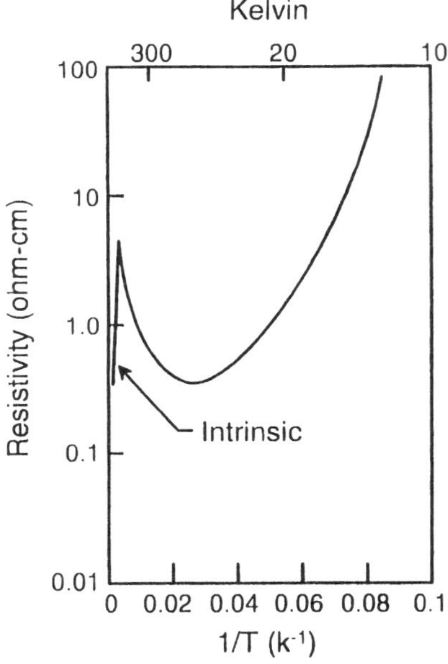

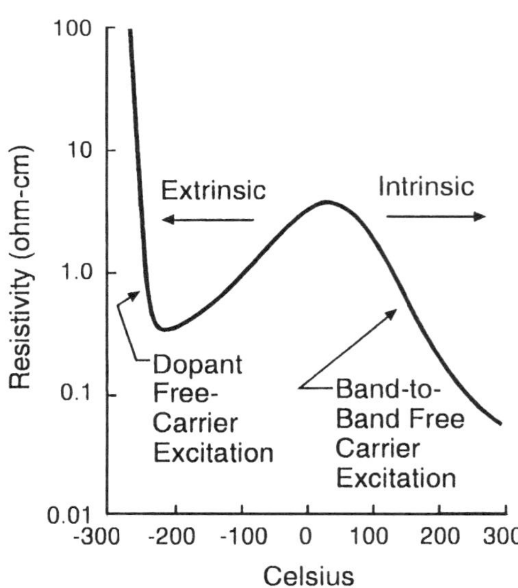

FIG. 1. Effect of temperature on resistivity. The portion of the curve between about −200 °C and 50 °C shifts up and down with the concentration of impurities in the semiconductor. These data are for Ge (following Teal *et al.*, 1985).

as the semiconductor energy gap increases. The mid-temperature, extrinsic range, where the impurities are fully ionized and before intrinsic conduction begins, is the region where most semiconductor devices operate. The resistivity in this region can readily be varied by the introduction of a properly selected dopant impurity concentration (Teal *et al.*, 1985). For example, the resistivity in silicon can be varied by seven orders of magnitude from 10^{-4} to 10^3 Ω cm (Irvin, 1962; Thurber *et al.*, 1980a, 1980b). This article reviews, in addition to the electrical properties, a wide range of material properties for the group-IVa crystalline semiconductors diamond, silicon, germanium, and gray tin, as well as silicon–germanium alloys and β-SiC, important semiconducting material combinations derived from the group-IVa elements. The variation and correlation of their structural, thermodynamic, electrical, thermal, mechanical, and optical properties are emphasized, rather than an exhaustive analysis of the data *per se*. Although these properties are dopant-type, dopant-density, and temperature dependent, the characteristics presented are mainly for lightly doped materials at room temperature.

The atomic structure of the group-IVa elements carbon, silicon, germanium, and tin is introduced in Sec. 1. This is followed by an examination of their condensed-state crystalline semiconducting properties including crystal structure and chemical bonding, electronic structure, and thermodynamic properties, as well as kinetic considerations where appropriate. Appropriate data will also be included for silicon–germanium alloys and β-SiC where available. Comprehension of the defect solid state and the consequences of deviations from perfect periodicity, central to the understanding of semiconductor material properties, is discussed in Sec. 2. Semiconductor fabrication processes have a critical, complementary influence on material microstructure and, thereby, material properties. The process–structure–property methodology is critical, in conjunction with device and integrated circuit (IC) fabrication, to facilitate tailoring the semiconductor material properties to ensure optimal device/IC performance, reliability, and yield. Accordingly, the variation of the significant electrical, thermal, mechanical, and optical properties of the group-IVa crystalline semiconductors in Sec. 3, silicon–

germanium alloys in Sec. 4, and β-SiC in Sec. 5 should be considered in that context. Section 6 discusses the emergence of silicon, and silicon-based materials such as silicon–germanium, as the premier materials driving the IC microelectronics revolution. Section 6 also cautions against undue optimism for diamond, which is often cited as the ultimate semiconducting device material. Section 7 presents several conclusions. Finally, a further reading list complements the list of works cited.

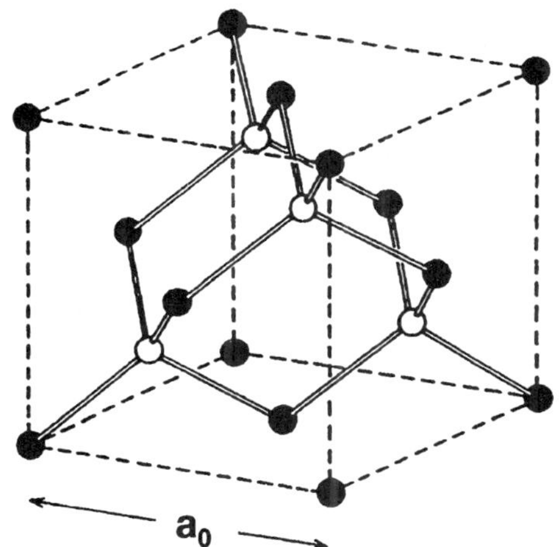

FIG. 2. Silicon crystal structure; the open circles indicate substitutional sites interior to the unit cell (following Kittel, 1996).

1. CHEMICAL BONDING AND SEMICONDUCTORS

1.1 Atomic Structure

Carbon, silicon, germanium, and tin are elements in group IVa of the periodic table (Lide, 1993–1994). Their atomic numbers, gram atomic weights, electronic configurations, atomic radii, and ionization energies are listed in Table 1. Each element has four valence electrons, two in *s* states and two in *p* states, for the ground-state configuration, which is designated by the spectroscopic notation $^3P_0$ (Landau and Lifshitz, 1958). The monotonic decrease of the ionization energy with increasing atomic number of the elements will be seen to exhibit analogs with the variation in a number of crystalline semiconducting material properties.

1.2 Group-IVa Crystalline Semiconductors

1.2.1 Crystal Structure The group-IVa semiconductors crystallize in the tetrahedrally coordinated diamond crystal structure. The diamond crystal structure is conveniently described as two interpenetrating face-centered cubic (fcc) crystals displaced along the $\langle 111 \rangle$ direction by one-fourth the diagonal length (see Fig. 2) (Harrison, 1980; Bullis and Huff, 1994). The unit cell of this structure contains eight sites at the points (0,0,0), $(0,\frac{1}{2},\frac{1}{2})$, $(\frac{1}{2},0,\frac{1}{2})$, $(\frac{1}{2},\frac{1}{2}0)$, $(\frac{1}{4},\frac{1}{4},\frac{1}{4})$, $(\frac{1}{4},\frac{3}{4},\frac{3}{4})$, $(\frac{3}{4},\frac{1}{4},\frac{3}{4})$, and $(\frac{3}{4},\frac{3}{4},\frac{1}{4})$. The length of the unit cell in the $\langle 100 \rangle$ direction, referred to as the lattice constant a_0, is the distance required to translate the unit cell to the position of the adjacent unit cell along the three $\langle 100 \rangle$ directions. This replication of the unit cell results in the crystal structure. A useful characteristic is the covalent radius, r_c, defined as one-half the distance between nearest neighbors in a crystal of the pure element, in terms of the cube edge a_0. Alternatively, the closest separation of atomic sites along the $\langle 111 \rangle$ direction or bond length d (twice the value of r_c) is conveniently utilized. The bond length is a critical parameter affecting a number of physical properties involving the elastic stiffness coefficients such as the bulk modulus (see Sec. 3.3). The bond energy is a useful

Table 1. Characteristics of several group-IVa elements.

Element	Atomic number	Gram atomic weight (g)	Electronic configuration	Atomic radius (nm)	Ionization energy (eV)
Carbon	6	12.01	$1s^22s^22p^2$	0.091	11.3
Silicon	14	28.09	$1s^22s^22p^63s^23p^2$	0.132	8.1
Germanium	32	72.59	$1s^22s^22p^63s^23p^63d^{10}4s^24p^2$	0.137	7.9
Tin	50	118.69	$1s^22s^22p^63s^23p^63d^{10}4s^24p^64d^{10}5s^25p^2$	0.162	7.3

Table 2. Variation of several structural/bonding characteristics for the group-IVa and β-silicon carbide semiconductors.

Element	Metallicity parameter (α_m)	Lattice constant a_0(nm)	Bond length d(nm)	Cohesive energy (eV/bond)
Diamond[a]	0.34	0.356683	0.154	3.68
β-Silicon carbide	0.45	0.43596	0.189	3.17
Silicon	0.66	0.543075	0.235	2.32
Germanium	0.81	0.565754	0.245	1.90
Gray tin[b]	0.87	0.64912	0.281	1.56

[a]Carbon crystallizes as diamond under conditions discussed in Sec. 1.2.3. p-type semiconducting diamond is referred to as type IIb (see DIAMOND AND DIAMONDLIKE CARBON).
[b]Also referred to as α-tin.

parameter denoting the strength of the chemical bond (Pauling, 1960). The cohesive energy (enthalpy of sublimation) is a measure of the energy per bond required to separate the solid into isolated atoms. It has been a useful parameter historically and is included for completeness, although it is not currently a fundamental parameter describing variations in material properties (Harrison, 1980). The values of a_0, d, and the cohesive energy for the group-IVa semiconductors are summarized in Table 2. The increase in a_0 and d and the concurrent decrease in cohesive energy from diamond to gray tin in Table 2 can be explained in terms of the decrease in the strength of the covalent bond from diamond through gray tin. β-SiC's parameters are between those of silicon and carbon, as expected. These trends may also be correlated with the decrease in the energy gap E_g and the increase in the metallicity parameter α_m, to be discussed in Sec. 1.2.2.

Each atomic site in the diamond crystal is surrounded by four nearest-neighbor atomic sites (coordination number 4), which reside at the points of a tetrahedron with bond angles $\theta = \arccos(-\frac{1}{3}) = 109.47°$. The relative positions of the four nearest neighbors have different directions for atoms in the two fcc structures. Therefore, the two fcc structures are not equivalent and the diamond crystal structure cannot be reduced to a simpler crystal structure. The covalent bond formed between nearest neighbors is described by sp^3 hybridization (see Sec. 1.2.2.) Each atom of the pair contributes one electron, which bond with opposite spins as described by the Pauli principle. As a result, each atom on an atomic site in a perfect crystal is surrounded by eight electrons: four of "its own" and one from each of its four nearest neighbors. This octet of electrons, similar to that of a noble gas, is chemically very stable (Pauling, 1960). No electrons are free to move through the crystal unless sufficient energy is applied to break the bond.

The sp^3 hybridized bond is highly directional and accounts for the diamond crystal structure exhibiting the smallest packing density (i.e., volume fraction of the atoms in the unit cell) among the cubic crystal classes, which include the simple cubic, body-centered cubic (bcc), and fcc structures (see Table 3). The variation of the packing density correlates with the geometric coeffi-

Table 3. Variation of structural parameters for the cubic crystal structures.

Crystal structure	Coordination number	Covalent radius (r_c)	Atoms per unit cube	Packing density	Geometric coefficient (α)
fcc	12	$(a_0/4)\sqrt{2}$	4	$(\pi/12)\sqrt{8} = 0.7405$	1.79
bcc	8	$(a_0/4)\sqrt{3}$	2	$(\pi/8)\sqrt{3} = 0.6802$	1.79
Simple cubic	6	$a_0/2$	1	$\pi/6 = 0.5236$	1.76
Diamond	4	$(a_0/8)\sqrt{3}$	8	$(\pi/16)\sqrt{3} = 0.3401$	1.67

cient α characterizing the electrostatic environment of the cubic crystal classes (Harrison, 1980). Alpha depends only on the geometry of the atomic arrangements; charge magnitudes and distances have been factored out. The diamond crystal structure exhibits the smallest geometric constant. Silicon and germanium can also exhibit a phase transformation at room temperature to a body-centered tetragonal crystal structure without macroscopic plastic deformation (Roos *et al.*, 1995a).

An equivalent but more fundamental description of the diamond crystal structure utilizes the concept of an fcc lattice with basis vectors (0,0,0) and $(\frac{1}{4},\frac{1}{4},\frac{1}{4})$. The symmetry operations required to describe the arrangement of atoms associated with the lattice (i.e., the space group) is $Fd3m$ using the International notation (Burns, 1985). The F refers to the fcc lattice and $d3m$ refers to the point symmetry, glide, and screw operations. The diamond crystal structure is an example of a nonsymmorphic space group, which requires a translation by a fraction of a unit cell length in order to specify the space group. The point group is $m3m$, since only lattice translations are involved (the lattice symbol F is ignored) and all translations by a fraction of a unit cell length are set equal to zero. β-SiC and the III–V compounds such as gallium arsenide (zinc-blende structure) are characterized by $F\bar{4}3m$ and $\bar{4}3m$ for the corresponding space and point groups, respectively. In particular, the zinc-blende structure does not have a screw axis or a center of symmetry inversion between the (0,0,0) and $(\frac{1}{4},\frac{1}{4},\frac{1}{4})$ positions because of the dissimilarity of the atoms. The origin of their polarity effects in the $\langle 111 \rangle$ directions is due to this lack of a center of symmetry inversion. Silicon–germanium alloys, however, are not ordered wherein every silicon and germanium atom, for example, is precisely locked onto a (0,0,0) and $(\frac{1}{4},\frac{1}{4},\frac{1}{4})$ lattice site, respectively (or vice versa). Accordingly, one does not expect to see a polarity effect in silicon–germanium alloys, although, in principle, $Si_{0.5}Ge_{0.5}$ should be ordered.

1.2.2 Electronic Structure Consider a representative group-IVa element arranged in the diamond crystal structure at 0 K. At large internuclear distances, the two s and two p valence electrons are still associated with each atom. As the atoms approach each other, using the conventional one-electron model, the discrete atomic energy s and p levels are each broadened into a band as a result of coupling between the neighboring atomic states, as described by the tight-binding theory (Harrison, 1980) (see Fig. 3). These bands may be regarded as a quasicontinuum of energy states propagating throughout the material. The s bands are completely filled with two electrons donated per atom. The p bands, however, can accommodate six electrons per atom and, therefore, are only partially filled with two electrons donated per atom. This partial filling of bands is characteristic of a metal. As the atoms are brought still closer together, the broadening bands reach each other. At this point, the s and p valence electron states reshuffle to form bonding and antibonding bands. This process has also been described as the formation of the prototypical sp^3 hybridized, tetrahedrally bonded purely covalent bonds (Harrison, 1980). This reshuffling, or sp^3 hybridization process, results in an energy gap, E_g, with four energy states below and four energy states above the energy gap, with no propagating energy states within the energy gap. The original four valence electrons of each atom completely fill the lower four states, which are collectively referred to as the bonding states or valence band. The four states above the energy gap are completely empty and are collectively referred to as the antibonding states or conduction band. Each of the four bonding states is doubly degenerate on account of the presence of one bonding state contributed by each of the four neighboring atoms, as described in Sec. 1.2.1. The valence band, therefore, is completely filled with a stable octet of electrons per atom. Although the original valence electrons belong to the crystal system, there is, nevertheless, a high concentration of electronic charge pileup in the regions between the atoms. This is interpreted as the covalent bond between adjacent atoms formed between two oppositely spinning electrons. The system has no free electronic carriers at 0 K and is an insulator. This description of the band model is consistent with the bond model for crystalline materials described in Sec. 1.2.1. Interesting theoretical challenges are posed, however, in reconciling the band and bond approaches in

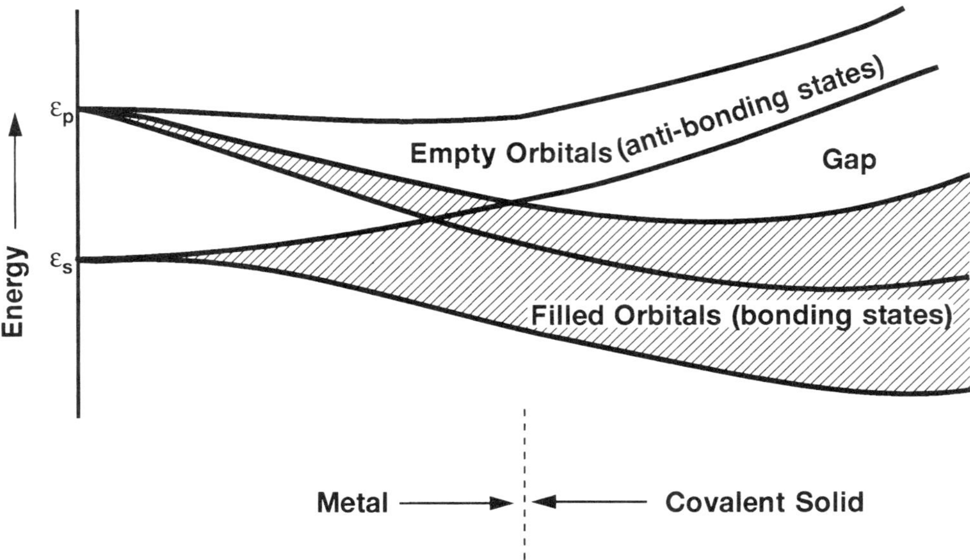

FIG. 3. Formation of energy bands in a homopolar tetrahedral semiconductor as the atoms are brought together. The internuclear distance decreases to the right. (Following Harrison, 1980.)

the absence of long-range order (Ziman, 1979).

The energy gap is the defining characteristic of an insulator and is typically greater than 5–6 eV. A pure semiconductor, however, has a sufficiently small value of E_g relative to kT to facilitate the thermal excitation of electrons from valence-band states to conduction-band states, resulting in free electrons and holes at finite temperatures (see Sec. 3.1). The value of E_g slightly decreases with increasing temperature (Lide, 1993–1994) for the group-IVa semiconductors since the increased temperature increases the amplitude of atomic vibrations, thereby increasing the effective internuclear distance. The value of E_g decreases (increases) slightly with increasing pressure for silicon (germanium) (Lide, 1993–1994). Germanium exhibits greater changes, compared to silicon, in E_g with changes in temperature and pressure because of its weaker covalent bonding, as noted in Sec. 1.2.1. The magnitude of E_g varies with different directions in the crystal and is generally expressed with respect to specific directions in reciprocal (**k**) space (Herman, 1955; Sze, 1981).

The fundamental parameter distinguishing the elemental semiconductors is the metallicity parameter α_m. This parameter is defined as the ratio of the metallic energy V_1 to the covalent energy V_2 within a factor of 1.1 (Harrison, 1980):

$$\alpha_m = 1.1(V_1/V_2). \tag{1}$$

The metallic energy relates to the band broadening and is determined principally by the separation in energy between the atomic p and s states $[\frac{1}{4}(\epsilon_p - \epsilon_s)]$. The sp splitting (and therefore V_1) does not change greatly as one proceeds from diamond to gray tin (i.e., with increasing atomic number) as seen in Table 4. The separation in energy between the valence and conduction bands arises from the coupling strength between states on neighboring atoms, which resulted in the bonding–antibonding splitting (Harrison, 1980). Inasmuch as the bond length significantly increases as one proceeds from diamond to gray tin, the covalent bond becomes weaker and, likewise, the covalent energy V_2 decreases relative to the band broadening. Accordingly, the energy gap E_g decreases and α_m increases as one proceeds from diamond to gray tin (see Table 4) (Harrison, 1980). β-SiC's energy gap is between diamond and silicon as expected. The energy gap for silicon–germanium alloys can be tuned to intermediate values between silicon and germa-

Table 4. Metallic energy, covalent energy, metallicity, and energy gap for the group-IVa and β-silicon carbide semiconductors.

Element	Metallic energy V_1 (eV)	Covalent energy V_2 (eV)	Metallicity α_m	Energy gap at 0 K (eV)	Energy gap at 300 K (eV)
Diamond	2.13	6.94	0.34	5.48	5.45
β-Silicon carbide	1.91	4.66	0.45	2.6	2.3
Silicon	1.76	2.98	0.66	1.17	1.12
Germanium	2.01	2.76	0.81	0.74	0.66
Gray tin	1.64	2.10	0.87	0.075	0.0

nium by varying the germanium content (see Sec. 4). When α_m approaches unity, semiconducting behavior disappears as the valence and conduction bands approach each other, and metallic behavior (i.e., partially filled bands) is approached. This indeed occurs for gray tin (Harrison, 1981), at least at the wave vector $\mathbf{k} = 0$, where E_g approaches zero. The metallicity parameter is the parameter *sine qua non* by which the distinct variations in the group-IVa crystalline semiconductor material characteristics may be correlated.

1.2.3 Thermodynamic Properties The first and second laws of thermodynamics are usefully expressed in terms of the Gibbs free energy ΔG of a chemical reaction at constant temperature and pressure (Swalin, 1972; Kubaschewski and Alcock, 1973):

$$\Delta G = \Delta H - T\Delta S, \tag{2}$$

where ΔH is the change in enthalpy and ΔS is the change in entropy associated with the formation of the product at absolute temperature T. The enthalpy of formation is a constant for every substance at a given temperature and pressure relative to the reference or standard state. The reference state usually selected for crystalline solids is the most stable form of the element at 298 K and one atmosphere pressure. Therefore, the enthalpy of formation for carbon, silicon, germanium, and white tin is zero. Several thermodynamic data for the group-IVa elements and β-SiC are tabulated in Table 5 (Kubaschewski and Alcock, 1973). Negative values for ΔH^0_{298} refer to an exothermic process whereas a positive value for ΔH^0_{298} refers to an endothermic process. The positive value of ΔH^0_{298} for diamond indicates it is not stable at room temperature in comparison to the graphitic form. The negative value of ΔH^0_{298} for gray tin, compared with white tin, might indicate that the former is the stable form of tin at 298 K. However, white tin is the stable form at 298 K (Swalin, 1972); the temperature at which gray tin transforms to white tin at one atmosphere pressure is 286 K (Swalin, 1972). Clearly, the enthalpy change is not a sufficient criterion to deduce the direction of a chemical reaction.

The complete description of a chemical reaction requires utilization of the entropy term in Eq. (2). Entropy may be interpreted

Table 5. Selected thermodynamic properties for the group-IVa and β-silicon carbide semiconductors at 298 K.

Element	ΔH^0_{298} (kJ/mole)	S^0_{298} (J/mole K)	ΔG^0_{298} (kJ/mole)	c_{p298} (J/mole K)
Carbon	0	5.69	−2.87	10.62
Diamond	1.90	2.44	—	7.56
β-Silicon carbide	−66.88	16.51	−71.80	29.54
Silicon	0	18.81	—	20.73
Germanium	0	42.22	—	26.07
White tin	0	51.41	−0.03	26.32
Gray tin	−1.96	44.73	—	25.28[a]

[a]From Lide (1993–1994).

as the decrease in atomic order of a system undergoing any process. It is quantitatively calculated by the enthalpy exchanged during an isothermal and isobaric reversible process divided by the absolute temperature at which the process takes place. The entropy of formation at a given temperature may be determined for every substance relative to a reference state. Since the third law of thermodynamics indicates that the entropy of any ordered crystalline solid approaches the same value at 0 K (Swalin, 1972; Kubaschewski and Alcock, 1973), the common practice is to select 0 K as the reference state and arbitrarily choose the value zero for the entropy at 0 K (and one atmosphere). Entropy, therefore, can be determined "absolutely," unlike enthalpy, which was seen to be determined relative to the value of enthalpy at 298 K. β-SiC's entropy is significantly closer to that of silicon, rather than simply between diamond and silicon's values (see Table 5). The entropy of fusion decreases as one proceeds from silicon to gray tin (with increasing metallicity) (Kubaschewski and Alcock, 1973). This is reasonable since the strength of the covalent bond and the Debye temperature decrease with increasing metallicity and the group-IVa semiconductors are metallic in the liquid state.

The Gibbs free energy of reaction favors white tin over gray tin at temperatures greater than 286 K (Swalin, 1972) and graphitic carbon over diamond at room temperature (Kubaschewski and Alcock, 1973) (see Table 5). The observation of diamond at room temperature, therefore, is indicative of its nonequilibrium thermodynamic status. Bulk crystalline diamond is thermodynamically stable with respect to graphite at high temperature (>1000 °C) and high pressure (6×10^{11} dyn/cm^2); the reaction rate is enhanced by the utilization of a catalyst such as nickel (see DIAMOND AND DIAMONDLIKE CARBON). Crystalline diamond has also been grown via nonequilibrium chemical vapor deposition (CVD) processes, most of which have in common the production of atomic hydrogen and hydrocarbon radicals in regimes where solid carbon is expected to be the stable product. Two mechanisms have been proposed whereby crystalline diamond, rather than graphite or vitreous carbon, is observed. The first suggests that graphite is etched by atomic hydrogen at a faster rate than diamond, and hence diamond is kinetically stable with respect to graphite. The second mechanism suggests that diamond surfaces are stabilized by termination with atomic hydrogen. This topic continues to be an important area of research, and the reader is referred to the literature for further details (Davis *et al.*, 1988; Yarbrough, 1990) (see CARBON MATERIALS; DIAMOND AND DIAMONDLIKE CARBON; GRAPHITE).

The molar specific heat c_{p298} is the energy required to raise the temperature of one molecular weight of the material by one degree at constant pressure. The classical Dulong–Petit law indicates that c_{p298} for one molecular weight of a monatomic solid is about 26.33 J/mole K (6.3 cal/mole K) at room temperature (De Launay, 1956). The specific heat at 300 K for silicon, and especially diamond, however, is significantly less than 26.33 J/mole K because the Debye temperature θ_D for these elements is significantly larger than 300 K (see Sec. 3.2 and Tables 5 and 14). That is, the Dulong–Petit law is valid only when the atoms are sufficiently loosely bound that the equipartition of energy is valid (T/θ_D approximately greater than 80%) (De Launay, 1956). β-SiC's specific heat of 29.54 J/mole K, however, is only about 6% less than the sum of c_{p298} for carbon and silicon. This additivity of the specific heat of the elements to obtain the specific heat of the compound in a chemical reaction is known as Neumann and Kopp's rule (Kubaschewski and Alcock, 1973).

2. THE DEFECT SOLID STATE

2.1 Point Defects

2.1.1 Vacancies and Interstitials At 0 K, silicon is a perfect insulator inasmuch as the isotopes Si^{28}, Si^{29}, and Si^{30} exhibit only zero-point (weakly coupled) atomic vibrations. At any nonzero temperature, however, thermodynamic requirements introduce a degree of disorder (entropy) into the crystal structure so as to minimize the Gibbs free energy [Eq. (2)] at thermal equilibrium. In this case, ΔH is the enthalpy, and ΔS is mainly the configurational (i.e., mixing) entropy associated with the formation of each type of point defect at absolute temperature T (Swalin, 1972; Seeger, 1976).

Atoms transfer off the substitutional sites onto interstitial sites creating Frenkel (vacancy–interstitial) point defects (see POINT AND EXTENDED DEFECTS IN SOLIDS). The interstitial is also referred to as a self-interstitial. Five equivalent tetrahedral interatomic voids, in which interstitial atoms are easily accommodated in silicon, are located along the body diagonals at the points $(\frac{1}{2},\frac{1}{2},\frac{1}{2})$, $(\frac{1}{4},\frac{1}{4},\frac{3}{4})$, $(\frac{1}{4},\frac{3}{4},\frac{1}{4})$, $(\frac{3}{4},\frac{1}{4},\frac{1}{4})$, and $(\frac{3}{4},\frac{3}{4},\frac{3}{4})$. A description of additional interstitial configurations has recently been summarized (Hu, 1994). Vacancies are also injected from the surface into the crystal, creating Schottky point defects. The Frenkel and Schottky defect systems are described by the Boltzmann statistics; the temperature dependence of the Gibbs free energy for these reactions, however, is still an active area of investigation (Brown *et al.*, 1994). The point-defect concentrations and the size of their clusters may be experimentally determined, however, by such techniques as diffuse x-ray scattering (Schulz, 1982), positron annihilation (Asoka-Kumar *et al.*, 1994), or the Bond x-ray method (Bond, 1960). The Bond method can precisely measure the change in lattice constant and deduce the point-defect concentrations. The concentrations are extremely low, probably $<10^5$ cm^{-3} at temperatures approximately 77 K for silicon. The point defects are in the range of 10^{13}–10^{16} cm^{-3} from approximately 800 to 1400 °C for silicon; their presence and concentration are deduced by comparing experimental dopant diffusion profiles in devices and ICs with model calculations. The temperature-dependent concentration and diffusion coefficient for vacancies and interstitials in silicon remains a significant theoretical challenge (Taylor *et al.*, 1993; Brown *et al.*, 1994).

The creation of a vacancy in an otherwise rigid crystal should increase the crystal volume by one atomic volume (Seeger, 1976). Likewise, the creation of one interstitial should decrease the crystal volume by one atomic volume. In reality, the crystal elastically deforms in response to the introduction of point defects, which results in additional changes of the crystal volume. The elastic deformation, however, affects the lattice constant and the macrosopic dimensions equally. Thermal expansion also affects the unit cell and the crystal volume equally. By subtracting the relative change of volume of the unit cell from the relative change of the crystal volume, one obtains the difference between the vacancy and the interstitial concentrations. For a cubic crystal, one may write (Seeger, 1976)

$$C_V - C_I = 3[(\Delta l/l_0) - (\Delta a/a_0)] + \ldots, \qquad (3)$$

where Δl refers to the length change relative to the sample length, l_0, in a low-temperature state essentially free of vacancies and interstitials and Δa and a_0 are the corresponding quantities for the lattice constant (Seeger, 1976).

A measure of the type of point defect present in metals such as molybdenum and tungsten and the group-IVa semiconductors is illustrated in Fig. 4 (Slack and Bartram, 1975). The linear thermal expansion coefficient α is represented by the continuous line and is related to $\Delta l/l_0$. The dashed line represents the linear extrapolation of α from approximately 70% of the absolute melting temperature and is related to $\Delta a/a_0$. Accordingly, the dominance of vacancies or interstitials may be determined from the sign of the deviation of the experimental quantities in Eq. (3). Vacancies are the dominant intrinsic point defect in metals such as molybdenum and tungsten. The evidence is not so clear for germanium and diamond, although self-diffusion data in germanium is compatible with the presence of vacancies (Seeger and Swanson, 1968; Holwigle *et al.*, 1985). There has been no evidence of interstitials in germanium (Gösele and Tan, 1991). The thermal expansion coefficients of silicon and SiC are significantly different from those of metals, germanium, and diamond. The rather constant value of the thermal expansion coefficient of silicon at temperatures greater than approximately 725 °C has been interpreted as evidence that interstitials are as significant as vacancies. Collaborative evidence supporting the importance of the interstitial in silicon at high temperature comes from self-diffusion data (Seeger and Swanson, 1968; Holwigle *et al.*, 1985; Gösele and Tan, 1991) and transmission electron microscopy (TEM) analysis, indicating that A-type swirl defects are interstitial-type microscopic dislocation loops (Foll *et al.*, 1977; Chikawa and Shirai, 1979) (see Sec. 2.2.1). Swirl is a generic term referring to a class of defects in silicon distributed in a swirl pat-

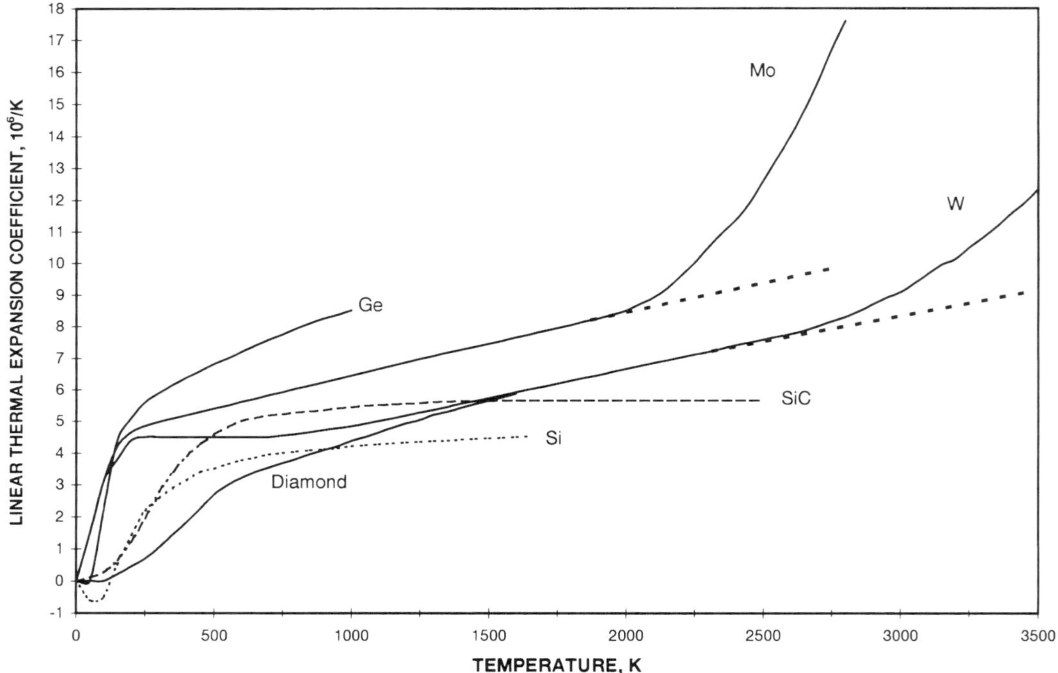

FIG. 4. Thermal expansion coefficients of several diamondlike solids (following Slack and Bartram, 1975).

tern reflecting melt convection, rotation of the crystal, and temperature fluctuations at the crystal–melt interface during crystal growth. The defects develop as a result of the condensation of point-defect complexes formed via solid-state reactions in the absence of readily available line and/or surface sinks. It should be emphasized, however, that the observed defects reflect conditions subsequent to the complete crystal growth process, resulting from the complicated set of microscopic defect reactions in silicon (Huff, 1992; Brown *et al.*, 1994), and require consideration of the initially solidifying silicon at the melt interface during crystal growth (Dornberger and von Ammon, 1996).

Defects such as vacancies and interstitials, therefore, are thermodynamically required at any nonzero temperature in order to minimize the Gibbs free energy ΔG in Eq. (2). On the other hand, one- and two-dimensional defects, such as dislocations (line defects) and stacking faults or surfaces (planar/interface defects), respectively, have high Gibbs free energies of formation because of the large number of atoms involved in the defects and a reduced configurational entropy due to the large number of defect sites. Consequently, line and planar defects are introduced only in environments that are far away from thermal equilibrium. Nonequilibrium defects can be minimized, in principle, by bulk reorganization and surface reconstruction processes during thermal annealing. Since diamond is not thermodynamically stable at room temperature and pressure, however, but is controlled by kinetic considerations, achieving relatively defect-free diamond is challenging.

Any deviation from perfect periodicity, furthermore, may be regarded as a defect. This includes free electronic carriers, particularly in semiconductors and insulators, phonons, impurities, isotopic variations, etc. Controlling point-defect impurities such as oxygen, carbon, hydrogen, and nitrogen are, in particular, of immense importance for silicon IC process technologies (Bullis and Huff, 1994). The wave functions describing defects are nonpropagating solutions to Schrödinger's equation; that is, they are spatially localized in the crystal and often exhibit localized (and deep) energy states within the energy gap. Defects that introduce localized energy levels within the energy gap may or may not influence the fabrication or performance of ICs. For example, fabrication of IC structures at reduced process temperatures (≤800 °C) may accentuate the effects of different charge states of localized defects through the promotion of unique solid-state defect reactions compared to high-temperature processing and/or significantly different rates for the same reactions. The resulting changes in the silicon microstructure will affect the subsequent device and IC performance (Huff, 1992).

2.2 Line Defects

2.2.1 Dislocations Dislocations are (nonequilibrium) line imperfections in the crystal that extend over many atomic spacings (see POINT AND EXTENDED DEFECTS IN SOLIDS). They can move (slip) under stress fields in the {111} planes (glide plane) and ⟨110⟩ directions (slip directions) in the diamond crystal, once they have been activated from the bound state. Their movement provides the mechanism for plastic deformation; that is, the permanent displacement of atomic planes past each other as the result of an external stress (Cottrell, 1961; Hirth and Lothe, 1982). The {111} planes in the silicon crystal are unequally spaced in the ratio 1:3 (see Fig. 2). Therefore, two sets of {111} glide planes, the *shuffle* and *glide*, exist. The *shuffle* set is observed if glide occurs between the widely spaced planes. The *glide* set is observed if glide occurs between the narrowly spaced planes.

When an extra plane of atoms occurs on one side of the slip plane, the line of atoms at the end of this extra half plane is called an edge dislocation (see Fig. 5a). The motion of edge dislocations in the glide plane occurs at kink sites (Hirth and Lothe, 1982) and is along the direction of the shearing force. The effect of having two atoms per lattice site in the group-IVa semiconductor is to double the number of atomic half planes associated with an edge dislocation (Hobstetter, 1960). Under some conditions the crystal may twist. In this case, there are no extra half planes of atoms. This type of dislocation, known as a screw dislocation, occurs along the line where the planes above and below the slip plane return to their original relationship. Screw dislocations move through the crystal in the direction perpendicular to the shearing force (Cottrell, 1961; Hirth and Lothe, 1982). The stress required to move dislocations decreases and their velocity increases exponentially with increasing temperature (Alexander and Haasen, 1968).

Dislocations are characterized by a net displacement, the Burgers vector **b**, and a unit vector **I** that is parallel to the dislocation line. Dislocations in real crystals may be pure edge or pure screw, where the dot

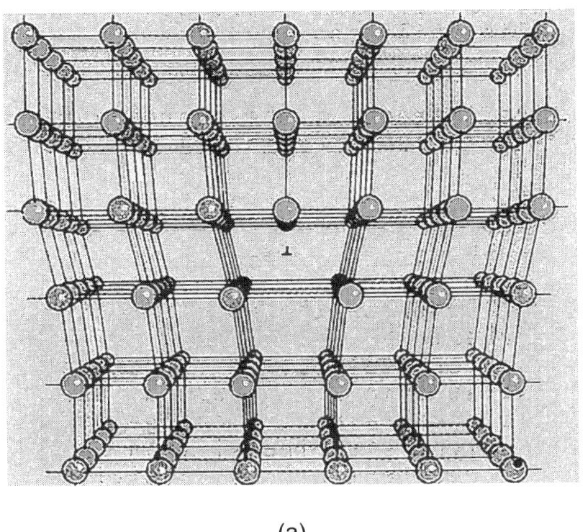

(a)

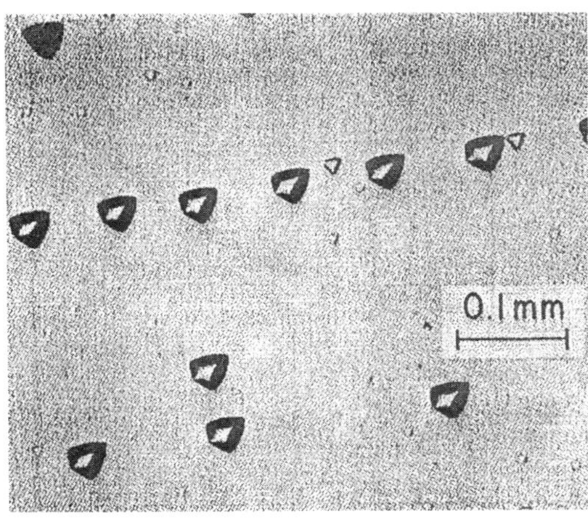

(b)

FIG. 5. (a) Atomistic representation of an edge dislocation (after Gay, 1960). (b) Photomicrograph of a preferentially etched (111) silicon wafer illustrating etch pits.

product of **b** and **I** equals zero and **b**, respectively, or more commonly, a combination of both configurations. In silicon and germanium, a more common configuration than the double half-plane edge dislocation occurs. This particular structure requires only one extra half plane and is referred to as the 60° dislocation (Hobstetter, 1960; Hirth and Lothe, 1982). In this case, the angle between the dislocation line and the Burgers vector is 60°. Electron and x-ray diffraction methods, which resolve the orientation of **b** relative to **I**, are routinely employed for structural determination. The motion of the edge and screw dislocations described above are confined to a glide plane that contains both **b** and **I**. Glide motion does not increase the dislocation line length. Motion normal to the glide plane is referred to as climb and is nonconservative. That is, climb requires the addition or subtraction of point defects and results in an increase of the dislocation line length.

The atomic arrangement and peculiar features of dislocations in silicon have been described (Hornstra, 1958; Hirth and Lothe, 1982). A dislocation cannot end within a single crystal except at another dislocation; thus, dislocations form closed loops or interconnecting networks. Unbound dislocations can multiply as a result of thermal stress fields during IC thermal processing, penetrate active regions of devices and ICs, and lead to electrical malfunctions (Lawrence and Huff, 1982; Huff and Shimura, 1985; Huff and Goodall, 1995). It was originally believed that dislocations could not form at low temperatures in brittle solids such as silicon (Allen, 1957). However, it has been found that (somewhat bound) dislocations could be produced in a pure cracking event (i.e., "crack dislocation") during room-temperature semiconductor material shaping processes (Allen, 1957; Wise *et al.*, 1994) and by nanoindentation at room temperature (Page *et al.*, 1992). These dislocations are not necessarily in a glide configuration (i.e., lying entirely in {111} planes), but at elevated temperatures such as during IC processing and with sufficient thermal stress, these crack dislocations can act as sources for extensive slip. That is, although silicon wafers utilized for the fabrication of ICs are macroscopically dislocation free, latent sources often exist at the wafer edges and back surface, reducing the critical resolved shear stress to activate dislocation movement. On the other hand, bound dislocations are incapable of glide, or they form loops too small to be activated into macroscopic slip (Wise *et al.*, 1994). Historical reviews of the effects of dislocations on semiconductor devices and IC performance and the efforts to obtain and maintain dislocation-free material during IC fabrication include extensive bibliographies (Hu, 1986; Huff and Goodall, 1995).

The correspondence between the intersection of an edge dislocation line with the surface and the etch pit formed by using a preferential etchant is extensively utilized to assess the crystallographic perfection of semiconductors (see Fig. 5b) (Dash, 1960; Gatos and Lavine, 1965). Substantiation of this correlation was obtained by the use of copper decoration and direct imaging of the dislocations (Dash, 1956). The presence of screw dislocations has also been deduced by etchants, although this correlation posed a slightly more complicated situation as regards their clear-cut etching characteristics.

The activation enthalpy required to move an edge dislocation to an adjacent configuration, neglecting dislocation charge effects and reconstruction at the dislocation core, is called the Peierls energy (Chen *et al.*, 1992). The nonmonotonic decrease of the Peierls energy with increasing metallicity, however, highlights the simplicity of neglecting dislocation core reconstruction effects and related structure-sensitive factors (see Table 6). The elementary process (i.e., kink energy U_K), by which an atom in an edge dislocation moves, has been correlated with the energy gap at the Tamm temperature (Siethoff, 1994), taken as 0.7 times the absolute melting temperature T_M. Although the resistance to motion of dislocations in covalent crystals has been suggested to decrease with excitation of electrons from the valence band to the conduction band, the effects of dislocation core reconstruction effects must also be considered. The activation enthalpy for vacancy self-diffusion, Q^{SD}, was observed to be a constant factor of 1.5 times U_K, reflecting the similarity of breaking and reconstructing bonds with the kink mechanism (Siethoff, 1994). The hardness—breaking and rearrangement of electronic bonds—of many materials scales with their atomic density; diamond, with the highest atomic density, is the

Table 6. Metallicity parameter, Peierls energy, kink energy, energy gap at 0.7 T_M, hardness, atomic density, and mass density for the group-IVa and β-silicon carbide semiconductors.

Element	Metallicity parameter α_m	Peierls energy (eV)	Kink energy U_K (eV)	Energy gap at 0.7 T_M (eV)	Hardness (kg/mm^2)	Atomic density (10^{23} atoms/cm^3)	Mass density (g/cm^3)
Diamond	0.34	11.35	10.7	5.35	8375	1.76	3.51525
β-SiC	0.45	—	—	—	3100–3475	0.97	3.21
Silicon	0.66	5.95	2.35	0.93	1125	0.50	2.3283
Germanium	0.81	6.67	1.89	0.44	750	0.44	5.3234
Gray tin	0.87	5.66	—	—	—	0.29	5.765

hardest material (see Sec. 3.3 for a related discussion of the bulk modulus, i.e., the resistance of atomic electron shells being pushed closer together). The general monotonic decreases in the theoretical Peierls energy, kink energy U_K, energy gap at $0.7T_M$, hardness, and atomic density with increasing metallicity are summarized in Table 6.

A fairly low concentration of point defects such as oxygen, (2–5) $\times$ 10^{17}/cm^3 (4–10 ppma) in silicon, where the oxygen concentration is based on the "old" ASTM calibration factor, F121-79, Annual Book of ASTM Standards, Part 43, Electronics (1979), can impede dislocation multiplication by dislocation locking (solution hardening) (Sumino, 1983). A somewhat higher concentration of oxygen facilitates the formation of small ($\leq$500 Å) dislocation loops and half loops as a result of the condensation of oxygen agglomerates into small silicon oxide precipitates SiO_x during IC fabrication. That is, as the wafer cools down during IC fabrication, the localized stress, caused by the difference in thermal expansion coefficients between SiO_x and silicon, induces dislocations and silicon interstitials to form around the oxide precipitates. The dislocations are bound to the precipitate, however, and are small enough not to catastrophically multiply. These dislocations and associated precipitate dislocation complexes (PDCs) (Tan *et al.*, 1977) for 20–30 ppma of oxygen also impede additional dislocation movement or recombination by work-hardening the wafer. Precipitate dislocation complexes can exhibit either a positive or a negative influence on IC electronic characteristics. If the PDCs are located sufficiently far away from the device active region (typically >30 μm below the active device area), they can provide sites, due to the localized crystal disruption, for gettering metallic impurities inadvertently introduced during IC fabrication. Metallic impurities become trapped in the strain fields around the PDCs and/or may form metallic silicides by reaction of the metals with silicon interstitials. Internal (intrinsic) gettering requires the controlled incorporation of oxygen within the crystal and the subsequent formation and growth of the PDCs within the wafer to achieve high-performance ICs (Tan *et al.*, 1977; Lawrence and Huff, 1982; Huff *et al.*, 1983; Shimura, 1994). The importance of reducing the carbon concentration to less than 1 $\times$ 10^{16}/cm^3 (0.2 ppma) to minimize its catalytic effects on oxygen precipitation continues to receive attention (Fukuda, 1994).

An oxygen concentration of approximately 1.5 $\times$ 10^{18}/cm^3 (>30 ppma), in conjunction with no bulk dislocations to assist in dispersing in the oxygen, results in large numbers of silicon oxygen precipitates (size approximately $\geq$2000 Å) during thermal processing, which serve as dislocation generation sources. Dislocation loops can expand beyond a critical stage during subsequent thermal processes resulting in catastrophic multiplication and can lead to a large fraction of ICs (especially bipolar ICs) fabricated in such material exhibiting nonfunctional electrical performance (Taylor *et al.*, 1960; Ravi, 1981; Lawrence and Huff, 1982; Kolbesen and Strunk, 1985). Control of dislocations and macroscopic slip during epitaxial layer deposition and IC fabrication has been extensively investigated (Dyer *et al.*, 1971; Rossi *et al.*, 1985; Wise *et al.*, 1994). The formation of dislocations is an extremely process- and structure-sensitive parameter, dependent on the wafer diameter, interstitial and precipitated oxygen content, carbon content, thermal history of the wafer, latent sources of damage such as periphery damage (e.g., edge chips, crack dislocations), and strain rate (rate of heating/cooling in the furnace), as well as the detailed IC thermal process conditions (and sequence) such as insert (idle) and final temperature, location of wafer support points, wafer spacing, and type of boat material during thermal processing.

The synergistic interaction of the oxygen concentration in the silicon wafer, the wafer-to-wafer spacing in the quartz boat during thermal processing, and the detailed thermal process conditions (including sequence), for example, significantly control the growth of the oxygen precipitates. Excessive oxygen precipitation can result in unacceptable warpage of the silicon wafer precluding effective lithographic printing and reducing the critical stress for plastic deformation or slip by the generation of internal stresses around the precipitates. Process engineers continue aggressively to pursue identification of the optimal oxygen concentration and associated IC thermal-process sequences to utilize internal gettering, perhaps in conjunction with back-surface external gettering, while concurrently preventing uncontrolled dislocation generation, propagation, and the attendant wafer warpage (Lawrence and Huff, 1982; Huff and Goodall, 1995).

The microscopic nature and electrical effect of the various dislocation structures (Kimerling and Patel, 1985) continue to form an important area of investigation. Plastic deformation in silicon at 700 °C was shown to introduce a deep generation–recombination (g–r) energy level, which generally anneals out at 900 °C, leaving behind energy levels much closer to both the conduction- and valence-band edges. Thermal annealing of "freshly formed" dislocations and associated point defects is clearly evidence of bulk reconstruction processes and is reflected in the change of the energy-level spectrum (Kimerling and Patel, 1979). It is now known that dislocation motion produces a whole range of deep energy levels, mostly residual defects from dislocation interactions and defective locations in the dislocation core. Here, also, thermal annealing simplifies the energy level spectrum (Kisielowski and Weber, 1991). Correlation of the structural and electronic properties of dislocations with their photoluminescence *D*-line spectra constitutes a major area of research (Sauer *et al.*, 1985; Weber and Alonso, 1990; Weber, 1994).

2.2.2 Misfit Dislocations. Misfit dislocations, resulting from the mismatched lattice constants between a (generally) lightly doped epitaxial layer and a heavily doped substrate, can partially relieve the misfit-induced strain. The misfit dislocations are localized at the epitaxial layer–substrate interface and are visible to the eye as a cross-hatched array. The cross-hatch lines appear at 90° angles to each other on {100} oriented silicon and at 60° angles on {111} oriented silicon (Wise *et al.*, 1994). Misfit dislocations can provide gettering sites very close to the device active regions, which may be useful in low-temperature IC processes (Bean *et al.*, 1988). However, if misfit dislocations are too close to or terminate at the surface, rather than the wafer edges, they might degrade IC performance by penetrating the device active regions.

2.3 Planar/Interface Defects

2.3.1 Stacking Faults Stacking faults are a local error in the order of the sequential layering of crystallographic planes. The diamond crystal structure normally exhibits an *ABCABCABC*, etc., stacking sequence in the direction of the {111} planes (Bullis and Huff, 1994). A stacking sequence *ABCABABC*, etc., however, does not change the nearest-neighbor (*AB*) configuration. Indeed, the localized energy level associated with the stacking fault lies close to the conduction-band edge in silicon (Kimerling *et al.*, 1977). The second-nearest-neighbor configuration, however, is changed. Stacking faults are surrounded by partial dislocations, which facilitate the faults joining smoothly with the nonfaulted regions within the crystal structure (Cottrell, 1961; Hirth and Lothe, 1982). The partial dislocations exhibit deep energy levels (Kimerling *et al.*, 1977). Decoration of the partial dislocations by metals can cause electrical degradation of the device if they terminate within *p-n* junctions (Huff and Goodall, 1995). Since the stacking-fault nuclei occur in the silicon crystal, significant effort has been expended to understand, control, and minimize the occurrence of these nuclei during crystal growth and subsequent epitaxial deposition.

2.3.2 Twins A twin is a portion of a crystal that contains atoms stacked in one order (e.g., *ABCAB*) on one side of a boundary (the twin plane) and in the mirror image order (e.g., *BACBA*) on the other. Although these structures can occur during Czochralski crystal growth (see Sec. 3.1.2), the crystals or parts of crystals containing them are not dislocation free and, therefore, are recycled or discarded. Twinned structures can also occur during epitaxial silicon deposition. With improvements in current silicon wafer preparation technology and contamination-free manufacturing methodologies, however, twin-plane defects rarely occur in epitaxial silicon (nor germanium) films (Wise *et al.*, 1994). The state of the art for diamond and α- and β-SiC (see Sec. 5), however, is not as advanced.

2.3.3 Surfaces Surfaces and interfaces are significant crystal defects, exhibiting unique physicochemical characteristics and phenomena. The surface energy γ is defined as one-half the energy required to rupture the bonds per unit area of surface at 0 K. Theoretical surface energies for the major crystallographic orientations are listed in Table 7 for the group-IVa semiconductors (Clarke, 1992). The experimental fracture surface energies F for the {111} orientation are comparable to the theoretical estimates (Clarke, 1992). The decrease in γ and fracture surface energy with increasing semiconductor metallicity (weaker covalent bonding), for a given surface orientation, is not unexpected. The variation in γ for the various surface orientations of silicon appears to correlate with the bond density (see Table 8), but more sophisticated models of surface reconstruction are required for quantitative comparison with experiments (Ligenza, 1961; Shimura and Huff, 1985; Shimura, 1989).

Table 7. Major crystallographic theoretical surface energies γ (J/m^2) and experimental fracture surface energy F (J/m^2) for the group-IVa semiconductors.

Element	$\gamma\{100\}$	$\gamma\{110\}$	$\gamma\{111\}$	$F\{111\}$
Diamond	9.82[a]	—	5.35[b]	~6.0
Silicon	2.13	1.51	1.46	1.23
Germanium	1.84	1.30	1.07	1.06
Gray tin	—	—	—	—

[a]Theoretical surface energy $\gamma\{100\}$ (Adamson, 1990).
[b]Alternative theoretical surface energy $\gamma\{111\}$ equals 5.65 J/m^2 (Adamson, 1990).

Table 8. Properties of the major crystallographic surfaces for silicon.

Property	{100}	{110}	{111}
Spacing (nm)	0.543	0.384	0.313
Atomic density ($10^{14}/cm^2$)	6.78	9.59	7.83
Bond density ($10^{14}/cm^2$)	13.56	9.59	7.83

Because of its importance to the semiconductor industry, the silicon surface has been extensively studied; both polished and epitaxial silicon surfaces exhibit a plethora of defects, which must be controlled to achieve high-yielding IC performance. These defects include particles, minute shallow pits and/or surface microroughness, residual surface chemical residues, or structural defects such as the occasional epitaxial stacking fault. The measurement, counting, composition, morphology, removal, and prevention of these defects is a state-of-the-art challenge in silicon wafer technology. Laser scanning instruments are widely utilized to monitor these defects (Huff *et al.*, 1995). An inspection technique that has also become widely used in the examination of polished wafers is collimated light. The light illuminates the wafer surface, is reflected to a light-receiving screen, and is transformed to a visual image on a television monitor. By adjusting the contrast and focal depth, a variety of surface anomalies can be qualitatively observed.

In particular, surface microroughness is an imperfection of both the polished and epitaxial silicon surface and may degrade metal oxide semiconductor (MOS) gate oxide integrity (Ohmi *et al.*, 1992; Rajagopalan *et al.*, 1993; Zhong *et al.*, 1996). The magnitude and spatial frequency of surface microroughness, as well as the configuration (i.e., bandwidth) of the measuring instrument used to collect the data (i.e., pixel density, image area), must be considered in characterizing surface microroughness (Bullis, 1994; Strausser *et al.*, 1994). Today's high-quality polished or epitaxial silicon surfaces typically exhibit root mean square (rms) microroughness values from approximately 0.05 to 0.5 nm in the spatial frequency range from approximately 0.1 to 10 μm^{-1}.

3. CRYSTALLINE MATERIAL PROPERTIES

A number of crystalline material properties may conveniently be described in terms of the flux generated in response to an externally applied field. The proportionality constant relating the field to the resulting flux is referred to as a constitutive material property. The isotropic or anisotropic behavior of the constitutive property can be predicted on the basis of the tensor rank of the property and the crystal symmetry (Nye, 1964; Kelly and Groves, 1970). The electrical conductivity, thermal conductivity, and dielectric constant, for example, are second-rank tensors relating two vectors. Each of these tensors is reduced to a single number, and the resulting material properties are isotropic, in the case of the diamond crystal structure. This result is especially important in the case of the electrical conductivity.

The elastic stiffness coefficients, however, form a fourth-rank tensor that relates two second-rank tensors. The elastic properties of the group-IVa semiconductors, therefore, are anisotropic since the elastic stiffness tensor is reduced to three principal (diagonal) coefficients for the diamond crystal structure. Micromachined silicon structures exhibit unique mechanical and electronic characteristics due to their anisotropic mechanical characteristics. A number of interesting phenomena are observed by examining the material properties in the presence of various external fields (see Sec. 3.1.1).

There are experimental difficulties, however, in determining the tensor rank of a number of material properties such as the yield stress, cleavage strength, refractive index (even though the tensor property of the dielectric constant is known), and a host of surface properties such as surface energy, surface hardness, chemical etch rate, and crystal growth rate (Nye, 1964). The diffusion coefficient is an especially interesting example. Although diffusion inside the diamond crystal is isotropic, the surface itself exhibits a potential barrier dependent on the surface orientation. Additionally, very thin regions of a crystal at interfaces or surfaces may exhibit two-dimensional properties and thus display a different symmetry from the bulk (three-dimensional) crystal. The electron mobility in silicon inversion layers, for example, is anisotropic, whereas the bulk mobility is isotropic.

3.1 Electronic Properties

3.1.1 Electrons and Holes Distributed within the vibrating system of atoms and vacancy–interstitial point defects at finite temperatures are free carriers generated by collisions between the atoms and phonons. Electrons excited out of the valence band with energies greater than E_g populate the multiple conduction-band minima at nonzero values of the wave vector **k** and are referred to as "intrinsic" electrons. Concurrently, an equal number of (light and heavy) holes are left resident at the valence-band maximum at the wave vector $\mathbf{k} = 0$ (i.e., zone center) along with holes resident at the split-off valence band, also at $\mathbf{k} = 0$. The total sum of these holes is referred to as "intrinsic" holes, although the split-off holes are often ignored. The offset in **k** space of the valence-band maximum and conduction-band minima accounts for the designation of the group-IVa semiconductors (except gray tin) as indirect energy-gap semiconductors.

The lowest energy of the bands above E_g and the highest energy of the bands below E_g as a function of the reduced wave vector **k** are plotted along the [100] (X) and [111] (L) directions in Fig. 6 for silicon (Chelikowsky and Cohen, 1976). The minima at the bottom of the lowest conduction band are located at about 80% of the distance between the center ($\mathbf{k} = 0$) and the edge of the Brillouin zone along the six ⟨100⟩ axes. The six equivalent minima, 1.12 eV above the valence-band maximum, are denoted by the Brillouin-zone symmetry point X (Harrison, 1980; Ridley, 1982). The surfaces of constant energy are ellipsoids of revolution, rather than spheres, as in the case of GaAs and InSb, so that at each (equivalent) minimum the band curvature varies with direction in **k** space. The effective mass is an important parameter characterizing the energy bands; it is inversely proportional to the curvature of the energy band. The longitudinal electron

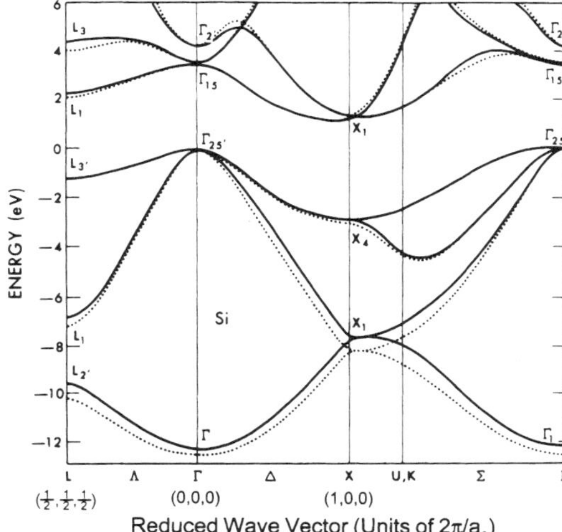

(a)

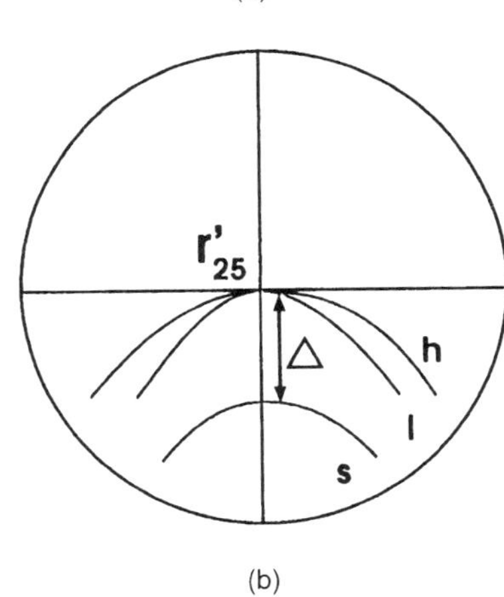

(b)

FIG. 6. Energy bands of silicon: (a) in the vicinity of the energy gap (following Chelikowsky and Cohen, 1976) and (b) near the valence-band maximum (following Anderson, 1989). The solid (dashed) line in (a) represents a nonlocal (local) pseudopotential analysis.

effective mass normalized by the free electron mass, m_l^*/m_0, describes the variation of energy with wave number measured from the minimum of X in the direction of Γ ($\mathbf{k}$ = 0 symmetry point). The normalized transverse electron effective mass, m_t^*/m_0, describes the variation of energy with wave number in the perpendicular $\mathbf{k}$-space direction. Typically, m_l^* is of the order of the free-electron mass, while m_t^* is of the order of a tenth of the free-electron mass (see Table 9).

The isotropic density-of-states electron effective mass is defined as (Smith, 1961; Green, 1990)

$$m_{nd}^*/m_0 = [(m_l^*/m_0) \times (m_t^*/m_0)^2]^{1/3}. \quad (4)$$

There are two additional conduction bands in silicon with lowest energy gaps at the Brillouin-zone symmetry points L and Γ. The former is located at the edge of the Brillouin zone along the eight ⟨111⟩ directions. The L and Γ energy gaps are 2.2 and 4.08 eV, respectively, above the valence-band maximum. Transitions between the valence and the L or Γ conduction bands can be observed under optical excitation (see Sec. 3.4). The energy gaps for silicon at the X, L, and Γ symmetry points are summarized in Table 10 (Pollak *et al.,* 1970; Chelikowsky and Cohen, 1976; Ridley, 1982).

In the case of germanium, the minima at the bottom of the lowest conduction band are located at the edge of the Brillouin-zone boundary along the eight ⟨111⟩ directions at the L symmetry point. This behavior in germanium, compared with silicon, is related to the increased lattice constant a_0 (i.e., decreased strength of the interatomic potential) and electronic core effects. The energy gaps for germanium at the X, L, and Γ symmetry points are also summarized in Table 10 (Ridley, 1982). The influence of d-band mixing in germanium reduces the L and Γ energy gaps, relative to Si, so that the L indirect gap is smaller than the X indirect gap. The influence of d-band mixing is even more dominant in the case of gray tin. In this case, Γ (direct-gap) undercuts the energy gap at the L symmetry point and, accordingly, gray tin is a direct-gap semiconductor (see Table 10) (Pollak *et al.,* 1970). The gray-tin band structure, however, is further complicated by crossing of the lowest conduction band and the upper valence band, resulting in a complicated set of effective masses (Pollak *et al.,* 1970). Fuller comprehension of the variation of these energy gaps requires an examination of the interatomic matrix elements.

The valence bands for the group-IVa elements have a single extremum at the center of the Brillouin zone ($\mathbf{k}$ = 0) (see Fig. 6) (Chelikowsky and Cohen, 1976; Anderson, 1989). The two upper valence bands are degenerate at the maximum and the constant-energy surfaces are warped spheres. The va-

Table 9. Several electronic characteristics for the group-IVa and β-silicon carbide semiconductors at 300 K.[a] See Sec. 4 for a discussion of the variation of the energy gap and mobility for $Si_{1-x}Ge_x$.

Element	m_n^*/m_0	m_p^*/m_0	μ_n[k] (cm^2/V s)	μ_p[k] (cm^2/V s)	n_i[o] (cm^{-3})	σ_i[o] $(\Omega\ cm)^{-1}$
Diamond	1.4[b] 0.36[c] 0.57[d]	0.7[e] 2.18[f] 1.06[g] 2.78[h]	2000	1600	—	$<10^{-15}$
β-Silicon carbide	0.65[b] 0.24[c] 0.33[d]	—	900	40	—	$<10^{-6}$
Silicon	0.92[b] 0.19[c] 0.32[d]	0.153[e] 0.537[f] 0.234[g] 0.590[h]	1450	500	1.07×10^{10}	3.3×10^{-6}
Germanium	1.588[b] 0.0815[c] 0.22[d]	0.043[e] 0.352[f] 0.095[g] 0.36[h]	3900	1800	2.33×10^{13}	2.0×10^{-2}
Gray tin	0.024[i] 0.21[j]	0.2–0.45[e]	1.2×10^5 [l] 3000[m]	3000[n]	—	$\sim 2 \times 10^{-4}$

[a]References: Lide, 1993–1994; Anderson, 1989; Madelung, 1991; EMIS Datareview series, 1988; DIAMOND AND DIAMONDLIKE CARBON; GERMANIUM; King, 1994; Davis, 1994; Sze, 1981; Green, 1990; Smith, 1961; Morkoc *et al.*, 1994.
[b]Longitudinal electron effective mass, m_l^*/m_0
[c]Transverse electron effective mass, m_t^*/m_0
[d]Isotropic density-of-states electron effective mass, m_{nd}^*/m_0; see Eq. 4
[e]Light-hole effective mass, m_{lh}^*/m_0
[f]Heavy-hole effective mass, m_{hh}^*/m_0
[g]Split-off hole effective mass, m_{so}^*/m_0
[h]Isotropic density-of-states hole effective mass, m_{pd}^*/m_0; see Eq. 5
[i]Light electron effective mass, m_{le}^*/m_0
[j]Heavy electron effective mass, m_{he}^*/m_0
[k]Mobilities are for lightly doped material.
[l]Light electron, 100 K.
[m]Heavy electron, 270 K.
[n]270 K.
[o]See discussion associated with Eqs. 6a through 9.

Table 10. Energy gaps for the group-IVa and β-silicon carbide semiconductors at L, X, and Γ symmetry points and split-off valence-band energy (Δ) at 300 K.

Element	L (eV)	X (eV)	Γ (eV)	Δ (eV)
Diamond	5.77[a]	5.45	13.04	0.006
β-Silicon carbide	—	2.2	7.75	—
Silicon	2.2[b]	1.12	4.08	0.044
Germanium	0.66	0.96	0.89	0.29
Gray tin	0.3	0.35[c]	0.082	0.64

[a]Theoretical calculation (Ridley, 1982).
[b]Theoretical calculation (Chelikowsky and Cohen, 1976).
[c]Pollak *et al.* (1970).

lence band exhibiting the smaller curvature is denoted as the "heavy-hole" band while the higher curvature valence band is the "light-hole" band. A third valence band is split off from the other two by relativistic spin–orbit interactions. The split-off valence band, Δ, is 0.044 eV below the maximum of the light- and heavy-hole bands for silicon. The curvature (and, therefore, effective mass) of this approximately spherical band is intermediate between the curvature at the maxima of the heavy- and light-hole bands for silicon and germanium. The light-hole effective mass m_{lh}^*/m_0, heavy-hole effective mass m_{hh}^*/m_0, and split-off hole effective mass m_{so}^*/m_0, are summarized in Table 9.

The isotropic density-of-states hole effective mass is defined as (Green, 1990)

$$m^*_{pd}/m_0 = \{(m^*_{lh}/m_0)^{3/2} + (m^*_{hh}/m_0)^{3/2} + [(m^*_{so}/m_0)\exp(-\Delta/kT)]^{3/2}\}^{2/3}. \quad (5)$$

The split-off valence-band energies (Ridley, 1982) are also summarized in Table 10.

The number of intrinsic holes (p_i) and electrons (n_i) per cm^3 at thermal equilibrium, for the nondegenerate case, are described by the Boltzmann statistics (Smith, 1961):

$$p_i = N_V \exp(-|E^V_{F,i}|/kT), \quad (6a)$$

$$n_i = N_C \exp(-|E^C_{F,i}|/kT), \quad (6b)$$

where N_v is the effective density of states in the valence band, equal to $2(2\pi m_0 kT/h^2)^{3/2} \times (m^*_{pd}/m_0)^{3/2}$; N_C is the effective density of states in the conduction band, equal to $2(2\pi m_0 kT/h^2)^{3/2}\ M(m^*_{nd}/m_0)^{3/2}$; M is the number of minima in the conduction band [i.e., six for Si and four for Ge (Smith, 1961)]; $|E^V_{F,i}|$, $|E^C_{F,i}|$ are the absolute values of the intrinsic Fermi energy relative to the valence-band edge and conduction-band edge, respectively; and m^*_{pd}, m^*_{nd}, and m_0 are the isotropic density-of-states hole effective mass, the isotropic density-of-states electron effective mass, and the free-electron mass, respectively.

The intrinsic electron–hole excitation phenomenon is described by the mass-action relationship (Smith, 1961):

$$n_i p_i = n_i^2. \quad (7)$$

Utilizing Eqs. (6a), (6b), and (7), the intrinsic electron and hole concentrations per cm^3 may be determined:

$$n_i = p_i = (N_C N_V)^{1/2} \exp(-E_g/2kT), \quad (8)$$

where E_g is the temperature-dependent activation enthalpy (i.e., the minimum energy gap) equal to $|E^V_{F,i}| + |E^C_{F,i}|$. The intrinsic concentration of electrons and holes, calculated for silicon and germanium at 300 K utilizing Eq. (8) and the electronic characteristics summarized in Tables 9 and 10, is a fundamental material electrical characteristic. The corresponding intrinsic conductivity is determined at 300 K using Table 9 and the relation

$$\sigma_i = p_i|e|\mu_p + n_i|e|\mu_n, \quad (9)$$

where μ_p and μ_n are the temperature-dependent hole and electron mobilities, respectively, in the case of phonon-limited scattering at 300 K (Sze, 1981).

The coefficients of electrical conductivity, σ_{kl}, a second-rank tensor, relate the electric field component E_l to the component of electric charge flow per unit area per second j_k (i.e., the current per unit area) in the kth direction. The electrical conductivity tensor σ_{kl} can be transformed to its principal axes and, therefore, principal conductivity coefficients, σ_1, σ_2, and σ_3, which are equal (isotropic) for the diamond crystal structure (see Eq. 9). Although the individual conduction bands for the group-IVa semiconductors are anisotropic, the macroscopic conductivity is isotropic as a result of the symmetry of the diamond crystal structure. An external mechanical stress or magnetic field, however, can lift the degeneracy of the conduction bands, and the conductivity will no longer be isotropic, leading to a number of useful electronic device applications.

3.1.2 Donors and Acceptors The fundamental property of a crystalline semiconductor that results in its technological importance is the easily controllable and spatially variable concentration and type of free carriers (Irvin, 1962; Thurber *et al.*, 1980a, 1980b). According to G. K. Teal (Teal, 1976; Teal *et al.*, 1985):

> "I reasoned that polycrystalline germanium, with its variations in resistivity and its randomly occurring grain boundaries, twins and crystal defects that acted as uncontrolled resistances, electron or hole emitters and traps would affect transistor operation in uncontrolled ways. . . . My general aims for the single crystal research were:
>
> **1.** to produce a conducting medium in which a high degree of crystal perfection, of uniformity of structure and of chemical purity is attained; and
> **2.** to build into this highly perfect medium in a controlled way the re-

quired resistivities and electrical boundaries to give a variety of device possibilities by control of the chemical composition (i.e., donor and acceptor concentration) along the direction of single crystal growth."

Significant perturbation of the intrinsic electron–hole concentrations is accomplished by the selective (extrinsic) doping of the semiconductor. Group-Va donor impurities such as phosphorus, arsenic, or antimony result in free electrons in the conduction band (*n*-type semiconductor) while the group-IIIa acceptor impurities such as boron result in free holes in the valence band (*p*-type semiconductor). These dopants occupy substitutional sites in the group-IVa semiconductors with shallow ground-state energy levels in the energy gap. The energy levels are described by the hydrogenic approximation in the effective-mass theory, taking into account the dielectric constant of the semiconductor and the effective mass of the free carrier (i.e., electron or hole in the semiconductor) (Harrison, 1980). The dopant wave functions are localized (i.e., nonpropagating) solutions to Schrödinger's equation for doping densities less than the isotropic density of states. The ground states are typically 40–50 meV from the appropriate band edge for silicon (donors below the conduction-band edge, acceptors above the valence-band edge) and are readily ionized at 300 K. Table 11 summarizes the energy levels of several common dopants in diamond, β-SiC, silicon, and germanium (Sze, 1981; Beadle *et al.*, 1985).

The dopant impurities are introduced into the growing silicon (or germanium) crystals during growth from the melt by the modified Czochralski (Teal–Little) technique (Teal *et al.*, 1985; Runyan and Huff, 1994). The concentration $C_s(g)$ along the direction of crystal growth, for the group-IIIa or group-Va dopant impurity, is described by the equation for normal freezing (Pfann, 1966):

$$C_s(g) = k_e C_1 (1 - g)^{k_e(d_l/d_s) - 1}, \qquad (10)$$

where g is the fraction of crystal solidified (seed end, $g = 0$; bottom end, g approaches 1); C_l is the initial dopant concentration in the melt $[C_s(0) = k_e C_l]$; k_e is the effective distribution coefficient, equal to the ratio of the dopant concentration in the crystal to the dopant concentration in the bulk melt (i.e., far away from the crystal–melt interface); and d_l/d_s is the ratio of the melt density to the crystal density. This ratio, about 1.1 and 1.05 for silicon and germanium, respectively, is typically taken as unity since the experimental uncertainty in k_e is generally greater than 10%.

The effective distribution coefficient k_e is dependent on the crystal growth rate and melt flow conditions at the crystal–melt interface and is related to the equilibrium distribution coefficient k_0 by the Burton–Prim–Slichter relationship (Burton *et al.*, 1953):

Table 11. Energy levels (meV) for several common dopants in group-IVa and β-silicon carbide semiconductors.[a]

Dopant	Diamond[b]	β-Silicon carbide[c]	Silicon	Germanium	Gray Tin
B	370	—	45	10	—
Al	—	200	67	10	—
Ga	—	—	72	11	—
P	—	—	44	12	—
As	—	—	49	13	—
Sb	—	—	39	10	—

[a]Energy levels are measured from the conduction-band edge for *n*-type dopants and the valence-band edge for *p*-type dopants.

[b]Nitrogen and phosphorus are *n*-type dopants. Phosphorus is believed to be too large to fit within the diamond crystal (DIAMOND AND DIAMONDLIKE CARBON), but donor energy levels about 1.7 and 4.0 eV below the conduction band edge have been identified (Davies, 1994; Collins, 1990).

[c]For nitrogen concentrations $\leq 10^{17}/cm^3$, the electrical ionization energy approximately equals 45 meV (Pensl and Troffer, 1995). For $10^{17}/cm^3$–$10^{19}/cm^3$, the electrical ionization energy decreases to as low as approximately 20 meV for $10^{19}/cm^3$ (Pensl and Choyke, 1993).

Table 12. Equilibrium distribution coefficients for several dopants at the melting points of silicon[a] and germanium.

Dopant	Silicon	Germanium
B	0.8	17
Al	0.002	0.073
Ga	0.008	0.087
Si	1	5.5
Ge	0.33	1
P	0.35	0.08
As	0.30	0.02
Sb	0.023	0.003

[a]The equilibrium distribution coefficient of oxygen, an important impurity in silicon, is about 0.25 (Lin and Hill, 1983).

$$k_e = k_0[k_0 + (1 - k_0)\exp(-f\delta/D)]^{-1}, \tag{11}$$

where f is the microscopic crystal growth rate, δ is the thickness of the stagnant boundary layer in the melt at the crystal–melt interface, and D is the dopant diffusivity in the bulk melt.

Equilibrium distribution-coefficient data for several dopants in silicon and germanium are summarized in Table 12 (Trumbore, 1960). It is of particular interest that k_0 for acceptor impurities is greater in germanium than for silicon while k_0 for donor impurities is greater in silicon than for germanium. These observations may correlate with the experimental observation that donor dopants diffuse faster in germanium and acceptor dopants diffuse faster in silicon (Kooi, 1991). The interpretation of Fig. 4 that interstitials are more predominant in silicon than in germanium may be consistent with this observation. That is, a dopant with a larger (smaller) value of k_0 might preferentially be (not be) incorporated within the crystal and, therefore, exhibit a smaller (larger) diffusion coefficient. These ideas are qualitatively summarized in Table 13, although detailed variations occur between each dopant for both silicon and germanium. The defect-mediated diffusion by point defects in Table 13 refers to the dopants in Table 12.

The donor (acceptor) impurity becomes positively (negatively) charged when the free electron (hole) resides in the conduction (valence) band. Charge neutrality requires

$$n + N_A^- = p + N_D^+, \tag{12}$$

where N_A^- and N_D^+ are the ionized acceptor and donor concentrations, respectively.

One may utilize the generalized mass-action relationship (applicable in the case of nondegenerate statistics at thermal equilibrium) (Smith, 1961),

$$np = n_i^2, \tag{13}$$

with the charge neutrality condition from Eq. (12) to relate the electron and hole concentrations.

For example, an (essentially) uncompensated p-type silicon wafer at 300 K with an acceptor concentration N_A of 1.3×10^{15} boron atoms per cm^3 (10 Ω cm) and an intrinsic concentration of 1.07×10^{10} cm^{-3} (Green, 1990) has hole and electron concentrations

$$p \sim N_A^- = 1.3 \times 10^{15}\,\text{cm}^{-3}, \tag{14a}$$

$$n = n_i^2/p = 8.81 \times 10^4\,\text{cm}^{-3}. \tag{14b}$$

The corresponding extrinsic conductivity σ equals 9.88×10^{-2} (Ω cm)$^{-1}$, calculated from Eqs. (9), (14a), and (14b) with p_i and n_i replaced by p and n, respectively, in Eq. (9). For this dopant concentration, the phonon-limited mobilities at 300 K are valid approximations to the usual dopant-density–dependent mobilities (Sze, 1981) (see Table 9).

Table 13. Qualitative correlation among equilibrium distribution coefficient, predominant point defect, and diffusion coefficient at high temperature in silicon and germanium.

Element	k_0 (p type)	k_0 (n type)	$[I]$[a]	$[V]$[a]	D (p type)	D (n type)
Silicon	Less[b]	More	More	Less	More	Less
Germanium	More[b]	Less	Less	More	Less	More

[a]Interstitial and vacancy concentrations, respectively.
[b]Less and more refer to a comparison between silicon and germanium.

In this case, the holes are referred to as majority carriers, and the electrons are minority carriers. Under thermal equilibrium conditions, the electron (minority carrier) contribution to the extrinsic conductivity is negligible in comparison to the hole (majority carrier) contribution. The hole and electron concentrations for the nondegenerate extrinsic case are related to the Fermi energy by analogy with Eqs. (6a) and (6b):

$$p = N_V \exp(-|E_F^V|/kT), \tag{15a}$$

$$n = N_C \exp(-|E_F^C|/kT), \tag{15b}$$

where $|E_F^V|$ and $|E_F^C|$ are the absolute values of the Fermi energy relative to the valence-band edge and conduction-band edge, respectively.

Alternatively, one may combine Eqs. (6a), (6b), (15a), and (15b) to yield

$$p = p_i \exp(+|\varphi_F|), \tag{16a}$$

$$n = p_i \exp(-|\varphi_F|), \tag{16b}$$

where $|\varphi_F|$ is the absolute value of the Fermi energy relative to the intrinsic Fermi energy normalized by kT. It should be noted that, although the Fermi energy, or alternatively the normalized Fermi energy, is spatially constant at thermal equilibrium, it may nevertheless be spatially variable relative to the band edges, even at thermal equilibrium. In this case, the free-carrier flux due to the spatially varying internal electric field set up by the energy-band bending [resulting in a space-charge region (SCR)] is exactly balanced by the free-carrier diffusion flux. The intrinsic Fermi energy, however, follows the energy-band bending. Similar concepts may be utilized to describe the surface energy-band bending due to surface states and the resulting surface space-charge region (SSCR) at thermal equilibrium.

3.1.3 Metal Impurities Transition metals in silicon exhibit a significant component of interstitial, as well as substitutional, solubility. This results in a mixed diffusion process, with a fast interstitial component. The 3*d* transition metals in silicon, such as iron and chromium, mainly occupy and diffuse via interstitial sites. They accordingly exhibit high diffusion coefficients. Cobalt, Ni, and Cu diffuse sufficiently fast, even at 300 K, that these elements transfer from interstitial sites onto dislocation sites as well as form clusters and precipitates. The 3*d* metals in germanium, on the other hand, occupy and diffuse via substitutional sites (vacancies are the dominant point defect in germanium). The 4*d* metals in silicon, such as Ag, exhibit a solubility on substitutional sites comparable to interstitial sites and, therefore, diffuse more slowly than the 3*d* transition metals. The 5*d* metals, such as Au and Pt, exhibit a dominant substitutional solubility. The 5*d* metals accordingly take longer until their substitutional solubility is attained throughout the wafer. The U-shaped diffusion profile of Au, however, is clear evidence that there is also a fast-diffusing interstitial component. The lack of sufficient interstitial sinks in dislocation-free silicon limits the kinetics of the transfer of Au onto substitutional sites (kick-out mechanism) during diffusion (Taylor *et al.*, 1993). The temperature-dependent solubility and diffusion coefficients, taking into account kinetic considerations during cool-down from high temperatures, are especially important for transition metals such as Au in determining their state of aggregation at room temperature. The properties of transition metals in silicon have been reviewed with particular emphasis on their diffusion, solubility, and electrical activity (Weber, 1988; Lemke, 1994). Size effects and electronegativity differences between the impurity and host atoms are also important considerations and must be taken into account for these analyses.

Although metallic impurities can affect the electrical resistivity of silicon if they are present in large enough concentration, they are more important as traps or generation–recombination (g–r) centers that can degrade the carrier lifetime (Schroder, 1990; Bullis and Huff, 1995). The carrier lifetime is an extremely important example of the structure sensitivity of material properties in semiconductors. Metallic elements exhibit deep energy levels (i.e., far from the band edges) in silicon with ground states typically 0.4–0.6 eV from the nearest band edge (Sze, 1981). In general, metals are considered undesirable in silicon even at levels approaching 1 part per trillion atomic (1 ppta = 5×10^{10} atoms/cm^3). Since most transition elements have very small distribution coefficients (10^{-5}) in silicon (Trumbore, 1960), their direct incorporation from the melt during crystal growth is unlikely if normal precautions are taken. Metallic contamination by Fe, Ni, and Cu in the range of 10^{11}–10^{13} cm^{-2}—for example, during ion implantation, dry etching, and plasma CVD—is much more likely to be introduced onto the wafer surface during wafer preparation or IC fabrication (Huff and Goodall, 1996a). The surface metals diffuse into the wafer bulk during subsequent thermal processing. The surface and bulk descriptions, however, cannot be simply transposed via the wafer thickness because of surface–bulk segregation during thermal processing. In view of this, prescriptions are under development to determine the equivalent bulk concentration from a given surface concentration during subsequent thermal processing (Helms, 1995). An extensive methodology has developed for the removal of surface metals by cleaning with suitable solutions of high-purity chemicals (Kern, 1993). The sensitivity of surface diagnostic techniques varies but ranges between $\approx 10^8$ and 10^{10} atoms/cm^2 for a variety of transition metals (Diebold *et al.*, 1992).

3.2 Thermal Properties

The Debye temperature θ_D is related to the maximum phonon (quantized normal mode of vibration) frequency ν_D that can propagate in the crystalline material at a given temperature (De Launay, 1956):

$$k\theta_D = h\nu_D. \tag{17}$$

The value for θ_D at 300 K decreases from diamond to gray tin, because of the decrease in the bond energy (cohesive energy) of the group-IVa semiconductors and the increase in the interatomic distance, making it easier to populate higher-energy phonon modes. β-SiC's Debye temperature is virtually intermediate between diamond and silicon's values. The Debye temperature is a key parameter in thermal characteristics, including the melting point, coefficient of linear thermal expansion α, and thermal conductivity κ. These data are listed in Table 14 for 300 K (De Launay, 1956; Slack and Bartram, 1975; Field, 1991).

3.2.1 Thermal Expansion The coefficients of thermal expansion α_{ij}, a second-rank tensor, relate a scalar ΔT to a second-rank strain tensor ϵ_{ij}. When a small temperature change ΔT is uniformly applied throughout the crystal, all the components of ϵ_{ij} are proportional to ΔT. The thermal expansion tensor α_{ij} can be transformed to its principal axes and, therefore, principal thermal expansion coefficients, α_1, α_2, and α_3. These coefficients are equal (isotropic) for the diamond crystal structure (see Table 14). In heavily doped silicon at 300 K, α is expected to increase above 2.57×10^{-6}/K (Bullis and Huff, 1994), but in all cases it is smaller than the value for germanium (see Fig. 4). The values of α for silicon and silicon carbide are significantly different from those of metals, germanium, and diamond. In relatively pure silicon, α is very small and positive below 18 K. It is negative between 18 K and about 120 K, reaching a minimum of about -0.7×10^{-6}/K at about 80 K. Above approximately 120 K, α becomes positive and increases monotonically to a nearly constant value beyond 1200 K of about 4.4

Table 14. Several thermal characteristics of the group-IVa and β-silicon carbide semiconductors at 300 K.

Element	Debye temperature (K)	Melting temperature (°C)	Coefficient of linear thermal expansion, α (10^{-6} K^{-1})	Thermal conductivity, κ (W/cm K)
Diamond	1862	3827[a]	1.05	20
β-Silicon carbide	1430	>1827[b]	2.77	5.0
Silicon	645	1412	2.57	1.5
Germanium	374	958	5.90	0.6
Gray tin[c]	260	—	~5.5	—

[a]Diamond-graphite-liquid eutectic at 1.23×10^5 atm.
[b]Sublimes at 1 atm pressure; melts at 2830 °C at 35 atm.
[c]Gray tin transforms to white tin at 13 °C; white tin melts at 232 °C.

$\times$ 10^{-6}/K (Slack and Bartram, 1975). At temperatures above 1000 K, silicon exhibits the smallest value of α for a number of diamondlike solids (Slack and Bartram, 1975). As noted in Sec. 2.1.1, this contributed to the interpretation that interstitials are as significant as vacancies in silicon. The high-temperature dependence of α for silicon carbide also appears consistent with a similar interpretation. Although α for SiC is similar to that of silicon at 300 K, its saturated value of 5.65 $\times$ 10^{-6}/K is larger than that for silicon.

The difference in α between silicon and various films such as silicon dioxide or silicon nitride, which are used in silicon IC technology, is particularly important. It is responsible for stresses that can generate and propagate dislocations in silicon resulting in plastic deformation and subsequent degradation of IC performance (Lawrence and Huff, 1982; Huff and Shimura, 1985; Huff and Goodall, 1995). These effects are particularly important at corners in windows of patterned films and trench structures (Kooi, 1991).

3.2.2 Thermal Conductivity The coefficients of thermal conductivity κ_{ij}, a second-rank tensor, relate the negative of the temperature gradient component $\partial T/\partial x_j$ to the component of heat flow per unit area per second (h_i) in the ith direction. The thermal conductivity tensor κ_{ij} can be transformed to its principal axes and, therefore, principal thermal conductivity coefficients, κ_1, κ_2, and κ_3. These coefficients are equal (isotropic) for the diamond crystal structure. The thermal transport mechanism in semiconductors is predominantly by phonons, rather than by free electrons (as in metals) or holes. Thus, the Wiedemann-Franz law, which relates the electronic thermal conductivity to the electrical conductivity in metals, does not apply to semiconductors.

Natural diamond consists of 98.892% of C^{12} and 1.108% C^{13}. Enriched diamond with 99.93% C^{12} has a large room-temperature value of κ of about 33 W/cm K (Field, 1991; DIAMOND AND DIAMONDLIKE CARBON). Silicon, germanium, and tin also exhibit a number of natural isotopic species (see Table 15). The isotope effect on the thermal conductivity for these elements, however, is significantly smaller compared to diamond because of their much larger atomic mass. For example, the mass differential between C^{12} and C^{13} is 8.3% while the mass differential between Si^{28} and Si^{29} is 3.6%. The room-temperature value of κ for silicon is 1.5 W/cm K (see Table 14). As the temperature increases above 300 K, κ decreases on account of increased phonon scattering until it reaches a value of about 0.22 W/cm K at the melting point of silicon (Runyan, 1965). The thermal conductivity increases for silicon below room temperature, reaching a maximum at about 20–30 K as a result of reduced phonon scattering. Below this temperature range, κ decreases because of increased boundary (umklapp) scattering. In this region, κ is also reduced by increased impurity concentration. The room-temperature value of κ for silicon is essentially independent of impurity concentration, except for concentrations greater than about $10^{18}/cm^3$ (Bullis and Huff, 1994). At these free-carrier concentrations, κ decreases as a result of increased phonon–electron (or hole) intravalley and intervalley scattering processes. The potential increase in κ due to electron (or hole) transport is apparently more than offset by these increased scattering processes (Runyan, 1965).

Table 15. Natural isotopic species and percent for the group-IVa elements.

Element	Isotopic species/Percent
Carbon	C^{12}, 98.892%; C^{13}, 1.108%
Silicon	Si^{28}, 92.23%; Si^{29}, 4.67%; Si^{30}, 3.10%
Germanium	Ge^{70}, 20.4%; Ge^{72}, 27.4%; Ge^{73}, 7.8%; Ge^{74}, 36.6%; Ge^{76}, 7.8%
Tin	Sn^{112}, 1.01%; Sn^{114}, 0.68%; Sn^{115}, 0.35%; Sn^{116}, 14.28%; Sn^{117}, 7.67%[a]; Sn^{118}, 23.84%; Sn^{119}, 8.68%[a]; Sn^{120}, 32.75%; Sn^{122}, 4.74%; Sn^{124}, 6.01%

[a]Mass number of "stable" radioactive isotopes.

3.3 Mechanical Properties

The variation of the group-IVa semiconductor mechanical properties generally correlates with the variation of the metallicity parameter α_m and Debye temperature θ_D. These properties include the hardness number (see Sec. 2.2.1), elastic stiffness coefficients, Young's modulus Y, bulk modulus K, shear modulus G, Poisson's ratio ν, and the theoretical fracture strength σ_{th}. The reciprocal of Young's modulus Y, which relates an applied stress σ to the resulting strain ϵ, [Hooke's law, $\epsilon = (1/Y)\sigma$], is an especially important material characteristic.

The mechanical properties are often expressed via an alternative, but equivalent, representation of the generalized form of Hooke's law:

$$\sigma_{ij} = c_{ijkl}\epsilon_{kl}, \tag{18}$$

where the elastic stiffness coefficients c_{ijkl}, composing a fourth-rank tensor, relate the strain ϵ_{kl} to the stress σ_{ij}. Both the strain and the stress are second-rank tensors. The symmetry of both the strain tensor and, in the absence of body torques, the stress tensor implies symmetry of the c_{ijkl} in the first two and last two suffixes, respectively. This reduces the 81 c_{ijkl} coefficients to 36 c_{mn} coefficients ($c_{11} \ldots c_{66}$), thereby facilitating utilization of matrix notation. That is, the first two suffixes are abbreviated into a single one running from 1 to 6, and the last two suffixes are abbreviated in the same way. Accordingly, the tensor notation 11, 22, 33, 23 and 32, 31 and 13, and 12 and 21 for σ_{ij} and ϵ_{kl} are reduced to 1, 2, 3, 4, 5, and 6, respectively (Nye, 1964). The assumption of a parabolic crystal potential (i.e., $c_{mn} = c_{nm}$) results in 21 independent coefficients. Finally, symmetry considerations for cubic crystals such as the diamond crystal structure (i.e., cyclic interchange of the x, y, and z coordinates) results in three independent elastic stiffness coefficients, c_{11}, c_{12}, and c_{44}, and anisotropic elastic properties (see Table 16) (McSkimin *et al.*, 1951; McSkimin, 1953; Nye, 1964; Runyan, 1965; Henisch and Roy, 1969; Kelly and Groves, 1970; Madelung, 1991; Chen *et al.*, 1992; Tu *et al.*, 1992). It should be noted that the c_{mn} are not the components of and, therefore, do not transform like the components of a second-rank tensor. To transform the c_{mn} to other axes, it is necessary to go back to tensor notation. The elastic properties form an important example in which the family of cubic crystal structures exhibit anisotropic behavior of a material property.

Inasmuch as the elastic stiffness coefficients are determined by the interatomic forces in the crystal, the general decrease in the elastic stiffness coefficients in Table 16 with increasing metallicity and increased lattice constant a_0 (see Table 2) is expected. The composition dependence of the elastic stiffness coefficients of the silicon–germanium alloys and superlattices is not well understood (see Table 16). However, the most reliable data suggest a linear interpolation between the elastic stiffness coefficients of silicon and germanium, despite the 4.2% difference in the bond lengths of silicon and germanium (see Table 2) (Mendik *et al.*, 1991; Baker and Arzt, 1995). A more systematic study of the elastic properties of silicon–germanium alloys is required (Roos *et al.*, 1995a, 1995b).

Although microscopic models are generally expressed in terms of the c_{ij} coefficients, computation of the engineering parameters Y, K, G, and ν is facilitated by utilizing the

Table 16. Elastic stiffness coefficients for the group-IVa, β-silicon carbide, and several silicon–germanium alloy and superlattice semiconductors at 300 K.

Element	c_{11} (10^{11} dyn/cm^2)	c_{12} (10^{11} dyn/cm^2)	c_{44} (10^{11} dyn/cm^2)
Diamond	107.640	12.520	57.740
β-Silicon carbide	35.23	14.04	23.29
Silicon	16.772	6.498	8.036
SiGe (5084 Å)[a]	14.7 ± 0.36	5.90 ± 0.42	7.30 ± 0.22
SiGe (6795 Å)[b]	14.29 ± 0.34	5.64 ± 0.41	6.81 ± 0.20
$Si_{0.49}Ge_{0.51}$[c]	13.77 ± 0.33	5.12 ± 0.4	6.66 ± 0.19
Germanium	13.112	4.923	6.816
Gray tin	6.90	2.93	3.62

[a]Material density is 2.79 g/cm^3; the Å data refer to the superlattice thickness (Mendik *et al.*, 1991).
[b]Material density is 3.27 g/cm^3; the Å data refer to the superlattice thickness (Mendik *et al.*, 1991).
[c]Material density is 3.85 g/cm^3 (Mendik *et al.*, 1991).

Table 17. Elastic compliance coefficients for the group-IVa, β-silicon carbide, and several silicon–germanium alloy and superlattice semiconductors at 300 K.

Element	s_{11} $(10^{13}$ dyn/cm$^2)^{-1}$	s_{12} $(10^{13}$ dyn/cm$^2)^{-1}$	s_{44} $(10^{13}$ dyn/cm$^2)^{-1}$
Diamond	0.952	−0.0992	1.73
β-Silicon carbide	3.67	−1.05	4.29
Silicon	7.61	−2.12	12.44
SiGe (5084 Å)[a]	8.83	−2.53	13.70
SiGe (6795 Å)[b]	8.95	−2.53	14.68
$Si_{0.49}Ge_{0.51}$[c]	9.05	−2.45	15.02
Germanium	9.59	−2.62	14.67
Gray tin	19.4	−5.78	27.62

[a]Material density is 2.79 g/cm^3; the Å data refer to the superlattice thickness (Mendik *et al.*, 1991).
[b]Material density is 3.27 g/cm^3; the Å data refer to the superlattice thickness (Mendik *et al.*, 1991).
[c]Material density is 3.85 g/cm^3 (Mendik *et al.*, 1991).

elastic compliance coefficients s_{ijkl}. The s_{ijkl} relate a sufficiently small applied stress σ_{kl} to strain ϵ_{ij} (Hooke's law). Note that the s_{ijkl} tensor is the reciprocal of the c_{ijkl} tensor. Here also, the 81 s_{ijkl} coefficients reduce to three s_{mn} coefficients for the diamond crystal structure. The s_{mn} coefficients are related to the c_{ij} by

$$s_{11} = (c_{11} + c_{12})/[(c_{11} - c_{12})(c_{11} + 2c_{12})], \tag{19a}$$

$$s_{12} = -c_{12}/[(c_{11} - c_{12})(c_{11} + 2c_{12})], \tag{19b}$$

$$s_{44} = 1/c_{44}, \tag{19c}$$

and are summarized in Table 17 (Runyan, 1965). The c_{mn} coefficients can be expressed in terms of the s_{mn} by interchanging the c's and s's of Eqs. (19) (Nye, 1964; Runyan, 1965).

The expression for Young's modulus Y is

$$[Y(l_1l_2l_3)]^{-1} = s_{11} - 2(s_{11} - s_{12} - 0.5s_{44}) \times [(l_1^2l_2^2 + l_2^2l_3^2 + l_3^2l_1^2)]. \tag{20}$$

Young's modulus is anisotropic and varies with direction via the expression in brackets containing the direction cosine terms, l_1, l_2, l_3, with respect to the unit vectors along the ⟨100⟩ axes, with indices 1, 2, and 3 corresponding to the x, y, and z (⟨100⟩) axes, respectively (Nye, 1964). The direction cosine term is zero for the ⟨100⟩ directions, 1/4 for the ⟨110⟩ directions, and 1/3 for the ⟨111⟩ directions. Accordingly, if $s_{11} - s_{12} - 0.5s_{44}$ is positive (as is the case for the group-IVa semiconductors), Young's modulus is a maximum in the ⟨111⟩ directions, a minimum in the ⟨100⟩ directions, and intermediate for the ⟨110⟩ directions. For an applied force lying in an arbitrary direction in the {100} plane, the value of Young's modulus is between the ⟨100⟩ and ⟨110⟩ values. For an applied force lying along an arbitrary direction in a {111} plane, Young's modulus is constant and equal to the value for the ⟨110⟩ direction (Riney, 1961; Runyan, 1965; Bullis and Huff, 1994). Experimental data determined for Y are summarized in Table 18 for the major crystallographic orientations for the group-IVa and β-SiC semiconductors (Tu *et al.*,

Table 18. Experimental Young's modulus data for the major crystallographic directions for the group-IVa and β-silicon carbide semiconductors compared to expected values (in parentheses) at 300 K.

Element	⟨100⟩ $(10^{12}$ dyn/cm$^2)$	⟨110⟩ $(10^{12}$ dyn/cm$^2)$	⟨111⟩ $(10^{12}$ dyn/cm$^2)$
Diamond	— (10.504)	— (11.64)	11.60 (12.077)
β-Silicon carbide	— (2.725)	— (3.85)	4.80 (4.46)
Silicon	1.302 (1.314)	1.689 (1.708)	1.875 (1.897)
Germanium	1.037 (1.043)	1.380 (1.398)	1.551 (1.569)
Gray tin	— (0.515)	— (0.729)	— (0.845)

1992; Wentorf, 1992; Schlichting and Riley, 1992; Bullis and Huff, 1994). These data compare rather favorably with the values calculated from Eqs. (20) (parenthesized data in Table 18) using the elastic compliance coefficients in Table 17. The decrease in Young's modulus, for a given direction, with increasing metallicity (weaker covalent bonding) appears reasonable.

The elastic stiffness coefficients in silicon have been noted to decrease fairly linearly with temperature between approximately 27 and 700 °C; the quantity $c_{ij}^{-1}\, dc_{ij}/dT$ varies as -9.4×10^{-5}/K, -9.8×10^{-5}/K, and -8.3×10^{-5}/K for ij equal to 11, 12, and 44, respectively (Nielsen, 1988). The temperature dependence of Young's modulus, in an isotropic approximation over approximately 800–1200 °C, has been fitted as $Y \sim 5.8 \times 10^{-2} \exp[(1\ \text{eV})/kT]$ dyn/cm^2 (Wijaranakula, 1993).

The reciprocal of the bulk modulus K (the reciprocal of the volume compressibility) is an isotropic parameter relating an applied three-dimensional unit hydrostatic pressure to the fractional change in volume. The reciprocal of the linear compressibility β relates an applied unit hydrostatic pressure to the fractional change in length in the same direction and, in general, depends on direction. The reciprocal of the shear modulus G relates an applied shear force to the fractional change in angular displacement and also varies with direction. Poisson's ratio ν is the negative ratio of the strain perpendicular to an applied stress to the strain in the direction of the applied stress. Poisson's ratio depends on both the direction of the applied stress and the measured perpendicular direction of contraction. These mechanical characteristics may be expressed in terms of the elastic compliance coefficients, for the diamond crystal structure, by (Riney, 1961; Nye, 1964)

$$K = \{3(s_{11} + 2s_{12})\}^{-1}, \tag{21}$$

$$\beta = s_{11} + 2s_{12}, \tag{22}$$

$$G = 1/s_{44}, \tag{23}$$

$$\nu = -s_{12}/s_{11}. \tag{24}$$

The values of K, Y, and ν are related in the isotropic case by (Nye, 1964)

$$K = Y\{3(1 - 2\nu)\}^{-1}. \tag{25}$$

The values of Y, K, G, and ν for an isotropic polycrystalline aggregate, computed from single-crystal data by a Green's-function method, are summarized in Table 19 (Ruoff, 1979; Field, 1991). Young's modulus and the bulk-modulus data compare rather favorably with the single-crystal values derived from Eqs. (20) (⟨110⟩ case for Young's modulus) and (21), respectively (see parenthesized data in Table 19), using the elastic compliance coefficients in Table 17. The decrease in K with increasing metallicity correlates with the similar trend in hardness observed in Table 6. Since the bulk modulus is an energy per unit volume and the energy varies inversely with the second power of the bond length d, K scales inversely with the fifth power of the bond length (Harrison, 1980).

Poisson's ratio varies considerably with direction from 0.01 to 0.2, 0.048 to 0.403,

Table 19. Experimental isotropic polycrystalline aggregate mechanical data compared to single-crystal values (in parentheses) for the group-IVa semiconductors at 300 K.

Element	Y^a (10^{12} dyn/cm^2)	$K^{b,c}$ (10^{12} dyn/cm^2)	G^d (10^{12} dyn/cm^2)	ν^e
Diamond	11.41 (11.64)	4.42 (4.42)	5.53 (5.78)	0.07 (0.104)
Silicon	1.629 (1.708)	0.979 (0.990)	0.666 (0.800)	0.223 (0.279)
Germanium	1.316 (1.398)	0.76 (0.766)	0.545 (0.680)	0.207 (0.273)
Gray tin	— (0.729)	— (0.426)	— (0.362)	— (0.298)

[a] ⟨110⟩ case.
[b] Additional bulk modulus data in (Harrison, 1980).
[c] Computed from Eq. (21).
[d] Computed from Eq. (23).
[e] Computed from Eq. (24).

and 0.022 to 0.403 for diamond (Field, 1991), silicon (Wortman and Evans, 1965), and germanium (Wortman and Evans, 1965), respectively. It usually has a value slightly greater than 0.25 and is 0.29 for any direction in a {111} plane for silicon (Bullis and Huff, 1994). Poisson's ratio for germanium has been reported to be 0.278 (McSkimin, 1953). The nonmonotonic behavior of G and ν in Table 19 may be due to the method of averaging (Ruoff, 1979; Field, 1991). The often poor agreement for the experimental G and ν values in Table 19, in comparison with the calculations using Eqs. (23) and (24), respectively, may arise because the equations are applicable for specific crystallographic directions, rather than for an isotropic polycrystalline aggregate.

As noted earlier, the elastic stiffness coefficients are determined by the interatomic forces in the crystal. A measure of the elastic anisotropy in group-IVa semiconductors may be assessed by determining the deviation from unity of the expression

$$s = (c_{11} - c_{12})/2c_{44}. \tag{26}$$

Values of s for diamond, β-SiC, silicon, germanium, and gray tin are 0.824, 0.455, 0.639, 0.601, and 0.548, respectively (Harrison, 1980). The elastic stiffness coefficients may be combined to yield the radial and angular rigidity force constants, C_0 and C_1, respectively (Harrison, 1980):

$$C_0 = (3a_0^3/16)(c_{11} + 2c_{12}), \tag{27}$$

$$C_1 = (a_0^3/32)(c_{11} - c_{12}). \tag{28}$$

The degree to which the group-IVa semiconductors are not described by central forces (i.e., ionic bonds), but rather by highly directional covalency forces, may be determined by examining the deviation of c_{11} from c_{44} (i.e., the Cauchy relation) (Harrison, 1980). The crystal is "softened" by those distortions that can be accommodated by the hybridized bonds; this "softening" is reflected in a lower value of c_{44}. The c_{44} elastic stiffness coefficient can be expressed in terms of the force constants C_0 and C_1 (Harrison, 1980):

$$c_{44} = (32/a_0^3)(C_0C_1/[C_0 + 8C_1]). \tag{29}$$

The three independent elastic stiffness constants reflect the three important physical effects summarized in the force constants C_0 and C_1 and the elastic stiffness coefficient c_{44}. Table 20 illustrates that the radial and angular rigidity force constants C_0 and C_1 and the Cauchy relation decrease with increasing metallicity as expected, using the experimental elastic stiffness coefficients from Table 16. Inclusion of the a_0^3 term in Eqs. (27) and (28) in Table 20 does not change this trend. The macroscopic force constants C_0 and C_1 derived from the experimental elastic stiffness coefficients can be compared with those derived from the characteristic vibrational frequencies (see Sec. 3.4). The speed of propagation of sound waves can be determined from the elastic stiffness coefficients and the material density

Table 20. Several elastic stiffness characteristics related to group-IVa, β-silicon carbide, and silicon–germanium alloy and superlattice semiconductors at 300 K.

Element	Metallicity parameter (α_m)	$(c_{11} + 2c_{12})$ (10^{12} dyn/cm^2)	$(c_{11} - c_{12})$ (10^{12} dyn/cm^2)	$(c_{11} - c_{44})$ (10^{12} dyn/cm^2)
Diamond	0.34	13.26	9.51	4.99
β-Silicon carbide	0.45	6.33	2.12	1.19
Silicon	0.66	2.98	1.03	0.87
SiGe (5084 Å)[a]	—	2.65	0.88	0.74
SiGe (6795 Å)[b]	—	2.56	0.87	0.75
$Si_{0.49}Ge_{0.51}$[c]	—	2.40	0.87	0.71
Germanium	0.81	2.30	0.82	0.63
Gray tin	0.87	—	0.44[d]	—

[a]Material density is 2.79 g/cm^3; the Å data refer to the superlattice thickness (Mendik *et al.*, 1991).
[b]Material density is 3.27 g/cm^3; the Å data refer to the superlattice thickness (Mendik *et al.*, 1991).
[c]Material density is 3.85 g/cm^3 (Mendik *et al.*, 1991).
[d]Theoretical data (Harrison, 1980).

(Boer, 1992). Finally, third-order elastic stiffness coefficients reflecting nonlinear effects such as thermal expansion and the interaction of phonons have also been summarized (Boer, 1992).

The theoretical fracture strength, σ_{th}, is expressed by (Davies, 1994)

$$\sigma_{th} = (Y\gamma/d)^{1/2}. \quad (30)$$

Utilizing the experimental Y {111} data from Table 18, the γ {111} data from Table 7, and the bond-length d data from Table 2, the values of σ_{th} decrease with increasing metallicity from diamond, silicon, and germanium in the order 2.01×10^{12}, 0.34×10^{12}, and 0.26×10^{12} dyn/cm^2, respectively. The theoretical fracture strength is approximately 0.2 the Young's modulus and approximately 0.4 the bulk modulus. Microscopic cracks and other material defects, however, reduce the fracture strength in real materials. Accordingly, the plastic deformation of semiconductors during device and IC fabrication is extremely structure sensitive. Extensive investigations on both semiconductor material preparation and the retention of the improved material microstructure during device and IC fabrication is an extremely active area of research.

3.4 Optical Properties

The coefficients of the dielectric constant ϵ_{ij}, a second-rank tensor, relate the electric field component E_j to the displacement component D_i in the ith direction. The dielectric constant tensor ϵ_{ij} can be transformed to its principal axes and, therefore, principal dielectric constant coefficients ϵ_1, ϵ_2, and ϵ_3, which are equal (isotropic) for the diamond crystal structure. The refractive index n, the ratio of the speed of light in a vacuum to the speed of light in a material, is related to the dielectric constant ϵ, at the same frequency, from Maxwell's equations:

$$n = (\epsilon\mu_m)^{1/2}, \quad (31)$$

where μ_m is the magnetic permeability. The extinction coefficient is sufficiently small in the long-wavelength (static) regime that the imaginary part of both the dielectric constant and the index of refraction can be neglected. The magnetic permeability may be taken as unity since the magnetic susceptibility χ is very small for the group-IVa semiconductors ($\mu_m = 1 + \chi$). The experimental data for ϵ, n, and χ are summarized in Table 21. The trend in the properties is consistent with previous material trends for the group-IVa semiconductors. For example, the increase in the index of refraction with increasing metallicity and decreasing Debye temperature (see Table 14) indicates that the semiconductor is becoming "less stiff" to the excitation and propagation of vibrational modes in the crystal. The optical excitation of electrons from the valence band to higher-order conduction-band minima, noted in Sec. 3.1.1, can be qualitatively described by the Phillips ionicity number (Phillips, 1973). The theoretical values of the Phillips ionicity are also summarized in Table 21.

Phonons in the group-IVa semiconductors facilitate indirect energy-gap electronic transitions via absorption or emission as appropriate. The phonons are required in order to conserve crystal momentum during optical transitions. The acoustic modes (i.e., sound waves) exhibit zero energy (other than zero-point motion) at zero wave number. Optical modes, however, can be excited into oscilla-

Table 21. Experimental data for the static dielectric constant, index of refraction, magnetic susceptibility, and Phillips ionicity number for the group-IVa and β-silicon carbide semiconductors.

Element	Static dielectric constant (ϵ)	Index of refraction (n)	Magnetic susceptibility (χ) (10^{-6}) (cgs)	Phillips ionicity number
Diamond	5.7	2.42	−5.9	2.50
β-Silicon carbide	9.72	3.12	−12.8	—
Silicon	11.9	3.42	−3.9	1.41
Germanium	16.0	4.00	−0.12	1.35
Gray tin	24.0	4.90[a]	−37.0	1.15

[a]Theoretical value (Harrison, 1980).

tory motion, even at zero wave number. Both acoustic and optical modes of vibration exhibit longitudinal and transverse motion. The transverse acoustical-mode frequencies at the Brillouin-zone boundary and the transverse optical-mode frequencies at $k = 0$ are summarized in Table 22 (Harrison, 1980); the reduction in the frequency of the transverse acoustical mode with increasing metallicity is observed. The force constants derived from these normal modes of vibration are not in agreement with the values derived from the elastic stiffness coefficients in Sec. 3.3, indicating the limitation of current models. An extensive listing of a number of phonon modes at crystal symmetry points has been summarized (Boer, 1992).

Finally, many semiconductors such as silicon are optically almost indistinguishable from metals because of their high reflectivity. For example, approximately 30% of the visible to far-infrared incident electromagnetic energy radiation is lost by reflection at each air–silicon interface (Smith, 1961; Bullis and Huff, 1994). The transmission of energy in silicon can be increased by the use of an antireflecting (AR) coating with a thickness equal to 1/4 of the incident radiation's wavelength and an index of refraction equal to $(n_{Si}n_{air})^{1/2}$ ($n_{air} = 1$) (Bullis and Huff, 1994). In the visible region of the spectrum, semiconductors, like metals, generally absorb strongly, having absorption coefficients approximately 10^4–10^5 cm^{-1} (Smith, 1961). Over the range 700 to 1040 nm, the absorption coefficient of silicon, α_{Si}, is given by the empirical relationship (Nartowitz and Goodman, 1985)

$$\alpha_{Si} = (8.4732 \times 10^{-3}\,\nu - 76.417)^2, \qquad (32)$$

where α_{Si} is in cm^{-1} and ν is the wave number in cm^{-1}. A significant increase in absorption occurs below about 400 nm due to the onset of direct transitions from the valence band to the higher-energy conduction band at $k = 0$.

On the other hand, pure diamond is a good "window" for all wavelengths longer than 227 nm. Diamond is unique in that it exhibits almost no optical absorption from extremely low frequencies, approaching dc, to energies just below the energy gap of 5.45 eV (227 nm) (see DIAMOND AND DIAMONDLIKE CARBON).

Table 22. Transverse acoustical mode frequencies at the Brillouin zone and transverse optical mode frequencies at zero wave number.

Element	Transverse acoustic mode frequency (Brillouin zone) (10^{13}/s)	Transverse optical mode frequency ($k = 0$) (10^{13}/s)
Diamond	2.42	0.392
Silicon	0.449	0.155
Germanium	0.0242	0.899
Gray tin	0.0126	0.060

4. SILICON–GERMANIUM

There are basically two classes of transistors: bipolar, in which both electrons and holes play an active role, and unipolar, such as MOS devices, based on the modulation of the conductance of a (surface) layer. The limitations of silicon electronic devices have warranted the development of compound semiconductors such as gallium arsenide for specialized, mainly high-frequency, electronic applications. The limited device integration possible with compound semiconductors, however, and the complexity of processing compound semiconductors have driven the extension of silicon technology by forming alloys of silicon with germanium (Kasper and Bean, 1988; Madelung, 1991; Meyerson, 1992). The silicon energy gap (E_g = 1.12 eV) at the X symmetry point can be monotonically decreased with the addition of germanium (E_g = 0.66 eV). The flexibility of tuning the energy gap in silicon–germanium alloys can improve the performance of specialized, high-performance electronic devices.

In some cases, the germanium content is changed abruptly at a certain distance from the surface of the semiconductor to form an abrupt "heterojunction," with a discontinuity in the energy gap. Because silicon and germanium have similar electron affinities, the energy required to remove an electron from the bottom of the conduction band to outside the solid, of 4.05 and 4.00 eV, respectively, the conduction-band edge is nearly continuous at the silicon–germanium boundary, while the valence-band edge has a discontinuity large enough to modify device performance (Kamins *et al.*, 1989; Kasper *et al.*, 1993). This valence-band discontinuity is especially useful in bipolar transistors. The gain of an *n-p-n* bipolar transistor is related to the ratio of the electron current injected across the emitter–base junction to the sum of the electron and hole currents crossing that junction. The valence-band discontinuity creates a barrier that decreases the injection of holes from the base into the emitter, thereby increasing the gain of the transistor. A further approach to improving bipolar transistor performance is accomplished by increasing the germanium content gradually with position in the base layer (Patton *et al.*, 1989; Cressler *et al.*, 1994). In this case, the energy gap varies as a function of position across the base region of the transistor. The varying energy gap creates an internal electric field that accelerates the free carriers (i.e., electrons in the case of an *n-p-n* transistor) across the base region, thereby increasing the maximum frequency response f_t of the bipolar transistor toward 100 Ghz.

A *p*-channel silicon MOS transistor has a small hole mobility (see Table 9). The hole mobility in silicon–germanium alloys, however, can be greater than in pure silicon (Bean, 1988). The increased carrier mobility enhances the transconductance of the *p*-channel MOS transistor. Furthermore, in silicon–germanium *p*-channel MOS transistors (Garone *et al.*, 1991), a thin layer of silicon–germanium, slightly separated from the oxide–semiconductor interface by a silicon layer, can spatially confine holes. These confined holes travel in the silicon–germanium layer and, accordingly, their velocity is not reduced by scattering from the oxide–semiconductor interface.

Optimal electrical properties are achieved by ensuring that the crystal bonding is uninterrupted across the silicon/silicon–germanium interface. Table 2, however, indicates that the lattice constant of germanium is about 4% greater than that of silicon. The lattice mismatch increases with increasing germanium content in silicon–germanium alloys. Modern growth techniques using CVD can form a thin layer of silicon–germanium alloy on silicon with a continuous crystal structure across the interface (i.e., the crystal structure is "commensurate"). The alloy layer, however, is compressed in the plane of the film. With increasing layer thickness, the energy associated with compressing the alloy layer to fit the silicon crystal increases. Above a "critical" thickness (Matthews and Blakeslee, 1974), the strain energy exceeds the energy necessary to form misfit dislocations, discussed in Sec. 2.2.2. As the bonds across the interface break to form the misfit dislocations, the strain in the system is reduced, thereby lowering the total energy. Accordingly, although silicon–germanium layers thicker than the "equilibrium" critical thickness can be formed, they are unstable (Matthews and Blakeslee, 1974). When these "metastable" layers are heated above the stability temperature appropriate to the layer thickness and germanium content, any irregularity already present can nucleate a misfit dislocation, taking into account kinetic considerations. The dislocation then propagates along the interface, creating a line of broken bonds. While misfit dislocations at the interface may be detrimental to device performance, interactions between misfit dislocations may more seriously degrade the electrical performance of devices built in the layer. The limited stable thickness of silicon–germanium, therefore, constrains the application of silicon–germanium layers. Since very high-performance devices require thin layers ($\leq$100 nm) to limit carrier transit times, however, this restriction is not serious. Thin silicon–germanium layers are indeed compatible with device requirements.

The strain in silicon–germanium layers also exhibits the beneficial effect of increasing the energy-gap difference relative to pure silicon (People, 1985). For example, while an unstrained silicon–germanium layer containing 20% germanium has an energy gap about 0.09 eV smaller than that of silicon, the energy gap of a strained layer with the same germanium content is about 0.17 eV smaller than that of silicon. The required electrical performance can, therefore, be obtained with a smaller germanium content in a strained, rather than an unstrained, layer.

Carrier mobilities in silicon–germanium layers are expected to differ from the silicon values for several reasons. First, carrier mobilities in germanium are greater than for silicon (see Table 9). Second, the random arrangement of germanium in the silicon crystal structure leads to a reduction in mobility due to an additional scattering mechanism, referred to as "alloy" scattering. Third, the strain in the silicon–germanium layer modifies the curvature of the energy bands, changing the carrier effective mass and, thereby, the carrier mobility (Smith, 1961). Finally, carrier lifetime is strongly reduced by impurities such as oxygen in the film. Since oxygen is more readily incorporated into silicon–germanium alloys than in silicon, a higher-purity ambient is required when forming silicon–germanium alloys compared to elemental silicon films.

While most applications of silicon–germanium focus on high-performance transistors, optical devices may also become important (Bean, 1988). Since the energy gap of silicon–germanium alloys is smaller than for silicon, the optical absorption edge corresponds to a longer wavelength (i.e., smaller energy), and silicon–germanium is useful as a photodetector at longer wavelengths compared to silicon.

5. β-SILICON CARBIDE

Silicon carbide can crystallize in the diamond crystal structure, where the silicon and carbon atoms are arranged at the (0,0,0) and $(\frac{1}{4},\frac{1}{4},\frac{1}{4})$ basis vector positions, respectively (or vice versa). This configuration is referred to as β-SiC. Silicon carbide can also occur in a variety of modifications that can be described as different stacking sequences of the closest-packed {111} layers (see Sec. 2.3.1). Since these polymorphs only differ in their stacking sequences, the term polytypes instead of polymorphs is used to describe them (Azaroff, 1960). These structures occur in the hexagonal and rhombohedral crystal structures and are collectively referred to as α-SiC. The most commonly observed polytypes are referred to as 4*H*, 6*H*, and 15*R*, where the numerical coefficient designates the number of double layers required to replicate the crystal structure. The letter *H* denotes that the hexagonal cell is primitive, and the letter *R* signifies a rhombohedral cell (i.e., a hexagonal cell that is centered). Additional structures such as 21*R*, 33*R*, and 51*R* are also observed. The reason only particular polytypes are observed is not yet fully comprehended, but it appears to be the rule rather than the exception and is determined by the conditions during their growth (Tairova and Tsvetkov, 1983). It has been observed, however, that 6*H* is most often formed in the absence of impurities, whereas polytypes such as 51*R*, 141*R*, and others may occur when the melt contains a relatively large number of impurities. The role of impurities has previously been suggested to aid in the formation of screw dislocations, which facilitate the formation of complex polytypes (Azaroff, 1960) and spurious nucleation effects (Davis, 1994). This topic continues to be an important area of research, and the reader is referred to the literature for further details (Davis *et al.*, 1988; Pirouz *et al.*, 1991; Morkoc *et al.*, 1994; Pensl and Troffer, 1995).

The physicochemical properties of α- and β-SiC are different. In particular, α-SiC exhibits significantly different properties as a result of the effects of differing stacking sequences (Henisch and Roy, 1969). Several structural, electrical, thermal, mechanical, and optical properties of β-SiC have been summarized in earlier tables. α-SiC, however, especially 4*H*, has been reported to exhibit superior device performance compared to β-SiC (Spencer, 1993; Itoh *et al.*, 1994) because of its high electron mobility, low anisotropy of the electron effective mass, small nitrogen donor electrical ionization energy (Pensl and Troffer, 1995), and low micropipe-density epitaxial layers. Nevertheless, the usefulness of β-SiC as a semiconducting device material (Davis *et al.*, 1990) is briefly discussed in Sec. 6 because of the emphasis in this review on semiconductor material configurations with the diamond crystal structure.

6. DEVICE/INTEGRATED-CIRCUIT IMPLICATIONS

Silicon rapidly replaced germanium in device technology on account of its larger energy gap, which allowed higher-temperature operation and lower reverse current. In addition, silicon's oxide is stable, whereas germanium's oxide is water soluble. Accordingly, silicon was favored for IC fabrication since the stable oxide facilitated its utilization in the planar process as

1. a diffusion mask for *p-n* junction fabrication,
2. passivation of the silicon surface and *p-n* junctions intersecting the surface, and
3. a dielectric layer for supporting and sep-

Table 23. Material and device characteristics utilized in figures of merit.

Material	κ (W/cm K)	E_{bd} (V/cm)	V_{sat} (cm/s)	ϵ
Diamond	20	1.0×10^7	2.7×10^7	5.7
β-Silicon carbide	5.0	4.0×10^6	2.5×10^7	9.72
Silicon	1.15	3.0×10^5	1.0×10^7	11.9
Gallium arsenide	0.46	3.5×10^7	2.0×10^7	13.1

arating metallic conductor overlayers (Adcock, 1969; Kooi, 1991; Huff, 1996).

Alternative materials to silicon that might exhibit "better" device parameters have, of course, been proposed. These materials include gallium arsenide, silicon carbide, and diamond (Collins, 1990; Davis *et al.*, 1990). Several important material and device characteristics utilized in figures of merit include the thermal conductivity κ, the breakdown electric field E_{bd}, and the saturation (limiting) electron velocity v_{sat}. These characteristics, as well as the dielectric constant ϵ, are summarized in Table 23.

The square of the product of breakdown electric field E_{bd} and the saturation (limiting) electron velocity v_{sat} has been proposed as a measure of the power handling capability (V/s) of a transistor (Johnson, 1965). This figure of merit is

$$J = (E_{bd}v_{sat}/2\pi)^2. \qquad (33)$$

An additional figure of merit relates to the thermally limited switching speed of transistors (Keyes, 1972) in ICs for computer logic applications. The ability of a semiconductor to dissipate the thermal energy generated by the transport of electrons or holes from the source to drain during operation of the MOS transistor is critically important for dense MOS circuitry. This figure of merit is

$$K = \kappa(v_{sat}/\epsilon)^{1/2}, \qquad (34)$$

where a smaller dielectric constant ϵ (i.e., reduced amount of stored charge) results in a faster switching device. These figures of merit are summarized in Table 24. Diamond exhibits the best characteristics, followed by β-SiC (Davis *et al.*, 1990). Gallium arsenide exhibits a poorer thermally limited switching characteristic compared with silicon, on account of its smaller thermal conductivity (Meindl, 1987). Although these figures of merit are useful for comparison, they are theoretical values that assume that the semiconductor material is being operated at its physical limits (Collins, 1990). In practice, carrier mobilities (see Table 9) and, especially, device geometry appear to be more influential on performance than might be inferred from figure-of-merit values (Collins, 1990). Diamond's high thermal conductivity makes it especially useful as a heat sink material to mitigate excessive temperature gradients during the operation of dense IC components.

The potential benefit of gallium arsenide due to its electron mobility of 8500 cm^2/V s has been extensively discussed. This benefit, however, may not be realized when one considers not only the thermally limited switching time as in Table 24 but ULSI system-level limitations such as interconnection resistance (Meindl, 1987). It has also been suggested that the only diamond device that appears to offer some promise is an enhancement-mode *p*-type MOSFET operating at room temperature (Collins, 1990). The challenge for the future is to fabricate such a device with submicron dimensions necessary for operation at microwave frequencies (Collins, 1990). β- and α-SiC will also continue to

Table 24. Summary of several device figures of merit.

Material	J $(V/s)^2$	Ratio to Si	K (W/K $cm^{1/2}$ $s^{1/2}$)	Ratio to Si
Diamond	1.85×10^{27}	8103	4.35×10^4	31.52
β-Silicon carbide	2.53×10^{26}	1107	8.03×10^3	5.82
Gallium arsenide	1.24×10^{24}	5.44	5.68×10^2	0.41
Silicon	2.28×10^{23}	1.0	1.38×10^3	1.0

draw attention. Nevertheless, crystalline silicon and silicon-based materials, such as silicon–germanium, are expected to continue, through sophisticated device design and processing technologies, to be the main engine driving the microelectronics revolution to the approximate regime of 50-nm design rules and, perhaps, beyond (Huff, 1993; Semiconductor Industry Association, 1994; Huff and Goodall, 1996). Improved control of the magnitude, tolerance, and uniformity of silicon material characteristics, coupled with an improved understanding of the interrelationships among IC performance, device parameters, IC fabrication processes, and silicon material characteristics, continue to be the key to superior IC product performance, reliability, and yield. The silicon material trends anticipated to ensure the required performance characteristics for gigabit-memory ICs are summarized in the National Technology Roadmap for Semiconductors (Semiconductor Industry Association, 1994).

7. CONCLUSIONS

The group-IVa semiconductors form the prototypical sp^3-hybridized, tetrahedrally bonded, purely covalent bonds. The bonds are highly directional in the diamond crystal structure, which is conveniently described as two interpenetrating face-centered cubic crystals displaced along the ⟨111⟩ direction by one-fourth the diagonal length. The diamond crystal structure was seen to exhibit the smallest packing fraction in the cubic crystal class. This large open space can accommodate a number of interstitial impurities and accounts for the extreme structure sensitivity of diverse material properties such as the yield stress and carrier lifetime.

The fundamental parameter distinguishing the group-IVa semiconductors was seen to be the metallicity parameter, the parameter *sine qua non* by which the distinct variations in the crystalline semiconductor characteristics may be correlated. The variation of the energy gap, lattice constant, Debye temperature, hardness, atomic density, bond energy, radial and angular force constants, and dielectric constant, for example, correlate with the metallicity parameter. β-SiC generally follows the correlation with the metallicity parameter. Although silicon–germanium also follows some of these trends, alloy effects complicate the analysis. The electrical conductivity, thermal conductivity, and dielectric constant are each second-rank tensors that can be reduced to a single number and, therefore, the resulting properties are isotropic. The elastic stiffness coefficients, however, form a fourth-rank tensor that can be reduced to three principal (diagonal) coefficients; therefore, most of the elastic properties of the group-IVa semiconductors were seen to be anisotropic.

Silicon–germanium alloys and β-SiC are important semiconducting material combinations derived from the group-IVa elements. Semiconductor properties and device considerations for these materials as well as diamond were briefly examined. Nevertheless, crystalline silicon and silicon-based materials, such as silicon–germanium, are expected to continue, through sophisticated device design and processing technologies, to be the main engine driving the microelectronics revolution to the approximate regime of 50-nm design rules and, perhaps, beyond.

Crystalline silicon and silicon-based materials will continue to be the most intensively studied material systems in the scientific literature. The opportunities for further silicon-based advancements appear boundless, certainly quite sufficient to support numerous research and development thrusts well into the 21st century (Huff and Goodall, 1996b). In conjunction with IC process/device/circuit innovations for cost-effective production of ICs (Ohmi, 1995; Koike, 1995), electronics will offer unsurpassed opportunities for an improved quality of life for the world's citizens.

ACKNOWLEDGMENTS

The critical reading of this manuscript by Dr. W. M. Bullis, Dr. R. K. Goodall, Dr. E. Kooi, Dr. B. S. Roos, Dr. W. R. Runyan, Prof. H. Berk, Prof. U. Gösele, Prof. W. A. Harrison, Prof. R. Helms, Prof. L. C. Kimerling, Prof. J. Meindl, Prof. F. H. Pollack, and Prof. E. Weber as well as selected portions by Dr. J. Hersener, Dr. G. Pensl, Dr. E. Wijaranakula, Prof. A. L. Ruoff, and Prof. E. A. Fitzgerald is appreciated. Dr. A. Burk supplied a number of references to β-SiC material characteristics. Dr. T. Kamins kindly contributed the section on silicon–germanium, and Mr. V. Bhat replicated several figures for the manuscript. Finally, I thank Linda Ofshe for compiling and formatting the manuscript.

GLOSSARY

Conduction Band: The antibonding states or band, where the electrons "belong" to the whole crystal.

Crystal: A periodic three-dimensional structure.

Defect: A deviation from perfect periodicity of the crystalline structure.

Diamond Crystal Structure: The group-IVa semiconductor crystal structure, often described as two interpenetrating face-centered cubic crystals displaced along the ⟨111⟩ direction by one-fourth the diagonal length.

Dopant: A group-Va (group-IIIa) element that resides on a substitutional site, introducing an electron (hole) into the conduction (valence) band.

Effective Distribution Coefficient: Ratio of the impurity concentration incorporated into a growing crystal at the crystal-melt interface to the impurity concentration in the bulk of the liquid.

Energy Gap: The minimum energy separating the valence-band maximum from the conduction-band minimum. When the energy extrema are located at different values in momentum (**k**) space, the energy gap is denoted as indirect.

Interstitial Atom: An atom residing on a crystal site other than a substitutional site.

Line Defect: A line imperfection in the crystal, such as a dislocation, that extends over many atomic spacings.

Metallicity Parameter: The ratio of the metallic energy to the covalent energy for a semiconductor.

Point Defect: A defect that extends only over a few atomic spacings such as a vacancy, interstitial atom, oxygen atom, or carbon atom.

Semiconductor: A material that exhibits a reduction in resistivity with increasing temperature over selected regions of temperature.

***sp^3* hybridization:** Interaction between s and p valence electrons to form bonding and antibonding states or bands in the diamond crystal structure.

Substitutional Site: A crystal site where the atoms of the crystal normally reside.

Vacancy: A crystal site where a substitutional atom has been removed.

Valence Band: The bonding states or band, where the electrons "belong" to the whole crystal, but statistically are spatially localized between the atoms.

Works Cited

Adamson, A. W. (1990), *Physical Chemistry of Surfaces,* 5th ed. New York: Wiley.

Adcock, W. (1969), in: R. R. Haberecht, E. L. Kern, (Eds.), *Semiconductor Silicon,* Pennington, NJ: The Electrochemical Society, pp. 36–54.

Alexander, H., Haasen, P. (1968), in: F. Seitz, D. Turnbul, H. Ehrenreich (Eds.), *Solid State Physics,* Vol. 22, New York: Academic, pp. 27–158.

Allen, J. W. (1957), *Philos. Mag.* **2,** 1475–1481.

Anderson, H. L. (Ed.) (1989), *A Physicist's Desk Reference,* 2nd ed., New York: American Institute of Physics.

Asoka-Kumar, P., Lynn, K. G., Welch, D. O. (1994), *J. Appl. Phys.* **76,** 4935–4982.

Azaroff, L. V. (1960), *Introduction to Solids,* New York: McGraw-Hill.

Baker, S. P., Arzt, E. (1995), in: E. Kasper (Ed.), *Properties of Strained and Relaxed Silicon Germanium,* Exeter, U.K.: INSPEC, pp. 67–69.

Beadle, W. E., Tsai, J. C. C., Plummer, R. D. (1985), *Quick Reference Manual for Silicon Integrated Technology,* New York: Wiley.

Bean, J. C. (1988), in: E. Kasper, J. C. Bean (Eds.), *Silicon-Molecular Beam Epitaxy,* Vol. 2, Pennington, NJ: The Electrochemical Society, pp. 65–110.

Bean, K., Lindberg, K., Rozgonyi, G. (1988), in: M. Scott, Y. Akasaka, R. Reif (Eds.), *Epitaxy in Advanced Materials for ULSI,* Pennington, NJ: The Electrochemical Society, pp. 3–24.

Boer, K. W. (1992), *Survey of Semiconductor Physics,* New York: Van Nostrand Reinhold.

Bond, W. L. (1960), *Acta Crystallogr.* **13,** 814–818.

Brown, R. A., Maroudas, D., Sinno, T. (1994), *J. Cryst. Growth* **137,** 12–25.

Bullis, W. M. (1994), in: H. R. Huff, W. Bergholz, K. Sumino (Eds.) *Semiconductor Silicon/94,* Pennington, NJ: The Electrochemical Society, pp. 1156–1169.

Bullis, W. M., Huff, H. R. (1994), in: D. Bloor, R. J. Brook, M. C. Flemings, S. Mahajan (Eds.), *The Encyclopedia of Advanced Materials,* London: Pergamon, pp. 2478–2496.

Bullis, W. M., Huff, H. R. (1996), *J. Electrochem. Soc.* **143,** 1399–1405.

Burns, G. (1985), *Solid State Physics,* New York: Academic.

Burton, J. A., Prim, R. C., Slichter, W. P. (1953), *J. Chem. Phys.* **21,** 1987–1991.

Chelikowsky, J. R., Cohen, M. L. (1976), *Phys. Rev. B* **14,** 556–581.

Chen, A. B., Sher, A., Yost, W. T. (1992), in: K. T. Faber, K. Malloy (Eds.) *The Mechanical Properties of Semiconductors,* New York: Academic, pp. 1–77.

Chikawa, J., Shirai, S. (1979), *Japan J. Appl. Phys.* **18** (Suppl. 18-1), 153–164.

Clarke, D. R. (1992), in: K. T. Faber, K. Malloy (Eds.), *The Mechanical Properties of Semiconductors,* New York: Academic, pp. 79–142.

Collins, A. T. (1990), in: J. T. Glass, R. Messier, N. Fujimori (Eds.), *Diamond, Silicon Carbide and Related Wide Bandgap Semiconductors,* MRS Symposium Proceedings Vol. 162, Pittsburgh, PA: Materials Research Society, pp. 3–14.

Cottrell, A. H. (1961), *Dislocations and Plastic Flow in Crystals,* Oxford: Oxford University Press.

Cressler, J. D., Harame, D., Comfort, J. H. (1994), in: *IEEE International Solid State Circuits Conference,* New York: IEEE, pp. 24–27.

Dash, W. C. (1956), *J. Appl. Phys.* **27,** 1193–1195.

Dash, W. C. (1960), in: H. C. Gatos, (Ed.), *Properties of Elemental and Compound Semiconductors,* Metallurgical Society Conferences, Vol. 5, New York: Interscience, pp. 195–209.

Davies, G. (Ed.) (1994), *Properties and Growth of Diamond,* EMIS Datareviews Series, London: INSPEC, p. 9.

Davis, R. F. (1994), in: D. Bloor, R. J. Brook, M. C. Flemings, S. Mahajan (Eds.), *The Encyclopedia of Advanced Materials,* London: Pergamon, pp. 2465–2469.

Davis, R. F., Palmour, J. W., Edmond, J. A. (1990), in: J. T. Glass, R. Messier, N. Fujimori (Eds.), *Diamond, Silicon Carbide and Related Wide Bandgap Semiconductors,* MRS Symposium Proceedings Vol. 162, Pittsburgh: Materials Research Society, 463–474.

Davis, R. F., Sitar, Z., Williams, B. E., Kong, H. S., Kim, H. J., Palmour, J. W., Edmond, J. A., Ryu, J., Glass, J. T., Carter, C. H., Jr. (1988), *Materials Science and Engineering, B1,* Lausanne: Elsevier, pp. 77–104.

De Launay, J. (1956), in: F. Seitz, D. Turnbull (Eds.), *Solid State Physics,* Vol. 2, New York: Academic, pp. 219–303.

Diebold, A. C., Maillot, P., Gordon, M., Baylis, J., Chacon, J., Witowski, R., Arlinghaus, H. F., Knapp, J. A., Doyle, B. L. (1992), *J. Vac. Sci. Technol. A.* **10,** 2945–2952.

Dornberger, E., von Ammon, W. (1996), *J. Electrochem. Soc.* **143,** 1648–1653.

Dyer, L. D., Huff, H. R., Boyd, W. W. (1971), *J. Appl. Phys.* **42,** 5680–5688.

EMIS Datareviews (1988), *Properties of Silicon,* EMIS Datareviews Series Vol. 4, London: INSPEC, p. 4.

Field, J. E. (1991), in: R. E. Clausing, L. L. Horton, J. C. Angus, P. Koidl (Eds.), *Diamond and Diamond-Like Films and Coatings,* NATO Advanced Study Institutes Series B, Physics, Vol. 266, New York: Plenum, pp. 17–35.

Foll, H., Gösele, U., Kolbesen, B. O. (1977), in: H. R. Huff, E. Sirtl (Eds.) *Semiconductor Silicon/77,* Pennington, NJ: The Electrochemical Society, pp. 565–574.

Fukuda, T. (1994), *Appl. Phys. Lett.* **65,** 1376–1378.

Garone, P. M., Venkataraman, V., Sturm, J. C. (1991), *IEEE Electron Device Lett.* **12,** 230–232.

Gatos, H. C., Lavine, M. C. (1965), in: A. F. Gibson, R. E. Burgess (Eds.) *Progress in Semiconductors,* Vol. 9, New York: Wiley, pp. 3–45.

Gay, A. G. (1960), *Elements of Physical Metallurgy,* Reading, MA: Addison-Wesley.

Gösele, U., Tan, T. Y. (1991), *MRS Bull.* **16** (11), 42–46.

Green, M. A. (1990), *J. Appl. Phys.* **67,** 2944–2954.

Harrison, W. A. (1980), *Electronic Structure and The Properties of Solids,* San Francisco: W. H. Freeman.

Harrison, W. A. (1981), *Phys. Rev. B* **24,** 5835–5843.

Helms, R. (1996), in: W. M. Bullis, D. Seiller, A. C. Diebold (Eds.), *Semiconductor Characterization: Present Status and Future Needs,* New York: American Institute of Physics, pp. 110–117.

Henisch, K. H., Roy, R. (Eds.) (1969), *Silicon Carbide—1968,* New York: Pergamon.

Herman, F. (1995), *Proc. IRE* **43,** 1703–1732.

Hirth, J. P., Lothe, J. (1982), *Theory of Dislocations,* 2nd ed., New York: Wiley.

Hobstetter, J. N. (1960), in: Hannay, N. B. (Ed.), *Effect of Imperfections on Germanium and Silicon in Semiconductors,* New York: Reinhold, pp. 508–540.

Holwigle, N. A., Frank, W., Hobel, J., Pearton, P., Haller, E. E. (1985), *J. Appl. Phys.* **57,** 5211–5219.

Hornstra, J. (1958), *J. Phys. Chem. Solids* **5,** 129–141.

Hu, S. M. (1994), in: F. Shimura (Ed.), *Oxygen in Silicon,* New York: Academic, pp. 153–190.

Huff, H. R. (1992), in: S. Mahajan, L. C. Kimerling (Eds.), *Concise Encyclopedia of Semiconducting Materials & Related Technologies,* New York: Pergamon, pp. 478–492.

Huff, H. R. (1993), in: G. K. Celler, E. Middlesworth, K. Hoh (Eds.), *Proceedings of the Fourth International Symposium on ULSI Science and Technology,* Pennington, NJ: The Electrochemical Society, pp. 103–132.

Huff, H. R. (1996), *Electrochem. Soc. Extended Abstracts,* **96-1,** 341–343.

Huff, H. R., Goodall, R. K. (1995), in: R. B. Fair, B. Lojek (Eds.), *RTP '95,* Rapid Thermal Processing Conference, Round Rock, TX: RTP '95, pp. 9–40.

Huff, H. R., Goodall, R. K. (1996a), in: W. M. Bullis, D. Seiller, A. C. Diebold (Eds.), *Semiconductor Characterization: Present Status and Future Needs,* New York: American Institute of Physics, pp. 67–96.

Huff, H. R., Goodall, R. K. (1996b), *Interface* **5** (2), 31–35.

Huff, H. R., Shimura, F. (March, 1985), *Solid State Tech.* **28** (3), 103–118.

Huff, H. R., Schaake, H. F., Robinson, J. T., Baber, S. C., Wong, D. (1983), *J. Electrochem. Soc.* **130,** 1551–1555.

Huff, H. R., Goodall, R. K., Williams, E., Woo, K.-S., Liu, B. Y. H., Warner, T., Hirleman, D., Gildersleeve, K., Bullis, W. M., Scheer, B. W., Stover J. (1995), in: E. M. Middlesworth, H. Z. Massoud (Eds.), *Proceedings of the Fifth International Symposium on ULSI Science and Technology,* Pennington, NJ: The Electrochemical Society, pp. 365–383.

Irvin, J. C. (1962), *Bell Syst. Tech. J.* **41,** 387–412.

Itoh, A., Akita, H., Kimoto, T., Matsunami, H. (1994), *Appl. Phys. Lett.* **65,** 1400–1402.

Johnson, E. O. (1965), *RCA Rev.* **26,** 163–177.

Kamins, T. I., Nauka, K., Camnitz, L. H., Kruger, J. B., Turner, J. E., Rosner, S. J., Scott, M. P., Hoyt, J. L., King, C. A., Noble, D. B., Gibbons, J. F. (1989), *International Electron Devices Meeting, Washington, D.C.,* Piscataway, NJ: IEEE, pp. 647–650.

Kasper, E., Bean, J. C. (Eds.) (1988), *Silicon-Molecular Beam Epitaxy,* Pennington, NJ: The Electrochemical Society.

Kasper, E., Gruhle, A., Kibbel, H. (1993), *International Electron Devices Meeting, Washington, D.C.,* Piscataway, NJ: IEEE, pp. 79–81.

Kelly, A., Groves, G. W. (1970), *Crystallography and Crystal Defects,* Reading, MA: Addison-Wesley.

Kern, W. (Ed.) (1993), *Handbook of Semiconductor Wafer Cleaning Technology,* Park Ridge, NJ: Noyes Publications.

Keyes, R. W. (1972), *Proc. IEEE* **60,** 225.

Kimerling, L. C., Patel, J. R. (1985), in: N. G. Einspruch, H. R. Huff (Eds.), *VLSI Electronics: Microstructure Science,* Vol. 12, New York: Academic, pp. 223–267.

Kimerling, L. C., Leamy, H. J., Patel, J. R. (1977), *Appl. Phys. Lett.* **30,** 217–219.

Kimerling, L. C., Patel, J. R. (1979), *Appl. Phys. Lett.* **38,** 73–75.

King, C. A. (1994), in: D. Bloor, R. J. Brook, M. C. Flemings, S. Mahajan (Eds.), *The Encyclopedia of Advanced Materials,* London: Pergamon, pp. 2496–2500.

Kisielowski, C., Weber, E. R. (1991), *Phys. Rev. B* **44,** 1600–1612.

Kittel, C. (1996), *Introduction to Solid State Physics,* 7th ed., New York: Wiley.

Koike, A., Shimoyashura, S., Kubota, K., Suzuki, Y., Kiguchi, A., Fujisawa, A., Takamatsu, A., Okabe, T. (1995), in: *International Symposium on Semiconductor Manufacturing,* Piscataway, NJ: IEEE, pp. 239–242.

Kolbesen, B. O., Strunk, H. P. (1985), in: N. G. Einspruch, H. R. Huff (Eds.), *VLSI Electronics: Microstructure Science,* Vol. 12, New York: Academic, pp. 143–222.

Kooi, E. (1991), *The Invention of LOCOS, IEEE Case Histories of Achievement in Science and Technology,* Vol. 1, Parsipanny, NJ: IEEE Press.

Kubaschewski, O., Alcock, C. B. (1973), *Metallurgical Thermochemistry,* 5th ed., New York: Pergamon.

Landau, L. D., Lifshitz, E. M. (1958), *Quantum Mechanics, Non-Relativistic Theory,* New York: Pergamon.

Lawrence, J. E., Huff, H. R. (1982), in: N. G. Einspruch (Ed.), *VLSI Electronics: Microstructure Science,* Vol. 5, New York: Academic, pp. 51–102.

Lemke, H. (1994), in: H. R. Huff, W. Bergholz, K. Sumino (Eds.), *Semiconductor Silicon/1994,* Pennington, NJ: The Electrochemical Society, pp. 695–710B.

Lide, D. R. (Ed.) (1993–1994), *CRC Handbook of Chemistry and Physics,* 74th ed., Boca Raton, FL: CRC Press.

Ligenza, J. R. (1961), *J. Phys. Chem.* **65,** 2011–2014.

Lin, W., Hill, D. W. (1983), *J. Appl. Phys.* **54,** 1082–1085.

Madelung, O. (Ed.) (1991), *Data in Science and Technology, Semiconductors, Group IV Elements and III-V Compounds,* Berlin: Springer-Verlag.

Matthews, J. W., Blakeslee, A. E. (1974), *J. Cryst. Growth* **27,** 118–125.

McSkimin, H. J. (1953), *J. Appl. Phys.* **24,** 988–997.

McSkimin, H. J., Bond, W. L., Buckler, E., Teal, G. K. (1951), *Phys. Rev.* **83,** 1080.

Meindl, J. (1987), in: S. Broydo and C. M. Osburn (Eds.), *ULSI Science and Technology/1987,* Pennington, NJ: The Electrochemical Society, pp. 3–17.

Mendik, M., Ospelt, M., Von Kanel, H., Wachter, P. (1991), *Appl. Surface Sci.* **50,** 303–307.

Meyerson, B. S. (1992), *Proc. IEEE* **80,** 1592–1608.

Morkoc, H., Strite, S., Gao, G. B., Lin, M. E., Sverdlov, B., Burns, M. (1994), *J. Appl. Phys.* **76,** 1363–1398.

Nartowitz, E. S., Goodman, A. M. (1985), *J. Electrochem. Soc.* **132,** 2992–2997.

Nielsen, O. H. (1988), in: *Properties of Silicon,* EMIS Datareviews Series Vol. 4, London: INSPEC, p. 15.

Nye, J. F. (1964), *Physical Properties of Crystals*, Oxford: Oxford University Press.

Ohmi, T., Miyashita, M., Itano, M., Imaoka, T., Kawanabe, I. (1992), *IEEE Trans. Electron Devices* **39,** 537–545.

Ohmi, T. (1995), *Ultra Clean Soc. Proc.* **26,** 91.

Page, T. F., Oliver, W. C., McHargue, C. J. (1992), *J. Mater. Res.* **7,** 450–473.

Patton, G. L., Harame, D. L., Stork, J. M. C., Meyerson, B. S., Scilla, G. J., Ganin, E. (1989), *IEEE Electron Device Lett.* **10,** 534–536.

Pauling, L. (1960) *The Nature of the Chemical Bond*, Ithaca, NY: Cornell Univ. Press.

Pensl, G., Choyke, W. J. (1993), *Physica B* **185,** 264–283.

Pensl, G., Troffer, Th. (1995), in: H. Richter, M. Kittler, C. Claeys (Eds.), *GADEST '95*, Zurich: Scitec Publications, pp. 115–126.

People, R. (1985), *Phys. Rev. B* **32,** 1405–1408.

Pfann, W. G. (1966), *Zone Melting*, 2nd ed. New York: Wiley.

Phillips, J. C. (1973), *Bonds and Bands in Semiconductors*, New York: Academic.

Pirouz, P., Yang, J. W., Powell, J. A., Ernst, F. (1991), in: A. G. Cullis, A. R. Long (Eds.), *Microscopy of Semiconducting Materials 1991*, IOP Conference Proceedings No. 117, Bristol, U.K.: IOP Publishing, Sec. 3, 149–154.

Pollak, F. H., Cardona, M., Higginbotham, C. W., Herman, F., Van Dyke, J. P. (1970), *Phys. Rev. B* **2,** 352–363.

Rajagopalan, S., Mitra, U., Pan, S., Gupta, K., Lin, C. M., Sery, G., Mittal, S., Hasserjian, K., Lo, W. J., Neubauer, G. (1993), in: *31st Annual Proceedings of Reliability Physics/1993*, New York: IEEE, pp. 28–31.

Ravi, K. V. (1981), *Imperfections and Impurities in Semiconductor Silicon*, New York: Wiley.

Ridley, B. K. (1982), *Quantum Processes in Semiconductors*, Oxford: Oxford University Press.

Riney, T. D. (1961), *J. Appl. Phys.* **32,** 454–460.

Roos, B. S., Richter, H., Wollweber, J. (1995a), in: H. Richter, M. Kittler, C. Clayes (Eds.), *GADEST '95*, Zurich: Scitec Publications, pp. 509–515.

Roos, B. S., Richter, H., Morgenstern, Tillack, B. (1995b) in: S. P. Baker, P. Borgesen, B. H. Townsend, C. A. Ross, C. A. Volkert (Eds.), *Thin Films: Stresses and Mechanical Properties V*, MRS Symposium Proceedings Vol. 356, Pittsburgh, PA: Materials Research Society, pp. 277–282.

Rossi, J. A., Dyson, W., Hellwig, L. G., Haley, T. M. (1985), *J. Appl. Phys.* **58,** 1798–1802.

Runyan, W. R. (1965), *Silicon Semiconductor Technology*, New York: McGraw-Hill.

Runyan, W. R., Huff, H. R. (1994), in: J. J. McKetta (Ed.), *Encyclopedia of Chemical Processing and Design*, Vol. 49, New York: Marcel Dekker, pp. 340–365.

Ruoff, A. L. (1979), in: K. D. Timmerhaus, M. S. Barber (Eds.), *High Pressure Sci. Technol.* **2,** 525–548.

Sauer, R., Weber, J., Stolz, J., Weber, E. R., Kusters, K.-H., Alexander, H. (1985), *Appl. Phys. A* **36,** 1–13.

Schlichting, J., Riley, F. L. (1992), in: S. Mahajan, L. C. Kimerling (Eds.), *Concise Encyclopedia of Semiconducting Materials & Related Technologies*, London: Pergamon, pp. 451–455.

Schroder, D. K. (1990), *Semiconductor Material and Device Characterization*, New York: Wiley.

Schulz, H. (1982), in: E. Kaldis (Ed.), *Current Topics in Materials Science*, Vol. 8, Amsterdam: North-Holland, pp. 276–379.

Seeger, A. (1976), in: J. Treusch (Ed.) *Festkörperprobleme XVI—Advances in Solid State Physics*, Braunschweig: Vieweg, pp. 149–178.

Seeger, A., Swanson, M. L. (1968) in: R. R. Hasiguti (Ed.), *Lattice Defects in Semiconductors*, Tokyo: Univ. of Tokyo Press, pp. 93–130.

Semiconductor Industry Association (November, 1994), *1994 National Technology Roadmap for Semiconductors*, San Jose, CA: Semiconductor Industry Association.

Shimura, F. (1989), *Semiconductor Silicon Crystal Technology*, New York: Academic.

Shimura, F., Huff, H. R. (1985), in N. G. Einspruch (Ed.), *VLSI Handbook*, New York: Academic, pp. 191–269.

Shimura, F. (Ed.) (1994), *Oxygen in Silicon*, New York: Academic.

Siethoff, H. (1994), *Appl. Phys. Lett.* **65,** 174–176.

Slack, G. A., Bartram, S. F. (1975), *J. Appl. Phys.* **46,** 89–98.

Smith, R. A. (1961), *Semiconductors*, Cambridge: Cambridge Univ. Press.

Spencer, M. G. (Ed.) (1993), *Proceedings Fifth International Conference on SiC and Related Materials*, IOP Conference Series No. 137, Bristol: IOP Publishing.

Strausser, Y. E., Doris, B., Diebold, A. C. Huff, H. R. (1994), *Electrochem. Soc. Extended Abstracts* **94-1,** 461–462.

Sumino, K. (1983), J. Nishizawa (Ed.), *Semiconductor Technologies* **8,** 23–44.

Swalin, R. A. (1972), *Thermodynamics of Solids*, 2nd ed. New York: Wiley.

Sze, S. M. (1981), *Physics of Semiconductor Devices*, 2nd ed. New York: Wiley.

Tairova, Y. M., Tsvetkov, V. F. (1983), in: P. Krishna (Ed.), *Growth and Characterization of Polytypic Crystals*, New York: Pergamon, pp. 111–162.

Tan, T. Y., Gardner, E. E., Tice, W. K. (1977), *Appl. Phys. Lett.* **30,** 175–176.

Taylor, W., Gösele, U., Tan, T. Y. (1993), in: G. R. Srinivasan, K. Taniguchi, C. S. Murthy (Eds.), *Process Physics and Modeling in Semiconductor*

Technology, Pennington, NJ: The Electrochemical Society, pp. 3–19.

Taylor, W. E., Dash, W. C., Miller, L. E., Mueller, C. W. (1960), H. C. Gatos (Ed.), *Properties of Elemental and Compound Semiconductors,* Metallurgical Society Conferences, New York: Interscience, Vol. 5, pp. 327–335.

Teal, G. K. (1976), *IEEE Trans. Electron Devices* **ED-23,** 621–639.

Teal, G. K., Runyan, W. R., Bean, K. E., Huff, H. R. (1985), in: J. F. Young, R. S. Shane (Eds.), *Materials and Processes,* 3rd ed., New York: Marcel Dekker, pp. 219–312.

Thurber, W. R., Mattis, R. L., Liu, Y. M., Filliben, J. J. (1980a), *J. Electrochem. Soc.* **127,** 2291–2294.

Thurber, W. R., Mattis, R. L., Liu, Y. M., Filliben, J. J. (1980b), *J. Electrochem. Soc.* **127,** 1807–1812.

Trumbore, F. A. (1960), *Bell Syst. Tech. J.* **39,** 205–233.

Tu, K.-N., Mayer, J. W., Feldman, L. C. (1992), *Electronic Thin Film Science,* New York: Macmillan.

Weber, E. R. (1988), *Properties of Silicon,* EMIS Datareviews Series Vol. 4, London: INSPEC, pp. 409–451.

Weber, J. (1994), *Diffus. Defect Data* **37–38,** 13–24.

Weber, J., Alonso, M. I., (1990), in: K. Sumino (Ed.), *Defect Control in Semiconductors, 1990,* Amsterdam: Elsevier, pp. 1453–1457.

Wentorf, R. H. (1992), in: S. Mahajan, L. C. Kimerling (Eds.), *Concise Encyclopedia of Semiconducting Materials & Related Technologies,* New York: Pergamon, pp. 91–94.

Wijaranakula, W., (1993), *J. Electrochem. Soc.* **140,** 3306–3316.

Wise, R. L., Lindberg, K. J., Dyer, L. D., Huff, H. R. (1994), in: D. Bloor, R. J. Brook, M. C. Flemings, S. Mahajan (Eds.) *The Encyclopedia of Advanced Materials,* London: Pergamon, 2469–2478.

Wortman, J. J., Evans, R. A. (1965), *J. Appl. Phys.* **36,** 153–156.

Yarbrough, W. A. (1990), in: J. T. Glass, R. Messier, N. Fujimori (Eds.), *Diamond, Silicon Carbide and Related Wide Bandgap Semiconductors,* MRS Symposium Proceedings No. 162, Pittsburgh, PA: Materials Research Society, pp. 75–83.

Ziman, J. M. (1979), *Models of Disorder,* London: Cambridge University Press.

Further Reading

Cerofolini, G. F., Meda, L. (1989), *Physical Chemistry Of, In and On Silicon,* Berlin: Springer-Verlag.

Graff, K. (1994), *Metal Impurities in Silicon-Device Fabrication,* Berlin: Springer-Verlag.

Harbeke, G., Schulz, M. J. (1989), *Semiconductor Silicon,* Berlin: Springer-Verlag.

Monch, W. (1993), *Semiconductor Surfaces and Interfaces,* Berlin: Springer-Verlag.

Runyan, W. R., Bean, K. E. (1990), *Semiconductor Integrated Circuit Processing Technology,* Reading, MA: Addison-Wesley.

Sze, S. M. (Ed.) (1988), *VLSI Technology,* 2nd ed., New York: McGraw-Hill.

Tarui, Y. (Ed.) (1986), *VLSI Technology: Fundamentals and Applications,* Berlin: Springer-Verlag.

Wolf, S., Tauber, R. N. (1986), *Silicon Processing for the VLSI Era,* Vol. 1, *Process Technology,* Sunset Beach, CA: Lattice Press.

Wolf, S. (1990), *Silicon Processing for the VLSI Era,* Vol. 2, *Process Integration,* Sunset Beach, CA: Lattice Press.

Wolf, S. (1995), *Silicon Processing for the VLSI Era,* Vol. 3, *The Submicron MOSFET,* Sunset Beach, CA: Lattice Press.

Yu, P. Y., Cardona, M. (1995), *Fundamentals of Semiconductors, Physics and Materials Properties,* Berlin: Springer-Verlag (1996).

Zangwill, A. (1988), *Physics at Surfaces,* Cambridge: Cambridge University Press.

SEMICONDUCTORS, ORGANIC—ELECTRONIC PROPERTIES

MARTIN POPE, *Chemistry Department, New York University, New York, New York, U.S.A.*
CHARLES E. SWENBERG, *Physics Department, George Mason University, Fairfax, Virginia, U.S.A.*

3-527-28139-8/96/$5.00 + .50

INTRODUCTION

Semiconductors play a vital role in society. They are at the base of all computers, all modern communication systems, all controls for modern transportation systems, and in fact all modern control systems in general. They constitute a multibillion-dollar industry. The vast majority of semiconductors (and as will be explained shortly, all photoconducting insulators are included in this group) are inorganic, being composed of such compounds as silicon, germanium, gallium arsenide, and others (see INSULATORS AND SEMICONDUCTORS, CONDUCTIVITY IN). These materials are used in both crystalline and amorphous forms. The theoretical treatment of the crystalline material is well developed and is capable of providing both a rationalization of observed behavior and a predictive capacity for as yet uncharted areas. The theory of amorphous inorganic semiconductors is not as well developed.

Alongside the vast electronics industry based on inorganic semiconductors is the huge polymer industry, based on organic compounds. The synthetic genius of chemists in this field is well known. The ability of nature to fashion biological structures capable of thought, sensation, and communication, and using light to grow food provides an existence proof of the feasibility of synthesizing polymers that can mimic these functions. There is sufficient evidence that this goal is not a dream. As will be described in Sec. 8, examples of such developments have been realized. Most electrographic copiers have polymeric films that capture and transfer the image to the page. The development of polymeric electronic materials, with all the benefits of magnitude of production capacity and product versatility, depends on an adequate theoretical understanding of the fundamental structural and electronic properties of organic solids. A brief description of progress towards the end will be presented here.

This article will be devoted entirely to organic molecular solids. An organic compound is one that contains carbon in combination with other elements. Of the more than two million known compounds, about 90% are organic. Some compounds such as metal carbonates are usually considered inorganic compounds. However, the number of carbon compounds considered as inorganic is relatively small, and a minimal error is committed by accepting the definition given above. A molecular solid is one in which the individual molecules are electrically neutral, but are attracted to neighboring molecules by the weak van der Waals forces (see COHESIVE ENERGY). Examples of molecular solids are the rare gases at sufficiently low temperatures, polymers like polyethylene, and many organic compounds, like anthracene. Typical intermolecular bond energies are $\leq$0.1 eV, but the intramolecular bond energies can be quite high; in polyethylene, the energy is about 3.4 eV. Most of the organic compounds of particular interest to this discussion are called conjugated hydrocarbons. Essentially all electronic phenomena that have been observed in inorganic compounds have also been found in organic compounds. The differences are in degree only.

As a further subdivision, the intramolecular bonding of the organic systems to be discussed will be conjugated in nature. These are the only compounds to date that have been extensively studied. Conjugated compounds have delocalized electronic orbitals, which lead to facile intramolecular carrier transport and to a system of lower-lying ionization levels than would be the case if the orbitals were highly localized. In addition, the overlap of orbitals on neighboring molecules facilitates electron mobility. The greater ease of generating carriers and the increased carrier mobility of conjugated compounds make them better semiconductors than other compounds such as sugars or plastics like polyethylene or Teflon. The description of a conjugated bond will be found in Sec. 1. Aside from polymeric systems, single-crystal solids will be reported on whenever possible. Finally, only those processes that involve the generation, transport, and recombination of free charges in organic semiconductors (OSC) will be discussed. Generally speaking, a semiconductor is a material that exhibits a thermally activated dark conductivity in the range 10^{-3} to 10^{-10} Ω^{-1} m^{-1}. This definition will be broadened herein to include molecular materials that can efficiently transport photoexcited or dark-injected carriers (in some cases with negligible loss to deep traps) even though these materials are insulators in the dark. This excludes a significant group of organic materials that exhibit metallic conductivity, and even superconductivity (Ramirez, 1994). It also excludes phenomena based on lattice polarization rather than current flow, such as pyroelectricity and piezoelectricity (Mort and Pfister, 1982), and ferroelectricity (Kepler and Anderson, 1992). In addition, recently developed organic paramagnetic solids will not be discussed (see Miller and Epstein, 1994).

The plan of this article is to introduce the reader to the types of molecular structures that will be encountered. This is followed by a brief description of the photoexcited states that appear in the more detailed exposition. Then comes a discussion of the various techniques that are available for generating the charge carriers that form the basis of all the anticipated functions of organic semiconductors (OSC). Following this comes a discussion of carrier recombination, which is one of the most characteristic features of OSC. Carrier recombination can be very efficient, and presents a problem if a high quantum efficiency of free-carrier generation is desired. If the carriers survive the recombination obstacle, they must move through the solid, and this process is not adequately understood; in fact, it represents a major unsolved problem.

An obstacle to carrier transport consists of chemical and physical defects in the solid that trap the carriers; this is discussed next. The following two sections deal with special techniques for probing the electronic structure of solids, and the last section describes several applications in which organic materials can seriously challenge inorganic materials.

Note: In view of the extensive referencing that will be made to the monographs authored by Pope and Swenberg (1982) and by Silinsh and Capek (1994), these works will be referred to by the initials of the authors and the relevant page number in each book, e.g., PS, p. 450 or SC, p. 265.

1. MOLECULAR STRUCTURE OF TYPICAL ORGANIC SEMICONDUCTORS

1.1 π-Electron Molecules

The electronic configuration of the ground state of the carbon atom is $1s^2 2s^2 2p^2$. There are four electrons in the outer electronic orbitals. The two *s* electrons are paired, and the two *p* electrons are unpaired in each of two of the three mutually perpendicular *p* orbitals. Another possible electronic configuration for the carbon atom may be realized by mixing the 2*s* and the three 2*p* orbitals to create a set of four equivalent degenerate orbitals, referred to as sp^3 hybrid orbitals. These orbitals are oriented so that each points in the direction of one corner of an equivalent tetrahedron. Methane, CH_4, provides an example of sp^3 bonding. It is also possible for one *s* and two *p* orbitals (e.g., p_x, p_y) to combine to form three sp^2 orbitals (sp^2 hybridization); these are referred to as planar trigonal orbitals. In this case, one of the three original mutually perpendicular atomic *p* orbitals, the p_z orbital, remains unaltered. The three sp^2 orbitals are all coplanar and directed about 120° apart from each other; the bonds formed from these orbitals are called σ bonds; the p_z is perpendicular to the plane of the sp^2 orbitals. The hybrid-orbital concept provides a convenient model for un-

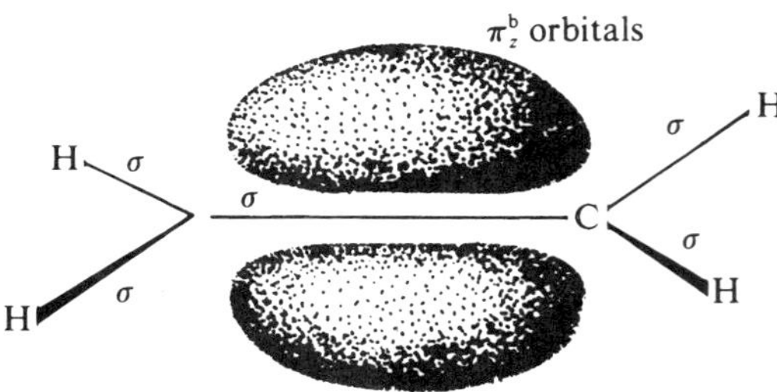

FIG. 1. Double-bond character of ethylene, C_2H_4. All σ bonds are represented by straight lines. The p_z atomic orbitals of the C atoms overlap to form the bonding molecular π_z^b orbital.

derstanding the formation of such double-bonded carbon structures as ethylene (see Fig. 1) and polyacetylene [see Fig. 2(a)]. More important for this work is the type of structure found in benzene, anthracene, and the entire family known as conjugated aromatic compounds. The p_z orbital of each carbon atom in the anthracene molecule, as an example, is perpendicular to the plane containing all the carbon atoms, and the spacing between the carbon atoms is such that there is an overlap between neighboring p_z orbitals, creating what is called a molecular π bond. The π bond establishes a delocalized electron density above and below the plane of the carbon atoms with zero π-electron density in the nodal plane coinciding with the plane of the molecule. The C–C interatomic distance in anthracene and other conjugated molecules is small enough so that the π-electron system is delocalized over the entire molecule. In contrast, the sp^2 orbitals generate highly localized electron density in the plane of the ring, and between the carbon and hydrogen atoms. The new form of carbon referred to as buckminsterfullerene, or fullerene, C_{60}, is neither a linear polyacene nor a hydrocarbon. Nevertheless, in its crystalline form, it behaves like a typical organic molecular solid, with electronic prop-

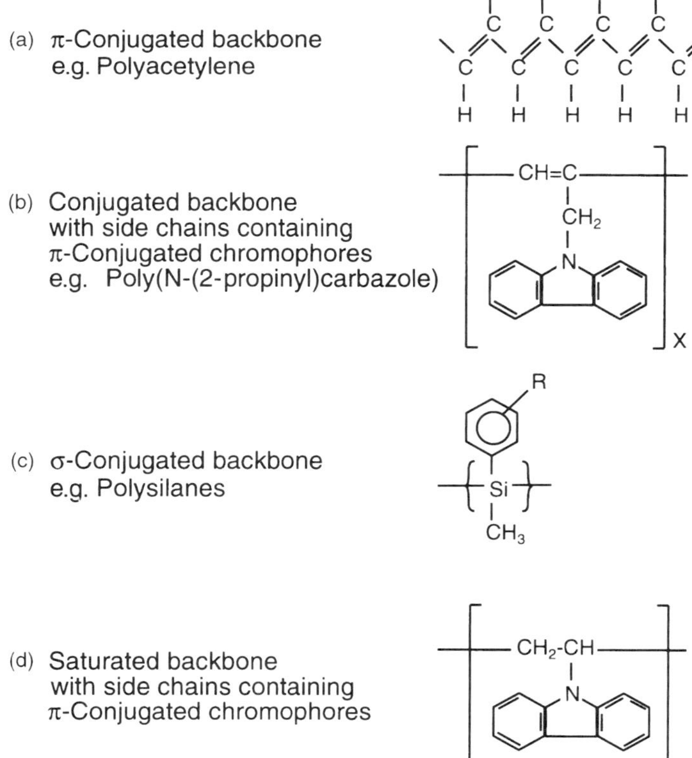

FIG. 2. Three characteristic groups of polymeric organic semiconductors. (a), (b) Polymers with completely π-conjugated backbones, and some with π-conjugated pendant groups as well; (c) polymers with σ-conjugated backbones like the polysilanes and some with σ-saturated backbones and π-conjugated pendant groups; (d) polymers with a saturated backbone with π-conjugated pendant groups, such as polyvinylcarbazole (PVK). (Redrawn from Nešpůrek, 1993.)

erties similar to those of anthracene (see Sec. 3.4.2.1). It is therefore included in the class of organic semiconductors.

In addition to the ringed aromatic compounds, there is the important class of conjugated polymeric compounds, an example of which is polyacetylene, the simplest linear conjugated polymer [see Fig. 2(a)]. In studying the electronic properties of these carbon-backbone conjugated compounds, it will be sufficient to focus on the properties of the π electrons, to the exclusion of the σ-electron network, since σ electrons are most tightly bound and thus do not participate in transport. The π electrons are in the highest-energy occupied molecular orbitals (HOMO) and are the most easily excited or ionized without disrupting the σ bonds that hold the molecule together. Since the lowest-energy unoccupied molecular orbitals (LUMO) are also π orbitals, the electronic transitions from the HOMO to the LUMO are referred to as π–π^* transitions. These transitions can result in singlet or triplet excited states, depending on whether the spin of the electron remains the same or changes orientation upon excitation.

1.2 π-Electron Solids

1.2.1 Organic Crystals As a crystal, anthracene is the most thoroughly studied conjugated aromatic hydrocarbon insofar as its electronic properties are concerned. This crystal is a typical molecular crystal; it is soft, easily deformed, and has a relatively low melting point. However, despite the weak interaction between the molecules in the crystal, there is sufficient interaction between neighboring molecules to create a band of crystal energy levels. This band of states is referred to as an exciton band (see EXCITONS) and is depicted graphically in Fig. 3. Anthracene has two inequivalent molecules per unit cell, and the interactions among these molecules produce a divided band of crystal states referred to as a Davydov splitting (see EXCITONS).

1.2.2 Polymers The polymers that are most important as semiconductors and photoconductors can be classified into three groups (Nešpůrek, 1993). These are shown in Fig. 2; in Figs. 2(a) and 2(b) are polymers with completely π-conjugated backbones,

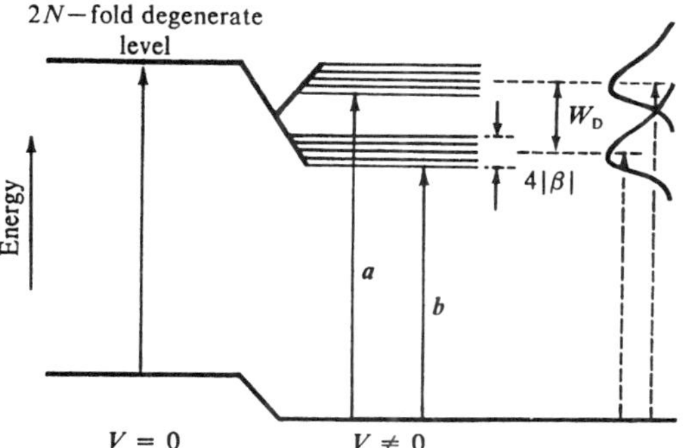

FIG. 3. Splitting of the $2N$-fold degenerate level for a crystal with two inequivalent molecules per unit cell into two distinct Davydov bands. The polarizations of each band are denoted by a and b. The energy separating the two $k = 0$ states is W_D, the Davydov splitting. Its magnitude arises from inequivalent intermolecular interactions only.

and some with π-conjugated ligands as well. These become highly conductive when doped with donors or acceptors. As photoconductors, they operate in the visible and near infrared. Shown in Fig. 2(c) are polymers with σ-conjugated backbones, such as the polysilanes. These are good photoconductors, and have hole mobilities of the order 10^{-8} m^2 V^{-1} s^{-1}. Polymers like poly(N-vinylcarbazole) with a saturated (σ-bonded) backbone and conjugated side chains [Fig. 2(d)] are normally photoconductive in the UV.

1.2.3 Molecularly Doped Polymers Polymers have highly desirable mechanical and chemical properties such as flexibility, strength, and chemical stability. It is difficult, however, to synthesize polymers at will with any of a list of desirable electronic properties. On the other hand, there are monomeric molecules with desirable electronic properties, but with none of the properties of polymers. By dissolving or dispersing short-chain molecules in a stable polymeric matrix, it is possible to combine some of the best properties of both polymer and dopant. The great advantage of the molecularly doped polymers is that they can be prepared in the form of thin films, and their mobilities can be adjusted from 10^{-13} to 10^{-7} m^2 V^{-1} s^{-1} by varying the concentration of the active species. More significantly, they can be rendered effectively trap free for the majority carrier by judicious choice of the dopant (see Sec. 5.3).

As a host polymer, a favorite is a polycarbonate (commercial name Lexan). The dopant can be an electron acceptor, an electron donor, or a combination of both. A typical example of a molecularly doped polymer is a dispersion of *N*-isopropyl carbazole (NIPCA) in Lexan. Lexan by itself is an excellent insulator. When NIPCA is dispersed in the Lexan in a concentration of $\sim 10^{26}$ molecules m^{-3}, then, under the action of light, hole generation and transport are observed; the photoexcited dopant ionizes into a hole and electron but only the hole is mobile. Hole transport takes place by hopping from molecule to molecule of NIPCA.

1.2.4 Langmuir–Blodgett (LB) Films The LB film technique makes possible the construction of thin films with a thickness varying from that of one or two molecules to many tens of molecules, such that each layer can possess carefully tailored properties (see Zasadzinski *et al.*, 1994). The production of thin films is of considerable technological importance (see THIN FILMS). By use of the LB technique, a multilayer structure can be assembled in which essentially two-dimensional (2-D) organic semiconductors separated by insulating spacers form a sequence of quantum wells (see COHESIVE ENERGY) that behave like those in inorganic systems (SC, p. 195; Roberts, 1990).

1.3 σ-Bonded Solids

1.3.1 Polysilanes Unlike the conjugated aromatic π-bonded organic compounds that constitute almost the entire group that possesses semiconducting properties, there is the group of polysilanes (and polygermanes). In the polysilanes, an example of which is shown in Fig. 2(c), the σ bonds along the Si–Si backbone are delocalized. These σ-bonded polymers exhibit many of the properties of the π-bonded compounds. These include a high fluorescence quantum efficiency (~ 0.5 in some compounds) and excellent photoconductivity. A simplified picture of the basis for delocalization is the following: In a polysilane polymer there are approximately sp^3 orbitals facing each other on neighboring Si atoms (called vicinal Si orbitals) and weaker interactions between approximately sp^3 orbitals on the same Si atom (called germinal Si orbitals). Ignore for the present the sp^3 orbitals that are attached to the ligands on each Si atom. The interacting vicinal orbitals split into strongly localized σ bonding and antibonding orbitals. The weaker geminal interactions connect the entire linear chain of localized bond orbitals with each other, creating delocalized orbitals over the entire Si backbone (see also Sec. 5.1.5) (Miller and Michl, 1989; Kepler and Soos, 1993).

1.4 Fullerenes, C_{60}

In 1985, a remarkable discovery was made of a new allotropic form of carbon, called buckminsterfullerene or fullerene, having the molecular formula C_{60}. It has a structure resembling a soccer ball; it is a truncated icosahedron consisting of 32 faces, of which 20 are hexagonal and 12 are pentagonal. The compound is aromatic in its chemical properties. The solid is a typical molecular crystal and the crystalline material is an organic semiconductor. The molecule has a large electron affinity, and so it can trap electrons when added as a dopant to other systems (see Sec. 3.4.2.1 for additional data). When alkali-metal atoms are diffused into a C_{60} film, ionic compounds are formed, some of which are superconductors at the relatively high temperature of 33 K and ambient pressure (e.g., $RbCs_2C_{60}$) (see Ramirez, 1994).

2. ELECTRONICALLY EXCITED STATES

A complete discussion of optically excited states in organic molecular crystals is outside the bounds of this article. However, an important characteristic feature of these solids is the ability of the optical excitation to migrate inside the crystal in the form of an exciton. The ground state of the molecule is usually a singlet state. Nevertheless, there is great interest in organic molecules that have triplet ground states, with the view of preparing ferromagnetic organic solids (Miller and Epstein, 1994).

2.1 Excited States of Aggregates of Molecules

2.2.1 Excitons Many, but not all, spectral properties of small aggregates of molecules and crystals are directly traceable to properties of individual molecules. Weak as the interaction between molecules may be, it imposes a communal response upon the aggregate; the collective response to electronic excitation is embodied in a quasiparticle called an exciton (see EXCITONS). In conjugated organic solids, exciton properties are greatly affected by the size of the exciton, defined as the average distance between the electron and hole. These limiting states are called Frenkel and Wannier excitons, with charge-transfer (CT) excitons occupying an intermediate position.

2.2.1.1. Frenkel Excitons. In a solid containing a high concentration of identical chromophores (see the Glossary), the electronic excitation of any particular chromophore (neglecting surface effects) produces the same electronic state of the solid. This final crystal state is called a Frenkel exciton. Frenkel excitons are electrically neutral and represent the limiting case where the electron–hole pairs are located primarily on the same molecule, moving as a unit from molecule to molecule. In the case of polymers, the molecule is so large that distinctions among the various excitonic types blur. When pairs are far apart, a Frenkel exciton can behave like a polaron pair (see POLARONS). The exciton can have singlet or triplet spin character, depending on the excited state of the molecule. The exciton bandwidth ($4|\beta|$ in Fig. 3) is determined by the intermolecular forces. The bandwidth of a singlet exciton in anthracene is about 200 cm^{-1}, whereas for the lowest triplet exciton, it is $\sim$20 cm^{-1}. These bandwidths are quite anisotropic as a result of the anisotropy of the lattice. Since they are mobile, these excitons are capable of dissociation at an electrode, a defect, or a surface to produce a charge carrier. They are thus important in photoconductivity. As for polymers, the backbone of the π-conjugated polymers such as substituted poly(p-phenylenevinylene) (PPV) and σ-bonded polysilanes also supports excitonic motion, as do the pendant π-conjugated groups, known as chromophores; fullerene, C_{60}, also exhibits excitonic properties (Vardeny and Rothberg, 1994, p. 187).

2.2.1.2 Wannier–Mott Excitons. In contrast to the Frenkel exciton, the Wannier–Mott exciton is not confined to an individual molecule, but consists of a widely separated hole and electron pair that are still bound to each other by the Coulomb interaction. The energy levels and corresponding absorption bands associated with the Wannier–Mott states resemble Rydberg transitions in the hydrogen atom. These modified Rydberg levels form an energy-level series (E_n) described by the equation

$$E_n = E_g - G/n^2 \quad n = 1, 2, \ldots,$$

where E_g is the band gap and G is the binding energy of the exciton. The dissociation energy of the exciton is given by the value $n = 1$. Wannier–Mott excitons are important in inorganic solids such as Si, Ge, II–VI compounds, and rare-gas solids; they may play a role in the polysilane polymeric semiconductors. For a discussion of excitons in inorganic solids, see Knox (1963).

2.2.1.3 Charge-Transfer (CT) Excitons. Charge-transfer excitons can be considered as intermediate in size between small (Frenkel) and large (Wannier-Mott) excitons. The hole and its paired electron are on different molecules, but are bound Coulombically to each other and move as a unit through the solid. CT excitons play an important role in carrier generation and recombination in organic semiconductors. These excitons also exhibit a series of excited states consisting of increased electron–hole separations. The energies of these states can be fitted by a Rydberg-type series for electron–hole separations $>$1.0 nm.

2.2.2 Polarons In addition to neutral quasiparticles, there are charged states that can differ significantly from that of a free hole or electron by virtue of their interaction with the lattice phonons (see PHONONS). This interaction results in the polarization of the neighboring electrons and nuclei, and produces a lattice distortion. If the distortion is confined to nearest neighbors, it is referred to as a small polaron (see POLARONS). If the polarization covers a region with a radius much larger than a lattice constant, it is called a large polaron. As discussed in Sec. 5, the lattice distortion that accompanies the carrier modifies the temperature dependence of the mobility.

2.2.3 Solitons In the context of organic semiconductors, a soliton is a mobile chemical structural defect. In *trans*-polyacetylene (*trans*-PA), an interchange of single and double bonds does not change the energy of the system. This is not true for *cis*-PA or with most conjugated polymers. There are therefore two degenerate forms of the ground state of *trans*-PA. In the process of forming *trans*-PA by isomerization of *cis*-PA, the transformation can take place at different points of the *cis*-PA molecule, and so it is possible for the two degenerate forms (A and B in Fig. 4) to meet, producing the defect shown in the figure. This defect consists of two adjoining single bonds with an unpaired electron at the point of junction, and is called a soliton. This is not a completely appropriate appellation, since unlike the rigorously defined soliton, these topological defects cannot pass through each other unchanged. The soliton can move along the chain almost isoenergetically, and it can be charged or uncharged. It can also hop from chain to chain (Heeger *et al.*, 1988). The mechanism of charge transport through other polymers lacking a degenerate ground state cannot depend upon charged soliton transport, or intersoliton hopping (see Sec. 5.1.4).

2.2.4 Bipolarons Another model for transport in polymers proposes the existence of like-charged solitons as pairs on a single chain that remain near each other despite the presence of Coulomb repulsion. The repulsive interaction is overcome by the energy gained by each charge as it polarizes the lattice in its vicinity. These charged pairs are referred to as bipolarons, and transport from one chain to another can take place by the transfer of a pair of charges (bipolaron) from a charged chain to a neutral chain, just as occurs for single charges in normal semiconductors (see Sec. 5.1.4) (Brédas and Street, 1985).

H H H H
| | | |
$[C=C-C=C]_n$

FIG. 4. Bond-alternation pattern of *trans*-polyacetylene showing the connection of the two degenerate forms (A and B) of the molecule. The soliton defect is the junction of the two single bonds. The defect is a radical (unpaired electron) and is neutral as shown. Hydrogen bonds are omitted for clarity.

3. CARRIER GENERATION

In conjugated organic solids, the intrinsic conductivities can vary continuously from that of excellent insulators to superconductors. The mechanism of carrier generation is therefore not general for these solids. The organic carrier-transport solids discussed herein consisting of a single component are essentially insulators in the dark. Unless the band gap is less than 0.5 eV, the density of free carriers produced thermally at room temperature will be less than the number of trapped carriers generated by impurities. This section is devoted to those solids in which the intrinsic dark conductivity is low, below $10^{-3}\ \Omega^{-1}\ m^{-1}$.

Carriers can be generated in organic solids by a variety of methods. It is possible to use special electrodes that will inject carriers of either sign into the solid in the dark. It is also possible to photogenerate carriers by creating excitons that dissociate at an impurity, or react with an electrode. In this case, free carriers of either sign can be generated inside the solid, but not both at the same time in the same place. The examples just cited describe what is referred to as extrinsic carrier generation. Intrinsic photogeneration of carriers in a molecular crystal requires light of energy at least equal to that of the band gap. In a single-component polymer, the individual polymer molecule must be ionized for intrinsic generation; in multicomponent polymers, photogeneration involves the participation of the dopant and is classified as extrinsic.

3.1 Band Theory

In typical inorganic semiconductors, where there is a crystalline lattice and intermolecular interactions are large, there is a well-defined forbidden energy gap E_g separating broad ($\gg kT$) valence and conduction bands. Carrier generation in these solids takes place when an absorbed photon energy exceeds E_g, the quantum efficiency of carrier generation is a weak function of the electric field and temperature, and the mobility μ varies as T^{-n}, $n > 1$ (see Sec. 5).

In typical organic molecular-crystal semiconductors at room temperature, the carrier bandwidths are narrow, in many cases $\ll kT$. This is a consequence of their weak intermolecular interaction energy and low dielectric constants. The Coulomb binding energy between the hole and electron is therefore large and long range. As a consequence, the quantum efficiency of carrier generation is a function of the electric field and temperature. However, Karl (1990) has shown that as the temperature is lowered to liquid He temperatures in ultrapure naphthalene, the one-electron band picture is valid (bandwidths $\gg kT$), at least insofar as the mobility is concerned (see sec. 5).

In linear conjugated-backbone polymers, broad 1-D carrier bands exist along the chain at room temperature. In *trans*-PA, the π bandwidth is ~10 eV. However in the direction perpendicular to the chain, the interaction energy is ~0.1 eV, and in view of the lack of long-range order, an intermolecular band theory is not tenable. The quantum efficiency of carrier generation is greatly decreased by geminate recombination (sec. 4.1). The product of the quantum efficiency of carrier production and the probability of escaping fast geminate recombination is 0.01 (Něspůrek, 1993). The effective generation efficiency therefore depends markedly on the electric field and temperature, as it does in other linear-chain polymers such as poly(*p*-phenylenevinylene) polymers (Gailberger and Bässler, 1991).

The question of whether carrier generation takes place as a result of a valence-to-conduction-band transition (referred to as a band-to-band or BB transition) in organic semiconductors is taken up in Sec. 3.4.2.2.

3.2 Energy Levels

3.2.1 Valence and Conduction Bands

In view of the weak intermolecular interactions, the ionic energy levels in a molecular crystal bear a close relationship to those in the isolated molecule, modified by the polarization energy gained by bringing the molecular ions into the solid phase. In its simplest form, the energy E_g required to produce a pair of free carriers in organic molecular crystals is given by

$$E_g = I_g - A_g - P^+ - P^-, \tag{1}$$

where I_g and A_g are respectively the ionization energy and electron affinity of the isolated molecule, and P^+ and P^- are the polarization energies of the positive and negative carriers inside the solid, respectively; they are typically between 1 and 2 eV (Seki, 1989). Although Eq. (1) is only an approximation, it nevertheless is still a useful guide.

Generally, the polarization energies for the positive and negative carriers have been taken as equal to each other. However because there is an asymmetry in the charge-quadrupole interactions of electrons and holes, the total effective polarization energy of the hole is greater than that of the electron, varying from 0.08 eV for benzene to 0.48 eV for pentacene (see SC, p. 116).

The adiabatic values of E_g for anthracene, tetracene, and pentacene are respectively 4.1, 3.1, and 2.5 eV. These are the energies of the relaxed electrons and holes with respect to each other. The optical band gaps are 4.4, 3.4, and 2.8 eV; these are the optical excitation energies required to produce the separated holes and electrons in their unrelaxed polarization states.

As for polymers, Eq. (1) is not useful. Carrier generation is observed at the absorption edge in the carbon-backbone polymers. However, in these systems, where the generation efficiency and carrier mobility are strong functions of the field and temperature, it is difficult and perhaps irrelevant to claim that a BB transition rather than an extrinsic process is responsible for the conductivity (see Sec. 3.4.2.2).

3.3 Single Positive or Negative Carriers in Organic Solids

3.3.1 Generation Mechanisms

3.3.1.1 Dark Injection from an Electrode. When an electrode is placed in contact with an organic semiconducting solid, there is a redistribution of charge that accompanies the equalization of the Fermi energies of the two phases. Let Φ_M be the work function of the electrode, Φ_I the work function of the insulator, and χ (also referred to as A_C) the electron affinity of the insulator, as is shown in Fig. 5. If the ionization energy of the solid is I_c, the energy required to move an electron from the insulator into the electrode is $I_c - \Phi_M$, whereas the energy required for the opposite process is $\Phi_M - \chi$. If $I_c - \Phi_M < \Phi_M - \chi$, then surface electrons will shift from the insulator into the electrode, making the electrode negatively charged with respect to the insulator.

The energy difference $\phi_0 = \Phi_M - \chi$ is the smallest barrier that must be overcome by an electron in the electrode if it is to be injected into the insulator. For holes, there is an analogous value, $\phi_0^+ = I_c - \Phi_M$. Electrons injected into the surface region of the insulator must still overcome a potential barrier in order to enter the bulk of the crystal; this barrier is field dependent, and is shown in Fig. 6 as $\phi_m - \phi_0$.

For anthracene, $E_g \approx 4$ eV, $I_c = 5.8$ eV,

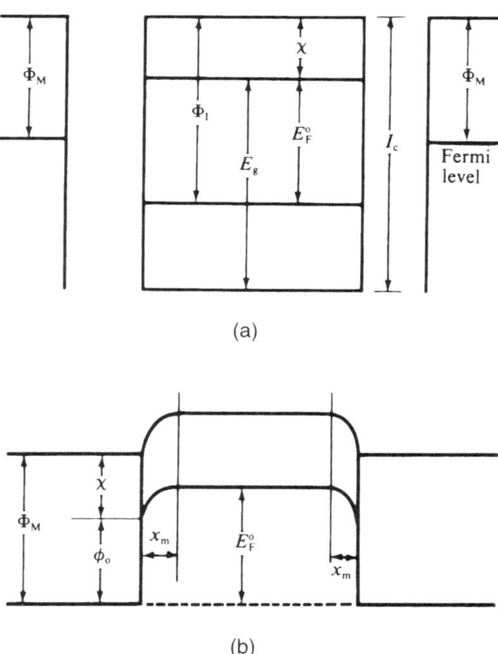

FIG. 5. Energy-level diagram for an insulator with identical metal contacts showing the barrier shape at zero bias (a) before the electrodes are attached to the insulator, (b) after the electrodes are attached. Symbols are defined in the text. (Redrawn from Frank and Simmons, 1967.)

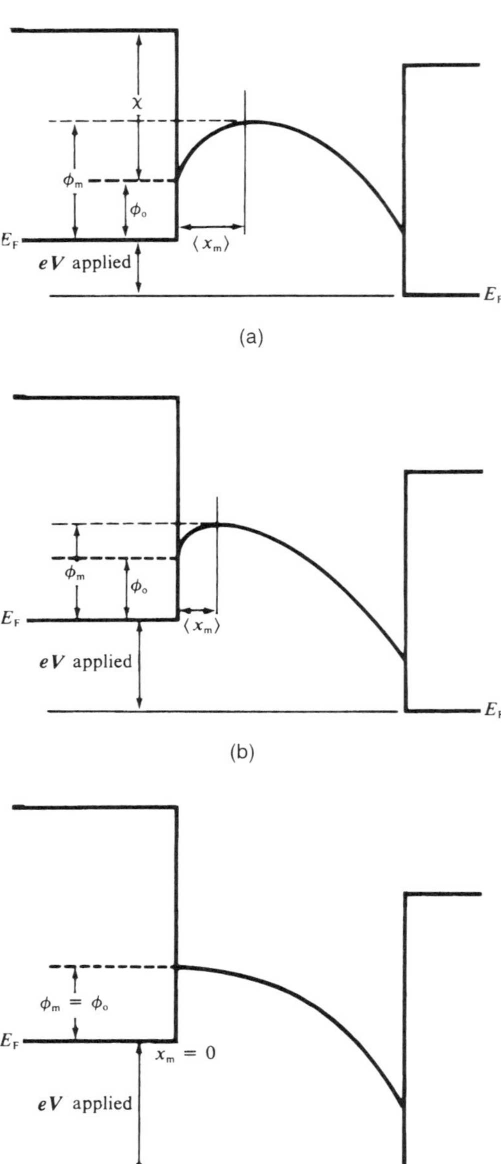

FIG. 6. Effect of increasing the electrical bias on the height and position of the barrier maximum at x_m; (a) small bias; (b) intermediate bias; (c) bias sufficient for the onset of emission-limited current. (Redrawn from Frank and Simmons, 1967.)

and $\chi(A_C) \sim 1.8$ eV; so Φ_I for anthracene should be ~3.8 eV. A good choice for an electrode for electron injection is one with a small work function Φ_M. The alkali metals satisfy this requirement. Sodium (Na) has a Φ_M of ~2.8 eV, and so the barrier to electron injection would be ~1 eV. This is further reduced by a chemical reaction between the Na and anthracene. For hole injection, large–work-function electrodes are required. The noble metals, such as gold (Au), are suitable. The Φ_M of Au is ~5.4 eV, providing a barrier of only 0.4 eV. By using both Au and Na as opposing electrodes on anthracene, electroluminescence is produced (see Sec. 4.3.1).

If the contacting electrode has a low work function, electrons will be injected thermionically into the surface region of the OSC. Equilibrium is established when the rate of injection equals the rate of surface recombination. Upon application of an external field, the Fermi levels of the electrodes are displaced by an amount equal to the external voltage V, and the potential barrier to electrons drops eventually from ϕ_m to ϕ_0 [see Fig. 6(c)]. The surface region of extent x_m serves as an excess-charge reservoir that supplies carriers upon demand. As the applied electric field increases, carriers are forced into the solid at the injecting contact to a maximum determined by the buildup of space charge inside the crystal. This space charge creates an electric field in the opposite direction to that of the external field at the injecting contact, reducing the field strength there virtually to zero. The current that flows under these conditions is called a space-charge–limited current (SCLC), and is the maximum current that can flow in the semiconductor at the applied external field and temperature. A contact that can inject this amount of space charge into the reservoir at the surface is referred to as an Ohmic contact (see Sec. 5.2.1).

In addition to metal electrodes, it has been found that oxidizing electrolyte solutions, such as Ce^{4+}, form excellent hole-injecting contacts (Pope *et al.*, 1962; see Gerischer and Willig, 1976, for a more complete discussion of electrolyte contacts). Currents as high as 1 A m^{-2} were injected in the dark into an anthracene crystal 15 μm thick with 40 V applied.

In addition to the Ohmic contacts just discussed, a contact can be non-Ohmic; i.e., it can be neutral or blocking. In a neutral contact, the energy bands in the insulator remain flat up to the point of contact with the electrode. The concentration of carriers is the same at the interface near the electrode as in the bulk. For a defect-free insulator without surface states, this occurs when $\Phi_M = \Phi_I$. A blocking contact that effectively prevents electron injection, as an example, would be created if in Fig. 5, $\Phi_I < \Phi_M$; in this case, electrons would spill over into the contact, charging the interface negative. This creates an added potential-energy barrier and results in a depletion of electrons in the solid at the interface, relative to the concentration of electrons deeper into the solid.

3.3.1.2 Injection by Optically Excited Electrodes. It follows from Fig. 6 that the barrier to injection of carriers into organic insulators can be overcome if the electrode in contact with the insulator is optically excited. Under these conditions, an electron in the electrode can be excited into the conduction level of the insulator, for electron conductivity, or an electron in the insulator can be excited into the Fermi level of the electrode, enabling hole conductivity in the insulator. The electrode can be a metal, an adsorbed dye, an electrolyte, or a solid mixture of a dye in a polymer.

The condition for electron injection is

$$E_{\text{ext}} + A_C - \chi_e > 0, \tag{2}$$

where E_{ext} is the externally supplied energy, A_C is the electron affinity of the crystal, and χ_e is the electron affinity of the electrode. For metals, χ_e is also the work function Φ_M. The condition for hole injection is

$$E_{\text{ext}} + \chi_e - I_C > 0, \tag{3}$$

where I_C is the ionization energy of the crystal.

If the electrode (e.g., a metal) is optically excited, photoemission of electrons from the metal into the organic solid may be induced in the same way as into a vacuum, except that much lower energies are required.

3.3.2 Exciton Interactions Excitons represent mobile excited states that transport the potential energy of optically excited molecular states. The singlet and triplet excitons are particularly well studied. In anthracene, the singlet exciton transports 3.1 eV and the triplet exciton transports 1.8 eV; hence, this amount of energy can be transferred in exciton dissociation at electrodes, impurities, and surface molecules (see Sec. 3.4.2.2). The same processes can take place in polymers that support exciton transport, which includes all the important single-component systems.

3.3.2.1 Exciton Dissociation at an Electrode. Exciton dissociation is a major mechanism for photoconduction when an electrode is placed in direct contact with an organic semiconductor. The energy-level scheme shown in Fig. 7 shows that both electron and hole injection are energetically possible in anthracene by the reaction of excitons with the electrode. The exciton can transfer its energy via a Förster dipole–dipole mechanism to an electron inside the metal. This can take place over a range of about 3 nm. For triplet excitons, energy can be transferred over a distance of 1.8 nm by electron exchange. Similar behavior is noted in most polymers that support exciton motion, as in polysilanes (Kepler and Soos, 1993).

3.3.2.2 Exciton Dissociation at Impurities. The presence of electron or hole acceptors at the surfaces provides additional sites for exciton dissociation. Oxygen is a well-known electron acceptor, and both singlet and triplet excitons can dissociate at these sites, delivering an electron to the O_2 and liberating a hole for transport. The exciton dissociation rate at an O_2 site on anthracene is about 10^{13} s^{-1}. Iodine is also effective in dissociating excitons at a free surface. The oxidation product anthraquinone on anthracene creates a 1-eV electron trap and provides a ready site for singlet-exciton dissociation to release a hole.

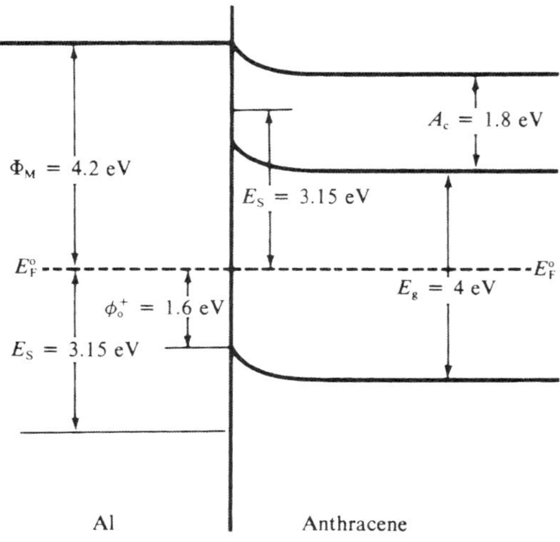

FIG. 7. Energy relationships at the interface of an Al electrode and an anthracene crystal. The singlet energy E_S is shown to be sufficient to cause the injection of either a hole or an electron.

3.3.2.3 Exciton Dissociation at Surface Molecules. Surface-adsorbed dyes are effective as hole injectors into anthracene when they react with singlet excitons. The injection efficiency of a dye depends on the degree of overlap between the singlet fluorescence spectrum and the absorption spectrum of the dye. This overlap determines the efficiency of Förster dipole–dipole transfer (Knox, 1963). The sensitization of charge injection into organic semiconductors finds application in photography and electrophotography. For a more complete discussion of dyes at surfaces, see Gerischer and Willig (1976).

3.3.3 Photovoltaic Effect Consider a system consisting of a layer of a semiconducting material or materials, sandwiched between two opposing electrodes, at least one of which is partially transparent to light. If light is shone through the transparent electrode in the absence of an external field and a voltage difference is produced thereby, then a photovoltaic effect has been demonstrated. One type of photovoltage can result from differences in the diffusivity of the carriers. This is known as the Dember effect and historically was the mechanism for the photovoltaic effect first observed in an anthracene crystal. More important are two other mechanisms for generating a photovoltage. In both cases, advantage is taken of a potential barrier that is present in the dark as a result of differences in the work functions of contacting materials. One barrier is called a Schottky barrier, and is set up between the electrode and the semiconductor (see SCHOTTKY BARRIERS). The other is a *p/n*-junction barrier (see LIGHT-EMITTING DIODES), set up between two semiconductors in which the majority carriers are holes and electrons respectively.

When all the contacts are made, the Fermi levels equalize and the band bending seen in Fig. 8 is produced. Light absorbed at the Schottky contact causes electrons to be excited into the conduction level, and the built-in potential forces the electrons into the Schottky contact and the holes into the Ohmic contact. The maximum photovoltage obtainable is the potential difference between the Fermi level and the conduction-band edge. In the *p*/*n* junction, metals A and B are Ohmic contacts, and the potential difference at the interface is caused by the depletion of majority carriers that flow into the regions of opposite majority-carrier conductivity. The maximum photovoltage in this case is the potential difference between the upper edge of the valence band and the lower edge of the conduction band.

Typical examples of normally *p*-type organic semiconductors are phthalocyanines and quinacridone pigments. For *n*-type materials, there are pyrylium dyes, and perylene–3,4,9,10 tetracarboxylic acid. The origin of the basic *p* or *n* character of the organic compounds is not well understood. By doping, it is a straightforward matter to create the required majority-carrier type. Unlike inorganic materials, *p*- and *n*-type organic semiconductors are not generated by the introduction of minute amounts of impurities into the pristine semiconductor. The doping is produced by adding mol% amounts of acceptor or donor molecules into the host matrix. Typical acceptor molecules are O_2, halogens, and tetracyanoquinodimethane. For donor molecules, there are NH_3 and phenothiazine (see Borsenberger and Weiss, 1993).

For a Schottky configuration, the cell would be glass/(Au, Ag, or ITO electrode)/(organic semiconductor)/(In or Al electrode). For *p*/*n* cells, the configuration would be glass/ITO/*p*-type/*n*-type)/(Au or Ag). Overall power conversion efficiencies of 2% have been achieved with *p*/*n* junctions and 1% with a Schottky-type cell. In inorganic systems, efficiencies of more than 20% have been attained.

3.4 Carrier Generation of Pairs in the Bulk

As previously stated, the selection of organic semiconductors to be discussed is based on an arbitrary demarcation level of conductivity of $10^{-3}\ \Omega^{-1}\ m^{-1}$. Thus, all the organic compounds with conductivities $>10^{-3}\ \Omega^{-1}\ m^{-1}$ would fall into the class of

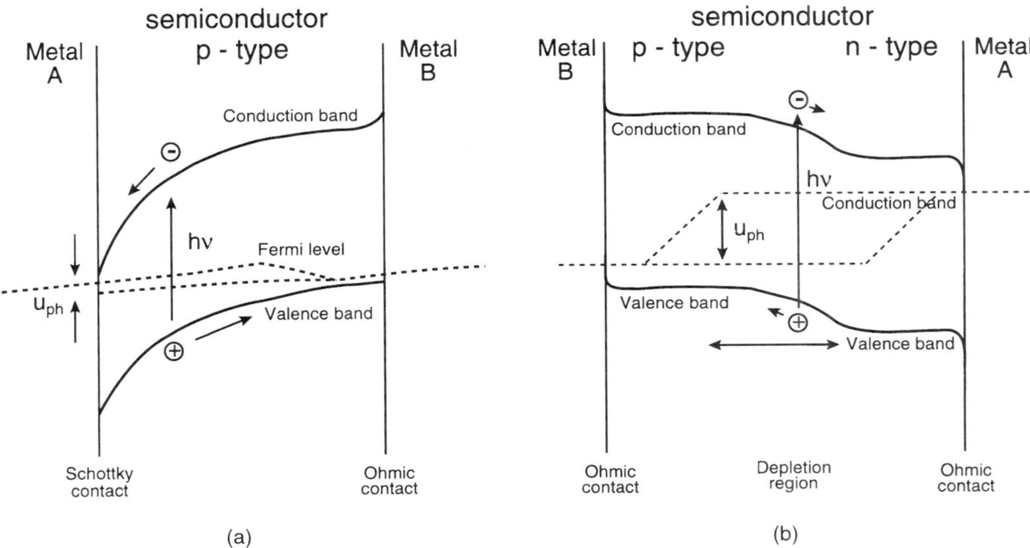

FIG. 8. Schematic of photovoltaic mechanisms. (a) A Schottky contact is formed between metal *A* of low work function and a *p*-type semiconductor. The metal *B* forms an Ohmic contact. (b) A typical *p*-*n* junction. Both metals form Ohmic contacts. The maximum photovoltage in each case is given by U_{ph}. (Redrawn from Wöhrle and Meissner, 1991.)

organic conductors and will not be discussed further. This class includes high-conductivity donor–acceptor complexes such as TTF–TCNQ and doped polymers such as I_2-doped polyacetylene.

3.4.1 Thermal Mechanism All the materials reported on in this section are essentially insulators in the dark, and so intrinsic thermal generation of carrier pairs is negligible.

3.4.2 Photonic Mechanisms

3.4.2.1 Autoionization. The mechanism of bulk photogeneration of carrier pairs in molecular organic crystals has been most extensively studied on anthracene. In the threshold region of photogeneration there is evidence for the direct generation of bound hole–electron pairs, which require thermal activation energy and/or a high electric field for complete separation into pairs of carriers. For photon energies larger than the band gap, a higher excited molecular state is formed, which autoionizes to create a pair of carriers that move apart from each other with excess kinetic energy. However, the efficiency of free-carrier generation is small, $\sim 10^{-4}$, because of efficient carrier recombination (see Sec. 4.1). In passing, it may be noted that the efficiency of carrier generation in crystalline fullerene, C_{60}, is also 10^{-4}. The band gap of this new form of carbon is ~2.2 eV; its hole mobility is 1.74×10^{-4} m^2 V^{-1} s^{-1} and its electron mobility is 0.52×10^{-4} in the same units. The mobility is independent of temperature from 46 to 250 K; this and its other electronic properties are similar to those of other organic molecular crystals.

In addition to carrier generation by the action of a single photon, it is also possible to generate carrier pairs by a multiphoton process involving low-energy photons. The final excited state produced by multiphoton excitation autoionizes to generate the carriers as in the case with single-photon excitation.

In polymers consisting of a saturated backbone and π-conjugated pendant chromophores, e.g., poly(*N*-vinylcarbazole), or a conjugated backbone with π-conjugated pendant chromophores, e.g., poly(*N*-2 propinyl) carbazole, photogeneration proceeds as with monomeric crystals, namely by autoionization followed by Onsager-type geminate recombination (see Sec. 4.1) (for a review, see Nešpůrek, 1993, and Borsenberger and Weiss, 1993).

3.4.2.2 Band-to-Band Transitions. In molecular crystals of the polyacene type, electron and hole bands in ultrapure materials have been demonstrated to exist at $T \sim 4$ K by Karl (1990). At higher temperatures, the thermal energy exceeds the bandwidth and band theory is not applicable. In some polymers, like *trans*-PA and polydiacetylenes (PDA), there is evidence of a direct valence-to-conduction-band transition. While the lowest π–π^* transition in *trans*-PA along the chain may be BB (Sariciftci and Heeger, 1994), in which an intrachain widely separated electron–hole pair is formed, the lowest π–π^* transition in PDA is excitonic. In *trans*-PA, in ≤ 1 ps, the carriers polarize their surroundings, becoming charged solitons, which neutralize and decay into phonons in <1 ps. In PDA, the oppositely charged species undergo geminate recombination in subpicosecond times. Since the lowest π–π^* excitation in polydiacetylene is excitonic, any conductivity that is associated with the absorption edge must be attributed to an external agency that supplies the binding energy of the singlet exciton, which is about 0.5 eV. This could be an impurity or a large electric field.

The lowest π–π^* transition in most polymers is not generally a BB transition (see Bässler *et al.*, 1995). In PPV and derivative polymers, the lowest π–π^* transition is a neutral intrachain exciton. This exciton can become an interchain exciton of CT character, dissociating into carriers under the action of an external field and thermal energy. Geminate recombination in all cases limits the carrier-generation efficiency (see Sec. 4).

In poly(methylphenylsilane), the lowest σ–σ^* transition at 3.5 eV generates a singlet exciton. There is no evidence of polaron formation. The threshold for direct carrier generation is at ~4.6 eV, producing holes and electrons. Only the holes are mobile, and the transport is nondispersive. The efficiency of carrier generation is $\sim 10^{-2}$ (Kepler and Soos, 1993).

There is as yet no unanimity in the field regarding the relative importance of BB (Sariciftci and Heeger, 1994) as opposed to excitonic mechanisms for carrier generation. However, if by a BB transition is meant the generation of an uncorrelated hole–electron pair that is free to dissociate even under the influence of a small external field, and with a weak temperature dependence, this has not yet been observed in organic semiconductors at room temperature.

3.4.2.3 Excitonic Mechanisms. When two singlet excitons come within a critical distance ~1 nm from each other (known as the Förster transfer distance), a transfer of energy from one exciton to the other can occur (see EXCITONS) The final state is a highly excited exciton that possesses sufficient energy to ionize. In anthracene, the singlet energy is 3.15 eV, so that the exciton fusion produces 6.3 eV. This is greater than the band gap of ~4 eV, and is even greater than the ionization energy of the crystal, 5.8 eV. The exciton–exciton fusion rate constant is $\sim 10^{-14}$ m^3 s^{-1} (diffusion-limited) and the probability of ionization is $\sim 10^{-4}$. A remarkably similar excitonic behavior has been observed in the polysilanes.

A singlet exciton can also excite a triplet exciton, yielding an energy of ~4.9 eV. The triplet–singlet collision rate constant is $\sim 5 \times 10^{-15}$ m^3 s^{-1}, with about the same probability of ionization as in the singlet–singlet case. Furthermore, photons can ionize either singlet or triplet excitons. The ionization cross section for the photon–singlet is $\sim 10^{-23}$ m^2, and for the photon–triplet is $\sim 10^{-25}$ m^2.

A direct two-photon ionization process can take place starting at a photon energy of 2 eV in anthracene. For two 3.1-eV photons, the cross section for carrier generation is $\sim 10^{-32}$ m s. These cross sections are typical of other organic molecular semiconductors.

4. CARRIER RECOMBINATION

In organic semiconductors, carrier recombination is a highly efficient process, and it limits the overall efficiency of carrier generation to a degree not approached in inorganic semiconductors. The reason for this is twofold. First, because of the generally low dielectric constant of most organic semiconductors, there is very little shielding of the Coulombic force between the electron and hole of the newly generated *e–h* pair. The attractive force is therefore potent at separation distances of as much as 17 nm in the *e–h* pair. Second, the intermolecular bandwidths for the carriers are narrow, implying low mobilities and frequent carrier-scattering events. This scattering facilitates energy loss in the initially created *e–h* pair with excess kinetic energy. The task of complete separation thus will depend on the external field strength and the temperature.

4.1 Geminate Recombination—Onsager Theory

Geminate recombination is the recombination of an electron with a hole, both of which came from the same molecule. Such recombination in OSCs dominates the recombination process for all but the highest concentration of *e–h* pairs. In contrast to bimolecular recombination between uncorrelated *e–h* pairs, geminate recombination is kinetically first order (see CHEMICAL KINETICS).

Geminate recombination was given a firm theoretical basis by Onsager (1938), who treated the problem as it existed in a liquid, and the theory was applied in a definitive manner to organic solids by Chance and Braun (1973). The Onsager theory gives the probability that a thermalized charge pair will escape geminate recombination and dissociate under the influence of an external electric field of strength E.

The Onsager theory requires a knowledge of the thermalization distance r_{th}, the temperature, and the dielectric constant of the medium. With these parameters and a knowledge of the initial quantum yield of electron–hole pairs, the field and temperature dependence of the free-carrier yield can be calculated. An important quantity is the Coulomb-capture distance r_c. This is the distance of separation of the *e–h* pair at which point the electrical potential energy is equal to the thermal energy. This distance is given by

$$r_c = e^2/4\pi\epsilon\epsilon_0 kT \equiv 2q. \tag{4}$$

The average thermalization distance $\langle r_{th} \rangle$ is defined as the mean distance traversed by the hot carrier as it comes to thermal equilibrium with its surroundings, and is usually not known. There is a distribution of thermalization distances and no single distribution function fits all occasions. The energy-loss mechanisms for a highly excited initial state vary. Ionization may take place from the initial state, in which case $\langle r_{th}\rangle$ will increase with the energy of excitation, until an energy-loss mechanism is encountered in the medium, such as a vibrational excitation. The initial state may undergo an internal conversion to some lower-lying state, which then dissociates, in which case $\langle r_{th}\rangle$ will be independent of the initial energy of excitation.

The Onsager relationship for the low-field case (applicable for most situations) is

$$f(r_{th},\theta) = [\exp(-A)](1 + AB), \tag{5a}$$

where $A = 2q/r_{th}$, $B = \beta\, r_{th}(1 + \cos\theta)$, $q \equiv e^2/8\pi\epsilon\epsilon_0 kT$, $\beta = eE/2kT$, θ is the angle between the radius vector $\mathbf{r}$ and the applied field vector $\mathbf{E}$, and $f(r_{th},\theta)$ is the probability that the charge pair will dissociate as a function of the separation distance and the angle between the applied field and the radius vector. For an isotropic medium of dielectric constant $\epsilon = 3.02$, and an applied field of 10^5 V m^{-1}, r_{th} would have to be 45 nm to have a probability of 0.7 for escaping geminate recombination. At 10^7 V m^{-1}, r_{th} could be 8 nm and still yield the same probability (0.7) for escape.

To calculate the quantum yield as a function of the external field $\Phi(E)$ using Eq. (5a), it is necessary to have a proper distribution function for r_{th}. If $g(r_{th},\theta)$ is the probability per unit volume of finding the ejected electron in a volume element $d\tau$ at (θ,r_{th}), then, assuming an isotropic medium and a spherically symmetric $g(r_{th},\theta)$, the quantum yield would be

$$\Phi(E) = 4\pi\Phi_0(1 + 2\beta q) \times \int_0^{\infty} [\exp(-2q/r_{th}]g(r_{th})r_{th}^2 dr_{th}, \tag{5b}$$

where Φ_0 is the primary quantum yield of carrier pairs per absorbed photon and is generally not known. It is possible to get an estimate of Φ_0, which can help distinguish among various functions for $g(r_{th})$ (see SC, p. 259, and PS, p. 494). The most used functions are the delta function, the Gaussian, and the exponential. For anthracene, in the photon-energy range 4.4–5.2 eV, the values of Φ_0 based on various assumptions in the use of the Onsager equation (see PS, p. 494) are 5×10^{-3} for a delta-function $g(r_{th})$, 10^{-2} for a Gaussian, and 1.4×10^{-1} for an exponential. For a photon range of 5.4–6.2 eV, the values of Φ_0 are 3.4×10^{-5} for the delta function, 5.2×10^{-3} for the Gaussian, and 3.4×10^{-2} for the exponential. The constancy of Φ_0 over a wide photon-energy range is indicative of a fixed lower-lying state that autoionizes to yield the carriers.

The Onsager theory predicts a value of the ratio of the slope to the intercept (S/I) of a curve such as that shown in Fig. 9. This is given by the expression

$$\begin{aligned} S/I &= 2\beta q/E = e^3/8\pi\epsilon\epsilon_0 k^2T^2 \\ &= 3.38 \times 10^{-7}(298/T)^2 \text{ m V}^{-1}. \end{aligned} \tag{6}$$

Taking $\epsilon = 3.23$, the experimental value of the S/I ratio for electrons is $(3.21 \pm 0.12) \times 10^{-7}$ m V^{-1}, and for holes it is $(3.02 \pm 0.08) \times 10^{-7}$ m V^{-1}. The temperature dependence of the value of S/I at a fixed excitation wavelength is in agreement with the Onsager prediction.

The evaluation of $\langle r_{th}\rangle$ was based on the model shown in Fig. 10 and an assumption of a delta function for $g(r_{th})$. As may be seen, when using light of energy less than E_g, it is possible to populate intermediate e–h pair states either directly or by autoionization from some higher-lying level. It follows then that there will be a thermal energy of activation for photogeneration defined as E_a^{ph} that must be supplied in order to produce free carriers. If measured as a function of photon energy, one obtains

$$\langle r_{th}\rangle(h\nu) = e^2/4\pi\epsilon\epsilon_0 E_a^{ph}(h\nu). \tag{7}$$

The results for anthracene are shown in Fig. 11. A more sophisticated modification of the Onsager theory was made by SC (1994), who included the effect of electric field and temperature on the thermalization distance. The Onsager theory, however, is still in wide use.

In dealing with geminate recombination in some polymers, it is found that another theory propounded by Onsager in 1934 is more appropriate (Braun, 1984). This theory gives the same low-field S/I ratio as the 1938 theory, but it predicts the increased dissociation efficiency of a CT state in the presence of an electric field. Such near-neighbor CT states are often the initial photogenerated

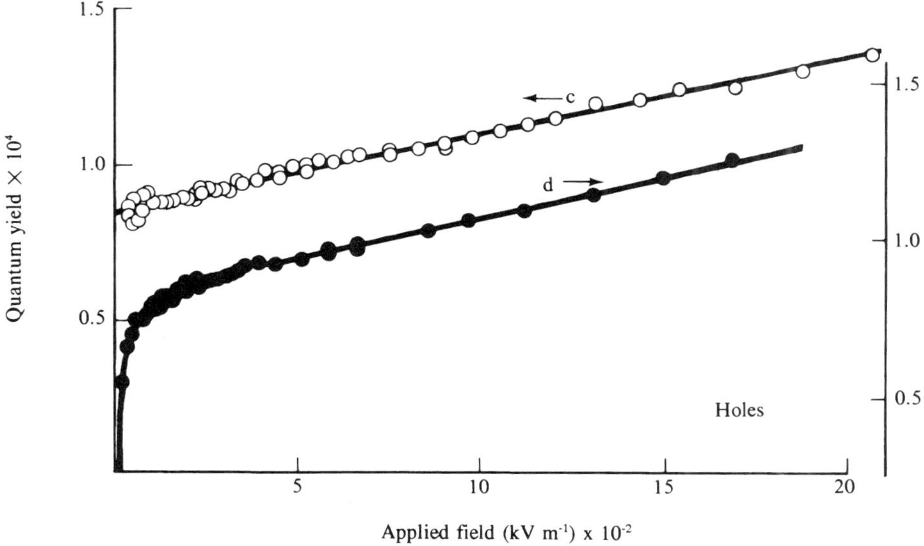

FIG. 9. Hole-carrier quantum yield at 255 nm as a function of the applied field. Curve *c* is for a previously unirradiated crystal sample; in curve *d*, the trapped electrons, left behind from the earlier curve-*c* experiments, reduce the low-field free-hole yield. Note that for clarity of presentation, the origins of curves *c* and *d* are different. (Redrawn from Chance and Braun, 1973.)

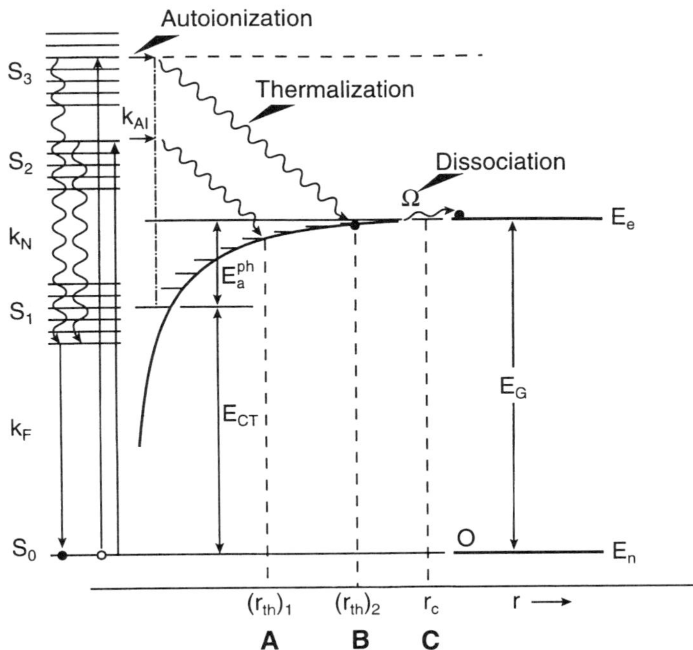

FIG. 10. Schematic diagram of the multistep photogeneration process in organic molecular crystals. Molecular levels in the crystal are shown at the left for scale purposes. The ground state S_0 is shown with one electron in some excited state S_n. The various processes of autoionization (AI), thermalization, and final dissociation of an electron–hole pair state are depicted. The curve depicts the Rydberg-type energy levels of the CT exciton. These levels can be populated directly in the subthreshold region, and by thermalization as shown. E_{CT} is the energy of the CT state, and E_a^{ph} is the activation energy required to dissociate the CT state. *A*, *B* represent various thermalization distances; *C* is at the Coulomb-capture distance. E_G is the band gap. (Adapted from SC, p. 256.)

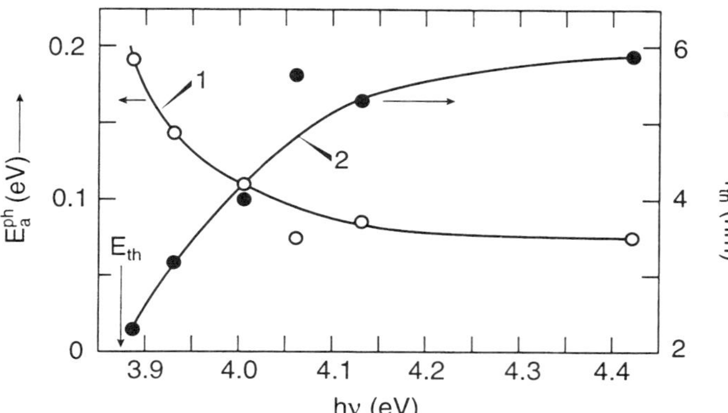

FIG. 11. Experimental $E_a^{ph}(h\nu)$ dependence in an anthracene crystal (curve 1) and the corresponding $r_{th}(h\nu)$ (curve 2) from Eq. (7) (redrawn from SC, p. 265).

states (see Borsenberger and Weiss, 1993, p. 91).

The original Onsager theory was based on a condensed-phase continuum. However, a crystal is not a continuum, and other models have been proposed (for references, see Borsenberger and Weiss, 1993, p. 93). In addition, the original Onsager theory was developed for 3D materials, whereas in polymers such as *t*-PA and PPV, the initial carrier generation takes place within a single 1D polymeric chain. The thermalization distance for the carriers in 1D may exceed the Coulomb capture radius and still the carriers will not escape geminate recombination. For such situations, a 1D Onsager theory has been derived (see Gailberger and Bässler, 1991, for discussion and references).

4.2 Geminate Recombination—Time-Dependent Theory

The Onsager (1938) theory of geminate recombination is a steady-state theory, and has as one of its boundary conditions that recombination occurs with unit efficiency when the distance separating the *e–h* pair reaches a specific minimum value. Both of these conditions were removed by Hong and Noolandi (1978; see also Noolandi, 1982), who derived an analytical time-dependent theory with variable recombination efficiencies at the origin. In addition to the analytical theory, there have been computer simulations of time-dependent geminate recombination by Ries *et al.* (1983).

4.3 Bimolecular Recombination

Bimolecular recombination following intrinsic carrier photogeneration in organic semiconductors is relatively unimportant compared with geminate recombination, except under conditions of extremely high light intensity ($\sim 10^{30}$ m^{-2} s^{-1}). However, bimolecular recombination takes place when electrons and holes are injected separately into the same organic semiconductor (double injection), and brought together by the electric field. The rate of the recombination is given by $\gamma_{eh}np$, where γ_{eh} is the bimolecular rate constant, and n, p are the concentrations of holes and electrons, respectively.

Bimolecular recombination was treated theoretically by Langevin under the restriction that the mean free path of the carriers be less than the Coulomb-capture radius r_c. This condition is amply met in organic semiconductors. The calculated value for γ_{eh} in anthracene is 10^{-12} m^3 s^{-1}, in good agreement with the experimental result, $(3 \pm 2) \times 10^{-12}$ m^3 s^{-1}.

4.3.1 Electroluminescence The most dramatic result of bimolecular recombination is the production of light. This radiative recombination is referred to as electroluminescence (EL) (see ELECTROLUMINESCENCE). The first observation of EL caused by radiative recombination in organic crystals was made by Pope *et al.* (PS, p. 93). Double-injection EL in anthracene was produced by Helfrich and Schneider (1965), who used solutions of negative and positive anthracene ions as cathode and anode respectively. The EL is essentially identical with the normal fluorescence.

In addition to the organic molecular crystals, EL has been observed in composite polymeric systems consisting of a dye (e.g., squarylene) dispersed in a plastic matrix (e.g., polycarbonate), and in homogeneous polymers (e.g., polyphenylenevinylene). An outstanding feature of many organic materials is their high fluorescence efficiencies. Charge-injecting electrodes are required to generate the *e–h* pair that will recombine to fluoresce. Electron-injecting electrodes are chosen from low–work-function materials (e.g., Na, Mg, Ca) and hole-injecting electrodes are chosen from high–work-function materials [e.g., Au, CuI, indium tin oxide (ITO)]. At least one transparent electrode is necessary to permit the transmission of the luminescence.

A dominant characteristic of *e–h* recombination in organic semiconductors is that the *e–h* pair passes through an intermediate exciton or other intergap state, which then decays radiatively into a singlet state, or nonradiatively into any one of three triplet states. Thus the maximum EL efficiency per injected carrier pair is 25%, which is a serious limitation (see also Sec. 7.2.)

5. CARRIER TRANSPORT

Transport in inorganic semiconductors generally is band-type; the electron and hole wave functions are delocalized; holes in Ge have a mean free path (at RT) before scattering of approximately 100 nm, a value considerably larger than nearest-neighbor lattice spacing of 0.25 nm. If the carrier wave functions are highly localized, carriers are strongly scattered and motion through the lattice is random-walk–like with short hopping distances.

For each extreme type of motion, the mobility, defined as the drift velocity per unit electric field, has a different magnitude and temperature dependence. In wide conduction bands, the mobility $\mu > 10^{-4}$ m^2 V^{-1} s^{-1} and $\mu \propto T^{-n}$, $n > 1$. For strongly localized carriers, $\mu \ll 10^{-4}$ m^2 V^{-1} s^{-1} and $\mu \propto \exp(-E/kT)$ where E is an activation energy; the band model stresses the collective nature of the conducting states whereas the hopping, or diffusive, model emphasizes the molecular character. Molecular crystals, such as the polyacenes, have mobilities between these limiting cases and exhibit transport characteristics influenced strongly by the anisotropy of the crystal lattice. In polyacenes near room temperature, $\mu \propto T^{-n}$, $n > 1$, whereas the carrier's mean free path λ inferred from both theory and experiment is only of the order of a lattice constant, a limit where temperature-activated hopping is required.

Only recently has a phenomenological model been available capable of quantitatively accounting for such a carrier dependence on temperatures (from 4 to 300 K) and for electric fields $\approx 10^5$ up to 10^8 V m^{-1}, thereby explaining both hot- and thermalized-carrier transport in polyacene crystals. In this model, the quasiparticle effective mass depends significantly on temperature (varying from a few times the electron's free mass m_e at near zero temperature to well over $400m_e$ at room temperature), the sign of the carrier, the direction of transport, and the electric field, at least in the high-field regime (SC, p. 318).

In most polymers, molecular glasses [such as amorphous films of tetracene and other polyacenes (Bässler, 1981)], and doped polymers, charge transport is described by a formalism based on disorder. Disorder is characterized by the terms *diagonal* and *off-diagonal,* as well as *static* and *dynamic.* Diagonal disorder owes its origin to the nonhomogeneous local environment around any given molecular site due, as examples, to chemical impurities, chain termination, or pendant-group orientations, and to thermally induced motion. This introduces a distribution of site energies. This situation in turn leads to a distribution of intersite distances, and therefore transfer energies J. Such effects are termed off-diagonal disorder. It is sometimes convenient to distinguish static (diagonal or off-diagonal) disorder, due to environmental inhomogeneities, from that produced by thermal fluctuations, which is referred to as dynamic (diagonal or off-diagonal) disorder. As a result of these types of disorder, carrier hopping occurs through a manifold of localized states having different positional disorder and different localized energies.

A fundamental distinction between polaron models and disorder formalism is that in polaron transport, the activation energy reflects the deformation energy (large electron–phonon coupling) whereas in the disorder formalism, the electron–phonon coupling is weak and the activation energy arises from transport between sites characterized by static disorder.

5.1 Types of Transport

5.1.1 Band Motion When the nearest-neighbor overlap energy J is large relative to other interactions for the carrier, the eigenfunctions are approximated by Bloch-band states. In this case, in the constant–relaxation-time ($\tau_{\rm rel}$) approximation,

$$\mu = e\tau_{\rm rel}/m^*, \tag{8}$$

where m^* is the constant effective mass and need not be equal to m_e because the lattice phonons renormalize the bare bands. In some solids this can enhance m^* relative to m_e as in Li metal where $m^*/m_e = 1.4$; and in others, it gives an $m^* < m_e$, as in InSb where m^* is isotropic and $\sim 0.014\ m_e$. The temperature dependence of μ is reflected primarily in those microscopic processes that determine $\tau_{\rm rel}$. For acoustic-phonon scattering, $\tau_{\rm rel} \propto T^{-n}$ $(n > 1)$, whereas for optical-phonon scattering $\tau_{\rm rel} \propto n^{-1}(\omega)$ where $n(\omega)$ is the Bose–Einstein distribution. In general, for band-type behavior, the temperature dependence of μ is of the form aT^{-n} (see PHONONS IN CRYSTAL LATTICES).

5.1.2 Electric-Field Effects Equation (8) predicts a carrier drift velocity linear in applied field E, provided m^* and $\tau_{\rm rel}$ are independent of E. However, low-temperature measurements in ultrapure polyacene crystals have shown a nonlinear field dependence for the drift velocity v_d. For $E < 10^6$ V m^{-1}, $v_d \propto \sqrt{E}$, whereas at higher field, v_d saturates. This initial field-dependent decrease in mobility can be rationalized as due to field-enhanced acoustic-phonon–carrier scattering. For band transport, Shockley has shown that

$$\tau_{\rm rel}^{-1}(E) = \tau_{\rm rel}^{-1}(0) \times \{\tfrac{1}{2} + [\tfrac{1}{4} + (3\pi/32)(\mu_0 E/\mu_1)^2]^{1/2}\}^{1/2}, \tag{9}$$

where μ_0 is the low-field mobility and μ_1 is the longitudinal sound velocity, typically of the order of (0.2 to 0.5) $\times$ 10^3 m s^{-1}. At higher field strengths, the energy the carrier acquires from the external field between successive scattering events can exceed the crystal optical-phonon energies, $\hbar\omega_{\rm ph}$; hence the emission of crystal phonons imposes a saturation velocity on the charge carriers,

$$(V_d)_s = (\hbar\omega_{ph}/2m^*)^{1/2}, \tag{10}$$

as has been demonstrated in polyacene crystals.

5.1.3 Hopping Motion When static disorder is negligible, and when J (the nearest-neighbor interaction energy) is comparable to the site-energy differences due to dynamic disorder, the effect of carrier–lattice interactions needs to be considered in forming the first-order transport states. Because of electron–lattice interactions (polarization), a carrier's neighboring atoms are displaced from their equilibrium positions, and these distortions form a potential well around the carrier in which it is (partially) self-trapped. In organic molecular solids, J is of the order of 10^{-2} eV and is small compared with the polarization energy. The composite entity consisting of the carrier and the local distortion is called a polaron (see POLARONS). The size of the polaron will be strongly reflected in the temperature dependence (and field dependence) of hopping models, and many polaron-transport models have been formulated and reported in the literature.

In the adiabatic limit, the effects of lattice polarization impart to the carrier an effective mass

$$m^* = m_e \exp[S(T)], \tag{11}$$

where S is a Huang–Rhys factor that depends on temperature and the coupling constants of the carrier to the phonons. $S(T)$ increases with increasing temperature, and correspondingly, the polaron bandwidth narrows. At room temperature, transport is more appropriately considered as a sequence of hops through the lattice; in this limit $\mu \propto \exp(-E/kT)$ where E depends on temperature and crystallographic direction, and is approximately one-half the polaron binding energy (E_p); i.e., $E \approx E_p/2 - J$. However, in crystalline molecular solids, it is frequently found that the mobility varies as T^{-n}, and in some molecular crystals such as naphthalene and anthracene, the electron mobility in the high-T region is practically T independent in the c' direction. This behavior is not consistent with band theory or polaron transport.

A rigorous microscopic theory for the dependence of μ on T is still lacking. However, the μ-vs-T behavior has been rationalized (SC, 1994) by considering the carrier as a "nearly small" polaron that moves via tunneling from one lattice site to another. In this picture, the activation energy decreases with increasing temperature, and the dependence of μ on T is modeled by the variation of m^* and τ_{rel} with T and E; m^* varies from $\sim 10m_e$ to $\sim 400m_e$ from $T \sim 40$ K to room temperature. The physical model for transport is therefore one of strong scattering ($\lambda \sim a$) coupled with tunneling through a highly anisotropic local-potential barrier.

5.1.4 Transport in Polymeric Materials

Polymers generally do not form crystals that are large enough to study as such, and exhibit transport properties considerably more diverse than single crystals. A notable exception is the class of polydiacetylenes. These polymers are prepared by forming a single crystal of the monomer, after which the crystal is polymerized to produce the final, essentially single-crystal polymer. The mobility of the carriers is typical of those measured in the polyacenes, namely, of the order of 10^{-4} m^2 V^{-1} s^{-1}.

Although small regions of polymers may be crystalline, the overall response of polymers is typical of an amorphous system having a manifold of localized site energies and intersite distances within a highly anisotropic electronic structure. The chainlike structure provides for strongly covalent bonding along the chains and considerably weaker van der Waals interactions between chains. For polyacetylene, the bandwidth of conducting states along the chain is of the order of 10 eV but is only one hundredth of this perpendicular to the chain. A distinct feature of most polymers is their lack of long-range order [static and dynamic diagonal and off-diagonal disorder (see Sec. 5.1.3)], small intermolecular interaction energies, and large polarization energies (~ 1 eV). Thus the charge-transporting sites (which can be either molecules or a polymer-chain segment) have site energies ϵ, and transfer energies J, that are characterized by distribution functions. Most theoretical studies of charge transport in disordered molecular solids and molecularly doped polymers have assumed that the inhomogeneous broadening of the hopping-site energies ϵ has the form

$$\rho(\epsilon) = (2\pi\sigma)^{-1/2} \exp(-\epsilon^2/2\sigma^2), \tag{12}$$

as this is the shape usually observed in the absorption (and fluorescence) bands of disordered organic solids. Here, ϵ is measured relative to the center of the distribution. Equation (12) neglects polaronic effects and assumes that adjacent site energies are uncorrelated. This distribution-function form for the transfer energies is admittedly oversimplified but contains a large measure of validity.

If site-energy fluctuations are sufficiently large that they overcome the delocalizing effects of the (average) transfer energy $\langle J \rangle$, then the ordinarily distinct boundaries of the valence and conduction bands become diffuse, and energy states appear as a tail extending into the forbidden gap. These gap states overlap each other weakly if at all, and injected charge becomes localized. The quantitative localization criterion (Duke, 1982) can be written as

$$\Delta > cz\langle J \rangle, \tag{13}$$

where Δ denotes the rms deviation of $\rho(\epsilon)$, z is the number of nearest neighbors, and c is a dimensionless constant of order unity that depends on the dimensionality of the system. For polymers and molecular glasses the static polarization-induced fluctuations in ϵ nearly always produce localized states. The term Fermi glass is often used to denote systems satisfying the criteria given by Eq. (13), i.e., molecular solids in which all electronic states are localized. Experimental evidence supporting the Fermi-glass description of polymeric systems comes from an analysis of the gaseous- and solid-phase valence photoemission spectra (see Duke, 1982).

A characteristic of molecular glasses at short times (Bässler, 1981) and molecularly doped polymers is the strong electric-field and temperature dependence of their carrier mobilities. Despite the wide diversity in both chemical and physical structure of Fermi glasses, most display a mobility μ for $T_0 > T$ varying as

$$\mu(E,T) = \mu_0 \exp[-(\Delta_0 - \beta E^{1/2})/kT_e], \qquad (14)$$

where Δ_0 is a low-field activation or barrier energy, reflecting the system static diagonal disorder, β is a constant, k is the Boltzmann constant, and $T_e^{-1} = T^{-1} - T_0^{-1}$; T_0 is a characteristic temperature, typically of the order of the glass-transition temperature.

It appears that in polysilanes (and polygermanes) the dependence of μ on E is insensitive to the nature of the pendant groups (either aromatic or aliphatic) as well as the molecular weight of the polymer. This observation has been taken to mean that only the long-chain backbone is involved in the transport. The observation of hopping behavior, despite the delocalized σ orbitals, indicates that the transport states do not extend over the entire chain, but consist of a population of short segments interspersed among longer segments. These short segments provide relatively localized sites. A fit of lnμ(hole) versus $E^{1/2}$ for films of poly(methylphenylsilylene) (PMPS), a saturated-backbone polymer with pendant groups (CH_3 and benzene), is shown in Fig. 12. (*Note:* The nomenclature in the literature concerning the Si-backbone polymers can be confusing. The above-named compounds are referred to by other authors as silanes instead of silylenes; the same equivalence exists between the germylenes and the germanes.) The square-root dependence of μ on E is also characteristic of hole mobilities in poly(methyl-cyclo-hexylsilylene), poly(methyl-*n*-propylsilylene), and poly(methyl-*p*-methoxy-silylene).

The functional dependence of ln μ on E and T given in Eq. (14), initially used for poly(*N*-vinylcarbazole) (PVK), is by no means universal, and there is a lack of consensus as to the origin of this field and temperature behavior. However, to discriminate unequivocally between square-root and linear field dependence requires a span of E measurements greater than one sees in the μ-vs-E plots. Similarly, in most studies, the temperature range investigated is narrow and thus the exp(T^{-1}) dependence can easily be interpreted as an exp(T^{-2}) dependence. In particular, the disorder formalism of Bässler (see Sec. 5.1.6) predicts a mobility $\mu \propto \exp(-T_0^2/T^2)$ for $T_0 < T < T_g$, where T_g is the glass-transition temperature and T_0 is the temperature associated with the onset of dispersive transport (see Sec. 5.1.5). In doped polymers, the range of temperatures is narrow, so that the distinction between T^{-1} and T^{-2} behavior is small. The temperature dependences of the slopes of lnμ vs $E^{1/2}$ for poly(methylphenylsilylene) (PMPS) plotted against T^{-1} are illustrated in Fig. 13. The intercept (at 408 K) is the characteristic temperature T_0 for PMPS.

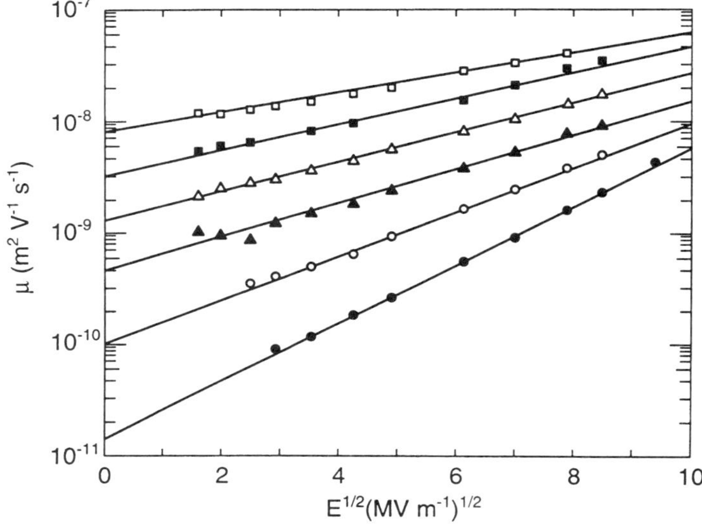

FIG. 12. Hole mobility vs $E^{1/2}$ in PMPS films at various temperatures: ●, 214 K; ⊙, 234 K; ▲, 254 K; △, 274 K; ■, 294 K; □, 314 K (redrawn from Abkowitz *et al.*, 1991).

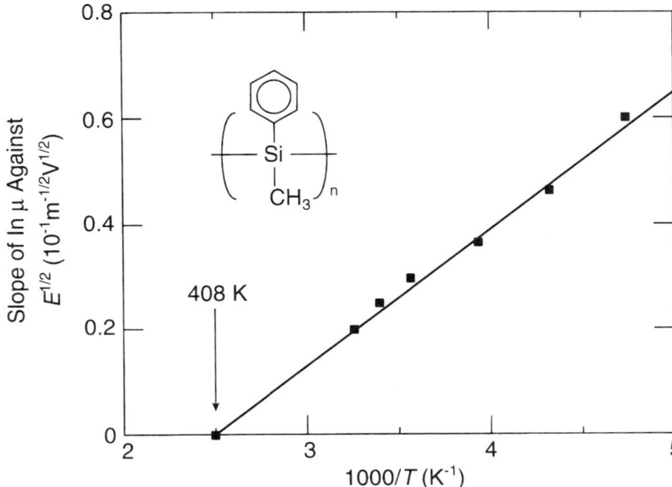

FIG. 13. Temperature dependence of the slopes of lnμ vs $E^{1/2}$. The intercept at 408 K is the characteristic temperature T_0. PMPS shown in insert. (Redrawn from Abkowitz *et al.*, 1991.)

There is as yet no satisfactory theory to rationalize the functional form $\Delta_0 - \beta E^{1/2}$ in Eq. (14), and this is a major unsolved problem in this field.

In contrast to the extensive data on hole transport, there are considerably fewer studies of electron transport because of the difficulties associated with trapping; O_2, an electron trap, is an almost unavoidable chemical impurity. In many systems, the field dependence is similar to that of hole mobilities, e.g., mixtures of 3,3′-dimethyl-5,3′-di-*t*-butyl diphenoquinone (DPQ) in a polymer host, such as polystyrene or polycarbonate.

In addition to holes and electrons, carriers in polymers can consist of charged solitons and bipolarons. In *trans*-PA, the neutral-soliton (see Sec. 2.1.1.2) diffusion coefficient is estimated to be 10^{-10} to 10^{-12} m^2 s^{-1}. It has a spatial extent or size (i.e., a range of lattice distortions associated with the defect) of the order of 10–14 lattice constants, and has an effective mass $\sim 5m_e$. The neutral soliton has an associated spin of $\frac{1}{2}$. Upon doping with donors or acceptors (residing between polymer chains) a soliton can become negatively or positively charged by the transfer of an electron to (or from) its unpaired orbital. The charged soliton is spinless. The literature values of the π–π^* band gap vary from 1.4 to 1.8 eV and the neutral-soliton level is situated in the middle of the gap. Figure 14(a) illustrates the mid-gap positions of a soliton in the neutral, positively, and negatively charged states. When Coulomb interactions are included, the charged-soliton level is no longer at the center of the gap.

In addition to solitons, both *trans*- and *cis*-PA can support polaron states when either electrons or holes are injected into the polymer. Figure 14(b) illustrates the position of polaron states relative to the bands as two states symmetrically disposed about the center of the gap. For the hole polaron, the

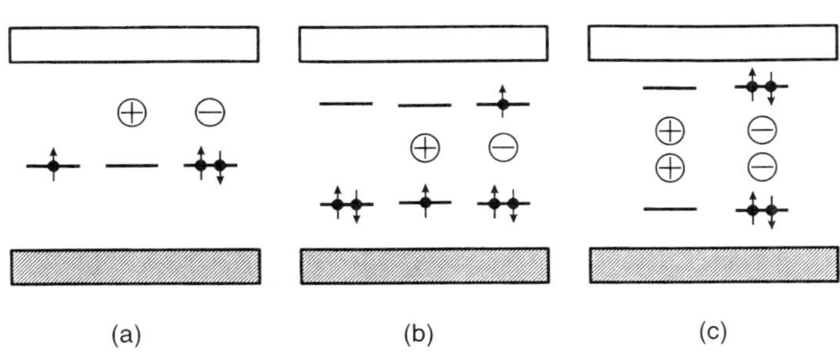

FIG. 14. Positions of the energy levels of (a) solitons in the neutral, positively, and negatively charged states; (b) polarons in the positively and negatively charged states; (c) bipolarons charged positively and negatively within the band gap.

lower level is singly occupied and the higher level is empty. For the electron polaron, the lower level is doubly occupied by antiparallel spins and the upper level is singly occupied. In *trans*-PA, it is energetically favorable for two neighboring polarons in the same chain to interact and form two like-charged solitons (kink and antikink charged states). However, in materials with a nondegenerate ground state, such as *cis*-PA, a widely separated kink–antikink pair is energetically unfavorable (Brédas and Street, 1985); instead, a correlated pair of polarons can form called a bipolaron. Their energies are located within the gap as illustrated in Fig. 14(c). The bipolaron is stabilized by the presence of the strong local lattice distortion, which screens the repulsive Coulomb force in a manner analogous to that exhibited by the Cooper electron pairs responsible for superconductivity (see SUPERCONDUCTIVITY, LOW-TEMPERATURE). In *trans*-PA, the interchain hopping of a single charged soliton necessitates surmounting a relatively large potential-energy barrier because of the large number of carbon atoms on each chain that must relax to new equilibrium positions. For this reason, a pair of closely spaced solitons can more easily hop from one chain to another chain since a smaller number of carbon atoms on each chain have to relax. This transport process is illustrated in Fig. 15.

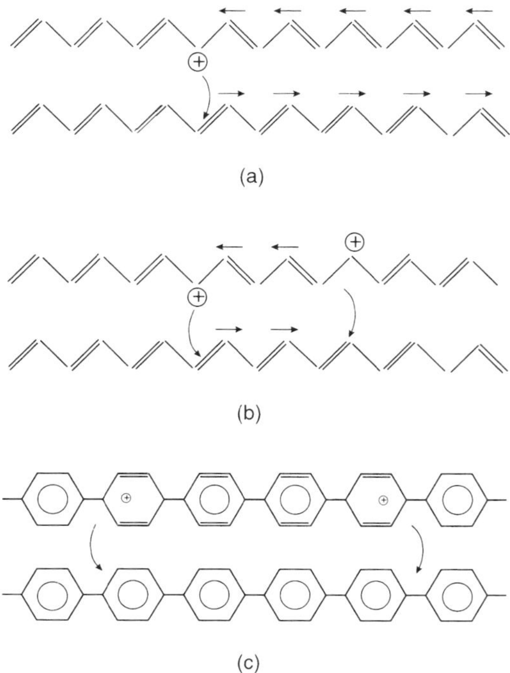

FIG. 15. Schematic of motion of a soliton and bipolaron between polymer chains. (a) Interchain hopping of a soliton in *trans*-polyacetylene; arrows represent those carbon atoms that move. (b) Interchain hopping of a bipolaron. (c) Interchain hopping of a bipolaron in poly(*p*-phenylene).

5.1.5 Dispersive Transport When a delta-function distribution of charge drifts across a sample under the influence of an electric field, its shape alters through thermal spreading and also as a result of the presence of molecular disorder. In the latter instance, the transition probabilities from site to site vary. The arrival of the leading edge of the charge packet at the collecting electrode results in a decrease in the measured current, and a plot of current vs time can exhibit a well-defined inflection point (see Fig. 16). Transport exhibiting this behavior is termed nondispersive. If, however, the time required for the mean velocity of the carriers to reach steady state is comparable to or exceeds the mean transit time, then the plot of current vs time does not show a well-defined inflection point. This transport behavior is called dispersive (see Borsenberger and Weiss, 1993). At sufficiently low temperature, all molecular glasses and doped polymers display dispersive transport on account of the enhanced importance of site-energy distributions and spatial-disorder energy variations relative to kT.

It has been found that by plotting the logarithm of the photocurrent vs the logarithm of the time, an inflection point can be distinguished that can be associated with the arrival time of a characteristic segment of the carrier packet. A few models have been proposed to rationalize this dispersive behavior, the most significant being those referred to as Scher–Montroll, multiple trapping, and Bässler formalism (see Borsenberger and Weiss, 1993). A brief description of the Bässler formalism is now presented.

5.1.6 Bässler Formalism Transport in disordered molecular solids and molecularly doped polymers is strongly influenced by the effects of diagonal and off-diagonal disorder.

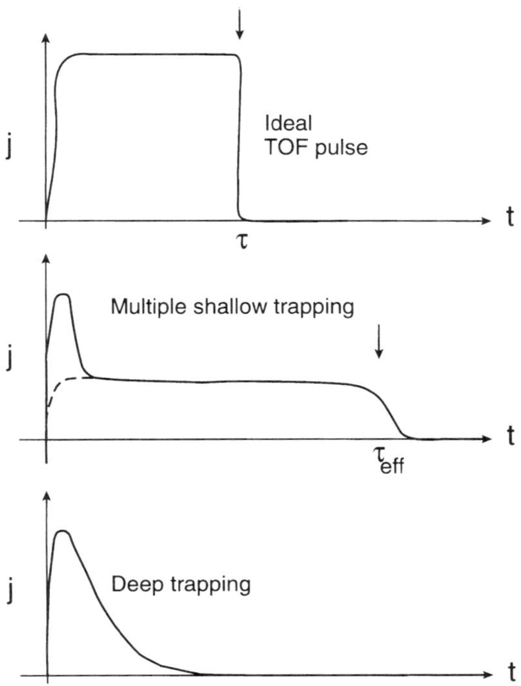

FIG. 16. Top curve: ideal time-of-flight (TOF) current-pulse shape; arrow shows the arrival (transit) time τ_t of the carriers at the collecting electrode. Middle curve: multiple–shallow-trapping case with the initial spike decay representing the trap-filling process, followed by the equilibrium trapping–detrapping process. Bottom curve: deep trapping, with no defined transit time. For all curves, the ordinate is the current and the abscissa is time, in arbitrary units. (Redrawn from Karl, 1990.)

Lacking analytical solutions, most investigators have resorted to Monte Carlo simulation (Bässler, 1993; see also Monte-Carlo Methods). In this approach, one adopts a Gaussian distribution for the hopping-site energies [as given by Eq. (12)] and assumes the jump rate between sites i and j separated by a distance R_{ij} with energies ϵ_i and ϵ_j to be of the form

$$\nu_{ij} = \nu_0 W_{ij} \times \begin{cases} \exp[-(\epsilon_j - \epsilon_i)/kT] & \text{for} \quad \epsilon_j \geq \epsilon_i, \\ 1 & \text{for} \quad \epsilon_j \leq \epsilon_i, \end{cases} \tag{15}$$

where ν_0 is a frequency factor. Implicit in Eq. (15) are the assumptions of weak electron–phonon coupling (i.e., polaronic effects are negligible) and that downward jumps in energy are unaffected by an electric field. Calculations assume that the variation in the intersite electronic wave-function overlap factor W_{ij} is exponential,

$$W_{ij} = \exp(-\gamma_{ij} R_{ij}/a), \tag{16}$$

where a is the lattice constant. The parameter γ_{ij} is a dimensionless random variable describing off-diagonal disorder effects (i.e., a measure of the inverse wave-function decay constant). Monte Carlo calculations (see Monte Carlo Methods) based on the assumption that γ_{ij} is a simple sum of uncorrelated Gaussian distributions, $\Gamma_i + \Gamma_j$, each characterized by a variance Σ, predict mobilities described well by the analytical expression

$$\mu = \begin{cases} \mu_0 \exp(-2\hat{\sigma}^2/3) \exp[C(\hat{\sigma}^2 - \Sigma^2)E^{1/2}] & \text{for} \quad \Sigma > 1.5, \\ \mu_0 \exp[C(\hat{\sigma}^2 - 2.25\Sigma^2)E^{1/2}] & \text{for} \quad \Sigma < 1.5, \end{cases} \tag{17}$$

where $\hat{\sigma} = \sigma/kT$; σ is the variance for the site-energy distribution function [Eq. (12)], $\hat{\sigma}$ is referred to as the disorder parameter, and C is a constant with the value 3×10^{-4} $m^{1/2}$ $V^{1/2}$. Figure 17 illustrates the good agreement between the calculated constant C and that inferred from the slope of β ($= C\hat{\sigma}^2$) vs $\hat{\sigma}^2$ for DPQ-doped polystyrene (PS) or polycarbonate (PC), C(experiment) $= 2.9 \times 10^{-5}$ $m^{1/2}$ $V^{1/2}$. The decrease in mobility of the carrier in PC compared with PS can be attributed to the increased disorder associated with the carbonyl groups in PC. The parameter $\hat{\sigma}$ is of central importance to this theory because together with a knowledge of the sample thickness, it can be used to predict the temperature at which the mobility transients become dispersive. The nondispersive transport regime is one in which the mean carrier-arrival time in a time-of-flight experiment is independent of the sample thickness; this is no longer true in the dispersive transport regime.

The assumption that $\gamma_{ij} = \Gamma_i + \Gamma_j$ implies

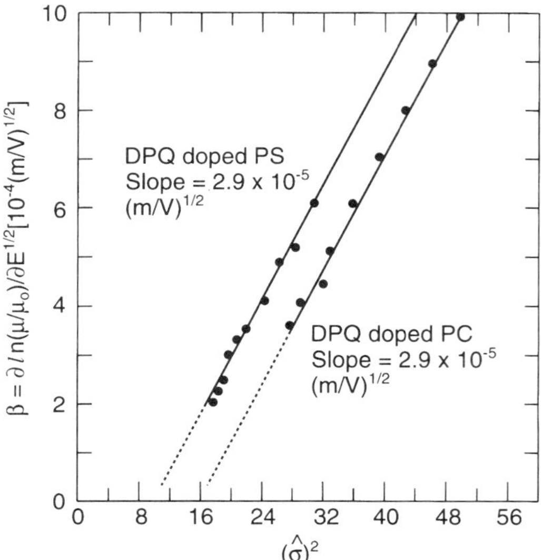

FIG. 17. β vs $\hat{\sigma}^2$ for a mixture of two asymmetrical diphenoquinone (DPQ) compounds in polystyrene (PS) and a polycarbonate (PC). The DPQ concentrations were 30%. Here, $\hat{\sigma} = \sigma/kT$ and $\beta = \partial \ln(\mu/\mu_0)/\partial E^{1/2}$. (Redrawn from Borsenberger and Rossi, 1992.)

a strong correlation between all transitions to and from a given site and omits factors of probable importance. The proper inclusion and estimate of the effects of off-diagonal disorder and, in addition, polaronic effects in carrier transport in disordered molecular solids and molecularly doped polymers constitute a major task (Gartstein and Conwell, 1994).

Nevertheless, the Bässler formalism does account for a large body of information on transport in polymers. In particular are the following:

1. its correct predictions of the field dependence of the mobility in the high-field regime;
2. the observed value of the universal constant C [see Eq. (17)], which is the slope of the plot of $\ln\mu/E^{1/2}$ vs $\hat{\sigma}^2$;
3. the temperature dependence of the mobility; and
4. the transition from nondispersive to dispersive transport (the shape of the log current vs log carrier transit time), thereby generalizing the Scher–Montroll formalism.

5.1.7 Positron Mobility As noted, the motion of an electron through a lattice involves interactions with local perturbations, including those produced by impurity molecules or lattice defects. A useful method for obtaining an estimate of the microscopic mobility in insulators with a high density of electron traps is provided by the mobility of positrons (PS, p. 369; Mills and Karl, 1993). That the positron drift mobility is an upper bound to the electron mobility can be rationalized as follows: For positrons, a chemical impurity does not act as a trap because there is no stable empty orbit that the positron can occupy to form a temporary ion, as would be the case with an electron. The chemical impurity would act as a scattering center because it distorts the local electric field. Just as is the case with electrons, positrons can become trapped in vacancies and structural imperfections if polarization energy is gained. Thus if the positron mobility is significantly larger than the electron mobility, it can be concluded that the electron mobility is chemical-impurity dominated; this is the most frequently encountered case in organic semiconductors.

That μ_p can be extremely large in aromatic solids was shown by the Mills and Karl (1993) measurements in high-purity monoclinic α-perylene single crystals. Over a temperature range 100–350 K, $\mu_p \propto T^{-1}$, a clear departure from the $T^{-3/2}$ dependence expected if the mobility was acoustic-phonon limited and relatively independent of temperature for $T < 50$ K. They obtained a value of $\sim$0.1 m^2 V^{-1} s^{-1}, quite similar to their measured value of 0.09 m^2 V^{-1} s^{-1} measured for the electron mobility in naphthalene at 10 K. The limit on μ_p was ascribed to the presence of residual impurities. In anthracene at 300 K in highly purified material, $\mu_p \sim 0.03$ m^2 V^{-1} s^{-1}, 300 times the room-temperature mobility. This large value of μ_p can be shown to imply that positrons move in extended states with a mean free path of $\sim$8.5 nm, approximately 15–20 times the lattice constant.

5.2 Steady Current Flow

5.2.1 Basic One-Carrier Theory A simplified theory of space-charge–limited currents (SCLC) was introduced into the field of organic crystals by Helfrich and Mark (1963).

The total current that flows in an insulator is composed of two components, one driven by the electric field and the other by carrier-concentration gradients. These components are referred to as drift and diffusion currents, respectively. In the simplified theory, consideration of diffusion currents is omitted. For a more complete theory including diffusion, the work of Bonham and Jarvis (1978) should be consulted.

The nature of the contact to the organic semiconductor (OSC) is of major importance in determining the qualitative and quantitative response of the current to an external field, both in the dark and in the presence of absorbed light in the contact region (see Sec. 3.3). A contact can change its characteristics under the influence of an electric field. At high field strength, an Ohmic contact can be exhausted, leading to an emission-limited contact (limited by the rate of injection from the electrode). The energetic description of an Ohmic contact and its depletion is shown in Fig. 6.

5.2.2 Space-Charge–Limited Currents (SCLC) By definition, the voltage dependence of the SCLC current in all materials provided with Ohmic contacts is the same. For a trap-free insulator (or semiconductor), or for one containing shallow traps (trap depth from the transport level $\leq$0.3 eV) in which only one type of carrier is mobile, and with neglect of diffusion current, this SCLC is

$$J = (9/8)\theta\mu\epsilon\epsilon_0 V^2/L^3, \qquad (18)$$

where the trapping ratio $\theta = n_f/(n_f + n_t) = 1$ for a trap-free solid.

If, as is usually the case, the carriers encounter deeper traps during their passage through the sample, then J is reduced, sometimes by factors of 10^{-8}. It is the aim of SCLC theory to be able to deduce, from the J–V response of the material as a function of its temperature and thickness, such parameters as the concentration of traps, and their energetic distribution in the forbidden gap.

By making specific assumptions about the energetic distribution of traps (still neglecting diffusion), it is possible to derive more complicated expressions for the SCLC, which reveal a more detailed microscopic picture of the material under study. By providing a solid with hole- and electron-injecting contacts on opposite sides, it is possible to inject SCLC for both carriers simultaneously. This is referred to as double injection. The presence of both types of carrier introduces a new process, namely carrier recombination. Carrier recombination plays a major role in limiting current in two-carrier injection systems, and is the physical basis for electroluminescence (see Sec. 8.2). A good description of double injection may be found in Lampert and Mark (1970) and in PS, pp. 413 *et seq.*

5.2.3 Carrier Trapping The transport of carriers through organic crystals is dominated by the presence and distribution of carrier-trapping sites. An excellent discussion of the effects of traps on carrier mobility is given by Silinsh (1980, p. 250), SC, p. 372, Karl (1990), and PS, pp. 227 *et seq.* Traps can be situated at one specific energy level within the forbidden gap. This would be the case for a chemical impurity. Traps can also be produced by structural defects, such as vacancies, interstitials, grain boundaries, and dislocations. In anthracene, there are about 2×10^{20} m^{-3} lattice vacancies at room temperature. The number of dislocations depends on crystal preparation and handling. For melt-grown crystals, the total density can reach 10^{25} m^{-3}. Some defects such as lattice expansions do not trap carriers, but do cause scattering; such defects are called antitraps.

Chemical impurities can provide sites of lower potential energy for a carrier, causing localization for periods of time determined by the trap depth from the appropriate conduction level; chemical impurities can also act as antitraps. A hole trap is produced when an impurity has a lower ionization energy than the host material. The hole-trap depth E_t^h is given to a first approximation by the relation

$$E_t^h = I_c^h - I_c^g, \qquad (19)$$

where I_c^h and I_c^g are the ionization energies of the host and guest, both in crystal form. If $E_t^h > 0$, a hole trap is formed. If $E_t^h < 0$, an antitrap is formed. An electron trap is produced when an impurity has a larger electron affinity than that of the host. The electron-trap depth E_t^e is given to a first approximation by the relation

$$E_t^e = A_c^g - A_c^h, \tag{20}$$

where A_c^g and A_c^h are the electron affinities of the host and guest, respectively, both in the crystal form. If $E_t^e > 0$, an electron is trapped, and for the inverse, an antitrap is formed.

As is the case with inorganic semiconductors, ultrapurification of organic crystals has produced not only quantitative increases in the mobility of carriers, but qualitative differences in the sense that at low temperature (≈5 K) it was possible to demonstrate band mobility as opposed to the hopping motion that dominates at room temperature, and also to detect hot conduction electrons and holes (Karl, 1990).

The chief technique for determining whether trapping is taking place, and whether the traps are shallow or deep, is to measure the drift mobility of the carriers in a TOF method. In this method, the sample is placed between two blocking electrodes and a strongly absorbed light pulse is used to generate carriers close to one of the electrodes in the presence of an external field. Forced by the external field, the carriers of one sign drift across the bulk of the sample until they reach the opposite surface. By reversing the direction of the external field, the drift mobility of the oppositely charged carrier may be determined. As the carriers drift across the sample, they induce a current flow in the external circuit, which provides the TOF curve.

In Fig. 16 are shown the shapes of TOF pulses and what they indicate about trapping. In (a) is shown an ideal TOF pulse indicative of no significant trapping. The arrow marks the arrival time of the sheet of injected carriers at the back surface of the sample. In (b) is shown the effect of multiple shallow trapping, thermally activated detrapping, and retrapping. There is still a distinct arrival time for the injected carriers but the transit time has been increased because of the time spent in traps by the carriers. In (c) is shown a TOF curve where deep trapping is encountered. Here, there is no definite arrival time because the spread in arrival times exceeds the experimental sampling time.

In Fig. 18 is an illustration of a classic example of trap-free TOF curves for anthracene under external fields varying from 0 to 4.5×10^5 V m^{-1}. The broadening of the arrival tail is due to the mutual repulsion of the electrons as they drift, and also to the finite width of the charge-generation region. For a more complete discussion of methods of measuring and interpreting drift mobility, see PS, pp. 709 *et seq.* In order to obtain data such as are shown in Fig. 18, it is necessary to employ ultrapure single crystals, or, as was shown by Abkowitz *et al.* (1993), to use materials that have small ionization energies, making it difficult for an impurity to trap a hole by ionizing into the host hole.

It is practically impossible to determine the complete trap constellation in an organic semiconductor using any single experimental technique. The dominant methods depend on filling the various traps with carriers, and then emptying the traps to produce a characteristic pattern of enhanced conductivity in

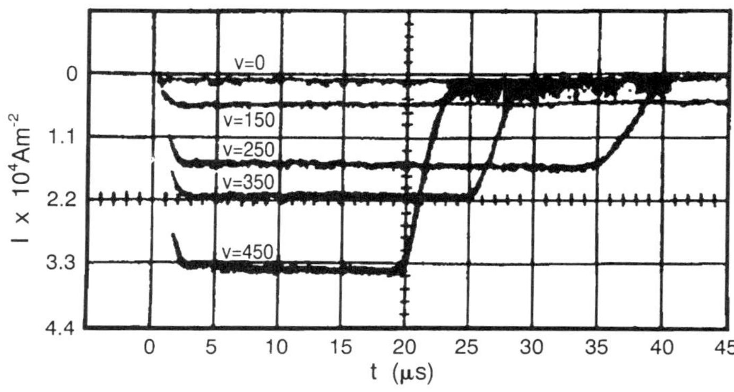

FIG. 18. Trap-free TOF curves for a 1.01-mm-thick anthracene crystal under applied voltages ranging from 0 to 450 V. The ordinate shows the current and the abscissa shows the time. The light pulse had a 1-μs halfwidth. (Redrawn from Karl *et al.*, 1970.)

the external circuit. These methods are mainly

1. thermally activated mobility;
2. thermal release, referred to as thermally stimulated currents (TSC);
3. photoexcitation of charged traps using specific wavelengths of light in order to deduce the depth of the traps;
4. excitation of trapped charge by energy transfer from excitons;
5. space-charge–limited current measurements as a function of temperature.

5.2.3.1 Thermally Activated Mobility. This method is most effective for the determination of shallow trap depths, if the concentration or the influence of deep traps is small enough to permit the carrier to traverse the crystal while interacting multiply only with shallow traps. When a single level of shallow trap at a density N_t and depth E_t is present then

$$\mu = \mu_0[1 + (N_t/N_0)\exp(E_t/kT)]^{-1}. \tag{21}$$

From this equation, it follows that at high temperature ($kT \gg E_t$), shallow trapping is virtually absent and $\mu \propto \mu_0$, whereas at lower temperature,

$$\mu \approx \mu_0(N_t/N_0)\exp(-E_t/kT). \tag{22}$$

In Fig. 19 is shown an Arrhenius plot of the electron and hole mobilities in an anthracene crystal doped with tetracene. Although the concentration of impurity was only 4 parts in 10^7, the effect on the mobility is considerable. The trap depths were calculated from these curves, using the above equations.

5.2.3.2 Thermally Stimulated Currents (TSC) and Isothermal Decay Currents (IDC). In the TSC technique, the traps in a solid are filled by passing a current through the solid at low temperatures; the voltage is then reversed and the temperature of the solid is increased linearly by means of a controlled heat input. As the traps empty, a current is measured and characteristic peaks are observed, indicative of traps lying at specific levels below the conduction band (in the case of electrons). In the IDC technique, the current through an insulator is abruptly increased, as with a pulse of exciting light, and the time decay of the current is measured at constant temperature. Both techniques involve observing current changes produced by the return of the system to thermal equilibrium after a nonequilibrium trap occupancy has been established. This initial trap loading can be due to either electrons or holes. It is possible to load the traps with carriers of only one sign if an injecting contact is used. The temperature range over which TSC and IDC experiments are performed determines the lower and upper limits to trap depths capable of being probed; generally traps shallower than 0.3 eV or deeper than 0.9 eV are not detectable by TSC techniques in the lower–melting-point polyacenes.

The analysis of TSC spectra is beset with difficulties stemming from overlapping peaks, an unknown spatial distribution of

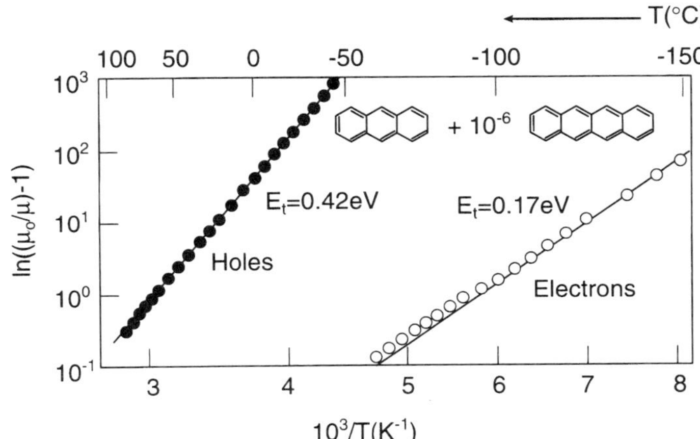

FIG. 19. Plot of $\ln[(\mu_0/\mu) - 1]$ vs $1/T$ for tetracene-doped anthracene, according to Eq. (21). Tetracene is a hole and electron trap and is present in a concentration of 0.4 ppm. Trap depths for holes and electrons are shown. (Redrawn from Karl, 1990.)

filled traps, and lack of temperature control across the sample. Also, the data are markedly affected by heating rate, applied electric field, efficiency of trap filling, and, in the case of polymers, the thickness of the sample (Stasiak *et al.*, 1980). In short, these experiments must be carried out with full awareness of the extensive literature that exists (see PS pp. 231 and 428 for references).

The lifetime of trapped carriers, τ_t, is related to the trap depth E_t (for the specific case of electrons) by

$$\tau_t^{-1} = \nu \exp[-(E_c - E_t)kT], \tag{23}$$

where ν is the escape-frequency factor representing the effective rate of lattice collisions with the trapped carrier. The parameter τ_t can be determined by measuring the IDC at several temperatures near the temperature peak in the TSC spectrum. The values inferred for the hole trap in anthracene are $E_t = 0.63$ eV and $\nu = 1.5 \times 10^9$ s^{-1}. The escape frequency is low compared with typical lattice frequencies of $\approx 10^{13}$ s^{-1}. A high retrapping rate would produce such a small escape frequency.

5.2.3.3 Optically Stimulated Current. The experimental arrangement for optically stimulated currents is the same as for TSC, except that the sample is not subjected to a rising temperature schedule. Instead, monochromatic light is focused on the sample and the photocurrent is measured as a function of the wavelength of light. The spectra show the electronic radical-ion transitions of the local trap. The electronic transitions of the radical ions can be identified from solution studies.

By combining optically stimulated currents with TSC, it becomes possible to identify the nature of the trap. This was done with tetracene-doped anthracene, and a hole trap at 0.45 eV was identified.

It is also possible to carry out a direct trap-to-band transition using light of the proper wavelength. Thus, for the case of tetracene in anthracene, it is possible to use IR light of energy close to the trap energy of 0.45 eV and detrap the trapped charge with an efficiency that outlines the vibrational states of the trap.

5.2.3.4 Detrapping by Excitons. Singlet and triplet excitons migrating through the solid can detrap trapped carriers when the exciton encounters the trap, and energy transfer to the trapped carrier is sufficient for detrapping. Neither the identity nor the depth of the trap can be determined by this method. However, important information about the spatial distribution of traps can be obtained.

The quenching of excitons by trapped carriers can, in addition, be used to distinguish between Ohmic and non-Ohmic contacts. When a contact is Ohmic and a SCLC is flowing through a solid, the total charge Q injected into the solid obeys the simple relationship $Q \approx CV$. In most crystals, Q consists mainly of trapped charge. At relatively low light intensity, the quenching of excitons depends linearly on the concentration of carriers (mostly trapped). Triplet excitons have a long natural lifetime, and so they are sensitive to relatively low concentrations of trapped carriers. The quenching of triplet excitons by trapped carriers reduces the exciton lifetime and concentration. The decrease in exciton concentration can be monitored by measuring the delayed singlet-exciton fluorescence produced by the fusion of two triplet excitons. It is thus possible to follow the quenching of the triplet excitons by trapped charge as a function of the voltage applied to the crystal. When an Ohmic contact is operative, then $Q \approx CV$ and the quenching is linear with applied voltage. With an Ohmic CuI contact applied to an anthracene crystal, the linear quenching curve persists up to about 1000 V on a 225-μm-thick crystal. When the contact is depleted, the photocurrent saturates and the quenching is essentially constant as the voltage is increased. An example of a contact that is Ohmic to begin with and then becomes emission-limited is provided by Ce^{4+} in aqueous solution (PS, p. 399).

5.2.3.5 Trap Distributions. Theoretically, SCLC and its temperature dependence can be used to deduce the concentration and energetic distribution of traps in a solid. As was discussed in Sec. 5.2.3, this can be validly accomplished only if the correct distribution function is chosen for the traps in the solid. An example of the use of an exponential distribution of traps is given by Bässler *et al.* (1969), who studied single-crystal tetracene. From the J-vs-V dependence and from measurements of the temperature dependence of J at various values of V, it was possible to determine the position of the quasi-Fermi level E_F, and hence the level of filled traps. It was concluded that in the energy range $0.6 < E_t < 0.9$ eV, the traps are distributed exponentially. It was also possible to determine the concentration of traps as $\sim 3 \times 10^{21}\ m^{-3}$.

For a Gaussian distribution of traps, the diagnostic criteria are completely different. The SCLC equations must be solved with a Gaussian distribution function

$$N(E_t) = (N_t/\sigma\sqrt{2\pi}) \exp[-(E - E_t)^2/2\sigma^2], \quad (24)$$

where E is the mean trap depth, N_t is the total trap density, and σ is a measure of dispersion of trap levels on either side of the mean. These equations can be solved analytically only by dividing the observed SCLC regime into about four sections. When this is done, the results for various materials are as shown in Table 1.

5.3 Transient Currents (Single Carrier)

5.3.1 Space-Charge–Free (SCF) Transients An important application of space-charge–free transients is to measure the time of flight (TOF) of a sheet of charge injected into the solid. From the shape of the transient pulse amplitude as a function of time, it is possible to deduce the mobility of the carrier, and whether it had encountered a shallow or deep trap en route to the collecting electrode. The shape of the response may be seen in Fig. 16. From the transit time and the electric field, the TOF can be calculated. In addition, the electric field is large enough to make the transit time of the carriers much shorter than the dielectric relaxation time of the solid, i.e., $\tau < \epsilon\epsilon_0/\sigma$. This is necessary in order to see a field effect on the transit time. In high-resistivity materials, this condition is easily realized.

In the ideal case, the arrival time of the charge at the collecting electrode is sharply delineated. In the case of deep trapping, there is no obvious transit time at all, as is shown in Fig. 16.

5.3.2 SCLC Transients A current transient brought about by the injection of an amount of charge $Q = CV$, where C is the capacitance of the sample, has been thoroughly analyzed theoretically, and is discussed in detail in Lampert and Mark (1970). A good example of the utility of trap-free SCLC transients (TFSCLC) is given by Abkowitz *et al.* (1993). They outlined a tech-

Table 1. Typical values of Gaussian trap-distribution parameters E_t, σ, N_t, and N_e for a number of aromatic crystals (anthracene, tetracene, pentacene, and perylene) in different crystalline states. N_e is the density of states at the conductivity level, and the other parameters are defined in the text. SC = single crystal; OL = oriented polycrystalline layers obtained by evaporation in vacuo; QA = quasiamorphous obtained by evaporation in vacuo on a cooled surface; PC = polycrystalline. (From Silinsh, 1980).

Compound	Crystalline state	E_t (eV)	σ (eV)	N_t ($10^6\ m^{-3}$)	N_e ($10^6\ m^{-3}$)
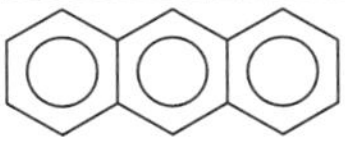	SC	0.5–0.8	0.03–0.13	2×10^{11}–8×10^{13}	10^{16}–10^{20}
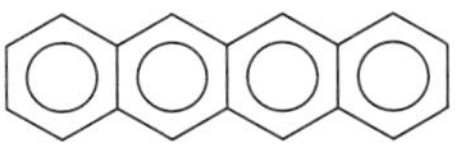	SC	0.6	0.05–0.14	10^{13}–10^{14}	10^{18}–10^{21}
	OL	0.3–0.4	0.05–0.14	6×10^{13}–2×10^{14}	10^{12}–10^{15}
		0.1	0.07–0.13	4×10^{14}–6×10^{15}	
	QA	0.26	0.11	2×10^{14}	
		0.1	0.05	4×10^{15}	
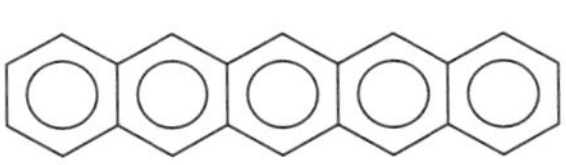	OL	0.35	0.11	8×10^{13}	10^{14}–10^{15}
		0.08	0.07	8.5×10^{14}	
	QA	0.35	0.06	8×10^{13}	10^{14}–10^{15}
		0.04	0.07	7×10^{15}	
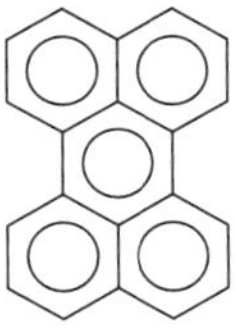	PC	0.4	0.03	4×10^{13}	10^{11}–10^{15}
		0.3	0.05–0.09	$(4–7) \times 10^{15}$	

nique for determining whether a contact is Ohmic, using trap-free polymers. The polymer was a thin film (9 μm) of poly-(tetraphenylbenzidine) (PTPB). This polymer is a hole conductor and has an ionization energy so low that there are very few impurities with lower ionization energies that can act as hole traps (see Sec. 5.2.3). The experimental configuration is shown in Fig. 20.

For a trap-free solid, the maximum current is given by the relation

$$J = (9/8)\mu\epsilon\epsilon_0 V^2/L^3 = (9/8)CV/At_{tr}, \qquad (25)$$

where C is the capacitance of the experimental configuration, t_{tr} is the transit time of the injected carrier, and A is the area of the contact. Thus, by measuring t_{tr} and C, it should be possible to calculate the steady-state SCLC at the same electric field, if the contact is Ohmic. A TFSCLC transient current plotted against time is shown as the top curve in Fig. 21(a). The time t_p at which there is a peak in the response curve, according to the theory of TFSCLC, is related to the transit time of the carriers simply as $t_{tr} = t_p/0.8$. The transient TFSCLC is generated by applying a positive field step of 10^7 V m^{-1} at T = 323 K. In a TOF experiment made on the same film at the same applied field, the bottom curve in Fig. 20(b) is obtained. The mean hole-transit time is taken at the point where the TOF signal has dropped to one-half of the plateau amplitude; within experimental accuracy, it is the same as is shown in the top curve of the same figure.

The steady-state TFSCLC is attained after the elapse of several transit times, and is shown in the top curve of Fig. 21(b). It is possible to calculate the steady-state TFSCLC, I_{SS}, using the expression

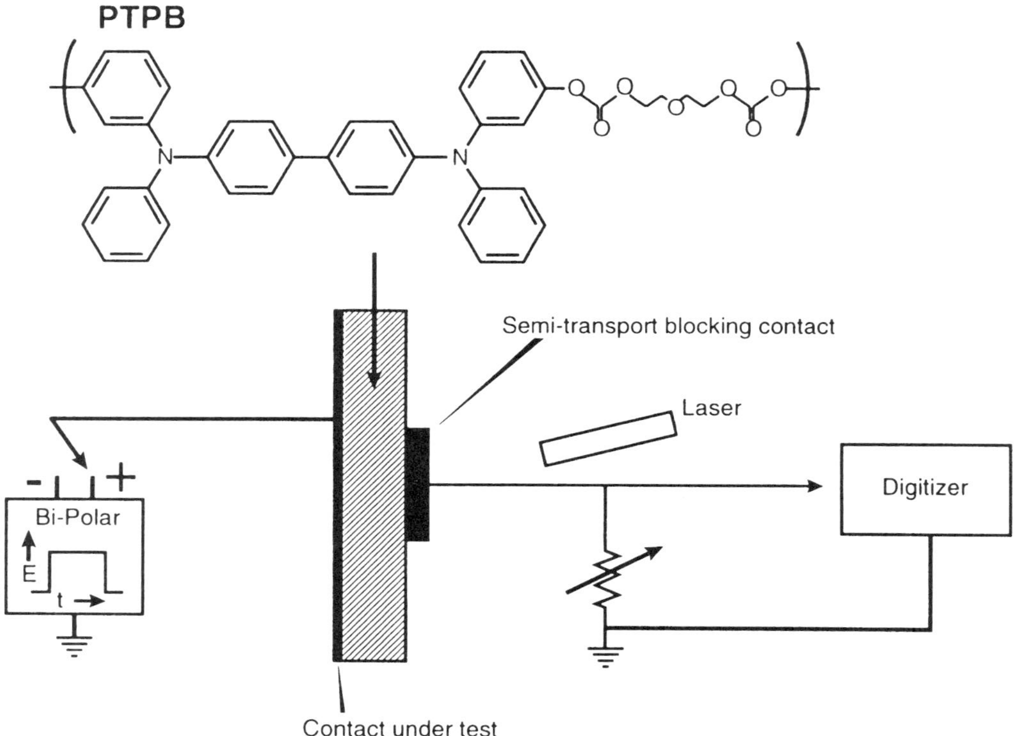

FIG. 20. Schematic of experimental configuration. Structure of the trap-free polymer PTPB is displayed. During TOF measurements, the semitransparent contact is positively biased and the field is applied 15 ms prior to the light flash. During the transient dark-injection measurements (see Fig. 21), the contact under test (substrate) is excited by a positive voltage pulse. (Redrawn from Abkowitz *et al.*, 1993.)

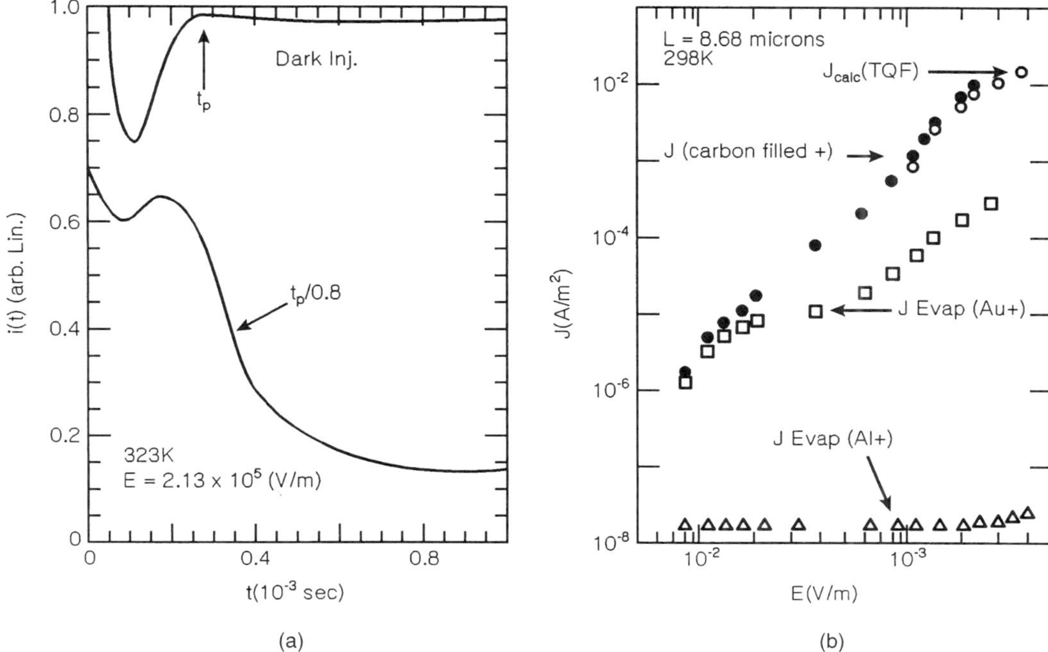

FIG. 21. (a) Comparison of a dark hole-injection transient and the TOF hole transient in the same PTPB film (T = 323 K, E = 2.13 × 10^7 V m^{-1}). (b) Steady-state J-vs-E plots on the same PTPB film fitted with various contacts. Solid circles are hole currents sustained by the carbon-filled polymer contact. Open squares are hole currents sustained by a gold contact. Triangles are hole currents injected from an evaporated aluminum contact. Open circles are steady-state dark currents computed from TOF transit times using TFSCLC theory. Departure of the slope from the theoretical value of 2 is due to field dependence of the mobility. (Redrawn from Abkowitz *et al.*, 1993.)

$$I_{SS} = (9/8)CV/t_{tr} \tag{26}$$

and the value of t_{tr} taken from the TOF measurement. If the contact is Ohmic, the calculated and measured I_{SS} should agree.

In Fig. 21(b) are shown the results for several different contacts. As is seen from the equivalence of the calculated and measured steady-state TFSCLC, the carbon-filled polymer contact is Ohmic. The evaporated Au contact approaches Ohmicity at low field strength, but becomes emission-limited at higher field strength. The Al contact is strictly emission-limited. The departure of the slope of the curve of ln$J$ vs ln$V$ from the theoretical value of 2 is due to the field dependence of the mobility that is characteristic of transport in molecularly doped polymers (see Sec. 5.2.3).

This rare demonstration of TFSCLC is made possible because the film is thin and the hole conductor has an ionization energy so low that there are few impurity species with a lower ionization energy. Thus there are no hole traps in the film, and the hole is the majority carrier.

6. PHOTOEMISSION

Photoemission takes place from and provides information about the surface region of a solid; this region is of the order of a few nanometers in depth. However, unlike typical inorganic materials, organic semiconductors do not exhibit dangling bonds at the surface, and so the surface properties do not differ markedly from those in the bulk.

Photoemission is used mainly to obtain such information as the ionization energy of the solid, the work function, the energy-level structure in the valence band, and the density of states in the valence band. By using multiphoton photoemission, in which two or more quanta of low-energy light add their energies and succeed in providing a total energy equal to or greater than the ionization energy of the solid, it is possible to measure exciton levels, the band gap, and trap levels.

The yield of photoelectrons per incident photon, which in the case of anthracene is about 10^{-4} for a photon energy about 1 eV above the threshold, is too low to be of practical interest.

6.1 Energy-Level Analysis

In Fig. 22 is shown a series of photoemission curves plotted against the binding energy for a series of polyacenes. Included in the figure are the photoemission data for the corresponding gas-phase molecules, shifted by an arbitrary energy interval to bring the peaks of the yields into coincidence. The binding energy represents the location of an energy level in the valence band of the compound, relative to the position of the Fermi level of the solid, E_F. Thus, for anthracene, the first peak appears about 1.8 eV below the Fermi level. This is the level of the highest occupied orbitals. In anthracene, using the known values for the band gap of 4.0 eV, and the ionization energy of the crystal, the Fermi level should lie at 2.0 eV above the highest occupied valence level. If the valence and conduction bands remained flat up to the surface (no surface states to pin the Fermi level at some other position) then in Fig. 22, the first peak should appear at 2.0 eV instead of 1.8 eV. Thus E_F is closer to the valence band, indicating the presence of charge between the Fermi level and the valence band, due perhaps to the presence of some negatively charged surface states.

As may also be seen in Fig. 22, there is a close correspondence between the vapor-phase and solid-phase spectra. This is a further manifestation of the weak intermolecular interaction in the van der Waals solids that allows the individual molecular properties to dominate the solid-state spectrum. The shift in energy required to put the vapor-phase and solid-phase spectra into coincidence is a measure of the cohesive-energy contribution to that transition. The broadening of each peak in the solid-phase spectrum can be traced to fluctuations in the intermolecular electronic polarization. This illuminates one of the most significant differences

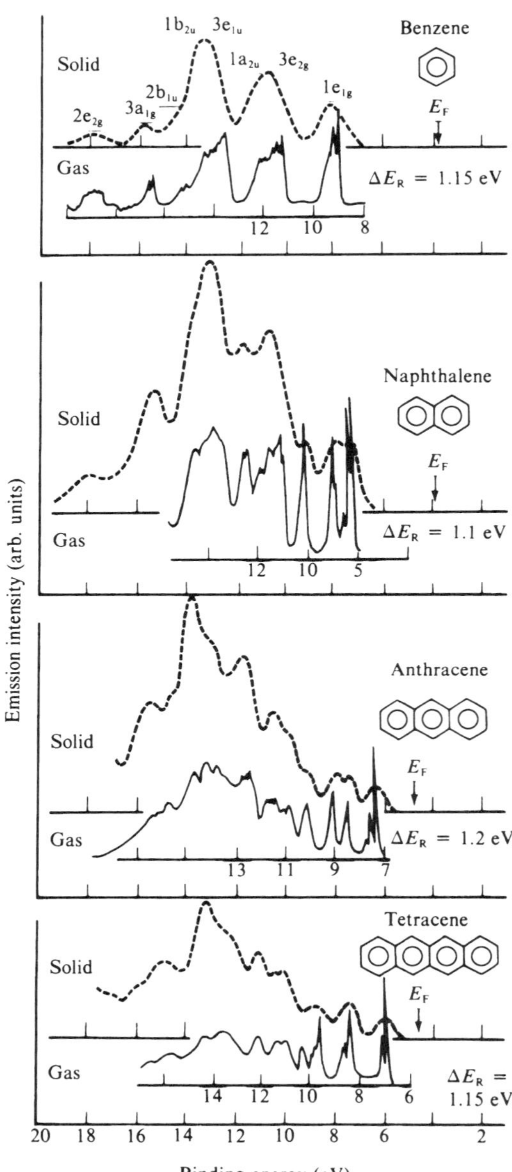

FIG. 22. Comparison of gas-phase photoemission spectra with photoemission yield from polycrystalline films using He I radiation ($h\nu$ = 21.22 eV). ΔE_R gives the shift resulting from the cohesive-energy contribution to the positions of the energy levels in going from the gaseous to the solid state in order to obtain coincidence for most of the prominent bands. The broadening of each peak in the solid phase is due to the polarization fluctuations caused by the presence of defects and phonon interactions. E_F denotes the position of the Fermi level. (Redrawn from Grobman and Koch, 1979.)

between organic crystals and inorganic materials, namely the importance of electron–phonon interactions. These fluctuations (0.1 to 0.5 eV) are caused by defects, termination of the crystal at the surface, and thermal vibrations. Electrons originating from monomolecular layers at different depths from the surface show slightly different energies, since the polarization energy is a function of the distance from the surface.

To obtain information about the density of states in the valence and conduction bands, accurate photoemission and optical data must be available. It is also based on other assumptions, most of which can be justified. For more details, see PS, pp. 552 *et seq.*

Since autoionization plays a large role in the photogeneration of carriers, it is not surprising that it also plays a role in photoemission. This is exemplified by the fact that absorption and photoemissive yields vary in the same way (though not proportionally) (PS, p. 568). The effects of AI are more evident in multiphoton photoemission.

6.2 Multiphoton Studies

The first multiquantum photoemission study of organic semiconductors was carried out on anthracene (Pope *et al.*, 1965), and was shown to be due to a fusion of two Frenkel singlet excitons, each at 3.15 eV, to generate an excited state at 6.3 eV, which autoionized to emit an electron. The bimolecular rate constant for this process was calculated to be $\sim 10^{-14}$ m^3 s^{-1}.

In addition to Frenkel exciton–exciton fusion reactions, it is also possible to determine a wide variety of lower-lying energy states, such as trapped-charge and *e–h* pair states, and also photon–photon photoemission (see PS, p. 477).

7. MAGNETIC-FIELD EFFECTS

7.1 Hall Mobility

Measurement of the Hall effect (PS, p. 374) allows the free-carrier density and trap-free mobility to be determined. In the aromatic solids anthracene and naphthalene, Hall mobilities are of the order of 10^{-4} m^2 V^{-1} s^{-1} at room temperature. In simple metals, $\mu_H = \mu_D$, while usually in semiconductors $\mu_H > \mu_D$, the exact value being determined by the energy dependence of the carrier-scattering mechanism. In general, dark Hall measurements on most organic semiconductors are too difficult to perform because their conductivity is too small at the low fields necessary to have Ohmic currents. In these cases, the Hall measurements are made using the photocurrents. The ratio of the photo-Hall hole mobility in the *b* direction to the drift mobility is about 8 in anthracene (PS, p. 376). Recently, however, the Hall effect for the organic semiconductor *bis* (1,2,5-thiadiazolo)-*p*-quinobis (1,3-dithiole) was found to be 4×10^{-4} m^2 V^{-1} s^{-1} at room temperature with the sign of the carrier as positive, suggestive of a band model for the holes, since $\mu_H > 10^{-4}$ m^2 V^{-1} s^{-1}, the lower bound for band mobility.

7.2 Magnetic-Field Effects on Spin–Spin Interactions

Magnetic-field effects in organic solids were initially observed in delayed fluorescence in anthracene single crystals due to the fusion of two mobile triplet excitons to create a final fluorescent singlet state (Merrifield, 1968). Subsequently, magnetic-field effects in organic semiconductors were observed in photoconductivity, electroluminescence, charge-detrapping processes, scintillation, fluorescence, and geminate recombination, in both crystalline and polymeric systems.

Figure 23 illustrates the high-field anisotropy of the photoenhanced bulk conductivity in anthracene as reported by Geacintov *et al.* (1970). The dips in the photocurrent owe their origin to a reduction in the efficiency of detrapping of holes by triplet excitons. The common theoretical elements of these field effects can be summarized as follows. Consider two species, *A* and *B*, initially spatially separated, which through their motion enter a correlated state (*AB*); these could represent triplet or singlet excitons, trapped or free carriers, or paramagnetic defects. This correlated state evolves in time according to a Hamiltonian H_S that does not include spatial variables and is therefore called a spin Hamiltonian. After a time of the order of nanoseconds, two or more final products, *C* and

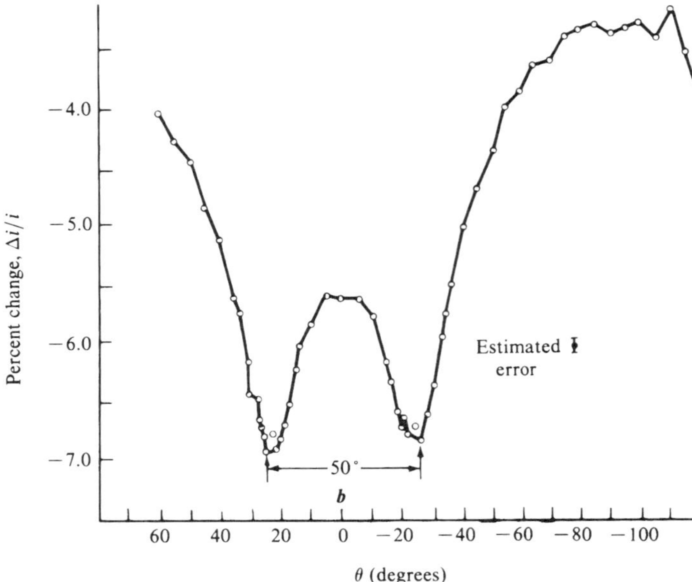

FIG. 23. Effect of magnetic field $H = 3400$ G on hole photocurrent i. H was rotated in the *ab* crystal plane. Photocurrents were measured with 24 V across the crystal; current with $H = 0$ was 1.3×10^{-5} A m^{-2} at 366 nm. The *b* axis is at 0 and the resonances (arrows) are calculated from theory. $T = 298$ K. (Redrawn from Geacintov *et al.*, 1970.)

D, are formed. Viewed as a chemical reaction, the overall process is usually written as

$$A + B \underset{k_{-1}}{\overset{k_1}{\leftrightarrows}} (AB) \xrightarrow{k_2} C + D, \tag{27}$$

where k_1 denotes the pair-formation rate constant and k_{-1} the pair-dissociation rate constant, and k_2 represents the transition to a new state of the system. The basic assumptions that allow the process shown in Eq. (27) to have a magnetic-field effect are that

1. the overall reaction is spin conserving, and
2. the thermalization times τ_t of the pair spin states are longer than the pair-state lifetimes τ_p.

Since H_S does not commute with the total spin of the pair state, the external magnetic field H can impart a different percentage of spin character to each of the pair states ψ_l. If the final states $C + D$ in Eq. (27) have a total spin S, and P_S denotes the S projection operator, then only pair states ψ_l for which $|\langle S|P_S\psi_l\rangle|^2 \neq 0$ contribute to the overall reaction since $k_2 \propto |\langle S|P_S\psi_l\rangle|^2$.

In geminate recombination, the relevant interactions are the electron and hole hyperfine terms; generally one includes only the contact term $\mathbf{I}\cdot\mathbf{S}$, the Zeeman interaction, and the exchange interaction $J(\mathbf{r})\mathbf{S}_e\cdot\mathbf{S}_h$. As an example of how field effects provide insight into fundamental processes, measurements by Frankevich and Lymarev (1992) on the magnetic-field effects on photocurrent in polymers PTS, PPV, PPPV, and PMOP-PPV have permitted estimates of the size of the geminate electron–hole (CT) state; these vary from 1.5 to 3.5 nm.

Triplet-exciton pairs can fuse to produce not only a fluorescent singlet state, but also free carriers provided that twice the triplet energy exceeds the band gap; and in such cases, the intrinsic current should be magnetic-field dependent. Evidence for exciton fusion as a mechanism for the intrinsic generation of free carriers in molecularly doped polymer consisting of solid solutions of the amine TTA [tri-(p-tolyl)amine] in the polymer Lexan was observed by Orlowski and Scher (1983). It was found that the photogeneration yield Y depended on light intensity as I^2 at low intensity and as I at high light intensity and at some intermediate value in between with I depending on the electric-field strength. The fusion of triplet excitons in this case resulted initially in a charge-separated state in which the polymer contributed an acceptor state and a hole was created on the TTA molecule. The overall rate of carrier generation was diffusion limited and had a rate constant of the order of

10^{-17} m^3 s^{-1}, which is typical of organic materials.

8. APPLICATIONS

8.1 Electrophotography

By far the greatest industrial application of organic semiconductors is in the field of electrophotography. A schematic of a conventional organic photoreceptor is shown in Fig. 24. It consists of a two-layer system: a charge-generation layer, between 1 and 3 μm in thickness, and a charge-transport layer, between 15 and 30 μm in thickness. Charge-transport layers are typically mixtures of low-molecular-weight donor or acceptor compounds in a polymer. The carrier moves by thermally assisted hopping along the path created by molecules in contact with each other. The charge-generation layer is chosen for its sensitivity to the light used for exposure, and for its quantum efficiency of charge generation. This layer plays the role of the photosensitive electrode, injecting charge into the charge-transport layer. The charge-transport layer is chosen for its superior carrier mobility, interlayer compatibility, absence of interfacial trapping, mechanical properties, and injected-carrier range.

The quantum efficiency of a two-layer system consisting of a metal-free phthalocyanine charge-generating layer and a hydrazone-doped polycarbonate charge-transport layer was measured, yielding a primary quantum yield of 0.5. See Borsenberger and Weiss (1993) and Schein (1988) for more complete details regarding electrophotography.

8.2 Electroluminescence (EL)

The industrial potential for electroluminescent organic materials, particularly polymers, is great. Not only does this provide the possibility of large-area, flexible illumination, with significant energy saving, but it makes possible flat video screens. EL was first reported in polymers by Burroughes *et al.* (1990) in *p*-phenylene vinylene (PPV). This thin film has a low turnon voltage (~3 V) and relatively high quantum efficiency (1% photons out/electrons in), and shows green and orange light using different polymers. They are bright, and can be seen in room lighting. In addition, a light-emitting diode (LED) made out of oriented polymers will emit polarized light; this opens another possibility for transferring information. Oriented photoluminescence has already been demonstrated (Heeger, 1992). Recent improvements in PPV involve preparing block copolymers consisting of alternating sequences of conjugated and nonconjugated segments (Holmes *et al.*, 1992). This mimics the GaAs quantum-well structure. By varying the degree of conjugation and copolymerization, the colors can be varied from green-blue to orange-red; the light is also brighter. More efficient LEDs can be made with transparent polyaniline as a hole-injecting electrode, instead of indium and tin oxide (Yang and Heeger, 1994).

The same device that emits EL can, under reverse bias, act as a photodetecting diode, with a yield of 0.2 electron/photon (Yu *et al.*, 1994). Electron injection is usually the difficult function to accomplish because it requires efficient chemical reducing power. Such materials are susceptible to attack by ubiquitous oxidants, such as air. Alloys of Mg and Ag were used by Tang and Van Slyke (1987), who ushered in the modern era of EL by introducing the scheme of using a bilayer structure, in which the functions of charge generation, transport, and luminescence were accomplished by using materials best suited for each function. This structure consists of a hole-transport layer and a luminescent layer. The hole-transport layer is an amorphous diamine film, and the luminescent layer consists of 8-hydroxyquinoline aluminum (Alq_3), which mostly transmits electrons. An intermediate layer of coumarine-doped Alq_3 improves EL by a factor of 2. A typical multilayer EL configuration is shown in Fig. 25.

The main problem with organic EL devices is operational stability. At present, 10 000 hours of continuous operation has been achieved.

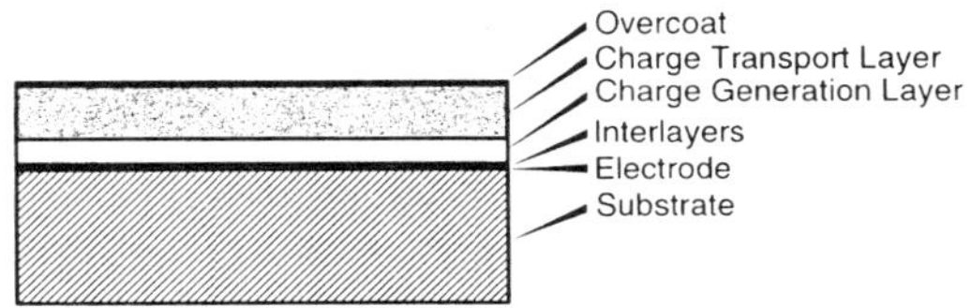

FIG. 24. A cross-section schematic of a conventional dual-layer photoreceptor (redrawn from Borsenberger and Weiss, 1993.)

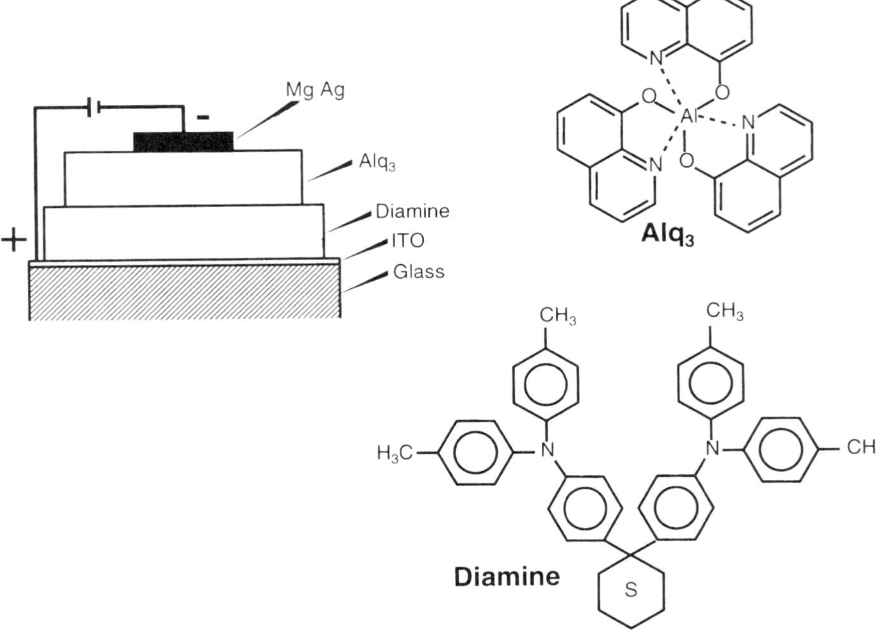

FIG. 25. Configuration of a dual-layer electroluminescent cell and molecular structures of the hole-transport layer (diamine) and the electron and luminescent layers (Alq_3). The top electrode is an electron injector and the ITO layer injects holes. This organic diode acts as a rectifier, with the forward bias represented by a positive voltage on the ITO. The unit can emit visible light at 2.5 V. (Redrawn from Tang and Slyke, 1987.)

8.3 Sensors

Considerable work has been carried out on the phthalocyanines with the goal of preparing chemically stable sensors for various gases. A reproducible sensor sensitive to nitrogen dioxide at relatively low temperature (313 K) was prepared from tetra-*t*-butyl–substituted copper phthalocyanine, deposited by the Langmuir–Blodgett (LB) technique on a glass substrate (Chyla *et al.*, 1993) and aged. The change in conductivity ΔJ of the LB film exposed to NO_2 relative to the conductivity J_0 of a film free of NO_2 varied from $\Delta J/J_0 \sim 0.2$ to $\Delta J/J_0 \sim 0.8$ as the NO_2 concentration was changed from 0.28 to 1.22 ppm.

Sensors based on conductive polymers, capable of detecting biologically important compounds such as glucose, are discussed by Heller (1992).

8.4 Semiconductor Devices

Three types of devices that have been fabricated are a Schottky-barrier diode, a metal-insulator–semiconductor (MIS) device, and a metal-insulator–semiconductor field-effect transistor (MISFET). The major advance in the use of organic semiconductor devices is the ability to prepare a solution of the precursor polymer that can be deposited in layers in the correct geometry, after which the precursor is transformed into the final conjugated polymer. Polyacetylene is one important polymer that can be used in this manner; it can be doped n type with Na and p type with I_2.

To prepare a Schottky-barrier diode, a layer of gold is evaporated on a substrate followed by a coating of polyacetylene, on top of which is placed the low-work-function layer of aluminum. The ratio of forward to reverse current is about 5×10^5 at a bias voltage of 1.5 V with a saturation current of about 10 A m^{-2}.

A MIS device has been prepared by depositing a layer of gold (20 nm) on a silica substrate, followed by a polyacetylene layer 20 nm thick, upon which in turn is deposited by spin-coating a layer of polymethylmethacrylate 150 nm thick, upon which is evaporated another 20-nm layer of gold. This device permits the creation of accumulation, depletion, and inversion layers in the polyacetylene, and thus a greater control of the current passing through the device.

In Fig. 26 is shown an example of a MISFET device. The source (emitter) and the drain (collector) are supplied with *n*-Si contacts, which are intended to make Ohmic contact to *n*-type polyacetylene film. The channel conductance changes by a factor of 10^5 when the gate-to-source voltage is changed from -40 to $+10$ V.

The MISFET device demonstrates that the charge at the polyacetylene/insulator interface in the MIS device is mobile, since, as is seen in Fig. 26, it is possible to pass current from source to drain and thus measure the conductance of the surface-charge layer. From these conductance measurements, it is possible to determine the carrier mobility, which comes out to 10^{-8} m^2 V^{-1} s^{-1}. Similar values were found for the extrinsic carriers in undoped polyacetylene.

8.5 Photorefractive Polymers

A photorefractive (PR) material is one in which a spatial modulation of its refractive index occurs as a result of a light-induced spatial modulation of an internal charge distribution. The material must therefore contain an optically nonlinear component. Such a material has potentially important applications, such as high-density optical storage, phase conjugation, and programmable optical interconnection. At present most if not all commercially available photorefractive materials are inorganic, such as $LiNbO_3$, $(Sr,Ba)Nb_2O_6$, GaAs, and $BaTiO_3$. The discovery of a photorefractive effect in polymers (see Moerner and Silence, 1994) is of great potential importance, since polymers are inexpensive, easy to process, and versatile.

The basic requirements for a PR material are

1. a photoionizable charge-generating component,
2. a charge-transporting medium,
3. trapping sites, and
4. a nonlinear optical (NLO) material whose index of refraction can be altered under the influence of a space-charge field.

The principle of operation is as follows: A polymer that contains all of the components listed above is prepared in the form of a thin film (~100 μm) sandwiched between transparent electrodes. A large electric field is applied to the sandwich, at a temperature appropriate for the alignment of the NLO chromophores with the external field. The optimal NLO optical materials are conjugated organic compounds with an asymmetric charge distribution as would be produced by having an electron-donating substituent at one end of the molecule and an electron-accepting molecule at the other end. Matters are improved if the distance between these groups is increased judiciously. The orientation of the NLO chromophores by means of an external electric field breaks the isotropy of the sample making possible a maximum electro-optic effect.

To write a hologram, two intersecting coherent beams of light are focused on the surface of the film, using a wavelength that addressed the charge-generating component only. The interference between the beams of

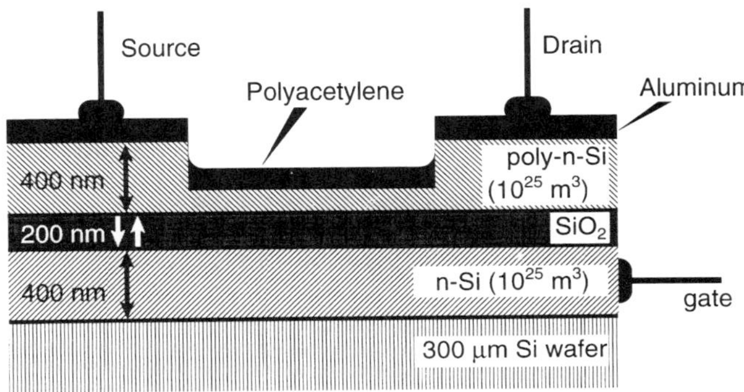

FIG. 26. Schematic diagram for a polyacetylene MISFET structure. Dimensions shown are to scale, except the channel width (20 μm) and length (1.5 $\times$ 10^{-2} m). (Redrawn from Burroughes *et al.*, 1990.)

light produces a stationary pattern of light and dark fringes. In the illuminated regions, charge generation takes place, and the external electric field produces charge separation. The charge-generation component is dispersed in a matrix that is chosen to support the movement of only one sign of carrier, usually the hole. The electron thus remains stationary and the positive charge is displaced into traps in the dark region. The light source is removed, leaving a pattern of displaced electrons and holes similar to the interference pattern, but the phase of the space-charge electric field is shifted in phase from the light-intensity pattern by about 90°. The displacement of the trapped holes from the immobile electrons produces an electrostatic restoring field that is high enough to alter the refractive index of the NLO chromophore. Interestingly, the presence of traps, which is the bane of photoconductors, is a basic necessity in the PR effect.

If the material has a linear electro-optic effect, the magnitude of the refractive-index modulation Δn is related to the space-charge field E_{sc} by the relation

$$\Delta n = -\tfrac{1}{2}E_{sc}n^3r_e, \tag{28}$$

where r_e is the effective electro-optic coefficient for the geometry under consideration (about 1–20 for organics). Using a figure of merit $Q = n^3r_c/\epsilon_r$ where ϵ_r is the dc dielectric constant, the value for the inorganic material is 19.3; for present polymeric materials it is 2.2 with a potential value of 31.1 in sight.

The phase-shifted refractive-index modulation forms a grating that can diffract light and permit the utilization of two-beam coupling whereby one of the writing beams gains energy at the expense of the other, a property that is made use of in photonic devices. A sample composition for a PR polymer is polyvinylcarbazole (PVK) functioning as the hole transporter, doped with a NLO chromophore compound, 3-fluoro-4-(*N*,*N*-diethylamino)-β nitrostyrene (FDEA-NST) and sensitized for charge generation with 2,4,7-trinitro-9-fluorene (TNF). Typical poling field strengths are about 10^7 V m^{-1}, and space-charge fields about half that. Carrier mobilities are in the range 10^{-11} to 10^{-10} m^2 V^{-1} s^{-1}, and quantum efficiencies of charge generation are $\sim 10^{-2}$. A grating efficiency η, defined as the intensity ratio of the diffracted beam to that of the incoming reading beam, of the order of 100% has been reported (Meerholz *et al.*, 1994).

Although the readout requires the use of light, which tends to destroy the information residing in the charge pattern, it is possible to minimize this effect (Silence *et al.*, 1994). This greatly increases the utility of the PR effect as an information-storage method. For a complete introduction to and survey of PR methodology, see Moerner and Silence (1994).

ACKNOWLEDGMENTS

We have benefited from the support of Department of Energy grant No. DE-FG02-86ER-60405 and from discussions with Heinz Bässler, Charles B. Duke, R. Glen Kepler, William E. Moerner, Lewis J. Rothberg, Lawrence B. Schein, Ching W. Tang, and Marek Zielinski, and especially with Martin A. Abkowitz and Nicholas E. Geacintov.

GLOSSARY

Aliphatic: Containing no aromatic rings.

Amphiphilic: Having both hydrophobic and hydrophilic substituents on the same molecule.

Aromatic Compound: An unsaturated hydrocarbon containing planar ring systems stabilized by delocalized electrons, as in benzene.

Autoionization: The spontaneous ionization of an electronically excited molecular state.

Bipolaron: A quasiparticle composed of two correlated like-charged polarons.

Chromophore: An atom or group of atoms that acts as a unit in light absorption.

cis: A molecular geometry; thus in the structure shown, *A* and *B* are *cis* with respect to the double bond, which does not permit independent rotation of *A* or *B* around the bond.

```
R         R
 \       /
  C  =  C
 /       \
A         B
```

Conjugated: A system of alternating covalent single and double bonds; electron delocalization ensues.

Coulomb-Capture Radius: The mean distance at which the Coulomb-interaction energy between oppositely charged particles equals the thermal energy.

DPQ: 3,3′-dimethyl-5,3′-di-*t*-butyldiphenoquinone.

Exciton: A mobile electronically excited neutral nonconducting state.

Fermi Glass: A molecular solid in which there are no bandlike states, i.e., all electronic states are localized.

Geminate Recombination: Fusion of a hole–electron pair formed from the same molecule.

Glass-Transition Temperature: A small range of temperature over which the viscosity of the melt increases steeply, viz., several orders of magnitude within 10 K.

HOMO: Highest occupied molecular orbital.

ITO: Indium–tin oxide transparent contact.

LUMO: Lowest unoccupied molecular orbital.

OSC: Organic semiconductor.

PA: Polyacetylene.

PC: Polycarbonate.

Peierls Instability: A lattice instability in a one-dimensional system due to electron–phonon (and/or Coulomb) interactions, having rational-fractionally filled orbitals.

Pendant Group: A group of atoms attached as a unit to a polymeric backbone.

π Bond: A chemical bond formed by the overlap of neighboring *p* orbitals that has one and only one nodal plane, which includes the internuclear axis.

PMPS: Poly(methylphenylsilylene).

Polaron: A quasiparticle consisting of an excess charge and its associated polarization field.

Primary Quantum Yield: The yield per photon of carriers before any recombination of any kind has occurred.

PS: Polystyrene.

PVK: Polyvinylcarbazole.

Schottky Barrier: A potential barrier formed by a space-charge double layer whose width and magnitude are dependent on the applied voltage.

Sigma (σ) Bond: A chemical bond made by the overlap of orbitals that have cylindrical symmetry around the bonding axis.

Soliton: A mobile bond-alternation defect in a π-conjugated chain; if charged, has a spin of zero, and if uncharged, has a spin of one-half.

Symbatic: In phase with the absorption spectrum.

Thermalization Distance: The distance traveled by an ionized and excited electron before it comes into thermal equilibrium with its surroundings. Distance is measured in a straight line between starting and final points.

TNF: 2,4,7-trinitro-9-fluorene.

TOF: Time of flight.

trans: A type of molecular geometry; thus in the structure shown, *A* and *B* are *trans* with respect to the double bond, which does not permit independent rotation of *A* or *B* around the bond.

```
R         B
 \       /
  C = C
 /       \
A         R
```

Works Cited

Abkowitz, M., Bässler, H., Stolka, M. (1991), *Philos. Mag. B* **63,** 201–220.

Abkowitz, M., Facci, J. S., Stolka, M. (1993), *Appl. Phys. Lett.* **63,** 1892–1894.

Bässler, H., Brandl, V., Deussen, M., Göbel, E. O., Kersting, R., Kurz, H., Lemmer, U., Mahrt, R. F., Ochse, A. (1995), *Pure Appl. Chem.* **67,** 377–385.

Bässler, H. (1993), *Phys. Status Solidi B* **175,** 15–56.

Bässler, H. (1981), *Phys. Status Solidi B* **107,** 9–54.

Bässler, H., Herrmann, G., Riehl, N., Vaubel, G. (1969), *J. Phys. Chem. Solids* **30,** 1579–1585.

Bonham, J. S., Jarvis, D. H. (1978), *Aust. J. Chem.* **31,** 2103–2115.

Borsenberger, P., Rossi, M. (1992), *J. Chem. Phys.* **96,** 2390–2394.

Borsenberger, P. M., Weiss, D. S. (1993), *Organic Photoreceptors for Imaging Systems,* New York: Dekker.

Braun, C. L. (1984), *J. Chem. Phys.* **80,** 4157–4161.

Brédas, J. L., Street, G. B. (1985), *Acc. Chem. Res.* **18,** 309–315.

Burroughes, J. H., Bradley, D. D. C., Brown, A. R., Marks, R. N., Mackey, K., Friend, R. H., Burns, P. L., Holmes, A. B. (1990), *Nature* **347,** 539–541.

Burroughes, J. H., Jones, C. A., Lawrence, R. A., Friend, R. H. (1990), in: J. L. Brédas, R. R.

Chance (Eds.), *Conjugated Polymeric Materials: Opportunities in Electronics, Optoelectronics, and Molecular Electronics,* Dordrecht, the Netherlands: Kluwer, p. 221.

Chance, R. R., Braun, C. L. (1973), *J. Chem. Phys.* **59,** 2269–2272.

Chyla, A., Sworakowski, J., Szczurek, A., Brynda, E., Nešpůrek, S. (1993), *Mol. Cryst. Liq. Cryst.* **230,** 1–6.

Duke, C. B. (1982), "The Electronic Structure of Semiconducting Polymers," in: Joel S. Miller (Ed.), *Extended Linear Chain Compounds,* Vol. 2, New York: Plenum, pp. 59–125.

Frank, R. I., Simmons, J. G. (1967), *J. Appl. Phys.* **38,** 832–840.

Frankevich, E. L., Lymarev, A. A. (1992), *Mol. Cryst. Liq. Cryst.* **218,** 103–108.

Gailberger, M., Bässler, H., (1991), *Phys. Rev. B.* **44,** 8643–8651.

Gartstein, Y. N., Conwell, E. M. (1994), *Chem. Phys. Lett.* **217,** 41–47.

Geacintov, N. E., Pope, M., Fox, S. (1970), *J. Phys. Chem. Solids* **31,** 1375–1379.

Gerischer, H., Willig, F. (1976), *Top. Curr. Chem.* **61,** 31–84.

Grobman, W. D., Koch, E. E. (1979), in: L. Ley, M. Cardona (Eds.), *Photoemission in Solids II,* Topics in Applied Physics Vol. 27, New York: Springer-Verlag.

Heeger, A. J., Kivelson, S., Schrieffer, J. R., Su, W.-P. (1988), *Rev. Mod. Phys.* **60,** 781–850.

Heeger, A. J. (1992), *Conjugated Polymers and Related Materials: Interconnection of Chemical and Electronic Structure,* Oxford, U.K.: Oxford Univ. Press.

Helfrich, W., Mark, P. (1963), *Z. Phys.* **171,** 527–536.

Helfrich, W., Schneider, W. G. (1965), *Phys. Rev. Lett.* **14,** 229–231.

Heller, A. (1992), *J. Phys. Chem.* **96,** 3579–3587.

Holmes, A. B., Burn, P. L., Kraft, A., Friend, R. H., Bradley, D. D. C. (1992), *Nature* **356,** 47–49.

Hong, K. M., Noolandi, J. (1978), *J. Chem. Phys.* **69,** 5026–5039.

Karl, N., Schmid, E., Seeger, M. (1970), *Z. Naturforsch.* **25a,** 382–391.

Karl, N. (1990), in: K. Sumino (Ed.), *Defect Control in Semiconductors,* North-Holland: Elsevier, pp. 1725–1746.

Kepler, R. G., Anderson, R. A. (1992), *Adv. Phys.* **42,** 1–57.

Kepler, R. G., Soos, Z. G. (1993), in: T. Kobayashi (Ed.), *Relaxation in Polymers,* Singapore: World Scientific, p. 100–133.

Knox, R. S. (1963), *Theory of Excitons,* New York: Academic.

Lampert, M. A., Mark, P. (1970), *Current Injection in Solids,* New York: Academic.

Meerholz, K., Volodin, B. L., Sandalphon, Kippelen, B., Peyghambarian, N. (1994), *Nature* **371,** 497–500.

Merrifield, R. E. (1968), *Acc. Chem. Res.* **1,** 129–135.

Miller, J. S., Epstein, A. J. (1994), *Angew. Chem. Int. Ed. Engl.* **33,** 385–415.

Miller, R. B., Michl, J. (1989), *Chem. Rev.* **89,** 1359–1410.

Mills, A. P., Jr., Karl, N. (1993), *Phys. Rev. B.* **48,** 7050–7056.

Moerner, W. E., Silence, S. M. (1994), *Chem. Rev.* **94,** 127–155.

Mort, J., Pfister, G. (1982) *Electronic Properties of Polymers.* New York: Wiley.

Nešpůrek, S. (1993), *Synth. Met.* **61,** 55–60.

Noolandi, J. (1982), *J. Electrostat.* **12,** 13–25.

Onsager, L. (1938), *Phys. Rev.* **54,** 554–557.

Orlowski, T. E., Scher, H. (1983), *Phys. Rev. B* **27,** 7691–7702.

Pope, M., Kallmann, H., Giachino, J. (1965), *J. Chem. Phys.* **42,** 2540–2543.

Pope, M., Kallmann, H. P., Chen, A., Gordon, P. (1962), *J. Chem. Phys.* **36,** 2486–2492.

(PS) Pope, M., Swenberg, C. E. (1982), *Electronic Processes Crystals,* New York: Oxford Univ. Press; (1997) *Electronic Processes in Organic Crystals and Polymers,* New York: Oxford Univ. Press.

Ramirez, A. P. (1994), *Superconductiv. Rev.* **1,** 1–101.

Ries, B., Schönherr, G., Bässler, H., Silver, M. (1983), *Philos. Mag.* **48,** 87–106.

Roberts, G. (1990), *Langmuir-Blodgett Films,* New York: Plenum.

Sariciftci, N. S., Heeger, A. J. (1994), *Int. J. Mod. Phys. B* **8,** 237–274.

Schein, L. B. (1988), *Electrophotography and Development Physics,* Heidelberg: Springer-Verlag.

Seki, K. (1989), *Mol. Cryst. Liq. Cryst.* **171,** 255–270.

Silence, S. M., Twieg, R. J., Bjorklund, G. C., Moerner, W. E. (1994), *Phys. Rev. Lett.* **73,** 2047–2050.

Silinsh, E. A. (1980), *Organic Molecular Crystals; Their Electronic State,* Berlin: Springer-Verlag.

(SC) Silinsh, E. A., Capek, V. (1994), *Organic Molecular Crystals: Interaction, Localization, Transport Phenomena,* New York: American Institute of Physics.

Stasiak, M., Jeszka, J. K., Zielinski, M., Plans, J., Kryszewski, M. (1980), *J. Phys. D.: Appl. Phys.* **13, L221–L224.**

Tang, C. W., Van Slyke, S. A. (1987), *Appl. Phys. Lett.* **51,** 913–915.

Vardeny, Z. V., Rothberg, L. J. (Eds.) (1994), Proceedings of 2nd International Conference on Optical Probes of Conjugated Polymers and Fullerenes, *Mol. Cryst. Liq. Cryst.* **256,** 1–932.

Wöhrle, D., Meissner, D. (1991), *Adv. Mater.* **3,** 129–138.

Yang, Y., Heeger, A. J. (1994), *Appl. Phys. Lett.* **64,** 1245–1247.

Yu, G., Zhang, C., Heeger, A. J. (1994), *Appl. Phys. Lett.* **64,** 1540–1542.

Zasadzinski, J. A., Viswanathan, R., Madsen, L., Garnaes, J., Schwartz, D. K. (1994), *Science* **263,** 1726–1733.

Further Reading

Borsenberger, P. M., Weiss, D. S. (1993), *Organic Photoreceptors for Imaging Systems,* New York: Dekker.

Brédas, J. L., Chance, R. R., (Eds.) (1990), *Conjugated Polymeric Materials: Opportunities in Electronics, Optoelectronics, and Molecular Electronics,* Dordrecht, the Netherlands: Kluwer.

Duke, C. B. (1982), "The Electronic Structure of Semiconducting Polymers," in: Joel S. Miller (Ed.), *Extended Linear Chain Compounds,* Vol. 2, New York: Plenum, pp. 59–125.

Gerischer, H., Willig, F. (1976), *Top. Curr. Chem.* **61,** 31.

Grobman, W. D., Koch, E. E. (1979), in: L. Ley, M. Cardona (Eds.), *Photoemission in Solids II,* Topics in Applied Physics Vol. 27, New York: Springer-Verlag.

Heeger, A. J. (1992), *Conjugated Polymers and Related Materials: Interconnection of Chemical and Electronic Structure,* Oxford, U.K.: Oxford Univ. Press.

Karl, N. (1990), in: K. Sumino (Ed.), *Defect Control in Semiconductors,* North-Holland: Elsevier, 1725–1746.

Knox, R. S. (1963), *Theory of Excitons,* New York: Academic.

Lampert, M. A., Mark, P. (1970), *Current Injection in Solids,* New York: Academic.

Miller, J. S., Epstein, A. J. (1994), *Angew. Chem. Int. Ed. Engl.* **33,** 385–415.

Moerner, W. E., Silence, S. M. (1994), *Chem. Rev.* **94,** 127–155.

Mort, J., Pfister, G. (1982), *Electronic Properties of Polymers,* New York: Wiley.

Mott, N. F., Davis, E. A. (1979), *Electronic Processes in Non-Crystalline Materials,* Oxford, U.K.: Clarendon.

Noolandi, J. (1987), "Stochastic Theory of Electron-Hole Transport and Recombination in Amorphous Materials," in: G. R. Freeman (Ed.), *Kinetics of Non-Homogeneous Processes,* New York: Wiley-Interscience, pp. 465–525.

Pope, M., Swenberg, C. E. (1984), *Annu. Rev. Phys. Chem.* **35,** 613–656.

(PS) Pope, M., Swenberg, C. E. (1982), *Electronic Processes in Organic Crystals,* New York: Oxford Univ. Press.

Schein, L. B. (1988), *Electrophotography and Development Physics,* Heidelberg: Springer-Verlag.

Silinsh, E. A. (1980), *Organic Molecular Crystals; Their Electronic State,* Berlin: Springer-Verlag.

(SC) Silinsh, E. A., Capek, V. (1994), *Organic Molecular Crystals: Interaction, Localization, Transport Phenomena,* New York: American Institute of Physics.

Vardeny, Z. V., Rothberg, L. J. (Eds.) (1994), Proceedings of 2nd International Conference on Optical Probes of Conjugated Polymers and Fullerenes, *Mol. Cryst. Liq. Cryst.* **256,** 1–932.

SENSORS, ACOUSTIC

LAWRENCE C. LYNNWORTH, *Panametrics, Inc., Waltham, Massachusetts, U.S.A.*

INTRODUCTION

Acoustic sensors may be understood to mean transducers, devices, or systems that generate, detect, or measure mechanical vibrations. These vibrations may be characterized in terms of frequency, bandwidth, information content, intensity, polarization, speed of propagation, and the rate at which they are attenuated. Mechanical vibrations or elastic waves interact with, influence, or are influenced by other elastic waves and also by other forms of energy (thermoacoustic or optoacoustic interactions), a material's elastic properties and stress, and metallurgical and chemical processes (ultrasonic welding, sonochemistry). Acoustic waves are influenced by gradients and inhomogeneities in, and by the motion of, the medium in which they are propagating (turbulent or particulate scattering, beam drift, Doppler shift), and by the boundaries of that medium, or boundary conditions.

Ultrasonic refers to elastic waves of frequency $f \geq 20$ kHz. Ultrasonic waves are thus vibrations at a frequency above human hearing. With respect to sensing and measurements, high frequency avoids interference from many audible, low-frequency noises due to wind, machinery, pumps, and vibration of large bodies. High frequency allows resolution of "the small" in both the temporal and spatial senses. Attenuation often imposes the upper limit on the maximum usable frequency. *Audible* frequencies, 16 Hz $< f <$ 20 kHz, are sometimes selected to overcome attenuation or to limit propagation to one mode in a plate. The parameter that is to be measured is called the *measurand.*

The *sensor* may be a system that includes one or more transducers and the medium itself or an existing internal or bounding structure. The sensor system may also be a foreign sensor, intrusive, added to the existing process to make the required measurement. The four main measurands in process control are flow, pressure, temperature, and level. (To the extent that flow measurements include mass flow rate and flow profile, the rheological properties of fluids such as density and viscosity may be viewed as falling occasionally under the "flow" umbrella.) To sense these measurands by ultrasonic pulse techniques, one can measure propagation in the medium itself (e.g., speed of sound in air

3-527-28139-8/96/$5.00 + .50

is proportional to the square root of the air's absolute temperature), or in a "sensor" that touches the medium. Rather than perturb the situation, one prefers to exploit an existing structural element as the sensor, or as a reflector that bounds the sensed medium at a known location. A pipe wall or an entire section of pipe, for example, can sometimes be used as a temperature sensor, a fluid-density sensor, or an acoustic window that allows one to interrogate the fluid within the pipe noninvasively. The fluid then serves as its own "sensor." On the other hand, to avoid uncertainties attributed to variability in existing structural elements not expressly designed nor quality controlled for sensing, one may need to introduce a foreign sensor [or foreign reflector(s)] for measurement purposes. This invasive sensor is often a resonator or waveguide—some form of probe—in which propagation is a unique and repeatable function of the measurand in the medium. The invasive or "wetted" sensor may also consist of one or more transducers that are each acoustically isolated from the medium's boundary, and hence from one another if two transducers are used, so that the medium can be interrogated or listened to without interference from boundary-borne noises.

The discussion of acoustic sensors is limited here to technical equipment and practical applications in the frequency range 10^3 to 10^9 Hz, and to "small" amplitudes (particle displacement $\ll$ wavelength). States of matter will include gas, liquid, and solid media and multiphase mixtures. Emphasis is placed on

1. two decades of the ultrasonic spectrum, between 50 kHz and 5 MHz;
2. measurement of the four principal measurands in the field of industrial process control, with emphasis on flow and liquid level in closed conduits and tanks, respectively; and
3. sensing of such measurands based on their influence on the propagation of ultrasound.

Waves of interest include longitudinal, transverse shear, extensional, torsional, and plate waves (Lamb, Rayleigh). It will be seen that appropriate utilization of transducers often allows one to interpret wave speed, reflections, transmission, and attenuation in terms of one or more measurands. "Appropriate utilization" includes selecting the proper piezoelectric material (Gualtieri *et al.*, 1994), vibration mode, transducer type and location, coupling means, probing means, electronic excitation, detection, and display or output means, etc.

Acoustical waves are often selected as the means to measure a particular parameter. One example is measuring the distance to a moving web or sheet. The sound waves can accomplish this task without contacting the web. Ultrasound often gets the job done with no penetration of the boundary (e.g., clamp-on transducers) or with minimal disturbance of the measurand even when the boundary must be penetrated (e.g., wall-mounted transducers for an across-the-stack flow measurement). Another example is downward-looking "air sonar" to measure the level of liquids in a tank or solids in a bin. Process-control engineers may select acoustical waves to sense process measurands when they need

- noninvasive, noncontact, or minimally invasive measurement;
- high accuracy (measurands usually transformed to time or frequency measurements);
- reliability (no moving parts, in the usual sense);
- fast response (can be <1 ms);
- remote sensing, sometimes with no physical contact;
- average reading over an extended region;
- profile information (point by point, or small-path average);
- computer compatibility of time, frequency, or, say, 8-bit amplitude data;
- low cost, especially for multiplexed and/or mass-produced sensors;
- small size, small mass;
- avoidance of problem(s) associated with competing (nonultrasonic) technologies;
- data or results unobtainable any other way; and
- useful sensing of two or more parameters simultaneously, e.g., flow velocity and composition, density and viscosity, or flow velocity and density.

The seven basic elements in a typical active acoustical sensing system may be represented as [Electrical Energy Source] → [Electronic Transmitter] → [Electroacoustic

Transmitting Transducer] → [Wave–Measurand Interaction Zone] → [Acoustoelectric Receiving Transducer] → [Electronic Receiver] → [Information Output]. These [items] or [functions] will become clear in relation to measuring specific measurands. An abbreviated representation of an active acoustical sensing system is [Transmit] → [Interact] → [Receive].

Much of the material in this article is condensed from Lynnworth (1989, 1994a). That book and chapter, respectively, or the references cited therein, may be referred to for more technical or historical details than could be included here. The present article, however, includes some new developments, new data, and references to works published in the period 1992 to early 1995. See also Papadakis (1997). In this article, space restrictions generally make it impractical to identify every concept with its inventor or author. Manufacturers are identified and specific products and model numbers for measuring flow, liquid level, temperature, pressure, etc., by ultrasonic as well as nonultrasonic sensors are indexed or reviewed annually in several periodicals, e.g., in the U.S. journals *m&c (Measurements & Control)* and *Sensors,* and in archival periodicals such as *Journal of the Acoustical Society of America, Journal of Sound and Vibration, IEEE Transactions on Ultrasonics, Ferroelectrics, and Frequency Control,* and *Acoustica.* Journals of interest that are published in the United Kingdom include *Measurement Science & Technology* and *Ultrasonics.*

1. PRINCIPAL MEASURANDS IN PROCESS CONTROL

1.1 Flow

All four principal process-control measurands have been analyzed and measured by acoustics, but acoustic measurement of flow is probably the subject of more studies than pressure, temperature, and level combined. A study of the issues involved in measuring flow by transmission and reflection methods provides a base for understanding many other acoustic/ultrasonic measurements.

1.1.1 Flow Profile Flow velocity is theoretically zero at the walls of a conduit and maximum at the center or at some other point. One is usually interested in *total* volumetric flow $Q = V_A A$ or total *mass* flow $M_f = \rho Q$, where V_A = area-averaged flow velocity, A = duct area, and ρ = fluid density. Now V_A can be thought of as the area-weighted average of V_i's in annuli or pixels of areas A_i. One can try to measure V_i in many small regions of area dA and then integrate over A, or one can measure over one or more paths and relate the path averages V_P to V_A. Range-gated Doppler measurement is one way to get to $Q = \int_A V_i dA$. Use of a meter factor K leads to $V_A = KV_P$. Special paths or quadrature methods yield V_A largely independent of flow profile. Examples include midradius paths and Gauss–Chebyshev paths. If the various paths that are to be interrogated differ in transit time by more than the ringdown time of the transducers, it is possible to interrogate a number of these paths simultaneously by connecting sets of transmitting or receiving transducers in parallel electrically (Lynnworth, 1994c, 1995). In some cases interrogation is possible over the entire area A all at once. Examples of recent European research on the influence of flow disturbances are Fleury (1994) and Holm (1995). If the flow is irrotational, the circulation Γ can be measured along a closed path such as the inscribed equilateral triangle in a round pipe (Smith, 1994; Smith *et al.,* 1995).

In general, flow velocity is not only a function of position but also a function of time t. Acoustic flowmeters tend to be linear and so yield the correct time-averaged value for $V(t)$, unlike square-law devices that yield the rms value. Acoustic flowmeters are also inherently faster in response than sensor elements having inertia. [See Hamidullin (1993).] Regarding clamp-on, see, for example, Baumoel (1994).

1.1.2 Single-Phase Fluids In the contrapropagation method, ultrasonic (or sometimes audible) waves are transmitted upstream and downstream. From the transit times t_1 and t_2 in each direction, and knowledge of the path and flow profile, the average flow velocity V_A is determined. A rather simple derivation of the basic flow-sensing equation is possible if one imagines a fluid of sound speed c flowing at a uniform velocity $V < c$ in a duct of cross-sectional area A, interrogated by two point sensors on the axis spaced a distance L apart. The transit times in the upstream and downstream directions, respectively, are

$$t_1 = L/(c - V) \quad \text{and} \quad t_2 = L/(c + V). \tag{1}$$

The reciprocals of these transit times, multiplied by L, are

$$L/t_1 = c - V \quad \text{and} \quad L/t_2 = c + V. \tag{2}$$

Accordingly,

$$V = \frac{L}{2}\left(\frac{1}{t_2} - \frac{1}{t_1}\right) = \frac{L}{2}\left(\frac{\Delta t}{t_1 t_2}\right) \quad \text{and}$$
$$c = \frac{L}{2}\left(\frac{1}{t_2} + \frac{1}{t_1}\right) = \frac{L}{2}\left(\frac{\Sigma t}{t_1 t_2}\right). \tag{3}$$

The upstream–downstream time difference can be obtained from Eq. (1) as $\Delta t = 2LV/(c^2 - V^2)$. This can be expressed in terms of the Mach number $M_s = V/c$ for $M_s \ll 1$:

$$\Delta t = \frac{2LV/c^2}{1 - M_s^2} = (2LV/c^2)(1 + M_s^2 + M_s^4 + \cdots). \tag{4}$$

At sufficiently small Mach numbers, the following approximations are valid:

$$V \approx c^2 \Delta t/2L \tag{5}$$

and

$$\Delta t \approx 2LV/c^2. \tag{6}$$

For liquids, V is usually $\ll c$, but for gases in the tall stacks of 100-MW fossil-fueled electric utilities (Matson and Davis, 1994) or in offshore flare lines (Folkestad and Mylvaganam, 1993), it is not uncommon to find $V/c > 0.1$ at the time that flow measurements are required. If the flow is interrogated obliquely, or if refraction or beam drift occurs, the above equations must be re-examined.

Differences between ultrasonic measuring systems for gas flow and liquid flow stem from differences in kinematic and acoustic properties of these fluids. The densities of air and water, for example, are 1.3 and 1000 kg/m^3, respectively. Examples of contrapropagation flowmeters for liquids and gases are shown in Figs. 1 through 3. (See Liu *et al.*, 1996.)

Prior to about 1988, contrapropagation was limited to single-phase fluids. However, the use of cross-correlation detection has extended this method to two-phase cases (Jacobson *et al.*, 1988; Hayward, 1994). Conversely, improvements in reflection methods have enabled such equipment to measure "clear" liquids in which the dominant scatterers might be only turbulent eddies. By 1994 contrapropagation had been applied to single-phase liquids and gases up to 260 °C and to cryofluids such as liquid N_2 down to approximately −200 °C. The flow of liquids, even at these temperature extremes, can be measured using the *clamp-on* method, if one has the appropriate couplant and coupling method. Soft-metal foil and high coupling pressure constitute one of the solutions at temperature extremes. Measuring the flow of gases, however, generally requires an opening in the pipe, unless the pipe is plastic, or the gas pressure is $\gtrsim$100 bar. Exceptions may occur for special pipe materials and geometries, or where opportunities exist to subtract electronically the pipe-borne acoustic cross talk that normally would overwhelm the receiver if one attempted to build a gas clamp-on flowmeter.

As explained by Pollard (1977, pp. 139–140), for signals in the form of a pulse, the cross-correlation coefficient is given by

$$R_{12}(\tau) = \int_{-\infty}^{\infty} f_1(t) f_2(t - \tau) dt. \tag{7}$$

The cross-correlation coefficient is a measure of the correlation or similarity between two (nonidentical) wave forms as one of the signals, $f_2(t)$, is shifted along the time axis past the other signal $f_1(t)$. To understand how cross-correlation can find a signal buried in noise, consider that if the form of the transmitted signal is known, then it may be cross-correlated with the received sum of signal and noise. Since the cross-correlation coefficient for random noise eventually averages out to zero, there remains only the signal in the form of its autocorrelation function. Furthermore, since phase information is retained in computing the cross-correlation coefficient, the value of τ for which $R_{12}(\tau)$ has its peak value is a measure of the transit time of the signal.

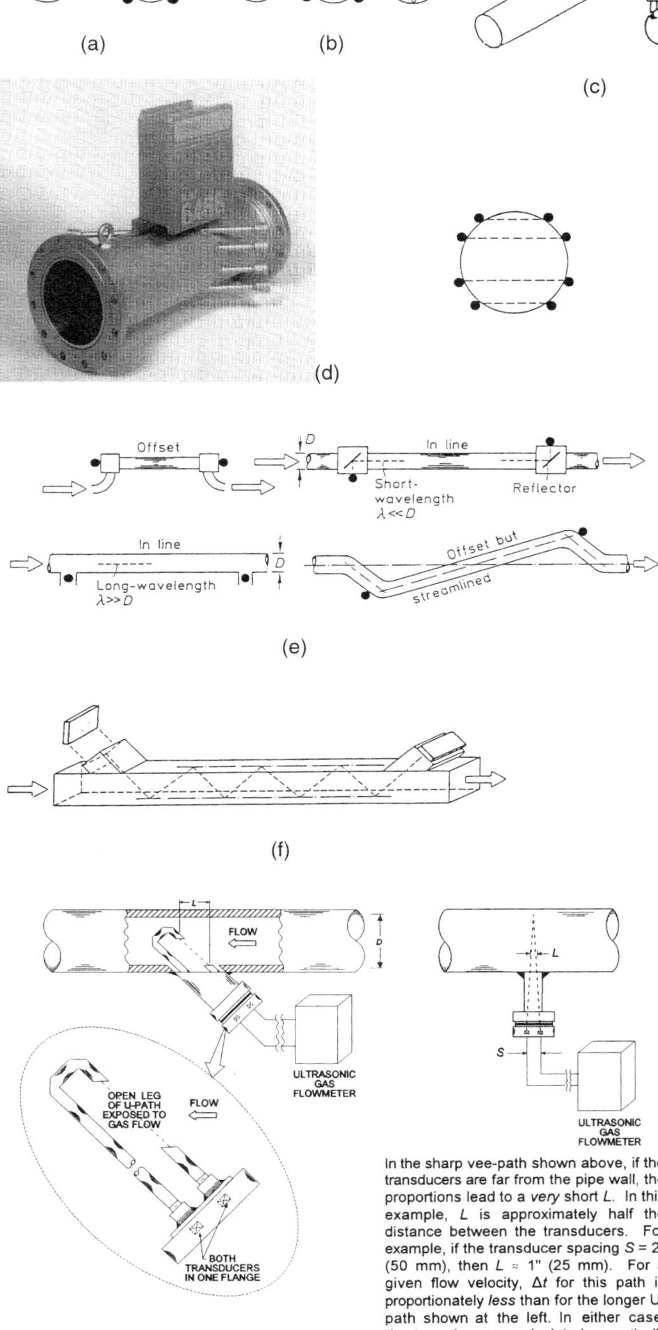

FIG. 1. Examples of transducer locations (●) and interrogation paths (---) for contrapropagation flow measurements along (a) tilted diameter(s), (b) one, two, or three midradius chords, (c) chord segment, (d) Gauss–Chebyshev chords, (e) axial paths that are offset, in-line, and offset but using an uninterrupted streamlined cell, (f) 45° zigzag interrogation of 100% of flowing cross section in a square conduit, and (g) paths requiring only one port. Copyright credits: (a), (b), (e), and (f) VCH, 1994; (c) Texas A&M University, 1984; (d) and (g) Panametrics, 1992 and 1994.

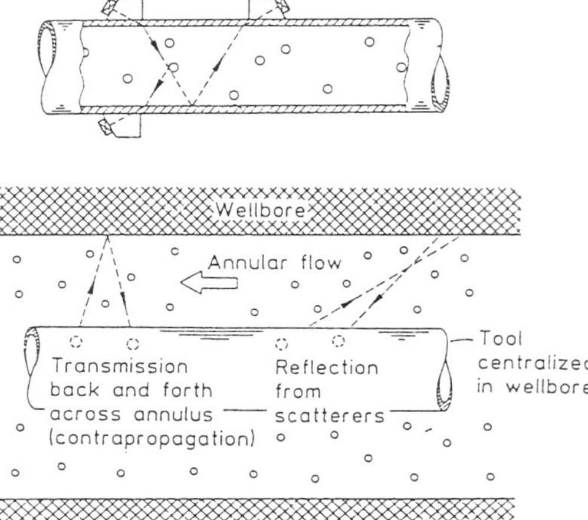

FIG. 2. Example of transducer (⊠,⊠) locations to measure the flow velocity of a fluid that is intermittently two-phase. Top, ordinary pipe. Bottom, downhole flow tool.

1.2 Level of Liquids

The level of liquids in closed vessels such as tanks or pipes can be measured with audible and ultrasonic elastic waves. The methods currently in use in industry may be divided into nonintrusive and intrusive categories, then divided according to whether the measurement is at one or more discrete levels or continuous, further divided according to the type of wave used (longitudinal, transverse shear, torsional, Lamb, Rayleigh, etc.), further divided according to whether propagation is measured in the liquid, in the liquid's container wall, or in a sensor immersed in the liquid, and lastly divided according to whether the effect of the liquid is primarily an attenuation effect or a transit-time or sound-speed effect. Some special solutions may not fit this categorization exactly; some of the following remarks may apply to measuring the level of solids too, and also to liquids in open channels.

Figures 4 and 5 compare a number of different liquid-level measuring approaches that are organized according to the above categories. As with flow, one would think that the nonintrusive measurement would be preferred over intrusive. The large number of intrusive liquid-level sensors in use, however, suggests that there must be important limitations on nonintrusive methods, or that there must be important advantages to intrusive sensors.

1.2.1 Nonintrusive Methods Nonintrusive methods (clamp-on, bond-on, press-on, weld-on) have the obvious apparent advantages of minimally interfering with normal operations in order to be used, and being transportable from one site to another so that one sensor can yield measurements in different places. For example, a fuel-truck driver could temporarily attach a magnetically clamped high-level sensor to each customer's steel tank that was to be filled, to avoid overfilling. On the other hand, a clamp-on device that is seemingly so simple to use might be misused or abused. Permanent intrusive sensors can be mass produced largely independent of the tank-wall details, as they are typically threaded into a standard pipe fitting.

The nonintrusive techniques that are well known include liquid-presence detecting at a discrete point, based on detecting an echo from the far wall; sensing the change in ring-down when a probe is slid up and down a vessel wall; and sending the sound wave vertically up and timing the round trip. A hybrid case occurs when the nonintrusive transducer is outside the vessel but a stillwell or reflectors are placed inside the vessel to improve the measurement. (A stillwell is a tube of relatively small diameter compared with length, so that sloshing and wavy action are smoothed out or stilled. The stillwell sometimes provides the further advantage of guiding the wave. In other words, the stillwell prevents attenuation caused by beam spread, and avoids spurious echoes from off-axis reflectors. A potential disadvantage is that the density of liquid inside the stillwell could end up being different from that outside the stillwell.)

Nonintrusive techniques that have been overlooked in numerous surveys of liquid-level sensors are those using plate waves and transverse shear waves. One plate-wave solution, due to Royer *et al.* (1992), is attenuation based and makes use of the leakage of symmetric plate waves. Another plate-wave or flexural-wave solution is sound-speed based and makes use of the retarding effect of a liquid adjacent to a tank wall (Liu and Lynnworth, 1993, 1995). This latter plate-

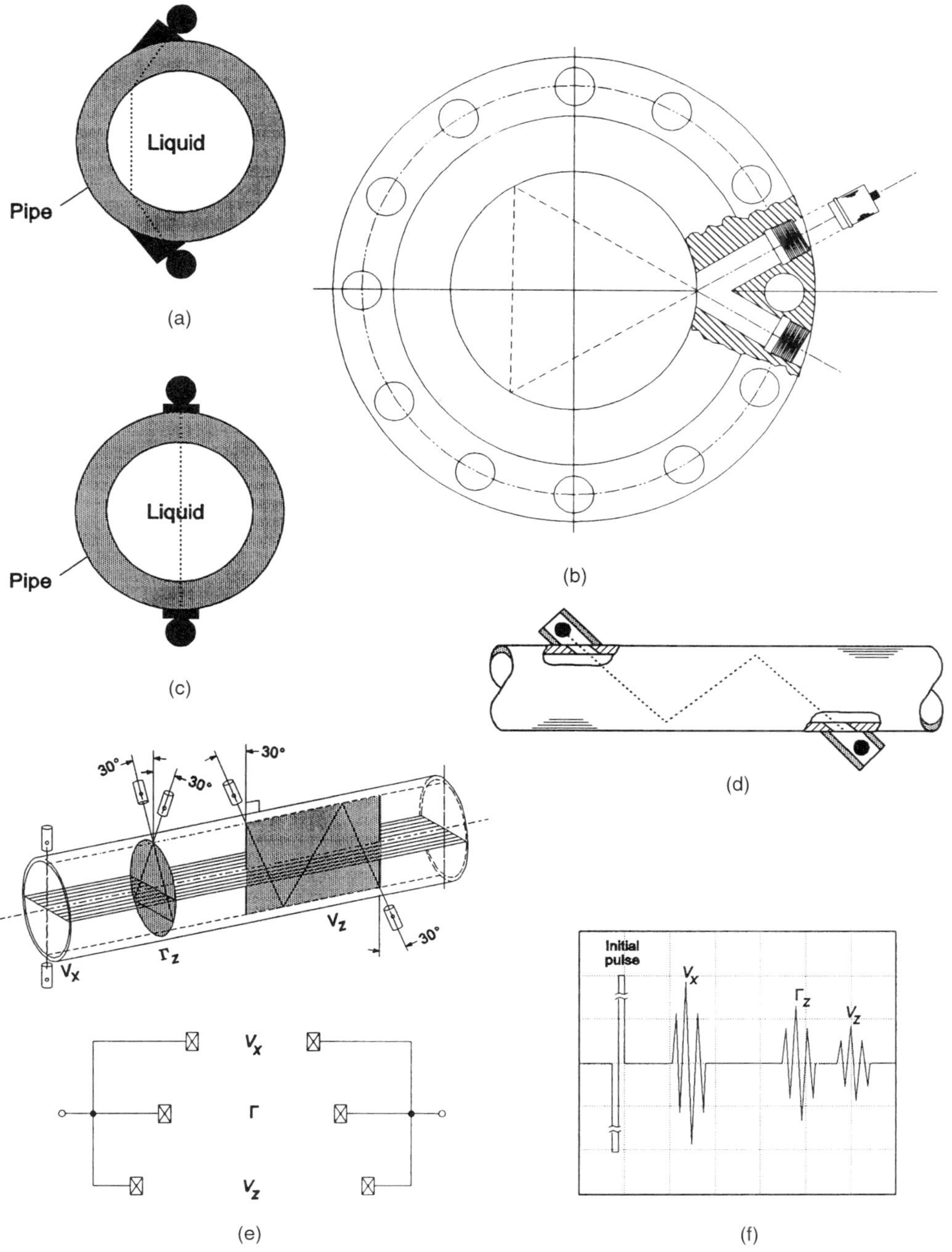

FIG. 3. Transducer locations (●) and paths (---) for measuring (a) liquid swirl presence by clamp-on, (b) circulation Γ_z of a vortex, (c) liquid cross flow V_x by clamp-on, (d) axial flow V_z, (e) the simultaneous measurement of Γ_z, V_x, and V_z in air, using pairs of electrically paralleled, acoustically isolated transducers and paths differing in transit time by more than the ringdown time of the transducers, (f) schematic of oscillogram corresponding to (e).

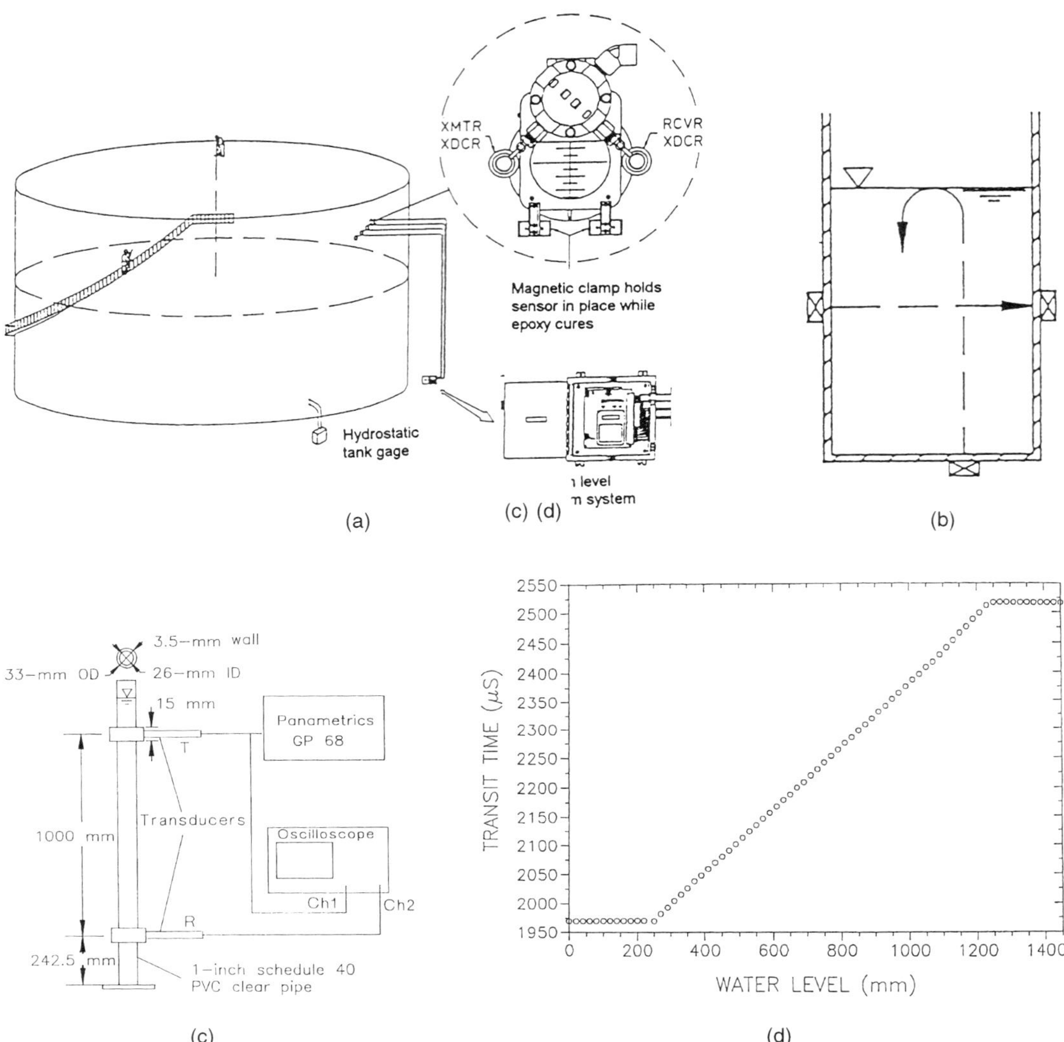

FIG. 4. Nonintrusive liquid-level techniques (a) for detecting liquid presence near the top of large fuel-storage tanks, and (b) for discrete and continuous measurements on a relatively small-diameter vessel whose bottom is accessible and not covered with deposits. Method in (c) uses flexural or bending waves, and yielded the data plotted in (d), after Liu and Lynnworth (1993, 1995). © 1993 Panametrics. Although such waves propagate dispersively (Brillouin, 1960) they nevertheless may prove practical.

wave method can be applied with the path horizontal or vertical. In some cases, the choice depends on the objective being discrete or continuous measurements.

1.2.2 Intrusive Methods In the situation shown in Fig. 5(a), the downward-looking sonars may operate free-field or in a multiple-notch stillwell. In either case the transducers do not contact the product, but a hole in the roof is necessary. The hole is sealed by a flange when the sensor or transducer is installed. Recent advances in transducers for use in air, as may be used in downward-looking sonars, are discussed or reviewed by Dabirikhah and Turner (1994), Hutchins (1994), and Mágori (1994). If the product beneath the vapor is a settled liquid that is clean and not bubble- nor residue-bearing, a wetted probe such as a torsional waveguide may be used (Kim *et al.,* 1993). [Some versions of this waveguide may be used to sense liquid density ρ and viscosity η. Whether this is better than alternative and especially noninvasive solutions based on the complex reflection coefficient (Sheen, 1994) or streaming (Hertz *et al.*, 1991) needs to be determined in each case.] Sometimes a magnetic float rides up and down a magnetostrictive waveguide, using the Wiedemann effect to achieve positional or liquid-level accuracy of a fraction of 1 mm over long spans (manufacturer: MTI Corp.).

Tuning forks and gap sensors are in wide use, installed by threading or flanging. Liquid dampens the tuning-fork vibration, an "attenuation" effect, but liquid increases the gap signal compared with the near-zero ultrasonic signal strength transmitted across the vapor. On the other hand, an airborne signal across a wall-to-wall "gap" will be interrupted (attenuated) by the presence of solids piling up in the path. SAW (surface acoustic wave) and plate-wave sensors are often proposed to sense the presence of particular substances (Wenzel *et al.*, 1994).

The upward-looking transducer runs the risk of being buried in sludge, but if cleaned or if used only in clean liquids, it can provide a reliable echo from the liquid–vapor interface. A stillwell stabilizes the interface and can contain reference reflectors to compensate for sound-speed gradients, the importance of which increases if a temperature gradient exists. This need for multiple reflectors is readily appreciated if one recognizes that, for water and some other liquids, sound speed changes by about $\frac{1}{4}$% per °C. A 45° internal (immersed) reflector sometimes allows a sidewall-mounted transducer to interrogate vertically. The difficult problem of measuring the level of molten aluminum is discussed by Nygaard and Mylvaganam (1993).

1.3 Temperature

This section covers average temperature, temperature profile, and tomographic reconstruction. In most cases, the temperature is derived from a measurement of sound speed [Fig. (6a)].

1.3.1 Average Temperature In contrast to thermocouples, resistance temperature detectors, integrated-circuit temperature sensors, and other small devices that sense temperature essentially at one point, by using acoustic or ultrasonic waves one can measure the speed of sound over an extended path to obtain an average temperature reading. At sufficiently low carrier frequency (57 Hz), and with a sufficiently intense (221 dB) coded source, this has been done over a distance of 18 000 km, using the ocean's 1- to 2-km-thick *SOFAR* channel, achieving a range that reached halfway around the world (Baggeroer and Munk, 1992; see also Holing, 1994). On a somewhat smaller scale, as in combustors where gas temperature may be of the order of 1000 °C (Sec. 1.3.3), the path lengths may be 10 to 20 m, and the frequency may be 1500 Hz (Kleppe, 1989, 1991; Yori, 1992). In smokestacks at somewhat milder conditions of only 200 °C and flow velocities up the stack of Mach 0.1, the frequency can be higher, reaching up to the lowest octave of ultrasound, 20 to 40 kHz. A frequency as high as 100 kHz was used to measure sound speed and flow velocity across a 3-m path in a smokestack duct at 150 °C (Matson and Davis, 1994).

If a *wire waveguide* is used as the temperature sensor, the frequency used in the past has typically been 100 kHz. A wire waveguide is simply a wire that guides acoustic or ultrasonic waves, something like the string in a "string telephone" stretched between paper cups. A wire can guide sound something like the way a fiber-optic waveguide guides light, or a microwave waveguide guides electromagnetic microwaves. Starting around 1960 various investigators proposed using wire-waveguide sensors as replacements for thermocouples (tc's) in order to reduce the number of sensor elements, to avoid tc errors attributed to insulator failure at very high temperature, to avoid tc diffusion errors, and to achieve a more rugged design. In some nuclear-fuel pin studies, some or all of these potential advantages were realized. However, at the time those studies were conducted, no practical way was found to extend the results to analogous industrial applications in an economic fashion. In recent times, however, the widespread use of microprocessors and of ultrasonic high-precision intervalometer instruments for other purposes, principally flow and liquid level, means that it may soon be practical to extend to temperature measurements the use of electronic instruments that

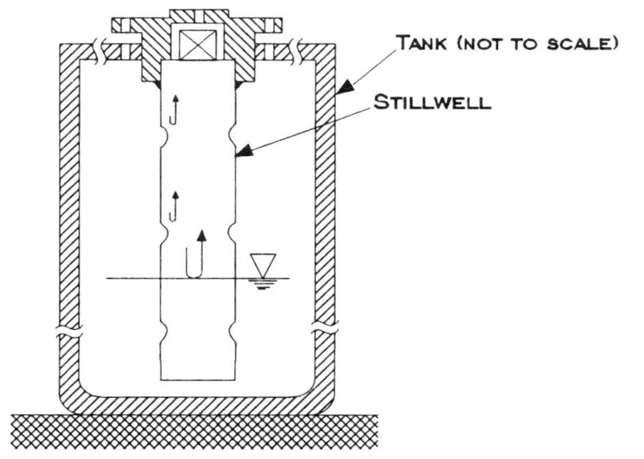

(I) NOTCHED STILLWELL GUIDES THE WAVES, IN A DESIGN MANUFACTURED BY SAAB TANK CONTROL

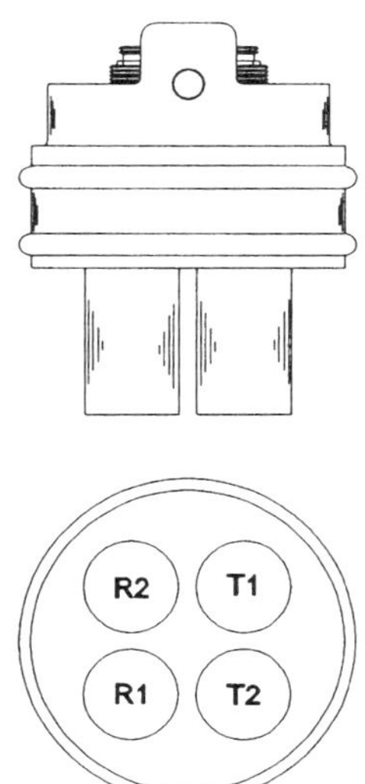

(II) TRANSDUCER ARRAY TRANSMITS UNGUIDED SOUND BEAM, THEN WAITS FOR ECHOES. THIS TYPE TRANSDUCER WAS ORIGINALLY DEVELOPED FOR WALL-TO-WALL FLOW MEASUREMENTS IN LARGE SMOKESTACKS (MATSON AND DAVIS, 1994).

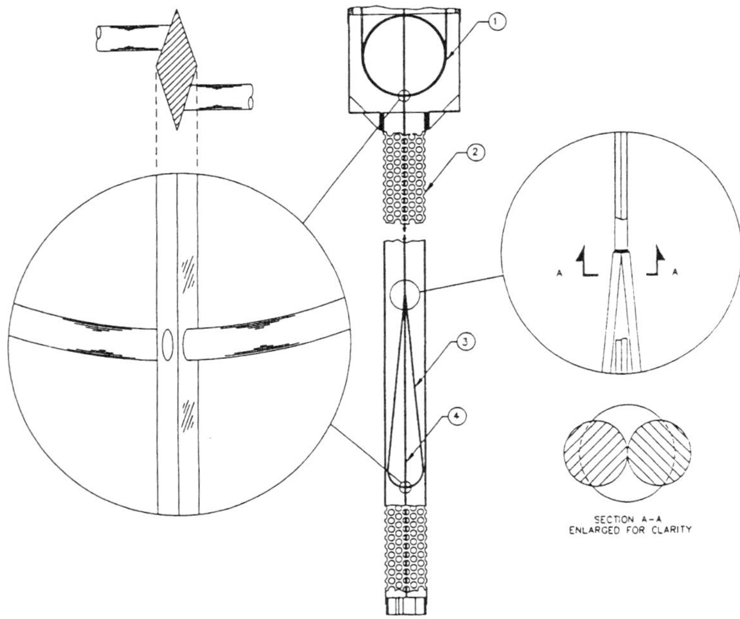

(III) TORSIONAL WAVEGUIDE, AFTER KIM ET AL. (1993). LEGEND: 1 -TOP LEAD-IN/OUT WIRES; 2 -PERFORATED STILLWELL; 3 -BOTTOM LEAD-IN/OUT WIRES; 4 -DIAMOND CROSS-SECTION WAVEGUIDE.

(a)

FIG. 5. Intrusive liquid-level techniques. (a) Downward-looking sonar using (i) guided or (ii) unguided waves, generated and detected using an array of four transducers operating in pairs, and (iii) torsional waveguide top-mounted sensors. (Details in Kim *et al.,* 1993.) [Regarding (i) an early downward-looking sonar with reference reflectors is due to Tomioka (1968).] (b) Side-mounted gap probe. (c) Upward-looking transducers with or without reflector.

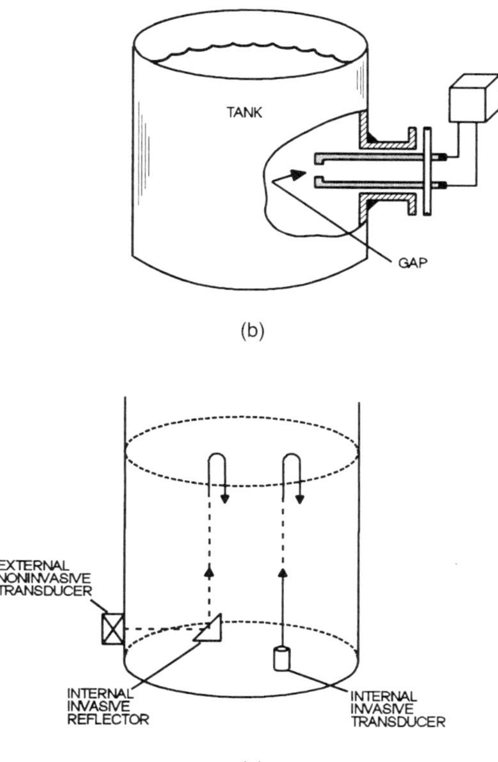

FIG. 5. Continued.

were originally developed for one of the other major process-control measurands. Sometimes the sensor is a waveguide added to the system; sometimes an existing structural element or process boundary can be the sensor [Figs. 6(b)–6(g)].

1.3.2 Temperature Profile along One Path Shortly after the wire waveguide was introduced into the United States as a candidate sensor to measure temperature in certain extreme environments, it was recognized that by intentionally creating a series of discontinuities along the waveguide, one could measure temperature profile. By 1970 this had been demonstrated in laboratory prototype sensors (Lynnworth and Patch, 1970). In other words, the "average" temperature measurement of Sec. 1.3.1, utilizing the sound-velocity–temperature data represented by the thin-wire curves in Figure 6(a), can be applied to the smaller path lengths of a multizone sensor such as that in Fig. 6(b). It is also possible to "tap" a multizone wire waveguide at n points to extract profile information (Kim *et al.*, 1993), somewhat analogous to picking up signals at stations along the 18 000-km *SOFAR* path mentioned above.

As an example of how an ultrasonic "flowmeter" can be applied to multizone and multisensor measurements of sound velocity in wire waveguides, see Collard (1990). While Collard's 1991 dissertation was concerned with determining elastic moduli at high temperature, as was Bell's pioneering work (1957), there are many similarities between such work and measurements of temperature. Such similarities, in fact, were recognized in the United Kingdom by 1960 and provided the stimulus for the earliest wire-waveguide (average) temperature-sensor developments by Bell, Hub, Thorne, Jacques, and colleagues in that decade. Those investigators, particularly Thorne, also demonstrated how melting-point indicators (e.g., solder in a small hole, or solder between the waveguide and a surrounding cylindrical mass) could be built into a waveguide, to provide a precise measurement of when a particular temperature was reached at a particular point along a waveguide. Fendrock and Varela (1995) describe a waveguide thermometer instrument used to measure furnace temperature profiles.

1.3.3 Tomographic Reconstruction of Profile from Averages Obtained over Many Scanned Paths Audible sound and ultrasound offer potential advantages over competing technologies in applications that require long-path averages and in temperature-profile determinations based on a multizone path or tomographic reconstruction. The "sensor" may be gas, liquid, or solid [Fig. 6(a)]. A number of multizone waveguide applications are reviewed in Lynnworth (1989). Many of these used a waveguide similar to that shown in Fig. 6(b). Methods of isolating the electrical part of the assembly or the sensor from the environment are shown in Figs. 6(c) and 6(d).

Since the mid-1980s, CAT scanning of large combustors and boilers has been proven practical. Early proposals from about 1970, as well as one of the first large-scale working examples, due to Green (1985), are shown in Fig. 7. In the first two schematics of Fig. 7(a), lines with arrows represent the

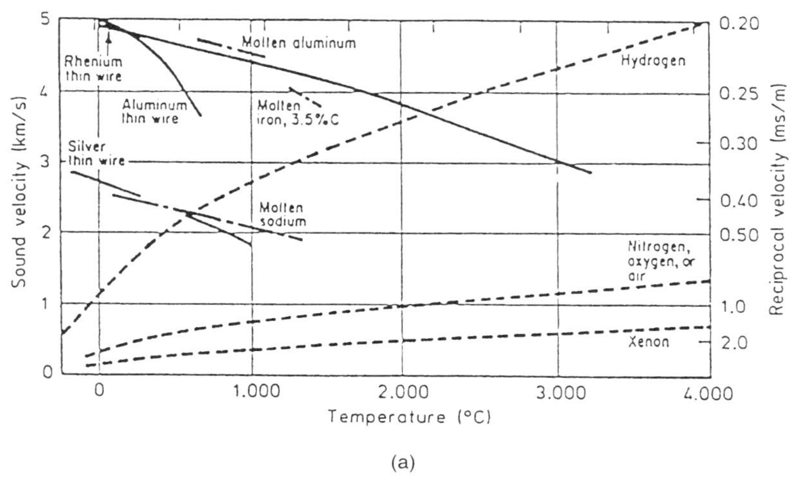

(a)

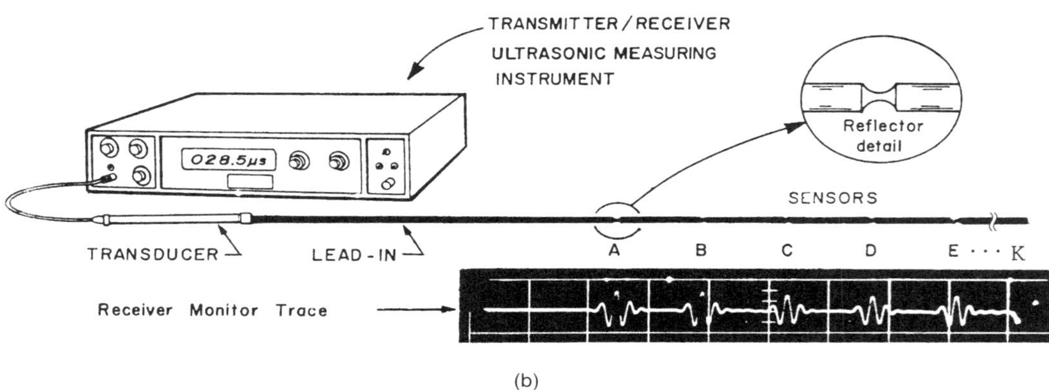

(b)

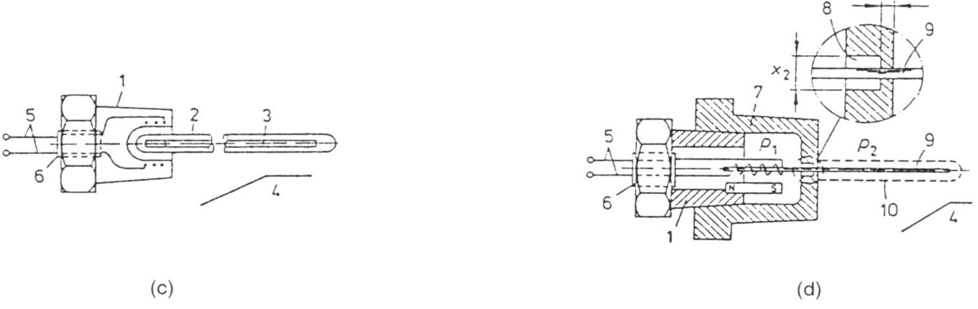

(c) (d)

FIG. 6. (a) Sound speed *c* as a function of temperature *T* in gases, liquids, and solids. (b) An ultrasonic *T* profiler waveguide comprised of ten zones, after Lynnworth and Patch (1970). Echoes are obtained from reflective grooves A, B, . . . , K. The time between echoes depends on sound speed in the segment from whose ends the echoes are generated. The data in (a) are used to convert sound speed measured in each zone to the average temperature in that zone. (c) Magnetic feedthrough: 1, secondary seal plug; 2, sealed all-metal sheath protecting sensor (3) from environment (4); 5, leadwires passing through glass-to-metal or other feedthrough (6). (d) Diaphragm feedthrough, after Lynnworth (1994a), © 1994 VCH. Legend: 7, plug contains thin-wall cavity (8) where x_1 and x_2 are chosen to withstand pressure difference $\Delta P = P_2 - P_1$ and other stresses; sensor (9) passes through diaphragm, to which it is welded or brazed; 10, optional sheath. The bar magnet (N–S) biases the magnetostrictive transducer material into its linear range (analogous to poling of ferroelectric material) and orients its domains so that the generated wave will be purely the extensional mode (no torsion). Thermometer designs proposed for bakery: (e) Moving hearth plate is its own sensor. (f) Rolling drum is its own sensor. (g) Multizone waveguide sensor. After Lynnworth (1992). © 1992 John Wiley & Sons.

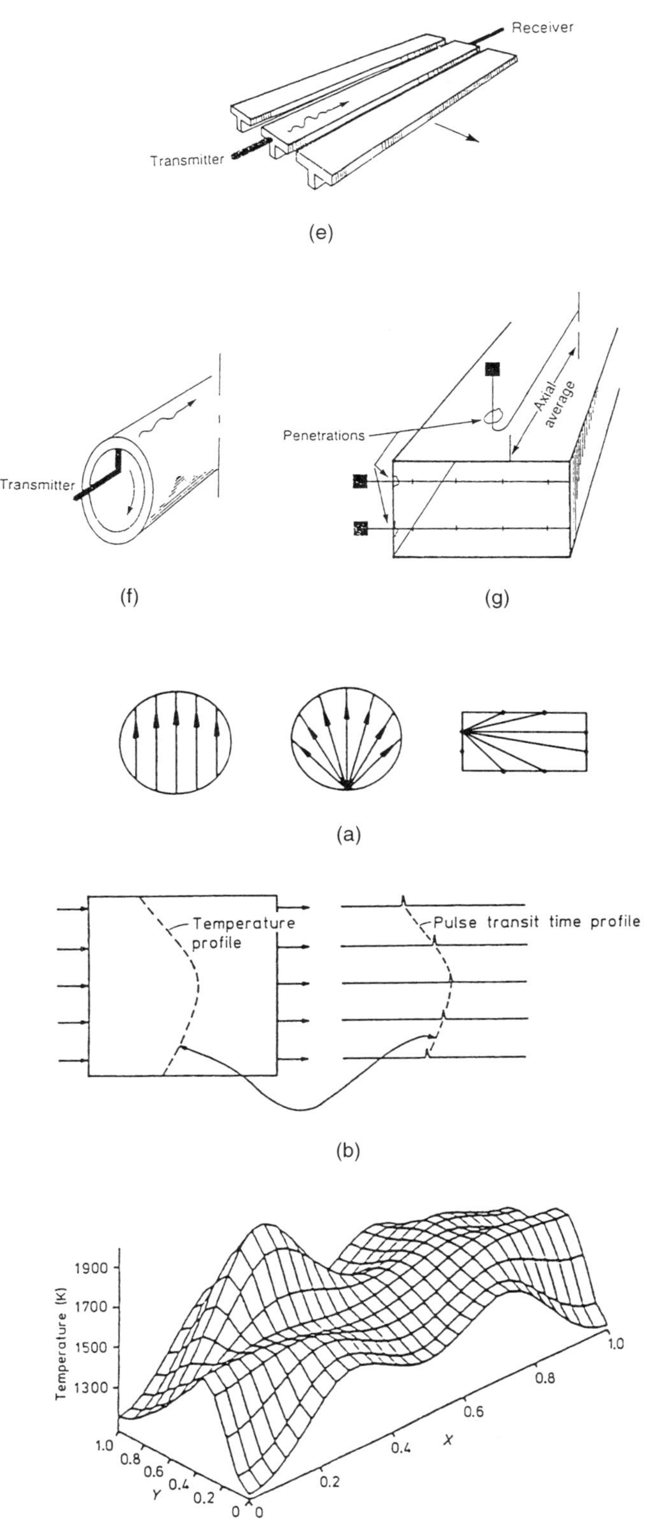

FIG. 6. Continued.

FIG. 7. (a) Tomographic thermometry proposed for gases by Lynnworth and Carnevale (1966) and for hot steel billets by Lynnworth and Patch (1970). In first and second drawings, lines with arrows represent the central rays of beams that sample the temperature profile in the circular ducts. In practice, each beam actually spreads; see text. Third drawing, ray paths launched from one transducer in one wall of scheme described in text for rectangular duct. (b) Ultrasonic determination of temperature distribution inside large metal billets, proposed by Lynnworth (1970). Measuring the transit time vs elapsed (real) time was also proposed, to monitor the approach to equilibrium. (c) Tomographically reconstructed temperature profiles in the exit plane of a large furnace, after Green (1985). © 1985 AIP; reproduced with permission.

central rays of beams that sample the temperature profile in the circular ducts. In practice, each beam actually spreads. Beam spread, or *diffraction*, depends mainly on the diameter d of the transducer compared with the wavelength λ transmitted into the fluid medium. For $d/\lambda = 10$, the most intense part of the beam's energy is confined to a cone having an included angle $\theta_{3\,\mathrm{dB}} = 0.1$ radian $= 5.7°$. Suppose that along each wall of a rectangular duct, two transducers are installed, such that each transducer communicates with all the other transducers except the one on its own wall; see the last schematic in Fig. 7(a). Then there are 24 independent paths. More recent work along these lines is reported by Kleppe (1989, 1991) and Basarab-Horwath and Dorozhevets (1994). To overcome attenuation in the hot, sooty, turbulent gas, Kleppe used intense waves at audible, not ultrasonic, frequencies, e.g., ~1.5 kHz for paths of 10 to 20 m.

1.4 Pressure

1.4.1 Fluids The speed of sound c generally increases as pressure P increases in liquids, but the effect is too small to be of much practical use for sensing P, especially as the P effect on c would probably be masked by effects of temperature T or composition. In gases near atmospheric pressure, the sign of dc/dP can be plus or minus, depending on the type of gas (Bhatia, 1967, p. 15). The increase in c caused by P up to approximately 300 bar in several inorganic gases is shown in Lynnworth (1989), p. 232, based on work by Carey *et al.* (1969). The attenuation coefficient α and density ρ do not change much with pressure in a liquid. But in a gas the effects of P on α and Z are large enough to provide a basis for using the gas as its own P sensor (Fig. 8).

Intrusive P-sensing probes are available commercially that utilize the change in resonant frequency of a quartz crystal. See Figs. 9(a) and 9(b), which include contributions from researchers at Quartztronics and Schlumberger, respectively [EerNisse *et al.* (1988); Gordon and Wiggins (1990); Besson *et al.* (1993)].

One way of measuring fluid pressure by exploiting pressure coupling across the dry interface between two solids contained in a sealed two-crystal transducer assembly is shown in Fig. 10, after a proposal by Lynnworth (1995). [For pressure-coupling data see Crecraft (1964).] In Fig. 10, the cylindrical object between gaps G1 and G2 is an insulated solid rod such as fused silica. If it is metallized, at least one end must be electrically insulated. The rod is supported by soft O-rings or by other means, in a way that allows electrical lead wires to pass from the transducers to the connectors at the right.

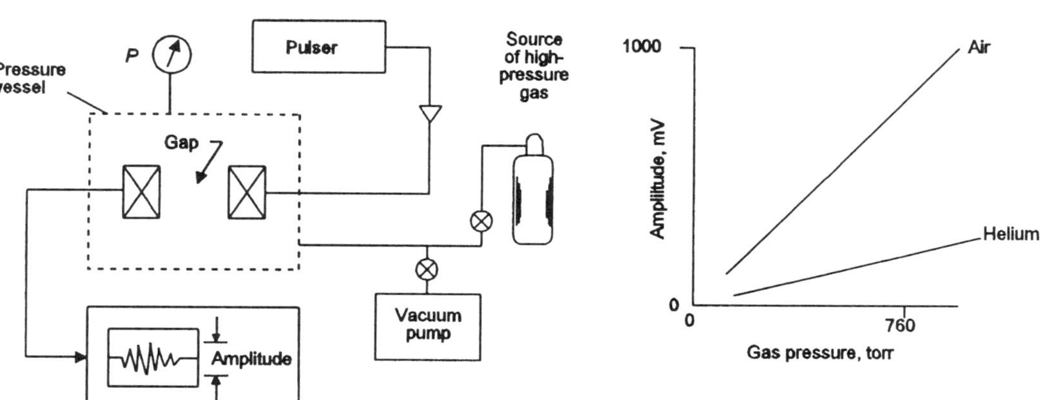

FIG. 8. Amplitude of an ultrasonic pulse transmitted across the gap is a nearly linear function of gas pressure up to ~10 bar for inorganic gases such as air or helium. The idealized graph covers the bottom portion of that range. The slope for air is greater than that for He by approximately the ratio $r = (M_{\mathrm{air}}\ \gamma_{\mathrm{air}}/M_{\mathrm{He}}\gamma_{\mathrm{He}})^{1/2} = 2.5$, where the M's are molecular weights and γ's are specific heat ratios. Beam spread and attenuation also need to be taken into account, in a more accurate calculation. Virial coefficients are used to quantify departures from ideal-gas behavior.

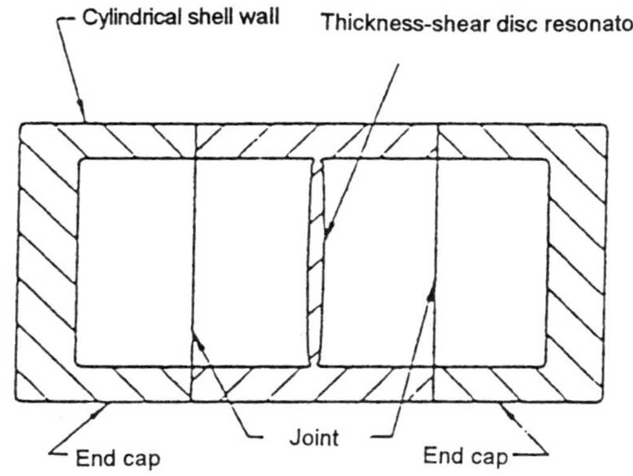

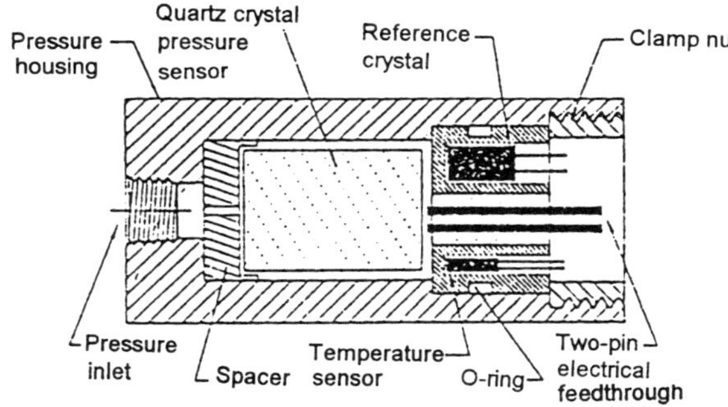

(a)

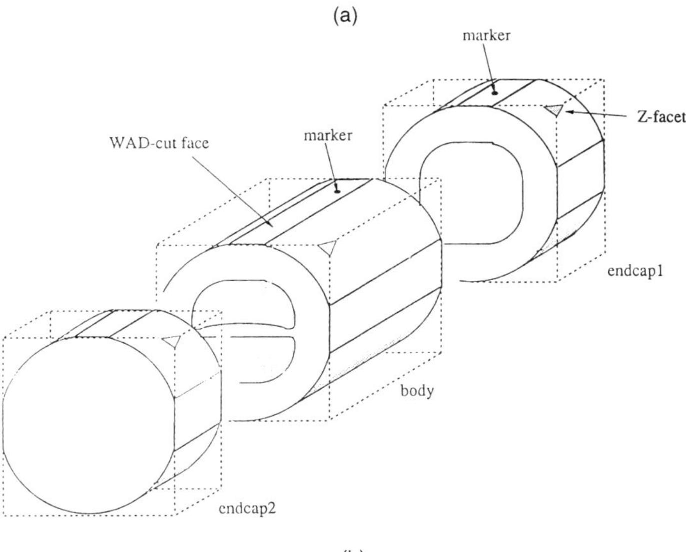

(b)

FIG. 9. Quartz-cystal *P* sensors. (a) Design manufactured by Quartztronics, Inc., after EerNisse *et al.* (1988) and Gordon and Wiggins (1990), courtesy of and © 1990 Quartztronics. (See also Clayton and EerNisse, 1992.) The upper part represents a zoomed enlargement of the quartz-crystal pressure sensor shown below it. (b) Schematic diagram of the three elements of a dual-mode thickness-shear SPA design, after Besson *et al.* (1993), courtesy of and © 1994 Schlumberger, Ltd. This design has been developed by Schlumberger, and is being manufactured by them for downhole pressure measurements in oil fields. It is referred to as CQG™ (Combinable Quartz Gauge) in the oil-field industry.

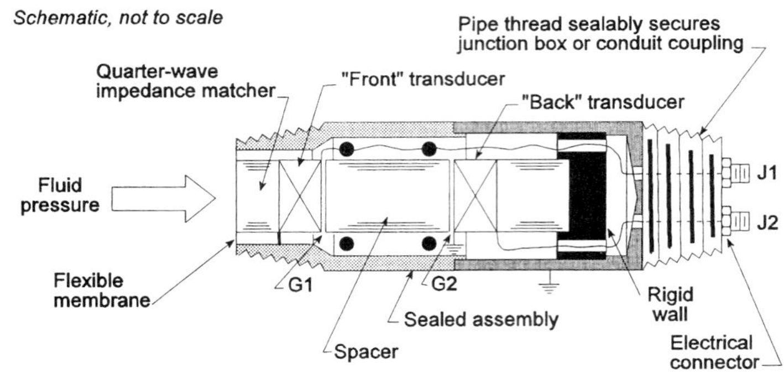

FIG. 10. Pressure coupling within a sealed transducer assembly. The strength of the signal internally coupled between "front" and "back" transducers is a function of fluid pressure bearing against the assembly. Gaps G1, G2, exaggerated in this schematic, close up and couple more strongly when fluid pressure P increases. Accordingly, a signal would be transmitted more efficiently from one transducer to the other as P increases. The front transducer is intended to be part of a trans-fluid interrogation system—for example, part of an ultrasonic flowmeter where transducers are diagonally positioned on opposite sides of a pipe as in Fig. 2 or in Fig. 3d. The two O-rings (●) symbolize one way of concentrically centralizing the spacer. The ungrounded lead wire from the front transducer passes through these O-rings, then through a small hole in the rigid wall, and then to electrical connector J1. The ungrounded lead from the back transducer passes through a second small hole in the rigid wall to J2. After a proposal by Lynnworth (1995).

Compared with a ceramic rod such as fused silica, a metal rod will normally exhibit a much bigger change in sound speed as temperature changes. In this case, the time delay in the metal can be interpreted, with the help of data like those in Fig. 6(a), in terms of the average temperature in the rod. To the extent that the rod is at the same temperature as the fluid, then the fluid *pressure* sensor also acts as a fluid *temperature* sensor. It should be noted that while "pressure coupling" provides a means to sense wide *ranges* in pressure, it has not yet (1995) been demonstrated to do so with high accuracy.

1.4.2 Solids under Stress Provided textural effects can be eliminated, the small change in sound speed that is associated with elastic nonlinearity or higher-order elastic constants can be used to measure residual stress. Measurements are made orthogonally or along crystal axes, preferably from the no-load to full-load condition to eliminate texture or anisotropy effects that could mask the stress effect that is sought. Recent work by Shaikh (1995) suggests that a surface-skimming longitudinal wave may be particularly sensitive and convenient for measuring *relative* stress.

2. OTHER MEASURANDS

2.1 Density of Gases; Density of Liquids

2.1.1 Gas Density For a one-component ideal gas at pressure $P < 10$ bar, the sound speed is $c = (\gamma RT/M_W)^{1/2}$ where $\gamma =$ specific heat ratio C_P/C_V, $C_P = C_V + R$, $R =$ universal gas constant, $T =$ absolute temperature, and $M_W =$ molecular weight. For a mixture of two ideal gases, c is calculated by weighting the specific heats and M_W's according to the mole fractions X_1 of the first gas and $X_2 = 1 - X_1$ of the second gas; see Polturak *et al.* (1986) or Valdes and Cadet (1991). This means replacing γ by the weighted average $\bar{\gamma}$ and similarly replacing M_W by $\bar{M}_W$. We now have for the binary mixture $c = (\bar{\gamma}RT\bar{M}_W)^{1/2}$ where

$$\bar{\gamma} = \frac{X_1C_{P1} + (1 - X_1)C_{P2}}{X_1C_{V1} + (1 - X_1)C_{V2}} = 1 + \left(\frac{X_1}{\gamma_1 - 1} + \frac{1 - X_1}{\gamma_2 - 1}\right)^{-1} \quad (8)$$

and

$$\bar{M}_W = X_1M_{W1} + (1 - X_1)M_{W2}. \quad (9)$$

Generalizing to a multicomponent gas yields

$$\bar{M}_W = \sum_{i=1}^{N} X_iM_{Wi}, \quad (10)$$

where

$$\sum_{i=1}^{N} X_i = 1. \quad (11)$$

If the specific heats and M_W's are known for the individual gases, measurement of c and T yields the average molecular weight $\bar{M}_W$ of the binary gas mixture. Additionally, if the pressure P is known, the gas mixture's density ρ_{mix} can be calculated. In units of g/L,

$$\rho_{mix} = \frac{\bar{M}_W}{22.4}\frac{P}{P_0}\frac{T_0}{T}, \quad (12)$$

where P_0 and T_0 are standard pressure and temperature, 760 mm and 273 K. If the gases are not ideal, one or more virial coefficients (or supercompressibility factors) must be introduced into the equations relating sound speed to molecular weight and density.

2.1.2 Liquid Density For special cases such as some pure liquids, liquid density ρ_{liq} can be determined from sound speed c_{liq} in the liquid (the "medium as its own sensor" method). For water, c_{liq} is not uniquely associated with T unless T is only above or only below 74 °C. Salinity, pressure, and contaminants also influence c_{liq}.

Two other approaches to determining ρ_{liq} are

1. use of a foreign sensor such as a torsional waveguide, provided the cross section is noncircular;
2. use of a tube, or duct wall, as the sensor, interrogated by flexural or bending waves.

2.2 Concentration Analysis of Binary Gas Mixtures

By using the equation for sound speed in an ideal gas, $c = (\gamma RT/M_W)^{1/2}$, one can readily calculate c if one knows γ, R, T, and M_W. Conversely, if c is measured, one can determine T (Mayer, 1873), γ (Sherratt and Griffiths, 1934), or M_W, provided the other terms are known. [In flare gases (Smalling *et al.*, 1986) and in some other multicomponent gas mixtures $\bar{M}_W$ and $\bar{\gamma}$ are not independent of one another. Therefore, c, T, and the theoretical or empirical relationship between $\bar{M}_W$ and $\bar{\gamma}$ suffice to determine the mixture's average $\bar{M}_W$. Another multicomponent case in which c and T measurements can yield $\bar{M}_W$ is the *pseudobinary* gas mixture. This is a mixture of two gases, one of which is itself a mixture, such as air, that has a relatively stable composition and presumably known M_W and γ.]

Although the concept of determining M_W from c is contained in the works of Herzfeld and Litovitz (1959) and Bhatia (1967), and demonstrated by data in their books or in other publications, the idea did not gain much attention until the 1980s. Since the 1980s, there has been a growing interest in determining M_W for binary mixtures of MOCVD (metal-organic chemical-vapor deposited) gases and some other binary mixtures; see Valdes and Cadet (1991) and Hallewell and Lynnworth (1994). This 1994 reference deals with the special case of two gases A and B of widely different molecular weights, say $M_A \gg M_B$, and where one of the gases, say A, is present at very low concentrations, <1%. In this special case X_A, the concentration of gas A can be computed using the approximation

$$X_A \approx \left[2\frac{M_B}{M_A}\right]\left[1 - \frac{c_{mixture}}{c_B}\right]. \quad (13)$$

As a numerical example, consider a dilute (<1%) solution of TMIn (trimethylindium) in H_2. Approximating the molecular weights of these gases by 160 and 2, respectively, and taking c in H_2 to be 1284 m/s at 20 °C, Eq. (13) yields

$$X_{TMI} \approx \frac{2 \times 2}{160}\left(1 - \frac{c_{mixture}}{1284}\right) = \frac{1}{40}\left[1 - \frac{c_{mixture}}{1284}\right]. \quad (14)$$

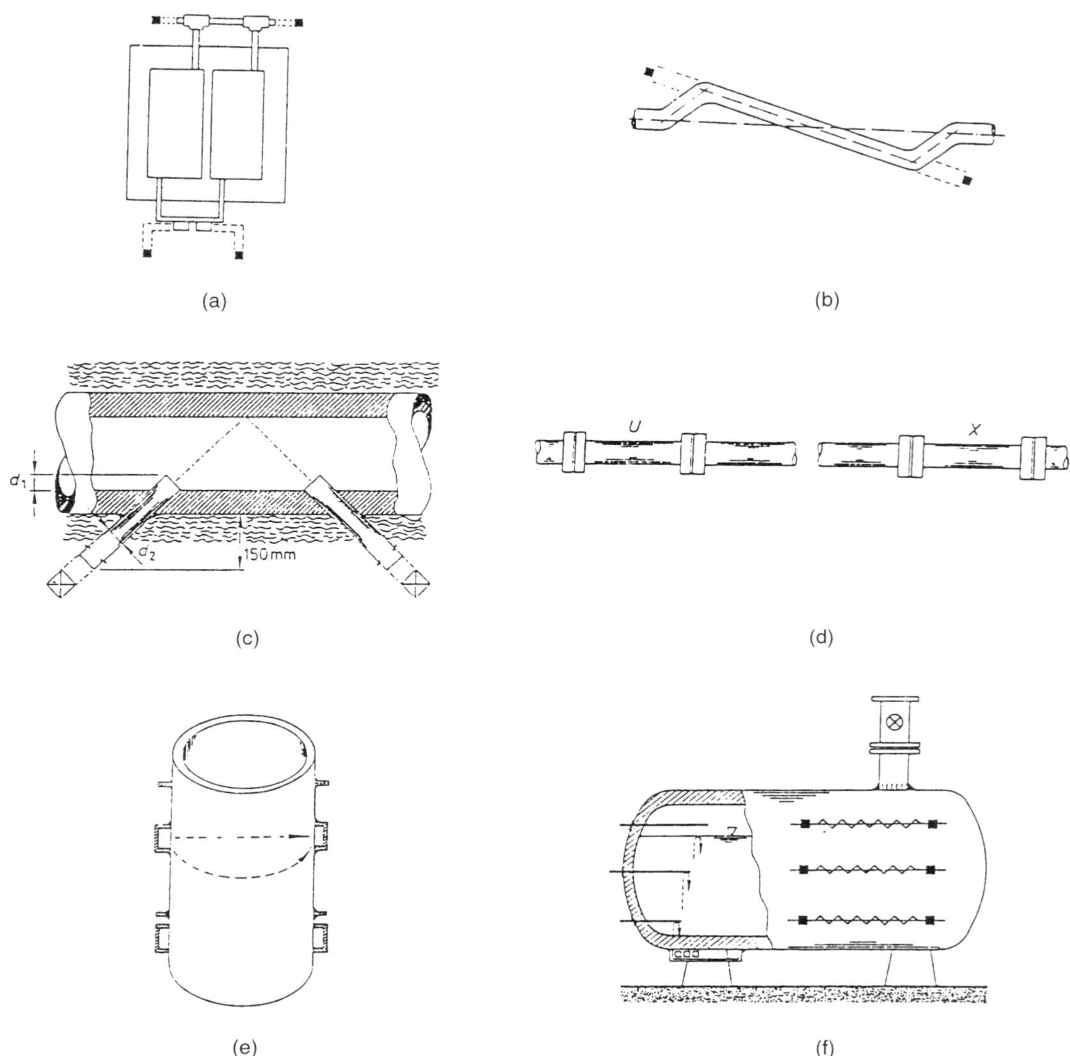

FIG. 11. Examples of adapting a process boundary at the time of manufacture to accommodate ultrasonic sensors that are to be installed later. (a) Clamps for clamp-on vee-path flowmeters; plugged tees for hybrid axial-path flow interrogation. (b) Tube bent smoothly, for clamp-on axial-path interrogation. (c) Buffer rods welded into a high-temperature insulated pipe. (d) In a long straight pipe, two similar or identical dummy spoolpieces U and X have been installed as spacers. Spoolpiece U is to be replaced later by a precision-bore ultrasonic flowmeter like that in Fig. 1(d) having one or more paths. Spoolpiece X is to be replaced by a nonultrasonic flowmeter [turbine, orifice plate, Venturi, magmeter (magnetic field flowmeter), or Coriolis sensor, for example] to establish the installed accuracy of U relative to X. Some designers would put U and X in a bypass line so that replacement can occur without shutting down the main process line. Spoolpiece X can also be thought of as representing a general-purpose sensor port to accommodate whatever sensor is "best" as a function of new requirements and ongoing improvements in sensor technology. (e) Clamps and threaded studs welded up along wall of a tall storage tank. (f) Clamps, reference reflectors, and isolation-valved port welded on horizontal-axis thick-walled storage tank. © 1992 Panametrics.

Note that γ and T do not appear in Eq. (13). Gas A is hardly present and so its γ is not important here. But T must be known, in order to enter the proper value for c_B in Eq. (13).

In practical applications, considerable care may be required to bring the gas to a stable temperature and, if pulse techniques are used, to measure over a path long enough to achieve adequate precision in c for a given instrument-limited resolution of transit time. Resonant cavities or interferometers provide a way to achieve high precision in a short path.

2.3 Designing Pipes and Tanks to Accommodate Sensors, Particularly Ultrasonic Sensors

Just as one essential design parameter of a welded structure or weldment is its inspectability, so too should be the *sensorability* of a process. Sensorability, the ability to be sensed, requires that the designer select the sensor generically or specifically, as part of the design process, or at least provide a sensor nozzle, port, or mounting bracket. If clamp-on sensors are to be used, there are cases where it will be more economical to weld the clamps or other sensor-mounting fixtures onto the pipe or tank when that pipe or tank is being fabricated (Fig. 11), rather than as an afterthought. At the design stage, rather than after the pipe or tank is in operation, if one can anticipate or select the sensors that will be used later, one can enhance their performance, and simplify their installation, e.g., avoid subsequent hot-tapping or plant shutdown merely to install a sensor. Sensor mounts and reference reflectors, for example, could be welded safely and easily prior to filling a pipe or tank with an explosive hydrocarbon. Once the container is filled, welding is forbidden in some areas (e.g., above or within ~1 m of the liquid level) and certainly is more complicated, even in areas where welding is allowed, than if the necessary provisions for sensors could have been created earlier.

Works Cited

Baggeroer, A., Munk, W. (1992), *Phys. Today* **45** (9), 22–30. See also Holing, D. (1994), *Amicus J.* **16** (3), 19–23.

Basarab-Horwath, I., Dorozhevets, M. M. (1994), "Measurement of the Temperature Distribution in Fluids Using Ultrasonic Tomography," in: M. Levy, S. C. Schneider, B. R. McAvoy (Eds.), *1994 IEEE Ultrasonics Proceedings,* New York: IEEE, pp. 1891–1894.

Baumoel, J. (1994), U.S. Patent No. 5,343,737.

Bell, J. F. W. (1957), *Philos. Mag.* **2,** 1113–1120.

Besson, R. J., Boy, J. J., Glotin, B., Jinzaki, Y., Sinha, B., Valdois, M. (1993). *IEEE Trans. Ultrason. Ferroelec. Freq. Control* **40,** 584–591.

Bhatia, A. B. (1967), *Ultrasonic Absorption,* London: Oxford Univ. Press.

Brillouin, L. (1960), *Wave Propagation and Group Velocity,* Pure and Applied Physics, Vol. 8, New York: Academic.

Carey, C. A., Carnevale, E. H., Uva, S., Marshall, T. (1969), *Experimental Determination of Gas Properties at High Temperatures and/or Pressures,* Arnold Engineering Development Center Report No. AEDC-TR-69-78, Springfield, VA: National Technical Information Service.

Clayton, L. D., EerNisse, E. P. (1992), *Sensors* **9** (11), 37–42.

Collard, S. M. (1990), *Scripta Metall. Mater.* **24,** 623–627. See also Collard, S. M. (1991), Ph.D. Diss., Rice University.

Crecraft, D. I. (1964), *J. Sound Vib.* **1,** 381–387.

Dabirikhah, H., Turner, C. W. (1994), *Electron. Lett.* **30** (18), 1549–1550.

Dogan, A., Yoshikawa, S., Uchino, K., Newnham, R. E. (1994), in: M. Levy, S. C. Schneider, B. R. McAvoy (Eds.), *1994 IEEE Ultrasonics Proceedings,* New York: IEEE, pp. 935–939.

EerNisse, E. P., Ward, R. W., Wiggins, R. B. (1988), *IEEE Trans. Utrason. Ferroelec. Freq. Control* **35,** 323–330.

Fendrock, C., Varela, D. W. (1995), *Sensors* **12** (1), 23–37.

Fleury, G. (1994), *Reconstruction de Profils de Vitesse à Partir de Données Lacunaires et Optimisation d'Instrument—Application à la Débitmétrie Ultrasonore,* Orsay, France: Université de Paris-Sud, École Supérieure d'Électricité.

Folkestad, T., Mylvaganam, K. S. (1993), *IEEE Trans. Ultrason. Ferroelec. Freq. Control* **40,** 193–215.

Gordon, J., Wiggins, R. B. (1990), *Sensors* **7** (8), 30–34.

Green, S. F. (1985), *J. Acoust. Soc. Am.* **77,** 759–763.

Gualtieri, J. G., Kosinski, J. A., Ballato, A. (1994), *IEEE Trans. Ultrason. Ferroelec. Freq. Control* **41,** 53–59.

Hallewell, G. D., Lynnworth, L. C. (1994), in: M. Levy, S. C. Schneider, B. R. McAvoy (Eds.), *1994 IEEE Ultrasonics Proceedings,* New York: IEEE, pp. 1311–1316.

Hamidullin, V. K. (1993), in: M. Levy, S. C. Schneider, B. R. McAvoy (Eds.), *1994 IEEE Ultrasonics Proceedings,* New York: IEEE, pp. 395–399.

Hayward, G. (Ed.) (1994), special issue on Correlation Techniques and Applications in Ultrasonics, *IEEE Trans. Ultrason. Ferroelec. Freq. Control* **41,** 577–663.

Hertz, T. G., Dymling, S. O., Lindström, K., Persson, H. W. (1991), *Rev. Sci. Instrum.* **62** (2), 457–462.

Herzfeld, K. F., Litovitz, T. A. (1959), *Absorption and Dispersion of Ultrasonic Waves,* New York: Academic.

Holm, M. (1995), *Simulation of Flowmeter Installation Effects,* LUTHMDN/TMVK—7019-SE, Lund, Sweden: Lund Institute of Technology.

Hutchins, D. A. (1994), in: M. Levy, S. C. Schneider, B. R. McAvoy (Eds.), *1994 IEEE Ultrasonics Proceedings,* New York: IEEE, Paper II-3.

Jacobson, S. A., Lynnworth, L. C., Korba, J. M. (1988), U.S. Patent No. 4,787,252.

Kim, J. O., Bau, H. H., Liu, Y., Lynnworth, L. C., Lynnworth, S. A., Hall, K. A., Jacobson, S. A., Korba, J. M., Murphy, R. J., Strauch, M. A., King, K. G. (1993), *IEEE Trans. Ultrason. Ferroelec. Freq. Control* **40** (5), 563–576.

Kleppe, J. A. (1989), *Engineering Applications of Acoustics,* Artech. See also Kleppe, J. A. (1991), *I & CS (Instruments & Control Systems)* **64** (6), 35–40; Yori, L. G. (1992), *I & CS (Instruments & Control Systems)* **65** (11), 61–62.

Lalande, F., Chaudhry, Z., Rogers, C. A. (1995), *IEEE Trans. Ultrason. Ferroelec. Freq. Control* **42,** 21–27.

Liu, Y., Lynnworth, L. C. (1993), in: *1993 IEEE Ultrasonics Symposium Proceedings,* pp. 385–390; see also Liu, Y., Lynnworth, L. C. (1995), U.S. Patent No. 5,456,114.

Liu, Yi, Lynnworth, L. C., Nguyen, T. H., Xiao, D., Walters, J. (1996), in: *Proc. International Pipeline Conference,* ASME.

Lynnworth, L. C. (1970), *Nav. Res. Rev.* **23** (7), 1–22, 32.

Lynnworth, L. C. (1989), *Ultrasonic Measurements for Process Control,* New York: Academic.

Lynnworth, L. C. (1992), in: P. H. Sydenham, R. Thorn (Eds.), *Handbook of Measurement Science,* Vol. 3, New York: Wiley, Chap. 38, 1655–1689.

Lynnworth, L. C. (1994a), in: H. H. Bau, N. F. deRooij, B. Kloeck (Eds.), *Mechanical Sensors,* Vol. 7, Weinheim: VCH, Chap. 8, pp. 285–329.

Lynnworth, L. C. (1994b), U.S. Patent No. 5,275,060.

Lynnworth, L. C. (1994c), in: M. Levy, S. C. Schneider, B. R. McAvoy (Eds.), *1994 IEEE Ultrasonics Proceedings,* New York: IEEE, pp. 1317–1321.

Lynnworth, L. C. (1995), U.S. Patent No. 5,437,194.

Lynnworth, L. C., Carnevale, E. H. (1966), *NASA CR-54979.*

Lynnworth, L. C., Patch, D. R. (1970), *MTRSA (Materials Research and Standards)* **10** (8), Cover, 6–11, 40.

Lynnworth, L. C., Hallewell, G. D., Bragg, M. I. (1994), in: *Proceedings of the "FLOMEKO '94" Conference on Flow Measurement in the Mid 90s,* Session 7, Paper #3, pp. 1–15.

Mágori, V. (1994), in: M. Levy, S. C. Schneider, B. R. McAvoy (Eds.), *1994 IEEE Ultrasonics Proceedings,* New York: IEEE, pp. 471–481.

Matson, J., Davis, R. (1994), *I&CS (Instruments & Control Systems)* **67** (2), 67–69.

Mayer, A. M. (1873), *Philos. Mag.* Ser. 4, **45,** 18–22.

Nygaard, O. G. H., Mylvaganam, K. S. (1993), *Technisches Messen* **60** (1), 4–14.

Papadakis, E. P. (1997) (Ed.), *Ultrasonic Instruments and Devices: Reference for Modern Instrumentation, Techniques and Technology,* Academic, to be published.

Pollard, H. F. (1977), *Sound Waves in Solids,* Pion, pp. 139–140.

Polturak, E., Garrett, S. L., Lipson, S. G. (1986), *Rev. Sci. Instrum.* **57,** 2837–2841.

Royer, D., Dieulesaint, E., Legras, O. (1992), *J. Phys. III* **2,** 145–168.

Shaikh, N. (1995), private communication.

Sheen, S. H. (1994), *Machine Design* **66** (23), 60 and 62.

Sheen, S. H., Chien, H.-T., Raptis, A. C. (1995), private communication.

Sherratt, G. G., Griffiths, E. (1934), *Proc. R. Soc. London, Ser. A* **147,** 292–308.

Smalling, J. W., Braswell, L. D., Lynnworth, L. C., Wallace, D. R. (1984), in: *Proceedings of the 39th Annual Symposium on Instrumentation for the Process Industries,* Research Triangle Park, NC: Instrument Society of America, pp. 27–38.

Smalling, J. W., Braswell, L. D., Lynnworth, L. C. (1986), U.S. Patent No. 4,596,133.

Smith, R. H., Jr. (1994), *The Direct Measurement of Circulation in Free Surface Vortices,* MS Thesis, Worcester Polytechnic Institute, Worcester, Massachusetts; see also Smith, R. H., Jr., Durgin, W. W., Johari, H. (1995), *Proceedings of AIAA Conference,* AIAA Paper 95-0104, American Institute of Aeronautics and Astronautics, Inc., 9 pp.

Sunthankar, Y., (1973), *IEEE Trans. Son. Ultrason.,* **SU-20,** 274–278.

Tomioka, G. (1968), U.S. Patent No. 3,394,589.

Valdes, J. L., Cadet, G. (1991), *Anal. Chem.* **63,** 366–369.

Wenzel, S. W., Costello, B. J., White, R. M. (1994), *Sensors* **11** (12), 47–49.

Further Reading

Babikov, O. I. (1960), *Ultrasonics and Its Industrial Applications,* Consultants Bureau, New York: Plenum.

Ballato, A. (1983), "Piezoelectric Resonators," in: B. Parzen (Ed.), *Design of Crystal and Other Harmonic Oscillators,* New York: Wiley, Chap. 3, pp. 66–122 and pp. 432–436.

Brekhovsikh, L. M. (1980), *Waves in Layered Media,* Applied Mathematics and Mechanics Vol. 16, New York: Academic.

Duncombe, E. (1984), *J. Phys. E.: Sci. Instrum.* **17,** 17–18.

Frederick, J. R. (1965), *Ultrasonic Engineering,* New York: Wiley.

Gerber, E. A., Ballato, A. (Eds.) (1985), in: *Precision Frequency Control,* Vol. 1, *Acoustic Resonators and Filters,* 448 pp.; Vol. 2, *Oscillators and Standards,* 480 pp.; New York: Academic.

Hamidullin, V. K. (1989), *Ultrasonic Control Instrumentation Systems* (in Russian), Leningrad (St. Petersburg) University, 248 pp.

Hamidullin, V. K. (1992), *Ultrasonic Transducers for Flow and Liquid Level Measurements* (in Russian), Leningrad (St. Petersburg) University, 81 pp.

Hayward, G. (Ed.) (1994), special issue on Correlation Techniques and Applications in Ultrasonics, *IEEE Trans. Ultrason. Ferroelec. Freq. Control.* **41,** 577–663.

Kažys, R.-J. (1986), *Ultrasonic Information Measuring Systems,* Vilnius Mokslas (in Russian).

Krautkrämer, J., Krautkrämer, H. (1983), *Ultrasonic Testing of Materials,* 3rd ed., New York: Springer.

Leighton, T. G. (1994), *The Acoustic Bubble,* New York: Academic, 613 pp.

Malghan, S. G. (Ed.) (1993), *Electroacoustics for Characterization of Particulates and Suspensions,* National Institute of Standards and Technology Special Publication No. 856, Washington, DC: U.S. GPO.

Mason, W. P. (1950), *Piezoelectric Crystals and Their Application to Ultrasonics,* New York: Van Nostrand.

Meyers, R. A. (Ed.) (1992), *Encyclopedia of Physical Science and Technology,* 18-volume set, New York: Academic, 13,788 pp.

Papadakis, E. P. (1997) (Ed.), *Ultrasonic Instruments and Devices: Reference for Modern Instrumentation, Techniques and Technology,* Academic, to be published.

Pierce, A. D. (1981), *Acoustics,* New York: McGraw-Hill, 642 pp.

Pierce, A. D., Thurston, R. N. (Eds.) (1993), *Physical Acoustics,* Vol. 22, New York: Academic, 384 pp.

Rosa, D. (Ed.), (1994), "1995 Buyer's Guide," *Sensors,* Vol. 11, No. 9, Part 2, Peterborough, NH: Helmers.

Schlichting, H. (1955), *Boundary Layer Theory,* London: Pergamon.

Viktorov, I. A. (1967), *Rayleigh and Lamb Waves,* New York: Plenum.

SENSORS, INFRARED

JOSEPH S. ACCETTA, *Industrial Technology Institute, Ann Arbor, Michigan, U.S.A.*

INTRODUCTION

The infrared (IR) region of the electromagnetic spectrum lies just above the visible region, as measured in wavelengths, and extends out to submillimeter wavelengths. Human vision is adapted to the spectral output of the sun and does not respond to infrared radiation. The sensation of heat is a familiar biological sensory response to infrared radiation common to most living species and is particularly well developed in some. An infrared sensor is defined as a device to transform infrared radiation in the spectral region of 0.75 to 1000 μm into interpretable information. As a result of specialization, modern infrared sensor technology is complex and sophisticated.

As described in a brief account of the history of infrared sensors given by Hudson (1969), IR sensing began with the discovery by Herschel in 1800 of "invisible rays." While measuring the heating effects of the refracted spectral components of sunlight with a prism, Herschel discovered that a portion of the spectrum beyond red that was invisible nevertheless elicited a response from the thermometer. This very first infrared detector, the thermometer, was followed by the thermocouple, the bolometer, and a number of other novel detection schemes including infrared-sensitized photographic film. In 1917 Case employed the first infrared detector based upon the photoconductive effect (Dereniak and Crowe, 1984). Although early infrared sensors were little more than primitive detectors, the characteristic exploitation of a number of related and enabling technologies such as the addition of optical elements and electronic amplifiers followed rap-

3-527-28139-8/96/$5.00 + .50

idly. As is frequently but sadly the case, much of the success of modern infrared sensors can be attributed to the massive investment of military research since WW II directed at night warfare and surveillance.

One of the most familiar examples of an infrared sensor is the thermal imager that converts radiation emitted from an object or scene into a visually interpretable image. The variations in gray scale or color frequently represent relative temperature differences. A typical example of the imagery from one of these devices is shown in Fig. 1. The relative contrast of the author's face is displayed as function of temperature. In this case the lighter regions correspond to higher temperature, although it could be presented in reverse as well. On occasion images such as these are coded in false color, each color representing a range of temperature. With certain assumptions about the material properties of the object, a temperature range can be assigned to the various regions. Versions of the thermal imager are widely used in applications ranging from astronomy to medicine. The greatest scientific and technological impact of the development of infrared sensing devices is in infrared astronomy, environmental remote sensing, and industrial process control. In addition infrared sensors have clearly played a revolutionary role in modern warfare. Complexity and usage range from the exotic, including satellite-borne astronomical, environmental, and military reconnaissance systems, to numerous commercial and industrial remote temperature-sensing applications.

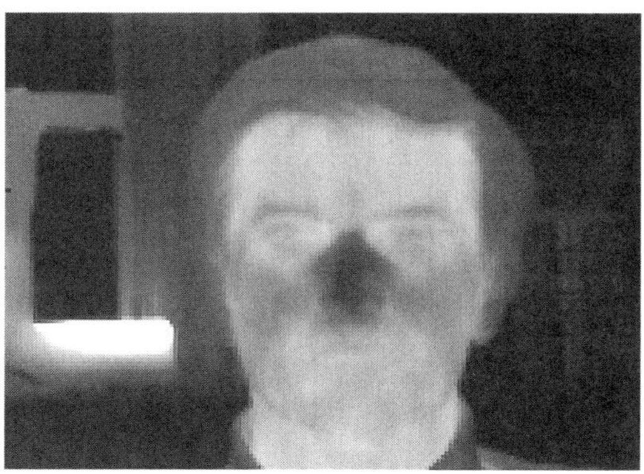

FIG. 1. Infrared image of the author taken with a Cincinnati Electronics IRC 160 infrared imager. Darker regions correspond to lower relative temperatures.

A generally accepted taxonomy of infrared sensors has yet to be established. However, Hudson (1969) made one of the earliest attempts at application-dependent classification (see Sec. 3). Characteristic components and technology afford yet another means of categorization. Distinctions can be made on the basis of imaging versus point detection, the means by which contiguous imagery is generated, and active versus passive systems. In the active case the employment of illuminators is the distinguishing feature, while in the passive case natural illumination or emission is relied on. A multitude of highly specific sensor configurations have been developed to meet the demand for specialized usage. To some extent categorization depends upon the accepted jargon and practices of specific applications.

In its most elementary form, an infrared sensor is a device that responds in a measurable way to incident infrared radiation. Although it is common practice to regard thermometers, thermocouples, and other infrared-sensitive devices as infrared sensors, they are more properly categorized as infrared detectors. In modern usage an infrared sensor is comprised of a thermal or photon solid-state detector element combined with other application-dependent components, as described below. Although sensor configurations vary widely, the basic building blocks are shown in Fig. 2 with additions or deletions as necessary to meet certain requirements.

1. PHYSICAL AND TECHNICAL PRINCIPLES

All objects at finite temperature and in thermal equilibrium radiate energy. The radiation per unit wavelength is given by the Planck radiation law (Reif, 1965):

$$M_e(\lambda,T) = \epsilon(\lambda)(2\pi hc^2/\lambda^5) \times \exp[hc/kT - 1]^{-1}\ (\mathrm{W/m^3})$$

where $\epsilon(\lambda)$ = spectral emissivity, c = speed of light in vacuum = 2.997925 × 10 m/s, h = Planck constant = 6.6260755 × 10^{-34} J s, k = Boltzmann constant = 1.38054

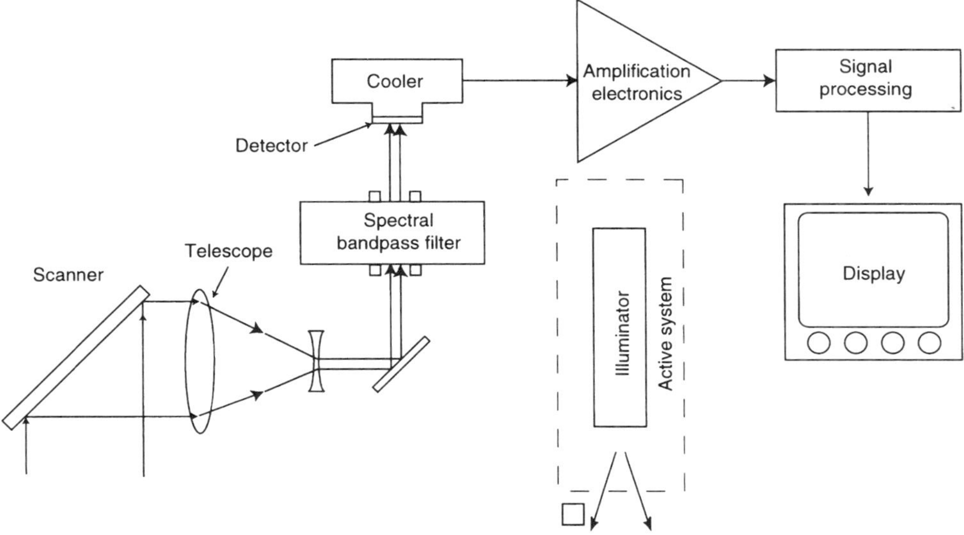

FIG. 2. Generic components of a modern, complex infrared sensor system. The laser is used as a source of monochromatic illumination in certain environmental and military radar applications.

$\times\ 10^{-23}$ J/K, T = absolute temperature in kelvins, and λ = radiation wavelength in meters.

This expression is shown graphically in Fig. 3. We note that for a given temperature, the emitted radiation has a maximum at a specific wavelength. As the temperature increases, the wavelength at maximum emission shifts progressively to shorter wavelengths, and the relative emission rises. In accordance with Wien's law, the wavelength at maximum emission varies inversely with temperature. For objects near room temperature, for example, the maximum occurs in

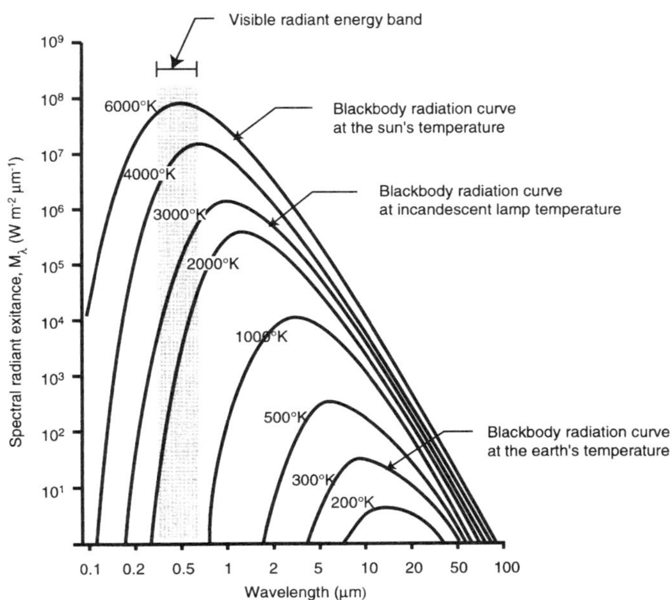

FIG. 3. Spectral distribution of energy emitted per unit wavelength radiated from blackbodies of various temperatures (Lillesand and Kiefer, 1994). As the temperature rises, the emission both increases and shifts to shorter wavelengths.

the long-wave IR region near 10 μm. In general the position of this maximum has a strong influence on the selection of the sensor wavelength band of operation.

Spectral emissivity $\epsilon(\lambda)$ is a property of macroscopic matter that is a measure of the efficiency of emission of absorption of radiation relative to a perfect blackbody, for which $\epsilon(\lambda) = 1.0$. For a "graybody," that is an object whose emissivity is independent of wavelength, only the magnitude (but not the relative spectral distribution) of the radiation is altered. For nonblackbody radiation where the emissivity is a function of wavelength, the distribution of the radiation may be radically altered from the Planck distribution. In some substances such as gases, vibrational-rotational molecular transitions give rise to characteristic emission and absorption lines in the infrared. These gases can thus be identified by analysis of their corresponding spectra. In general, because of energy-level broadening in the solid state, narrow-line emission or reflectance spectra are not usually observed in solid materials. Relatively broad spectral features are much more common. High-temperature line broadening produces the same effect in the radiation spectra of gases at high temperature. It is the analysis of such spectral features of materials and even living organisms that allows remote discrimination and identification of various materials constituting the observed scene. This technique forms the basis of the science of multispectral remote sensing (see Sec. 4.4). Integrating the Planck expression over all wavelengths yields the Stefan–Boltzmann equation, which, under certain assumptions about the object's emissivity, enables an inference of object temperature. Remote infrared temperature-sensing devices use this principle to measure object temperature indirectly. The accuracy of this technique depends largely upon on how well the emissivity is known.

1.1 Principles of Operation

A modern infrared sensor comprises an infrared-sensitive element, or detector, to convert the collected radiation to electrical signals and at least one or more of the following:

1. optical elements to collect and focus radiation;
2. a scanning device to image the scene of interest;
3. a spectral filter to eliminate extraneous radiation; and
4. some form of display device or subsequent signal processing to convert the electrical signal into useful information.

In some cases an illuminator may be used to provide a controlled source of radiation. Additional components could include cryogenics for detector cooling, internal temperature references, mechanical gimbals, servo control for pointing and tracking, and a host of other components in specialized applications. The physical and technical principles of the typical major components of an infrared sensor are discussed elsewhere in this work. In many instances strong physical analogies exist between sensors operating in the optical region of the spectrum and the infrared region, and some devices are capable of operating in both.

Complex infrared sensors contain a composite of diverse technologies including solid-state detectors, optical design and fabrication techniques, optical materials, mechanical scanning devices, electronics, and display and signal-processing electronics. These are briefly discussed below. For an extended discussion of the components of infrared systems reviewed below see Accetta and Shumaker (1993), Vol. 3.

1.1.1 Infrared Detectors and Arrays

The responsive element in all modern infrared sensors is the detector that converts the incoming radiation into electrical signals by one of several possible physical mechanisms. These signals are subsequently amplified by accompanying electronics. The physical mechanisms that have been employed in detectors include bolometric, photoconductive, photoelectromagnetic, photovoltaic, pyroelectric, thermopneumatic, and thermovoltaic processes. The majority of modern infrared sensors depend upon photon detectors using photoconductive and photovoltaic processes and thermal detectors using bolometric and pyroelectric processes. For an extended discussion of detectors see RADIATION DETECTORS, INFRARED.

In general, photon detectors employ some form of cooling to approach their ultimate limits of sensitivity. In contrast, thermal detectors are generally not cooled and are used in applications requiring modest levels of sensitivity. (A prominent exception is the infrared bolometer employed in astronomy, which is sometimes cooled to liquid helium temperatures to achieve very high sensitivity and wide spectral response). Figure 4 shows several specific detector types and the spectral ranges commonly employed. The parameter commonly used to describe the detector figure of merit is referred to as D* ("D star") and is widely employed in sensor performance predictions (Dereniak and Crowe, 1984).

Detectors are often fabricated in linear, rectangular, or square arrays as an alternative to creating images by mechanical scanning. These focal-plane arrays (FPAs) spatially sample the optical image in the focal plane of the sensor to create nearly contiguous imagery. Each individual detector in the FPA senses the radiation falling within its active boundaries. The detector element and its corresponding contribution to the image is referred to as a "pixel" (short for picture element). FPAs are employed in some imaging sensors and possess certain technical advantages over systems that mechanically scan the field of view over a single detector. For example, rapid scanning over the sensor scene of interest may have mechanical limitations that may be circumvented with the use of FPAs. The choices here depend on the particular application.

Modern FPAs may consist of thousands of individual detector elements. An associated amplifier and output wires from each detector are thus impractical. A means for amplifying and multiplexing (handling many detector outputs over a few wires) the signals was required. To solve this problem the FPA is now integrated with multiplexing electronics on a single chip and commonly referred to as a charge-coupled device (CCD) or infrared CCD (IRCCD). The term CCD really comes from the specific means by which the outputs are multiplexed but has become generic in usage.

The majority of infrared detector elements in use are uncooled single-element types. Detectors suitable for this purpose include pyroelectrics, which create a temperature-dependent change in effective capacitance, and bolometers, whose resistance is temperature dependent. There are many industrial applications where sufficient signal to overcome inherent detector noise is readily available, especially those employing active illuminators or intense thermal sources. Thermal detectors are the standard of usage in many industrial applications. Because of the attendant advantages in simplicity and cost, a great deal of recent work in detector development has concentrated on high-sensitivity uncooled operation (Rogalski, 1994). See also RADIATION DETECTORS, INFRARED for an extended discussion of modern infrared detector technology.

1.1.2 Collection Optics The optical design principles used in the design of IR sensors are strongly analogous to those used in

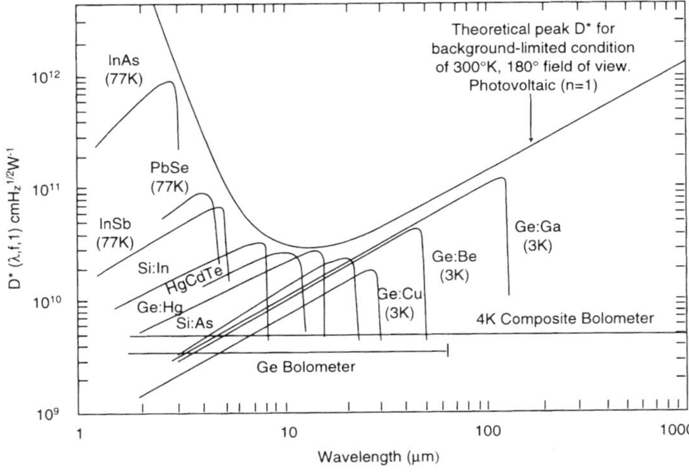

FIG. 4. The spectral response for various types of detectors expressed in terms of detectivity D* (Dereniak and Crowe, 1984). D* is a commonly used parameter for the figure of merit that reflects the measured signal-to-noise ratio per watt of incident flux under specific conditions including the incident wavelength, spectral bandwidth, and temporal frequency. The measurement is further normalized by the detector area and electrical bandwidth of the measurement.

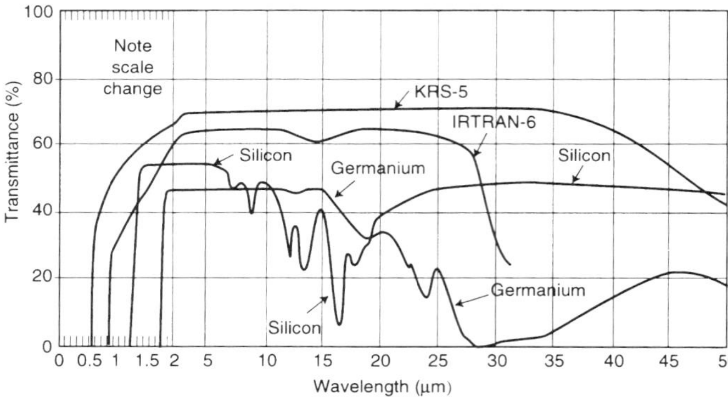

FIG. 5. Relative transmittance of various materials used in infrared refractive elements (2.0 mm thickness) (Spiro and Schlessinger, 1995).

the visible region of the spectrum. Departures from conventional techniques in high-sensitivity applications include the necessity of accounting for thermal radiation from the optical elements themselves. The collection optics for a typical infrared sensor function in the same manner as elements in the optical region of the spectrum, i.e., to collect and focus the radiation on the detector. To avoid unacceptable signal attenuation, highly transmissive refractive elements and highly reflective mirrors are required. Although these considerations are seemingly elementary, the practicalities of achieving these properties with conventional optical materials are often problematic. In some cases exotic materials have been developed that have the necessary mechanical and thermal properties as well. The spectral transmittance of some typical IR optical materials are shown in Fig. 5. Reflective elements are typically metal substrates with protective coatings to achieve environmental stability. The spectral reflectance of various metals used for reflective elements are shown in Fig. 6. See OPTICS, LINEAR; OPTICS, GEOMETRICAL.

1.1.3 Filters Spectral filters serve to limit the spectral range of the IR radiation falling on the detector. Limiting the observable spectral bandpass pays large dividends in signal detectivity by restricting the extent of undesirable contributions from extraneous backgrounds. Filters may employ interference coatings or transmissive elements exhibiting selective absorption. Other selectively reflective devices and gratings can be used as well (Wolfe and Zissis, 1985). These devices include Christiansen filters, made up of small, closely packed particles of an infrared-transparent substance suspended in a liquid or a gas; wavelength-selective reflection filters made up of various crystalline materials; refraction filters that depend on the

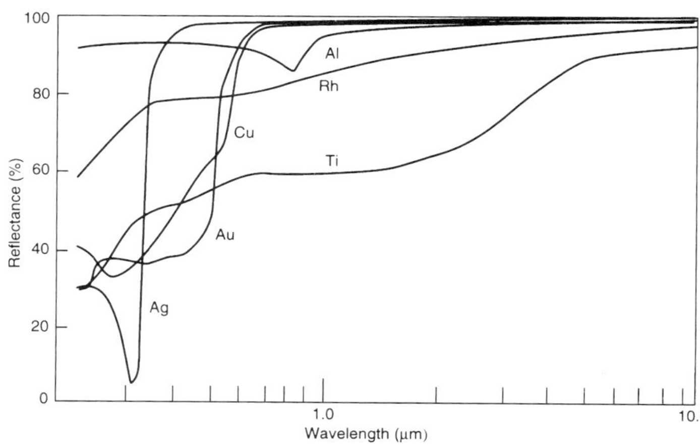

FIG. 6. Reflectance of various metallic films used in infrared reflective elements (Wolfe and Zissis, 1985).

wavelength dependence of the index of refraction to focus radiation selectively at different points along the optical axis; and polarization interference filters.

1.1.4 Cryogenics To achieve high sensitivity and in some cases simply to effect operation over a given spectral region, detectors must be cooled well below room temperature. Cooling reduces inherent detector and preamplifier noise and is nearly essential in most astronomical, environmental, and military applications. In applications beyond 10 μm, additional cooling of the optics, filter, and other components is necessary to reduce internally generated background radiation. These devices are sometimes called coolers but are more properly termed cryostats. Cryostats may use electrical means, such as the thermoelectric effect, or compressed gases, as in the Joule–Thompson effect. Other types include liquid or solid cryogens as the refrigerant such as liquid nitrogen, liquid helium, or solid CO_2. Cryostats must be well insulated from the environment, holding the detector temperature constant within a very narrow range. They must be highly reliable and compact, especially for spaceborne applications, and must possess a considerable degree of mechanical integrity under harsh environmental conditions. A typical cryostat is shown in Fig. 7.

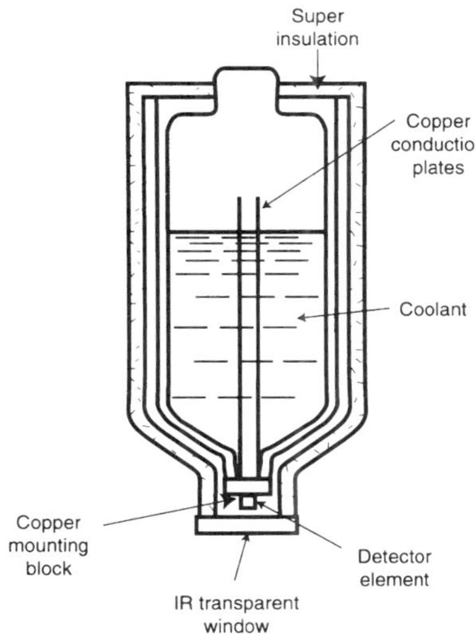

FIG. 7. A conventional Dewar flask used for holding liquid cryogens to cool the detector. The detector is in physical contact with a copper block immersed in the coolant. Typical coolants are liquid nitrogen at 77 K or helium at 4 K. (Wolfe and Zissis, 1985.)

1.1.5 Electronics Modern IR sensors employ sensitive electronics to amplify the extremely weak signals generated by the IR detectors. Such IR-detector electronics are quite similar to the circuitry used with visible detectors. The desirable characteristics are low noise, high gain, and impedance matched to the detector for efficient power transfer. Often the preamplifier electronics are cooled along with the detector to reduce noise. The idealized electronics design is such that the overall detector/preamplifier performance is detector limited. High-impedance detectors are very sensitive to electromagnetic interference and external noise, and amplifiers must be located in close physical proximity to the detectors.

In early sensor designs, preamplifiers built from discrete components were placed in close proximity to the detectors on the cold focal-plane mounting. The advent of detector arrays with thousands of separate detectors requires a different approach that integrates the amplifier and readout electronics with the detector. The CCD or change-coupled device integrates all of these functions on a single chip and has become the most prevalent device used for infrared imaging. A typical CCD is shown in Fig. 8.

1.1.6 Scanners To create images of infrared scenes, a means must be provided to scan the detector field of view over a much larger area than that addressed by a single detector. As described above, one way to achieve this is with focal-plane arrays imaging simultaneously the entire scene of interest on the detector array. In single-detector designs, electromechanical scanning devices are employed to move one or more optical elements physically to achieve the desired effect. Regular and precise motion is achieved by optical-component rotation or oscillation or by various electromagnetic means. A typical mechanical scanner is shown in Fig. 9. Piezoelectric, electro-optical, and acousto-optical effects are used to effect motion of the optical axis. Finally, motion of a spacecraft or aircraft platform may be used to create contiguous imagery and is often employed when such motion is regular and predictable.

1.1.7 Displays and Signal Processing The electrical signals from the sensor may ultimately be used to drive various cathode-

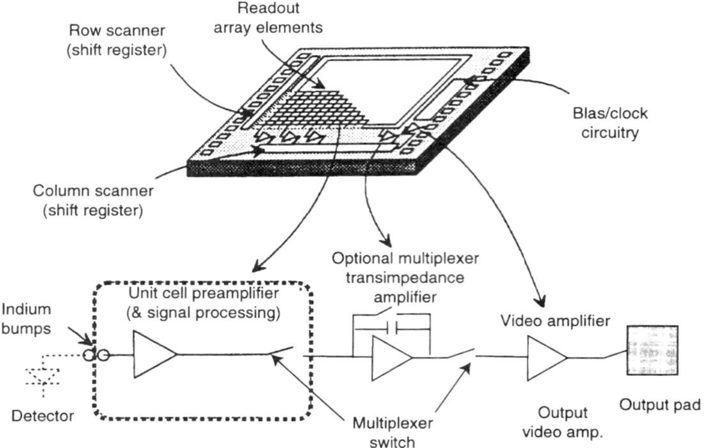

FIG. 8. Focal-plane array and readout electronics integrated onto a single chip (Accetta and Shumaker, 1993). Charge is transferred synchronously from the individual detector elements to shift registers and subsequently transformed into electrical impulses by the video amplifiers. This technique reduces the number of output leads in an $n \times n$ array from n^2 to $2n$.

ray tubes (CRTs) or solid-state displays for visual interpretation, transferred to photographic film for archival purposes, or processed for additional information. Further processing may take the form of noise or background reduction or call upon a multitude of image-processing techniques for feature enhancement and detection. The nature of this additional information and the processing required to extract it is application specific and has a myriad of possibilities. These include electrical signals indicating the presence or absence of a particular object, utilization of infrared spectral information in the assignment of various biological or geological forms as in remote sensing, and the location and nature of emitting objects as in infrared astronomy. The spatial distribution

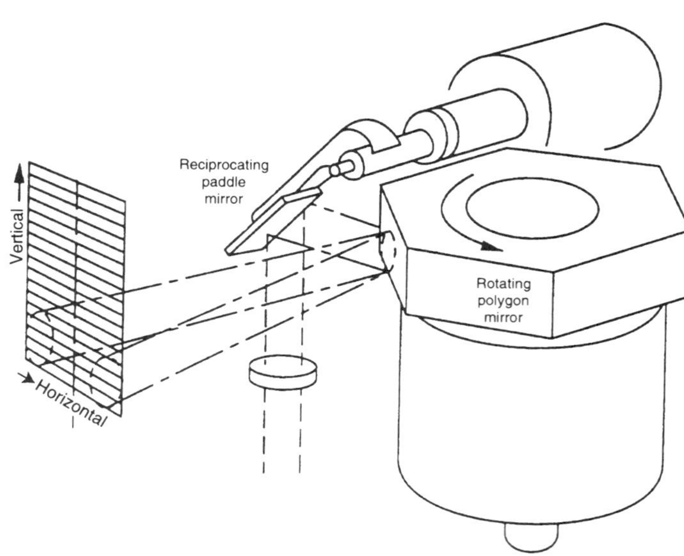

FIG. 9. Rotating-polygon and nodding-mirror scanner configured for rectangular scanning (Accetta and Shumaker, 1993). The reciprocating paddle mirror moves the field of view in the vertical direction while the rotating polygon mirror creates motion in the horizontal direction.

of temperature is used in diagnostic medicine, process control, and industrial inspection. Other applications include surveillance and guidance and control as in military and comparable industrial operations.

2. DESIGN, FUNCTION, AND OPERATION

2.1 Sensor Figures of Merit

The figure of merit of an IR sensor often depends on the specific application. In industrial applications the minimum detectable temperature change may serve as a figure of merit. In certain military applications the figure of merit is quite frequently the probability of detecting a given target in a given background, given a certain tolerable false alarm rate. In imaging applications spatial resolution in combination with sensitivity is often the determining factor, i.e., the degree to which fine detail can be resolved by a given sensor. In an infrared radiometer (discussed below) the absolute radiometric accuracy of the sensor is the overriding concern. Some formalisms commonly employed in the assessment of infrared sensor performance include noise equivalent irradiance (NEI), noise equivalent temperature difference (NETD), and minimum resolvable temperature difference (MRTD). In complex cases such as forward-looking infrared systems (FLIRs) discussed below, a human operator is involved. Thus psychophysical factors are an intrinsic part of the figure of merit, invoking the statistical performance of a large number of human operators. In environmental systems minimum detectable concentration of a given molecular species can be a figure of merit.

As implied by the foregoing, figures of merit are quite specific to a given sensor type. However, in almost all applications the signal-to-noise ratio is an underlying consideration in determining sensor performance. The probability of detection of an infrared signal is a strong function of the signal-to-noise ratio. In imaging applications it affects the quality of the displayed image. Noise may result from the detector, preamplifier, or background, and noise mitigation is the major preoccupation of the sensor designer. Information extraction in noisy conditions constitutes a voluminous portion of the modern literature on signal processing.

2.2 Scanning Systems

Historically, the first infrared sensors were simple point detectors with the field of view limited by the detector geometry. To create infrared scenes or maps, the point detectors were mechanically scanned over the scene to create infrared images. The essence of the scanning system is that a single detector element may be used to form an entire image if the optics in the system scans over the object point by point. An infrared line scanner is an imaging device that forms images, usually by successive scans of a rotating mirror in a specific direction. The perpendicular scan direction needed for a two-dimensional image is provided by forward motion of the sensor platform or alternatively by motion of the scene of interest across the field of the sensor as in Fig. 10. In environmental remote-sensing applications, the platform could be an aircraft or satellite. Early "forward-looking infrared systems" (FLIRS), as they were called (described below), were adaptations of line scanners with the addition of a mechanical means to do the scanning in the other dimension. A modern version of this system is the satellite-based LANDSAT and SPOT sensors described in Lillesand and Kiefer (1994).

2.3 Staring Systems

Staring systems, as the name implies, simply stare. There is a strong analogy to the human eye. The entire scene is imaged onto a two-dimensional array of detectors (FPA) as previously discussed. Because of the small but finite spacing of the individual detector elements, the image is essentially sampled at regularly spaced intervals. The ultimate spatial resolution of the image so obtained is determined by the individual detector geometry, the array geometry, the resolving capability of the optics, and other factors. In the visible region of the spectrum, the largest commercial and military applications of staring systems are conventional TV cameras that operate over a very wide range of lighting conditions. With the availability of infrared focal-plane arrays, the TV-camera con-

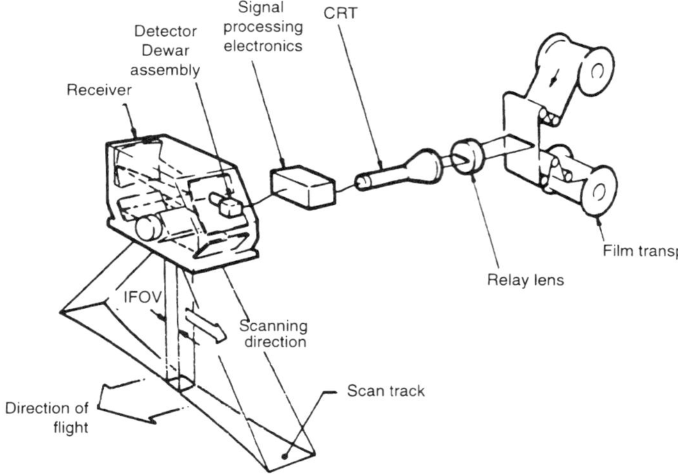

FIG. 10. Typical IR line scanner used for environmental and military data gathering from a moving platform (Accetta and Shumaker, 1993). The image is recorded in synchronism with the motion of the platform and with the scanning device. Generally, the imagery must be further corrected for small perturbations in the flight path of the aircraft.

cept has been extended into the infrared regions of the spectrum as with the thermal imager discussed previously.

2.4 Infrared Phenomenology

It is the combined phenomenology of the source, background, and atmosphere that will drive the design of a sensor. Other factors such as economics, the environment, and available technology may play a significant role in the choice as well. A proper understanding of the phenomenological relationships between these factors is essential to achieving satisfactory sensor performance.

As shown in Fig. 11, the radiation from a typical object of interest embedded in a terrestrial background is actually a superposition of a number of sources of radiation including the background radiation from the terrain within the sensor field of view, the background radiation reflected from the object into the field of view, the radiation from the object itself, and the radiation from the intervening atmospheric path. These sources are in turn affected by the atmospheric absorption along the path from the object to the sensor. In the general case the sensor-design process requires a detailed understanding of all relevant sources of radiation. The determination of these effects is often difficult and highly dependent on local environmental conditions. The proper choice of the spectral band that optimizes sensor performance demands a consideration of the above factors. The degree of optimization of a specific sensor design depends on the extent to which these factors are known and understood.

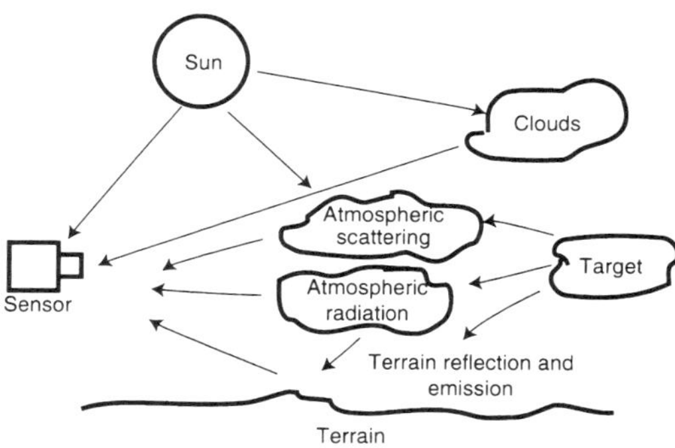

FIG. 11. Naturally occurring phenomenology associated with infrared sensing. There are multiple sources of direct and reflected natural radiation that interfere with the target signal.

2.4.1 Object Radiation The essential infrared characteristics of an object are temperature, spectral reflectance, and emissivity. From these quantities, the emitted and reflected radiation from an object or scene in a given environmental geometry can be determined. In the general case object radiation is not blackbody in nature, and thus, its spectral properties have to be accounted for. Reflectance spectra from some natural materials are shown in Fig. 12. The spectral characteristics of the materials constituting the scene or object of interest will again af-

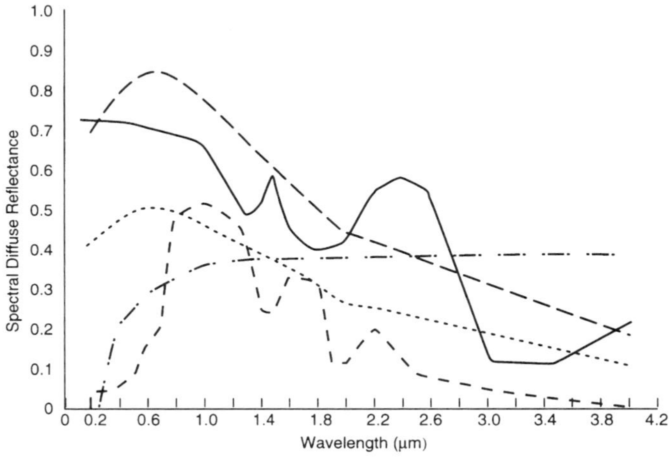

——— Clouds. Data are directional reflectance of a middle layer cloud.

— — — Winter, Snow and Ice. Data are directional reflectance of dry snow.

······· Summer Ice. Data are directional reflectance of summer Arctic ice.

·—· ·– Soil and Rocks. Data represent the average value of the bidirectional reflectance, ρ_λ (45°, 0, 0, 0), of gravel, wet clay, dry clay, tuff bedrock, and sandy loam.

– – – Vegetation. Data represent the average value of directional reflectance of many types of vegetation (from ERIM Data Files).

FIG. 12. Diffuse reflectance of earth-atmosphere components (Wolfe and Zissis, 1985). The spectral dependence of these components may be used to discriminate against undesirable effects on sensor performance.

fect the choice of sensor spectral band of operation.

2.4.2 Backgrounds Backgrounds represent an undesirable component of incident radiation and are a serious source of interference in the detection problem. In Fig. 11 there are several sources of radiation including reflected solar and emitted radiation from surrounding terrain or clouds; thermal radiation from the atmosphere itself; and solar, lunar, and stellar sources. The distinction between objects and background is arbitrary depending on the interests of the observer. Sensors are generally designed to maximize object signal and minimize background interference by selection of the spectral band that optimizes the ratio of signal power to background power.

2.4.3 Atmospherics The atmosphere can play a significant role in the absorption of IR radiation over long observation distances. The absorption is both wavelength and distance dependent; thus sensors operating over long distances must use regions of the spectrum with relatively low absorption to minimize signal loss. Most infrared absorption in the atmosphere is due to water vapor and CO_2 with a minor dependence on other gaseous species. Scattering also plays a role as a potential signal attenuator. A typical atmospheric transmission curve is shown in Fig. 13. Note that the "windows" of relatively high transmission occur in the 3.0–5.5- and 8.0–14.0-μm regions of the spectrum. In certain astronomical applications observation at wavelengths outside of these windows requires exoatmospheric-based instrumentation using balloons, high-altitude aircraft, satellites, or sounding rockets as platforms.

3. MANUFACTURING AND ECONOMICS

Because of the diversity in both technology and complexity of infrared sensors, the economic and manufacturing factors are not readily discussed in detail. In the simplest cases—e.g., an industrial sensor employing an uncooled detector—the major components of the sensor including the detector element, electronics, and housing can be readily manufactured in-house for very low cost. Optics, if required, are rarely manufactured

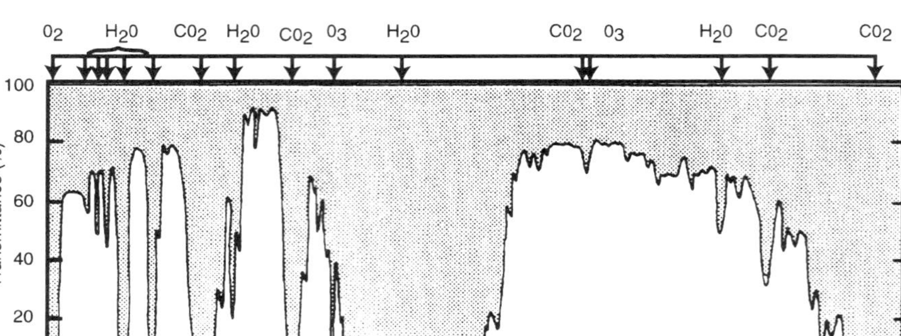

FIG. 13. Spectral transmittance of the atmosphere from 0.0 to 15.0 μm. Note the presence of "windows" at 3.0–5.5 μm and 8.0–14.0 μm. (Lillesand and Kiefer, 1994).

in-house. In contrast, most manufacturers of complex sensors typically buy required specialized and sophisticated components when possible rather than attempting to manufacture them. "Build or buy" decisions generally lean in the "buy" direction. The potential reduction in revenue resulting from nonvertically integrated manufacturing is offset by relieving the manufacturer of the necessity of keeping up with the state of the art in a number of diverse technologies. Other expenses are reduced as well.

It is not uncommon to purchase detectors, cryogenics, and electronic and optical components from several outside sources. Manufacturers typically rely on in-house sources for mechanical design, electronic design, final assembly, and testing. It would be extremely rare to find all the necessary technology under one roof. Complex sensor manufacturers are thus highly dependent on external sources of supply.

As might be expected, the overall economic aspects depend upon the attributes of the sensor components and volume produced. Infrared detectors and arrays that depend upon semiconductor technology and cryostats are usually much more expensive than their uncooled pyroelectric and thermopile equivalents. Infrared optical materials and components are also more expensive than their visible counterparts. For a comparison with components that operate in the visible region of the spectrum, a modern visible TV camera employing an uncooled silicon detector array can be currently purchased for a few hundred dollars in quantity. However, a sensitive infrared equivalent will likely cost tens of thousands of dollars and could exceed millions in some extremely specialized cases. As usage increases, prices will drop accordingly, especially in those cases highly amenable to mass production and the use of inexpensive and uncooled detectors or FPAs, plastic optics, and application-specific integrated circuits. Significant production quantities should strongly drive price reductions.

Although we might conjecture that many additional commercial and industrial activities could benefit greatly from the mass application of infrared sensors, the overall cost–benefit relationship must be understood by the user before such large-scale application would occur. There is a great deal of promise that over the near term the cost of sensitive, rugged, high-quality infrared sensors will be significantly reduced.

4. MAJOR APPLICATIONS

There are a multitude of scientific, industrial, and military applications for IR sensors. Some of the major applications are shown in Table 1.

Table 1. Major infrared applications (adapted from Hudson, 1969).

Military	Industrial	Medical	Scientific
Intrusion detection	Fire detection	Obstacle detection for the blind	Satellite detection
Bomber defense	Fuel ignition monitor	Skin-temperature measurement	Space-vehicle navigation and flight control
Missile guidance	Law enforcement	Cancer detection	Horizon sensors
Navigation and flight control	Remote dimensioning	Monitoring of healing and infection	Solar trackers
Ship, aircraft, ICBM, mine detection	Process control	Remote biosensors	Optical structure of horizon
Fire control	Temperature measurement of brake linings, power lines, cutting tools, welding and soldering operation, ingots	Studies of skin heating	Lunar, planetary, and stellar temperatures
Proximity fuses	Turbulence detection	Detection and monitoring of pollution	Remote sensing of weather conditions
Aircraft collision warning	Analysis of organic chemicals and gases	CO_2 levels in blood and breath	Heat transfer in plants
Target signatures	Blood-alcohol levels	Localization of placental sites	Earth heat balance
Terrain analysis	Pipeline leaks	Clothing efficiency	Detection of extraterrestrial life
Poison gas detection	Water contaminants and pollution	Diagnosis of incipient stroke	Terrain analysis
Fuel vapor detection	Display of heating patterns	Measurement of pupillary diameter	Monitor spacecraft atmospheres
Detection of contaminants	Efficiency of thermal insulation	Location of vein and arterial blockages	Zero-*g* liquid-level gage
Reconnaissance	Crime prevention	Monitoring eye movements	Magnetic field measurement
Surveillance	Crop-disease detection	Nocturnal habits of animals	Earth resource survey
Thermal mapping	Agricultural assessments	Corneal opacities	Mapping ocean flow
Underground facilities detection	Automatic camera focus	Heat therapy	Forest fire detection
Damage assessment	Intrusion detection		Water pollution studies
Night driving	Auto-collision prevention		Location of crevasses
Weapon location	Traffic monitoring		Sea-ice reconnaisance
Camouflage detection	Radiant heating and drying		Petroleum exploration
Station keeping	Data and voice communications		Forgery detection
Docking and loading	Intervehicle speed sensing		Epitaxial film thickness
Communications	Aircraft landing aids		Planetary constituents
Countermeasures	Material bonding inspection		Gem identification
Range finding	Forgery detection		Water quality analysis
Command links	Personal identification		Detection of diseased crops
			Animal communications

4.1 Infrared Radiation Pyrometry

Infrared radiation pyrometry is a generic term that is often used to describe remote temperature measurement by observation of emitted thermal flux in one or more infrared spectral bands. An infrared pyrometer is analogous to an optical pyrometer whose complexity ranges from a simple single-detector and amplifier combination to more complex units such as a hand-held optically sighted device operating in multiple spectral bands. Except for the hand-held variety, fields of view and directions of observation are generally fixed. Pyrometers generally require modest sensitivity, simplicity, rugged construction, and low cost. The characteristics of pyroelectric and thermopile detectors are consistent with these requirements. Pyrometers may be calibrated to measure temperature or relative flux. Numerous examples of relevant applications are cited in Table 1, for many of which detailed discussions are given in Hudson (1969) and Kaplan (1993). Of particular importance are industrial applications such as the monitoring and control of processes requiring continuous remote measurement of temperature. An illustrative example is the temperature control of moving steel in a rolling mill or extrusion line. Another is the monitoring of transformer and insulator temperature in power-distribution systems. The general inaccessibility of these components requires portable remote measurement systems. There are several models of this type of sensor in current production.

Infrared radiation pyrometry can be applied to many other industrial problems such as intrusion alarms or appliance control where the measured temperature may cross certain pre-established thresholds triggering a control signal or other event.

4.2 Search and Tracking Systems

An infrared search and tracking system (IRST) is designed to detect reliably, to locate, and continuously to track infrared-emitting objects and targets in the presence of background radiation and other disturbances. Principally employed by the military, IRST systems are usually confined to unresolved point-source objects in highly structured backgrounds and at ranges beyond 5 km. They generally use mechanical scanning in both dimensions to create a two-dimensional image or "frame." The spectral bands employed are the usual atmospheric windows at 3.0–5.5 and 8–12 μm. Psychophysical factors are not a consideration in these systems since the search and track processes are highly automated. Most of the current research activity in IRST systems is in the signal-processing domain to extract targets from severe background clutter. A typical military IRST is shown in Fig. 14.

4.3 Imaging Systems

4.3.1 Forward-Looking Infrared (FLIR) Systems An FLIR is an infrared analog of a visible TV camera. The term FLIR is an acronym for "forward-looking infrared" that originated from the modification of an early 1960s U.S. Air Force downward-looking infrared mapper to a forward-looking real-time infrared imaging system. The name has remained, although inappropriate, and is usually used in a military context. Thermal or infrared imager would be a better name. The FLIR systems are operated nearly exclusively in the 3.0–5.5- or 8.0–14.0-μm region of the spectrum corresponding to atmospheric windows of high transmissivity. The FLIR systems are found in both scanning and staring configurations and generally do not rely on platform motion for the creation of an image frame. High-performance military systems are used in targeting, navigation, and surveillance applications for the imaging of extended targets. A block diagram of a typical FLIR system is shown in Fig. 15.

4.3.2 Medical and Industrial Thermal Imagers Medical and industrial thermal imagers use the same basic technology as FLIR systems. In these applications infrared imaging and analysis is often referred to as thermography. The instruments used for this

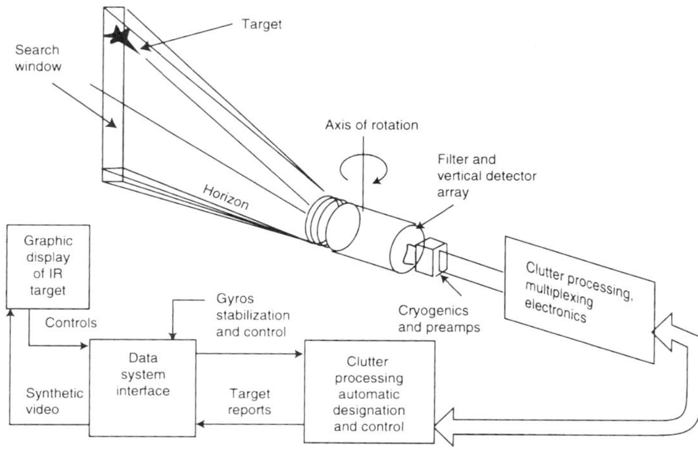

FIG. 14. Block diagram of a typical infrared search and track system (IRST) (Accetta and Shumaker, 1993). The sensor axis is continuously swept about the horizon. When a target detection occurs, the system reverts to a tracking mode of operation maintaining continuous surveillance of the target.

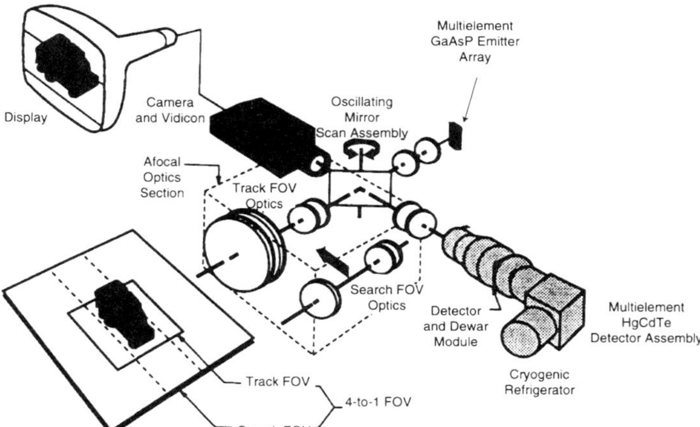

FIG. 15. First-generation forward looking infrared system (FLIR) (Accetta and Shumaker, 1993). Rather than driving the display directly in this modular system, the electrical signals from the detector array drive a solid-state GaAsP light-emitter array. The output of the emitter array is used as a strong optical excitation source for the Vidicon. "FOV" stands for field of view.

purpose are generally less expensive than their military counterparts. The infrared image of the author in Fig. 1 is characteristic of medical infrared imagery. Temperature anomalies may be indicative of certain pathological conditions such as breast tumors, abnormal arterial and venous functions, and infections. Industrial applications of IR imagers include the remote inspection of industrial facilities, and process monitoring and control. For example, utility companies often use mobile industrial thermography units to locate overheating components on inaccessible utility poles. Electronics companies use imagers to detect overheating components, and biologists use these systems to gauge cellular metabolism.

4.3.3 Astronomical Imagers The importance of infrared sensors in astronomy lies in their ability to detect the emissions of otherwise invisible dust grains and compact interstellar sources heated by stellar sources or shock waves. Prototypical stars at their earliest stages of formation and not yet at the threshold of visible emission are also detectable (Zeilik and Smith, 1987). Additional advantages provided by observation in the infrared regions of the spectrum are less attenuation by interstellar dust than in the visible region and the ability to observe in daylight. Analogous to their visible photometry counterparts, the earliest infrared astronomical applications simply focused the radiation from a stellar source onto a calibrated lead sulfide infrared detector. The emitted infrared intensity from stellar sources was computed after correcting for atmospheric and interstellar absorption and the detector spectral response.

Modern infrared astronomical sensors range from sensitive bolometers consisting of a tiny chip of doped germanium operated at very low temperature to the employment of sophisticated InSb (indium antimonide) or HgCdTe (mercury-cadmium-telluride, also known as MCT) infrared focal-plane arrays to produce detailed imagery. Infrared astronomy has become a major subdiscipline of observational astronomy.

A vast majority of optical telescopes have been used for infrared observations at one time or another. Early adaptations used infrared-sensitive photometers. There are now a number of astronomical telescopes specifically designed for infrared use and equipped with sophisticated focal-plane arrays (McLean, 1994). One of the many major triumphs of infrared astronomy is the Infrared Astronomical Satellite (IRAS). Designed to circumvent the limitations of endoatmospheric observation due to extinction and turbulence (also known as "seeing"), this satellite-borne infrared imager generated extensive survey maps revealing a multitude of previously unknown IR celestial sources (Beichman *et al.*, 1988). The telescope, shown in Fig. 16, was a two-mirror Ritchey–Chretien configuration mounted inside a superfluid helium tank mounted within an evacuated main shell. The liquid helium supply cooled the entire telescope including the focal-plane array to achieve extremely sensitive performance. The major specifications of

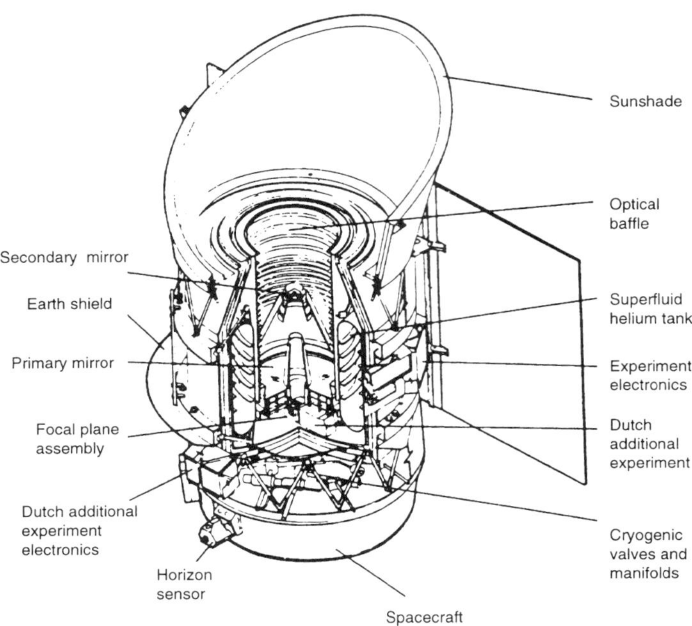

FIG. 16. The Infrared Astronomical Satellite (IRAS) infrared sensor configuration (Beichman *et al.*, 1988). This instrument is a successful example of a modern, extremely complex infrared sensor.

this remarkable testimony to modern infrared-sensor technology include a collecting aperture of 57.0 cm, a focal-plane operating temperature of 2.6 K, eight infrared and two visible detector arrays, and operation in the 12-, 25-, 60-, and 100-μm wavelength bands.

4.4 Remote Sensing and Environmental Applications

4.4.1 Airborne and Spaceborne Spectral Systems The earliest environmental applications of infrared sensors used photographic film sensitized in the near-infrared region. As sensor technology evolved, the spectral region of observance was extended to longer wavelengths and multiple spectral bands. Analogous to its counterpart in the visible region of the spectrum, infrared multispectral sensing is a relatively new airborne or spaceborne technique to exploit the information content of infrared images by intercomparisons of identical imagery in multiple spectral bands. Images are collected simultaneously in multiple spectral bands, and good image registration is critical to achieve optimum performance. One means of collection is line scanning, as shown in Fig. 10. Useful information arises from differences in spectral reflectance properties of the terrain that can be correlated to various material types after correcting for source temperatures, background radiation sources, path radiance, and transmission factors. The basic technique described is employed in satellites and aircraft sensor platforms for pollution detection and monitoring, land-use assessment, treaty monitoring, change detection, soil and moisture content, crop inventories, snow cover, bathymetry, mineral exploration, and disaster-damage assessment (Lillesand and Kiefer, 1994).

The principle behind multispectral analysis is that the ratio of the intensities in two or more spectral bands is nearly constant and unique for a given material. Objects within a given scene can be classified on the basis of the observed ratio. The spectral bands that optimize separation between various materials are carefully chosen from large trial data sets for consistency. This category of infrared sensing is further subdivided into multispectral, hyperspectral, and ultraspectral, depending on the spectral resolution used. Multispectral sensors operate with spectral bandwidths on the order of 10% of the center wavelength and typically use from 5 to 20 distinct bands. Hyperspectral systems operate at about 1% of the center wavelength and may use 100–200 bands. Ultraspectral sensors have bandwidths on the order of 0.1% of the center wavelength and may have up to 1000 separate spectral bands. The LANDSAT series is perhaps the most well known of the spaceborne multispectral sensor systems. Depending on the configuration, several different sensor configurations were employed, including both infrared and visible sensors operating in multiple bands. The French SPOT satellite offers similar capabilities.

Thermal scanners operate in the longwave infrared region (LWIR) of the spectrum and can be used to infer surface-temperature differences. A good example of the use of this technique is to monitor surface water temperature near nuclear power plants. Heating and cooling rates over diurnal periods offer information about the type, condition, and properties of observed materials as well. The NASA Thermal Infrared Multispectral Scanner operates exclusively in the thermal infrared region using six spectral bands.

4.4.2 Infrared Remote Sensing of the Atmosphere A special class of infrared instruments has arisen in the context of atmospheric remote sensing for either pollutant or natural composition studies. The primary properties of the atmosphere usually observed for this purpose are absorption, aerosol scattering, and fluorescence. A given gaseous species has a unique absorption-band structure, and thus with some modification the techniques of molecular spectroscopy are applicable to infrared remote sensing of the atmosphere.

Rayleigh or elastic molecular scattering by gases involves no shift in the incident wavelength and is strongly wavelength dependent. Mie scattering, also elastic, is an observable aerosol characteristic and may exhibit wavelength dependence depending on the size of the particle. Inelastic scattering involves a shift between incident and scattered radiation. Raman scattering is an example of inelastic scattering at the molecular level. By observation of the scattering and absorption properties of the atmosphere with infrared instruments, significant information can be obtained regarding the molecular and aerosol species and concentration. Scattering is often investigated with atmospheric light detection and ranging (LIDAR) systems (see Sec. 4.7).

A prominent instrument and technique for the observation of the absorptive properties of the atmosphere is the Fourier-transform spectrometer (FTS). Operating in the infrared regions of the spectrum, this application is usually referred to as an FT-IR (Simpson, 1994). In its elementary form an infrared spectrometer is an infrared sensor with a highly selective wavelength-filtering device attached. (See SPECTROMETERS, INFRARED.) The wavelength-selection device used in the typical FT-IR is based upon the Michelson interferometer and has superior sensitivity over conventional spectrometers for the observation of relatively weak spectral features (see INTERFEROMETERS AND INTERFEROMETRY). The FT-IR is usually employed in the long-path configuration, as shown in Fig. 17, using a hot background source as the source of radiation and observing molecular absorption lines from the intervening atmosphere. The preferable configuration for this instrument is termed "monostatic" where a wide-banded infrared illumination source is collocated with the receiver. The "bistatic" mode may be used over extremely long paths to maximize signal return. In this configuration the illuminator and receiver are located at opposite ends of the region under observation.

FT-IR systems have been employed in a

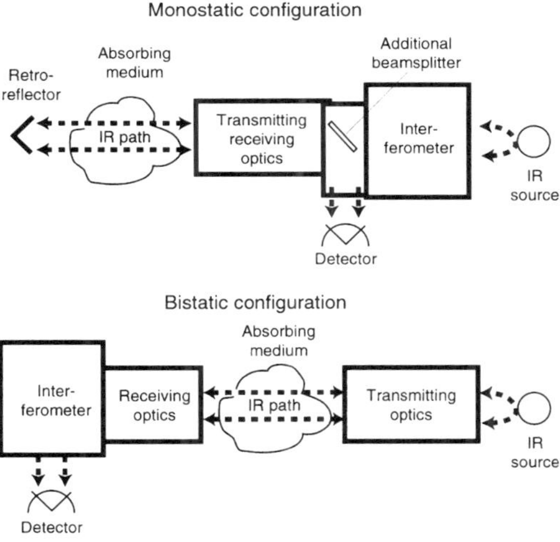

FIG. 17. Monostatic and bistatic FT-IR long-path configurations (Russwurrm and Childers, 1995). Because of superior noise canceling ability, the monostatic configuration is preferred.

variety of data-gathering experiments, including limb sounding of atmospheric absorption from satellites and balloon platforms, absorption of the atmosphere using the moon as a source, and earthbound measurements of absorption against the sun. In the context of pollution monitoring, FT-IR instruments have been used for stratospheric emission measurements from aircraft and balloons, smokestack emission monitoring at a distance, and analysis of forest fire emissions. Recently, FT-IR techniques have emerged as a technique for the generation of spectrally resolved images for use in medical, industrial, and astronomical applications.

4.5 Radiometers

Radiometers are calibrated infrared sensors and are used as laboratory or field instruments for quantitative measurement of emitted infrared flux from sources of interest (Wolfe and Zissis, 1985). There is a direct and quantifiable relationship between the observed signal level and the apparent radiance of the source. In addition to wide-scale research applications, such as the measurement of terrestrial heat transfer and solar output, there are a number of commercial applications, including plastic, glass, and thin-film process monitoring, and laser power meters. A number of special- and general-purpose commercial units are available. More sophisticated devices are spectroradiometric insofar as they may use a number of discrete spectral bands or be continuously tunable over a given region of the infrared spectrum and are thus akin to the conventional laboratory infrared spectrometer. Some radiometers employ focal-plane arrays to yield calibrated imagery.

4.6 Infrared Laser Radar

Infrared laser radar in its simplest form is simply the optical analog of the conventional microwave radar. It is an example of an active infrared sensor that employs its own source of (generally narrow-band) illumination and has the additional capability of measuring the range to an object of interest. More complex versions may be employed in imaging modes. Laser radar instruments are used to measure range-dependent quantities such as aerosol concentration and molecular composition (see Sec. 4.7). Laser radars may utilize repetitively pulsed, amplitude-modulated or frequency-modulated lasers to determine range as in conventional radars. Laser radar is not confined to the infrared region but is conveniently employed there because of the fortuitous existence of both atmospheric windows and high-power lasers operating at 0.8, 1.06, and 10.6 μm using GaAs, Nd-YAG, and CO_2 lasers, respectively. (See LASERS, GAS; LASERS, SOLID STATE; LASER TECHNOLOGY.) Increased levels of sophistication include heterodyne detection and synthetic-aperture imaging techniques. Infrared laser radar is employed in both military and industrial applications for tasks ranging from the elementary, such as proximity detectors and communications systems, to the highly complex, such as target detection, robotic guidance, and inspection of manufactured parts.

4.7 LIDAR and DIAL

Light detection and ranging (LIDAR) is an example of a relatively sophisticated active infrared sensor. Although the following definition is not universally accepted, LIDAR is a variant of laser radar used by the environmental science community to glean atmospheric information by analysis of backscattered range-dependent radiation from atmospheric molecular or aerosol constituents (Measures, 1985). The objects of interest for these systems are atmospheric particulates, aerosols, and absorptive species. The observable is the range-resolved backscattered radiation in which the intensity of the backscattered component is a function of the aerosol concentration. The LIDAR systems operate in the IR region but are also employed at the shorter wavelengths where the scattering effects are greater. Frequency shifts in the backscattered component of high-power visible or UV lasers due to Raman scattering can be characteristic of molecular rotational-vibrational transitions. Thus, they can be used to identify the presence of certain gases including those which, by homonuclear symmetry, do not radiate in the infrared.

Differential absorption LIDAR (DIAL) refers to a subset of LIDAR systems that use differences in atmospheric backscatter be-

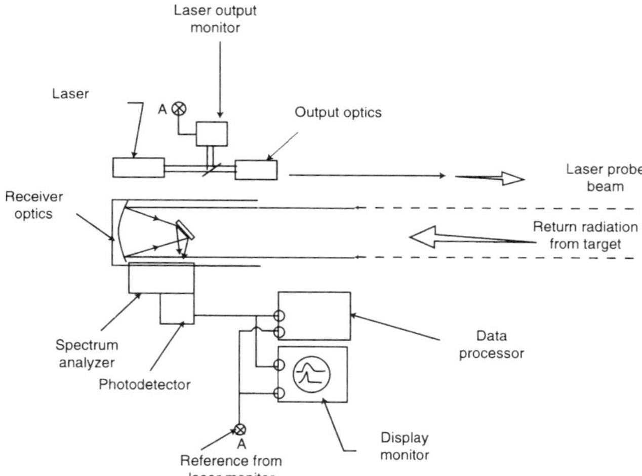

FIG. 18. A block diagram of a typical laser environmental sensor. The laser may operate in a comparative mode at multiple frequencies, as in the case of DIAL.

tween two closely spaced transmitted laser wavelengths to deduce concentrations of atmospheric components. The wavelengths are chosen so that one lies within the absorption line of the molecule of interest and the other lies just off the line. The range-resolved difference in intensity between the two returns is a measure of the species concentration along the path. Figure 18 illustrates a typical infrared LIDAR.

5. FURTHER RESEARCH

Infrared sensors could benefit greatly from further research in several associated technology areas. Clearly there is a fairly rapid migration from military applications to civilian and industrial uses. The reader is referred to the applicable sections in this Encyclopedia for discussions of recent developments in infrared sensor components. Of particular importance is the development of sensitive and rugged detector elements free from the encumbrances of cryogenics or thermoelectric cooling. Advances here could substantially reduce cost and complexity of IR detectors and imaging arrays, thus yielding wider applications in industrial, transportation, and medical applications. Increased yields in the semiconductor detector manufacturing process would be a contributing factor to lower cost.

In the imaging domain a number of new techniques under development such as aperture-plane interferometry, passive interferometric range-angle imaging, and range-Doppler imaging (Accetta and Shumaker, 1993, Vol. 8) are expected to expand the versatility of imaging processes useful in the infrared.

The environmental sciences would benefit greatly from the availability of low-cost, high-power tunable laser diodes for atmospheric investigations (see LASERS). More spectral versatility in laser sources for DIAL and LIDAR systems would increase the general utility of these instruments.

Although many of the early advances in infrared technology were heavily driven by military funding, current commercial interests have sparked a considerable amount of civilian research and development that will undoubtedly lead to further usage of infrared sensors in many new applications.

GLOSSARY

Bathymetry: Mapping of the sea bottom.

Bolometer: A very sensitive, wide-spectral-band detector based on an element whose resistance varies as a function of absorbed radiation.

CCD: Charge-coupled device. A generic term for a focal-plane array device to multiplex outputs of the many elements over a few data lines.

Clutter: Highly inhomogeneous infrared backgrounds.

Cryostat: A cooling device.

DIAL: Differential absorption LIDAR. A laser radar system used for atmospheric investigations. The system transmits multiple laser frequencies near an absorption-line edge and measures the differential backscattered radiation between on-line and off-line measurements to deduce species concentration.

FLIR: Forward-looking infrared. Military term for an infrared imaging system used in navigation and targeting applications. Civilian versions are in use in industrial and medical applications.

FPA: Focal-plane array. A square or rectangular array of detectors located in the focal plane of an optical system.

FT-IR: Fourier-transform interferometer operating in the infrared.

IRAS: Infrared imaging astronomical satellite, launched by NASA in 1983.

IRCCD: A CCD operating in the infrared region of the spectrum.

IRST: Infrared search and track system. A passive infrared system primarily used by the military to locate and track infrared emitting targets.

LIDAR: Light detection and ranging. A research laser radar system used for atmospheric investigations using backscattered radiation from aerosols.

Long-Wave Infrared (LWIR or Thermal): 8.0 to 14.0 μm.

Mid-Wave Infrared (MWIR): 3.0 to 5.5 μm.

Near Infrared (NIR): 0.7 to 1.1 μm.

Short-Wave Infrared (SWIR): 1.1 to 2.5 μm.

Thermography: A term for infrared imaging and analysis.

Thermopile: A detector comprised of a number of thermocouples connected in series.

Works Cited

Accetta, J. S., Shumaker, D. L. (Eds.) (1993), *Infrared and Electro-Optical Systems Handbook,* Ann Arbor, MI: Environmental Research Institute of Michigan (ERIM), and Bellingham, WA: SPIE Press.

Beichman, C. A. *et al.* (1988), *Infrared Astronomical Satellite (IRAS),* NASA RP-1190, Vol. 1.

Dereniak, E. L., Crowe, D. G. (1984), *Optical Radiation Detectors,* New York: Wiley.

Hudson, R. D. Jr. (1969), *Infrared System Engineering,* New York: Wiley.

Kaplan, H. (1993), *Practical Applications of Infrared Thermal Sensing and Imaging Equipment,* Bellingham, WA: SPIE Press.

Lillesand, T. M., Kiefer, R. W. (1994), *Remote Sensing and Image Interpretation,* New York: Wiley.

McLean, I. S. (Ed.) (1994), *Infrared Astronomy with Arrays,* Boston: Kluwer Academic Publishers.

Measures, R. M. (1985), *Laser Remote Sensing,* New York: Wiley.

Reif, F. (1965), *Fundamentals of Statistical and Thermal Physics,* New York: McGraw-Hill, p. 381.

Rogalski, A. (1994), *Infrared Physics and Technol.* **40** (1), 1–218.

Russwurrm, G. M., Childers, J. W. (1995), "Environmental Sensing," *Laser Focus World* **31** (4), 79–84.

Simpson, O. A. (Chair) (1994), *Proceedings of the Symposium on Optical Sensing for Environmental Monitoring, (Sp-89),* Pittsburgh: Air And Waste Management Association.

Wolfe, W. L., Zissis, G. J. (Eds.) (1985) *The Infrared Handbook,* Ann Arbor: Environmental Research Institute of Michigan (ERIM).

Zeilik, M., Smith, E. v. P. (1987), *Introductory Astronomy and Astrophysics,* Philadelphia: Saunders College Publishing.

Further Reading

Barnes, N. P. (1995), "Laser Remote Sensors," *Laser Focus World* **31** (4), 87–94.

Killinger, D. K., Mooradian A., *Optical and Laser Remote Sensing,* Berlin: Springer-Verlag.

Lloyd, M. (1975), *Thermal Imaging Systems,* New York: Plenum Press.

Russwurrm, G. M., Childers, J. W. (1995), "Environmental Sensing," *Laser Focus World* **31** (4), 79–84.

Seyrafi, K. (1985), *Electro-Optical Systems Design,* Los Angeles: Electro-Optical Research Company.

Spiro, I. J., Schlessinger, M. (1995), *Infrared Technology Fundamentals,* New York: Marcel Dekker.

SENSORS, OPTICAL

Elmar Wagner, Roland Grisar, and Maurus Tacke, *Fraunhofer Institut für Physikalische Meßtechnik (Fraunhofer Institute of Physical Measurement Techniques), Freiburg im Breisgau, Germany.*

Christian Rückauer, *Dettenheim, Germany.*

INTRODUCTION

Around the turn of the century, optical components and measurement techniques became available with high quality. This spurred the development of precision instruments; for most of them, however, an operator is needed for signal evaluation. The size of these instruments usually is so large that the term "sensor" is not appropriate.

Together with the revolution of microelectronics and digital signal processing, there also proceeded a less obvious revolution for optical instruments. New semiconductor elements here too allowed designs with higher lifetime, less power consumption, higher stability, and smaller size. These features, together with the improved electronics, allowed the design of compact systems that are self-contained and fully automated, these systems truly being sensors.

Since the technical realizations of optical sensors are quite different depending on the application, we want rather to stress the principles. Specific sensors mentioned should be taken as examples; no completeness of sensor products and suppliers is claimed.

The following section deals with the components and techniques essential for optical sensors. It explains the features of the mostly miniaturized components that made optical-sensor design possible.

We then turn to the sensor principles. Overviewing the contents from threshold sensors to fiber sensors, it becomes evident that the underlying techniques are used for quite different applications: at first sight the measurement of turbidity (like in diesel exhaust) does not have much in common with a

3-527-28139-8/96/$5.00 + .50

safety sensor for an automatic coach door. Yet both can be accounted for by sensing the transmission of an optical path.

Optical sensors have many special features that make them important for today's technology, including their ability to sense without physical contact. The control of robots and production lines would be an insolvable task without optical sensors. Optical sensors will keep and enlarge their importance in the future.

1. COMPONENTS AND TECHNIQUES

1.1 Detectors

The detectors most frequently used for optical sensors are semiconductor detectors. Among the semiconductor materials for optical detectors, silicon is used preferentially, because this material is best understood from the development of electronic components like integrated circuits, and it therefore can be processed with high yield and at low cost. For special applications, however, other materials like gallium arsenide, gallium phosphide, lead sulfide, or other compound materials are chosen.

As a characteristic of pure semiconductors, there is the "forbidden" gap in the energy scale of electrons without permitted electron states. At very low temperature the region below this gap—the valence band—is filled completely with electrons while the region above—the conduction band—is empty. Electrons can be lifted from the valence to the conduction band—e.g., by elevating the temperature, but also by energy transfer from particles. These processes generate mobile electrons in the conduction band and mobile "holes" in the valence band. Both types of particles contribute to the electrical conductivity of the material.

The effect using electromagnetic radiation, i.e., photons, for this process of mobile charge-carrier generation is called the "internal photoeffect," the base of all semiconductor radiation detectors. Photons can only generate free carriers if they have energies equal to or greater than the energy gap of the semiconductor. This fact causes a sharp edge of the sensitivity of semiconductor detectors vs the energy of photons. Typical spectral sensitivity curves in terms of current vs incident wavelength for some detector materials are given in Fig. 1. After the irradiation is turned off, photoconductivity decays until the charge carriers have recombined, with the charge-carrier lifetime as decay time constant.

The detector devices used in optical measurement systems vary with regard to the design and the internal gain achieved. The

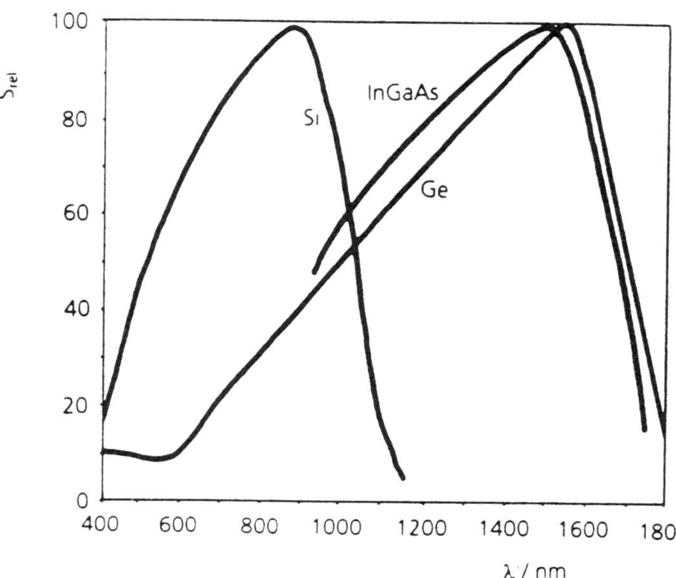

FIG. 1. Relative spectral sensitivities of detectors made of Si, Ge, and InGaAs as functions of the wavelength.

most common components are photoconductors, *pn* and *pin* photodiodes, position-sensitive devices, phototransistors, avalanche photodiodes, and charge-coupled devices.

Photoconductive elements are conduction elements of semiconductor material with electrical connections at both ends. External voltage is applied to the electrodes, and current is detected in the external circuit. In the dark, the conductivity is due to the thermally excited electron–hole pairs. If the element is exposed to radiation, conductivity increases because of the photogenerated charge carriers. If the carriers travel only a portion κ of the distance between the electrodes until they recombine, $1/\kappa$ photons are needed to detect one charge carrier as a signal in the outer electrical circuit. However, in very pure materials the lifetime of charge carriers can be very long (Si: milliseconds), and therefore the carriers can drift over distances much longer than the electrode separation. In this case κ is greater than unity and can achieve values up to 10^5. This means that one photon is able to cause a flux of 10^5 electrons, or the gain is 10^5. On the other hand, devices with high gain have slow response because charge-carrier lifetime determines the maximum frequency. Thus photoconductive elements are used where high gain is needed and slow events in the Hz to 1 kHz range have to be measured.

All other semiconductor devices take advantage of the possibility of *p*-type and *n*-type doping, with acceptors and donors, respectively, and of the electric field at *pn* junctions. The width w of the space-charge layer of a *pn* junction is given by (Sze, 1981)

$$w^2 = 2\epsilon(V_{bi} + V)/qN, \qquad (1)$$

with ϵ the dielectric constant, V_{bi} the built-in voltage equivalent to the band-gap energy, V the applied voltage, q the electron charge, and N the doping concentration. The current–voltage characteristic of a *pn* diode is shown in Fig. 2. Irradiation shifts the curves to higher reverse currents. The generation of charge-carrier pairs is not affected by the space charge; i.e., the density of charge carriers generated through photon absorption is proportional to $\exp(-\alpha d)$, where α is the absorption constant of the semiconductor material at the wavelength of irradiation and d is the depth underneath the surface. The charge carriers generated inside the space-charge region w drift in the electric field: electrons toward the *n* region, and holes toward the *p* region. This response of the device toward instantaneous turnon or shutoff of irradiation is fast because it is not limited by the charge-carrier lifetime. Nevertheless, there are also charge carriers generated in the field-free parts of the device, which diffuse randomly and if approaching the space-charge region are pulled by the electric field, thus contributing to the photocurrent. This latter portion of the response is again carrier-lifetime limited and therefore slow.

In order to achieve components with good efficiency and fast temporal response, it is therefore necessary to extend the depth w of the space-charge regions, thereby reducing the contribution from the field-free regions. This is accomplished in *pin* photodiodes: a lightly (unintentionally) doped "intrinsic" region is located between the *p* and *n* regions, thus extending the space-charge region and the volume where charge carriers are collected by the electric field (Fig. 3). High doping levels in the *p* and *n* regions keep the

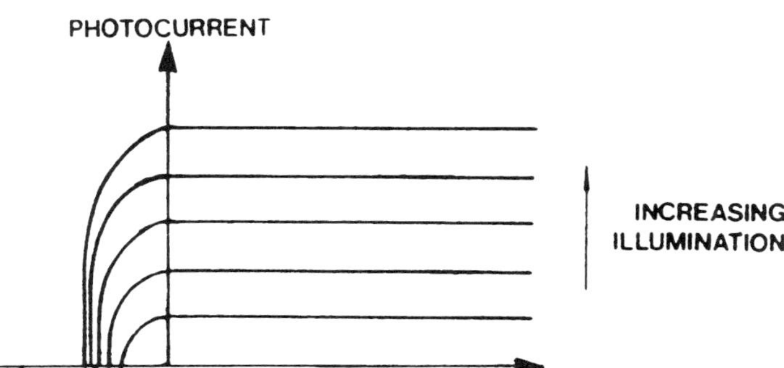

FIG. 2. Current vs voltage characteristic of detector diodes under dark condition and under illumination.

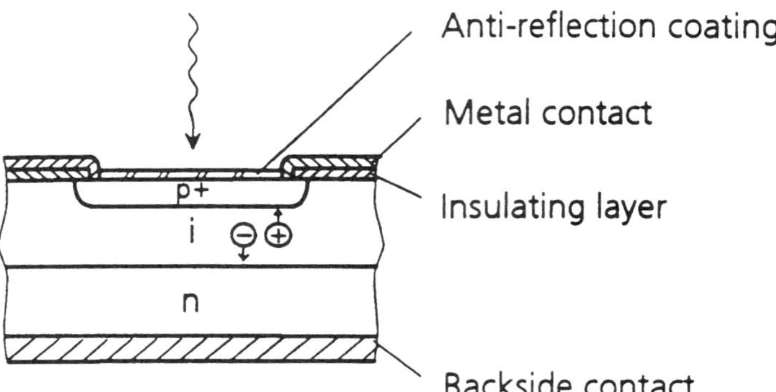

FIG. 3. Cross section of a *p-i-n* detector diode.

lifetimes of the generated carriers short, thus also improving the high-frequency response. Time resolutions in the nanosecond range can be achieved. As the performance of *pin* photodiodes is so much superior to that of normal *pn* diodes, *pin* diodes are almost exclusively used as discrete devices.

Position-sensitive devices (PSDs) are special versions of *pin* diodes. Their design allows locating the position of impact of photons in one or two dimensions of the active area. There are electrodes at either end of the sensitive stripe in the one-dimensional version. The photocurrent generated by an optical beam splits into two partial currents through the electrodes (Fig. 4). The ratio between these two currents is inversely proportional to the ratio of the distances of the illuminated spot from the respective electrodes. The beams do not have to be well focused for such measurements, since the measurement yields the center of the distribution of the radiation. The accuracy of these measurements can be better than 0.1% with regard to linearity and 1 μm in terms of the absolute resolution if proper optical focusing is provided. The time response is that of *pin* photodiodes and goes up to 100 MHz. Two-dimensional detection is achieved by additionally having two backside electrodes oriented orthogonally to the top electrodes.

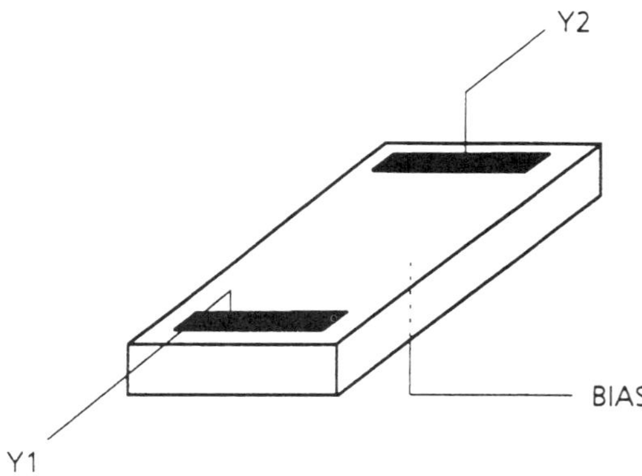

FIG. 4. Schematic of a position-sensitive device (PSD).

Phototransistors have principally the same structure as electrical transistors, consisting of collector, base, and emitter. Just the lateral design is different to keep the percentage of radiation shaded by the contacts as low as possible. The *pn* junction between the collector (substrate) and the base is used as the source of current, which is amplified by the transistor structure. Amplification, i.e., gain, between some 100 and 1000 is common for phototransistors. The temporal response is relatively slow, ranging around 100 kHz. Therefore, phototransistors are mainly used for the detection of mechanical movements like in interrupter switches, where their frequency response is sufficient and the advantage of amplified signals and thereby safer transmission is important. In all commercial devices, the base, usually open, can be contacted to adjust the point of operation for improving the high-frequency response at the expense of the gain.

Avalanche photodiodes use the internal multiplication of charge carriers due to the generation of electron–hole pairs by charge carriers with high kinetic energy. Such kinetic energies can only be gained in a region of a sufficiently high electric field and long free paths, i.e., in pure semiconductor material. Avalanche diodes usually combine the structure of a *pin* diode, resulting in a good collection efficiency with quantum yields exceeding 90%, with a region of an electric field exceeding the avalanche field strength. Silicon avalanche diodes exhibit gains of 200 at frequencies of 1 GHz approximately and are the photodetectors with the highest value of the gain–bandwidth product. Because of the high gradient at the operating point, the voltage has to be well stabilized and individually adjusted for each specimen. This causes high costs and prevents avalanche photodiodes from being used in consumer applications.

Charge-coupled devices (*q.v.*) (CCDs) allow the recording of one- or two-dimensional intensity distributions. They are used in consumer TV cameras as well as in industrial applications for production control, quality assurance, etc. A CCD image sensor works with four steps of operation:

1. During exposure radiation is converted into electrical charge through the internal photoeffect and electrical potentials built into the device. Both photodiodes and metal-oxide-semiconductor (MOS) structures can be used. The saturation charge is 200 000 electrons/pixel, approximately.
2. Photocharges from each photoelement are stored in a MOS capacitor,
3. The photocharges are transferred sequentially along the MOS capacitors, the MOS capacitors acting like a shift register. This charge transfer is achieved by periodically changing the potential of the electrodes. There are two-, three-, and four-phase structures, i.e., each second, third, or fourth electrode has the same potential (Fig. 5).
4. For read-out, each photocharge arriving at the charge detector is converted into a voltage.

The clock rate can be increased beyond that needed for video frequencies, up to 20 MHz depending on the bandwidth of the amplifier. The charge transfer is usually not the limiting process. On the other hand, in applications there is a minimum read-out frequency since charge carriers are lost through the dark current. The spectral response of CCDs is similar to that of normal photodiodes, MOS structures with sensitive regions very close to the surface having a lower sensitivity for short wavelengths in the blue. Two-dimensional CCD arrays are available with up to 2000 × 2000 pixels approximately, the element dimensions ranging between 20 × 20 and 5 × 5 mm², approximately.

1.2 Light Sources

The choice of a light source for a given sensor application depends on application, size, sensitivity, and signal-processing

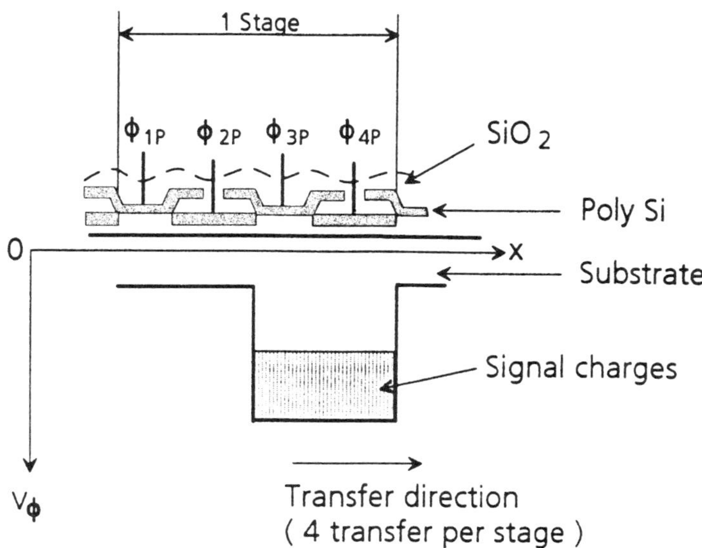

FIG. 5. Principle of the charge transfer in a four-phase CCD structure.

scheme. Besides the light-emission characteristics—emitting-element size, emitted power, spectral distribution, useful radiation, coherence, etc.—there are electrical parameters such as operating voltage and current, modulatability, mechanical features like total size and weight, and thermal properties such as operating temperature and heat dissipation. Various types of light sources represent extremely differing combinations of these parameters. We focus in the following on four predominant light-source types in sensor applications. A more detailed survey can be found in Grisar (1992). The characteristic numbers given in the following are taken from information sheets of commercial suppliers of light sources.

1.2.1 Incandescent Lamps Incandescent lamps consist of electrically heated filaments or rods, which emit thermal radiation. The most commonly used type is the tungsten-filament lamp. The emitted spectral power is essentially determined by Planck's blackbody radiation law, modified by the emission factor $E < 1$ of the material at the operating conditions.

Figure 6 gives the calculated spectral emission of a tungsten filament in the visible spectrum, where E can be approximated as 0.45, at typical operating temperatures of light-bulb filaments. The strong increase of the output versus operation temperature unfortunately comes along with a decreased lifetime, which varies approximately as the inverse 13th power of the operating current.

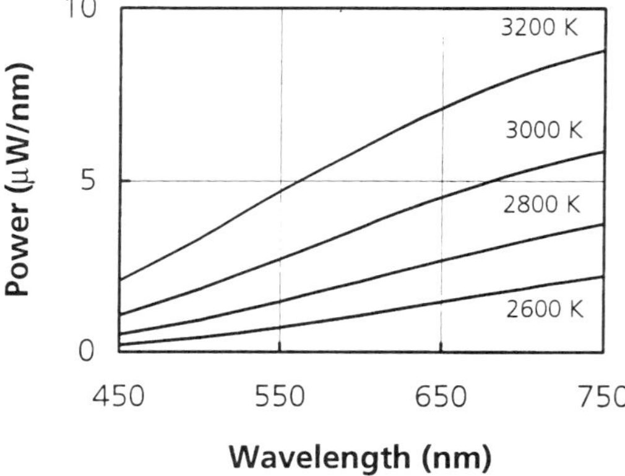

FIG. 6. Power emitted by a tungsten filament of 1 mm^2 in a f/5 aperture at various temperatures.

A larger filament size will increase the total useful power only if the sensor optics can take up a larger entrance spot. The filament shape, however, should ensure an optimum coverage of the optics entrance spot that sometimes can be achieved by tilting the bulb.

Tungsten-filament lamps generally have higher output at near-infrared (NIR) wavelength than in the visible, since the blackbody law overcompensates the drop in emissivity (to ≈0.2 at 2000 nm) and in glass-bulb transmission (to ≈0.8 at 4000 nm and 150 μm thickness). Lifetime and degradation of light bulbs are determined by evaporation from the filament and deposition on the inner surface of the glass bulb. In contrast to evacuated lamps, this effect is reduced in krypton or argon noble-gas–filled lamps. A more efficient way is the halogen cycle, where a chemical reaction and a vapor-phase transport efficiently prohibit a loss of filament material. The process requires compact quartz glass bulbs, and high filament temperatures, >3000 K, leading to high visible output (Fig. 6). Table 1 lists some data on commercial incandescent lamps.

1.2.2 Discharge Lamps Discharge lamps, which are nonthermal sources, make use of radiation emitted from an electrical discharge in a noble-gas atmosphere with addition of metal or nonmetal atoms or molecules. In most cases the emission is restricted to discrete spectral ranges, resulting in high efficiency for respective applications. Low-pressure lamps in general emit intense pure line spectra, whereas high-pressure lamps in addition exhibit spectral bands. The emission is generally anisotropic and directed perpendicular to the electrodes, the specific radiation profile depending on their shape. The size of the emitting area is ≈1 mm^2 or more. Parameters of some species are listed in Table 2 and commented on in the following.

The deuterium lamp, although of the low-pressure type, is a continuum source for the ultraviolet >150 nm with strongly decreasing efficiency toward the visible spectrum. Low-pressure mercury lamps emit about a dozen extremely narrow spectral lines between 250 and 580 nm. High-pressure discharge lamps mostly use xenon or krypton with addition of mercury or metal halides. An example is

Table 1. Typical characteristics of incandescent lamps.

Lamp type	Voltage (V)	Current (A)	Input power (W)	Color Temperature (K)	Average lifetime (h)
Evacuated	2.5	0.4	1.0	2710	10
	2.5	0.35	0.9	2230	27 000
	6.0	0.3	1.6	2350	5000
	8.0	0.2	1.6	2510	400
Noble-gas filled	2.5	0.45	1.1	2870	15
	3.5	0.45	1.6	2270	50 000
	4.0	1.1	4.3	2760	150
	10.0	0.95	9.3	2670	1500
Halogen cycle	2.5	0.425	1.06	3250	10
	2.6	0.85	2.21	2840	4000
	5.0	0.97	4.85	2800	10 000
	6.27	1.44	9.03	3335	60

Table 2. Typical characteristics of discharge lamps.

Lamp type	Voltage (V)	Current (A)	Emitting area (mm^2)	Emission angle (deg)	Average lifetime (h)
Deuterium	75–95	0.2–0.6	1	~30	1500
	60–90	0.3	1	~20	1000
Mercury short arc	22	2.3	0.2–0.35	~80 axial, 360 radial	200
	61	3.6	0.6–2.2	~80 axial, 360 radial	200
Xenon short arc	14	5.4	0.28–0.54	~80 axial	200
	20	50	1.1–2.8	~80 axial	1500
	29	70	1.3–4.8	~80 axial	2000
Metal halide	45	6	0.75–2.35	not stated	250

the short-arc mercury lamp emitting continuum and superimposed pressure-broadened mercury lines between 200 and 500 nm. Metal-halide lamps are characterized by a spectral distribution close to that of daylight. Xenon short-arc lamps cover the visible and the near infrared with maximum emission in the 800–1000-nm range. Long-arc krypton lamps emit a group of intense lines between 750 and 900 nm.

1.2.3 Light-Emitting Diodes Light-emitting diodes (*q.v.*) (LEDs) are semiconductor *pn*-junction devices (Fig. 7) using free or bound electron–hole recombination. This light-generation process is the inverse of the previously mentioned internal photoeffect of semiconductor detectors. On application of a forward bias, narrow-bandwidth light with the wavelength depending on the band gap is emitted from the top surface. The divergence angle depends on packaging and included lenses. The emitting area is 0.1–10 mm^2. Smaller spots of some 10 μm^2 with better focussing properties, but lower power levels, are obtained by edge emitters. Standard visible and NIR LEDs are listed in Table 3. Operating voltages are a few volts, electrical input powers are a few milliwatts, and lifetimes are a few times 10 000 h.

1.2.4 Lasers Mainly two types of lasers are of interest for sensors at the time of printing: HeNe and diode lasers (Fig. 8). The

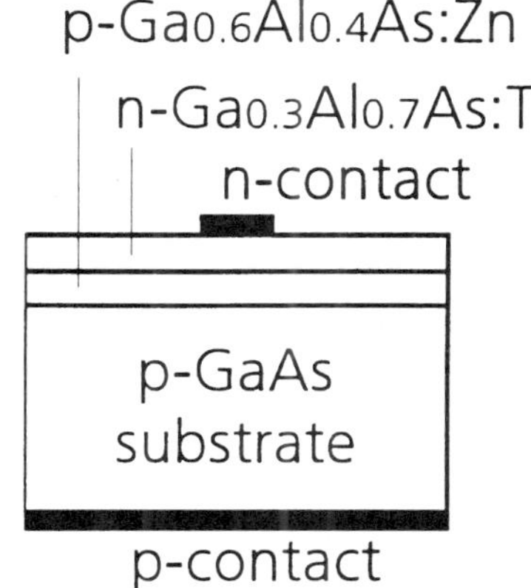

FIG. 7. Schematic of a light-emitting diode (LED).

Table 3. Typical characteristics of light-emitting diodes.

LED type	Emission wavelength (nm)	Operating current (mA)	Output power (mW)	Bandwidth (nm)	Emitting area (mm^2)	Emission angle (deg)
Visible	480	7.5–45	0.02	90	1–25	15–140
	565		0.1	25		
	590		0.04	50		
	635		0.2	45		
	655		1.1	35		
Near-infrared	870	20–100	5	40	0.1–0.25	12–130
	880		10–25	80		
	950		3–16	55		
	1060	75	0.12	80	not stated	not stated
	1300		0.12	120		
	1550		0.1	170		
	1950		0.05	250		

laser principle requires a resonator for light generated in the active medium by recombination in a semiconductor *pn* junction or by luminescence in a gas discharge with a He–Ne mixture. The cavity consists of cleaved crystal faces for diode and of discrete mirrors for gas lasers. Optical feedback and amplification by stimulated emission lead to a coherent directed beam of radiation.

Typical diode-laser dimensions are 0.3 mm in a package of 1 cm. With an emitting area of only a few square microns, diffraction determines the output beam divergence of a few times 10°, which differs in parallel and perpendicular directions to the junction. Diffraction-limited focussing or collimation by suitable optical elements is possible. The operating voltage is 0.8–2 V, the electrical input power is 10–500 mW, and the lifetime under normal conditions >10 000 h. The wavelength depends on the semiconducting material and composition; available values (Table 4) are those required in compact-disk players, optical communication, and optically pumped solid-state lasers. Other wavelengths can in principle be supplied by specialized companies in larger quantities. Individual lasers tune by typically ~0.06 nm/°C and ≈0.006 nm/mA with temperature and current, respectively.

HeNe lasers generate a well-collimated parallel beam. The discharge operating voltage is typically 1 kV at a few milliamperes current. The mechanical size ranges from 15 to 50 cm in length with ~5-cm diameter. HeNe lasers are available with different fixed wavelengths (Table 4).

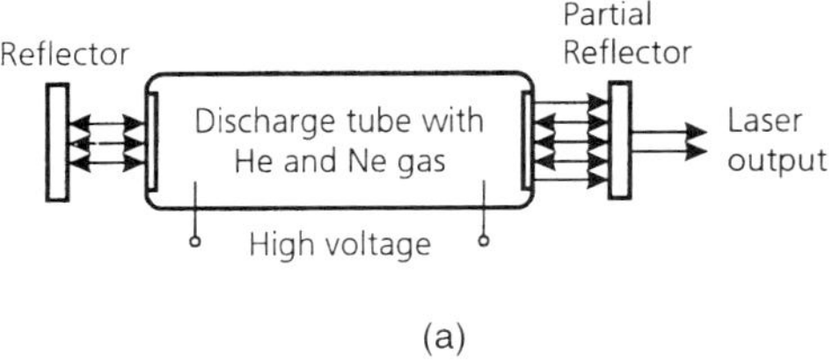

(a)

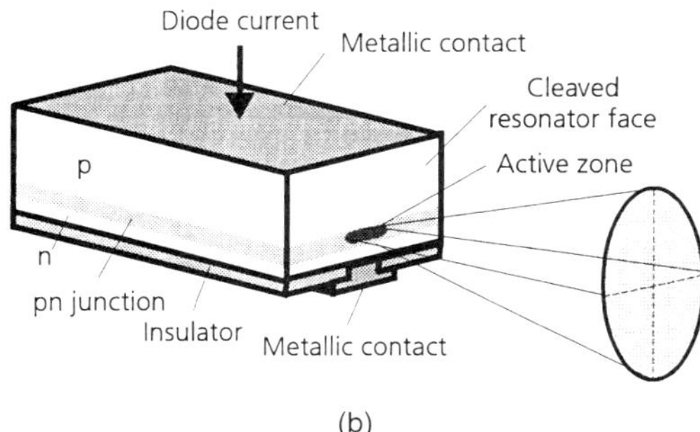

(b)

FIG. 8. (a) Schematic of a He–Ne laser; (b) schematic of a diode laser.

1.3 Spectral Filters

Spectral filtering of optical radiation is employed in many sensors for generating or enhancing specific sensor signals, for reduction of system noise, and for obtaining special spectral response. The last application is usually restricted to exposure-meter type applications, where the sensor is used to detect the intensity of optical radiation related to some process that depends on spectral intensity. In this case the spectral transmission of the filter will generally vary slowly within a

Table 4. Typical characteristics of diode and HeNe lasers.

Type	Emission wavelength (nm)	Operating current (mA)	Output power (mW)	Beam diameter (mm)	Divergence (mrad)	Divergence (deg)
Diode laser	670	85	3	—	—	10 ∥, 40 ⊥
	750	65	5	—	—	10 ∥, 38 ⊥
	780	65–90	5–15	—	—	11 ∥, 35 ⊥
	830	100–300	40–100	—	—	11 ∥, 33 ⊥
	1200	30	5	—	—	30 ∥, 40 ⊥
	1310	30	5–10	—	—	30 ∥, 40 ⊥
	1550	30	5	—	—	30 ∥, 40 ⊥
HeNe laser	543	—	0.2–25	0.84	0.9	—
	594	—	1	0.84–1.4	0.9–3.5	—
	612	—	0.5–1	0.84	0.9	—
	633	—	0.5–50	0.5–6	0.4–4	—
	1152	—	1	0.84	0.9	—
	1523	—	1	0.84	0.9	—

wide wavelength range, according to the technical goal. More frequently, however, typical filter applications ask for bandpass or high-low-pass characteristics. By using such color filters, a sensor can, for instance, distinguish different objects that naturally have or are marked by specific colors. This is an example of system selectivity due to spectral filtering.

The noise-reduction potential of filters is another topic for optical sensors. It gets obvious if one regards the typical wide-band responsivity of optical detectors and wide-band spectra of background natural and artificial radiation that is picked up by the sensor and leads to unwanted signals or even blinding of the sensor. By band-pass filtering of the system of radiation source and detector, most of the background radiation can be rejected, and hence the signal-to-noise ratio is improved. These two applications give an impression of the importance of spectral filters for optical sensors.

Depending on the application, different filter types are used: Historically, glass color filters were the first obtainable in optical quality. Color filters with dyes embedded in glass or plastic sheets have slowly varying filter spectral transmission. Filters without spectral selectivity are important too. These so-called neutral density filters have constant transmission and are used, for instance, for detector response calibration or for adjustment of detector sensitivity. Interference filters made of specially designed thin multilayer stacks of optical materials typically are used for band-pass filters. Low-pass filters can be made of semiconducting material with specially chosen or designed band-gap energy. This energy marks the high-energy/short-wavelength cut off. For technical details on filters we refer to the article OPTICAL FILTERS.

Two more optical filters that are spectrally neutral should still be mentioned: diffusers and polarizers. Diffusers are usually made of optically transparent windows with one surface polished for optimum transmission, the other one, however, ground or etched. This surface will then scatter incident radiation. One usually tries to prepare the surfaces such that they scatter equal amounts into all angles. Hence a light ray incident at any angle will give the same response of a detector located behind the diffuser. This wide-angle property is important for illumination sensors.

Polarizers are used to select the polarization states of light. Electromagnetic radiation is composed of electric fields that in optically transparent material oscillate perpendicular to the propagation direction. There are two possible orthogonal linear polarization states of light, and the corresponding parts of the intensity can be analyzed and selected by polarizers. A popular type of polarizer is made of oblong absorbing molecules imbedded parallel to each other in otherwise transparent plastic films. These molecules absorb radiation polarized parallel to the molecules, but much less or no radiation polarized perpendicular to the molecule axes.

Such filters can be used to reject unwanted radiation reflected from surfaces: This radiation is quite often "polarized"—it contains predominantly one specific polarization. These filters also may serve as tunable neutral density filters, a filter type that is made by adding two filters, the first transmitting only one polarization and the second rotatable around the optical axis relative to the first. If aligned in parallel, transmission is at maximum, if perpendicular, transmission is at minimum. Any transmission in between is accessible by rotation. Further, polarization filters are the key components of sensors that detect stress in transparent objects. Under stress, transparent material can change the polarization. The objects are placed between polarizers; the second polarizer is then rotated until the transmission is minimum. If now stress is applied, polarization changes brighten the image, and those parts of the object with most stress are easily identified.

1.4 Noise-Reduction Techniques

Various spurious unwanted signals are superimposed on or mixed in other ways with the desired sensor signal, thus giving rise to system noise. As long as the noise signals are smaller than systematic basic measurement errors or at least induce only small and acceptable errors, they need not be specifically addressed. In general, however, one has to reduce the influence of possible noise sources by special sensor design in order to achieve good sensor performance. System noise, for instance, limits sensor sensitivity for small signals.

System noise originates from all system components; generally, however, only a few noise sources dominate. Which do, and how their influence is reduced in detail, depend on the sensor type and design. Hence we can only give examples and restrict this discussion to noise that originates from optical radiation, and to lock-in techniques, which are mentioned as a powerful electronic means of noise reduction for optical sensors.

As shown schematically in Fig. 9, aside from some wanted optical signal that in general originates from an artificial optical source, which is part of the sensor, unwanted radiation will fall on the sensor radiation detector. This illumination may be very intense, like, for instance, direct or reflected sunlight. Furthermore, it may be strongly modulated, for instance, by the moving shadow cast by a sensor operator. This radiation frequently is of higher intensity than the intended optical signal and thus

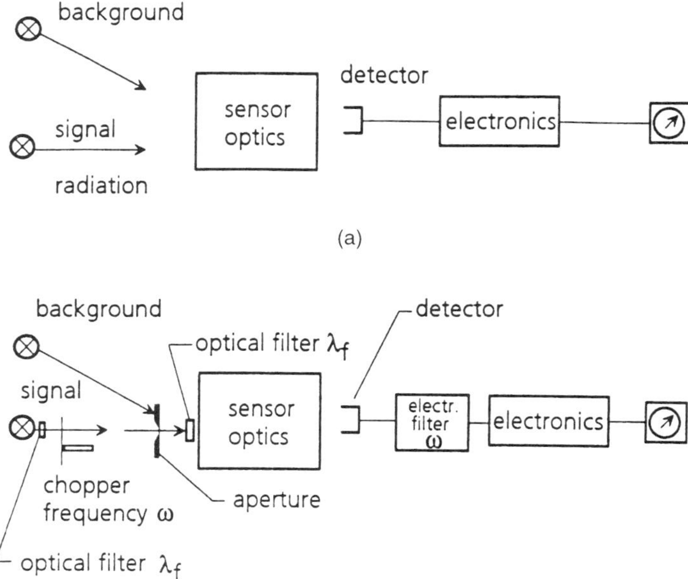

FIG. 9. Sources of sensor noise and reduction means.

cannot be allowed to enter the sensor without reduction to an acceptably low level.

The most simple shielding consists of a completely closed housing made from metal or other opaque material. Quite often, this is not possible, or the sensor light beam has to be led in an open path—for instance, for length measurements. In this case, the sensor optics have to be designed so that the detector optics has field stops that limit the detector field of view to the sensor radiation source area. Typically, one uses slender beams that allow detector optics with a small aperture. If then the detector is housed with an entrance pupil that is equivalent or equal to the aperture, and the radiation source has matching optics, the possibility of unwanted radiation hitting the detector is minimized. The optics housing is usually blackened inside in order to reduce stray light that enters through any necessary housing openings.

Unwanted radiation can also or additionally be reduced by spectral filtering. Artificial sources like LEDs or lasers have a limited or even very narrow spectral bandwidth. Hence a spectral filter in front of the detector can be used to transmit the signal radiation but block all unwanted radiation outside the signal band. With lasers as signal source and white or gray background light, interference filters can reduce background levels by many orders of magnitude. A filtering of artificial broad-band radiation does not help in all cases, since it reduces the signal just as it does the background. It is helpful only if the sensor effect is spectrally dependent, so that a reduction of the spectral bandwidth optimizes the signal-to-noise ratio.

The radiation source itself is a noise source as well. Fluctuations of signal are caused by instabilities of the driving current or temperature fluctuations. In general, stabilization of the source operation parameters can suppress such noise sufficiently. Another source of radiation noise is the quantum nature of light. A given power detected by a detector is equivalent to a given number of quanta. The number of quanta being statistically defined, the intensity of any radiation source must vary in time in a random manner with given statistical deviations from the mean intensity. Whenever light sources are faint, this radiation noise gives the ultimate minimum system noise limit.

A very efficient means of noise rejection is modulating the signal at an early stage with fixed frequency and then filtering the signal for only this signal frequency and thus reducing all signals that are not modulated. If possible, the light source is modulated, or at least its radiation is modulated by a chopper, a segmented rotating blade that periodically blocks the beam. Aside from the noise reduction potential, this technique, yielding ac signals, also is preferential to permanent and constant illumination, since ac electronics inherently have a better signal stability than dc electronics.

The signal narrow-band filtering preferentially is performed by "phase-sensitive rectification" with so-called lock-in amplifiers. The function of these filters can be explained by a simple analog shown in Fig. 10 that his-

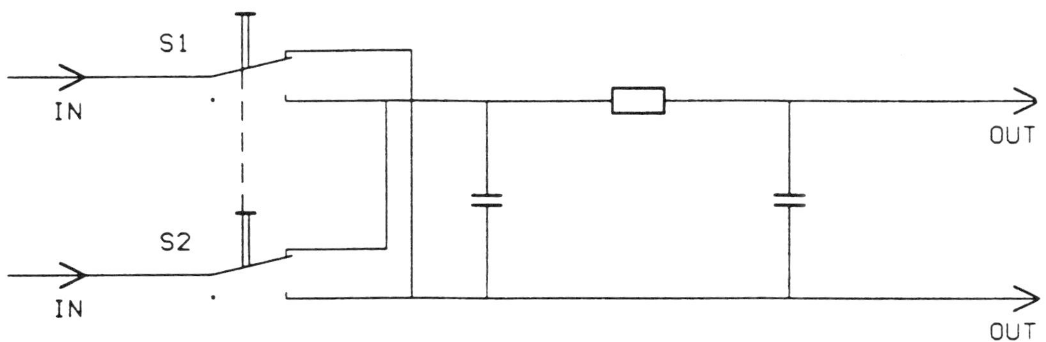

FIG. 10. The function principle of lock-in amplifiers. The incoming signal is inverted by a switch that is reversed periodically with the same frequency and phase as the signal modulation, thus it is rectified. A low-pass filter passes the effective voltage of the signal.

torically was used for early lock-in techniques. The incoming periodically modulated signal is fed into an inverter switch that is operated at the modulation frequency and with fixed phase relative to the modulation. Historically, this was a mechanical switch connected to the chopper. Nowadays this function is taken by electronics driven by a reference signal of fixed phase relative to the signal modulation. If the phase is chosen properly, the output of the switch is a rectified signal, and the following low-pass filter yields the effective voltage of the input as signal.

Figure 11 shows a set of input, reference, and inverter output signals. If the frequency of input and reference are chosen to be different, the inverter output is not always positive but rather varies like a beating between reference and signal. The output of the low-pass filter is an oscillating signal with its frequency equal to the difference of the signal and reference frequencies. The amplitude of this output is reduced with increasing frequency, so that output is generated only for input frequencies close to the reference frequency—hence the band-pass filter characteristic of the phase-sensitive detector.

As the time constant of the low-pass filter is increased, the filter bandwidth is decreased. This is desirable for noise reduction, since the filter passes less of the usually wide-band noise and thus increases the signal-to-noise ratio. On the other hand, the sensor output response time is limited by the time constant, so that there is a direct trade-off between low sensor noise and sensor speed.

Various types of lock-in amplifiers exist that combine signal amplification and the phase-sensitive filtering, ranging from integrated-circuit dimensions for standard sensors to complete and versatile laboratory equipment for highest sensor demands. At present there is a trend to include the lock-in function into digital signal processing. This can, for instance, be done by multiplying the digitized signal by a rectangular periodic reference, as shown in Fig. 11, and then averaging numerically.

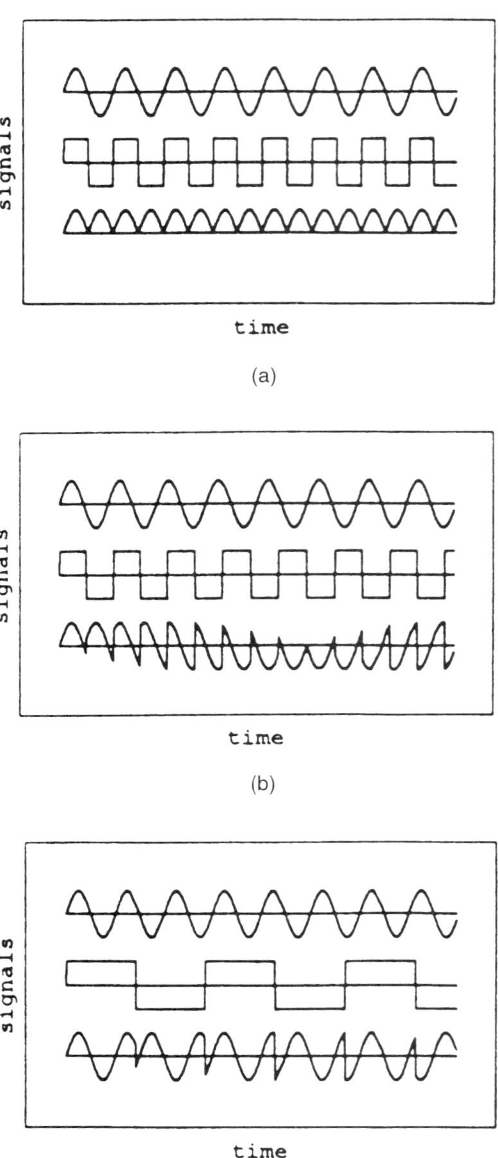

FIG. 11. Examples of modulated signals (upper trace), switching sequence (middle), and resulting signal (lower) before the low-pass filter. The reference (switching) frequency is equal to the signal frequency in (a), and differs for the other examples.

2. OPTICAL SENSORS

Optical sensors use radiation emitted, transmitted, or reflected from an object, or they use the influence of a physical or chemical quantity onto the optical properties of a sensor. One differentiates between optical sensors using the natural emission of the object and those depending on special illumination.

Emission from the object under test is utilized in all areas that are based on the visibility of objects. For cost reasons video components from the consumer market are used in many cases as detectors. If manufactured for the visible, the spectral sensitivity of these components matches the eye sensitivity V_λ, which is shown in Fig. 12. The radiometric (physical) quantities and the photometric quantities are linked by the equation

$$1\ \mathrm{W} = 673\ \mathrm{lm}\ @\ \lambda = 555\ \mathrm{nm}. \quad (2)$$

As most common detectors have sensitivities extending into the near infrared, IR radiation is frequently used for detection. Such sensors can be applied in surveillance, feature recognition, flame control, etc. Detectors for the middle and the far infrared are used to monitor the thermal radiation of objects for monitoring the presence of men, hot parts, vehicles, etc. (see SENSORS, INFRARED).

The majority of applications need optical sensors with their own matching radiation source. This allows the measurement of nonvisible properties of the measurands. The advantage is the use of light or radiation of particular wavelength, structure, focus, etc. Semiconductor emitters are used in many cases because of long lifetime, high stability, low weight, etc. Their radiation is commonly in the visible or the near infrared. The incorporation of optical emitters allows also the operation of sensors that are based on optical read-out principles, like fiber-optical or integrated-optical sensors.

The following sections describe some important principles used by optical sensing.

2.1 Threshold Sensors

Threshold sensors are probably the most frequently used optical sensor type. They are usually active sensors that include a system light source, so that the sensor does not depend on natural lighting conditions. Light barriers are used, e.g., to test whether an object is at a certain location. The light from the sensor light source falls onto the sensor detector and its intensity is dependent on the location of the object under test. If a certain threshold intensity level is reached, the sensor will give a signal that indicates the object location.

The most simple position switch just needs the light source and detector without any further optical elements, as shown in Fig. 13. Typically, LEDs are employed as light source; they already include collimating optics as a result of their transparent-housing surface shape.

If larger separations are to be bridged, like for safety sensors of automatic doors, optical elements have to be included in order to collimate the light beam sufficiently for optimum detector optics illumination and in

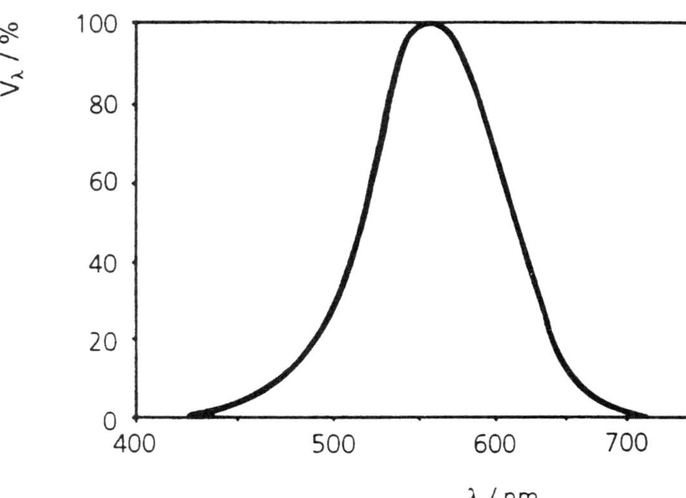

FIG. 12. Relative sensitivity V_λ of the human eye.

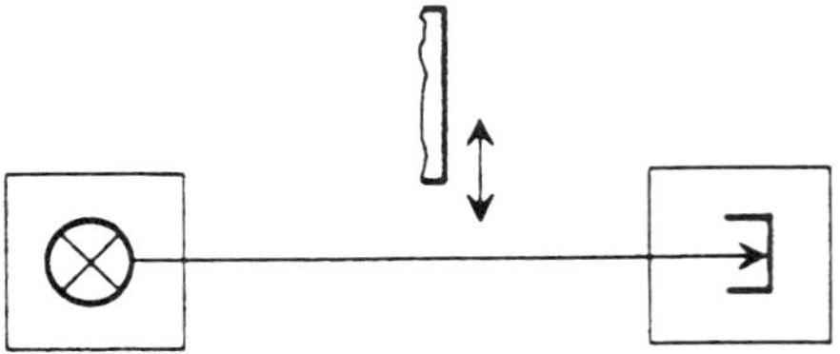

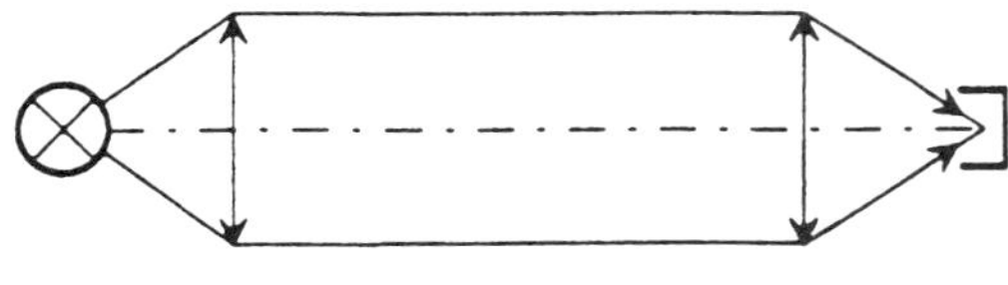

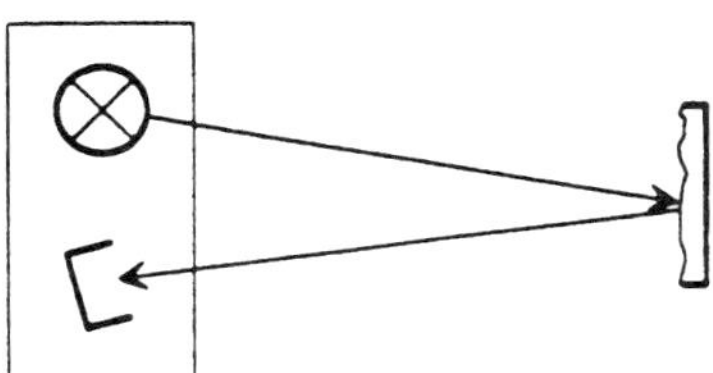

FIG. 13. (a) Transmission sensor for detection of objects by reduction of the detector illumination. (b) Reflection sensor for detection of objects or marked surfaces by their reflection.

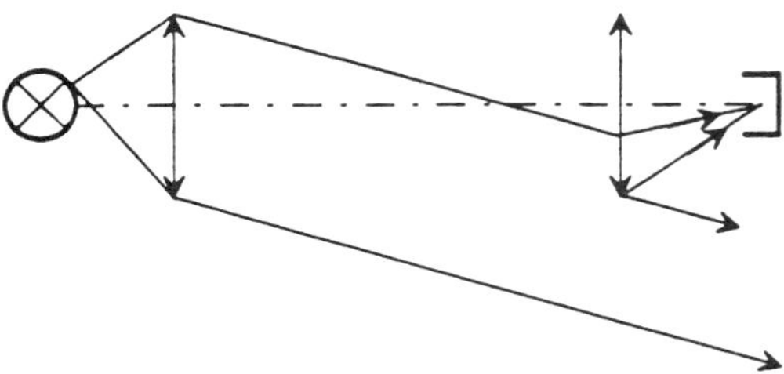

FIG. 14. The maximum separation between light source and light sensor is limited by the light-source size (which should be small) and lens size (which should be large). The upper part shows the perfect transmission of light generated on the optical axis, and the lower part, the reduction of detected intensity by vignetting radiation that originates from noncentral source areas.

order to concentrate the light beam onto the detector element. Figure 14 shows the schematics of a corresponding sensor optics.

The total optical system consisting of a collimating and a collecting lens gives an image of the light source on the detector. Because of the finite size of the light source, there will be vignetting of the light path. As indicated in Fig. 14, part of the radiation of noncentral source areas will not hit the detector optics and hence is lost. This loss increases with increasing separation of the two sensor parts and has to be counteracted either by increasing the collecting optics diameter or by reduction of the light source area.

With LEDs as light source, and lens diameters of a few centimeters, large open sensing paths of up to 100 m are realized in commercial systems. The optimum geometrical performance in terms of ratio of beamwidth minimum diameter to path length is obtained by using lasers as light source. Single–spatial-mode laser radiation has the theoretical limiting minimal radiation source area. Use of lasers for threshold sensors, however, is not straightforward. The monochromatic radiation gives rise to interference patterns that strongly modulate the sensor signal. While the sensor signal should depend on a smooth and predictable pattern on the object location to enable good threshold switching, such interferences can cause erratic multiple threshold response. Using lasers as light source, care must be taken to suppress interference—e.g., by spatial averaging of the received light.

The so-called direct-scan sensor design of Fig. 13, upper part, with separated light source and light sensor, has disadvantages for large separation of the two, since in general only one of the two components is located close to the sensor electronics. Hence this arrangement (called bistatic) is not favorable for position switches that have to allow for relative movements of the sensor optical components. For these cases it is preferable to use a monostatic design with light source and detector combined in a single housing and the open path folded by a retroreflector—e.g., a cat's-eye reflector sheet. This way, the function of Fig. 13, upper part, is obtained with an optical arrangement analog to the reflective scan of Fig. 13, lower part, in retroreflective scan mode.

The contrast of the optical signal intensity for "on" and "off" and the position dependence is increased by special sensor designs. An example is given in Fig. 14 for the detection of small objects. Objects with dimensions small as compared to the optical beam will only block part of the beam (giving small contrast) and will give a constant signal while passing the beam. In this case one employs intermediate focus positions, as shown in Fig. 15, that give the desired high contrast and local resolution.

The threshold-position switch principle is also the basis of position sensors that use the encoder technique. The location of an object is determined by mechanically connecting it to a linear optically coded strip (like a ruler), or to a disc if rotation is to be sensed. The code can consist of black and white segments for reflection scanning, or of opaque and transparent segments for direct scan. As the object is moved, the threshold sensor counts the sectors moving by. Encoders with equal periodic segments are termed incremental sensors, since the absolute position cannot be evaluated. Absolute encoders use sector groups, e.g., by splitting a single-sector area into a discrete set of subsectors that have varying size and position. Separate position switches for each subsector give then a binary code that identifies the sector field position (for code types, see, e.g., Norton, 1982).

Threshold sensors are also the key component of bar-code readers. The most elementary bar-code reader consists of a reflection-scan sensor that is mounted such that different objects are moved past them, e.g., in a production line. These objects are equipped with an object-type–specific bar-code field, e.g., by stickers with a black and white bar code. The typical sequence of threshold signals is then used to identify the object.

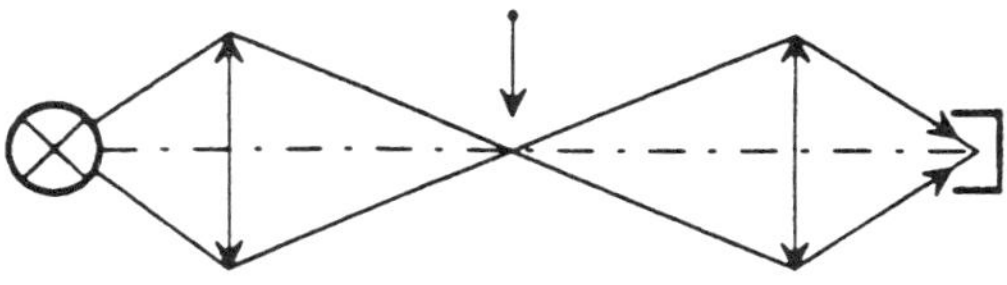

FIG. 15. Optical layout for a position sensor for small objects.

A more versatile bar-code reader design employs scanned beams (Marshall, 1980), as shown schematically in Fig. 16. If an area with a complete bar code is in the scanned region, the corresponding position switch signals are generated, and the sensor electronics can recognize and identify the bar code.

The scanning principle is used not only in bar-code readers but also for extending the concept of optical barriers from line surveillance to two-dimensional optical curtains. Such sensors are usually either reflection-scan sensors for sensing objects by their reflection, or use retroreflectors, thus being made for surveillance of the scanned area between the sensor head and retroreflector.

2.2 Turbidity Measurement

Some examples for turbidity measurements are monitoring of visibility in traffic control, characterization of particle density

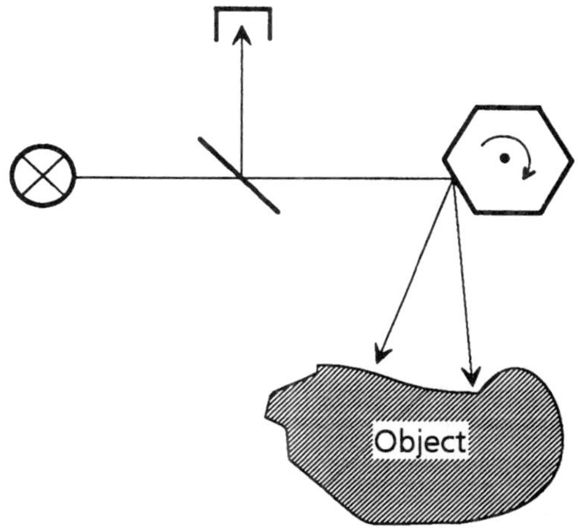

FIG. 16. Schematic of a scanning threshold sensor. The output is scanned by a rotating mirror (as indicated in the figure for a polygon mirror), or a vibrating mirror. Shown is a reflection sensor that employs the same scanning device for the reflected radiation split off with a beam splitter. Necessary collimating and focussing optics are omitted.

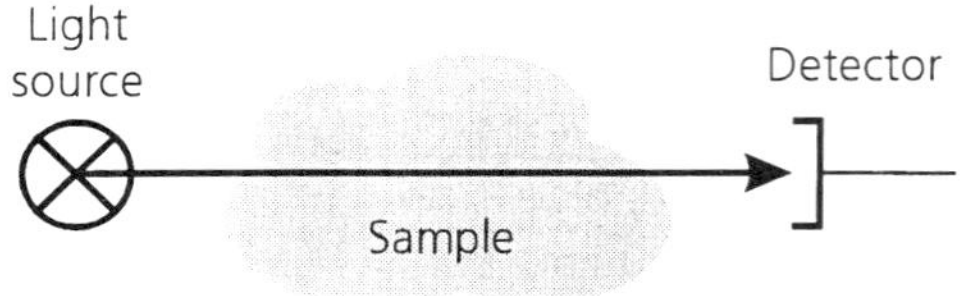

FIG. 17. Schematic of a turbidity sensor.

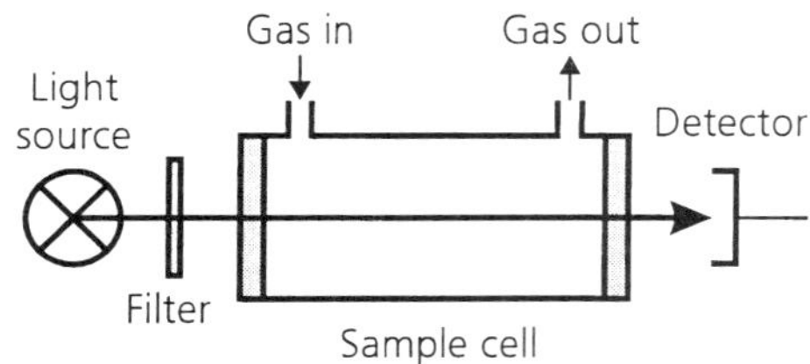

FIG. 19. Principle of a particle-emission sensor.

in exhaust gas, or determination of solid material in liquid process control. The common feature in these applications is the physical relation between the light attenuation by a medium and the concentration of solid materials, particles, or droplets. In general, no specific spectral features of the attenuation are measured. The basic optical setup therefore is as simple as shown schematically in Fig. 17. It consists of a broad-band light source, the sample to be investigated within an open path or within a cell, and a broadband radiation detector. The measured quantity is the transmission T of the optical path, i.e., the detector signal in the presence of a turbid medium normalized to the signal without turbid medium. In some cases also scattered light in a simple setup is taken as a measure of turbidity.

2.2.1 Visibility Sensors Turbidity sensors are commonly applied to visibility monitoring. An example from automobile or aircraft traffic is fog warning. In that case, the attenuation of light is caused by water droplets. Figure 18 shows the principle of the optics of a commercial transmittometer (Erwin Sick GmbH, 1995) performing a single-ended measurement by means of an autocollimation reflector. The sensor also features a reference channel and a zero-point correction by a self-test procedure as well as a sophisticated dust protection for the optics. The detection limit of this sensor is a few mg/m^3 of particles or aerosols. The output signal directly gives the meteorological visibility. In a different version with a light-emitting diode as a radiation source, the dust concentration is determined, e.g., for monitoring of emitted particles in tunnels.

2.2.2 Vehicle Particle-Emission Sensors The limits for particle emission by diesel engines in vehicles are regulated in most industrial countries. For a control, opacimeters are applied. Opacity in that context is defined as the fraction of light directed from a source through a smoke-obscured path that is prevented from reaching the observer or the instrument's receiver.

Figure 19 gives a schematic of the setup for particle-emission measurement. We here quote some of the German regulations for diesel-engine measurement systems. Compa-

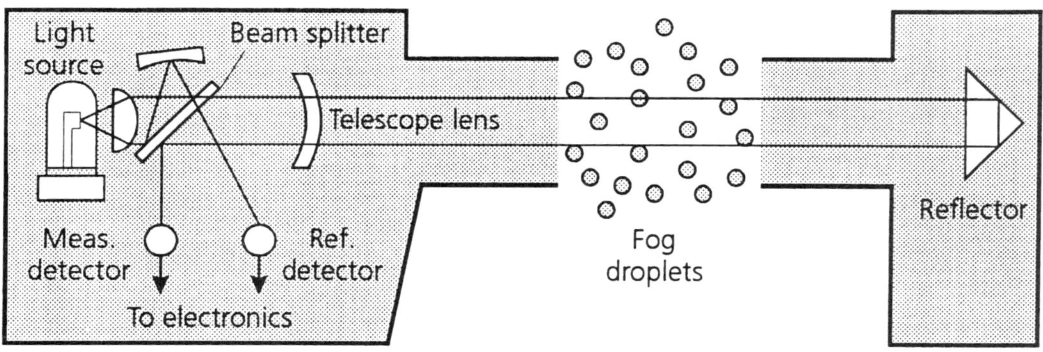

FIG. 18. Optics of a transmissometer for fog alarm monitoring.

rable regulations hold in other countries. The spectral characteristics of the transmission measurement are required to match the normalized spectral eye sensitivity characteristics with the highest sensitivity in the range of 550 to 570 nm and less than 4% of this value below 430 nm and above 680 nm. This is achieved by insertion of a corresponding spectral band-pass filter. The light source has to be an incandescent lamp with a color temperature between 2800 and 3250 K. A green-light–emitting diode with a spectral peak between 550 and 570 nm may also be taken without further filtering. Further required are nonreflective inner walls of the measurement cell. The optical path length is 0.43 m, and the resulting lower detection limit amounts to approximately 5% turbidity.

Modern low-emission diesel engines require higher sensitivities and hence longer optical path lengths. An instrument with 2-m optical path length and a more refined transmission measurement at different wavelengths in combination with a data processing derived from the Mie scattering theory (Bohren and Huffman, 1993) was shown to yield distinctly higher sensitivities. In addition, it enables a determination of the average particle diameter and the particle density.

2.2.3 Smoke Detectors For the case of particle measurement with a view toward smoke detection, light-scattering detectors are applied. For completeness, the optical setup of a commercial smoke detector (Hekatron GmbH, 1991) is given in Fig. 20. It consists of a black chamber, a diode-laser source, and a photodiode detector. A 90° scattering geometry is applied. In the absence of smoke particles, no signal is observed. For low concentrations, where no multiple scattering is present, the observed signal is proportional to the smoke density.

2.2.4 Turbidity Measurement in Liquids Turbidity measurements in liquids are more and more of interest—e.g., in sewerage plants or for characterization of sludge, in chemical processes, for filtration control, and in biomass determination. As an example, a simple immersion probe for turbidity measurements without optical elements besides lamps, windows, and detectors (EUR-Control, 1984) is schematically drawn in Fig. 21. This probe is open on two sides. Two incandescent lamps are alternately turned on and produce signals corresponding to two different path lengths on the two detectors. The four single signals are processed so as to yield the extinction modulus that is proportional to the particle concentration. All degradation effects of lamps and windows are compensated for. The device has to be cleaned only when the signals decrease below a given value.

Also applied to liquid analysis are light-scattering techniques, which are somewhat similar to that illustrated in Fig. 20. Diode lasers and light-emitting diodes enable the application of fiber transmitters and hence measurements under adverse ambient conditions. One commercial version (Mettler-Toledo AG, 1995) uses an infrared LED emitting at 880 nm in combination with backward scattering.

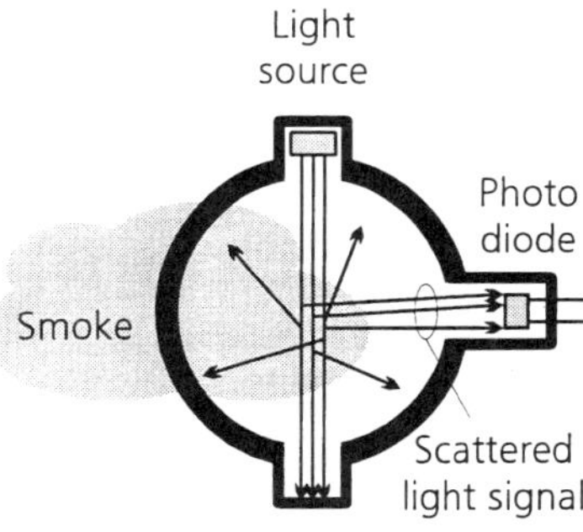

FIG. 20. Principle of a light scattering smoke detector.

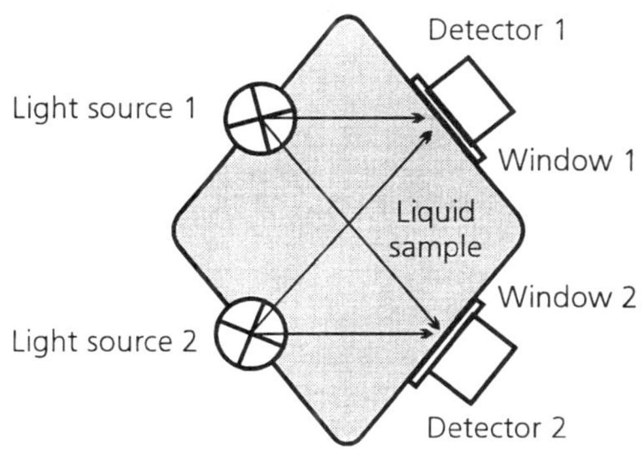

FIG. 21. Schematic of a turbidity sensor for liquids.

2.3 Colorimetry

There are two aspects to the definition of color of a given object. One regards physical or technical topics, the other is related to physiology and to the human perception of color, which is not connected by deterministic laws to physical properties. Color sensors must bridge this gap by allowing objective measurements and a physiological interpretation thereof (see, e.g., Gobrecht, 1978; Wyszecki, 1978; Gerlinger, 1992; MacAdam, 1981).

The physical part can be described by stating the spectral intensity $S(\lambda)$ of radiation. A given intensity distribution is interpreted as a specific color by an observer. Hence the task of colorimetry, recognition and quantification of color, can be performed by taking spectra of optical radiation and analyzing them with empirical physiological spectral-response data of human observers. We discuss the guidelines and reasoning of doing colorimetry and regard related sensor (colorimeter) designs.

The intensity distribution $S(\lambda)$ as a function of wavelength fully defines the color. This color hence is an attribute of radiation reaching the eye. Any type of light source has its own emission intensity distribution $S_e(\lambda)$. Usually for lighting purposes, lamps are chosen that have a "white" appearance. An observer will generally interpret the color of the intensity $S_g(\lambda)$ reaching his eye and attribute his perception to the objects that are origin of the light rays imaged on the retina. These objects modify the illuminating spectrum $S_e(\lambda)$, which can be stated by

$$S_g(\lambda) = M(\lambda)S_e(\lambda). \tag{3}$$

The "color" information of an object is thus contained in the function $M(\lambda)$, which can stand for spectral transmission (of, e.g., color glass), specular reflection (of, e.g., interference filters), diffuse reflection (of, e.g., textiles or wall paintings), scattering (like heaven's blue), or some mixture of these. Usually $M(\lambda)$ is taken not to depend on the light intensity. This assumption is reasonable for typical objects of colorimetry.

The different physical origins of M make it necessary not only to determine the illuminating spectrum $S_e(\lambda)$ and the observable spectrum $S_g(\lambda)$ but also for the general case to scan all possible illumination and receiving geometries. This is a formidable task and very rarely performed. The illumination and receiving angular dependence of color is, however, gaining in technical importance on account of the increased use of special-effect varnishes like "metallic" colors.

While a complete determination of the angular dependence of color spectra technically is not reasonable, spectroscopic sensors for colorimetry must allow for some characteristic illumination and receiving geometries. Typically one selects representative beam angles of 0°, 30°, or 45° to the surface normal, and/or diffuse radiation, as shown schematically in Fig. 22. The sensor-head geometry of Fig. 22(a) uses direct fiber illumination and pickup, and thus, in general, it covers a relatively large surface area. By use of imaging optics, a well-defined inspection area can be chosen. This is shown in Fig. 22(b) in a different configuration. With recent designs of miniature spectrometers equipped with detector arrays (Noll *et al.*, 1995), mobile sensor heads containing all optical parts are possible. In the geometry shown schematically in Fig. 22(b), the image of the detector elements through the input slit determines the surface area under inspection.

Another geometry of practical importance is represented in Fig. 22(c) and makes use of an integrating sphere (*Ulbrichtsche Kugel;* Gobrecht, 1978). The integrating (also called diffusing) sphere consists of a hollow sphere with a white, diffusely reflecting inner surface. Because of this surface, light rays that enter the sphere through an orifice get multiply scattered and thus get a homogeneous angular and local distribution. If one introduces light through a small orifice and brings the sample under test close to another small orifice that is not hit directly by the illumination radiation, the sample will thus be lighted by diffuse radiation. One can then pick up the radiation emerging from the sample through a further aperture at a given angle from the normal. Alternatively, one can illuminate the sample with directed light and detect the light reflected into all angles by reversing the setup.

The described spectroscopic techniques are a basis for colorimetry with their ability to record complete spectra with good spectral resolution, be they usually limited to representative geometries. In practice, most

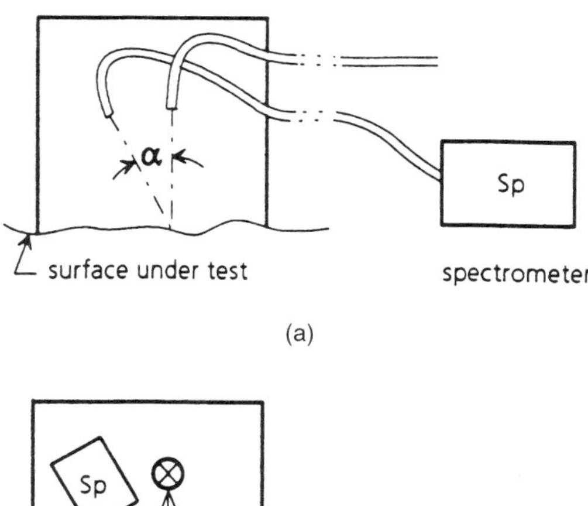

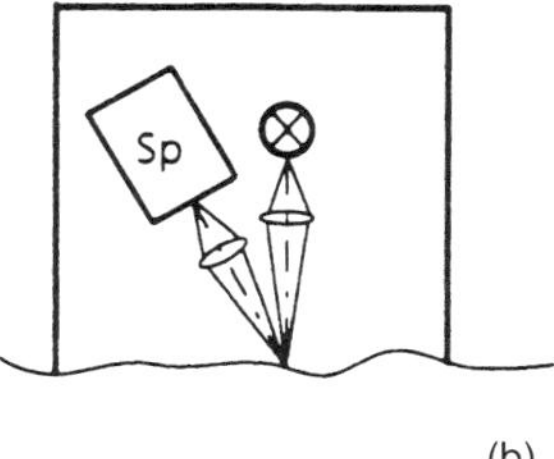

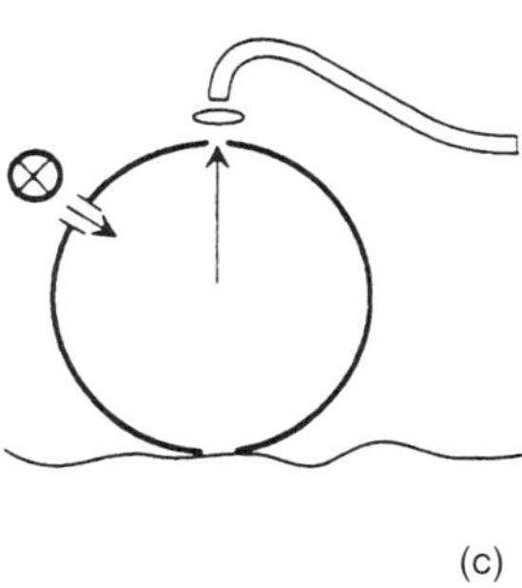

FIG. 22. (a) Fiber-optic sensor head and (b) miniature spectrometer sensor for color measurement. The fiber-optic link allows mobility of the sensor head while using conventional spectroscopic equipment. The sensor heads must allow for specific measurement geometries like choice of the angles of illumination and received light. With recent designs of miniature spectrometers, the whole optical system can be integrated into a handy sensor head. (c) The special geometry of a sensor head with integrating sphere for diffuse light.

colorimetric sensors use reduced spectral resolution, too. The validity of this approach stems from the physiological properties of human vision.

The eye has receptors of three different spectral sensitivity types. It is nearly impossible quantitatively to determine exactly their spectral sensitivity; moreover, it differs individually. Hence representative spectra have been gained empirically and are shown in Fig. 23. As a consequence of this special spectral analysis by the eye, colors can be mixed from a small number of elementary colors.

With three well-selected colors, one can mix spectra that generate all colors necessary for color prints (Kaase, 1992). One further important result is that the perception of color does not depend on the illumination intensity as long as the eye is adapted to daylight brightness. Thus, if all single colors are attenuated or amplified in the same ratio, the object will not change color, but will just appear dark or bright.

While an artist mixes colors on the basis of his perception, this is not acceptable for technical color reproduction or color control, since individual color reception differences prohibit an objective color determination. Technically, color is usually measured using three elementary sensitivity spectra. The shapes of these spectra are standardized for intercomparison. Figure 24 shows a set of standard spectra. If a detector is equipped with a filter, e.g., with transmission according to $\bar{x}$, then its signal R is due to an integral of the product of $\bar{x}$ and the observed spectral density. Analogous equations hold for G and B, the letters standing for red, yellow (German *gelb*), and blue.

Since color is independent of intensity, it is sufficient to state the relative intensity for

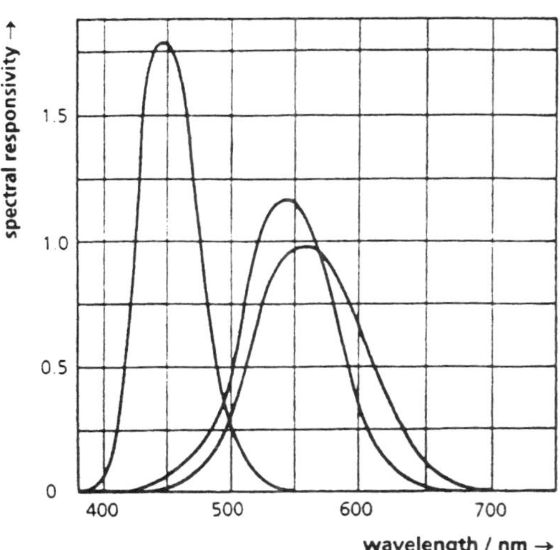

FIG. 23. Spectral response of the three receptor types of the human eye [after Kaase (1992)].

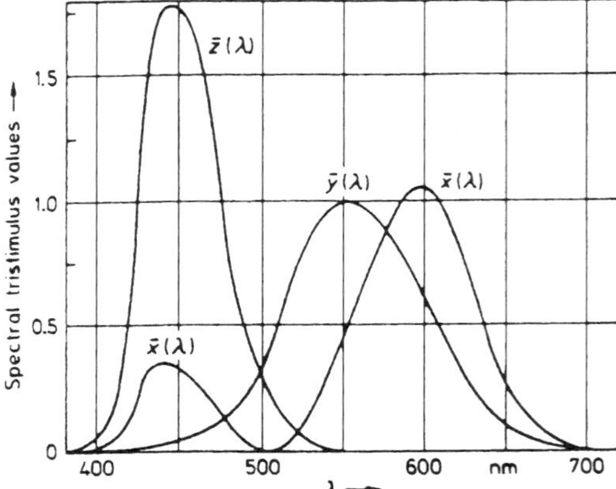

FIG. 24. Spectral shape of the three CIE (International Commission on Illumination) standardized spectra for analysis and synthesis of colors. The so-called spectral tristimulus values are denoted $\bar{x}$, $\bar{y}$, and $\bar{z}$.

colorimetric purposes. The corresponding chromaticity values x, y, z are gained from R, G, B by three equations of the type

$$x = R/(R + G + B). \quad (4)$$

Since the sum $x + y + z$ is equal to 1, a color is fully defined by a set of two values like x and y, and by the total intensity $R + G + B$. All colors can thus be represented in a color chart like the one shown in Fig. 25 for the standard colors of Fig. 24. Spectrally pure monochromatic colors give rise to x,y values on a curve that has a shape dependent on the standard filter spectra. This curve is open at one side and closed by the purple line, which results from mixing monochromatic light from the two extrema of the visible spectrum. Since all colors must be a mixture of the spectrally pure colors, the area within the boundary thus described represents all colors.

In technical colorimetry one hence uses sensors that illuminate sample surfaces with a light source that has a spectrum close to white light (one should rather use the term "colorless" light). Either the light source has to be filtered by special color filters to have a "white" character, or its spectral characteristic has to be normalized by standard "white" samples. The sample radiation is then analyzed by detectors with filters that match the standard *RGB* filter curves. The result of the measurement is a quantitative determination of the color-diagram coordinates of the color under test and can thus be used to identify, test, or regulate colors.

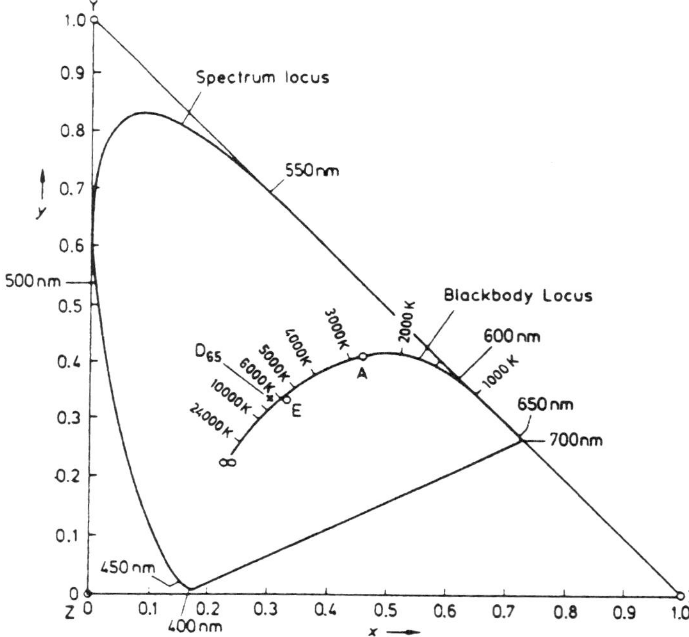

FIG. 25. Diagram of the chromaticities x, y, and z as defined by the CIE. The spectrum locus line represents monochromatic colors; some corresponding wavelengths are noted. The blackbody locus connects values that would be observed with blackbody radiation sources of the indicated temperatures. The points D_{65}, E, and A correspond to CIE-defined special light sources; see, e.g., Sec. 3.4.6.

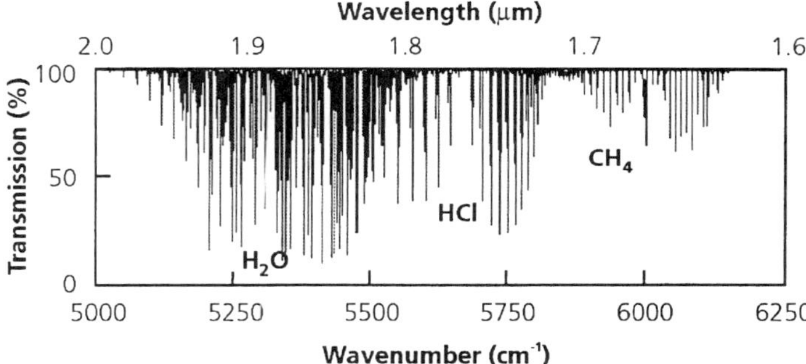

FIG. 26. Near-infrared transmission spectrum of a gas mixture.

2.4 Spectral Gas Sensing

All molecules—except diatomic symmetric ones such as O_2 or N_2—exhibit specific absorption structures in the infrared (IR) range due to molecular vibrational-rotational transitions. The strong fundamental absorption in the 3–20-μm wavelength region is the basis of various gas sensors. The harmonic transitions in the 1–2-μm NIR region cause weaker absorption features, which can also be taken for gas analysis. As an example, Fig. 26 gives a transmission spectrum of H_2O, HCl, and CH_4, at 10-cm optical path and 10% concentration each. Typical in the complex structure are overlapping bands of absorption. Some ranges, however, are dominated by single gas-species absorption, enabling selective spectroscopic gas analysis. For spectrally fully resolved measurement, the transmission is given by the Beer–Lambert law:

$$T = e^{-\alpha(\lambda)lc}, \tag{5}$$

where $\alpha(\lambda)$ is the specific absorption coefficient, l is the optical path length, and c is the molecule density, which according to Eq. (5) can be solved for c as

$$c = -\frac{\ln T}{\alpha(\lambda)l}. \tag{6}$$

The detection limit of gas sensors based on a transmission measurement is given by the lowest detectable transmission change and according to Eq. (6) increases linearly with optical path length and the absorption strength. For spectrally unresolved measurement, the relation becomes nonlinear, and the corresponding relation is obtained generally by calibration. For high signal stability, special signal-processing procedures such as the two-wavelength method are applied. If one takes the transmission ratio within and adjacent to a specific molecular absorption band, any attenuation of light by degradation of optical components will not affect the result, provided these effects are independent of the wavelength.

2.4.1 Filter Gas Analyzers Figure 27 is a schematic of an infrared gas sensor based on narrow-band filtering. The beam of a thermal light source passes a chopper wheel with two or more narrow-band filters, traverses a sample gas cell, and is incident on an infrared detector. For a stable measurement of one component, two filters may be chosen as indicated in the lower part of Fig. 27. For multicomponent analysis, a corresponding number of filter pairs is taken. Dielectric filters for that purpose have bandwidths in the 10–100-nm range at peak transmissions >70%. The detector signals are processed by lock-in type amplifiers for

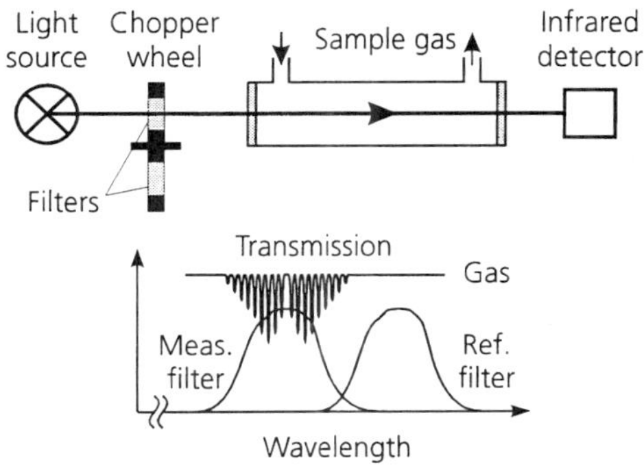

FIG. 27. Schematic of a filter gas sensor.

synchronous detection and background-light suppression. In a modified version without mechanically moving parts, the lamp emission is modulated directly by the current. Again, one or two optical channels with respective filters and detectors and subsequent lock-in amplification are applied.

2.4.2 Light-Emitting–Diode Gas Sensors This type of sensor uses NIR radiation of comparable bandwidth to the filter sensors and a NIR LED type source. Typical emission bandwidths are 100–250 nm. The power per bandwidth is comparable to the combination of thermal sources and narrowband filters. Thus, LED-based sensors are similar in their setup and their properties to filter sensors with the exception of higher potential modulation frequencies. An appealing additional feature is the potential coupling to optical fibers with a view toward remote measurements and multiplexing of several measurement cells at different locations.

2.4.3 Wavelength-Scanning Near Infrared Gas Sensors By continuous wavelength scanning, a larger amount of spectral absorption information can be gathered at the expense of shorter effective data collection times at specific wavelength. Scanning is obtained, e.g., with a grating or prism monochromator. Also in use are circular variable dielectric filters, where the transmitted wavelength is a function of the position. Acousto-optical modulators render possible sensors without mechanically moving parts using diffraction of light by an acoustic wave of variable wavelength. The large amount of spectral information in scanning devices allows for sophisticated data processing with possibility of multicomponent analysis by correlation techniques or factor analysis with efficient cross-interference suppression.

2.4.4 Optical Multichannel Analyzers Linear-array detectors enable spectroscopic devices with simultaneous measurement in a large number of optical channels, as schematically shown in Fig. 28. They use grating or prism monochromators and array detectors in the focal plane. Each detector pixel hence corresponds to a different wavelength interval. The remainder of the gas-sensor setup in Fig. 28 is similar to that for filter sensors. Data-processing schemes are similar

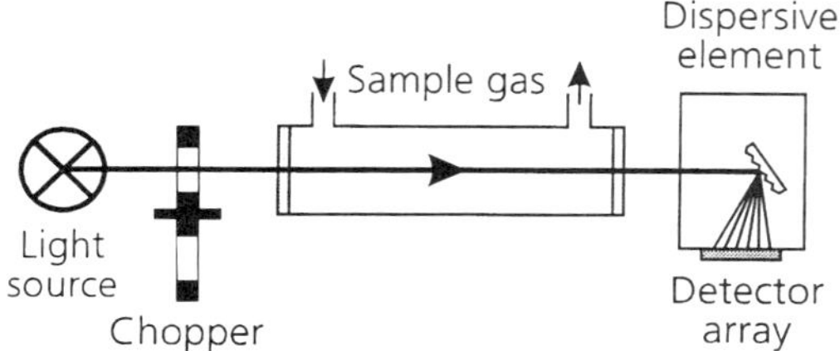

FIG. 28. Schematic of an optical multichannel analyzer (OMA)-based gas sensor.

to those applied in wavelength-scanning systems but with better signal-to-noise ratios due to the generally longer averaging times at specific wavelengths.

2.4.5 Diode-Laser Gas Sensors A special type of scanning spectroscopic system uses diode lasers as a source that is tuned by the operating current over typically 0.2 nm. Figure 29 gives a schematic of a fiber-coupled device that enables remote sensing. The narrow laser bandwidth allows for extremely high selectivity even in complex gas mixtures by tuning over one single absorption line of typically 0.2 nm in width, as indicated in the inset. Since diode lasers also tune with temperature, a control to within <0.01 K is required, whereas the operating current stability should be in the order of some 0.001 mA. The optical isolator in Fig. 29 is for suppression of optical feedback into the diode laser, which would affect the tuning behavior in an

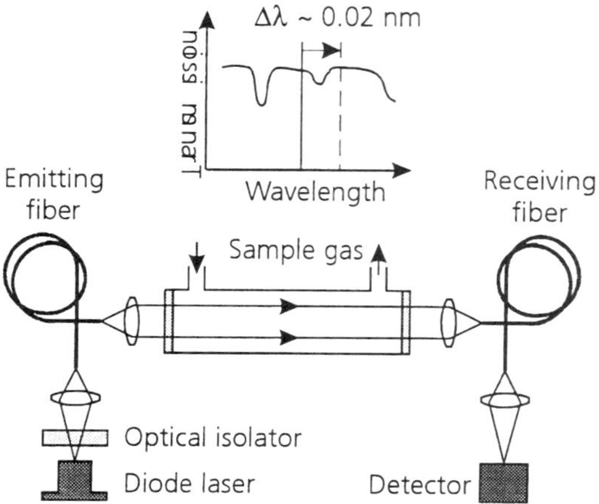

FIG. 29. Schematic of a fiber-coupled diode-laser gas sensor.

uncontrolled way. Diode-laser gas analyzers are yet not much in use. Since their sensitivity and selectivity, however, exceed those of present optical sensors, they will meet the requirements of decreasing concentrations to be detected in the future.

2.5 Optical Surface Analysis

Surface analysis is important for many technical, technological and scientific applications. Probably the overwhelming majority of these tasks is performed by inspection with the unaided eye or by use of special enlarging optics like microscopes. The need for skilled operators is disadvantageous, hence the importance of surface-analysis sensors that deliver signals that can be calibrated in an objective fashion. We describe three sensor principles that make use of scattering for surface roughness detection, of autofocus techniques for surface profile measurements, and of near-field optics that allow subwavelength resolution.

While a perfectly parallel surface reflects an incoming parallel beam only specularly, any imperfections like scratches or other surface irregularities will give rise to scattering. With θ_i defining the input angle (see Fig. 30), one will find output not only at the same angle θ_i but also at different angles. One way to make use of this is to employ a parallel input beam and then to focus the output. At the focal point a beam stop is located, so that only scattered light will get beyond the focal plane and can be collected by a detector. The intensity signal of this detector is hence due to surface roughness.

The angular distribution of the scattered light depends on the microscopic surface structure. Its influence can be explained easily in a semiquantitative manner. The incident beam with wave vector $\mathbf{k}_i$ is assumed to have its component in the surface parallel to the x axis. It gives rise to an electric field in the plane, which travels in the x direction (see Fig. 31). This surface field has a wave vector k_{ix} that is given by

$$\mathrm{k}_{ix} = k \sin\theta_i,$$
$$\lambda_{ix} = \lambda/\sin\theta_i, \tag{7}$$

where the surface wavelength λ_{ix} corresponds to the inverse surface wave vector. In a perfect boundary layer only outgoing fields with the same surface wavelength can match the incoming field, and hence only specularly reflected or transmitted radiation can exist. This situation is drastically changed if the surface is not homogeneous. One can show that, if the boundary conditions of the wave equation are periodic in the x direction (with periodicity length Λ; see Fig. 32), a general solution of the wave equation must have the form

$$\phi(x) = \sum_{l=-\infty}^{\infty} a_l \exp[i(k_{ix} + lq)]x. \tag{8}$$

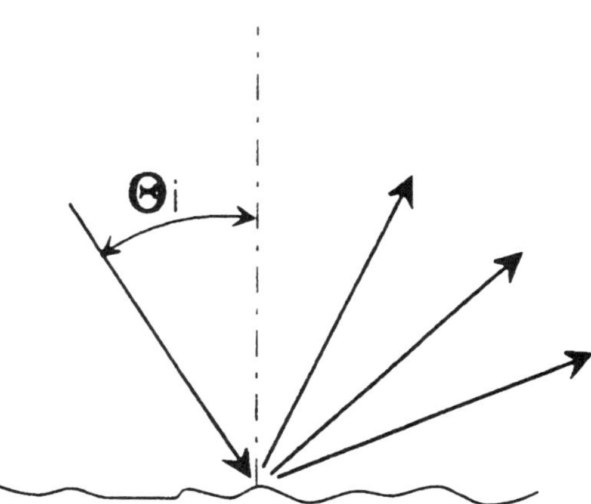

FIG. 30. Roughness sensing by scattering: an illumination beam with angle of incidence θ_i is scattered into nonspecular reflection output angles. The intensity of scattered radiation depends on the surface roughness.

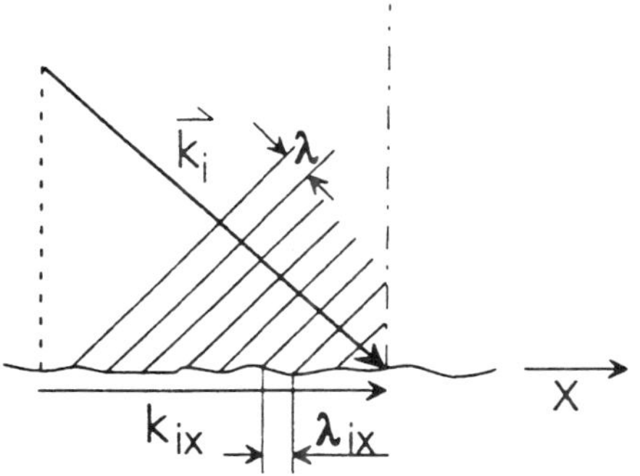

FIG. 31. An incoming beam with wave vector $\mathbf{k}_i$, $\mathbf{k}_i = 2\pi/\lambda$, gives rise to electric fields in the surface plane with wave vector $\mathbf{k}_{ix}$, wavelength λ_{ix}.

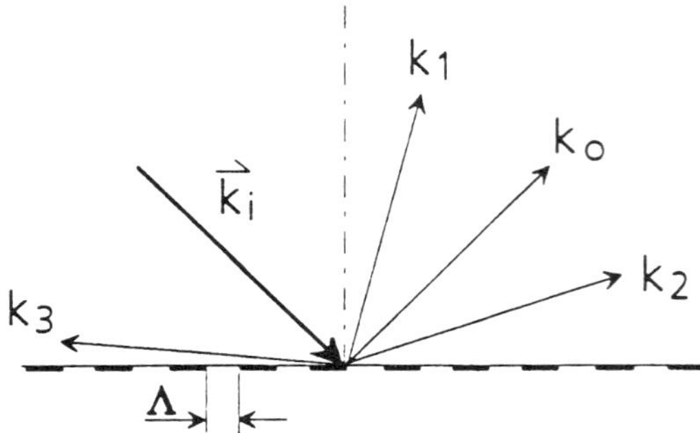

FIG. 32. A periodic surface (like a diffraction grating) scatters into well-defined output angles that depend on the input angle and the periodicity length Λ.

The grating vector $q = 2\pi/\Lambda$ enters into this equation. Hence, a surface excitation exists that is a sum of periodic functions. Any of these spatial components can couple to freely propagating waves, if an output wave vector k_l and output angle θ_l match the resulting wave vectors:

$$k_l \sin\theta_l = k_{ix} + lq. \tag{9}$$

Equation (9) is just a different way of stating the well-known grating equation for diffraction gratings.

The intensity profile is contained in the coefficients a_i, which depend on the geometry of the surface profile. A specific case is a surface that is sinusoidally corrugated. If the amplitude of this surface ripple is small, then only the first coefficient a_1 will give a significant contribution apart from a_0, the specularly reflected beam. Hence for small surface variations, any surface profile can be expanded into sine variations, and the angular intensity distribution can be calculated. While this approximation of the scattering angular distribution gives a good qualitative image, it cannot be applied directly to irregular wave surfaces in general. A numerical calculation of the scattering profile is usually quite involved, and hindered by the fact that microscopic realistic surface profiles are scarcely known. Hence, one usually tries to correlate statistical properties of the surface profile theoretically with the intensity distribution. Technically, one rather tries to have standard surfaces, and then compares the scattering distribution of samples under test to these standards.

A helpful numerical approximation connects the integral radiation intensity scattered into all angles except the specular reflection l_s to the rms height δ of surface irregularities (Elson *et al.*, 1979):

$$l_s \approx (4\pi\delta\theta_i/\lambda)^2. \tag{10}$$

This approximation holds for $\delta < \lambda$ and Gaussian height distribution. Typically, a sensor head for roughness measurements consists of a light source like a LED, which is focused onto the sample surface; the same focusing optics are then used to collimate the scattered radiation, which is separated from the input by a beam splitter and led to a detector array. Parameters like the total detected intensity width of the scattering distribution and its variance are then used as the roughness signal.

While scattering usually gives statistical information on the surface, autofocus sensors can be used to give a realistic surface profile. This technique is limited to a lateral resolution of the order of the wavelength. Autofocus sensors make use of confocal optics, as shown in Fig. 33. The radiation of a laser is focused onto the surface under test and reflected by the surface back into the focusing optics. These optics will then yield an image at the laser position that is separated from the input beam by a beam splitter and led to a detector aperture. If the optical system is in focus at the surface, all of the reflected radiation will fall through the aperture onto the detector. If not in focus, as shown in Fig. 33, some or most of this radiation will hit the boundary of the aperture, and the signal at the detector will be reduced. The separation between the optical unit and the surface is kept constant with a servo loop using the detector signal. As the sensor is scanned across the surface, the resulting deviation signal d gives the surface profile.

As a result of diffraction, the laser beam at the focal point will have a minimum waist width w_0. Its intensity profile is usually chosen to be Gaussian. If so, most of the intensity is contained in a spot with a diameter $3w_0$ (Yariv, 1975) (see Fig. 34). This waist width is connected to the far-field angle of aperture of the beam δ by the equation

$$w_0 = \lambda/\pi \sin\delta. \tag{11}$$

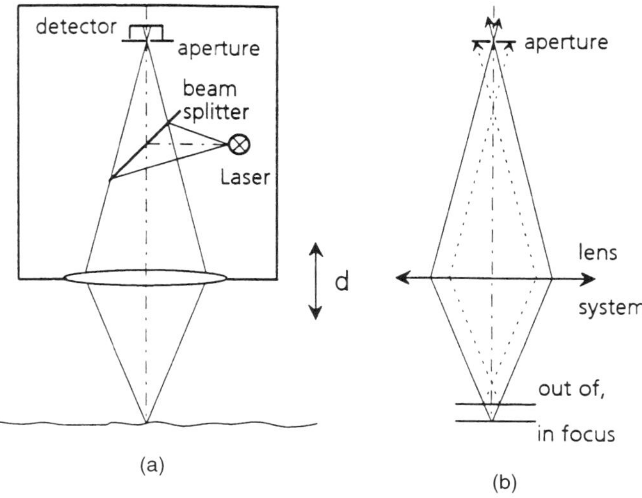

FIG. 33. (a) Schematic of an autofocus sensor. As the sensor-body separation from the surface is changed, the image of the illumination spot gets out of focus, as shown in (b). Usually, only the lens is moved by a servo for maximum detector signal.

One can see that the angle of aperture δ should be chosen as large as possible to get a small width and hence good lateral resolution. [The interested reader may note that in the denominator of Eq. (11), a sine appears instead of the tangent, which follows from the approximation of slowly varying distributions. The sine results from a direct evaluation of the far field of a Gaussian beam waist, which holds for a very small beam waist.] Corresponding optics of high quality are available, since they are used in conventional compact-disk players. At a separation z_0 from the beam waist, where

$$z_0 = \lambda/\pi \tan\delta, \tag{12}$$

the beam waist has been widened by a factor of $\sqrt{2}$. As can be seen from the equation, this distance is of the order of the wavelength as well and gives the depth resolution of such a

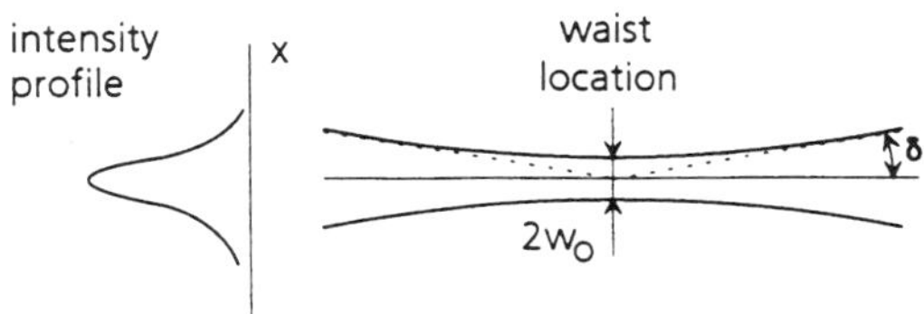

FIG. 34. The contours of a Gaussian laser beam showing waist minimum width w_0 at a focus. The width is defined as the radial 1// field-strength position. The far-field angle of aperture δ is also indicated.

system if it is used to measure the surface position absolutely. In the autofocus mode that is used for getting the profile of a surface, the relative maximum intensity separation can be determined much more exactly, down to a few nanometers. Hence, such autofocus systems can give accurate surface profiles across large dimensions; at the same time, the measurement range is quite large, of the order of a few millimeters.

As can be seen from the reasoning for the lateral resolution, it is diffraction limited like optical microscopes, albeit with somewhat better performance due to the confocal technique. Optical sensors of the near-field type allow scanning sensors that have increased resolution. These optical sensors are an optical analog of the scanning tunneling microscope (see TUNNELING SCANNING MICROSCOPY). In scanning near-field optical microscopy (SNOM), a very small source of light or detector of light, with dimensions small compared to the wavelength of light, is used. As an example, shown schematically in Fig. 35, a monomode optical fiber with core dimensions of the order of a wavelength is pulled to yield a tiny tip. Most of the radiation will leave the fiber at this tip if additional shielding by metal layers inhibits radiation from the conical face. The fiber is then brought next to the surface under test having microscopic structures. Because of this surface structure, the radiation from the tip is either transmitted through the sample or attenuated or reflected locally differently.

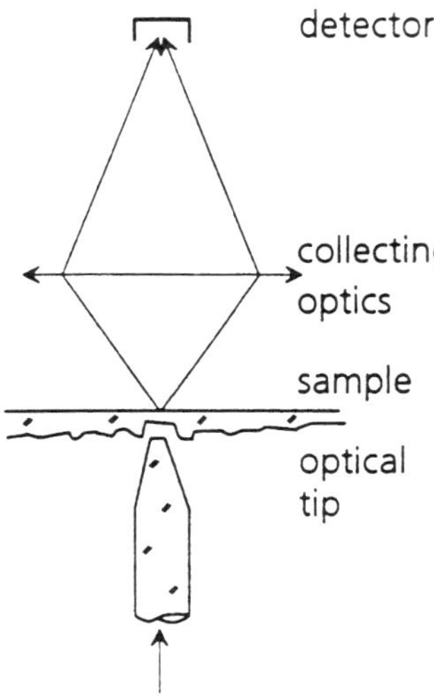

FIG. 35. Schematic of a SNOM near-field sensor. Laser radiation is fed in this example into an optical orifice with dimensions small compared to a wavelength (here a fiber tip). By scanning a surface and collecting scattered light, a high-resolution optical image is generated.

The transmitted radiation is collected by a detector. The fiber-tip separation relative to the surface can be controlled—for instance, for constant detected intensity. If the fiber-tip sensor is scanned across the surface, the vertical position will be a surface microscopic image. Since the point source can have very small dimensions compared to the wavelength, the resolution of such a surface sensor is high and reaches the order of 15–100 nm. Using the same principle, systems can be made that have wide-angle illumination and fiber-tip detector optics, or systems can be made that work in the reflection mode—for instance, by combining a fiber-tip illumination and collection optics on the same sample side. At the time of writing, the first commercial SNOM systems are under development.

Optical surface sensors can be quite rugged and simple, like a scattering sensor that uses integrated scattered intensity, or they may have to be embodied into a system with very high mechanical quality and matching electronics and computer evaluation, like the near-field surface optical rastering sensors. Because of the noncontact optical technique, they can be used quite universally—e.g., for soft surfaces like human tissue samples.

2.6 Optical Distance Metering

In optical distance-metering devices, generally, a spot of the object to be measured is illuminated, and a signal is derived from the directly or diffusely reflected light. Both coaxial and noncoaxial configurations are used with emitted and received beams collinear or not (Fig. 36). With noncollinear systems one should be careful to suppress shadow effects, as indicated in Fig. 36, by a suitable beam geometry. Tightly focussed illumination or receiver beams lead to a good lateral resolution but low depth of field and vice versa. A further point is the signal-versus-distance characteristic, which determines the sensitivity and which is not necessarily linear. Four types of measurement are of main practical importance, making use of geometrical, transit time, interferometric, and focussing principles.

2.6.1 Triangulation The principle of distance measurement by triangulation as illustrated in Fig. 37 has been in use in geodesy for a long time. It is clearly a noncollinear method. As seen from the base b, the object appears at a certain angle a. The object is illuminated by a light source and a lens. The reflected or scattered light is focussed by imaging optics onto a position-sensitive, a linear-array, or a differential photodiode detector. The location of the focus on the detector plane is related to the object distance by trigonometric formulas for similar triangles. The sensitivity decreases with increasing distance, as indicated in Fig. 37,

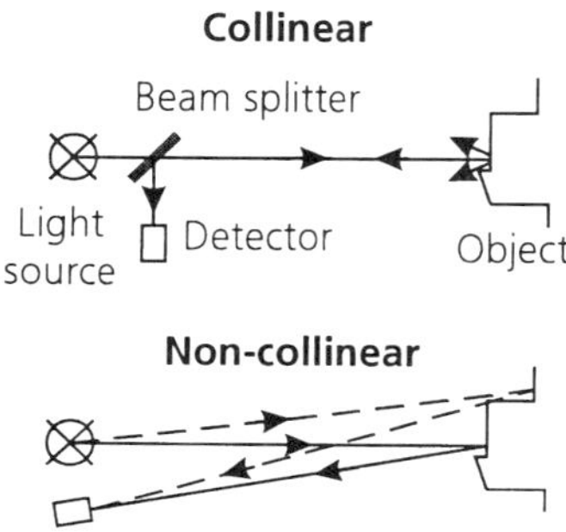

FIG. 36. Distance-metering configurations.

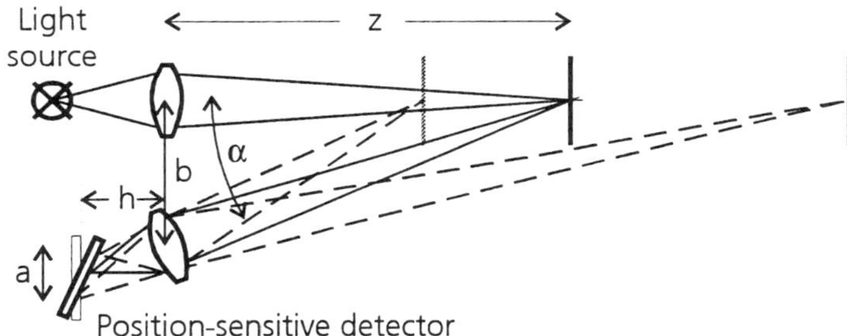

FIG. 37. Principle of triangulation.

by a hyperbolic law. The technique thus is mainly used for shorter distances. Sharp focussing on the detector of light reflected from different planes is achieved by tilting the detector. The range of accessible distances is given by the detector size. For the case of a nontilted detector, the distance is given by

$$z = bh/a. \tag{13}$$

2.6.2 Pulse–Transit-Time Measurement This technique uses the finite velocity of light, which causes a delay time of a back-reflected or scattered light pulse. The setup often is similar to that of Fig. 37; in other versions, emitter and receiver optics are collinear. If the detector signal is delayed by Δt with respect to the emitted signal, the object distance is given by

$$d = \Delta t\, c/2n. \tag{14}$$

Under normal conditions, light travels 30 cm in 1 ns, which corresponds to a distance of 15 cm. Hence, an extremely high system time resolution is required, and the spatial resolution of sensors generally amounts to a few centimeters.

2.6.3 Phase Measurement The transit time may also be determined by the phase of the reflected light. For that purpose, the emitting laser is operated in continuous mode with amplitude modulation of frequency f. The object distance can be calculated from the measured phase shift φ according to

$$d = \varphi c/4\pi n f. \tag{15}$$

Modulation frequencies can be as high as 20 MHz. The determined distance is unambiguous in a range of 2π. Figure 38 shows a compact distance-measurement system ("laser radar") based on phase-shift determination (Müller and Wölfelschneider, 1995) that is characterized by a collinear arrangement. A diode laser ($\lambda = 670$ nm, $P < 1$ mW) with an integrated collimator is used as emitter

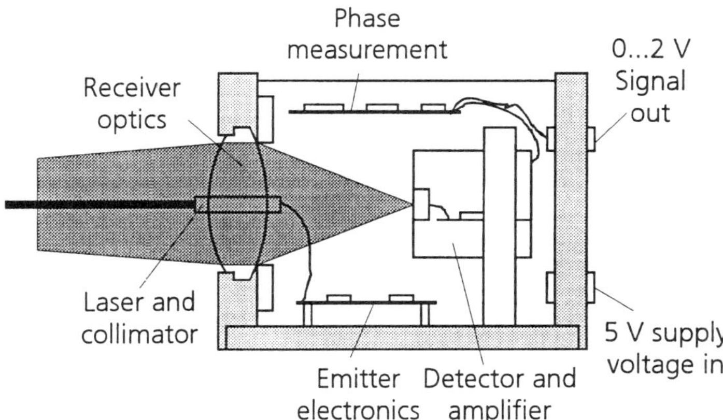

FIG. 38. Laser radar distance measurement system.

and collinear receiver optics. The modulation frequency is 12 MHz. The accessible range of distances is 0.3 to 6 m, the resolution is 2 mm, and the precision is ±5 mm.

2.6.4 Interferometric Measurement Figure 39 shows the principle of a Michelson interferometer, which can be applied to precision distance-measurement devices. A coherent laser beam passes a beam splitter, forming a reference beam (intensity I_1) and a measurement beam (intensity I_2). After reflection, the two beams are recombined. By interference, the resulting beam I_{res} depends on the object position z, as indicated in the inset, as

$$I_{res} = I_1 + I_2 + 2\sqrt{I_1 I_2}\cos\varphi, \tag{16}$$

where the phase difference φ between the two waves is given by

$$\varphi = n \times 2z \times 2\pi/\lambda. \tag{17}$$

Here n is the refractive index within the interferometer. A single-valued distance determination by the periodic signal is not possible from a single measurement but can be obtained by counting fringes from an initial position. Relative measurements within $\lambda/2$ are performed by interpolation of the periodic signal. Light-modulation or phase-shift techniques lead to higher single-valued measurement ranges.

In practical cases, a frequency-stabilized HeNe laser is used as a source with the distance defined by the speed of light. An alternative is offered by single-mode diode lasers, which individually have slightly differing wavelengths. A stable wavelength requires highly stabilized temperature and current with typically $\Delta\lambda = 5 \times 10^{-4}$ nm and knowledge of n to within 10^{-6} for a measurement error of 1 μm/m. For the measurement reflector, high optical quality is required. Optical feedback into the laser even at extremely low levels will lead to wavelength uncertainties and has to be suppressed. In high-precision distance sensors, a fixed-length reference interferometer within the same medium as the measurement beam can be used to compensate for changes in n by temperature and humidity.

2.6.5 Autofocus Systems The principle of these sensors has been derived from the focussing device in compact-disk players and is mentioned in Sec. 2.5. In one example (Fig. 40), the test beam is focussed onto the surface of the object with an intentional astigmatism. Focussing close to the optimum is controlled by imaging the focus on a quadrant detector and setting the minimum astigmatism in both directions. The four detector signals are processed in such a way that deviations from the optimum focussing produce a distance signal that is linear close to the optimum. The resolution is between 1 and 100 μm within ranges from 2 to 200 mm.

2.7 Laser Velocimetry

The laser Doppler velocimeter (LDV), in literature frequently called laser Doppler anemometer, has been introduced by Yeh and Cummins (1964), Lehmann (1968), and Rudd (1969). Various manufacturers devel-

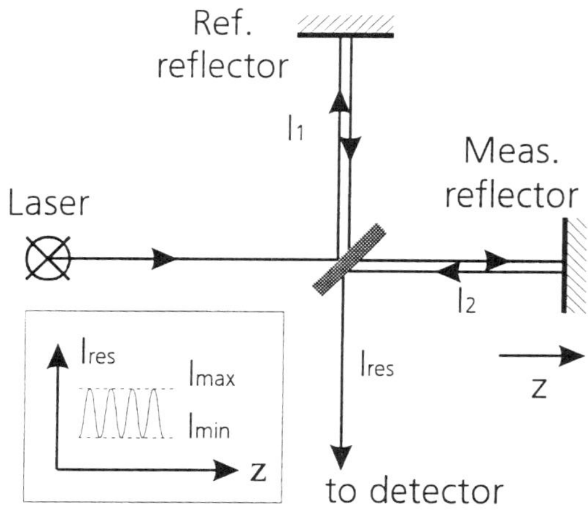

FIG. 39. Schematic of a Michelson interferometer.

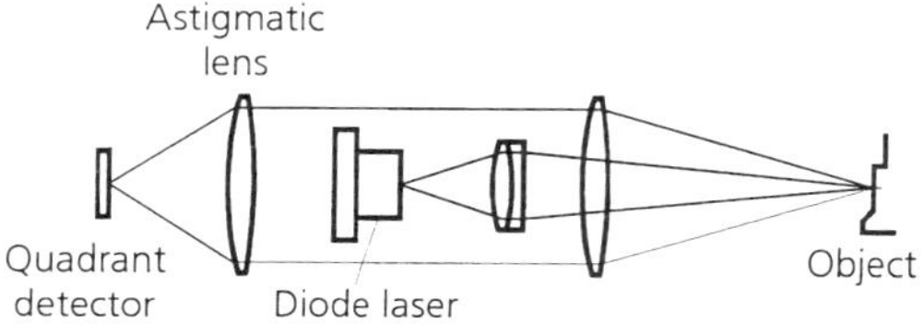

FIG. 40. Schematic of an autofocus system.

oped the fringe or differential method with two beams that today is almost exclusively in use.

The measuring method allows time-resolved and selective measurements of velocity and direction of stationary and nonstationary flows in gases and liquids. It measures without contact in free flows and without disturbing probes. The correlation between the flow velocity and the output signal is linear. There is no need of any calibration or test flow. Measurement errors may only be induced by changes of temperature, density, or chemical reactions. The measuring volume is of small physical dimension. Therefore laser velocimetry allows measurements in boundary layers and near to walls. In addition, the distance between the front lens and the measuring volume can be up to several meters.

The measurement principle is based on laser light scattered by small particles in the flow. These particles may be small crystals, dust, air bubbles, or water or oil droplets and are supposed to be small enough to follow each instantaneous value of the flow velocity without slippage. For enhancing the data rate, particles are often added intentionally. For that purpose fine powders of nontoxic titanium dioxide, small hollow glass bubbles, or small droplets of DEHS (a nontoxic fluid for pulmonary test aerosols in medicine) are used.

In the differential mode the laser beam is split into two partial beams of equal intensities, which are combined and focused by a lens on the crossing point at an angle θ. Thus the smallest possible diameter and therefore highest possible light intensity occurs at the crossing point, which is the measuring spot. On a particle crossing the measuring volume, the two partial beams generate a scattered light field. On a photodetector located at any place around the measuring volume, an output signal with the Doppler frequency f_D is generated by superheterodyning:

$$f_D = \frac{nn}{\lambda_0} |\mathbf{V}| \, 2 \sin \frac{\theta}{2} \cos \delta. \tag{18}$$

The Doppler frequency is strictly proportional to the particle velocity **V**. The calibration constant only includes the known laser wavelength λ_0, the refraction index nn, and the beam crossing angle θ, which easily can be determined. The component of the velocity vector is measured that is in the plane of the crossing incident beams and perpendicular to the bisector of the angle θ (see Fig. 41). Angle δ takes into account the vector component in the direction of the bisector.

The lens for focusing the incident beams can simultaneously be used for collecting the scattered light. The optical efficiency is less favorable for backward scattering than for forward scattering, but in practical use all instruments industrially produced are designed for backward scattering because of the simplicity of optical adjustments.

Modern LDV lens systems commonly make use of a beam expander. The greater the two incident-beam diameters in front of the focusing lens are, the smaller are the diameter and the depth of the measuring volume. This way the light intensity in the measuring volume is increased, and the signal quality is improved appreciably.

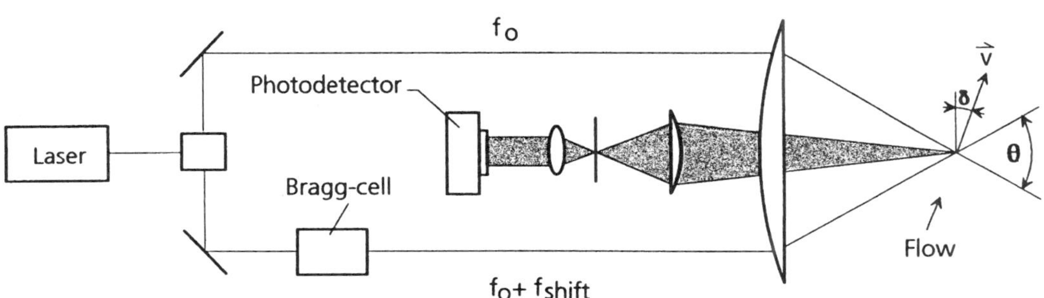

FIG. 41. Principle of the laser Doppler velocimeter with Bragg cell for optical frequency shift and backward scattering.

In order to distinguish the flow direction, an offset is generated by a Bragg cell introduced into one of the incident beams. With this electroacoustical transformer, the laser frequency is slightly shifted. A particle at zero velocity therefore generates a signal with the acoustical frequency. In motion the Doppler frequency is added or substracted to the acoustical frequency, thus providing information on the direction of the flow. Under certain adjustment conditions a Bragg cell can generate a frequency-shifted and -nonshifted beam simultaneously. Herewith, a beam splitter may be saved and the efficiency increased.

For the measurement of two or all three components of the velocity vector, one mostly uses different colors for the channel separation. If an Ar-ion laser with a multiline mirror is used, one simultaneously can get the colors 514.5 nm (green), 488.3 nm (blue), and 476.5 nm (deep blue). The two components of the velocity perpendicular to the optical axis can be measured with one front lens. For the measurement of all three components, at least two front lenses with three pairs of crossing beams are needed.

For simplification and ease of use, the bulky and heavy parts of the LDV, i.e., the laser, the Bragg cell, the color separator, the photomultiplier, and the signal processor, are separated from the relatively light-weight front lens system and are connected by optical fibers. A separate fiber is used for each of the incident beams. To maintain the coherence of the beams, only polarization-maintaining single-mode fibers can be used. Multimode gradient-index or step-index fibers are sufficient for the transmission of the scattered light to the photodetector.

Single-mode fibers have a very thin fiber core with a diameter of about 3 μm to permit the transmission of the fundamental mode only. Polarization-maintaining properties are achieved by means of mechanical stress. The launching of light into such fibers needs a high-precision optical and mechanical coupling because under certain circumstances light power densities of several MW/cm^2 can be reached at the coupling spot. These may cause burning of the fiber termination.

The Doppler signals are evaluated with especially designed signal processors. The two most commonly used processing methods are based on correlation functions or on the discrete Fourier transform function. They have replaced the simpler but noise-sensitive counting processors used formerly. The signal processors can analyze more than 100 000 events per second, making LDV suited for the investigation of turbulent processes. With the same instrument one can measure the lowest velocities like convectional flows as well as the highest velocities like supersonic or plasma flows. The Doppler frequency ranges between 100 Hz and 120 MHz.

In flows with high density or temperature gradients like in flames or mixing flows, the crossing point of the two incident beams may insignificantly migrate, caused by optical schlieren. This may prevent the two beams from crossing for a short time, which decreases the data rate. However, such events have only a small influence on the Doppler frequency, and so correct readings are achieved even under difficult conditions. An additional encoding allows measurements in rotating systems. Examples are the investigation of flows in rotating impellers of hydraulic pumps (Radke and Siekmann, 1992) or inside the cylinder of fired piston engines (Lorenz and Prescher, 1990).

Recently LDV was supplemented by the particle image velocimeter (Adrian, 1991) in an ideal manner, although this is not based on the Doppler effect. The entire flow field or part of it is illuminated by intensive laser double-pulse flashes. A photo or video camera registers the light scattered by particles spread out over the flow field as pairs of light spots. From their distances the local velocities can be calculated by the correlation function. From the graph of the flow velocities, complete information can be achieved of the flow distributions, e.g., the turbulent flows behind obstacles or in mixing processes.

Based on laser Doppler velocimetry, the phase Doppler method was developed for simultaneously measuring the size and the velocity of small droplets in gases or gas bubbles in liquids. The optical arrangement for the crossing incident beams as well as the use of a Bragg cell is identical with that of the LDV. However, the scattered light from the single particles is received not only from one but simultaneously from two or more photodetectors located at different positions.

They all receive the same Doppler signal but with a phase shift depending on the particle diameter d_P.

With careful determination of the position of the photodetectors to ensure that the scattered light is received either from reflection or from refraction conditions, the phase shift is proportional to the particle diameter. In the expressions (Naqui *et al.*, 1991) for the phase difference, only the laser wavelength, the optical refractive index, and three angles—the incident-beam crossing angle, the scattering angle, and the elevation angle—enter.

It should be noted that the expressions for the phase shift contain no calibration constants, and no calibration with test particles is needed. The measuring principle was originally developed for cavitation experiments. An important field of application is measurements in fuel sprays in direct-injection diesel engines and in oil combustion systems.

The laser Doppler vibrometer (Buchhave, 1995) is used for noncontact measurements of oscillations and vibrations on solid surfaces. This principle is closely related to the laser Doppler velocimeter and was developed from it.

2.8 Fiber- and Integrated-Optical Sensors

Fibers are widely used in communication applications for guiding optical signals. The attenuation of fibers is extremely low as compared with that of electrical cables for comparable frequencies, and the transmission is immune against electromagnetic fields. For sensing, fibers can be used as transportation means for radiation equally well from a source to an optical sensor and back to a detector. Beyond that, an optical sensor can be a piece of fiber itself since the optical properties of fibers are affected by external influences like temperature, stress, etc. Two types are commonly called fiber sensors: extrinsic fiber sensors if radiation is launched from the fiber into a sensor and intrinsic fiber sensors if radiation is kept guided within the fiber.

The sensors can be of transmissive or reflective type, i.e., either two fibers are needed for coupling with the source and the detector, or one fiber only is used with a fiber coupler for the forth-and-back transmission. A schematic of a reflective sensor is shown in Fig. 42.

The measurement signals that can be used are

- amplitude (attenuation);
- phase (optical or phase of a modulation frequency);
- polarization;
- spectral shift or attenuation; and
- temporal behavior (pulse duration, transient).

It is the goal of most development efforts to overcome shortcomings of the first—in most cases the easiest—method, that of attenuation measurements, since the resulting power amplitude is not free from effects of the transmission line to and from the sensors.

The type of fiber to be used—monomode or multimode, polarization maintaining or not, etc.—is up to the sensing principle. Attenuation measurements, e.g., can be executed with both monomode and multimode techniques, while phase-sensitive interferometry requires monomode as well as, in many cases, polarization-maintaining fibers.

Many physical parameters can be converted into one of the measuring signals given above. A few examples are

- displacement, velocity, vibration;
- stress and strain;
- reflectivity, refractivity, absorptivity, turbidity, fluorescence;
- filling level, flux;
- temperature; and
- electric field, magnetic field, electric current, etc.

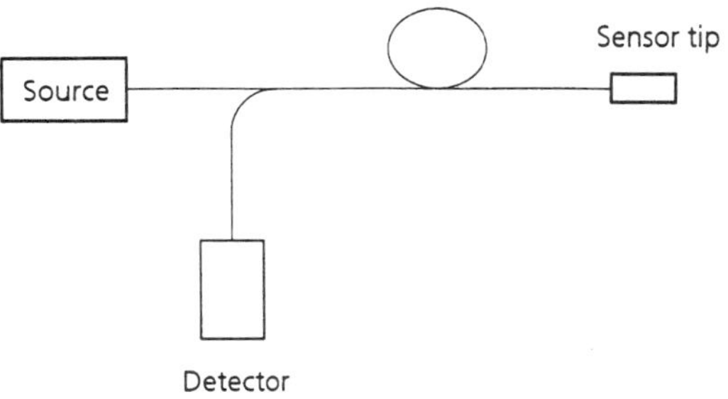

FIG. 42. Schematic of a reflective-type fiber-optical sensor.

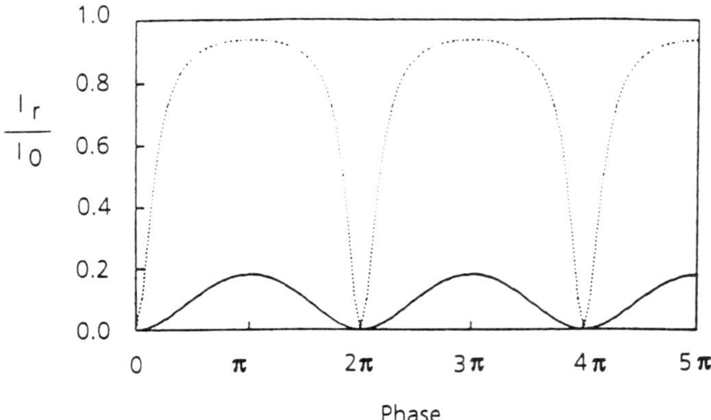

FIG. 43. Fringes from Fabry–Pérot resonators (in transmission or reflection) for two different reflectivities of the resonator end faces.

Interferometric sensors take advantage of the extreme sensitivity that can be achieved by splitting the radiation into two branches, allowing only one to be affected by the external force, and finally recombining the branches. The external force—the measurand—changes the optical path length of the measuring branch by $\Delta(nL)$ and thereby changes the relative optical phase φ of the two beams by

$$\Delta\varphi = 2\pi\Delta(nL)/\lambda, \tag{19}$$

with λ the wavelength of the radiation used. Like in conventional optics, there are several types of interferometers used in fiber optics: Michelson interferometer, Mach–Zehnder interferometer, Sagnac interferometer, and Fabry–Pérot resonator.

Michelson and Mach–Zehnder interferometers can be used to measure any effect that changes the difference *nd* between optical path lengths of the branches. The optical intensity $I(\Delta\varphi)$ resulting after interference of the beams I_1 and I_2 with a phase difference of $\Delta\varphi = nd \times 2\pi/\lambda$ is

$$I(\Delta\varphi) = \tfrac{1}{2}(I_1 + I_2) + (I_1 I_2)^{1/2} \cos(\Delta\varphi). \tag{20}$$

The *Sagnac interferometer* is similar to the Mach–Zehnder type. It splits the radiation into two parts counterpropagating in a fiber coil. Thus the mechanical length to be passed is the same for both beams. The relativistic Sagnac effect, however, causes a phase shift if the system is accelerated, i.e., is rotating:

$$\Delta\varphi = 4\pi LR\Omega/\lambda c, \tag{21}$$

where Ω is the angular velocity and R and L are the radius and the fiber length of the coil, respectively. With Sagnac-type interferometers, drifts and resolutions better than 0.1 deg/h can be achieved.

Fabry–Pérot sensors are resonators with semitransparent end faces. Because of the multiple passing of the radiation inside the resonator, a mode selection occurs (like in a laser resonator) that is stronger the lower the attenuation is inside the resonator. The shape of the fringes observed in reflection or transmission depends on the reflectivity of the resonator end faces (cf. Fig. 43).

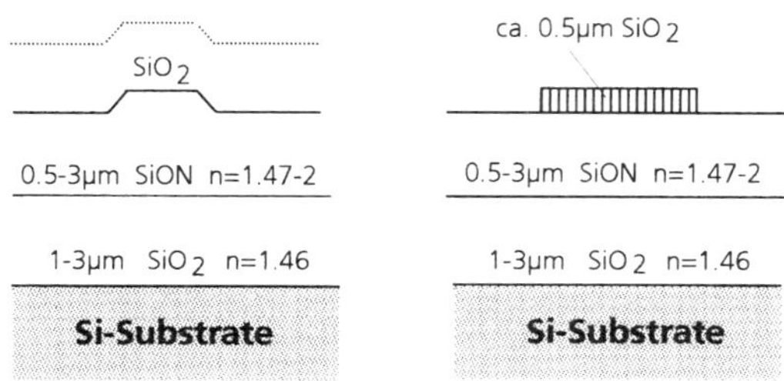

FIG. 44. Two typical cross sections of silicon oxynitride waveguides on silicon substrates.

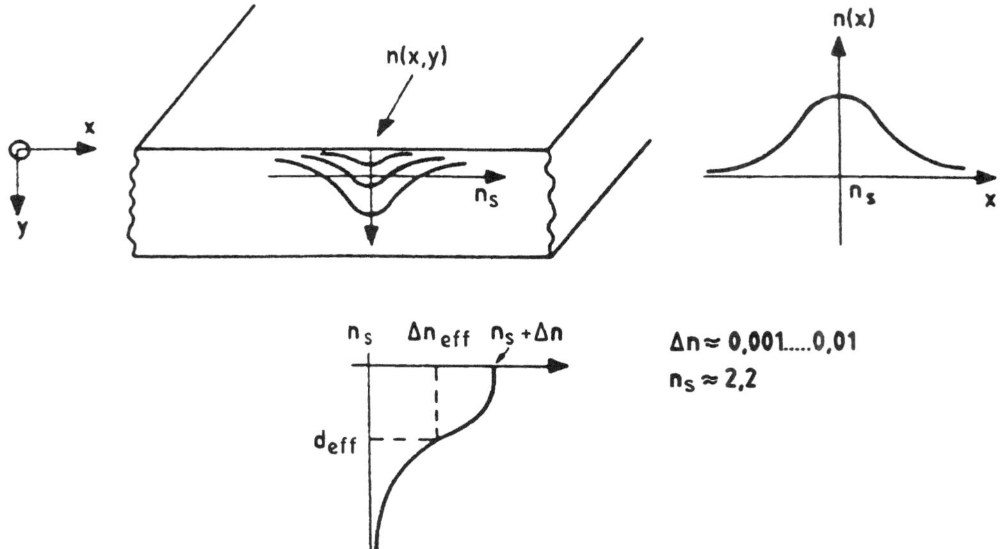

FIG. 45. Cross section and refrective-index profiles of channel waveguides made by diffusion of Ti in lithium niobate, $LiNbO_3$.

The effects that can be measured with monomode fibers at wavelengths of 850 nm approximately are

- stress: $\Delta\varphi/\Delta(\epsilon L) = 10^7$ rad/m;
- force K: $\Delta\varphi/\Delta(KL) = 2 \times 10^{-4}$ rad/(N·m);
- pressure p: $\Delta\varphi/\Delta(pL) = 5 \times 10^{-5}$ rad/(Pa·m); and
- temperature T: $\Delta\varphi/\Delta(TL) = 100$ rad/(K·m).

Integrated optical sensors are planar-type components incorporating waveguide structures. As compared with fiber optical sensors, they can be produced with submicron resolution using techniques and means of semiconductor production. Depending on the material used, the wave guides are formed by various techniques:

- integrated optics in glass (for passive components): waveguides are formed by ion exchange;
- integrated optics on silicon (for passive components): silicon is used as substrate only, the waveguides formed by layers of SiO_2 and silicon oxynitride (cf. Fig. 44);
- integrated optics in lithium niobate, $LiNbO_3$ (for optically active components): waveguides are formed either by diffusion of titanium (cf. Fig. 45) or by proton exchange; and

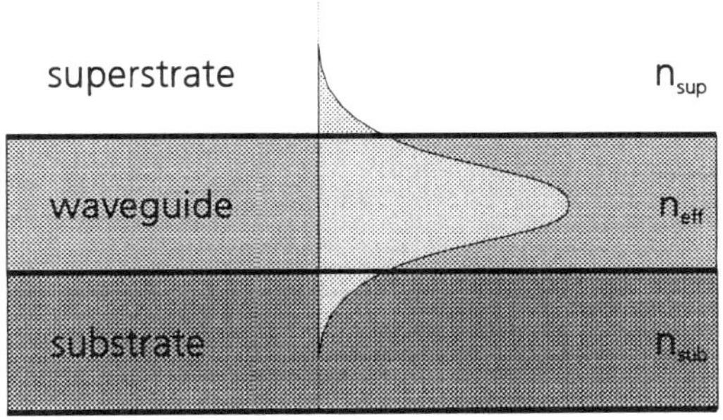

FIG. 46. Schematic of the influence of the superstate on the effective refractive index of the waveguide through the evanescent wave.

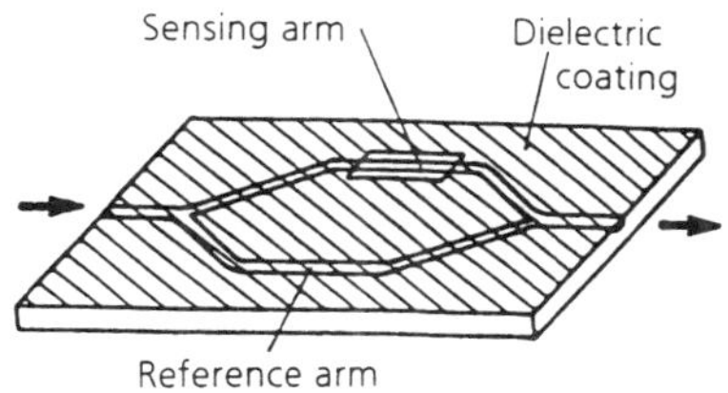

FIG. 47. Schematic of an integrated-optical sensor outlined as a Mach–Zehnder interferometer.

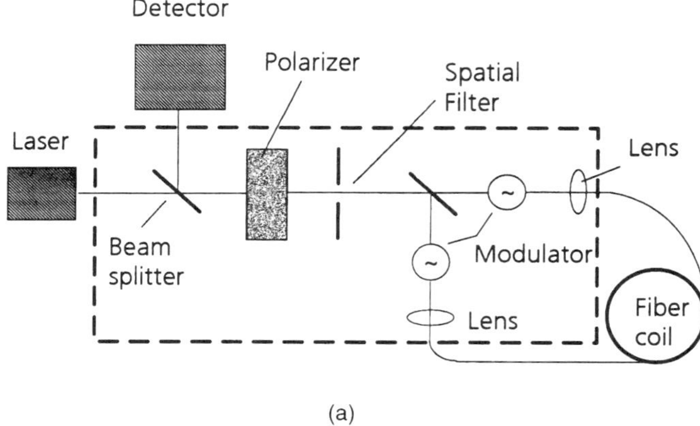

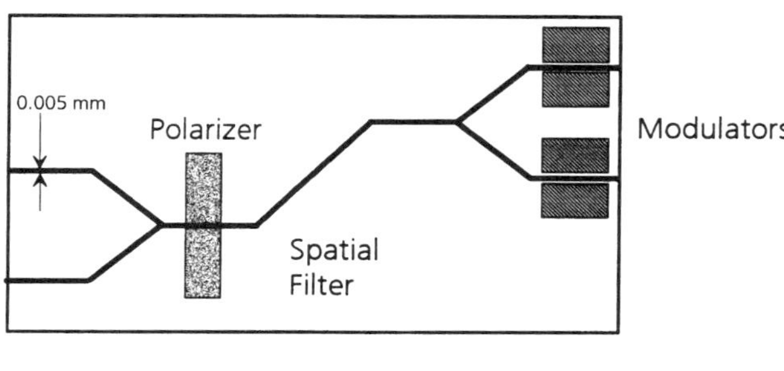

FIG. 48. (a) Schematic of a Sagnac interferometer. (b) Integrated optical layout with polarizer, Y branches, spatial filters, and optical phase modulators.

- integrated optics in polymers (for passive and for active components): waveguides are mostly formed by rib structures.

Integrated optics allows the integration of active and passive optical components. The monolithic integration enhances accuracy and reproducibility as well as reducing the assembling costs incurred with the use of discrete elements. There are numerous approaches to the use of integrated optics for sensing. Many of them are based on the use of the evanescent wave—the radiation leaking from the waveguide into the vacuum or into a sensitive layer deposited onto the integrated-optical chip (the superstrate) (cf. Fig. 46). If the superstrate changes its optical refractive index through a reaction with the gas atmosphere, the effective refractive index of the waveguide is affected. This small variation can be detected, e.g., by using a Mach–Zehnder interferometric structure on the chip (cf. Fig. 47). Lithium niobate allows the incorporation of active components like phase modulators needed for a dynamic technique for Sagnac interferometers (cf. Fig. 48).

Works Cited

Adrian, R. J. (1991), "Particle-Imaging Techniques for Experimental Fluid Mechanics," *Annu. Rev. Fluid Mech.* **23,** 261–304.

Bohren, C. F., Huffman, D. R. (1993), *Absorption and Scattering of Light by Small Particles,* New York: Wiley.

Buchhave, P. (1995), "Laser Doppler Vibration Measurement Using Variable Frequency Shift," *DISA Information* No. 18, 15–20.

Elson, J. M., Bennet, H. E., Bennet, J. M. (1979), "Scattering from Optical Surfaces," in: R. R. Shannon, J. C. Wyant (Eds.), *Applied Optics and Optical Engineering,* Vol. 7, New York: Academic.

Erwin Sick GmbH (1995), Information brochure on Visibility System VISIC 400 SM.

EUR-Control (1984), Information brochure on MEX 2 Process Photometer.

Gerlinger, H. (Ed.) (1992), Special issue on Colorimetry, *Tech. Messen* **59** (5).

Gobrecht, H. (1978), *Bergmann-Schaefer Lehrbuch der Optik,* Band III, Hawthorne, NY: Walter de Gruyter.

Grisar, R. (1992), in: Sensors, Vol. 6, *Optical Sensors,* Weinheim: VCH.

Hekatron GmbH (1991), Information brochure on smoke switching systems.

Kaase, H. (1992), "Fundamentals and Limitations of Optical Radiation Measurements," in: E. Wagner *et al.* (Eds.), *Sensors,* Vol. 6, *Optical Sensors,* Weinheim: VCH.

Lehmann, B. (1968), "Geschwindigkeitsmessung mit dem Laser-Dopplerverfahren," *Wiss. Ber. Telefunken AG,* pp. 141 ff.

Lorenz, M., Prescher, K. (1990), "Cycle Resolved LDF Measurements on a Fired SI-Engine at High Data Rates Using a Conventional Modular LDV System," SAE Technical Paper Series 900054.

MacAdam, D. L. (1981), *Color Measurement,* Heidelberg: Springer.

Marshall, G. F. (1980), "Scanning Devices and Systems," in: R. Kingslake, B. J. Thompson (Eds.), *Applied Optics and Optical Engineering,* Vol. 6, New York: Academic.

Mettler-Toledo AG (1995), Information brochure on in-line turbidity measurement with FSC402.

Müller, R., Wölfelschneider, H. (1995), "Low Cost Distance Metering System," in: *Achievements and Results, IPM Annual Report,* Freiburg: Institut für Physikalische Messtechnik, pp. 18–19.

Naqui, A., Durst, F., Liu, X. (1991), "Extended Phase-Doppler System for Characterization of Multiphase Flows," *Part. Syst. Charact.* **8,** 16–22.

Noll, G., Rogher, M., Dörr, J. (1995), *Chem. Anl. Verf.* **5/95,** 30.

Norton, Harry N. (1982), *Sensor and Analyzer Handbook,* Englewood Cliffs, NJ: Prentice-Hall.

Radke, M., Siekmann, H. E. (1992), "Performance of Axial-Flow Pump Stage with Variable Geometry Guide Vanes," in: J. H. Kim, W.-J. Yang (Eds.), *Proceedings of the Fourth Symposium on Transport Phenomena and Dynamics of Rotating Machinery (ISROMAC4),* Honolulu, Hawaii, Vol. A, Washington: Hemisphere Publ. Corp., pp. 14–23.

Rudd, M. J. (1969), "A New Model for the Laser Dopplermeter," *J. Sci. Instrum.* **2,** 55–58.

Sze, S. M. (1981), *Physics of Semiconductor Devices,* New York: Wiley.

Wyszecki, Gunter (1978), "Colorimetry," in: W. G. Driscoll, W. Vaugham (Eds.), *Handbook of Optics,* New York: McGraw-Hill.

Yariv, A. (1975), *Quantum Electronics,* New York: Wiley.

Yeh, Y., Cummins, H. Z. (1964), "Localized Fluid Flow Measurement with an He-Ne Laser Spectrometer," *Appl. Phys. Lett.* **4,** 176–178.

Further Reading

Bohren, C. F., Huffmann, D. R. (1993), *Absorption and Scattering of Light by Small Particles,* New York: Wiley.

Dakin, J., Culshaw, B. (1988), *Optical Fiber Sensors,* Vol. 1, *Principles and Components,* Vol. 2, *Systems and Applications,* Boston: Artech House.

Gobrecht, H. (Ed.) (1978), *Bergmann-Schaefer Lehrbuch der Optik,* Band III, Hawthorne, NY: Walter de Gruyter.

MacAdam, D. L. (1981), *Color Measurement,* Heidelberg: Springer.

Marshall, G. F. (1980), "Scanning Devices and Systems," in: R. Kingslake, B. J. Thompson (Eds.), *Applied Optics and Optical Engineering,* Vol. 6, New York: Academic.

Norton, Harry N. (1982), *Sensor and Analyzer Handbook,* Englewood Cliffs, NJ: Prentice-Hall.

Saleh, B. E. A., Teich, M. C. (1991), *Fundamentals of Photonics,* New York: Wiley.

Sze, S. M. (1981), *Physics of Semiconductor Devices,* New York: Wiley.

Wyszecki, Gunter (1978), in: W. G. Driscoll, W. Vaugham (Eds.), *Handbook of Optics,* New York: McGraw-Hill.

Yariv, A. (1974), *Quantum Electronics,* New York: Wiley.

SEPARATION PROCESSES

JAMES ZHOU, *Research and Technology, Allied Signal Inc., Des Plaines, Illinois, U.S.A.*

W. S. WINSTON HO, *Corporate Research, Exxon Research and Engineering Company, Annandale, New Jersey, U.S.A.*

NORMAN N. LI, *Research and Technology, Allied Signal Inc., Des Plaines, Illinois, U.S.A.*

3-527-28139-8/96/$5.00 + .50

INTRODUCTION

Separation processes are widely used for the recovery and purification of desired products in manufacturing operations in various industrial sectors including the chemical, petrochemical, petroleum, pharmaceutical, medical, food, dairy, beverage, paper, textile, metallurgical, and electronic industries, as well as for the removal of pollutants in pollution control and environmental protection. This article covers the major separation techniques employed in these processes, namely, distillation, absorption, adsorption, ion exchange, extraction and leaching, crystallization, filtration, sedimentation, and membrane separations. Included for the description of each separation technique are introduction and process description, process equipment, process operations, and important industrial applications. While in-depth description is not possible because of the scope of the article, references are given for further reading.

1. DISTILLATION

1.1 Introduction and Process Description

Distillation is the most commonly used method for separating liquid mixtures in chemical industries and the most studied and understood method (Fair, 1987; Seader, 1984). Distillation effects separation of the liquid mixtures by the relative difference of the effective vapor pressure, or volatility, of liquid components. Thus, a large volatility difference between components in a liquid mixture means that they will be easily separated via distillation. When such difference does not exist, or is very small, separation by distillation becomes impossible or costly. On the basis of the mode of operation, distillation can be classified as continuous or batch. The term fractional distillation is often used interchangeably with distillation.

The overall separation efficiency of the distillation process is determined by the relative volatility, the number of contacting stages (trays), and the liquid and vapor flow rates.

1.2 Distillation-Process Equipment

The equipment used to carry out distillation can be divided into two large categories, i.e., tray-type columns and packed-bed–type columns. The former provides a stagewise contacting, whereas the latter provides a countercurrent differential contacting between the vapor and the liquid. A tray-type distillation column is shown schematically in Fig. 1, and a packed-bed–type distillation column is shown in Fig. 2.

The feed, which is to be separated into fractions, is introduced into the column at one or more places along the column. Because of density difference, the liquid flows downward, while the vapor flows upward. Equilibrium is assumed to exist between the liquid and vapor streams on all trays. Liquid reaching the bottom is partially vaporized in the reboiler, and the vapor is pushed upward. The heavier fraction left at the bottom is withdrawn as the bottom product. The vapor reaching the top is condensed in the overhead condenser. Part of the condensate is returned to the column as reflux to provide liquid overflow, and the remainder is the top product.

The trays in tray columns generally contain some type of vapor dispersion device such as liftable valves, bubble caps, or simple perforations. The packed-type column can have either random packing or ordered packing. Figure 3 shows some of these random packings used in the industry. The choice of which packing to use depends on the efficiency of contact required for the separation and on the pressure drop from top to bottom of the column (Strigle, 1994; Fair, 1987). Ordered packing, also known as struc-

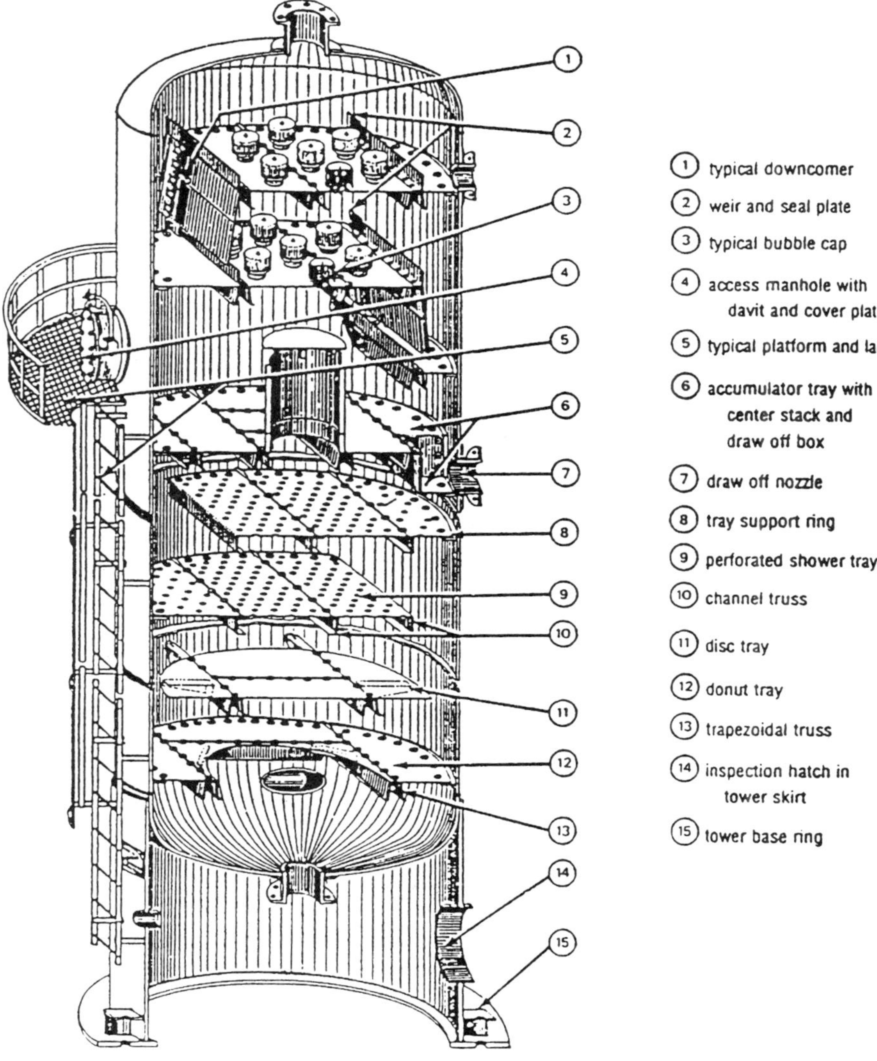

FIG. 1. Schematic diagram of a typical tray column [reprinted from Paruit (1982) with permission].

tured packing, provides high mass-transfer efficiency but also at a higher cost of manufacturing.

1.3 Distillation-Process Design and Operations

The basic design equations for the distillation process are obtained from mass and energy balances and from thermodynamic principles. For a generalized equilibrium stage shown in Fig. 4, the mass-balance equation is

$$F^n z_i^n + L^{n+1} x_i^{n+1} + V^{n-1} y_i^{n-1} = (V^n + V^{n'}) y_i^n + (L^n + L^{n'}) x_i^n. \quad (1)$$

The summation of mole fractions must be unity:

$$\sum_i x_i^n = \sum_i y_i^n = \sum_i z_i^n = 1. \quad (2)$$

Since each stream is also associated with an energy (enthalpy), thus, the energy balance equation of the stages is

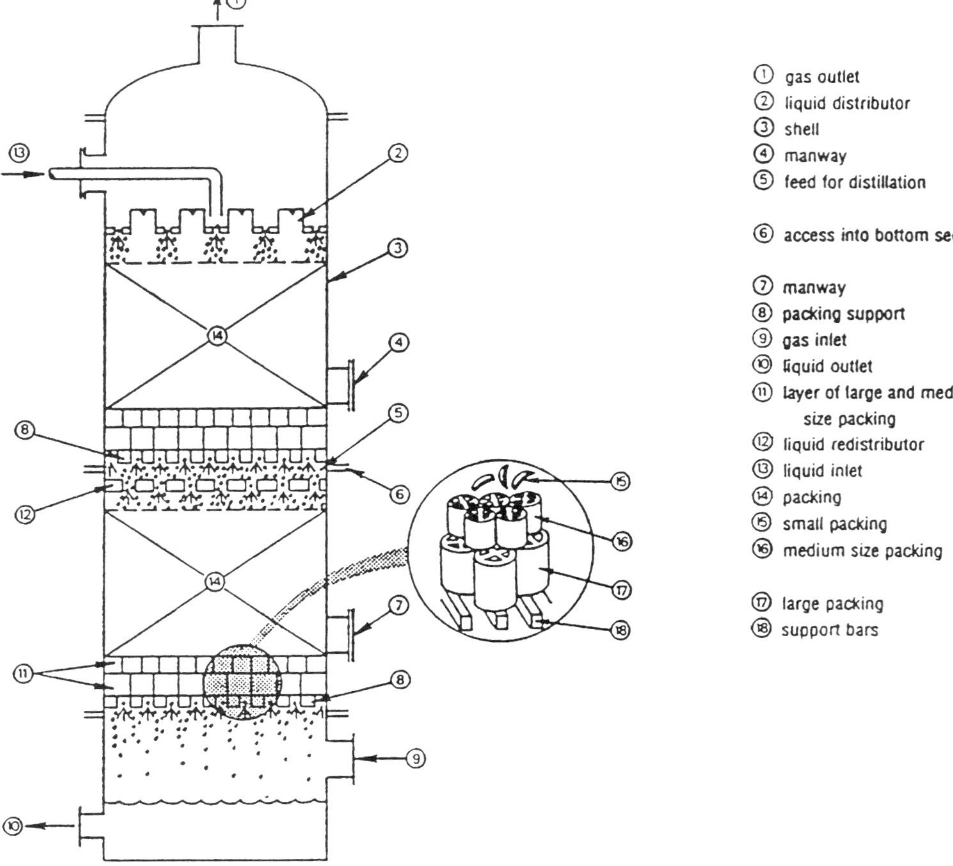

FIG. 2. Schematic diagram of a typical packed column [reprinted from Paruit (1982) with permission].

$$Q^n + F^n H_f^n + L^{n+1} h^{n+1} + V^{n-1} H^{n-1} = (V^n + V^{n'}) H^n + (L^n + L^{n'}) h^n, \quad (3)$$

where Q^n is the heat input into stage n, and the h's and H's are the liquid and vapor enthalpies of the streams, respectively.

The other important equation is supplied by equilibrium thermodynamics, which relates the liquid and vapor phase compositions by

$$K_j^n = y_i^n / x_i^n, \quad (4)$$

where K_i^n is the relative volatility.

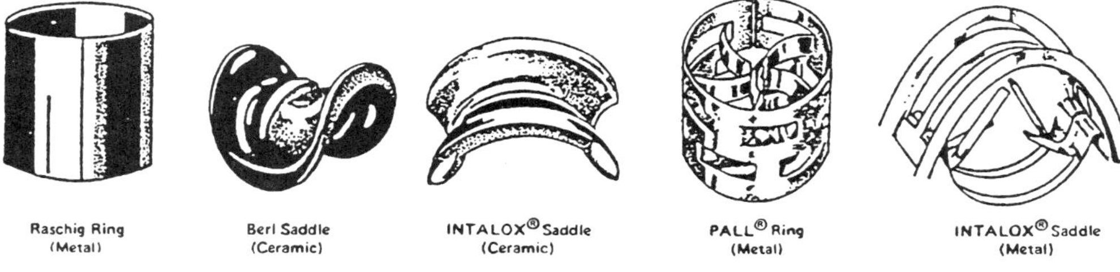

FIG. 3. Some of the random packings used in packed columns [reprinted from Fair (1987) with permission].

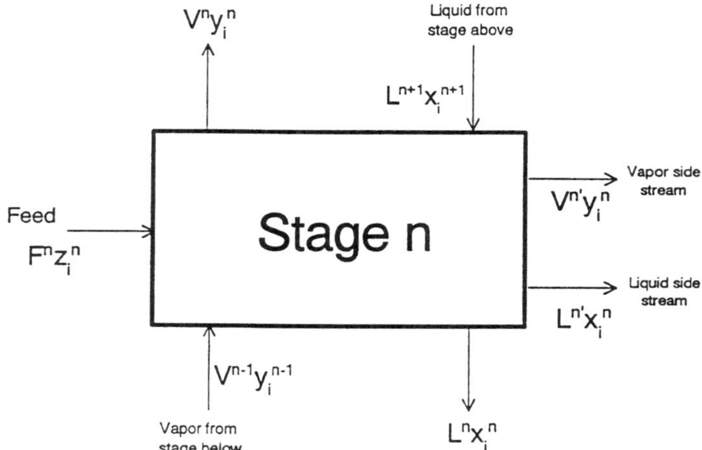

FIG. 4. Generalized equilibrium stage.

The above relations have been incorporated into many process simulation software packages available from several sources, for example, ASPEN-PLUS (1994) and ChemCAD (1993). These software packages also contain a large amount of thermodynamic data for pure substances and for mixtures. Furthermore, they can also be used to estimate the vapor–liquid equilibrium (VLE) behavior based on certain theoretical models such as UNIFAC.

For a given separation, the above relationships are used to determine the key operating parameters such as the number of theoretical stages required, the reflux ratio, the maximum operational capacity, the minimum operating liquid or vapor rate, the efficiency as compared with the theoretical value, and, most importantly, the operating pressure of the column. The choice of operating pressure can generally be made on the basis of the relative volatility between the components.

1.4 Important Industrial Applications

The single largest application of the distillation process is in oil refining where crude oil is separated into many fractions before and after catalytic reforming. The other applications are numerous in the chemical and food industries. It is safe to say that if there is a chemical plant, there must be one or more distillation columns. In addition to the application in refining, the following is a short list of some of the most important industrial applications:

1. Production of styrene monomer: vacuum distillation of styrene monomer from a mixture of ethylbenzene, water, benzene, and toluene using packed columns.
2. Ethylene glycol (EG) purification: vacuum distillation of EG using tray columns.
3. Ethanol production: azeotropic distillation of ethanol from water using benzene as an entrainer (a third component added to the system to effect the separation of an otherwise inseparable mixture by distillation).
4. Ethylene/ethane separation: pressure distillation of ethylene from ethane and propylene using packed or tray columns.
5. Cryogenic fractional distillation: a special case of separating gaseous mixtures by fractional distillation of liquid air to produce high-purity nitrogen, oxygen, and argon.

2. ABSORPTION

Absorption and adsorption can be generally termed sorption, which refers to processes of selective transfer of one or more solutes from a fluid phase to a *liquid phase* where absorption occurs, or to a batch of *solid particles* where adsorption occurs. The absorption process involves contacting a gas phase with a liquid absorbent, whereas the adsorption process consists of contacting either a gas or a liquid phase with solid particles known as adsorbent. Both of these processes have grown into essential unit operations and are widely practiced in chemical, petrochemical, and many other industries. We briefly summarize both of these processes separately in the following sections.

2.1 Introduction and Process Description

Absorption can be classified into three general types based on the interaction between the absorbent liquid and the absorbed components (absorbates) of the gas phase (Zarzycki, 1993; Kohl, 1987).

1. *Reversible reaction:* A reversible chemical reaction occurs between the absorbates and a component of the absorbent. In this case, the absorbates are taken up by the absorbent under one set of conditions and released under another set of conditions. This type of absorption is the most important in consequence of large-scale applications in natural-gas sweetening processes (removal of acid gases such as CO_2 and H_2S from light hydrocarbons) (Campbell, 1976; Kohl and Riesenfeld, 1979).
2. *Physical solution:* The absorbates simply dissolve in the liquid absorbent. Separation of components from the gas stream occurs as a result of the difference in solubility between components; the absorbate having a higher solubility preferentially dissolves in the absorbent.
3. *Irreversible reaction:* An irreversible chemical reaction occurs between the absorbates and a component of the absorbent. Since the absorbates are irreversibly captured by the absorbent, the rate of mass transfer of this type of process tends to be quite high as compared with the other two types of absorption processes.

The inverse operation, called stripping or desorption, involves the transfer of dissolved volatile components from the liquid to the gas phase. The same basic principles apply to both absorption and desorption operations.

2.2 Process Equipment

Absorption processes are similar to distillation processes; both involve a gas–liquid contacting, and both are equilibrium processes. Therefore, the process equipment used in these two processes is also similar. The three major types of equipment used in absorption are tray-type, packed, and spray columns (Kohl, 1987; Edwards, 1984). The first two types of columns are identical to those used for distillation. The spray column is unique to the absorption process.

Tray columns are well suited for clean, noncorrosive, and nonfoaming liquid absorbents, and low to medium absorbent flow rates. The advantages of tray columns are their simple design, capability of internal cooling, and, most importantly, large number of mass-transfer units due to the absence of gas channeling. The main disadvantages are that the tray columns are difficult to operate, and cannot adapt easily to changing operating conditions such as feed flow-rate change.

The packed columns have either random packing or structured packing. The random packings used are shown in Fig. 3. These packings are dumped into the column and can generally be changed easily when column operating parameters change. Packed columns are used for corrosive liquid, under foaming conditions, at high liquid-to-gas ratios, and at low pressure drops. Structured packing offers higher column efficiency but at an added cost.

Spray columns are used when solids are present in either the liquid or the gas stream and when a higher pressure drop cannot be tolerated. Spray columns offer a very limited number of mass-transfer units (theoretical number of trays) and hence are not suitable for difficult separations. The main application of spray columns is for the absorption of SO_2, H_2S, and other impurities from flue gas where low pressure drop is essential.

One other type of gas–liquid contacting device is the bubble column, where the gas is bubbled through a column filled with the liquid absorbent. This type of column has very low efficiency and is most often used as a reactor.

2.3 Process Design and Operations

Design of absorption processes includes the choice of absorbents, the choice of liquid–gas contactors, and, finally, operating conditions. The operation of the absorption is generally straightforward. The gas stream is introduced from the bottom of the tray, packed, or spray column and allowed to flow upward. The liquid stream is introduced from the top of the column and flows downward.

Selection of absorbents is primarily based on experimental data and on economic considerations. Thus, the absorbent should have a high solubility for the absorbates, and a low vapor pressure for reduced vapor loss and effluent contaminations, and to the extent possible it should be inexpensive, stable, noncorrosive, nonfoaming, and nonflammable.

The next step in the design process is the determination of the VLE behavior of the absorbent–absorbate system. The VLE represents the limiting conditions of the system, yielding the minimum liquid-to-gas ratio and the maximum gas purity. The VLE data are generally presented as the solubility of absorbates in the absorbents. When this solubility is very low, it can usually be expressed by the well-known Henry's law:

$$p_A = Hx_A,$$
$$p_A = H'c_A, \tag{5}$$

where p_A is the partial pressure of component A in the gas phase; x_A and c_A are mole fraction and molar concentration, respectively, of A in the liquid phase; and H and H' are Henry's-law constants in appropriate units.

Vapor–liquid equilibrium data for many gas–liquid systems are available in *Perry's Chemical Engineers' Handbook* (Perry and Green, 1984), and in other engineering data books (Ohe, 1990, 1989; Cheremisinoff, 1995; Gas Processors Suppliers Association, 1987). When experimental data are not available, prediction can be made using several techniques, such as those discussed in Sec. 1. Comprehensive discussion on this subject is given by Reid *et al.* (1987).

Selection of the mass transfer equipment is based on the operating requirements, economics, and the advantages and disadvantages of each type of equipment, as discussed in Sec. 2.2. Once the selection is made, detailed mass and energy balances in conjunction with the mass-transfer efficiency considerations determine the number of trays needed or the height of the packing and the physical size of the column. We do not discuss the computational methods in detail here, but the readers are referred to the discussion in Secs. 14 and 18 of *Perry's Chemical Engineers' Handbook* (Perry and Green, 1984). Computer simulations of absorption processes are also available in ASPEN-PLUS (1994), ChemCAD (1993), etc.

2.4 Important Industrial Applications

Large-scale industrial applications of absorption processes include the absorption of SO_3 with sulfuric acid to make fuming sulfuric acid, and the absorption of HCl with water to produce hydrochloric acid. More recently, the use of lime/limestone slurries for the removal of SO_2 from flue gases has developed. Many other applications using absorption processes are given in the literature (Kohl, 1987).

As mentioned earlier, the largest industrial application of absorption processes is in the natural- and synthetic-gas sweetening processes, where acid gases such as H_2S and CO_2 are removed from gaseous hydrocarbons consisting mostly of methane and small amounts of ethane and higher hydrocarbons (Kohl and Riesenfeld, 1979; Campbell, 1976). This process is based on reversible chemical reactions of the acid gases with the basic liquid absorbents such as monoethanolamine, diglycolamines, diethanolamine, methyldiethanolamine, and sterically hindered amines (Sartori *et al.*, 1987), as well as mixtures of two or more of the amines. The absorbents react with CO_2 and H_2S either selectively or nonselectively under absorption conditions and release the absorbed species under desorption/stripping conditions (usually higher temperature, vacuum, or use of stripping gas). The regenerated absorbents are returned to the absorber, and the cycle continues.

3. ADSORPTION

3.1 Introduction and Process Description

Adsorption occurs when a gas or liquid stream is brought into contact with a batch of solid particles (adsorbent). Specific component(s) of the fluid stream can then be concentrated on the adsorbent surface and thus separated from the remaining components. Adsorption is heterogenous in nature and classified by the interaction between the adsorbents and adsorbates. Three types of adsorption exist based on the interactions, which are physical, chemical, and specific (Suzuki, 1990; Keller *et al.*, 1987; Vermeulen *et al.*, 1984).

1. *Physical adsorption* occurs through weak van der Waals forces between the adsorbents and the adsorbates. These forces include the dispersion force and classical electrostatic forces. Physical adsorption constitutes the most important type of adsorption because this weak interaction makes regeneration of adsorbents relatively straightforward.
2. *Chemical adsorption* involves chemical reactions between adsorbents and adsorbates, resulting in new species being formed on the adsorbent surface. The chemical bond formed is generally much stronger than the van der Waals interaction. Chemical adsorption is typically identified with monolayer coverage.
3. When adsorbates interact with specific functional groups on the adsorbent surface without chemical transformation, the resulting adsorption process is termed *specific adsorption*. The forces associated with these specific interactions range from values common to physical adsorption to the stronger forces common to chemical adsorption.

We concentrate our discussion on physical adsorption because of its importance to industrial separations. Ion-exchange processes are discussed in more detail in a later section. A related process known as chromatography is not discussed here. The interested reader is referred to Brown and Grushka (1993) for chromatographic analysis and to Ganetsos and Barker (1993) for industrial applications; see also CHROMATOGRAPHY.

The most often used process equipment for adsorption is the packed column, otherwise known as the fixed-bed adsorber. Its design and operation are governed by the adsorption equilibria. Therefore, an understanding of adsorption equilibria provides one with the tool for understanding adsorption processes.

3.2 Adsorption Equilibria

The distribution of the adsorbates in the fluid phase and in the adsorbent solid phase at thermodynamic equilibrium is described by the adsorption isotherm, which relates fluid-phase adsorbate concentration to the amount of adsorbate per unit weight of adsorbent. A wide variety of equilibrium isotherms exists. The most common two types are presented here.

3.2.1 Langmuir Isotherm This isotherm was derived by Langmuir on the basis of the assumptions that at most a monolayer of adsorbate could cover the solid surface and that the adsorption energy is constant and independent of surface coverage. For gas–solid systems, the Langmuir isotherm is expressed as

$$Q = Q_{\mathrm{max}} K_a P/(1 + K_a P), \tag{6}$$

where Q is the adsorbate loading (wt. of adsorbate/wt. of adsorbent), Q_{max} is the maximum loading at monolayer coverage, K_a is the adsorption equilibrium constant, and P is the partial pressure of the adsorbate in the gas phase at equilibrium. For liquid–solid systems, P is replaced by concentration C of the adsorbate in the liquid phase at equilibrium.

3.2.2 BET Isotherm The Brunauer-Emmett-Teller (Brunauer *et al.*, 1938) isotherm is based on the assumptions that the adsorption is multilayer and that the energy associated with adsorption beyond the first layer is the same as for condensation. The simplest form of the BET isotherm is

$$\frac{Q}{Q_{\mathrm{mono}}} = \frac{KP/P^s}{[1 + (K - 1)P/P^s](1 - P/P^s)}, \tag{7}$$

where Q_{mono} is the adsorbate loading for monolayer coverage, P^s is the vapor pressure of pure adsorbate at the adsorption temperature, and K is a constant related to the energy of adsorption. For liquid systems, P^s is replaced by the saturation concentration C^s

of the pure adsorbate in the liquid phase, and P by concentration C.

The adsorption of many gas–solid and liquid–solid systems can be fitted by the Langmuir or the BET isotherm. However, these isotherms are phenomenological in nature, and their representation of the adsorption data does not constitute positive proof that the assumptions used to derive these isotherms are operative in the specific adsorption process in question.

3.3 Process Design and Operations

Keys to the design of adsorption processes are the choice of adsorbent and choice of adsorption and regeneration procedures. In general, it is desirable to use adsorbents with as small a particle size as conditions for efficient operation permit, because the rate of adsorption is controlled by the transport of adsorbate molecules on the surface or within the pores of adsorbents. The pressure drop through a packed column of adsorbent particles is usually inversely proportional to the particle size. Therefore, tradeoffs must be made sometimes between the mass-transfer rate and pressure drop. To avoid such tradeoffs, compound-size columns packed with two different particle sizes, usually arranged in series, are often employed.

The adsorbents of commercial importance are zeolite molecular sieves (ZMSs), activated alumina, silica gel, and activated carbon. The ZMS is crystalline, and the adsorption takes place within the pores formed by the crystal lattice. The ZMS differentiates species by size, shape, and polarity. Activated alumina and silica gel adsorbents are amorphous Al_2O_3 and SiO_2, respectively. Both types of adsorbent are used extensively for drying gases and liquids. Activated carbon adsorbents are microcrystalline, nongraphitic carbon with internal pores and large surface area ranging from 300 to 2500 m^2/g. Activated carbon adsorbents are most often used for removing organic compounds from liquid and gas streams.

Adsorption processes most often used in industrial applications are pressure-swing adsorption for gas purification, temperature-swing adsorption for gas or liquid purification, and inert-purge adsorption for gas or liquid purification. These three different processes are shown schematically in Fig. 5.

In the *pressure-swing adsorption* (PSA) cycle, a stream containing a small amount of an adsorbate at partial pressure P_1 is passed through the adsorbent column. The equilibrium loading X_1 is expressed in units of weight or moles of adsorbate per unit weight or volume of adsorbent. After equilibrium is reached, more feed is passed through the column (usually from the opposite direction) at an adsorbate partial pressure P_2. Desorption occurs and a new equilibrium loading, X_2 is established. The difference between X_1 and X_2 is known as the delta loading that determines the maximum removal capacity of the adsorbent column. After desorption, new feed is introduced at P_1 and the cycle continues. The time required to load, depressurize, regenerate, and repressurize a packed column is usually from a few seconds to a few minutes.

The *temperature-swing* cycle uses temperature to lower the equilibrium partial pressure of the adsorbate, thus regenerating the column. The time required to heat, desorb, and cool the adsorbent column for regeneration can range from a few hours to a few days. The long regeneration time makes the temperature-swing cycle unsuitable for removing high concentrations of adsorbates.

The *inert-purge* cycle uses an inert (nonadsorbing) fluid containing little or no adsorbate to lower the partial pressure or concentration of the adsorbate in the column. If enough inert fluid is passed through the column, the adsorbent can be depleted of adsorbate and completely regenerated. Thus, the equilibrium loading is X_1 at pressure P_1.

The selection of adsorption processes for a given task is often based on process economics as compared to various schemes of adsorption processes and competing processes such as distillation, absorption, and membrane separation.

3.4 Important Industrial Applications

Adsorption process is widely used in gas and liquid purification where small amounts of impurities are removed by adsorbents from feed streams. Drying of gas and liquid streams using adsorbents (ZMS, activated alumina, silica gel) can be found in almost all sectors of chemical, petrochemical, petroleum, natural gas, and pharmaceutical indus-

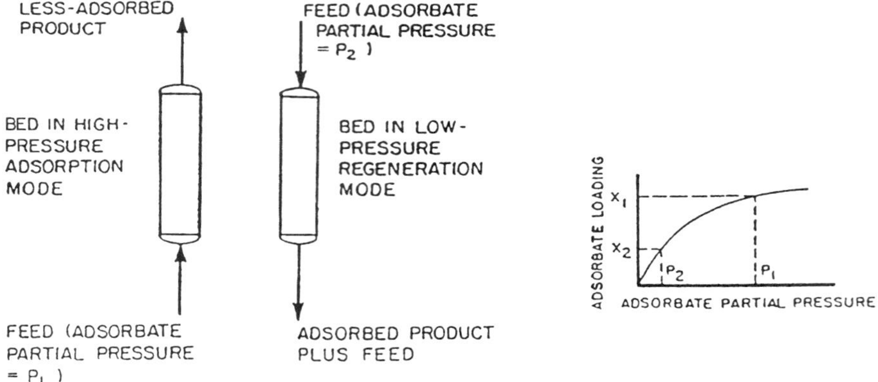

Pressure-swing adsorption cycle

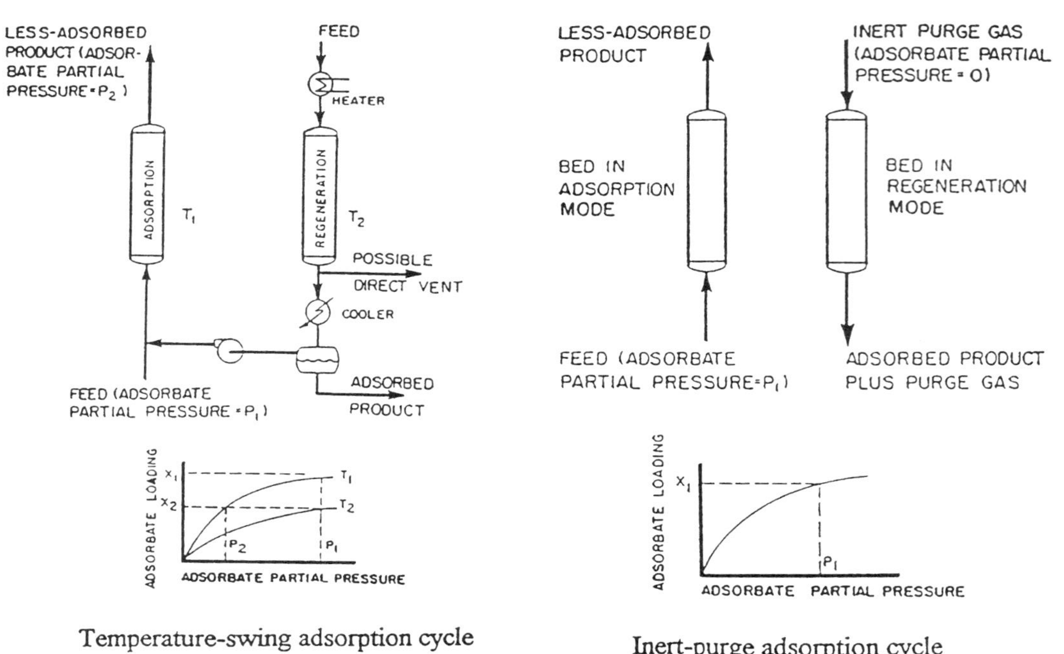

Temperature-swing adsorption cycle

Inert-purge adsorption cycle

FIG. 5. Typical adsorption process schemes [reprinted from Keller *et al.* (1987) with permission].

tries. Other large-scale applications include removal of volatile organic compounds (VOC) from air streams, removing SO_2 from effluent gases, and removing color and odor from water and organic liquid streams (Noll *et al.,* 1991).

A large-scale PSA process is used for O_2 and N_2 production using ZMS or using a type of adsorbent known as carbon molecular sieve. Examples of liquid bulk separation using adsorbents include separation of fructose and glucose, normal paraffin from aromatics and isoparaffins, and *p*-xylene from other C_8 aromatics (Ruthven, 1984).

4. ION EXCHANGE

4.1 Introduction and Process Description

Ion exchange is intimately related to adsorption in that both involve a fluid phase and a solid particulate adsorbent phase and both use the same types of process equipment (Streat and Cloete, 1987; Vermeulen *et al.*, 1984). Moreover, the basic fundamentals involved in liquid adsorption and ion exchange are similar. The adsorbents used in the ion-exchange process are often referred to as ion-exchange resins because these adsorbents are often made of cross-linked polymeric materials, such as polystyrene cross-linked with divinyl benzene (DVB) and polyacrylic acid cross-linked with DVB.

The processes used are batch process, fixed-bed (column) process, continuous process, and fluidized-bed (slurry-bed) process (Dorfner, 1991; Streat and Cloete, 1987). In all these processes, the electrolyte solution is brought into contact with the ion exchanger, and certain ions in the solution are replaced by ions from the ion-exchange material. Similarly, all processes involve not only an ion take-up step (also known as a service step) but also a regeneration step.

On account of this similarity between adsorption and ion exchange, we do not discuss the process equipment used for ion exchange but concentrate our discussion on ion-exchange resins and on applications of ion-exchange processes.

4.2 Ion-Exchange Resins

Ion-exchange resins are naturally divided into two categories, cation- and anion-exchange resins. These resins are microporous and have either cations or anions chemically bonded to the resin backbone. For cation-exchange resin, the anions are bound to the solids. For anion-exchange resins, the cations are bound to the solids. Since electric neutrality is always preserved, ions having the opposite charge to the fixed ions, known as counter ions, must be present in sufficient quantity. These counter ions are free to move in and out of the micropores of the ion-exchange resin either by diffusion or under the influence of an electrical field. They can be replaced by other ions of the same charge from solution in the ion-exchange process.

The concentration of fixed charges on the ion-exchange resin defines the ion-exchange capacity, which is often measured in milliequivalents (meq.) of charge per unit weight of dry resin. The ion-exchange process is always stoichiometric to preserve electroneutrality. The ion-exchange capacity is independent of the nature of the counter ions; e.g., a divalent counter ion will replace two univalent ions. The rate of the ion-exchange process is governed by the diffusion rate of the counter ions in the micropores and in the stagnant layer adjacent to the ion-exchange resin.

4.2.1 Polymeric Ion-Exchange Resins Polymeric ion-exchange resins are widely used in the industry. These resins essentially come from either cross-linked polystyrene or cross-linked polyacrylic acid with the desired functional groups attached to the polymer backbone for ion exchange.

Among cation-exchange resins, polystyrene sulfonic acid (PSSA) is by far the most common cation-exchange resin. It is made by sulfonating DVB cross-linked polystyrene (PS) in chlorosulfuric or concentrated sulfuric acid at about 80 °C. The sulfonic acid group may be placed in one of three positions on the benzene ring of PS, the most probable being meta or para. The resulting resin acidity depends on the position of the sulfonic acid group and on the matrix structure of the resin. The total capacity of the PSSA resin is about 4.0 to 5.5 meq/dry gram. Polystyrene sulfonic acid is a strong acid ion-exchange resin.

Polyacrylic acid cation-exchange resin is the second most commonly used resin. It is made by hydrolysis of cross-linked polyacrylates or polyacrylonitrile. The polyacrylic acid resin is characterized by higher capacity (about 8.0 to 13.0 meq/dry gram).

Anion-exchange resins are generally based on an amine or ammonium group. Strong base anion-exchange resins use quaternary ammonium compounds, whereas weak base products employ tertiary amine groups. The backbone polymers used to make the anion-exchange resins are exactly the same as those used for the cation exchangers. The PS polymer is made into the anion-exchange resin in two steps: cloromethylation and subsequent addition reaction with ammonium or primary, secondary, or tertiary amine. The common strong base anion-exchange resins are made from trimethylamine or diethanolamine. The available amine groups are numerous; therefore, a large number of anion-exchange resins can be made to suit the special requirements.

4.2.2 Inorganic Ion-Exchange Materials Inorganic ion-exchange materials differ from polymeric ion-exchange resins in that they may have extremely high selectivity for one ion or a group of similar ions, that they may be used at temperatures much higher than the operating temperature limit of the polymeric resins, and that they are stable under high radiation dose (Dorfner, 1991). The most commonly used inorganic ion-exchange materials are natural and synthetic zeolite; hydrous oxides—for example, zirconia and stannic oxide—; insoluble salts—for example, molyphosphate and zirconium phosphate—; ferrocyanides; and clays and pillared clays.

4.2.3 Other Ion-Exchange Materials There are other types of ion-exchange materials that do not fit into the above-mentioned two large categories—for example, liquid ion-exchange materials and ion-exchange membranes (Dorfner, 1991).

Liquid ion exchangers are formed by attaching ion-exchange groups to oligomers, uncross-linked polymers, or organic molecules. These materials are most often in the liquid state and insoluble in aqueous solutions but soluble in organic solvents. Liquid ion exchangers can be used where solid ion exchangers can be used. The largest industrial application for liquid ion exchangers is in the production of citric acid (Code of Federal Regulations, 1977).

4.3 Ion-Exchange Operation

As mentioned earlier, the ion-exchange operation can be divided into two steps: the exchange step and the regeneration step. There are four basic techniques used for the ion-exchange process, i.e., batch, fixed bed, continuous, and fluidized bed.

Batch operation involves contacting the ion-exchange material with the electrolyte solution in any suitable vessel with agitation until equilibrium. The ion exchanger is then separated from the solution by filtration, centrifugation, or settling. The separated ion exchanger is regenerated by mixing it with a suitable regeneration chemical.

The fixed-bed or column process involves packing the solid ion exchanger into a column and flowing the electrolyte solution in a downward or upward manner. The ions that are to be exchanged always contact the fresh ion exchanger until the ion-exchange capacity of the column is fully utilized and breakthrough occurs. At that point, the column is washed free of the feed electrolyte mixture with water, and elution and regeneration take place subsequently.

The continuous ion-exchange process employs either fixed-bed or fluidized-bed operations. When fixed-bed operation is used, a bank of ion-exchange beds is used, and each bed in the bank is in a different process step at any given time because feed solution to the bed changes with time.

The fluidized-bed process is similar to the batch process, except that the exchanger material is continuously cycled in and out of the main process step to be washed and regenerated.

4.4 Important Industrial Applications

Ion-exchange processes find their largest applications in water treatment, i.e., water deionization and water softening. More recently, ion-exchange processes have also been used for pollution control. Other important applications include sugar processing, noble-metal recovery from spent catalysts, leach solutions, uranium recovery from low-grade mineral leach solutions, and purification and recovery of pharmaceutical products.

4.4.1 Water Treatment Water deionization basically removes cation and anion species such as Ca^{2+}, Mg^{2+}, K^+, Na^+, and other metal ions using H^+ from a cation-exchange resin and Cl^-, SO_4^{2-}, NO_3^-, etc., using OH^- from an anion-exchange resin. The process generally uses a fixed bed of intimately mixed cation- and anion-exchange resins in the hydrogen and hydroxide forms, respectively. Ultrapure water produced by this process is used in electronic, chemical, and pharmaceutical industries.

Water-softening processes essentially remove Ca^{2+} and Mg^{2+} ions from water using Na^{+} from cation-exchange resins. Most cation-exchange resins can be used for this process because the selectivity for divalent ions is much higher. Regeneration of the resin is accomplished using concentrated sodium chloride solution.

Water desalination using ion-exchange membranes in an electrodialysis process is also practiced industrially in large scale.

4.4.2 Metals Recovery Ion-exchange processes have been used to recover various metals from dilute solutions. These metals include noble metals such as gold, silver, platinum, and palladium. Uranium ore upgrade from low-grade ore is practiced on a large scale using an ion-exchange process in conjunction with a liquid–liquid extraction process. In general, the ion-exchange process is highly suited for recovering high-value products from dilute streams.

4.4.3 Other Applications Ion-exchange processes are routinely used in effluent treatment to remove and recover pollutants. Typical examples include chromium removal from an electroplating effluent, ammonia and nitrate removal from groundwater, and cesium and strontium removal from radioactive wastewater. Ion-exchange processes are also used in sugar processing plants to soften sugar juices by removing scale-forming elements such as Ca^{2+} before concentration, to decolorize, to catalyze inversion of sucrose to fructose and glucose, and finally to separate fructose from glucose. Ion-exchange processes are also used extensively in the pharmaceutical industry for separation and analysis.

5. EXTRACTION AND LEACHING

5.1 Introduction and Process Description

The extraction and leaching processes are among the most utilized ones for separating one or more constituents from a mixture. This separation is based on the solubility difference of the components in a second phase formed by the extraction or leaching solvents (Skelland and Tedder, 1987; Robbins, 1984). The term extraction generally refers to the separation of components (solutes) from a liquid mixture by a second liquid (solvent) phase immiscible or only partially miscible with the original liquid mixture, whereas leaching refers to the separation from a solid mixture.

The extraction and leaching processes can be classified into physical and chemical processes. The physical process is based on difference in solubility as expressed by the partition coefficient. The chemical process depends on chemical changes or chemical interaction between the solute and the extraction solvent to increase the solute solubility in the solvent. The solute extracted can be separated from the solvent by distillation, evaporation, or back-extraction (stripping) with a different solvent. This allows the solvent to be recycled to the extraction/leaching process. Industrial success of extraction/leaching hinges on the economics of solvent and solute recovery.

5.2 Principles and Selection of Solvents

Extraction and leaching are also equilibrium-based processes. The thermodynamic principles suggest that in order for phases to be in equilibrium, the chemical potential of each species in all phases must be uniform. The partition coefficient m of a solute in the two liquid phases at equilibrium can be expressed as

$$m = y_A/x_A, \tag{8}$$

where y_A and x_A are the molar (or mass) fractions of species A in the two liquid phases.

Liquid–liquid extraction in practice rarely involves only three components. For such systems, if the two liquid phases are immiscible, the linear distribution law can be applied to solutes 1, 2, . . . i, separately:

$$m_i = y_i/x_i. \tag{9}$$

The ease of separation is represented by the selectivity or separation factor α, defined as the ratio of the partition coefficients:

$$\alpha_{AB} = m_A/m_B = y_A x_B/x_A y_B. \tag{10}$$

The chemical extraction process depends on a chemical change to effect mass transfer from the mixture to the chemical extractant. The transferring species can associate, dissociate, or react with the solvent. However, chemical and physical extraction processes have in common the need to maintain a high interfacial area for ultimate contact between the two phases and a quick and clean phase separation after the transfer of species has been completed. Chemical extraction finds many applications, especially in the separation of metals (Chapman, 1987). The metal ion of interest is converted selectively to an organometallic compound or complex soluble in an organic carrier solvent.

The selection of solvent depends on equilibrium considerations, physical properties, cost, ease of recovery, and safety and environmental concerns. Solvent selectivity is determined by physical and chemical equilibria and is the determining factor to the purity of the recovered products. A higher selectivity can reduce processing costs substantially by reducing solvent and solute recovery costs.

Physical properties such as viscosity, density, boiling point, and surface tension determine the extent of phase dispersion and separation, the solvent and solute diffusion rate, the wetting of solid surfaces, and recovery conditions. In chemical extraction and leaching, reaction conditions such as pH and redox potential must also be considered.

5.3 Mass-Transfer Coefficients

Equilibrium considerations define the ultimate extent of separation. However, the rate of the processes is determined by the mass-transfer considerations (Rydberg *et al.*, 1992; Skelland and Tedder, 1987). Accordingly, the mass-transfer rate depends on the interfacial contact area. For both liquid–liquid extraction and liquid–solid extraction, the mass flux at the interface is given via Fick's first law by the following equation derived on the basis of the film model:

$$J = D^*(C_s - C)/\delta, \tag{11}$$

where D^* is a diffusion coefficient; C_s and C are concentrations of solute at the interface and in the bulk liquid, respectively; and δ is the effective thickness of the liquid film at the interface.

The mass-transfer rate based on molecular diffusion through the interfacial film can be expressed phenomenologically by mass-transfer coefficients k_x and k_y and the concentration differences for each phase:

$$J = k_x(x_A - x'_A) = k_y(y'_A - y_A), \tag{12}$$

where x_A and y_A are the bulk concentrations of species A and x'_A and y'_A are concentrations of A adjacent to the interface. The concentration adjacent to the interface, in turn, is usually represented by the equilibrium partition relationship $y'_A = mx'_A$. Thus, an overall mass-transfer coefficient K_A can be defined and the mass flux expressed as

$$J = K_x(x_A - x^*_A), \tag{13}$$

where

$$\frac{1}{K_x} = \frac{1}{k_x} + \frac{1}{mk_y}; \tag{14}$$

x^*_A is the concentration of A at equilibrium with y_A. Equation (14) defines the overall mass-transfer coefficient and is known as the resistance additivity rule; i.e., the overall resistance to mass transfer is the sum of the resistances from the two phases. Furthermore, K_x is also a function of the partition coefficient m, which could be large depending on the nature of the solvent.

The rate of mass transfer depends strongly on agitation and temperature, and on the degree of mobility of the droplet surface. When two immiscible liquids are mixed together by agitation, a dispersion of droplets is created to increase the contact area. Increasing the degree of agitation results in smaller droplets, hence larger contact area, and higher mass-transfer rate.

5.4 Process Equipment

Continuous countercurrent extraction is usually carried out in a vertical column. The contactor is designed to disperse one phase into droplets and achieve the desired mass transfer in a given number of stages. The most commonly used contactor devices are mixer-settler, packed column, perforated-plate column, and various rotary agitated columns (Robbins, 1984; Skelland and Tedder, 1987).

The various packed and perforated-plate columns are similar to the ones discussed in previous sections, for distillation, absorption, adsorption, and ion exchange. For the packed column, the packing may be different for liquid–liquid extraction than for other purposes discussed earlier.

In addition, there are several types of mechanically agitated columns. The agitation can be provided by a centrifugal device, a stator-rotor type design, or a reciprocating-plate design. For example, the rotating-disk contactor (RDC) and the Karr reciprocating-plate column (RPC) are shown in Fig. 6.

Leaching is carried out in equipment designed to handle solid particles. The often used types of equipment are moving-bed percolation extractor and rotary-tray extractor (Wadsworth, 1987; Schwartzberg, 1987; Miller, 1984).

5.5 Extraction and Leaching Operations

There are three basic methods in extraction and leaching processes for contacting solvent(s) with the feed: concurrent, crosscurrent, and countercurrent.

Concurrent stagewise contact is not usually necessary because equilibrium is established or approached in stage 1. Therefore, a single-stage contactor is generally used, and the two phases are mixed to establish equilibrium and allowed to settle, after which the extracted phase containing solvent plus extracted solute and the raffinate phase composed of feed depleted of solute are separated. Single-stage leaching only differs slightly from single-stage liquid–liquid extraction in its operation. The extent of separation by a single-stage contactor is limited by equilibrium attainable under the operating conditions. To enhance the degree of separation, multistage contacting in a crosscurrent or countercurrent manner becomes necessary.

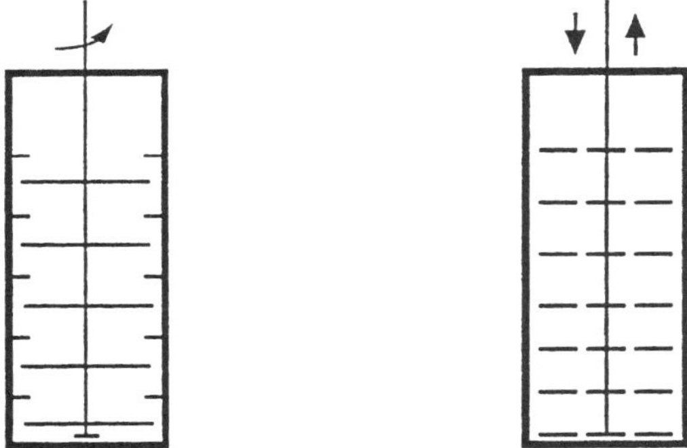

FIG. 6. Two types of mechanically agitated columns.

The crosscurrent arrangement in which fresh solvent is added at each stage enhances to a certain extent the separation obtainable for a given solvent-to-feed ratio over the single-stage arrangement. To increase further the degree of separation, the solvent-to-feed ratio has to be increased.

Countercurrent extraction represents the most efficient process for a given solvent-to-feed ratio. The feed and solvent flow countercurrently and emerge at opposite ends of the contacting stages (cascade). The maximum concentration driving force is maintained during the course of countercurrent flow. The feed having the highest solute concentration, entering stage 1, contacts solvent that has already passed through all other stages, whereas the raffinate leaving the last stage has been contacted with fresh solvent. Because of the economic advantages, continuous countercurrent operation is the choice for commercial-scale operations whenever possible.

5.6 Important Industrial Applications

Liquid–liquid extraction finds its largest applications in the petroleum industry for the separation of aromatic and aliphatic hydrocarbons, for removing aromatics from lubricating oils and jet fuels, and for desulfurization of crude oil. It is also used in large-scale production of anhydrous acetic acid and for polymer-grade ϵ-caprolactam.

Liquid–liquid extraction is also used as a major separation process in the pharmaceutical industry for the recovery of antibiotics and extraction of vitamins from natural products. For example, penicillin is extracted from fermentation broth using a batch extraction process and amyl acetate or n-butyl acetate as a solvent at low pH's from 2 to 2.5 and then back-extracted using an aqueous buffer solution at a pH of about 7 to 7.5.

The extraction of metals is also a large area of application for solvent extraction. Nuclear fuel processing is almost exclusively carried out using solvent extraction. For example, uranium is obtained from ore leach liquors by extraction with tributyl phosphate. Spent nuclear fuel is also reprocessed using the same process.

Leaching finds its largest application in food- and metal-processing industries—for example, the extraction of coffee from ground roasted coffee beans, oil from oil seeds, and natural flavor and color from nuts, leaves, and flowers, and in the metal-processing industry, the mining of gold by leaching with sodium cyanide solutions.

6. CRYSTALLIZATION

6.1 Introduction and Process Description

Crystallization is the process of producing crystals in the conversion of a chemical species from an amorphous solid, liquid, or gaseous state to the crystalline state. Crystallization is important as a thermal separation process because a large number of highly purified species, which are and can be marketed in the form of crystals, can be obtained from relatively impure solutions, melts, or the gaseous phase in a single processing step. Crystallization from a solution involving a solvent is called solution crystallization, whereas crystallization from a melt in the absence of a solvent is termed melt crystallization. Solution crystallization involves most industrial applications of the crystallization operation, and it is the subject of this section. For melt crystallization, the reader is referred to the literature (Moyers, 1984; Toyokura *et al.*, 1994).

Solution crystallization involves two distinct steps: crystal formation (nucleation) and crystal growth. The crystals must first form and then grow. The formation of a new solid phase in the bulk of a fluid phase without the involvement of a solid–fluid interface is called homogeneous nucleation; in the presence of surfaces, other than that of the crystal itself, such as those of the container, agitator, or foreign particles, is called heterogeneous nucleation; and in the presence of crystals of the crystallizing species itself is termed secondary nucleation. Crystal growth is the increase in size of the nucleus with a layer-by-layer addition of solute.

Both nucleation and crystal growth require that the solution be supersaturated. Supersaturation is a measure of the quantity of solute present in the solution compared with the quantity that would be present if the solution were kept for a very long period of time with the solid phase of the solute in contact with the solution. The latter quantity is the equilibrium solubility at the temperature and pressure under consideration. In general, supersaturation is expressed as a coefficient:

$$S = \frac{\text{parts solute/100 parts solvent}}{\text{parts solute at equilibrium/100 parts solvent}} \geq 1.0. \tag{15}$$

Generally, the rate of nucleation is directly proportional to the supersaturation coefficient, and it has been modeled (Randolph and Larson, 1971; Singh, 1979; Bennett, 1984; Mersmann, 1944b). Crystal growth involves two steps:

1. diffusion of the solute to the solution-crystal interface and
2. surface reaction to absorb the solute into the crystal lattice.

The crystal growth rate has also been modeled (Singh, 1979; Bennett, 1984; Mersmann, 1994b).

Supersaturation can be generated by the following five methods (Singh, 1979; Mersmann, 1994a):

1. evaporation: by evaporating the solvent from the solution;
2. cooling: by cooling the solution via heat exchange;
3. vacuum cooling: by vacuum flashing the solution adiabatically to a lower temperature, leading to inducing crystallization by simultaneous cooling of the solution and evaporation of the solvent;
4. reaction: by chemical reaction (e.g., molecular-sieve zeolites); and
5. salting out: by adding a third species to change the solubility relationship.

6.2 Process Equipment

According to the means of suspending the growing crystals, four basic types of crystallizers (Singh, 1979; Bennett, 1984; Mersmann, 1994a) may be classified:

1. Circulating magna crystallizers: All growing crystals are circulated through the zone of the crystallizer where supersaturation is generated. Circulating magna (slurry) crystallizers are the most important crystallizers in use today, and they include the forced-circulation (FC) and draft-tube-baffle (DTB) types. In the FC crystallizer shown in Fig. 7, the feed solution enters the circulating pipe below the product discharge at a point sufficiently below the free liquid surface to prevent flushing. The slurry of crystals together with the feed is circulated through a heat exchanger to raise its temperature by about 2 to 6 °C. The heated slurry enters the crystallizer slightly below the liquid surface to cause boiling. The vaporization of the solvent and the consequent cooling result in the supersaturation needed for nucleation and crystal growth. The DTB crystallizer is shown in Fig. 8. It has an inner baffle forming a partitioned settling area in a closed vessel. Inside the settling area is a top-entering agitator located near the bottom. The agitator operates at low speeds and gives axial flow. The draft tube prevents body swirl and minimizes turbulence in the circulating slurry.
2. Circulating-liquor crystallizers: In this type of crystallizer, only the liquor is circulated while the crystals are retained in the suspension vessel by the upflowing solution. By changing the liquor flow rate, the output and crystal-size distribution are controlled.
3. Scraped-surface crystallizers: The supersaturation in this type of crystallizer is generated by direct-contact heat exchange between the slurry and a jacket or double wall containing a cooling medium. The cooling medium circulates through the shell side, while an internal agitator fitted

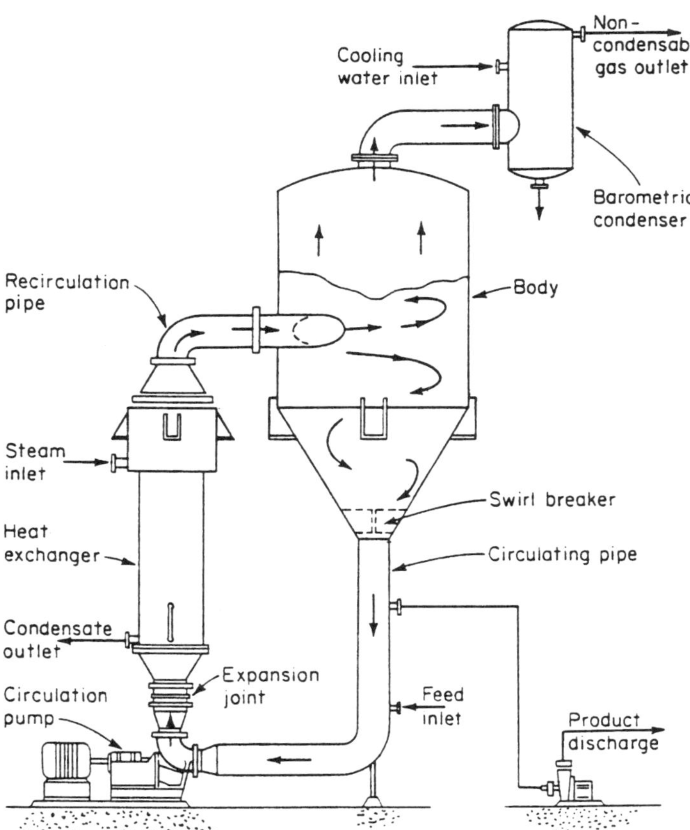

FIG. 7. Forced-circulation crystallizer [Swenson Division, Whiting Corporation, reprinted from Bennett (1984) with permission].

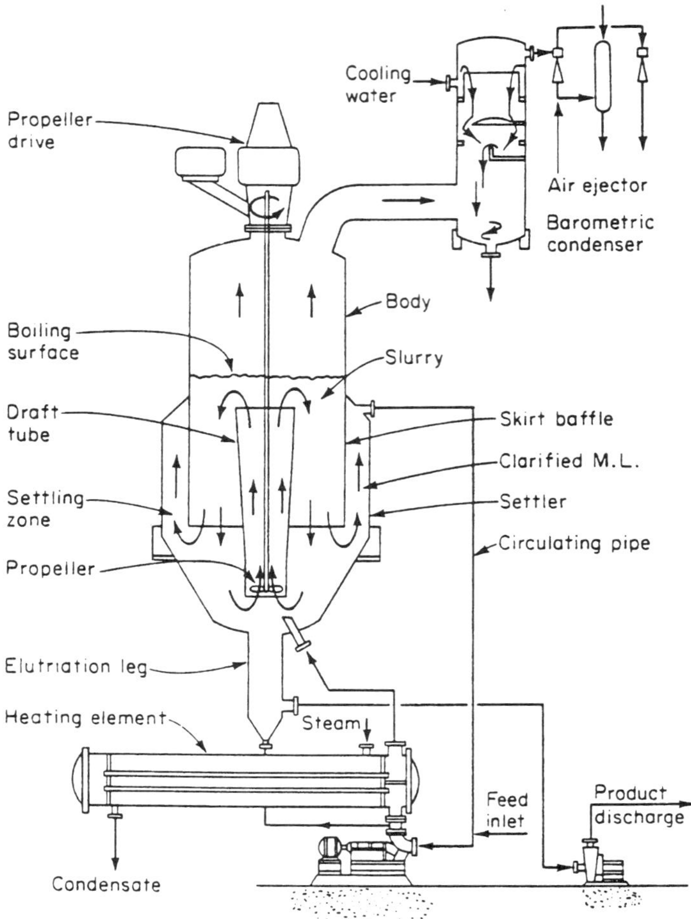

FIG. 8. Draft-tube-baffle crystallizer [Swenson Division, Whiting Corporation, reprinted from Bennett (1984) with permission].

with spring-loaded scrapers scrapes the heat-transfer surface to prevent solids buildup. These crystallizers are best suited for handling high-viscosity solutions.

4. Tank crystallizers: The hot feed solution with the solute(s) of normal solubility is pumped into the crystallizer and allowed to cool for supersaturation to occur either by natural convection and radiation or by surface cooling through coils in the tank or in a jacket on the outside of the tank. This type of crystallizer is generally used for small-capacity applications including fine chemicals, pharmaceuticals, and food products in which the liquor, not the crystals, is generally the desired product.

6.3 Crystallizer Operations

The key operations are maintaining crystallizer stability, controlling crystal-size distribution, minimizing fouling, and separating solids from the liquor (Singh, 1979; Mersmann, 1994a). Steadiness of operation is much more important in crystallization by maintaining crystallizer stability than in other separation processes. Crystal growth is a layer-by-layer process, and the retention time required in most commercial crystallizers to produce crystals of the size normally desired is on the order of 2 to 6 h. On the other hand, nucleation can be generated in a fraction of a second. The influence of any upsets in operating conditions, in terms of the excess nuclei produced, is very short term in comparison with the total growth period of the crystal product. In general, four to six retention periods will pass before the effects of any upset will be damped out. Thus, the recovery period may be from 8 to 36 h.

The keys to controlling crystal-size distribution are the rate of nucleation and slurry density. The rate of nucleation required to produce crystals of a given size decreases exponentially as the crystal size increases. For example, 1 lb of 150-mesh seed crystals is enough to generate 1 ton of 14-mesh product crystals. Thus, extreme care must be exercised to prevent prenucleation in the incoming feed stream or unwanted nuclei being introduced through recycle streams. The slurry density is primarily controlled by regulating the flow rate through the crystallizer.

Fouling (Krause, 1993; Mersmann, 1994b) can be controlled to a great extent by determining the optimum slurry density and circulation rates. Variations in liquor purity affect fouling and must be controlled.

The product stream from crystallizers generally contains less than 30 wt % solids and is usually thickened to 50 wt % or greater prior to the solid–liquid separation step, for which either conical settlers or hydrocyclones can be used. In general, the final solid–liquid separation is achieved by filters or centrifuges. Filters are preferred for high-capacity applications, and a separate washing step is needed, whereas centrifuges are used for a more thorough separation of solids and liquor.

6.4 Important Industrial Applications

Important industrial applications include the manufacture of the following products:

1. Industrial chemicals: Ammonium sulfate, sodium and potassium chlorides, potassium nitrate, trisodium phosphate, sodium ferrocyanide, and sodium tetraborate are produced via crystallization. For these products, circulating-magna crystallizers are typically used for the production of large crystals. For ammonium sulfate, circulating-liquor crystallizers are sometimes employed for the production of small, uniform crystals (Singh, 1979; Mersmann, 1994a).
2. Inorganic chemicals: Gypsum, sodium nitrate, ammonium nitrate, silver nitrate, urea, sodium sulfate, inorganic chlorates, dichromates and sulfites are produced in small, uniform crystals by the use of circulating-liquor crystallizers (Singh, 1979; Mersmann, 1994a).
3. Organic compounds: Organic dyes, *p*-xylene, chlorobenzene, and organic acids (benzoic, propionic, butyric, adipic, sorbic, etc.) are processed via scraped-surface crystallizers (Singh, 1979; Mersmann 1994a).
4. Food products: Vegetable oils, sugar, and fatty acids are prepared in batch operations via tank crystallizers (Singh, 1979; Mersmann, 1994a).
5. Molecular-sieve zeolites: The zeolites are processed via circulating-liquor or tank crystallizers (Singh, 1979).

7. FILTRATION

7.1 Introduction and Process Description

Filtration is the process of removing particles from a fluid by forcing the fluid through a filtering medium (septum), on which the particles are deposited. The fluid passing through the septum is called filtrate, and it may be a liquid or a gas. The solid particles may be coarse or fine, rigid or soft (or slimy), separate individuals or aggregates. The valuable product is either the fluid, the solid, or both.

The operation of filtration depends on the mechanism for arrest and accumulation of particles, and it can be divided into two classes, deep-bed filtration and cake filtration (Jacobs, 1984; Chiang and He, 1993). Deep-bed filtration is usually preferred when the solids content of the slurry is very low (about 0.1%). A deep bed of porous media such as diatomaceous earth (precoat) or fibrous material supported on a coarse filter is used to remove fine particles from the slurry. The particles to be removed are often considerably smaller than the pores of the filter medium and can penetrate a considerable depth before they are captured. The particles can be captured via the following mechanisms: sieving, impingement, interception at solid–liquid interfaces, gravity settling, Brownian diffusion, and electrostatic forces

(Grace, 1956; Tien and Payatakes, 1979; Chiang and He, 1993).

Most liquid filtration processes operate on the mechanism of cake filtration. The filtered solids are deposited on the surface of the filter medium, and the solid cake thus deposited quickly becomes the actual filter medium.

For the relationship between pressure drop and filtration rate in cake filtration and in deep-bed filtration through a precoat filter, the Kozeny–Carman equation can be used in the following form (Donald, 1958; Chiang and He, 1993; Pinheiro and Cabral, 1993a):

$$\frac{\Delta p}{L} = \frac{k\mu u(1-\epsilon)^2(s_p/V_p)^2}{\epsilon^3}, \tag{16}$$

where Δp is the pressure drop across the cake at thickness L, μ is the viscosity of the filtrate, u is the velocity of the filtrate based on the filter area (the volumetric flux of the filtrate), ϵ is the porosity of the cake, s_p is the surface area of a single particle, V_p is the volume of a single particle, and k is a constant. The constant k has a value of 5 for uniform, spherical particles, and it ranges between 3.5 and 5.5 for other particles. Wide particle size distributions tend to decrease k, whereas high porosity ($\epsilon > 0.8$) tends to increase it.

7.2 Process Equipment

As shown in Eq. (16), liquid flows through a filter medium via a pressure differential across the medium. The pressure differential is achieved by application of a pressure on the slurry side or a vacuum on the filtrate side. Most industrial filters are either pressure filters or vacuum filters. The filters include the following types (McCabe and Smith, 1956; Jacobs, 1984; Pinheiro and Cabral, 1993a).

1. Filter press: A filter press is a discontinuous pressure filter, and it is commonly in the plate-and-frame type shown in Fig. 9. The filter press consists of alternating plates and frames mounted on two parallel horizontal bars and clamped together. The filter cloths are stretched against the plates, which have channeled surfaces to allow filtrate drainage. The slurry material under pressure is fed into the hollow frame space, and the filtrate passing through the filter cloth drains via the channels on the plate. The cake is accumulated inside the frame until it fills the frame space. Washing the cake can be done by introducing the washing fluid through washing plates. The washing fluid flows through the entire cake thickness. The cake is then discharged by separating the plates and frames.
2. Rotary-drum filter: This is the most common type of continuous vacuum filter, shown schematically in Fig. 10 (Davis and Grant, 1992). A horizontal cylindrical drum (with a radius r_d) covered by a filtering medium (e.g., canvas cloth) turns at a constant angular velocity Ω (0.1 to 2 rpm), and it is partially submerged in an agitated slurry trough containing the suspension (slurry) to be filtered. To the interior space of the drum, a vacuum is applied. As the drum dips under the surface of the suspension (with an angle Θ_f subtended by the portion of the drum's surface that is in contact with the suspension), the vacuum sucks permeate perpendicularly through the drum surface. A layer of solids, i.e., the cake, builds up on the filter medium as the filtrate is sucked through the filter medium. As the cake leaves the suspension, it enters the washing and drying zones, in which the applied vacuum draws wash liquid and air through the cake. When the cake leaves the drying zone, the cake is discharged normally by scraping off with a doctor blade. Once the cake is removed, that part of the drum reenters the suspension and the operation cycle repeats. In addition, applying pressure on the suspension can be achieved by enclosing the filter in a shell.
3. Precoat filter: This is a modified rotary-drum filter with a layer of filter aid, e.g., diatomaceous earth, applied to the filter surface. The cake discharge mechanism is an advancing scraper knife to remove a thin layer of the precoat along with the cake in order to present a clean surface to the slurry. This filter is particularly useful for filtering small amounts of fine or gelatinous solids and deformable flocs, which normally plug a filter medium.
4. Cartridge filter: This type consists of a re-

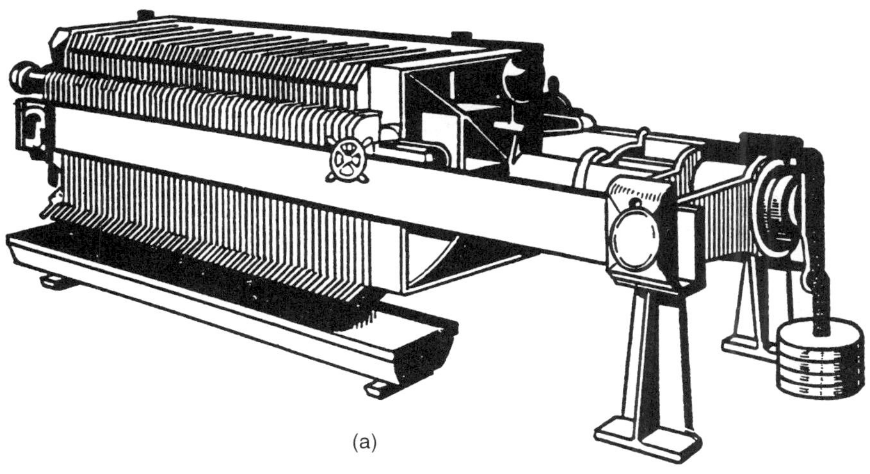

(a)

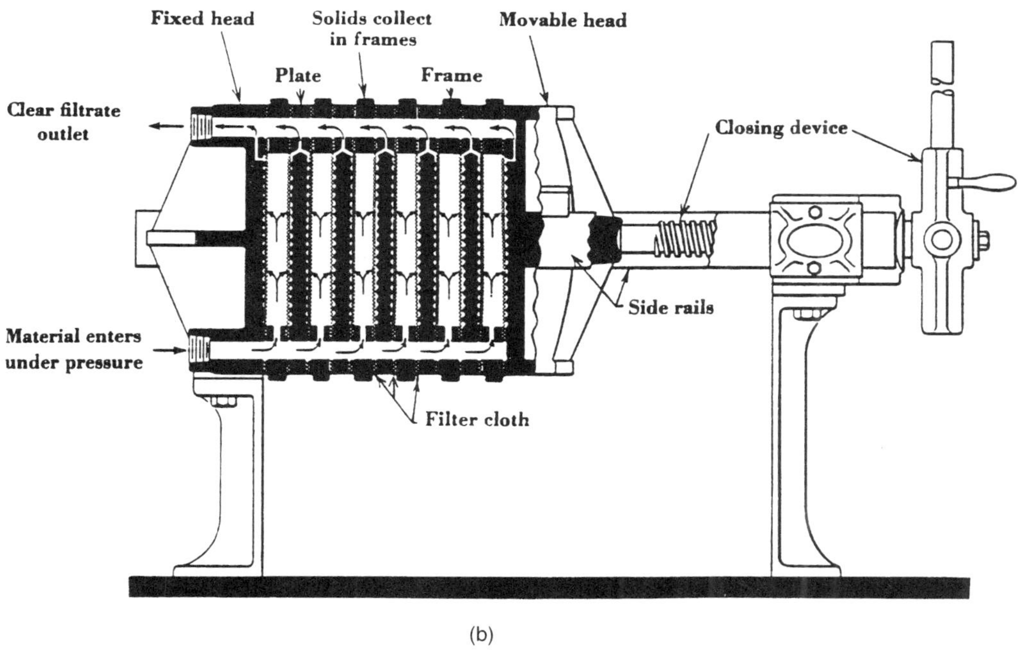

(b)

FIG. 9. Plate-and-frame filter presses: (a) outside view [Eimco Process Equipment Co., reprinted from Jacobs (1984) with permission] and (b) inside view [T. Shriver and Co., Inc., reprinted from Bennett and Meyers (1982) with permission].

newable filter element of fiber, resin-impregnated filter paper, or porous stainless steel of controlled porosity. The filter is usually used for removing small amounts of solids from process fluids, and it is placed in a line carrying the fluid to be clarified; i.e., clarification occurs while the fluid is in transit.

In all filters, filtering media (septa) are used to effect fluid-particle separations. In industrial applications, the most common fil-

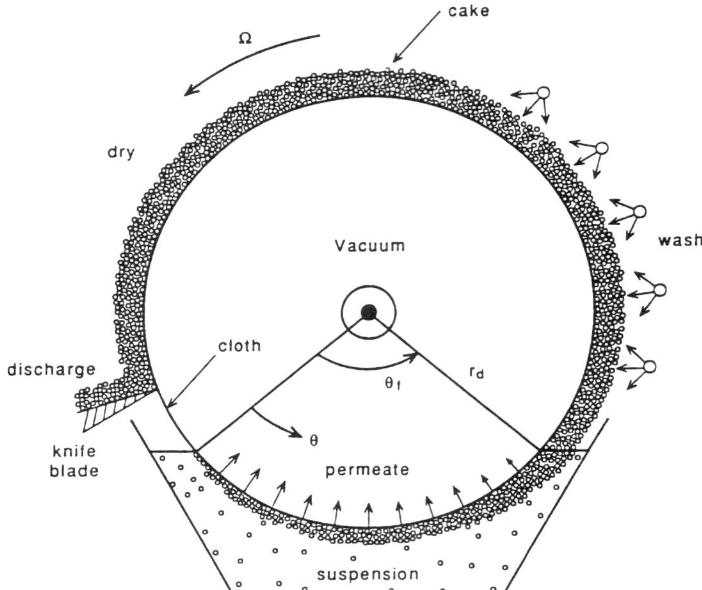

FIG. 10. Rotary-drum filter [reprinted from Davis and Grant (1992) with permission].

tering medium is canvas cloth. Other filtering media such as stainless steel cloth, monel metal cloth, glass cloth, and paper are used for corrosive fluids. Some synthetic fabrics, e.g., polyvinylidenechloride, nylon, and polyester, are also highly chemically resistant. Materials used in rigid porous media include sintered stainless steel and other metals, alumina, silica, porcelain, graphite, and some plastics. These media offer a wide range of chemical and temperature resistances, and they are mainly used for clarification.

Filter aids, e.g., diatomaceous earth, perlite (aluminum alkali silicate), and cellulosic fibers, are frequently used for filtration operations with problems of slow filtration rate and rapid medium plugging. The filter aid is mixed with the slurry before filtration, and it forms a highly permeable filter that can trap very fine or gelatinous particles and deformable flocs.

7.3 Filtration Operations

Filtration operations may be classified into two categories: batch and continuous. In batch operations, filter presses and cartridge filters are generally used, whereas in continuous operations, rotary-vacuum filters and precoat filters are commonly employed. The operations may include the following three modes:

1. constant-pressure filtration is the most frequent mode of operation, in which compressed air or vacuum is used as the driving force,
2. constant-rate filtration is usually achieved by the use of positive displacement pumps, and
3. variable-pressure, variable-rate filtration is a result of using centrifugal pumps (McCabe and Smith, 1956; Jacobs, 1984; Pinheiro and Cabral, 1993a).

As shown in Eq. (16), the rate of filtration can be improved by

1. decreasing the viscosity of filtrate: this can be achieved by heating the slurry or diluting it with a fluid of lower viscosity;
2. increasing the pressure drop across the filter cake, but there is an upper limit on the pressure drop in vacuum filtration, and the pressure drop can increase the filter cake resistance; and
3. decreasing filter cake resistance, i.e., increasing the porosity of the cake and decreasing the thickness of the cake: increasing the porosity of the cake can be

achieved by the use of filter aids or by coagulation via using flocculants or changing the pH of the slurry, whereas decreasing the cake thickness can be done by employing moving blades to remove the filter cake mechanically (Chiang and He, 1993).

The filtrate retained in the pores of the filter cake is removed via washing. The maximum wash ratio between wash fluid volume and cake void volume is usually 1.5–2.0 for removing about 90% of the retained material (Pinheiro and Cabral, 1993a).

7.4 Important Industrial Applications

The industrial applications include the following (McCabe and Smith, 1956; Gregor *et al.*, 1992; Pinheiro and Cabral, 1993a; Combest and Corporation, 1994):

1. Filtration and washing of pigments and dyes: Plate-and-frame filter presses are usually used for these applications in batch operation.
2. Dewaxing of petroleum: This is carried out by the use of plate-and-frame filter presses.
3. Pharmaceutical processes for the recovery of fermentation broths, ascorbic acid, and antibiotics (albamycin, erythromycin, penicillin, streptomycin, etc.): For these processes, vacuum precoat filters are typically employed in continuous operation.
4. Dewatering of sludges: This is conducted via rotary-drum pressure filters in continuous operation.
5. Filtration and washing of fine crystals: Rotary-drum vacuum filters are usually employed in continuous operation.
6. Clarification of process liquids and cooling water: Cartridge filters are typically used in batch operation.
7. Automotive and truck filters: Cartridge filters are generally employed for fuel injection and air intake.

8. Sedimentation

8.1 Introduction and Process Description

Sedimentation, similar to filtration, is the process of separating a fluid from particles suspended in it by allowing the particles to settle out from the fluid. The fluid above the bed of solid particles is called supernatant, and it can then be removed from the particles. In the beginning of the sedimentation process, the particles in a slurry (suspension) behave independently. But as the particles fall closer to the bottom of the settling vessel, the influences of neighboring particles become pronounced, which is called hindered settling. To remove relatively large, heavy particles such as coarse sands, which have reasonably high settling velocities, gravity sedimentation under free or hindered settling is satisfactory. However, to remove fine particles, flocculation to increase the effective diameter of the settling particle and centrifugation to replace the force of gravity can be used to enhance the fluid-particle separation (Baczek *et al.*, 1988; McCabe and Smith, 1956; Pinheiro and Cabral, 1993b).

Sedimentation may be divided into two functional operations: thickening and clarification, both operations taking place simultaneously in sedimentation. However, the primary objective of thickening is to increase the concentration of suspended particles in a slurry stream, whereas the main purpose of clarification is to remove relatively small amounts of particles and produce a clear effluent. In general, thickeners are designed for the heavy-duty requirements imposed by highly concentrated slurry, whereas clarifiers have features to ensure essentially complete removal of particles, including greater depth, special provision for coagulation and flocculation, and greater overflow-weir length (Dahlstrom *et al.*, 1984).

8.2 Process Equipment

As described above, sedimentation equipment includes thickeners and clarifiers. Industrially important thickeners and clarifiers are for continuous operations.

8.2.1 Thickeners There are two types of continuous thickeners: conventional and high-rate, both with circular horizontal cross section since the rectangular alternative tends to give higher maintenance costs and poor performance for high underflow concentrations (Baczek *et al.*, 1988; Dahlstrom *et al.*, 1984; Pinheiro and Cabral, 1993b). A conventional thickener is shown in Fig. 11. The slurry feed is introduced to the center of

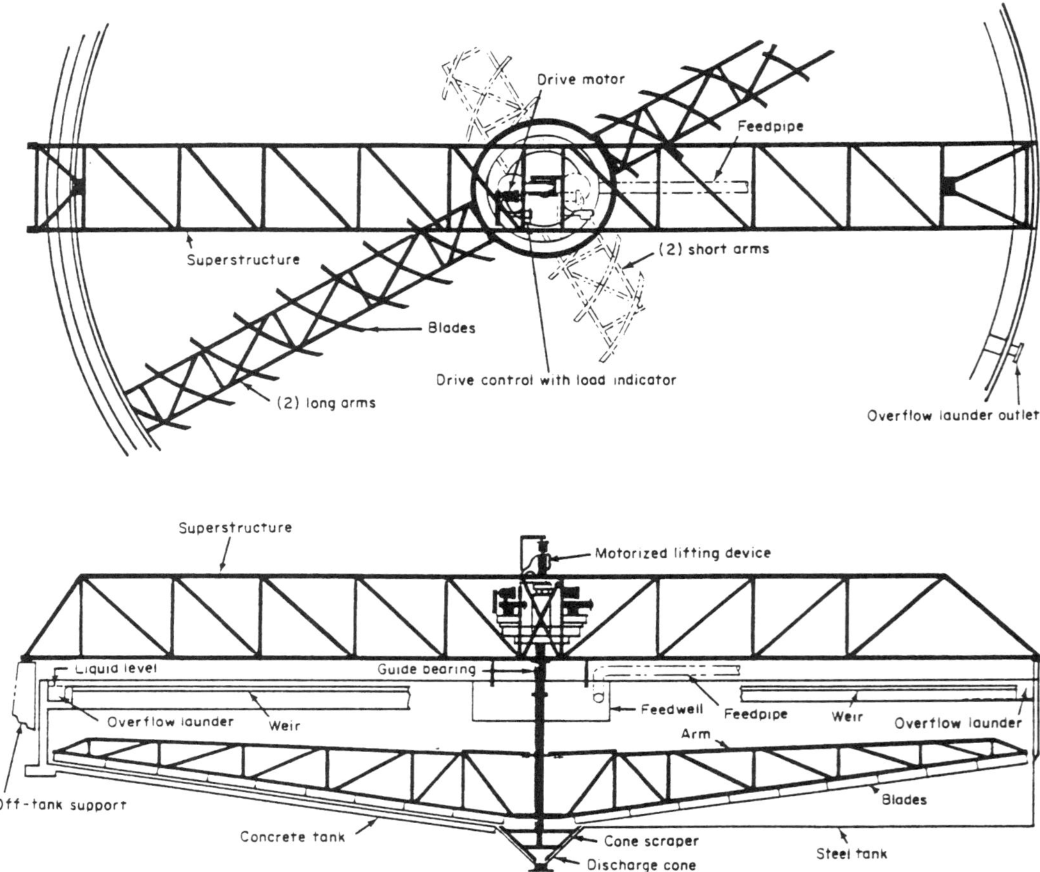

FIG. 11. Conventional thickener [Eimco Process Equipment Co., reprinted from Dahlstrom *et al.* (1984) with permission].

the thickener in a pipe or an open launder and enters the feed well. The feed well is designed to minimize the turbulence caused by the feed entry and to introduce the feed below the clear-liquid surface. The solid particles settle to the bottom where they are directed to the underflow part by a revolving raking mechanism, while clear liquid is removed from the top of the thickener in a peripheral launder. The conventional thickener can be used with or without flocculants. If flocculants are utilized, they are usually added to the feed launder or in the feed well.

A high-rate thickener is designed specifically to maximize the efficiency of flocculation. As shown in Fig. 12, the feed stream enters the feed well, and the flocculant is added also at the feed well but at various locations to optimize mixing with the feed stream. A mechanical mixer is used to disperse the flocculant. A deaeration step must be employed to remove entrained air from the feed stream so that the turbulence in the thickener can be minimized and the flotation of solid particles will not occur. This thickener can decrease the equipment area and volume by a factor of about 5 in comparison with the conventional thickener.

8.2.2 Clarifier Clarifiers are basically identical to thickeners in design and layout except that the former equipment uses a structure of lighter construction and a drive head with a lower torque capability (Baczek *et al.*, 1988; Dahlstrom *et al.*, 1984; Pinheiro and Cabral, 1993b; Springer, 1993). These

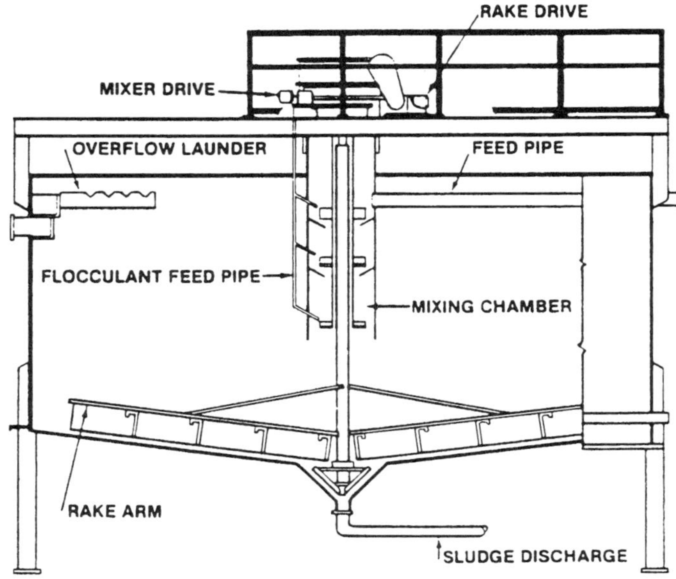

FIG. 12. High-rate thickener [Eimco Process Equipment Co., reprinted from Dahlstrom *et al.* (1984) with permission].

differences are allowed because in clarifiers the thickened pulp (concentrated solid particles) produced is considerably less in volume and lower in concentration. Both circular and rectangular horizontal cross-section geometries are used. As shown in Fig. 13 for a rectangular clarifier, the influent (feed) enters along one of the short sides, and the clear liquid overflows along the opposite side. The chain-type drag moves the deposited pulp to a sludge hopper located at the influent end by means of scrapers fixed to the chains. During their return, the scrapers travel near the liquid level and thus act as skimming devices for removing surface scum.

8.3 Sedimentation Operations

Batch sedimentation operation allows the slurry (suspension) to settle for the required period of time, after which the sludge and

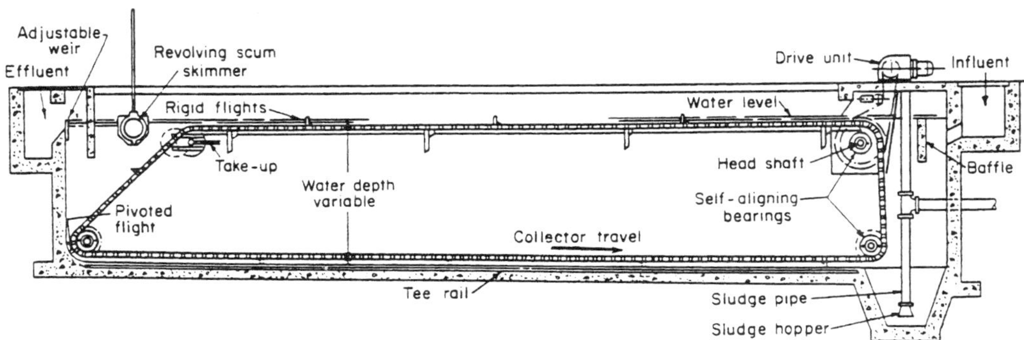

FIG. 13. Rectangular clarifier [Rexnord Inc., reprinted from Dahlstrom *et al.* (1984) with permission].

supernatant are removed separately. In this operation, the retention time is the major design parameter. But this operation is expensive in labor requirements and limited in capacity. Thus, industrial applications are mainly in continuous operation (Dahlstrom *et al.*, 1984; Pinheiro and Cabral, 1993b).

In continuous operation, a solids vertical concentration profile in the thickener or clarifier is kept constant to ensure that both thickening and clarification occur (Dahlstrom *et al.* 1984; Pinheiro and Cabral, 1993b). Such a concentration profile is shown schematically in Fig. 14 (Pinheiro and Cabral, 1993b). In the local thickening (compression) zone, a scraper or a rake is used to direct the sludge to the bottom outlet and to aid in solids compression. The solids content and the degree of thickening depend on the time of retention of the solids in the thickening zone. The retention time is, in turn, proportional to the depth of this zone, which can be increased only at the expense of a deeper and more costly thickener. A clear overflow effluent can be obtained if the upward flow velocity of the liquid in the dilute zone is less than the minimum terminal velocity of the solid particle. The liquid velocity is directly proportional to the overflow rate. If the overflow rate is too high, a cloudy overflow liquid will result. Thus, the overflow rate needs to be controlled in order to obtain a clean overflow effluent (McCabe and Smith, 1956; Pinheiro and Cabral, 1993b).

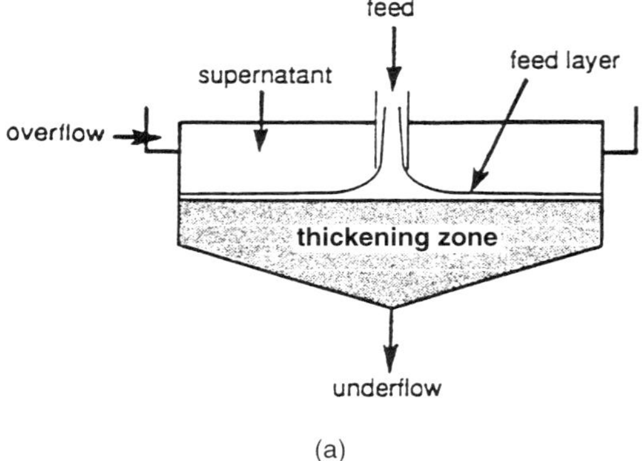

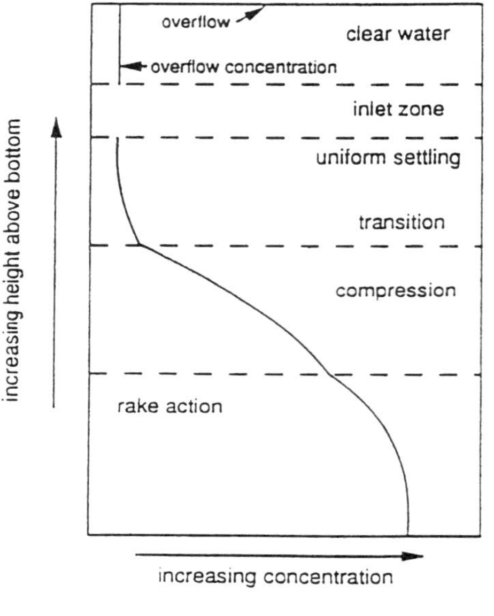

FIG. 14. Schematic of a solids vertical concentration profile: (a) thickener and (b) solids vertical concentration profile [reprinted from Pinheiro and Cabral (1993) with permission].

8.4 Important Industrial Applications

The applications include the following (Dahlstrom *et al.*, 1984; Pinheiro and Cabral, 1993b):

1. Municipal wastewater treatment: Both thickeners and clarifiers are used for these applications. Thickeners are employed for primary sludge, waste-activated sludge, and anaerobically digested sludge.
2. Drinking-water treatment: Clarification is normally carried out after 30-min. flocculation. It is also conducted with softening treatment using lime.
3. Metallurgical operations: Sedimentation is used in processing the concentrates and tailings of ores including copper, iron, lead, nickel, zinc, etc.
4. Pulp and paper: Clarifiers are used for processing green and white liquors. Thickeners are employed for handling kraft waste, deinking waste, and paper-mill waste.
5. Flue-gas desulfurization sludge: For treating the sludge, sedimentation is used.
6. Oil–water separations in refineries: For preliminary oil–water separations in refineries, rectangular clarifiers are usually employed.

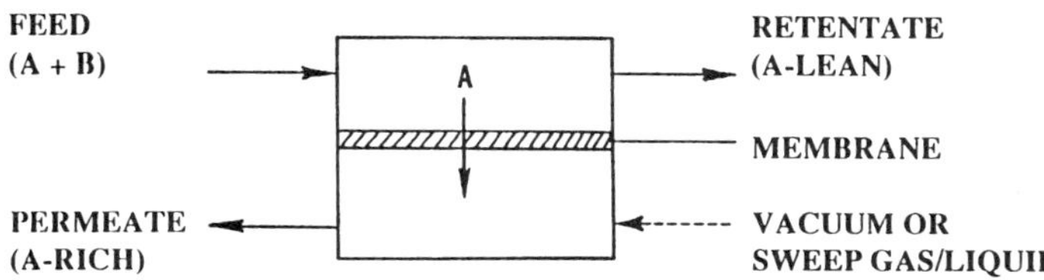

FIG. 15. Schematic of a membrane process.

9. MEMBRANE SEPARATIONS

9.1 Introduction and Process Description

A membrane process, shown schematically in Fig. 15, requires two bulk phases physically separated by a third phase, the membrane. In all membrane processes, the feed is separated into a stream that goes through the membrane, the permeate, and a portion of the feed that is retained by the membrane, the retentate (the reject or the concentrate). In some processes [e.g., pervaporation and vapor (gas) permeation], a vacuum is applied in the permeate side, whereas in a few processes (e.g., dialysis, electrodialysis, and emulsion liquid membrane), a sweep, wash, or strip stream is needed on the permeate side. The membrane controls the transport of mass between the two bulk phases. One of the species in the feed mixture, i.e., *A* in this figure, is allowed to pass through the membrane in preference to others, that is, the membrane is selective to species *A*. The permeate phase is enriched in the species *A*, while the retentate phase is depleted of it (Ho and Li, 1984; Ho and Sirkar, 1992a, 1992b).

The membrane is either a homogeneous phase or a heterogeneous collection of phases, and it may be any one or a combination of the following: nonporous solid, microporous or macroporous solid with a fluid (liquid or gas) in the pores, a liquid phase with or without a second phase, or a gel. The membrane phase is almost always thin when compared with the dimensions of the bulk phases in at least two other directions. Hollow fibers and emulsion liquid membranes provide exceptions where the membrane thickness is of the order of the dimensions of one of the bulk phases (Ho and Sirkar, 1992b; Lonsdale, 1989). In general, a commercial membrane consists of a thin, selective, active layer or skin (about 0.1–5 μm) on a porous support layer (about 30–200 μm), which provides mechanical strength. If the active and support layers are formed in a single operation from a given material, the membrane is referred to as having an integrally skinned structure. If the active layer is a coating on the support layer, or if a coating is put on the active layer, the membrane is said to have a composite structure. Two geometries of membranes are usually used, namely, flat films and hollow fibers (outside diameters of about 50–1000 μm and walls of about 10–100 μm thick). Figure 16 shows an integrally skinned hollow-fiber membrane (Kesting, 1985).

The transport of any species across the membrane is caused by one or more driving forces. These driving forces arise from a gradient of chemical potential or electrical potential. A gradient in chemical potential may be due to concentration gradient or pressure gradient or both. The transmembrane flux of

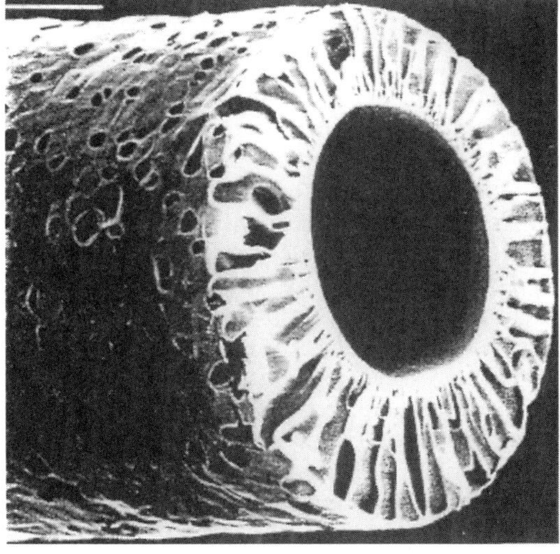

FIG. 16. Integrally skinned hollow-fiber membrane with a dense, active layer on a porous support structure [Amicon Corporation, reprinted from Kesting (1985) with permission].

any species per unit driving force is proportional to the permeability of the species. If the driving force is described by the use of a partial pressure difference (Δp_i) or a concentration difference (Δc_i) across the membrane for species i (if there is an osmotic pressure difference $\Delta\pi$, the pressure gradient driving force is $\Delta p_i - \Delta\pi$), then

$$\text{Transmembrane Flux of Species } i = \left(\frac{\text{Permeability of Species } i}{\text{Effective Membrane Thickness}}\right)(\Delta p_i \text{ or } \Delta c_i). \tag{17}$$

The ratio of permeability of species i to effective membrane thickness is sometimes called the permeance (which is also called permeability in some special processes, e.g., dialysis and microfiltration). The membrane selectivity between any two species can be defined in a number of ways. A common definition called separation factor α_{ij} for two species i and j is

$$\alpha_{ij} = \frac{c_i''/c_j''}{c_i'/c_j'} \tag{18}$$

where the prime and double prime superscripts denote the upstream bulk phase (feed or retentate) and the downstream bulk phase (permeate), respectively. The separation factor is equal to the ratio of the permeabilities of the two species under conditions where the downstream pressure or concentration is negligible in comparison with the upstream pressure or concentration. Such selectivities are reduced or destroyed if the membrane has gross defects through which nonselective hydrodynamic flow of one bulk phase into the other bulk phase occurs.

There have been eight commercialized membrane separation processes; they are gas permeation, pervaporation, dialysis, electrodialysis, reverse osmosis, ultrafiltration, microfiltration, and emulsion liquid membrane (the nanofiltration process is included in reverse osmosis) (Ho and Sirkar, 1992a, 1992b; Belfort, 1984; Bungay *et al.*, 1986; Rautenbach and Albrecht, 1989; Porter, 1990). Gas permeation involves gases on the high-pressure side of the membrane permeating through the membrane to its low-pressure side. The membranes currently used in most commercial gas separations are polymeric. The transport mechanism through polymeric membranes is solution-diffusion, and it consists of three steps:

1. solution of the permeating molecules from the feed at the upstream side of the membrane,
2. diffusion of these molecules through the membrane, and
3. desorption at the downstream side of the membrane to become the permeate (Ho and Li, 1984; Zolandz and Fleming, 1992; Koros and Pinnau, 1994; Paul and Yampol'skii, 1994).

In a pervaporation process, a liquid feed contacts one side of a membrane while a vacuum is drawn on the other side of the membrane to produce a permeate vapor, which is subsequently condensed to a liquid. The pervaporation membranes currently employed in most commercial applications are polymeric, and the transport mechanism through the membranes is solution-diffusion. The separation of feed components at given operating conditions is based on the selectivity of the membrane employed. In turn, the membrane selectivity comes from the different solubilities and diffusivities of the components in the membrane (Huang, 1991; Fleming and Slater, 1992; Wijmans and Baker, 1993).

Dialysis is the transfer of solute molecules across a membrane by diffusion from a concentrated solution to a dilute solution. A simultaneous diffusion of solvent molecules through the membrane occurs in the opposite direction—a phenomenon called osmosis. The ratio of mass of water transported to mass of dissolved solute dialyzed is defined as the water-transport number. The membranes used can be either porous or nonporous. The remaining solution on the feed side is called retentate, and the solution on the other side to receive the solutes to be removed from the feed is called dialyzate. Separation of solutes in dialysis is due to differences in their diffusion rates. The process of mass transfer from the feed compartment to the dialyzate compartment involves diffusion through a liquid film on each side of the membrane as well as through the membrane itself. Diffusion of solute molecules is governed by a number of factors, such as the inherent mobility of the molecules and the restrictive effect or drag exerted by the membrane pores on both solute and solvent molecules (Ho and Li, 1984; Kessler and Klein, 1992).

The electrodialysis process uses an electrodialysis stack such as shown schematically in Fig. 17 consisting of a series of anion- and cation-exchange membranes arranged in an alternating pattern between an anode and a cathode to form individual cells (Strathmann, 1992). A cell is composed of a volume with two adjacent membranes (about 0.5–2 mm between two membranes), i.e., anion- and cation-exchange membranes. If an ionic solution, e.g., an aqueous salt solution, is pumped through these cells and an electrical potential established between the anode and cathode, the negatively charged anions move toward the anode while the positively charged cations migrate toward the cathode. The anions pass easily through the positively charged anion-exchange membrane but are rejected by the negatively charged cation-exchange membrane. Similarly, the positively charged cations migrate through the cation-exchange membrane but are retained by the anion-exchange membrane. The overall result is an increase in the ion concentration in alternate compartments, while the ion concentration in the other compartments decreases. The solution with the reduced ion concentration is usually called the diluate, and the concentrated solution is the brine or the concentrate. The driving force for the ion transport is the applied electrical potential between the anode and cathode.

Reverse osmosis separates the solvent from a solution by forcing it to pass through a membrane by applying a pressure greater than the normal osmotic pressure. The solvent molecules are of about the same size as that of solute molecules. The solute rejection efficiency R_S is defined as

$$R_s = 1 - c''_s/c'_s, \quad (19)$$

where c'_S is the solute concentration in the upstream retentate and c''_S is the solute concentration in the downstream permeate (Ho and Li, 1984). Reverse osmosis processes may be classified into three types:

1. high-pressure reverse osmosis (5.6–10.5 MPa, i.e., 813–1524 psi), e.g., seawater desalination with very high rejection of inorganics (95–99.9% NaCl rejection);
2. low-pressure reverse osmosis (1.4–4.2 MPa, i.e., 203–610 psi), e.g., brackish water desalination and moderate to high rejection of low molecular weight organics; and

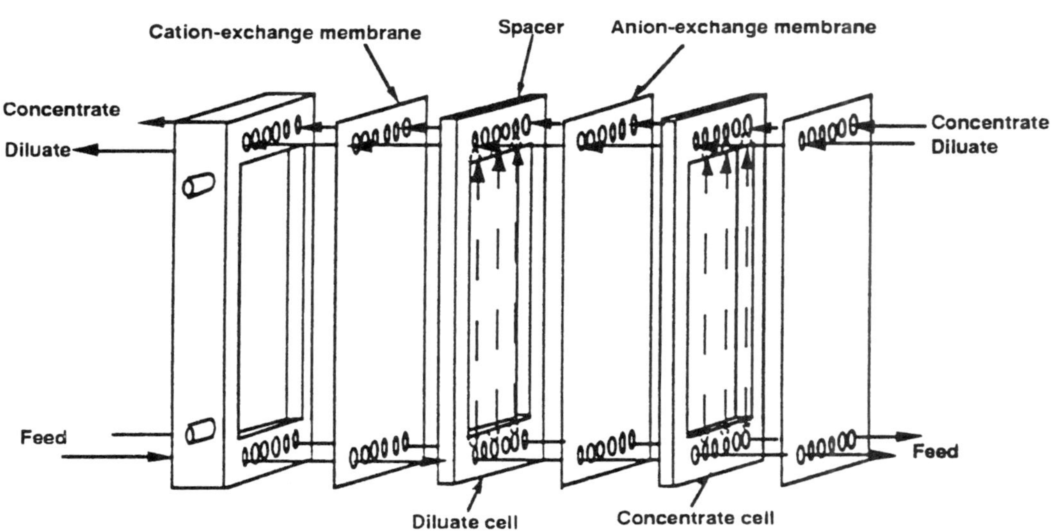

FIG. 17. Exploded view of components in an electrodialysis stack [reprinted from Strathmann (1992) with permission].

3. nanofiltration or loose reverse osmosis (0.3–1.4 MPa, i.e., 44–203 psi), e.g., divalent ion removal, partial demineralization, or 0–20% NaCl rejection.

Most transport models for reverse osmosis assume the solution-diffusion mechanism, the sieving mechanism, and the surface force–pore flow mechanism (Bhattacharyya *et al.*, 1992; Ho and Li, 1984; Sourirajan and Matsuura, 1985).

Similar to reverse osmosis, ultrafiltration is a pressure-driven membrane process capable of separating solution components on the basis of molecular size and shape. Under an applied pressure difference across an ultrafiltration membrane, solvent and small solute species pass through the membrane and are collected as permeate while larger solute species are retained by the membrane and recovered as a concentrated retentate. Ultrafiltration involves solutes whose molecular dimensions are 10 or more times larger than those of the solvent and are usually below 0.02 μm (200 Å) in size. The solutes or the materials to be separated usually have molecular weights greater than 500, such as macromolecules (proteins, polymers, starches, natural gums, enzymes, etc.), colloidal dispersions (clays, pigments, minerals, latex particles, and microorganisms), and emulsions (grease–detergent and oil–water emulsions). As a consequence of relatively high molecular weight, the osmotic pressures of the solutes are usually low. Thus, the operating pressures of ultrafiltration range only from 34 to 689 kPa (5 to 100 psi). The lower pressures reduce the energy requirement for pumping and compression and the equipment cost by a considerable margin over reverse osmosis. Ultrafiltration can be used to concentrate, purify, and fractionate macrosolutes or materials in the feed liquid. The solute rejection efficiency for ultrafiltration is defined in the same way as that for reverse osmosis, Eq. (19). Transport in ultrafiltration is through the sieving mechanism. When solvent migrates toward the membrane surface, it carries solute, which is rejected at the membrane surface, resulting in an accumulation of solute on the membrane. This accumulation usually leads to the formation of a gel layer, i.e., secondary membrane (Ho and Li, 1984; Cheryan, 1986; Kulkarni *et al.*, 1992; Sourirajan and Matsuura, 1985).

Similar to reverse osmosis and ultrafiltration, microfiltration is a pressure-driven membrane process, and it separates micron-sized particulates from a fluid. In general, microfiltration is defined as the filtration of a suspension containing colloidal or fine particulates with linear dimensions in the approximate range of 0.02 to 10 μm. These particulates are usually larger than the solutes separated by reverse osmosis and ultrafiltration. Thus, the osmotic pressure for microfiltration is negligible, and the transmembrane pressure drop is relatively small (typically, 7 to 345 kPa, i.e., 1 to 50 psi). The membrane pore size and permeate flux for microfiltration are generally larger than those for reverse osmosis and ultrafiltration. Similar to ultrafiltration, microfiltration is based on the transport through the sieving mechanism (Davis and Grant, 1992).

The emulsion liquid membrane (ELM) process is unique and different from the membrane processes discussed above. The membrane is a liquid phase involving an emulsion configuration. Emulsion liquid membranes, also called surfactant liquid membranes or liquid surfactant membranes, are essentially double emulsions, i.e., water/oil/water (W/O/W) systems or oil/water/oil (O/W/O) systems. For the W/O/W systems, the oil phase separating the two aqueous phases is the liquid membrane. For the O/W/O systems, the liquid membrane is the water phase that is between the two oil phases. Figure 18 shows a schematic of a continuous ELM process. This process includes four steps:

1. emulsification;
2. dispersion of the emulsion in contact with the external, continuous phase for extraction;
3. settling to separate the emulsion from the external phase, which is the raffinate if the internal phase becomes the extract; and
4. breaking the emulsion to recover the internal phase as the extract and the membrane phase for recycle.

The ELMs are usually prepared by first forming an emulsion between two immiscible phases and then dispersing the emulsion in a third (continuous) phase by agitation for extraction. The membrane phase is the liquid phase that separates the encapsulated,

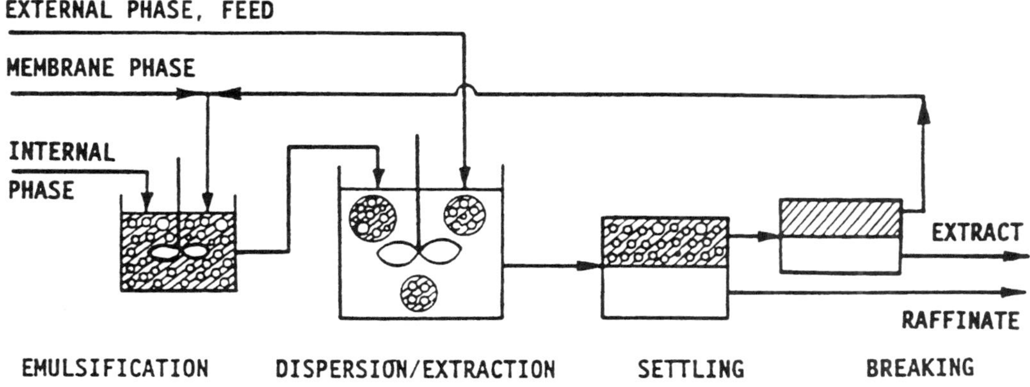

FIG. 18. Schematic of a continuous emulsion liquid membrane process [reprinted from Ho and Li (1992) with permission].

internal droplets in the emulsion from the external, continuous phase, as shown schematically in Fig. 19. In general, the internal, encapsulated phase and the external, continuous phase are miscible. However, the membrane phase must not be miscible with either of these two phases in order to be stable. To maintain the integrity of the emulsion during the extraction process, the membrane phase generally contains some surfactant(s) and additive(s) as stabilizing agents, and it also contains a base material that is solvent for all the other ingredients.

Separation of mixtures is via the solution-diffusion mechanism by selective transport of one component through the membrane phase into the receiving phase of lower equivalent concentration. Surfactant(s), additive(s), and extractant(s) included in the membrane phase can control the selectivity and permeability of the membrane. An individual component can be trapped and concentrated in the internal phase for later disposal or recovery. Once separation is achieved, the emulsion and external, continuous phases are separated, usually by settling as in conventional solvent extraction. The extracted component can be recovered from the "loaded" internal phase of the emulsion by breaking the emulsion, usually by the use of an electrostatic coalescer. From the broken emulsion, the membrane phase recovered can then be recycled to the emulsification step for the preparation of the emulsion with a regenerated or fresh internal reagent phase. Since their discovery by Li (1968), emulsion liquid membranes have demonstrated considerable potential as effective tools for a wide variety of separations (Ho and Li, 1992; Gu *et al.,* 1992).

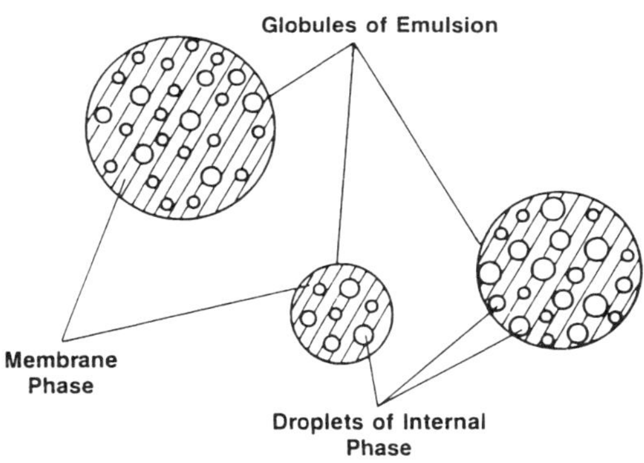

FIG. 19. Schematic of an emulsion liquid membrane system [reprinted from Ho and Li (1992) with permission].

Large-scale or widespread commercial uses were developed earliest for microfiltration, dialysis (hemodialysis), electrodialysis, and reverse osmosis. Ultrafiltration was adopted for commercial uses immediately afterward. The first large-scale gas-separation modules were developed by Du Pont as early as 1970; but the first successful commercial membrane gas-separation processes were announced by Monsanto in the late 1970s. Many more have appeared since then. Significant commercialization of pervaporation separation began during the period 1985–1990. Commercialization of emulsion liquid membranes for the removal of zinc and phenol from wastewaters was also achieved in Austria and China, respectively, around 1986.

All commercialized membrane separation processes, except gas permeation and microfiltration of gaseous feeds, are capable of separating liquid solutions or liquid-phase feeds, and virtually all of them are principally used for separating aqueous solutions. In the aqueous solutions, the solutes may be microsolutes (inorganic or organic), macrosolutes, proteins, polymers, colloids, whole cells, precipitates, dirt particles, etc. An exception is pervaporation, whose principal commercial use at this time is for removing small amounts of water from organic liquids such as alcohols. Also, a new pervaporation technology based on a polyimide–polyester copolymer membrane (Ho *et al.*, 1990, 1991) has been developed recently to separate heavy catalytically cracked naphtha into an aromatics-rich permeate and an aromatics-lean retentate (Exxon, 1994). The higher aromatic content of the permeate improves its octane and therefore makes it an excellent gasoline blendstock, while the retentate becomes a much better distillate (jet fuel and diesel) blend component because of its reduced aromatics level. In addition, ultrafiltration has been used for treating waste aqueous streams with a very high volume fraction of organics, and microfiltration separation of organic liquids is practiced.

In general, staging in membrane processes is economically unfavorable. Thus, membrane processes are usually best suited for bulk separation rather than for purification. Exceptions include membrane separations with very high selectivities, e.g., reverse osmosis for desalination, pervaporation for dehydration of ethanol and isopropanol, pervaporation for removal of sparingly soluble organics from wastewater and some ultrafiltration and microfiltration applications, and the emulsion liquid membrane process where staging and facilitated transport mechanisms with chemical reactions (Li, 1978; Ho and Li, 1992; Gu *et al.*, 1992) can be incorporated (Ho and Sirkar, 1992a).

Although most membranes used in the commercialized processes are polymeric, other membranes, e.g., ceramic, metal, etc., are also used or being studied. However, a few common features of membranes are dictated by the mechanism for transport/selectivity, e.g., only nonporous or solvent membranes can give the solution-diffusion mechanism. Therefore, the selective layer in gas permeation and pervaporation membranes is solventlike, i.e., nonporous. Sieving by a membrane presupposes the existence of pores, and thus membranes in dialysis, ultrafiltration, and microfiltration have pores. Membranes for ultrafiltration and microfiltration are generally identified as microporous (Ho and Sirkar, 1992b).

9.2 Process Equipment

In commercial processes, membranes must be packaged into modules with as high surface area per unit volume as possible but still allow good flow distribution and efficient contact of the feed with the membrane. Module housings are normally fabricated from standard-sized pipe, and they range from about 4 to 12 in. (10 to 30 cm) in diameter and from 4 to 20 ft (1.2 to 6.1 m) in length. The housing materials depend on applications, and carbon steel and aluminum are usually suitable. Membrane geometries lead to different module designs. Membrane modules include the following:

1. Spiral-wound modules: A spiral-wound module constructed from flat-film membranes is shown schematically in Fig. 20 (Zolandz and Fleming, 1992). It is fabricated by taking two membrane sheets with a permeate spacer in between and sealing them together with glue at three edges to form a membrane leaf. The open end of the leaf is glued to a perforated tube. The membrane leaf and a feed spacer are then rolled around the perforated tube and glued with an outer wrap to form the module. When the module is used, the feed is brought to the feed spacer and in contact with outer surfaces of the membrane leaf. Permeant passes through the walls of the membrane leaf and then flows along the permeate spacer to the perforated, collection tube. The modules typically contain about 1000 ft^2 (93 m^2) of membrane area per cubic foot (0.028 m^3) volume, i.e., 3281 m^2/m^3, for

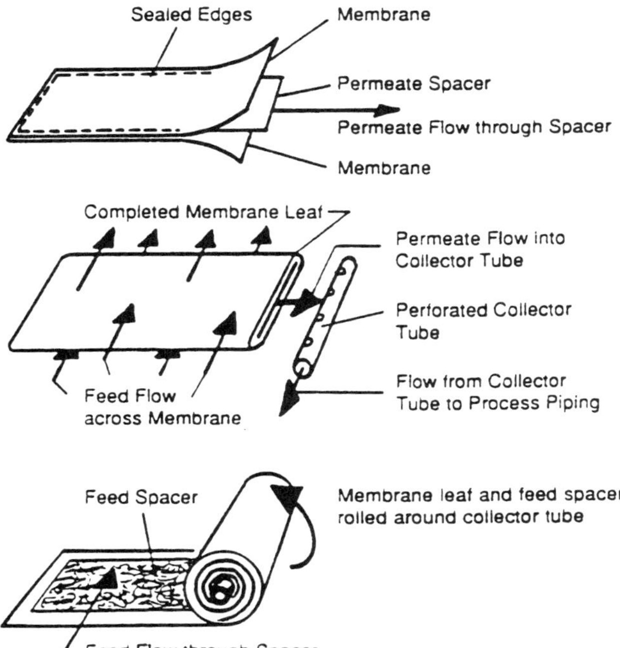

FIG. 20. Schematic of a spiral-wound module assembly [reprinted from Zolandz and Fleming (1992) with permission].

gas permeation and about 300 ft^2/ft^3 (984 m^2/m^3) for liquid separations.

2. Hollow-fiber modules: The modules are fabricated in a similar manner as shell-and-tube heat exchangers. The fibers are arranged parallel to one another and glued together to form tube sheets at either one or both ends of the module. There is a seal between the tube-sheet exterior and the pressure vessel in order to isolate the feed from the permeate. Figure 21 shows a schematic of a cross-flow feed hollow-fiber module. Hollow-fiber modules can provide the highest membrane area per unit volume among all membrane modules. The hollow-fiber modules for gas permeation can have about 3000 ft^2 per cubic foot volume, i.e., 9843 m^2/m^3 (Zolandz and Fleming, 1992).
3. Tubular modules: Similar to hollow-fiber modules, tubular models are like shell-and-tube heat exchangers. But the tube diameters are generally 2.5 to 25 mm, and they are larger than hollow-fiber diameters. In general, the feed enters the tubes at one end of the module, is concentrated, and exits at the other end, while the permeate comes out from the shell side. Recently, ceramic modules have become available. Figure 22 shows the Ceraflo™ ceramic module containing nineteen 3-mm-i.d. tubes. The feed flows through the tubes while the permeate passes through the microporous walls of the module into the surrounding shell for withdrawal (Kulkarni *et al.,* 1992; Mir *et al.,* 1992). The ceramic modules are particularly suited for high-temperature applications (about 200 °C and higher).
4. Plate-and-frame modules: Flat-film membranes can be assembled in a plate-and-frame module similar to the filter presses shown in Fig. 9. A new variation, i.e., flat-stack membrane module, has been developed recently and is shown in Fig. 23 (Ohlrogge *et al.,* 1995). In this new module, two round, flat-film membranes are thermally welded at the cutting edges forming a membrane envelope. The envelopes are placed on a perforated, central tube, and the stack of the envelopes is divided into asymmetrical compartments by means of baffle plates. In each compartment, the number of the envelopes is reduced progressively in the direction of the feed flow with the decrease of feed vol-

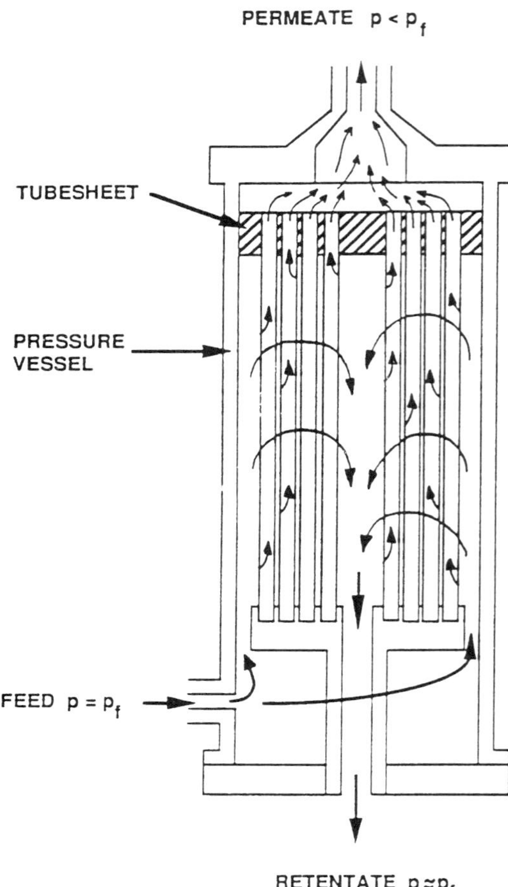

FIG. 21. Schematic of a cross-flow feed hollow-fiber module [reprinted from Zolandz and Fleming (1992) with permission].

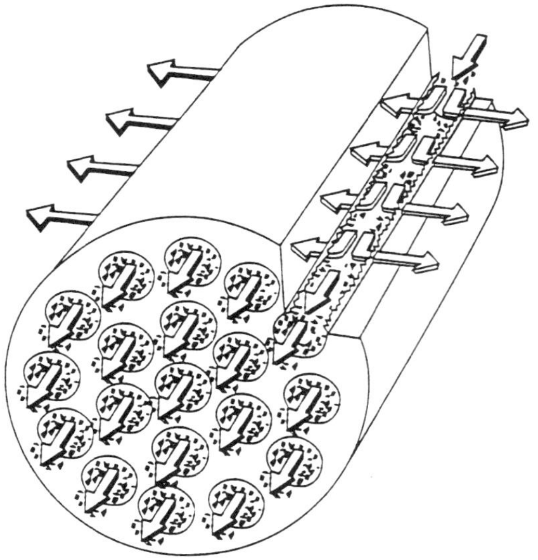

FIG. 22. Ceraflo™ ceramic module (courtesy Millipore Corporation, Literature No SD117, May 1989).

ume flow caused by permeation through the membranes. The space between the membranes ranges from 0.25 to 2.5 mm (Mir *et al.*, 1992). As described earlier, an electrodialysis stack similar to the plate-and-frame arrangement but using rectangular membranes is shown in Fig. 17.

5. Rotary-drum filters: Rotary-drum vacuum filters are widely used for difficult filtrations including microfiltrations for the removal of microbial cells and other particulates that form compressible cakes with large resistances (Davis and Grant, 1992). As described earlier, the filter is shown schematically in Fig. 10.

For gas permeation, pervaporation, and reverse osmosis, both spiral-wound and hollow-fiber modules are used. Plate-and-frame modules are sometimes used for gas permeation and pervaporation. Dialysis processes usually employ hollow-fiber and plate-and-frame modules. In ultrafiltration and microfiltration, tubular, spiral-wound, and plate-and-frame modules are normally used. As mentioned above, electrodialysis stacks are employed for electrodialysis, and rotary-drum filters are sometimes used for microfiltration. For emulsion liquid membranes, the contacting devices are the same as those for solvent extraction, including mixer settlers and mechanically agitated column extractors (Gu *et al.*, 1992).

9.3 Membrane Separation Operations

Membrane process operations with compact modular systems are usually the simplest among all separation processes. As long as the driving force for the membrane process exists, the membrane separation can take place. The driving force for gas permeation, reverse osmosis, ultrafiltration, and microfiltration processes comes from the pressure differential across the membrane for the species transported through the membrane. Thus, these processes are called

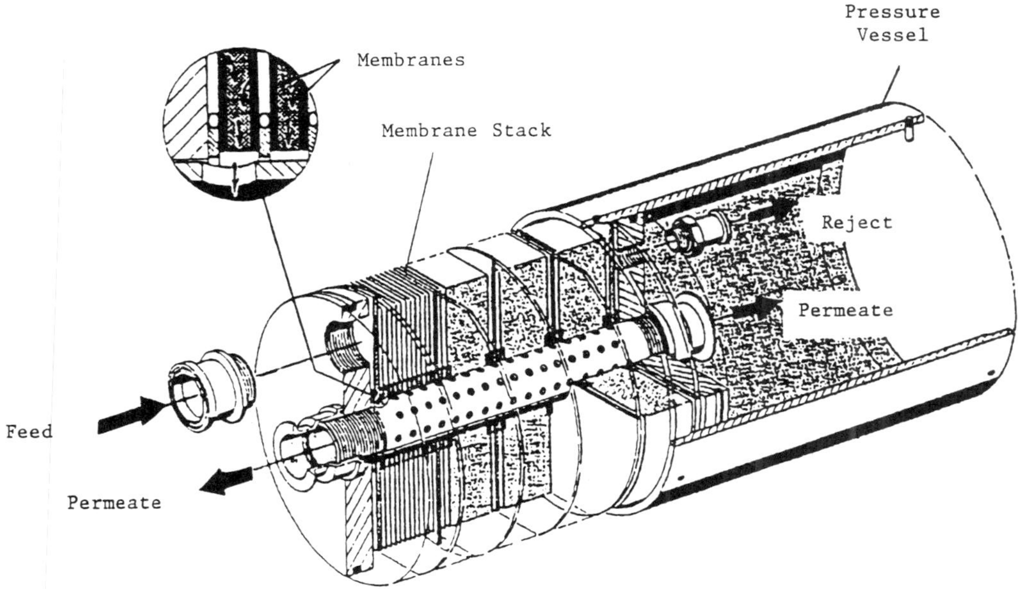

FIG. 23. Flat-stack membrane module [reprinted from Ohlrogge *et al.* (1995) with permission].

pressure-driven processes. The driving force for the concentration-driven processes, namely, pervaporation, dialysis, and emulsion liquid membrane, is the concentration difference across the membrane for the species diffusing through membrane. For pervaporation, temperature can also be used to increase the concentration differential for a given vacuum in the permeate side. As mentioned earlier, the driving force for ion transport in electrodialysis is the applied electrical potential between the anode and cathode. Industrial membrane processes are generally continuous in operation.

Membrane modules are designed to operate within a range of feed flow rates, and it is a good practice to stay within the range. Operating the module below the minimum design flow rate can give poor distribution of the feed over the available membrane area, whereas operating above the design maximum flow rate can result in excessive pressure drop for the feed stream flowing through this module. Operation with poor distribution of the feed results in both poorer product (e.g., permeate) purity and lower module productivity. In a given membrane stage, the proper feed rate is achieved by piping modules in parallel and dividing the feed stream equally among the modules. If the feed rate will change with time, a membrane stage can consist of parallel modules, and the required number of modules is then employed as needed (Zolandz and Fleming, 1992).

The purity of the permeate stream, e.g., the product stream in gas permeation, reduces from its maximum at a stage cut (the ratio of the permeate flow rate to the feed flow rate) of zero as the recovery of the permeate increases. The trade-off between purity and recovery is intrinsic to membrane processes. To increase product purity beyond what can be achieved by once-through operation, the membrane process must be operated with recycle. Recycle of permeate to the feed side increases the component concentration in the feed side, resulting in higher product purity at a given recovery. However, recompression of the recycled permeate in a pressure-driven process, e.g., gas permeation, adds to both the investment and operating costs (Zolandz and Fleming, 1992).

In pervaporation, the process is mainly driven by condensation of the permeate, resulting in a significant vacuum and lower temperatures on the permeate site of the membrane. Even though a vacuum pump is still needed for startup, it is normally not used in steady-state operation, and it is small. The enthalpy of vaporization for the feed is consumed, and interstage heaters to reheat the feed are employed (Fleming and Slater, 1992).

In dialysis (Kessler and Klein, 1992), advantage should be taken of natural convection currents in dialysis cells in order to maintain a maximum concentration gradient across the membrane. The membrane may be mounted vertically with the feed and solvent phases flowing countercurrently on opposite sides of the membrane. The directions of flow of these two phases should be determined by their densities. For example, if the feed has a higher density than the solvent, it should be introduced at the bottom of the feed compartment and the solvent should be introduced at the top of the dialyzate compartment. The solvent in the feed compartment, entering by osmosis through the membrane, will produce a rising stream of low-density weak liquor, whereas the diffusing solute in the dialyzate compartment will form a stream of high-density solution flowing down the membrane surface. This density-streaming effect can reduce greatly the thickness of the liquid films at both sides of the membrane and thus lower the overall resistance to dialysis (Ho and Li, 1984).

In electrodialysis, concentration polarization effects result in the depletion of ions in the laminar boundary layer at the membrane surfaces in the cell containing the diluate flow stream and in the increase of ions in the laminar boundary layer at the membrane surfaces in the cell containing the brine solution. Concentration polarization decreases separation efficiency. The magnitude of concentration polarization can be controlled by the electrical current density, the spacer design to decrease the boundary-layer thickness, and the flow velocities of the diluate and brine solutions (Strathmann, 1992).

In reverse osmosis, ultrafiltration, and microfiltration, concentration polarization decreases fluxes. When solvent passes through a membrane, the solute (or particulates in microfiltration) left behind concentrates in a solution next to the membrane surface. Accompanying the increase in solute concentration at the interface are an increase in the osmotic pressure of the solution at the interface and the formation of a secondary membrane, resulting in flux reduction. The solute layer built up near the membrane surface may reach a constant thickness as a result of a balance of two opposing factors, i.e., the connective transport of the solute toward the membrane by the bulk motion of the solvent and the back diffusion of the solute away from the membrane surface due to the concentration gradient established near the phase boundary. Concentration polarization can be minimized by increasing the fluid shear rate to decrease the boundary-layer thickness at the membrane surface and to promote the mass transport of the solute away from the membrane surface. Turbulence promoters (e.g., feed spacers in spiral-wound modules and static mixers in tubular modules) have been used to decrease concentration polarization (Ho and Li, 1984; Bhattacharyya *et al.*, 1992; Kulkarni *et al.*, 1992; Davis and Grant, 1992).

As mentioned earlier, an emulsion liquid membrane process consists of four steps:

1. emulsification,
2. dispersion of the emulsion in contact with the external, continuous phase for extraction,
3. settling to separate the emulsion from the external phase, and
4. breaking the emulsion to recover the internal phase and the membrane phase.

Emulsification is carried out by the use of colloidal mills or static homogenizers. Dispersion and settling of the emulsion in contact with the external phase are conducted with mixer settlers and mechanically agitated column extractors followed by settlers, which are the same as those used conventionally for solvent extraction. An electrostatic coalescer is usually employed to break the emulsion (Ho and Li, 1992; Gu *et al.*, 1992).

Pretreatment of feed, such as the use of micron-sized filtration (about 0.5–5 μm), is usually needed to minimize solids buildup in membrane modules. Fouled membranes are normally cleaned by clean-in-place procedures that minimize downtime. Periodic reversal in flow direction may help to prevent particulates from buildup in membrane modules. Cleaning solutions, e.g., detergent and surfactant solutions and some solvents (e.g., isopropanol), are generally employed. A typical cleaning cycle may use 1 to 3 module volumes of the cleaning solution at an elevated temperature (about 50–80 °C) to circulate through the membrane system. The cycle may be carried out once every 1–4 weeks (Fleming and Slater, 1992; Ho and Sirkar, 1992a).

9.4 Important Industrial Applications

Applications include the following for eight commercialized membrane processes.

9.4.1 Gas Permeation Gas separation applications include hydrogen recovery, air (oxygen and nitrogen) separation, acid gas removal, and helium recovery. Hydrogen is recovered from hydroprocessing purge streams in refineries, from ammonia-plant purge streams and other petrochemical plant streams, and from synthesis-gas (H_2/CO) ratio adjustment. Air is separated into oxygen- and nitrogen-enriched streams. Oxygen-enriched air has usually been limited to medical uses, whereas commercial-scale uses (e.g., enhanced combustion) have yet to reach their full potential. The purity of the nitrogen-enriched stream ranges from 95 to 99% nitrogen, which is suitable for use as inert gas. Acid gases, e.g., CO_2 and H_2S, are removed from natural gas. A potentially large market for CO_2 separation membrane systems is the recovery of CO_2 from gas produced in oil fields where enhanced oil recovery via CO_2 injection is used. Helium is recovered from natural gas (Zolandz and Fleming, 1992).

9.4.2 Pervaporation Its industrial uses are primarily for the dehydration of organics such as ethanol and isopropanol and the removal of organics—e.g., ethanol from dealcoholization of beers, wines, and liquors, and trichloroethane from groundwaters and wastewaters (Fleming and Slater, 1992). As mentioned earlier, a new pervaporation technology has been developed recently to separate heavy catalytically cracked naphtha into an aromatics-rich permeate for mogas blending and an aromatics-lean retentate for distillate blending (Ho *et al.*, 1990, 1991; Exxon, 1994).

9.4.3 Dialysis The artificial kidney, hemodialysis, for purifying human blood is the major application of dialysis in the medical field. The principal solutes to be removed for chronic patients are urea, uric acid, creatinine, phosphates, and excess amounts of chloride (Ho and Li, 1984; Kessler and Klein, 1992). Industrial applications include the recovery of caustic soda in rayon processing and the recovery of spent acids from metallurgical liquors (Ho and Li, 1984).

9.4.4 Electrodialysis Its main applications include (Strathmann, 1992) these:

1. Desalination of brackish water. This application competes directly with reverse osmosis and multistage flash evaporation. For brackish water with relatively low salt concentration (less than 5000 ppm), electrodialysis is usually the most economic process.
2. Production of table salt from seawater. This application to concentrate sodium chloride up to 200 g/L prior to evaporation is used nearly exclusively in Japan.
3. Wastewater treatment. The main application is in processing rinse waters from the electroplating industry.

9.4.5 Reverse Osmosis Its applications include the following (Ho and Li, 1984; Bhattacharyya *et al.*, 1992):

1. Desalination of sea and brackish waters. Reverse-osmosis units have successfully desalted sea and brackish waters to potable quality (Ko and Guy, 1988).
2. Demineralization. Reverse-osmosis units have been integrated into the ultrapure water systems that supply water for electronics-manufacturing processes (Rautenbach and Albrecht, 1989).
3. Recycle operations in electroplating and electrocoating plants. This application not only recovers valuable materials, such as nickel, chromium, and cyanide, but at the same time eliminates water-discharge problems.
4. Processing wheys, juices, and effluents in the food industry. Cheese and cottage-cheese wheys are processed by the use of reverse osmosis, which is combined with ultrafiltration to separate the solutes into fractions containing mostly protein, lactose, and lactic acid as well as to concentrate each of these fractions. Reverse osmosis is also used to concentrate juices, milk, and syrup.
5. Wastewater treatment. This includes applications to treat dilute processing

streams in the pulp and paper and textile industries for recycling water and decreasing color, biochemical oxygen demand, and other objectionable components in effluents.

6. Water treatment in the power-generation industry. The primary uses are to produce boiler-quality water for steam generation and to treat wastewater effluents from cooling-water blowdown (Patra *et al.*, 1987).

9.4.6 Ultrafiltration Its applications include the following (Ho and Li, 1984; Cheryan, 1986; Kulkarni *et al.*, 1992):

1. Electrophoretic paint recovery. Ultrafiltration is used to process the paint by retaining the polymer resins and pigmented solids while allowing inorganic salts, water, and solvent to permeate through the membrane. The retained species are returned to the electropaint tank. The permeate is then used to rinse the freshly painted components as they emerge from the paint and to recover the drag-out excess paint (Michaels, 1981).
2. Protein extraction in food and dairy industries. Large protein molecules from cheese, casein whey, or skim milk are concentrated.
3. Juice clarification. Haze components are removed from fruit juices, e.g., apple juice (Blanck and Eykamp, 1986).
4. Waste-machining and metal-rolling emulsions. Ultrafiltration is used to concentrate emulsified machining and metal-rolling oils from an initial 2 to 10% to a level of 25 to 50%. The concentrated emulsion can be disposed of by burning and the water permeate discharged to the drain.

9.4.7 Microfiltration Its applications include the following:

1. Sterilization of pharmaceuticals: Microfiltration is used to sterilize pharmaceuticals, including water for injection, ophthalmic solutions, antibiotics, vaccines, etc. (Goel *et al.*, 1992).
2. Stabilization of beers and wines: Microfiltration is used increasingly in the beverage industry mainly for clarification and biological stabilization of beers and wines (Goel *et al.*, 1992).
3. Purification of process fluids in semiconductor manufacture: Both gaseous and liquid process fluids used to manufacture semiconductors and integrated circuits are purified by microfiltration to remove particles (Goel *et al.*, 1992).
4. Clarification of antibiotics: Microfiltration is used for the recovery and purification of antibiotics from fermentation broths (Mir *et al.*, 1992).

9.4.8 Emulsion Liquid Membrane Applications for ELM processes include these (Ho and Li, 1992):

1. Zinc removal from wastewater in the viscose fiber industry: This was first commercialized at Lenzing AG in Austria with a wastewater feed rate of 75 000 L/h (11 300 bbl/d), and zinc was removed from 200 to 0.3 mg/L (Marr and Draxler, 1992).
2. Phenol removal from wastewater in the chemical industry: This was commercialized at the Nanchung Plastic Factory in Guangzhou, China with a feed rate of 250 L/h, and phenol was reduced from 1000 to 0.5 mg/L (Marr and Draxler, 1992).
3. Cyanide removal from waste liquors in gold processing: This is being commercialized at the Huang-hua Mountain Gold Plant, near Tian-jin, China with a feed rate of about 600 L/h, and cyanide can be reduced from 130 to 0.5 mg/L (Jin and Zhang, 1990; Jin *et al.*, 1994).
4. Well control fluid for preventing well blowout and sealing loss zones in wells: The well control fluid is a pumpable water-in-oil emulsion containing clay particles in the oil phase. High shear at drill-bit nozzles breaks up the oil film and allows clay particles to contact water droplets, resulting in thickening the emulsion to very viscous paste for controlling the well in the drilling operation.

GLOSSARY

Absorbate: The solute component that is absorbed.

Absorbent: The liquid that absorbs solute components from the feed.

Adsorbate: The solute component that is adsorbed.

Adsorbent: The solid that adsorbs solute components from the feed.

Adsorption Isotherm: The isotherm that relates the fluid-phase adsorbate concentration to the amount of the adsorbate adsorbed per unit weight of the adsorbent.

BET: Abbreviation for the Brunauer-Emmett-Teller thermodynamic isotherm.

Boundary Layer: The region near a solid surface, e.g., a membrane, where fluid motion is affected by the surface.

Cake Filtration: Filtration in which the filtered solids are deposited on the surface of the filter medium, and the solid cake thus deposited quickly becomes the actual filter medium.

Chemical Potential: A measure of how the Gibbs free energy for a system changes when a specific amount of a component is added to the system. For membrane separation, the driving forces for transport through a membrane arise from a gradient of chemical potential or electrical potential. A gradient in chemical potential may be due to concentration gradient or pressure gradient or both.

Clarification: Process that removes relatively small amounts of particles and produces a clear effluent.

Composite Structure: Membrane structure in which the active layer is a coating on the support layer, or a coating is put on the active layer.

Concentrate: The portion of the feed solution that is retained on the high-pressure or -concentration side of the membrane. The terms "concentrate" and "retentate" are used interchangeably.

Concentration Polarization: When solvent transports through a membrane, the solute (or particulates, in microfiltration) left behind concentrates in a solution next to the membrane surface. Accompanying the increase in solute concentration at the interface are an increase in the osmotic pressure of the solution at the interface and the formation of a secondary membrane, resulting in flux reduction.

Counter Ions: The ions having the opposite charge to the fixed ions of an ion-exchange resin.

Deep-Bed Filtration: A deep bed of porous media such as diatomaceous earth (precoat) or fibrous material supported on a coarse filter is used to remove fine particles from a slurry. The particles to be removed are often considerably smaller than the pores of the filter medium and can penetrate a considerable depth before they are captured.

Dialyzate: The solution on the side of the membrane to receive the solutes to be removed from the feed on the other side of the membrane in dialysis.

Diffusion Coefficient (Diffusivity): A measure of the diffusive mobility of a solute in a solvent [see Eq. (11)].

Entrainer: A third component added to the system to effect the separation of an otherwise inseparable mixture by distillation.

Equilibrium: A state from which there is no tendency to depart spontaneously. An equilibrium state still has certain permissible changes or processes, e.g., heat transfer, work of volume displacement, and mass transfer across a phase boundary.

Extract: The phase comprising a solvent and the solute extracted from the feed.

Fluidized Bed: A bed in which the particles, e.g., ion-exchange resin beads, are held in suspension by the gas phase in contact with them.

Flux: Amount of fluid passing through the membrane.

Fouling: Phenomenon in which the membrane or equipment adsorbs or interacts in some manner with solutes in the feed stream, resulting in a decrease in performance (reduction in the membrane flux or equipment productivity). This phenomenon is usually irreversible and time dependent. In general, fouling effects can only be remedied by shutting down the system and cleaning the membrane or equipment by chemical means.

Integrally Skinned Structure: Membrane structure in which the active and support layers are formed in a single operation from a given material.

Ion-Exchange Capacity: Milliequivalents (meq.) of charge in an ion-exchange resin per unit weight (gram) of the dry resin.

Mass-Transfer Coefficient: A measure of the mobility of a solute, which is defined as the ratio of the solute flux to the solute concentration difference [see Eq. (12)].

Nucleation: Crystal formation initiated by the presence of a foreign body in the fluid or an irregularity in the wall of the container. The formation of a new solid phase in the bulk of a fluid phase without the involvement of a solid–fluid interface is called homogeneous nucleation; in the presence of surfaces, other than that of the crystal itself, such as those of the container, agitator, or foreign particles is called heterogeneous nucleation; or in the presence of the crystals of the crystallizing species itself is termed secondary nucleation.

Packed Column: Mass-transfer device filled with ceramic, metal, or plastic packings to facilitate mass transfer by increasing the contacting area. The packings are classified as random packings or structured packings. Random packings are dumped into the column onto a support plate whereas structured packings are shaped in the same size as the column with a regular structure.

Partition Coefficient: The ratio of the mole (or mass) fraction of the solute in the extract phase to its mole (or mass) fraction in the raffinate phase [see Eq. (8)].

Permeability: The ratio of the flux through the membrane to the concentration or pressure gradient across the membrane [see Eq. (17)].

Permeate: The portion of the feed that passes through the membrane.

PSA: Abbreviation for pressure-swing adsorption.

PSSA: Abbreviation for polystyrene sulfonic acid.

Reflux Ratio: The ratio of the condensate returned to the distillation column to the total condensate.

Rejection Efficiency: A measure of how well a membrane retains or allows passage of a solute [see Eq. (19)].

Relative Volatility: The ratio of the vapor pressure of two components.

Retentate: The portion of the feed that is retained on the high pressure or concentration side of the membrane. The terms "retentate" and "concentrate" are used interchangeably.

Separation Factor: The selectivity between species i and j is defined as the ratio of species i and j mole or mass fractions (or concentrations) in the extract (for solvent extraction) or in the permeate (for membrane separation), divided by their corresponding ratio of mole or mass fractions (or concentrations) in the raffinate, retentate, or feed [see Eqs. (10) and (18)].

Stage Cut: The ratio of the permeate flow rate to the feed flow rate.

Supersaturation: A measure of the quantity of solute present in the solution compared with the quantity, i.e., equilibrium solubility, which would be present if the solution were kept for a very long period of time with the solid phase of the solute in contact with the solution.

Theoretical Stage: A theoretical stage (defined at the existing or intended reflux rate) is considered as the unit stage that will produce a stage of equilibrium between the vapor and the liquid leaving the given stage.

Thickening: Process that increases the concentration of suspended particles in a slurry stream.

Tray Column: Gas–liquid contacting device, also known as mass-transfer device (usually in cylindrical shape) equipped with trays (plates). The tray columns are usually operated in countercurrent mode.

UNIFAC: Abbreviation for UNIQUAC (universal quasi-chemical activity coefficient) functional group activity coefficient, which is a thermodynamic group contribution method.

VLE: Abbreviation for vapor–liquid equilibrium.

VOC: Abbreviation for volatile organic compound.

Water-Transport Number: The ratio of mass of water transported per mass of dissolved solute dialyzed.

ZMS: Abbreviation for zeolite molecular sieve.

Works Cited

ASPEN-PLUS Reference Manual (1994).

Baczek, F. A., Emmett, R. C., Kominek, E. G. (1988), in: P. A. Schweitzer (Ed.), *Handbook of Separation Technologies for Chemical Engineers,* 2nd ed., New York: McGraw-Hill, p. 4-121.

Belfort, G. (1984), *Synthetic Membrane Processes,* New York: Academic.

Bennett, C. O., Meyers, J. E. (1982), *Momentum, Heat and Mass Transfer,* 3rd ed., New York: McGraw-Hill, pp. 224–246.

Bennett, R. C. (1984), in: R. H. Perry, D. W. Green (Eds.), *Perry's Chemical Engineers' Handbook,* 6th ed., New York: McGraw-Hill, pp. 19-24 to 19-40.

Bhattacharyya, D., Williams, M. E., Ray, R. J., McCray, S. B. (1992), in: W. S. Ho, K. K. Sirkar (Eds.), *Membrane Handbook,* New York: Chapman & Hall, pp. 263–390.

Blanck, R. G., Eykamp, W. (1986), *AIChE Symp. Ser.* **82** (250), 59.

Brown, P. R., Grushka, E. (1993), *Advances in Chromatography,* New York: Marcel Dekker.

Brunauer, S., Emmett, P. H., Teller, E. (1938), *J. Am. Chem. Soc.* **60,** 309.

Bungay, P. M., Lonsdale, H. K., dePinho, M. N. (1986), *Synthetic Membranes: Science, Engineering and Applications,* Dordrecht, Holland: D. Reidel Publishing Co.

Campbell, J. M. (1976), *Gas Conditioning and Processing: Absorption and Fractionation, Pumping, Compression and Expansion, Refrigeration, Hydrate Inhibition, Dehydration and Process Control,* Vols. 1 and 2, Norman OK: Campbell Petroleum Series.

Chapman, T. W. (1987), in: R. W. Rousseau (Ed.), *Handbook of Separation Process Technology,* New York: Wiley.

ChemCAD III Users' Guide (1993).

Cheremisinoff, P. N. (1995), *Processing Engineering Data Book,* Lancaster, PA: Technomic Publishing Co.

Cheryan, M. (1986), *Ultrafiltration Handbook,* Lancaster, PA: Technomic Publishing Co.

Chiang, S.-H., He, D. (1993), *Fluid/Particle Sep. J.* **6,** 64.

Code of Federal Regulations (1977), Title 21, §173.280, Washington, DC: U.S. Government Printing Office.

Combest, J., Corporation, K. (1994), *Fluid/Particle Sep. J.* **7,** 131.

Dahlstrom, D. A., Emmett, R. C., Klepper, R. P., Silverblatt, C. E. (1984), in: R. H. Perry, D. W. Green (Ed.), *Perry's Chemical Engineers' Handbook,* 6th ed., New York: McGraw-Hill, p. 19–53.

Davis, R. H., Grant, D. C. (1992), in: W. S. Ho, K. K. Sirkar (Eds.), *Membrane Handbook,* New York: Chapman & Hall, pp. 455–505.

Donald, M. B. (1958), *Chem. Eng. Practice* **6,** 473.

Dorfner, K. (Ed.) (1991), *Ion Exchangers,* Berlin: Walter de Gruyter.

Edwards, W. M. (1984), in: R. H. Perry, D. W. Green (Eds.), *Perry's Chemical Engineers' Handbook,* 6th ed., New York: McGraw-Hill, p. 14-1.

Exxon Research & Engineering Co. (1994), The ER&E Membrane Process, Refining Technology Brochure.

Fair, J. R. (1987), in: R. W. Rousseau (Ed.), *Handbook of Separation Process Technology,* New York: Wiley.

Fleming, H. L., Slater, C. S. (1992), in: W. S. Ho, K. K. Sirkar (Eds.), *Membrane Handbook,* New York: Chapman & Hall, pp. 103–159.

Ganetsos, G., Barker, P. E. (1993), *Preparative and Production Scale Chromatography,* New York: Marcel Dekker.

Gas Processors Suppliers Association (1987), *Engineering Data Book,* Tulsa, OK: Gas Processors Suppliers Association.

Goel, V., Accomazzo, M. A., DiLeo, A. J., Meier, P., Pitt, A., Pluskal, M., Kaiser, R. (1992), in: W. S. Ho, K. K. Sirkar (Eds.), *Membrane Handbook,* New York: Chapman & Hall, pp. 506–570.

Grace, H. P. (1956), *AIChE J.* **2,** 316.

Gregor, E. C., Walt, C., Mollet, J. R. (1992), *Fluid/Particle Sep. J.* **5,** 163.

Gu, Z. M., Ho, W. S., Li, N. N. (1992), in: W. S. Ho, K. K. Sirkar (Eds.), *Membrane Handbook,* New York: Chapman & Hall, pp. 656–700.

Ho, W. S., Li, N. N. (1984), in: R. H. Perry, D. W. Green (Eds.), *Perry's Chemical Engineers' Handbook,* 6th ed., New York: McGraw-Hill, pp. 17-14 to 17-35.

Ho, W. S., Li, N. N. (1992), in: W. S. Ho, K. K. Sirkar (Eds.), *Membrane Handbook,* New York: Chapman & Hall, pp. 595–655.

Ho, W. S., Sartori, G., Thaler, W. A., Dalrymple, D. C. (1990), Polyimide/Aliphatic Polyester Copolymers, U. S. Patent 4,944,880.

Ho, W. S., Sartori, G., Thaler, W. A., Dalrymple, D. C. (1991), Polyimide/Aliphatic Polyester Copolymers, U. S. Patent 4,990,275.

Ho, W. S., Sirkar, K. K. (Eds.) (1992a), *Membrane Handbook,* New York: Chapman & Hall.

Ho, W. S., Sirkar, K. K. (1992b), in: W. S. Ho, K. K. Sirkar (Eds.), *Membrane Handbook,* New York: Chapman & Hall, pp. 3–15.

Huang, R. Y. M. (Ed.) (1991), *Pervaporation Membrane Separation Processes,* Amsterdam: Elsevier.

Jacobs, L. J. (1984), in: R. H. Perry, D. W. Green (Eds.), *Perry's Chemical Engineers' Handbook,* 6th ed., New York: McGraw-Hill, p. 19-65.

Jin, M., Wen, T., Lin, L., Liu, F., Liu, L., Zhang, Y., Zhang, C., Deng, P., Song, Z. (1994), *Mo Kexue Yu Jishu* **14,** 16.

Jin, M., Zhang, Y. (1990), *Proc. International Congress on Membranes and Membrane Processes,* Chicago, IL, August 20–24, 1990, Vol. I, pp. 676–678.

Keller, G. E., II, Anderson, R. A., Yon, C. M. (1987) in: R. W. Rousseau (Ed.), *Handbook of Separation Process Technology,* New York: Wiley, pp. 644–696.

Kessler, S. B., Klein, E. (1992), in: W. S. Ho, K. K. Sirkar (Eds.), *Membrane Handbook,* New York: Chapman & Hall, pp. 161–215.

Kesting, R. E. (1985), *Synthetic Polymeric Membranes: A Structural Perspective,* 2nd ed., New York: Wiley, pp. 1–21.

Ko, A., Guy, D. (1988), in: B. Parekh (Ed.), *Reverse Osmosis Technology,* New York: Marcel Dekker, Chap. 5.

Kohl, A. L. (1987), in: R. W. Rousseau (Ed.), *Handbook of Separation Process Technology,* New York: Wiley.

Kohl, A. L., Riesenfeld, F. C., (1979), *Gas Purification,* 3rd ed., Houston: Gulf Publishing Co.

Koros, W. J., Pinnau, I. (1994), in: D. R. Paul, Y. Yampol'skii (Eds.), *Polymeric Gas Separation Membranes,* Boca Raton, FL: CRC Press, pp. 209–271.

Krause, S. (1993), *Int. Chem. Eng.* **33,** 355.

Kulkarni, S. S., Funk, E. W., Li, N. N. (1992), in: W. S. Ho, K. K. Sirkar (Eds.), *Membrane Handbook,* New York: Chapman & Hall, pp. 391–453.

Li, N. N. (1968), Separating Hydrocarbons with Liquid Membranes, U. S. Patent 3,410,794.

Li, N. N. (1978), *J. Membr. Sci.* **3,** 265.

Lonsdale, H. K. (1989), *J. Membr. Sci.* **43,** 1.

Marr, R. J., Draxler, J. (1992), in: W. S. Ho, K. K. Sirkar (Eds.), *Membrane Handbook,* New York: Chapman & Hall, pp. 701–724.

McCabe, W. L., Smith, J. C. (1956), *Unit Operations of Chemical Engineering,* New York: McGraw-Hill, pp. 324–376.

Mersmann, A. (1994a), *Crystallization Technology Handbook,* New York: Marcel Dekker.

Mersmann, A. (1994b), in: A. Mersmann (Ed.), *Crystallization Technology Handbook,* New York: Marcel Dekker, p. 79.

Michaels, A. S. (1981), *Chemtech* **(January),** 36.

Miller, S. A. (1984), in: R. H. Perry, D. W. Green (Eds.), *Perry's Chemical Engineers' Handbook,* 6th ed., New York: McGraw-Hill, p. 19-1.

Mir, L., Michaels, S. L., Goel, V., Kaiser, R. (1992), in: W. S. Ho, K. K. Sirkar (Eds.), *Membrane Handbook,* New York: Chapman & Hall, pp. 571–594.

Moyers, G. C. (1984), in: R. H. Perry, D. W. Green (Eds.), *Perry's Chemical Engineers' Handbook,* 6th ed., New York: McGraw-Hill, p. 17-3.

Noll, K. E., Gounaris, V., Hou, W. S. (1991), *Adsorption Technology for Air and Water Pollution Control,* Chelsea, MI: Lewis Publishers.

Ohe, S. (1989), *Vapor-Liquid Equilibrium Data,* New York: Elsevier.

Ohe, S. (1990), *Vapor-Liquid Equilibrium Data at High Pressure,* New York: Elsevier.

Ohlrogge, K., Wind, J., Behling, R.-D. (1995), *Sep. Sci. Technol.* **30,** 1625.

Paruit, B. H. (Ed.) (1982), *Illustrated Glossary of Process Equipment,* Houston, TX: Gulf Publishing Co.

Patra, R., Prabhakar, S., Misra, B., Ramani, M. (1987), *Desalination* **67,** 507.

Paul, D. R., Yampol'skii, Y. (Eds.) (1994), *Polymeric Gas Separation Membranes,* Boca Raton, FL: CRC Press.

Perry, R. H., Green, D. W. (Eds.) (1984), *Perry's Chemical Engineers' Handbook,* 6th ed., New York: McGraw-Hill.

Pinheiro, H., Cabral, J. M. S. (1993a), in: J. F. Kennedy, J. M. S. Cabral (Eds.), *Recovery Processes for Biological Materials,* New York: Wiley, pp. 67–95.

Pinheiro, H., Cabral, J. M. S. (1993b), in: J. F. Kennedy, J. M. S. Cabral (Eds.), *Recovery Processes for Biological Materials,* New York: Wiley, pp. 97–131.

Porter, M. C. (1990), *Handbook of Industrial Membrane Technology,* Park Ridge, NJ: Noyes Publications.

Randolph, A. D., Larson, M. A. (1971), *Theory of Particulate Processes,* New York: Academic.

Rautenbach, R., Albrecht, R. (1989), *Membrane Processes,* New York: Wiley.

Reid, R. C., Prausnitz, J. M., Polling B. E. (1987), *The Properties of Gases and Liquids,* New York: McGraw-Hill.

Robbins, L. A. (1984), in: R. H. Perry, D. W. Green (Eds.), *Perry's Chemical Engineers' Handbook,* 6th ed., New York: McGraw-Hill Book Co., p. 15–1.

Ruthven, D. M. (1984), *Principles of Adsorption and Adsorption Processes,* New York: Wiley-Interscience.

Rydberg, J., Musikas, C., Choppin, G. R. (1992), *Principles and Practices of Solvent Extraction,* New York: Marcel Dekker.

Sartori, G., Ho, W. S., Savage, D. W., Chludzinski, G. R., Wiechert, S. (1987), *Separ. Purif. Methods* **16,** 171.

Schwartzberg, H. G. (1987), in: R. W. Rousseau (Ed.), *Handbook of Separation Process Technology,* New York: Wiley.

Seader, J. D. (1984), in: R. H. Perry, D. W. Green (Eds.), *Perry's Chemical Engineers' Handbook,* 6th ed., New York: McGraw-Hill, p. 13-1.

Singh, G. (1979), in: P. A. Schweitzer (Ed.), *Handbook of Separation Technologies for Chemical Engineers,* New York: McGraw-Hill, p. 2-151.

Skelland, A. H. P., Tedder, D. W. (1987), in: R. W. Rousseau (Ed.), *Handbook of Separation Process Technology,* New York: Wiley.

Sourirajan, S., Matsuura, T. (1985), *Reverse Osmosis/Ultrafiltration Process Principles,* Ottawa, Canada: National Research Council Canada Publications.

Springer, A. M. (1993), in: A. M. Spriger (Ed.), *Industrial Environmental Control,* 2nd ed., Atlanta, GA: TAPPI, pp. 344–360.

Strathmann, H. (1992), in: W. S. Ho, K. K. Sirkar (Eds.), *Membrane Handbook,* New York: Chapman & Hall, pp. 217–262.

Streat, M., Cloete, F. L. D. (1987), in: R. W. Rousseau (Ed.), *Handbook of Separation Process Technology,* New York: Wiley.

Suzuki, M. (1990), *Adsorption Engineering,* New York: Elsevier.

Tien, C., Payatakes, A. C. (1979), *AIChE J.* **25,** 737.

Toyokura, K., Wintermantel, K., Hirasawa, I., Wellinghoff, G. (1994), in: A. Mersmann (Ed.), *Crystallization Technology Handbook,* New York: Marcel Dekker, p. 459.

Vermeulen, T., LeVan, M. D., Hiester, N. K., Klein, G. (1984), in: R. H. Perry, D. W. Green (Eds.), *Perry's Chemical Engineers' Handbook,* 6th ed., New York: McGraw-Hill, p. 16-1.

Wadsworth, E. M. (1987), in: R. W. Rousseau

(Ed.), *Handbook of Separation Process Technology,* New York: Wiley.

Wijmans, J. G., Baker, R. W. (1993), *J. Membr. Sci.* **79,** 101.

Zarzycki, R. (1993), *Absorption: Fundamentals & Applications,* New York: Pergamon Press.

Zolandz, R. R., Fleming, G. K. (1992), in: W. S. Ho, K. K. Sirkar (Eds.), *Membrane Handbook,* New York: Chapman & Hall, pp. 17–101.

CONTENTS OF PREVIOUS VOLUMES

Volume 1

Volume 2

Volume 3

Volume 4

Volume 5

Volume 6

Volume 7

Volume 8

Volume 9

Volume 10

Volume 11

Volume 12

Volume 13

Volume 14

Volume 15

Volume 16

LIST OF RECOMMENDED UNITS AND SYMBOLS

RECOMMENDED UNITS AND CONVERSION FACTORS

The SI system provides six basic units: meter m, kilogram kg, second s, ampere A, kelvin K, candela cd, and mole mol.

Some important derived units are also allowed and bear special names, e.g.:

1 N (newton) = 1 kg m s^{-2}
1 J (joule) = 1 N m = 1 kg m^2 s^{-2}
1 W (watt) = 1 J s^{-1} = 1 kg m^2 s^{-3}
1 Pa (pascal) = 1 N m^{-2} = 1 kg m^{-1} s^{-2}

For mass, the gram g, or the metric ton t which equals 1000 kg, may be used instead of kilogram kg.

The so-called "long ton" (UK) and "short ton" (US) have been abandoned and will not be used in the *Encyclopedia of Applied Physics*.

For pressure, the bar (name and symbol alike) may be used, and for temperature, the degree celsius °C.

From all of these, decimal multiples or fractions can be derived:

Power of ten	Prefix	Symbol	Power of ten	Prefix	Symbol
10	deca	da	10^{-1}	deci	d
10^2	hecto	h	10^{-2}	centi	c
10^3	kilo	k	10^{-3}	milli	m
10^6	mega	M	10^{-6}	micro	μ
10^9	giga	G	10^{-9}	nano	n
10^{12}	tera	T	10^{-12}	pico	p
10^{15}	peta	P	10^{-15}	femto	f
10^{18}	exa	E	10^{-18}	atto	a

For mass, multiples or fractions are derived from g, not from kg (since the latter already contains a prefix), e.g., mg.

Units with the prefixes are considered as one entity and can, therefore, be raised to any power, e.g., cm^3.

The liter is now considered as synonymous with dm^3 (which does not hold in the older literature!). Please use the capital letter L as unit symbol (following a recent IUPAC recommendation). The use of the Ångström unit is discouraged; it should be replaced by fractional meters (1 Å = 100 pm = 0.1 nm).

SELECTED QUANTITIES, UNITS, AND SYMBOLS

Name of quantity[a]	Symbol[b]	SI unit[c]	Name of unit	Other units
Space and time				
length*	l	m	meter[d]	
breadth, width	b	m		
height	h	m		
radius	r	m		
thickness	d	m		
area	A, S	m^2		a (are) h (hectare)
volume	V	m^3		L, l(liter)[c]
plane angle	$\alpha, \beta, \gamma,$ ϑ, φ	1, rad	radian	° (degree) ′ (minute) ″ (second)
solid angle	ω, Ω	1, sr	steradian	
wavelength	λ	m		
wave number	σ, ν	m^{-1}		
time*	t	s	second	min (minute) h (hour) d (day)
frequency	ν, f	s^{-1}		Hz (hertz)[f]
relaxation time	τ	s		
velocity	u, v	$m\ s^{-1}$		km/h
acceleration	a	$m\ s^{-2}$		
Mechanics				
mass*	m	kg	kilogram	g (gram) t (tonne)[g]
(mass) density	ρ	$kg\ m^{-3}$		g/cm^3
momentum	$\mathbf{p}$	$kg\ m\ s^{-1}$		
angular momentum	$\mathbf{L}$	$kg\ m^2\ s^{-1}$		
force	$\mathbf{F}$	N	newton[f]	
moment of force	$\mathbf{M}$	N m		
weight	G, W	N		
pressure	p	Pa	pascal	bar (bar)[h]
energy	E, W	J	joule	W h (watt hour)[i] eV (electron volt)[j]
work	W, A	J		
power	P	W	watt	J/s, V A[k]
Molecular Physics and Thermodynamics				
thermodynamic temperature*	T	K	kelvin	°C (degrees Celsius)
Celsius temperature	ϑ, t			°C
number of entities	N			
Avogadro constant[l]	N_A, L	mol^{-1}	(particles) per mole	
Boltzmann constant[m]	k, k_B	$J\ K^{-1}$		
Planck constant[n]	h	J s		
(molar) gas constant[o]	R	$J\ mol^{-1}\ K^{-1}$		
(quantity of) heat	Q	J		
entropy[p]	S	$J\ K^{-1}$		
internal energy[p]	U	J		
Helmholtz function,[p] (Helmholtz) free energy, Helmholtz energy	F, A	J		
enthalpy[p]	H	J		
Gibbs function,[p] (Gibbs) free energy, Gibbs energy	G	J		
heat capacity[p]	C_p, C_V	$J\ K^{-1}$		

Name of quantity[a]	Symbol[b]	SI unit[c]	Name of unit	Other units
Chemical Physics				
amount of substance*	n	mol	mole	
relative atomic mass	A_r	1		
relative molecular mass	M_r	1		
atomic mass constant	m_u	kg		u (atomic mass unit)[q]
mass of a portion (of substance B)	$m_B, m(B)$	kg		g (gram)
molar mass (of substance B)	$M_B, M(B)$	$kg\ mol^{-1}$		
concentration (of substance B)	$c_B, c(B)$	$mol\ m^{-3}$		mol/L
mole fraction[r] (of substance B)	$\kappa_B, \kappa(B)$	1		
mass fraction (of substance B)	$\omega_B, \omega(B)$	1		%,‰,ppm,ppb
volume fraction (of substance B)	$\varphi_B, \varphi(B)$ $\phi_B, \phi(B)$	1		%,‰,ppm,ppb
mass concentration	ρ	$kg\ m^{-3}$		g/L
molality		$mol\ kg^{-1}$		mmol/kg
volume concentration[s]	σ	1		
molar volume	V_m	$m^3\ mol^{-1}$		L/mol
molar heat capacity	C_m	$J\ mol^{-1}\ K^{-1}$		
molar conductivity	Λ_m	$S\ m^2\ mol^{-1}$	(S: siemens)	
Faraday constant[t]	F	$C\ mol^{-1}$		
Electricity and Magnetism				
quantity of electricity[u]	Q	C	coulomb	
charge density	ρ	$C\ m^{-3}$		
electric potential	ϕ, V	V	volt	
electric potential difference, voltage	$U, \Delta\phi, \Delta V$	V		
electric dipole moment	$\mathbf{p}, \mathbf{p}_c$	C m		
electric current*	I	A	ampere	
electric current density	j	$A\ m^{-2}$		
electric field strength	$\mathbf{E}$	$V\ m^{-1}$		
electric displacement	$\mathbf{D}$	$C\ m^{-2}$		
capacitance	C	F	farad	
permittivity	ε	$F\ m^{-1}$		
relative permittivity	ε	1		
dielectric polarization	$\mathbf{P}$	$C\ m^{-2}$		
electric susceptibility	χ_e	1		
polarization (of a particle)	α	$m^2\ C\ V^{-1}$		
magnetic flux	Φ	Wb	weber	
magnetic flux density	$\mathbf{B}$	T	tesla	
magnetic field strength	$\mathbf{H}$	$A\ m^{-1}$		
permeability	μ	$H\ m^{-1}$, $N\ A^{-2}$	(H: henry)	
relative permeability	μ_r	1		
magnetization	M	$A\ m^{-1}$		
magnetic susceptibility	χ	1		
molar magnetic susceptibility	χ_m	$m^3\ mol^{-1}$		
(electrical) resistance	R	Ω	ohm	
(electrical) conductance	G	S	siemens	
(electrical) resistivity	ρ	Ω m		
(electrical) conductivity	κ,σ	$S\ m^{-1}$		
self-inductance	L	H	henry	
Radiation				
radiant energy	Q, W, Q_e	J		
luminous intensity*	I	cd	candela	
radiant intensity	I_e	$W\ sr^{-1}$, W		
emissivity, emittance	ε	1		

Name of quantity[a]	Symbol[b]	SI unit[c]	Name of unit	Other units
absorptance	α	1		
reflectance	ρ, R	1		
transmittance	τ	1		
absorption coefficient:				
linear (decadic)	a	m^{-1}		
molar (decadic)	ε	$m^2\ mol^{-1}$		
refractive index	n	1		
molar refraction	R_m	$m^3\ mol^{-1}$		
angle of optical rotation	α	1, rad		
Transport Properties				
flux of quantity X	J_X, J	(varies)		
mass flow rate	q_m, $\dot{m}$	$kg\ s^{-1}$		
volume flow rate	q_V, $\dot{V}$	$m^3\ s^{-1}$		
heat flow rate	Φ	W		
thermal conductivity	κ, k, λ	$W\ m^{-1}\ K^{-1}$		
coefficient of heat transfer	h	$W\ m^{-2}\ K^{-1}$		
thermal diffusivity	a	$m^2\ s^{-1}$		
diffusion coefficient	D	$m^2\ s^{-1}$		
thermal diffusion coefficient	D_T	$m^2\ s^{-1}$		
viscosity	η, μ	Pa s		
kinematic viscosity	ν	$m^2\ s^{-1}$		

[a]SI base quantities are marked by asterisks (*)
[b]Recommended by IUPAC
[c]SI base units as well as derived and supplementary units are listed; all are to be used with prefixes as needed
[d]do not use "metre"
[e]do not use "litre"; $1\ L = 10^{-3}\ m^3$
[f]$1\ Hz = 1\ s^{-1}$; $1\ N = 1\ kg\ m\ s^{-2}$
[g]Formerly metric ton; $1\ t = 10^3\ kg$
[h]$1\ bar = 10^5\ Pa$
[i]$1\ W\ h = 3.6 \times 10^3\ J$
[j]$1\ eV = 1.602\ 189 \times 10^{-19}\ J$
[k]$1\ W = 1\ J/s = 1\ VA$
[l]$N_A = 6.022\ 136\ 7 \times 10^{23}\ mol^{-1}$
[m]$k = 1.380\ 658 \times 10^{-23}\ J\ K^{-1}$
[n]$h = 6.626\ 075\ 5 \times 10^{-34}\ J\ s$
[o]$R = 8.314\ 510\ J\ mol^{-1}\ K^{-1}$
[p]Molar quantities can be distinguished from the quantity of a system by adding the subscript m; e.g., molar internal energy U_m, in $J\ mol^{-1}$
[q]$1\ u = 1.660\ 565\ 5 \times 10^{-27}\ kg$
[r]A more accurate, but rather uncommon, name is "amount-of-substance fraction"
[s]σ refers to the total volume of a mixture, whereas the volume fraction φ relates the volume of a substance to the volume of several components before mixing
[t]$F = 9.648\ 530\ 9 \times 10^4\ C\ mol^{-1}$
[u]Also called electric charge